ENGINEERING SOLID MECHANICS

Fundamentals and Applications

ENGINEERING SOLID MECHANICS

Fundamentals and Applications

Abdel-Rahman Ragab
Salah Eldin Bayoumi

CRC Press
Boca Raton London New York Washington, D.C.

Acquiring Editor: Cindy Carelli
Project Editor: Maggie Mogck
Marketing Manager: Jane Stark
Cover design: Dawn Boyd

Library of Congress Cataloging-in-Publication Data

Ragab, Abdel-Rahman A. F.
Engineering solid mechanics : fundamentals and applications / Abdel-Rahman A. F. Ragab, Salah Eldin A. Bayoumi.
p. cm.
Includes bibliographical references and index.
ISBN 0-8493-1607-3 (alk. paper)
1. Deformations (Mechanics). 2. Mechanics, Applied. I. Bayoumi, Salah E. A. II. Title.
TA417.6.R34 1998
620.1 ' 05—dc21

98-7492
CIP

International Standard Book Number 0-8493-1607-3
Library of Congress Card Number 98-7492
Printed in the United States of America 1 2 3 4 5 6 7 8 9 0
Printed on acid-free paper

Preface

The subject of the mechanics of solids deals with the deformations set up in a solid as a result of a system of forces. The deformations can be elastic, plastic, or elastic–plastic depending upon the material stress–strain relation and the severity of loading. In the present text, rate-dependent behavior, such as creep and superplasticity, have been included.

The contents of this text are an outgrowth of lecture notes for courses delivered by the authors to undergraduate and graduate university students during the past 25 years. Although primarily written for mechanical engineering students, the text serves students in other fields of engineering, such as aerospace, civil, and material engineering, and is a useful reference for professional engineers interested in advancing their knowledge in the subject.

The text is intended to bridge the gap between elementary strength-of-materials texts and the more-advanced and specialized texts on the subject. An engineering analytical approach is adopted with particular emphasis on problem formulation. Analytical treatment is a necessary foundation for the student to be capable of handling modern numerical methods, as well as assess and create software packages. The potential of analytical solutions is unlimited since it provides reliable results that have universal application. The educational benefits gained from investing teaching hours in developing analytical solutions are obvious and merit the rigorous effort demanded from both teacher and student.

The material presented has been chosen and arranged to serve three requisites: first, to provide the student with a basic understanding of the fundamentals of elasticity and plasticity, as in Chapters 1, 2, 3, 4, and 10; second, to apply these fundamentals to solve analytically some problems of practical interest, as presented in the rest of the chapters; third, to get the student introduced to advanced topics of mechanics of materials, such as fracture, creep, superplasticity, fiber-reinforced composites, and powder compacts.

Chapters 1 and 2 develop the concepts of stress and strain, equilibrium and compatibility to provide a sound basis for elastic and plastic analysis. Chapter 3 covers elastic stress–strain relations. Chapter 4 discusses the solution of the elastic problem and develops the stress function approach to solve plane elastic problems. Some applications of the stress function solution in Cartesian and polar coordinates are given in Chapters 5 and 6, respectively. Chapters 7 and 8 present solutions to some elastic problems of rods, plates, and shells through formulating a strain compatibility function and also by applying energy methods. Chapter 9 is an application to linear-elastic fracture mechanics and safe-life prediction in the presence of fatigue crack propagation. The next three chapters deal with inelastic deformation and its applications. In Chapter 10 the fundamentals of plasticity are set out and explained in detail including anisotropic behavior, the kinematic hardening concept, and the effect of voids on plastic deformation. Applications to plastic instability, superplasticity, and mechanics of creep deformation are dealt with in Chapter 11. In Chapter 12, solutions to some engineering problems involving elastic–plastic deformation are developed.

Throughout the text the material is presented in such a manner to make it self-instructive and suitable for individual study. This has been accomplished by adopting an attitude of generous explanation, systematic derivations, and detailed discussions supplemented by a vast number and variety of solved examples, more than 200 devised mostly by the authors. Another feature is the variety of topics and applications divided into a relatively large number of sections in each chapter, which simplifies the assimilation of the material by the student and allows teachers to make their own selection of course contents by omitting some sections, hopefully without upsetting the continuity of the presentation.

At the end of each chapter, a variety of problems is provided. A solutions manual may be ordered.

The authors acknowledge the generous help and valuable suggestions received from their colleagues, particularly Professor M. M. Megahed and Dr. M. Abou-Hamda. They also acknowledge the patience and devotion of Mr. B. El-Hadidi in the preparation of the drawings and Mr. A. M. Nasr in typing the manuscript.

Abdel-Rahman A. F. Ragab
Salah Eldin A. Bayoumi

Authors

Abdel-Rahman A. F. Ragab graduated with honors (B. Sc.) in Mechanical Engineering from Cairo University, Egypt, and received his M. Sc. from Sherbrooke University and his Ph.D. from McMaster University in Canada. Since that time he has been on the Faculty of Engineering at Cairo University and has occupied visiting professorship posts at both the Universite de Metz in France (1974 and 1977) and Kuwait University (1983 to 1987). For the past decade he has been the director of the Center of Postgraduate Studies and Research in Engineering at Cairo University.

Professor Ragab's primary research interest lies in the area of mechanics of materials applied to material testing, plasticity, superplasticity, metal forming, and plastics engineering. He has published and presented over 45 research papers in international and national periodicals and at conferences. He also is a consultant for several large industrial firms and is the holder of the State Decoration for Sciences and Arts – First Grade, 1985.

Salah Eldin A. Bayoumi received his Ph.D. (1951) from Manchester University in the U.K. and joined the Engineering Laboratory at Cambridge University as a research worker. He headed research and development in the Egyptian General Organization for metallurgical industries and was later appointed to the post as professor of Mechanics of Solids (1966) and Head of the Department of Mechanical Design and Production (1971 to 1978) and is now Professor Emeritus at the Faculty of Engineering at Cairo University.

Professor Bayoumi has published several scientific papers in experimental stress analysis, mechanical systems, metal cutting, metal forming, and polymer mechanics. He is a member of several national educational and engineering societies and committees, and is a consultant on mechanical engineering projects.

Nomenclature

A, ΔA	area, change in area
a	crack length
B	creep constant in the law $\dot{\varepsilon}_c = B\sigma^N$
b	beam width
C_v	void volume fraction
C	complementary elastic strain energy
D	flexural rigidity of a plate
d	diameter
E	modulus of elasticity
e	distance locating neutral axis or shear center
f	shear flow in thin-walled sections
F	body force
g	acceleration of gravity
G	modulus of rigidity or shear modulus
h	thickness, beam height
I, (I_1, I_2, I_3)	second moment of area, or stress invariants
(J_1, J_2, J_3)	strain invariants
J	contour integral around a crack
k, k_t, k_s	shear yield strength, stress concentration factor, and stiffness, respectively
K	bulk modulus of elasticity, or stress intensity factor, strength coefficient in the law $\sigma = K(\varepsilon^p)^n$
l, m, n	direction cosines
L, ΔL	length and change in length
m	strain-rate-sensitivity exponent in th law $\sigma = C\dot{\varepsilon}^m$ or exponent in fatigue crack growth law $da/dN = C(\Delta K)^m$
M, M_t	bending moment and twisting moment, respectively
n	Strain-hardening-sensitivity exponent in the law $\sigma = K(\varepsilon^p)^n$
N	creep exponent in the law $\dot{\varepsilon}_c = B\sigma^N$
p	pressure
P	load
q	distributed load
Q	shear force
r	radius
R	radius of curvature, or anisotropic strain ratio
s	distance measured along perimeter
S, S^*	stress resultant vector and first moment of area, respectively
t	time
T	temperature
u, v, w	displacement components
U	elastic strain energy
V	potential function, volume, or linear velocity
W	work, weight, or width
X_s, Y_s, Z_s	surface tractions
Y	yield strength

α	coefficient of thermal expansion, or angle of twist per unit length
δ	displacement, deflection, or crack tip opening displacement
Δ	radial interference
ε	strain
ϕ	stress function
γ	engineering shear strain
ψ	warping function in torsion of bars
μ, λ	Lame's elastic constants
λ	extension ratio, or radii ratio
ν	Poisson's ratio
ρ	mass density, or radius of curvature
σ	normal stress
τ	shear stress
ω	angular speed
Ω	potential energy, or creep damage parameter
x, y, z	Cartesian coordinates
r, θ, z	polar cylindrical coordinates
r, θ, ϕ	polar spherical coordinates
α, β, γ	curvilinear coordinates
CCF	configuration correction factor for the stress intensity around a crack
F.S.	factor of safety

Contents

Chapter 1 Analysis of Stress

1.1 Rigid and Deformable Bodies 1
1.2 Body Forces and Surface Tractions 1
1.3 Concept of Stress and Strain 2
1.4 The State of Stress at a Point 2
1.5 Cartesian Stress Components 6
1.6 Some Special States of Stress 8
1.6.1 Plane Stress 8
1.6.2 Plane Strain 9
1.6.3 Axial Symmetry 9
1.6.4 Free Torsion 9
1.7 Stress Equations of Equilibrium 9
1.7.1 Cartesian Coordinates 10
1.7.2 Cylindrical Polar Coordinates 14
1.7.3 Spherical Polar Coordinates 17
1.7.4 Curvilinear Coordinates 18
1.8 Stress Transformation Law 21
1.9 Plane Stress Transformation — Mohr's Circle of Stress 27
1.10 Principal Stresses 29
1.11 Maximum Shear Stresses 34
1.12 Octahedral Shear Stress — Pure Shear 38
1.13 Mean (Hydrostatic) Stress and Deviatoric Stresses 39
1.14 A Note on the Stress Equations 41
Problems 41
References 48

Chapter 2 Analysis of Strain

2.1 Infinitesimal Strains 49
2.1.1 Normal Strain 49
2.1.2 Shear Strain 50
2.1.3 Volumetric Strain 52
2.2 Infinitesimal Strain–Displacement Relations 54
2.2.1 Cartesian Coordinates 54
2.2.2 Cylindrical Polar Coordinates 59
2.2.3 Spherical Polar Coordinates 61
2.3 Strain Compatibility Conditions 62
2.3.1 Cartesian Coordinates 62
2.3.2 Cylindrical Polar Coordinates 64
2.3.3 Spherical Polar Coordinates 65
2.4 Strain Tensor 67
2.5 Some Special States of Strain 71
2.5.1 Plane Strain 72
2.5.2 Plane Stress 72
2.5.3 Axial Symmetry 72

2.5.4 Free Torsion 72
2.6 Principal Strains — Maximum and Octahedral Shear Strains 74
2.7 Mean Strain Dilatation and Strain Deviations 76
2.8 Mohr's Circle of Strain 78
2.9 Strain Gauge Rosettes 80
2.10 Notes on Finite Strains 82
2.11 Strain Rate–Velocity Relations 87
Problems 90
References 99

Chapter 3 Elastic Stress–Strain Relations
3.1 Introduction 101
3.2 Basic Assumptions: Elasticity, Homogeneity, and Isotropy 101
3.2.1 Elasticity 101
3.2.2 Homogeneity 102
3.2.3 Isotropy 103
3.3 Hooke's Law for Homogeneous Isotropic Materials 103
3.3.1 Simple Loading 103
3.3.2 Triaxial Loading 104
3.4 Relations Among the Elastic Constants 108
3.5 Inverse Form of Hooke's Law 110
3.6 Dilatation and Distortion 112
3.7 Thermoelastic Stress–Strain Relations 114
3.8 Strain Energy for an Elastic Isotropic Solid 116
3.9 Strain Energy for a Solid Obeying Hooke's Law 122
3.10 Some Elastic Energy Theorems 128
3.10.1 Principle of Work 128
3.10.2 Principle of Virtual Work 129
3.10.3 Principle of Stationary Potential Energy 131
3.10.4 Castigliano's Theorems 132
3.11 Generalized Hooke's Law 135
3.11.1 Anisotropic Elasticity 135
3.11.2 Application to Fiber-Reinforced Composites 140
3.12 Note on Composite Elastic Constants 146
3.13 Stress–Strain Relations for Large Elastic Deformation 146
Problems 150
References 154

Chapter 4 Solution of the Elastic Problem
4.1 The Elastic Problem 155
4.2 Boundary Conditions 156
4.3 Saint-Venant's Principle 159
4.4 Uniqueness and Semi-Inverse Method of Elastic Solution 160
4.5 Example of Solution in Terms of Stress: Pressurized Thick-Walled Sphere 161
4.6 The Elastic Plane Problem 164
4.6.1 Plane Strain Formulation 165
4.6.2 Plane Stress Formulation 167
4.6.3 Deduction of Plane Stress Equations from Plane Strain Equations 169
4.7 Stress Function Formulation for Plane Elastic Problems 170
4.8 Governing Equations in Terms of a Stress Function in Cartesian Coordinates 171

4.8.1 Plane Strain 171
4.8.2 Plane Stress 174
4.8.3 Thermoelastic Plane Problem 174
4.8.3.1 Thermoelastic Plane Strain 174
4.8.3.2 Thermoelastic Plane Stress 175
4.8.4 Finding a Stress Function in Cartesian Coordinates 182
4.9 Governing Equations in Terms of a Stress Function in Polar Coordinates 183
4.9.1 Plane Strain 183
4.9.2 Plane Stress 185
4.9.3 Axisymmetric Plane Problems 187
4.9.3.1 Axisymmetric Problems without Body Forces 187
4.9.3.2 Axisymmetric Problems with Centrifugal Body Forces 189
4.9.3.3 Axisymmetric Problems with Radial Temperature Gradient 191
4.9.4 A Note on Finding a Stress Function in Polar Coordinates 195
4.10 A Glossary of Stress Functions for Some Plane Problems 195
4.10.1 Cartesian Coordinates 195
4.10.2 Polar Coordinates 197
Problems 200
References 203

Chapter 5 Elastic Plane Problems in Cartesian Coordinates

5.1 Introduction 205
5.2 Problems Solved in Terms of Algebraic Polynomials 205
5.2.1 Retaining Wall Subjected to Hydrostatic Pressure 209
5.2.2 Simply Supported Beam under Uniformly Distributed Load 213
5.2.3 Cantilever Beam Subjected to an End Load 220
5.2.3.1 Stresses 220
5.2.3.2 Displacements 222
5.3 Problems Solved in Terms of Trigonometric Stress Functions 229
5.3.1 Simply Supported Beam under Laterally Distributed Sinusoidal Load on Both Sides 230
5.3.2 Simply Supported Beam under Two Equal Lateral Loads at the Middle of the Span 232
5.3.3 Bar subjected to Two Equal and Opposite Axial Loads 233
5.4 A Note on Some Other Forms of Stress Functions 234
Problems 237
References 241

Chapter 6 Elastic Plane Problems in Polar Coordinates

6.1 Introduction 243
6.2 Axisymmetric Problems 243
6.2.1 Thick-Walled Cylinder Subjected to Uniform Internal and/or External Pressure 243
6.2.1.1 Cylinder Subjected to Internal Pressure Only 246
6.2.1.2 Cylinder Subjected to External Pressure Only 247
6.2.2 Thick-Walled Cylinder Subjected to Steady-State Radial Thermal Gradient 250
6.2.2.1 Plane Strain 250
6.2.2.2 Plane Stress 253
6.2.2.3 Other End Conditions 254

6.2.3 Cylinder Compounding by Shrink Fit 255
6.2.4 Rotating Disk of Uniform Thickness 262
6.2.4.1 Annular Rotating Disk of Constant Thickness 262
6.2.4.2 Solid Rotating Disk of Constant Thickness 264
6.2.5 Rotating Solid Disk of Uniform Strength (De Laval Disk) 266
6.2.6 Rotating Drums and Rotors 269
6.2.7 Rotating Disks and Rotors Subjected to Radial Thermal Gradients 270
6.3 Axially Nonsymmetric Problems 273
6.3.1 Bending of a Circularly Curved Beam 274
6.3.1.1 Beam Subjected to an End Shearing Force 274
6.3.1.2 Beam Subjected to Pure Bending 280
6.3.1.3 Beam Subjected to an End Moment and a Normal Force 282
6.3.1.4 Beam Subjected to an Inclined End Force 282
6.3.2 Thermal Stresses in Curved Beams 283
6.3.3 Wedge Subjected to a Concentrated Load at its Vertex 286
6.3.3.1 Force Acting Along a Wedge Axis 286
6.3.3.2 Force Perpendicular to the Wedge Axis 289
6.3.3.3 Force Inclined to the Wedge Axis 290
6.3.3.4 Bending Moment Acting at the Vertex 292
6.3.4 Concentrated Line Load Acting on the Edge of a Straight Boundary 295
6.3.4.1 Force Acting Normal to the Boundary 295
6.3.4.2 Force Acting Along the Boundary 296
6.3.4.3 Force Acting Inclined to the Boundary 297
6.3.5 Uniformly Distributed Line Load Acting on the Edge of a Straight Boundary 297
6.3.6 Circular Solid Disk Subjected to Two Equal and Opposite Diametral Loads 299
6.3.7 Concentrated Load Acting on a Rectangular Beam 301
6.4 Stresses Concentration Around a Small Circular Hole 303
Problems 311
References 318

Chapter 7 Elastic Rods Subjected to General Loading

7.1 Introduction 319
7.2 Stress Resultants 319
7.2.1 Note on Sign Convention for Stress Resultants 321
7.3 Bending of Rods 322
7.3.1 Bending Stresses 322
7.3.2 Elastic Curve in Bending 326
7.3.3 Bending of Curved Beams 330
7.3.3.1 Determination of the Location of the Neutral Axis 334
7.3.3.2 Approximate Determination of the Neutral Axis 336
7.3.3.3 Maximum Stresses 337
7.3.3.4 Bending of a Curved Beam by Lateral Forces Acting in the Plane of Its Axis 337
7.3.3.5 Strain Energy in Curved Beams 338
7.3.3.6 Comparison with Exact and Other Solutions 339
7.3.4 Thermoelastic Bending of Straight Bars 341
7.4 Shear Stresses in Rods 345
7.4.1 Rectangular Solid Section 347

7.4.2 Circular Solid Section 349
7.4.3 Thin-Walled Open Sections 352
7.4.3.1 Shear Center 355
7.4.4 Thin-Walled Closed Sections 357
7.5 Torsion of Bars 359
7.5.1 Saint-Venant's Free Torsion 360
7.5.2 Solid Circular Section 363
7.5.3 Solid Elliptical Section 365
7.5.4 Solid Rectangular Section 366
7.5.5 Thin-Walled Open Sections 368
7.5.6 Thin-Walled Closed Sections 370
7.5.7 Effect of Internal Stiffening Webs 372
7.5.8 Effect of End Constraint 373
7.5.8.1 Solid Sections 374
7.5.8.2 Thin-Walled Sections 375
7.6 Displacements in Rods — Energy Approach 376
7.6.1 Application of Castigliano's Theorem 376
7.6.2 Mohr's Unit Load Method 384
7.6.3 A Note on the Deflection of Curved Beams 387
7.6.4 Application to Springs 391
7.6.4.1 Helical Compression Spring 391
7.6.4.2 Spiral Helical Compression Spring 393
7.6.4.3 Flat Compression Spring 394
7.6.4.4 Flat Torsion Spring 395
7.7 Buckling of Rods 398
7.7.1 Buckling of Columns 398
7.7.1.1 Equilibrium Approach 398
7.7.1.2 Minimum Potential Energy Solution: Rayleigh–Ritz Method 403
7.7.2 Beam–Columns 410
7.7.3 Lateral Buckling of Beams 412
7.8 Beams on Elastic Foundation 415
7.8.1 Infinitely Long Beams 416
7.8.1.1 Concentrated Force 416
7.8.1.2 Concentrated Moment 418
7.8.1.3 Uniform Load 420
7.8.2 Semi-Infinite Beams 422
7.8.3 Short Beams 425
Problems 425
References 439

Chapter 8 Some Problems of Elastic Plates and Shells

8.1 Introduction 441
8.2 State of Stress in Plates and Shells 441
8.3 Plate Equations in Cartesian Coordinates 442
8.3.1 Deformation Pattern 442
8.3.2 Stress Resultants 444
8.3.3 Equations of Equilibrium 445
8.3.4 Method of Solution: Pure Bending of a Plate 448
8.3.5 Effect of Thermal Gradient Throughout Plate Thickness 449
8.3.5.1 A Plate with Free Edges 449
8.3.5.2 A Plate with Clamped Edges 450

8.3.5.3 A Plate with Simply Supported Edges 451
8.4 Bending of Rectangular Plates — Energy Approach 452
8.4.1 Uniformly Loaded Rectangular Plate Simply Supported Along Its Four Edges 452
8.4.2 Uniformly Loaded Rectangular Plate Clamped Along Its Four Edges 458
8.4.3 An Approximate Strip Method for Rectangular Plates 465
8.5 Axisymmetric Bending of Flat, Circular Plates 466
8.5.1 Solid Circular Plates 468
8.5.1.1 Simply Supported Plate Subjected to Uniform Pressure 468
8.5.1.2 All-Around Clamped Plate Subjected to Uniform Pressure 469
8.5.1.3 All-Around Clamped Plate Subjected to a Concentrated Force at the Center 472
8.5.1.4 Simply Supported Plate Subjected to a Concentrated Force at the Center 474
8.5.2 Annular Circular Plates 475
8.5.2.1 Simply Supported Annular Plate Subjected to Edge Moments 475
8.5.2.2 Simply Supported Annular Circular Plate Subjected to a Shearing Force at the Inner Edge 477
8.5.3 Other Loadings and Edge Conditions 478
8.5.4 Thermal Stresses in Circular Plates 479
8.5.4.1 Temperature Gradient Across the Thickness of a Disk with Free Edges 479
8.5.4.2 Temperature Gradient Across the Thickness of a Disk with All-Around Clamped Edges 480
8.5.4.3 Axisymmetric Radial Temperature Gradient 480
8.5.5 Comments on the Deflection of Circular Plates 482
8.5.5.1 Deflection Due to Shear 482
8.5.5.2 Large Deflection 483
8.6 Membrane Stresses in Axisymmetric Shells 486
8.6.1 Axisymmetric Shells Subjected to Uniform Pressure 486
8.6.2 Applications to Pressurized Containers 490
8.6.2.1 Spherical Shell 490
8.6.2.2 Circular Cylindrical Shell 491
8.6.2.3 Conical Shell 491
8.6.2.4 Toroidal Shell 492
8.6.3 Displacement in Axisymmetric Shells 493
8.6.4 Axisymmetric Shells Subjected to Gravity Loading 499
8.6.4.1 Hemispherical Liquid Container Freely Supported at Its Top Edge 499
8.6.4.2 Conical Liquid Container Freely Supported at Its Top Edge 500
8.6.4.3 Spherical Container on a Skirt Support 504
8.7 Bending of Thin-Walled Cylinders Subjected to Axisymmetric Loading 507
8.7.1 Problem Formulation 507
8.7.2 Long, Thin-Walled Pressurized Pipe with a Rigid Flange at its End 513
8.7.3 Short, Thin-Walled Pressurized Pipe with Two Rigid Flanges at Both Ends 517
8.7.4 Long, Thin-Walled Pipe Subjected to Uniform Radial Compression Along a Circular Section at its Middle Length 518
8.7.5 Long, Thin-Walled Pipe Subjected to a Uniform Circumferential Load Along a Finite Length 521
8.7.6 Cylindrical Pressure Vessels With End Closures 523
8.7.6.1 Case of a Flat End 524

8.7.6.2 Case of a Curved End 527
8.7.6.3 Case of a Hemispherical End 529
8.7.7 Cylindrical Storage Tanks 534
8.7.8 Effect of Thermal Gradient 537
8.8 Elastic Buckling of Plates and Shells 540
8.8.1 Buckling of Uniformly Compressed Rectangular Plate 540
8.8.1.1 All-Around Clamped Rectangular Plate 542
8.8.1.2 Rectangular Plates with Other Boundary Conditions 543
8.8.2 Axisymmetric Buckling of Circular Plates 544
8.8.3 Buckling of Thin-Walled Cylinders Under External Uniform Pressure 546
8.8.3.1 Effect of Out-of-Roundness, Cylinder Length, and End Constraints 550
Problems 550
References 559

Chapter 9 Applications to Fracture Mechanics

9.1 Introduction 561
9.2 Griffith Energy Criterion 562
9.3 Stress Concentration Around Elliptical Holes 565
9.4 The Elastic Stress Field at the Crack Tip 566
9.5 The Stress Intensity Factor and Fracture Toughness 570
9.6 Stress Intensity Factors for Various Configurations 574
9.6.1 Plates under Tensile Loading 575
9.6.2 Cracks Emanating from Circular Holes in Infinite Plates 577
9.6.3 Plates under Bending 579
9.6.4 Circular Rods and Tubes 579
9.6.5 Pressurized Thick-Walled Cylinders 581
9.6.6 Rotating Solid Disks and Drums 582
9.7 Superposition under Combined Loading 587
9.8 Mixed-Mode Loading 589
9.9 Plastic Zone Geometry at Crack Tip 590
9.10 Notes on Fracture Toughness Testing 595
9.11 Fracture Due to Crack Growth 598
9.11.1 Fatigue Crack Propagation 598
9.11.1.1 Region (i) of Nonpropagating Cracks 599
9.11.1.2 Region (ii) of Steady Crack Propagation 599
9.11.1.3 Region (iii) of Unstable Crack Growth Rate 601
9.11.2 Safe-Life Prediction 603
9.11.3 Comments on Safe-Life Predictions 609
9.11.3.1 Margin of Safety 609
9.11.3.2 Variable Amplitude Loading 609
9.11.3.3 Mixed-Mode Crack Growth 611
9.11.3.4 Correlation with *S–N* Curves 611
9.11.3.5 Growth of Physically Short Cracks 611
9.11.3.6 Crack Closure 612
9.12 Stress Corrosion Cracking 613
9.13 Elastic–Plastic Fracture Mechanics 616
9.13.1 J Integral 616
9.13.2 Experimental Determination of J 619
9.13.3 A Scheme for Fracture Estimation Using J_{Ic} 623
9.13.4 Crack Opening Displacement 627

9.13.5 Experimental Determination of COD 629
9.13.6 Application of CTOD to Structural Design 630
Problems 632
References 639

Chapter 10 Plastic Deformation

10.1 Introduction 641
10.2 Basic Assumptions 642
10.3 Definition of Large Plastic Strains 644
10.4 Strain Hardening in Simple Tension 646
10.5 Empirical Relations for Stress–Strain Curves 647
10.6 Idealized Stress–Strain Curves 652
10.7 Yield Criteria 653
10.7.1 von Mises Yield Criterion 654
10.7.2 Comments on the von Mises Criterion 656
10.7.3 Tresca Yield Criterion 658
10.7.4 Geometrical Representation of von Mises and Tresca Criteria 659
10.7.5 Experimental Verification of Yield Criteria 662
10.8 Plastic Stress–Strain Relations — Flow Rule 664
10.9 Principle of Normality and Plastic Potential 667
10.10 Plastic Work, Effective Stress, and Effective Strain Increment 669
10.11 Experimental Determination of the Flow Curve 674
10.12 Isotropic Hardening 678
10.13 Uniqueness and Path Dependence 679
10.14 Complete Elastic–Plastic Stress–Strain Relations 683
10.15 Plastic Deformation of Anisotropic Materials 686
10.15.1 A Yield Criterion for Anisotropic Materials 687
10.15.2 A Flow Rule for Anisotropic Materials 688
10.15.3 Measurement of Anisotropic Parameters 688
10.15.4 Normal Anisotropy 689
10.15.5 Effective Stress and Effective Plastic Strain Increment 691
10.15.6 A Special Case: Rotational Symmetry (Planar Isotropy) 692
10.15.7 A Modified Nonquadratic Criterion for Planar Isotropy 697
10.16 Kinematic Hardening 700
10.16.1 Uniaxial Behavior under Cyclic Loading 701
10.16.2 Triaxial Behavior — Yield Function and Flow Rule 709
10.17 Plastic Deformation of Porous Solids 715
10.17.1 Yield Function 716
10.17.2 Flow Rule 719
10.17.3 Void Growth Characteristics 720
10.17.4 Application to Metal Powder Compacts 722
Problems 723
References 731

Chapter 11 Plastic Instability, Superplasticity and Creep

11.1 Introduction 733
11.2 Unstable Plastic Deformation 733
11.2.1 Necking of a Tensile Bar 734
11.2.2 Local Necking of a Wide Strip 738
11.2.3 Limit Tensile Strain for a Bar with an Imperfection 740

11.2.4 Stresses in the Neck of a Tensile Bar....741
11.2.4.1 Round Bar....742
11.2.4.2 Wide Strip....745
11.2.5 Biaxial Stretching — Flat and Bulged Circular Sheets....746
11.2.5.1 Flat Sheet....746
11.2.5.2 Bulging of a Circular Sheet....749
11.2.6 Pressurized Axisymmetric Thin-Walled Containers....754
11.2.6.1 Thin-Walled Sphere....754
11.2.6.2 Thin-Walled Cylinder....756
11.3 Strain-Rate Dependent Plastic Behavior — Application to Superplasticity....760
11.3.1 Neck-Free Elongations....763
11.3.2 Limit Tensile Strains for a Bar of Strain-Rate-Dependent Material....764
11.3.3 Forming Time for a Bulged Circular Sheet of Rate-Dependent Material....765
11.4 Creep Deformation....767
11.4.1 Creep Testing and Data....767
11.4.2 Empirical Creep Equation of State....771
11.4.2.1 Uniaxial Behavior....771
11.4.2.2 Multiaxial Behavior....773
11.4.3 Steady Creep of Beams under Bending....773
11.4.4 Steady Creep of Thin-Walled Pressurized Cylinders....777
11.4.5 Steady Creep of Thick-Walled Pressurized Cylinders....780
11.4.6 Steady Creep in Rotating Disks....785
11.4.7 Steady Creep of Circular Shafts Under Torsion....785
11.4.8 Creep Buckling of Columns....788
11.4.9 The Reference Stress Method....791
11.4.10 Stress Relaxation....796
11.4.11 Creep under Variable Loading: Time Hardening vs. Strain Hardening....798
11.4.12 Creep Rupture and Damage Concept....802
11.4.12.1 Ductile Creep Rupture under Uniaxial Stress....803
11.4.12.2 Creep Damage Concept....805
11.4.12.3 Brittle Creep Rupture under Uniaxial Stress....807
Problems....809
References....816

Chapter 12 Some Elastic–Plastic Problems

12.1 Introduction....819
12.2 Plane Strain Bending of Plates....820
12.2.1 Elastic State....820
12.2.2 Initial Yielding....822
12.2.3 Partial and Full Yielding — Shape Factor....822
12.2.4 Unloading: Residual Stresses and Springback....825
12.3 Plane Stress Bending of Beams....828
12.3.1 Initial Yielding, Full Yielding, and Springback....828
12.3.2 Combined Bending and Tension....831
12.3.2.1 Elastic State....831
12.3.2.2 Elastic–Plastic State....83
12.3.2.3 Unloading and Residual Stresses....
12.3.3 Plastic Collapse of Beams — Plastic Hinges....

12.3.4 Deflection and Shear Stresses 837
12.3.5 Effect of Strain Hardening 839
12.4 Biaxial Bending of Flat Plates 843
12.4.1 Rectangular Plates 843
12.4.2 Circular Plates 847
12.5 Bending of Circularly Curved Beams 849
12.6 Buckling of Bars Under Axial Compression 853
12.6.1 Tangent Modulus Formula 854
12.6.2 Double-Modulus Formula 854
12.7 Bars Subjected to Torsion 856
12.7.1 Circular Solid and Hollow Sections 857
12.7.1.1 Solid Circular Section 857
12.7.1.2 Hollow Circular Section 858
12.7.2 Thin-Walled Tubular Sections 861
12.7.2.1 Uniform Wall Thickness 861
12.7.2.2 Nonuniform Wall Thickness 861
12.7.3 Combined Torsion and Tension 861
12.7.3.1 Solid Circular Section 861
12.7.3.2 Hollow Circular Sections 862
12.7.3.3 Thin-Walled Cylinder of Uniform Thickness 863
12.7.3.4 Remarks 865
12.8 Pressurized Thick-Walled Cylinders 865
12.8.1 Initial and Partial Yielding 866
12.8.1.1 Stresses in the Elastic Region $r_p \le r \le r_o$ 870
12.8.1.2 Stresses in the Plastic Region $r_i \le r \le r_p$ 870
12.8.1.3 Radial Displacements in Partially Yielded Cylinders 873
12.8.2 Full Yielding and Plastic Expansion Process 874
12.8.2.1 Full Yielding 874
12.8.2.2 Plastic Expansion Process 875
12.8.3 Residual Stresses — The Autofrettage Process 880
12.8.4 Effect of Strain Hardening and Temperature Gradient 884
12.8.4.1 Strain Hardening 884
12.8.4.2 Radial Temperature Gradient 885
12.9 Annular Rotating Disks of Uniform Thickness 886
12.9.1 Initial Yielding 886
12.9.1.1 Tresca Yield Criterion 886
12.9.1.2 von Mises Yield Criterion 887
12.9.2 Partial and full Yielding 887
12.9.2.1 Stresses in the Plastic Region $r_i \le r \le r_p$ 887
12.9.2.2 Stress in the Elastic Region $r_p \le r \le r_o$ 888
12.9.3 Residual Stresses at Stoppage 890
12.9.4 Shrink-Fitted Disks 890
12.10 Solid Rotating Disks of Uniform Thickness 892
12.10.1 Initial, Partial, and Full Yielding 892
12.10.1.1 Initial Yielding 892
12.10.1.2 Partial Yielding 893
12.10.1.3 Full Yielding 893
12.10.2 Residual Stresses at Stoppage 895

12.11 Shakedown Limit: Application to Pressurized Cylinders 896
Problems 899
References 904

Index 905

Dedication

to Samia

A.R. Ragab

to Fatma and Karam

S.E. Bayoumi

1 Analysis of Stress

The concept of stress is developed by considering the equilibrium of a solid subjected to general loading. It is shown that six independent stress components define the state of stress at a point. The equations of equilibrium are obtained in terms of the variation of stress from one point to another in Cartesian and cylindrical polar coordinates. The stress components at a point change when changing the reference coordinates; the stress components and the directions of the axes are related by the stress transformation law. On one set of orthogonal planes through a point, the shear stresses vanish and the state of stress is defined by three principal stress components. Expressions for the maximum shear stress, octahedral shear stress, and the hydrostatic and deviatoric stresses are obtained. These stress expressions find application in the analysis of plastic deformation.

1.1 RIGID AND DEFORMABLE BODIES

Mechanics is concerned with the motion of particles or bodies under the action of a system of forces. A particle has a mass that is concentrated in one point whereas a body is an aggregate of constrained particles that are geometrically distributed within a certain boundary.

If the particles within a body are constrained such that the distance between any two in the aggregate remains unchanged, the body form will remain unaltered and this defines a rigid body. If the distance between any two particles changes, there will be a change of form and the body is deformable.

The displacement of any particle in a body may be considered to be the displacement resulting from the motion of the body as a whole, called rigid body displacement, and the displacement due to change of form, called deformation. The two displacements are independent of each other.

All bodies are essentially deformable. A rigid body is, in fact, an idealized concept adopted when studying the mechanics of rigid bodies. The subject of mechanics of deformable bodies is concerned with determining deformations set up in a body as a result of a system of applied forces. According to their behavior under load, deformable bodies are classified into solids, fluids, or semifluids. This book is concerned with solids.

1.2 BODY FORCES AND SURFACE TRACTIONS

At this stage, distinction may be made between two systems of forces which may act on a solid. The first one represents forces acting on the volume of the solid, called body forces, and is denoted per unit volume. Examples of these forces are gravity forces, magnetic forces, and centrifugal forces. The second system of forces acts on the external surface (boundary) of the body. Surface forces — or surface tractions — are expressed per unit area of the boundary surface of the solid. These forces arise as a result of the interaction of the solid under consideration with other solids or fluids surrounding it. Reactive forces, contact forces, fluid pressure, and friction are examples of surface forces. It is explicitly assumed that under the action of both body and surface forces, the solid is in equilibrium.

1.3 CONCEPT OF STRESS AND STRAIN

Consider a straight rod of uniform cross-sectional area A_0 and length L_0. Let the rod cross section be subjected to a tensile force P. From experience, the rod will stretch by an amount that increases as the force is increased. The force P vs. extension ΔL may be plotted as in plot a of Figure 1.1. The area under this curve represents the work done during deformation.

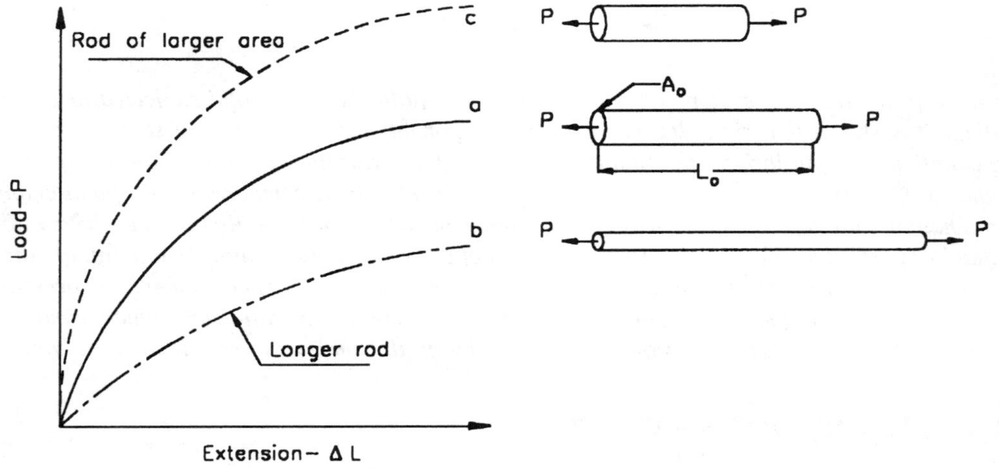

FIGURE 1.1 Load–extension curves for bars of different cross-sectional areas and lengths.

Now consider other rods of the same material but having different dimensions and plot the force–extension relation. Different curves will be obtained as in plots b and c of Figure1.1 for rods of larger cross-sectional area or longer length, respectively. This means that there are unlimited number of P–ΔL curves according to rod dimensions. Now if these curves are replotted in terms of PA_0 and $\Delta L/L_0$, a one single curve for all rods of the same material is shown to be obtained. This curve expresses the mechanical characteristics of the rod material, and the area under this curve represents the work done per unit volume in uniaxial tension.

The quantity P/A_0 is the stress σ, and the ratio $\Delta L/L_0$ is the strain ε. The idea of using σ and ε has provided a unifying concept to describe the mechanical behavior of solids, which corresponds to using pressure p and specific volume v to describe the behavior of gases in thermodynamics. Note that the original bar dimensions A_0 and L_0 have been arbitrarily chosen in defining σ and ε Another choice based on the current cross-sectional area A and elongated length L could have been made. For small changes in rod geometry, i.e., small deformations occurring due to force application, both choices will give virtually the same results. For large deformations other definitions of σ and ε are used.

In general, a structural member or a machine element will not possess uniform geometry of shape or size proportions as the rod dealt with above. The applied forces will also be more complex than a simple tensile force. For such general cases, the concepts of stress and strain at a point are developed in the following sections of this chapter for stress and in Chapter 2 for strain.

1.4 THE STATE OF STRESS AT A POINT

The study of the state of stress at a point of a loaded solid starts by considering its free-body diagram indicating all forces acting on its boundaries and maintaining its static equilibrium* as illustrated in Figure 1.2. In order to investigate the internal forces set up within this body at a point

* All bodies can be thought of as being in a state of static equilibrium, by applying the d'Alembert principle.

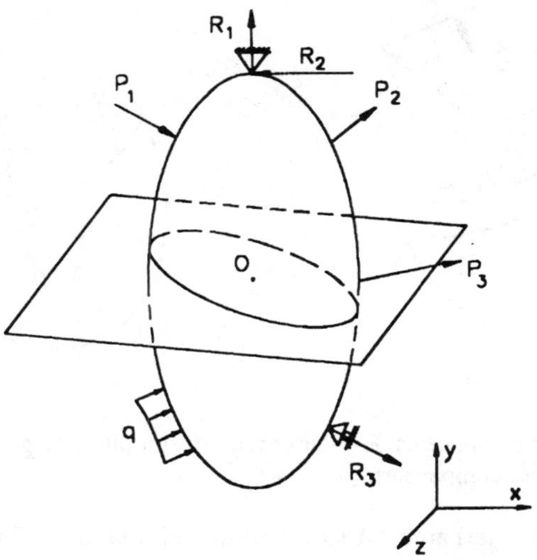

FIGURE 1.2 External loads acting on a solid in static equilibrium.

O, which balance the externally applied forces, an arbitrary cutting plane is imagined to pass through this point as shown in Figure 1.2.

The result of this is to separate the body into two distinct parts. If the body as a whole is in equilibrium, any part of it must also be in equilibrium. For the two parts of the body, Figure 1.2, some internal forces must act at the two cut sections to maintain equilibrium as shown in the free-body diagrams of Figure 1.3. The internal forces acting on the first cut section are equal in magnitude and opposite in direction to those acting on the second section. This is obtained by considering each force on one section to be an action and the corresponding force on the other section to be its reaction. In general, the internal forces F_1, F_2, F_3, ... etc. acting on the cut section are of varying magnitudes and directions.

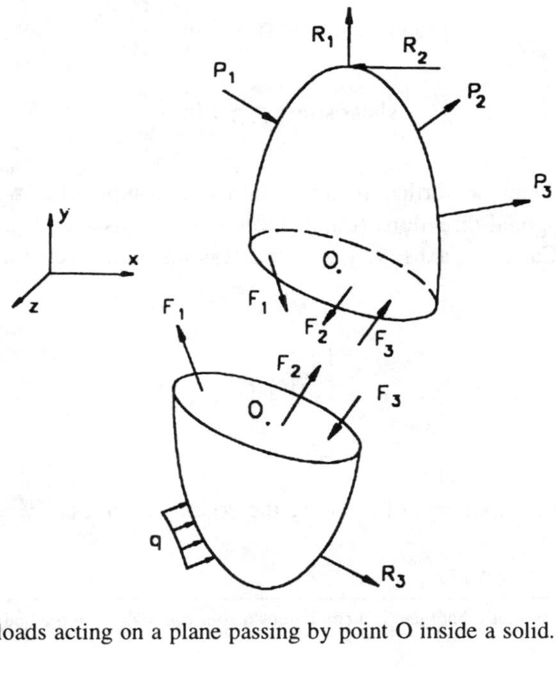

FIGURE 1.3 Internal loads acting on a plane passing by point O inside a solid.

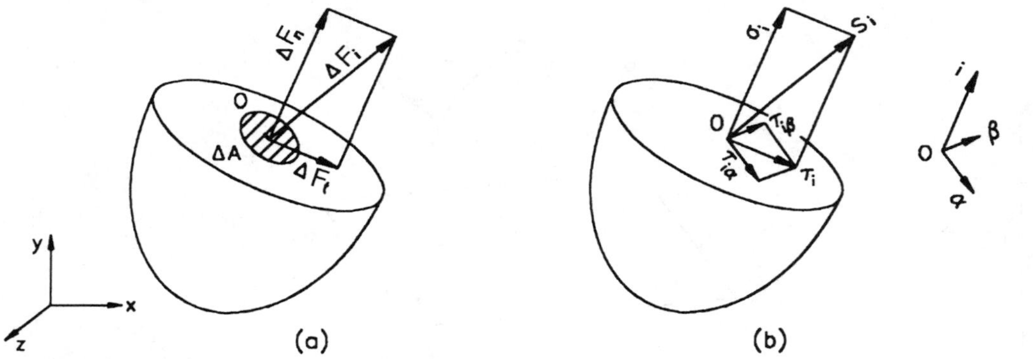

FIGURE 1.4 Stress vector components at point O related to a plane of outward normal *i* through O, (a) load components and (b) stress components.

Now, by considering a small area ΔA surrounding point O as in Figure 1.4a, the internal force vector acting on ΔA will be ΔF_i. The cutting plane through has an outward normal *i*. The internal force ΔF_i acts along a direction which is not necessarily along *i*. Applying the concept of stress to the area ΔA defines the stress vector as

$$\operatorname*{Lim}_{\Delta A \to 0} \frac{\Delta \mathbf{F}_i}{\Delta A} = \mathbf{S}_i$$

The vector $\mathbf{S}_i$ is the stress resultant vector acting at point O lying in a plane with an outward normal *i* as shown in Figure 1.4b

The internal force $\Delta \mathbf{F}_i$ may be resolved to two components: a normal component $\Delta \mathbf{F}_n$ and an in-plane one $\Delta \mathbf{F}_t$. For a continuum,* taking the limits will define the normal and shear stress components as:

$$\text{normal stress } \sigma_i = \operatorname*{Lim}_{\Delta A \to 0} \frac{\Delta \mathbf{F}_n}{\Delta A}$$

$$\text{shear stress } \tau_i = \operatorname*{Lim}_{\Delta A \to 0} \frac{\Delta \mathbf{F}_t}{\Delta A}$$

The shear stress τ_i can be further resolved into two components $\tau_{i\alpha}$ and $\tau_{i\beta}$ along the two arbitrary in-plane orthogonal directions α and β, respectively, as shown in Figure 1.4b.

With reference to Cartesian axes (x, y, z), the stress resultant vector may be expressed by the vector:

$$\mathbf{S}_i = \begin{Bmatrix} S_x \\ S_y \\ S_z \end{Bmatrix} = \{S_x S_y S_z\}^T \tag{1.1a}$$

where S_x, S_y, S_z are the components of $\mathbf{S}_i$ along the coordinate axes x, y, and z, respectively. Also $\mathbf{S}_i$ may be represented by

* The continuum hypothesis assumes that there is a continuous matter everywhere in the body and neglects any microscale effects.

$$\mathbf{S}_i = \{\sigma_i \tau_{i\alpha} \tau_{i\beta}\}^T \tag{1.1b}$$

The following relations thus hold for $\mathbf{S}_i$ and its components:

$$\begin{aligned} \mathbf{S}_i^2 &= S_x^2 + S_y^2 + S_z^2 \\ &= \sigma_i^2 + \tau_i^2 \\ &= \sigma_i^2 + \left(\tau_{i\alpha}^2 + \tau_{i\beta}^2\right) \end{aligned} \tag{1.2}$$

Obviously, at any other point lying in the same cut plane, the stresses σ and τ will be different as the resultant stress vector **S** takes different values from point to point. Also at the same point O, the stresses σ, τ depend on the orientation of the cutting plane.

Example 1.1:

Find the state of stress at point O in a circular rod of 335 mm diameter as shown in Figure 1.5 by considering two cut planes:

a. *Plane* A: *perpendicular to rod axis*
b. *Plane* B: *making an angle of 30° to rod axis*

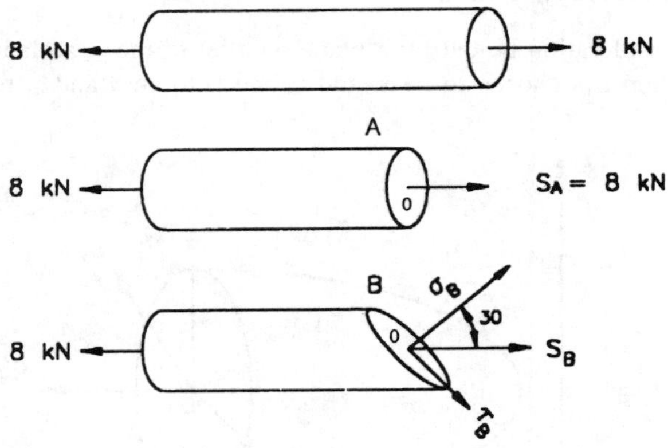

FIGURE 1.5 Example 1.1.

Solution:

a. After cutting by plane A, equilibrium requires:

$$\sigma_x = \frac{S_A}{A_A} = \frac{8000}{\pi\,(0.035)^2\!/\!4} = 8.319 \text{ MPa}$$

where σ_x is considered to be uniformly distributed over the cut plane A.

b. Considering equilibrium after cutting by plane B inclined 30° to the x axis:

$$S_B A_B = 8 \text{ kN}$$

where S_B is the stress vector acting on the area A_B of the cut plane B. The area of the ellipse A_B is

$$A_B = \frac{\pi}{4}(0.035)\,(0.035/\cos\ 30°) = 0.00111\ \text{m}^2$$

Hence,

$$S_B = \frac{8000}{0.00111} = 7.205\ \text{MPa}$$

Resolving into normal and shear stresses σ_B and τ_B, respectively, and assuming uniform stress distribution gives:

$$\sigma_B = S_B \cos\ 30° = 6.239\ \text{MPa}$$

$$\tau_B = S_B sin\ 30° = \sqrt{S_B^2 - \sigma_B^2} = 3.603\ \text{MPa}$$

This example, although simple, illustrates the following:

1. To define the stress resultant at a point, one should specify the orientation of the plane on which it acts.
2. The description of the state of stress at a point varies with the choice of coordinates.

1.5 CARTESIAN STRESS COMPONENTS

Let the body under consideration be cut by a plane parallel to the y–z plane as in Figure 1.6. In other words, the direction i is chosen to be x, and α and β to be y and z, respectively. In such a

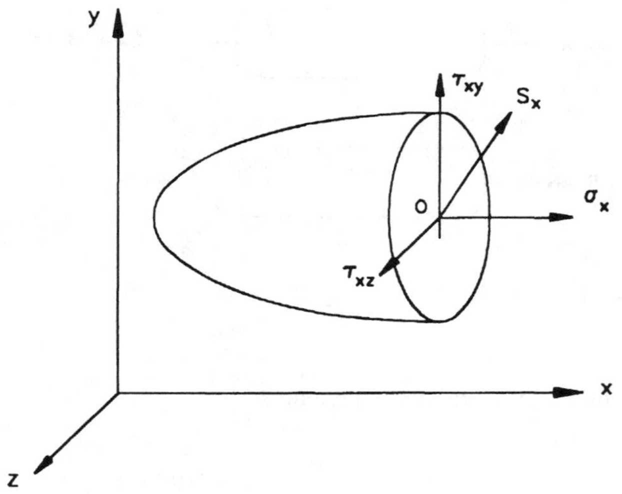

FIGURE 1.6 Stress vector components at point O related to a plane normal to the x-direction.

case, the stress vector S_x at point O may be used to define the following stress components according to Equation 1.1b as

$$\mathbf{S}_x - \{\sigma_{xx}\tau_{xy}\tau_{xz}\}^T \tag{1.3}$$

In the above stress designation, the first subscript x indicates the normal to the plane on which the stresses are acting, and the second subscript x, y, or z indicates the direction of the stress component. For instance, τ_{xz} designates the shear stress at point O acting on a plane perpendicular to the x-axis and directed in the z-direction. Since for normal stress the two subscripts are always identical, only one will be used, i.e., σ_x instead of σ_{xx}.

At any point O in the body of Figure 1.6, consider three mutually perpendicular cutting planes, the x-plane, y-plane, and the z-plane, to isolate an infinitesimally small cube element whose centroid is the point O, as shown in Figure 1.7. The stress components similar to Equation 1.3 are written as

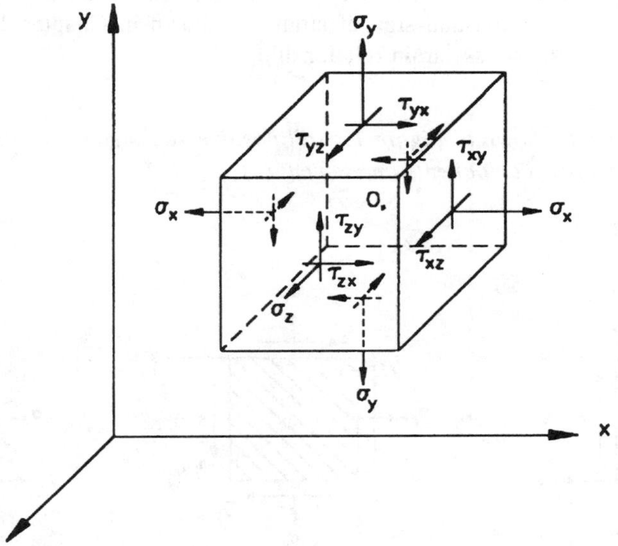

FIGURE 1.7 Stress vectors components at point O related to planes normal to the x, y, and z directions, respectively.

$$S_x = \{\sigma_x \tau_{xy} \tau_{xz}\}^T$$
$$S_y = \{\tau_{yx} \sigma_y \tau_{yz}\}^T \tag{1.4a}$$
$$S_z = \{\tau_{zx} \tau_{zy} \sigma_z\}^T$$

This may be expressed in a matrix form as

$$[\sigma] = \begin{bmatrix} S_x \\ S_y \\ S_z \end{bmatrix} = \begin{bmatrix} \sigma_x & \tau_{xy} & \tau_{xz} \\ \tau_{yz} & \sigma_y & \tau_{yz} \\ \tau_{zx} & \tau_{zy} & \sigma_z \end{bmatrix} \tag{1.4b}$$

where $[\sigma]$ designates the 3×3 stress matrix or stress tensor,* involving nine stress components; three normal stresses and six shear stresses. It will be shown in Section 1.7 that $[\sigma]$ is a symmetric tensor so that $\tau_{xy} = \tau_{yx}$, $\tau_{xz} = \tau_{zx}$, and $\tau_{yz} = \tau_{zy}$.

* In indicial notation, the stress tensor is expressed by $[\sigma_{ij}]$, where $i, j = 1, 2, 3$. Hence, for instance, for $i = j$, σ_{11}, σ_{22}, and σ_{33} correspond to normal stress σ_{xx}, σ_{yy}, σ_{zz} or, simply, σ_x, σ_y, and σ_z, respectively, when $i \neq j$; σ_{12}, σ_{13}, and σ_{23} designate τ_{xy}, τ_{xz}, and τ_{yz}, respectively.

On any plane, three components of stress are acting: one normal and two shear stresses. The normal stress components σ_x, σ_y, σ_z are taken to be positive when having an outward direction, i.e., when they produce tension. The shear stress components are taken to be positive when their direction has the same sense of the normal stress acting on their plane. For example, if a positive normal stress is in the negative direction of one of the axes, then a positive shear component will have the negative direction of the other two axes. Following this convention, the stresses shown in Figure 1.7 are all positive. It may be noted that the sign convention adopted for the stresses is based, in fact, on the definition and sign of strain, as shown in Chapter 2, which constitutes a necessary condition for the stress–strain relationship.

Example 1.2:

Indicate on the elements shown in Figure 1.8: all positive stress components on elements (a) and (b) and all negative stress components on element (c).

Solution:

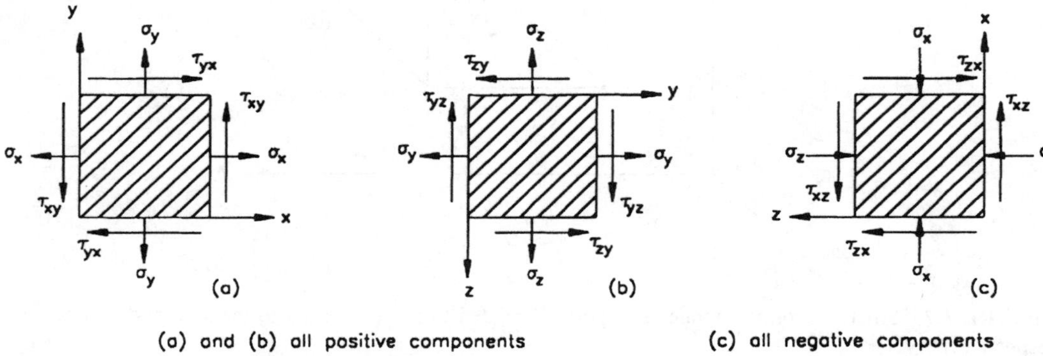

FIGURE 1.8 Example 1.2 and solution: (a) and (b) all positive components; (c) all negative components.

1.6 SOME SPECIAL STATES OF STRESS

There are states of stress that often occur in several engineering applications. Some of these states are given below.

1.6.1 Plane stress

A state of plane stress exists on a plane parallel to *x–y*, when $\sigma_z = \tau_{zx} = \tau_{zy} = 0$ and, hence, is determined by three components namely; $(\sigma_x, \sigma_y, \tau_{xy})$.

For cylindrical polar coordinates (r, θ, z), plane stress, exists when $\sigma_z = \tau_{zr} = \tau_{z\theta} = 0$ and is determined by $(\sigma_r, \sigma_\theta, \tau_{r\theta})$. The stress tensor is

$$[\sigma] = \begin{bmatrix} \sigma_x & \tau_{xy} & 0 \\ \tau_{yx} & \sigma_y & 0 \\ 0 & 0 & 0 \end{bmatrix} \text{ or } \begin{bmatrix} \sigma_r & \tau_{r\theta} & 0 \\ \tau_{\theta r} & \sigma_\theta & 0 \\ 0 & 0 & 0 \end{bmatrix} \tag{1.5a}$$

Thin plates and sheets subjected to loads acting in the middle planes (*x–y*) are typical examples of plane stress states.

1.6.2 Plane Strain

In some applications, a body is constrained against lateral deformation in a direction normal to the plane of loading (x–y). This superimposes a normal stress σ_z on the state of plane stress to give the stress tensor:

$$[\sigma]=\begin{bmatrix}\sigma_x & \tau_{xy} & 0\\ \tau_{yx} & \sigma_y & 0\\ 0 & 0 & \sigma_z\end{bmatrix} \tag{1.5b}$$

which is known as a state of plane strain.

1.6.3 Axial symmetry

If a geometrically axisymmetric body is loaded symmetrically, the state of stress is simplified. Employing polar cylindrical coordinates (r, θ, z) where symmetry is around the z-axis, then, because of this complete axial symmetry,

$$\tau_{r\theta} = \tau_{\theta z} = 0$$

The stress tensor is given by:

$$[\sigma]=\begin{bmatrix}\sigma_r & 0 & \tau_{rz}\\ 0 & \sigma_\theta & 0\\ \tau_{zr} & 0 & \sigma_z\end{bmatrix} \tag{1.5c}$$

A cylinder under internal pressure or a rotating circular disk are typical cases of axial symmetry.

1.6.4 Free Torsion

Free torsion of a rod having its axis along the z-axis, produces a state of stress defined by

$$\sigma_x = \sigma_y = \sigma_z = \tau_{xy} = 0$$

The stress tensor is thus given by

$$[\sigma]=\begin{bmatrix}0 & 0 & \tau_{zx}\\ 0 & 0 & \tau_{zy}\\ \tau_{xz} & \tau_{yz} & 0\end{bmatrix} \tag{1.5d}$$

In the case of a rod with a circular cross section, the only nonzero stress in cylindrical polar coordinates is $\tau_{z\theta}$.

1.7 STRESS EQUATIONS OF EQUILIBRIUM

So far the state of stress has been considered at a point in the solid. In general, the state of stress varies from one point to another. If the solid as a whole is in a state of equilibrium, then an infinitesimal element, such as that in Figure 1.7, will also be in equilibrium. The conditions of static equilibrium, i.e., zero force and moment resultants with respect to the three coordinate directions, must be satisfied for the stress components acting on the six faces of the element.

1.7.1 Cartesian Coordinates

Figure 1.9 shows the stresses acting on a parallelepiped having sides *dx*, *dy*, and *dz*. Consider the stress component σ_x undergoing a change from face *x* of the element to the other face $x + dx$. If

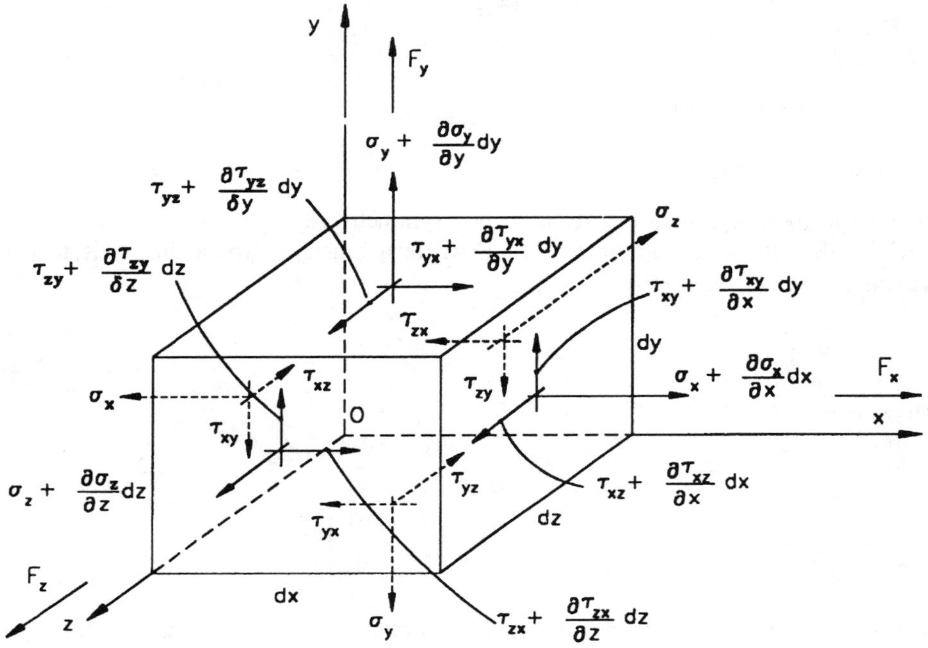

FIGURE 1.9 Static equilibrium of an infinitesimal element subjected to a general state of stress in Cartesian coordinates.

the stress and its derivatives in space are continuous, this variation is expressed by

$$\sigma_{(x+dx)} = \sigma_x + \left(\frac{\partial \sigma_x}{\partial x}\right)dx + \frac{1}{2!}\left(\frac{\partial^2 \sigma_x}{\partial x^2}\right)dx^2 + \frac{1}{3!}\cdots \text{etc.}$$

As the volume of the element shrinks to zero, higher orders of *dx* are neglected and hence,

$$\sigma_{(x+dx)} = \sigma_x + \left(\frac{\partial \sigma_x}{\partial x}\right)dx$$

Consequently, the stresses acting on the element are as follows:
For the face of ordinate *x*:

$$\sigma_x, \quad \tau_{xy}, \quad \tau_{xz}$$

For the face of ordinate $x + dx$:

$$\left(\sigma_x + \frac{\partial \sigma_x}{\partial x} dx\right), \quad \left(\tau_{xy} + \frac{\partial \tau_{xy}}{\partial x} dx\right), \quad \left(\tau_{xz} + \frac{\partial \tau_{xz}}{\partial x} dx\right)$$

For the face of ordinate y:

$$\sigma_y, \quad \tau_{yx}, \quad \tau_{yz}$$

For the face of ordinate $y + dy$:

$$\left(\sigma_y + \frac{\partial \sigma_y}{\partial y} dy\right), \quad \left(\tau_{yx} + \frac{\partial \tau_{yx}}{\partial y} dy\right), \quad \left(\tau_{yz} + \frac{\partial \tau_{yz}}{\partial y} dy\right)$$

For the face of ordinate z:

$$\sigma_z, \quad \tau_{zx}, \quad \tau_{zy}$$

For the face of ordinate $z + dz$:

$$\left(\sigma_z + \frac{\partial \sigma_z}{\partial z} dz\right), \quad \left(\tau_{zx} + \frac{\partial \tau_{zx}}{\partial z} dz\right), \quad \left(\tau_{zy} + \frac{\partial \tau_{zx}}{\partial z} dz\right)$$

The components of body force per unit volume are F_x, F_y, and F_z in the x-, y-, z-directions, respectively.

Consider now the static equilibrium* condition of the element. In the x-direction the sum of the forces is equal to zero.

$$\left(\sigma_x + \frac{\partial \sigma_x}{\partial x} dx\right) dydz - \sigma_x dydz + \left(\tau_{yx} + \frac{\partial \tau_{yx}}{\partial y} dy\right) dzdx - \tau_{yx} dzdx$$

$$+\left(\tau_{yx} + \frac{\partial \tau_{yx}}{\partial y} dy\right) dzdx - \tau_{yx} dzdx + \left(\tau_{zx} + \frac{\partial \tau_{zx}}{\partial z} dz\right) dxdy - \tau_{zx} dxdy + F_x dxdydz = 0$$

Reducing and dividing by ($dxdydz$) gives

$$\frac{\partial \sigma_x}{\partial x} + \frac{\partial \tau_{yx}}{\partial y} + \frac{\partial \tau_{zx}}{\partial z} + F_x = 0 \tag{1.6a}$$

Similarly, equating forces in the y- and z-directions, respectively, to zero gives two more equations, namely,

$$\frac{\partial \sigma_y}{\partial y} + \frac{\partial \tau_{zy}}{\partial z} + \frac{\partial \tau_{xy}}{\partial x} + F_y = 0 \tag{1.6b}$$

and

* As mentioned before, this can be readily extended to take into account dynamic effects by applying d'Alembert's principle.

$$\frac{\partial \sigma_z}{\partial z} + \frac{\partial \tau_{xz}}{\partial x} + \frac{\partial \tau_{yz}}{\partial y} + F_z = 0 \tag{1.6c}$$

The three equations (Equations 1.6) are the stress differential equations of equilibrium in Cartesian coordinates. Now, the moment equilibrium should also be satisfied. Considering the moments of forces about the *z*-axis.

For the two faces normal to the *x*-axis:

$$\left(\sigma_x + \frac{\partial \sigma_x}{\partial x} dx - \sigma_x\right)(dydz)\frac{dy}{2} - \left(\tau_{yx} + \frac{\partial \tau_{xy}}{\partial x}\right)(dydz)dx$$

For the two faces normal to the *y*-axis:

$$-\left(\sigma_y + \frac{\partial \sigma_y}{\partial y} dy - \sigma_y\right)(dxdz)\frac{dx}{2} + \left(\tau_{yx} + \frac{\partial \tau_{yx}}{\partial y}\right)(dxdz)dy$$

For the planes normal to the *z*-axis:

$$\left(\tau_{zx} + \frac{\partial \tau_{zx}}{\partial z} dz - \tau_{zx}\right)(dxdy)\frac{dy}{2} - \left(\tau_{zy} + \frac{\partial \tau_{zy}}{\partial z} - \tau_{zy}\right)(dxdy)\frac{dx}{2}$$

Equating the sum of these moments to zero, dividing by (*dxdydz*) and neglecting terms containing *dx* or *dy* as the element converges to a point gives

$$\tau_{xy} = \tau_{yx} \tag{1.7a}$$

Similarly, by summing moments about the *y*- and *x*-directions, respectively, and equating to zero results in

$$\tau_{xz} = \tau_{zx} \tag{1.7b}$$

and

$$\tau_{yz} = \tau_{zx} \tag{1.7c}$$

Equations 1.7 state that there are only six independent components of stress at a point. The stress matrix of Equation 1.4b is thus a symmetric tensor. Note that there are only three equilibrium equations in six stress components. The problem is, thus, statically indeterminate and the equilibrium equations are not sufficient to obtain a solution.

Example 1.3:

The state of stress in a solid is given by

$$\sigma_x = C_1 x^2 yz, \qquad \sigma_y = C_2 xyz^3,$$

$$\sigma_z = 2\left(x^3 + y^3 - 2yz\right), \quad \tau_{xy} = -3xy^2 z,$$

$$\tau_{yz} = C_3\left(6y^2 - 5xz^2\right)z^2 + 8\left(x^2 + y^2\right), \textit{ and}$$

$$\tau_{zx} = -3xyz^2$$

Find the value of the constants C_1, C_2, *and* C_3.

Solution:

The state of stress must satisfy the equilibrium equations (Equations 1.6). Substitution in these equations gives, receptively,

Equation 1.6a: $2C_1xyz - 6xyz - 6xyz = 0$; hence, $C_1 = 6$
Equation 1.6b: $C_2xz^3 + C_3(12y^2z - 20xz^3) - 3y^2z = 0$; hence, $C_2 - 20C_3 = 0$ and $12C_3 - 3 = 0$ or $C_2 = 5$ and $C_3 = {}^1/_4$
Equation 1.6c: $-4y - 3yz^2 + 12C_3yz^2 + 16C_3y = 0$, which is satisfied for $C_3 = {}^1/_4$

Example 1.4:

The strength of materials solution gives the stresses in an elastic cantilever beam under end load P, *as shown in Figure 1.10a by the expressions:*

$$\sigma_x = \frac{My}{I_z}, \quad \sigma_y = \sigma_z = \tau_{xz} = \tau_{yz} = O,$$

$$\tau_{xy} = -Q\frac{\left(r_o^2 - y^2\right)}{3I_z} \quad \textit{for circular cross section, or}$$

$$\tau_{xy} = -Q\frac{\left(r_o^2 - y^2\right)}{2I_z} \quad \textit{for rectangular cross section,}$$

where M *is the bending moment,* Q *is the shearing force, and* I_z *is the second moment of area of the cross section about the* z *axis. Which of these expressions satisfy the equilibrium equations? And if they do not, why? Neglect body forces.*

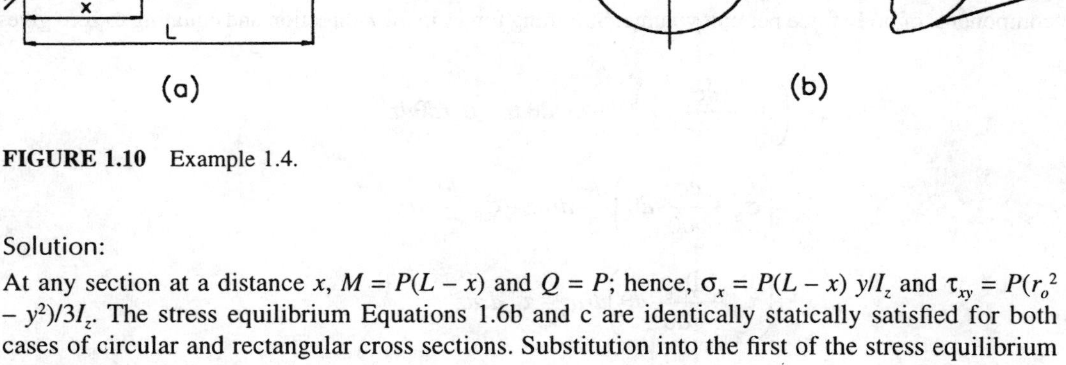

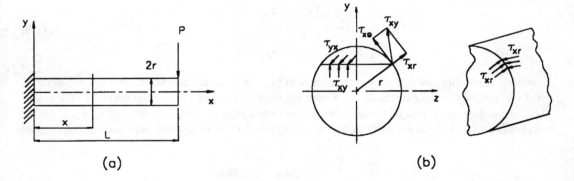

FIGURE 1.10 Example 1.4.

Solution:

At any section at a distance x, $M = P(L - x)$ and $Q = P$; hence, $\sigma_x = P(L - x)\, y/I_z$ and $\tau_{xy} = P(r_o^2 - y^2)/3I_z$. The stress equilibrium Equations 1.6b and c are identically statically satisfied for both cases of circular and rectangular cross sections. Substitution into the first of the stress equilibrium equations (1.6a) gives

a. Circular cross section:

$$\frac{\partial \sigma_x}{\partial x} + \frac{\partial \tau_{yx}}{\partial y} + \frac{\partial \tau_{zx}}{\partial z} = 0$$

$$\frac{-Py}{I_z} + \frac{2Py}{3I_z} + 0 \neq 0$$

b. Rectangular cross section:

$$\frac{-Py}{I_z} + \frac{2Py}{3I_z} + 0 = 0$$

The given stress expressions for the circular cross section do not satisfy equilibrium because they suffer from physical inconsistency. The shear stress τ_{xy} at any level y acts vertically around the periphery. Hence, it can be resolved into two components $\tau_{x\theta}$ and τ_{xr} as shown in Figure 1.10b thus resulting in a non-zero value normal to the peripheral surface of the beam. This is in contradiction with the condition of a free surface (i.e., with no load acting on the surface). Although this solution is fairly accurate for engineering analysis, a more rigorous solution based on a shear stress distribution tangent to the periphery can be obtained.[1] Obviously, the shear stress distribution given for the rectangular cross section does not suffer from this inconsistency since the direction of shear stresses is parallel to the y-axis, which is the direction of free surface.

1.7.2 Cylindrical Polar Coordinates

The expressions of the stress equations of equilibrium depend on the coordinate axes. For cylindrical polar coordinates (r, θ, z); θ is measured round the z-axis from the positive x-axis in a right-handed sense, the element will have sides dr, dz, $rd\theta$, and $(r + dr)d\theta$ as shown in Figure 1.11. The stress tensor is thus given by

$$[\sigma] = \begin{bmatrix} \sigma_r & \tau_{r\theta} & \tau_{rz} \\ \tau_{\theta r} & \sigma_\theta & \tau_{\theta z} \\ \tau_{zr} & \tau_{z\theta} & \sigma_z \end{bmatrix} \tag{1.8}$$

Considering equilibrium in the r-direction, it is to be noted that the stresses σ_θ and $(\sigma_\theta + \partial\sigma_\theta/\sigma_\theta d\theta)$ have components in the r-direction since they are each at an angle $d\theta/2$ to the normal to the r-axis. Also, the difference in area of the faces normal to r must be taken into account. Let F_r, F_θ, F_z be the components of body force per unit volume. Summing forces in the r-direction and equating to zero gives

$$\left(\sigma_r + \frac{\partial \sigma_r}{\partial r} dr\right)(r + dr)d\theta dz - \sigma_r r d\theta dz$$

$$-\left(\sigma_\theta + \frac{\partial \sigma_\theta}{\partial \theta} d\theta\right)\frac{d\theta}{2} drdz - \sigma_\theta \frac{d\theta}{2} drdz$$

$$+\left(\tau_{\theta r} + \frac{\partial \tau_{\theta r}}{\partial \theta} d\theta\right) drdz - \tau_{\theta r} drdz$$

$$+\left(\tau_{zr} + \frac{\partial \tau_{zr}}{z} dz\right) drrd\theta - \tau_{zr} drrd\theta + F_r drrd\theta dz = 0$$

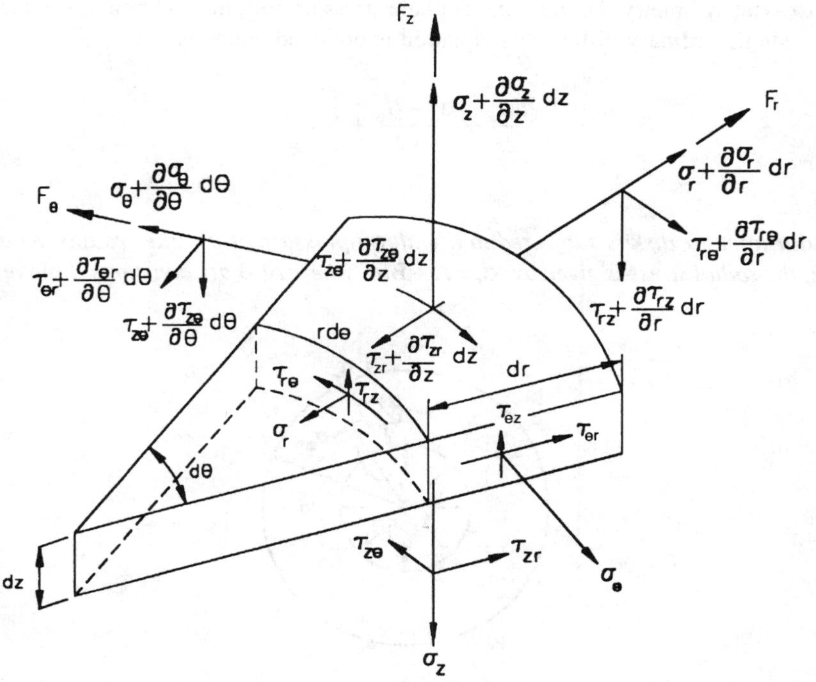

FIGURE 1.11 Static equilibrium of an infinitesimal element subjected to a general state of stress in cylindrical polar coordinates.

Reducing terms and dividing by $dr d\theta dz$ yields

$$\frac{\partial \sigma_r}{\partial r} + \frac{\partial \tau_{zr}}{\partial z} + \frac{1}{r}\frac{\partial \tau_{r\theta}}{\partial \theta} + \frac{\sigma_r - \sigma_\theta}{r} + F_r = 0 \tag{1.9a}$$

Similarly, equilibrium in the θ- and z-directions gives, respectively,

$$\frac{\partial \tau_{r\theta}}{\partial r} + \frac{1}{r}\frac{\partial \sigma_\theta}{\partial \theta} + \frac{\partial \tau_{\theta z}}{\partial z} + \frac{2\tau_{r\theta}}{r} + F_\theta = 0 \tag{1.9b}$$

and

$$\frac{\partial \tau_{rz}}{\partial r} + \frac{1}{r}\frac{\partial \sigma_{z\theta}}{\partial \theta} + \frac{\partial \sigma_z}{\partial z} + \frac{\tau_{rz}}{r} + F_z = 0 \tag{1.9c}$$

In most cases axial symmetry exists and stresses are independent of θ; $\tau_{r\theta} = \tau_{\theta z} = 0$. Equations 1.9 thus reduce to

$$\frac{\partial \sigma_r}{\partial r} + \frac{\partial \tau_{zr}}{\partial z} + \frac{\sigma_r - \sigma_\theta}{r} + F_r = 0 \tag{1.10a}$$

and

$$\frac{\partial \tau_{rz}}{\partial r} + \frac{\partial \sigma_z}{\partial z} + \frac{\tau_{rz}}{r} + F_z = 0 \tag{1.10b}$$

Added to axial symmetry if conditions of plane stress prevail, the components in the z-direction vanish and a single ordinary differential equation is obtained, namely,

$$\frac{d\sigma_r}{dr} + \frac{\sigma_r - \sigma_\theta}{r} + F_r = 0 \tag{1.10c}$$

Example 1.5:

A thin circular annular disk is subjected to a uniform pressure at its inner radius. At any radius r*, Figure 1.12, the radial stress is given by:* $\sigma_r = A + B/r^2$*, where* A, B *are constants. Find an expression for* σ_θ.

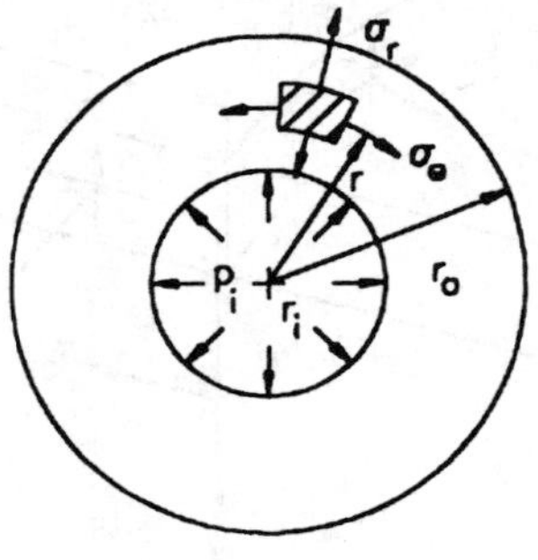

FIGURE 1.12 Example 1.5.

Solution:

This is a case of plane stress with axial symmetry and no body forces. The equations of equilibrium reduce to

$$\frac{d\sigma_r}{dr} + \frac{\sigma_r - \sigma_\theta}{r} = 0$$

Hence,

$$\sigma_\theta = r\frac{d\sigma_r}{dr} + \sigma_r$$

$$= -\frac{2B}{r^2} + A + \frac{B}{r^2} = A - \frac{B}{r^2}$$

Example 1.6:

The stresses in a rotating circular solid disk, as shown in Figure 1.13, are given by $\sigma_r = A - B/r^2$*, and* $\sigma_\theta = A - Cr^2$*, where* A, B *are constants. Show that* $(3B - C) = \rho\omega^2$*, where* ρ *= mass density and* ω *= angular velocity of rotation.*

Solution:

Again this is a case of plane stress with axial symmetry, but with body forces F_r. The equation of equilibrium is

$$\frac{d\sigma_r}{dr} + \frac{\sigma_r - \sigma_\theta}{r} + F_r = 0$$

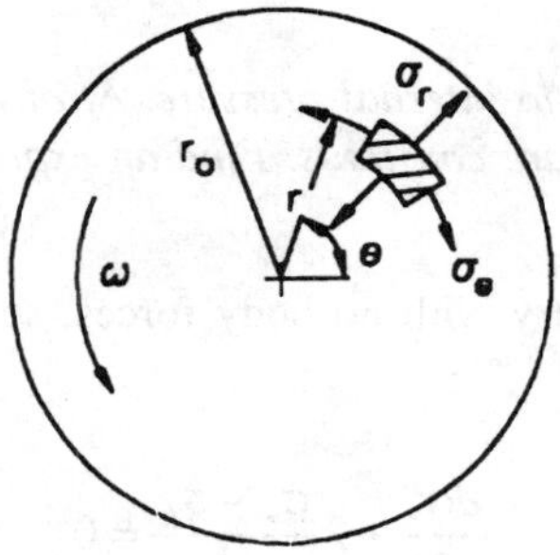

FIGURE 1.13 Example 1.6.

Substituting σ_r and σ_θ in the equation of equilibrium yields

$$-2Br - (B - C)r + F_r = 0$$

Hence,

$$F_r = (3B - C)r = \rho\omega^2 r$$

or

$$3B - C = \rho\omega^2$$

1.7.3 Spherical Polar Coordinates

Spherical coordinates (r, θ, ϕ) may be advantageously employed to analyze some shell problems. The equilibrium equations referred to (r, θ, and ϕ) will be stated here* without derivation and are given by

$$\frac{\partial \sigma_r}{\partial r} + \frac{1}{r\sin\phi}\frac{\partial \tau_{r\theta}}{\partial \theta} + \frac{1}{r}\frac{\partial \tau_{\phi r}}{\partial \phi} + \frac{2\sigma_r - \sigma_\theta - \sigma_\phi + \tau_{\phi r}\cot\phi}{r} + F_r = 0$$

$$\frac{\partial \tau_{r\theta}}{\partial r} + \frac{1}{r\sin\phi}\frac{\partial \sigma_\theta}{\partial \theta} + \frac{1}{r}\frac{\partial \tau_{\theta\phi}}{\partial \phi} + \frac{3\tau_{r\theta} + 2\tau_{\theta\phi}\cot\phi}{r} + F_\theta = 0 \tag{1.11a}$$

$$\frac{\partial \tau_{\theta r}}{\partial r} + \frac{1}{r\sin\phi}\frac{\partial \tau_{\theta\phi}}{\partial \theta} + \frac{1}{r}\frac{\partial \sigma_\phi}{\partial \phi} + \frac{3\tau_{\phi r} + (\sigma_\phi - \sigma_\theta)\cot\phi}{r} + F_\phi = 0$$

where F_r, F_θ, and F_ϕ are the body force components per unit volume along r, θ, ϕ, respectively; θ is the same as in Section 1.7.2 and ϕ is the angle between a radius r and the positive z-axis.

For the simple case of complete spherical symmetry, stresses are independent of θ and ϕ. Shear stresses vanish and $\sigma_\theta = \sigma_\phi$. Equations 1.11 reduce to

$$\frac{d\sigma_r}{dr} + 2\frac{\sigma_r - \sigma_\theta}{r} + F_r = 0 \tag{1.11b}$$

* For a detailed derivation of the stress equilibrium equations in spherical coordinates, see Ford.[2]

Example 1.7:

A thick-walled sphere is subjected to internal pressure. At any radius r *the radial stress is given by:* $\sigma_r = A + B/r^3$, *where* A *and* B *are constants. Find an expression for* σ_θ *and* σ_ϕ.

Solution:

This is a case of spherical symmetry with no body forces, so that $\sigma_\theta = \sigma_\phi$ and the equations of equilibrium reduce to

$$\frac{d\sigma_r}{dr} + 2\frac{\sigma_r - \sigma_\theta}{r} = 0$$

Hence,

$$\sigma_\theta = \sigma_\phi = \frac{r}{2}\frac{d\sigma_r}{dr} + \sigma_r$$

$$= -\frac{3B}{2r^3} + A + \frac{B}{r^3} = A - \frac{B}{2r^3}$$

1.7.4 Curvilinear Coordinates

In curvilinear coordinates, the space axes and coordinates are all curved and the curvature from one point to another is not usually constant. They are the most general form of coordinates, and it will be observed that Cartesian, cylindrical, and spherical coordinates are special cases of curvilinear coordinates. The space coordinates consist of three families of curves, α, β, and γ, which correspond to certain characteristics in the body. In orthogonal curvilinear coordinates these three families of curves meet mutually at right angles. Orthogonal curvilinear coordinates are utilized in some solid mechanics problems such as thin-walled shells and slip-line field analysis.

The equations of equilibrium in orthogonal curvilinear coordinates will be stated here without derivation.[2] They are

$$\frac{\partial\sigma_\alpha}{\partial\alpha} + \frac{\partial\tau_{\beta\alpha}}{\partial\beta} + \frac{\partial\tau_{\gamma\alpha}}{\partial\gamma} + \frac{\sigma_\alpha - \sigma_\beta}{r_{\beta\alpha}} - \frac{\sigma_\alpha - \sigma_\gamma}{r_{\gamma\alpha}} - \tau_{\alpha\beta}\left(\frac{2}{r_{\alpha\beta}} - \frac{1}{r_{\gamma\beta}}\right) + \tau_{\alpha\gamma}\left(\frac{2}{r_{\alpha\gamma}} - \frac{1}{r_{\beta\gamma}}\right) + F_\alpha = 0$$

$$\frac{\partial\sigma_\beta}{\partial\beta} + \frac{\partial\tau_{\gamma\beta}}{\partial\gamma} + \frac{\partial\tau_{\alpha\beta}}{\partial\alpha} + \frac{\sigma_\beta - \sigma_\gamma}{r_{\gamma\beta}} - \frac{\sigma_\beta - \sigma_\alpha}{r_{\alpha\beta}} - \tau_{\beta\gamma}\left(\frac{2}{r_{\beta\gamma}} - \frac{1}{r_{\alpha\gamma}}\right) + \tau_{\beta\alpha}\left(\frac{2}{r_{\beta\alpha}} - \frac{1}{r_{\gamma\alpha}}\right) + F_\beta = 0 \quad (1.12a)$$

$$\frac{\partial\sigma_\gamma}{\partial\gamma} + \frac{\partial\tau_{\alpha\gamma}}{\partial\alpha} + \frac{\partial\tau_{\beta\gamma}}{\partial\beta} + \frac{\sigma_\gamma - \sigma_\alpha}{r_{\alpha\gamma}} - \frac{\sigma_\gamma - \sigma_\beta}{r_{\beta\gamma}} - \tau_{\gamma\alpha}\left(\frac{2}{r_{\gamma\alpha}} - \frac{1}{r_{\beta\alpha}}\right) + \tau_{\gamma\beta}\left(\frac{2}{r_{\gamma\beta}} - \frac{1}{r_{\alpha\beta}}\right) + F_\gamma = 0$$

where

$r_{\alpha\beta}$ = radius of curvature of the α-line with respect to the α–β plane
$r_{\alpha\gamma}$ = radius of curvature of the α-line with respect to the α–γ plane
$r_{\beta\alpha}$ = radius of curvature of the β-line with respect to the β–α plane
$r_{\beta\gamma}$ = radius of curvature of the β-line with respect to the β–γ plane
$r_{\gamma\alpha}$ = radius of curvature of the γ-line with respect to the γ–α plane
$r_{\gamma\beta}$ = radius of curvature of the γ-line with respect to the γ–β plane

The sign convention for r_α, r_β, and r_γ follows the same sign for the curvature related to x, y, and z coordinates.

If α, β, and γ are principal directions (1,2,3), i.e., corresponding to a state of stress for which all shear stress components vanish* then Equations 1.12a reduce to

$$\frac{\partial\sigma_1}{\partial\alpha}+\frac{\sigma_1-\sigma_2}{r_{21}}-\frac{\sigma_1-\sigma_3}{r_{31}}+F_1=0$$

$$\frac{\partial\sigma_2}{\partial\beta}+\frac{\sigma_2-\sigma_3}{r_{32}}-\frac{\sigma_2-\sigma_1}{r_{12}}+F_2=0 \tag{1.12b}$$

$$\frac{\partial\sigma_3}{\partial\gamma}+\frac{\sigma_3-\sigma_1}{r_{13}}-\frac{\sigma_3-\sigma_2}{r_{23}}+F_3=0$$

For the case of axial symmetry, the β-lines will be coaxial circles and the γ-lines will be radial straight lines. The α-lines are not curved in the α–β plane; hence,

$$\frac{1}{r_{31}}=\frac{1}{r_{32}}=\frac{1}{r_{12}}=0,\ \text{also}\ \frac{\partial\sigma_\beta}{\partial\beta}=0$$

So that Equations 1.12b reduce to**

$$\frac{\partial\sigma_1}{\partial\alpha}+\frac{\sigma_1-\sigma_2}{r_{21}}+F_1=0$$

$$\frac{\partial\sigma_3}{\partial\gamma}+\frac{\sigma_3-\sigma_1}{r_{13}}-\frac{\sigma_3-\sigma_2}{r_{23}}+F_3=0 \tag{1.12c}$$

For the case of plane stress, Equations 1.12a reduce to

$$\frac{\partial\sigma_\alpha}{\partial\alpha}+\frac{\partial\tau_{\beta\alpha}}{\partial\beta}+\frac{\sigma_\alpha-\sigma_\beta}{r_\beta}-\frac{2\tau_{\alpha\beta}}{r_\alpha}+F_\alpha=0$$

$$\frac{\partial\sigma_\beta}{\partial\beta}+\frac{\partial\tau_{\alpha\beta}}{\partial\alpha}+\frac{\sigma_\alpha-\sigma_\beta}{r_\alpha}-\frac{2\tau_{\beta\alpha}}{r_\beta}+F_\beta=0 \tag{1.12d}$$

where

r_α = radius of curvature of the α-line
r_β = radius of curvature of the β-line

---

* See Section 1.10.

** If α, β are chosen to be the directions of the maximum shear stress, for a constant maximum shear stress k and $\sigma_\alpha = \sigma_\beta = -p$. With no body forces, Equations 1.12d reduce to

$$\frac{\partial p}{\partial\alpha}+2k\frac{\partial\phi}{\partial\alpha}=0\ \text{and}\ \frac{\partial p}{\partial\beta}+2k\frac{\partial\phi}{\partial\beta}=0$$

Such equations are used in slip-line field solution of plane strain rigid–plastic problems.

It may be easily shown that Equations 1.12a will reduce to Equations 1.6 by substituting x, y, z for α, β, γ, respectively, and $1/r_\alpha = 1/r_\beta = 1/r_\gamma = 0$. Similarly, Equations 1.12a will reduce to Equations 1.9 and 1.11 by applying appropriate substitutions.

Example 1.8:

Apply the equations of equilibrium in curvilinear coordinates to the case of a thin-walled axisymmetric pressure container having a continuous surface.

Solution:

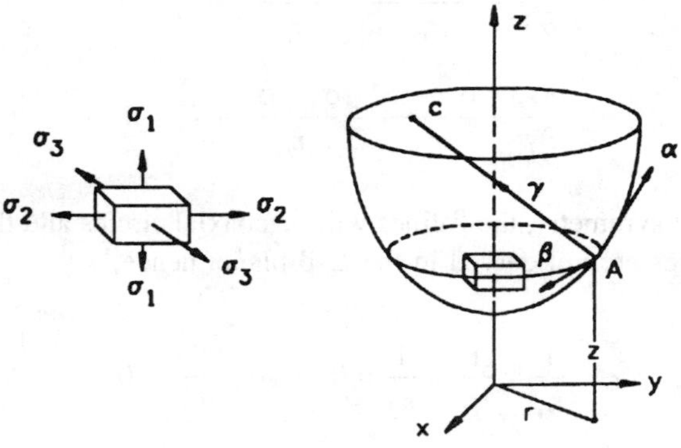

FIGURE 1.14 Example 1.8.

Consider a point $A(r, z)$ on the container wall, Figure 1.14, and take orthogonal directions α, β, γ passing through A, such that

α is along the meridian,
β is along the circumference,
γ normal to the wall surface.

From Figure 1.14:

AC = radius of curvature of the meridian line with respect to α–γ plane $= -r_{\alpha\gamma} = r_{13} = r_1$
AB = radius of curvature of the circumference line with respect to β–γ plane $= r_{\beta\gamma} = r_{23} = r_2$
$\sigma_\alpha = \sigma_1$, $\sigma_\beta = \sigma_2$ and $\sigma_\gamma = \sigma_3$

Let p be the internal pressure acting on inner surface and h the wall thickness. Now $\sigma_3 = -p$ at the inner surface and is zero at the outer surface. For a thin-walled container the variation of σ_3 along the thickness may be assumed linear so that $\partial\sigma_3/\partial\gamma = -p/h$ and the average value of $\sigma_3 = -p/2$. Substituting in the second equation of Equations 1.12c yields

$$\frac{\sigma_1}{r_1} + \frac{\sigma_2}{r_2} = \frac{p}{h}\left(1 + \frac{h}{2r_1} + \frac{h}{2r_2}\right)$$

For thin-walled containers $h/2r_1$ and $h/2r_2$ are extremely small and may be neglected for design purposes and, hence, the equation of equilibrium reduces to

$$\frac{\sigma_1}{r_1} + \frac{\sigma_2}{r_2} = \frac{p}{h}$$

This equation is used extensively to determine membrane stresses in thin shells subjected to internal pressure, as in Chapter 8.

Example 1.9:

A thin strip of a uniform thickness having a curved cross section is subjected to only an axial stress σ_z and shear stress τ in its cross section as in Figure 1.15. Show, from the equations of equilibrium in curvilinear coordinates, that if τ does not vary along z, then $(\partial\sigma_z/\partial z) + (\partial\tau/\partial s) = 0$.

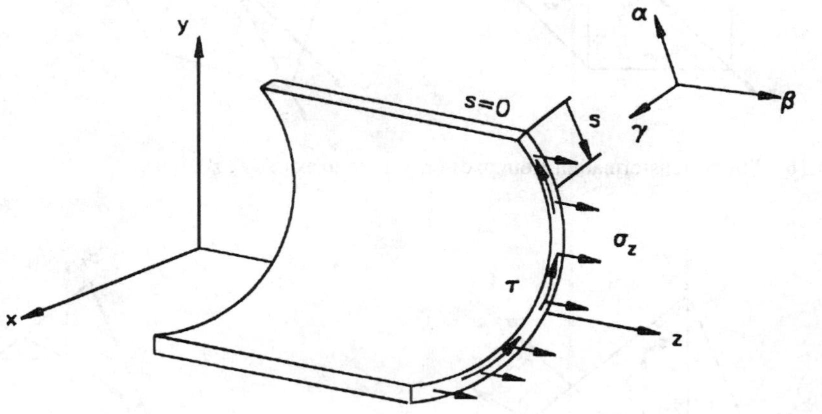

FIGURE 1.15 Example 1.9.

Solution:

Take curvilinear coordinates α along s, β along z, and γ normal to the strip surface. The state of stress referred to α, β, γ is given by $\alpha_\beta = \alpha_z$, $\tau_{\beta\alpha} = \tau$, $\sigma_\alpha = \sigma_\gamma = \tau_{\alpha\gamma} = \tau_{\beta\gamma} = 0$ and $1/r_{\beta\alpha} = 1/r_{\gamma\alpha} = 1/r_{\gamma\beta} = 1/r_{\alpha\beta} = 1/r_{\beta\gamma} = 0$. Substituting in the second equation of equilibrium of Equations 1.12a gives

$$\frac{\partial\sigma_z}{\partial z} + \frac{\partial\tau}{\partial s} = 0$$

which is used to analyze the shear flow in rods as in Chapter 7.

1.8 STRESS TRANSFORMATION LAW

It has been shown in Section 1.5 that the state of stress at any point O in a body is completely determined when the values of six independent stress components are specified at that point. These stress components (σ_x, σ_y, σ_z, τ_{xy}, τ_{xz}, τ_{yz}) have been specified in Section 1.5 with respect to three mutually perpendicular planes (x–y, y–z, z–x) passing through that point. It is demonstrated below that, by starting with these stress components, it is possible to specify the stresses on any other three orthogonal planes (say x'–y', y'–z', z'–x') passing through the same point O, as shown in Figure 1.16.

To this end consider an infinitesimal tetrahedron element having three orthogonal surfaces perpendicular to the axes x, y, and z and the fourth inclined face having its normal to x', as shown in Figure 1.17. The relation between the two sets of coordinates (x, y, z) and (x', y', z') are defined by the direction cosines of the latter set with respect to the former as given in Table 1.1. Note that since the transformation of coordinates is orthogonal, the following trigonometric relations hold.

$$l^2 + m^2 + n^2 = 1, \qquad l'^2 + m'^2 + n'^2 = 0 \ \ldots \text{ etc.}$$

$$ll' + mm' + nn' = 0, \quad l'l'' + m'm'' + n'n'' = 0 \ \ldots \text{ etc.}$$

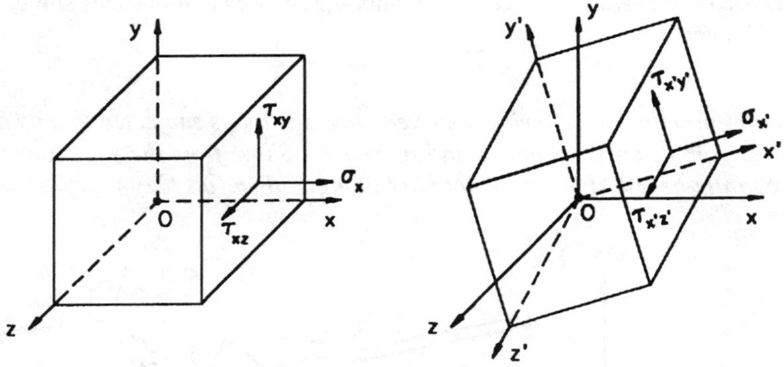

FIGURE 1.16 Stress transformation from axes (x, y, z) to axes (x', y', z').

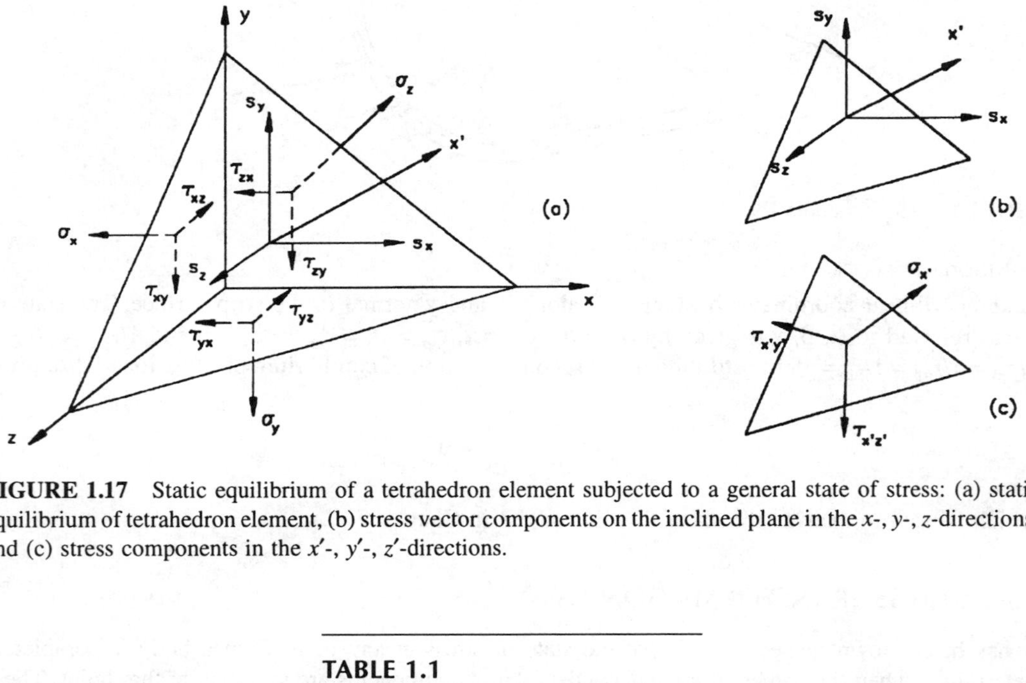

FIGURE 1.17 Static equilibrium of a tetrahedron element subjected to a general state of stress: (a) static equilibrium of tetrahedron element, (b) stress vector components on the inclined plane in the x-, y-, z-directions, and (c) stress components in the x'-, y'-, z'-directions.

TABLE 1.1
Direction Cosines between Two Sets of Coordinate Axes

Direction Cosine	of x'	of y'	of z'
With x	l	l'	l''
With y	m	m'	m''
With z	n	n'	n''

If the surface area of the inclined plane is $A_{x'}$, the surface areas of the coordinate planes will be given, respectively, by

$$A_x = A_{x'}\, l, \quad A_y = A_{x'}\, m, \quad \text{and} \quad A_z = A_{x'}\, n$$

With reference to Figure 1.17a, the stress components acting on the coordinate planes are σ_x, σ_y, σ_z, τ_{xy}, τ_{yz}, and τ_{zx}, while the stress resultant vector on the inclined plane has the components S_x, S_y, and S_z in the x-, y-, and z-directions, respectively, Figure 1.17b. Considering the equilibrium of the element in the x-direction gives*

$$S_x A_{x'} - \sigma_x A_{x'}\, l - \tau_{yx} A_{x'}\, m - \tau_{zx} A_{x'}\, n = 0$$

Repeating by considering equilibrium in y' and z'–directions and simplifying gives

$$\begin{aligned} S_x &= \sigma_x l + \tau_{yx} m + \tau_{zx} n \\ S_y &= \tau_{xy} l + \sigma_y m + \tau_{zy} n \\ S_x &= \tau_{xz} l + \tau_{yz} m + \sigma_z n \end{aligned} \tag{1.13a}$$

These equations can be put in a matrix form as

$$\begin{bmatrix} S_x \\ S_y \\ S_x \end{bmatrix} = \begin{bmatrix} \sigma_x & \tau_{xy} & \tau_{xz} \\ \tau_{yx} & \sigma_y & \tau_{yz} \\ \tau_{zx} & \tau_{zy} & \sigma_z \end{bmatrix} \begin{bmatrix} l \\ m \\ n \end{bmatrix} = [\sigma] \begin{bmatrix} l \\ m \\ n \end{bmatrix} \tag{1.13b}$$

Equations 1.13 are known as Cauchy's stress formulas.

From Figure 1.17c, the stress resultant on the inclined plane may be resolved into one normal stress $\sigma_{x'}$, and two shear stresses $\tau_{x'y'}$ and $\tau_{x'z'}$. By projection, they are related to the components S_x, S_y, and S_z according to

$$\begin{aligned} \sigma_{x'} &= S_x l + S_y m + S_z n \\ \tau_{x'y'} &= S_x l' + S_y m' + S_z n' \\ \tau_{x'z'} &= S_x l'' + S_y m'' + S_z n'' \end{aligned} \tag{1.14a}$$

or in a matrix form as

$$\begin{bmatrix} \sigma_{x'} \\ \tau_{x'y'} \\ \tau_{x'z'} \end{bmatrix} = \begin{bmatrix} l & m & n \\ l' & m' & n' \\ l'' & m'' & n'' \end{bmatrix} \begin{bmatrix} S_x \\ S_y \\ S_z \end{bmatrix} \tag{114b}$$

Combining Equations 1.13a and 1.14a gives for the three stress components

$$\begin{aligned} \sigma_{x'} &= \sigma_x l^2 + \sigma_y m^2 + \sigma_z n^2 + 2\tau_{xy} lm + 2\tau_{yz} mn + 2\tau_{zx} nl \\ \tau_{x'y'} &= \sigma_x ll' + \sigma_y mm' + \sigma_z nn' + \tau_{xy}(lm' + l'm) + \tau_{yz}(mn' + nm') + \tau_{zx}(nl' + n'l) \\ \tau_{x'z'} &= \sigma_x ll'' + \sigma_y mm'' + \sigma_z nn'' + \tau_{xy}(lm'' + l''m) + \tau_{yz}(mn'' + nm'') + \tau_{zx}(nl'' + n''l) \end{aligned} \tag{1.15a}$$

* The body force component is neglected since the element volume shrinks to zero in the limit.

These can be obtained in a matrix form by substituting Equation 1.13b into Equation 1.14b as

$$\begin{bmatrix} \sigma_{x'} \\ \tau_{x'y'} \\ \tau_{x'z'} \end{bmatrix} = \begin{bmatrix} l & m & n \\ l' & m' & n' \\ l'' & m'' & n'' \end{bmatrix} \begin{bmatrix} \sigma_x & \tau_{xy} & \tau_{xz} \\ \tau_{yx} & \sigma_y & \tau_{yz} \\ \tau_{zx} & \tau_{zy} & \sigma_z \end{bmatrix} \begin{bmatrix} l \\ m \\ n \end{bmatrix} \tag{1.15b}$$

or, simply,

$$\begin{bmatrix} \sigma_{x'} \\ \tau_{x'y'} \\ \tau_{x'z'} \end{bmatrix} = [\alpha]^T[\sigma] \begin{bmatrix} l \\ m \\ n \end{bmatrix} \tag{1.15c}$$

$[\alpha]$ expresses the direction cosines in the form of a 3×3 matrix as follows:

$$[\alpha] = \begin{bmatrix} l & l' & l'' \\ m & m' & m'' \\ n & n' & n'' \end{bmatrix} \text{ and } [\alpha]^T = \begin{bmatrix} l & m & n \\ l' & m' & n' \\ l'' & m'' & n'' \end{bmatrix} \tag{1.16}$$

The stress components $\sigma_{y'}$, $\sigma_{z'}$, and $\tau_{y'z'}$ can be obtained by considering two more elements having their inclined planes perpendicular to the y' and z' axes, respectively. Repeating the above procedure gives

$$\begin{bmatrix} \tau_{y'x'} \\ \sigma_{y'} \\ \tau_{y'z'} \end{bmatrix} = [\alpha]^T[\sigma] \begin{bmatrix} l' \\ m' \\ n' \end{bmatrix} \tag{1.17}$$

$$\begin{bmatrix} \tau_{z'x'} \\ \tau_{z'y'} \\ \sigma_{z'} \end{bmatrix} = [\alpha]^T[\sigma] \begin{bmatrix} l'' \\ m'' \\ n'' \end{bmatrix} \tag{1.18}$$

Combining Equations 1.15c, 1.17, and 1.18 gives

$$[\sigma]_{x'y'z'} = [\alpha]^T \, [\sigma]_{xyz} \, [\alpha] \tag{1.19}$$

where $[\sigma]_{x'y'z'}$ and $[\sigma]_{xyz}$ express the stress tensor with respect to the original axes (x, y, z) and the new axes (x', y', z') respectively.

Equation 1.19 represents the stress transformation law, from one set of orthogonal coordinates to another. Any quantity that transforms according to Equation 1.19 is called a tensor of the second rank.

Example 1.10:

At a point in a body, only the stresses $\tau_{xy} = 15$ *MPa,* $\tau_{yz} = 2$ *MPa and* $\tau_{yz} = 37$ *MPa are known;* σ_x, σ_y, and σ_z *being unknown. If the stress resultant at the same point is 140 MPa and acts in a direction making angles 43°, 75° with the* x-, y-axes, *respectively, find:*

a. *The normal and shear stresses on an oblique plane whose normal makes angles 67° 13′, 30° with* x-, y-axes, *respectively;*
b. *The stress components* σ_x, σ_y, *and* σ_z.

Solution:

a. The angle that the stress vector makes with the z-axis is

$$\cos^{-1}\left[\sqrt{1-\left(\cos^2 43^0+\cos^2 75\right)^0}\right]=50.878°$$

Hence, the stress vector components are

$$S_x = 140\cos 43° \quad = 102.5 \text{ MPa}$$

$$S_y = 140\cos 75° \quad = 36.2 \text{ MPa}$$

$$S_z = 140\cos 50.878° = 88.2 \text{ MPa}$$

Now, an oblique plane making angles 67° 13′ and 30° with the x-, y-axes, respectively, makes an angle:

$$\cos^{-1}\left[\sqrt{1-\left(\cos^2 67°13' + \cos^2 30°\right)}\right]=71°$$

with the z-axis.

On such an oblique plane, the normal stress is obtained from Equations 1.14a as

$$\sigma_{x'} = S_x l + S_y m + S_z n$$

$$= 102.5 \;\cos\; 67°13' + 36.2 \;\cos\; 30° + 88.2 \;\cos\; 71°$$

$$\sigma_{x'} = 98.2 \text{ MPa}$$

The resultant shear stress in this plane is

$$\tau = \sqrt{\left(S_x + S_y + S_z\right)^2 - (98.2)^2} = \sqrt{140^2 - (98.2)^2}$$

$$\tau = 99.5 \text{ MPa}$$

b. Applying Equation 1.13b,

$$\begin{bmatrix} S_x \\ S_y \\ S_x \end{bmatrix} = [\sigma]_{xyz} \begin{bmatrix} l \\ m \\ n \end{bmatrix}$$

or

$$\begin{bmatrix} 102.5 \\ 36.2 \\ 88 \end{bmatrix} = \begin{bmatrix} \sigma_x & 15 & 37 \\ 15 & \sigma_y & -2 \\ 37 & -2 & \sigma_z \end{bmatrix} \begin{bmatrix} \cos\; 67°13' \\ \cos\; 30° \\ \cos\; 71° \end{bmatrix}$$

This gives

$$102.5 = 0.388\sigma_x + 0.866(15) + 0.315(37)$$

or

$$\sigma_x = 199 \text{ MPa}$$

Similarly,

$$\sigma_y = 35.8 \text{ MPa and } \sigma_z = 24 \text{ MPa}$$

Example 1.11:
A state of stress at a point is given by

$$[\sigma] = \begin{bmatrix} 10 & 6 & -8 \\ 6 & 20 & -4 \\ -8 & -4 & 10 \end{bmatrix} MPa$$

with respect to a set of axes (x, y, z). *Find the state of stress for a new set of axes rotated about the* x*-axis through an angle 45°.*

Solution:
The new axes (x', y', z'), Figure 1.18, are related to the original ones by the direction cosines of Table 1.2. Note that in determining the direction cosines, positive angle is measured from a new to an original axis following a counterclockwise direction.

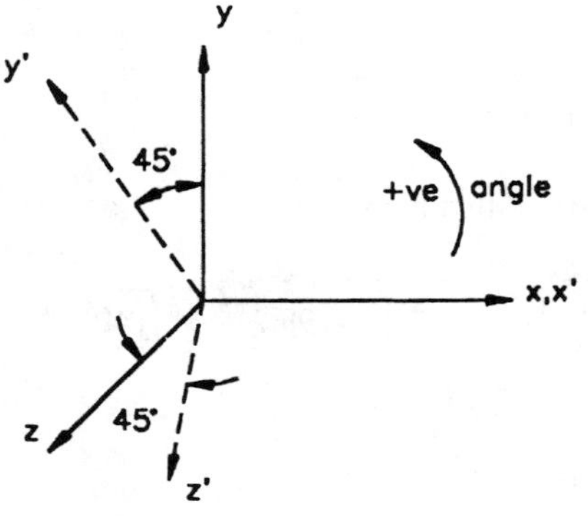

FIGURE 1.18 Example 1.11.

Applying the stress transformation law, Equations 1.19 gives

$$\begin{bmatrix} \sigma_{x'} & \tau_{x'y'} & \tau_{x'z'} \\ \tau_{y'x'} & \sigma_{y'} & \tau_{y'z'} \\ \tau_{z'x'} & \tau_{z'y'} & \sigma_{z'} \end{bmatrix} = \begin{bmatrix} 1 & 0 & 0 \\ 0 & 1/\sqrt{2} & 1/\sqrt{2} \\ 0 & -1/\sqrt{2} & 1/\sqrt{2} \end{bmatrix} \begin{bmatrix} 10 & 6 & -8 \\ 6 & 20 & -4 \\ -8 & -4 & 10 \end{bmatrix} \begin{bmatrix} 1 & 0 & 0 \\ 0 & 1/\sqrt{2} & -1/\sqrt{2} \\ 0 & 1/\sqrt{2} & 1/\sqrt{2} \end{bmatrix} \text{MPa}$$

This yields

TABLE 1.2
Direction Cosines Between Two Sets of Coordinate Axes

Direction Cosine	of x'	of y'	of z'
With x	1	0	0
With y	0	$1/\sqrt{2}$	$-1/\sqrt{2}$
With z	0	$1/\sqrt{2}$	$1/\sqrt{2}$

$$[\sigma]_{x'y'z'} = \begin{bmatrix} 10 & 6 & -8 \\ -1.414 & 11.321 & 4.242 \\ -9.898 & -16.968 & 9.898 \end{bmatrix} \begin{bmatrix} 1 & 0 & 0 \\ 0 & 1/\sqrt{2} & -1/\sqrt{2} \\ 0 & 1/\sqrt{2} & 1/\sqrt{2} \end{bmatrix} \text{ MPa}$$

Hence,

$$[\sigma]_{x'y'z'} = \begin{bmatrix} 10 & -1.414 & -9.9 \\ -1.414 & 11 & -5 \\ -9.9 & -5 & 19 \end{bmatrix} \text{ MPa}$$

1.9 PLANE STRESS TRANSFORMATION — MOHR'S CIRCLE OF STRESS

Consider a case of plane stress for which $\sigma_z = \tau_{xz} = \tau_{yz} = 0$. This state of stress is represented in Figure 1.19a and b for two sets of axes (x, y) and (x', y'). If θ denotes the angle between the new x' axis and the original x, the direction cosines of the new system (x', y', z') are as given in Table 1.3. Substituting into the stress transformation equations (Equations 1.15a) gives

$$\begin{aligned} \sigma_{x'} &= \sigma_x \cos^2\theta + \sigma_y \sin^2\theta + 2\tau_{xy} \sin\theta\cos\theta \\ \sigma_{y'} &= \sigma_x \sin^2\theta + \sigma_y \cos^2\theta - 2\tau_{xy} \sin\theta\cos\theta \\ \tau_{x'y'} &= (\sigma_x - \sigma_y)\sin\theta\cos\theta + \tau_{xy}(\cos^2\theta - \sin^2\theta) \end{aligned} \tag{1.20a}$$

which reduce to

$$\begin{aligned} \sigma_{x'} &= \frac{\sigma_x + \sigma_y}{2} + \frac{\sigma_x - \sigma_y}{2}\cos 2\theta + \tau_{xy}\sin 2\theta \\ \sigma_{y'} &= \frac{\sigma_x + \sigma_y}{2} - \frac{\sigma_x - \sigma_y}{2}\cos 2\theta - \tau_{xy}\sin 2\theta \\ \tau_{x'y'} &= -\frac{\sigma_x - \sigma_y}{2}\sin 2\theta + \tau_{xy}\cos 2\theta \end{aligned} \tag{1.20b}$$

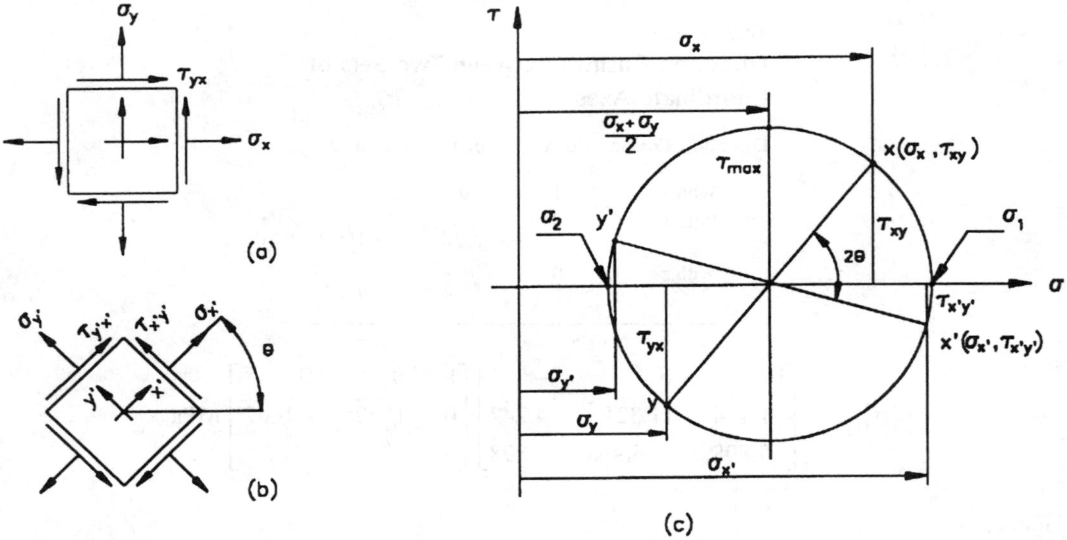

FIGURE 1.19 Mohr's circle for plane stress.

TABLE 1.3
Direction Cosines Between Two Sets of Coordinate Axes

Direction Cosine	of x'	of y'	of z'
With x	cos θ	–sin θ	0
With y	sin θ	cos θ	0
With z	0	0	1

Equations 1.20 are parametric equations of a circle. Squaring and adding gives

$$\left(\sigma_{x'} - \frac{\sigma_x + \sigma_y}{2}\right)^2 + \tau_{x'y'}^2 = \left(\frac{\sigma_x - \sigma_y}{2}\right)^2 + \tau_{xy}^2$$

which represents a circle in the σ–τ plane with its center at $(\sigma_x + \sigma_y)/2$ and a radius equal to the square root of $[((\sigma_x + \sigma_y)/2)^2 + \sigma_{xy}^2\,]$ as shown in Figure 1.19c. The circle represents the locus of all pairs of stress components (σ, τ) defining the state of plane stress at a point for any set of axes (x', y') and is known as Mohr's circle, presented in books of strength of materials* as a graphical means for stress transformation.

A three-dimensional stress state may be graphically represented by three touching Mohr's circles, the largest being an envelope to the other two.[2] Such construction is too elaborate and is not often used.

* For details about the construction of Mohr's circle and the sign convention, refer to Popov.[3]

1.10 PRINCIPAL STRESSES

For any plane passing through a point, the stress resultant possesses generally a normal stress component and two shear stress components. It is very useful in design and failure analysis problems to define planes on which only normal stresses act without any shear stresses. These normal stresses are known as principal stresses. Assume that there is a plane, Figure 1.20, with direction cosines.

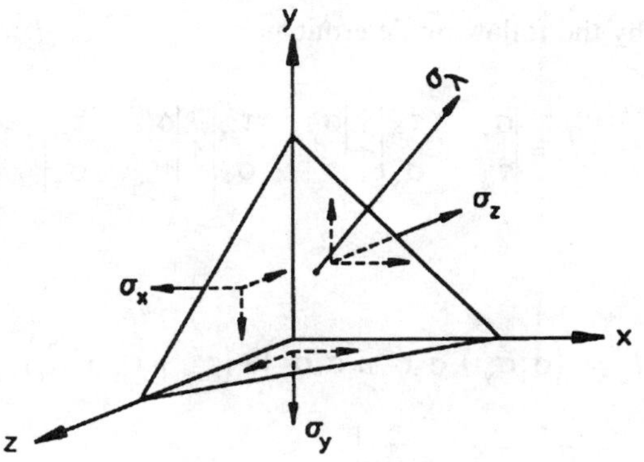

FIGURE 1.20 The stress vector is normal to the principal plane, which is free of shear stresses.

$(l_\lambda, m_\lambda, n_\lambda)$ on which the stress vector is normal and has the magnitude σ_λ. On such a plane the shear stress components vanish and Equations 1.13a give

$$S_x = \sigma_\lambda l_\lambda, \quad S_y = \sigma_\lambda m_\lambda, \quad \text{and} \quad S_z = \sigma_\lambda n_\lambda$$

Hence,

$$\begin{aligned} \sigma_\lambda l_\lambda &= \sigma_x l_\lambda + \tau_{yx} m_\lambda + \tau_{zx} n_\lambda \\ \sigma_\lambda m_\lambda &= \tau_{xy} l_\lambda + \sigma_y m_\lambda + \tau_{zy} n_\lambda \\ \sigma_\lambda n_\lambda &= \tau_{xz} l_\lambda + \tau_{yz} m_\lambda + \sigma_z n_\lambda \end{aligned} \tag{1.21}$$

Rearranging in a matrix form results in

$$\begin{bmatrix} \sigma_x - \sigma_\lambda & \tau_{xy} & \tau_{xz} \\ \tau_{yx} & \sigma_y - \sigma_\lambda & \tau_{yz} \\ \tau_{zx} & \tau_{zy} & \sigma_z - \sigma_\lambda \end{bmatrix} \begin{bmatrix} l_\lambda \\ m_\lambda \\ n_\lambda \end{bmatrix} = 0$$

For a nontrivial solution for l_λ, m_λ, and n_λ, the characteristic determinant must vanish, hence,

$$\begin{vmatrix} \sigma_x - \sigma_\lambda & \tau_{xy} & \tau_{xz} \\ \tau_{yx} & \sigma_y - \sigma_\lambda & \tau_{yz} \\ \tau_{zx} & \tau_{zy} & \sigma_z - \sigma_\lambda \end{vmatrix} = 0$$

which upon expansion leads to the cubic equation:

$$\sigma_\lambda^3 - I_1\sigma_\lambda^2 - I_2\sigma_\lambda - I_3 = 0 \tag{1.22}$$

where

$$I_1 = \sigma_x + \sigma_y + \sigma_z \tag{1.23a}$$

and I_2 and I_3 are defined by the following determinants:

$$-I_2 = \begin{vmatrix} \sigma_x & \tau_{xy} \\ \tau_{yx} & \sigma_y \end{vmatrix} + \begin{vmatrix} \sigma_y & \tau_{yz} \\ \tau_{zy} & \sigma_z \end{vmatrix} + \begin{vmatrix} \sigma_z & \tau_{zx} \\ \tau_{xz} & \sigma_x \end{vmatrix} \tag{1.23b}$$

or

$$I_2 = -\left(\sigma_x\sigma_y + \sigma_y\sigma_z + \sigma_z\sigma_x\right) + \left(\tau_{xy}^2 + \tau_{yz}^2 + \tau_{zx}^2\right)$$

$$I_3 = \begin{vmatrix} \sigma_x & \tau_{xy} & \tau_{xz} \\ \tau_{yx} & \sigma_y & \tau_{yz} \\ \tau_{zx} & \tau_{zy} & \sigma_z \end{vmatrix} \tag{1.23c}$$

$$= \sigma_x\sigma_y\sigma_z + 2\tau_{xy}\tau_{yz}\tau_{zx} - \sigma_x\tau_{yz}^2 - \sigma_y\tau_{zx}^2 - \sigma_z\tau_{xy}^2$$

Equation 1.22 is called the characteristic equation of $[\sigma]$ and will always possess three real roots,* namely, σ_1, σ_2, and σ_3. Since they act on planes where shear stresses vanish, they are called principal stresses and arranged such that $\sigma_1 \geq \sigma_2 \geq \sigma_3$. They represent stationary values (maximum or minimum) for the normal stresses.** The three roots of the characteristic equation may be obtained analytically by using either the algebraic method or Newton's approximation procedure, as shown in Example 1.12.

Corresponding to the principal stresses, there are three sets of direction cosines. Each set defines the direction of a principal axis, i.e., the normal to the principal plane, and is obtained by substituting into Equation 1.21 the corresponding value of σ_λ (λ = 1, 2, or 3) together with the relation:

$$l^2 + m^2 + n^2 = 1$$

The principal axes are mutually perpendicular. If the values of σ_1, σ_2, and σ_3 are different, then the principal axes are distinct. If, say, $\sigma_1 = \sigma_2$, then the third principal axis is distinct and every direction perpendicular to it is a principal direction associated with $\sigma_1 = \sigma_2$. If $\sigma_1 = \sigma_2 = \sigma_3$, then every direction is a principal direction, as will be shown later.

The principal stresses and the principal directions are, respectively, the eigenvalues and eigenvectors of the stress tensor $[\sigma]$. The principal stresses will diagonalize the stress tensor so that

$$[\sigma] = \begin{bmatrix} \sigma_1 & 0 & 0 \\ 0 & \sigma_2 & 0 \\ 0 & 0 & \sigma_3 \end{bmatrix} \tag{1.24}$$

---

* To prove this, see Srinath,[4] p. 18.

** To prove this, see Venkatraman and Patel.[5]

The principal stresses at a point depend only on the state of stress at that point and not on the coordinate system of axes. Hence, if (x, y, z) and (x', y', z') are two orthogonal sets of axes at that point, then the following two characteristic equations hold:

$$\sigma_\lambda^3 - I_1\sigma_\lambda^2 - I_2\sigma_\lambda - I_3 = 0$$

and

$$\sigma_\lambda^3 - I_1^*\sigma_\lambda^2 - I_2^*\sigma_\lambda - I_3^* = 0$$

and both equations must give the same solution for σ_λ. Since the two systems of axes are arbitrary, the coefficients I_1, I_2, I_3, and I_1^*, I_2^*, I_3^* must be correspondingly identical and these coefficients as defined by Expressions 1.23 are called stress invariants and in terms of the principal stresses become

$$\begin{aligned} I_1 &= \sigma_1 + \sigma_2 + \sigma_3 \\ I_2 &= -(\sigma_1\sigma_2 + \sigma_2\sigma_3 + \sigma_3\sigma_1) \\ I_3 &= \sigma_1\sigma_2\sigma_3 \end{aligned} \tag{1.25}$$

For the case of plane stress, where $\sigma_z = \tau_{zx} = \tau_{zy} = 0$, z is a principal direction. The stress invariants become

$$I_1 = \sigma_x + \sigma_y, \quad I_2 = -\sigma_x\sigma_y + \tau_{xy}^2, \quad \text{and} \quad I_3 = 0$$

Taking $\sigma_1 = \sigma_x$ and $\sigma_2 = \sigma_y$ gives $I_1 = \sigma_1 + \sigma_2$, $I_2 = -\sigma_1\sigma_2$. The characteristic Equation 1.22 thus reduces to a quadratic one as

$$\sigma_\lambda^2 - (\sigma_x + \sigma_y)\sigma_\lambda + (\sigma_x\sigma_y - \tau_{xy}^2) = 0$$

with the roots

$$\left.\begin{matrix}\sigma_1 \\ \sigma_2\end{matrix}\right\} = \frac{\sigma_x + \sigma_y}{2} \pm \sqrt{\left(\frac{\sigma_x - \sigma_y}{2}\right)^2 + \tau_{xy}^2} \tag{1.26a}$$

These are the principal stresses that are obtained from the construction of Mohr's circle for plane stress as shown in Figure 1.19c. The principal directions occur at right angles given by

$$\tan 2\theta = \frac{2\tau_{xy}}{\sigma_x - \sigma_y} \tag{1.26b}$$

which yields θ and $(\theta + \pi/2)$.

Example 1.12:

Determine the magnitude and direction of the principal stresses for the given stress tensor:

$$[\sigma] = \begin{bmatrix} 8 & -3 & -3 \\ -3 & 8 & -3 \\ -3 & -3 & 8 \end{bmatrix} MPa$$

Solution:

The stress invariants are:

$$I_1 = 24, \quad I_2 = -165, \quad \text{and} \quad I_3 = 242$$

The characteristic equation is

$$\sigma_\lambda^3 - 24\sigma_\lambda^2 + 165\sigma_y - 242 = 0$$

The algebraic method of solving cubic equations gives the real roots as

$$\sigma_1 = \frac{I_1}{3} + 2\sqrt{-\frac{a}{3}} \cos\frac{\beta}{3},$$

$$\sigma_2 = \frac{I_1}{3} + 2\sqrt{-\frac{a}{3}} \cos\left(\frac{\beta}{3} + 120°\right),$$

$$\sigma_3 = \frac{I_1}{3} + 2\sqrt{-\frac{a}{3}} \cos\left(\frac{\beta}{3} + 240°\right)$$

where

$$\cos\beta = -b/2\sqrt{-a^3/27},$$

$$a = \frac{1}{3}\left(3I_2 - I_1^2\right), \quad \text{and}$$

$$b = \frac{1}{27}\left(-2I_1^3 + 9I_1I_2 - 27I_3\right)$$

which has roots

$$\sigma_1 = \sigma_2 = 11 \text{ MPa} \quad \text{and} \quad \sigma_3 = 2 \text{ MPa}$$

From Equation 1.21, for $\sigma_\lambda = \sigma_3 = 2$:

$$6l_3 - 3m_3 - 3n_3 = 0$$

$$-3l_3 + 6m_3 - 3n_3 = 0$$

which yields

$$l_3 = m_3 = n_3 = 1/\sqrt{3}$$

Since $\sigma_1 = \sigma_2$, then all directions are principal directions in the plane (1–2).

Another method of trial and error based on Newton's approximation may be used to solve the cubic characteristic equation.

Let σ^* be an initial guess for a root; then a more accurate root is

$$\sigma = \sigma^* - \frac{f(\sigma^*)}{f'(\sigma^*)}$$

where $f(\sigma^*)$ and $f'(\sigma^*)$ are the values of the left-hand side of the characteristic equation and its derivative at $\sigma_\lambda = \sigma^*$. Applying Newton's approximation method and starting by $\sigma_\lambda = \sigma^* = 1$ gives:

$$\sigma = \left[\sigma^* - \frac{\sigma^{*3} - 24\sigma^{*2} + 165\sigma^* - 242}{3\sigma^{*2} - 48\sigma^* + 165}\right]_{\sigma^*=1}$$

$$= 1.833$$

Repeating the same procedure for $\sigma^* = 1.833$ gives $\sigma^* = 1.994$. A third trial results in $\sigma = 1.999992$ which is a root ($\sigma_3 = 2$) as above. Division determines the other two roots $\sigma_1 = \sigma_2 = 11$ MPa.

Example 1.13:

Determine the magnitude and direction of the principal stresses for the given stress tensor.

$$[\sigma] = \begin{bmatrix} 0 & 10 & -10 \\ 10 & 0 & 20 \\ -10 & 20 & 0 \end{bmatrix} MPa$$

Solution:

For the given stress tensor, from Expressions 1.23, the stress invariants are

$$I_1 = 0, \quad I_2 = 600, \quad \text{and} \quad I_3 = -4000$$

The characteristic Equation 1.22 is

$$\sigma_\lambda^3 - 600\sigma_\lambda + 4000 = 0$$

which has the roots:

$$\sigma_1 = 20, \quad \sigma_2 = 7.32, \quad \text{and} \quad \sigma_3 = -27.32$$

To determine the first principal direction, Equation 1.21 is used with $\sigma_\lambda = \sigma_1 = 20$; hence, only two independent equations are obtained:

$$-20l_1 + 10m_1 - 10n_1 = 0$$

$$10l_1 - 20m_1 + 20n_1 = 0$$

which yields

$$l_1 = 0 \quad \text{and} \quad m_1 = n_1$$

And since

$$l_1^2 + m_1^2 + n_1^2 = 1$$

then

$$m_1 = n_1 = 1/\sqrt{2}$$

Repeating for $\sigma_\lambda = \sigma_2$ and $\sigma_2 3$, respectively, yields the second and third principal directions, respectively, as

$$(l_2, m_2, n_2) = (0.887, 0.324, -0.324)$$

$$(l_3, m_3, n_3) = (0.459, -0.627, 0.627)$$

1.11 MAXIMUM SHEAR STRESSES

Shear stresses vanish on planes normal to the principal directions. This is the minimum absolute value for the shear stresses. It is required now to determine the magnitude and direction of the maximum shear stresses.

For simplicity, take the coordinate axes to be the principal directions as shown in Figure 1.21. Consider an inclined plane whose normal i has direction cosines (l_i, m_i, n_i) with the coordinate axes. Let the maximum value of the in-plane component of the stress resultant vector **S** acting on the inclined plane, be τ_i. Equations 1.3 thus gives

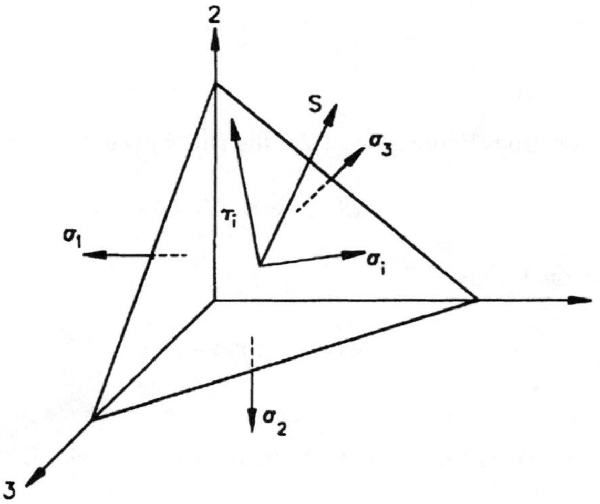

FIGURE 1.21 Static equilibrium of principal planes tetrahedron element.

$$S_1^2 + S_2^2 + S_3^2 = \sigma_i^2 + \tau_i^2$$

where S_1, S_2, and S_3 are the components of the stress vector **S** with respect to the principal coordinate axes (1, 2, 3). Hence,

$$\tau_i^2 + S_1^2 + S_2^2 + S_3^2 - \sigma_i^2 \tag{1.27}$$

And from Equations 1.13a

$$S_1 = \sigma_1 l_i, \quad S_2 = \sigma_2 m_i, \quad \text{and} \quad S_3 = \sigma_3 n_i \tag{1.28}$$

Applying the stress transformation law results in

$$\sigma_i = \sigma_1 l_i^2 + \sigma_2 m_i^2 + \sigma_3 n_i^2 \tag{1.29}$$

Therefore, substituting from Equations 1.28 and 1.29 into Equation 1.27 gives

$$\tau_i^2 = \sigma_1^2 l_i^2 + \sigma_2^2 m_i^2 + \sigma_3^2 n_i^2 - \left(\sigma_1 l_i^2 + \sigma_2 m_i^2 + \sigma_3 n_i^2\right)^2 \tag{1.30}$$

For maximum values of τ_i, the conditions

$$\frac{\partial \tau_i}{\partial l_i} = 0 \text{ and } \frac{\partial \tau_i}{\partial m_i} = 0$$

are applied noting that $l_i^2 + m_i^2 + n_i^2 = 1$. These two conditions are simultaneously satisfied for the following values of l_i, m_i, and n_i:

$$\begin{aligned} &\text{a. } l_i^2 = n_i^2 = \frac{1}{2}, \quad m_i^2 = 0 \text{ which gives } \tau_1 = \frac{\sigma_1 - \sigma_3}{2} \\ &\text{b. } m_i^2 = l_i^2 = \frac{1}{2}, \quad n_i^2 = 0 \text{ which gives } \tau_2 = \frac{\sigma_1 - \sigma_2}{2} \\ &\text{c. } n_i^2 = m_i^2 = \frac{1}{2}, \quad l_i^2 = 0 \text{ which gives } \tau_3 = \frac{\sigma_2 - \sigma_3}{2} \end{aligned} \tag{1.31}$$

For $\sigma_1 > \sigma_2 > \sigma_3$, the maximum shear stresses τ_1, τ_2, and τ_3 are positive and $\tau_1 = (\sigma_1 - \sigma_3)/2$ is the maximum shear stress that occurs on the plane $l = n = \pm 1/\sqrt{2}$ and $m = 0$. This means that there are two planes (on which τ_1 occurs) making angles of 45° and 135° with σ_1 and σ_3 planes and passing by one of the principal axes σ_2 (since $m_1 = 0$, for this case). Unlike the principal stresses with which zero shear stresses are associated, there are nonzero normal stresses acting on the planes of maximum shear. Their values are obtained from Equation 1.29 by substituting the values of maximum shear from Equations 1.31. These normal stresses are $(\sigma_1 + \sigma_3)/2$, $(\sigma_1 + \sigma_2)/2$ and $(\sigma_2 + \sigma_3)/2$ associated with τ_1, τ_2, and τ_3, respectively. Figure 1.22 shows the planes on which τ_3 and the associated σ_1 are acting drawn with respect to the principal directions.

Note that for plane stress the maximum shear is equal to $(\sigma_1 - \sigma_2)/2$, if $\sigma_2 < 0$, and acts on planes making 45° and 135° with the first principal direction. The associated normal stress is $(\sigma_1 + \sigma_2)/2$. These values can be read from Mohr's circle representation of Figure 1.19c.

The maximum shear stresses are important in studying the failure of loaded engineering components. Yielding may occur where the maximum shear stress attains a critical value (yield shear stress) as postulated by the Tresca yield condition.* Simply, since $\sigma_1 > \sigma_2 > \sigma_3$, the maximum shear stress theory is stated as

* See Section 10.7.

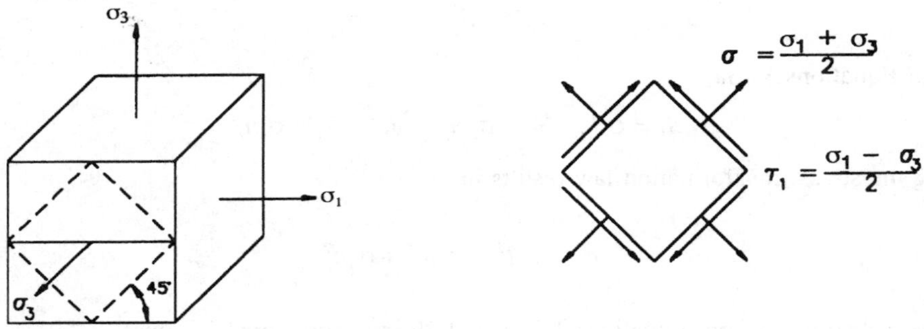

FIGURE 1.22 Plane of maximum shear stresses makes 45° with the principal plane.

$$\tau_1 = {}^1/_2\,(\sigma_1 - \sigma_3) \le k \tag{1.32a}$$

where k is the yield shear stress of the material under consideration. Taking $k = Y/2$, where Y is the tensile yield strength, the maximum shear stress condition is stated as

$$(\sigma_1 - \sigma_3) \le Y \tag{1.32b}$$

Example 1.14:

Select a material for a circular shaft subjected to a combined loading consisting of an axial tension P *and a twisting moment* M_t. *The axial tension is kept constant at a value that produces a tensile stress of 0.4*Y, *where* Y *is the yield strength of the shaft material. The twisting moment* M_t *is increased gradually until failure occurs. The shaft material could be selected (based on maximum torque capacity) from a ductile alloy if the maximum shear stress attains a value equal to* Y/2, *or a brittle alloy if the maximum normal stress attains a value equal to* Y *(for brittle materials the yield strength may be considered close to the fracture stress point). Indicate the plane on which fracture may occur in both cases.*

Solution:

For combined loading, the principal stresses are calculated. An element on the shaft surface is shown in Figure 1.23. For a two-dimensional state of stress, Mohr's circle or Equations 1.26a may be used:

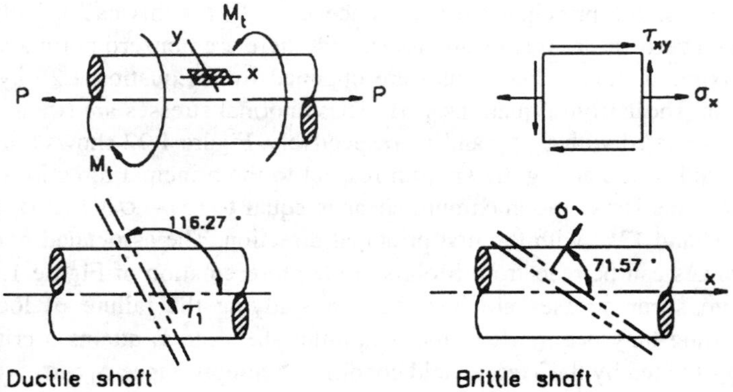

FIGURE 1.23 Example 1.14.

$$\sigma_1 = 0.2Y + \sqrt{(0.2Y)^2 + \tau_{xy}^2}$$

$$\sigma_3 = 0.2Y - \sqrt{(0.2Y)^2 + \tau_{xy}^2}$$

while $\sigma_2 = 0$.

The maximum shear stress from Equations 1.31 is

$$\tau_1 = \frac{\sigma_1 - \sigma_3}{2} = \sqrt{(0.2Y)^2 + \tau_{xy}^2}$$

a. For the ductile alloy, failure occurs according to Equation 1.32a when

$$\tau_1 = \frac{Y}{2} = \sqrt{(0.2Y)^2 + \tau_{xy}^2}$$

giving

$$\tau_{xy} = 0.4583Y$$

b. For the brittle alloy,

$$\sigma_1 = Y = 0.2Y + \sqrt{(0.2Y)^2 + \tau_{xy}^2}$$

giving

$$\tau_{xy} = 0.7746Y$$

Since τ_{xy} is directly related to the applied twisting moment, the shaft made of a ductile alloy will support lower moment at failure.

c. Fracture in a brittle alloy such as cast iron starts along a plane normal to the largest principal stress σ_1, i.e., at an angle θ_b determined from Equation 1.26b as

$$\tan\theta_b = \frac{2\tau_{xy}}{\sigma_x} = \frac{1.2Y}{0.4Y}$$

giving

$$\theta_b = 71.57°$$

Fracture in steel starts along a plane of maximum shear which is inclined 45° to the principal direction as given by Equation 1.31, i.e.,

$$\theta_d = 71.57° + 45° = 116.57°$$

These fracture surfaces are shown in Figure 1.23.

1.12 OCTAHEDRAL SHEAR STRESS — PURE SHEAR

Consider now a tetrahedron element similar to that of Figure 1.21 with a plane equally inclined to the principal axes. Hence, its normal has the direction cosines:

$$1 = m = n = \pm 1/\sqrt{3} \tag{1.33}$$

There are eight planes that satisfy this condition, as shown in Figure 1.24, and hence are called octahedral planes. The shear stress τ_{oct} acting on these planes is called the octahedral shear stress and is obtained by substituting Equation 1.32 into Equation 1.30 to give

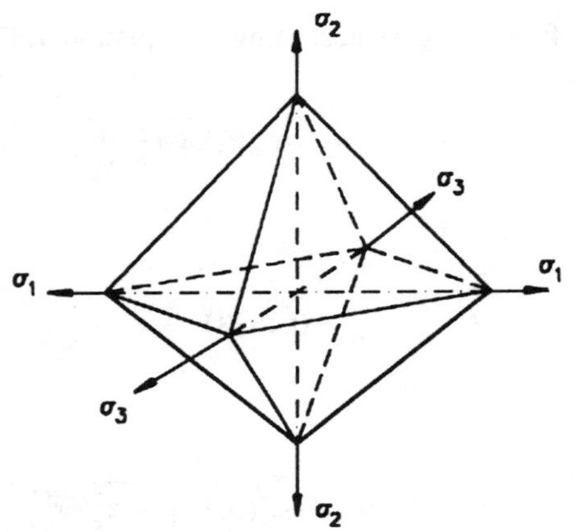

FIGURE 1.24 Eight octahedral planes equally inclined to the principal axes.

$$\begin{aligned}\tau_{oct}^2 &= \tfrac{1}{9}\left[(\sigma_1-\sigma_2)^2+(\sigma_2-\sigma_3)^2+(\sigma_3-\sigma_1)^2\right]\\ &= \tfrac{2}{9}(\sigma_1+\sigma_2+\sigma_3)^2-\tfrac{2}{3}(\sigma_1\sigma_2+\sigma_2\sigma_3+\sigma_3\sigma_1)\\ &= \tfrac{2}{9}\left(I_1^2-3I_2\right)\end{aligned} \tag{1.34a}$$

The normal stress acting on the octahedral plane is obtained from Equation 1.29 as

$$\sigma_{oct} = \tfrac{1}{3}(\sigma_1+\sigma_2+\sigma_3) = \tfrac{1}{3}I_1 \tag{1.34b}$$

If in any state of stress the first stress invariant I_1 is zero, then the normal stresses on the octahedral planes will be zero and only the octahedral shear stress will act. This defines a state of pure shear namely, when $I_1 = 0$.*

The definition of τ_{oct} and σ_{oct} referred to coordinate axes x, y, z gives:

$$\begin{aligned}\tau_{oct}^2 &= \tfrac{1}{9}\left[(\sigma_x-\sigma_y)^2+(\sigma_y-\sigma_z)^2+(\sigma_z-\sigma_x)^2+6\left(\tau_{xy}^2+\tau_{yz}^2+\tau_{zx}^2\right)\right]\\ \sigma_{oct} &= \tfrac{1}{3}(\sigma_x+\sigma_y+\sigma_z)\end{aligned} \tag{1.35}$$

* For a proof, refer to Srinath,[4] p. 58.

The octahedral shear τ_{oct} and normal stresses σ_{oct} play a basic role in formulating some failure theories (e.g., the von Mises yield condition).*

1.13 MEAN (HYDROSTATIC) STRESS AND DEVIATORIC STRESSES

For any state of stress at a point (σ_x, σ_y, σ_z, τ_{xy}, τ_{yx}, τ_{zx}), the mean normal stress is given by

$$\sigma_m = {}^1\!/_3(\sigma_x + \sigma_y + \sigma_z) = {}^1\!/_3(\sigma_1 + \sigma_2 + \sigma_3) = {}^1\!/_3 I_1 \tag{1.36}$$

If $\sigma_x = \sigma_y = \sigma_z$, the stress is uniform in all directions and is equal to σ_m. This is the case of a hydrostatic pressure and hence σ_m is sometimes called the hydrostatic stress. A solid would experience such a stress when immersed in a pressurized fluid.

Generally, the normal stress components are decomposed into a hydrostatic part σ_m and a deviatoric part called the stress deviation σ_x', σ_y', and σ_z', such that

$$\sigma_x' = \sigma_x - \sigma_m, \quad \sigma_y' = \sigma_y - \sigma_m, \quad \sigma_z' = \sigma_z - \sigma_m \tag{1.37}$$

The stress tensor $[\sigma]$ may be thus written in the form:

$$[\sigma] = [\sigma'] + [\sigma_m]$$

or

$$[\sigma] = \begin{bmatrix} \sigma_m & 0 & 0 \\ 0 & \sigma_m & 0 \\ 0 & 0 & \sigma_m \end{bmatrix} + \begin{bmatrix} \sigma_x' & \tau_{xy} & \tau_{xz} \\ \tau_{yx} & \sigma_y' & \tau_{yz} \\ \tau_{zx} & \tau_{zx} & \sigma_z' \end{bmatrix}$$

The corresponding stress invariants are obtained from Equation 1.25 as

$$\begin{aligned} &\text{for } [\sigma_m]: I_{1m} = 3\sigma_m, \ I_{2m} = 3\sigma_m^2, \ I_{3m} = \sigma_m^3 \\ &\text{for } [\sigma']: I_1' = 0, I_2' = -(\sigma_1'\sigma_2' + \sigma_2'\sigma_3' + \sigma_3'\sigma_1') = I_2 - I_1^2/3, I_3' = \sigma_1'\sigma_2'\sigma_3' \end{aligned} \tag{1.38}$$

Since for the deviatoric stress matrix $[\sigma']$, the first stress invariant $I_1' = 0$, then it represents a state of pure shear. It will be shown later** that the hydrostatic part σ_m produces change in volume while the deviatoric part $[\sigma']$ produces change in shape.

Example 1.15:

For the stress matrix:

$$[\sigma] = \begin{bmatrix} 0 & a & a \\ a & 0 & a \\ a & a & 0 \end{bmatrix}$$

Determining:

a. *The principal stresses and principal directions.*

---

* See Section 10.7.

** See Section 3.6.

b. *The maximum shear stresses.*
c. *The hydrostatic and deviatoric stresses.*
d. *The octahedral shear and normal stresses.*

Solution:

a. According to Expressions 1.23, the stress invariants are

$$I_1 = 0, \quad I_2 = 3a^2, \quad \text{and} \quad I_3 = 2a^3$$

The characteristic equation is

$$\sigma_\lambda^3 - 3a^2\sigma_\lambda - 2a^3 = 0$$

By inspection, a root of this equation is $\sigma_\lambda = -a$. By division, the above equation reduces to

$$(\sigma_\lambda - a)(\sigma_\lambda^2 - a\sigma_\lambda - 2a^2) = (\sigma_\lambda - a)(\sigma_\lambda - 2a)(\sigma_\lambda - a) = 0$$

which gives

$$\sigma_\lambda = -a \quad \text{and} \quad \sigma_\lambda = 2a$$

Hence,

$$\sigma_1 = 2a \quad \text{and} \quad \sigma_2 = \sigma_3 = -a$$

Since $\sigma_2 = \sigma_3$, then all directions in the (2–3) plane are principal directions. The first principal direction is obtained from Equations 1.21 by putting $\sigma_\lambda = 2a$, i.e.,

$$2al_1 - am_1 - an_1 = 0$$

and

$$al_1 - 2am_1 + an_1 = 0$$

which gives

$$l_1 = m_1 = n_1 = 1/\sqrt{3}$$

b. The maximum shear stresses are

$$\tau_1 = \frac{\sigma_1 - \sigma_3}{2} = \frac{2a + a}{2} = \frac{3a}{2}$$

$$\tau_2 = \frac{\sigma_1 - \sigma_2}{2} = \frac{2a + a}{2} = \frac{3a}{2}$$

$$\tau_3 = \frac{\sigma_2 - \sigma_3}{2} = \frac{-a + a}{2} = 0$$

c. The hydrostatic component $\sigma_m = 0$ and the deviatoric components are

$$\sigma_1' = \sigma_1 - \sigma_m = 2a, \quad \sigma_2' = -a, \quad \sigma_3' = -a$$

Note that the deviatoric stress components are the same as those of the principal stress matrix since $\sigma_m = 0$. In fact, the given state of stress is a state of pure shear.

d. The octahedral stresses are obtained from Equation 1.34a as

$$\tau_{\text{oct}} = \sqrt{2}\, a$$

1.14 A NOTE ON THE STRESS EQUATIONS

In developing the stress equations comprising stress equilibrium, stress transformation law, principal stresses, etc., the derivations are obviously independent of material behavior. In fact, these equations are strict expressions for equlibrium and therefore are valid for any body disregarding its material behavior under load.

It should also be noted that the stress equations of equilibrium are strictly satisfied in terms of the current deformed configuration.* Nonlinearities introduced by changes of geometry due to large deformations produce a significant effect on the derived forms of the equations of equilibrium.**

PROBLEMS

Stress at a Point and Stress Equilibrium

1.1 Represent all positive stress components on the elements shown below.

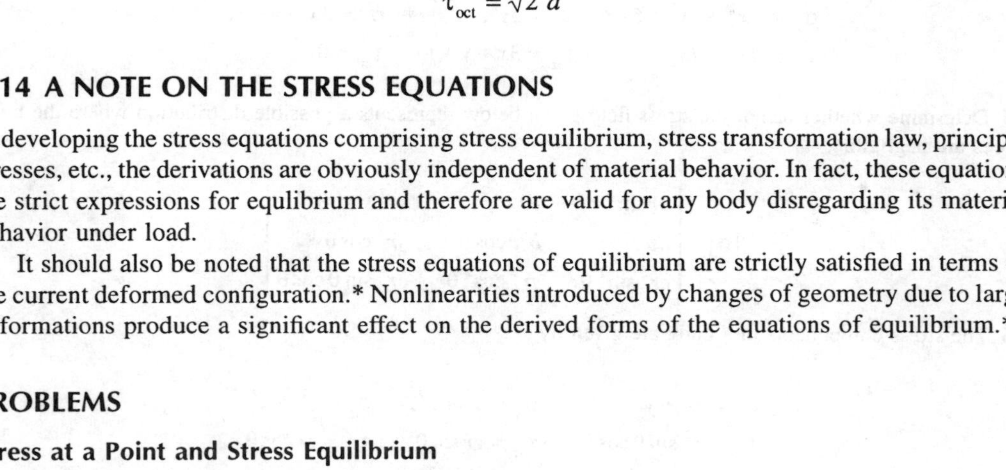

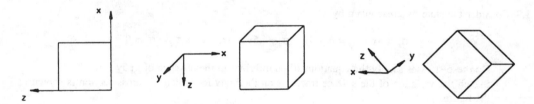

Problem 1.1

1.2 In the absence of body forces, does the following state of stress at a point in a body satisfy the equations of equilibrium?

$$\sigma_x = 3x^2y, \quad \sigma_y = 4x^2y^2, \quad \sigma_z = 7xz^3$$
$$\tau_{xy} = 8x^2y^2, \quad \tau_{yz} = 9y^2z^2, \quad \tau_{zx} = 10x^2y^2$$

---

* Stresses defined with respect to the deformed configuration form a Cauchy stress tensor. In the original undeformed configuration, the stresses represent Kirchoff's stress tensor. For this, see Fung.[6]

** To illustrate this, consider a simple example of a pressurized thin-walled sphere for which equilibrium yields the stresses $\sigma_\theta = \sigma_\phi = p_i r_m / 2h$, where p_i is the internal pressure. Strictly speaking, r_m and h are the deformed mean radius and wall thickness. However, for a sphere (balloon) made of rubbery material undergoing deformation, the condition of constant volume applies. Hence, $4\pi r_{mo}^2 h_o = 4\pi r_m^2 h$ and equilibrium stresses in the original configurations are expressed as $\sigma_\theta = \sigma_\phi = (p_i r_m / 2h_o)\,(r_m / r_{mo})^3$. The nonlinearity is self-evident. Another example is the equilibrium of a flexible cantilever where large deflections change the lever arm for bending moment by a significant amount.

1.3 Determine the restrictions that must be imposed on the constants a_1, a_2, a_3, a_4, a_5, a_6, and a_7 for the following stresses and body forces to satisfy the equilibrium equations:

$$\sigma_x = a_1 x^2 y^2 z, \quad \sigma_y = a_2 x^3 y^2, \quad \sigma_z = a_3 y^2 z^3$$
$$\tau_{xy} = a_4 y^3 z^2, \quad \tau_{yz} = a_5 x^3 yz, \quad \tau_{zx} = a_6 xy^2 z^2$$
$$F_x = 0, \quad F_y = 0, \quad F_z = a_7 x^3 z$$

1.4 Obtain expressions for the body forces acting on a solid having the following stress field:

$$\sigma_x = -x^2 + 3y^2 - 5z, \quad \sigma_y = 2y^2, \quad \sigma_z = 3x + y + 3z - 5$$
$$\tau_{xy} = 4xy - 7, \quad \tau_{yz} = 3x + y + 1, \quad \tau_{zx} = 0$$

1.5 Determine whether or not the stress field given below represents a possible distribution where the body forces are negligible.

$$[\sigma] = \begin{bmatrix} ar\sin\theta z & dz^2 & er^2\sin^2\theta \\ dz^2 & bzr\cos\theta & fr^2\cos\theta \\ er^2\sin^2\theta & fr^2\cos^2\theta & cr^2\sin\theta\cos\theta \end{bmatrix}$$

1.6 The stress components in a body are given by

$$\sigma_x = x^2 + y^2, \quad \sigma_y = y^2 + z^2, \quad \sigma_z = z^2 + x^2$$
$$\tau_{xy} = r^2\sin\theta\cos\theta, \quad \tau_{yz} = zr\sin\theta, \quad \tau_{zr} = zr\cos\theta$$

What must the body forces be in order to satisfy the conditions of equilibrium?

1.7 Consider the state of stress given by

$$\sigma_x = 4x \quad \tau_{xy} = 6 - 4y, \quad \sigma_y = -6y + 4x, \quad \tau_{xy} = \tau_{yz} = \sigma_z = 0$$

a. Do these stresses satisfy the equations of equilibrium in the absence of body forces?
b. Sketch the variation of the surface tractions on the body for which the cross section is shown in the figure.

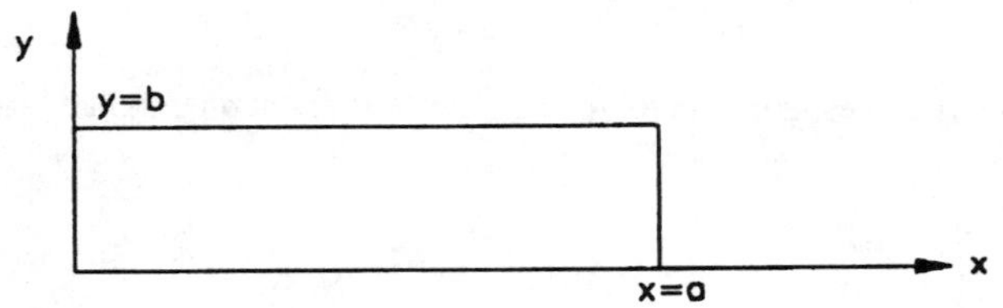

Problem 1.7

1.8 Suppose that the body force vector is $F_r = F_\theta = 0$ and $F_z = -g$, where g is a constant. Consider the following stress tensor:

$$[\sigma]_{r\theta z} = \alpha \begin{bmatrix} r\sin\theta & -z & 0 \\ -z & 0 & -r\sin\theta \\ 0 & -r\sin\theta & \sigma_z \end{bmatrix}$$

where α is a constant, and find an expression for σ_z such that $[\sigma]$ satisfies the equations of equilibrium.

1.9 For a state of plane stress given by

$$[\sigma]_{xyz} = \begin{bmatrix} \sigma_x & \tau_{yx} & 0 \\ \tau_{xy} & \sigma_y & 0 \\ 0 & 0 & 0 \end{bmatrix}$$

a. Derive the equilibrium equations in this special case.
b. If the stress components are expressed in terms of a function $\phi(x, y)$, such that

$$\sigma_x = \frac{\partial^2 \phi}{\partial y^2}, \quad \sigma_y = \frac{\partial^2 \phi}{\partial x^2}, \quad \tau_{xy} = -\frac{\partial^2 \phi}{\partial x \partial y}$$

will this stress distribution be in equilibrium at zero body forces?

1.10 For each of the following stress distributions, determine whether or not the conditions of equilibrium are satisfied in the absence of body forces. If the stress distribution does not satisfy the equilibrium conditions, calculate the body forces required to achieve equilibrium.

a. $\sigma_x = x \sin \frac{2\pi y}{L}, \quad \tau_{xy} = \frac{L}{2\pi} \cos \frac{2\pi y}{L} + z$

$\sigma_y = -y \sin \frac{2\pi x}{L}, \quad \tau_{yz} = 0, \; \tau_{zx} = 0, \; \tau_{zx} = 0$

b. $\sigma_x = c_1 xz, \; \sigma_y = 0, \; \sigma_z = (c_2 z + c_3)(x + c_4),$

$\tau_{xy} = 0, \; \tau_{yx} = 0, \; \tau_{zx} = -\frac{c_1}{2} z^2 - \frac{c_2}{2} x^2 - c_2 c_4 x + c_5$

c. $\sigma_x = -ay, \; \tau_{xy} = 0, \; \sigma_y = -ay, \; \tau_{yz} = 0,$

$\sigma_z = -ay, \; \tau_{zx} = 0.$

Stress Transformation

1.11 Given $\sigma_x = 30$, $\sigma_y = 0$, $\sigma_z = -20\backslash 3$, $\tau_{xy} = 40$, $\tau_{yz} = 20$, and $\tau_{zx} = 50$ (MPa), calculate for a plane (l, m, n), where $l = 0.4$ and $m = 0.6$, the resultant shear stress.

1.12 On a plane whose normal makes 60° with the x-axis and 45° with the y-axis, the stress vector at a point has a magnitude = 200 MPa and a direction that makes 45° with the x-axis and 60° with the z-axis. If $\sigma_y = \tau_{zx} = \tau_{yz} = 0$, determine the state of stress at this point.

1.13 The state of stress at a point O of a deformable solid is characterized by the matrix:

$$[\sigma]_{xyz} = \begin{bmatrix} \sigma_x & 20 & 10 \\ 20 & 0 & 20 \\ 10 & 20 & 0 \end{bmatrix} \text{MPa}$$

Determine σ_x such that there is at least one plane passing through the point O in such a way that the stress resultant vector on that plane is zero. Determine the direction cosines of the normal to that plane.

1.14 The matrix of Cartesian stress components at a point is shown below in MPa. Determine a direction **i** such that the stress resultant vector on a plane normal to **i** has $S_x = S_x = 0$, and determine S_z on that plane.

$$[\sigma]_{xyz} = \begin{bmatrix} 20 & 0 & 40 \\ 0 & 30 & 60 \\ 40 & 60 & 0 \end{bmatrix} \text{MPa}$$

1.15 For a plane whose normal has 40° and 70° with x and y, respectively, calculate the resultant normal and vectors shear stress on the plane given that $\sigma_x = 20$, $\sigma_y = 30$, $\sigma_z = -10$, $\tau_{xy} = 50$, and $\tau_{zx} = \tau_{zy} = 0$ (MPa).

1.16 The matrix of Cartesian components of stress at a point is shown below. Determine τ_{xy} so that there will be a stress-free plane through the point and determine the direction cosines of the normal **i** of the stress-free plane.

$$[\sigma] = \begin{bmatrix} 10 & \tau_{yz} & 10 \\ \tau_{xz} & 0 & 20 \\ 10 & 20 & 0 \end{bmatrix} \text{MPa}$$

1.17 Determine the stress vector and the resultant shear stress on a plane whose normal is equally inclined to the axes for the stress tensor:

$$[\sigma] = \begin{bmatrix} a & 0 & d \\ 0 & b & e \\ d & e & c \end{bmatrix}$$

1.18 The stresses acting at a point are (units in MPa)

$$\sigma_x = 40,\ \sigma_y \text{ is unknown},\ \sigma_z = 0,\ \tau_{yz} = 40,$$

$$\tau_{zx} = 20,\ \tau_{xy} = 30$$

Calculate the direction cosines of the plane on which there is no stress and the value of σ_y.

1.19 Using the transformation equations of plane stress, determine σ_x and $\tau_{x'y'}$ for the x'-axis rotated 30° clockwise from the x-axis. The nonzero stress components in MPa are $\sigma_x = 200$, $\sigma_y = 100$, and $\sigma_x = -50$.

1.20 Using the transformation equations of plane stress, determine σ_x and $\tau_{x'y'}$ for the x'-axis oriented 10° counterclockwise from the x-axis. The nonzero stress components in MPa are $\sigma_x = -90$, $\sigma_y = 50$, and $\tau_{xy} = 60$.

1.21 Consider a state of stress in which the nonzero stress components are σ_x, σ_y, σ_z, and τ_{xy}. Note that this is not a state of plane stress since $\sigma_z \neq 0$. Consider another set of coordinate axes (x', y', z') with the z'-axis coinciding with the z-axis and with the x'-axis located counterclockwise through an angle θ from the x'-axis. Show that the transformation equations for this state of stress are identical with that for plane stress.

1.22 A state of plane stress is specified by the following stress components (in MPa): $\sigma_x = 90$, $\sigma_y = -10$, $\tau_{xy} = 40$. The state of stress is referred to axes (x', y', z'), where the x'-axis lies in the x–y plane at $\theta = \pi/6$ clockwise from the x-axis and the z'-axis coincides with the z-axis. Determine the normal and shear stresses on a plane perpendicular to the x-axis.

1.23 A tie-bar of steel has a rectangular cross section 150 × 20 mm, and carries a tensile load of 200 kN. Find the stress normal to a plane making an angle of 30° with the cross section and the shear stress on this plane. Assume uniform stress distribution.

1.24 At a point in a vertical cross section of a beam, there is a resultant stress of 75 MPa which is inclined upward at 35° to the horizontal. On the horizontal plane through that point there is only shear stress. Find, in magnitude and direction, the resultant stress vector on the plane that is inclined at 45° to the *x*-axis and 60° to the *y*-axis.

1.25 A rectangular beam is subjected to a pure bending moment *M*. The cross section of the beam is rectangular. Using the elementary bending formula $\sigma = My/I$, determine the normal and shear stresses at point (x, y) on a plane inclined 45° to the horizontal axis.

1.26 A round rod of diameter *d* made of parallel fiberglass filaments embedded in epoxy resin is subjected to an axial tensile force *P*. The number of filaments per unit area normal to the laying direction of the filaments is *n*. Determine the tensile force carried by each filament assuming uniform stress distribution and neglecting the load-carrying capacity of the epoxy resin when

a. The filaments are along the direction of *P*.
b. The filaments are inclined at an angle α to *P*.

1.27 Consider a spherical shell subjected to internal pressure. Let σ_r, σ_θ, and σ_ϕ be the normal stresses and $\tau_{r\theta}$, $\tau_{\theta\phi}$, and $\tau_{\theta r}$ the shear stresses at a point (x, y, z). Show that

a. $\sigma_\theta = \sigma_\phi$ and $\tau_{r\theta} = \tau_{\theta\phi} = \tau_{\theta r} = 0$.
b. For a point on the outer surface find the stress components σ_x, σ_y, and σ_z in terms of the spherical stress components.

Principal Stresses — Maximum Shear Stresses

1.28 The nonzero stress components in MPa are $\sigma_x = 200$, $\sigma_y = 100$, and $\tau_{xy} = -50$. Determine the principal stresses and the maximum shear stress. Determine the angle between the x'-axis and the *x*-axis when the x'-axis is in the direction of the maximum principal stress.

1.29 A rivet is under the action of a shear stress of 60 MPa and a tensile stress, due to contraction, of 45 MPa. Determine the magnitude and direction of the greatest tensile and shear stresses in the rivet.

1.30 A propeller shaft is subjected to an end thrust producing a stress of 90 MPa, and the maximum shear stress arising from torsion is 60 MPa. Calculate the magnitude of the maximum tensile and shear stresses.

1.31 Two states of stress are given in MPa by the two tensors below. Calculate the magnitudes and directions of the principal stresses for each state.

$$\begin{bmatrix} 50 & -100 & -100 \\ -100 & 150 & 0 \\ -100 & 0 & 200 \end{bmatrix} \text{ and } \begin{bmatrix} 50 & 100 & 100 \\ 100 & 150 & 0 \\ 100 & 0 & 200 \end{bmatrix} \text{ MPa}$$

Calculate the deviatoric stress components for each state. Compare the results of the two cases.

1.32 Determine the principal stresses when all six components of the state of stress are equal. Show that this state of stress is a simple tension.

1.33 The state of stress at a point is given in MPa by $\sigma_x = -120$, $\sigma_y = 140$, $\sigma_z = 66$, $\tau_{xy} = 45$, $\tau_{yx} = -65$, and $\tau_{zx} = 25$. Determine the three principal stresses and their directions.

1.34 The state of stress at a point is given in MPa by $\sigma_x = 0$, $\sigma_y = 100$, $\sigma_z = 0$, $\tau_{xy} = -60$, $\tau_{yz} = 35$, and $\tau_{zx} = 50$. Determine the three principal stresses and their directions.

1.35 A state of stress is given by the tensor

$$[\sigma]_{xyz} = \begin{bmatrix} 0 & \tau & 0 \\ \tau & 0 & \tau \\ 0 & \tau & 0 \end{bmatrix}$$

Calculate the magnitude and direction of the principal stresses.

1.36 An element in a body in a state of plane stress is subjected to the following stresses in MPa: $\sigma_x = 100$, $\sigma_y = -60$, and $\tau_{xy} = 60$. Find the principal stresses and their directions.

1.37 The components of stresses at a point in MPa are $\sigma_x = 120$, $\sigma_y = -80$, $\sigma_z = -40$, $\tau_{xy} = 30$, $\tau_{xz} = 0$, and $\tau_{yz} = 20$. Determine the principal stresses and their directions with respect to the x, y, z axes.

1.38 The states of stress at two points in a loaded beam are given in MPa by

$$\sigma_x = 150,\ \ \tau_{xy} = 100,\ \ \sigma_y = 0 \ \text{ and } \ \sigma_x = 100,$$

$$\sigma_y = 50,\ \ \tau_{xy} = 60.$$

Determine for each point:

a. The magnitude of the maximum and minimum principal stresses and the maximum shear stress and the orientation of the principal and maximum shear planes.
b. Draw Mohr's circle for the state of stress at each point and sketch the results on properly oriented elements.

1.39 For the case of plane stress σ_x is unknown, while $\sigma_y = 30$ MPa and $\tau_{xy} = -80$ MPa.

a. Determine in terms of σ_x the stress invariants I_1, I_2, and I_3.
b. If the smallest principal stress σ_3 is -10 MPa, find σ_x.
c. What are the values of the three principal stresses?
d. Determine the direction cosines of the first principal axis.
f. Determine the principal stress deviations.

1.40 The state of stress at a point is specified by the following stress components in MPa: $\sigma_x = \sigma_y = \sigma_z = 0$, $\tau_{xy} = -75$, $\tau_{yz} = 65$, and $\tau_{zx} = -55$. Determine the principal stresses, the direction cosines for the three principal stress directions, and the maximum shear stress.

1.41 The state of stress at a point is specified by the following stress components in MPa: $\sigma_x = 11$, $\sigma_y = -86$, $\sigma_z = 55$, $\tau_{xy} = 60$, and $\tau_{yz} = \tau_{zx} = 0$. Determine the principal stresses, the direction cosines of the principal stress directions, and the maximum shear stress.

1.42 The nonzero stress components in MPa are $\sigma_x = 80$, $\sigma_z = -60$, and $\tau_{xy} = 30$. Determine the principal stresses and the maximum shear stress. Determine the angle between the x'- and the x-axes when the x'-axis is in the direction of the principal stress with largest absolute magnitude.

1.43 The state of stress at a point is given by the following components in MPa: $\sigma_x = 123$, $\sigma_y = 89$, $\sigma_z = 43$, $\tau_{xy} = 42$, $\tau_{yz} = 53$, and $\sigma_{zx} = 8$. Find the value of the principal stresses and their directions.

1.44 Let the state of stress at a point be given in MPa by $\sigma_x = 120$, $\sigma_y = -55$, $\sigma_z = -85$, $\tau_{xy} = -55$, $\tau_{yx} = 33$, and $\tau_{zx} = -75$. Determine (a) the three principal stresses, (b) the maximum shear stress.

1.45 Determine the principal stresses and their directions for the state of stress given by the following stress tensor:

$$[\sigma]_{xyz} = \begin{bmatrix} 18 & 0 & 24 \\ 0 & -50 & 0 \\ 24 & 0 & 32 \end{bmatrix} \text{MPa}$$

1.46 At a point in a body, the state of stress is expressed by

$$[\sigma]_{xyz} = \begin{bmatrix} 10 & 20 & 10 \\ 20 & -20 & -30 \\ 10 & -30 & 40 \end{bmatrix} \text{MPa}$$

Find the directions of the principal axes and the values of the corresponding principal stresses.

1.47 A three-dimensional state of stress is given in MPa by

$$\sigma_x = 50, \quad \sigma_y = 0, \quad \sigma_z = 0,$$
$$\tau_{xy} = 30, \quad \tau_{yz} = 20, \quad \tau_{zx} = -30$$

a. Show that one principal stress is 20 MPa, and find its direction.
b. Find the value of the other two principal stresses.
c. Determine the principal stress deviations.

1.48 The state of stress at a point in a loaded element is specified by

$$[\sigma]_{xyz} = \sigma_o \begin{bmatrix} 1 & a & b \\ a & 1 & 0 \\ b & 0 & c \end{bmatrix}$$

Here σ_o is a known value of the stress, and *a*, *b*, and *c* are constants. Find the constants such that the octahedral plane is free of stress.

1.49 Let $\sigma_x = -5C$, $\sigma_y = C$, $\sigma_z = C$, $\tau_{xy} = -C$, $\tau_{yz} = \tau_{zx} = 0$ where C is a constant. Determine the principal stresses, principal stress deviations, principal axes, greatest shear stress, and octahedral stress.

1.50 In a metal hot-forming operation, the state of stress is given by

$$[\sigma]_{r\theta z} \begin{bmatrix} -80 & -60 & 0 \\ -60 & -80 & 0 \\ 0 & 0 & -80 \end{bmatrix} \text{MPa}$$

Find the hydrostatic pressure and the maximum shear stresses. Also obtain the principal stress deviations.

1.51 A long, thin-walled pipe is subjected to an internal pressure p_i and an axial tensile force $P/p_i = 1000$. The radius and the thickness of the pipe are $r_o = 180$ mm and $h = 2$ mm. The axial and tangential stresses under internal pressure are $\sigma_z = pr_o/2h$ and $\sigma_\theta = \sigma_z$. Determine the values of P and p and find the stress components on a plane making 30° with the pipe axis. The maximum allowable normal stress is 100 MPa.

1.52 A thin-walled, cylindrical pressure vessel of 300 mm radius and 2.5 mm wall thickness has a welded seam at an angle of 30° with the axial direction. The vessel is subjected to an internal pressure p MPa and an axial load of 20 kN applied through rigid end plates. Find the allowable value of p if the normal and shear stresses acting simultaneously in the plane of welding are limited to 30 and 10 MPa, respectively.

1.53 A rod of gray cast iron of uniform circular cross section is subjected to combined loading consisting of a tensile load P and a twisting moment M_t. The load P remains at such a value that it produces a tensile stress $\sigma_u/2$ on any cross section where σ_u is the ultimate tensile strength of cast iron. The torque T is increased gradually until fracture occurs on some inclined surface. Assuming that fracture takes place when the maximum principal stress σ_1 reaches the ultimate strength σ_u, determine the magnitude of the torsional shear stress produced by M_t at fracture and determine the orientation of the fracture surface.

1.54 A circular pipe made of a brittle material has a 100-mm inside diameter and a 150-mm outside diameter. It is subjected to a twisting moment $M_t = 70$ kN · m and a compressive axial force P kN. The material is known to have an ultimate compressive strength $\sigma_c = 250$ MPa and an ultimate tensile strength $\sigma_u = 100$ MPa. Determine the maximum value of the compressive force P so that the pipe does not fail in either tension or compression.

REFERENCES

1. Love, A. E., *Mathematical Theory of Elasticity*, 4th ed., Dover, New York, 1944, 348.
2. Ford, H., *Advanced Mechanics of Materials*, Longman, London, 1963, 51, 79.
3. Popov, E. P., *Mechanics of Materials*, 2nd ed., Prentice-Hall, Englewood Cliffs, NJ, 1987, 248.
4. Srinath, L. S., *Advanced Mechanics of Solids*, Tata McGraw-Hill, New Delhi, 1980, 18.
5. Venkatraman, B. and Patel, S. A., *Structural Mechanics with Introductions to Elasticity and Plasticity*, McGraw-Hill, New York, 1970, 30.
6. Fung, Y. C., *Foundations of Solid Mechanics*, Prentice-Hall, Englewood Cliffs, NJ, 1965.

2 Analysis of Strain

Infinitesimal strain at a point is defined by considering the undeformed and deformed geometrical configurations of the solid. Strain–displacement relations are derived in Cartesian and cylindrical polar coordinates. Conditions for strain compatibility are established. It is shown that strain, similar to stress, is a symmetric tensor composed of six components that transform from one set of orthogonal axes to another following the same transformation law of stress. Application to strain gauge rosettes is demonstrated. Expressions for the principal strains, maximum shear strain, octahedral, dilatation, and deviatoric strains are obtained. Strain rate–velocity relations, which find application in the analysis of plastic flow, are also derived and a note on large strain definition is included.

2.1 INFINITESIMAL STRAINS

In Chapter 1, it has been shown that the material mechanical behavior of a straight rod of uniform cross-sectional area is identified by the stress–strain relationship. The strain $\varepsilon = \Delta L/L_0$ in such a case of simple geometry and loading has been defined as the ratio of rod extension ΔL to the original length L_0. Following the same approach adopted for the generalization of the concept of stress, the concept of strain will be developed in this chapter from the simple case of a straight rod in tension to the general case of a deformable solid.

2.1.1 Normal Strain

Consider a cube element in the body, as shown in Figure 1.7, having its sides oriented to the x-, y-, and z-directions. For simplicity the cube is assumed to deform only in the x- and y-directions, while there is no deformation in the z-direction. Such a case of deformation is called a state of plane strain. It will be further assumed that the deformation in the x- and y-directions are small and uniform, thus presenting a state of infinitesimal homogeneous strain.

Consider now the deformation in the x- and y-directions. Referring to Figures 2.1 and 2.2 the undeformed element is shown by the rectangle *ABDC* in which *A* is fixed in position to exclude rigid-body linear displacement. It will be noted that such fixation does not eliminate rigid body rotation about *A*. The deformed element is indicated by *AB′D′C′*. Two modes of deformation are identified which may occur either separately or simultaneously. The first mode is shown in Figure 2.1 in which the element remains rectangular and the deformation is merely a linear displacement along each side. The strain along *AB* is *BB′*/*AB*, which equals the strain along *CD* for homogeneous conditions. Also the strain along *AC* is *CC′*/*AC*, which again equals the strain along *BD* but not necessarily equal the strain along *AB*. These strains are called normal strains in analogy to normal stresses and are sometimes called linear or direct strains. Since *AB* and *AC* are parallel to the x- and y-directions, respectively, then the normal strain components in the x-, y-, and z-directions for the state of plane strain are expressed by

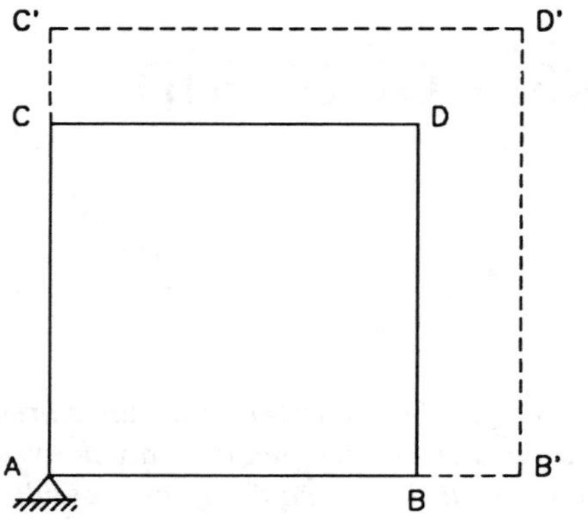

FIGURE 2.1 Normal homogeneous strains resulting from linear homogeneous displacements with no change in geometry.

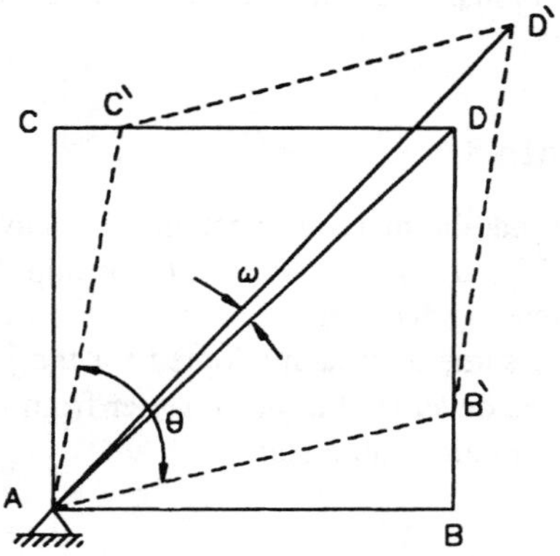

FIGURE 2.2 Shear strain resulting from relative angular displacements; a rectangle changes shape to a parallelogram.

$$\varepsilon_x = \frac{BB'}{AB}, \quad \varepsilon_y = \frac{CC'}{AC}, \quad \varepsilon_z = 0 \tag{2.1}$$

The normal strains indicated in Figure 2.1 are positive since BB' and CC' are positive displacements in the x- and y-directions, respectively. This means that tensile strains are positive while compressive strains are negative, which is the basis of the sign convention adopted for normal stresses.

2.1.2 Shear Strain

The second mode of deformation is shown in Figure 2.2 in which the element changes from a rectangle into a parallelogram of sides equal to the original sides of the element. In this deformation mode the displacements BB' and CC' have been normal to AB and AC, respectively. The angle between AB and AC is no longer a right angle as a result of deformation. Such deformation is expected to occur when the side surfaces of the rectangle element are subjected to shear forces. The shear strain for an orthogonal set of coordinates is thus defined by the deviation from orthogonality and is expressed for small strain by

$$\gamma_{xy} = \frac{\pi}{2} - \theta \tag{2.2a}$$

where θ is the angle between AB' and AC' given for small deviations by the relation:

$$\theta = \frac{\pi}{2} - \left[\tan^{-1}\left(\frac{BB'}{AB}\right) + \tan^{-1}\left(\frac{CC'}{AC}\right)\right]$$

Substitution in Expression 2.2a, noting that for small angles the tangent is approximately equal to the angle itself, yields

$$\gamma_{xy} = \left(\frac{BB'}{AB}\right) + \left(\frac{CC'}{AC}\right) \tag{2.2b}$$

The shear strain indicated in Figure 2.2 is positive since BB' and CC' are positive displacements in the y- and x-directions, respectively. The rule is that the shear strain is positive when the angle θ is an acute angle. In fact, the sign convention for the stresses in Section 1.5 is based on this sign convention of the strain.

By referring to Figure 2.2, it will be noted that diagonal AD has rotated an angle ω, which means that the shear strain is accompanied by a rigid-body rotation of the element about the fixed point A. From simple geometry, it can be shown that for infinitesimal displacements the angle of rotation is given by

$$\omega = \frac{1}{2}\left(\frac{BB'}{AB} - \frac{CC'}{AC}\right) \tag{2.3}$$

Expression 2.3 shows that the shear strain is independent of element rotation and can thus occur in different configurations having the same angle θ as shown in Figure 2.3. The rotation is determined by the displacement field. Cases in Figure 2.3a and b apply to the boundary interface of a deformable solid in contact with a rigid body. The case in Figure 2.3c applies to uniform stretching of a strip in the direction of the diagonal AD.

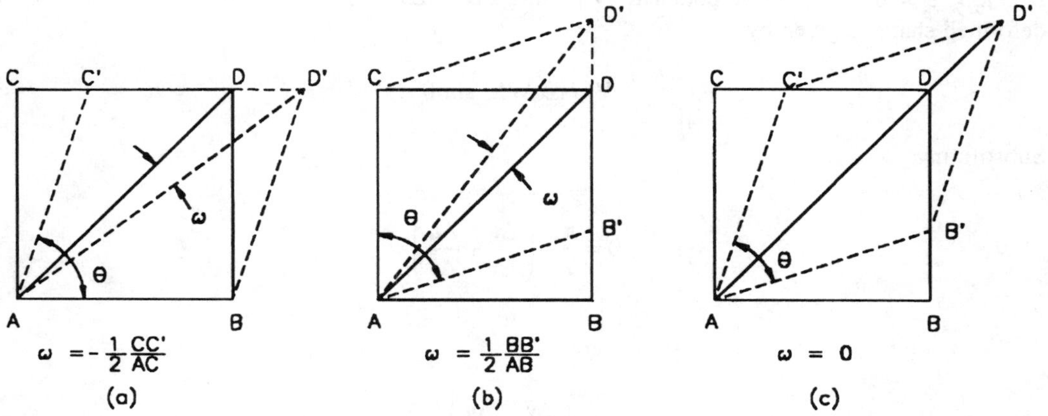

FIGURE 2.3 Shear strain and rigid-body rotation.

2.1.3 Volumetric Strain

Another type of strain is yet to be considered. The element $ABDC$ of Figure 2.1 has a new volume in the deformed shape. Let $AB = CD = \Delta x$, $AC = BD = \Delta y$, thickness $= \Delta z$, $BB' = \Delta u$, $CC' = \Delta v$ and the change in thickness is Δw. The volume of the deformed element V is given by

$$V = (\Delta x + \Delta u)\,(\Delta y + \Delta v)\,(\Delta z + \Delta w)$$

$$= \Delta x \Delta y \Delta z \left(1 + \frac{\Delta u}{\Delta x}\right)\left(1 + \frac{\Delta v}{\Delta y}\right)\left(1 + \frac{\Delta w}{\Delta z}\right)$$

$$= \Delta x \Delta y \Delta z \left(1 + \frac{\Delta u}{\Delta x} + \frac{\Delta v}{\Delta y} + \frac{\Delta w}{\Delta z} + \text{higher powers}\ldots\right)$$

For infinitesimal strains, higher powers of $\Delta u/\Delta x$, $\Delta v/\Delta y$, and $\Delta w/\Delta z$ may be neglected. A volumetric strain ε_v may be thus defined as the ratio of the change in volume to the original volume. This gives

$$\varepsilon_v = \left[\Delta x \Delta y \Delta z \left(1 + \frac{\Delta u}{\Delta x} + \frac{\Delta v}{\Delta y} + \frac{\Delta w}{\Delta z}\right) - \Delta x \Delta y \Delta z\right] \bigg/ \Delta x \Delta y \Delta z$$

which simplifies to

$$\varepsilon_v = \frac{\Delta u}{\Delta x} + \frac{\Delta v}{\Delta y} + \frac{\Delta w}{\Delta z}$$

or by virtue of Relations 2.1:

$$\varepsilon_v = \varepsilon_x + \varepsilon_y + \varepsilon_z \tag{2.4}$$

Thus, the volumetric infinitesimal strain is the sum of the three normal strains.

Applying Expression 2.4 to the element of Figure 2.2 gives $\varepsilon_v = 0$ since in this particular case $\varepsilon_x = \varepsilon_y = \varepsilon_z = 0$. This can be obtained by putting $BB' = \Delta v$, $CC' = \Delta u$ so that the volume of the deformed shape is given by

$$V = \Delta x \Delta y \Delta z \sin\theta$$

Substituting

$$\theta = \frac{\pi}{2} - \left(\frac{\Delta v}{\Delta x} + \frac{\Delta u}{\Delta y}\right)$$

gives

$$\sin\theta = \cos\left(\frac{\Delta v}{\Delta x} + \frac{\Delta u}{\Delta y}\right) = \cos\gamma_{xy}$$

and, since γ_{xy} is infinitesimal, then $\cos \gamma_{xy} \cong 1$ and, hence,

$$V = \Delta x \cdot \Delta y \cdot \Delta z = V$$

i.e., there is no volume change in this case.

The above expressions of strain apply when the strains are infinitesimal, as in most engineering structures, and hence are called engineering strains. For large strains, such as those encountered in rubber deformation or forming processes, other strain expressions are employed, as will be shown in Section 2.10.

Example 2.1:

Defining strain as the ratio of change in length to original length, derive expressions for the strain components and element rotation in

a. *A thick-walled cylinder expanding uniformly under internal pressure as in Figure 2.4a;*
b. *A straight beam subjected to bending and bent to a radius* R *as in Figure 2.4b;*
c. *Torsion of a bar of circular cross section, Figure 2.4 c.*

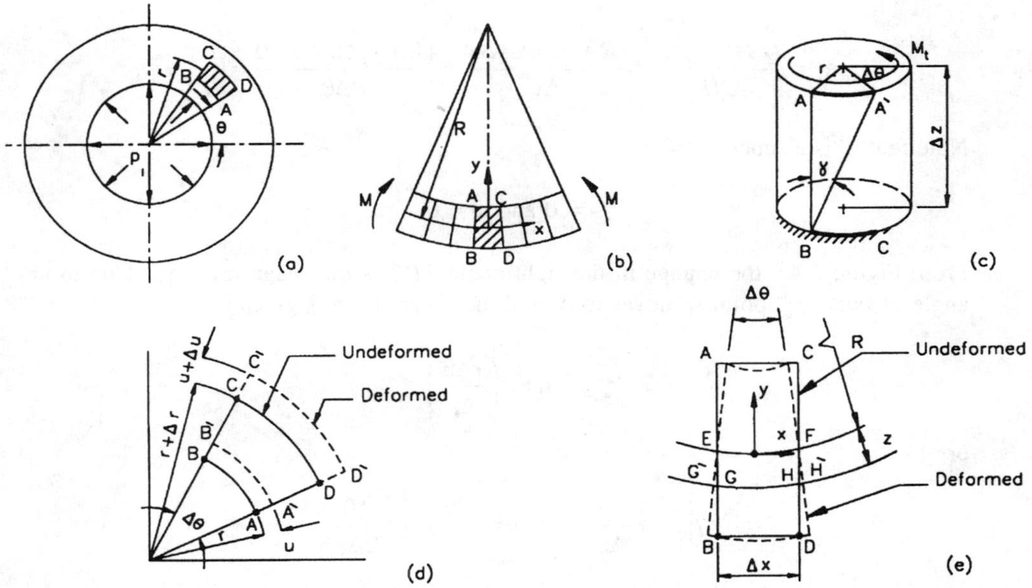

FIGURE 2.4 Example 2.1.

Solution:

a. Take an element as shown in Figure2.4a. From axial symmetry the cylinder is expanding only radially as in Figure2.4d. The strain in the radial direction is

$$\varepsilon_r = \frac{B'C' - BC}{BC} = \frac{(r + \Delta r + u + \Delta u - r - u) - (r + \Delta r - r)}{(r + \Delta r - r)} = \frac{\Delta u}{\Delta r}$$

Strain in the tangential (hoop) direction is

$$\varepsilon_\theta = \frac{A'B' - AB}{Ab} = \frac{(r+u)\Delta\theta - r\Delta\theta}{r\Delta\theta} = \frac{u}{r}$$

$$\gamma_{r\theta} = \text{change in the right angle } ABC = 0$$

$$\omega = 0$$

Note that ε_r and ε_θ are functions of r. Also, although the displacement is only radial, there is a tangential strain ε_θ in the θ-direction.

b. The undeformed and deformed beam elements are shown in Figure 2.4e. It is assumed that plane sections taken normal to its axis (e.g., AB and CD) remain plane after bending. Under the action of moments M, the top fibers are compressed while the bottom are stretched. Obviously, a certain fiber in between (e.g., EF) remains unchanged. This is the neutral fiber. Hence,

$$\Delta_x = R\Delta_\theta$$

The strain at any fiber GH at distance y from the neutral fiber is

$$\varepsilon_x = \frac{G'H' - GH}{GH} = \frac{(R+y)\Delta\theta - \Delta x}{\Delta x} = \frac{(R+y)\Delta\theta - R\Delta\theta}{R\Delta\theta} = \frac{y}{R}$$

Note that ε_x is a function of y.

$$\gamma_{r\theta} = 0 \text{ and } \omega = 0$$

c. From Figure 2.4c, the change in the right angle ABC is the shear strain $\gamma_{z\theta}$. Due to an angle of twist $\Delta\theta$, point A moves to A' and the shear strain is given by

$$\gamma_{z\theta} = \tan^{-1}\left(\frac{r\Delta\theta}{\Delta z}\right)$$

or

$$\gamma_{z\theta} = r\frac{\Delta\theta}{\Delta z} \quad \text{and} \quad \omega = -\frac{\gamma}{2} = -r\frac{\Delta\theta}{2\Delta z}$$

Note that in deriving this shear strain, circular cross sections are assumed to remain plane and circular while they rotate with respect to the base.

In solving this problem, the geometry of the deformed element is always defined according to some practically acceptable assumptions* (e.g., plane sections remain plane in bending).

2.2 INFINITESIMAL STRAIN–DISPLACEMENT RELATIONS

2.2.1 Cartesian Coordinates

In the preceding section, the amount of stretching, contraction, and change in right angles has been visualized by comparing the deformed and undeformed elements together. This is consistent with

* Assumptions fulfilling certain deformed geometry are in essence related to strain compatibility conditions, as will be explained in Section 2.3. Solutions available in strength of materials textbooks are based on initially assumed geometry of deformation.

the definition of a deformable solid in which the distance between any two particles changes upon application of forces. In other words, what matters in defining strains is the displacement of one point relative to another, excluding any rigid-body motion (translation and/or rotation) the solid may undergo as a whole.

Each particle (or point) will generally undergo a displacement of components (u, v, w) in the x-, y-, z-directions, respectively, which varies from one point to another within the solid. This describes a state of nonhomogeneous deformation.

For simplicity, a one-dimensional case is first considered. Figure 2.5 shows a strand of material positioned along the x-axis and fixed at one end while being stretched at the other end. Two points

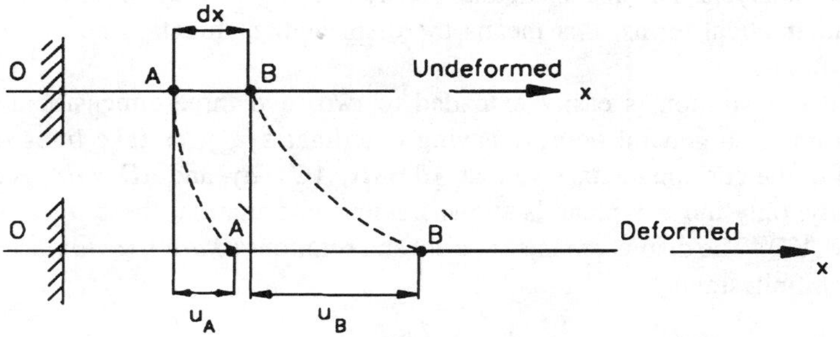

FIGURE 2.5 Normal strain in terms of linear displacement.

A and B, positioned apart a small distance dx on the undeformed strand, will assume the new positions A' and B' in the deformed state. Obviously, the displacement $u(x)$ of any particle will be a function of its position x measured from point 0 where the displacement is zero. The displacement of point B with respect to that of point A is defined according to Taylor series expansion as

$$u_B = u_A + \left(\frac{du}{dx}\right)_A (dx) + \frac{1}{2!}\left(\frac{du}{dx^2}\right)_A^2 (dx)^2 + \frac{1}{3!}\left(\frac{d^3u}{dx^3}\right) dx^3 + \cdots$$

For infinitesimal length element (dx), higher powers of dx are negligibly small and hence

$$u_B = u_A + \left(\frac{du}{dx}\right)_A dx$$

Applying the definition of normal strain to the element AB gives

$$\varepsilon_x = \frac{A'B' - AB}{AB}$$

where

$$A'B' = dx + u_B - u_A = dx + \left(\frac{du}{dx}\right)_A dx = dx\left[1 + \left(\frac{du}{dx}\right)_A\right]$$

Hence,

$$\varepsilon_x = \left(\frac{du}{dx}\right)_A$$

or generally

$$\varepsilon_x = \frac{du}{dx} \tag{2.5}$$

Thus, (du/dx) defines the normal infinitesimal strain at a generic point in the strand in terms of the displacement function $u(x)$. Expression 2.5, which bears similarity to Relation 2.1, applies to nonhomogeneous states of deformation.

The assumption that the displacement of points lying on the strand is continuous is implicitly made in the above analysis. This means that no separation or gaps may occur between two adjacent particles. In mathematical terms, this means the displacement function and its derivatives (e.g., du/dx) are continuous.

The above demonstration is easily extended to two- and three-dimensional states of strain. Consider the strains at a general point A having coordinates x, y, z. Take lines AB, AC, and AD initially parallel to the coordinate axes so that $AB = dx$, $AC = dy$, and $AD = dz$, as shown in Figure 2.6 (for simplicity, only the x–y plane is shown). After deformation, these lines will be displaced to $A'B'$, $A'C'$, and $A'D'$. The displacements (u, v, w) and rotations $(\partial u/\partial y)$, $(\partial v/\partial x)$..., etc. are assumed, as before, to be infinitesimal.

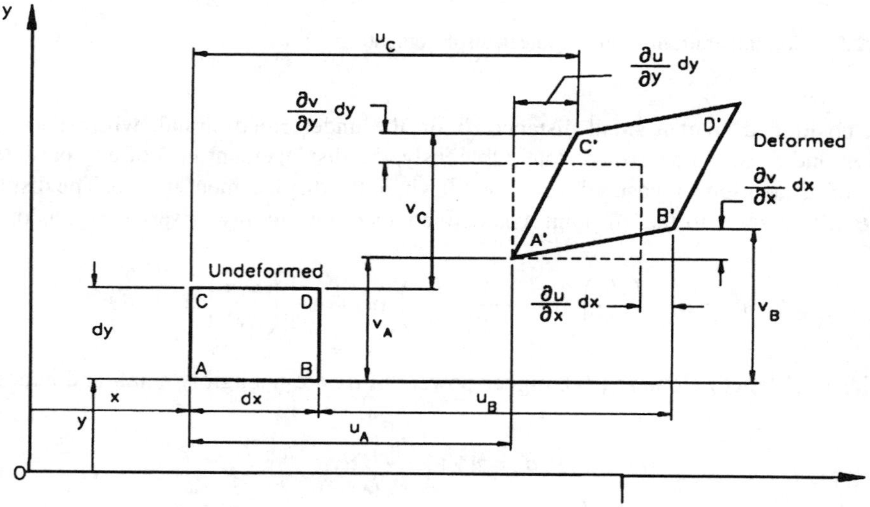

FIGURE 2.6 Two-dimensional strain in terms of two-dimensional displacement components in Cartesian coordinates.

The change in length for AB is $(A'B' - AB) = u_B - u_A$. But

$$u_B = u_A + \frac{\partial u}{\partial x} dx$$

Hence,

$$\varepsilon_x = \frac{A'B' - AB}{AB} = \frac{\left(u_A + \frac{\partial u}{\partial x} dx\right) - u_A}{dx} = \frac{\partial u}{\partial x}$$

Partial derivatives are used since the displacement components (u, v, w) are each a function of (x, y, z). Also

$$\varepsilon_y = \frac{A'C' - AC}{AC} = \frac{\left(v_A + \frac{\partial v}{\partial y}dy\right) - v_A}{dx} = \frac{\partial v}{\partial y}$$

The change in the right angle BAC defines the shear as

$$\gamma_{xy} = BAC - B'C'A' = \tan^{-1}\left(\frac{v_B - v_A}{\Delta x}\right) + \left(\frac{u_c - u_A}{\Delta y}\right)$$

For small angles and referring to Figure 2.6, this becomes

$$\gamma_{xy} \cong \frac{v_B - v_A}{dx} + \frac{u_c - u_A}{dy} \cong \frac{1}{dx}\left(\frac{\partial v}{\partial x}dx\right) + \frac{1}{dy}\left(\frac{\partial u}{\partial x}\right)dy$$

Hence,

$$\gamma_{xy} = \frac{\partial v}{\partial x} + \frac{\partial u}{\partial y}$$

Considering the projections in the y–z and z–x planes, gives

$$\varepsilon_z = \frac{\partial w}{\partial z}, \quad \gamma_{yz} = \frac{\partial v}{\partial z} + \frac{\partial w}{\partial y}, \quad \gamma_{zx} = \frac{\partial w}{\partial x} + \frac{\partial u}{\partial z}$$

Hence, the state of strain at a point in a deformable solid in Cartesian coordinates is determined by the following six infinitesimal strain–displacement relations:

$$\varepsilon_x = \frac{\partial u}{\partial x}, \quad \varepsilon_y = \frac{\partial v}{\partial y}, \quad \varepsilon_z = \frac{\partial w}{\partial z}$$

$$\gamma_{xy} = \frac{\partial v}{\partial x} + \frac{\partial u}{\partial y}, \quad \gamma_{yz} = \frac{\partial w}{\partial y} + \frac{\partial v}{\partial z}, \quad \gamma_{zx} = \frac{\partial u}{\partial z} + \frac{\partial w}{\partial x} \tag{2.6}$$

$$\varepsilon_v = \varepsilon_x + \varepsilon_y + \varepsilon_z = \frac{\partial u}{\partial x} + \frac{\partial v}{\partial y} + \frac{\partial w}{\partial z}$$

It should be indicated that rigid-body displacement does not contribute to strain components. For instance, rigid-body translation expressed by constant terms in the displacement components (u, v, w) do not obviously produce strains, since their derivates are zero. Also, rigid-body rotations do not produce any strain.

The rotation of the diagonal AD is

$$\omega_{AD} = \frac{1}{2}\left(\frac{\Delta v}{\Delta x} - \frac{\Delta u}{\Delta y}\right) \quad \text{or} \quad \omega_z = \frac{1}{2}\left(\frac{\partial v}{\partial x} - \frac{\partial u}{\partial y}\right) \tag{2.7a}$$

Similarly,

$$\omega_x = \frac{1}{2}\left(\frac{\partial w}{\partial y} - \frac{\partial v}{\partial z}\right) \text{ and } \omega_y = \frac{1}{2}\left(\frac{\partial u}{\partial z} - \frac{\partial w}{\partial x}\right) \tag{2.7b}$$

These rotations expressed by derivatives of displacements according to Relation 2.7 are rigid-body rotations.

Example 2.2:

Given the displacement field: u = axy, v = by², *where* a, b, *and* c *are constants, determine the strain components and the volumetric strain at point (1,2). Starting from the strain–displacement relation, would it be possible to determine the originally given displacement field? Comment on the result.*

Solution:

The strains are determined from Relations 2.6 as

$$\varepsilon_x = \frac{\partial u}{\partial x} = ay, \quad \varepsilon_y = \frac{\partial v}{\partial y} = 2by$$

$$\gamma_{xy} = \frac{\partial v}{\partial x} + \frac{\partial u}{\partial y} = ax$$

At point (1,2)

$$\varepsilon_x = 2a, \quad \varepsilon_y = 4b, \text{ and } \gamma_{xy} = a$$

From Relation 2.4,

$$\varepsilon_v = \varepsilon_x + \varepsilon_y = 2a + 4b$$

In order to solve the inverse problem of determining the displacement functions from strains, integrations are performed according to

$$u^* = \int \frac{\partial u}{\partial x} dx, \quad v^* = \int \frac{\partial v}{\partial y} dy$$

where u^* and v^* are the displacement functions (they should reduce to the same function u and v as originally given above). Hence,

$$u = \int (ay)dx = ayx + f(y)$$

$$v = \int (2by)dy = by^2 + g(x)$$

Note that the constants of integration are unknown functions of the other variable. To determine these functions, the condition that u and v must provide the correct expression for γ_{xy} has to be satisfied. Hence,

$$\gamma_{xy} = ax = \frac{\partial v^*}{\partial x} + \frac{\partial u^*}{ay} = \frac{dg}{dx} + ax + \frac{df}{dy}$$

or

$$\frac{f}{y} + \frac{g}{x} = 0$$

Since f is a function of y only and g is a function of x only, the above condition cannot be satisfied unless

$$\frac{df}{dy} = -\frac{dg}{dx} = \text{constant}$$

say = c_1. Hence, $f = c_1y + c_2$ and $g = -c_1x + c_3$, where c_1, c_2, and c_3 are unknown constants. The displacement components are, thus,

$$u^* = axy + (c_1y + c_2)$$

$$v^* = by^2 - (c_1x - c_3)$$

Comparing these expressions for u* and v* with the originally given ones for u, v shows that they give the same strain field but not the same displacement field. In fact, the last two terms of u^* and v^* represent rigid-body motions (c_2 and c_3: rigid-body translation and c_1y and $-c_1x$ rigid-body rotations). This becomes obvious by applying Equations 2.7 for the rigid-body rotation to give

$$\omega_z = \frac{1}{2}\left(\frac{\partial v^*}{\partial x} - \frac{\partial u^*}{\partial y}\right) = \frac{1}{2}(-c_1 - c_1) = -c_1$$

2.2.2 Cylindrical Polar Coordinates

The strain–displacement relations in cylindrical polar coordinates (r, θ, z) can be derived by considering the undeformed and deformed elements and applying the previous strain definitions. The derivation is simple for the case of axial symmetry and is already obtained in Example 2.1. For the general case the derivation is rather lengthy.[1] A simple way to derive these relations is to consider the relation between Cartesian and cylindrical polar coordinates together with the strain-displacement relations (Equations 2.6).

Let (u, v, w) be the displacement components at point 0 in Cartesian coordinates (x, y, z) respectively. Similarly (u_p, v_p, w_p) are the displacement components at the same point in cylindrical polar coordinates (r, θ, z), respectively. The relation between the two sets of components is given by Figure 2.7.

$$u = u_p \cos\theta - v_p \sin\theta$$

$$v = u_p \sin\theta + v_p \cos\theta$$

$$w = w_p$$

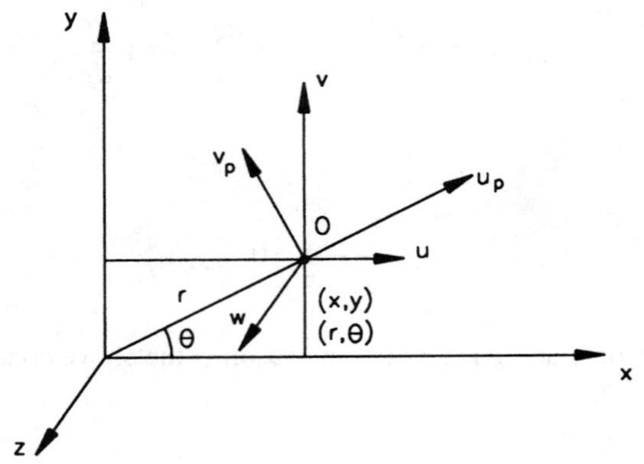

FIGURE 2.7 Transformation of displacements from Cartesian to cylindrical polar coordinates.

The relation between (r, θ) and (x, y) are

$$r^2 = x^2 + y^2, \quad x = r\cos\theta, \quad y = r\sin\theta, \quad \tan\theta = y/x$$

From Relation 2.6,

$$\varepsilon_x = \frac{\partial u}{\partial x} = \frac{\partial u}{\partial r}\frac{\partial r}{\partial x} + \frac{\partial u}{\partial \theta}\frac{\partial \theta}{\partial x}$$

Differentiation gives

$$\frac{\partial u}{\partial r} = \frac{\partial u_p}{\partial r}\cos\theta - \frac{\partial v_p}{\partial r}\sin\theta$$

$$\frac{\partial u}{\partial \theta} = \frac{\partial u_p}{\partial \theta}\cos\theta - u_p\sin\theta - \frac{\partial v_p}{\partial \theta}\sin\theta - v_p\cos\theta$$

$$\frac{\partial r}{\partial x} = \cos\theta \text{ and } \frac{\partial \theta}{\partial x} = -(\sin\theta)/r$$

Hence, upon substitution,

$$\varepsilon_x = \frac{\partial u_p}{\partial r}\cos^2\theta - \frac{\partial v_p}{\partial r}\cos\theta\sin\theta - \frac{1}{r}\left(\frac{\partial u_p}{\partial \theta}\cos\theta\sin\theta - u_p\sin^2\theta - \frac{\partial v_p}{\partial \theta}\sin^2\theta - v_p\cos\theta\sin\theta\right)$$

The above expression yields ε_r and ε_θ by putting

$$\varepsilon_r = (\varepsilon_x)_{\theta=0} \text{ and } \varepsilon_\theta = (\varepsilon_x)_{\theta=\pi/2}$$

$$\varepsilon_r = \frac{\partial u_p}{\partial r} \text{ and } \varepsilon_\theta = \frac{1}{R}\frac{\partial v_p}{\partial \theta} + \frac{u_p}{r}$$

In a similar way $\gamma_{r\theta}$, $\gamma_{\theta z}$ are obtained from the relation

$$\gamma_{xy} = \frac{\partial u}{\partial y} + \frac{\partial v}{\partial x}$$

observing that $\gamma_{r\theta} = (\gamma_{xy})_{\theta=0}$ and $\gamma_{\theta z} = (\gamma_{yz})_{\theta=0}$.

Finally, in terms of displacement components (u, v, w) in the cylindrical polar coordinates, the strain–displacement relations are given by

$$\begin{aligned} &\varepsilon_r = \frac{\partial u}{\partial r}, \quad \varepsilon_\theta = \frac{1}{r}\frac{\partial v}{\partial \theta} + \frac{u}{r}, \quad \varepsilon_z = \frac{\partial w}{\partial z} \\ &\gamma_{r\theta} = \frac{1}{r}\frac{\partial u}{\partial \theta} + \frac{\partial v}{\partial r} - \frac{v}{r}, \quad \gamma_{\theta z} = \frac{\partial v}{\partial z} + \frac{1}{r}\frac{\partial w}{\partial \theta}, \quad \gamma_{zr} = \frac{\partial u}{\partial z} + \frac{\partial w}{\partial r} \end{aligned} \tag{2.8}$$

and

$$\varepsilon_v = \varepsilon_r + \varepsilon_\theta + \varepsilon_z = \frac{\partial u}{\partial r} + \frac{1}{r}\frac{\partial v}{\partial \theta} + \frac{u}{r} + \frac{\partial w}{\partial z}$$

For complete axial symmetry (for both geometry of the body and its loading), conditions are independent of θ and $v = 0$; hence Relations 2.8 reduce to

$$\begin{aligned} &\varepsilon_r = \frac{\partial u}{\partial r}, \quad \varepsilon_\theta = \frac{u}{r}, \quad \varepsilon_z = \frac{\partial w}{\partial z} \\ &\gamma_{r\theta} = \gamma_{\theta z} = 0 \ \text{ and } \ \gamma_{zr} = \frac{\partial u}{\partial z} + \frac{\partial w}{\partial r} \end{aligned} \tag{2.9a}$$

For an axisymmetric plane problem in (r, θ), $\varepsilon_z = \gamma_{zr} = 0$ and the only nonvanishing strains are

$$\varepsilon_r = \frac{du}{dr} \ \text{ and } \ \varepsilon_\theta = \frac{u}{r} \tag{2.9b}$$

Note that the same expessions have been previously obtained in Example 2.1a.

2.2.3 Spherical Polar Coordinates

The strain–displacement relations in spherical polar coordinates (r, θ, ϕ) will not be derived here;* they are only presented for application to problems of spherical geometry.

Let u, v, w be the displacement components in the r, θ, ϕ coordinates, respectively. The strain components are given by

* See Ford,[1] p, 165.

$$
\begin{aligned}
\varepsilon_r &= \frac{\partial u}{\partial r}, \varepsilon_\theta = \frac{1}{r}\frac{\partial v}{\partial \theta} + \frac{u}{r} \\
\varepsilon_\phi &= \frac{1}{r\sin\theta}\left(\frac{\partial w}{\partial \phi} + u\sin\theta + v\cos\theta\right) \\
\gamma_{\theta\phi} &= \frac{1}{r\sin\theta}\left(\sin\theta\frac{\partial w}{\partial \theta} - w\cos\theta + \frac{\partial v}{\partial \phi}\right) \\
\gamma_{r\theta} &= \frac{1}{r\sin\theta}\frac{\partial u}{\partial \phi} + \frac{\partial w}{\partial r} - \frac{w}{r} \\
\gamma_{\phi r} &= \frac{\partial v}{\partial r} - \frac{v}{r} + \frac{1}{r}\frac{\partial u}{\partial \theta}
\end{aligned}
\tag{2.10}
$$

For the case of complete spherical symmetry (e.g., hollow sphere subjected to internal pressure), the strains in terms of the only nonvanishing displacement u are given by

$$
\varepsilon_r = \frac{du}{dr}, \quad \varepsilon_\theta = \varepsilon_\phi = \frac{u}{r}, \quad \gamma_{r\theta} = \gamma_{\theta\phi} = \gamma_{\phi r} = 0 \tag{2.11}
$$

2.3 STRAIN COMPATIBILITY CONDITIONS

The strain–displacement relations involve three independent displacement functions $u(x, y, z)$, $v(x, y, z)$, and $w(x, y, z)$. If these functions are known, the six independent strain components are determined by differentiation as from Relations 2.6. However, in many situations, strain data are available either through experimental measurements or numerical analysis. If they are given as functions of (x, y, z), it will be required to determine the displacement functions by integration of Equations 2.6. Here a problem arises since six equations are to be used to determine three unknown displacement functions. If the strains are arbitrarily described, the six Equations 2.6 are not expected, in general, to yield single-valued continuous solutions for u, v, and w. Hence, certain restrictions must be imposed on strains. Strain fields for which a single-valued displacement solution exists are called compatible strain fields. To this end, the additional conditions to be satisfied to limit the arbitrariness of strain fields are known as strain compatibility equations.

2.3.1 Cartesian Coordinates

Consider first a state of two-dimensional strain, i.e.,

$$
\varepsilon_x = \frac{\partial u}{\partial x}, \quad \varepsilon_y = \frac{\partial v}{\partial y}, \quad \text{and} \quad \gamma_{xy} = \frac{\partial u}{\partial y} + \frac{\partial v}{\partial x}
$$

where the other strain components $\varepsilon_z = \gamma_{zx} = \gamma_{zy} = 0$ and the displacement component $w = 0$. Differentiation of these equations gives

$$
\frac{\partial^2 \varepsilon_x}{\partial y^2} = \frac{\partial^3 u}{\partial x \partial y^2}, \quad \frac{\partial^2 \varepsilon_y}{\partial x^2} = \frac{\partial^3 v}{\partial y \partial x^2}, \quad \frac{\partial^2 \gamma_{xy}}{\partial x \partial y} = \frac{\partial^3 u}{\partial x \partial y^2} + \frac{\partial^3 v}{\partial y \partial x^2}
$$

This results in a condition to be satisfied by the three strains, namely,

$$\frac{\partial^2 \varepsilon_x}{\partial y^2} + \frac{\partial^2 \varepsilon_y}{\partial x^2} = \frac{\partial^2 \gamma_{xy}}{\partial y \partial x}$$

This is the first equation in a set of six compatibility equations for infinitesimal strains, which may be derived in a similar way to give

$$\frac{\partial^2 \varepsilon_x}{\partial y^2} + \frac{\partial^2 \varepsilon_y}{\partial x^2} = \frac{\partial^2 \gamma_{xy}}{\partial x \partial y}$$

$$\frac{\partial^2 \varepsilon_y}{\partial z^2} + \frac{\partial^2 \varepsilon_z}{\partial y^2} = \frac{\partial^2 \gamma_{yz}}{\partial y \partial z}$$

$$\frac{\partial^2 \varepsilon_z}{\partial x^2} + \frac{\partial^2 \varepsilon_x}{\partial z^2} = \frac{\partial^2 \gamma_{zx}}{\partial z \partial x}$$

$$\frac{\partial}{\partial y}\left(\frac{\partial \gamma_{xy}}{\partial z} + \frac{\partial \gamma_{yz}}{\partial x} - \frac{\partial \gamma_{zx}}{\partial y}\right) = 2\frac{\partial^2 \varepsilon_y}{\partial z \partial x} \tag{2.12}$$

$$\frac{\partial}{\partial z}\left(\frac{\partial \gamma_{yz}}{\partial x} + \frac{\partial \gamma_{zx}}{\partial y} - \frac{\partial \gamma_{xy}}{\partial z}\right) = 2\frac{\partial^2 \varepsilon_z}{\partial x \partial y}$$

$$\frac{\partial}{\partial x}\left(\frac{\partial \gamma_{zx}}{\partial y} + \frac{\partial \gamma_{xy}}{\partial z} - \frac{\partial \gamma_{yz}}{\partial x}\right) = 2\frac{\partial^2 \varepsilon_x}{\partial y \partial z}$$

The preceding explanation illustrating the necessity* of satisfying the compatibility conditions by the infinitesimal strain components is essentially mathematical. Another explanation may be based on considering geometries of the undeformed and deformed bodies. First, the undeformed body is imagined to consist of a large number of cubic elements. Without any restriction on the strain field, each one of these elements will deform differently, and parallelepiped elements of arbitrary shapes will result. It will be generally impossible to fit these deformed parallelepiped elements together to form a continuous body without gaps, Figure 2.8a. Thus, the strain field imposed on the body cannot be arbitrary. Physical limitations in this regard seem obvious. For instance, it is unrealistic to have a displacement or strain fields predicting that two different elements will occupy the same place in the deformed body, Figure 2.8b. Also, a group of neighboring elements cannot end up all over the place, as shown in Figure 2.8c. The mathematical conditions which rule out such physically nonacceptable (or incompatible) deformations are the strain compatibility** conditions; Equations 2.12.

Example 2.3:

Determine the values of the constants a_1, a_2, a_3, a_4, a_5, *and* a_6 *so that the following state of strain be possible:*

$$\varepsilon_x = 3x^2y^2 + a_1y^3z^2, \qquad \varepsilon_y = 4a_2y^3z + 2a_3x^2yz^2$$
$$\varepsilon_z = 3a_4xyz^2 + 2x^3y^2, \qquad \gamma_{yz} = 3y^4 + 4x^2y^2z + 2a_6x^3yz$$
$$\gamma_{xy} = 2a_5x^3y + xy^2z^2, \qquad \gamma_{zx} = a_4yz^3 + 3a_6x^2y^2z - 2xy^3z$$

---

* A proof for the compatibility conditions being also sufficient to assure the existence of a single–valued continuous displacement solution, is found in the literature. For instance, see: Fung,[2] p. 101.

** Compatibility conditions are also referred to as conditions of integrability or conditions of strain continuity.

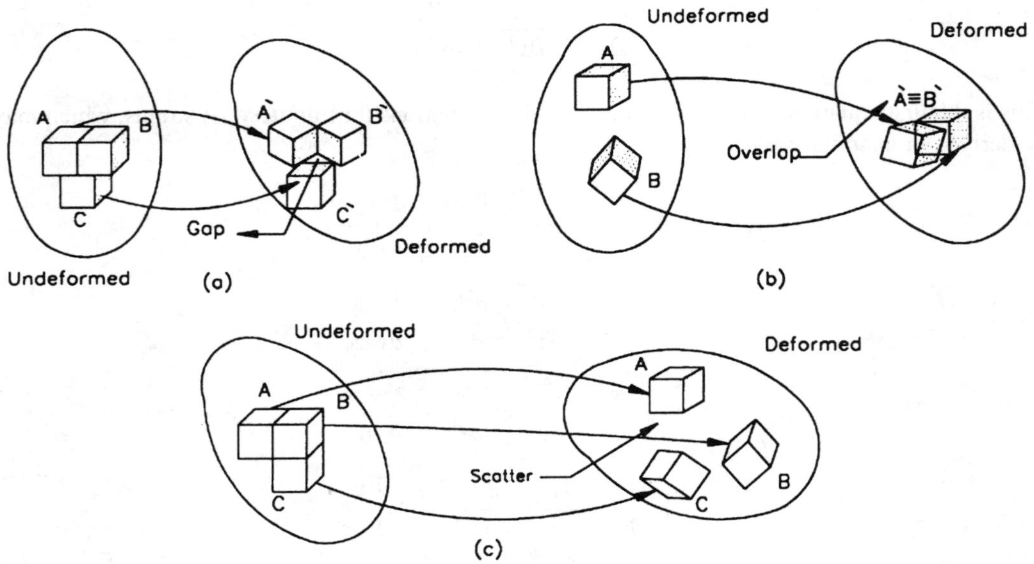

FIGURE 2.8 Physical illustration indicating that the strain field cannot be arbitrary but must be compatible; (a) incompatibility deformation due to gaps, (b) incompatible deformation due to material overlap, and (c) completely discontinous material.

Solution:
Differentiation and substituting into the first three equations of the compatibility conditions (Equations 2.12) gives

$$6x^2 + (6a_1 + 4a_3)yz^2 = 6a_5x^2 + 2yz^2$$

$$4a_3x^2y + 4x^3 = 8x^2y + 2a_6x^3$$

$$12xy^2 + 2a_1y^3 = 6a_6xy^2 - 2y^3$$

To satisfy these equations for any values of (x, y, z)

$$a_5 = 1, \quad 6a_1 + 4a_3 = 2$$

$$a_3 = 2, \quad a_6 = 2$$

$$a_1 = -1 \text{ and hence, } a_3 = 2$$

The fourth and fifth Equation 2.12 are identically satisfied, while the sixth requires $a_4 = 0$. Since all six equations are satisfied irrespective of a_2, it may have any arbitrary value.

2.3.2 Cylindrical Polar Coordinates

Analogous to Equations 2.12, the compatibility conditions are derived by appropriate differentiation of the strain–displacement relations in cylindrical polar coordinates, Equations 2.8, to give:

$$r\frac{\partial \varepsilon_r}{\partial r}-\frac{\partial^2 \varepsilon_r}{\partial \theta^2}-\frac{\partial}{\partial r}\left[r\left(r\frac{\partial \varepsilon_\theta}{\partial r}-\frac{\partial \gamma_{r\theta}}{\partial \theta}\right)\right]=0$$

$$r\frac{\partial}{\partial z}\left[2\varepsilon_r-2\frac{\partial}{\partial r}(r\varepsilon_\theta)+\frac{\partial \gamma_{r\theta}}{\partial \theta}\right]+\frac{\partial}{\partial \theta}\left[\frac{\partial}{\partial r}(r\gamma_{\theta z})-\frac{\partial \gamma_{zr}}{\partial \theta}\right]=0$$

$$\frac{\partial}{\partial z}\left[2r\frac{\partial \varepsilon_r}{\partial \theta}-\frac{\partial}{\partial r}(r\gamma_{r\theta})\right]+r^2\frac{\partial}{\partial r}\left[\frac{1}{r}\left[\frac{\partial}{\partial r}(r\gamma_{\theta z})-\frac{\partial \gamma_{zr}}{\partial \theta}\right]\right]=0$$

$$r^2\frac{\partial^2 \varepsilon_\theta}{\partial z^2}+r\frac{\partial \varepsilon_z}{\partial r}+\frac{\partial^2 \varepsilon_z}{\partial \theta^2}-r\frac{\partial}{\partial z}\left(\frac{\partial \gamma_{\theta z}}{\partial \theta}+\gamma_{zr}\right)=0 \qquad (2.13a)$$

$$2\frac{\partial}{\partial \theta}\left(\frac{1}{r}\frac{\partial \varepsilon_z}{\partial \theta}\right)+\frac{\partial}{\partial z}\left[\frac{\partial \gamma_{r\theta}}{\partial \theta}-r\frac{\partial}{\partial r}\left(\frac{\gamma_{\theta z}}{r}\right)-\frac{1}{r}\frac{\partial \gamma_{zr}}{\partial \theta}\right]=0$$

$$\frac{\partial^2 \varepsilon_r}{\partial z^2}+\frac{\partial^2 \varepsilon_z}{\partial r^2}+\frac{\partial^2 \gamma_{zr}}{\partial r \partial z}=0$$

The above equations reduce for the case of axial symmetry ($v=0$, $\gamma_{r\theta}=\gamma_{\theta z}=0$) and being independent of θ to

$$\frac{\partial \varepsilon_r}{\partial r}-\frac{\partial^2(r\varepsilon_\theta)}{\partial r^2}=0$$

$$r\frac{\partial^2 \varepsilon_\theta}{\partial z^2}+\frac{\partial \varepsilon_z}{\partial r}=\frac{\partial \gamma_{zr}}{\partial z} \qquad (2.13b)$$

$$\frac{\partial^2 \varepsilon_r}{\partial z^2}+\frac{\partial^2 \varepsilon_z}{\partial r^2}=\frac{\partial^2 \gamma_{zr}}{\partial r \partial z}$$

For an axisymmetric plane state of strain $\varepsilon_z=\gamma_{zr}=0$ (or $\gamma_z=$ constant); hence, the only compatibility condition to be satisfied is

$$\frac{d(r\varepsilon_\theta)}{dr}-\varepsilon_r=0 \qquad (2.13c)$$

2.3.3 Spherical Polar Coordinates

Compatibility conditions are expressed in spherical polar coordinates[3] (r, θ, ϕ) for the special case when the shear strains vanish, i.e., in terms of principal strains ε_r, ε_θ and ε_ϕ as

$$2\frac{\partial}{\partial\theta}\left(\frac{\partial\varepsilon_r}{\partial\theta}-\varepsilon_r\cot\phi\right)=0$$

$$2r\frac{\partial}{\partial\theta}\left(\varepsilon_r-r\frac{\partial\varepsilon_\phi}{\partial r}\right)=0$$

$$2\sin^2\phi\frac{\partial\varepsilon_r}{\partial\phi}-2r\sin^2\phi\frac{\partial}{\partial r}\left[\frac{\partial\varepsilon_\theta}{\partial\phi}+\left(\varepsilon_\theta-\varepsilon_\phi\right)\cot\phi\right]=0$$

(2.14a)

$$r\frac{\partial\varepsilon_r}{\partial r}-\frac{\partial^2\varepsilon_r}{\partial\phi^2}-\frac{\partial}{\partial r}\left(r^2\frac{\partial\varepsilon_\phi}{\partial r}\right)=0$$

$$\frac{\partial^2\varepsilon_\phi}{\partial\theta^2}+\frac{\partial}{\partial\phi}\left(\sin^2\phi\frac{\partial\varepsilon_\theta}{\partial\phi}\right)-\sin\phi\cos\phi\left(\frac{\partial\varepsilon_\phi}{\partial\phi}\right)+\sin^2\phi\left[r\frac{\partial}{\partial r}\left(\varepsilon_\theta+\varepsilon_\phi\right)+2\left(\varepsilon_\phi-\varepsilon_r\right)\right]=0$$

$$r\sin^2\phi\frac{\partial\varepsilon_r}{\partial r}-\frac{\partial^2\varepsilon_r}{\partial\theta^2}-\sin^2\phi\frac{\partial}{\partial r}\left(r^2\frac{\partial\varepsilon_\theta}{\partial r}\right)-\sin\phi\cos\phi\left(\frac{\partial\varepsilon_r}{\partial\phi}\right)=0$$

When complete spherical symmetry exists, relations 2.14a reduce to

$$\frac{d\varepsilon_\phi}{dr}+\frac{\varepsilon_\phi-\varepsilon_r}{r}=0 \tag{2.14b}$$

Relation 2.14b may be easily derived from Equation 2.10.

Example 2.4:
A hollow circular cylinder is deformed such that the nonzero strains are given by

$$\varepsilon_r = A - B/r^2,\ \varepsilon_\theta = A + B/r^2$$

a. *Determine the radial and tangential displacement components;*
b. *Check the compatibility of these strains;*
c. *Determine the radial change at the bore for a cylinder of* $r_0 = 2r_i$, *knowing that the radial and hoop strains measured at the mean radius are 0.002 and 0.001, respectively.*

Solution:

a. Integrating the strain–displacement Relation 2.8 gives

$$u=\int\varepsilon_r dr=\int\frac{\partial u}{\partial r}dr=\int\left(A-B/r^2\right)dr=Ar+B/r$$

Hence,

$$\varepsilon_\theta=A+B/r^2=\frac{1}{r}\frac{\partial v}{\partial\theta}+\frac{u}{r}=\frac{1}{r}\frac{\partial v}{\partial\theta}+\frac{1}{r}(Ar+B/r)$$

or

$$\frac{\partial v}{\partial \theta} = r\left(A + B/r^2\right) - \left(Ar + B/r\right) = 0$$

i.e., $v = 0$. This is a case of axial symmetry.

b. The only compatibility condition to be satisfied is given by Equation 2.13b as

$$\frac{d\varepsilon_r}{dr} - \frac{d^2\left(r\varepsilon_\theta\right)}{dr^2} = 0$$

or by Equation 2.13c as

$$\frac{d\left(r\varepsilon_\theta\right)}{dr} - \varepsilon_r = 0$$

Hence,

$$\frac{d}{dr}\left(Ar + B/r\right) - \left(A - B/r^2\right) = 0$$

i.e., compatible.

c. To determine the constants A and B, the radial and hoop strains measured at the mean radius $r_m = 3r_i/2$ are used as

$$\left(\varepsilon_r\right)_{r=r_m} = 2\times10^{-3} = A - B/r_m^2 = A - 4B/9r_i^2$$

$$\left(\varepsilon_\theta\right)_{r=r_m} = 1\times10^{-3} = A + B/r_m^2 = A + 4B/9r_i^2$$

Solving for A and B gives

$$A = 1.5\times10^{-3} \text{ and } B = -1.125\times10^{-3}\,r_i^2$$

Hence, at the inner radius r_i

$$(u)_{r=r_i} = \left(Ar + B/r\right)_{r=r_i} = 1.5\times10^{-3}\,r_i - 1.125\times10^{-3}\,r_i^2/r_i$$

$$(u)_{r=r_i} = 0.375\times10^{-3}\,r_i$$

i.e., 0.0375% radial change at the bore.

2.4 STRAIN TENSOR

The state of strain at a point has been defined by six components (ε_x, ε_y, ε_{xy}, γ_{xy}, γ_{yz}, γ_{zx}) referred to the coordinate axes (x, y, z). Provided that these strain components are given, it is required to calculate the strain in any direction x' passing through this point.

Consider a line PQ in the undeformed state having an infinitesimal length ds_o, Figure 2.9, and direction x' determined by direction cosines (l, m, n) with the coordinate axes (x, y, z) as

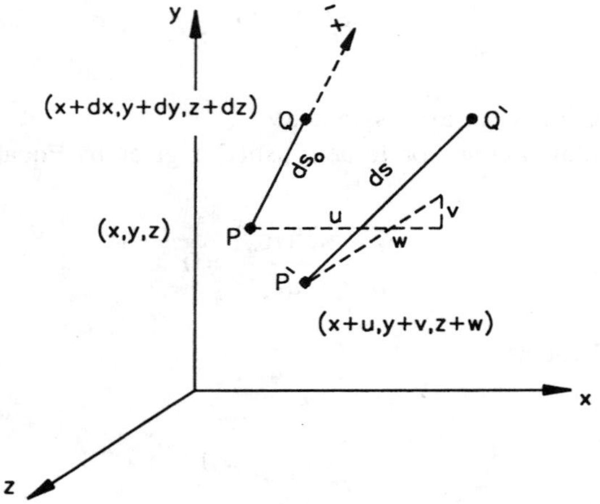

FIGURE 2.9 Transformation of linear strain from one set of axes to another.

$$l, m, n = \frac{dx}{ds_o}, \frac{dy}{ds_o}, \frac{dz}{ds_o} \tag{2.15}$$

where dx, dy, dz are the projections of length along the coordinate axes respectively. The length ds_o is thus

$$ds_o^2 = dx^2 + dy^2 + dz^2 \tag{2.16}$$

In the deformed state, point P and Q are displaced to P' and Q', respectively. Point P has undergone a displacement (u, v, w) while Q is displaced ($u + du$, $v + dv$, $w + dw$). Hence, the deformed length ds is given by

$$ds^2 = (dx + du)^2 + (dy + dv)^2 + (dz + dw)^2 \tag{2.17}$$

The strain occurring along the line PQ will be

$$\varepsilon_{x'} = \frac{ds - ds_o}{ds_o} = \frac{ds}{ds_o} - 1$$

Hence,

$$\left(\frac{ds}{ds_o}\right)^2 = (\varepsilon_{x'} + 1)^2 = \varepsilon_{x'}^2 + 2\varepsilon_{x'} + 1$$

Neglecting second-order powers of the infinitesimal strain ε_x gives

$$\varepsilon_{x'} = \frac{1}{2}\frac{ds^2 - ds_o^2}{ds_o^2} \tag{2.18}$$

Equations 2.16 and 2.17, neglecting second-order powers of du, dv, and dw, yield

$$ds^2 - ds_o^2 = 2(dudx + dvdy + dwdz) \tag{2.19}$$

Since u, v, and w are functions of (x, y, z), differentiation gives

$$\begin{aligned} du &= \frac{\partial u}{\partial x}dx + \frac{\partial u}{\partial y}dy + \frac{\partial u}{\partial z}dz \\ dv &= \frac{\partial v}{\partial x}dx + \frac{\partial v}{\partial y}dy + \frac{\partial v}{\partial z}dz \\ dw &= \frac{\partial w}{\partial x}dx + \frac{\partial w}{\partial y}dy + \frac{\partial w}{\partial z}dz \end{aligned} \tag{2.20}$$

Substituting from Equation 2.20 into Equation 2.19 yields

$$\begin{aligned} ds^2 - ds_o^2 = 2\left(\frac{\partial u}{\partial x}dx^2 + \frac{\partial v}{\partial y}dy^2 + \frac{\partial w}{\partial z}dz^2\right) + 2\left(\frac{\partial u}{\partial y} + \frac{\partial v}{\partial x}\right)dxdy \\ +2\left(\frac{\partial v}{\partial z} + \frac{\partial w}{\partial y}\right)dydz + 2\left(\frac{\partial w}{\partial x} + \frac{\partial u}{\partial z}\right)dzdx \end{aligned} \tag{2.21a}$$

Equation 2.21a by virtue of Equations 2.15, 2.16, and 2.18 reduces to

$$\varepsilon_{x'} = \frac{\partial u}{\partial x}l^2 + \frac{\partial v}{\partial y}m^2 + \frac{\partial w}{\partial z}n^2 + \left(\frac{\partial u}{\partial y} + \frac{\partial v}{\partial x}\right)lm + \left(\frac{\partial v}{\partial z} + \frac{\partial w}{\partial y}\right)mn + \left(\frac{\partial w}{\partial x} + \frac{\partial u}{\partial z}\right)nl \tag{2.21b}$$

Using the strain–displacement relations (Equations 2.6) gives

$$\varepsilon_{x'} = \varepsilon_x l^2 + \varepsilon_y m^2 + \varepsilon_z n^2 + \gamma_{xy} lm + \gamma_{yz} mn + \gamma_{zx} nl \tag{2.22a}$$

Expression 2.22a determines the normal strain at point P along the x' direction in terms of the strain components referred to coordinates (x, y, z). In other words, Equation 2.22a expresses the transformation of normal infinitesimal strain from one set of axes to another.

The transformation of shear strain may be also considered in a similar way. In such a case, the change in the right angle between two line elements (PQ and PR) emanating from point P is to be considered as shown in Figure 2.10, which results in*

$$\gamma_{x'y'} = 2(ll'\varepsilon_x + mm'\varepsilon_y + nn'\varepsilon_z) + (lm' + ml')\,\gamma_{xy} + (mn' + nm')\,\gamma_{yz} + (nl' + ln')\,\gamma_{zx} \tag{2.22b}$$

where (l, m, n) and (l', m', n') are the direction cosines of the x' axis and y' axis with respect to the coordinate axes (x, y, z), respectively.

Expressions 2.22a and b are similar to Expressions 1.15a for transformation of stress. Hence, the state of strain at a point will have the same mathematical properties as the state of stress. More explicitly, the strain is a tensor of the second rank.

* See Boresi and Sidebottom,[4] p. 45.

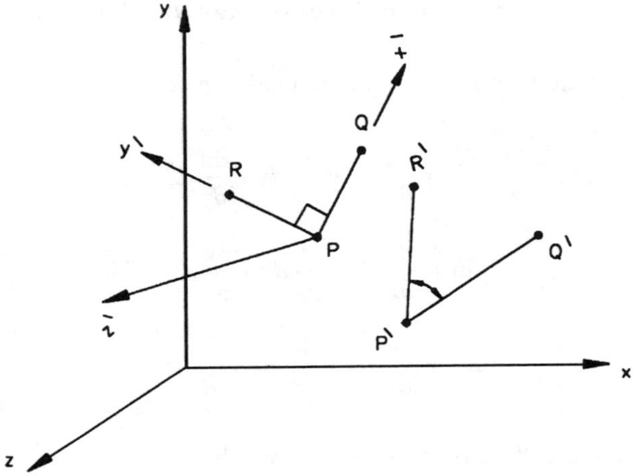

FIGURE 2.10 Transformation of shear strain from one set of axes to another.

By comparing the expression of ε_x with that of $\sigma_{x'}$, it is observed that the coefficients of the terms *lm*, *mn*, *ln* in the strain expression is equal to $^1/_2$ those in the stress expression. Hence, the shear components in the strain tensor is half the engineering shear strain, i.e., $\gamma_{xy}/2$, ... etc.

The strain is thus expressed by a symmetric tensor as

$$[\varepsilon]=\begin{bmatrix}\varepsilon_x & \frac{1}{2}\gamma_{xy} & \frac{1}{2}\gamma_{xy}\\ \frac{1}{2}\gamma_{yx} & \varepsilon_y & \frac{1}{2}\gamma_{yz}\\ \frac{1}{2}\gamma_{zx} & \frac{1}{2}\gamma_{zy} & \varepsilon_z\end{bmatrix} \tag{2.23}$$

Often ε_{xy}, ε_{yz}, and ε_{zx} are used to denote $\gamma_{xy}/2$, $\gamma_{zy}/2$, and $\gamma_{zx}/2$, respectively.

The transformation law for the infinitesimal strain components at a point is thus

$$[\varepsilon]_{x'y'z'} = [\alpha]^T\,[\varepsilon]_{xyz}\,[\alpha] \tag{2.24}$$

where $[\alpha]$ is the direction cosine matrix as defined by Table 1.1, $[\varepsilon]_{x'y'z'}$ and $[\varepsilon]_{xyz}$ are the strain tensors referred to the original coordinates (x, y, z) and the transformed ones (x', y', z') respectively.

Example 2.5:

The deformation of a body is defined by the displacement components:

$$u = a(3x^2 + y),\quad v = a(2y^2 + z),\quad w = a(4z^2 + x)$$

a. *Define the state of strain at point (2,1,1).*
b. *Determine the change of length in a line segment joining the origin (0,0,0) and point (2,1,1) using Equation 2.22a. Take* $a = 0.001$.

Solution:

a. The strains are obtained by differentiating the displacement functions according to Equations 2.6:

$$\varepsilon_x = 6ax,\quad \varepsilon_y = 4ay,\quad \varepsilon_z = 8az$$

$$\gamma_{xy} = a,\qquad \gamma_{yz} = a,\qquad \gamma_{zx} = a$$

This is a nonhomogenous strain field. At point (2,1,1)

$$\varepsilon_x = 12a, \quad \varepsilon_y = 4a, \quad \varepsilon_z = 8a, \quad \gamma_{xy} = \gamma_{yz} = \gamma_{zx} = a$$

b. The undeformed length is $L_o = \sqrt{2^2 + 1^2 + 1^2} = \sqrt{6}$ units along the direction $l = 2/\sqrt{6}$ and $m = n = 1/\sqrt{6}$. The strain along this direction (say, x') is calculated from Equation 2.22a as

$$\varepsilon_{x'} = \varepsilon_x l^2 + \varepsilon_y m^2 + \varepsilon_z n^2 + \gamma_{xy} lm + \gamma_{yz} mn + \gamma_{zx} nl$$

$$\varepsilon_{x'} = 12a(4/6) + 4a(1/6) + 8a(1/6) + a(2/6) + a(2/6) = 65a/6$$

For $a = 0.001$ and $\varepsilon_{x'} = 0.001083$. If the strain along the line is assumed to be homogeneous, the change in length is thus

$$\Delta L = \varepsilon_{x'} \; L_o = 0.001083\sqrt{6} = 0.02653 \text{ units}$$

This assumption is not true since a nonhomogeneous state of strain exists along this length. For instance at point (0,0,0), $\varepsilon_x = \varepsilon_y = \varepsilon_z = 0$. An approximation is to calculate an average value for $\varepsilon_{x'}$ along the x'-direction at the midpoint $(1,{}^1/_2,{}^1/_2)$. Hence, at this point

$$\varepsilon_x = 6a, \quad \varepsilon_y = 2a, \quad \varepsilon_z = 4a, \quad \gamma_{xy} = \gamma_{yz} = \gamma_{zx} = a$$

and

$$\left(\varepsilon_{x'}\right)_{\text{average}} = 6a(4/6) + 2a(1/6) + 4a(1/6) + a(2/6) + a(1/6) + a(2/6)$$

$$= 35a/6$$

The change in length of the line segment between points (0,0,0) and (2,1,1) is thus

$$\Delta L = \left(\varepsilon_{x'}\right)_{\text{average}} L_o = \frac{35(0.001)}{6}\sqrt{6} = 0.014289$$

The actual deformed length is obtained from the calculated displacements of point (2,1,1), namely, (0.013, 0.003, 0.006), giving a change in length of

$$\Delta L = \sqrt{(2.013)^2 + (1.003)^2 + (1.006)^2} - \sqrt{6} = 0.01429$$

which is nearly the same result as obtained by assuming an average value of the strain at the midpoint.

2.5 SOME SPECIAL STATES OF STRAIN

There are certain states of strain that often occur in several engineering applications. Some of these states are given below.

2.5.1 Plane Strain

In some applications, a body is constrained against lateral deformation in a direction normal to the plane of loading x–y, this is known as a state of plane stain.

A state of plane strain exists on a plane parallel to x–y, when $\varepsilon_z = \gamma_{zx} = \gamma_{zy} = 0$ and, hence, is determined by three components, namely, $(\varepsilon_x, \varepsilon_y, \gamma_{xy})$.

For cylindrical polar coordinates (r, θ, z), plane strain exists when $\varepsilon_z = \gamma_{zr} = \gamma_{z\theta} = 0$ and is determined by $(\varepsilon_r, \varepsilon_\theta, \gamma_{r\theta})$. The strain tensor is

$$[\varepsilon] = \begin{bmatrix} \varepsilon_x & \tfrac{1}{2}\gamma_{xy} & 0 \\ \tfrac{1}{2}\gamma_{yx} & \varepsilon_y & 0 \\ 0 & 0 & 0 \end{bmatrix} \text{ or } \begin{bmatrix} \varepsilon_r & \tfrac{1}{2}\gamma_{r\theta} & 0 \\ \tfrac{1}{2}\gamma_{\theta r} & \varepsilon_\theta & 0 \\ 0 & 0 & 0 \end{bmatrix} \tag{2.25a}$$

Thick plates loaded by forces in the x–y planes are typical examples of plane strain states in the z-direction.

2.5.2 Plane Stress

In a state of plane stress $\gamma_{zx} = \gamma_{zy} = 0$, but ε_z does not vanish and, hence, the stress tensor is given by

$$[\varepsilon] = \begin{bmatrix} \varepsilon_x & \tfrac{1}{2}\gamma_{xy} & 0 \\ \tfrac{1}{2}\gamma_{yx} & \varepsilon_y & 0 \\ 0 & 0 & \varepsilon_z \end{bmatrix} \text{ or } \begin{bmatrix} \varepsilon_r & \tfrac{1}{2}\gamma_{r\theta} & 0 \\ \tfrac{1}{2}\gamma_{\theta r} & \varepsilon_\theta & 0 \\ 0 & 0 & \varepsilon_z \end{bmatrix} \tag{2.25b}$$

2.5.3 Axial Symmetry

If a geometrically axisymmetric body is loaded symmetrically, the state of strain is simplified. By employing cylindrical polar coordinates (r, θ, z) where symmetry is around the z-axis, then due to this symmetry

$$\gamma_{r\theta} = \gamma_{\theta z} = 0$$

The strain tensor is given by

$$[\varepsilon] = \begin{bmatrix} \varepsilon_r & 0 & \tfrac{1}{2}\gamma_{rz} \\ 0 & \varepsilon_\theta & 0 \\ \tfrac{1}{2}\gamma_{zr} & 0 & \varepsilon_z \end{bmatrix} \tag{2.25c}$$

A cylinder under internal pressure or a rotating circular disk strains, $\gamma_{r\theta} = \gamma_{\theta z} = \gamma_{zr} = 0$, are typical states of complete axial symmetry where all shear stresses vanish.

2.5.4 Free Torsion

Free torsion of a rod having its axis along the z-axis produces a state of strain defined by

$$\varepsilon_x = \varepsilon_y = \varepsilon_z = \gamma_{xy} = 0$$

The stress tensor is thus given by

$$[\varepsilon]=\begin{bmatrix} 0 & 0 & \tfrac{1}{2}\gamma_{xz} \\ 0 & 0 & \tfrac{1}{2}\gamma_{yz} \\ \tfrac{1}{2}\gamma_{zx} & \tfrac{1}{2}\gamma_{yz} & 0 \end{bmatrix} \quad (2.25d)$$

In the case of a rod with a circular cross section, the only nonzero stress in cylindrical polar coordinates is $\gamma_{z\theta}$.

Example 2.6:

A finite square of a unit side length drawn on a sheet metal deforms without change in thickness to the dimensions shown in Figure 2.11.

a. *Define the state of strain for this square.*
b. *Determine the change in length of* OB; *apply strain transformation.*
c. *Determine the change in angle* COA; *apply strain transformation.*

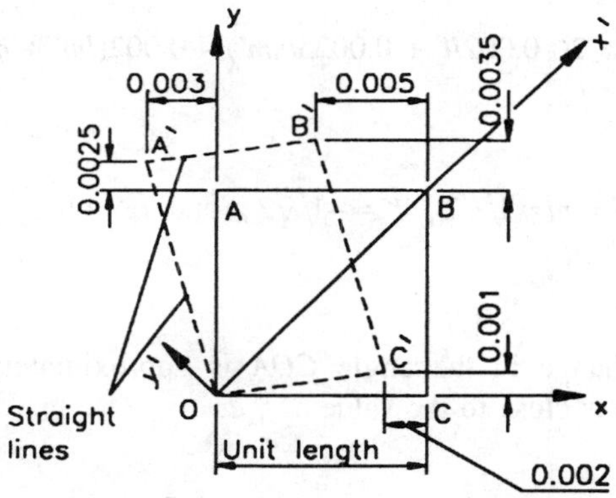

FIGURE 2.11 Example 2.6.

Solution:

a. This is a state of plane strain where $\varepsilon_z = \gamma_{zx} = \gamma_{zy} = 0$. Since the sides of the deformed element remain straight lines, the displacement function is of the form:

$$u = a_1x + b_1y, \quad v = a_2x + b_2y$$

The constants a_1, b_1, a_2, and b_2 are determined from the known displacements of the element corner points; hence

$$u = -0.002x - 0.003y, \quad v = 0.001x + 0.0025y$$

The strains, using Relations 2.6 are

$$\varepsilon_x = -0.002, \quad \varepsilon_y = 0.0025, \quad \text{and} \quad \gamma_{xy} = -0.002$$

which are independent of point location, i.e., homogeneous strain field.

b. The strain along OB, direction x', is calculated from Equation 2.22a as

$$\varepsilon_{OB} = \varepsilon_{x'} = -0.002l^2 + 0.0025m^2 - 0.002lm$$

Since $l = m = 1/\sqrt{2}$, then

$$\varepsilon_{OB} = -7.5 \times 10^{-4} \text{ and } \Delta_{OB} = OB \quad \varepsilon_{OB} = -1.06066 \times 10^{-3}$$

The same result may be obtained from geometry:

$$OB' = \sqrt{(1-0.005)^2 + (1+0.0035)^2} = 1.413166$$

Hence,

$$\Delta OB = -1.04768 \times 10^{-3}$$

The error (about 1.2%) in using the strain transformation law (Equations 2.22a) is due to neglecting higher orders of strains and displacements in deriving this law.

c. The shear strain with respect to orthogonal axes x' and y' is calculated from Equation 2.22b:

$$\gamma_{x'y'} = 2(-0.002ll' + 0.0025mm') - 0.002(lm' + ml')$$

where

$$l = m = 1/\sqrt{2}, \quad l' = -1/\sqrt{2}, \text{ and } m' = 1/\sqrt{2}$$

Hence, $\gamma_{x'y'} = 0.0045$.

From geometry, the change in the angle COA is approximately equal to $\tan^{-1}(0.001) + \tan^{-1}(0.003) \cong 0.004$, which is close to the value of $\gamma_{x'y'}$.

2.6 PRINCIPAL STRAINS — MAXIMUM AND OCTAHEDRAL SHEAR STRAINS

The similarity between stress and strain tensors makes it possible to apply the analysis performed on the stress tensor in Chapter 1 to the strain tensor without any need for further proofs.

The same eigenvalue problem could be described by the strain tensor to determine the principal strains and their directions. On the principal planes $\gamma_{xy} = \gamma_{yz} = \gamma_{zx} = 0$. The principal strains ε_1, ε_2, ε_3 are obtained from the characteristic equation:

$$\varepsilon_\lambda^3 - J_1\varepsilon_\lambda^2 - J_2\varepsilon_\lambda - J_3 = 0 \tag{2.26a}$$

where

$$J_1 = \varepsilon_x + \varepsilon_y + \varepsilon_z = \varepsilon_1 + \varepsilon_2 + \varepsilon_3$$

$$\begin{aligned} J_2 &= -\left(\varepsilon_x\varepsilon_y + \varepsilon_y\varepsilon_z + \varepsilon_z\varepsilon_x\right) + \tfrac{1}{4}\left(\gamma_{xy}^2 + \gamma_{yz}^2 + \gamma_{zx}^2\right) \\ &= \varepsilon_1\varepsilon_2 + \varepsilon_2\varepsilon_3 + \varepsilon_3\varepsilon_1 \end{aligned} \tag{2.26b}$$

$$\begin{aligned} J_3 &= \varepsilon_x\varepsilon_y\varepsilon_z + \tfrac{1}{4}\left(\gamma_{xy}\gamma_{yz}\gamma_{zx} - \varepsilon_x\gamma_{yz}^2 - \varepsilon_y\gamma_{zx}^2 - \varepsilon_z\gamma_{xy}^2\right) \\ &= \varepsilon_1\varepsilon_2\varepsilon_3 \end{aligned}$$

The quantities J_1, J_2, and J_3 are expressed in terms of ε_1, ε_2, and ε_3 in the same way as for the stress invariants I_1, I_2, and I_3 in terms of σ_1, σ_2, and σ_3. Again J_1, J_2, and J_3 are invariants of the strain tensor. They are independent of the orientation of coordinate axes.

The principal directions of strain,* which are mutually orthogonal, are determined in the same way as for the principal directions of stress by using equations similar to Equations 1.21 and replacing σ_λ by ε_λ, i.e.,

$$\begin{aligned} \varepsilon_\lambda l_\lambda &= \varepsilon_x l_\lambda + \tau_{yx} m_\lambda + \tau_{zx} n_\lambda \\ \varepsilon_\lambda m_\lambda &= \tau_{xy} l_\lambda + \sigma_y m_\lambda + \tau_{zy} n_\lambda \\ \varepsilon_\lambda n_\lambda &= \tau_{xz} l_\lambda + \tau_{yz} m_\lambda + \sigma_z n_\lambda \end{aligned} \tag{2.26c}$$

The maximum shear strains in terms of the principal strains are given by

$$\begin{aligned} \gamma_1 &= \varepsilon_1 - \varepsilon_3 \\ \gamma_2 &= \varepsilon_1 - \varepsilon_2 \\ \gamma_3 &= \varepsilon_2 - \varepsilon_3 \end{aligned} \tag{2.27}$$

The maximum shear strains occur on three planes each passing by one of the principal axes and making 45° with the other two axes. The normal strains on the planes of maximum shear strains are, respectively, $(\varepsilon_1 + \varepsilon_3)/2$, $(\varepsilon_1 + \varepsilon_2)/2$ and $(\varepsilon_2 + \varepsilon_3)/2$.

An octahedral shear strain occurring on a plane equally inclined to the principal directions of strain may also be derived as

$$\begin{aligned} \gamma_{oct}^2 &= \tfrac{4}{9}\left[(\varepsilon_1 - \varepsilon_2)^2 + (\varepsilon_2 - \varepsilon_3)^2 + (\varepsilon_3 - \varepsilon_1)^2\right] \\ &= \tfrac{4}{9}\left(\gamma_1^2 + \gamma_2^2 + \gamma_3^2\right) = \tfrac{4}{9}\left(2J_1^2 - 6J_2\right) \end{aligned} \tag{2.28a}$$

together with

$$\varepsilon_{oct} = \tfrac{1}{3}(\varepsilon_1 + \varepsilon_2 + \varepsilon_3) = J_1/3 \tag{2.28b}$$

Example 2.7:

A circular disk undergoes a deformation: u = Br *cos 2θ,* v = Ar + Br *sin 2θ,* w = *0, where* A *and* B *are positive constants. Calculate the locations and magnitude of the largest positive strains, maximum shear strain, and volumetric strain.*

Solution:

Equations 2.8 relating strains to displacements in cylindrical polar coordinates yields:

---

* Later when considering the behavior of elastic solids, it will be shown from Hooke's law that principal directions of stress and strain coincide. For isotropic plastic solids it is postulated that principal directions of stress and strain increment coincide; see Levy–Mises flow rule, Chapter 10.

$$\varepsilon_r = \frac{\partial u}{\partial r} = B\cos 2\theta$$

$$\varepsilon_v = \frac{1}{r}\frac{\partial v}{\partial \theta} + \frac{u}{r} = \frac{1}{r}(2Br\cos 2\theta) + B\cos 2\theta$$

$$= 3B\cos 2\theta$$

$$\gamma_{r\theta} = \frac{1}{r}\frac{\partial u}{\partial \theta} + \frac{\partial v}{\partial r} - \frac{v}{r}$$

$$= \frac{1}{r}(-2Br\sin 2\theta) + (A + B\sin 2\theta) - (A + B\sin 2\theta)$$

$$= -2B\sin 2\theta$$

Other strain components ε_z, $\varepsilon_{\theta z}$, and γ_{zr} are zero, i.e., this is a state of plane strain. Principal strains are obtained from Equations 2.26 according to

$$J_1 = 4B\cos 2\theta,\quad J_2 = -2B^2\cos^2 2\theta,\quad J_3 = 0$$

and

$$\varepsilon_\lambda^3 - (4B\cos 2\theta)\varepsilon_\lambda^2 + \left(2B^2\cos^2 2\theta\right)\varepsilon_\lambda = 0$$

or

$$\varepsilon_\lambda^2 - (4B\cos 2\theta)\varepsilon_\lambda + \left(2B^2\cos^2 2\theta\right) = 0$$

which gives $\varepsilon_1 = (2 + \sqrt{2}\,)$ B cos 2θ and $\varepsilon_2 = (2 - \sqrt{2}\,)$ B cos 2θ.

The maximum positive normal strain ε_1 occurs at $\theta = 0$, $\pi/2$, and $3\pi/2$. The maximum shear strain is $\gamma_3 = \varepsilon_1 - \varepsilon_2 = 4B\cos 2\theta$. The volumetric strain $\varepsilon_v = 4B\cos 2\theta$.

2.7 MEAN STRAIN DILATATION AND STRAIN DEVIATIONS

The volumetric infinitesimal strain ε_v is defined by Equation 2.4 as

$$\varepsilon_v = \varepsilon_x + \varepsilon_y + \varepsilon_z$$

or

$$\varepsilon_v = \varepsilon_1 + \varepsilon_2 + \varepsilon_3 = J_1 \tag{2.29a}$$

This means that the volumetric strain is independent of coordinate axes, i.e., invariant of the strain tensor.

The mean strain ε_m is

$$\varepsilon_m = {}^1\!/_3(\varepsilon_x + \varepsilon_y + \varepsilon_z) = {}^1\!/_3\varepsilon_v = {}^1\!/_3 J_1 \tag{2.29b}$$

which, similarly to the mean stress, corresponds to the strain under hydrostatic pressure. For constant-volume deformation, $\varepsilon_v = 0$.

The strain matrix may thus be divided into two matrices as

$$[\varepsilon]=\begin{bmatrix}\varepsilon_m & 0 & 0\\ 0 & \varepsilon_m & 0\\ 0 & 0 & \varepsilon_m\end{bmatrix}+\begin{bmatrix}\varepsilon'_x & \tfrac{1}{2}\gamma_{yx} & \tfrac{1}{2}\gamma_{zx}\\ \tfrac{1}{2}\gamma_{xy} & \varepsilon'_y & \tfrac{1}{2}\gamma_{zy}\\ \tfrac{1}{2}\gamma_{xz} & \tfrac{1}{2}\gamma_{yz} & \varepsilon'_z\end{bmatrix}$$

where the strain deviation components ε_x', ε_y', and ε_z' are defined as: $\varepsilon_x' = \varepsilon_x - \varepsilon_m$ ... etc. Equations 2.29 are simply written as

$$[\varepsilon] = [\varepsilon_m] + [\varepsilon']$$

where $[\varepsilon_m]$ and $[\varepsilon']$ are 3×3 matrices called mean and deviatoric strain matrices, respectively. Their invariants are

$$J_{1m} = 3\varepsilon_m,\ J_{2m} = 3\varepsilon_m^2,\ J_{3m} = \varepsilon_m^3 \tag{2.30a}$$

and

$$J_1' = 0,\ J_2' = -\left(\varepsilon_1'\varepsilon_2' + \varepsilon_2'\varepsilon_3' + \varepsilon_3'\varepsilon_1'\right),\ J_3' = \varepsilon_1'\varepsilon_2'\varepsilon_3' \tag{2.30b}$$

Since $J_1' = 0$ for the strain deviations, it represents a state of no volume change. Deformation is only a pure shear when $J_1 = 0$.

Example 2.8:

A state of free torsion is given by the strain tensor:

$$[\varepsilon]=\begin{bmatrix}0 & a & 0\\ a & 0 & b\\ 0 & b & 0\end{bmatrix}$$

Determine the dilatation, strain deviations, the magnitude, and first direction of principal strains.

Solution:

Dilatation is given by Expression 2.4, namely; $\varepsilon_v = \varepsilon_x + \varepsilon_y + \varepsilon_z = 0$ and hence, $\varepsilon_m = \varepsilon_v/3 = 0$. The strain deviations are then $\varepsilon_x' = \varepsilon_y' = \varepsilon_z' = 0$.

The principal strains are obtained by solving Equation 2.26a, where $J_1 = 0$, $J_2 = (a^2 + b^2)/4$, $J_1 = 0$, i.e.,

$$\varepsilon_\lambda^3 - \tfrac{1}{4}\left(a^2 + b^2\right)\varepsilon_\lambda = 0$$

which gives

$$\varepsilon_1 = \tfrac{1}{2}\sqrt{a^2 + b^2},\ \varepsilon_2 = 0 \text{ and } \varepsilon_3 = -\tfrac{1}{2}\sqrt{a^2 + b^2}$$

This is obviously a state of plane strain along the second principal direction.

Substituting the first principal strain ε_1 into Equations 2.26c gives for the direction cosines of the first principal direction of strain (l_1, m_1, n_1):

$$\tfrac{1}{2}\sqrt{a^2+b^2}\,l_1 = am_1$$

$$\tfrac{1}{2}\sqrt{a^2+b^2}\,m_1 = al_1 + bn_1$$

$$\tfrac{1}{2}\sqrt{a^2+b^2}\,n_1 = bm_1$$

Solving for l_1, m_1, and n_1 knowing that $l_1^2 + m_1^2 + n_1^2 = 1$ yields:

$$l_1 = 2a\Big/\sqrt{5(a^2+b^2)},\quad m_1 = 1/\sqrt{5},\quad \text{and}\quad n_1 = 2b\Big/\sqrt{5(a^2+b^2)}$$

2.8 MOHR'S CIRCLE OF STRAIN

Both the stress and strain tensors obey the same law of transformation. Recalling that Mohr's circle of plane stress is merely a graphical representation of this transformation law, there is also a Mohr's circle of plane strain.*

By analogy with stress, Mohr's circle for strain will have ε as abscissa and $\gamma/2$ as ordinate as shown in Figure 2.12. Note that the strain quantities represented in Figure 2.12 correspond to the stress quantities of Figure 1.13.

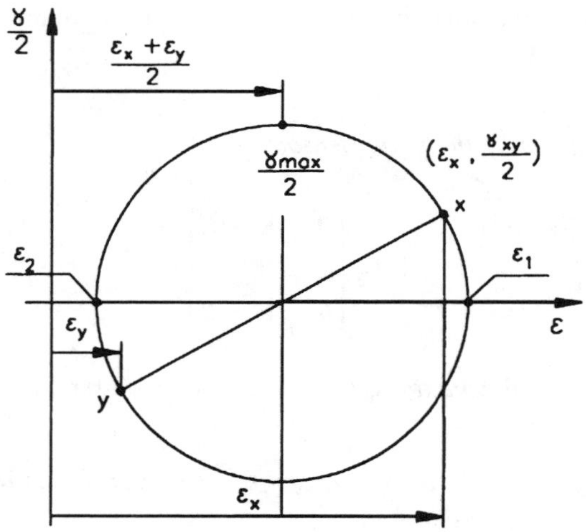

FIGURE 2.12 Mohr's circle for plane strain.

Note also that Equations 1.26 hold for strains by replacing σ by ε and τ by $\gamma/2$. Thus, the principal strains are given by

$$\frac{\varepsilon_1}{\varepsilon_2} = \left(\frac{\varepsilon_x + \varepsilon_y}{2}\right) \pm \sqrt{\left(\frac{\varepsilon_x - \varepsilon_y}{2}\right)^2 + \left(\frac{\gamma_{xy}}{2}\right)^2} \tag{2.31a}$$

at angles with x–y directions obtained from

$$\tan\theta = \gamma_{xy}/(\varepsilon_x - \varepsilon_y) \tag{2.31b}$$

which gives θ and $[\theta + (\pi/2)]$ for the two principal directions.

* Popov,[5] p. 258.

An appropriate sign convention must be adopted in constructing Mohr's circle. Here, a positive shear strain (causing a counterclockwise rotation about any point in the physical element) is plotted above the horizontal axis of Mohr's circle. Also, an angle θ on the physical element is represented on Mohr's circle by an angle 2θ measured in the opposite direction.

Example 2.9:

A flat, thin rubber band, Figure 2.13a is subjected to a uniaxial uniform stress σ_x. *Assume that the two lateral strains,* ε_y *and* ε_z, *are equal and the deformation occurs at constant volume. Determine the direction in the* x–y *plane along which zero extension will occur. What are the stresses acting on a cut-plane along this direction. Verify the results graphically using Mohr's circles.*

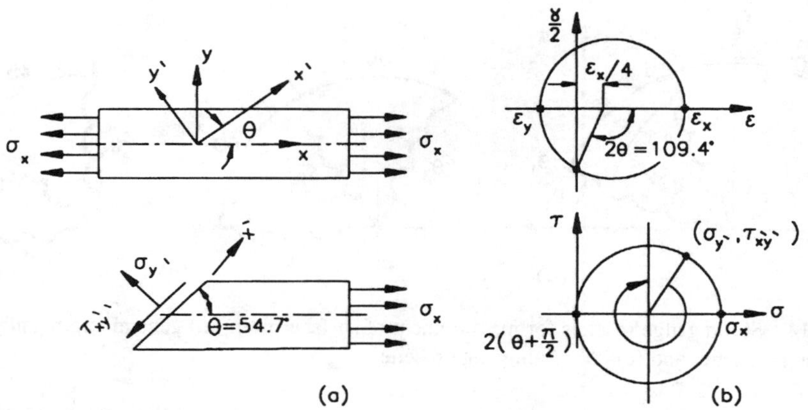

FIGURE 2.13 Example 2.9.

Solution:

For constant-volume condition and $\varepsilon_y = \varepsilon_z$, Equation 2.4 reduces to

$$\varepsilon_v = 0 = \varepsilon_x + \varepsilon_y + \varepsilon_z = \varepsilon_x + 2\varepsilon_y$$

hence $\varepsilon_y = \varepsilon_z = -{}^1\!/_2\varepsilon_x$.

Applying the strain transformation law, Equation 2.22a, assuming that the x'-direction is that of zero strain, gives

$$\varepsilon_{x'} = 0 = \varepsilon_x l^2 + \varepsilon_y m^2 + \varepsilon_z n^2 + \gamma_{xy} lm + \gamma_{yz} mn + \gamma_{zx} nl$$

$$0 = \varepsilon_x \cos^2\theta - \tfrac{1}{2}\varepsilon_x \sin^2\theta - \tfrac{1}{2}\varepsilon_x \cos^2 90 + (0)\sin\theta\cos\theta + 0 + 0$$

which gives $\tan\theta = \sqrt{2}$ or $\theta_1 = 54.736°$ and $\theta_2 = 234.736°$.

The stresses acting on a plane with a normal y' are obtained from the stress transformation law for plane stress ($\sigma_z = \gamma_{zx} = \gamma_{zy} = 0$), Equations 1.20, as

$$\sigma_{y'} = \frac{\sigma_x}{2} + \frac{\sigma_x}{2}\cos 2(\theta_1 + \pi/2) = 0.667\sigma_x$$

$$\tau_{y'x'} = -\frac{\sigma_x}{2}\sin 2(\theta_1 + \pi/2) = 0.471\sigma_x$$

Mohr's circles for strain and stress, Figures 2.13b, verify the above results.

2.9 STRAIN GAUGE ROSETTES

Strains that occur in most engineering structures and machine elements are very small. For instance, for a metallic tie-rod, the maximum allowable axial strain will be less than the offset strain of 0.2%.

A widely used method of strain measurement is based on the electric resistance strain gauge.* These gauges measure normal (longitudinal) strain whether extension or contraction at a point on the surface of a deforming solid. Shear strain cannot be directly measured by one single strain gauge. However, an arrangement of three gauges mounted at the same point is used to define the state of strain at this point completely. This gauge arrangement is known as a strain gauge rosette. Gauge arrangements are made to several configurations as shown schematically in Figure 2.14.

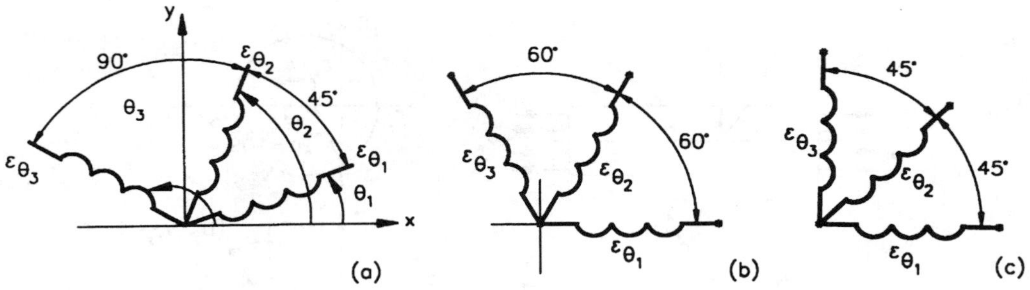

FIGURE 2.14 Strain gauge rosettes for measurement of surface strain; (a) general strain gauge rosette, (b) 60° strain gauge rosette, and (c) 45° strain gauge rosette.

Consider the general case where the three gauges are cemented to the surface at angles θ_1, θ_2, and θ_3 from the *x*-direction, Figure 2.14a. If the strain readings are $\varepsilon_{\theta 1}$, $\varepsilon_{\theta 2}$, and $\varepsilon_{\theta 3}$, the transformation law of normal strain, Equation 2.22a is applied for the *x–y* plane as

$$\begin{aligned} \varepsilon_{\theta_1} &= \varepsilon_x \cos^2\theta_1 + \varepsilon_y \sin^2\theta_1 + \gamma_{xy} \sin\theta_1 \cos\theta_1 \\ \varepsilon_{\theta_2} &= \varepsilon_x \cos^2\theta_2 + \varepsilon_y \sin^2\theta_2 + \gamma_{xy} \sin\theta_2 \cos\theta_2 \\ \varepsilon_{\theta_3} &= \varepsilon_x \cos^2\theta_3 + \varepsilon_y \sin^2\theta_3 + \gamma_{xy} \sin\theta_3 \cos\theta_3 \end{aligned} \tag{2.32}$$

This set of equations can be solved for ε_x, ε_y, and γ_{xy} defining the state of strain at the surface point. For instance, for the gauge rosette of Figure 2.14c $\theta_1 = 0$, $\theta_2 = 45°$, and $\theta_3 = 90°$, thus yielding

$$\varepsilon_{\theta_1} = \varepsilon_{0°} = \varepsilon_x, \quad \varepsilon_{\theta_3} = \varepsilon_{90°} = \varepsilon_y$$

and

$$\varepsilon_{\theta_2} = \varepsilon_{45°} = \tfrac{1}{2}\left(\varepsilon_x + \varepsilon_y + \gamma_{xy}\right)$$

or

* Electric resistance strain gauges consist basically of thin wire filaments folded back and forth to increase the sensitivity while confining the gauge to a small region. A strain gauge is mounted by carefully cementing it to a free surface of the deforming solid. If the gauge wire filament is strained, a change in its electrical resistance occurs. By appropriate calibration and using a Wheatstone bridge, this change of resistance is monitored giving normal strain reading along the wire filament direction.

$$\gamma_{xy} = 2\varepsilon_{45^\circ} - \left(\varepsilon_{0^\circ} + \varepsilon_{90^\circ}\right)$$

Mohr's circle of strain may be used and graphical solutions for the various rosettes are available.*

Example 2.10:

Three strain gauges are symmetrically arranged at 120° on the free surface of a machine apart as shown in Figure 2.15. The gauge readings gave strains: $\varepsilon_{\theta 1} = 0.001$, $\varepsilon_{\theta 2} = 0.002$, and $\varepsilon_{\theta 3} = 0.003$. The strain normal to the surface is $\varepsilon_z = -0.00156$. Determine the magnitude of the principal and volumetric strains.

Solution:

Take the x-, y-axes as shown in Figure 2.15. Applying strain transformation, Equations 2.32, for $\theta_1 = 0$, $\theta_2 = 120°$ and $\theta_3 = 240°$:

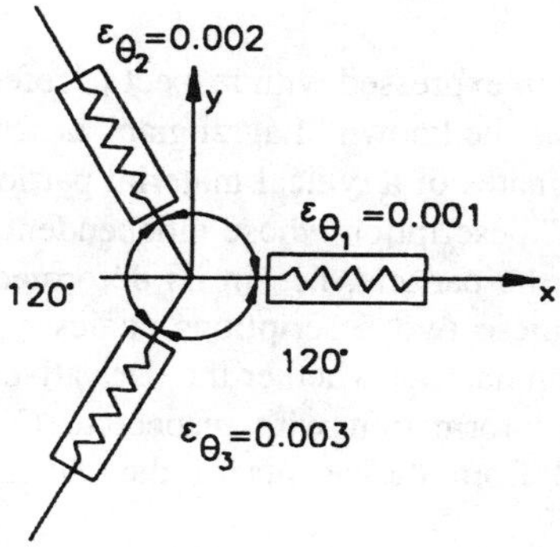

FIGURE 2.15 Example 2.10.

$$\varepsilon_{\theta_1} = 0.001 = \varepsilon_x$$

$$\varepsilon_{\theta_2} = 0.002 = 0.25\varepsilon_x + 0.75\varepsilon_y - 0.433\gamma_{xy}$$

$$\varepsilon_{\theta_3} = 0.003 = 0.25\varepsilon_x + 0.75\varepsilon_y + 0.433\gamma_{xy}$$

Solving gives $\varepsilon_x = 0.001$, $\varepsilon_y = 0.003$, and $\gamma_{xy} = 0.001154$. Note that although this state of strain is not two dimensional since $\varepsilon_z = -0.00156$, Equations 2.31 may be used to determine the principal strains since, for a free surface, z is a principal direction. In general, the strain matrix invariants are given by Equations 2.26 as

$$J_1 = 0.00244, \quad J_2 = -3.5729 \times 10^{-6}, \text{ and } J_3 = -4.1606 \times 10^{-9}$$

The characteristic equation (Equaton 2.25) is

$$\varepsilon_\lambda^3 - 2.44 \times 10^{-3}\ \varepsilon_\lambda^2 + 3.5729 \times 10^{-6}\ \varepsilon_\lambda + 4.1606 \times 10^{-9} = 0$$

In this case $\gamma_{xy} = \gamma_{zy} = 0$, and z is a principal direction; hence, $\varepsilon_3 = \varepsilon_z = -1.560 \times 10^{-3}$. By division, the resulting equation gives two principal strains: $\varepsilon_1 = 3.155 \times 10^{-3}$ and $\varepsilon_2 = 0.845 \times 10^{-3}$. The volumetric strain is given by Expression 2.29a as

$$\varepsilon_v = J_1 = \varepsilon_x + \varepsilon_y + \varepsilon_z = \varepsilon_1 + \varepsilon_2 + \varepsilon_3 = 2.44 \times 10^{-3}$$

* See Dally and Riley,[6] p. 422.

2.10 NOTES ON FINITE STRAINS

In the preceding sections, deformation and strain are analyzed assuming that they are small. Indeed, infinitesimal displacements and strains describe deformation of most engineering structures. However, there are certain applications, such as metal forming processes, ductile fracture, rubber elasticity, thin shells and plates, buckling and stability problems, etc., where large or finite deformations are encountered. Strains thus should be analyzed without the restriction of being small.

Referring to Section 2.4, where the deformed length ds of a line segment ds_o has been calculated, it is found that

$$\begin{aligned} ds^2 - ds_o^2 &= \left[(dX+du)^2 + (dY+dv)^2 + (dZ+dw)^2\right] - \left[dX^2 + dY^2 + dZ^2\right] \\ ds^2 - ds_o^2 &= 2(dudX + dvdY + dwdZ) + \left(du^2 + dv^2 + dw^2\right) \end{aligned} \tag{2.33}$$

The above calculation has been expressed with respect to reference coordinates (X, Y, Z) which are fixed in space. This constitutes the known "Lagrangian" description for which the independent variables (X, Y, Z), are the coordinates of a typical material particle in its initial undeformed state. This is opposed to the "Eulerian" description whose independent variables (x, y, z) are the present or current position occupied by the particle, i.e., in its deformed state. When displacements and deformations are infinitesimal, these two descriptions coalesce without any need to distinguish between them. It thus becomes immaterial whether the derivatives of displacements are calculated at the position before or after deformation with respect to (X, Y, Z) or (x, z, y) respectively. Substituting into Equation 2.33 from the results of the chain rule of differentiation, namely, Equations 2.20 gives

$$\begin{aligned} ds^2 - ds_o^2 = {} & 2\left[\left(\frac{\partial u}{\partial X}\right) + \frac{1}{2}\left[\left(\frac{\partial u}{\partial X}\right)^2 + \left(\frac{\partial v}{\partial X}\right)^2 + \left(\frac{\partial w}{\partial X}\right)^2\right]\right] dX^2 \\ & + 2\left[\left(\frac{\partial v}{\partial Y}\right) + \frac{1}{2}\left[\left(\frac{\partial u}{\partial Y}\right)^2 + \left(\frac{\partial v}{\partial Y}\right)^2 + \left(\frac{\partial w}{\partial Y}\right)^2\right]\right] dY^2 \\ & + 2\left[\left(\frac{\partial w}{\partial Z}\right) + \frac{1}{2}\left[\left(\frac{\partial u}{\partial Z}\right)^2 + \left(\frac{\partial v}{\partial Z}\right)^2 + \left(\frac{\partial w}{\partial Z}\right)^2\right]\right] dZ^2 \\ & + 2\left[\left(\frac{\partial v}{\partial X} + \frac{\partial u}{\partial Y}\right) + \frac{\partial u}{\partial X}\frac{\partial u}{\partial Y} + \frac{\partial v}{\partial X}\frac{\partial v}{\partial Y} + \frac{\partial w}{\partial X}\frac{\partial w}{\partial Y}\right] dXdY \\ & + 2\left[\left(\frac{\partial w}{\partial X} + \frac{\partial u}{\partial Z}\right) + \frac{\partial u}{\partial X}\frac{\partial u}{\partial Z} + \frac{\partial v}{\partial X}\frac{\partial v}{\partial Z} + \frac{\partial w}{\partial X}\frac{\partial w}{\partial Z}\right] dXdZ \\ & + 2\left[\left(\frac{\partial w}{\partial Y} + \frac{\partial v}{\partial Z}\right) + \frac{\partial u}{\partial Y}\frac{\partial u}{\partial Z} + \frac{\partial v}{\partial Y}\frac{\partial v}{\partial Z} + \frac{\partial w}{\partial Y}\frac{\partial w}{\partial Z}\right] dXdY \end{aligned} \tag{2.34}$$

Now, one must remember that the definition of strain is rather arbitrary as far as it describes the change in the relative position between points of a deformable solid. Hence, the quantity ($ds^2 - ds_o^2$) could provide a convenient means for describing deformation if Expression 2.34 is rewritten as

$$ds^2 - ds_o^2 = 2\varepsilon_X^F dX^2 + 2\varepsilon_Y^F dY^2 + 2\varepsilon_Z^F dZ^2 + 2\gamma_{XY}^F dXdY + 2\gamma_{XZ}^F dXdZ + 2\gamma_{YZ}^F dYdZ \tag{2.35}$$

where

$$\begin{aligned}
\varepsilon_X^F &= \left(\frac{\partial u}{\partial X}\right) + \frac{1}{2}\left[\left(\frac{\partial u}{\partial X}\right)^2 + \left(\frac{\partial v}{\partial X}\right)^2 + \left(\frac{\partial w}{\partial X}\right)^2\right] \\
\varepsilon_Y^F &= \left(\frac{\partial v}{\partial Y}\right) + \frac{1}{2}\left[\left(\frac{\partial u}{\partial Y}\right)^2 + \left(\frac{\partial v}{\partial Y}\right)^2 + \left(\frac{\partial w}{\partial Y}\right)^2\right] \\
\varepsilon_Z^F &= \left(\frac{\partial w}{\partial Z}\right) + \frac{1}{2}\left[\left(\frac{\partial u}{\partial Z}\right)^2 + \left(\frac{\partial v}{\partial Z}\right)^2 + \left(\frac{\partial w}{\partial Z}\right)^2\right] \\
\gamma_{XY}^F &= \left(\frac{\partial v}{\partial X} + \frac{\partial u}{\partial Y}\right) + \left(\frac{\partial u}{\partial X}\frac{\partial u}{\partial Y} + \frac{\partial v}{\partial X}\frac{\partial v}{\partial Y} + \frac{\partial w}{\partial X}\frac{\partial w}{\partial Y}\right) \\
\gamma_{XZ}^F &= \left(\frac{\partial w}{\partial X} + \frac{\partial u}{\partial Z}\right) + \left(\frac{\partial u}{\partial X}\frac{\partial u}{\partial Z} + \frac{\partial v}{\partial X}\frac{\partial v}{\partial Z} + \frac{\partial w}{\partial X}\frac{\partial w}{\partial Z}\right) \\
\gamma_{YZ}^F &= \left(\frac{\partial w}{\partial Y} + \frac{\partial v}{\partial Z}\right) + \left(\frac{\partial u}{\partial Y}\frac{\partial u}{\partial Z} + \frac{\partial v}{\partial Y}\frac{\partial v}{\partial Z} + \frac{\partial w}{\partial Y}\frac{\partial w}{\partial Z}\right)
\end{aligned} \tag{2.36}$$

Expressions 2.36 constitute the strain–displacement relations for large deformations. The superscript F stands for "finite" and the strain components ε_X^F, …, γ_{XY}^F are known as Green strains.*

For small deformations, the derivatives of displacements are infinitesimal and their products or powers of higher orders could be neglected. If this is done, expressions (2.36) reduce to relations (2.6) for infinitesimal strains. The geometrical interpretation of finite strains is not always as simple as infinitesimal strains. Consider a line segment ds_o that lies parallel to the x-axis, then $l = 1$, $m = n = 0$ and $ds_o = dX$. Substitution into expressions (2.35) gives

$$ds^2 - ds_o^2 = 2\varepsilon_X^F ds_o^2$$

Hence,

$$\varepsilon_X^F = \frac{1}{2}\left[\left(\frac{ds}{ds_o}\right)^2 - 1\right] \tag{2.37}$$

which relates the deformed length to the undeformed length. Similarly, it can be demonstrated that the finite shear strain γ_{XY}^F expresses the change in angle between two orthogonal lines.**

---

* "Green" strains are defined with reference to the underformed geometry, hence they represent a Lagrangian description. An Eulerian description defines "Almansi" strains with reference to the deformed Geometry as

$$\varepsilon_x^F = \frac{\partial u}{\partial x} - \frac{1}{2}\left[\left(\frac{\partial u}{\partial x}\right)^2 + \left(\frac{\partial v}{\partial x}\right)^2 + \left(\frac{\partial w}{\partial x}\right)^2\right] \cdots \text{ and}$$

$$\gamma_{xy}^F = \left(\frac{\partial v}{\partial x} + \frac{\partial u}{\partial y}\right) - \left[\frac{\partial u \partial u}{\partial x \partial y} + \frac{\partial v \partial v}{\partial x \partial y} + \frac{\partial w \partial w}{\partial x \partial y}\right] \cdots$$

** See Fung,[2] p. 98.

The relation between the engineering strain ε_x, defined as $(ds - ds_o)/ds_o$ and finite strain is easily found from Equation 2.37 as

$$\varepsilon_X = \frac{ds - ds_o}{ds_o} = \sqrt{1 + 2\varepsilon_X^F} - 1 \tag{2.38a}$$

or

$$\varepsilon_X^F = \varepsilon_X + \tfrac{1}{2}\varepsilon_X^2$$

which reduces to $\varepsilon_X \cong \varepsilon_X^F$ for infinitesimal strains. As regards the shear strain, consider two elements oriented in their undeformed state along the *X*- and *Y*-directions. After deformation, these two lines undergo finite strains ε_X^F and ε_Y^F respectively, together with a change γ_{XY} in the right angle originally included between them. It can be shown that this change is given by

$$\sin \gamma_{XY} = \frac{\gamma_{XY}^F}{\sqrt{1 + 2\varepsilon_X^F}\sqrt{1 + 2\varepsilon_Y^E}} \tag{2.38b}$$

It is seen that for finite strains, shear deformation as defined by angle change depends not only on γ_{XY}^F but also on the strain components ε_X^F and ε_Y^F. For infinitesimal strains $\sin \gamma_{XY} \cong \gamma_{XY}$ and the above definition reduces to $\gamma_{XY} = \gamma_{XY}^F$ as expressed before in the strain matrix (Equation 2.23).

There are other definitions used to describe large deformation of solids. For the case of simple extension or contraction the following definitions apply:

Extension Ratio (λ). This definition is particularly used in problems of large deformation of rubbery materials* as:

$$\lambda = (L/L_o) \tag{2.39}$$

where L and L_o are the initial and current lengths respectively.

Natural (Logarithmic) Strain (ε^p). This is used in analyzing problems of plastic deformation and in particular metal-forming problems. It is defined as**

$$\varepsilon^p = \ln(L/L_o) = \ln(\lambda) \tag{2.40a}$$

For large strains, the volumetric strain ε_v cannot be longer expressed by Expression 2.4 or 2.29a. This may be shown by considering a volume element of initial dimension $(ds_o)_1$, $(ds_o)_2$, and $(ds_o)_3$, where 1, 2, and 3 designate the principal strain directions. After large deformation, the volume element dimensions become $(ds)_1$, $(ds)_2$, and $(ds)_3$, respectively and the change of volume $\Delta V/V_o$ is thus

$$\begin{aligned}\frac{\Delta V}{V_o} &= \left(\frac{ds}{ds_o}\right)_1 \left(\frac{ds}{ds_o}\right)_2 \left(\frac{ds}{ds_o}\right)_3 - 1 \\ &= (1 + \varepsilon_1)(1 + \varepsilon_2)(1 + \varepsilon_3) - 1 \\ &= \varepsilon_1 + \varepsilon_2 + \varepsilon_3 + \varepsilon_1\varepsilon_2 + \varepsilon_2\varepsilon_3 + \varepsilon_3\varepsilon_1 = J_1 + J_2 + J_3\end{aligned}$$

---

* See Williams,[7] p. 30.

** See Chapter 10. For convenience, ln is used hereafter to signify $\log_e$.

where the second powers of ε cannot be neglected for large strains. Employing natural strains, however gives

$$\frac{\Delta V}{V_o} = \exp(\varepsilon_1^p) + \exp(\varepsilon_2^p) + \exp(\varepsilon_3^p) - 1 = \exp(\varepsilon_1^p + \varepsilon_1^p + \varepsilon_1^p) - 1$$

Defining a natural volumetric strain ε_v^p as

$$\varepsilon_v^p = \ln\left(\frac{V}{V_o}\right) = \ln\left(1 + \frac{\Delta V}{V_o}\right)$$

Hence,

$$\varepsilon_v^p = \varepsilon_1^p + \varepsilon_2^p + \varepsilon_3^p \tag{2.40b}$$

Large plastic deformation occurs at constant volume and, hence, $\varepsilon_v^p = 0$.

The above finite strain definitions are related to each other, and strain–displacement relations could be easily derived for any of them. However, it must be recalled that the transformation laws (Equations 2.22 and 2.24) apply strictly for infinitesimal strains. Other derivations are required for the transformation of finite or natural strains.*

Example 2.11:
Derive the relation between the engineering strain and each of

a. *Green strain,* ε^F;
b. *Extension ratio,* λ;
c. *Natural strain,* ε^P;

for the case of simple tension. Then calculate the various strains for rod of initial length L_o *pulled axially* ΔL = *5, 50%, or 100%.*

Solution:
Relations are as follows

a. Engineering strain to Green strain: $\varepsilon_x = (\Delta L/L_o) = \sqrt{1 + 2\varepsilon_X^F} - 1$
b. Engineering strain to extension ratio: $\varepsilon_x = (\Delta L/L_o) = \lambda - 1$
c. Engineering strain to natural strain: $\varepsilon_x = (\Delta L/L_o) = (\exp \varepsilon_X^p) - 1$

For the case of a uniaxially pulled rods, the results are given by the following table:

$\varepsilon_x = (\Delta L/L_o)$	ε_x^F	λ	ε_x^P
0.05	0.05125	1.05	0.05127
0.5	0.625	1.5	0.6487
1.0	1.5	2.0	0.6931

* Hoffman and Sachs,[8] p. 32.

Obviously, as deformation becomes larger, the difference between the engineering strain and both Green and natural strain becomes more appreciable.

Example 2.12:

The displacement field for a solid is given by the components (in length units) by

$$u = (X^2 + Y^2 + 2)\,10^{-2},\quad v = (3X + 4Y^2)\,10^{-2},\quad w = (2X^3 + 4Z^2)\,10^{-2}$$

Determine the strain components at a point originally at (1,2,3) using both infinitesimal strain–displacement relations and finite strain–displacement relations (Green strains). Also determine the change in length of a line joining the two points originally at (0, 0, 0) and (1, 2, 3) using both infinitesimal and finite strain definitions.

Solution:

Employing Equations 2.6 for infinitesimal strains gives

$$\varepsilon_X = \frac{\partial u}{\partial x} = (2X)10^{-2},\quad \varepsilon_Y = \frac{\partial v}{\partial Y} = (8Y)10^{-2},$$

$$\varepsilon_Z = \frac{\partial w}{\partial Z} = (8Z)10^{-2},\quad \gamma_{XY} = \frac{\partial v}{\partial X} + \frac{\partial u}{\partial Y} = (3+2Y)10^{-2}$$

$$\gamma_{XY} = \frac{\partial v}{\partial Z} + \frac{\partial w}{\partial X} = \left(6X^2\right)10^{-2},\quad \gamma_{YZ} = \frac{\partial w}{\partial Y} + \frac{\partial v}{\partial Z} = 0$$

For a point originally at (1, 2, 3), these strain components are

$$\varepsilon_X = 0.02,\quad \varepsilon_Y = 0.16,\quad \varepsilon_Z = 0.24$$

$$\gamma_{XY} = 0.07,\ \gamma_{XZ} = 0.06,\quad \gamma_{YZ} = 0$$

The change in length of a line segment originally joining the two points (0, 0, 0) and (1, 2, 3) is obtained from Equation 2.21 for $dX = 1$, $dY = 2$, and $dZ = 3$, as

$$ds^2 - ds_o^2 = 2\left(\varepsilon_X dX^2 + \varepsilon_Y dY^2 + \varepsilon_Z dZ^2\right) + 2\left(\gamma_{XY} dXdY + \gamma_{XZ} dXdZ + \gamma_{YZ} dYdZ\right) = 6.28$$

Knowing that the original length is

$$ds_o = \sqrt{dx^2 + dY^2 + dZ^2} = 3.7417$$

then $ds = 4.5033$. This corresponds to an elongation of about 20%.

For finite strains, Equations 2.36 apply as

$$\varepsilon_X^F = (2X)10^{-2} + ½\left[(2X)^2 + (3)^2 + \left(6X^2\right)\right]10^{-4}$$

$$\varepsilon_Y^F = (8Y)10^{-2} + ½\left[(2Y)^2 + (8Y)^2 + (0)\right]10^{-4}$$

$$\varepsilon_Z^F = (8Z)10^{-2} + ½\left[(0) + (0) + (8Z)^2\right]10^{-4}$$

$$\gamma_{XY}^F = (3 + 2Y)10^{-2} + (2X)(2Y)10^{-4} + (3)(8Y)10^{-4} + \left(6X^2\right)(0)10^{-4}$$

$$\gamma_{XZ}^F = \left(6X^2\right)10^{-2} + (2X)(0)10^{-4} + (3)(0)10^{-4} + \left(6X^2\right)(8Z)10^{-4}$$

$$\gamma_{YZ}^F = 0 + (2Y)(0)10^{-4} + (8Y)(0)10^{-4} + (0)(8Z)10^{-4}$$

This gives for the point originally at (1, 2, 3)

$$\varepsilon_X^F = 0.02245, \quad \varepsilon_Y^E = 0.1736, \quad \varepsilon_Z^F = 0.2688$$

$$\gamma_{XY}^F = 0.0762, \quad \gamma_{XZ}^F = 0.0744, \quad \gamma_{YZ}^F = 0$$

The change in length of the line segment is obtained by applying Equation 2.35 as

$$\left(ds^2\right)^F - ds_o^2 = 2\left(\varepsilon_X^F dX^2 + \varepsilon_Y^F dY^2 + \varepsilon_Z^F dZ^2\right) + 2\left(\gamma_{XY}^F dXdY + \gamma_{XZ}^F dXdZ + \gamma_{YZ}^F dYdZ\right) = 7.0233$$

Hence,

$$(ds)^F = 4.5851$$

with a difference of only 1.8% compared with infinitesimal strain calculations. Such small error is to be expected for moderate elongations as 20%.

2.11 STRAIN RATE — VELOCITY RELATIONS

The mechanical behavior of many materials is highly influenced by the strain rate during deformation. Typical examples are polymers, superplastic materials, and materials subjected to impact or creep loading and during hot forming processes. The deformation changes with time and the material flows.

If u, v, w are the displacement components of a point at any time t in the x-, y-, z-directions, respectively, the velocity components of the point are expressed by

$$\dot{u} = \frac{\partial u}{\partial t}, \quad \dot{v} = \frac{\partial v}{\partial t}, \quad \dot{w} = \frac{\partial w}{\partial t} \tag{2.41}$$

The strain-rate components are defined as

$$\dot{\varepsilon}_x = \frac{\partial \varepsilon_x}{\partial t} = \frac{\partial}{\partial t}\left(\frac{\partial u}{\partial x}\right) = \frac{\partial}{\partial x}\left(\frac{\partial u}{\partial t}\right) = \frac{\partial \dot{u}}{\partial x}$$

Similarly

$$\dot{\varepsilon}_y = \frac{\partial \dot{v}}{\partial y}, \quad \dot{\varepsilon}_z = \frac{\partial \dot{w}}{\partial z}, \quad \dot{\gamma}_{xy} = \frac{\partial \dot{v}}{\partial x} + \frac{\partial \dot{u}}{\partial y}, \quad \dot{\gamma}_{yz} = \frac{\partial \dot{w}}{\partial y} + \frac{\partial \dot{v}}{\partial z}, \quad \dot{\gamma}_{zx} = \frac{\partial \dot{u}}{\partial z} + \frac{\partial \dot{w}}{\partial x} \tag{2.42}$$

The strain-rate tensor expressed by Equations 2.42, simply as the time rate of strain,* transforms according to the same transformation law as for the infinitesimal strain tensor.

In the analysis of large plastic deformation, the logarithmic or natural strain ε^p is encountered. However, it is the increments of this, $d\varepsilon^p$, which appear generally in the governing equations as given in Chapter 10. Dividing $d\varepsilon^p$ by dt defines the plastic strain-rate tensor [$\dot{\varepsilon}^p$]. As a simple illustration, for uniaxial deformation $\varepsilon^p = \ln(L/L_0)$, hence, $\dot{\varepsilon}^p = d\varepsilon^p/dt = (dL/dt)/L = \dot{u}/L$ where $\dot{u}$ is the velocity component along the L-direction.

Example 2.13:

A bar is drawn through a die to reduce its cross section by 20%; the die is tapered with a 10° semiangle, Figure 2.16. Assuming uniform velocity on the bar cross section throughout the deformation zone, obtain expressions for the strain rate for no change of volume and an exit speed of 5 m/s if

a. *The bar is a strip of width 200 mm and initial thickness 40 mm and the width is maintained constant through the die;*
b. *The bar is round of initial diameter 40 mm.*

Solution:

For an area reduction of 20% and exit speed of 5 m/s, the entry speed at constant volume is given by

$$A_1\dot{u}_1 = A_2\dot{u}_2$$

Substituting $A_2 = 0.8A_1$, $\dot{u}_2 = 5$ m/s yields $\dot{u}_1 = 4$ m/s.

a. For a strip of constant thickness, the velocity at any plane distance x from the entry plane is given by

$$\dot{u} = \dot{u}_1 \frac{A_1}{A} = \dot{u}_1 \frac{h_1}{h}$$

---

* Distinction must be made between the time rate of strain $\dot{\varepsilon}_{ij} = \partial \varepsilon_{ij}/\partial t$ and the deformation or in the material strain-rate

$$\frac{d\varepsilon_{ij}}{dt} = \dot{D}_{ij} = \dot{\varepsilon}_{ij} + \left(\partial \varepsilon_{ij}/\partial x_k\right)\dot{u}_k$$

or explicitly

$$\dot{D}_x = \dot{\varepsilon}_x + \left(\frac{\partial \varepsilon_x}{\partial x_1}\dot{u} + \frac{\partial \varepsilon_x}{\partial y}\dot{v} + \frac{\partial \varepsilon_x}{\partial z}\dot{w}\right)\cdots \text{ etc.}$$

where the last term involves product of the strain gradient $(\partial\varepsilon_{ij}/\partial x_k)$ and the material velocity $\dot{u}_k$. Only for homogeneous strain, quasi-state conditions, or small displacement, $\dot{D}_{ij} = \dot{\varepsilon}_{ij}$ and hence no distinction among them is needed. See Drucker,[9] p. 263.

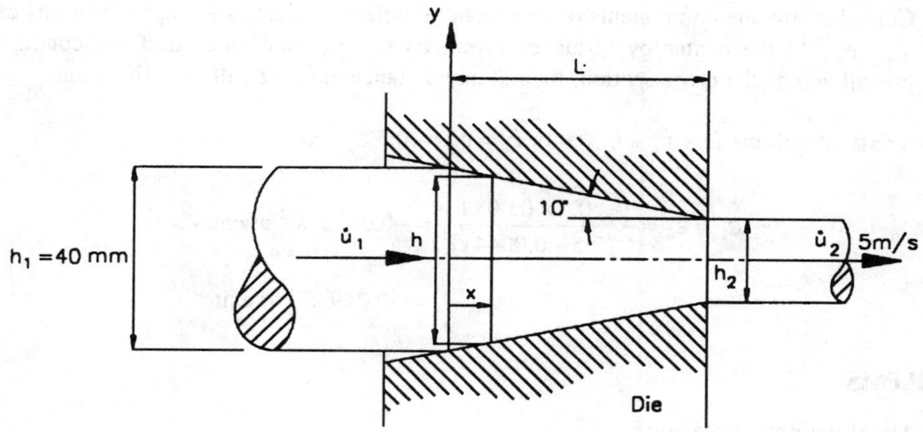

FIGURE 2.16 Example 2.13.

where $h = h_1 - 2x \tan 10° = 40 - 0.3526x$.
Hence,

$$\dot{u} = \frac{4 \times 40}{40 - 0.3526x} = \frac{1}{0.25 - 0.0022x}$$

$$\dot{\varepsilon}_x = \frac{\partial \dot{u}}{\partial x} = \frac{0.0022}{(0.25 - 0.0022x)^2} \quad s^{-1}$$

Since the strip thickness remains constant, then $\varepsilon_z = \dot{\varepsilon}_z = 0$ and for constant volume $\dot{\varepsilon}_x = -\dot{\varepsilon}_y$.

At entry $x = 0$: $\dot{\varepsilon}_{x_1} = 0.0352\ s^{-1}$; $\dot{\varepsilon}_{y_1} = -0.0352\ s^{-1}$

At exit: $h_2 = 0.8h_2 = 32$ mm and $x = L = \dfrac{40 - 32}{0.3526} = 22.6885$ mm

Hence: $\dot{\varepsilon}_{x_2} = 0.055\ s^{-1}$, $\dot{\varepsilon}_{y_2} = -0.055\ s^{-1}$

b. For a round bar of diameter 40 mm, the velocity at any plane is

$$\dot{u} = \dot{u}_1 \frac{A_1}{A} = \dot{u}_1 \frac{h_1^2}{h^2}$$

$$= \frac{4 \times 40^2}{(40 - 0.3526x)^2} = \frac{1}{(0.5 - 0.0044x)^2}$$

$$\dot{\varepsilon}_x = \frac{\partial \dot{u}}{\partial x} = \frac{0.0088}{(0.5 - 0.0044x)^3} \quad s^{-1}$$

At entry: $\dot{\varepsilon}_{x_1} = 0.704\ s^{-1}$, $\dot{\varepsilon}_{y_1} = -0.704\ s^{-1}$

At exit: $h_2 = \sqrt{0.8}\ h_1 = 35.777$ mm and $L = \dfrac{40 - 35.777}{0.3526} = 11.977$ mm

Hence: $\dot{\varepsilon}_{x_2} = 0.0983\ s^{-1}$ $\dot{\varepsilon}_{y_2} = -0.0983\ s^{-1}$

Consider now the components of strain rate in polar coordinates $\dot{\varepsilon}_r$, $\dot{\varepsilon}_\theta$, $\dot{\varepsilon}_z$; in this case $\dot{\varepsilon}_z = \dot{\varepsilon}_x$. At the center by virtue of symmetry $\dot{\varepsilon}_r = \dot{\varepsilon}_\theta$, and since uniform conditions prevail across the cross section, then at any distance x, $\dot{\varepsilon}_r = \dot{\varepsilon}_\theta$ all over the plane.

For constant volume $\dot{\varepsilon}_r + \dot{\varepsilon}_\theta + \dot{\varepsilon}_z = 0$ or $\dot{\varepsilon}_r = \dot{\varepsilon}_\theta = -0.5\,\dot{\varepsilon}_x$. Hence,

$$\dot{\varepsilon}_r = \dot{\varepsilon}_\theta = -\frac{0.5 \times 0.0088}{(0.5 - 0.0044x)^3} = -0.0352\ \text{s}^{-1} \text{ at entry}$$

$$= -0.049\ \text{s}^{-1} \text{ at exit}$$

PROBLEMS

Strain–Displacement Relations

2.1 Given the displacement field:

$$u = \left(3x^4 + 2x^2y^2 + x + y + z^3 + 3\right) \times 10^{-3}$$

$$v = \left(3xy + y^3 + y^2z + z^2 + 1\right) \times 10^{-3}$$

$$w = \left(x^2 + xy + yz + zx + y^2 + z^2 + 2\right) \times 10^{-3}$$

Determine the associated strains at point (1, 2, 3).

2.2 Given the displacement field

$$u = axy, \quad v = by^2, \quad w = cxz$$

where a, b, and c are constants, determine the strain tensor components at a point with coordinates (1, 2, 3).

2.3 Given the displacement field

$$u = ae^{-x}, \quad v = ae^{-y}, \quad w = ae^{-z}$$

where a is constant, determine:

a. The strain tensor components at a point with coordinates (2, 2, 2).
b. The volumetric strain at (0, 0, 0).

2.4 Given the displacement field

$$u = ax^2 + bxy, \quad v = ay^2 + bxy, \quad w = cxyz$$

where a, b, and c are constants, determine the strain components at the point (1, 2, 1). What is the volumetric strain?

2.5 A displacement field is given by

$$u = (x^2 + 10) \times 10^{-2},\ v = (2yz) \times 10^{-2}, \text{ and } w = (z^2 - xy) \times 10^{-2}$$

Determine the state of strain of an infinitesimal element positioned at (0, 1, 2).

2.6 Find the strain components for a body whose deformation is described by the displacements:

$$u = M\,[\nu\,(x^2 - yz) + z^2]/2EI,\quad v = M\nu xy/EI \text{ and } w = -Mxz/EI$$

2.7 The strain in the longitudinal direction of a thin, square rubber sheet of 500 mm side long is $\varepsilon_x = 0.004 + 0.0001\,x$, where x is the distance in mm from the end of the rod. Determine the change in thickness for constant volume deformation.

2.8 A square plate *ABCD* has sides of 20 mm length. The x-axis is along side *AB* and the y-axis is along side *AD*. The plate is subjected to the shear strain $\gamma_{xy} = 0.004x + 0.002y$, where x and y are in mm. If side *AB* is fixed and lines parallel to *AB* remain parallel to it, determine the displacement of point *D* in the x-direction. Determine the change in the angle *CDA*.

2.9 A hollow circular cylinder with inner and outer radii r_i and r_o, respectively, undergoes a radial displacement:

$$u = ar + b/r\,,\ v = 0, \text{ and } w = 0$$

Determine (a) the radial and tangential strains at the inner and outer walls of the cylinder, taking $r_o/r_i = 3$. (b) The relation between a and b such that ε_r (at $r = r_i$) $= 2\varepsilon_\theta$ (at $r = r_o$).

2.10 A circular rod undergoes a radial displacement:

$$u = ar(\sin^2\theta - {}^1/_2),\ v = 0, \text{ and } w = 0$$

where a is a constant. Determine the strains at $r = 1$ for $\theta = 0$, $\pi/4$, and $\pi/2$.

Strain Compatibility Conditions

2.11 Derive the strain compatibility equations for the following states of strain:

a. Plane strain
b. Plane stress

2.12 Determine whether the following strain fields (a) and (b) are compatible:

a.
$$\begin{aligned}
\varepsilon_x &= 2x^2 + 3y^2 + z + 1,\\
\varepsilon_y &= 2y^2 + x^2 + 3z + 2,\\
\varepsilon_z &= 3x + 2y + z^2 + 1,\\
\gamma_{xy} &= 8xy,\\
\gamma_{yz} &= 0,\\
\gamma_{zx} &= 0
\end{aligned}$$

b.
$$\begin{aligned}
\varepsilon_x &= 3y^2 + xy,\\
\varepsilon_y &= 2y + 4z + 3,\\
\varepsilon_z &= 3zx + 2xy + 3yz + 2,\\
\gamma_{xy} &= 6xy,\\
\gamma_{yz} &= 2x,\\
\gamma_{zx} &= 2y
\end{aligned}$$

2.13 For the following plane strain distribution, verify whether or not the compatibility condition is satisfied:

$$\varepsilon_x = 3x^2y,\ \varepsilon_y = 4y^2x \text{ and } \gamma_{xy} = 2xy + 2x^3$$

2.14 Verify whether or not the following strain field satisfies the equations of compatibility (a is a constant):

$$\varepsilon_x = ay,\ \varepsilon_x y = ax,\ \varepsilon_z = 2a\,(x + y)\ \gamma_{xy} = a(x + y),\ \gamma_{yz} = 2az, \text{ and } \gamma_{zx} = 2az$$

2.15 Under what conditions are the following displacement field and the shear strain at a point compatible?

$$u = ax^2y^2 + bxy^2 + cx^2y,$$

$$v = ax^2 + bxy,$$

$$\gamma_{xy} = dx^2y + cxy + dx^2 + ey$$

2.16 Given the strain field:

$$\varepsilon_x = ay^2 + bxy,$$

$$\varepsilon_y = a(x^2 + y^2) + by,$$

$$\gamma_{xy} = bxy,$$

$$\varepsilon_z = \gamma_{xz} = \gamma_{yz} = 0$$

determine the condition that these strains become compatible.

2.17 Determine whether or not the following displacement field is possible in a continuous material:

$$\begin{bmatrix} u \\ v \\ w \end{bmatrix} = \begin{bmatrix} 0.001 & 0 & -0.003 \\ 0.0005 & 0.002 & 0 \\ 0 & 0.001 & -0.005 \end{bmatrix} \begin{bmatrix} x \\ y \\ z \end{bmatrix}$$

a Calculate the displacement of the point (1, 2, 1).

b Let $A(2, 0, 0)$ and $B(0, 1, 3)$ represent two points in the undeformed geometry. What displacement occurs between the two points?

2.18 Given the strain field:

$$\varepsilon_x = ay,\ \gamma_{xy} = \gamma_{yz} = \gamma_{zx} = 0,\ \varepsilon_y = by, \text{ and } \varepsilon_z = by$$

Determine the displacement fields.

2.19 A body is in a state of plane strain in the y–z plane. Given that

$$\varepsilon_y = 2ay + bz,\quad \varepsilon_z = 0,\quad \text{and}\quad \gamma_{yz} = by + \frac{c\pi}{L}\cos\frac{\pi y}{L}$$

determine the displacements v and w.

2.20 A body is in a state of plane strain in the x–y plane, such that

$$\varepsilon_x = ay,\ \varepsilon_y = bx,\ \gamma_{xy} = ax + by$$

Find the displacements $u(x, y)$ and $v(x, y)$ given that

$$u(0, 0) = 0,\ v(0, 0) = 0, \text{ and } u(0, 1) = a$$

2.21 A body is in a state of plane strain in the y–z plane, such that

$$\varepsilon_y = ae^{-z}, \ \varepsilon_z = by, \text{ and } \gamma_{yz} = -aye^{-z}$$

where a and b are constants. Determine the displacements $v(y, z)$ and $w(y, z)$ given that

$$v(0, 0) = 0, \ w(0, 0) = 0, \ w(1, 0) = 0$$

2.22 Determine whether or not the following strain fields are possible in a continuous material:

a. $$[\varepsilon] = \begin{bmatrix} c(x^2 + y^2) & 2cxy & 0 \\ 2cxy & y^2 & 0 \\ 0 & 0 & 0 \end{bmatrix}$$

b. $$[\varepsilon] = \begin{bmatrix} cz(z^2 + y^2) & 2cxyz & 0 \\ 2cxyz & y^2 z & 0 \\ 0 & 0 & 0 \end{bmatrix}$$

where c is a constant.

2.23 Consider the strain given by

$$\varepsilon_{z\theta} = b/(4\pi r)$$

$$\varepsilon_r = \varepsilon_\theta = \varepsilon_z = \varepsilon_{zr} = \varepsilon_{r\theta} = 0$$

Determine the displacement that these strains represent and show that the displacement field is not admissible for a solid cylinder with an axis along the z-direction.

Strain Transformation

2.24 The displacement components in a strained body are $u = 0.01\,x + 0.002y^2$ mm, $v = 0.02x^2 + 0.002\,z^3$mm, and $w = 0.001\,x + 0.005$ mm. Determine the change in distance between two points which, before deformation, have coordinates (3, 2, 0 mm) and (–1, 4, 5 mm).

2.25 The square plate in the figure is loaded so that the plate is in a state of plane strain ($\varepsilon_z = \varepsilon_{zx} = \varepsilon_{zy} = 0$).

a. Determine the displacements for the plate given the deformations shown and also the strain components for the (x, y) coordinate axes.
b. Determine the strain components for the (x', y') axes.

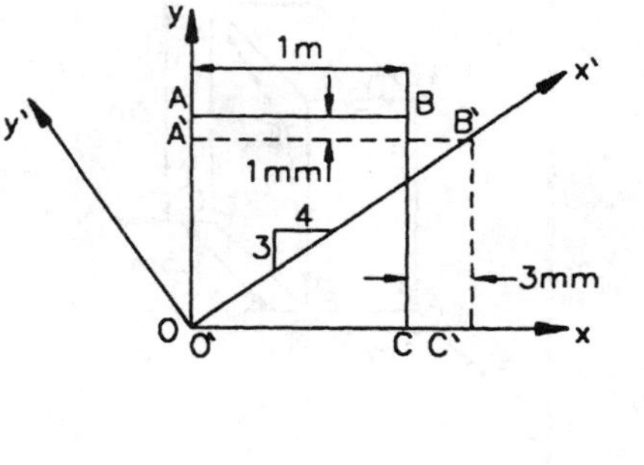

Problem 2.25

2.26 The plate shown in the figure is loaded so that a state of plane strain ($\varepsilon_z = \varepsilon_{zx} = \varepsilon_{zy} = 0$) exists.

a. Determine the displacement field for the plate and the strain components at point B.
b. Let the x'-axis extend from point 0 through point B. Determine $\varepsilon_{x'}$ at point B.

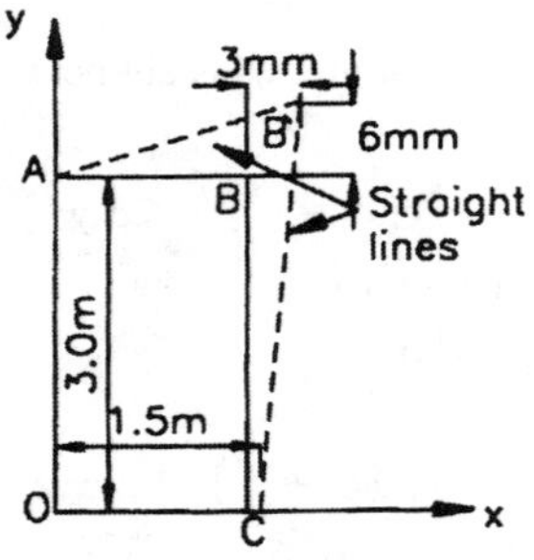

Problem 2.26

2.27 In plane strain conditions, line elements in the x- and y-directions increase by 1.5 and 2.4%, respectively, and decrease their right angle by 8°. What is the percentage increase in a line element originally at 45° to x- and y-axes and by how much has this line rotated?

2.28 If the displacement field is given by

$$u = kxy,\ v = kxy,\ \text{and}\ w = 2k(x + y)z$$

where k is a small constant,

a. Write down the infinitesimal strain matrix;
b. Determine the strain in the direction $l = m = n = 1\ \sqrt{3}$.

2.29 The parallelepiped in the figure is deformed into the shape indicated by dashed straight lines. The small displacements are given by the following relations:

$$u = c_1xyz,\ v = c_2xyz,\ \text{and}\ w = c_3xyz$$

a. Determine the state of strain at point E when the coordinates of point E' for the deformed body are (1.503, 1.001, 1.997).
b. Determine the normal strain at E in the direction EA.

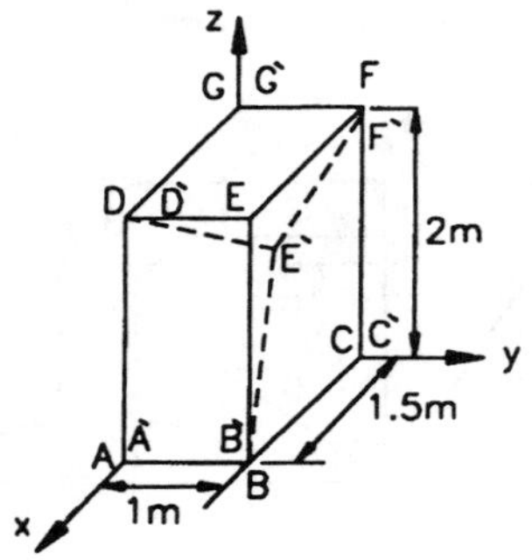

Problem 2.29

Principal Strains

2.30 The displacements u, v, and w of a material point are defined by $u = 2ay$ and $v = w = 0$, where a is a constant. Determine the corresponding strain components, strain tensor, strain invariants, and the principal strains.

2.31 The state of strain is given by

$$[\varepsilon] = \begin{bmatrix} a & a/2 & a/2 \\ a/2 & a & a/2 \\ a/2 & a/2 & a \end{bmatrix}$$

Determine the magnitude of the principal strains.

2.32 Given a major principal strain of $\varepsilon_1 = 600 \times 10^{-6}$ and strain invariants: $J_1 = 500 \times 10^{-6}$ and $J_2 = -4900 \times 10^{-6}$, (a) find the remaining principal strains; (b) find, in magnitude and direction, the normal and shear strains for the octahedral plane.

2.33 Determine the principal strains and the principal directions for the following states of strain:

a. $\varepsilon_x = 0.0008$, $\varepsilon_y = 0.0004$, $\gamma_{xy} = 0.0002$, and $\varepsilon_z = \gamma_{xz} = \gamma_{yz} = 0$.
b. $\gamma_{xy} = 0.0005$, $\gamma_{xz} = 0.0004$, $\gamma_{zy} = 0.002$, and $\varepsilon_x = \varepsilon_y = \varepsilon_z = 0$.
c. $\gamma_{xy} = 0.0024$ and $\varepsilon_x = \varepsilon_y = \varepsilon_z = \gamma_{xz} = \gamma_{yz} = 0$.

2.34 The state of strain in a cylindrical pressure vessel is given by

$$[\varepsilon] = \begin{bmatrix} 0.0005 & 0 & -0.0005 \\ 0 & 0.002 & 0 \\ -0.0005 & 0 & -0.001 \end{bmatrix}$$

Determine the magnitude and direction of the principal strains.

2.35 A state of strain is given by the tensor:

$$[\varepsilon] = \begin{bmatrix} 0.002 & 0.001 & 0 \\ 0.001 & 0.003 & 0 \\ 0 & 0 & 0.004 \end{bmatrix}$$

(a) Calculate the magnitude and direction for the principal strains. (b) Determine the dilatation and strain principal deviations.

2.36 The state of strain in a body is given by

$$[\varepsilon] = \begin{bmatrix} 0.005 & 0.0025 & 0 \\ 0.0025 & 0.005 & 0 \\ 0 & 0 & 0.005 \end{bmatrix}$$

Calculate:

a. Dilatation and deviatoric strains;
b. Principal strains; and
c. For the original dimensions of the body equal to $x_o = 20$ mm, $y_o = 30$ mm, and $z_o = 40$ mm, find the final dimensions.

2.37 The nonzero strain components at a point in a loaded member are $\varepsilon_x = 0.002$, $\varepsilon_y = -0.001$, and $\gamma_{xy} = -0.002$. Using Mohr's circle of strain, determine the principal strains and their directions.

2.38 The state of strain at a point on a steel plate is given by $\varepsilon_x = 550\varepsilon_x$, $\varepsilon_y = 140 \times 10^{-6}$, and $\gamma_{xy} = 360 \times 10^{-6}$. Determine, using Mohr's circle of strain,

a. The components of strain associated with axes x', y', which make an angle $\theta = 45$ with the axes x, y.
b. The principal strains and directions of the principal axes.
c. The maximum shear strains and associated normal strains.

2.39 The strains at a point in a body are given as $\varepsilon_x = k$, $\varepsilon_y = -k$, $\varepsilon_z = 0$, $\gamma_{xy} = 0$, $\gamma_{yz} = 0$, and $\gamma_{zx} = 2k$. Determine the principal strains and the maximum shear strain.

2.40 A 400 × 600 mm rectangular plate $OABC$ is deformed into a shape $O'A'B'C'$ shown in the figure, determine

a. The strain components ε_x, ε_y, γ_{xy};
b. The principal strains and the direction of the principal axes.

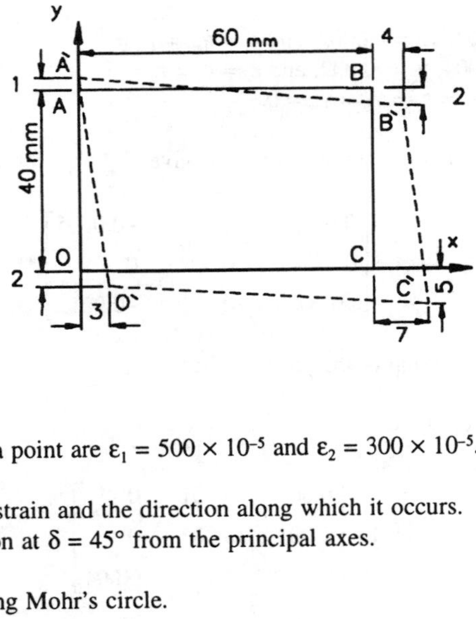

Problem 2.40

2.41 The principal strains at a point are $\varepsilon_1 = 500 \times 10^{-5}$ and $\varepsilon_2 = 300 \times 10^{-5}$. Determine

a. The maximum shear strain and the direction along which it occurs.
b. The strains in direction at $\delta = 45°$ from the principal axes.

Check the results by employing Mohr's circle.

2.42 In a state of two-dimensional strain, the principal strains at a point are given as $\varepsilon_1 = 0.0018$ and $\varepsilon_2 = -0.0009$. Determine the direction of the x'-axis such that the shear strain $\gamma_{x'y'}$ 0.0006, and the corresponding normal strains $\varepsilon_{x'}$ and $\varepsilon_{y'}$. Solve the problem analytically and graphically using Mohr's circle of strain.

Strain Gauge Rosettes

2.43 The data for a strain gauge rosette attached to a stressed steel member are $\varepsilon_{0°} = -0.00032$, $\varepsilon_{45°} = 0.00016$, and $\varepsilon_{90°} = 0.00025$. Determine the principal strains and their directions.

2.44 The strain at a point on the surface of a machine part has been measured in three directions by using three strain gauges. Obtain the magnitude of ε_x, ε_y, and γ_{xy} in terms of the three gauge readings if the angles between the gauges are equal and have the values of 60°. Apply the result to the gauge readings: $\varepsilon_{0°} = \varepsilon_{60°} = 0.0006$ and $\varepsilon_{120°} = -0.0008$.

2.45 A resistance strain gauge rosette has arms *a*, *b*, and *c*. Measurements taken on the surface of a thin plate subjected to a two-dimensional stress system are

$$\varepsilon_a,\ \varepsilon_b = 2\varepsilon_a \text{ and } \varepsilon_c = -\varepsilon_a$$

Determine the maximum shear strain at this point. The angle between *a* and *b* is 30° and between *b* and *c* is 60°.

2.46 A circular shaft is subjected to a bending moment and at the same time a twisting moment. A 60° equiangular resistance strain gauge rosette is fixed to the shaft surface to monitor the torque. The following readings are taken:

$$\varepsilon_0 = -10 \times 10^{-5},\ \varepsilon_{60} = 5 \times 10^{-5},\ \text{and } \varepsilon_{120} = 7 \times 10^{-5}$$

Determine the shear strain due to the twisting moment and the principal strains.

2.47 A 45° equiangular rosette is fixed to a plate structure and the following readings are taken:

$$\varepsilon_0 = 8 \times 10^{-4} \text{ and } \varepsilon_{90} = 3 \times 10^{-4}$$

ε_{45} cannot be read because its gauge is damaged. It is known, however, that at the point of attachment of the rosette there is a thickness strain $\varepsilon_z = -4 \times 10^{-4}$. Determine the maximum shear strain at the point and its direction in relation to gauge *a*.

2.48 A strain gauge records 950×10^{-6} when it is bonded in the direction of the major principal strain for a 25-mm-diameter shaft under torsion. What will be the change in this strain when a 150-mm length of this bar is simultaneously subjected to an axial force causing a 0.25-mm change in length and a 0.0125-mm diameter reduction?

2.49 A strain gauge rosette is bonded to the cylindrical surface of a 40-mm driveshaft of a motor. There are three gauges *a*, *b*, and *c* and the centerline of gauge a lies at 45° to the centerline of the shaft. The angles between gauges *a* and *b* and between *b* and *c* are both 60°, all angles measured counterclockwise. When the motor runs and transmits constant torque M_t, the readings of gauges *a*, *b*, and *c* are found to be 7.25×10^{-4}, -3.625×10^{-4}, and -3.625×10^{-4}, respectively. If $E = 208$ GPa and $\nu = 0.3$, calculate M_t, the principal stresses and the maximum shear in the shaft.

Finite Strain

2.50 The displacement field for a body is given by

$$u = \tfrac{1}{4}(X^2 + Y^3 + 2)10^{-1},\ v = \tfrac{1}{2}(3X + Y^2)10^{-1},\ \text{and } w = (X^{3/3} + 2Z)10^{-1}$$

What is the deformed position of a point originally at (1, 1, 2)? Determine

a. The strain components at (1, 1, 2), using only linear terms;
b. The strain components if nonlinear terms are also included using the Green strain definition.

2.51 If $\varepsilon_x = 0.002$, $\varepsilon_y = 0.001$, $\gamma_{xy} = 0.008$, and $\gamma_{yz} = \gamma_{zx} = 0.007$, (a) determine the change in length along a line having direction cosines $l = 0.4$, $m = 0.75$ using infinitesmal and finite strain definitions, i.e., Equations 2.21 and 2.35, respectively. (b) If all these strains were 50 times as large. Comment on the results.

2.52 Express the finite strain tensor E_{ij} given the spatial coordinates as

$$x = (1 + A)X + BY, \quad y = (1 + B) + CX$$
$$u = AX + BY, \quad v = BY + CX$$

Strain Rate–Velocity Relations

2.53 For a given velocity field:

$$\dot{u} = 2axy, \quad \dot{v} = -a\left(x^2 - y^2\right), \quad \text{and} \quad \dot{w} = 0$$

determine the components of the strain rate tensor, the deviatoric strain rate tensor, and the principal strain rates $\dot{\varepsilon}_1$, $\dot{\varepsilon}_2$, and $\dot{\varepsilon}_3$.

2.54 For the case of uniform forging at constant speed $2S_o$ as shown in the figure, satisfying constant volume deformation the following velocity components may be assumed:

a. Long billets with rectangular cross section, $\dot{w} = -2S_o z/H_o$,
b. Solid disk with circular cross section, $\dot{u} = S_o r/H_o$.

Determine the other remaining components of the velocity field and then the strain rate for each case.

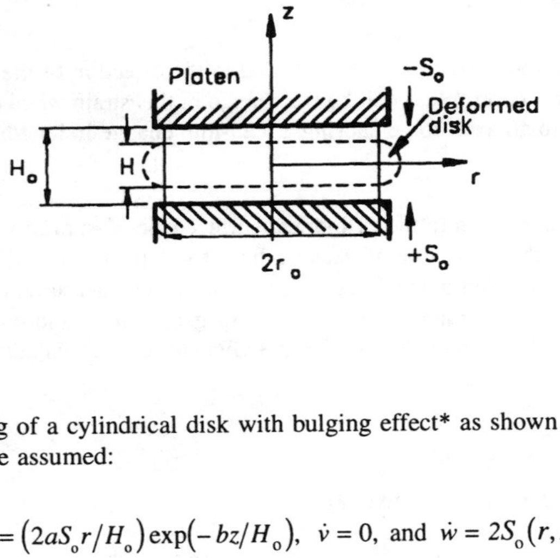

Problem 2.54

2.55 For the case of forging of a cylindrical disk with bulging effect* as shown in the figure, the following velocity components may be assumed:

$$\dot{u} = \left(2aS_o r/H_o\right)\exp\left(-bz/H_o\right), \quad \dot{v} = 0, \quad \text{and} \quad \dot{w} = 2S_o(r, z)$$

where a and b are constants.

a. Assuming constant volume deformation, determine the velocity component along the z-direction.
b. Determine the strain rate components.
c. By considering symmetry, express the constant a in terms of b. Discuss the case when the constant $b = 0$.

* See Avitzur,[10] p. 103.

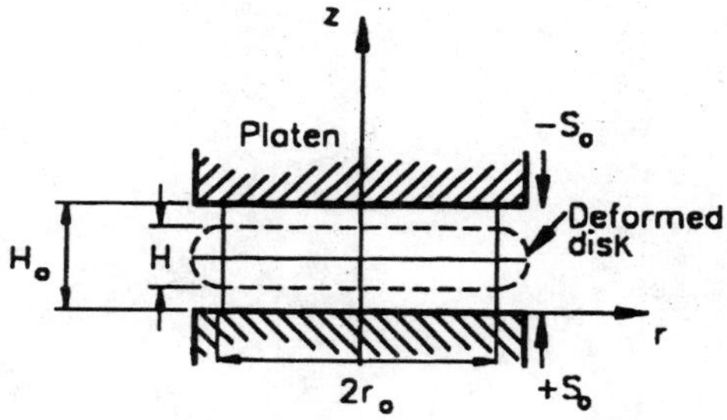

Problem 2.55

REFERENCES

1. Ford, H., *Advanced Mechanics of Materials, Longman*, London, 1972.
2. Fung, Y. C., *Foundations of Solid Mechanics,* Prentice-Hall, Englewood Cliffs, NJ, 1965.
3. Venkatraman, B. and Patel, S. A., *Structural Mechanics with Introductions to Elasticity and Plasticity,* McGraw-Hill, New York, 1970.
4. Boresi, A. P. and Sidebottom, O. M., *Advanced Mechanics of Materials,* 4th ed., John Wiley & Sons, New York, 1985.
5. Popov, E. P., *Mechanics of Materials,* 2nd ed., Prentice-Hall, Englewood Cliffs, NJ, 1987.
6. Dally, J. W. and Riley, W. F., *Experimental Stress Analysis,* McGraw-Hill, New York, 1965.
7. Williams, J. G., *Stress Analysis of Polymers,* Longman, London, 1973.
8. Hoffman, O. and Sachs, G., *The Theory of Plasticity for Engineers,* McGraw-Hill, New York, 1953.
9. Drucker, D. C., *Introduction to Mechanics of Deformable Solids,* McGraw-Hill, New York, 1967.
10. Avitzur, B., *Metal Forming: Processes and Analysis,* McGraw-Hill, New York, 1968.

3 Elastic Stress–Strain Relations

Hooke's law expressing the linear stress–strain relations for small deformations of an elastic homogeneous isotropic solid is obtained in both direct and inverse forms. These relations are applied to derive expressions of the strain energy for such a solid. Some energy theorems are discussed with application to simple bar problems. A generalization of Hooke's law describing linear elastic behavior of an anisotropic solid is formulated including an application to fiber-reinforced composites. Furthermore, nonlinear stress–strain relations for isotropic materials undergoing large elastic deformation, such as rubbers and polymers, are obtained by assuming a finite strain energy density function.

3.1 INTRODUCTION

In the preceding two chapters the concepts of stress and strain at a point are founded independently without any reference to the material of the solid dealt with. This simply means that these concepts are valid for any material under any conditions of loading or geometry.

Physical observations, however, indicate that there is a relation between the external loads applied to any solid and the deformation occurring due to this loading. Such a relation depends on the geometry of the loaded solid as well as the material from which it is made. For instance, calculating the deflection of a simply supported beam requires knowledge of the applied loads and beam dimensions as well as some properties of the material. A pressure vessel will certainly possess a load (pressure)–deformation (diametral expansion) relationship different from that of a simply supported beam even if both are made of the same material.

In order to determine the load–deformation behavior of a solid of a certain geometry due to a system of loads, the material stress–strain relation must be known. Such relation often termed as the material *constitutive law.* may be a quite general description of the material behavior under any conditions. It is aimed here to establish the relation between stress and strain for a state of material behavior defined as *elastic behavior.*

3.2 BASIC ASSUMPTIONS: ELASTICITY, HOMOGENEITY, AND ISOTROPY

In order to formulate the stress–strain relations describing any material behavior, some assumptions are made. These assumptions which simplify mathematical description, must not, however, obscure the basic material behavior observed experimentally.

3.2.1 ELASTICITY

A solid bar subjected to a monotonically increasing uniaxial stress σ may manifest a uniaxial stress–strain behavior as shown by curve a or b of Figure 3.1. Such behavior will be instantaneous, i.e., occurring immediately upon stress application. The linear behavior of Figure 3.1 corresponds

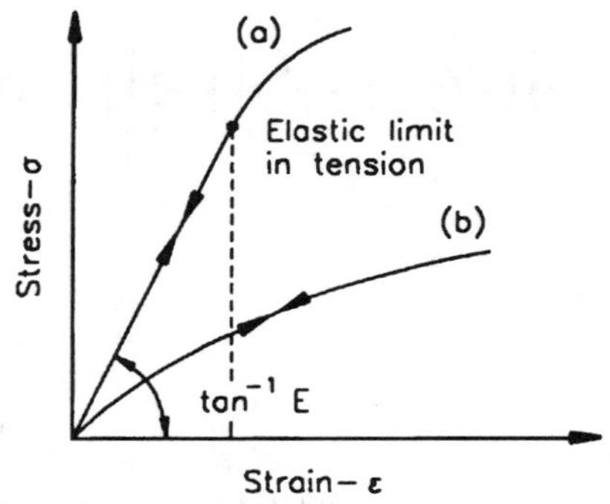

FIGURE 3.1 Idealized elastic behavior up to the elastic limit.

to most metallic materials, while the nonlinear one (curve b) describes behavior of materials such as rubbers. A common feature between these two behaviors is that, upon complete load removal, strain disappears. Furthermore, it is often assumed that loading and unloading take place along the same path. This means that the work done while loading is stored in the material as strain energy, which is fully recovered with no losses during unloading. This describes an idealized elastic behavior. Notably, this perfect elastic behavior is observed up to a certain value of stress, beyond which the relation displayed by Figure 3.1 no longer holds. Such limiting value of σ is known as the *proportional limit* and practically taken to be coincident with the elastic limit for many materials. The elastic limit is thus defined as the greatest stress that can be applied without producing any permanent strain upon unloading.

The first formulation of a relation between stress and strain is due to Robert Hooke (1678), who experimentally observed the proportionality between the elongation of a bar and the tensile force producing this elongation. The constant of proportionality is determined experimentally as the slope of the straight line representing σ–ε and is known as Young's modulus.*

3.2.2 Homogeneity

Consider a bar of uniform circular cross-sectional area A_0 and length L_0 subjected to a tensile force P uniformly distributed over its cross section, as shown in Figure 3.2. If the bar material is assumed "homogeneous," i.e., as possessing the same properties in the axial directions at all points, the bar would undergo a uniform elongation ΔL. More explicitly, if the bar is imagined to be an assembly of longitudinal fibers, homogeneity means that all fibers possess the same material properties and all particles in each fiber are identical. Hence, each fiber undergoes the same elongation ΔL when subjected to the same load P. Thus Hooke's law for such a homogeneous linear elastic bar of length L is

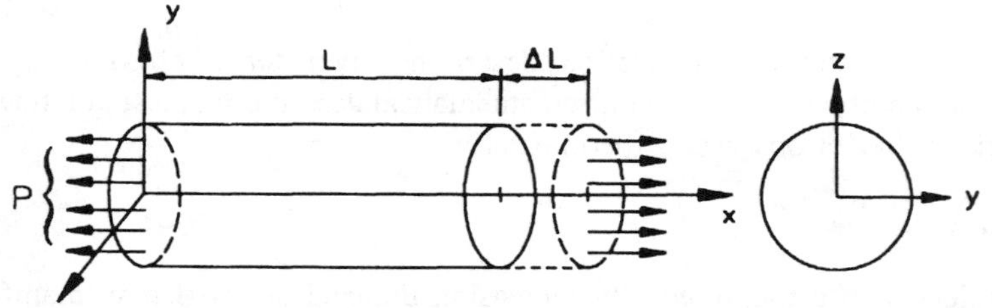

FIGURE 3.2 Homogeneous stress and strain in a bar under axial tension.

* Named after Thomas Young (1773–1829).

$$(\Delta L/L_o) = \frac{1}{E}(P/A_o) \tag{3.1}$$

where E is the material Young's modulus. Equation 3.1 is simply written as*

$$\varepsilon_x = \frac{\sigma_x}{E} \tag{3.2}$$

where ε_x and σ_x are the normal strain and normal stress, respectively, in the x-direction taken along the bar axis.

3.2.3 Isotropy

Observations indicate that as the bar extends longitudinally it also contracts laterally, resulting in diameteral decrease. Obviously, the two diameters along the y- and z-directions will not contract equal amounts unless the material properties along these two directions are exactly the same. If the case is so, the lateral strains ε_y and ε_z will be equal and, for infinitesimal strains, proportional to the longitudinal strain ε_x. This case is expressed by

$$\varepsilon_y = \varepsilon_z = -\nu\varepsilon_x = -\frac{\nu}{E}\sigma_x \tag{3.3}$$

Signs indicate that ε_x is extension (positive) while ε_y and ε_z are contraction (negative). The proportionality factor ν is known as Poisson's ratio** and is determined from a tensile experiment according to Equation 3.3. If Equation 3.3 holds for all directions, i.e., diameters, the bar material is said to be isotropic. Generally, isotropy means that material properties are independent of the direction along which they are measured. The assumption of isotropy holds for many engineering materials and conditions of elastic loading.

3.3 HOOKE'S LAW FOR HOMOGENEOUS ISOTROPIC MATERIALS

3.3.1 Simple Loading

Equations 3.2 and 3.3 constitute Hooke's law for uniaxial tension along the x-direction. Hooke's law may be also formulated for pure shear loading of homogeneous isotropic materials. As shown in Figure 3.3, the twisting moment M_t applied to a thin-walled circular tube of a mean radius r_m, produces an almost uniform shear stress τ in the tube wall given by $\tau = M_t/2\pi r_m^2 h$, where h is the tube wall thickness. This shear stress is accompanied by a shear strain γ, which is uniform throughout the tube thickness and attains a value $\gamma \equiv \tan\gamma = AA'/L$.

Experimentally, a plot between τ and γ follows a linear relationship (up to a certain limit) as shown in Figure 3.3b. In this case Hooke's law is expressed by

$$\gamma = \frac{1}{G}\tau \tag{3.4}$$

The proportionality constant G is known as the shear modulus or the modulus of rigidity.

* The small magnitude of elastic deformations is sensed by considering simple tension of, say, an aluminum rod up to its elastic limit (i.e., yield point) of 100 MPa. Hence, knowing that for aluminum E = 70 GPa gives $\varepsilon_x = \sigma_x/E$ = 100/70,000 = 0.00143.

** Named after S. D. Poisson (1781–1840).

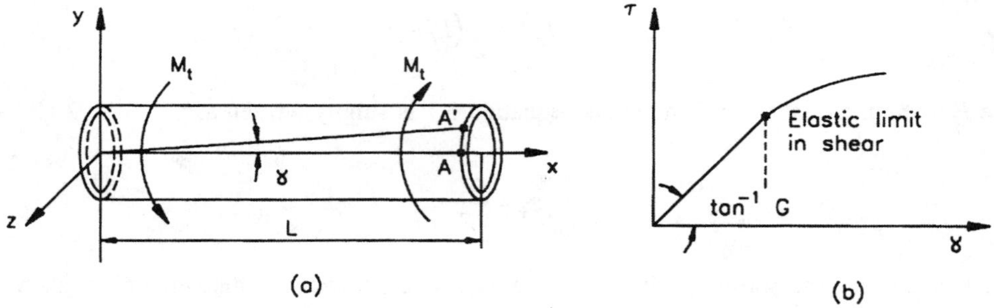

FIGURE 3.3 Thin-walled circular tube in torsion: (a) deformation in torsion and (b) shear stress–strain curve.

3.3.2 Triaxial Loading

Consider a parallelepiped element of a homogeneous isotropic solid obeying Hooke's law and subjected to three normal stresses σ_x, σ_y, and σ_z, as shown in Figure 3.4a. To express the relation among the applied three normal stresses and the resulting strains, each stress will be considered separately. Thus, according to Equations 3.2 and 3.3,

$$\text{due to } \sigma_x \text{ only}: \ \varepsilon_x = \frac{\sigma_x}{E} \ \text{ and } \ \varepsilon_y = \varepsilon_z = -\frac{\nu}{E}\sigma_x \tag{3.5a}$$

$$\text{due to } \sigma_y \text{ only}: \ \varepsilon_y = \frac{\sigma_y}{E} \ \text{ and } \ \varepsilon_x = \varepsilon_z = -\frac{\nu}{E}\sigma_y \tag{3.5b}$$

$$\text{due to } \sigma_z \text{ only}: \ \varepsilon_z = \frac{\sigma_z}{E} \ \text{ and } \ \varepsilon_x = \varepsilon_y = -\frac{\nu}{E}\sigma_z \tag{3.5c}$$

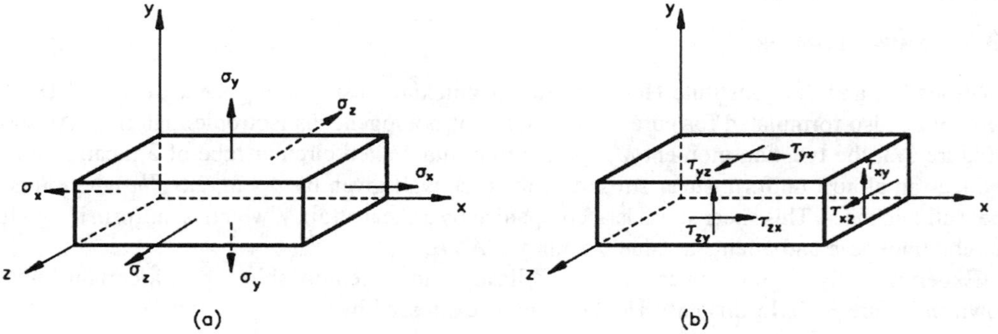

FIGURE 3.4 Normal and shear stresses acting on a parallelepiped element: (a) normal stresses and (b) shear stresses.

The total strains due to the simultaneous application of σ_x, σ_y, and σ_z are obtained by summing Equations 3.5a, b, and c as

$$\varepsilon_x = \frac{1}{E}\left[\sigma_x - \nu\left(\sigma_y + \sigma_z\right)\right]$$

$$\varepsilon_y = \frac{1}{E}\left[\sigma_y - \nu\left(\sigma_x + \sigma_z\right)\right] \quad (3.6a)$$

$$\varepsilon_z = \frac{1}{E}\left[\sigma_z - \nu\left(\sigma_x + \sigma_y\right)\right]$$

The validity of the procedure stems from the application of the principal of "superposition." This principal states that the resultant stress or strain in a system due to several forces is the algebraic sum of their effects when separately applied. This is only true if the effect is linearly related to the force causing it. For a material obeying Hooke's law undergoing small deformations, the principle of superposition holds.

If the parallelpiped element of Figure 3.4a is subjected instead of normal stresses to the shear stresses τ_{xy}, τ_{xz}, and τ_{yz} as shown in Figure 3.4b, it will, respectively, undergo shear strains γ_{xy}, γ_{xz}, and γ_{yz}. The stress–strain relations in this case are expressed by Equation 3.4 as

$$\gamma_{xy} = \tau_{xy}/G$$

$$\gamma_{yz} = \tau_{yz}/G \quad (3.6b)$$

$$\gamma_{zx} = \tau_{zx}/G$$

Obviously, for an isotropic material, the shear modulus G has the same value in all directions.

In the general case, when all the six stress components are present, Hooke's law is given by both Equations 3.6a and b. These Equations state that the normal strains are functions only of the normal stresses and that each shear strain is a function only of the corresponding shear stress. This is a consequence of the assumption of isotropy as demonstrated in Example 3.2. Furthermore, since the shear stresses vanish with respect to the principal axes of stress, it follows from Equation 3.6b that shear strains also vanish with respect to these axes. Hence, the principal axes of stress are indeed principal axes of strain.

For numerical computations, the six Equations expressing Hooke's law may be preferably written in matrix form as

$$\begin{bmatrix} \varepsilon_x \\ \varepsilon_y \\ \varepsilon_z \\ \gamma_{xy} \\ \gamma_{yz} \\ \gamma_{zx} \end{bmatrix} = \begin{bmatrix} 1/E & -\nu/E & -\nu/E & 0 & 0 & 0 \\ -\nu/E & 1/E & -\nu/E & 0 & 0 & 0 \\ -\nu/E & -\nu/E & 1/E & 0 & 0 & 0 \\ 0 & 0 & 0 & 1/G & 0 & 0 \\ 0 & 0 & 0 & 0 & 1/G & 0 \\ 0 & 0 & 0 & 0 & 0 & 1/G \end{bmatrix} \begin{bmatrix} \sigma_x \\ \sigma_y \\ \sigma_z \\ \tau_{xy} \\ \tau_{yz} \\ \tau_{zx} \end{bmatrix} \quad (3.7)$$

Example 3.1:

A square, thin plate (800 × 800 × 10mm) having sides parallel to the x–y *axes is made of aluminium alloy with E = 72 GPa and ν = 0.33. The plate is subjected to a uniform state of plane stress such that $\sigma_x = \sigma_1$ = 360 MPa and $\varepsilon_x = 2\varepsilon_y$. Determine the final dimensions of the plate under load.*

Solution:

A state of plane stress implies that $\sigma_3 = \sigma_2 = 0$. Hence, by applying Hooke's law,

$$\varepsilon_1 = \varepsilon_x = \frac{1}{E}[\sigma_1 - \nu\sigma_2] \qquad \text{(a)}$$
$$\varepsilon_2 = \varepsilon_y = \frac{1}{2}\varepsilon_x = \frac{1}{E}[\sigma_2 - \nu\sigma_1] \qquad \text{(b)}$$

Solving for σ_1 and σ_2 gives for $\nu = 0.33$ and $\sigma_1 = 360$ MPa:

$$\sigma_2 = (1 + 2\nu)\,\sigma_1/(2 + \nu) = 256.48 \text{ MPa}$$

Substituting back in Expression (a) gives:

$$\varepsilon_1 = \frac{10^6}{72 \times 10^9}[360 - 0.33 \times 256.48] = 3.824 \times 10^{-3}$$
$$\varepsilon_2 = \frac{1}{2}\varepsilon_1 = 1.912 \times 10^{-3}$$

Hence,

$$\varepsilon_3 = \frac{1}{E}[\sigma_3 - \nu(\sigma_1 + \sigma_2)], \text{ where } \sigma_3 = 0, \text{ i.e.,}$$
$$\varepsilon_3 = \frac{10^6}{72 \times 10^6}[0 - 0.33\,(256.48 + 360)] = -2.826 \times 10^{-3}$$

The change in plate dimensions are

$$\Delta x = \varepsilon_1 x_o = 3.824 \times 10^{-3} \times 800 = 3.0592 \text{ mm}$$
$$\Delta y = \Delta x/2 = 1.5296 \text{ mm}$$
$$\Delta z = \varepsilon_3 z_o = -2.826 \times 10^{-3} \times 10 = -0.02826 \text{ mm}$$

The final plate dimensions are 803.06 × 801.53 × 9.972 mm.

Example 3.2:

An isotropic material in the form of a cube is subjected to the three principal stresses σ_1, σ_2, and σ_3. If the material obeys Hooke's law, show that for any set of axes:

a. The normal strains are functions only of the normal stresses and the shear strain is a function only of its corresponding shear stress.
b. The relation among the elastic constants is given by $G = E/2(1 + \nu)$

Solution:

a. Referring to principal axes, Hooke's law provides

$$\varepsilon_1 = \frac{1}{E}\left[\sigma_1 - \nu(\sigma_2 + \sigma_3)\right]$$

$$\varepsilon_2 = \frac{1}{E}\left[\sigma_2 - \nu(\sigma_1 + \sigma_3)\right] \tag{3.8}$$

$$\varepsilon_3 = \frac{1}{E}\left[\sigma_3 - \nu(\sigma_1 + \sigma_2)\right]$$

Consider new axes x', y', z' of direction cosines (l_1, m_1, n_1), (l_2, m_2, n_2), and (l_3, m_3, n_3) with respect to the principal directions. Substituting into the strain transformation law given by Equation 2.22a yields

$$\varepsilon_{x'} = \varepsilon_1 l_1^2 + \varepsilon_2 m_1^2 + \varepsilon_3 n_1^2$$

and

$$\varepsilon_{y'} = \varepsilon_1 l_2^2 + \varepsilon_2 m_2^2 + \varepsilon_3 n_2^2$$

$$\gamma_{x'y'} = 2(l_1 l_2 \varepsilon_1 + m_1 m_2 \varepsilon_2 + n_1 n_2 \varepsilon_3)$$

Substitution from Hooke's law gives

$$\varepsilon_{x'} = \frac{1+\nu}{E}(l_1^2\sigma_1 + m_1^2\sigma_2 + n_1^2\sigma_3) - \frac{\nu}{E}(\sigma_1 + \sigma_2 + \sigma_3)$$

and

$$\gamma_{x'y'} = \frac{2(1+\nu)}{E}(l_1 l_2 \sigma_1 + m_1 m_2 \sigma_2 + n_1 n_2 \sigma_3)$$

b. In view of the stress transformation Equations 1.15a, and noting that $\sigma_1 + \sigma_2 + \sigma_3 = \sigma_{x'} + \sigma_{y'} + \sigma_{z'}$, the above equations reduce to

$$\varepsilon_{x'} = \frac{1+\nu}{E}\left[\sigma_{x'} - \frac{\nu}{E}(\sigma_{x'} + \sigma_{y'} + \sigma_{z'})\right]$$

$$= \frac{1}{E}\left[\sigma_{x'} - \nu(\sigma_{y'} + \sigma_{z'})\right] \tag{a}$$

and

$$\gamma_{x'y'} = \frac{2(1+\nu)}{E}\tau_{x'y'} = \frac{1}{G}\tau_{x'y'}$$

Similar expressions for each of the strain components $\varepsilon_{y'}$, $\varepsilon_{z'}$, $\gamma_{y'z'}$, and $\gamma_{z'x'}$ in the rotated axes can be easily obtained.

Hooke's law (Equations 3.8) and Equations (a) confirm that irrespective of the coordinate system, normal strains are functions only of normal stresses. Also each shear strain is a function only of the corresponding shear stress. Finally, from the second of Equations (a), a relation is obtained among the three elastic constants E, ν, and G as

$$G = E/[2(1 + \nu)] \tag{3.9}$$

3.4 RELATIONS AMONG THE ELASTIC CONSTANTS

Equation 3.9 expresses a relation between the modulus of elasticity E, Poisson's ratio ν, and the shear modulus G. This means that only two of these three constants are required to describe the stress–strain relations in linear isotropic elasticity.

Another elastic modulus, which is often used, is the bulk modulus K. The relation among K, E, and ν is obtained by studying the volumetric change during elastic deformation. From Hooke's law, Equations 3.6,

$$\varepsilon_x + \varepsilon_y + \varepsilon_z = \frac{1-2\nu}{E}\left(\sigma_x + \sigma_y + \sigma_z\right)$$

Noting that $\varepsilon_v = \varepsilon_x + \varepsilon_y + \varepsilon_z$ and $\varepsilon_m = (\sigma_x + \sigma_y + \sigma_z)/3$, hence

$$\varepsilon_v = \frac{3(1-2\nu)}{E}\sigma_m = \frac{1}{K}\sigma_m \tag{3.10}$$

Equation 3.10 expresses the volumetric strain ε_v in terms of the mean stress (often called hydrostatic stress) and the bulk modulus of elasticity K, namely,

$$K = \frac{E}{3(1-2\nu)} \tag{3.11}$$

Since for all materials a hydrostatic pressure tends to diminish volume, it is clear that K is positive. Hence, from Equation 3.11 the value of $(1 - 2\nu)$ is always positive with a limiting value of $\nu = 0.5$. This limiting value is attained when $K = \infty$, i.e., when no volume change occurs due to any applied hydrostatic stress. A lower limit on ν is set from Relation 3.9. In order that E and G be always positive, ν must be greater than -1. This sets up the limits $-1 < \nu \le 0.5$.

Poisson's ratio for most metallic alloys ranges from $\nu = 0.25$ to $\nu = 0.33$. Table 3.1 lists the modulus of elasticity and Poisson's ratio for some engineering materials.

Example 3.3:

A rectangular aluminum thin plate of the dimensions shown in Figure 3.5 is subjected to the stresses $\sigma_x = -100$ MPa, $\sigma_y = 115$ MPa, and $\tau_{xy} = -20$ MPa. If the elastic constants of the plate material are: E = 77,500 MPa and G = 29,500 MPa, determine the change in length of the diagonal AC.

Solution:

Since $G = E/2(1 + \nu)$, then

$$\nu = \frac{77{,}500}{2 \times 29{,}500} - 1 = 0.31$$

TABLE 3.1
Elastic Constants E and ν for Some Engineering Materials

Material	Modulus of Elasticity E = GPa	Poisson's Ratio, ν
Aluminum and its alloys	69–79	0.31–0.34
Copper and its alloys	105–150	0.33
Cast irons	105–150	0.21–0.03
Steels	190–210	0.28–0.33
Lead and alloys	14	0.43
Beryllium	272–300	0.01–0.06
Rubbers	0.01–0.1	0.5
Acrylics and nylons	1.4–3.4	0.32–0.4
Concrete	20–35	0.24

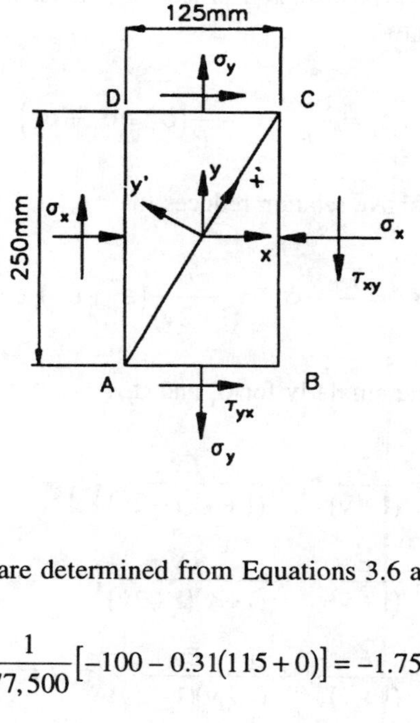

FIGURE 3.5 Example 3.3.

Noting that $\sigma_z = 0$, the strains are determined from Equations 3.6 as

$$\varepsilon_x = \frac{1}{77{,}500}\left[-100 - 0.31(115 + 0)\right] = -1.75 \times 10^{-3}$$

$$\varepsilon_y = \frac{1}{77{,}500}\left[0 - 0.31(-0 - 100)\right] = +1.88 \times 10^{-3}$$

$$\varepsilon_z = \frac{1}{77{,}500}\left[0 - 0.31(-100 + 115)\right] = 6 \times 10^{-5}$$

$$\gamma_{xy} = \frac{-20}{29{,}500} = -6.78 \times 10^{-4}$$

To determine the normal strain along the direction AC, i.e., $\varepsilon_{x'}$, the strain transformation law, Equation 2.22a is used, namely,

$$\varepsilon_{x'} = \varepsilon_x l^2 + \varepsilon_y m^2 + \varepsilon_z n^2 + \gamma_{xy} lm + \gamma_{yz} mn + \gamma_{zx} nl$$

where

$$l = 0.475, \; m = 0.88 \; \text{and} \; n = 0$$

Hence,

$$\varepsilon_{x'} = -1.75 \times 10^{-3}\,(0.475)^2 + 1.88 \times 10^{-3}\,(0.88)^2 - 6.78 \times 10^{-4}\,(0.475 \times 0.88) = 0.779 \times 10^{-3}$$

The original length of AC is 279.51 mm, and hence the change in its length is 0.218 mm.

3.5 INVERSE FORM OF HOOKE'S LAW

It would be more convenient to have expressions for the stresses in terms of the strain. This is often needed in determining stresses from strain gauge readings.

For numerical computations, this may be realized by manipulating the matrix form of Hooke's law, Equations 3.7, through matrix inversion procedures. It is, however, instructive to consider rewriting any of Equations 3.6 as

$$\varepsilon_x = \frac{1+\nu}{E}\sigma_x - \frac{\nu}{E}\left(\sigma_x + \sigma_y + \sigma_z\right)$$

In view of Equation 3.10, the above relation reduces to

$$\varepsilon_x = \frac{1+\nu}{E}\sigma_x - \frac{\nu}{(1-2\nu)}\left(\varepsilon_x + \varepsilon_y + \varepsilon_z\right)$$

Hence, solving for σ_x gives (and similarly for σ_y and σ_z)

$$\begin{aligned}
\sigma_x &= \frac{E}{(1+\nu)}\varepsilon_x + \frac{E\nu}{(1+\nu)(1-2\nu)}\left(\varepsilon_x + \varepsilon_y + \varepsilon_z\right)\\
\sigma_y &= \frac{E}{(1+\nu)}\varepsilon_y + \frac{E\nu}{(1+\nu)(1-2\nu)}\left(\varepsilon_x + \varepsilon_y + \varepsilon_z\right)\\
\sigma_z &= \frac{E}{(1+\nu)}\varepsilon_z + \frac{E\nu}{(1+\nu)(1-2\nu)}\left(\varepsilon_x + \varepsilon_y + \varepsilon_z\right)
\end{aligned} \tag{3.12a}$$

For shear stresses,

$$\begin{aligned}
\tau_{xy} &= G\gamma_{xy}\\
\tau_{yz} &= G\gamma_{yz}\\
\tau_{zx} &= G\gamma_{zx}
\end{aligned} \tag{3.12b}$$

Equations 3.12a and 3.12b are often written as

$$\sigma_x = 2\mu\varepsilon_x + \lambda\varepsilon_v \ldots$$
$$\tau_{xy} = \mu\gamma_{xy} \ldots \tag{3.13}$$

where μ and λ are known as Lame's constants.* Comparison between Equation 3.13 and the first of Equations 3.12a yields:

$$\mu = \frac{E}{2(1+\nu)} = G \tag{3.14a}$$

and

$$\lambda = \frac{E\nu}{(1+\nu)(1-2\nu)} \tag{3.14b}$$

Example 3.4:

The deformation of a solid is defined by the displacement components: u = a$(3x^2 + \sqrt{2}\, y)$, v = a$(2y^2 + 3\sqrt{2}\, x)$, w = *0, where* a *is a constant.*

a. *Find the principal strains at point (1,1).*
b. *Determine the principal stresses at the same point for* ν = *1/3.*

Solution:

a. The strain components are obtained from Equation 2.6 as

$$\varepsilon_x = \frac{\partial u}{\partial x} = 6ax, \varepsilon_y = \frac{\partial v}{\partial y} = 4ay$$

$$\gamma_{xy} = \frac{\partial u}{\partial y} + \frac{\partial v}{\partial x} = \sqrt{2}\, a + 3\sqrt{2}\, a = 4\sqrt{2}\, a$$

while the other strain components vanish. This is a plane state of strain.

At point (1,1), $\varepsilon_x = 6a$, $\varepsilon_y = 4a$, $\gamma_{xy} = 4\sqrt{2}\, a$. Using Expression 2.31a for the principal strains

$$\varepsilon_2^1 = \frac{6a+4a}{2} \pm \sqrt{\left(\frac{6a+4a}{2}\right)^2 + \left(\frac{4\sqrt{2}\, a}{2}\right)^2}$$

gives $\varepsilon_1 = 8a$ and $\varepsilon_2 = 2a$.

b. Since the principal directions of stress and strain are coincident, the principal stresses are obtained from Hooke's law in its inverse form, Equation (3.12), as

* Named after G. Lame (1798–1870).

$$\sigma_1 = \frac{E}{1+\nu}\varepsilon_1 + \frac{E\nu}{(1+\nu)(1-2\nu)}(\varepsilon_1 + \varepsilon_2 + \varepsilon_3) = 27\,Ea/2$$

$$\sigma_2 = \frac{E}{1+\nu}\varepsilon_2 + \frac{E\nu}{(1+\nu)(1-2\nu)}(\varepsilon_1 + \varepsilon_2 + \varepsilon_3) = 9Ea$$

$$\sigma_3 = \frac{E}{1+\nu}\varepsilon_3 + \frac{E\nu}{(1+\nu)(1-2\nu)}(\varepsilon_1 + \varepsilon_2 + \varepsilon_3) = 15\,Ea/2$$

Note that although $\varepsilon_3 = 0$, σ_3 has got a value.

3.6 DILATATION AND DISTORTION

In the preceding two chapters, it is shown that stress and strain may be expressed as the sum of their mean and deviatoric components. The relation between strain deviation components (say, ε_x') and the stress deviations is obtained from Equations 3.10 and 3.13 as

$$\sigma_x = \sigma_x' + \sigma_m = \sigma_x' + K\varepsilon_v = 2\mu\varepsilon_x + \lambda\varepsilon_v$$

hence,

$$\sigma_x' = 2\mu\varepsilon_x + (\lambda - K)\varepsilon_v$$

Substitution from Equations 3.13 and 3.14 gives

$$\sigma_x' = 2G\varepsilon_x + \left[\frac{E\nu}{(1+\nu)(1-2\nu)} - \frac{E}{3(1-2\nu)}\right]\varepsilon_v$$

or using Equation 3.9 yields

$$\sigma_x' = 2G\varepsilon_x - \frac{E}{3(1+\nu)}\varepsilon_v = 2G(\varepsilon_x - \varepsilon_m)$$

and, similarly, for σ_y and σ_z. This results in

$$\sigma_x' = 2G\varepsilon_x'$$
$$\sigma_y' = 2G\varepsilon_y' \tag{3.15}$$
$$\sigma_z' = 2G\varepsilon_z'$$

Equation 3.9 may also be rewritten as

$$\sigma_m = 3K\varepsilon_m \tag{3.16}$$

The physical interpretation of Equations 3.15 and 3.16 is that, for any state of strain [ε], its part representing the change in volume [ε_m] (dilatation) is related only to the mean (hydrostatic) stress $[\sigma_m]$ and the strain deviation part (distortion) $[\varepsilon']$ is related only to the stress deviation $[\sigma']$. Equations 3.15 and 3.16, thus offer another representation for Hooke's law as

$$\sigma_x = \sigma'_x + \sigma_m = 2G\varepsilon'_x + 3K\varepsilon_m \ldots \tag{3.17a}$$

In matrix form $[\sigma] = 2G\,[\varepsilon'] + 3K\,[\varepsilon_m]$ or, explicitly,

$$\begin{bmatrix} \sigma_x & \tau_{yx} & \tau_{zx} \\ \tau_{xy} & \sigma_y & \tau_{zy} \\ \tau_{xz} & \tau_{yz} & \sigma_z \end{bmatrix} = 2G \begin{bmatrix} \varepsilon_x - \varepsilon_m & \tfrac{1}{2}\gamma_{yx} & \tfrac{1}{2}\gamma_{zx} \\ \tfrac{1}{2}\gamma_{xy} & \varepsilon_y - \varepsilon_m & \tfrac{1}{2}\gamma_{zy} \\ \tfrac{1}{2}\gamma_{xz} & \tfrac{1}{2}\gamma_{yz} & \varepsilon_z - \varepsilon_m \end{bmatrix} + 3K \begin{bmatrix} \varepsilon_m & 0 & 0 \\ 0 & \varepsilon_m & 0 \\ 0 & 0 & \varepsilon_m \end{bmatrix} \tag{3.17b}$$

Example 3.5:
A rubber cylinder is compressed in a thin, steel tube by an axial stress σ_z, as shown in Figure 3.6. Find the pressure between the rubber and the steel tube:

a. Assuming that the steel tube is absolutely rigid;
b. Considering the deformation in the steel tube.

Take the elastic constants (E_R, ν_R) and (E_S, ν_S) for the rubber and steel, respectively. Neglect friction between steel and rubber and assume that both obey Hooke's law.

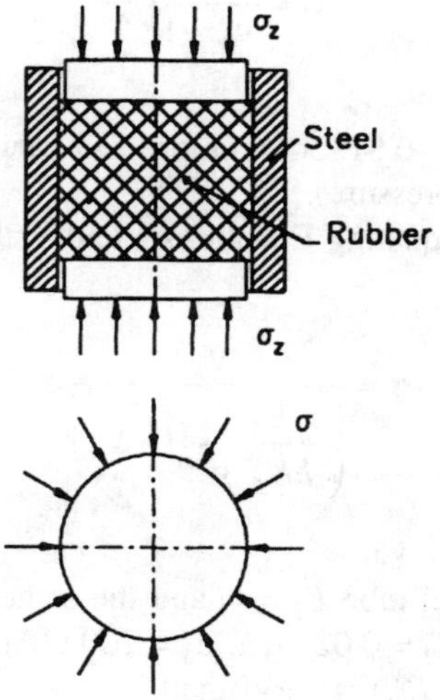

FIGURE 3.6 Example 3.5.

Solution:
For the rubber cylinder, due to symmetry $\tau_{r\theta} = \tau_{zr} = \tau_{z\theta} = 0$ and $\sigma_r = \sigma_\theta = \sigma$.*

Since rubber and steel remain always in contact, their circumferential strains are equal. Hence,

$$(\varepsilon_\theta)_R = (\varepsilon_\theta)_S$$

Applying Hooke's law for the rubber cylinder gives

* This can be easily proved from the equilibrium of an arbitrary slice-element plane across the solid. Also see Chapter 6.

$$\left(\varepsilon_\theta\right)_R = \frac{1}{E_R}\left[\sigma - \nu_R\left(\sigma - \sigma_z\right)\right] \tag{a}$$

By considering the steel tube to be subjected to a radial internal pressure σ, then from Hooke's law:

$$\left(\varepsilon_\theta\right)_S = \frac{1}{E_S}\left[\frac{\sigma r_m}{h} - \nu_S\left(\frac{\sigma}{2} + 0\right)\right] \cong \frac{\sigma r_m}{E_S h} \tag{b}$$

where r_i and h are the steel tube mean radius and thickness, respectively.

a. Rigid steel tube ($E_S = \infty$)

$$\left(\varepsilon_\theta\right)_S = \left(\varepsilon_\theta\right)_R = 0 = \frac{1}{E_R}\left[\sigma - \nu_R\left(\sigma - \sigma_z\right)\right]$$

Hence,

$$\sigma = \frac{\nu_R}{\nu_R - 1}\sigma_z \tag{a}$$

For $\nu_R = 0.45$, $\sigma_r = \sigma_\theta = \sigma = -0.8182\sigma_z$ (compressive) and, for $\nu_R = 0.5$, (incompressible); $\sigma = -\sigma_z$ (all-around equal pressure).

b. Deformable steel tube — equating Equations a and b, then solving for σ, gives

$$\sigma = \frac{\nu_R}{\left(\dfrac{r_m E_R}{h E_S}\right) - \left(1 - \nu_R\right)}\sigma_z$$

For an absolutely rigid steel tube $E_S = \infty$ and the same expression as a is obtained for σ. For $\nu_R = 0.5$, $\nu_S = 0.3$, $E_R = 0.02$ GPa, $E_S = 200$ GPa, and $\sigma = -\sigma_z$; the same result is obtained as in a since $(r_m E_R / E_S)$ is very small.

3.7 THERMOELASTIC STRESS–STRAIN RELATIONS

A solid subjected to a temperature increase T undergoes longitudinal thermal strains given by αT, where α is the coefficient of linear thermal expansion. For a homogeneous isotropic solid, α is the same in all directions and hence the strain αT is also equal in all directions. This means that the solid undergoes volumetric change without any change in shape, i.e., without distortion. Thus, for a state of stress σ_x, σ_y, σ_z, τ_{xy}, τ_{yz}, and τ_{zx} at a temperature increase T, the strains are

$$\varepsilon_x = \frac{1}{E}\left[\sigma_x - \nu\left(\sigma_y + \sigma_z\right)\right] + \alpha T$$

$$\varepsilon_y = \frac{1}{E}\left[\sigma_y - \nu\left(\sigma_x + \sigma_z\right)\right] + \alpha T$$

$$\varepsilon_z = \frac{1}{E}\left[\sigma_z - \nu\left(\sigma_x + \sigma_y\right)\right] + \alpha T \tag{3.18}$$

$$\gamma_{xy} = \tau_{xy}/G$$

$$\gamma_{xz} = \tau_{xz}/G$$

$$\gamma_{zy} = \tau_{zy}/G$$

Note that in the presence of temperature changes, Equations 3.16 will remain the same as they exclude any change of volume. Equation 3.16, however, becomes

$$\sigma_m = 3K(\varepsilon_m - \alpha T) \tag{3.19}$$

In writing the thermoelastic stress–strain relations, it must be remembered that E, ν, and α are dependent on temperature. However, for most materials the change is very small over an appreciable range of temperature. Use of average values for E, ν, and α over the temperature range of interest is thus justifiable for most engineering calculations.

Example 3.6:

A brass sheet 20 mm × 30 mm × 2 mm is clamped in a very rigid frame whose coefficient of thermal expansion is almost zero, Figure 3.7. Given that the temperature drops by 100°C, calculate the resulting stresses in the sheet. For brass, E = *120 GPa,* ν = *0.33, and* α = *16* × *10*$^{-6}$ *°C*$^{-1}$*.*

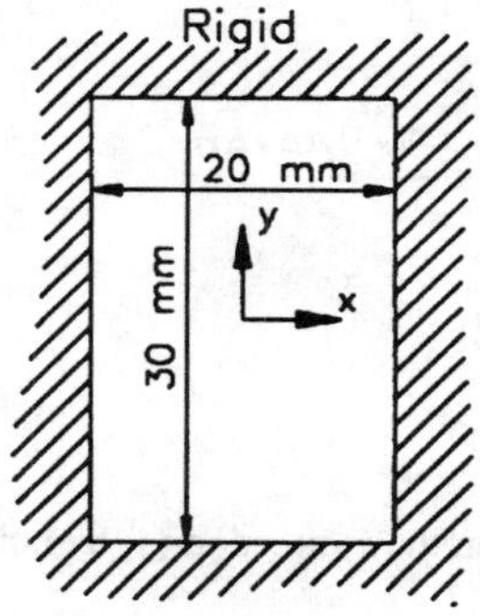

FIGURE 3.7 Example 3.6.

Solution:

Clamping to a very rigid frame imposes

$$\varepsilon_x = \varepsilon_y = 0$$

Noting that $\varepsilon_z = 0$, the thermoelastic stress–strain relations, Equations 3.18, yield

$$\varepsilon_x = 0 = \frac{1}{120\times10^9}\left[\sigma_x - 0.33\sigma_y\right] - 16\times10^{-6}\times100$$

$$\varepsilon_y = 0 = \frac{1}{120\times10^9}\left[\sigma_y - 0.33\sigma_x\right] - 16\times10^{-6}\times100$$

Solving for σ_x and σ_y gives

$$\sigma_x = \sigma_y = 286.6 \text{ MPa}$$

3.8 STRAIN ENERGY FOR AN ELASTIC ISOTROPIC SOLID

When an elastic solid is subjected to gradually increasing external loads, work W is done during deforming this solid from its initial state to its final state. On the assumption that equilibrium conditions prevail and neither heat exchange nor temperature rise takes place, it is valid to state that the work done by external loads is stored in the body as elastic strain energy U. Hence, this elastic strain energy is completely recovered upon load removal and the solid returns to its initial state. This concept is the basis of several strain energy theorems, which are employed effectively to solve elastic problems. Energy consideration is also used to derive stress–strain relations for large elastic deformation and in failure analysis.

It is now required to obtain expressions for the strain energy. For simplicity, an expression for the strain energy in a homogeneous isotropic elastic (but not necessarily linearly elastic) solid subjected to normal stresses will be developed first.

Consider a small parallelepiped of volume $\Delta V = \Delta x \Delta y \Delta z$, as shown in Figure 3.8a, cut from a stressed elastic solid. Let the corner o of this parallelepiped be immovable.

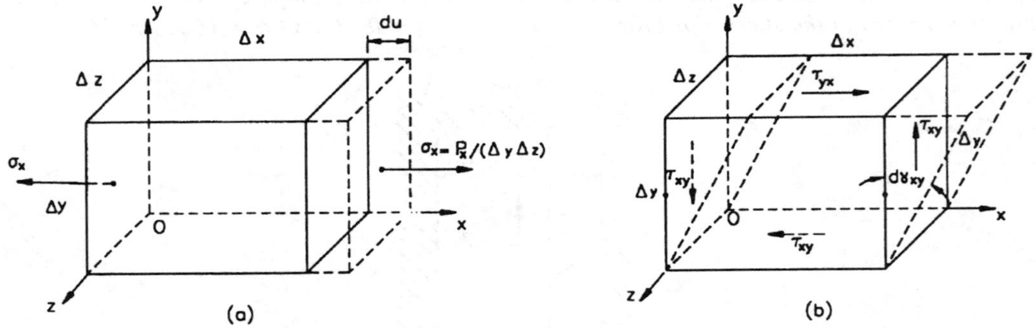

FIGURE 3.8 Increment of work done in the x-direction (a) by normal stress σ_x and (b) by shear stress $-\tau_{xy}$.

The increment of work done in the x-direction is given by $dW_x = P_x$, similarly, $dW_y = P_y dv$ and $dW_z = P_z dW$.

The increment of elastic strain energy; being equal to the work done by the external forces P_x, P_y and P_z, are expressed as

$$d(\Delta U) = P_x du + P_y dv + P_z dw \tag{3.20}$$

where du, dv, and dw are the increments of displacement in the x, y, and z-directions, respectively. For the faces $x = + \Delta x/2$, $y = + \Delta y/2$, and $z = + \Delta z/2$, the forces are given by

$$P_x = \sigma_x \Delta y \Delta z$$
$$P_y = \sigma_y \Delta x \Delta z \tag{3.21}$$
$$P_z = \sigma_z \Delta x \Delta y$$

The incremental displacements for these faces in the x, y, and z-directions are related to the strain components by

$$du = d\varepsilon_x \left(\frac{\Delta x}{2}\right)$$
$$dv = d\varepsilon_y \Delta y \tag{3.22}$$
$$dw = d\varepsilon_z \Delta z$$

Substitution from Equations 3.21 and 3.22 into Expression 3.20 gives

$$d(\Delta U) = \Delta x \Delta y \Delta z\left(\sigma_x d\varepsilon_x + \sigma_y d\varepsilon_y + \sigma_z d\varepsilon_z\right) \tag{3.23}$$

Hence, $d(\Delta U)$ for the deformed parallelepiped as a whole is

$$d(\Delta U) = \Delta V(\sigma_x d\varepsilon_x + \sigma_y d\varepsilon_y + \sigma_z d\varepsilon_z) \tag{3.24}$$

The total strain energy stored in the parallelepiped volume ΔV when the stresses and strains reach their final values is obtained by integration, i.e.,

$$\Delta U = \Delta V \int \left(\sigma_x d\varepsilon_x + \sigma_y d\varepsilon_y + \sigma_z d\varepsilon_z\right) \tag{3.25}$$

Attention is next directed toward finding an expression of the elastic strain energy stored in a homogeneous elastic solid due to shear stresses only. Let the parallelepiped of Figure 3.8b be subjected to a shear stress τ_{xy} only, which produces an increment of shear strain $d\gamma_{xy}$. Therefore, the increment of work done gives an expression for the increment of shear strain energy as

$$d(\Delta U) = \left(\tau_{yx} \Delta x \Delta z\right)\left(d\gamma_{yx} \Delta y\right)$$

or

$$d(\Delta U) = \Delta V \tau_{yx} d\gamma_{yx} \tag{3.26}$$

Note that the forces $\tau_{xy} \Delta y \Delta z$ acting on the planes 0, Δx do no work since for infinitesimal deformations the displacements may be assumed to be perpendicular to these forces. Since for an isotropic solid, each shear stress gives rise to only its corresponding shear strain,* the increment

* This is only valid within the framework of infinitesimal strains, where geometric changes do not result in changes in stress values arising from the applied loads.

of strain energy when the three shear stresses τ_{xy}, τ_{yz}, and τ_{zx} are all applied is simply obtained by adding three expressions similar to Equation 3.26, thus giving

$$d(\Delta U) = \Delta V(\tau_{xy}d\gamma_{xy} + \tau_{yz}d\gamma_{yz} + \tau_{zx}d\gamma_{zx}) \tag{3.27}$$

The total strain energy stored in the parallelepiped volume ΔV when the shear stresses and strains reach their final values is

$$\Delta U = \Delta V \int \left(\tau_{xy}d\gamma_{xy} + \tau_{yz}d\gamma_{yz} + \tau_{zx}d\gamma_{zx}\right) \tag{3.28}$$

As mentioned before* a consequence of isotropy is that normal strains are produced only from normal stresses and, hence, shear stresses do not contribute to the normal strain energy expressed by Equation 3.25. Similarly, normal stresses do not produce any shear strains and, hence, do not contribute to shear strain energy. Thus, for the general three-dimensional case, when all six components of stresses are acting, the strain energy is obtained by adding Equations 3.25 and 3.28 as

$$\Delta U = \int \left(\sigma_x d\varepsilon_x + \sigma_y d\varepsilon_y + \sigma_z d\varepsilon_z + \tau_{xy}d\gamma_{xy} + \tau_{yz}d\gamma_{yz} + \tau_{zx}d\gamma_{zx}\right) \Delta V \tag{3.29a}$$

where the integration is to be carried out from the initial to the final state of deformation of the solid. Expression 3.29a is written as

$$\Delta U = U_o \Delta V \tag{3.29b}$$

where

$$U_o = \int \left(\sigma_x d\varepsilon_x + \sigma_y d\varepsilon_y + \sigma_z d\varepsilon_z + \tau_{xy}d\gamma_{xy} + \tau_{yz}d\gamma_{yz} + \tau_{zx}d\gamma_{zx}\right) \tag{3.29c}$$

U_o is, by definition, the elastic strain energy stored in a unit volume of the material known as the "strain energy density."** Integration of this density over the volume V of the solid yields the total strain energy stored in the elastic solid as

$$U = \int_V U_o dV \tag{3.30}$$

Again, it is worth emphasizing that the above expressions are valid for all elastically deforming solids without any reference to their particular behavior as described by the stress–strain relations. The only limitation imposed on Equations 3.29 to 3.30 is the condition of infinitesimal deformation. Also note that for uniaxial loading, the strain energy density U_o is represented by the area under the stress–strain curve as shown in Figure 3.9.

Since dU is an exact differential, then

* See also the generalized Hooke's law (Equation 3.45) for anisotropic solids in Section 3.11.

** A strain energy density function U_o does not exist for all types of materials because the deformation is not, in general, independent of the path.

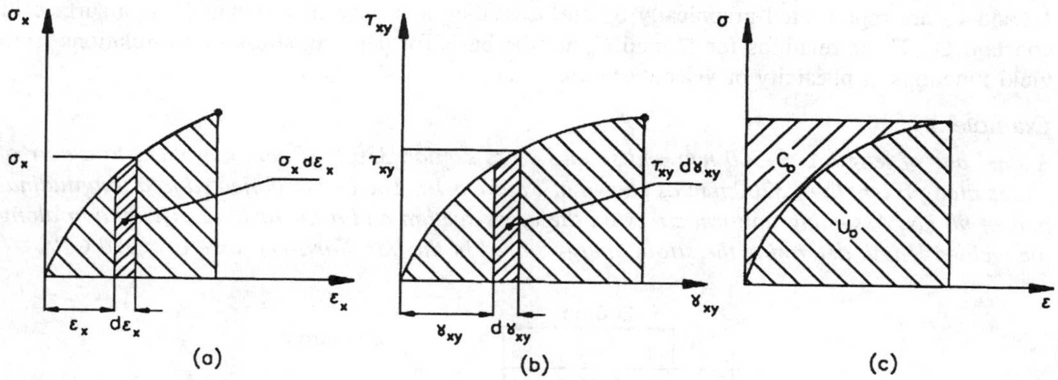

FIGURE 3.9 Strain energy and complementary energy (a) due to σ_x, (b) due to shear stress τ_{xy}, and (c) complementary energy.

$$dU_o = \frac{\partial U_o}{\partial \varepsilon_x} d\varepsilon_x + \frac{\partial U_o}{\partial \varepsilon_y} d\varepsilon_y + \frac{\partial U_o}{\partial \varepsilon_z} d\varepsilon_z + \frac{\partial U_o}{\partial \gamma_{xy}} d\gamma_{xy} + \frac{\partial U_o}{\partial \gamma_{yz}} d\gamma_{yz} + \frac{\partial U_o}{\partial \gamma_{zx}} d\gamma_{zx} \tag{3.31}$$

Thus, comparing Equations 3.29c and 3.31 results in

$$\sigma_x = \frac{\partial U_o}{\partial \varepsilon_x},\ \sigma_y = \frac{\partial U_o}{\partial \varepsilon_y},\ \sigma_z = \frac{\partial U_o}{\partial \varepsilon_z},\ \tau_{xy} = \frac{\partial U_o}{\partial \gamma_{xy}},\ \gamma_{yz} = \frac{\partial U_o}{\partial \gamma_{yz}},\ \gamma_{zx} = \frac{\partial U_o}{\partial \gamma_{zx}} \tag{3.32}$$

For emphasis, Equations 3.32 are valid for any elastic solid possessing a strain energy density without restriction on nonhomogeneity, anisotropy, or nonlinearity. Often, the strain energy density function U_o is postulated and the stress–strain relations are then derived using Equations 3.32, which are valid for all orthogonal axes such as Cartesian, cylindrical polar, and spherical polar coordinates.

Solution of some elasticity problems requires the definition of another energy function, which is called *complementary energy*. Referring to Figure 3.9c, the complementary energy density is represented by the area designated by C_o as opposed to U_o for the strain energy density. The complementary energy density for the general three-dimensional states of stress and strain is given by

$$C_o = \int \left(\varepsilon_x d\sigma_x + \varepsilon_y d\sigma_y + \varepsilon_z d\sigma_z + \gamma_{xy} d\tau_{xy} + \gamma_{yz} d\tau_{yz} + \gamma_{zx} d\tau_{zx} \right) \tag{3.33a}$$

Again, similar to dU_o, dC_o is an exact differential so that the strain components are expressed in terms of C_o by similar expressions to Equations 3.32 as

$$\varepsilon_x = \frac{\partial C_o}{\partial \sigma_x},\ \varepsilon_y = \frac{\partial C_o}{\partial \sigma_y},\ \varepsilon_z = \frac{\partial C_o}{\partial \sigma_z},\ \gamma_{xy} = \frac{\partial C_o}{\partial \tau_{xy}},\ \gamma_{yz} = \frac{\partial C_o}{\partial \tau_{yz}},\ \gamma_{zx} = \frac{\partial C_o}{\partial \tau_{zx}} \tag{3.33b}$$

U_o and C_o are fundamentally different quantities. U_o is always considered as a function of strain, whereas C_o is a function of stress no matter how they are algebraically expressed. The ratio U_o/C_o depends upon the material constitutive law. For a material with linear stress–strain relation $U_o = C_o$. For a stress–strain relation expressed by a single power term, namely, $\sigma = K\varepsilon^n$, $U_o/C_o = 1/n$. If

U_o and C_o are represented graphically by surfaces, then a surface of constant U_o is a surface of constant C_o. These relations for U_o and C_o are the basis for deriving stress–strain relations from yield functions in plasticity or viscoplasticity.

Example 3.7:

A steel bar of length L = *500 mm and square cross section 330 × 30 mm is welded to a pair of plates along a length of 3200 mm as shown in Figure 3.10. The bar is subjected to a longitudinal pull of 90 kN. Assuming uniform stress on the cross section and a linear load distribution along the welded length determine the strain energy stored in the bar. For steel, take* E = *210 GPa.*

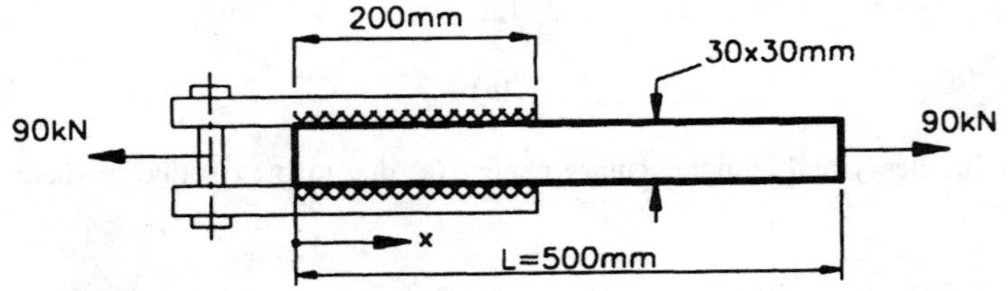

FIGURE 3.10 Example 3.7.

Solution:

From $x = 0$ to $x = 200$ mm, the stress σ_x is

$$\sigma_x = \frac{90 \times 10^3}{200 \times 30 \times 30} x = 0.5x \text{ MPa}$$

From $x = 200$ mm to $x = 500$ mm, the stress σ_x is constant = 10 MPa
For uniaxial stress and linear elastic material, the strain energy density is given by

$$U_o = \int \sigma_x d\varepsilon_x = \int \sigma_x \frac{d\sigma_x}{E} = \frac{\sigma_x^2}{2E}$$

From $x = 0$ to $x = 200$ mm, the strain energy density is

$$U_o = \frac{(0.5x)^2}{2 \times 210 \times 10^3} = 0.595 \times 10^{-6} x^2 \text{ MPa}$$

and

$$U_{\text{total}} = \int U_o dV = \left(0.595 \times 10^{-6}\right)\left(900 \int_0^{200} \left(x^2 dx\right)\right)$$

$$= 1.428 \text{ N} \cdot \text{m}$$

From $x = 200$ to $x = 300$ mm, the strain energy density is uniform:

$$U_o = 23.8 \times 10^{-3} \text{ MPa}$$

and $U_{total} = 6.426$ N · m.

Total energy stored = 1.428 + 6.426 = 7.854 N · m.

Example 3.8:

A linear elastic material possesses a strain energy density in terms of the strain invariants as

$$U_o = a_1 J_1^2 - a_2 J_2$$

where J_1, J_2 ***are the first and second invariants of the strain tensor, respectively. Derive the stress–strain relations for this material.***

Solution:

From Equation 2.26b

$$J_1 = \varepsilon_x + \varepsilon_y + \varepsilon_z$$

and

$$J_2 = -\left(\varepsilon_x\varepsilon_y + \varepsilon_y\varepsilon_z + \varepsilon_z\varepsilon_x\right) + \tfrac{1}{4}\left(\gamma_{xy}^2 + \gamma_{yz}^2 + \gamma_{zx}^2\right)$$

Applying Equations 3.32 yields

$$\sigma_x = \frac{\partial U_o}{\partial \varepsilon_x} = \frac{\partial U_o}{\partial J_1}\frac{\partial J_1}{\partial \varepsilon_x} + \frac{\partial U_o}{\partial J_2}\frac{\partial J_2}{\partial \varepsilon_x}$$

$$= 2a_1 J_1 + a_2\left(\varepsilon_y + \varepsilon_z\right)$$

$$= 2a_1\left(\varepsilon_x + \varepsilon_y + \varepsilon_z\right) + a_2\left(\varepsilon_y + \varepsilon_z\right)$$

$$\sigma_x = \left(2a_1 + a_2\right)\varepsilon_v - a_2\varepsilon_x \tag{a}$$

$$\tau_{xy} = \frac{\partial U_o}{\partial \gamma_{xy}} = \frac{\partial U_o}{\partial J_1}\frac{\partial J_1}{\partial \gamma_{xy}} + \frac{\partial U_o}{\partial J_2}\frac{\partial J_2}{\partial \gamma_{xy}}$$

$$\tau_{xy} = -\frac{a_2}{2}\gamma_{xy}$$

Similar expressions for σ_y, σ_z, τ_{yz}, and τ_{zx} are obtained by appropriate differentiation of U_o according to Equations 3.32. Comparing Equations a with the inverse form of Hooke's law given by Equations 3.13 indicates that they are identical provided that

$$2a_1 + a_2 = \lambda \qquad a_2 = -2\mu = -2G$$

$$a_1 = \tfrac{1}{2}(\lambda + 2\mu)$$

where λ and μ are Lame's constants. The strain energy density function U_o is thus given by

$$U_o = \left(\mu + \frac{\lambda}{2}\right)J_1^2 + 2\mu J_2 = G\left(\frac{1-\nu}{1-2\nu}J_1^2 + 2J_2\right)$$

using Equations 3.14.

This expression will be derived in the next section, Section 3.9, for a solid obeying Hooke's law.

3.9 STRAIN ENERGY FOR A SOLID OBEYING HOOKE'S LAW

For simplicity, consider first the case of a solid obeying Hooke's law and subjected to normal stresses σ_x, σ_y, and σ_z only. Hence, the strain energy density $(U_o)_\sigma$ is given by Expression 3.29c as

$$(U_o)_\sigma = \int\left(\sigma_x d\varepsilon_x + \sigma_y d\varepsilon_y + \sigma_z d\varepsilon_z\right)$$

Substituting from Hooke's law, Equations 3.6a, for $d\varepsilon_x$ and $d\varepsilon_z$ in terms of the stress increments $d\sigma_x$, $d\sigma_y$, and $d\sigma_z$ results in

$$\begin{aligned}(U_o)_\sigma = \int & \frac{\sigma_x}{E}\left[d\sigma_x - \nu\left(d\sigma_y + d\sigma_z\right)\right] \\ & + \frac{\sigma_y}{E}\left[d\sigma_y - \nu\left(d\sigma_z + d\sigma_x\right)\right] \\ & + \frac{\sigma_z}{E}\left[d\sigma_z - \nu\left(d\sigma_x + d\sigma_y\right)\right]\end{aligned}$$

Rearranging gives

$$\begin{aligned}(U_o)_\sigma = \frac{1}{E}\int & \left[\sigma_x - \nu\left(\sigma_y + \sigma_z\right)\right] d\sigma_x \\ & + \left[\sigma_y - \nu\left(\sigma_z + \sigma_x\right)\right] d\sigma_y \\ & + \left[\sigma_z - \nu\left(\sigma_x + \sigma_y\right)\right] d\sigma_z\end{aligned}$$

and after integration

$$(U_o)_\sigma = \frac{1}{2E}\left(\sigma_x^2 + \sigma_y^2 + \sigma_z^2\right) - \frac{2\nu}{E}\left(\sigma_x\sigma_y + \sigma_y\sigma_z + \sigma_z\sigma_x\right)$$

Hence, the following expression is obtained:

$$(U_o)_\sigma = \frac{\sigma_x}{2E}\left[\sigma_x - \nu\left(\sigma_y + \sigma_z\right)\right] + \frac{\sigma_y}{2E}\left[\sigma_y - \nu\left(\sigma_z + \sigma_x\right)\right] + \frac{\sigma_z}{2E}\left[\sigma_z - \nu\left(\sigma_x + \sigma_y\right)\right]$$

In view of Hooke's law, Equations 3.6, this reduces to

$$(U_o)_\sigma = {}^1\!/_2[\sigma_x\varepsilon_x + \sigma_y\varepsilon_y + \sigma_z\varepsilon_z]$$

The strain energy density for a solid obeying Hooke's law and subjected to three shear stresses τ_{xy}, τ_{yz}, and τ_{zx} only is given by Expression 3.29c as

$$(U_o)_\tau = \int \left(\tau_{xy} d\gamma_{xy} + \tau_{yz} d\gamma_{yz} + \tau_{zx} d\gamma_{zx}\right)$$

Substituting from Hooke's law and integrating gives

$$(U_o)_\tau = \frac{1}{2G}\left[\tau_{xy}^2 + \tau_{yz}^2 + \tau_{zx}^2\right]$$

In view of Hooke's law, Equations 3.6b, this reduces to

$$(U_o)_\tau = {}^1/_2\,[\tau_{xy}\gamma_{xy} + \tau_{yz}\gamma_{yz} + \tau_{zx}\gamma_{zx}]$$

The total strain energy density for a solid obeying Hooke's law and subjected to any stress system is obtained by addition as

$$U_o = {}^1/_2\,[\sigma_x\varepsilon_x + \sigma_y\varepsilon_y + \sigma_z\varepsilon_z + \tau_{xy}\gamma_{xy} + \tau_{yz}\gamma_{yz} + \tau_{zx}\gamma_{zx}] \tag{3.34}$$

Obviously, in view of the applicability of the superposition principle, Equation 3.34 is independent of the order of applying the stress components. It is also independent of how these stresses reach their final values, i.e., path independent. Also note that Expression 3.34 could have been simply obtained from Equation 3.29b considering the area under the curves representing linear stress–strain relations, i.e., straight-line plots analogous to that in Figure 3.9.

The strain energy density may be expressed in alternative forms. For instance, Equation 3.34 may be written as

$$U_o = \frac{1}{2E}\left(\sigma_x + \sigma_y + \sigma_z\right)^2 - \frac{2(1+\nu)}{E}\left[\sigma_x\sigma_y + \sigma_y\sigma_z + \sigma_z\sigma_x - \left(\tau_{xy}^2 + \tau_{yz}^2 + \tau_{zx}^2\right)\right] \tag{3.35a}$$

In terms of the stress invariants given by Equations 1.23, this reduces after algebraic manipulation to

$$U_o = \frac{I_1^2}{2E} + \frac{I_2}{2G} \tag{3.35b}$$

Similarly, the strain energy density may be expressed in terms of the strain invariants given by Equations 2.26b as

$$U_o = G\left(\frac{1-\nu}{1-2\nu} J_1^2 + 2J_2\right) \tag{3.35c}$$

Hence, the strain energy density (which is a scalar quantity) is, as expected, invariant to orthogonal rotations of coordinates. It can be expressed solely in terms of the principal stresses or the principal strains.

Expression 3.35 may be rearranged in view of Equations 1.25, 1.33, and 1.38 as

$$I_1 = 3\sigma_m$$

$$I_2 = -(\sigma_1\sigma_2 + \sigma_2\sigma_3 + \sigma_3\sigma_1) = I_2' - \tfrac{1}{3}I_1^2$$

or

$$I_2 = \frac{3}{2}\tau_{oct}^2 - 3\sigma_m^2$$

hence,*

$$U_o = \frac{1-2\nu}{6E}I_1^2 - \frac{1+\nu}{E}I_2' = \frac{\sigma_m^2}{2K} + \frac{3}{4G}\tau_{oct}^2 \tag{3.36a}$$

or

$$U_o = U_{ov} + U_{os}$$

This expression suggests dividing the strain energy density into two parts, namely, energy that causes change of volume U_{ov} (dilatational energy) and energy that causes change of shape U_{os} (distortion energy). Such division is important in studying plastic deformation, where the volume remains constant while yielding and plastic flow is governed by the distortion energy, as will be shown in Chapter 10.

Equation 3.36 may be written more explicitly in terms of the principal stresses as

$$U_o = \frac{1}{18K}(\sigma_1 + \sigma_2 + \sigma_3)^2 + \frac{1}{12G}\left[(\sigma_1 - \sigma_2)^2 + (\sigma_2 - \sigma_3)^2 + (\sigma_3 - \sigma_1)^2\right] \tag{3.36b}$$

Again since the elastic constants K and G are always positive, it follows that the strain energy density U_o is also positive for all stress systems. This property of U_o is important in establishing uniqueness of solution in linear elasticity problems.

The complementary strain energy density C_o given by Expression 3.33a is, obviously, for a solid obeying Hooke's law, equal to the strain energy density U_o and all above expressions for U_o apply similtaneously to C_o. Explicitly in terms of strains, it may be shown that C_o is expressed by

$$\begin{aligned} C_o = {} & \frac{E}{2(1+\nu)(1-2\nu)} \\ & \times\left\{(1-\nu)(\varepsilon_x + \varepsilon_y + \varepsilon_z)^2 - 2(1-2\nu)\left[\varepsilon_x\varepsilon_y + \varepsilon_y\varepsilon_z + \varepsilon_z\varepsilon_x - \tfrac{1}{4}(\gamma_{xy}^2 + \gamma_{yz}^2 + \gamma_{zx}^2)\right]\right\} \end{aligned} \tag{3.37a}$$

In terms of the strain invariants J_1, J_2, Equation 3.37a for C_o and, hence, U_o is given by

$$C_o = \frac{E(1-\nu)}{2(1+\nu)(1-2\nu)}J_1^2 + 2GJ_2 = (\lambda/2 + G)J_1^2 + 2GJ_2 \tag{3.37b}$$

which is the same expression as derived in Example 3.8.

* This expression for the strain energy density indicates that U_o is always positive, except at the instance when all stress and strain components vanish.

Example 3.9:

A closed, thin-walled pressure vessel of mean radius r_m *= 3 m and length* L *= 10 m, is used for chemical processing at 320°C. If the vessel wall thickness* h *is 24 mm, determine the diametral expansion and elongation caused by an operating pressure of* p_i *= 0.8 MPa. Knowing that this pressure produces hoop and axial stresses given by* $\sigma_\theta = p_i r_m/h$ *and* $\sigma_z = p_i r_m/2h$, *respectively,* what is the strain energy stored in the vessel wall material? Take* E *= 200 GPa,* ν *= 0.3, and* $\alpha = 11 \times 10^{-6}$ *°C*$^{-1}$.

Solution:

The stresses in the vessel wall are: $\sigma_r \cong 0$,

$$\sigma_\theta = \frac{p_i r_m}{h} = \frac{0.8\times10^6\times3}{0.024} = 100 \text{ MPa}$$

$$\sigma_z = \frac{p_i r_m}{2h} = \frac{0.8\times10^6\times3}{2\times0.024} = 50 \text{ MPa}$$

From the thermoelastic stress–strain relations, Equations 3.18,

$$\varepsilon_\theta = \frac{1}{E}\left[\sigma_\theta - \nu\sigma_z\right] + \alpha T$$

$$\varepsilon_\theta = \frac{1}{200\times10^9}\left[100\times10^6 - 0.3\times50\times10^6\right] + 11\times10^{-6}(320-20)$$

$$\varepsilon_\theta = 3.725\times10^{-3}$$

Similarly,

$$\varepsilon_z = \frac{1}{E}\left[\sigma_z - \nu\sigma_\theta\right] + \alpha T = 3.4\times10^{-3}$$

$$\varepsilon_r = \frac{1}{E}\left[0 - \nu(\sigma_\theta + \sigma_z)\right] + \alpha T = 3.5\times10^{-3}$$

The diametral expansion = $2\varepsilon_\theta r_m$ = 11.175 mm, the total elongation = $\varepsilon_z L$ = 34 mm, and the strain energy stored in the vessel wall is

$$U = 2\pi r_m hL\left(\sigma_r\varepsilon_r + \sigma_\theta\varepsilon_\theta + \sigma_z\varepsilon_z\right)/2$$

$$= \pi(3)\,(0.024)\,(10)\left[(0)\,\left(3.5\times10^{-3}\right) + (100)\,\left(3.725\times10^{-3}\right) + (50)\,\left(3.4\times10^{-3}\right)\right]$$

$$U = 1.2265 \text{ MN}\cdot\text{m}$$

---

* These stresses are found by considering the equilibrium of the free-body diagrams of the top half of the vessel and the right-hand part of the vessel separately and assuming uniform stress across the wall thickness. A proof of this is given in Chapter 7.

Example 3.10:

For a shaft, find an expression for the elastic strain energy in terms of the applied bending moment M, *twisting moment* M_t, *and thrust* P.

Solution:

The general expression for the elastic energy density is given by Equation 3.34 as

$$U_o = {}^1/_2\,(\sigma_x\varepsilon_x + \sigma_y\varepsilon_y + \sigma_z\varepsilon_z + \tau_{xy}\gamma_{xy} + \tau_{yz}\gamma_{yz} + \tau_{zx}\gamma_{zx})$$

If the shaft axis is denoted by x, the stresses developed in the shaft according to the strength of materials approach are

Due to bending: $\sigma_x = \dfrac{My}{I}$

Due to torsion: $\tau_{xy} = \dfrac{M_t r}{I_o}$

Due to thrust: $\sigma_x = \dfrac{P}{A}$

Hence, the strain energy density is given by

$$U_o = {}^1/_2\,(\sigma_x\varepsilon_x + \tau_{xy}\gamma_{xy})$$

From Hooke's law for the nonzero stresses, the strains are

$$\varepsilon_x = \frac{\sigma_x}{E} \quad \text{and} \quad \gamma_{xy} = \frac{\tau_{xy}}{G}$$

and, thus,

$$U_o = \frac{\sigma_x^2}{2E} + \frac{\tau_{xy}^2}{2G}$$

The total strain energy is

$$U = \int_V U_o dV = \int \left[\int_A U_o dA\right] dx$$

Substitution yields

$$U = \int \left[\int_A \frac{M^2 y^2}{2EI^2} dA + \int_A \frac{M_t^2 r^2}{2GI_o^2} dA + \int_A \frac{P^2}{2EA^2} dA\right] dx$$

Noting that $I = y^2 dA$, $I_o = r^2 dA$, the strain energy for the shaft reduces to

$$U = \int \frac{M^2}{2EI} dx + \int \frac{M_t^2}{2GI_o} dx + \int \frac{P^2}{2EA} dx$$

where the integration is taken along the shaft length. For a shaft with uniform cross section along its length L, the strain energy due to each one of the applied loads is

Thrust P: $U = P^2L/2EA$,
Moment M: $U = M^2L/2EI$, and
Twisting moment M_t: $U = M_t^2L/2GI_o$.

The above expressions are the same as derived in the textbooks of strength of materials.

Example 3.11:

A 45° equiangular strain gauge rosette is cemented on the surface of a circular shaft of diameter d = *0.05 m to monitor a constant transmitted twisting moment* M_t. *The shaft is also subjected to pure bending moment* M. *The strain gauge readings at the position of maximum negative bending stress were:* $\varepsilon_o = -10 \times 10^{-5}$, $\varepsilon_{45°} = 5 \times 10^{-5}$, *and* $\varepsilon_{90°} = 3.33 \times 10^{-5}$, *where the 0°-direction is taken along the shaft axis.*

a. *Determine the value of Possion's ratio for this material.*
b. *Determine the values of* M_t *and* M.
c. *Determine the elastic strain energy stored in the shaft due to this loading in terms of its modulus of elasticity* E *(in Pa).*

Solution:

a. An element on the shaft surface undergoes the following strains according to Equations 2.32 for $\theta_1 = 0°$, $\theta_2 = 45°$, and $\theta_3 = 90°$:

$$\varepsilon_{0°} = \varepsilon_x = -10 \times 10^{-5}$$

$$\varepsilon_{45°} = \tfrac{1}{2}\left(\varepsilon_x + \varepsilon_y + \gamma_{xy}\right) = 5 \times 10^{-5}$$

$$\varepsilon_{90°} = \varepsilon_y = 3.33 \times 10^{-5}$$

Applying Hooke's law to obtain σ_x, τ_{xy} yields for E (in Pa):

$$\varepsilon_x = -10 \times 10^{-5} = \frac{\sigma_x}{E} \text{ giving } \sigma_x = -E \times 10^{-4} \text{ Pa}$$

$$\varepsilon_y = 3.3 \times 10^{-5} = -\frac{\nu\sigma_x}{E} \text{ giving } \nu = 0.333$$

Hence,

$$\tau_{xy} = G\gamma_{xy} = \frac{E\gamma_{xy}}{2(1+\nu)} = \frac{3E}{8} 16.67 \times 10^{-5} = 6.25E \times 10^{-5} \text{ Pa}$$

b. In terms of the shaft loading for $d = 0.05$ m:

$$\sigma_x = -E \times 10^{-4} = \frac{My}{I} = \frac{M(d/2)}{\pi d^4/64} = \frac{32M}{\pi d^3}$$

giving

$$M = -1.2266E \times 10^{-9} \text{ N} \cdot \text{m}$$

$$\tau_{xy} = 6.25E \times 10^{-5} = \frac{M_t r}{I_o} = \frac{M_t (d/2)}{\pi d^4/32} = \frac{16 M_t}{\pi d^3}$$

giving $M_t = 1.5332E \times 10^{-9}$ N · m.

c. The strain energy density is calculated from the expression for $G = 3E/8$ as

$$U_o = \frac{1}{A}\left(\frac{M^2}{2EI} + \frac{M_t^2}{2GI_o}\right) = 3.854E \times 10^{-9} \text{ Pa}$$

3.10 SOME ELASTIC ENERGY THEOREMS

Some energy theorems applicable to elastic solids will be now presented. These theorems are usually expressed in terms of load–displacement and hence can be directly applied to determine the displacements in a solid due to applied loads. They offer powerful means to solve many structural mechanics problems and constitute the basis of numerical methods of solution, such as the finite-element method.

3.10.1 Principle of Work

The principle of work states that "For an elastic solid subjected to boundary loads in equilibrium, the total strain energy U is equal to the work W done by the external boundary loads." This is a statement of the energy conservation law and applies to linear and nonlinear elastic solids. Hence,

$$U = W \tag{3.38}$$

To illustrate the application of this principle, consider a bar of cross-sectional area A, length L, which is fixed at O and subjected at the free end to a gradually increasing axial load attaining a final value of P as shown in Figure 3.11a. Hence, the work done by the force P is

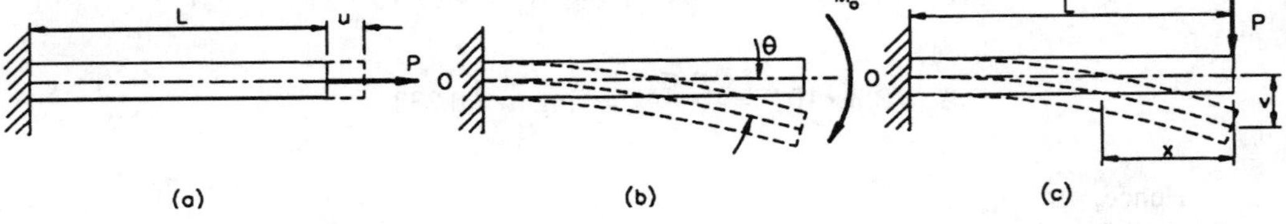

FIGURE 3.11 Displacements determined by applying the principle of work: (a) bar subjected to axial load, (b) cantilever beam subjected to edge bending moment, and (c) cantilever beam subjected to edge lateral force.

$$W = Pu/2$$

The total strain energy stored in the bar as obtained in Example 3.10 is

$$U = P^2L/2EA$$

Since $U = W$, the displacement due to a load P is obtained as

$$u = PL/EA$$

Consider now the bar to be subjected at the free end to a gradually increasing bending moment attaining a final value of M_o, as shown in Figure 3.11b; hence,

$$W = M_o\theta/2$$

Equating this work to the strain energy stored in the bar as expressed by the results of Example 3.10 is

$$U = M_o^2 L/2EI$$

gives $\theta = M_o L/EI$.

3.10.2 Principle of Virtual Work

The principle of virtual work states that "if a solid is in equilibrium, then the virtual work done in any virtual displacement is zero." What is meant by a virtual displacement in this statement is that one during which the applied loads remain constant in magnitude and direction. This principle applies to linear and nonlinear elastic solids.

Consider a rigid body subjected to a system of boundary loads P_1, P_2, P_3, P_n. If (X_i, Y_i, Z_i) are the components of any load P_i in the (x, y, z) directions, respectively, and (u_i, v_i, w_i) are the respective virtual displacement components, then

$$W_{\text{virtual}} = \sum_{i=1}^{i=n} \left(X_i u_i + Y_i v_i + Z_i w_i\right) \tag{3.39}$$

Now, since u_i, v_i, and w_i are arbitrary and can take any value, then choosing the same value of u_i, v_i, and w_i for all the applied loads gives

$$\sum X_i u_i = u_i \sum X_i,\quad \sum Y_i v_i = v_i \sum Y_i,\ \text{and}\ \sum Z_i w_i = w_i \sum Z_i$$

From the condition of equilibrium,

$$\sum X_i = \sum Y_i = \sum Z_i = 0$$

Hence, substituting in Equation 3.39 yields

$$W_{\text{virtual}} = 0$$

It will be noted that virtual displacements should not violate the conditions of constraint of the body, which means that they should be kinematically admissible. In the case of a rigid body there are only three independent virtual displacements and only external virtual work is done, Equation 3.39. However, in a deformable body, points within the body may move relative to one another and both external and internal forces will do virtual work. The internal virtual work may be expressed by

$$W_{\text{virtual (int)}} = \int_V \left(\sigma_x\varepsilon_x + \sigma_y\varepsilon_y + \sigma_z\varepsilon_z + \tau_{xy}\gamma_{xy} + \tau_{xz}\gamma_{xz} + \tau_{yz}\gamma_{yz}\right) dV \tag{3.40}$$

where V is the volume of the body. For a deformable body in equilibrium, the internal virtual work is equal to the external virtual work so that

$$W_{\text{virtual (int)}} = W_{\text{virtual (ext)}}$$

or

$$W_{\text{virtual (ext)}} - W_{\text{virtual (int)}} = 0 \tag{3.41}$$

which is stated as "in a deformable body in equilibrium, the total external virtual work is equal to the total internal virtual work for every virtual displacement consistent with the constraints."

To illustrate the application of the principle of virtual work, consider first the bar of Figure 3.11a.

$$\text{The internal virtual work} = \int_V \sigma_x\varepsilon_x dV$$

Hence, knowing that

$$\sigma_x = \frac{P}{A}, \quad \varepsilon_x \text{ (constant along } L) = \frac{u}{L}, \quad \text{and } V = AL$$

gives

$$W_{\text{virtual (int)}} = \frac{P}{A}\frac{u}{L}AL = Pu$$

This is equal to

$$W_{\text{virtual (ext)}} = Pu$$

Consider also the bar of Figure 3.11b with an axis along the x-axis and a rectangular cross section $b \times h$. Since each cross section is subjected to M_o, then the angular deflection per unit length at any distance x is the same and equals to θ/L. For a plane cross section that remains plane, the axial strain is given by $\varepsilon_e = y\theta/L$. Putting $dV = bdydx$, the internal virtual work is expressed by

$$W_{\text{virtual (int)}} = \int_V \sigma_x\varepsilon_x dV$$

or

$$W_{\text{virtual (int)}} = \frac{\theta}{L}\int_o^L \int_{-h/2}^{h/2} b\sigma_x ydydx$$

Since, from equilibrium,

$$\int_{-h/2}^{h/2} b\sigma_x y dy dx = M_o$$

Hence,

$$W_{\text{virtual (int)}} = \frac{\theta}{L} M_o L = M_o \theta$$

which is equal to the external virtual work:

$$W_{\text{virtual (ext)}} = M_o \theta$$

Note that no material stress–strain relation has been used in the above derivations since the principle of virtual work is in fact an expression for the Equations of equilibrium. However, the choice of virtual displacements has to satisfy the body constraints.

3.10.3 Principle of Stationary Potential Energy

The principle of stationary potential energy is another expression of the principle of virtual work. In this respect, the potential energy of a system is an arbitrary reference or datum for the capacity of the system to do work.

Consider a deformable body subjected to boundary loads P_1, P_2, P_3, where P can be either a force or a moment, which are in equilibrium. Assume that these loads have moved virtual incremental displacements δu_1, δu_2, δu_3, corresponding to P_1, P_2, P_3, respectively. The increment of virtual work δW done by the boundary loads is given by

$$\delta W = P_1 \delta u_1 + P_2 \delta u_2 + P_3 \delta u_3 + \ldots$$

where the subscript virtual belonging to δW has been omitted for simplicity. This increment of work done will increase the strain energy by an increment $\delta U = \delta W$. If the potential energy is Ω, then Ω will decrease by an amount equals to δW and increase by an amount δU, so that

$$\delta \Omega = \delta U - \delta W$$

and since $\delta U = \delta W$,

$$\delta \Omega = \delta U - \delta W = 0 \tag{3.42}$$

This is the principle of stationary potential energy which is stated as "if a structural system is in static equilibrium, the total potential energy of the system has a stationary value."

For stable equilibrium, Ω should be minimum. This is known as the principle of minimum potential energy expressed as "of all the displacements which satisfy the boundary conditions of a structural system, those corresponding to configurations of stable equilibrium make the total potential energy a relative minimum."

Note that the principle of minimum potential energy and the principle of virtual work deal with imaginary displacements. They have no physical meaning and are merely mathematical devices, which express equilibrium in terms of strain energy. A powerful approximate method of solving problems by applying the principle of minimum potential energy is the Raleigh–Ritz method. This is applied later in Chapter 7 for column buckling and in Chapter 8 for rectangular plates.

Example 3.12:

An elastic bar of length* L *is clamped at one end. Determine the vertical displacement* v *of the free end when subjected to a transverse force* P *as shown in Figure 3.11c.

Solution:

From Example 3.8

$$U = \int_0^L \frac{M^2}{2EI} dx$$

Hence, substituting $M = Px$ and $M = 0$ at $x = 0$ yields

$$U = \frac{P^2 L^3}{6EI}$$

Applying the principle of minimum potential energy gives

$$\delta\Omega = 0 \text{ or } \frac{\partial \Omega}{\partial P} = 0$$

where $\Omega = U - Pv$, hence,

$$\frac{\partial \Omega}{\partial P} = \frac{\partial U}{\partial P} - v = 0$$

This determine the vertical displacement v as

$$v = \frac{\partial}{\partial P}\left(\frac{P^2 L^3}{6EI}\right) = \frac{PL^3}{3EI}$$

3.10.4 Castigliano's Theorems*

If an elastic solid has a linear stress–strain relation, then the load–displacement relation is also linear. The loads that maintain equilibrium can be a system of forces P_i and/or moments M_i, and the corresponding displacement is either linear u_i or angular θ_i. In this case the strain energy U, which is the area below the straight line representing the load–displacement relation, equals the complementary energy C, which is the area above the straight line as shown in Figure 3.12. This means that

$$dU = Pdu$$

$$dC = vdP = dU$$

and hence

$$dU = vdP \tag{3.43a}$$

* Named after A. Castigliano (1847–1884).

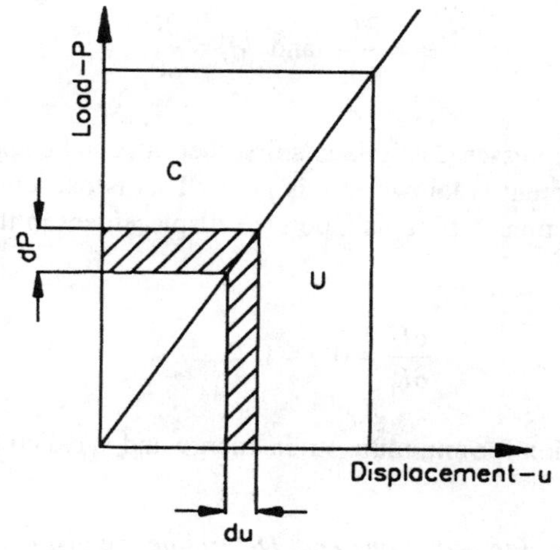

FIGURE 3.12 In a linear elastic solid, the increment of strain energy equals the increment of complementary energy.

If P_1, P_2, P_3, ... are the boundary loads that are in equilibrium and u_1, u_2, u_3, ... are the actual displacements of the points of application of the loads in the direction of these loads, then Equation 3.42 becomes

$$dU = u_1 dP_1 + u_2 dP_2 + u_3 dP_3 + \dots \tag{3.43b}$$

Writing

$$dU = \frac{\partial U}{\partial P_1} dP_1 + \frac{\partial U}{\partial P_2} dP_2 + \frac{\partial U}{\partial P_3} dP_3 + \cdots$$

and substituting this into Equation 3.43 yields

$$\left(\frac{\partial U}{\partial P_1} - u_1\right) dP_1 + \left(\frac{\partial U}{\partial P_2} - u_2\right) dP_2 + \cdots = 0$$

Since dP_1, dP_2, ... are arbitrary under the condition of equilibrium, then this requires that

$$u_1 = \frac{\partial U}{\partial P_1}, \quad u_2 = \frac{\partial U}{\partial P_2}, \quad \dots$$

or, in general,

$$u_i = \frac{\partial U}{\partial P_i} \text{ and } \theta_i = \frac{\partial U}{\partial M_i} \tag{3.44}$$

Equation 3.44 is Castigliano's second theorem, which can be stated as "the displacement corresponding to any one of a system of forces in equilibrium equals the partial derivative of the total elastic strain energy with respect to this force."

Similarly, from consideration of complementary energy, it is shown that

$$P_i = \frac{\partial U}{\partial u_i} \text{ and } M_i = \frac{\partial U}{\partial \theta_i}. \tag{3.45}$$

Equation 3.45, which expresses Castigliano's first theorm, can be applied to determine the load at any point if the displacement is known at this point. This is particularly useful in determining statically indeterminate reactions.* If at a support the displacement in the direction of the reaction R_i is zero, then

$$\frac{\partial U}{\partial R_i} = 0 \quad i = 1, 2, 3, \ldots \tag{3.46}$$

Equation 3.46 is an expression for minimum strain energy and is called the principle of least work.

Example 3.13:

An elastic bar of length L *is clamped at one end. Determine the displacement of the free end when subjected to*

a. Transverse end force P, *Figure 3.11c,*
b. End bending moment M_o, *Figure 3.11b.*

Solution:

a. For an end force P, the strain enegry is

$$U = \int_0^L \frac{M^2}{2EI} dx$$

Substituting $M = Px$ and $M = 0$ at $x = 0$ gives

$$U = \frac{P^2 L^3}{6EI}$$

Applying Castigliano's theorem results in

$$v = \frac{\partial}{\partial P}\left(\frac{P^2 L^3}{6EI}\right) = \frac{PL^3}{3EI}$$

which is the same result as obtained in Example 3.12.

b. For an end bending moment M_o, the strain energy is

$$U = \int_0^L \frac{M_o^2}{2EI} dz = \frac{M_o^2 L}{2EI}$$

Hence,

$$\theta = \frac{\partial U}{\partial M_o} = \frac{M_o L}{EI}$$

which is the same result as obtained in Section 3.10.1.

* See Chapter 7.

3.11 GENERALIZED HOOKE'S LAW

3.11.1 Anisotropic Elasticity

Hook's law for homogeneous isotropic linear materials is given by Equations 3.6. In many cases, the assumption of homogeneity and isotropy is not fulfilled and a more general formulation is required. Keeping only the linear proportionality between the components of the stress and strain tensors, a general relation may be written as

$$\begin{bmatrix} \sigma_x \\ \sigma_y \\ \sigma_z \\ \tau_{xy} \\ \tau_{xz} \\ \tau_{yz} \end{bmatrix} = \begin{bmatrix} a_{11} & a_{12} & a_{13} & a_{14} & a_{15} & a_{16} \\ a_{21} & a_{22} & a_{23} & a_{24} & a_{25} & a_{26} \\ a_{31} & a_{32} & a_{33} & a_{34} & a_{35} & a_{36} \\ a_{41} & a_{42} & a_{43} & a_{44} & a_{45} & a_{46} \\ a_{51} & a_{52} & a_{53} & a_{54} & a_{55} & a_{56} \\ a_{61} & a_{62} & a_{63} & a_{64} & a_{65} & a_{66} \end{bmatrix} \begin{bmatrix} \varepsilon_x \\ \varepsilon_y \\ \varepsilon_z \\ \gamma_{xy} \\ \gamma_{xz} \\ \gamma_{yz} \end{bmatrix} \tag{3.47}$$

These relations are called generalized Hooke's law referred to the strain. In reality, Equation 3.47 is merely an approximation that is valid for infinitesimal strains since any continuous function is approximately linear in a sufficiently small range of its variables. Note that in expressing the generalized Hooke's law as given by Equations 3.47, normal stresses are related to both normal strains and shear strains. This also applies to shear stresses unlike to Hooke's law for isotropic materials, Equation 3.6, in which normal and shear stresses are related to only normal and shear strains respectively. In general, the 36 coefficients (a_{11}, a_{12}, ...) are constants that depend upon the location in the body as well as on time and temperature. However, for isothermal conditions of a homogeneous material, the entire 36 coefficients will be constants.

The 36 constants of Equations 3.47 are shown not to be all independent by considering the existence of a strain energy density function. Equations 3.47 together with Equation 3.32 yield

$$\frac{\partial U_o}{\partial \varepsilon_x} = \sigma_x = a_{11}\varepsilon_x + a_{12}\varepsilon_y + a_{13}\varepsilon_z + a_{14}\gamma_{xy} + a_{15}\gamma_{xz} + a_{16}\gamma_{yz}$$

$$\frac{\partial U_o}{\partial \varepsilon_y} = \sigma_y = a_{21}\varepsilon_x + a_{22}\varepsilon_y + a_{23}\varepsilon_z + a_{24}\gamma_{xy} + a_{25}\gamma_{xz} + a_{26}\gamma_{yz} \cdots$$

and

$$\frac{\partial U_o}{\partial \gamma_{xy}} = \tau_{xy} = a_{41}\varepsilon_x + a_{42}\varepsilon_y + a_{43}\varepsilon_z + a_{44}\gamma_{xy} + a_{45}\gamma_{xz} + a_{46}\gamma_{yz}$$

$$\frac{\partial U_o}{\partial \gamma_{yz}} = \tau_{yz} = a_{61}\varepsilon_x + a_{62}\varepsilon_y + a_{63}\varepsilon_z + a_{64}\gamma_{xy} + a_{65}\gamma_{xz} + a_{66}\gamma_{yz} \cdots$$

Further differentiations yield

$$\frac{\partial^2 U_o}{\partial \varepsilon_x \partial \varepsilon_y} = a_{12} = a_{21}, \quad \frac{\partial^2 U_o}{\partial \varepsilon_x \partial \varepsilon_z} = a_{13} = a_{31}, \cdots$$

$$\frac{\partial^2 U_o}{\partial \gamma_{xy} \partial \gamma_{yz}} = a_{46} = a_{64}, \quad \frac{\partial^2 U_o}{\partial \gamma_{xy} \partial \gamma_{xz}} = a_{45} = a_{54}$$

These relations show that $a_{12} = a_{21}$, $a_{13} = a_{31}$, …, that is the elastic constants are symmetric. In other words, the general anisotropic linear elastic material has 21 elastic constants. Such a case is important in studying crystals.

If the material is elastically symmetric in certain directions, then the number of independent elastic constants is further reduced. These include symmetry with respect to three mutual perpendicular planes. For such cases of symmetry, some elastic constants remain invariant under a given orthogonal transformation of coordinates.[1] Thus, for one plane of symmetry, the number of independent elastic constants reduces to only 13. For symmetry with respect to three mutually perpendicular planes, this number becomes only 9. The latter case describes orthotropic materials, such as wood and some composites. The generalized Hooke's law is thus given by

$$\begin{aligned} \sigma_x &= a_{11}\varepsilon_x + a_{12}\varepsilon_y + a_{13}\varepsilon_z \\ \sigma_y &= a_{21}\varepsilon_x + a_{22}\varepsilon_y + a_{23}\varepsilon_z \\ \sigma_z &= a_{31}\varepsilon_x + a_{32}\varepsilon_y + a_{33}\varepsilon_z \\ \tau_{xy} &= a_{44}\gamma_{xy} \\ \tau_{xz} &= a_{55}\gamma_{xz} \\ \tau_{yz} &= a_{66}\gamma_{yx} \end{aligned} \tag{3.48a}$$

where symmetry is satisfied among the elastic constants, i.e.,

$$a_{12} = a_{21}, \quad a_{23} = a_{32}, \quad \text{and} \quad a_{31} = a_{13} \tag{3.48b}$$

Furthermore, if the material is fully isotropic, the above nine constants reduce to only two independent constants, as previously given by Equations 3.14; known as Lame's constants (see Example 3.14). Most structural materials as metallic alloys are formed of crystalline structure and, hence, very small portions of such materials cannot be regarded as being isotropic. Nevertheless, the assumption of isotropy and homogeneity can be accurately applied to the entire body whose dimensions are much larger than the dimensions of the randomly distributed single crystals. Sometimes, however, cast, drawn, and rolled metals possess a definite orientation of crystals and have to be treated as anisotropic solids.

Example 3.14:

For a fully isotropic solid, show that Equations 3.48 reduce to Hooke's law with two elastic constants only.

Solution:

Consider first three separate tensile bars; each is cut from the isotropic solid along directions x, y, and z, respectively. Hence, Equations 3.48a should satisfy the requirement of full isotropy, that is, along the loading direction the same strain is produced for the same loading. Hence,

$$a_{11} = a_{22} = a_{33}$$

Along two mutually orthogonal directions lying in a plane normal to the loading direction, the same strains are produced; hence,

$$a_{12} = a_{13}, \quad a_{21} = a_{23}, \quad \text{and} \quad a_{31} = a_{32}$$

By virtue of Equation 3.48b, it is found that

$$a_{12} = a_{21} = a_{13} = a_{31} = a_{23} = a_{32}$$

Therefore,

$$\sigma_x = a_{11}\varepsilon_x + a_{12}\left(\varepsilon_y + \varepsilon_z\right)$$

$$= \left(a_{11} - a_{12}\right)\varepsilon_x + a_{12}\left(\varepsilon_x + \varepsilon_y + \varepsilon_z\right) \tag{a}$$

$$\sigma_x = \left(a_{11} - a_{12}\right)\varepsilon_x + a_{12}\varepsilon_v$$

Similarly,

$$\sigma_y = \left(a_{22} - a_{21}\right)\varepsilon_y + a_{21}\varepsilon_v$$

$$\sigma_y = \left(a_{11} - a_{12}\right)\varepsilon_x + a_{21}\varepsilon_v \tag{b}$$

and

$$\sigma_z = \left(a_{11} - a_{12}\right)\varepsilon_z + a_{12}\varepsilon_v \tag{c}$$

Comparing these expressions with Hooke's law in its inverse form, Equation 3.12, reveals that

$$a_{11} - a_{12} \equiv 2\mu = 2G \quad \text{and} \quad a_{12} \equiv \lambda$$

where μ and λ are the two Lame's constants. It is also obvious that isotropy implies for shear loading along directions x, y, and z separately:

$$a_{44} = a_{55} = a_{66}$$

Now it remains to relate a_{44} to μ and λ. Considering simple shear in the x–y plane gives

$$\tau_{xy} = a_{44}\gamma_{xy} \tag{d}$$

Applying stress and strain transformation to this type of loading, namely, Equations 1.15a and 2.22a, gives along axes (x', y') rotated 45° with respect to x, y:

$$\sigma_{x'} = \tau_{xy}, \quad \sigma_{y'} = -\tau_{xy}, \text{ and } \tau_{x'y'} = 0$$

$$\varepsilon_{x'} = \gamma_{xy}/2, \ \varepsilon_{y'} = -\gamma_{xy}/2, \text{ and } \gamma_{x'y'} = 0$$

Substituting into Equations a and b for the (x', y') axes and making use of Equation d yields

$$\sigma_{x'} = \tau_{xy} = (a_{11} - a_{12})(\gamma_{xy}/2) = a_{44}\gamma_{xy}$$

$$\sigma_{y'} = -\tau_{xy} = (a_{11} - a_{12})(-\gamma_{xy}/2) = -a_{44}\gamma_{xy}$$

or $(a_{11} - a_{12})/2 = a_{44} = \mu = G$. Therefore, only two constants are required to describe a homogeneous linear isotropic elastic solid.

Example 3.15:

*A plate of a composite material is fabricated from boron fiber-reinforced epoxy as shown in Figure 3.13. For this orthotropic material, the generalized Hooke's law reduces to Equations 3.48 with nine elastic constants determined as**

$$a_{11} = 209, \ a_{22} = a_{33} = 29.4, \ a_{12} = a_{13} = 24.6$$

$$a_{23} = 13.7, \ a_{44} = 5.9, \text{ and } a_{55} = a_{66} = 8.14 \text{ GPa}$$

The plate is loaded in plane stress conditions, i.e., in the x–y plane by stresses σ_x and σ_y uniformly distributed over the edges such that

$$\varepsilon_x = 6.562 \times 10^{-4} \text{ and } \varepsilon_y = -59.055 \times 10^{-4}$$

Calculate the values of the applied stresses and the thickness change.

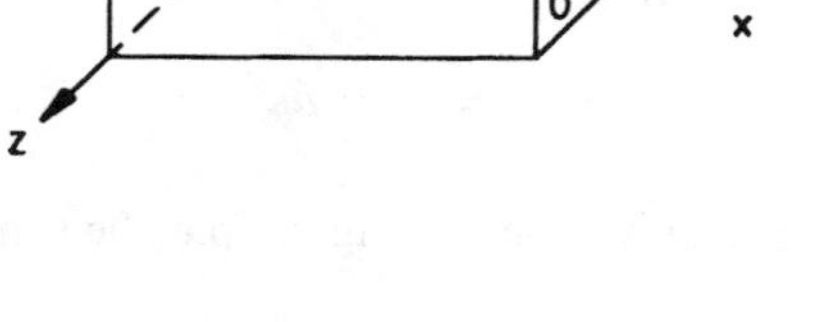

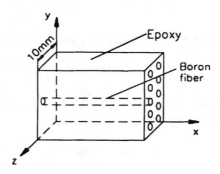

FIGURE 3.13 Example 3.15.

Solution:

Obviously the shear strains and stresses are zero and the other stresses are given by Equation 3.47 as

* See Tauchert and Guzelsu.[2]

$$\sigma_x = 209\left(6.562\times10^{-4}\right) + 24.6\left(-59.055\times10^{-4}\right) + 24.6\varepsilon_z \text{ GPa}$$

$$\sigma_y = 24.6\left(6.562\times10^{-4}\right) + 29.4\left(-59.055\times10^{-4}\right) + 13.7\varepsilon_z \text{ GPa}$$

$$\sigma_z = 24.6\left(6.562\times10^{-4}\right) + 13.7\left(-59.055\times10^{-4}\right) + 29.4\varepsilon_z \text{ GPa}$$

since $\sigma_z = 0$, the third equation gives

$$\varepsilon_z = 22.049 \times 10^{-4}$$

and the change in thickness is 0.022 mm. The applied stresses σ_x and σ_y are thus found to be

$$\sigma_x = 46.1 \text{ MPa}$$

$$\sigma_y = -127.3 \text{ MPa (compressive)}$$

Example 3.16:

Calculate the strain ratios ($\varepsilon_x/\varepsilon_y$) and ($\varepsilon_y/\varepsilon_x$) of the boron–epoxy plate of Example 3.15 in the two cases:

a. *The plate is subjected to σ_x = 75 MPa only.*
b. *The plate is subjected to σ_y = 75 MPa only.*

Comment on the result.

Solution:

a. Substitution in the generalized Hooke's law for this material, Equations 3.46, and using data of Example 3.15 gives

$$75\times10^{-3} = 209\varepsilon_x + 24.6\varepsilon_y + 24.6\varepsilon_z$$

$$0 = 24.6\varepsilon_x + 29.4\varepsilon_y + 13.7\varepsilon_z$$

$$0 = 24.6\varepsilon_x + 13.7\varepsilon_y + 29.4\varepsilon_z$$

Solving for strains gives

$$\varepsilon_x = 4.146 \times 10^{-4}, \quad \varepsilon_y = \varepsilon_z = -2.366 \times 10^{-4}$$

Hence, $|\varepsilon_x/\varepsilon_y| = |\varepsilon_x/\varepsilon_z| = 1.752$ and $|\varepsilon_y/\varepsilon_x| = \nu_f = 0.571$.

b. Repeating the same procedure for σ_y = 75 MPa gives

$$\varepsilon_x = -3.292 \times 10^{-4}, \ \varepsilon_y = 28.53 \times 10^{-4}, \text{ and } \varepsilon_z = -0.563 \times 10^{-4}$$

Hence, $|\varepsilon_y/\varepsilon_x| = 8.67$ and $|\varepsilon_x/\varepsilon_y| = \nu_n = 0.115$.

From the above results it is seen that such fiber-reinforced plate is much stiffer in the x-direction than in the y-direction. For the same magnitude of applied stress, 75 MPa, the strain ratio defined by

$$\varepsilon \text{ (loading direction)}/\varepsilon \text{ (transverse direction)}$$

is much higher if the plate is loaded in a direction perpendicular to the reinforcing fibers. Meanwhile, the ratio

$$\varepsilon \text{ (transverse direction)}/\varepsilon \text{ (loading direction)}$$

which signifies Poisson's ratio of the lamina is different for the two considered cases. The ratio $\nu_f > 0.5$ is expected for orthotropic plates, as will be discussed later.

3.11.2 Application to Fiber-Reinforced Composites

Composites represent a group of synthesized combinations of materials possessing specific properties that cannot be met by one single material. Such materials have wide engineering applications; for example, several structural and machine elements are made of composites, where high strength and rigidity at low weight are needed, such as in the aerospace and transportation industries. The mechanical properties of composites depend upon the individual mechanical properties of the constituent materials, their volume fraction, distribution, physical and chemical interactions.

In the following analysis a fiber-reinforced composite consisting of matrix and fiber reinforcement both being isotropic, is considered. The fibers are uniform in properties and cross section, continuous and parallel throughout the matrix. It is assumed that perfect bonding exists between fibers and matrix with no voids. The composite shape under consideration is a lamina subjected to in-plane loading only, i.e., the stresses normal to the plane of the lamina are zero. The x-axis is taken along the direction of fibers and the y- and z-axes are the directions of width and thickness, respectively, as shown in Figure 3.13. The material is orthotropic and, hence, the strains in the plane of the lamina are expressed in terms of the stresses, noting that $\sigma_z = \tau_{zx} = \tau_{zy} = 0$, as

$$\varepsilon_x = \frac{\sigma_x}{E_f} - \frac{\nu_n \sigma_y}{E_n}, \quad \varepsilon_y = \frac{\sigma_y}{E_n} - \frac{\nu_f \sigma_x}{E_f}, \quad \gamma_{xy} = \frac{\tau_{xy}}{G_{fn}} \tag{3.49}$$

where E_f, ν_f and E_n, ν_n are Young's moduli and corresponding Poisson's ratios for the lamina in the x- and y-directions, which are the directions along and normal to the fibers, respectively, and G_{fn} is the modulus of rigidity.*

Rearranging Equations 3.49 to express the stresses in terms of the strains yields

$$\begin{aligned} \sigma_x &= \left(\varepsilon_x + \nu_n \varepsilon_y\right) \frac{E_f}{1 - \nu_f \nu_y} \\ \sigma_y &= \left(\varepsilon_y + \nu_f \varepsilon_x\right) \frac{E_n}{1 - \nu_f \nu_n} \\ \tau_{xy} &= \gamma_{xy} G_{fn} \end{aligned} \tag{3.50}$$

Comparing Equations 3.50 with Equations 3.48a and b gives

* Textbooks on composites use the designation ν_{ij} as Poisson's ratio for transverse strain in the j-direction when stressed in the i-direction. Here, only one subscript is used for ν_f or ν_n designating the strain direction along the fiber and normal to it.

$$a_{11} = \frac{E_f}{1-\nu_f\nu_n}, \quad a_{22} = \frac{E_n}{1-\nu_f\nu_n}, \quad a_{44} = G_{fn}$$
$$a_{12} = a_{21} \frac{\nu_f E_n}{1-\nu_f\nu_n} = \frac{\nu_n E_f}{1-\nu_f\nu_n} \tag{3.51}$$

Note that the condition $a_{12} = a_{21}$ states that $\nu_f E_n = \nu_n E_f$ or $\nu_f/E_f = \nu_n/E_n$, which means that the nine constants of Equation 3.48a reduce in the case of plane stress to only four independent elastic constants, namely E_f, E_n, G_{fn}, ν_f noting that ν_n is obtained as $\nu_f\, E_n/E_f$. From Equations 3.51, hence

$$E_f = a_{11}\left(1 - \frac{a_{12}^2}{a_{11}a_{22}}\right), \quad \nu_f = a_{12}/a_{22}$$
$$E_n = a_{22}\left(1 - \frac{a_{12}^2}{a_{11}a_{22}}\right), \quad \nu_n = a_{12}/a_{11} \tag{3.52}$$

and $G_{fn} = a_{44}$.

The use of a lamina of the above construction is feasible only when it is expected to be subjected to a load along the fibers direction, since the properties in the width direction are not improved to allow in-plane general loading. This is a result of using a matrix of a lower strength as compared with the fibers strength. The lamina is therefore constructed of multilayers with different orientations of fibers in each layer. In this case the load share of each layer depends upon the orientation of its fibers with respect to the direction of loading as shown in Example 3.17.

Example 3.17:

A plate is constructed from two glass/epxoy identical fiber-reinforced layers arranged such that the fibers of one layer are oriented perpendicular to the fibers of the other layer along the directions of length and width, respectively. This is called a symmetric cross-ply structure. The elastic constants are $E_f = 40$ *GPa,* $E_n = 8$ *GPa,* $G_{fn} = 4$ *GPa, and* $\nu_f = 0.26$ *of each layer. If the plate width is 300 mm and the total thickness is 6 mm and the fibers are along the length and width directions, determine the maximum stress and the in-plane strains when the plate is subjected to a load of 60 kN along the direction of fibers of the first set of layers.*

Solution:

By taking the direction of length to be along the x-axis and the width along the y-axis designating the first set of layers by subscript 1 and the second set of layers by subscript 2, the directions of fibers for set 1 are along the x-axis and for set 2 are along the y-axis, as shown in Figure 3.14.

For perfect bond: $\varepsilon_x = \varepsilon_{x1} = \varepsilon_{x2}$ and $\varepsilon_y = \varepsilon_{y1} = \varepsilon_{y2}$. For a uniaxial load P in the longitudinal direction, x,

$$P = A_1\sigma_{x1} + A_2\sigma_{x2} = \frac{A}{2}(\sigma_{x1} + \sigma_{x2})$$
$$0 = A_1\sigma_{y1} + A_2\sigma_{y2} \text{ or } \sigma_{y1} = -\sigma_{y2} \tag{a}$$

From Equations 3.49 and 3.51

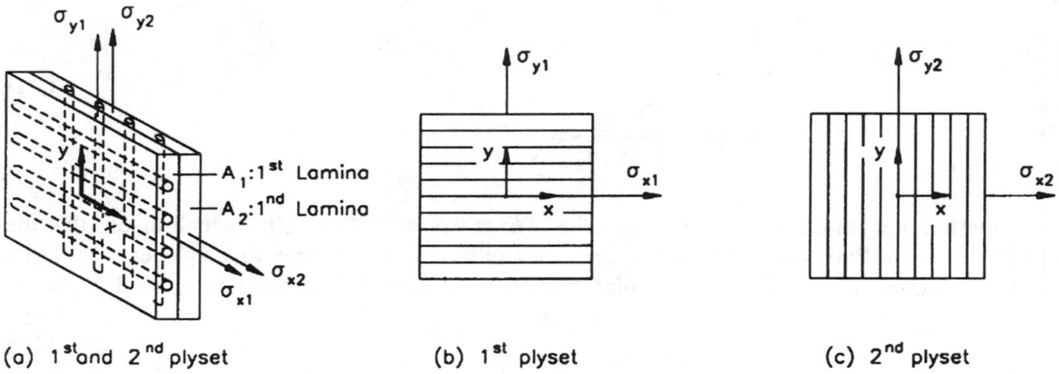

FIGURE 3.14 Example 3.17.

$$\varepsilon_x = \frac{\sigma_{x1}}{E_f} - \frac{\nu_n \sigma_{y1}}{E_n} = \frac{\sigma_{x2}}{E_n} - \frac{\nu_f \sigma_{y2}}{E_f} \quad \text{or} \quad \frac{\sigma_{x1}}{E_f} - \frac{\sigma_{x2}}{E_n} = 2\frac{\nu_n}{E_n}\sigma_{y1} \tag{b}$$

$$\varepsilon_y = \frac{\sigma_{y1}}{E_n} - \frac{\nu_f \sigma_{x1}}{E_f} = \frac{\sigma_{y2}}{E_f} - \frac{\nu_n \sigma_{x2}}{E_n} \quad \text{or} \quad \frac{\nu_f}{E_f}(\sigma_{x1} - \sigma_{x2}) = \left(\frac{1}{E_n} + \frac{1}{E_f}\right)\sigma_{y1} \tag{c}$$

Substituting

$$E_f = 40 \text{ GPa}, \ E_n = 8 \text{ GPa}, \ \nu_f = 0.26, \ \nu_n = 0.052$$

in Equations a, b, and c knowing that $A = 300 \times 6 = 1800 \text{ mm}^2$, yields

$$\sigma_{x1} + \sigma_{x2} = \frac{60 \times 10^3 \times 2}{1800} = 66.66 \text{ MPa} \tag{d}$$

$$\sigma_{x1} - 5\sigma_{x2} = 0.52\sigma_{y1} \tag{e}$$

$$\sigma_{x1} - \sigma_{x2} = 23.1\sigma_{y1} \tag{f}$$

Dividing Equation e by Equation f gives $\sigma_{x1} = 5.1\sigma_{x2}$. Substituting in Equation d results in

$$\sigma_{x1} = 55.73 \text{ MPa}, \ \sigma_{x2} = 10.93 \text{ MPa}$$

Note that without fiber reinforcement $\sigma_{x1} = \sigma_{x2} = 60000/1800 = 33.33$ MPa compared with σ_{x2} $=10.93$ MPa in the matrix material with reinforcement. Further calculations results in

$$\sigma_{y1} = 2.06 \text{ MPa}, \qquad \sigma_{y2} = 2.06 \text{ MPa}$$

$$\varepsilon_x = 1.38 \times 10^{-3}, \ \text{and} \ \varepsilon_y = -0.105 \times 10^{-3}$$

Example 3.18:

Consider a tube of a mean radius 51 mm and thickness 6 mm fabricated from symmetric layers of a composite material of the same properties as that of the plate in Example 3.17. The tube ends are closed and subjected to an internal pressure p_i *of 11 MPa. Determine the stresses in the tube wall for the following two cases of fibers orientation, Figure 33.15a and b:*

a. Fibers along and normal to the tube axis, and
b. Fibers at ±45° to the tube axis.

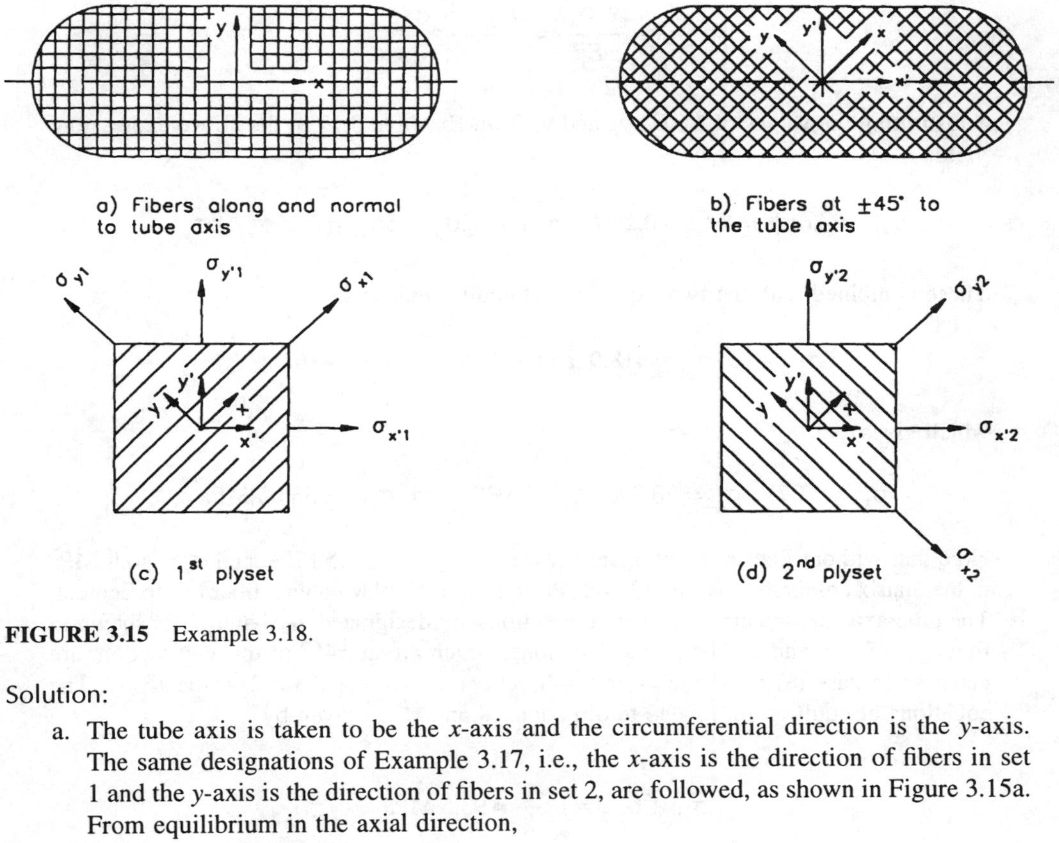

FIGURE 3.15 Example 3.18.

Solution:

a. The tube axis is taken to be the x-axis and the circumferential direction is the y-axis. The same designations of Example 3.17, i.e., the x-axis is the direction of fibers in set 1 and the y-axis is the direction of fibers in set 2, are followed, as shown in Figure 3.15a. From equilibrium in the axial direction,

$$2\pi r_m \frac{h}{2}\left(\sigma_{x1}+\sigma_{x2}\right)=\pi r_i^2 p_i$$

or

$$\sigma_{x1}+\sigma_{x2}=\frac{p_i r_m}{h}=93.5 \text{ MPa}$$

From equilibrium in the circumferential direction,

$$\sigma_{y1}+\sigma_{y2}=p_i\frac{2r_m}{h}=187 \text{ MPa}$$

By neglecting p_i with respect to σ_x or σ_y, then Equations 3.49 apply.

Now, since $\varepsilon_{x1} = \varepsilon_{x2}$ and $\varepsilon_{y1} = \varepsilon_{y2}$, the following compatibility conditions are obtained:

$$\frac{\sigma_{x1}}{E_f} - \frac{\nu_n \sigma_{y1}}{E_n} = \frac{\sigma_{x2}}{E_n} - \frac{\nu_f \sigma_{y2}}{E_f}$$

and

$$\frac{\sigma_{y1}}{E_n} - \frac{\nu_f \sigma_{x1}}{E_f} = \frac{\sigma_{y2}}{E_f} - \frac{\nu_n \sigma_{x2}}{E_n}$$

Substituting the values of E_f, E_n, ν_f, and ν_n from Example 3.17 in the above expressions yields

$$\sigma_{x1} - 0.26\sigma_{y1} = 5\sigma_{x2} - 0.26\sigma_{y2} \quad \text{and} \quad 0.26\sigma_{x1} - 5\sigma_{y1} = 0.26\sigma_{x2} - \sigma_{y2}$$

These combined with the two Equations of equilibrium yield

$$6\sigma_{x1} - 0.52\sigma_{y1} = 418.9 \quad \text{and} \quad 0.52\sigma_{x1} - 6\sigma_{y1} = -162.69$$

which gives

$$\sigma_{x1} = 72.71, \quad \sigma_{x2} = 20.79, \quad \sigma_{y1} = 33.38, \quad \text{and} \quad \sigma_{y2} = 153.62 \text{ MPa}$$

Note that without fiber reinforcement $\sigma_{y1} = \sigma_\theta = p_i r_m/h = 93.5$ MPa, and $\sigma_x = 46.8$ MPa in the matrix compared to $\sigma_{y1} = 33.38$ MPa and $\sigma_{x2} = 20.79$ MPa with fiber reinforcement.

b. The tube axis and the circumferential directions are designated as x' and y', as shown in Figure 3.15b, c, and d. The fibers directions, which are at ±45° to the x'-direction, are taken as in case (a) to be the x- and y-directions of set 1 and set 2, respectively. The equations of equilibrium related to directions x' and y' are given by

$$\sigma_{x'1} + \sigma_{x'2} = p_i \frac{r_m}{h} = 93.5 \text{ MPa}$$

$$\sigma_{y'1} + \sigma_{y'2} = p_i \frac{2r_m}{h} = 187 \text{ MPa}$$

The stresses σ_{x1}, σ_{y1}, σ_{x2}, and σ_{y2} are obtained from $\sigma_{x'1}$, $\sigma_{y'1}$, $\sigma_{x'2}$, and $\sigma_{y'2}$ by transformation of axes from x', y' to x, y using Equation 1.20a as

$$\sigma_{x1} = \sigma_{x'1}\cos^2 45° + \sigma_{y'1}\sin^2 45° + 2\tau_{x'y'1}\sin 45°\cos 45°$$

$$\sigma_{y1} = \sigma_{x'1}\sin^2 45° + \sigma_{y'1}\cos^2 45° - 2\tau_{x'y'1}\sin 45°\cos 45°$$

Hence,

$$\sigma_{x1} = {}^1/_2(\sigma_{x'1} + \sigma_{y'1}) + \tau_{x'y'1}, \quad \sigma_{y1} = {}^1/_2(\sigma_{x'1} + \sigma_{y'1}) - \tau_{x'y'1}$$

and

$$\sigma_{x1} + \sigma_{y1} = \sigma_{x'1} + \sigma_{y'1} \tag{a}$$

Similarly,

$$\sigma_{x2} = {}^1/_2(\sigma_{x'2} + \sigma_{y'2}) + \tau_{x'y'2}, \quad \sigma_{y2} = {}^1/_2(\sigma_{x'2} + \sigma_{y'2}) - \tau_{x'y'2}$$

and from the definition of the first stress invariant with $\sigma_z = 0$:

$$\sigma_{x2} + \sigma_{y2} = \sigma_{x'2} + \sigma_{y'2} \tag{b}$$

Adding Equations a and b and noting from symmetry that $\sigma_{x1} = \sigma_{y2}$ and $\sigma_{x2} = \sigma_{y1}$ yields

$$\sigma_{x1} + \sigma_{x2} = \sigma_{y1} + \sigma_{y2} = {}^1/_2(187 + 93.5) = 140.25 \text{ MPa}$$

and

$$\tau_{x'y'1} = -\tau_{x'y'2} \tag{c}$$

Equations c combined with the two equations of compatibility of $\varepsilon_{x1} = \varepsilon_{x2}$ and $\varepsilon_{y1} = \varepsilon_{y2}$ yield

$$\sigma_{x1} - 0.26\sigma_{y1} = 5(140.25 - \sigma_{x1}) - 0.26(140.25 - \sigma_{y1})$$

or

$$6\sigma_{x1} - 0.52\sigma_{y1} = 664.78$$

and

$$0.52\sigma_{x1} - 6\sigma_{y1} = -103.38$$

This gives

$$\sigma_{x1} = 113.11, \quad \sigma_{x2} = 27.14, \quad \sigma_{y1} = 27.14, \quad \sigma_{y2} = 113.11 \text{ MPa}$$

and

$$\tau_{x'y'1} = -\tau_{x'y'2} = 43 \text{ MPa}$$

The results show equal stress sharing along the fibers compared to case (a) together with the same reduced stresses in the matrix material. This fiber configuration is recommended for pressure vessels manufactured through winding the fibers on a cylindrical mandrel to form what is known as a filament-wound composite using several symmetric fiber orientations designed to give optimum stresses according to the constituent materials.

3.12 NOTE ON COMPOSITE ELASTIC CONSTANTS

The four elastic constants of the composite lamina, in Equations 3.49 and 3.50, namely, E_f, E_n, ν_f, and G_{fn}, can be obtained from the elastic constants of the fiber and matrix, namely, E_F, ν_f and E_M, ν_M, respectively, together with the volume fractions V_F and V_M from the rule of mixtures[3] to give the following simple practical expressions. These expressions are obtained by equating the strains along and normal to fibers combined with the Equations of equilibrium as applied in the solutions of Examples 3.17 and 3.18.

$$\begin{aligned}
E_f &= E_F V_F + E_M V_M = E_F V_F + E_M (1 - V_F) \\
E_n &= E_F E_M / (E_F V_M + E_M V_F) \\
\nu_f &= \nu_F V_F + \nu_M V_M \\
\nu_n &= E_y \nu_x / E_x \\
G_{fn} &= G_F G_M / (G_F V_M + G_M V_F)
\end{aligned} \tag{3.53}$$

in which the x-axis is the direction of the fibers, the y-axis is the direction of width, and $\sigma_z = 0$. There is a restriction on the value of Poisson's ratio to ensure a positive strain energy density. This restriction, in the case of an isotropic material, is given by $\nu \le 0.5$. For an orthotropic material this restriction can be obtained by referring to Equation 3.51 where the denominator must be positive and, hence, $\nu_f \nu_n \le 1$ or $\nu_f^2 \le E_f/E_n$ and $\nu_n^2 \le E_n/E_f$. It may be noted that values of ν_f higher than 0.5 have been experimentally* obtained. Table 3.2 gives the mechanical characteristics at room temperature of some fibers and matrix materials commonly used in the manufacture of fiber-reinforced composites and the elastic constants of some fiber-reinforced composites.

3.13 STRESS–STRAIN RELATIONS FOR LARGE ELASTIC DEFORMATION

There is a class of engineering materials that exhibit large elastic elongations up to several hundreds of percent. The elastic stress–strain curves of these materials is generally nonlinear. Rubbers and polymers above their glass transition temperature are good examples of these materials. The assumptions and definitions employed in infinitesimal strain elasticity are no longer applicable in this case. However, it is not intended here to derive a general theory for large elastic deformation, since this involves a rather complex formulation.[5] Only an illustration of this theory is given here by restricting attention to normal stresses and strains (no shear deformations).

In Chapter 2 it has been stated that a reasonable definition for strain in large deformation problems can be based on the stretch or extension ratio λ defined as the ratio of the deformed length L to the undeformed original length L_o, i.e.,

$$\lambda_i = (L_i/L_o)_i \; \ldots \quad i = 1, 2, 3$$

where i denotes a first principal direction. Note that λ is related to the finite strain** ε^F by Equation 2.41 according to

* See Jones,[3] p. 44.

** The strain here is generally finite and may be defined by Equations 2.36, i.e.,

$$\varepsilon_x^F = \frac{1}{2}\left[\left(1 + \frac{\partial u}{\partial x}\right)^2 + \left(\frac{\partial v}{\partial x}\right)^2 + \left(\frac{\partial w}{\partial x}\right)^2 - 1\right] \cdots$$

TABLE 3.2
Average Mechanical Characteristics of Some Composites

Constituent Material	Specific Gravity	Modulus of Elasticity (GPa)	Poisson's Ratio	Tensile Strength (GPa)
Fiber		E_F	ν_F	
Carbon	1.8	150–500	—	1.5–5.5
Aramid (Kevlar 49)	1.4	124	—	3.5
E-glass	2.5	72	0.24	3.5
Boron	2.6	400–450	0.21	3–3.5
Aluminum oxide	3.2	380	0.25	3.98
Silicon carbide	3.21	425	0.20	3.21
Molybdenum	10.2	360	0.32	1.4
Tungsten	19.3	400	0.27	4.3
Matrix		E_M	ν_M	
Epoxy	1.25	2.4	0.34	—
Polyester	1.35	3.2	0.36	0.050
Nylon 6.6	—	2.8	0.36	0.076
Polycarbonate	—	2.4	—	0.055
Composite lamina	E_f	E_n	ν_f	G_{fn}
Fiber/matrix (V_f)	(GPa)	(GPa)		(GPa)
Graphite/epoxy (0.65)	131	10.3	0.22	6.9
Aramid/epoxy (0.65)	76	5.5	0.34	2.3
E-glass/epoxy (0.45)	39	8.3	0.26	4.14
Boron/epoxy (0.65)	204	18.5	0.23	5.6

Data collected from Metal-Matrix Composites, UNIDO, MONITOR, Issue No. 17, Feb. 1990.[4]

$$\lambda^2 = 1 + 2\varepsilon^F$$

The strain invariants for large deformations are written in terms of the principal extension ratios λ_1, λ_2, and λ_3 as[5]

$$\begin{aligned} J_1^F &= \lambda_1^2 + \lambda_2^2 + \lambda_3^2 \\ J_2^F &= \lambda_1^2\lambda_2^2 + \lambda_2^2\lambda_3^2 + \lambda_3^2\lambda_1^2 \\ J_3^F &= \lambda_1^2\lambda_2^2\lambda_3^2 \end{aligned} \tag{3.54}$$

where the superscript F denotes finite strains.

Considering a unit cube of an initially isotropic homogeneous rubbery material undergoing large elastic deformation, the strain energy density is easily shown to be

$$dU_o = \sigma_1\lambda_2\lambda_3 d\lambda_1 + \sigma_2\lambda_3\lambda_1 d\lambda_2 + \sigma_3\lambda_1\lambda_2 d\lambda_3 \tag{3.55}$$

The term $\sigma_1\lambda_2\lambda_3$, for example, is the force acting in the first direction and $d\lambda_1$ is the displacement in the direction of this force. Note that the strain energy density U_o of Equation 3.55 may be generally expressed in terms of the strain invariants J_1^F, J_2^F, and J_3^F as in linear infinitesimal elasticity. The stresses are derived from U_o using expressions similar to Equations 3.32, namely,

$$\sigma_1 = \frac{1}{\lambda_2\lambda_3}\frac{\partial U_o}{\partial \lambda_1}, \quad \sigma_2 = \frac{1}{\lambda_1\lambda_3}\frac{\partial U_o}{\partial \lambda_2}$$

and

$$\sigma_3 = \frac{1}{\lambda_1\lambda_2}\frac{\partial U_o}{\partial \lambda_3} \tag{3.56}$$

For rubberlike materials, it is accurate enough to assume that deformation occurs at constant volume ($\nu = 0.46 - 0.5$ for rubber). Hence, for the unit cube,

$$\lambda_1\lambda_2\lambda_3 = 1$$

and the strain invariants become

$$J_1^F = \lambda_1^2 + \lambda_2^2 + \lambda_3^2$$

$$J_2^F = \frac{1}{\lambda_1^2} + \frac{1}{\lambda_2^2} + \frac{1}{\lambda_3^2} \tag{3.57}$$

$$J_3^F = 1 \text{ (incompressible material)}$$

This restriction of volume constancy alters the form of Expressions 3.56 to

$$\sigma_1 = \lambda_1\frac{\partial U_o}{\partial \lambda_1} + \sigma_p, \ldots\ldots \tag{3.58}$$

where σ_p is an arbitrary hydrostatic pressure.[6] Obviously, a hydrostatic stress acting on an incompressible material does no work, and hence energy consideration leaves σ_p arbitrary.

Several forms for the strain energy density functions are found in literature as reasonable representations for the behavior of rubberlike materials.* One of these forms is given by[7]

$$U_o = \frac{A_1}{2}\left(J_1^F - 3\right) + \frac{A_2}{2}\left(J_2^F - 3\right) \tag{3.59}$$

where A_1 and A_2 are constants and J_1^F and J_2^F are given by Equations 3.57. Applying Equations 3.58 to the function of Equation 3.59 according to

$$\sigma_1 = \frac{1}{\lambda_2\lambda_3}\left(\frac{\partial U_o}{\partial J_1^F}\frac{\partial J_1^F}{\partial \lambda_1} + \frac{\partial U_o}{\partial J_2^F}\frac{\partial J_2^F}{\partial \lambda_1}\right)\ldots$$

gives the stress–strain relations as

* Equation 3.52 is a description for a Mooney–Rivlin rubberlike material. If $A_2 = 0$, the equation describes a Neo-Hookian rubberlike material.

$$\sigma_1 = A_1\lambda_1^2 - \frac{A_2}{\lambda_1^2} + \sigma_p$$

$$\sigma_2 = A_1\lambda_2^2 - \frac{A_2}{\lambda_2^2} + \sigma_p \tag{3.60}$$

$$\sigma_3 = A_1\lambda_3^2 - \frac{A_2}{\lambda_3^2} + \sigma_p$$

The constants A_1 and A_2 are considered as material parameters and are determined experimentally.

Example 3.19:

For an acrylic sheet material, deforming elastically above its glass transition temperature, $A_2 \cong 0$ *in Expressions 3.59, plot the uniaxial tensile stress vs. the extension ratio and compare with linear Hooke's law.*

Solution:

For simple tension:

$$\sigma_1 = \sigma \text{ and } \sigma_2 = \sigma_3 = 0$$

Also for isotropic constant volume deformation,

$$J_3^F = \lambda_1\lambda_2\lambda_3 \text{ and } \lambda_2 = \lambda_3$$

Hence,

$$\lambda_1 = \lambda \text{ and } \lambda_2 = \lambda_3 = 1/\sqrt{\lambda}$$

Substitution into the stress–strain relations (Equations 3.60) gives

$$\sigma_1 = \sigma = A_1\lambda^2 + \sigma_p$$

$$\sigma_2 = \sigma_3 = 0 = (A_1/\lambda) + \sigma_p$$

Hence, $\sigma_p = -A_1/\lambda$ and

$$\sigma = A_1\lambda^2(1 - 1/\lambda^3) \tag{a}$$

According to Equation 2.40a, λ is expressed in terms of the logarithmic strain ε^p as $\lambda = \exp(\varepsilon^p)$. Hence, Equation a becomes

$$\sigma = A_1[\exp(2\varepsilon^p) - \exp(-\varepsilon^p)] \tag{b}$$

which is plotted in Figure 3.16 for both tension and compression.

However expressing λ in terms of the engineering strain ε according to $\lambda = (1 + \varepsilon)$ and neglecting powers of ε higher than unity for infinitesimal strains gives

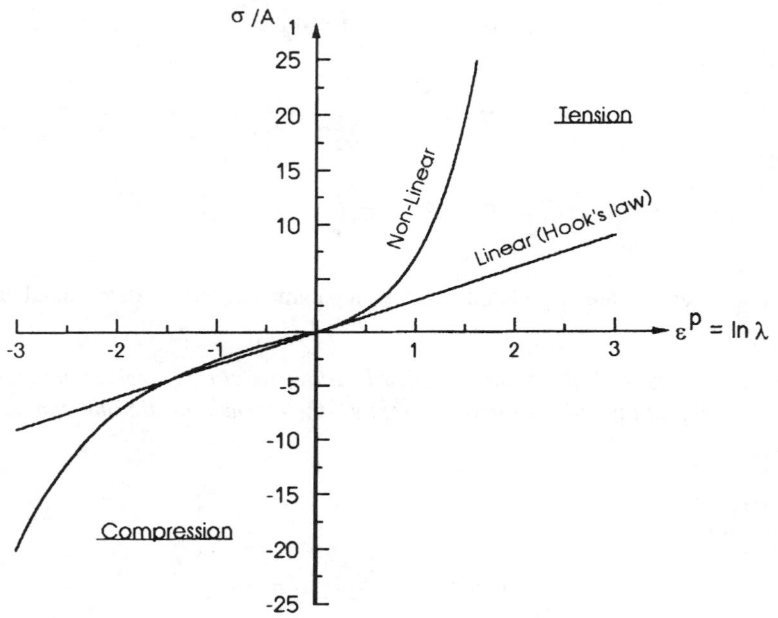

FIGURE 3.16 Example 3.19.

$$\sigma = 3A_1\varepsilon \tag{c}$$

Comparing this with Hooke's law in simple tension shows that $A_1 = E/3$ or, alternatively, $A_1 = G$ (for $\nu = 0.5$). For uniaxial compression where $\lambda < 1$ and $\varepsilon^p < 0$, Equations a, b, and c remain valid.

A plot of Equation c in Figure 3.16 shows that the error due to use of the linear relation (Equation b) is pronounced in both tension and large strain compression. This figure also indicates that Hooke's law gives a good approximation over the range, say, $-1 < \varepsilon^p < 0.5$. At higher strain values a nonlinear stress–strain relations as given by Equation 3.60 must be used.

PROBLEMS

Hooke's Law

3.1 In a flat, thin steel plate that is loaded in the x–y plane, it is known that $\sigma_x = 150$ MPa, $\tau_{xy} = 50$ MPa, and $\varepsilon_z = -0.0004$. What is the value of γ_{max}? Take $E = 200$ GPa and $\nu = 0.33$.

3.2 A bar of square cross section 100 × 100 mm parallel to the x- and y-axes has length $L = 1000$ mm and is made of an isotropic steel ($E = 200$ GPa and $\nu = 0.29$). The bar is subjected to a uniform state of stress. Considering a state of plane strain, determine the final dimensions of the bar assuming $\sigma_x = \sigma_y = 100$ MPa while all shear stresses vanish.

3.3 A member is made of an isotropic linearly elastic aluminium alloy ($E = 72$ GPa and $\nu = 0.33$). Consider a point on the free surface that is tangent to the x–y plane. If $\sigma_x = 110$ MPa, $\sigma_y = -40$ MPa, and $\tau_{xy} = -90$ MPa, determine the directions for strain gauges at this point to measure two of the principal strains. What are the magnitudes of these principal strains?

3.4 A member made of isotropic bronze ($E = 140$ GPa and $\nu = 0.33$) is subjected to a state of plane strain ($\varepsilon_z = \varepsilon_{zx} = \varepsilon_{zy} = 0$). Determine σ_z, ε_x, ε_y, and γ_{xy}, if $\sigma_x = 100$ MPa, $\sigma_y = -60$ MPa, and $\tau_{xy} = 50$ MPa.

3.5 A strain gauge is placed on a free surface making an angle α with the first principal direction. Obtain the value of α in order that the gauge reading = $A\sigma_1$, where A is a material constant and σ_1 is a principal stress.

3.6 A plate is subjected to $\sigma_x = -60$ MPa, $\sigma_y = 50$ MPa, and $\tau_{xy} = 35$ MPa. If the plate is subjected to a uniform temperature of 75°C, find the principal strains taking $E = 200$ GPa, $\nu = 0.3$, and $\alpha = 12 \times 10^{-6}$/°C.

3.7 A steel plate 150 × 250 mm and 8 mm thick is subjected to uniformly distributed loads along its edges.

a. If $P_x = 100$ kN and $P_y = 200$ kN, what change in thickness occurs due to the application of these loads?
b. To cause the same change in thickness as in (a) by P_x alone, what must be its magnitude? Let $E = 200$ GPa and $\nu = 0.3$.

3.8 A rectangular steel block has the following dimensions 50, 75, and 100 mm in the x-, y-, and z-directions, respectively. The faces of this block are subjected to uniformly distributed forces $P_x = 180$ kN, $P_y = 200$ kN, and $P_z = -240$ kN. Determine the magnitude of a single force acting only in the y-direction that would cause the same deformation in the y-direction as the initial forces. Let E = 200 GPa and $\nu = 0.25$.

3.9 In a steel member, normal strains of 650×10^{-6}, -420×10^{-6}, and 350×10^{-6} are measured in directions of 0°, 60°, and 120° anticlockwise with respect to the horizontal direction. Find the strains and the magnitude and direction of principal stresses knowing that these measurements are made at –50°C. Take $E = 207$ GPa, $\nu = 0.3$, and $\alpha = 12 \times 10^{-6}$/°C.

3.10 A long, thick-wall cylinder is subjected to an internal pressure of 120 MPa and a temperature rise of 200°C. The stresses at the inner surface are $\sigma_r = -120$ MPa and $\sigma_\theta = -350$ MPa. If the bore diameter is 150 mm and the length is 1000 mm, determine the change in bore diameter. Take $E = 210$ GPa, $\nu = 0.26$, and $\alpha = 1.25 \times 10^{-5}$/°C.

3.11 A strain gauge rosette is bonded to the surface of a 20-mm-thick plate. For the three gauges a, b, and c, gauge b is at 60° to gauge a and gauge c is at 120° to gauge a, both measured in the same direction. When the plate is loaded, the strain values from a, b, and c are 1.5×10^{-4}, 7×10^{-4} and 5×10^{-4}, respectively. Find the principal stresses and the change in plate thickness at the rosette point. Take $E = 200$ GPa and $\nu = 0.3$.

3.12 A rosette of three strain gauge elements spaced 120° apart is bonded to the web of a loaded beam. The elements record tensile strains of 2.75×10^{-4}, 5×10^{-4}, and 1.25×10^{-4}. Determine the magnitude of the principal strains and stresses and the direction of the maximum principal stress relative to the direction of the maximum strain reading. Take $E = 207$ GPa and $\nu = 0.28$.

3.13 A certain rubbery material is considered to be linear elastic with a Young's modulus of 0.05 GPa and a Poisson's ratio of 0.5. A thin sheet of the material is marked in the undeformed condition with three separate lines making the angles 0°, 30°, and 45°, respectively, relative to the horizontal x-axis. The lines are initially of unit length. Determine the lengths of these lines and the angle between them when an equibiaxial stress of 20 MPa is applied in the plane of the sheet in the x- and y-directions.

3.14 Consider an aluminum rectangular parallelapiped of a finite size whose sides have the initial lengths 450, 225, and 150 mm along x, y, and z, respectively. The faces of the parallelepiped are subjected to the stresses $\sigma_x = 115$ MPa, $\sigma_y = 0$, $\sigma_z = 68$ MPa, $\sigma_{xy} = 27$ MPa, $\tau_{yz} = -13.6$ MPa, and $\tau_{zx} = 0$. Determine the increase in length of its diagonal. Take $E = 75$ GPa and $\nu = 0.3$.

3.15 A circle of radius 20 mm is inscribed on a rectangular aluminum plate of dimensions 300 × 150 mm. The plate is subjected to the stresses $\sigma_x = -80$ MPa, $\sigma_y = 125$ MPa, and $\tau_{xy} = -30$ MPa. Determine the lengths and directions of the semiaxes of the ellipse formed from the circle after deformation. What is the change in area of a unit square whose sides are parallel to the x- and y-coordinates? Take $E = 70$ GPa and $\nu = 0.3$.

3.16 Determine the decrease in the volume of a solid steel sphere of 350 mm diameter submitted to a uniform hydrostatic pressure of 120 MPa. Take $E = 200$ GPa and $\nu = 0.3$.

3.17 An element is subjected to a stress σ_z. The element is free in the x-direction and restricted in the y-direction. Show that the apparent Young's modulus is given by $E' = E/(1 - \nu^2)$.

3.18 A flat steel plate 200 × 400 × 20 mm is compressed by forces in the plane of the plate so that the new lateral dimensions are 199.98 × 399.975 mm. Assuming that the plate is free in the thickness direction and that it is uniformly stressed (take $E = 200$ GPa and $\nu = 0.3$), determine

a. The change in thickness;
b. If the plate thickness was constrained to remain constant, what stress would be applied in the thickness direction?

Strain Energy for Elastic Solids

3.19 A hypothetical material obeys the following stress–strain relations in two dimensions:

$$\varepsilon_x = \frac{\sigma_x^2}{A} - \nu\frac{\sigma_y}{B}, \quad \varepsilon_y = \frac{\sigma_y^2}{A} - \nu\frac{\sigma_x}{B}, \quad \text{and } \gamma_{xy} = \frac{\tau_{xy}^2}{C}$$

where A, B, and C are elastic constants. Derive an expression for the strain energy density U_o and the complementary energy density C_o in a two-dimensional state of stress.

3.20 The stress–strain relations for a certain two-dimensional anisotropic body are

$$\sigma_x = A_1\varepsilon_x + A_2\varepsilon_y, \quad \sigma_y = (A_2 + A_3)\varepsilon_y; \quad \text{and} \quad \tau_{xy} = A_4\gamma_{xy}$$

where A_1, A_2, A_3, and A_4 are material constants. Derive expressions for the strain energy density U_o in terms of stress and the complementary strain energy density C_o in terms of strain for this material.

3.21 Derive the stress–strain relations for a material having the strain energy density expressed by

$$U_o = A_1(J_1 - 3) + A_2(J_2 - 3) + \sigma_p\varepsilon_v$$

where $J_1 = 3 + 2(\varepsilon_x + \varepsilon_y + \varepsilon_z) = 3 + 2\varepsilon_v$ and $J_2 = 3 + 4\varepsilon_v + 4(\varepsilon_x\varepsilon_y + \varepsilon_y\varepsilon_z + \varepsilon_z\varepsilon_x) - 4(\gamma_{xy}^2 + \gamma_{yz}^2 + \gamma_{yx}^2)$ and σ_p is an arbitrary hydrostatic pressure.

Strain Energy for a Solid Obeying Hooke's Law

3.22 A sphere of solid aluminum 500 mm in diameter is submerged in the ocean at a depth of 3 km. Considering that the sphere remains elastic, determine the volumetric strain. The density of seawater is about 1030 kg/m^3. Also calculate the strain energy of the sphere taking $E = 70$ GPa and $\nu = 0.3$

3.23 A bar of square cross section 35 × 35 mm is subjected to a longitudinal pull of 120 kN. The sides of the square contract by 0.005 mm and the bar exhibits an elongation of 1.5 mm. If $E = 200$ GPa, determine its Poisson's ratio and initial length. Also determine the strain energy stored in the bar.

3.24 A plate with dimensions 300 × 500 × 3 mm in the x-, y-, and z-directions, respectively, is placed between two rigid lubricated walls such that deformation is restrained in the z-direction. Given that the plate is subjected to the uniform stresses $\sigma_x = -25$ MPa and $\sigma_y = -15$ MPa, determine

a. The pressure exerted on the rigid walls,
b. The change in plate dimensions,

c. The strain energy stored in the plate,
d. The change in dimensions of the plate if it is free to expand in the z-direction, and
e. The strain energy stored in the plate.

Take E = 200 GPa and $\nu = 0.3$.

3.25 Three strain gauges are symmetrically arranged at 120° on the free surface of a machine part. The gauge readings gave strains 0.001, 0.002, and 0.003. Determine magnitude and direction of principal stresses and the strain energy density. Take E = 210 GPa and $\nu = 0.28$.

3.26 A thin-walled gas storage tank 10 m long with outer and inner diameters equal to 2.6 and 2.5 m, respectively, is subjected to an internal pressure of 3 MPa. Determine

a. The maximum normal and maximum shear stresses in the wall of the tank;
b. The normal and shearing stresses in the wall of the tank on a plane inclined to the axis of the tank through a 30° angle;
c. The change in tank length and diameter, knowing that E = 200 GPa and $\nu = 0.3$;
d. The strain energy density using both results of (a) and (b).

3.27 A linear elastic solid is in a state of plane strain in the x–y plane such that $\varepsilon_x = ay$, $\varepsilon_y = by$, and $\varepsilon_{xy} = ax + by$, where a and b are known constants. Determine the elastic strain energy in a cube bounded by the planes $x = y = z = 1$. Take $v = 0.25$.

3.28 Given the strain field, $\varepsilon_x = ay$, $\varepsilon_y = by$, and $\varepsilon_z = by$ and $\gamma_{xy} = \gamma_{yz} = \gamma_{zx} = 0$ for a linear elastic solid, find the stress field. What physical problem does this strain field represent? Determine the strain energy density U_o and the complementary energy density C_o?

Generalized Hooke's Law

3.29 The elastic constants of an anisotropic elastic solid that possesses symmetry about the x–y plane are specified by considering the generalized Hooke's law, Equations 3.45, with 21 independent constants. Show that

$$a_{14} = a_{24} = a_{34} = a_{44} = 0, \quad a_{15}\ (= a_{51}) = a_{25}\ (= a_{52}) = a_{35}\ (= a_{53})\ a_{45} = 0$$

$$\text{and} \quad a_{16} = a_{26} = a_{36} = a_{56} = a_{66} = 0$$

This can be demonstrated by applying the stress and strain transformation for the new set of axes: $x \equiv x'$, $y \equiv y'$, and $z \equiv -z'$ and imposing the condition that the elastic constants are invariant under this transformation.

3.30 A cube made of an anisotropic solid that possesses symmetry around the x–y plane such that its generalized Hooke's law is given by

$$\sigma_x = a_{11}\varepsilon_x + a_{12}\varepsilon_y + a_{13}\varepsilon_z + a_{16}\gamma_{yz},$$

$$\sigma_y = a_{21}\varepsilon_x + a_{22}\varepsilon_y + a_{23}\varepsilon_z + a_{26}\gamma_{yz},$$

$$\sigma_z = a_{31}\varepsilon_x + a_{32}\varepsilon_y + a_{33}\varepsilon_z + a_{36}\gamma_{yz},$$

$$\tau_{xy} = a_{41}\varepsilon_x + a_{42}\varepsilon_y + a_{43}\varepsilon_z + a_{46}\gamma_{yz},$$

$$\tau_{xz} = a_{54}\gamma_{xy} + a_{55}\gamma_{xz}, \quad \text{and}$$

$$\tau_{yz} = a_{64}\gamma_{xy} + a_{65}\gamma_{xz}$$

If the cube is deformed, such that $\varepsilon_x = \varepsilon_1 = \varepsilon$ and $\varepsilon_y = \varepsilon_2 = \varepsilon$, while all other strain components vanish,

a. Determine the components of the stress matrix producing this deformation;
b. Find the principal stresses and their directions;
c. Determine the angle between the principal directions of stress and those of the principal strains.

3.31 A closed pressure vessel of mean radius r_m, length L, and wall thickness h, such that $L/r_m = 3$ and $r_m/t = 10$, is fabricated from a composite material. The reinforcing fibers are aligned along the vessel circumference axis. The composite material elastic constants of Equation 3.48, are related by $a_{11} = A$, $a_{22} = a_{33} = 0.2A$, $a_{13} = a_{12} = a_{23} = 0.1A$, and $a_{44} = a_{55} = a_{66} = 0.05A$, where A is a constant. The vessel is subjected to an internal pressure p such that its dimensional changes do not exceed 0.2%. Determine the maximum value of this internal pressure in terms of the constant A.

3.32 A filament-wound cylindrical pressure vessel of mean diameter 1.5 m and wall thickness 30 mm is subjected to an internal pressure of 3 MPa. The filament winding angle is 60° to the vessel longitudinal axis. The material is glass/epoxy having $E_1 = 40$ GPa, $E_2 = 8$ GPa, $G_{12} = 4$, and $\nu_{12} = 0.26$. Determine the stresses and strains along and normal to the fiber direction.

3.33 Solve the above problem if the pressure vessel is manufactured of two filaments wound at 60° and –60° to the longitudinal axis. The vessel has the same mean diameter and thickness. What internal pressure can the vessel sustain at the stress values of the above case?

Stress–Strain Relations for Large Elastic Deformations

3.34 For a sheet of a rubbery polymeric material, the strain energy density is given by

$$U_o = \frac{A_1}{2}\left(J_1^F - 3\right) + \frac{A_2}{2}\left(J_2^F - 3\right)$$

where A_1 and A_2 are constants. Plot the uniaxial stress σ vs. the engineering strain ε in both of tension and compression. Compare these plots with linear Hooke's law for the cases $A_2 = 0$ and $A_2 = 0.25\,A_1$.

REFERENCES

1. Sokolnikoff, I. S., *Mathematical Theory of Elasticity*, 2nd ed., McGraw-Hill, New York, 1956.
2. Tauchert, T. R. and Guzelsu, A. N., *J. Appl. Mech.*, 39, 1972.
3. Jones, R. M., *Mechanics of Composite Materials*, McGraw-Hill, New York, 1975.
4. *Metal-Matrix Composites*, UNIDO, MONITOR, No. 17, Feb. 1990.
5. Rivlin, R. S., *Philos. Trans. R. Soc.*, A240, 57, 1948; A241, 379, 1948.
6. Green, A. E. and Zerna, M., *Theoretical Elasticity*, 2nd ed., Oxford University Press, New York, 1968.
7. Ragab, A. R. and Khorshied, S. A., *Plas. Rubber Proc. App.*, 6, 21–27, 1966.

4 Solution of the Elastic Problem

The boundary value problem of elastically deforming solids is defined. It is shown that the equations of equilibrium and strain compatibility together with the stress–strain relations and boundary conditions provide a complete solution of the problem. Simplifying the exact boundary conditions based on Saint-Venant's principle is demonstrated. The plane elastic problem is defined and the solution is formulated in terms of a stress function in Cartesian and polar coordinates taking into consideration thermoelastic strain and body forces. A glossary of stress functions for some engineering problems is included.

4.1 THE ELASTIC PROBLEM

The subject of mechanics of deformable solids is concerned with finding stresses, strains, and displacements within solids due to externally prescribed loads (or displacements). If deformations are infinitesimal and material behavior is linear elastic, the problem is a classical elastic problem. Complete solutions to these problems must satisfy sets of basic equations, namely; equations of equilibrium, equations of compatibility, and elastic stress–strain relations (Hooke's law).

The stresses and strains provided by the solution must satisfy the external loads and displacements prescribed on the boundary surfaces of the deformed solid. In other words, the solution to the elastic problem must satisfy the boundary conditions and the problem is generally described as an "elastic boundary-value problem." To find a solution to the general three-dimensional elastic boundary-value problem, the number of available independent equations has to be equal to the number of unknowns, as follows:

	Unknowns	
	3	Displacement components (u, v, w)
	6	Stress components (σ_x, σ_y, σ_z, τ_{xy}, τ_{yz}, τ_{zx})
	6	Strain components (ε_x, ε_y, ε_z, γ_{xy}, γ_{yz}, γ_{zx})
Total	15	Unknowns

In order to solve for the above-mentioned 15 unknown quantities, 15 independent equations are required. They are

	Equations:	
	3	Stress equations of equilibrium
	6	Strain–displacement relations
	6	Stress–strain relations (Hooke's law)
Total	15	Equations

Hence, it is possible to obtain a solution to the general elasticity problem provided that the body force distribution and boundary conditions are known.

The 15 equations can be reduced by appropriate substitutions to a set of three differential equations in the three unknown displacements.* They can be then solved in terms of displacements for given boundary conditions. In such a case, the strain–compatibility equations need not be considered since they are implicitly satisfied.

Alternatively, it is possible to solve the problem in terms of stress components. The six strain-compatibility equations must be included in the analysis, thus raising the number of equations to 21. These equations are not, however, independent. They can be reduced to six partial differential equations in terms of the stresses.** Their solution must satisfy the boundary conditions for stresses.

Analytical solutions to three-dimensional elasticity problems are generally quite difficult to obtain as they require the solution of partial differential equations. The number of problems that must be solved in an exact fashion is surprisingly small.

4.2 BOUNDARY CONDITIONS

Three types of boundary conditions are encountered in mechanics of solids boundary-value problems. In the first type, displacements are prescribed over the entire boundary of the body. In the second type, loads (forces, moments, stresses) are prescribed over the entire boundary of the body. In the third type, displacements are prescribed over a part of the boundary, while loads are prescribed over the rest of the boundary. This is known as a mixed boundary-value problem.

The boundary conditions cannot be arbitrarily prescribed and have to be carefully specified. The loads should satisfy the conditions of equilibrium, while the displacements have to be continuous.

For the first type of boundary condition, if a point on the surface is required to undergo a given displacement whose components in the x-, y-, and z-directions are, respectively, u_s, v_s, and w_s, the displacement solution u, v, and w must satisfy at that point the conditions:

$$u = u_s,\ v = v_s,\ \text{and}\ w = w_s$$

For the second type of boundary-value problems, the stresses furnished by the solution must produce tractions that are in equilibrium with the loads being applied to the boundary of the body. Thus, if at a boundary point of the body X_s, Y_s, and Z_s are surface forces per unit area (tractions) in the directions x, y, and z, equilibrium necessitates that Equation 1.13a must be satisfied, i.e.,

$$X_s = \sigma_x l_s + \tau_{yx} m_s + \tau_{zx} n_s$$

$$Y_s = \tau_{xy} l_s + \sigma_y m_s + \tau_{zy} n_s$$

$$Z_s = \tau_{xz} l_s + \tau_{yz} m_s + \sigma_z n_s$$

where l_s, m_s, and n_s are the direction cosines of the normal to the boundary surface at this point. The direction pointing outward perpendicular to the surface is considered positive.

Difficulties may arise in modeling the problem, i.e., translating the physical problem to a mathematical formulation. For engineering purposes, approximations are often made as a compromise with a difficult-to-obtain formal solution. Often, these approximations are concerned with boundary conditions.

Example 4.1:

Write the boundary conditions for the following problems:

---

* These are known as Navier equations. Named after C. L. M. H. Navier (1785–1836).

** These are known as Beltrami–Michell stress–compatibility equations; see Venkatraman and Patel.[1]

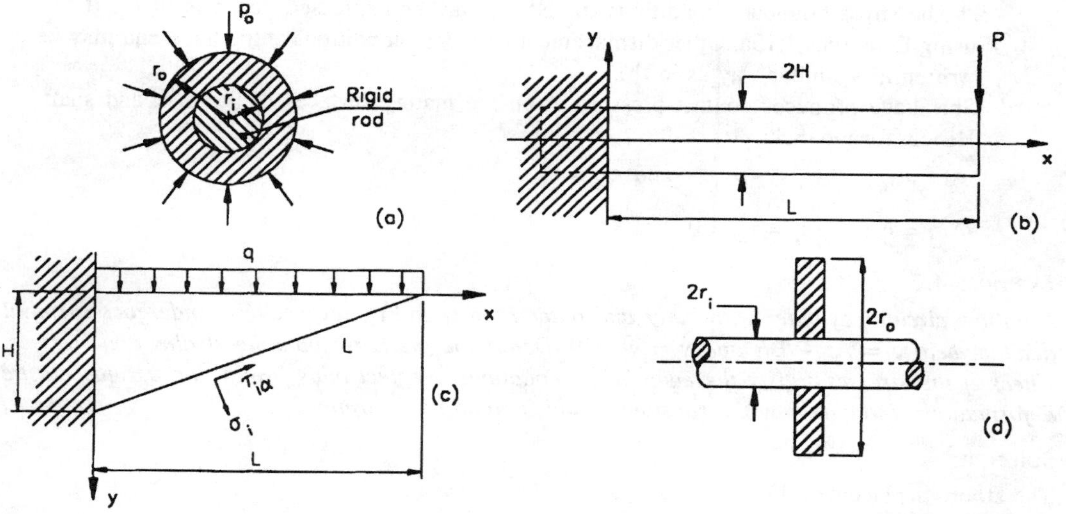

FIGURE 4.1 Example 4.1.

a. *Thick-walled cylinder — Figure 4.1a;*
b. *Cantilever beam of unit width — Figure 4.1b;*
c. *Tapered cantilever wedge of unit width — Figure 4.1c;*
d. *Rotating disk shrunk on a shaft — Figure 4.1d.*

Solution:

a. At $r = r_i$: $u = 0$
   At $r = r_o$: $\sigma_r = -p_o$
   i.e., mixed boundary conditions.
b. At $y = \pm H$: $\sigma_y = \tau_{yx} = 0$

   At $x = y = 0$: $u = v = 0$ and $\dfrac{\partial v}{\partial x} = 0$

   also

$$\int_{-H}^{H} \tau_{xy} dy = -P \text{ and } \int_{-H}^{H} \sigma_x dy \equiv PL$$

   At $x = L$: $\int_{-H}^{H} \tau_{xy} dy = -P$ and $\sigma_x = 0$

   Again these are mixed boundary conditions.

   Note that $\partial v/\partial x = 0$ at $x = y = 0$ is a condition on the slope of the centroidal axis of the beam to satisfy the cantilever fixed end. The concentrated end load P is mathematically difficult to describe and a statically equivalent and simple description is to consider that, at $x = L$, the load P equals the resultant shear stresses.
c. At $y = 0$: $\sigma_y = -q$ and $\tau_{yx} = 0$

   Along $y = H - H/L\,x$: $\sigma_i = 0$ and $\tau_{i\alpha} = 0$
   where the direction pointing outward of the surface is considered positive.

The stress boundary conditions σ_i and $\tau_{i\alpha}$ may be expressed in terms of σ_x, σ_y, τ_{xy} using Equations 1.15a. Other displacement boundary conditions at the fixed end may be written in a similar way as in (b).

d. Shrinkage produces contact pressure p_c on the mating surfaces of the disk and shaft. Hence, for the disk:

At $r = r_i$: $\quad \sigma_r = -p_c \quad$ and $\tau_{r\theta} = 0$

At $r = r_o$: $\quad \sigma_r = \tau_{r\theta} = 0$

Example 4.2:

A hollow circular cylinder with inner and outer radii r_i *and* r_o, *respectively, undergoes a radial displacement* u = Ar + B/r *and* v = w = 0. *Determine the corresponding strains and stresses. Check if the stresses satisfy the equilibrium equations. Neglect body forces and assume elastic deformations. Indicate how the constant* A *and* B *could be determined.*

Solution:

The strain-displacement Equations 2.9 give

$$\varepsilon_r = \frac{\partial u}{\partial r} = A - \frac{B}{r^2}, \quad \varepsilon_\theta = A + \frac{B}{r^2}, \quad \text{and } \varepsilon_z = 0$$

while all shear strain components vanish.

Note that these strains implicitly satisfy the compatibility condition Equation 2.13c, since strains are derived from given displacement fields.

The stresses are obtained from Hooke's law, Equations 3.6, as

$$\begin{aligned}
\varepsilon_r &= A - \frac{B}{r^2} = \frac{1}{E}\left[\sigma_r - \nu(\sigma_\theta + \sigma_z)\right] \\
\varepsilon_\theta &= A + \frac{B}{r^2} = \frac{1}{E}\left[\sigma_\theta - \nu(\sigma_r + \sigma_z)\right] \\
\varepsilon_z &= 0 = \frac{1}{E}\left[\sigma_z - \nu(\sigma_r + \sigma_\theta)\right]
\end{aligned} \tag{a}$$

All shear stresses vanish.

Solving for σ_r, σ_θ, and σ_z gives

$$\begin{aligned}
\sigma_r &= \frac{E}{(1+\nu)}\left(A - \frac{B}{r^2}\right) + \frac{E\nu}{3(1+\nu)(1-2\nu)}(2A) \\
\sigma_\theta &= \frac{E}{(1+\nu)}\left(A + \frac{B}{r^2}\right) + \frac{E\nu}{3(1+\nu)(1-2\nu)}(2A) \\
\sigma_z &= \frac{E}{3(1+\nu)(1-2\nu)}(2A)
\end{aligned} \tag{b}$$

Note that the above stresses could have been obtained directly from the inverse form of Hooke's law, Equations 3.12. For this stress field, the only equilibrium equation (Equation 1.10c)

$$\frac{d\sigma_r}{dr} + \frac{\sigma_r - \sigma_\theta}{r} = 0$$

is satisfied identically.

To determine the constants A and B, two further conditions are required. These are the boundary conditions, which may be expressed in displacement or stress form. For instance, if the cylinder is subjected to an internal pressure p_i, then

$$\text{At } r = r_i: \quad \sigma_r = -p_i$$

$$\text{At } r = r_o: \quad \sigma_r = 0$$

Imposing these two conditions on the expression of σ_r gives, after algebraic manipulations,

$$A = \frac{(1+\nu)(1-2\nu)}{E}\left(\frac{r_i^2}{r_o^2 - r_i^2}\right)p_i$$

$$B = \frac{(1+\nu)}{E}\left(\frac{r_i^2 r_o^2}{r_o^2 - r_i^2}\right)p_i \qquad \text{(c)}$$

Hence, Expressions a, b, and c define strains and stress completely.

4.3 SAINT-VENANT'S PRINCIPLE*

Generally, the formulation and solution of an elasticity boundary-value problem may become difficult because of complications in describing the boundary conditions. The task is often rendered more simple if, for instance, the surface loads are replaced by an alternative statically equivalent system. Obviously, the replacement has to be in terms of a simple system of loads rather than the exact distribution (often unknown) of the original boundary tractions. The two systems should have the same resultants.

Although such simplification in boundary conditions will make the problem mathematically tractable, it does not provide an exact solution at the boundary. However, this is always acceptable and sufficiently accurate for most engineering applications.

The first to propose such modification in boundary conditions was Saint-Venant in his principle stated as

> *Statically equivalent systems of surface tractions applied to a given small portion of the surface enclosing an elastic body, are elastically equivalent except in the immediate neighourhood of the region where these surface tractions are applied.*

Elastic equivalence means that the two systems of surface tractions produce the same stress–strain displacement distributions within the body.

To illustrate this principle, consider a simple compression test where a rod of unit thickness is subjected to a compressive load P, as in Figure 4.2. The distribution of P over the rod ends can take infinite forms of resultants P passing through the rod axis. If a uniform distribution of an intensity p per unit width as in Figure 4.2a is applied, the stress distribution will be uniform at any section all over the rod. This stress is given by $\sigma = pb/A$, where A is the rod cross-sectional area,

* Named after B. de Saint-Venant (1797–1886).

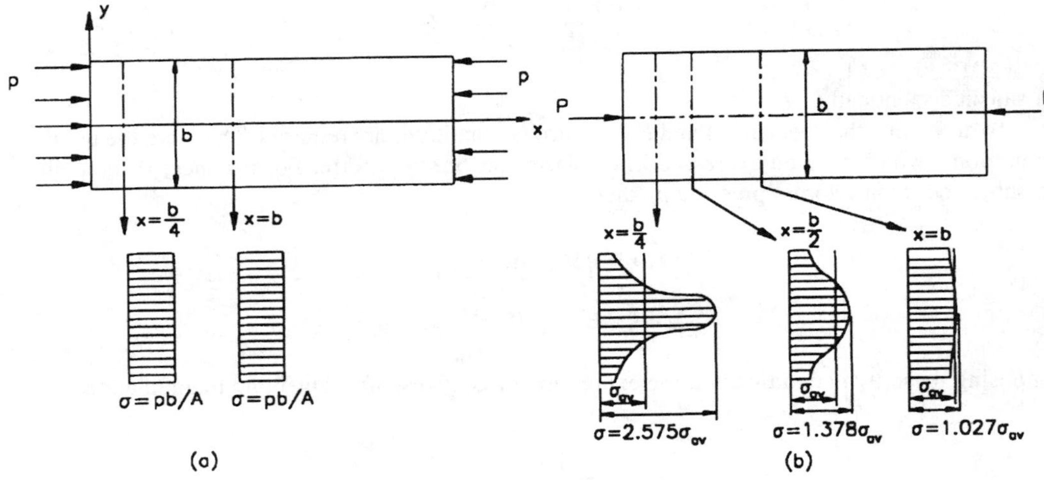

FIGURE 4.2 Saint-Venant's principle applied to a strip in axial compression: (a) uniform end pressure and (b) concentrated axial load.

Figure 4.2a. If instead, a statically equivalent concentrated load $P = pb$ acts on the rod, the stress distributions acting at various sections $x = b/4$, $x = b/2$, and $x = b$ will be as shown in Figure 4.2b.* Notably at a distance almost equal to the rod width b, the stress distribution becomes almost uniform as in the case of a uniformly distributed external compressive load. Hence, the effect of a specific distribution of the surface traction dies out at distances far enough from the loading surface.

Such illustration should not be considered as a proof to Saint-Venant's principle. In fact, it has been argued that this principle is not too convincing in several cases by many authors.[2] The principle is also inherently vague about the extent of the region within which local effects have to be considered. Nevertheless, application of Saint-Venant's principle to most engineering elasticity problems provides practically accepted solutions.

4.4 UNIQUENESS AND SEMI-INVERSE METHOD OF ELASTIC SOLUTION

On the premise that a solution to the elasticity boundary value problem can be found,** ambiguity will be avoided if it can be shown that this solution is unique. Although such a proof is beyond the present text, the positive definiteness of elastic strain energy density function, as noted in Section 3.9, is the basis of this proof for infinitesimal strains in linear elasticity.**

Uniqueness simply means that if any solution to the elastic boundary-value problem is found, it will be the only solution that exists for this problem. Many solutions to difficult problems benefit from this fact by simply applying intuition and assumptions as regard the form of the solution. This is often known as the semi-inverse method, where a general form of the solution in terms of stresses, strains, or displacements is assumed and yet leaves enough freedom in the quantities involved to satisfy the equations of equilibrium, compatibility, and boundary conditions. Finally, uniqueness assures that the obtained solution is the only one to exist.

---

* This problem is solved in Section 5.3.3 using the stress function approach.

** A general proof for the existence of a solution is based on the theory of integral equations (see Sokolnikoff[2]). A proof of the theorem of uniqueness is due to G. Kirchhoff (1824–1887).

4.5 EXAMPLE OF SOLUTION IN TERMS OF STRESS: PRESSURIZED THICK-WALLED SPHERE

A thick-walled sphere subjected to uniform internal pressure is chosen to demonstrate a method of solution in terms of stress.

Consider a thick-walled sphere whose inside and outside radii are r_i and r_o, respectively, subjected to a uniform internal pressure p_i. It is required to determine the stresses and displacements at any point within the sphere wall.

This is a boundary-value problem of the second type, where the boundary conditions are prescribed in terms of stresses as

$$\text{At} \quad r = r_i: \quad \sigma_r = -p_i$$

and

$$\text{At} \quad r = r_o: \quad \sigma_r = 0 \tag{4.1}$$

Obviously, the use of spherical coordinates r, θ, and ϕ offers an advantageous description to this problem. Moreover, a sphere possesses complete symmetry and variation in stresses, strains, and displacements as functions of only the coordinate r. The only nonvanishing stresses are σ_r and $\sigma_\theta = \sigma_\phi$, while all shear stresses $\tau_{r\theta} = \tau_{\theta\phi} = \tau_{r\phi}$ are zero by virtue of symmetry.

The only Equation of equilibrium that has to be satisfied is the first of Equations 1.11b, namely,

$$\frac{d\sigma_r}{dr} + \frac{2(\sigma_r - \sigma_\phi)}{r} = 0$$

where the body forces are taken zero. All other equilibrium Equations 1.11a are identically satisfied.

The displacement components in directions r, θ, and ϕ are u, v, and w, respectively. Obviously, due to symmetry $v = w = 0$ and u is a function of the coordinate r only. Hence, the strain–displacement relations (Equations 2.10) reduce to

$$\varepsilon_r = \frac{du}{dr} \quad \text{and} \quad \varepsilon_\theta = \varepsilon_\phi = \frac{u}{r}$$

In terms of these strains, the strain–compatibility relations (Equations 2.14) are satisfied if

$$\frac{d\varepsilon_\phi}{dr} = \frac{1}{r}(\varepsilon_r - \varepsilon_\phi)$$

Note that this compatibility condition may be simply derived from the strain–displacement relations, Equation 2.11, as

$$\frac{d\varepsilon_\phi}{dr} = \frac{1}{r}\frac{du}{dr} - \frac{u}{r^2} = \frac{1}{r}(\varepsilon_r - \varepsilon_\phi) \tag{4.2}$$

Hooke's law in terms of the above strains and stresses is given by

$$\varepsilon_r = \frac{1}{E}\left[\left(\sigma_r - 2\nu\sigma_\phi\right)\right]$$

$$\varepsilon_\phi = \frac{1}{E}\left[(1-\nu)\sigma_\phi - \nu\sigma_r\right] \tag{4.3}$$

The above equations, involving the unknowns σ_r, σ_ϕ, ε_r, ε_ϕ, and u, are sufficient to solve the problem provided that the boundary conditions (Equations 4.1) are satisfied.

Combining Equations 4.2 and 4.3 yields

$$(1-\nu)\frac{d\sigma_\phi}{dr} - \nu\frac{d\sigma_r}{dr} + (1+\nu)\frac{\sigma_\phi - \sigma_r}{r} = 0$$

This equation may be then combined with the equilibrium equation (Equations 1.11b) to eliminate σ_ϕ and, thus, obtain

$$\frac{d^2\sigma_r}{dr^2} + \frac{4}{r}\frac{d\sigma_r}{dr} = 0$$

Rearranging the above equation results in

$$\frac{d}{dr}\left[\frac{1}{r^2}\left(\frac{d}{dr}r^3\sigma_r\right)\right] = 0$$

which can be solved by successive integration to give

$$\sigma_r = \frac{A}{3} + \frac{B}{r^3}$$

The constants A and B are easily determined from the two stress boundary conditions (Equations 4.1). The solution is thus obtained after algebraic manipulation as

$$\sigma_r = -\frac{r_o^3/r^3 - 1}{r_o^3/r_i^3 - 1}p_i$$

$$\sigma_\phi = \sigma_\theta = \frac{r_o^3/2r^3 + 1}{r_o^3/r_i^3 - 1}p_i \tag{4.4}$$

$$u = \frac{(1-2\nu)r + (1+\nu)\left(r_o^3/2r^2\right)}{r_o^3/r_i^3 - 1}\frac{p_i}{E}$$

A plot of these stresses is shown in Figure 4.3 for a spherical shell of $r_o/r_i = 2$.

For a sphere subjected to an internal pressure p_i and an external pressure p_o, the boundary conditions are

At $r = r_i$: $\sigma_r = -p_i$
At $r = r_o$ $\sigma_r = -p_o$

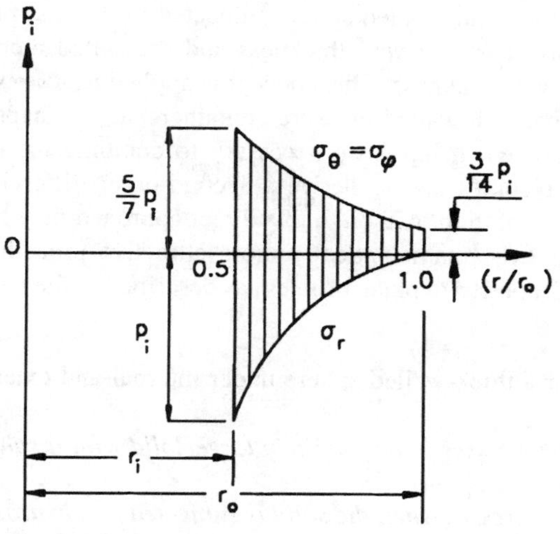

FIGURE 4.3 Stresses in a thick-walled pressurized sphere.

This results in the stresses:

$$\sigma_r = \frac{1}{r_o^3/r_i^3 - 1}\left[p_o\left(\frac{r_o^3}{r^3} - \frac{r_o^3}{r_i^3}\right) + p_i\left(1 - \frac{r_o^3}{r^3}\right)\right]$$

$$\sigma_\phi = \sigma_\theta = \frac{1}{r_o^3/r_i^3 - 1}\left[-p_o\left(\frac{r_o^3}{2r^3} + \frac{r_o^3}{r_i^3}\right) + p_i\left(1 + \frac{r_o^3}{2r^3}\right)\right] \tag{4.5}$$

Consider now the case when the wall thickness is small compared with the sphere radius. Let

r_m = mean radius = $(r_o + r_i)/2$
h = wall thickness = $(r_o - r_i)$

The outer and inner radii may be expressed in terms of r_m and h as

$$r_o = r_m\left(1 + \frac{h}{2r_m}\right) \quad \text{and} \quad r_i = r_m\left(1 - \frac{h}{2r_m}\right)$$

which, upon substitution in Equations 4.5 noting that $r/r_m \cong 1$ and neglecting second and third orders of $h/2r_m$, yields

$$\sigma_\theta = \sigma_\phi = \frac{r_m}{2h} p_i$$

$$\sigma_r = -\frac{r_o - r}{h} p_i \tag{4.6}$$

$$= -p_i/2 \text{ at } r = r_m$$

Equations 4.6 show that for a thin-walled sphere subjected to internal pressure, σ_θ and σ_ϕ can be considered to be uniform across the wall thickness and are called membrane stresses, while σ_r varies linearly across the wall thickness. This concept is applied as a stress compatibility condition to obtain a solution in other thin-walled pressure containers, as in Chapter 8.

In the problem just solved, it has been necessary to combine the equations of equilibrium, strain compatibility and Hooke's law to derive a second-order differential equation in stresses, which finally furnishes the solution. Although this direct approach may be followed to solve other elasticity problems, there exist mathematically more attractive procedures. One of these is the "stress function approach" applied to plane problem as described in the next sections of this chapter.

Example 4.3:

Using Expressions 4.5 for a thick-walled sphere under internal and external pressures, find

a. *Expressions for the stresses in an infinitely large solid with a spherical cavity subjected to an internal pressure* p_i *only.*
b. *Expressions for the stresses when the solid is subjected to a hydrostatic external pressure* p_o *only. Plot these stresses vs. the cavity radius.*

Solution:

a. For an infinitely large solid with a spherical cavity of radius r_i, $r_o/r_i \to \infty$. Hence, Equations 4.5 for $p_o = 0$ and $r_o/r_i = r_o/r = \infty$ give

$$\sigma_r = p_i \frac{\left(1 - r^3/r_o^3\right)}{\left(1 - r_i^3/r_o^3\right)} \frac{\left(r_o^3/r^3\right)}{\left(r_o^3/r_i^3\right)} = -p_i\, r_i^3/r^3$$

$$\sigma_\phi = \sigma_\theta = p_i \frac{\left(1 + 2r^3/r_o^3\right)}{\left(1 - r_i^3/r_o^3\right)} \frac{\left(r_o^3/2r^3\right)}{r_o^3/r_i^3} = p_i\, r_i^3/2r^3$$

which indicates that the tensile tangential stress σ_θ is half the compressive radial stress for a spherical cavity in an infinite solid.

b. If an external pressure is applied to the solid, Equations 4.5 for $p_i = 0$ and $r_o/r_i = r_o/r = \infty$, give

$$\sigma_r = -\frac{r_i^3/r^3 - 1}{r_i^3/r_o^3 - 1} p_o = p_o\left(r_i^3/r^3 - 1\right)$$

$$\sigma_\phi = \sigma_\theta = \frac{r_i^3/2r^3 + 1}{r_i^3/r_o^3 - 1} p_o = -p_o\left(r_i^3/2r^3 + 1\right)$$

which indicates that there always exist compressive stresses $\sigma_\phi = \sigma_\theta$ higher than p_o at the inside of the sphere with a maximum value of $-3p_o/2$ at the inside radius r_i. Note that if $r_i = 0$, then $\sigma_r = \sigma_\theta = \sigma_\phi = -p_o$ all over the spherical solid. Figure 4.4 shows plots of stresses along the radius for both cases (a) and (b).

4.6 THE ELASTIC PLANE PROBLEM

Many engineering stress analysis problems may be considered as plane problems. This means that the problem is reducible to a two-dimensional one, i.e., the stress or the strain components at every

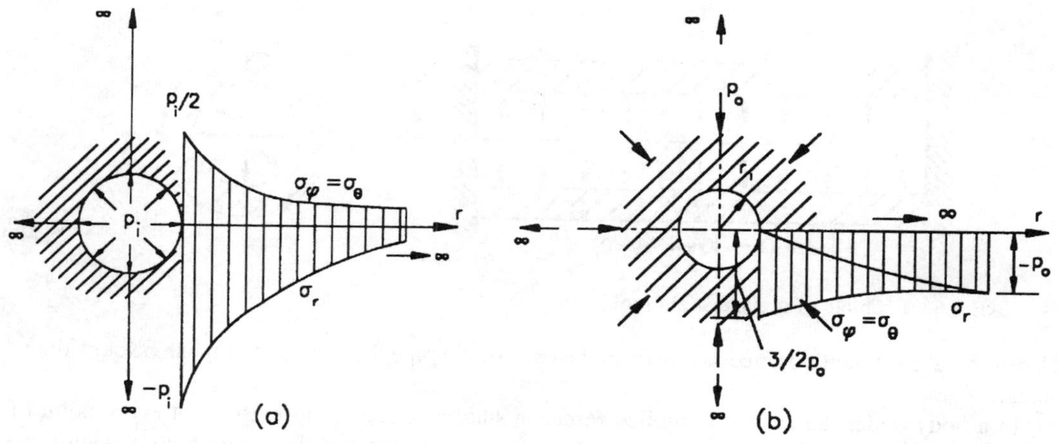

FIGURE 4.4 Example 4.3.

point in the body are functions only of the reference coordinates parallel to that plane. Such problems are encountered in a number of engineering applications, as shown in Figure 4.5. To name a few, a thin circular rotating disk, a long thick-walled cylinder subjected to uniform radial pressure, a retaining wall with lateral pressure, and a cylindrical roll compressed by loads in a diametral plane, may be all analyzed as plane problems.

For the purpose of analysis, the plane elastic problem can be formulated as either "plane stress" or "plane strain." Solution for these two types of problems will be developed in terms of a stress function.

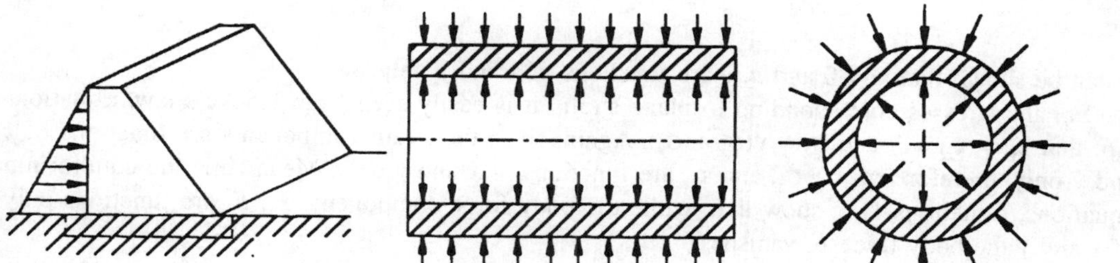

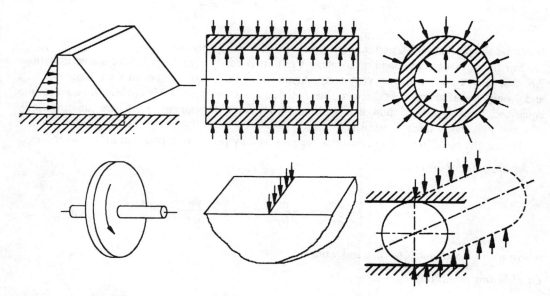

FIGURE 4.5 Examples of plane problems.

4.6.1 Plane Strain Formulation

A long pipe that is axially restrained against deformation at both ends and subjected to uniform internal pressure as shown in Figure 4.6 gives an illustration of a body in a state of plane strain.

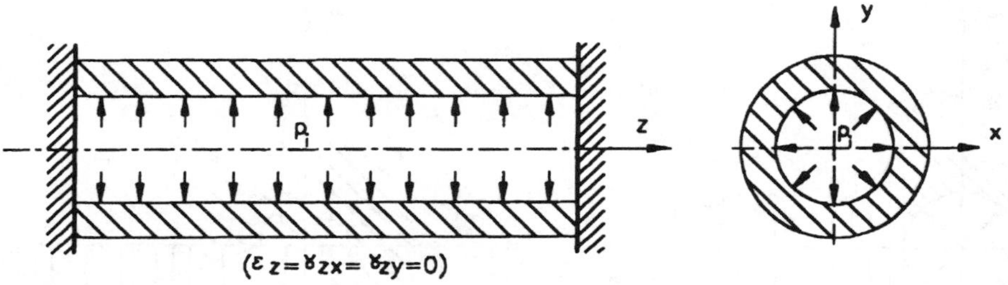

FIGURE 4.6 Example of a plane strain problem.

Therefore, at sufficient distances away from the ends of the pipe the state of strain can be considered planar.

In a body under the action of applied forces, a state of plane strain exists if at every point of the reference plane x–y, the strains ε_z, γ_{yz}, and γ_{zx} are zero and the remaining strain components are functions of x and y only. The strain tensor is thus

$$[\varepsilon] = \begin{bmatrix} \varepsilon_x & \frac{1}{2}\gamma_{yx} & 0 \\ \frac{1}{2}\gamma_{xy} & \varepsilon_y & 0 \\ 0 & 0 & 0 \end{bmatrix}$$

Hence, from the strain–displacement relations (Equations 2.6):

$$\varepsilon_z = \frac{\partial w}{\partial z} = 0, \quad \gamma_{yz} = \frac{\partial v}{\partial z} + \frac{\partial w}{\partial y} = 0, \quad \gamma_{zx} = \frac{\partial w}{\partial x} + \frac{\partial u}{\partial z} = 0$$

It can be shown that $w = 0$ and u, v are functions of x and y only.

For the stresses corresponding to plane strain, it is easily seen from Hooke's law, Equations 3.6, that $\tau_{yz} = \tau_{xz} = 0$ and $\sigma_z = \nu(\sigma_x + \sigma_y)$. Again, since the strain components are functions of x and y only, therefore σ_x, σ_y, σ_z, and τ_{xy} are functions of x and y only. Meanwhile the equilibrium equations, Equations 1. 6, show that, while the body force components F_x, F_y are functions only of x and y the body force F_z vanishes.

Hence, the basic governing equations for an elasticity problem in plane strain are as follows:

Strain-displacement relations

$$\varepsilon_x = \frac{\partial u}{\partial x}, \quad \varepsilon_z = \frac{\partial v}{\partial y}, \quad \gamma_{xy} = \frac{\partial v}{\partial x} + \frac{\partial u}{\partial y}$$

where u and v are functions of x and y only.

Equilibrium equations

$$\frac{\partial \sigma_x}{\partial x} + \frac{\partial \tau_{xy}}{\partial y} + F_x = 0, \quad \frac{\partial \tau_{xy}}{\partial x} + \frac{\partial \sigma_y}{\partial y} + F_y = 0$$

where σ_x, σ_y, τ_{xy}, F_x, and F_y are functions of x and y only.

Hooke's law

$$\varepsilon_z = \frac{1}{E}\left[\sigma_z - \nu\left(\sigma_x + \sigma_y\right)\right] = 0$$

Substituting σ_z in the stress–strain relations for ε_x and ε_y yields

$$\varepsilon_x = \frac{1}{E}\left[\sigma_x - \nu\left(\sigma_y + \sigma_z\right)\right] = \frac{1+\nu}{E}\left[\sigma_x - \nu\left(\sigma_x + \sigma_y\right)\right]$$

$$= \frac{1}{2G}\left[\sigma_x - \nu\left(\sigma_x + \sigma_y\right)\right], \text{ and}$$

$$\varepsilon_y = \frac{1}{2G}\left[\sigma_y - \nu\left(\sigma_x + \sigma_y\right)\right]$$

$$\gamma_{xy} = \tau_{xy}/G$$

Compatibility equations

Since the nonvanishing strains ε_x, ε_y, and γ_{xy} are functions of x and y only, it is readily seen that except for

$$\frac{\partial^2 \varepsilon_x}{\partial y^2} + \frac{\partial^2 \varepsilon_y}{\partial x^2} = \frac{\partial^2 \gamma_{xy}}{\partial x \partial y}$$

the other five compatibility conditions, Equations 2.12, are identically satisfied.

A minor extension of the definition of plane strain can be formulated by requiring that ε_z be a constant instead of being zero with $\gamma_{yz} = \gamma_{zx} = 0$. Therefore, the displacement w becomes a linear function of z. This state of strain may be called "generalized plane strain"[3] and exists in a thick-walled cylindrical tube subjected to uniform radial pressures without axial restraint. This state of generalized plane strain is also encountered in the analysis of rotating long cylinders.

4.6.2 Plane Stress Formulation

A plate loaded by forces in the x–y plane as shown in Figure 4.7a has the stress components σ_z, τ_{xy}, and τ_{zx} equal to zero on the lateral surfaces. It can be assumed with negligible error that for a sufficiently thin plate of Figure 4.7b, where $h/L \to 0$, the stress components σ_z, τ_{yz}, and τ_{zx} are zero throughout. In this case an element of the plate has stresses σ_x, σ_y, and τ_{xy}, as shown in Figure 4.7c, which may be assumed constant throughout the thickness h; that is, the values of σ_x, σ_y, and τ_{xy} are assumed independent of z.

The state of plane stress in a body under the action of applied forces is defined if the stress components σ_z, τ_{yz}, and τ_{xz} are zero and the remaining components σ_x, σ_y, and τ_{xy} are functions of x and y only.

The stress tensor for plane stress conditions is

$$[\sigma] = \begin{bmatrix} \sigma_x & \tau_{yx} & 0 \\ \tau_{xy} & \sigma_y & 0 \\ 0 & 0 & 0 \end{bmatrix}$$

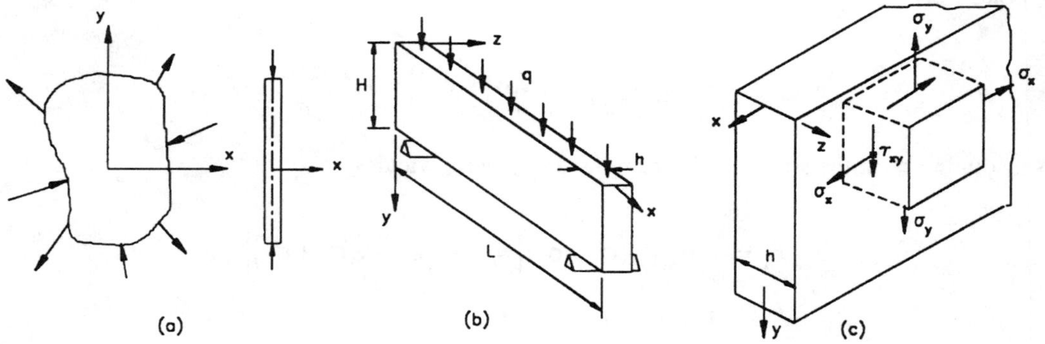

FIGURE 4.7 Example of a plane stress (a) state of plane stress problem, (b) load diagram, and (c) stress components for a laterally loaded beam in plane stress.

The basic governing equations of an elasticity problem in plane stress are as follows:

Equilibrium equations

$$\frac{\partial \sigma_x}{\partial x} + \frac{\partial \tau_{xy}}{\partial y} + F_x = 0$$

$$\frac{\partial \tau_{yx}}{\partial x} + \frac{\partial \sigma_y}{\partial y} + F_y = 0$$

where σ_x, σ_y, τ_{xy}, F_x, and F_y are functions of x and y only.

The third of the stress equilibrium equations (Equation 1.6c) in the z-direction shows that F_z vanishes.

Strain–displacement relations

Since $\tau_{xy} = \tau_{yz} = 0$ by definition, therefore, the shear strains $\gamma_{xz} = \gamma_{yz}$ are also equal to zero. The remaining components of strain are

$$\varepsilon_x = \frac{\partial u}{\partial x}, \quad \varepsilon_y = \frac{\partial v}{\partial y}, \quad \varepsilon_z = \frac{\partial w}{\partial z}, \quad \gamma_{xy} = \frac{\partial v}{\partial x} + \frac{\partial u}{\partial y}$$

where u and v are functions of x and y only.

Hooke's law

$$\varepsilon_x = \frac{1}{E}\left[\sigma_x - \nu\sigma_y\right],$$

$$\varepsilon_y = \frac{1}{E}\left[\sigma_y - \nu\sigma_x\right],$$

$$\varepsilon_z = -\frac{\nu}{E}\left[\sigma_x + \nu\sigma_y\right],$$

$$\gamma_{xy} = \tau_{xy}/G$$

Compatibility equations

Since the stresses σ_x, σ_y, and τ_{xy} are, by definition, functions of x and y only, therefore from Hooke's law ε_x, ε_y, ε_z, and γ_{xy} are also functions of x and y only. The six compatibility equations of strain (Equation 2.12) thus reduce to

$$\frac{\partial^2 \varepsilon_x}{\partial y^2} + \frac{\partial^2 \varepsilon_y}{\partial x^2} = \frac{\partial^2 \gamma_{xy}}{\partial x \partial y}$$

$$\frac{\partial^2 \varepsilon_z}{\partial x^2} = 0, \quad \frac{\partial^2 \varepsilon_z}{\partial y^2} = 0, \quad \frac{\partial^2 \varepsilon_z}{\partial x \partial y} = 0$$

Integration of the last three equations shows that ε_z must be a linear function in x and y. Since $\varepsilon_z = -(\nu/E)(\sigma_x + \sigma_y)$, therefore, $(\sigma_x + \sigma_y)$ is also a linear function of x and y. This condition is too restrictive and is a special case rather than the rule. To remove this restriction and obtain a solution satisfying all compatibility equations, the initial assumption made for plane stress must be modified. One modification is to assume that the stress components σ_x, σ_y, and τ_{xy} are no longer independent of z while σ_z, τ_{xz}, and τ_{yz} still remain zero. This modification yields a solution in which σ_x, σ_y, and τ_{xy} are parabolically distributed through the thickness,* i.e., of the form $f(x, y) + g(x, y)z^2$. As explained before, z is usually very small for bodies that can be analyzed under the plane stress assumption, and hence it can be assumed that the part proportional to z^2 is very small compared with the first term for sufficiently thin plates.

Therefore, the original assumption of plane stress problems, namely, σ_x, σ_y, and τ_{xy} being independent of z, is a good approximation for thin plates subjected to forces that are uniformly distributed over the thickness of its boundary and are parallel to its middle plane as shown in Figure 4.7c. The rotating disk of Figure 4.5 is a practical example of a body in which the stress distribution is very well approximated by plane stress conditions. However, for such a problem a mathematically rigorous analysis could be obtained by working with average values of u, v, σ_x, σ_y, τ_{xy}, F_x, and F_y along the z-direction, which thus defines a state of "generalized plane stress."**

4.6.3 Deduction of Plane Stress Equations from Plane Strain Equations

The stress–strain relations for the plane strain problem have been previously given by

$$\varepsilon_x = \frac{1+\nu}{E}\left[(1-\nu)\sigma_x - \nu\sigma_y\right]$$

$$\varepsilon_y = \frac{1+\nu}{E}\left[(1-\nu)\sigma_y - \nu\sigma_x\right]$$

$$\gamma_{xy} = \tau_{xy}/G$$

or, alternatively,

$$\sigma_x = \frac{E}{(1+\nu)(1-2\nu)}\left[(1-\nu)\varepsilon_x + \nu\varepsilon_y\right]$$

$$\sigma_y = \frac{E}{(1+\nu)(1-2\nu)}\left[(1-\nu)\varepsilon_y + \nu\varepsilon_x\right] \tag{4.7}$$

$$\tau_{xy} = G\gamma_{xy}$$

* See Timoshenko and Goodier,[4] p. 241.

** Sokolnikoff,[2] p. 254.

Substitution of Equations 4.7 into the equilibrium equations and making use of the strain-displacement relations gives

$$\frac{\partial}{\partial x}\left(\frac{\partial u}{\partial x}+\frac{\partial v}{\partial y}\right)+(1-2\nu)\nabla^2 u+\frac{2(1+\nu)(1-2\nu)}{E}F_x=0$$

where

$$\nabla^2=\frac{\partial^2}{\partial^2 x}+\frac{\partial^2}{\partial y^2}$$

Simplifying yields

$$\begin{aligned}G\nabla^2 u+\frac{G}{(1-2\nu)}\frac{\partial}{\partial x}\left(\frac{\partial u}{\partial x}+\frac{\partial v}{\partial y}\right)+F_x=0\\ G\nabla^2 v+\frac{G}{(1-2\nu)}\frac{\partial}{\partial y}\left(\frac{\partial u}{\partial x}+\frac{\partial v}{\partial y}\right)+F_y=0\end{aligned} \tag{4.8a}$$

The above two equations are Navier equations for plane strain. Similar equations can be obtained for plane stress problems, namely,

$$\begin{aligned}G\nabla^2 u+G\frac{1+\nu}{1-\nu}\frac{\partial}{\partial x}\left(\frac{\partial u}{\partial x}+\frac{\partial v}{\partial y}\right)+F_x=0\\ G\nabla^2 v+G\frac{1+\nu}{1-\nu}\frac{\partial}{\partial y}\left(\frac{\partial u}{\partial x}+\frac{\partial v}{\partial y}\right)+F_y=0\end{aligned} \tag{4.8b}$$

Equations 4.8a compared with Equations 4.8b indicate that plane strain equations reduce to those for plane stress if ν in the former is replaced by $\nu/(1+\nu)$ and E by $E(1+2\nu)/(1+\nu)^2$. This leaves G unchanged. Also, the plane stress equations reduce to plane strain equations if in the former, ν is replaced by $\nu/(1-\nu)$ and leaving G unchanged. These substitutions refer to Navier's equations (Equations 4.8a and 4.8b) as well as to the stress compatibility equations as will be shown later. These substitutions do not apply, however, to the boundary conditions or the stress-strain relations.

4.7 STRESS FUNCTION FORMULATION FOR PLANE ELASTIC PROBLEMS

For plane problems, it is observed that ε_z for plane stress and σ_z for plane strain are linear functions of $(\sigma_x+\sigma_y)$. If a plane problem is solved in terms of displacements, the number of unknowns reduces to eight, namely, u, v, ε_x, ε_y, γ_{xy}, σ_x, σ_y, and τ_{xy}. Meanwhile, the number of equations is eight, namely, three strain–displacement relations, two equations of equilibrium, and three stress–strain relations. If, however, a solution is sought in terms of stress, the strains must then satisfy the compatibility condition. In this case, the number of unknowns is six, namely, σ_x, σ_y, τ_{xy}, ε_x, ε_y, and γ_{xy}. The number of equations is also six, namely, two equations of equilibrium, three stress–strain relations, and one compatibility equation.

The problem of plane strain or plane stress can be formulated in terms of a single function called the stress function ϕ, whose appropriate derivatives define the stresses σ_x, σ_y, and τ_{xy}. To this end the following procedure is adopted:

1. The compatibility condition is obtained in terms of the stress components by substituting the stress–strain relations in the strain compatibility condition. This yields the stress compatibility equation.
2. The stress compatibility equation together with the two stress equations of equilibrium will provide three expressions for the three unknown stress components.
3. A function in terms of the stress components satisfying the three stress expressions together with the boundary conditions has to be found.

This procedure is illustrated by the diagram of Figure 4.8.

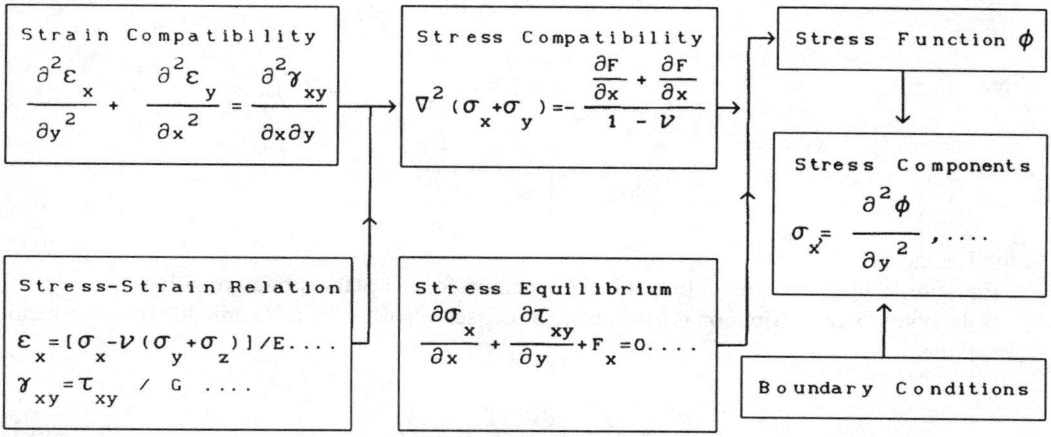

FIGURE 4.8 Block diagram for the stress function formulation of plane elastic problems.

In the general case, it is possible to express the three-dimensional stress components in terms of three independent stress functions satisfying both equilibrium and compatibility. This general case is extremely complicated and of limited practical value.[5]

4.8 GOVERNING EQUATIONS IN TERMS OF A STRESS FUNCTION IN CARTESIAN COORDINATES

The above procedure is adopted to derive the equations governing the solution of plane elastic problems in Cartesian coordinates using a stress function ϕ whose derivatives define the stress components σ_x, σ_y, and τ_{xy}. A separate formulation is made for each of the cases of plane strain, plane stress, and thermoelastic plane problems.

4.8.1 Plane Strain

Substitution of Hooke's law into the strain compatibility equation gives

$$\frac{\partial^2\sigma_x}{\partial y^2} - \nu\frac{\partial^2}{\partial y^2}\left(\sigma_x + \sigma_y\right) + \frac{\partial^2\sigma_y}{\partial x^2} - \nu\frac{\partial^2}{\partial x^2}\left(\sigma_x + \sigma_y\right) = 2\frac{\partial^2\tau_{xy}}{\partial x\partial y} \tag{4.9}$$

Differentiation of the equilibrium equations, the first with respect to x and the second with respect to y, and then adding yields

$$\frac{\partial^2\sigma_x}{\partial x^2}+\frac{\partial^2\sigma_y}{\partial y^2}+\frac{\partial F_x}{\partial x}+\frac{\partial F_y}{\partial y}=-2\frac{\partial^2\tau_{xy}}{\partial x\partial y} \tag{4.10}$$

Comparing Equation 4.9 with Equation 4.10 gives

$$\frac{\partial^2}{\partial x^2}\left(\sigma_x+\sigma_y\right)+\frac{\partial^2}{\partial y^2}\left(\sigma_x+\sigma_y\right)=-\frac{1}{1-\nu}\left(\frac{\partial F_x}{\partial x}+\frac{\partial F_y}{\partial y}\right)$$

or, alternatively,

$$\nabla^2\left(\sigma_x+\sigma_y\right)=-\frac{1}{1-\nu}\left(\frac{\partial F_x}{\partial x}+\frac{\partial F_y}{\partial y}\right) \tag{4.11}$$

where

$$\nabla^2=\left(\frac{\partial^2}{\partial x^2}+\frac{\partial^2}{\partial y^2}\right)$$

is the Laplacian.

Equation 4.11 is the stress compatibility equation for the plane strain condition.

If the body force distribution is assumed to be conservative so that it is derivable from a potential $V(x, y)$, then

$$F_x=-\frac{\partial V}{\partial x}\quad F_y=-\frac{\partial V}{\partial y} \tag{4.12}$$

Substituting into the equilibrium equation yields

$$\begin{aligned}\frac{\partial}{\partial x}\left(\sigma_x-V\right)+\frac{\partial\tau_{xy}}{\partial y}=0\\ \frac{\partial}{\partial y}\left(\sigma_y-V\right)+\frac{\partial\tau_{xy}}{\partial x}=0\end{aligned} \tag{4.13}$$

Hence. if the stresses are defined as

$$\sigma_x-V=\frac{\partial^2\phi}{\partial y^2},\sigma_y-V=\frac{\partial^2\phi}{\partial x^2},\ \text{and}\ \tau_{xy}=-\frac{\partial^2\phi}{\partial x\partial y} \tag{4.14}$$

it is seen that the stress equilibrium conditions (Equations 4.13) are identically satisfied. The function $\phi(x, y)$ is known as the Airy stress function.*

Therefore, by virtue of Equations 4.12 and 4.14, the stress compatibility equation (Equation 4.11) becomes

* The stress function of Equation 4.14 was introduced by the Astronomer G. B. Airy in 1862.

$$\nabla^2\left[\frac{\partial^2\phi}{\partial y^2}+V+\frac{\partial^2\phi}{\partial x^2}+V\right]=-\frac{1}{1-\nu}\left(-\frac{\partial^2 V}{\partial x^2}-\frac{\partial^2 V}{\partial y^2}\right),\ \nabla^2\left[\nabla^2\phi+2V\right]=\frac{1}{1-\nu}\nabla^2 V$$

or

$$\nabla^2\nabla^2\phi=-\frac{1-2\nu}{1-\nu}\nabla^2 V \tag{4.15}$$

If the body forces have a potential V, such that

$$\nabla^2 V = 0$$

i.e., V is harmonic, then Equation 4.15 reduces to

$$\nabla^2\nabla^2\phi = 0 \tag{4.16}$$

where ϕ is obviously a biharmonic function as

$$\nabla^2\nabla^2=\frac{\partial^4}{\partial x^4}+2\frac{\partial^4}{\partial x^2\partial y^2}+\frac{\partial^4}{\partial y^4}$$

A typical example for which $\nabla^2 V = 0$ is the case of gravity forces for which $F_x = 0$ and $F_y = \rho g$; therefore, $V = -\rho g y$.

If the body forces are neglected, i.e., $F_x = F_y = 0$, the governing stress function Equation 4.15 is then reduced to

$$\nabla^2\nabla^2\phi = 0$$

and the stress components are given by

$$\sigma_x=\frac{\partial^2\phi}{\partial y^2},\ \sigma_y=\frac{\partial^2\phi}{\partial x^2},\ \text{and}\ \tau_{xy}=-\frac{\partial^2\phi}{\partial x\partial y} \tag{4.17}$$

Example 4.4:
Given the functions:

a. $\phi = ax^3y^2$ *and*
b. $\phi = bxy^4 + cx^3y^2$

find whether these functions can be stress functions for plane strain problems without body forces.

Solution:
For plane strain conditions, Equation 4.15 requires that

$$\nabla^2\nabla^2\phi=-\frac{1-2\nu}{1-\nu}\nabla^2 V$$

Neglecting body forces, this equation reduces to

$$\nabla^2\nabla^2\phi = \frac{\partial^4\phi}{\partial x^4} + 2\frac{\partial^4\phi}{\partial x^2\partial y^2} + \frac{\partial^4\phi}{\partial y^4} = 0$$

a. Applying this condition to the function $\phi = ax^3y^2$ gives $24ax = 0$. This condition is satisfied for any value of x only when $a = 0$, which means that $\phi = 0$. Therefore, the function $\phi = ax^3y^2$ is not an appropriate stress function.
b. Differentiation of the function $\phi = bxy^4 + cx^3y^2$ and substituting into $\nabla^2\nabla^2\phi = 0$ gives

$$12cx - 24bx = 0$$

For any value of x, ϕ is a stress function if

$$c = 2b$$

4.8.2 Plane Stress

The governing equation for the Airy stress function for plane stress is obtained by replacing ν by $\nu/(1 + \nu)$ in Equation 4.15, thus yielding

$$\nabla^2\nabla^2\phi = -(1 - \nu)\nabla^2 V \tag{4.18}$$

Equation 4.18 can be obtained from a detailed analysis following the procedure of the previous section step by step. The stresses in the case of plane stress are obtained in terms of the stress function ϕ from either Equations 4.14 or 4.17.

4.8.3 Thermoelastic Plane Problem

In a thermoelastic plane problem the temperature rise T is a function of x and y only. A separate formulation is made for each of plane strain and plane stress conditions.

4.8.3.1 Thermoelastic Plane Strain

The stress–strain relations in this case are given by Equations 3.18 as

$$\varepsilon_x = \frac{1}{E}\left[\sigma_x - \nu\left(\sigma_y + \sigma_z\right)\right] + \alpha T$$

$$\varepsilon_y = \frac{1}{E}\left[\sigma_y - \nu\left(\sigma_z + \sigma_x\right)\right] + \alpha T$$

$$\varepsilon_z = \frac{1}{E}\left[\sigma_z - \nu\left(\sigma_x + \sigma_y\right)\right] + \alpha T$$

$$\gamma_{xy} = \tau_{xy}/G$$

Substituting $\varepsilon_z = 0$ results in $\sigma_z = \nu(\sigma_x + \sigma_y) - \alpha ET$ and, hence,

$$\varepsilon_x = \frac{1+\nu}{E}\left[\sigma_x - \nu\left(\sigma_x + \sigma_y\right) + \alpha ET\right]$$

$$\varepsilon_x = \frac{1}{2G}\left[\sigma_x - \nu\left(\sigma_x + \sigma_y\right) + \alpha ET\right]$$

$$\varepsilon_y = \frac{1}{2G}\left[\sigma_y - \nu\left(\sigma_x + \sigma_y\right) + \alpha ET\right]$$

$$\gamma_{xy} = \tau_{xy}/G$$

Substituting these stress–strain relations in the strain compatibility condition and combining the result with the equations of equilibrium yield the stress compatibility equation as

$$\nabla^2\left(\sigma_x + \sigma_y\right) = -\frac{\alpha E}{1-\nu}\nabla^2 T - \frac{1}{1-\nu}\left(\frac{\partial F_x}{\partial x} + \frac{\partial F_y}{\partial y}\right) \tag{4.19}$$

In plane strain thermoelastic problems with body forces having a potential V, the stress components are given by

$$\sigma_x = \frac{\partial^2\phi}{\partial y^2} + V, \quad \sigma_y = \frac{\partial^2\phi}{\partial x^2} + V, \quad \text{and} \quad \tau_{xy} = -\frac{\partial^2\phi}{\partial x \partial y} \tag{4.20}$$

Therefore, the stress compatibility Equation 4.19 reduces to

$$\nabla^2\nabla^2\phi = -\nabla^2\left(\frac{1-2\nu}{1-\nu}V + \frac{\alpha E}{1-\nu}T\right) \tag{4.21}$$

where ϕ is the stress function. It is seen that considering temperature effects is mathematically similar to considering the potential of body forces.

For steady-state conduction $\nabla^2 T = 0$, hence Equation 4.21 gives

$$\nabla^2\nabla^2\phi = -\nabla^2\left(\frac{1-2\nu}{1-\nu}V\right)$$

In the absence of body forces, the stress function satisfies again the biharmonic equation (Equation 4.16):

$$\nabla^2\nabla^2\phi = 0$$

4.8.3.2 Thermoelastic Plane Stress

The thermoelastic stress–strain relations for plane stress are given by

$$\varepsilon_x = \frac{1}{E}\left[\sigma_x - \nu\sigma_y\right] + \alpha T$$

$$\varepsilon_y = \frac{1}{E}\left[\sigma_y - \nu\sigma_x\right] + \alpha T$$

$$\gamma_{xy} = \frac{\tau_{xy}}{G} = \frac{2(1+\nu)}{E}\tau_{xy}$$

By substituting in the strain compatibility equation and combining the result with the equations of equilibrium, the stress compatibility relation is obtained as

$$\begin{aligned} \nabla^2\left(\sigma_x + \sigma_y\right) &= -\alpha E \nabla^2 T - (1-\nu)\nabla^2 V \\ \nabla^2\nabla^2\phi &= -\nabla^2\left[(1-\nu)V + \alpha E T\right] \end{aligned} \tag{4.22}$$

Again, for steady-state conduction $\nabla^2 T = 0$, hence, Equation 4.22 gives

$$\nabla^2\nabla^2\phi = -\nabla^2(1-\nu)V$$

In the absence of body forces, the stress function then satisfies

$$\nabla^2\nabla^2\phi = 0$$

The stresses are obtained from Equations 4.17.

Example 4.5:

In a thin, square plate whose sides are parallel to the Ox *and* Oy *axes, stresses are expressed* σ_x = Cy, σ_y = Cx, *and possibly some shearing stresses* τ_{xy}, *where* C *is a constant as shown in Figure 4.9.*

a. *Find the stress function by integration and then the most general shear stress that can be associated with the given stresses* σ_x, σ_y. *Neglect body forces.*
b. *Obtain expressions for the displacements* u *and* v *by integration of the strains.*

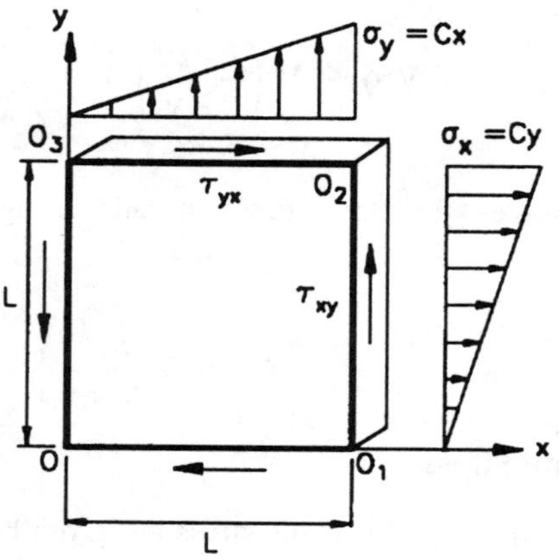

FIGURE 4.9 Example 4.5.

Solution:

a. From Equations 4.17 for a thin plate under plane stress conditions:

$$\sigma_x = \frac{\partial^2 \phi_1}{\partial y^2} = Cy$$

thus giving

$$\phi_1 = \frac{c}{6} y^3 + y f_1(x) + f_2(x)$$

where f_1 and f_2 are functions of x only. Similarly,

$$\sigma_y = \frac{\partial^2 \phi_2}{\partial x^2} = Cx$$

giving $\phi_2 = (C/6)\, x^3 + x f_3(y) + f_4(y)$, where f_3 and f_4 are functions of y only. Obviously the shear stress derived from either ϕ_1 or ϕ_2 must be the same; therefore,

$$\tau_{xy} = -\frac{\partial^2 \phi_1}{\partial x \partial y} = -\frac{\partial^2 \phi_2}{\partial x \partial y}$$

or

$$\frac{df_1(x)}{dx} = \frac{df_3(y)}{dy}$$

This implies that

$$\frac{df_1(x)}{dx} = \frac{df_3(y)}{dy} = \text{constant } C_1$$

Integrating gives

$$f_1(x) = C_1 x + C_2$$

$$f_3(y) = C_1 y + C_3$$

The stress function ϕ_1 and ϕ_2 are thus

$$\phi_1 = \frac{C}{6} y^3 + y(C_1 x + C_2) + f_2(x)$$

and

$$\phi_2 = \frac{C}{6}x^3 + x(C_1 y + C_2) + f_4(y)$$

Evidently, the stress functions ϕ_1 and ϕ_2 must yield the same original expressions for both of σ_x and σ_y, namely,

$$\sigma_x = \partial^2\phi_1/\partial y^2 = \partial^2\phi_2/\partial y^2 = Cy$$

and

$$\sigma_y = \partial^2\phi_1/\partial x^2 = \partial^2\phi_2/\partial x^2 = Cx$$

which yields

$$f_4(y) = \frac{Cy^3}{6} + C_4 y + C_5$$

and

$$f_2(x) = \frac{Cx^3}{6} + C_6 x + C_7$$

Therefore the stress functions are

$$\phi_1 = \frac{C}{6}y^3 + y(C_1 x + C_2) + \frac{C}{6}x^3 + C_6 x + C_7$$

$$\phi_2 = \frac{C}{6}x^3 + x(C_1 y + C_3) + \frac{C}{6}y^3 + C_4 y + C_5$$

These two stress functions are identical, as expected, and the final most general form of the stress function becomes

$$\phi = \frac{C}{6}(x^3 + y^3) + C_1 xy$$

It must be noted that constant terms as well as first-degree terms in x and y do not contribute to the stresses since they are obtained from the second derivatives of ϕ.

Having obtained the stress function, the stresses are

$$\sigma_x = Cy, \quad \sigma_y = Cx, \quad \text{and} \quad \tau_{xy} = -C_1$$

b. The strains are obtained by applying Hooke's law, Equations 3.6, for plane stress as

$$\varepsilon_x = \frac{1}{E}(\sigma_x - \nu\sigma_y) = \frac{C}{E}(y - \nu x)$$

$$\varepsilon_y = \frac{1}{E}(\sigma_y - \nu\sigma_x) = \frac{C}{E}(x - \nu y)$$

$$\varepsilon_z = -\frac{\nu}{E}(\sigma_x + \sigma_y) = -\frac{C\nu}{E}(x + y)$$

$$\gamma_{xy} = \tau_{xy}/G = -C_1/G$$

To obtain the displacement components u and v, Relations 2.6

$$\varepsilon_x = \frac{\partial u}{\partial x}, \quad \varepsilon_y = \frac{\partial v}{\partial y}, \quad \gamma_{xy} = \frac{\partial u}{\partial y} + \frac{\partial v}{\partial x}$$

have to be integrated.

Therefore,

$$u = \frac{C}{E}\left(yx - \frac{\nu}{2}x^2\right) + g_1(y)$$

and

$$v = \frac{C}{E}\left(xy - \frac{\nu}{2}y^2\right) + g_2(x)$$

where g_1 is a function of y only and g_2 is a function of x only. Substitution of the above expressions for u and v into the relation for γ_{xy} gives

$$\gamma_{xy} = \frac{C_1}{G} = \frac{C}{E}x + \frac{dg_1(y)}{dy} + \frac{C}{E}y + \frac{dg_2(x)}{dx}$$

By inspection, to reduce the right-hand side to a constant value (equal to the left-hand side), yields

$$\frac{dg_1(y)}{dy} = -\frac{C}{E}y + D_1 \quad \text{and} \quad \frac{dg_2(x)}{dx} = -\frac{C}{E}x + D_2$$

where D_1 and D_2 are constants satisfying the condition $D_1 + D_2 = -C_1/G$. Therefore, by integration,

$$g_1 = -\frac{C}{2E}y^2 + D_1 y$$

and

$$g_2 = -\frac{C}{2E}x^2 + D_2 x$$

The displacement u, v are thus given by

$$u = \frac{C}{E}\left(xy - \frac{\nu}{2}x^2 - \frac{y^2}{2}\right) + D_1 y$$

$$v = \frac{C}{E}\left(xy - \frac{\nu}{2}y^2 - \frac{x^2}{2}\right) + D_2 x$$

Example 4.6:

The stress function $\phi = a(x^2 + y^2) + bxy$ *provides the stresses in a thin square plate of side length* L, *with respect to Cartesian axes* x *and* y *whose origin* O *is located as shown in Figure 4.10. Knowing that the diagonal* OO_2 *does not undergo any change in length, find:*

a. *Expressions for the displacement components* u, v;
b. *Use the relation between* a *and* b *which makes the length of diagonal* OO_2 *unchanged to express* u, v;
c. *An expression for the deformed length of the diagonal* O_1O_3. *If this diagonal has elongated 0.2%, calculate the loads acting on the plate taking* $\nu = 1/3$.

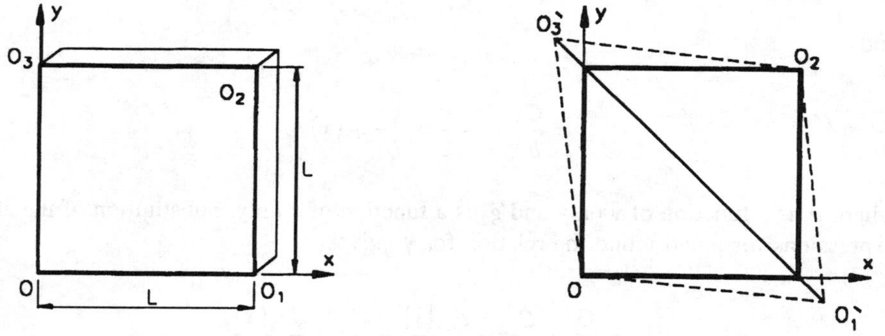

FIGURE 4.10 Example 4.6.

Solution:

a. The stresses are derived from the stress function ϕ as

$$\sigma_x = \frac{\partial^2 \phi}{\partial y^2} = 2a$$

$$\sigma_y = \frac{\partial^2 \phi}{\partial x^2} = 2a$$

$$\tau_{xy} = -\frac{\partial^2 \phi}{\partial x \partial y} = -b$$

Application of Hooke's law for plane stress conditions to determine the strain components and hence displacements results in

$$u = \int \varepsilon_x dx = \frac{1}{E}\int \left[\sigma_x - \nu\sigma_y\right] dx = \frac{1}{E}\int \left[2a - \nu(2a)\right] dx$$

$$v = \int \varepsilon_y dy = \frac{1}{E}\int \left[\sigma_y - \nu\sigma_x\right] dy = \frac{1}{E}\int \left[2a - \nu(2a)\right] dy$$

or

$$u = \frac{2(1-\nu)a}{E}x + f_1(y)$$

$$v = \frac{2(1-\nu)a}{E}y + f_2(x)$$

where f_1 (y) and f_2 (x) are determined to satisfy the condition:

$$\gamma_{xy} = \tau_{xy}/G = -b/G = \frac{\partial u}{\partial y} + \frac{\partial v}{\partial x} = \frac{\partial f_1(y)}{\partial y} + \frac{\partial f_2(x)}{\partial x}$$

This requires that

$$\frac{\partial f_1(y)}{\partial y} = \text{constant} = D_1 \text{ and } \frac{\partial f_2(x)}{\partial x} = \text{constant} = D_2$$

i.e., f_1 $(y) = D_1 y + D_3$ and f_2 $(x) = D_2 x + D_4$.
Expressions for displacements are thus

$$u = \frac{2(1-\nu)a}{E}x + D_1 y + D_3$$

$$v = \frac{2(1-\nu)a}{E}y + D_2 x + D_4$$

b. The boundary conditions at the origin O are such that $u = v = 0$, hence, $D_3 = D_4 = 0$. The other two constants D_1 and D_2 are determined by imposing the condition that at $x = y = L$, $u = v = 0$ since there is no change in length OO_2, which yields

$$0 = \frac{2(1-\nu)a}{E}L + D_1 L$$

and

$$0 = \frac{2(1-\nu)a}{E}L + D_2 L$$

Solving gives

$$D_1 = D_2 = -\frac{2(1-\nu)a}{E}$$

and, hence,

$$u = \frac{2(1-\nu)a}{E}(x - y)$$

and

$$v = \frac{2(1-\nu)a}{E}(y-x)$$

The deformed shape of the plate is shown in Figure 4.10b. A condition must be imposed on the constant b, such that

$$D_1 + D_2 = -b/G$$

giving

$$b = \frac{2(1-\nu)}{1+\nu}a$$

since

$$G = \frac{E}{2(1+\nu)}$$

The displacement components (u_1, v_1) and (u_3, v_3) at points O_1 and O_3, respectively, are

$$u_1 = 2(1-\nu)a\,L/E, \quad v_1 = -2(1-\nu)a\,L/E$$

$$u_3 = -2(1-\nu)a\,L/E, \quad v_3 = 2(1-\nu)a\,L/E$$

c. The deformed length of the diagonal O_1O_3 is

$$O_1'O_3' = \sqrt{\left[(L+u_1)^2 + v_1^2\right] + \left[(L+v_3)^2 + u_3^2\right]}$$

Substituting for u_1, v_1, u_3 and neglecting higher orders of displacements gives

$$O_1'O_3' = \sqrt{2}\,L\sqrt{1+2(1-\nu)\,a/E}$$

Since the elongation in the diagonal is $0.002 = (O_1'O_3' - O_1O_3)/O_1'O_3'$, where $O_1O_3 = \sqrt{2}$ L, hence

$$(1.002)^2 = 1 + 2(1-\nu)a/E$$

Hence, $a = 0.00075E = 0.002G$. For $\nu = 1/3$, the stresses applied to the plate are

$$\sigma_x = \sigma_y = 2a = 0.008G \text{ and } \tau_{xy} = -b = -0.004G$$

4.8.4 Finding a Stress Function in Cartesian Coordinates

If a function ϕ can be found to satisfy the biharmonic equation $\nabla^2\nabla^2\phi = 0$ in the interior of the deforming body and such that the stresses derived from ϕ according to Equations 4.14 satisfy the prescribed boundary conditions, then that function ϕ furnishes the solution to the plane elastic problem. Such a solution is unique as long as the analysis, is within the framework of the small deformation linear theory of elasticity as discussed in Section 4.4.

In many problems it is often found convenient to employ the semi-inverse method, specifying in advance the form of the stress function. In cases where specifying a stress function in advance becomes less obvious, a powerful and elegant method based on representing a biharmonic function by means of two analytic functions of a complex variable is proposed. The principal tool for this general approach in determining the stress function is the conformal mapping transformation. The details of this method are found in more-advanced texts.[6]

In Chapter 5, solutions for plane elastic problems are obtained by assuming simple forms of the Cartesian stress function such as algebraic polynomials (Section 5.2) or trigonometric series (Section 5.3).

4.9 GOVERNING EQUATIONS IN TERMS OF A STRESS FUNCTION IN POLAR COORDINATES

4.9.1 Plane Strain

The basic governing equations for the plane strain problem in polar coordinates (r, θ) are as follows:

Strain–displacement relations

$$\varepsilon_r = \frac{\partial u}{\partial r}, \quad \varepsilon_\theta = \frac{u}{r} + \frac{1}{r}\frac{\partial v}{\partial \theta}, \quad \gamma_{r\theta} = \frac{1}{r}\frac{\partial u}{\partial \theta} + \frac{\partial v}{\partial r} - \frac{v}{r}$$

where (u, v) are the displacement components in the radial and tangential directions (r, θ), respectively. By definition, the other strain components ε_z, γ_{zr}, and $\gamma_{z\theta}$ vanish.

Equilibrium equations

$$\frac{\partial \sigma_r}{\partial r} + \frac{1}{r}\frac{\partial \tau_{r\theta}}{\partial \theta} + \frac{\sigma_r - \sigma_\theta}{r} + F_r = 0$$

$$\frac{\partial \tau_{r\theta}}{\partial r} + \frac{1}{r}\frac{\partial \sigma_\theta}{\partial \theta} + \frac{2\tau_{r\theta}}{r} + F_\theta = 0$$

where F_r and F_θ are the body force components per unit volume in the radial and tangential directions, respectively. Note that F_r and F_θ are functions of r and θ only.

Hooke's law

For plane strain conditions $\varepsilon_z = 0 = (1/E)\,[\sigma_z - \nu(\sigma_r + \sigma_\theta)]$ and therefore,

$$\sigma_z = \nu(\sigma_r + \sigma_\theta)$$

Substituting σ_z into the stress–strain relation yields

$$\varepsilon_r = \frac{1}{E}\left[\sigma_r - \nu\left(\sigma_\theta + \sigma_z\right)\right] = \frac{1}{E}\left[\sigma_r - \nu\left(\sigma_\theta + \nu\left(\sigma_r + \sigma_\theta\right)\right)\right]$$

or

$$\varepsilon_r = \frac{1}{2G}\left[(1-\nu)\sigma_r - \nu\sigma_\theta\right]$$

Similarly,

$$\varepsilon_\theta = \frac{1}{2G}\left[(1-\nu)\sigma_\theta - \nu\sigma_r\right]$$

Also,

$$\gamma_{r\theta} = \tau_{r\theta}/G$$

Compatibility equations

$$r\frac{\partial \varepsilon_r}{\partial r} - \frac{\partial^2 \varepsilon_r}{\partial \theta^2} - \frac{\partial}{\partial r}\left[r\left(r\frac{\partial \varepsilon_\theta}{\partial r} - \frac{\partial \gamma_{r\theta}}{\partial \theta}\right)\right] = 0$$

Under plane strain conditions, the five other strain compatibility conditions of Equation 2.13 are identically satisfied.

Substitution of Hooke's law into the compatibility condition yields the stress compatibility equation. Differentiating the first of the equilibrium equations with respect to r and the second with respect to θ, adding the two equations and then comparing with the stress compatibility equation yields

$$\left(\frac{\partial^2}{\partial r^2} + \frac{1}{r}\frac{\partial}{\partial r} + \frac{1}{r^2}\frac{\partial^2}{\partial \theta^2}\right)(\sigma_r + \sigma_\theta) = -\frac{1}{1-\nu}\left(\frac{F_r}{r} + \frac{\partial F_r}{\partial r} + \frac{1}{r}\frac{\partial F_\theta}{\partial \theta}\right) \tag{4.23}$$

The differential operator on the left-hand side is the Laplacian in terms of the polar coordinates (r, θ); therefore, Equation 4.23 may be written as

$$\nabla^2(\sigma_r + \sigma_\theta) = -\frac{1}{1-\nu}\left(\frac{F_r}{r} + \frac{\partial F_r}{\partial r} + \frac{\partial F_\theta}{\partial \theta}\right) \tag{4.24}$$

As for the formulation of the problem in terms of a stress function $\phi(r, \theta)$ and a potential $V(r, \theta)$, the following expressions are introduced:

$$F_r = -\partial V/\partial r, F_\theta = -\partial V/\partial \theta$$

$$\sigma_r = \frac{1}{r}\left(\frac{\partial \phi}{\partial r} + \frac{1}{r}\frac{\partial^2 \phi}{\partial \theta^2}\right) + V \tag{4.25}$$

$$\sigma_\theta = \frac{\partial^2 \phi}{\partial r^2} + V \text{ and } \tau_{r\theta} = -\frac{\partial}{\partial r}\left(\frac{1}{r}\frac{\partial \phi}{\partial \theta}\right)$$

so that the stress–equilibrium equations are identically satisfied. Hence, substitution of Expressions 4.25 for σ_r and σ_θ in Equation 4.24 gives

$$\nabla^4\phi = \left(\frac{\partial^2}{\partial r^2} + \frac{1}{r}\frac{\partial}{\partial r} + \frac{1}{r^2}\frac{\partial^2}{\partial \theta^2}\right)\left(\frac{\partial^2\phi}{\partial r^2} + \frac{1}{r}\frac{\partial\phi}{\partial r} + \frac{1}{r^2}\frac{\partial^2\phi}{\partial \theta^2}\right) = -\frac{1-2\nu}{1-\nu}\nabla^2 V \tag{4.26}$$

Again, if V is harmonic or the body forces are neglected, Equation 4.26 reduces to

$$\nabla^2\nabla^2\phi = 0 \tag{4.27}$$

4.9.2 Plane Stress

In formulating the plane stress problem in polar coordinates, the stress components $\sigma_z = \tau_{z\theta} = \tau_{zr} = 0$ and the other stress components σ_r, σ_θ, and $\tau_{r\theta}$ as well as the body force components are all functions of r and θ only.

In a procedure similar to that followed in the plane strain condition, it can be shown that the basic governing equation satisfied by the stress function for conditions of plane stress is

$$\nabla^2\nabla^2\phi = -(1-\nu)\nabla^2 V \tag{4.28}$$

which reduces to $\nabla^2\nabla^2\phi = 0$, if either V is a harmonic function or the body forces are neglected. Note that Equation 4.28 could be deduced from Equation 4.26 by replacing ν in the latter equations by $\nu/(1 + \nu)$ as indicated in Section 4.6.3. Note again that Equations 4.26 and 4.28 are identical to those of Cartesian coordinates, Equations 4.15 and 4.18, respectively, considering that the definition of the operator ∇^2 in polar coordinates as given in Equation 4.23.

It is also emphasized that ϕ in polar coordinates is identical to that in Cartesian coordinates. It is the same scalar function merely expressed in two different coordinate systems, as demonstrated in Example 4.7.

Example 4.7:

The governing equation for the stress function in Cartestian coordinates neglecting body forces is given by

$$\nabla^2\nabla^2 f = \frac{\partial^4\phi}{\partial x} + 2\frac{\partial^4\phi}{\partial x^2\partial y^2} + \frac{\partial^4\phi}{\partial y^4} = 0$$

Express this equation in polar coordinates (r, θ) *and then find expressions for* σ_r, σ_θ *and* $\tau_{r\theta}$ *in terms of* $\phi(r, \theta)$.

Solution:

The relation between polar and Cartesian coordinates is given by

$$r^2 = x^2 + y^2, \quad \theta = \tan^{-1}\left(\frac{y}{x}\right)$$

from which

$$\frac{\partial r}{\partial x} = \frac{x}{r} = \cos\theta, \quad \frac{\partial r}{\partial y} = \frac{y}{r} = \sin\theta$$

$$\frac{\partial \theta}{\partial x} = -\frac{y}{r^2} = -\frac{\sin\theta}{r}, \quad \frac{\partial \theta}{\partial y} = \frac{x}{r^2} = \frac{\cos\theta}{r} \tag{a}$$

Using these relations, and considering ϕ as a function of r and θ only yields

$$\frac{\partial\phi}{\partial x} = \frac{\partial\phi}{\partial r}\frac{\partial r}{\partial x} + \frac{\partial\phi}{\partial\theta}\frac{\partial\theta}{\partial x} = \frac{d\phi}{dr}\cos\theta - \frac{1}{r}\frac{\partial\phi}{\partial\theta}\sin\theta$$

Differentiating again with respect to x yields

$$\frac{\partial^2\phi}{\partial x^2} = \frac{\partial^2\phi}{\partial r^2}\cos^2\theta - \frac{\partial^2\phi}{\partial\theta\partial r}\frac{\sin 2\theta}{r} + \frac{\partial\phi}{\partial r}\frac{\sin^2\theta}{r} + \frac{\partial\phi}{\partial\theta}\frac{\sin 2\theta}{r^2} + \frac{\partial^2\phi}{\partial\theta^2}\frac{\sin^2\theta}{r^2} \tag{b}$$

In the same manner,

$$\frac{\partial^2\phi}{\partial y^2} = \frac{\partial^2\phi}{\partial r^2}\sin^2\theta + \frac{\partial^2\phi}{\partial\theta\partial r}\frac{\sin 2\theta}{r} + \frac{\partial\phi}{\partial r}\frac{\cos^2\theta}{r} - \frac{\partial\phi}{\partial\theta}\frac{\sin 2\theta}{r^2} + \frac{\partial^2\phi}{\partial\theta^2}\frac{\cos^2\theta}{r^2} \tag{c}$$

$$\frac{\partial^2\phi}{\partial x\partial y} = \frac{\partial^2\phi}{\partial r^2}\sin\theta\cos\theta + \frac{1}{r}\frac{\partial^2\phi}{\partial r\partial\theta}(\cos 2\theta) - \frac{1}{r^2}\frac{\partial\phi}{\partial\theta}(\cos 2\theta) - \frac{1}{r}\frac{\partial\phi}{\partial r}\sin\theta\cos\theta - \frac{1}{r^2}\frac{\partial^2\phi}{\partial\theta^2}\left(\sin^2\theta\cos^2\theta\right) \tag{d}$$

Adding Equations b and c gives

$$\nabla^2\phi = \frac{\partial^2\phi}{\partial x^2} + \frac{\partial^2\phi}{\partial y^2} = \frac{\partial^2\phi}{\partial r^2} + \frac{1}{r}\frac{\partial\phi}{\partial r} + \frac{1}{r^2} + \frac{\partial^2\phi}{\partial^2\theta}$$

Therefore, the stress function Equation becomes

$$\nabla^2\nabla^2\phi = \left(\frac{\partial^2}{\partial r^2} + \frac{1}{r}\frac{\partial}{\partial r} + \frac{1}{r^2}\frac{\partial^2}{\partial\theta^2}\right)\left(\frac{\partial^2\phi}{\partial r^2} + \frac{1}{r}\frac{\partial\phi}{\partial r} + \frac{1}{r}\frac{\partial^2\phi}{\partial\theta^2}\right) = 0$$

which is identical to Equation 4.26 without body forces. Let the x-axis correspond to the direction $\theta = 0$; hence,

$$\sigma_r = (\sigma_x)_{\theta=0} = \left(\frac{\partial^2\phi}{\partial y^2}\right)_{\theta=0} = \frac{1}{r}\frac{\partial\phi}{\partial r} + \frac{1}{r^2}\frac{\partial^2\phi}{\partial\theta^2}$$

$$\sigma_\theta = (\sigma_y)_{\theta=0} = \left(\frac{\partial^2\phi}{\partial x^2}\right)_{\theta=0} = \frac{\partial^2\phi}{\partial r^2}$$

and

$$\tau_{r\theta} = (\tau_{xy})_{\theta=0} = -\left(\frac{\partial^2\phi}{\partial x\partial y}\right)_{\theta=0} = -\frac{\partial}{\partial r}\left(\frac{1}{r}\frac{\partial\phi}{\partial\theta}\right)$$

which are identical to Equations 4.25 without body forces.

4.9.3 Axisymmetric Plane Problems

Many engineering problems are concerned with solids of revolution, such as cylinders, disks, and spherical shells. Due to symmetry about the axis of revolution, taken as the z-axis, the displacements, strains, and stresses are independent of θ; hence,

$$v = 0 \text{ and } \gamma_{r\theta} = \gamma_{z\theta} = \tau_{r\theta} = \tau_{z\theta} = 0$$

Also, according to Equations 4.25, the stresses are given by

$$\sigma_r = \frac{1}{r}\frac{d\phi}{dr} + V \text{ and } \sigma_\theta = \frac{d^2\phi}{dr^2} \tag{4.29}$$

The solution is thus much simplified and the Laplacian reduces to

$$\nabla^2 = \frac{d^2}{dr^2} + \frac{1}{r}\frac{dr}{dr}$$

Equations 4.26 and 4.28 become for plane strain:

$$\frac{d^4\phi}{dr^4} + \frac{2}{r}\frac{d^3\phi}{dr^3} - \frac{1}{r^2}\frac{d^2\phi}{dr^2} + \frac{1}{r^3}\frac{d\phi}{dr} = -\frac{1-2\nu}{1-\nu}\nabla^2 V \tag{4.30a}$$

and for plane stress:

$$\frac{d^4\phi}{dr^4} + \frac{2}{r}\frac{d^3\phi}{dr^3} - \frac{1}{r^2}\frac{d^2\phi}{dr^2} + \frac{1}{r^3}\frac{d\phi}{dr} = -(1-\nu)\nabla^2 V \tag{4.30b}$$

where $V = V(r)$.

4.9.3.1 Axisymmetric Problems without Body Forces

In this case, both Equations 4.30 reduce to the homogeneous differential equation:

$$\frac{d^4\phi}{dr^4} + \frac{2}{r}\frac{d^3\phi}{dr^3} - \frac{1}{r^2}\frac{d^2\phi}{dr^2} + \frac{1}{r^3}\frac{d\phi}{dr} = 0$$

This can be integrated to give the following form for the stress function:

$$\phi = A \ln r + Br^2 \ln r + Cr^2 + D \tag{4.31}$$

where A, B, C, and D are constants of integration, which are to be determined from the boundary conditions of the specific problem under consideration.

The stress components corresponding to this stress function and according to Equations 4.29 are

$$\begin{aligned}
\sigma_r &= \frac{1}{r}\frac{d\phi}{dr} = \frac{A}{r^2} + B(1 + 2\ln r) + 2C \\
\sigma_\theta &= \frac{d^2\phi}{dr^2} = -\frac{A}{r^2} + B(3 + 2\ln r) + 2C \\
\tau_{r\theta} &= 0
\end{aligned} \tag{4.32}$$

For axisymmetric problems, the displacement component v vanishes and the displacement u is a function of r only; therefore, Equations 2.9b yield

$$\varepsilon_r = \frac{du}{dr}, \quad \varepsilon_\theta = \frac{u}{r}$$

or

$$\varepsilon_r - \varepsilon_\theta = r\frac{d\varepsilon_\theta}{dr} \tag{4.33}$$

which is the strain compatibility equation. Substituting Hooke's law in the compatibility equation, together with Equations 4.32, yields* $B = 0$, and since the stresses are the second derivatives of the stress function, the constant D in Expression 4.31 becomes trivial. Therefore, the stress function for plane axisymmetric problems without body forces is

$$\phi = A \ln r + Cr^2 \tag{4.34}$$

and the stress components are

$$\begin{aligned}
\sigma_r &= \frac{1}{r}\frac{d\phi}{dr} = \frac{A}{r^2} + 2C \\
\sigma_\theta &= \frac{d^2\phi}{dr^2} = -\frac{A}{r^2} + 2C \\
\tau_{r\theta} &= 0
\end{aligned} \tag{4.35}$$

* To prove that $B = 0$, since $\varepsilon_r - \varepsilon_\theta = r(\partial\varepsilon_\theta/\partial r)$, hence, from Hooke's law

$$\varepsilon_r - \varepsilon_\theta = \frac{1+\nu}{E}(\sigma_r - \sigma_\theta)$$

Substituting σ_r, σ_θ from Equation 4.32 gives

$$\varepsilon_r - \varepsilon_\theta = \frac{2(1+\nu)}{E}\left(\frac{A}{r_2} - B\right) \tag{a}$$

For plane stress $\varepsilon_\theta = (1/E)(\sigma_\theta - \nu\sigma_r)$, hence

$$r\frac{\partial\varepsilon_\theta}{\partial r} = \frac{r}{E}\left(\frac{\partial\sigma_\theta}{\partial r} - \nu\frac{\partial\sigma_r}{\partial r}\right) = \frac{2(1+\nu)}{E}\frac{A}{r^2} - \frac{2(1-\nu)}{E}B \tag{b}$$

For plane strain $\varepsilon_\theta = [(1+\nu)/E][(1-\nu)\sigma_\theta - \nu\sigma_r]$

$$r\frac{\partial\varepsilon_\theta}{\partial r} = \frac{2(1+\nu)}{E}\frac{A}{r^2} + 2\frac{(1+\nu)(1-\nu)}{E}B \tag{c}$$

Comparing Equations b and c with Equation a gives $B = 0$.

4.9.3.2 Axisymmetric Problems with Centrifugal Body Forces

A case of practical interest is where the body force is the centrifugal force resulting from a constant angular velocity ω. In this case, the only nonzero body force component is given by

$$F_r = -\frac{dV}{dr} = \rho\omega^2 r$$

where ρ is the material mass density. The potential V is then given by

$$V = -\, {}^1\!/_2\, \rho\omega^2 r$$

Considering conditions of plane stress, e.g., case of a rotating thin disk, Equation 4.30 becomes

$$\frac{d^4\phi}{dr^4} + \frac{2}{r}\frac{d^3\phi}{dr^3} - \frac{1}{r^2}\frac{d^2\phi}{dr^2} + \frac{1}{r^3}\frac{d\phi}{dr} = 2(1-\nu)\rho\omega^2$$

Solving this differential Equation gives

$$\phi = A\ln r + Br^2 \ln r + Cr^2 + D + \frac{1-\nu}{32}\rho\omega^2 r^4$$

Omitting the constant D and recalling that $B = 0$ for axisymmetry problems, the stress function ϕ for the plane stress condition with centrifugal body forces is given by

$$\phi = A\ln r + Cr^2 + \frac{1-\nu}{32}\rho\omega^2 r^4 \tag{4.36}$$

The stress components are thus

$$\begin{aligned} \sigma_r &= \frac{1}{r}\frac{d\phi}{dr} + V = \frac{A}{r^2} + 2C - \frac{3+\nu}{8}\rho\omega^2 r^2 \\ \sigma_\theta &= \frac{d^2\phi}{dr_2} + V = -\frac{A}{r_2} + 2C - \frac{1+3\nu}{8}\rho\omega^2 r^2 \end{aligned} \tag{4.37}$$

In a procedure similar to that of the case of plane stress, it can be shown that a stress function for plane strain condition with centrifugal body forces, e.g., the case of a rotating long cylinder or drum is given by

$$\phi = A\ln r + Cr^2 + \frac{1-2\nu}{32(1-\nu)}\rho\omega^2 r^4 \tag{4.38}$$

The stress components are given by

$$\begin{aligned} \sigma_r &= \frac{1}{r}\frac{d\phi}{dr} + V = \frac{A}{r^2} + 2C - \frac{3-2\nu}{8(1-\nu)}\rho\omega^2 r^2 \\ \sigma_\theta &= \frac{d^2\phi}{dr^2} + V = -\frac{A}{r^2} + 2C - \frac{1+2\nu}{8(1-\nu)}\rho\omega^2 r^2 \end{aligned} \tag{4.39}$$

Example 4.8:

A thin circular solid disk of radius r_o *and thickness* h *is an integral part of a solid shaft of diameter* $2r_i$ *which transmits a twisting moment* M_t *to the disk. The disk is subjected to an outside radial pressure* p_o *and a shear traction* $\tau_o = \mu p_o$ *all around its outer circumference, Figure 4.11. The function* $\phi = C_1 r^2 + C_2\theta$ *is a stress function that determines the stresses in the disk. Determine the constants* C_1 *and* C_2 *and obtain expressions for the stresses in the disk.* μ *is Coulomb's coefficient of friction and the disk is shrink-fit at its outer circumference.*

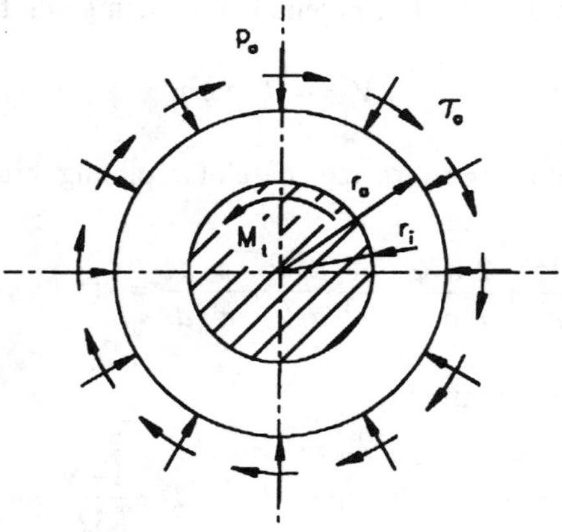

FIGURE 4.11 Example 4.8.

Solution:

The stresses are obtained from the stress function ϕ as

$$\sigma_r = \frac{1}{r}\left(\frac{\partial\phi}{\partial r} + \frac{1}{r}\frac{\partial^2\phi}{\partial\theta^2}\right) = 2C_1$$

$$\sigma_\theta = \frac{\partial^2\phi}{\partial r^2} = 2C_1$$

$$\tau_{r\theta} = -\frac{\partial}{\partial r}\left(\frac{1}{r}\frac{\partial\phi}{\partial\theta}\right) = \frac{C_2}{r^2}$$

The constants C_1 and C_2 are obtained from boundary conditions by imposing the condition that at the outer surface $\sigma_r = -p_o$; hence,

$$C_1 = -p_o/2$$

From the equilibrium of the disk, the twisting moment is

$$M_t = 2\pi r_o^2 h\tau_o$$

or

$$\tau_o = \frac{M_t}{2\pi h r_o^2},$$

hence,

$$C_2 = \frac{M_t}{2\pi h}$$

Hence, the stresses in the disk are

$$\sigma_r = \sigma_\theta = -p_o \text{ and } \tau_{r\theta} = \frac{M_t}{2\pi h r^2}$$

and at

$$r = r_i : \sigma_r = \sigma_\theta = -p_o \text{ and } \tau_{r\theta} = \frac{M_t}{2\pi h r_i^2}.$$

4.9.3.3 Axisymmetric Problems with Radial Temperature Gradient

Consider an axisymmetric plane problem with a temperature gradient such that the temperature rise distribution is a function of r only, as

$$T = T(r)$$

The governing equation for plane strain is obtained from Equation 4.21 by putting $\nabla^2 V = 0$ (case of zero body forces), i.e.,

$$\nabla^2\nabla^2\phi = -\frac{\alpha E}{1-\nu}\nabla^2 T \tag{4.40a}$$

Writing

$$\nabla^2 = \frac{1}{r}\frac{d}{dr}\left(r\frac{d}{dr}\right)$$

Equation (4.40a) takes the form:

$$\frac{1}{r}\frac{d}{dr}\left(r\frac{d}{dr}\right)\left(\nabla^2\phi\right) = -\frac{\alpha E}{1-\nu}\frac{1}{r}\frac{d}{dr}\left(r\frac{dT}{dr}\right)$$

or

$$r\frac{d}{dr}\left(\nabla^2\phi\right) = -\frac{\alpha E}{1-\nu}\int\left[r\frac{d^2T}{dr^2} + \frac{dT}{dr}\right]dr$$

Integrating by parts twice gives

$$\nabla^2\phi = -\frac{\alpha E}{1-\nu}\left(T + C_1 \ln r + C_2\right) \tag{4.40b}$$

where C_1 and C_2 are constants of integration. Further integration of Equation 4.40b gives the stress function, which yields the stresses

$$\begin{aligned}\sigma_r &= \frac{1}{r}\frac{d\phi}{dr} = \frac{A}{r^2} + B(1+2\ln r) + 2C - \frac{\alpha E}{1-\nu}\frac{1}{r^2}\int Trdr \\ \sigma_\theta &= \frac{d^2\phi}{dr^2} = -\frac{A}{r^2} + B(3+2\ln r) + 2C - \frac{\alpha E}{1-\nu}\left(T - \frac{1}{r^2}\int Trdr\right)\end{aligned} \tag{4.41}$$

where A, B, and C are constants of integration. Recalling that $B = 0$, for axisymmetric problems these expressions reduce to

$$\begin{aligned}\sigma_r &= \frac{A}{r^2} + 2C - \frac{\alpha E}{1-\nu}\frac{1}{r^2}\int Trdr \\ \sigma_\theta &= -\frac{A}{r^2} + 2C - \frac{\alpha E}{1-\nu}\left[T - \frac{1}{r^2}\int Trdr\right]\end{aligned} \tag{4.42}$$

Following a similar procedure, the stresses for the case of plane stress conditions with temperature gradient $T = T(r)$ could be obtained from the governing equation:

$$\nabla^2\nabla^2\phi = -\alpha E\nabla^2 T \tag{4.43}$$

or

$$\nabla^2\phi = -\alpha E(T + C_1 \ln r + C_2) \tag{4.44}$$

which yields the following stresses:

$$\begin{aligned}\sigma_r &= \frac{A}{r^2} + 2C - \frac{\alpha E}{r^2}\int Trdr \\ \sigma_\theta &= -\frac{A}{r^2} + 2C - \alpha E\left(T - \frac{1}{r^2}\int Trdr\right)\end{aligned} \tag{4.45}$$

These equations apply to steady and unsteady heat flow. For steady heat flow, $\nabla^2 T = 0$, i.e.,

$$\nabla^2 T = \frac{d^2T}{dr^2} + \frac{1}{r}\frac{dT}{dr} = \frac{1}{r}\frac{d}{dr}\left(r\frac{dT}{dr}\right) = 0 \tag{4.46a}$$

which gives $r\,dT/dr = K$, where K is a constant. If at $r = r_i$, $T = T_i$ and, at $r = r_o$, $T = T_o$, the temperature change from T_o at any radius is thus obtained by integration as

$$T = \frac{T_i - T_o}{\ln(r_o/r_i)}\ln(r_o/r) \tag{4.46b}$$

Example 4.9:

A solid disk of radius r_o *is subjected to a temperature gradient given by* $T = Kr^2$, *where* T *is the temperature rise,* K *being a constant.*

a. *Derive expressions for the stresses developed in the disk.*
b. *For a disk of 100 mm radius made of steel for which* $\alpha = 12 \times 10^{-6}/°C$, $E = 200$ *GPa and* $\nu = 0.3$, *what is the maximum permissible temperature rise if the allowable maximum shear stress at any point,* $^1/_2(\sigma_1 - \sigma_3)$, *in the disk should not exceed 48 MPa?*

Solution:

a. A solid, thin disk subjected to thermal gradient represents a plane stress problem (σ_z = 0). Hence, from Equation 4.45

$$\sigma_r = \frac{A}{r^2} + 2C - \frac{\alpha E}{r^2}\int Trdr$$

and

$$\sigma_\theta = -\frac{A}{r^2} + 2C - \alpha E\left[T - \frac{1}{r^2}\int Trdr\right]$$

To have finite stresses at the center $r = 0$, the constant A must be equal to zero. The boundary condition to be satisfied is

$$\text{at } r = r_o\text{: } \sigma_r = 0$$

This determines the constant C as

$$0 = 2C - \frac{\alpha E}{r_o^2}\int_0^{r_o} Kr^3 dr$$

or

$$2C = \alpha E K r_o^2/4$$

Hence, the stresses are given by

$$\sigma_r = \frac{\alpha E K r_o^2}{4} - \frac{\alpha E}{r^2}\int_o^r Kr^3 dr$$

and

$$\sigma_\theta = \frac{\alpha E K r_o^2}{4} - \alpha E\left[Kr^2 - \frac{1}{r^2}\int_o^r Kr^3 dr\right]$$

Finally, this yields

$$\sigma_r = (\alpha EK/4)\left(r_o^2 - r^2\right)$$

and

$$\sigma_\theta = (\alpha EK/4)\left(r_o^2 - 3r^2\right)$$

The radial displacement u is determined from the thermoelastic stress–strain Equations 3.19 for plane stress conditions as

$$\varepsilon_\theta = \frac{u}{r} = \frac{1}{E}\left[\sigma_\theta - \nu\sigma_r\right] + \alpha T$$

Hence, the radial displacement is

$$u = \frac{\alpha T}{4r}\left[(1-\nu)r_o^2 + (1+\nu)r^2\right]$$

The stress distribution as shown in Figure 4.12 indicates an equibiaxial tensile stress system as

$$\sigma_r = \sigma_\theta = \alpha EKr_o^2/4$$

at the center of the disk and a compressive stress

$$\sigma_\theta = -\alpha EKr_o^2/2$$

at the periphery of the disk.

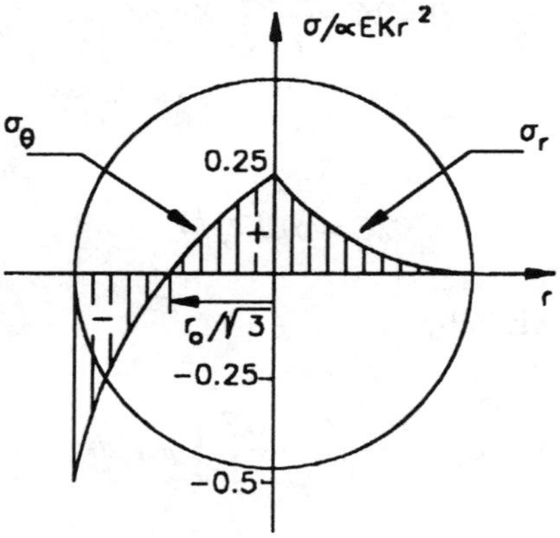

FIGURE 4.12 Example 4.9.

b. For $\alpha = 12 \times 10^{-6}$/°C, $E = 200$ GPa and a maximum shear stress at the center of 48 MPa ($\sigma_r = \sigma_\theta$, $\sigma_z = 0$) using Equations 1.31:

$$\tfrac{1}{2}\left(\sigma_\theta - \sigma_z\right) = \tfrac{1}{2}\alpha EKr_o^2/4 = 48 \text{ MPa}$$

Hence,

$$\tfrac{1}{2}\left(12\times10^{-6}\right)\left(200\times10^{9}\right)\left(T_o/r_o^2\right)\left(r_o^2/4\right) = 48\times10^{6}$$

giving $T_o = 160°C$ and $T = 16000r^2$ (r in m). The radial displacement at the periphery is given by

$$u = \frac{12 \times 10^{-6} \times 160}{4 \times 100}\left[(1-0.3)100^2 + (1+0.3)100^2\right]$$
$$= 0.096 \text{ mm}$$

If the disk as a whole is at uniform temperature of 160°C, then all the stress components vanish and the radial displacement at the periphery is equal to $\alpha r_o T_o$, i.e., o.192 mm.

4.9.4 A Note on Finding a Stress Function in Polar Coordinates

The biharmonic equation as expressed in polar coordinates, by Equation 4.26, is a fourth-order partial differential equation. It can be solved using a separation of variables method in the form:

$$\phi = f_1(r) \cdot f_2(\theta)$$

where $f_2(\theta)$ takes the form of a sine or cosine function. The resulting differential equation is an Euler type which upon solution yields four different stress functions.[7] One of these stress functions is that obtained in Equation 4.31 to solve axisymmetric problems without body forces. In such cases, the stress function is totally defined up to a certain number of constants, which are to be determined using the appropriate boundary conditions of the specific problem under consideration.

4.10 A Glossary of Stress Functions for Some Plane Problems

Numerous stress functions are already available in the literature to solve elastic problems. A glossary of some of these functions, which may be used to solve some engineering problems, are tabulated below for Cartesian and polar coordinates. Superposing any number of these stress functions together with the prevailing boundary conditions is obviously applicable and helps in solving problems with compound boundary conditions.

4.10.1 Cartesian Coordinates

Stress Function	Problem Description	Illustration
Algebraic Polynomials: $\sigma_x = \frac{\partial^2 \phi}{\partial x^2}$, $\sigma_y = \frac{\partial^2 \phi}{\partial y^2}$, $\tau_{xy} = -\frac{\partial^2 \phi}{\partial x \partial y}$		
1. $\phi = ay^2$	Plate under simple tension	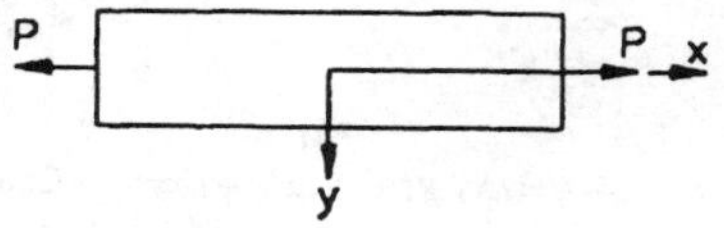
2. $\phi = axy$	Plate subjected to pure shear	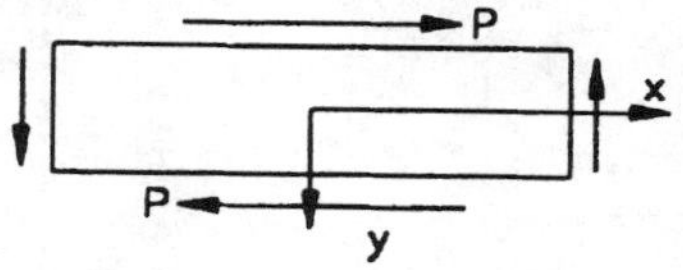

3. $\phi = ay^3$

Pure bending of beams (or cantilever beam under end moment).

Also

Retaining wall subjected to equal hydrostatic pressure at both sides

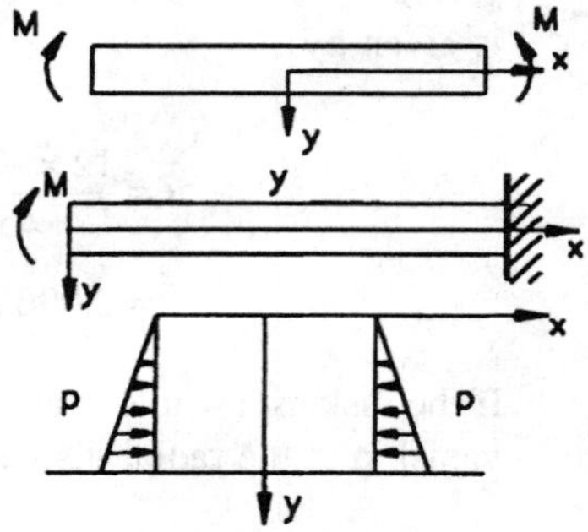

4. $\phi = ax^3 + bx^2y + cy^3$

Retaining wall subjected to hydrostatic pressure

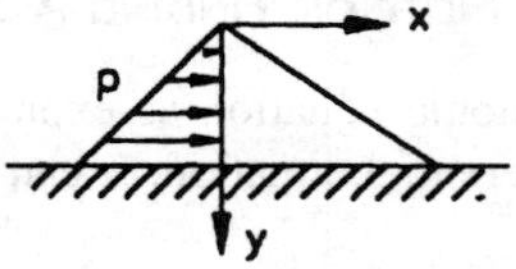

5. $\phi = axy + bxy^3$

Cantilever beam subjected to an end force

Also

Simply supported beam subjected to a concentrated vertical load (treat the load as a shear stress distribution on each half of the beam)

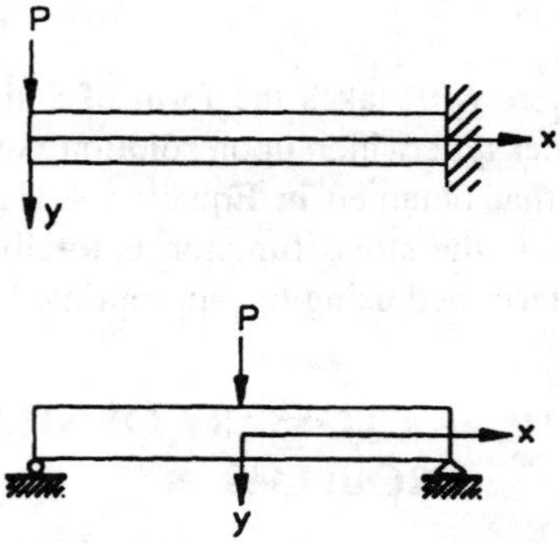

6. $\phi = ax^2 + bx^2y + cy^3 + dx^2y^3 + ey^5$

Simply supported beam subjected to vertical uniformly distributed load

Also

Cantilever beam subjected to uniformly distributed vertical load

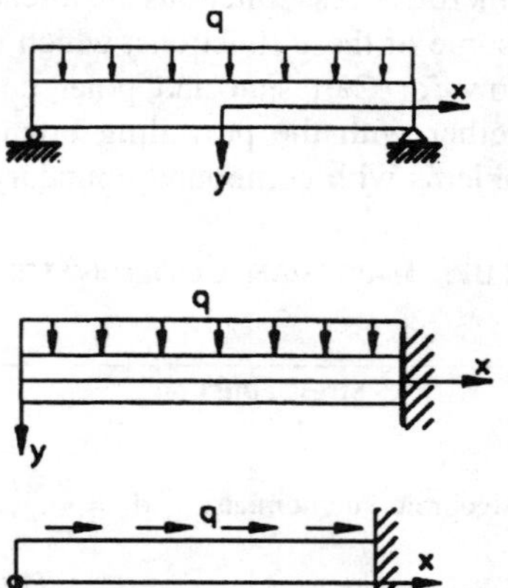

7. $\phi = axy + bxy^2 + cxy^3 + dy^2 + ey^3$

Cantilever beam subjected to uniformly distributed shear over its upper edge

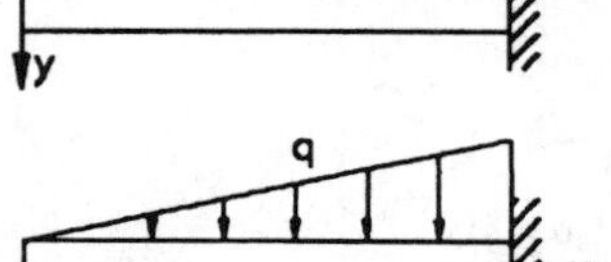

8. $\phi = axy + bx^3 + cx^3y + dxy^3 + ex^3y^3 + fxy^5$

Cantilever beam subjected to a vertical linearly distributed load

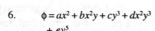

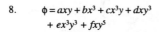

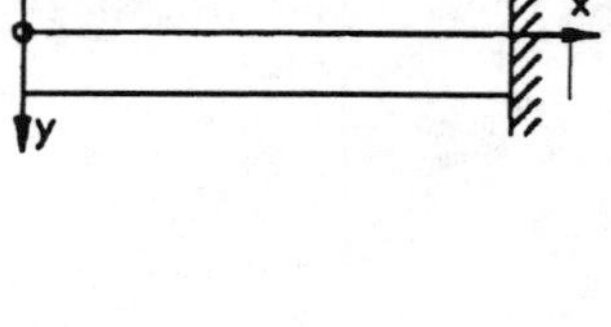

9. $\phi = axy + bx^2 + cx^3 + dx^2y + ey^3 + fxy^3 + gx^3y + hx^2y^3 + ky^5 + lxy^5 + mx^3y^3$ — Simply supported beam subjected to vertical triangular distributed load

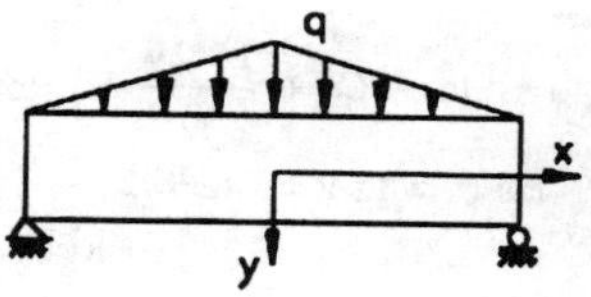

Trigonametric Functions: $\sigma_x = \frac{\partial^2\phi}{\partial y^2},\ \sigma_y = \frac{\partial^2\phi}{\partial x^2},\ \tau_{xy} = -\frac{\partial^2\phi}{\partial x\partial y}$

10. $\phi = \sin\alpha x(C_1\cosh\alpha y + C_2\sinh\alpha y + C_3y\cosh\alpha y + C_4y\sinh\alpha y)$ — Simply supported beam subjected along both edges to continuously distributed vertical loads of intensity $A\sin\alpha x$ and $B\sin\alpha x$

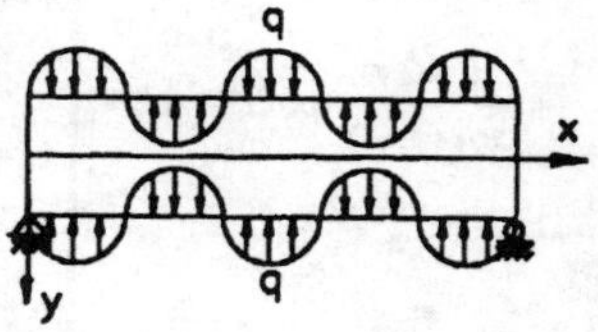

4.10.2 Polar Coordinates

Stress Function	Problem Description	Illustration

Axisymmetric without body forces: $\sigma_r = \frac{1}{r}\frac{d\phi}{dr},\ \sigma_\theta = \frac{d^2\phi}{dr^2},\ \tau_{r\theta} = 0$

1. $\phi = Cr^2$ — Circular solid cylinder or disk under peripheral uniformly distributed load

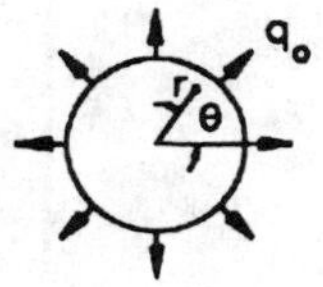

2. $\phi = A\ln r + Cr^2$ — Pressurized thick-walled cylinders or disks

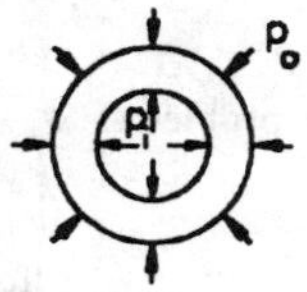

3. $\phi = A\ln r + Br^2\ln r + Cr^2$ — Pure bending of curved bars

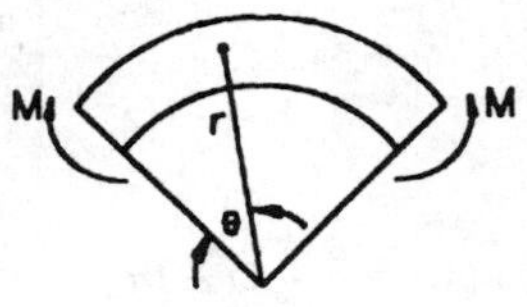

Axisymmetric with body forces: $\sigma_r = \frac{1}{r}\frac{d\phi}{dr} + V,\ \sigma_\theta = \frac{d^2\phi}{dr^2} + V,\ \tau_{r\theta} = 0$

4. $\phi = A \ln r + Cr^2 + \dfrac{1-2\nu}{32}\pi\omega^2 r^4$, with $V = -\pi\omega^2 r^2/2$ — Rotating thin disks

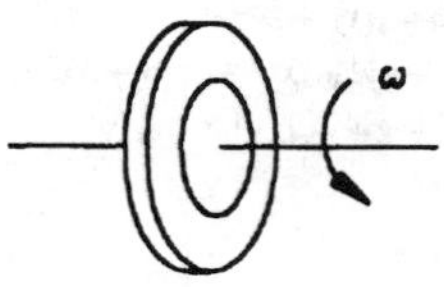

5 $\phi = A \ln r + Cr^2 + \dfrac{1-2\nu}{32(1-\nu)}\rho\omega^2 r^4$, with $V = -\rho\omega^2 r^2/2$ — Rotating long cylinders

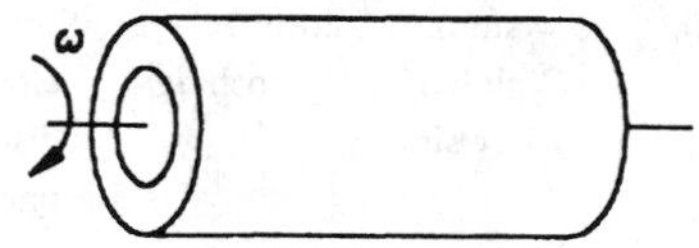

Axisymmetric with radial temperature gradiant: $\sigma_r = \dfrac{1}{r}\dfrac{d\phi}{dr}, \sigma_\theta = \dfrac{d^2\phi}{dr^2}, \tau_{r\theta} = 0$

6. $\phi = A \ln r + Cr^2 - \dfrac{\alpha E}{1-\nu}\displaystyle\int\left[\frac{1}{r}\int Trdr\right]dr$ — Plane strain conditions, e.g., long thick-walled cylinder

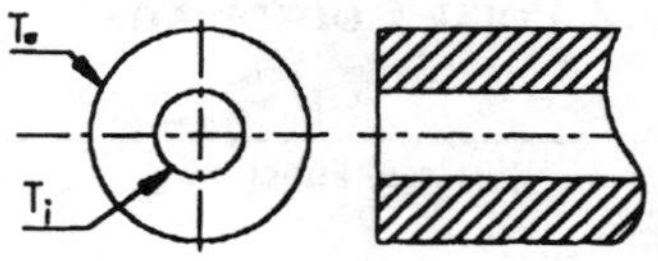

7. $\phi = A \ln r + Cr^2 - \alpha E \displaystyle\int\left[\frac{1}{r}\int Trdr\right]dr$ — Plane stress conditions, e.g., thin annular disk

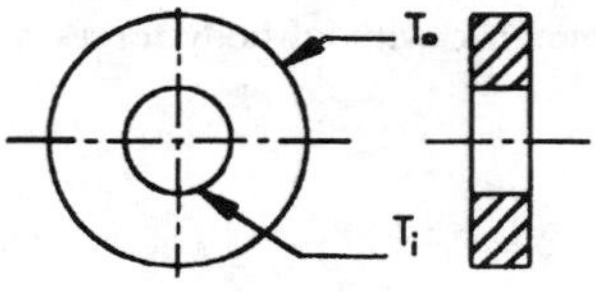

Axially nonsymmetric problems: $\sigma_r = \dfrac{1}{r}\left(\dfrac{\partial\phi}{\partial r} + \dfrac{1}{r}\dfrac{\partial^2\phi}{\partial\theta^2}\right) + V,\ \sigma_\theta = \dfrac{\partial^2\phi}{\partial r^2} + V,\ \tau_{r\theta} = -\dfrac{\partial}{\partial r}\left(\dfrac{1}{r}\dfrac{\partial\phi}{\partial\theta}\right)$

8. $\phi = C\theta$ — Torsion of thin disk by a twisting moment

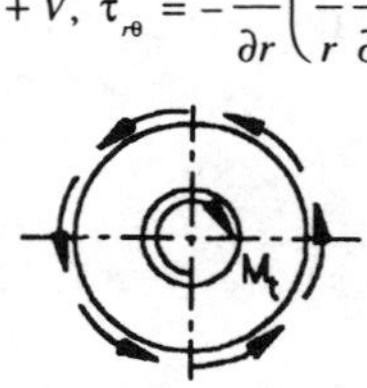

9. $\phi = [Ar^3 + \dfrac{B}{r} + Cr + Dr \ln r]\sin\theta$ — Bending of a curved bar by a force at the end

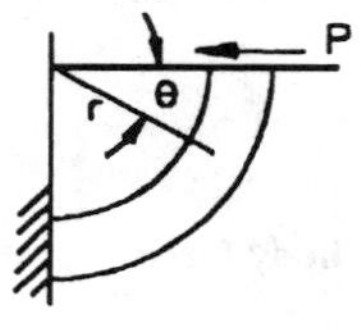

10. $\phi = [Ar^3 + \dfrac{B}{r} + Cr + \ln r]\sin\theta$ — Bending of a curved bar by two symmetrical end radial forces

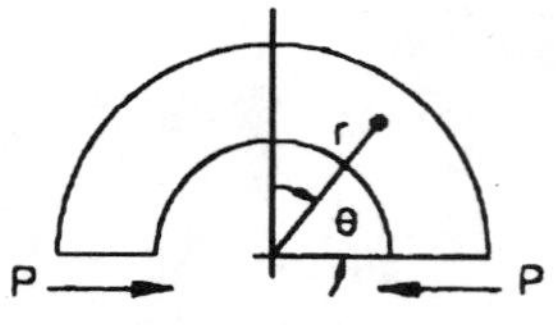

11. $\phi = Cr\,\theta \sin \theta$

Wedge subjected to an axial force at its vertex
Also
Semi-infinite plate subjected to a concentrated force perpendicular to its boundary

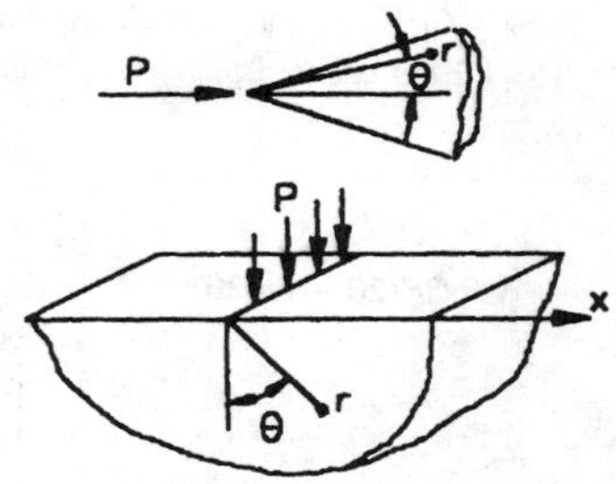

12. $\phi = Cr\,\theta \cos \theta$

Wedge subjected to a perpendicular force at its vertex
Also
Semi-infinite plate subjected to a concentrated horizontal force along its boundary

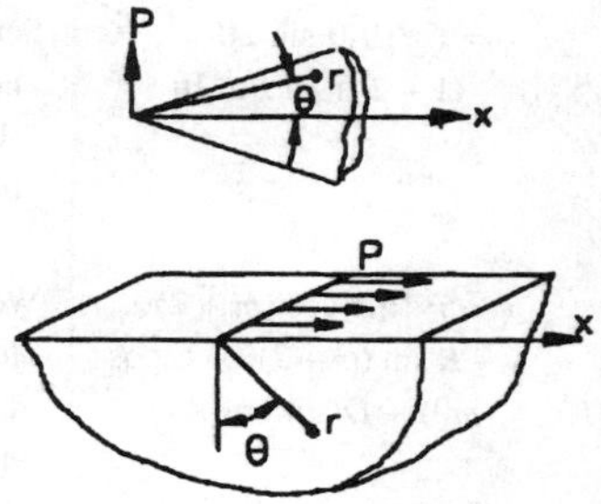

13. $\phi = A\theta + B \sin 2\theta$

Wedge subjected to a bending moment at the vertex
Also
Semi-infinite plate subjected to a moment at a point of its boundary

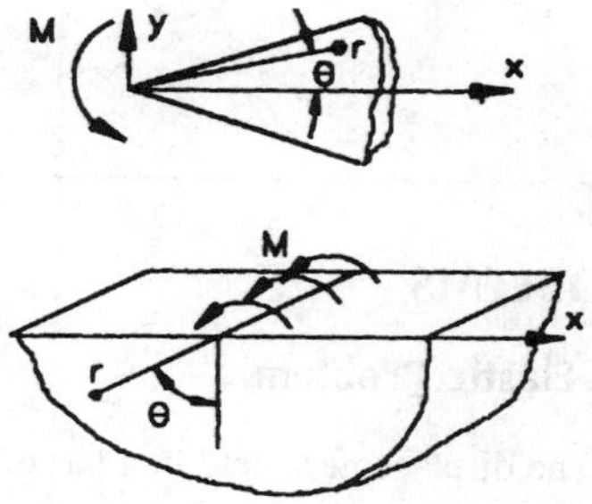

14. $\phi = C_1 r^2 \ln r + C_2 r^2 + C_3 \ln r + C_4 + (C_5 r^4 + C_6 r^2 + C_7 + C_8/r^2)\cos 2\theta$

Stress concentration around a circular hole in a loaded infinite thin plate

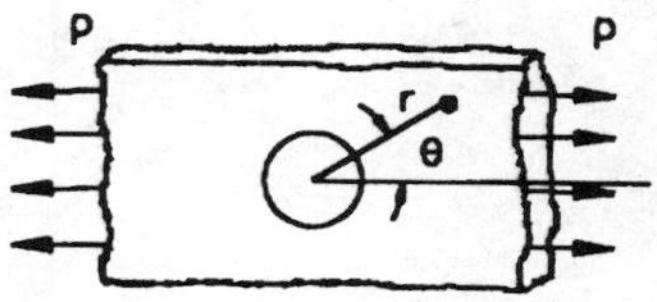

15. $\phi = r^2(C_1 \cos 2\theta + C_2 \sin 2\theta + C_3\theta + C_4)$

Wedge subjected to a uniform normal load along one edge

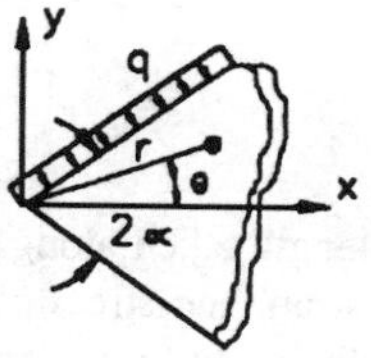

16. $\phi = r^2(C_1 \cos 2\theta + C_2)$

Wedge subjected to a uniform shear load on both edges

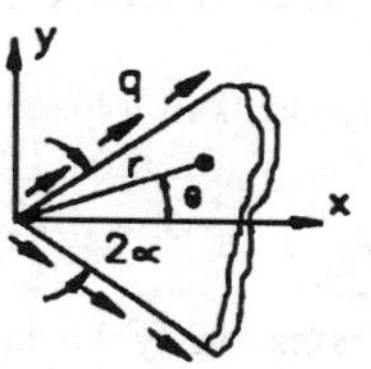

17. $\phi = A(r^2\theta - r_1^2\theta_1)$

Semi-infinite plate subjected to a uniform load over a portion of its boundary

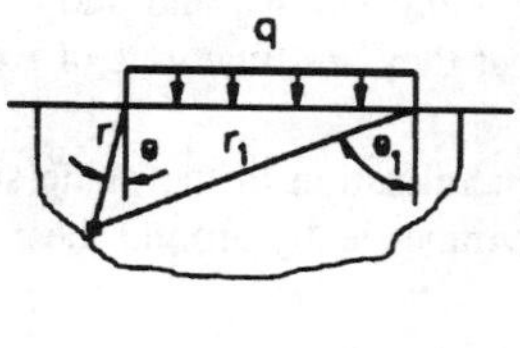

18.	$\phi = Cr^2[2(\alpha - \theta) + \sin 2\theta - 2 \tan \alpha \cos^2 \theta]$	Wedge subjected to a uniform load on the face $\theta = 0$	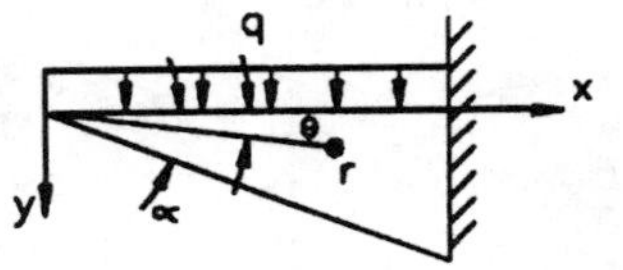
19.	$\phi = Cr^2(2\theta - \sin 2\theta)$	Semi-infinite plate with its boundary subjected to a uniform perpendicular load extending from the origin indefinitely to the left	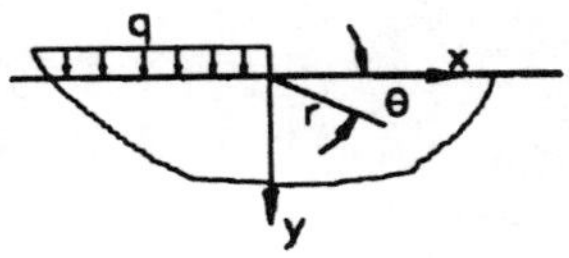
20.	$\phi = Cr^2[{}^1\!/_2\theta \sin 2\theta - (1 - 2 \ln r) \sin^2 \theta]$	Semi-infinite plate with its boundary subjected to a uniform shear load extending from the origin indefinitely to the left	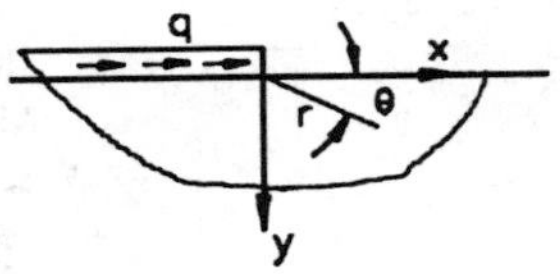
21.	$\phi = r^{m+2}[A \cos (m + 2)\theta + B \sin (m + 2)\theta + C \cos (m\theta) + D \sin (m\theta)]$	Wedge subjected to a distributed load of intensity $q = q_o r^m$ along the face $\theta = 0$ (with $m = 0$, the case no. 18 is obtained)	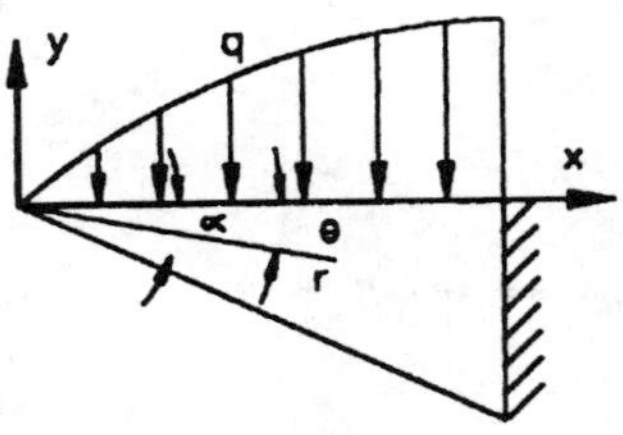

PROBLEMS

The Elastic Problem

4.1 The displacement field in a bar of uniform cross section deformed by its own weight is given by

$$u = \frac{\nu \rho g}{E} x(L - y)$$

$$v = \frac{-\rho g}{E}\left[\left(Ly - \tfrac{1}{2} y^2\right)\right]$$

$$w = \frac{\nu \rho g}{E}(L - y)z$$

where L is the bar length taken along the y-axis and ρ is its density.

Determine the strain and stress components and then check that both compatibility and equilibrium are satisfied. Note that the body force components are given by $F_x = F_z = 0$ and $F_y = -7\rho g$.

4.2 The stress components in a loaded elastic solid are given as

$$\sigma_r = A/r^3, \quad \sigma_\theta = \sigma_\phi = -A/2r^3, \quad \text{and} \quad \tau_{r\theta} = \tau_{\theta\phi} = \tau_\phi = 0$$

Show that these stresses satisfy the equilibrium equations, assuming that the body forces are zero. Determine the corresponding strains, and show that they satisfy the compatibility equations. Find the displacements, assuming that they are functions of r only.

4.3 In a generalization to the plane stress function, consider the stress components in a solid of revolution deformed symmetrically around the z-axis as

$$\sigma_r = \frac{\partial}{\partial z}\left(\nu\nabla^2\phi - \frac{\partial^2\phi}{\partial r^2}\right)$$

$$\sigma_\theta = \frac{\partial}{\partial z}\left(\nu\nabla^2\phi - \frac{1}{r}\frac{\partial\phi}{\partial r}\right)$$

$$\sigma_z = \frac{\partial}{\partial z}\left[(2-\nu)\nabla^2\phi - \frac{\partial^2\phi}{\partial z^2}\right]$$

$$\tau_{rz} = \frac{\partial}{\partial r}\left[(1-\nu)\nabla^2\phi - \frac{\partial^2\phi}{\partial z^2}\right]$$

a. Find the condition for which these stresses satisfy the equilibrium equations.
b. Using a stress function,

$$\phi = a_4\,(8z^4 - 24\,r^2z^2 + 3r^4) + b_4\,(2z^4 + r^2z^2 - r^4)$$

solve the problem of pure bending of a solid circular plate.

4.4 The displacement components in spherical polar coordinates (r, θ, ϕ) in a loaded elastic solid are given by

$$u = (2A\cos\phi + B)/r,\quad v = 0 \quad\text{and}\quad w = (C\cos^2\phi + B\cos\phi - A)/r$$

Derive expression for the strain and stress components and check if the compatibility and equilibrium equations are satisfied. Then, determine the relation among the constants A, B, and C. Neglect body forces.

Pressurized Thick-Walled Spheres

4.5 A steel spherical pressure vessel having an inside diameter of 650 mm is made of two halves welded together. If the efficiency of the welded joint is 75% and the allowable stress in tension is 130 MPa, determine the wall thickness of the vessel needed in order to sustain a pressure of 5 MPa.

4.6 A steel hollow sphere of 500 mm inside diameter is to withstand an external fluid pressure of 50 MPa. Find the necessary thickness if the maximum compressive stress in the vessel is limited to 250 MPa. Find also the diameter and thickness changes under this pressure.

4.7 A steel, thick-walled spherical shell is subjected to a uniform inner and outer pressure p. Find the change in the inner and outer diameters.

4.8 A thick-walled spherical pressure vessel has an inner and outer radii r_i and r_o, respectively. The allowable maximum shear stress of the vessel material is k. Taking $r_o/r_i = 3$ and $\nu = 1/3$, determine:

a. The maximum internal pressure p_i that acts alone and the corresponding displacement at both the inner and outer radii;
b. The maximum external pressure p_o that acts alone;
c. The maximum internal pressure p_i that acts simultaneously with an external pressure $p_o = \alpha p_i/2$, where α is a constant less than one.

Cartesian Stress Function

4.9 Derive the stress compatibility equations in Cartesian coordinates for (a) plane stress and (b) plane strain. Under what conditions do these equations become $\nabla^2(\sigma_x + \sigma_y) = 0$?

4.10 Find the conditions for which the following functions become valid stress functions:

a. $\phi = ax^2 + bxy + cy^2$;
b. $\phi = (C_1 e^{\alpha y} + C_2 e^{-\alpha y} + C_3 y e^{\alpha y} + C_4 y e^{-\alpha y}) \sin \alpha x$;

4.11 A rectangular strip bounded by $x = \pm a$, $y = \pm b$. For what value of c will the expression $\phi = ax^2 + by^3 + cx^2y^2$ be a stress function (neglect body forces)? With the proper expression for ϕ, determine the displacement along the boundary of the strip and make a sketch showing the deformed strip.

4.12 Derive the stresses associated with the stress function:

$$\phi = ax^2 + bxy + cy^2$$

Represent the distribution of these stresses on the boundaries of the rectangular plate shown in the figure.

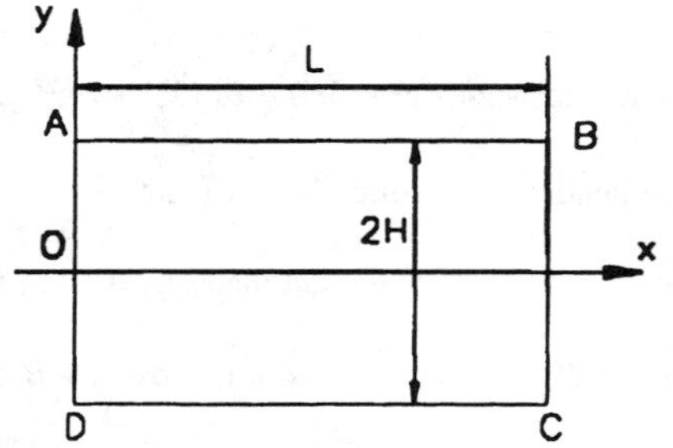

Problems 4.12 and 4.13

4.13 Show that $\phi = axy^3 + bxy$ is a valid stress function. Apply this to a rectangular plate, as shown in the figure accompanying Problem 4.12 and hence find the displacement u and v for points A, B, C, and D given that at O

$$u = v = \frac{\partial u}{\partial y} = 0$$

Polar Stress Function

4.14 Use the stress function listed in Section 4.10.2, stress function no. 21, taking $m = 1$ to derive the stress and strain components. Check that the equilibrium and compatibility equations are satisfied. Neglect body forces.

4.15 Use the stress function $\phi = (-P/\pi)r\theta \sin \theta$ to derive expressions for the stresses (σ_r, σ_θ, $\sigma_{r\theta}$) in a semi-infinite plate loaded with a vertical concentrated line load P per unit thickness.

a. Transform these stresses to Cartesian coordinates and plot them at a constant depth H below the surface.
b. Starting from the stresses in Cartesian coordinates, derive a stress function $\phi(x, y)$ and compare it with $\phi(r, \theta)$.

Thermoelastic Plane Problems

4.16 A square, thin plate is clamped all around its periphery and subjected to the temperature gradients shown in the figure. Suggest a suitable stress function in Cartesian coordinates and derive expressions for the thermal stresses in the plate. If a plate of side length H = 500 mm is made of copper alloy, E = 150 GPa, ν = 0.3 and $\alpha = 16 \times 10^{-6}$/°C, calculate the maximum shear stress value in the plate.

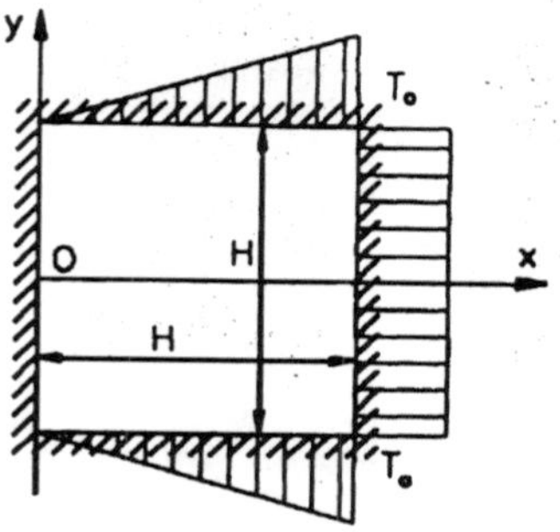

Problem 4.16

4.17 A thin, rectangular strip of length $4H$ and height $2H$ is subjected to a temperature field given by $T = K(x^2 + y^2)$, where x and y are the axes of symmetry for the strip and K is a given constant. If the four edges of the strip are held between rigid supports and given the polynomial

$$\phi = ax^2 + by^2 + c(x^4 + y^4)$$

a. Show that ϕ is a possible stress function for this problem and hence find the value of c.
b. By considering the boundary conditions, determine the values of the remaining constants a and b.
c. Obtain the final expressions for the stress components and sketch their distributions along the edges of the strip.
d. Show that $\varepsilon_y = 4\varepsilon_x$ at the center of the strip.

4.18 Derive expressions for the stress components in a thick-walled sphere under steady state conditions of heat transfer according to

$$T = \left(T_i - T_o\right)\left(\frac{r_o/r - 1}{r_o/r_i - 1}\right)$$

where T is the temperature difference between any radius r and the the outer radius r_o.

4.19 Use the stress function,

$$\phi = A\ln\left(\frac{r}{r_i}\right) + B\left(\frac{r}{r_i}\right)^2 \ln\left(\frac{r}{r_i}\right) + C\left(\frac{r}{r_i}\right)^2 + D\alpha E\int_{r_i}^{r}\left[\frac{1}{r}\int_{r_i}^{r} Trdr\right]dr$$

to derive expressions for the thermal stresses induced in a curved bar with radial temperature gradient $T(r)$.

REFERENCES

1. Venkatraman, B. and Patel, S. A., *Structural Mechanics with Introductions to Elasticity and Plasticity*, McGraw-Hill, New York, 1970.
2. Sokolnikoff, I. S., *Mathematical Theory of Elasticity*, McGraw-Hill, New York, 1956.
3. Wang, C.-T., *Applied Elasticity*, McGraw-Hill, New York, 1953, 44.
4. Timoshenko, S. and Goodier, J. N., *Theory of Elasticity*, 2nd ed., McGraw-Hill, New York, 1951, 241.
5. Ford, H., *Advanced Mechanics of Materials*, Longman, London.
6. Muskhelishili, N. I, *Some Basic Problems of the Theory of Elasticity*, Engl. trans. by Radok, J. R. M., Groningen, Noordhoff, 1956.
7. Dally, J. W. and Riley, W. F., *Experimental Stress Analysis*, McGraw-Hill, New York, 1965, 71–74.

5 Elastic Plane Problems in Cartesian Coordinates

Solutions to some engineering problems in Cartesian coordinates are obtained by finding appropriate stress functions for plane strain and plane stress conditions. Stress functions in the form of algebraic polynomials are chosen to solve the problems of a retaining wall subjected to hydrostatic pressure, a simply supported beam subjected to uniformly distributed load, and a cantilever beam under end force. The results are compared with those obtained from elementary strength-of-materials solutions. Use of stress functions in the form of trigonometric series is demonstrated by solving plane problems of bars subjected to two equal and opposite forces in the lateral and axial directions.

5.1 INTRODUCTION

In Chapter 4 the plane elastic problem is defined and a solution is formulated in terms of a stress function whose appropriate derivatives define the stress components. It is shown that the solution can be obtained by finding a biharmonic function, in which the body forces and thermoelastic effects are included, that satisfies boundary conditions. This approach will be adopted in the following sections to solve some engineering problems in Cartesian coordinates for which the stress functions can be expressed in terms of simple forms of algebraic polynomials and trigonometric series. The solutions are obtained by employing the semi-inverse method.

5.2 PROBLEMS SOLVED IN TERMS OF ALGEBRAIC POLYNOMIALS

As mentioned before, the semi-inverse method of solving elasticity problems consists of specifying in advance the form of stress function. For instance, for continuous boundary loads, such as distributed loads in the case of beam problems, the stress function ϕ may be generally expressed in the form of a polynomial in x and y. The problem is thus reduced to the determination of the polynomial coefficients that satisfy the biharmonic equation and the prescribed boundary conditions.

As a first step, consider the use of the simple constant term and first-degree polynomials:

$$\phi_o = a_o \quad \text{and} \quad \phi_1 = a_1 x + b_1 y$$

where a_o, a_1, and b_1 are constants. Both functions ϕ_o and ϕ_1 satisfy the biharmonic equation $\nabla^2\nabla^2\phi = 0$ identically and applying Equations 4.17 will lead to trivial stress solutions: $\sigma_x = \sigma_y = \tau_{xy} = 0$ over the entire elastic body. Therefore, the constant and first-degree terms are not included in the stress functions since they vanish upon differentiation and do not contribute to the stress components.

As a next step, consider the use of the second and third degree polynomials:

$$\phi_2 = a_2 x^2 + b_2 xy + c_2 y^2$$

$$\phi_3 = a_3 x^3 + b_3 x^2 y + c_3 xy^2 + d_3 y^3$$

Both these polynomials again satisfy the biharmonic equation $\nabla^2\nabla^2\phi = 0$ identically leading respectively, to the stress solutions, in the absence of body forces as

$$\sigma_x = \frac{\partial^2 \phi_2}{\partial y^2} = 2c_2, \quad \sigma_y = \frac{\partial^2 \phi_2}{\partial x^2} = 2a_2, \quad \tau_{xy} = -\frac{\partial^2 \phi_2}{\partial x \partial y} = -b_2,$$

$$\sigma_x = \frac{\partial^2 \phi_3}{\partial y^2} = 2(c_2 x + 3d_3 y), \quad \sigma_y = \frac{\partial^2 \phi_3}{\partial x^2} = 2(a_3 x + b_3 y), \quad \tau_{xy} = -2(b_3 x + c_3 y)$$

While the constants associated with these stresses must be determined from the boundary conditions of the given problem, the biharmonic equation does not impose any restrictions on the coefficients a_2, b_2, ... etc. Note that a second-degree polynomial yields uniform stress distribution, while the third-degree polynomial gives a linear stress distribution.

On the other hand, consider the use of the fourth- and fifth-degree polynomials:

$$\phi_4 = a_4 x^4 + b_4 x^3 y + c_4 x^2 y^2 + d_4 xy^3 + e_4 y^4$$

$$\phi_5 = a_5 x^5 + b_5 x^4 y + c_5 x^3 y^2 + d_5 x^2 y^3 + e_5 xy^4 + f_5 y^5$$

Substitution of these functions in the biharmonic equation shows that it will be satisfied if

$$e_4 = -(a_4 + c_4/3), \quad e_5 = -(5a_5 + c_5), \quad f_5 = -(b_5 + d_5)/5$$

The coefficients a_4, b_4, c_4, d_4 and a_5, b_5, c_5, d_5 must be then determined by the boundary conditions of the given problem as demonstrated by the following two examples.

Note that in the above analysis the selection of the origin and the positive directions of axes x and y is arbitrary and governed by the nature of the problem. However, in expressing boundary conditions the sign convention stated in the previous chapters for the displacement, strain, stress, and stress resultants must be strictly adhered to in relation to the selected directions of the axes.

Polynomials of higher degree as generally expressed by

$$\phi = \sum_{m=o}^{\infty} \sum_{n=o}^{\infty} x^m y^n$$

may be used to solve plane problems. It is then necessary to choose the coefficients so that the biharmonic equation is satisfied. A systematic way of doing this has been proposed[1] to eliminate the unwanted terms without resort to guessing.

Example 5.1:

A thin, rectangular plate of length L, *height* H, *and unit thickness is free of body forces and subjected to a loading around its four edges such that the stress function is found to be* $\phi = cxy^3/3$, *where* c *is an arbitrary constant. Show that this stress function satisfies the biharmonic equation* $\nabla^2\nabla^2\phi = 0$ *and derive the expressions for the stress components* σ_x, σ_y, *and* τ_{xy}. *Sketch the stress distribution along the four edges of the plate and verify that the equations of equilibrium for the plate as a whole are satisfied.*

Solution:

By inspection, it is obvious that the given stress function $\phi = cxy^3/3$ satisfies the biharmonic equation:

$$\nabla^2\nabla^2\phi = 0$$

The stress components are thus

$$\sigma_x = \frac{\partial^2\phi_2}{\partial y^2} = 2c_2xy, \quad \sigma_y = \frac{\partial^2\phi_2}{\partial x^2} = 0, \quad \tau_{xy} = -\frac{\partial^2\phi_2}{\partial x\partial y} = cy^2$$

The distribution of the normal stresses and shear stresses along the four edges of the plate are shown in Figure 5.1. It is obvious that the resultant force on the plate as a whole is zero.

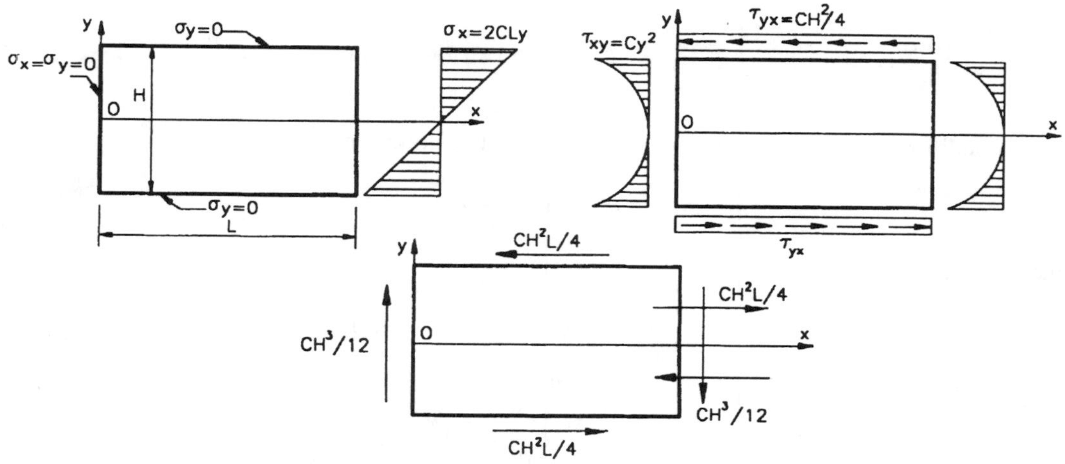

FIGURE 5.1 Example 5.1.

Taking moments about the origin O it is seen that

$$\frac{cH^2L}{4}\frac{2H}{3} + \frac{cH^3L}{12} - \frac{cH^2L}{4} = 0$$

Hence, the plate as a whole is in equilibrium.

Example 5.2:

A rectangular strip of thickness h *and width* 2H, *as shown in Figure 5.2 is subjected to a temperature gradient given by* $T = T_o - ky^2$, *the strip axis is the* x-*axis. If the edges* $y = \pm H$ *are subjected to a uniformly distributed compressive load,* q *per unit length, and the ends are constrained to prevent axial displacement, determine the stresses.*

Solution:

Due to symmetry about x and y, ϕ will contain only even powers for x and y. For a thin strip, plane stress conditions prevail. Neglecting body forces, the governing Equation 4.22 is $\nabla^2\nabla^2\phi = -\alpha E\nabla^2 T = 2\alpha EK$ = constant.

Therefore, ϕ is a fourth-degree polynomial, i.e.,

$$\phi = ax^2 + by^2 + cx^4 + dx^2y^2 + ey^4$$

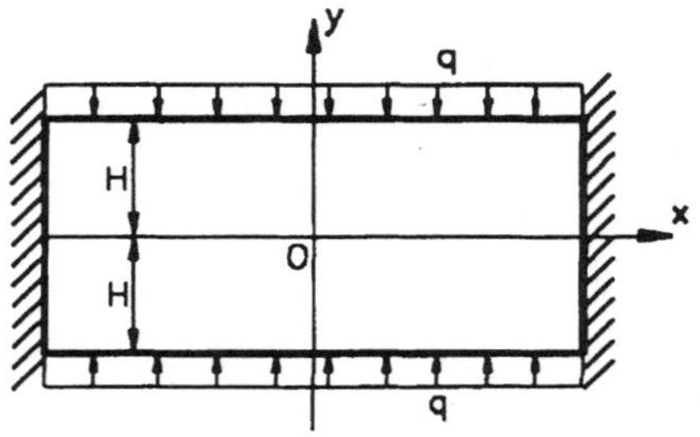

FIGURE 5.2 Example 5.2.

and the stresses are

$$\sigma_x = \frac{\partial^2\phi}{\partial y^2} = 2b + 2dx^2 + 12ey^2, \ \sigma_y = \frac{\partial^2\phi}{\partial x^2} = 2a + 12cx^2 + 2dy^2, \ \tau_{xy} = -\frac{\partial^2\phi}{\partial x \partial y} = -4dxy$$

Since conditions are independent of x, $c = d = 0$ and the stress function reduces to

$$\phi = ax^2 + by^2 + ey^4$$

and, hence,

$$\sigma_x = 2b + 12ey^2, \ \ \sigma_y = 2a, \ \ \tau_{xy} = 0$$

At $y = \pm H$, $\sigma_y - q/h$, thus giving

$$a = -q/2h$$

Also, $\nabla^2\nabla^2\phi = 2\alpha EK = 24e$, which yields

$$e = \alpha EK/12$$

By applying the condition that $u = \varepsilon_x = 0$,

$$\varepsilon_x = 0 = \frac{1}{E}\left[\sigma_x - \nu\left(\sigma_y + \sigma_z\right)\right] + \alpha T$$

$$2b + \alpha EKy^2 = -\nu q/h + E\alpha\left(T_o - Ky^2\right)$$

hence, $2b = -\nu q/h + \alpha ET_o - 2\alpha ETy^2$.

The stress components are thus

$$\sigma_x = -\nu q/h + \alpha ET_o - 2\alpha EKy^2 = -\nu q/h + \alpha E\left(T_o - Ky^2\right)$$

$$\sigma_y = -q/h \text{ and } \tau_{xy} = 0$$

Examples of some plane problems that can be solved in Cartesian coordinates using stress functions in the form of algebraic polynomials are given below.

5.2.1 Retaining Wall Subjected to Hydrostatic Pressure

The problem of a dam or a retaining wall subjected to pressure due to water or loose material could be considered to be a plane strain problem.

The pressure that is a result of gravity is linearly increasing with depth as shown in Figure 5.3a, and thus boundary conditions are on the face *OB* (i.e., for $x = 0$)

$$\sigma_x = -ky, \quad \tau_{xy} = 0 \tag{5.1}$$

and at the face *OA* (i.e., for $x = y \tan \beta$)

$$\sigma_i = 0, \quad \tau_i = 0 \tag{5.2}$$

where σ_i and τ_i are the normal and shear stress components.

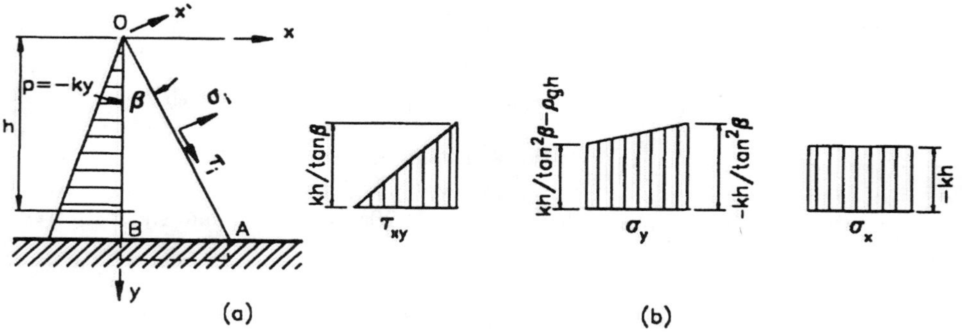

FIGURE 5.3 Stresses in a retaining wall subjected to hydrostatic pressure: (a) load diagram and (b) stress components at the cross section of height h.

The positive direction of *y* is taken to be downward to compare the results with those obtained from the strength of materials approach.

Since the number of boundary conditions are four, an algebraic polynomial with four coefficients would be a suitable choice for the stress function. Hence, ϕ is given by the expression:

$$\phi = a_3x^3 + b_3x^2y + c_3xy^2 + d_3y^3$$

From Equations 4.14 the stresses are obtained from

$$\sigma_x - V = \frac{\partial^2\phi}{\partial y^2}, \quad \sigma_y - V = \frac{\partial^2\phi}{\partial x^2}, \quad \tau_{xy} = -\frac{\partial^2\phi}{\partial x \partial y}$$

In calculating the stresses in retaining walls the weight of the wall must be taken into consideration, since it has a basic role in the wall equilibrium and therefore the potential *V* is given by $V = -\rho gy$, where ρg is the weight per unit volume of the wall material.

Therefore, employing Equations 4.14 the stresses are

$$\sigma_x = \frac{\partial^2 \phi}{\partial y^2} + V = 2c_3 x + 6d_3 y - \rho g y$$

$$\sigma_y = \frac{\partial^2 \phi}{\partial x^2} + V = 6a_3 x + 2b_3 y - \rho g y$$

$$\tau_{xy} = -\frac{\partial^2 \phi}{\partial x \partial y} = -2b_3 x - 2c_3 y$$

The boundary conditions, Equation 5.1, gives

$$c_3 = 0 \text{ and } d_3 = (\rho g - k)/6$$

and hence the stresses are

$$\sigma_x = -ky, \; \sigma_y = 6a_3 x + 2b_3 y - \rho g y, \; \tau_{xy} = -2b_3 x$$

The direction cosines of the normal to the face $0A$ is given by

$$l_i = \cos \beta \text{ and } m_i = -\sin \beta$$

Applying the stress transformation law, Equation 1.20a, the normal and shear stresses on the face OA are

$$\begin{aligned} \sigma_i &= \sigma_x \cos^2 \beta + \sigma_y \sin^2 \beta - \tau_{xy} \sin 2\beta \\ \tau_i &= \tau_{xy} \cos 2\beta + \left[\left(\sigma_x - \sigma_y\right)/2\right] \sin 2\beta \end{aligned} \tag{5.3}$$

or

$$\sigma_i = -ky \cos^2 \beta + \left(6a_3 x + 2b_3 y - \rho g y\right) \sin^2 \beta + 2b_3 x \sin 2\beta$$

$$\tau_i = -b_3 x \cos 2\beta + \left[\left(-ky - 6a_3 x - 2b_3 y + \rho g y\right)/2\right] \sin 2\beta$$

Applying the boundary condition (Equation 5.2) by setting $x = y \tan \beta$ into Equations 5.3 and equating to zero yields after simplification:

$$6a_3 \tan \beta + 6b_3 = \rho g + k/\tan^2 \beta$$

$$6a_3 \tan \beta + 2b_3 \left(2 \tan \beta / \tan 2\beta + 1\right) = \rho g - k$$

Solving gives

$$b_3 = \frac{k}{2 \tan^2 \beta}, \quad a_3 = \frac{\rho g}{6 \tan \beta} - \frac{k}{3 \tan^3 \beta}$$

The stresses thus take the form

$$\sigma_x = ky,$$

$$\sigma_y = \left(\frac{\rho g}{\tan\beta} - \frac{2k}{\tan^3\beta}\right)x + \left(\frac{k}{\tan^2\beta} - \rho g\right)y \tag{5.4}$$

$$\tau_{xy} = -\frac{kx}{\tan^2\beta}$$

and $\sigma_z = \nu(\sigma_x + \sigma_y)$.

The distribution of stresses σ_x, σ_y, and τ_{xy} over the horizontal section $y = H$ are shown in Figure 5.3 b.

Solving this problem by the elementary approach of strength of materials would give the same expression as in Equation 5.4 for only the stress σ_y.

The distribution of the shear stress τ_{xy} as obtained from the elementary approach of strength of materials is parabolic, which is obviously different from that shown in Figure 5.3b. The other stress components σ_x and σ_z are zero by the elementary approach.

By increasing the degree of polynomials representing the stress function ϕ, it is possible to find solutions for more-complicated problems. For instance, using a stress function of the sixth degree will give the solution for the case of a rectangular dam subjected to hydrostatic pressure, as shown in Figure 5.4.

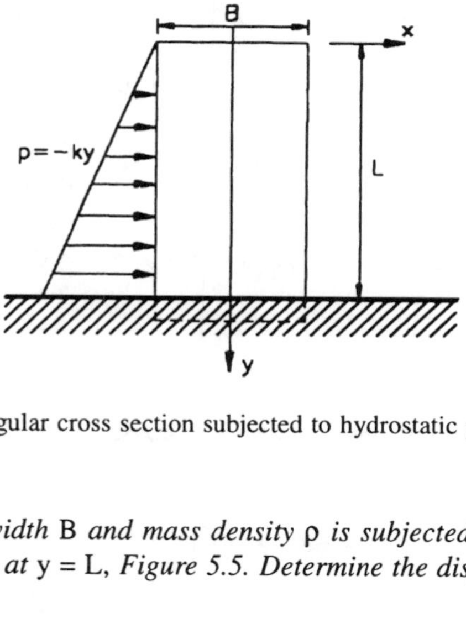

FIGURE 5.4 Wall of a rectangular cross section subjected to hydrostatic pressure.

Example 5.3:

A rectangular thin plate of width B *and mass density* ρ *is subjected to surface pressure* p = −ky *at* x = ±B/2 *and is supported at* y = L, *Figure 5.5. Determine the displacement components* (u, v) *at any point* (x, y)*.*

Solution:

The positive direction of y is taken to be downward. The body force components per unit volume are

$$F_x = 0, \quad F_y = \rho g = -\frac{\partial V}{\partial y}$$

hence,

$$V = -\rho g y \quad \text{and} \quad \nabla^2 V = 0$$

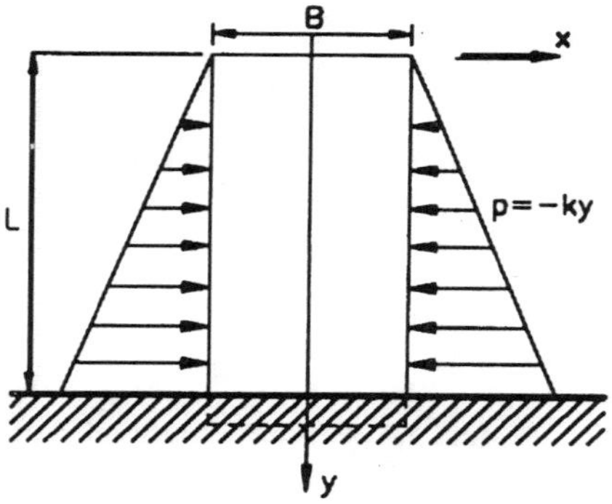

FIGURE 5.5 Example 5.3.

Therefore, the governing equation for plane stress conditions is

$$\nabla^2\nabla^2\phi = 0$$

and the stresses are given by Equations 4.14 as

$$\sigma_x = \frac{\partial^2\phi}{\partial y^2} + V, \quad \sigma_y = \frac{\partial^2\phi}{\partial x^2} + V, \quad \tau_{xy} = -\frac{\partial^2\phi}{\partial x \partial y}$$

Since the problem involves stress boundary conditions of linear form, hence ϕ may be assumed of the third degree, i.e.,

$$\phi = ax^2 + bxy + cy^2 + dx^3 + ex^2y + fy^2x + hy^3$$

Due to symmetry, the y-axis is a principal axis and hence $\tau_{xy} = 0$. Therefore, ϕ does not contain mixed terms of xy and it reduces to

$$\phi = ax^2 + cy^2 + dx^3 + hy^3$$

Furthermore, for the same reason of symmetry about y-axis, ϕ should contain only even powers of x, i.e.,

$$\phi = ax^2 + cy^2 + hy^3$$

The stresses are thus

$$\sigma_x = \frac{\partial^2\phi}{\partial y^2} + V = 2c + 6hy - \rho gy$$

$$\sigma_y = \frac{\partial^2\phi}{\partial x^2} + V = 2a - \rho gy \text{ and } \tau_{xy} = 0$$

The stress boundary conditions are such that

At $y = 0$: $\tau_{xy} = \sigma_y = 0$
At $x = \pm B/2$: $\sigma_x = -ky$ and $\tau_{xy} = 0$

This gives

$$a = c = 0 \text{ and } h = (\rho g - k)/6$$

Therefore,

$$\phi = (\rho g - k)y^3/6$$

and

$$\sigma_x = -ky, \ \ \sigma_y = -\rho g y, \ \ \tau_{xy} = 0$$

To obtain the displacement components, apply Hooke's law, Equations 3.6, for the plane stress condition:

$$\varepsilon_x = \frac{\partial u}{\partial x} = \frac{1}{E}\left(\sigma_x - \nu\sigma_y\right) = -\frac{1}{E}(k - \nu\rho g)y$$

$$\varepsilon_y = \frac{\partial v}{\partial y} = \frac{1}{E}\left(\sigma_y - \nu\sigma_x\right) = -\frac{1}{E}(\rho g - \nu k)y$$

Integration yields

$$u = f(y) - \frac{1}{E}(k - \nu\rho g)xy$$

$$v = f_1(x) - \frac{1}{2E}(\rho g - \nu k)y^2$$

where $f(y)$ is a function of y only and f_1 (x) is a function of x only. To determine these two functions, apply displacement boundary conditions:

At $x = 0$: $u = 0$
At $y = L$: $v = 0$

This yields the displacement components:

$$u = \frac{1}{E}(k - \nu\rho g)xy$$

$$v = \frac{1}{2E}(\rho g - \nu k)\left(L^2 - y^2\right)$$

5.2.2 Simply Supported Beam under Uniformly Distributed Load

Laterally loaded beams are basic components of engineering structures. As illustrated in Figure 5.6, the applied loads on a typical beam are in one single plane, say, the x–y plane, and consist of a uniformly distributed load per unit length. The thickness h of the beam normal to the x–y plane is relatively small compared with its other dimensions, H and L. It follows then that such a beam is essentially in a state of plane stress. Again, the positive direction of y is taken to be downward.

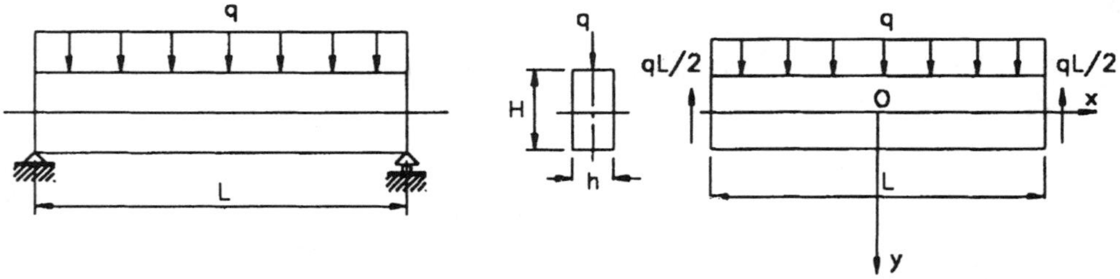

FIGURE 5.6 Simply supported beam under uniformly distributed load.

Referring to Figure 5.6, the reactions at the end supports can be looked upon as a shear force, which is distributed in the depth direction in the same way as the shear stress τ_{xy}. Therefore, the boundary conditions that apply to this problem are

At $x = \pm L/2$:

1. Shear force condition: $\int_{-H/2}^{H/2} \tau_{xy} h dy = \mp \frac{qL}{2}$
2. Condition of no resultant normal force: $\int_{-H/2}^{H/2} \sigma_x h dy = 0$
3. Condition of no bending moment: $\int_{-H/2}^{H/2} \sigma_x h y dy = 0$

At any value of x:

4. At $y = \pm H/2$: $\tau_{xy} = 0$
5. At $y = -H/2$: $\sigma_y = -q/h$
6. At $y = +H/2$: $\sigma_y = 0$

These boundary conditions have to be enforced on the stress solution obtained from a suitable stress function.

Since conditions are symmetrical about the y-axis, i.e., the same values are obtained for $\pm x$, ϕ will contain x of even powers only, as

$$\phi = ax^2 + by^2 + cx^2y + dy^3 + ex^4 + fx^2y^2 + gy^4 + kx^2y^3 + lx^4y + my^5 + nx^6 + \ldots$$

By neglecting the body forces, the stresses are thus

$$\sigma_x = \frac{\partial^2\phi}{\partial y^2} = 2b + 6dy + 2fx^2 + 12gy^2 + 6kx^2y + 20my^3 + \cdots$$

$$\sigma_y = \frac{\partial^2\phi}{\partial x^2} = 2a + 2cy + 12ex^2 + 2fy^2 + 2ky^3 + 12lx^2y + 30nx^4 + \cdots$$

$$\tau_{xy} = -\frac{\partial^2\phi}{\partial x \partial y} = 2cx + 4fxy + 6kxy^2 + 4lx^3 + \cdots$$

The boundary conditions 5 and 6 indicate that σ_y has a constant value, which is independent of x; therefore, ϕ will not contain terms in x higher than x^2. Hence,

$$e = l = n = \text{etc.} = 0$$

The boundary condition (2):

$$\int_{-H/2}^{H/2} \sigma_x h dy = 0$$

is valid for all values of x; hence, σ_x must contain only odd powers of y. Then,

$$b = f = g \ldots = 0$$

Therefore, ϕ is reduced to a simple form, namely,

$$\phi = ax^2 + cx^2y + dy^3 + kx^2y^3 + my^5$$

The requirement that this stress function satisfy the biharmonic equation

$$\nabla^2\nabla^2\phi = 0$$

imposes the restriction such that $k = -5l$ and, thus,

$$\phi = ax^2 + cx^2y + dy^3 + k(x^2y^3 - y^5/5) \tag{5.5}$$

This stress function yields the following expressions for the stresses:

$$\sigma_x = 6dy + 6kx^2y - 4ky^3$$

$$\sigma_y = 2a + 2cy + 2ky^3$$

$$\tau_{xy} = -2cx - 6kxy^2$$

The coefficients a, c, d, and k are determined from the boundary conditions (1), (3), (4), (5), and (6) such that

$$cH + kH^3/4 = q/2h$$

$$d + kL^2/4 - kH^3/10 = 0$$

$$2\left(c + 3kH^2/4\right) = 0$$

$$2a - cH - kH^3/4 = -q/h$$

$$2a + c/H + kH^3/4 = 0$$

These equations yield

$$a = -q/4h, \quad c = 3q/4hH, \quad k = -q/hH^3, \quad d = \frac{q}{hH^3}\frac{H^2}{10} - \frac{L^2}{4}$$

and thus lead to the following stress distribution:

$$\begin{aligned}
\sigma_x &= \frac{qy}{2I}\left(\frac{L^2}{4} - x^2\right) + \frac{qy}{2I}\left(\frac{2y^2}{3} - \frac{H^2}{10}\right) \\
\sigma_y &= -\frac{q}{2I}\left(\frac{y^3}{3} - \frac{H^2 y}{4} + \frac{H^3}{12}\right) \\
\tau_{xy} &= -\frac{qx}{2I}\left(\frac{H^2}{4} - y^2\right)
\end{aligned} \tag{5.6a}$$

Here $I = hH^3/12$ is the second moment of area taken around a line through the centroid.

Since the ends must be free of longitudinal forces, there being no applied force at $x = \pm L/2$, it would appear reasonable to assume that $\sigma_x = 0$ at the ends. However, substitution into the expression for σ_x in Equations 5.6a, shows that σ_x at $x = \pm L/2$ does not vanish and is given by

$$[\sigma_x]_{x=\pm L/2} = \frac{qy}{2I}\left(\frac{2y^2}{3} - \frac{H^2}{10}\right)$$

This distribution cannot exist, as no forces act at the ends. From Saint-Venant's principle it may be concluded, however, that the solution does predict the correct stresses throughout the beam except near the supports. The boundary condition $\sigma_x = 0$ at $x = +L/2$ is not satisfied exactly; however, it is satisfied on the average by the boundary condition 2.

The longitudinal normal stress derived from elementary beam theory is

$$\sigma_x = \frac{My}{I} = \frac{qy}{2I}\left(\frac{L^2}{4} - x^2\right)$$

This is equivalent to the first term of the expression for σ_x in Equations 5.6a. The second term is then the difference between the longitudinal stress given by the stress function approach and the elementary beam theory. The magnitude of this difference is appreciated if the ratio of this second term to σ_x, as given by the elementary theory at $x = 0$, where the bending moment is a maximum, is considered. Substituting $y = H/2$ for the condition of maximum stress gives

$$\frac{(qH/4I)\left(H^2/6 - H^2/10\right)}{(qH/4I)\left(L^2/4\right)} = \frac{4}{15}\left(\frac{H}{L}\right)^2$$

For a beam of (H/L), say, equal to 1/20, the ratio, being equal to 1/1500, is very small. However, for a short beam $(H/L) = 1/2$, the ratio is 1/15. It may be concluded that the elementary beam theory provides a result of sufficient accuracy for beams of ordinary proportions, i.e., when the beam length is large compared with its depth.

As for σ_y, this stress is neglected in the elementary theory. At the plane $y = 0$, the expression for σ_y in Equations 5.6a is given by

$$[\sigma_y]_{y=0} = -q/2h$$

This plane is assumed in the elementary beam theory as the neutral plane, i.e., free of stress. In fact, in deriving the solution in the elementary beam theory, it is assumed that the longitudinal fibers of the beam are in a condition of simple tension and compression along the x-axis. The stress function, however, does not confirm this assumption.

The result for τ_{xy} is, on the other hand, the same as that of the elementary beam theory, namely, a parabolic distribution through the depth H.

Rewriting Equation 5.6a as

$$\sigma_x = \frac{My}{I} + \frac{qy}{2I}\left(\frac{2y^2}{3} - \frac{H^2}{10}\right)$$

$$\sigma_y = -\frac{q}{2I}\left(\frac{y^3}{3} - \frac{H^2 y}{4} + \frac{H^3}{12}\right) \tag{5.6b}$$

$$\tau_{xy} = -\frac{Q}{2I}\left(\frac{H^2}{4} - y^2\right)$$

provides approximately the stress components for the general case of a beam subjected to a continuous load q of any distribution. In Equations 5.6b, q, Q, and M are the values of load intensity, shearing force, and bending moment, respectively, at the cross section under consideration.

The distributions of the stresses σ_y and τ_{xy} at the section $x = L/2$ are shown in Figure 5.7.

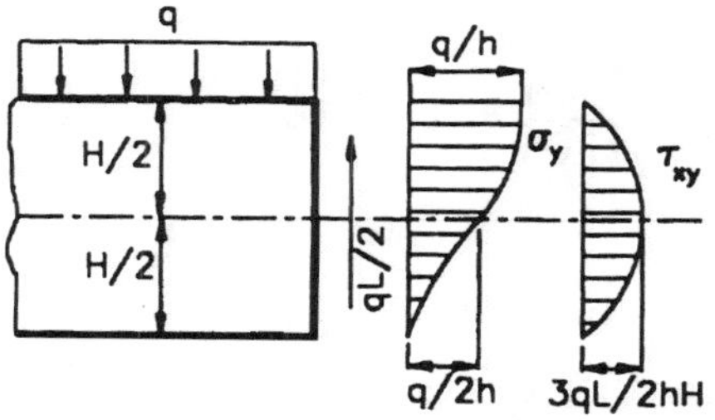

FIGURE 5.7 Distribution of σ_y and τ_{xy} along beam depth at end supports for the beam of Figure 5.6.

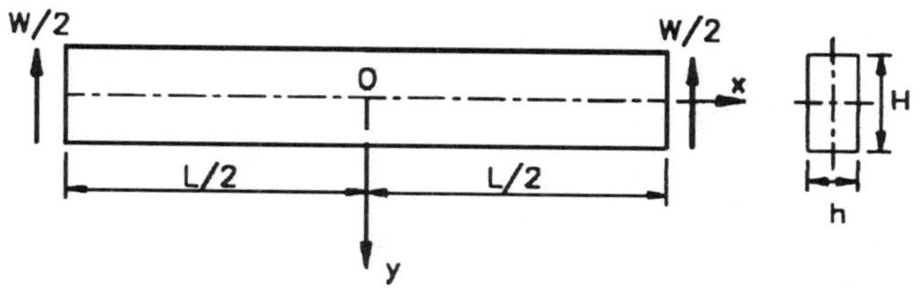

FIGURE 5.8 Example 5.4.

Example 5.4:

Using an appropriate stress function, obtain the stress components for a simply supported beam having a rectangular cross section and subjected to its own weight only, as shown in Figure 5.8.

Solution:

In this problem the only existing force is the body force. Since the total weight is $W = \rho gLHh$, where ρ is the density, the body components are

$$F_x = 0, \quad F_y = \frac{\rho gLHh}{LHh} = \rho g$$

From Equations 4.12, $F_y = -\partial V/\partial y$; hence, $V = -\rho gy$ and $\nabla^2 V = 0$. The stress compatibility condition for plane stress conditions with body forces, Equation 4.18, is

$$\nabla^4\phi = -(1-\nu)\nabla^2 V$$

Since the problem presents the same conditions of symmetry as in the case of a beam subjected to uniformly distributed external load, the stress function is given by Equation 5.5, as

$$\phi = ax^2 + cx^2y + dy^3 + k(x^2y^2 - y^5/5)$$

And the stresses are thus

$$\sigma_x = \frac{\partial^2\phi}{\partial y^2} + V = 6dy + 6kx^2y - 4ky^3 - \rho gy$$

$$\sigma_y = \frac{\partial^2\phi}{\partial x^2} + V = 2a + 2cy - 2ky^3 - \rho gy$$

$$\tau_{xy} = -\frac{\partial^2\phi}{\partial x\partial y} = -2cx - 6kxy^2$$

The boundary conditions are

1. Shear force at $x = \pm L/2$: $\int_{-H/2}^{H/2} \tau_{xy} h\,dy = \mp\frac{W}{2} = \mp\frac{\rho gLHh}{2}$
2. Normal force at $x = \pm L/2$: $\int_{-H/2}^{H/2} \sigma_x h\,dy = 0$
3. At $y = \pm H/2$: $\sigma_y = \tau_{xy} = 0$

Applying the boundary conditions yields

$$a = 0, \quad c = \frac{3\rho g}{4}, \quad k = -\frac{\rho g}{H^2 h}, \quad \text{and } d = \frac{\rho g}{h}\left(\frac{L^2}{4H^2} - \frac{H^2}{10} - \frac{1}{6}\right)$$

The stresses are expressed by

$$\sigma_x = \frac{\rho g y H}{2I}\left(\frac{L^2}{4} - x^2\right) + \frac{\rho g y H}{2I}\left(\frac{2y^2}{3} - \frac{H^2}{10}\right)$$

$$\sigma_y = \frac{\rho g H}{2I}\left(\frac{y^3}{3} - \frac{H^2 y}{12}\right)$$

and

$$\tau_{xy} = \frac{\rho g H}{I}\left(\frac{H^2}{4} - y^2\right)x$$

where $I = hH^3/12$.

Example 5.5:

The stress function for a straight rod of a constant cross section is given by $\phi = ay^2 + by^3$, *the rod axis being the* x-*axis, as shown in Figure 5.9. Determine the stresses in the beam and the external loading. Show that the rod cross section remains plane.*

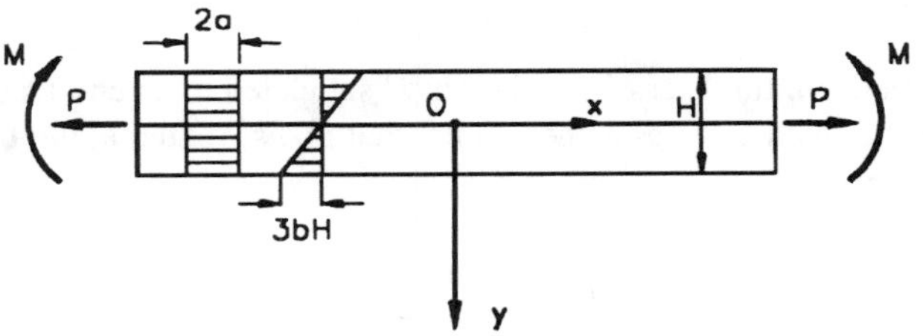

FIGURE 5.9 Example 5.5.

Solution:

Considering a plane stress problem without body forces, the stresses are

$$\sigma_x = \frac{\partial^2\phi}{\partial y^2} = 2a + 6by, \quad \sigma_y = \frac{\partial^2\phi}{\partial x^2} = 0, \quad \text{and} \quad \tau_{xy} = -\frac{\partial^2\phi}{\partial x \partial y} = 0$$

It is readily seen that σ_x is composed of two terms; the first is constant, while the second is linear in *y*. Moreover, considering that $\sigma_y = \tau_{xy} = 0$ suggests that the rod is subjected to axial loading *P* along *x* and pure bending *M* in the plane *x–y* as shown in Figure 5.9, thus resulting in a stress distribution:

$$\sigma_x = \frac{P}{A} + \frac{M}{I}y$$

where *A* and *I* are the cross-sectional area and second moment of area, respectively.

To prove that the rod cross section remains plane under this loading examine the displacement in the *x*-direction, as given by

$$u = \int \varepsilon_x dx = \int \frac{1}{E}\left(\sigma_x - \nu\sigma_y\right)dx$$

where $\sigma_z = 0$ for plane stress. Hence,

$$u = \frac{1}{E}\int (2a + 6by)dx = \frac{1}{E}(2a + 6by)x + f(y)$$

It can be shown that the function $f(y)$ reduces to a constant and hence the displacement u becomes linear in y. Therefore, at any value of x, the rod cross section is displaced, such that the cross section remains plane.

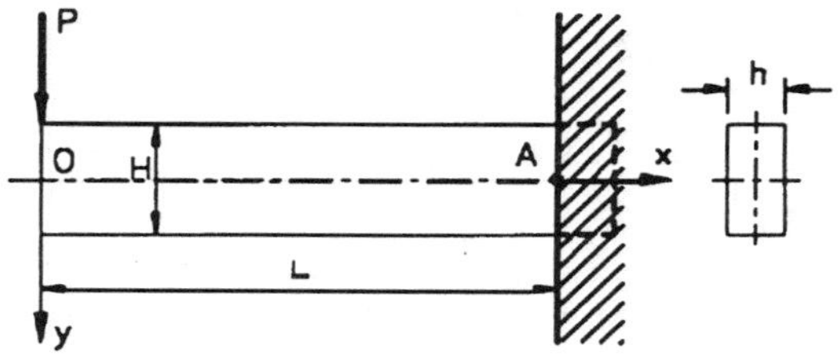

FIGURE 5.10 Cantilever beam subjected to an end force.

5.2.3 Cantilever Beam Subjected to an End Load

A cantilever beam of uniform thickness h and height H subjected to an end load is shown in Figure 5.10. The end load P is looked upon as a shear force that is distributed along the end cross section as a shear stress given by

$$\int_{-H/2}^{H/2} \tau_{xy} h dy = -P$$

Note that according to the sign convention of τ_{xy}, the shear force is negative.

5.2.4.1 Stresses

Consider the fourth-degree polynomial

$$\phi = ax^2 + bxy + cy^2 + dx^3 + ex^2y + fxy^2 + gy^3 + kx^4 + lx^3y + mx^2y^2 + nxy^3 + sy^4$$

Applying the compatibility condition, neglecting the body forces, $\nabla^2\nabla^2\phi = 0$ requires that

$$-3k - m = 3s$$

The boundary condition that $\sigma_y = 0$ at $y = \pm H/2$ for any value of x requires that

$$a = d = l = k = 0$$

and $m = e = 0$; therefore, $s = 0$.

Hence, ϕ reduces to $\phi = bxy + cy^2 + fxy^2 + gy^3 + nxy^3$ and the stresses are

$$\sigma_x = 2c + 6gy + 6nxy + 2fx$$

$$\sigma_y = 0$$

$$\tau_{xy} = -b - 2fy - 3ny^2$$

The shear stress must be zero at the top and bottom surfaces of the beam, i.e.,

$$\tau_{xy} = 0 \text{ at } y = \pm\frac{H}{2} \text{ for any value of } x$$

Therefore,

$$f = 0 \text{ and } b = -3nH^2/4$$

Applying the condition that at any value of x:

$$\int_{-H/2}^{H/2} \sigma_x h dy = 0$$

yields $c = 0$.

To satisfy the condition $\sigma_x = 0$ at $x = 0$, the coefficient g must be equal to zero. Again, at the loaded end, i.e., at $x = 0$,

$$\int_{-H/2}^{H/2} \tau_{xy} h dy = -P$$

Hence,

$$n = -2P/(hH^3)$$

The stress function is therefore given by

$$\phi = \frac{3P}{2hH}xy - \frac{2P}{hH^3}xy^3$$

and the stresses are

$$\sigma_x = -\frac{Pxy}{I}, \quad \sigma_y = 0, \text{ and } \tau_{xy} = -\frac{P}{2I}\left(\frac{H}{4} - y^2\right) \tag{5.7}$$

where $I = hH^3/12$ is the second moment of area of the cantilever cross section about the centroidal axis. This stress distribution is exactly the same as that given by the simple beam theory if the end load is distributed parabolically over the end section of the cantilever as required by the shear stress distribution τ_{xy}. If, however, this is not the case for the end load, Saint-Venant's principle permits one to accept the result as quite accurate for sections away from the end load.

5.2.3.2 Displacements

It is required to obtain the displacements corresponding to the stresses given by Equations 5.7. Applying Hooke's law, Equations 3.6, yields

$$\varepsilon_x = \frac{\partial u}{\partial x} = \frac{\sigma_x}{E} = -\frac{Pxy}{EI} \tag{5.8a}$$

$$\varepsilon_y = \frac{\partial v}{\partial y} = -\frac{\nu\sigma_x}{E} = -\frac{\nu Pxy}{EI} \tag{5.8b}$$

$$\tau_{xy} = \frac{\partial u}{\partial y} + \frac{\partial v}{\partial x} = \frac{\tau_{xy}}{G} = -\frac{P}{2IG}\left(\frac{H^2}{4} - y^2\right) \tag{5.8c}$$

The components u and v of the displacements are obtained by integrating Equations 5.8a and 5.8b; therefore,

$$u = -\frac{P}{2EI}x^2y + f(y)$$

$$v = \frac{\nu P}{2EI}xy^2 + f_1(x)$$

in which $f(y)$ and $f_1(x)$ are as yet unknown functions. Substituting these values of u in v in Equation 5.8c as

$$-\frac{Px^2}{2EI} + \frac{df(y)}{dy} + \frac{\nu Py^2}{2EI} + \frac{df_1(x)}{dx} = -\frac{P}{2IG}\left(\frac{H^2}{4} - y^2\right)$$

and rearranging yields

$$\left[\frac{df_1(x)}{dx} - \frac{Px^2}{2EI}\right] + \left[\frac{df(y)}{dy} + \frac{\nu Py^2}{2EI} - \frac{Py^2}{2IG}\right] = -\frac{P}{2IG}\frac{H^2}{4}$$

This expression shows that the first bracket on the left-hand side represents a function of x only and the second bracket is a function of y only, while the right-hand side is constant. Therefore, the two brackets of the left-hand side must individually be equal to a certain constant. Then, it follows that

$$df_1(x)/dx = Px^2/2EI + C_1$$

and

$$df_2(y)/dy = -\nu Py^2/2EI + Py^2/2IG + C_2$$

where the two constants C_1 and C_2 satisfy

$$C_1 + C_2 = -\frac{P}{2IG}\frac{H^2}{4} \tag{5.9}$$

The functions $f_1(x)$ and $f(y)$ are then obtained by integration as

$$f_1(x) = Px^3/6EI + C_1x + C_3$$

$$f(y) = -\nu Py^3/6EI + Py^3/6IG + C_2y + C_4$$

Substituting in Expressions 5.8c for u and v gives

$$\begin{aligned} u &= -Px^2y/2EI - Py^3/6IG + C_2x + C_4 \\ v &= \nu Pxy^2/2EI + Px^3/6EI + C_1x + C_3 \end{aligned} \tag{5.10}$$

The constants C_1, C_2, C_3, and C_4 may be now determined from Equations 5.9 and 5.10 and from the three conditions of constraint which are necessary to prevent the beam from moving as a rigid body in the x–y plane. Assume that the point A in Figure 5.11a is fixed; then,

$$\text{at } x = L \text{ and } y = 0: \quad u = v = 0$$

Substituting into Equations 5.10 yields

$$C_4 = 0, \quad C_3 = -\frac{PL^3}{6EI} - C_1L$$

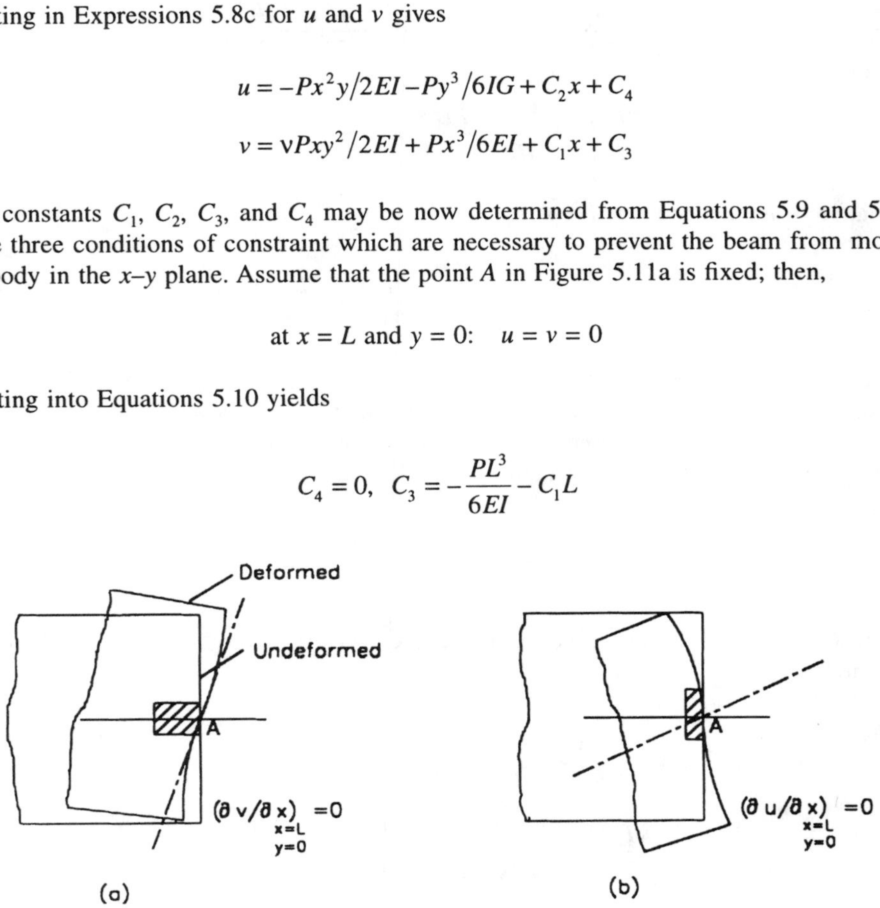

FIGURE 5.11 Displacement boundary conditions at the clamped end of the cantilever beam of Figure 5.10: (a) fixed A horizontal element and (b) fixed A vertical element.

For determining the constant C_1, a third condition of constraint has to be imposed by eliminating the possibility of rotation of the beam in the x–y plane about the point A. This constraint can be realized by fixing a horizontal element of the beam axis at point A as shown in Figure 5.11a. Then, the condition of constraint is

$$\text{at } x = L \text{ and } y = 0: \quad \frac{\partial v}{\partial x} = 0$$

which gives from Equations 5.9 and 5.10

$$C_1 = -\frac{PL^2}{2EI} \quad \text{and} \quad C_2 = \frac{PL^2}{2EI} - \frac{P(H^2/4)}{2IG}$$

Therefore, the expressions for displacements are

$$\begin{aligned} u &= -\frac{Px^2y}{2EI} - \frac{\nu Py^3}{6EI} + \frac{Py^3}{6IG} + \left[\frac{PL^2}{2EI} - \frac{P(H^2/4)}{2IG}\right]Y \\ v &= \frac{\nu Pxy^2}{2EI} + \frac{Px^3}{6EI} - \frac{PL^2x}{2EI} + \frac{PL^3}{3EI} \end{aligned} \tag{5.11}$$

The equation of the deflection curve is

$$(v)_{y=0} = \frac{Px^3}{6EI} - \frac{PL^2x}{2EI} + \frac{PL^3}{3EI} \tag{5.12}$$

which gives for the deflection at the loaded end, $x = 0$, the value $PL^3/3EI$. This coincides with the value usually derived in the elementary solution of strength of materials.

Another way of imposing a constraint condition on the beam is to fix a vertical element of the cross section at point A as shown in Figure 5.11b. Hence, by applying the condition,

$$\text{at } x = L \text{ and } y = o: \quad \frac{\partial u}{\partial y} = 0$$

and proceeding in a similar procedure as above, the following expressions for the displacements are obtained:

$$\begin{aligned} u &= -\frac{Px^2y}{2EI} - \frac{\nu Py^3}{6EI} + \frac{Py^3}{6IG} + \frac{PL^2}{2EI}Y \\ v &= \frac{\nu Pxy^2}{2EI} + \frac{Px^3}{6EI} - \frac{PL^2x}{2EI} + \frac{PL^3}{3EI} + \frac{PH^2/4}{2IG}(L-x) \end{aligned} \tag{5.13}$$

The deflection curve is thus

$$(v)_{y=0} = \frac{Px^3}{6EI} - \frac{PL^2x}{2EI} + \frac{PL^3}{3EI} + \frac{PH^2/4}{2IG}(L-x) \tag{5.14}$$

Comparing the deflection curves of Expressions 5.12 and 5.14 shows that the deflection of the beam is increased by the amount:

$$\frac{PH^2/4}{2IG}(L-x) = \frac{3P}{2hHG}(L-x)$$

This is the effect of shearing force on the beam deflection and is due to the rotation of the end of the beam axis at point A in Figure 5.11b.

At the loading point, $x = y = 0$, the deflection according to Expression 5.14 is

$$(v)_{x=y=0} = \frac{PL^3}{3EI} + \frac{PH^2/4}{2IG} L \tag{5.15}$$

Again the first term on the right-hand side of Expression 5.15 represents the deflection due to bending, while the second term on the same side is due to shearing.

To compare the total vertical deflection at this point with the deflection derived in the elementary theory of beams, namely, $PL^3/3EI$, consider the quotient:

$$\frac{PLH^2/4}{2IG} \Big/ \frac{PL^3}{3EI} = \left(\frac{H}{L}\right)^2$$

It is obvious that this ratio is proportional to the length-to-height ratio of the beam. If, for example $L/H = 10$, the above quotient is only about 0.01 for $\nu = {}^1/_3$. Table 5.1 gives the ratio of the deflection due to shear to the deflection due to bending. It is seen that for a slender beam $H << L$ the deflection is mainly due to bending.

TABLE 5.1
Effect of Beam Length-to-Height Ratio on Beam Deflection Due to Shear

L/H	15	10	5	2
v(shear)/v(bending)	0.004	0.01	0.04	0.25

Upon examining either set of Expressions 5.11 or 5.13 for the displacement components, it is seen that u and v include terms of y^2 and y^3 and, hence, plane sections do not remain plane after deformation as usually assumed in the elementary solution of strength of materials. A plane section, in fact, undergoes a constant amount of warping when subjected to a constant shearing force throughout the length of the beam. However, it is easily verified that this warping is negligibly small and plane sections can be considered to remain plane after deformation. This result, in fact, justifies concluding that the elementary solution of strength of materials gives very good approximate results for the solution of the problem of laterally loaded beams provided that the solution applies only at distances sufficiently remote from supports and points of load applications. This conclusion is applied as a strain compatibility to solve problems of rods and plates in Chapters 7 and 8.

Example 5.6:

A cantilever beam of unit thickness, depth 2H, *and length* L, *as shown in Figure 5.12 is subjected to a uniformly distributed load* q *per unit length. A stress function* ϕ *is suggested in the form:*

$$\phi = \frac{M}{I}\left(y^3/6 - H^2y/2 - H^3/3\right) - \frac{1}{I}\frac{d^2M}{dx^2}\left(y^5/60 - H^2y^3/30 + H^4y/60\right)$$

to solve this problem where M *is the bending moment at any section and* $I = 2H^3/3$,

a. *Show that* ϕ *is a valid stress function and then derive the stress components;*
b. *Calculate the values of the stress components at a point (0, 10) knowing that* q = 20 N/mm, L = 50 mm, *and* H = 25 mm;
c. *Compare the results with those obtained from simple beam formulas. Determine the principal stresses and maximum shear and estimate the difference between the values obtained from both methods.*

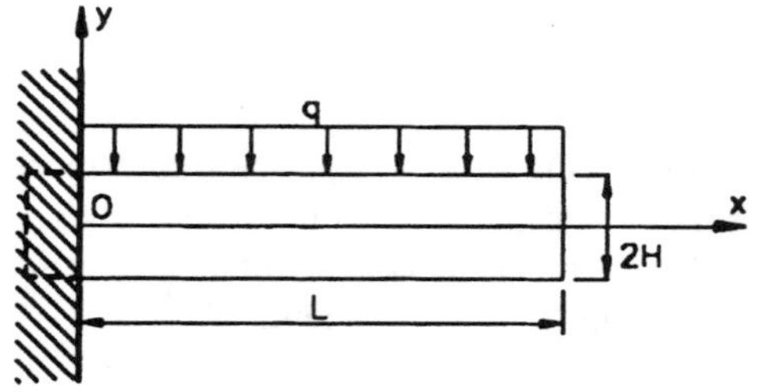

FIGURE 5.12 Example 5.6.

Solution:

a. The bending moment M at any distance x along the beam is given by

$$M = \frac{q}{2}(L-x)^2 \quad \text{where} \quad \frac{\partial M}{\partial x} = -q(L-x) \quad \text{and} \quad \frac{\partial^2 M}{\partial x^2} = q$$

Using this with the given expression for ϕ, the biharmonic equation is satisfied according to

$$\nabla^4\phi = \frac{\partial^4\phi}{\partial x^4} + 2\frac{\partial^4\phi}{\partial x^2\partial y^2} + \frac{\partial^4\phi}{\partial y^4}$$

$$= 0 + \frac{2}{I}\frac{\partial^2}{\partial y^2}\left[qy^3/6 - qH^2y/2 - qH^3/3\right] - \frac{2q}{I}y = 0$$

The stresses are obtained as

$$\sigma_x = \frac{\partial^2\phi}{\partial y^2} = \frac{My}{I} - \frac{1}{1}\frac{d^2M}{dx^2}\left(y^3/3 - H^2y/5\right)$$

$$= \frac{q(L-x)^2}{2I}y - \frac{q}{I}\left(y^3/3 - H^2y/5\right)$$

$$\sigma_y = \frac{\partial^2\phi}{\partial x^2} = \frac{1}{I}\frac{\partial^2 M}{\partial x^2}\left(y^3/6 - H^2y/2 - H^3/3\right)$$

$$= \frac{q}{I}\left(y^3/6 - H^2y/2 - H^3/3\right)$$

$$\tau_{xy} = -\frac{\partial^2\phi}{\partial x\partial y} = -\frac{\partial}{\partial y}\left[\frac{1}{I}\frac{\partial M}{\partial x}\left(y^3/6 - H^2y/2 - H^3/3\right)\right]$$

$$= -\frac{1}{I}\frac{\partial M}{\partial x}\left(y^2/2 - H^2/2\right)$$

$$= \frac{q(L-x)}{2I}\left(y^2 - H^2\right)$$

b. For $q = 20$ N/mm, $L = 50$ mm, $H = 25$ mm, $x = 0$ mm, $y = 10$ mm, and $I = 2\,(25)^3/3 = 10416.7$ mm^4, the stresses become

$$\sigma_x = \frac{20(50-0)^2 \times 10}{2 \times 10416.7} - \frac{20}{10416.7}\left(10^3/3 - 25^2 \times 10/5\right) = 25.76 \text{ MPa}$$

$$\sigma_y = \frac{20}{1046.7}\left(10^3/6 - 25^2 \times 10/2 - 25^3/3\right) = -15.36 \text{ MPa}$$

$$\tau_{xy} = \frac{20(50-0)}{2 \times 10416.7}\left(10^2 - 25^2\right) = -25.2 \text{ MPa}$$

c. The stresses based on the simple beam formulas are

$$\sigma_x = \frac{My}{I} = \frac{q(L-x)^2 y}{2I} = \frac{20(50-0)^2 \times 10}{10416.7} = 24 \text{ MPa}$$

$$\sigma_y = 0$$

(compared with – 15.36 MPa using the stress function method)

$$\tau_{xy} = -\frac{3}{2}\frac{Q}{2H^3}\left[y^2 - H^2\right] = -\frac{3(20 \times 50)}{2(2 \times 25)}\left[10^2 - 25^2\right] = -25.2 \text{ MPa}$$

which is the same value as determined using the stress function method.
The principal stresses according to Equation 1.26a are obtained as
Stress funtion method: $\sigma_1 = 37.72$, $\sigma_2 = 0$, $\sigma_3 = -27.32$ MPa
Simple beam formulas: $\sigma_1 = 39.91$, $\sigma_2 = 0$, $\sigma_3 = -15.91$ MPa
which results in a difference of (–16.5%) in the maximum shear value. Such difference is to be expected in stubby beams of $L/H \cong 1$.

Example 5.7:

A cantilever beam of a narrow rectangular cross section of unit width is subjected to a constant shear stress τ_{xy} (x, H) = q, *as shown in Figure 5.13.*

a. *Derive the stresses using a stress function of the form* $\phi = axy + bxy^2 + cxy^3 + dy^2 + ey^3$, *where* a, b, c, d, *and* e *are constants.*
b. *Write down the boundary conditions and use them to determine the constants* a, b, c, d, *and* e.
c. *Outline the procedure that may be used to determine the displacement components. What are the boundary conditions needed in this case?*

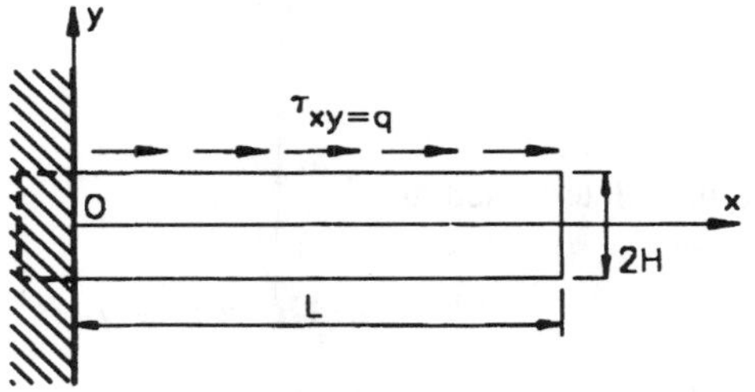

FIGURE 5.13 Example 5.7.

Solution:

a. Differentiation of the stress function yields the stresses as

$$\sigma_x = \frac{\partial^2\phi}{\partial y^2} = 2bx + 6cxy + 2d + 6ey, \quad \sigma_y = \frac{\partial^2\phi}{\partial x^2} = 0, \text{ and}$$

$$\tau_{xy} = -\frac{\partial^2\phi}{\partial x \partial y} = -\left(a + 2by + 3cy^2\right)$$

b. The boundary conditions for this problem are

1. At $y = \pm H$: $\sigma_y = 0$ for any x,
2. At $y = +H$ $\tau_{xy} = q$ for any x,
3. At $y = -H$: $\tau_{xy} = 0$ for any x,
4. At $x = L$: $\sigma_x = 0$ for any y,
5. At $x = L$: $\int_{-H}^{H} \tau_{xy} dy = 0$

These boundary conditions are sufficient to determine the five constants a, b, c, d, and e of the stress function. The first boundary condition is satisfied identically. Applying the second and third boundary conditions gives

$$-(a - 2bh + 3cH^2) = 0$$

$$-(a - 2bh + 3cH^2) = \text{q}$$

Hence,

$$a = 3cH^2 - q/2 \text{ and } b = -q/4H$$

The fourth boundary condition results in

$$2(bL + d) + 6(cL + e)\, y = 0$$

therefore,

$$d = -bL \text{ and } c = -e/L$$

The fifth boundary condition yields, after integration,

$$\left[ay + by^2 + cy^3\right]_{-H}^{+H} = 0$$

and, hence, $a = -cH^2$.

The constants are thus determined as

$$a = q/4, \qquad b = -q/4H, \quad c = -q/4H^2$$

$$d = qL/4H, \quad e = qL/4H^2$$

The stress components are

$$\sigma_x = \frac{q(L-x)}{2H}\left(1+\frac{3y}{H}\right),\ \sigma_y = 0,\ \tau_{xy} = \frac{q}{4}\left(\frac{3y^2}{H^2}+\frac{2y}{H}-1\right)$$

It is worth noting that the shear stress distribution does not satisfy the boundary condition $\tau_{xy}(L, y) = 0$ exactly, but instead it satisfies this boundary condition on the average, namely,

$$\int_{-H}^{H} \tau_{xy}(L, y)dy = 0$$

c. The displacement components u and v are obtained from stresses by the following procedure:
   1. Apply Hooke's law to obtain the strains ε_x, ε_y, and γ_{xy}.
   2. Integrate the expressions for the strains ε_x and ε_y, thus determining u and v by using the definition that

$$\gamma_{xy} = \frac{\partial u}{\partial y} + \frac{\partial v}{\partial x}$$

together with the boundary conditions at the clamped end, i.e.,

$$u(0,0) = v(0,0) = \frac{\partial v}{\partial x}(0,0) = 0$$

to determine the two unknown functions of x and y as constants of integration.

5.3 PROBLEMS SOLVED IN TERMS OF TRIGONOMETRIC STRESS FUNCTIONS

Stress functions in the form of algebraic polynomials have limited applications and are particularly suitable for problems where the load is continuous and of simple distribution along the boundaries. On the other hand, stress functions in the form of trigonometric series are more appropriate for solving problems where the boundary loads are of complicated distribution. By using Fourier's series expansions, trigonometric stress functions could handle discontinuous boundary loading conveniently.

Consider a stress function ϕ in the form:

$$\phi = \sin(\alpha x) \cdot f(y)$$

in which $f(y)$ is a function of y only and α is a constant. Substituting the expression for ϕ into the biharmonic equation

$$\nabla^2\nabla^2\phi = 0$$

yields the following differential equation for determining $f(y)$:

$$\alpha^4 f - 2\alpha^2 \frac{d^2 f}{dy^2} + \frac{d^4 f}{dy^2} = 0$$

The general solution of this equation is

$$f(y) = C_1 \cosh \alpha y + C_2 \sinh \alpha y + C_3 y \cosh \alpha y + C_4 y \sinh \alpha y \tag{5.16}$$

The stress function then is

$$\phi = \sin \alpha x \, [C_1 \cosh \alpha y + C_2 \sinh \alpha y + C_3 y \cosh \alpha y + C_4 y \sinh \alpha y] \tag{5.17}$$

and the corresponding stress components, neglecting body forces are

$$\begin{aligned}
\sigma_x &= \frac{\partial^2 \phi}{\partial y^2} = \sin \alpha x \big[C_1 \alpha^2 \cosh \alpha y + C_2 \alpha^2 \sinh \alpha y \\
&\quad + C_3 \alpha (2 \sinh \alpha y + \alpha y \cosh \alpha y) + C_4 \alpha (2 \cosh \alpha y + \alpha y \sinh \alpha y) \big] \\
\sigma_y &= \frac{\partial^2 \phi}{\partial x^2} = -\alpha^2 \sin \alpha x \big[C_1 \alpha \cosh \alpha y + C_2 \sinh \alpha y \\
&\quad + C_3 \alpha \cosh \alpha y + C_4 y \sinh \alpha y \big] \\
\tau_{xy} &= -\frac{\partial^2 \phi}{\partial x \partial y} = -\alpha \cos \alpha x \big[C_1 \alpha \sinh \alpha y + C_2 \alpha \cosh \alpha y \\
&\quad + C_3 (\cosh \alpha y + \alpha y \sinh \alpha y) + C_4 (\sinh \alpha y + \alpha y \cosh \alpha y) \big]
\end{aligned} \tag{5.18}$$

The coefficients C_1, C_2, C_3, and C_4 in Equations 5.18 are to be determined from the boundary conditions of the particular problem under consideration.

5.3.1 Simply Supported Beam Under Laterally Distributed Sinusoidal Loads on Both Sides

Consider a particular case of a rectangular thin beam of unit thickness supported at the ends and subjected along the upper and lower edges to continuously distributed vertical forces of intensity $A \sin \alpha x$. Figure 5.14a shows the case where $\alpha = m\pi/L$ for $m = 4$. The stress distribution for this case can be obtained from Equations 5.18. The boundary conditions are thus:*

$$\text{At } y = +H\text{: } \tau_{xy} = 0 \text{ and } \sigma_y = -A \sin \alpha x$$

$$\text{At } y = -H\text{: } \tau_{xy} = 0 \text{ and } \sigma_y = -A \sin \alpha x$$

Applying these boundary conditions to the stresses given by Equations 5.18 yields the following expressions for the constants C_1, C_2, C_3, and C_4:

* When the load intensity on the upper and lower edges are unequal, see Timoshenko and Goodier,[2] p. 47

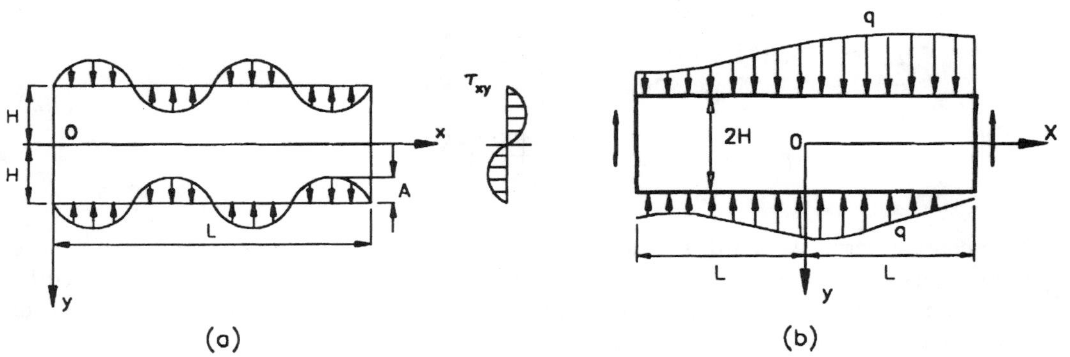

FIGURE 5.14 Simply supported beam under laterally distributed loads acting on both sides: (a) sinusoidal distribution and (b) general distribution.

$$C_1 = \frac{2A}{\alpha^2} \frac{\sinh \alpha H + \alpha H \cos \alpha H}{\sinh 2\alpha H + 2\,\alpha H}$$

$$C_2 = C_3 = 0$$

$$C_4 = -\frac{2A}{\alpha^2} \frac{\alpha \sinh \alpha H}{\sinh 2\alpha H + 2\,\alpha H}$$

Substituting into Equations 5.18 gives the expressions for the stress components as

$$\begin{aligned}
\sigma_x &= 2A \sin \alpha x \left[\frac{(\alpha H \cosh \alpha H - \sinh \alpha H)\cosh \alpha y - \alpha y \sinh \alpha y \sinh \alpha H}{\sinh 2\alpha H + 2\alpha H} \right] \\
\sigma_y &= -2A \sin \alpha x \left[\frac{(\alpha H \cosh \alpha H + \sinh \alpha H)\cosh \alpha y - \alpha y \sinh \alpha y \sinh \alpha H}{\sinh 2\alpha H + 2\alpha H} \right] \qquad (5.19) \\
\tau_{xy} &= -2A \cos \alpha x \left[\frac{\alpha H \cosh \alpha H \sinh \alpha y - \alpha y \cosh \alpha y \sinh \alpha H}{\sinh 2\alpha H + 2\alpha H} \right]
\end{aligned}$$

Recalling that $\alpha = 4\pi/L$, it is seen that at the ends of the beam, $x = 0$ and $x = L$, the stress σ_x is zero. The shear stress τ_{xy}, however, exists at these ends and has the distribution shown in Figure 5.14. Obviously, the resultant of this shear stress for a beam of unit thickness as given by

$$\int_{-H}^{H} \tau_{xy} dy$$

vanishes, and thus the reactions at the ends are nil.

For beams with a general boundary loading, Figure 5.14b, a more general representation for these loads could take the form of a Fourier's series (with $\alpha = m\pi/L$) for the upper and lower edges, q_u and q_l, respectively, as

$$q_u(x) = A_0 + \sum_{m=1}^{\infty} A_m \sin\frac{m\pi x}{L} + \sum_{m=1}^{\infty} A'_m \cos\frac{m\pi x}{L}$$
$$q_l(x) = B_0 + \sum_{m=1}^{\infty} B_m \sin\frac{m\pi x}{L} + \sum_{m=1}^{\infty} B'_m \cos\frac{m\pi x}{L} \tag{5.20}$$

The constant terms A_0 and B_0 represent uniform loading on the beam. The stresses produced by terms containing sin $m\pi x/L$ are directly given by Equations 5.19, while stresses produced by terms containing cos $m\pi x/L$ are easily obtained from Equation 5.19 by exchanging (sin αx) for (cos αx), and vice versa, as well as changing the sign of τ_{xy}.

5.3.2 Simply Supported Beam Under Two Equal Lateral Loads at the Middle of the Span

Consider first the case of a rectangular beam of unit thickness subjected to a load of intensity q, which is distributed over a small portion of width $2a$ of the upper and lower edges of the beam as shown in Figure 5.15a.

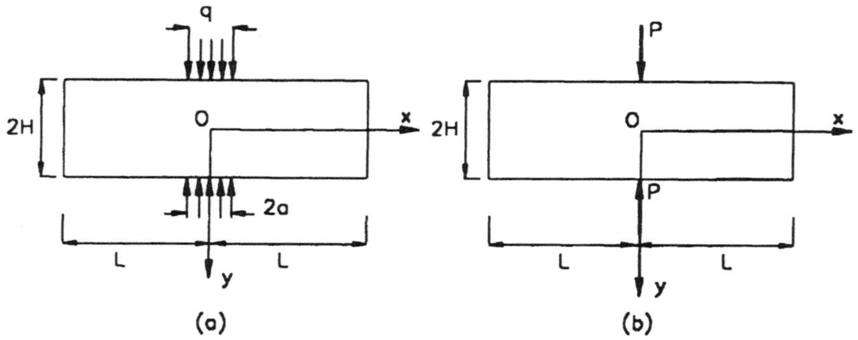

FIGURE 5.15 Beam under two equal lateral loads at midspan: (a) load uniformly distributed along a finite length and (b) concentrated load.

For this case of symmetrical loading, the terms with sin mx/L vanish from Expression 5.20 and the coefficients A_0 and A_m' are obtained in the usual manner. To determine A_0, both sides of Equation 5.20 are integrated between the proper limits, i.e.,

$$\int_{-a}^{a} q dx = \int_{-L}^{L} \left[A_0 + \sum_{m=1}^{\infty} A'_m \cos\frac{m\pi x}{L} \right] dx$$

which yields $A_0 = B_0 = qa/L$.

The coefficients A_m' and B_m' are obtained by multiplying both sides of Equation 5.20 by (cos $m\pi x/L$) and then integrating to give

$$A'_m = B'_m = \frac{2q \sin(m\pi a/L)}{m\pi}$$

The stresses produced by the trigonometric terms are calculated from Equation 5.19 by exchanging (sin αx) for (cos αx) and changing the sign of τ_{xy}. The term A_0 represents a uniform compression in the y-direction equal to $-qa/L$. Thus, for instance, the stress σ_y is given by

$$\sigma_y = -\frac{qa}{L} - \frac{4q}{\pi}\sum_{m=1}^{\infty}\frac{1}{m}\sin\frac{m\pi a}{L}\cos\frac{m\pi x}{L}$$

$$\times\frac{\left(\frac{m\pi H}{L}\cosh\frac{m\pi H}{L}+\sinh\frac{m\pi H}{L}\right)\cosh\frac{m\pi y}{L}-\frac{m\pi y}{L}\sinh\frac{m\pi y}{L}\sinh\frac{m\pi H}{L}}{\sinh\frac{2m\pi H}{L}+\frac{2m\pi H}{L}} \tag{5.21}$$

Considering the middle plane at $y = 0$, Equation 5.21 becomes

$$\left(\sigma_y\right)_{y=0} = -\frac{qa}{L} - \frac{4q}{\pi}\sum_{m=1}^{\infty}\frac{1}{m}\sin\frac{m\pi a}{L}\cos\frac{m\pi x}{L}$$

$$\times\frac{\left(\frac{m\pi H}{L}\cosh\frac{m\pi H}{L}+\sinh\frac{m\pi H}{L}\right)}{\sinh\frac{2m\pi H}{L}+\frac{2m\pi H}{L}} \tag{5.22}$$

For the case of an infinitely long beam (H/L very small) loaded by two opposite concentrated forces at the middle of the span, i.e., $2qa = P$, Figure 5.15b, the above expression is used to evaluate σ_y. The results[3] indicate that σ_y diminishes very rapidly far from the applied load. At a value of $(x/H) = 1.35$, it becomes zero compared with a value approximately equal to $3P/\pi H$ at $(x/H) = 0$.

5.3.3 Bar Subjected to Two Equal and Opposite Axial Loads

A thin plate of unit thickness subjected to two opposing forces P is shown in Figure 5.16a. The solution of this problem is obtained from Equation 5.22.

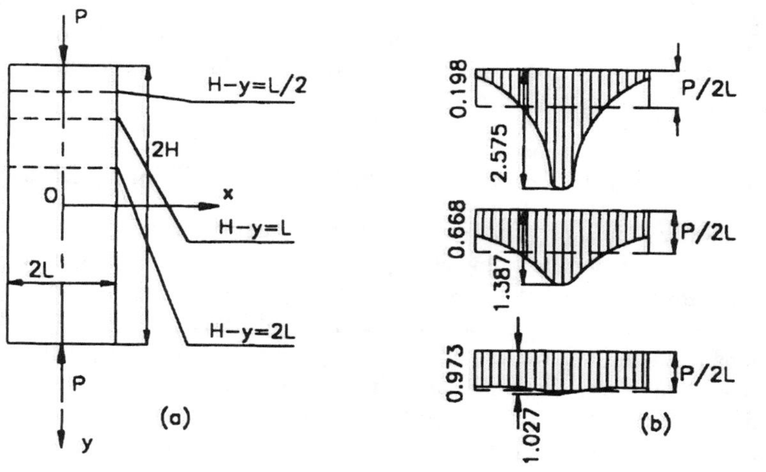

FIGURE 5.16 Distribution of compressive stress σ_y in a bar subjected to two equal and opposite axial loads.

If L is small in comparison to H, $(m\pi H/L)$ is a large quantity and, hence, for cross sections at large distances from the middle of the plate it is feasible to write:

$$\sinh m\pi\left(\frac{y}{L}\right) \cong \cosh m\pi\left(\frac{y}{L}\right) \cong \frac{1}{2}\exp\left(m\pi\frac{y}{L}\right)$$

Considering these approximations while neglecting $(m\pi H/L)$ in comparison with sinh $(m\pi H/L)$ in Equations 5.21 in which $P = 2qa$ yields

$$\sigma_y = -\frac{P}{2L} - \frac{P}{a\pi}\sum_{m=1}^{\infty}\frac{1}{m}\sin\frac{m\pi a}{L}\cos\frac{m\pi x}{L}$$
$$\times\left[\left(\frac{m\pi H}{L}+1\right)\exp\left(\frac{m\pi}{L}(y-H)-\frac{m\pi}{L}y\exp\left(\frac{m\pi}{L}(y-H)\right)\right)\right]$$

If $(H - y)$ is not very small, this series converges very rapidly. For instance, let $H - y = L$ and sin $(m\pi a/L) \cong m\pi a/L$; therefore,

$$\sigma_y = \frac{P}{2L} - \frac{P}{L}\sum_{m=1}^{\infty}(m\pi+1)\exp(-m\pi)\cos\frac{m\pi x}{L}$$

or

$$\sigma_y = -\frac{P}{2L} - \frac{P}{L}\left[(\pi+1)e^{-\pi}\cos\frac{\pi x}{L} + (2\pi+1)e^{-2\pi}\cos\frac{2\pi x}{L}\right.$$
$$\left. + (3\pi+1)e^{-3\pi}\cos\frac{3\pi x}{L} + \cdots\right] \tag{5.23}$$

For such series the consideration of the first three terms is sufficient to give a good accuracy in calculating σ_y. The stress distribution is shown in Figure 5.16b for $y = (H - L)$ and two other values of y. It is evident that at a distance from the end equal to the width of the strip the stress distribution is practically uniform which confirms the conclusion made on the basis of Saint-Venant's principle in Section 4.3.

5.4 A NOTE ON SOME OTHER FORMS OF STRESS FUNCTIONS

In this chapter deriving the appropriate stress function for any of the problems considered has been carried out systematically despite being lengthy. This has been possible because the stress function is expressed in terms of an algebraic polynomial or a trigonometric series. Such a procedure is not usually accessible when handling more elaborate forms of functions. Typical examples are biharmonic functions involving logarithmic or inverse trigonometric functions, such as: $\ln(x^2 + y^2)$, $e^{\alpha x}$ sin αy, $e^{\alpha x}$ cos αy, $\tan^{-1} y/x$, ... and any of their products by x, y, or $(x^2 + y^2)$.

The semi-inverse method of specifying the form of a stress function in advance is not easily applicable to many problems. For such problems, another approach based on complex variable functions is adopted,[4] as mentioned in Chapter 4.

It is worth noting that some problems in Cartesian coordinates become much easier to solve if expressed in polar coordinates. Typical examples are problems of semi-infinite plates subjected to edge loading, triangular cantilever plates, and circularly curved beams, as will be elaborated in Chapter 6.

Example 5.8:

Figure 5.17 shows a triangular beam of unit thickness subjected to a uniform load q *per unit area.*

a. *Derive expressions for the stresses in the beam based on the following stress function:*

$$\phi = \frac{q}{2(\tan\alpha - \alpha)}\left[-x^2\tan\alpha + xy + \left(x^2 + y^2\right)\left(\alpha - \tan^{-1}\left(\frac{y}{x}\right)\right)\right]$$

b. *Check that the boundary conditions are satisfied along the surfaces* OA *and* OB.
c. *Determine the stress distributions at the built-in end AB and represent them graphically for* $\alpha = 45°$. *Compare the distribution of* σ_x *with the simple beam formula* $\sigma_x = My/I$ *where* y *is measured vertically from the centroidal axis and* M *is the bending moment.*
d. *Calculate the stress* σ_x *at points* A *and* B *for* $\alpha = 5°$ *and comment on the results.*

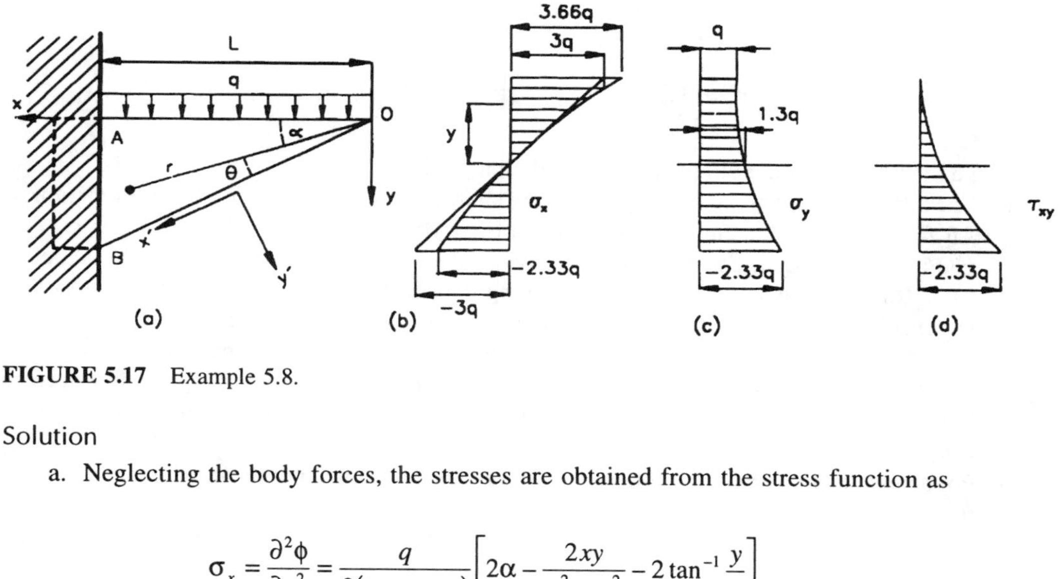

FIGURE 5.17 Example 5.8.

Solution

a. Neglecting the body forces, the stresses are obtained from the stress function as

$$\sigma_x = \frac{\partial^2\phi}{\partial y^2} = \frac{q}{2(\tan\alpha - \alpha)}\left[2\alpha - \frac{2xy}{x^2+y^2} - 2\tan^{-1}\frac{y}{x}\right]$$

$$\sigma_y = \frac{\partial^2\phi}{\partial x^2} = \frac{q}{2(\tan\alpha - \alpha)}\left[2(\alpha - \tan\alpha) + \frac{2xy}{x^2+y^2} - 2\tan^{-1}\frac{y}{x}\right]$$

$$\tau_{xy} = -\frac{\partial^2\phi}{\partial x\partial y} = \frac{-q}{2(\tan\alpha - \alpha)}\left[\frac{2y^2}{x^2+y^2}\right]$$

b. Along the surface OA, the boundary conditions are, at $y = 0$ for any x, $\sigma_y = -q$ and $\tau_{xy} = 0$, which are obviously satisfied. Along the surface OB, the boundary conditions referred to the transformed axes x', y' are such that, at $y' = 0$ for any x', $\sigma_{y'} = \tau_{x'y'} = 0$. The expressions for the stresses σ_y and τ_{xy} along the surface OB are simplified by putting

$$\tan\alpha = y/x, \quad \sin\alpha = y\Big/\sqrt{x^2+y^2}, \quad \cos\alpha = x\Big/\sqrt{x^2+y^2}$$

Hence,

$$\sigma_x = \frac{q\sin\alpha\cos\alpha}{(\tan\alpha-\alpha)}, \quad \sigma_y = \frac{q(\sin\alpha\cos\alpha-\tan\alpha)}{(\tan\alpha-\alpha)}$$

and

$$\tau_{xy} = \frac{-q\sin^2\alpha}{(\tan\alpha-\alpha)}$$

Applying the stress transformation equations (Equations 1.20a) for the set of axes (x, y) and (x', y') gives

$$\sigma_{y'} = \sigma_x\sin^2\alpha + \sigma_y\cos^2\alpha - \tau_{xy}\sin 2\alpha = 0$$

$$\tau_{x'y'} = \tau_{xy}\cos 2\alpha + \frac{\sigma_x-\sigma_y}{2}\sin 2\alpha = 0$$

i.e., the boundary conditions are satisfied along *OB*.

c. The distribution of stresses σ_x, σ_y, and τ_{xy} are shown in Figure 5.17b over the cross section *AB* for $\alpha = 45°$. The simple beam formula $\sigma_x = My/I = qx^2y/4I$ gives the stresses $\sigma_x = \pm 3q/\tan^2\alpha = \pm 3q$ at points *A* and *B* respectively, which are underestimated by about 21%. It may be also seen that the line joining the centroids of the cross sections, which is considered by the simple beam formula as the neutral fiber, is not a stress-free line. Moreover, simple beam analysis is not capable of predicting the distribution of the stress component σ_y, as shown in Figure 5.17c. The shear stress distribution τ_{xy}, as shown in Figure 5.17d, is obviously different from the parabolic distribution obtained by the simple beam theory.
d. For a beam angle $\alpha = 5°$, the following stresses are obtained at points *A* and *B*, according to:

Stress function method:

$$\left(\sigma_x\right)_A = 393.69q, \ \left(\sigma_y\right)_A = -q, \ \left(\tau_{xy}\right)_A = 0$$

$$\left(\sigma_x\right)_B = 390.81p, \ \left(\sigma_y\right)_B = -2.98q, \ \left(\tau_{xy}\right)_B = -34.2q$$

Simple beam theory:

$$(\sigma_x)_{A,B} = \pm 391.9q, \ (\sigma_y)_{A,B} = 0, \ (\tau_{xy})_{A,B} = 0$$

This indicates a very small error (about 0.5%) in estimating σ_x for beams with a small taper angle of 5°.

Note that this problem* could be solved more conveniently by means of the stress function expressed in polar coordinates, as shown later in Problem 6.52. The stress function in terms of r and θ is given by

$$\phi = \frac{-qr^2}{4(\alpha - \tan\alpha)}\left[2(\alpha - \theta) + \sin 2\theta - 2\tan\alpha\cos^2\theta\right]$$

PROBLEMS

5.1 A concrete retaining wall of an angle $\beta = 45°$ is subjected to a pressure $p = -ky$ as shown in the figure. Assuming plane strain conditions,

a. Write down the stress boundary conditions along the face OA and the face OB expressed in terms of σ_x, σ_y, and τ_{xy};
b. Given the stress function

$$\phi = ax^3 + bx^2y + cxy^2 + dy^3$$

and taking the body forces into consideration, find expressions for the stresses σ_x, σ_y, σ_z, and τ_{xy};
c. Considering an isolated element OCD, show that this element is in equilibrium and find the strains at the centroid of the section CD; taking $\nu = 1/6$ and assuming isotropy.
d. Find the stresses using the formula $\sigma = My/I$ and compare this result with that obtained in (b) and (c);
e. Outline the procedure one may follow to find the displacement components u and v.

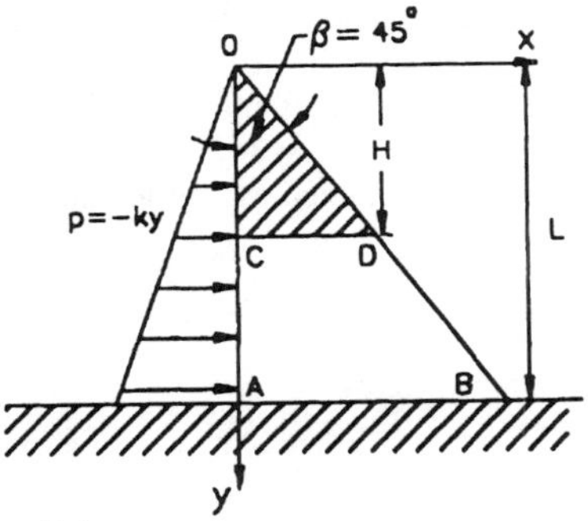

Problem 5.1

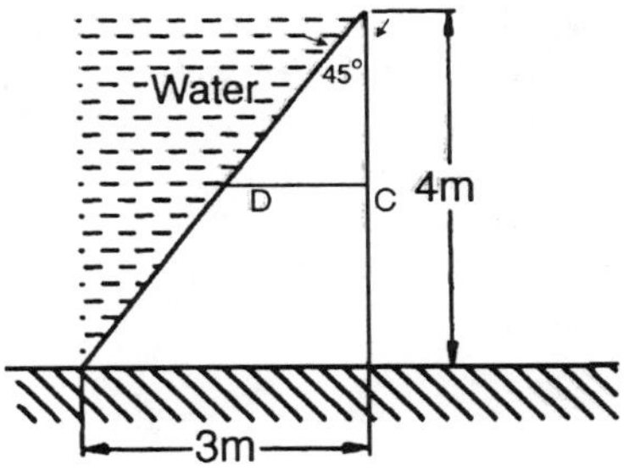

Problem 5.2

5.2 A small concrete dam retaining water is shown in the figure. Determine the stresses acting on a horizontal section 3 m below the top. Assume specific weights of water and concrete to be 10 kN/m^3 and 23 kN/m^3, respectively. Compare the results with those obtained from the elementary strength-of-materials formulas.

5.3 For a cantilever beam of rectangular cross section loaded by an end moment M, derive expressions for the stresses and the displacement components u and v. Use $\phi = ay^3$.

5.4 For the simply supported beam shown in the figure show that the stress function:

$$\phi = axy + bxy^3$$

satisfies the boundary conditions. Consider the concentrated load as a shear stress suitably distributed on each half of the beam, such that

* In particular the conditions at the tip of the wedge are of interest in investigating the problem of the sharp leading edge of a aeroplane wing. See Fung.[5]

$$\int_{-H}^{+H} \tau_{xy}\, dy = -\frac{P}{2}$$

Then derive expressions for σ_x, σ_y, = 0, and τ_{xy}.

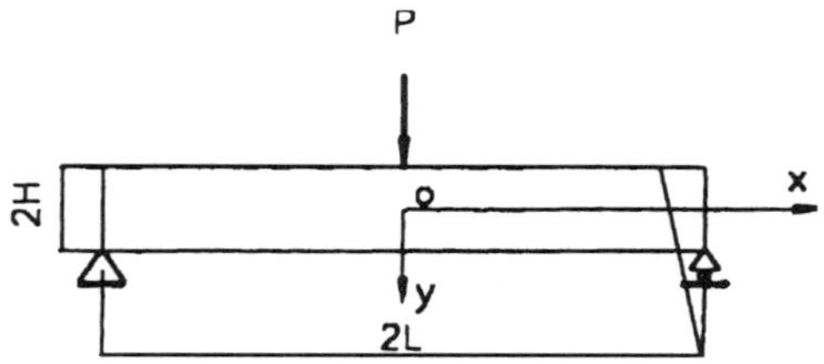

Problem 5.4

5.5 A beam of unit thickness is subjected to the uniformly varying normal load $q = q_o\, x/L$ as shown in the figure. Neglecting body forces and given the polynomial:

$$\phi = ax^3 + bx^3y^3 - \frac{q_o}{4H^3L}\left[\frac{xy^5}{10} + \frac{H^2x^3y}{2} - \left(\frac{H^2}{5} - \frac{L^2}{6}\right)xy^3 - \left(\frac{H^2L^2}{2} - \frac{H^4}{10}\right)xy\right]$$

a. Show that ϕ is a possible stress function for this problem and, hence, find the value of the constant b;
b. Write down all the stress boundary conditions for this problem;
c. Use one of these boundary conditions to determine the constant a, and then find the expression for the stresses σ_x, σ_y, and τ_{xy};
d. Show that only four boundary conditions are satisfied exactly, while the others could be satisfied on the average;
e. Compare the expression for stress σ_x as obtained above with that obtained using the strength-of-materials formula ($\sigma = My/I$).

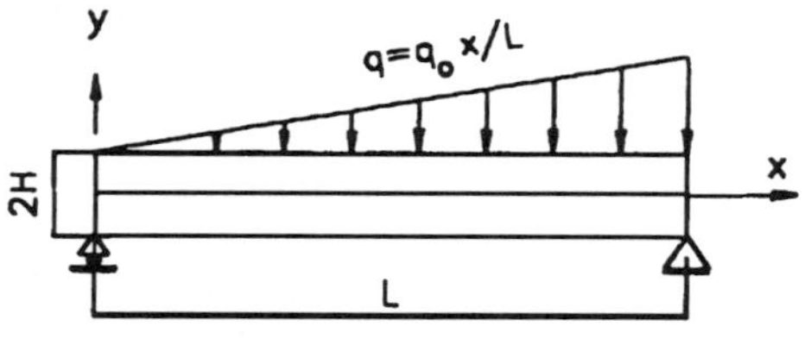

Problem 5.5

5.6 For a cantilever beam subjected to an end moment M and an axial pull P, suggest a suitable stress function to obtain the stresses. Derive expressions for the displacement components u and v. (use ϕ from Section 4.10.1 and apply superposition.)

5.7 Two simply supported beams of unit thickness as shown in the figure are subjected to a distributed load of q (N/m). (a) Use the results of Section 5.2.3 to determine the stresses in both beams at midspan at outer fibers. (b) Find the strains on the centroidal axis $y = 0$. Comment on the results. Take $\nu = {}^1/_3$. (c) Determine the ratios between the maximum τ_{xy} and σ_x for both beams.

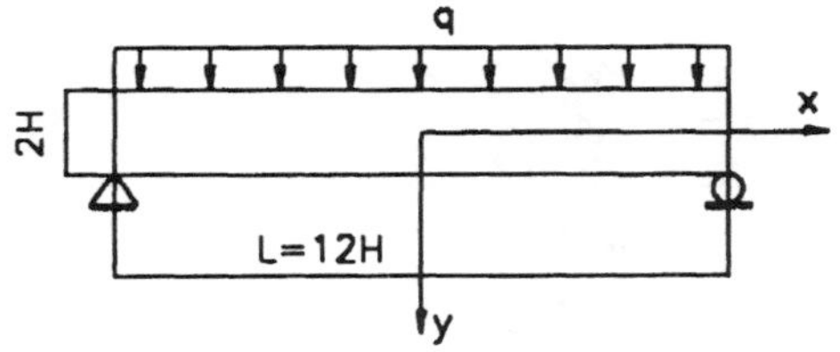

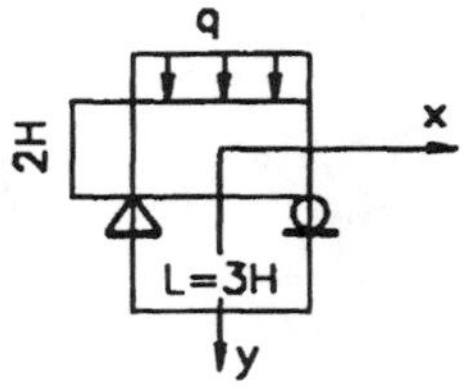

Problem 5.7

5.8 A cantilever beam of length L, depth H, and width h, is loaded by a downward vertical end force P.

a. Find the necessary conditions such that

$$f = axy + by^2 + cy^3 + dx^3y + ex^2y^2 + fxy^3$$

becomes a stress function.

b. Write down the boundary conditions at $y = \pm H/2$ and the average (integral) boundary conditions at $x = L$.
c. Determine the constants, a to f, using the results of (a) and (b) and, hence, the stress.
d. Let $h = 20$ mm, $H = 60$ mm, $L = 1000$ mm, and $P = 80$ kN.

Determine the magnitude and directions of the principal strains at the point (600 mm, 15 mm). Take $E = 210$ GPa and $\nu = 0.3$.

5.9 A cantilever beam of length L; rectangular cross section of depth H, and unit thickness is subjected to an end force P.

a. Using Castigliano's theorem, derive an expression for the deflection v^* at the free end.
b. Compare this deflection with that calculated from Equation 5.13, using the stress function solution for two beams of $L/H = 2$, 8 and $L/H = 32$, respectively. Comment on the different results obtained by the two methods. Take $\nu = {}^1/_3$.
c. Using the results of the stress function method, plot the deformed shape of the beam cross section lying in the plane $x = 0$. How does this compare with the assumption of plane cross sections adopted in the solutions of strength of materials?

5.10 A bar of square section (H × H) whose axis is taken along z-direction. The bar is laterally loaded by uniform load q_o per unit length. Derive the stresses and displacements assuming plain strain conditions such that $\varepsilon_z = 0$. Use the stress function $\phi = ax^3$.

5.11 A cantilever beam of unit thickness is loaded as shown in the figure. Using the stress function:

$$\phi = axy + by^3 + cxy^3$$

(a) derive expressions for the stress components and apply the boundary conditions to determine the coefficients a, b, and c. (b) Is the boundary surface x = 0 stress free?

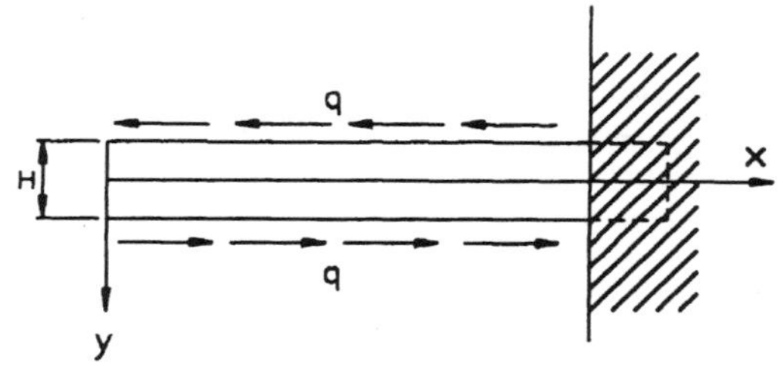

Problem 5.11

5.12 For the thin cantilever of unit thickness shown in the figure, the stress function is given by

$$\phi = -axy + bx^3/6 - cx^3y/6 - dxy^3/6 - ex^3y^3/9 - fxy^5/20$$

a. Write down the boundary conditions and use them to determine the coefficients a, b, c, d, e, and f.
b. Find the stress distribution σ_x.
c. Compare $(\sigma_x)_{max}$, as obtained from the stress function solution, with that (σ_x^*) max of the simple beam theory for L/H = and 10.

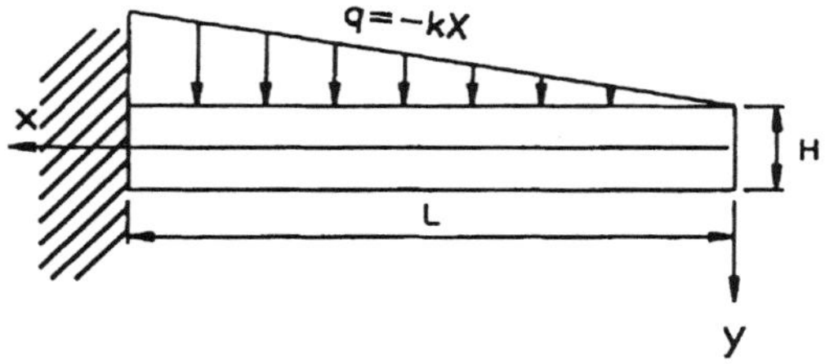

Problem 5.12

5.13 A rectangular strip of length L and height H is compressed along its length between smooth, frictionless plates. The pressure is given by $q = q_o \cos \pi x/L$, at $y = \pm H/2$, x and y being axes of symmetry. Assuming a function of the form $\phi = f(y) \cos \pi x/L$. Determine the stresses in the strip for $L/H = 2$. Determine their values along $y = 0$ for $x = 0$, $L/2$ and L.

5.14 The stress function: $\phi = a[(x^2 + y^2) \tan^{-1} (y/x) - xy]$, is used to solve the problem of a semi-infinite plate subjected to an edge load indefinitely extending on half of its boundary as shown in the figure.

a. Derive the stress components and find the constant a, using the appropriate boundary condition.
b. By applying superposition, determine the stresses in a semi-infinite plate subjected to the uniform load q on the segment $-L \leq x \leq L$ as shown in the figure. (*Note*: this problem may be solved with less algebraic effort using polar stress functions as shown in Chapter 6.)

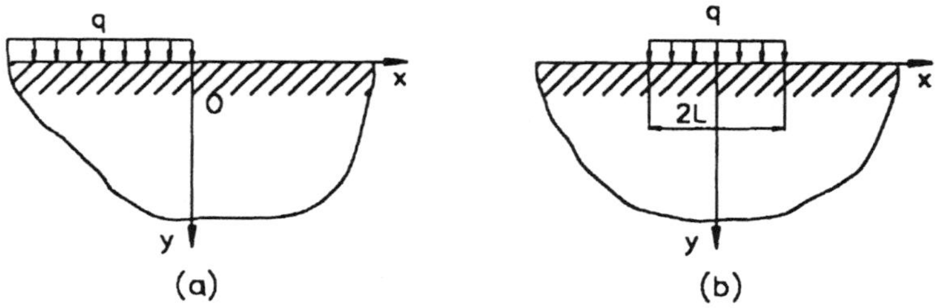

Problem 5.14

REFERENCES

1. Neou, C.-Y., *J. Appl. Mech.,* 24(3), 387–390, 1957.
2. Timoshenko S. and Goodier, J. N., *Theory of Elasticity,* 2nd ed., McGraw-Hill, New York, 1951.
3. Filon, L. N. G., *Trans. R. Soc.,* (London), A, 201, 67, 1903.
4. Muschelishvili, N. I., *Some Basic Problems of the Theory of Elasticity,* Groningen, Noordhoff, 1956.
5. Fung, Y. C., *J. Aeronaut. Sci.*, 20, 9, 1953.

6 Elastic Plane Problems in Polar Coordinates

Solutions are obtained to some engineering axisymmetric and axially nonsymmetric plane problems in polar coordinates by applying the governing stress function equations developed in Chapter 4. Axisymmetric problems comprise thick-walled cylinders subjected to internal and/or external pressure, shrink fits, rotating disks, and drums, taking into consideration the effect of radial thermal gradient. Axially nonsymmetric problems comprise curved beams subjected to bending, a wedge subjected to a concentrated load at its vertex, a concentrated load acting on a straight boundary, and stress concentration around a small circular hole in a plate.

6.1 INTRODUCTION

In many engineering problems, the geometry and boundary loading of the solid are more conveniently expressed in terms of cylindrical coordinates (r, θ, z). Typical examples are shown in Figure 6.1, which include cylinders subjected to internal and/or external pressure, rotating disks, bending of a circularly curved beam, … etc. In general, these problems can be treated as plane problems in (r, θ) by taking the axis of the prismatical solid to be the z-axis. Solutions are obtained by applying the stress function equations developed in Chapter 4 and substituting relevant boundary conditions.

The problems treated below comprise two groups, namely, axisymmetric problems and axially nonsymmetric problems. For the first group of problems, the stress functions and the stress components are readily obtained from Chapter 4. For the second group, the appropriate stress function is sought for each problem and the stresses are again obtained by applying the stress equations of Chapter 4.

6.2 AXISYMMETRIC PROBLEMS

In the following sections some axisymmetric problems that have engineering applications are treated. These problems comprise thick-walled cylinders subjected to internal and/or external pressures with and without radial thermal gradient, shrink fits, rotating disks, and drums with and without radial thermal gradient.

6.2.1 Thick-Walled Cylinder Subjected to Uniform Internal and/or External Pressure

The problem of a thick-walled cylinder subjected to internal and/or external pressures occurs in several types of engineering equipment, such as hydraulic presses, compressors, high-pressure boilers, extrusion containers, drawing dies, gun barrels, and high-pressure reactors. The solution of this problem is the basis of designing this equipment.

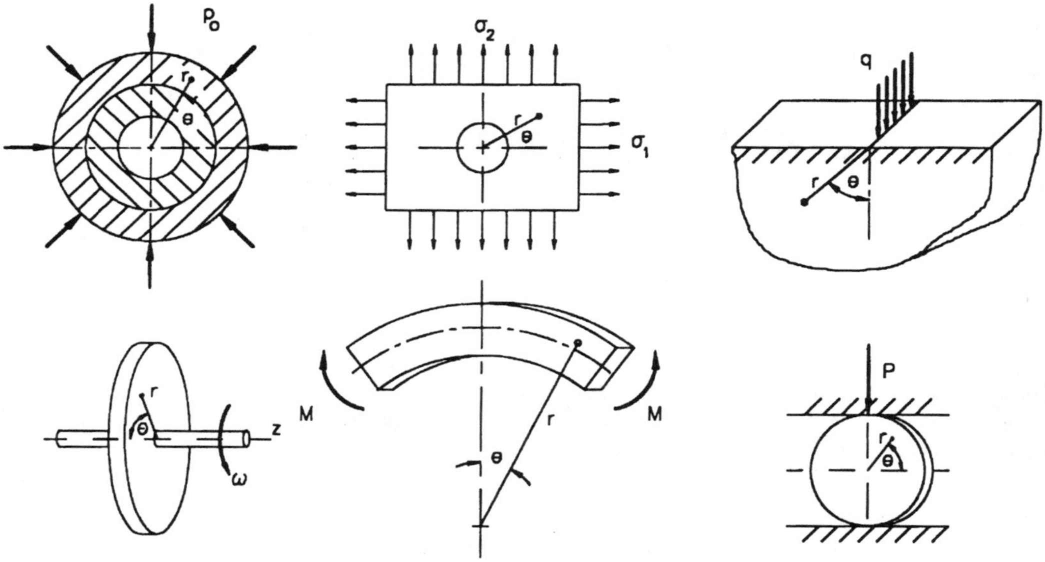

FIGURE 6.1 Examples of elastic plane problems in polar coordinates.

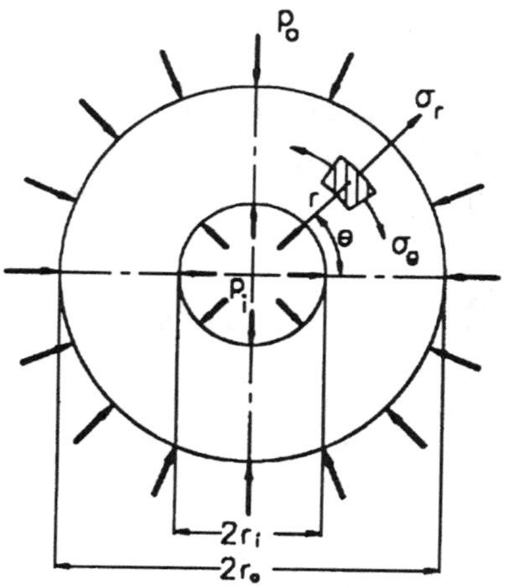

FIGURE 6.2 Thick-walled cylinder subjected to uniform internal and/or external pressure.

Consider a thick-walled cylinder of internal radius r_i and external radius r_o, which is subjected to internal pressure p_i and external pressure p_o, as shown in Figure 6.2.

The problem is axisymmetric with no body forces. The stress function and the stresses are given by Equations 4.34 and 4.35, namely,

$$\phi = A \ln r + Cr^2$$

$$\sigma_r = \frac{1}{r}\frac{d\phi}{dr} = \frac{A}{r^2} + 2C$$

$$\sigma_\theta = \frac{d^2\phi}{dr^2} = -\frac{A}{r^2} + 2C$$

The constants A and C are determined from the boundary conditions:

$$\sigma_r = -p_i \quad \text{at} \quad r = r_i$$

$$\sigma_r = -p_o \quad \text{at} \quad r = r_o$$

thus yielding

$$A = \frac{r_i^2 r_o^2 (p_o - p_i)}{r_o^2 - r_i^2}$$

$$2C = \frac{r_i^2 p_i - r_o^2 p_o}{r_o^2 - r_i^2}$$

Let $r_o/r_i = \lambda > 1$; the expressions for the stresses at any radius r are thus given by

$$\sigma_r = \frac{1}{\lambda^2 - 1}\left[p_i\left(1 - \frac{r_o^2}{r^2}\right) - p_o\left(\lambda^2 - \frac{r_o^2}{r^2}\right)\right]$$

$$\sigma_\theta = \frac{1}{\lambda^2 - 1}\left[p_i\left(1 + \frac{r_o^2}{r^2}\right) - p_o\left(\lambda^2 + \frac{r_o^2}{r^2}\right)\right] \tag{6.1a}$$

The value of the third normal stress component σ_z, depends on the end conditions. For plane strain conditions (a case of a cylinder with its ends axially constrained), it follows from Hooke's law that

$$\sigma_z = \nu(\sigma_r + \sigma_\theta) = \frac{2\nu}{\lambda^2 - 1}\left(p_i - p_o\lambda^2\right) \tag{6.1b}$$

On the other hand, for plane stress conditions $\sigma_z = 0$. However, in many engineering applications the case of a cylinder with free ends is encountered. In this case the appropriate boundary condition is not that $\sigma_z = 0$, but, rather, the integral

$$\int_{r_i}^{r_o} 2\pi\sigma_z r\,dr = 0$$

Due to symmetry in both geometry and loading, the nonzero displacements are the component, u for plane strain, and u and w for plane stress. These components are easily obtained by applying Hooke's law, namely,

For plane strain: $u = r\varepsilon_\theta = \frac{r}{E}\left[\sigma_\theta - \nu(\sigma_r + \sigma_z)\right], \quad w = 0$

For plane stress: $u = \frac{r}{E}\left[\sigma_\theta - \nu\sigma_r\right], \quad w = -\frac{\nu}{E}\int_o^z (\sigma_5 + \sigma_\theta)dz$

Substituting for the stresses from Equation 6.1 gives the radial displacement u in plane strain conditions as

$$u = \frac{(1+\nu)r}{E(\lambda^2 - 1)}\left\{\left[(1-2\nu) + \frac{\lambda^2}{(r/r_i)^2}\right]p_i - \left[(1-2\nu) + \frac{1}{(r/r_i)^2}\right]\lambda^2 p_o\right\} \tag{6.1c}$$

Special cases follow.

6.2.1.1 Cylinder Subjected to Internal Pressure Only

The stresses are obtained from Equations 6.1a by substituting $p_o = 0$; therefore,

$$\sigma_r = \frac{p_i}{\lambda^2 - 1}\left(1 - \frac{r_o^2}{r^2}\right)$$
$$\sigma_\theta = \frac{p_i}{\lambda^2 - 1}\left(1 + \frac{r_o^2}{r^2}\right) \tag{6.2}$$

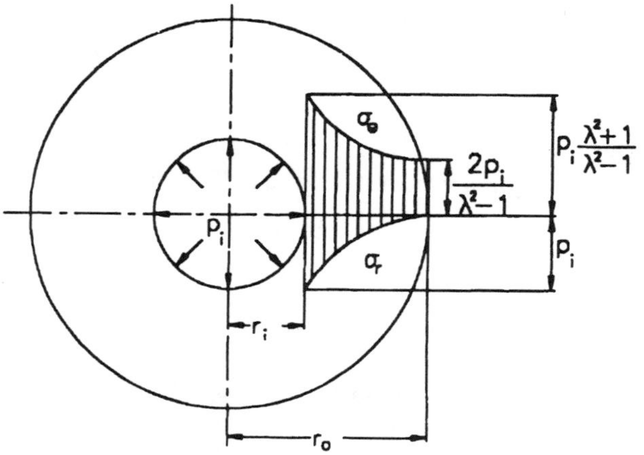

FIGURE 6.3 Stress distribution through wall thickness for a cylinder subjected to internal pressure.

The stress distribution in this case is shown in Figure 6.3. When $r_o = \infty$, i.e., the case of a uniform internal pressure applied to a circular hole in an infinite elastic plate, the stress distribution is given by

$$\sigma_r = -p_i r_i^2/r^2 \text{ and } \sigma_\theta = p_i r_i^2/r^2$$

It is seen that $\sigma_r = -\sigma_\theta$, thus representing a state of pure shear referred to axes inclined 45° to any radial directions.

For a thin-walled tube $r_o \cong r_i \cong r$ and $\lambda^2 = = (r_o/r_i)^2 \cong 1 + (2h/r_i)$, where h is the tube thickness. Therefore, from Equations 6.2 $\sigma_r \cong 0$ and $\sigma_\theta = p_i r_i/h$ as given by the elementary equations of thin-walled pressure vessels.

6.2.1.2 Cylinder Subjected to External Pressure Only

The stresses are obtained from Equations 6.1a by substituting $p_i = 0$; therefore,

$$\sigma_r = \frac{p_o}{\lambda^2 - 1}\left(\lambda^2 - \frac{r_o^2}{r^2}\right)$$

$$\sigma_\theta = -\frac{p_o}{\lambda^2 - 1}\left(\lambda^2 + \frac{r_o^2}{r^2}\right) \tag{6.3}$$

The stress distribution for this case is shown in Figure 6.4.

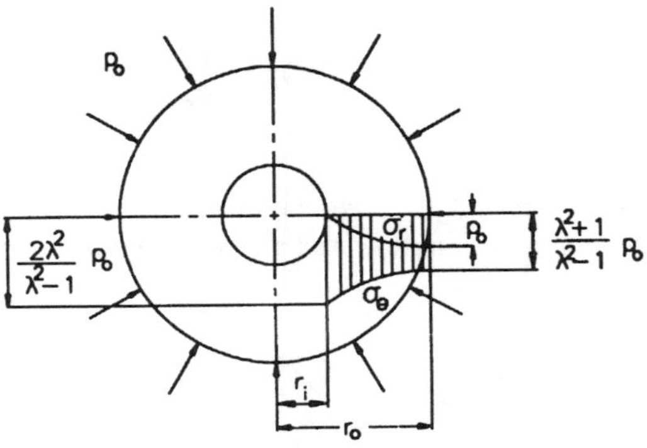

FIGURE 6.4 Stress distribution through wall thickness for a cylinder subjected to external pressure.

When $r_i = 0$, i.e., the case of a solid cylinder subjected to external uniform pressure, the stresses are given by

$$\sigma_r = -p_o \quad \text{and} \quad \sigma_\theta = -p_o$$

The stresses at all points in any direction are equal to $-p_o$, which represents a state of homogeneous stress all over.

Example 6.1:

Consider a thick-walled cylinder of inner and outer radii r_i and r_o, *respectively. When the cylinder is subjected to a uniform external pressure* p_o, *the circumferential stress* σ_θ *at* r_i *has the magnitude* σ. *It is required to reduce the magnitude of this stress to 0.8* σ *by the simultaneous application of a uniform internal pressure* p_i, *determine the relation of* p_o *to* p_i.

Solution:

For a cylinder subjected to external pressure only, Equation 6.3 gives

$$\sigma_\theta = -\frac{p_o}{\lambda^2 - 1}\left(\lambda^2 + \frac{r_o^2}{r^2}\right)$$

where

$$\lambda = r_o/r_i$$

At the inner radius $\sigma_\theta = \sigma$, therefore

$$\sigma = -\frac{p_o}{\lambda^2 - 1}\left(\lambda^2 + \lambda^2\right) = \frac{-2\lambda^2}{\lambda^2 - 1} p_o \tag{a}$$

By applying an internal pressure p_i while maintaining the external pressure p_o, the stress σ_θ is reduced to 0.8; hence, the second of Equations 6.1 gives at $r = r_i$:

$$0.8\sigma = \frac{1}{\lambda^2 - 1}\left[p_i\left(1 + \lambda^2\right) - 2\lambda^2 p_o\right] \tag{b}$$

Substituting Equation a into b yields the relation between p_i and p_o as

$$p_o / p_i = 5\left(1 + r_i^2 / r_o^2\right)/2$$

Example 6.2:

A long, thick-walled cylinder is embedded in a rigid wall and subjected to uniform internal pressure as shown in Figure 6.5. Determine the stresses and displacement in the cylinder for $r_o/r_i = 2.5$ *and* $\nu = 1/3$. *Illustrate the results graphically.*

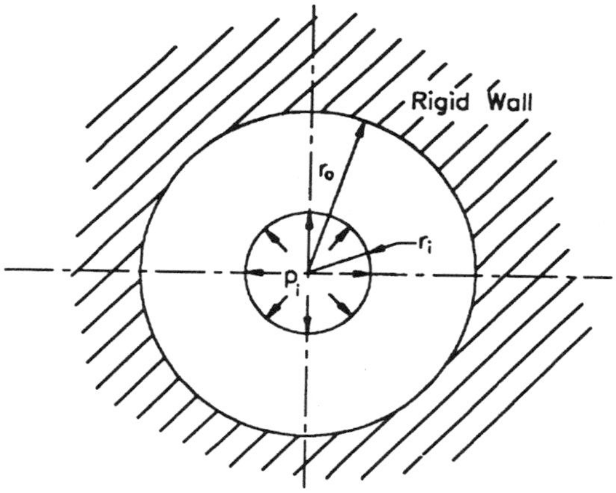

FIGURE 6.5 Example 6.2.

Solution:

The stresses are given by Equation 6.1. The pressure p_o is determined from the boundary condition:

At $r = r_o$: $u = 0$

where

$$u = r\varepsilon_\theta = \frac{r}{E}\left[\sigma_\theta - \nu\left(\sigma_r + \sigma_z\right)\right]$$

For a long cylinder there is no axial displacement, so that plane strain conditions prevail; hence,

$$\sigma_z = \nu(\sigma_r + \sigma_\theta)$$

and

$$u = \frac{r}{E}\left[\sigma_\theta - \nu(\sigma_r + \nu)(\sigma_r + \sigma_\theta)\right]$$

$$u = \frac{r}{E}\left[(1-\nu^2)\sigma_\theta - \nu(1+\nu)\sigma_r\right]$$

Substituting $\nu = {}^1/_3$ yields:

$$u = \frac{4r}{9E}(2\sigma_\theta - \sigma_r)$$

At $r = r_o$: $u = 0, \ \sigma_r = p_o$, and $\sigma_\theta = \frac{1}{\lambda^2 - 1}\left[2p_i - p_o(\lambda^2 + 1)\right]$

Hence, substituting these into the above expression for u results in

$$0 = \frac{2}{\lambda^2 - 1}\left[2p_i - p_o(\lambda^2 + 1)\right] + p_o$$

or

$$p_o = 4p_i/(\lambda^2 + 3)$$

Substituting $\lambda = 2.5$ yields $p_o = 16p_i/37$, which when substituted in Equations 6.1 gives the following expressions for the stresses and displacement:

$$\sigma_r = -\frac{4p_i}{37}\left[3 + \left(\frac{r_o}{r}\right)^2\right]$$

$$\sigma_\theta = -\frac{4p_i}{37}\left[3 - \left(\frac{r_o}{r}\right)^2\right]$$

$$\sigma_z = -\frac{8p_i}{37}$$

$$u = -\frac{16r_o p_i}{111E}\left[\frac{r_o}{r} - \frac{r}{r_o}\right]$$

The stress distribution, Figure 6.6, shows that σ_θ changes from tensile to compressive at $r = r_o/\sqrt{3}$.

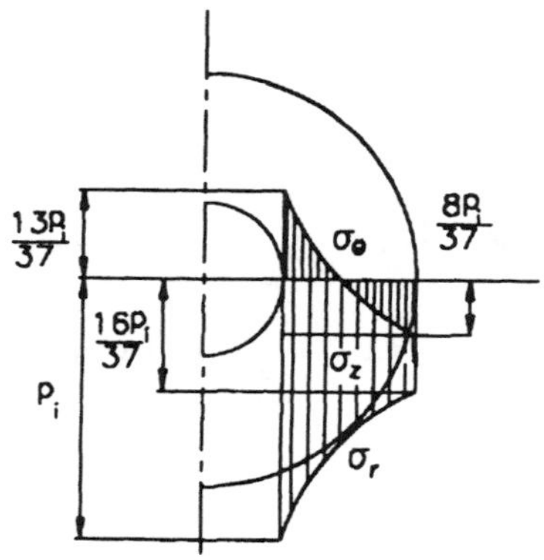

FIGURE 6.6 Example 6.2.

6.2.2 Thick-Walled Cylinder Subjected to a Steady-State Radial Thermal Gradient

For thick-walled cylinders subjected to radial thermal gradient $T = T(r)$, the stresses are obtained from Equations 4.42 and 4.45 for conditions of plane strain and plane stress, respectively. Note that if $T = T(r, z)$ the problem is a three-dimensional one.

6.2.2.1 Plane Strain

Considering the case of plane strain, $\varepsilon_z = 0$, the stresses are given by Equations 4.42 as

$$\sigma_r = \frac{A}{r^2} + 2C - \frac{\alpha E}{1-\nu}\frac{1}{r^2}\int Trdr$$

$$\sigma_\theta = -\frac{A}{r^2} + 2C - \frac{\alpha E}{1-\nu}\left[T - \frac{1}{r^2}\int Trdr\right]$$

and, from the thermoelastic stress–strain relation (Equation 3.18),

$$\sigma_z = \nu(\sigma_r + \sigma_\theta) - \alpha ET = 4\nu C - \frac{\alpha ET}{1-\nu}$$

The constants A and C are determined from the boundary conditions:

$$\sigma_r = 0 \text{ at } r = r_i, \text{ and } \sigma_r = 0 \text{ at } r = r_o$$

Applying these boundary conditions to Equations 4.42 yields, after algebraic manipulation, the following stress expressions:

$$\sigma_r = \frac{\alpha E}{1-\nu}\frac{1}{r^2}\left[\frac{(r/r_i)^2 - 1}{\lambda^2 - 1}\int_{r_i}^{r_o} Trdr - \int_{r_i}^{r} Trdr\right]$$

$$\sigma_\theta = \frac{\alpha E}{1-\nu}\frac{1}{r^2}\left[\frac{(r/r_i)^2 + 1}{\lambda^2 - 1}\int_{r_i}^{r_o} Trdr + \int_{r_i}^{r} Trdr - Tr^2\right] \tag{6.4}$$

$$\sigma_z = \frac{\alpha E}{1-\nu}\left[\frac{2\nu}{\lambda^2 - 1}\frac{1}{r_i^2}\int_{r_i}^{r_o} Trdr - T\right]$$

The last of Equations 6.4 represents the distribution of the axial normal stress σ_z, which must be applied in order to fulfill the plane strain condition: $\varepsilon_z = 0$. If, however, the tube is allowed to deform longitudinally, such that every cross section remains plane,[1] a state of "generalized plane strain" is developed. The stress components σ_r and σ_θ remain unchanged and are given by the first and second of Equations 6.4, respectively. The value of σ_z is modified by imposing a constant strain ε_o in the z-direction, such that

$$\sigma_z = \frac{\alpha E}{1-\nu}\left[\frac{2\nu}{\lambda^2 - 1}\frac{1}{r_i^2}\int_{r_i}^{r_o} Trdr - T\right] + E\varepsilon_o \tag{6.5}$$

The constant strain ε_o is arbitrary.

Example 6.3:

Find the expressions of σ_r, σ_θ, and σ_z in a long, thick-walled cylinder under plane strain conditions, which conducts heat in a steady state according to Equation 4.46b, namely,

$$\mathrm{T} = \mathrm{K}\,ln(\mathrm{r_o}/\mathrm{r}) \quad where \quad \mathrm{K} = \frac{\mathrm{T_i} - \mathrm{T_o}}{ln(\mathrm{r_o}/\mathrm{r_i})}$$

T *is the temperature difference at any radius* r, *and* $\mathrm{T_i}$ *and* $\mathrm{T_o}$ *are the temperatures at* $\mathrm{r} = \mathrm{r_o}$, *respectively. Plot the stress distribution for $\lambda = 3$ and $\nu = 0.3$.*

Solution

The stresses in this case are given by Equations 6.4 where the integrals

$$\int_{r_i}^{r_o} Trdr \quad \text{and} \quad \int_{r_i}^{r} Trdr$$

are obtained from the temperature distribution given above as

$$\int_{r_i}^{r} Trdr = K\left[\frac{1}{2}r^2 \ln(r_o/r) - \frac{1}{2}r_i^2 \ln\lambda + \frac{1}{4}(r^2 - r_i^2)\right]$$

$$\int_{r_i}^{r_o} Trdr = \int_{r_i}^{r_o} K\ln(r_o/r)rdr = \frac{Kr_i^2}{2}\left[\frac{1}{2}(\lambda^2 - 1) - \ln\lambda\right]$$

Substituting into Equations 6.4 yields

$$\sigma_r = \frac{\alpha E K}{2(1-\nu)}\left[-\ln\left(\frac{r_o}{r}\right) - \frac{\ln\lambda}{\lambda^2-1}\left(1-\frac{r_o^2}{r^2}\right)\right]$$

$$\sigma_\theta = \frac{\alpha E K}{2(1-\nu)}\left[1-\ln\left(\frac{r_o}{r}\right) - \frac{\ln\lambda}{\lambda^2-1}\left(1+\frac{r_o^2}{r^2}\right)\right] \tag{6.6}$$

$$\sigma_z = \frac{\alpha E K}{2(1-\nu)}\left[\nu-2\ln\left(\frac{r_o}{r}\right) - \frac{2\nu\ln\lambda}{\lambda^2-1}\right]$$

Figure 6.7 shows the distribution of stresses across the wall of a tube of $\lambda = 3$ and $\nu = 0.3$. If $T_i - T_o$ is positive, the stresses are compressive at the inner surface and tensile at the outer surface. In case of materials, such as cast iron, concrete, or brick, which are weak in tension, cracks are likely to start on the outer surface of the cylinder under the above conditions.

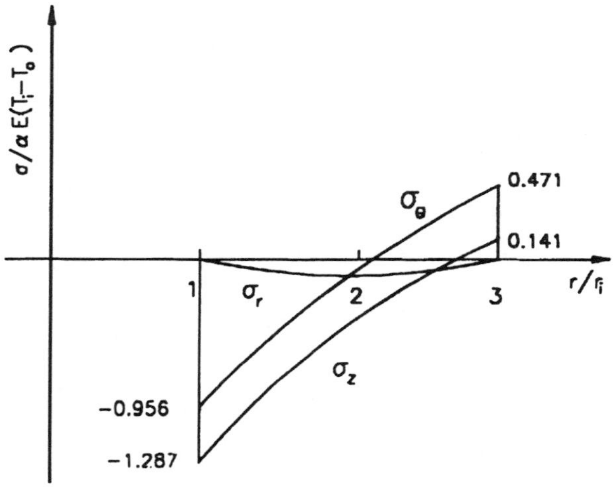

FIGURE 6.7 Example 6.3.

Example 6.4:

A long, thick-walled cast iron tube of radius ratio $\lambda = 3$ is used to transport a hot fluid at temperature T_i and pressure p_i. Considering plane strain conditions,

a. *Determine the stress at $r = r_i$ and $r = r_o$ knowing that $p_i = 5$ MPa. $E = 150$ GPa, $\nu = 0.3$, and $\alpha = 12 \times 10^{-6}/°C$.*
b. *If the tube fails when the maximum normal stress reaches either the value of the ultimate strength in tension or compression, what is the maximum temperature T_i that the tube can withstand taking a factor of safety F.S. = 2.5? Take ultimate tensile strength = 250 MPa and ultimate compressive strength = 750 MPa.*

Solution:

a. The results of Example 6.3, specifically Figure 6.7, may be used to determine the stresses at $r = r_i$ and $r = r_o$ due to the thermal gradient, as given in the following table:

Cause	Stress	$r = r_i$	$r = r_o$
Thermal gradient	σ_r	0	0
(T_i)	σ_θ	$-0.957\alpha ET_i$	$0.471\alpha ET_i$
Equations 6.6	σ_z	$-1.287\alpha ET_i$	$0.141\alpha ET_i$
Internal pressure	σ_r	$-p_i$	0
p_i	σ_θ	$p_i(\lambda^2 + 1)/(\lambda^2 - 1)$	$2p_i/(\lambda^2 - 1)$
Equations 6.2	σ_z	$2\nu p_i/(\lambda^2 - 1)$	$2\nu p_i/(\lambda^2 - 1)$

Hence, for $p_i = 5$ MPa, $\nu = 0.3$, $E = 150$ GPa, and $\alpha = 12 \times 10^{-6}/°C$, the total stresses are obtained by superposition as

At $r = r_i$ (in MPa):	At $r = r_o$ (in MPa):
$\sigma_r = -5$	$\sigma_r = 0$
$\sigma_\theta = 6.25 - 1.723T_i$	$\sigma_\theta = 1.25 + 0.848T_i$
$\sigma_z = 0.375 - 2.317T_i$	$\sigma_z = 0.375 + 0.254T$

b. The maximum tensile stress σ_θ occurs at $r = r_o$, such that

$$\sigma_\theta = 1.25 + 0.848T_i \leq (250/\text{F.S.})$$

giving $T_i \leq 116.45°C$.

On the other hand, the maximum compressive stress σ_θ occurs at the bore, such that $\sigma_z = 0.375 - 2.317T_i \leq (750/\text{F.S.})$ giving $T_i \leq 129.3°C$. Hence, the maximum permissible temperature is 116.5°C.

6.2.2.2 Plane Stress

For the case of plane stress,$\sigma_z = 0$, the stresses are given by Equations 4.45:

$$\sigma_r = \frac{A}{r^2} + 2C - \frac{\alpha E}{r^2}\int Trdr$$

$$\sigma_\theta = -\frac{A}{r^2} + 2C - \alpha E\left[T - \frac{1}{r^2}\int Trdr\right]$$

This is the case of a disk centrally cooled or externally heated at the outer rim, with its two faces insulated to prevent any thermal gradient across the disk thickness.

The constants A and C are determined from the boundary conditions:

$$\sigma_r = 0 \text{ at } r = r_i, \text{ and } \sigma_r = 0 \text{ at } r = r_o$$

which yields the following stress expressions:

$$\sigma_r = \frac{\alpha E}{r^2}\left[\frac{(r/r_i)^2 - 1}{\lambda^2 - 1}\int_{r_i}^{r_o} Trdr - \int_{r_i}^{r} Trdr\right]$$

$$\sigma_\theta = \frac{\alpha E}{r^2}\left[\frac{(r/r_i)^2 + 1}{\lambda^2 - 1}\int_{r_i}^{r_o} Trdr + \int_{r_i}^{r} Trdr - Tr^2\right] \tag{6.7a}$$

6.2.2.3 Other End Conditions

Other end conditions may be considered such as those for cylinders with open or closed ends. For the former condition, i.e., cylinder with free ends, the resultant axial force must vanish such that $\int_{r_i}^{r_o} \sigma_z 2\pi\, rdr = 0$. Integrating the third of the thermoelastic stress-strain Equations 3.18 gives

$$\int_{r_i}^{r_o} \varepsilon_z 2\pi\, rdr = \int_{r_i}^{r_o}\left\{\frac{1}{E}\left[\sigma_z - v(\sigma_r + \sigma_\theta)\right] + \alpha T\right\}2\pi\, rdr$$

For sections far removed from the ends, ε_z is constant.* Hence, substituting from the condition for zero axial force and using Expressions 6.4 for $(\sigma_r + \sigma_\theta)$ yields

$$\varepsilon_z = \frac{2\alpha}{\lambda^2 - 1}\frac{1}{r_1^2}\int_{r_i}^{r_o} Trdr$$

The stress σ_z is thus given by

$$\sigma_z = \frac{E\alpha}{1 - v}\left[\frac{2}{\lambda^2 - 1}\frac{1}{r_i^2}\int_{r_i}^{r_o} Trdr - T\right] \tag{6.7b}$$

For a pressurized cylinder with closed ends, the resultant axial force must be equal to $p_i\, \pi\, r_i^2$. Following a similar procedure, results in ε_z and σ_z as

$$\varepsilon_z = \frac{1 - 2v}{E}\frac{p_i}{\lambda^2 - 1} + \frac{2\alpha}{\lambda^2 - 1}\frac{1}{r_i^2}\int_{r_i}^{r_o} Trdr$$

$$\sigma_z = \frac{p_i}{\lambda^2 - 1} + \frac{E\alpha}{1 - v}\left[\frac{2}{\lambda^2 - 1}\frac{1}{r_i^2}\int_{r_i}^{r_o} Trdr - T\right] \tag{6.7c}$$

Example 6.5:

In example 6.3, replace the long, thick-walled cylinder under plane strain conditions by a thin disk that has the same r_i, r_o, v, T_i, T_o and the same radial thermal gradient.

Solution:

Substituting in Equations 6.7 yields:

---

* In this analysis it is assumed tht the cylinder is very long. Near the ends, local bending stress develop due to moments pertaining to the particular distribution of σ_z. Also see Timoshenko and Goodier.[3]

$$\sigma_r = \frac{\alpha EK}{2}\left[-\ln\left(\frac{r_o}{r}\right) - \frac{\ln\lambda}{\lambda^2 - 1}\left(1 - \frac{r_o^2}{r^2}\right)\right]$$

$$\sigma_\theta = \frac{\alpha EK}{2}\left[1 - \ln\left(\frac{r_o}{r}\right) - \frac{\ln\lambda}{\lambda^2 - 1}\left(1 + \frac{r_o^2}{r^2}\right)\right] \qquad (6.7d)$$

The stress distribution pattern for σ_r and σ_θ is the same as Figure 6.7, while $\sigma_z = 0$. The stress values are obtained by multiplying σ_r and σ_θ by $(1 - \nu)$, i.e., 0.7, and making $\sigma_z = 0$ for plane stress conditions.

6.2.3 Cylinders Compounding by Shrink Fit

Cylinders subjected to extremely high pressures, e.g., gun barrels and metal-forming containers, are usually prestressed to fulfill strength requirements that cannot be practically obtained by increasing the cylinder wall thickness. One of the methods used is to build up the cylinder from two or more cylinders shrunk together. In this case of assembling two cylinders, the outer cylinder has an internal diameter slightly less than the external diameter of the inner cylinder. For heavy shrink fits, the outer cylinder is expanded by heating or the inner cylinder is chilled and the two cylinders are assembled together. When the assembly attains the same temperature, the outside and inside cylinders become prestressed circumferentially in tension and in compression, respectively. In this way, by building a "prestressed compound cylinder," it is possible to reduce the stresses inside the cylinder when an internal pressure acts.

Shrink fits are also used in other applications, for example, to transmit torque between rotating shafts and rotors such as flywheels and pinions. Railway wheels have to be shrunk fit on their axles to avoid fretting fatigue. The axle is slightly tapered and the wheel is assembled by using a heavy press.

In solving the problem of cylinders assembled by shrink fits, the equations obtained for stresses in a thick-walled cylinder subjected to internal and external pressures are applied.

Consider a compound cylinder made of two cylinders. Let r_i, r_1 be the radii of the inside cylinder before fitting, where $\lambda_1 = r_1/r_i$ and r_2 ($<r_1$); and r_o be the radii of the outside cylinder before fitting, where $\lambda_2 = r_o/r_2$. The radial interference is $\Delta = r_1 - r_2$. After assembly, the two cylinders are in contact at a common radius r_c, Figure 6.8. The resulting contact pressure p_o acting equally on both cylinders at $r = r_c$ causes a radial displacement u_1 at the outer surface of the inside cylinder and a radial displacement u_2 at the inner surface of the outside cylinder, such that

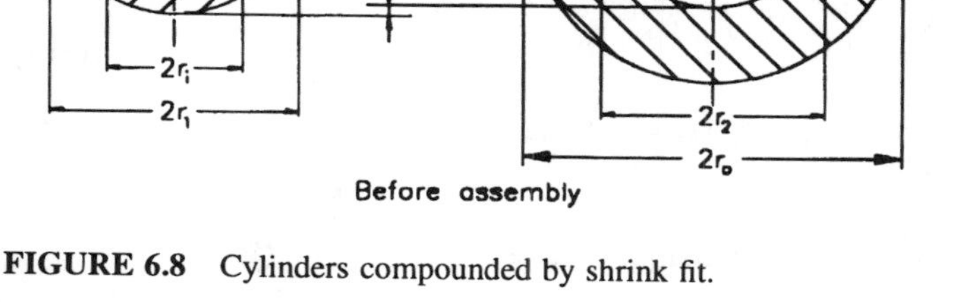

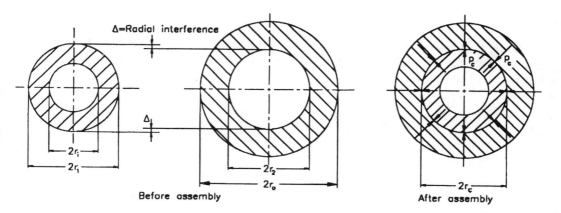

FIGURE 6.8 Cylinders compounded by shrink fit.

$$r_c = r_1 + u_1 = r_2 + u_2$$

and, hence,

$$u_2 - u_1 = r_1 - r_2$$

The assembly condition is, therefore,

$$\Delta = u_2 - u_1$$

The circumferential strain components ε_θ at the contact surface for the two cylinders are given by

$$\varepsilon_{\theta_2} = \frac{u_2}{r_c}$$

for the outside cylinder and

$$\varepsilon_{\theta_1} = \frac{u_1}{r_c}$$

for the inside cylinder. Therefore,

$$\frac{\Delta}{r_c} = \varepsilon_{\theta_2} - \varepsilon_{\theta_1} \tag{6.8}$$

For plane stress conditions, $\sigma_z = 0$ and, hence,

$$\frac{\Delta}{r_c} = \frac{1}{E_2}\left(\sigma_{\theta_2} - \nu_2\sigma_{r_2}\right) - \frac{1}{E_1}\left(\sigma_{\theta_1} - \nu_1\sigma_{r_1}\right) \tag{6.9}$$

where (E_1, ν_1) and (E_2, ν_2) are the elastic constants of the inside and outside cylinders, respectively. The radial stresses at the common radius r_o are

$$\sigma_{r_1} = \sigma_{r_2} = -p_c \tag{6.10}$$

Since, after assembly, the outside and inside cylinders are subjected to internal and external pressures, respectively, the circumferential stresses at the common radius r_c are thus given by Equations 6.2 and 6.3 as

$$\begin{aligned} \sigma_{\theta 2} &= \frac{p_c}{\lambda_2^2 - 1}\left(1 + \frac{r_o^2}{r_c^2}\right) = \frac{p_c}{\lambda_2^2 - 1}\left(1 + \lambda_2^2\right) \\ \sigma_{\theta 1} &= \frac{-p_c}{\lambda_1^2 - 1}\left(\lambda_1^2 + \frac{r_o^2}{r_c^2}\right) = -\frac{p_c}{\lambda_1^2 - 1}\left(1 + \lambda_1^2\right) \end{aligned} \tag{6.11}$$

Substitution of Equations 6.10 and 6.11 into Equation 6.9 yields

$$\frac{\Delta}{r_c} = p_c\left[\frac{1}{E_z}\left(\frac{\lambda_2^2+1}{\lambda_2^2-1}+\gamma_2\right)+\frac{1}{E_1}\left(\frac{\lambda_1^2+1}{\lambda_1^2-1}-\gamma_1\right)\right] \tag{6.12}$$

Equation 6.12 provides a relation between the radial interference Δ and the contact pressure p_c, which results from assembly.

For a compound cylinder built up from two cylinders made of the same material, i.e., $E_1 = E_2 = E$ and $\nu_1 = \nu_2$, Equation 6.12 reduces to

$$\frac{\Delta}{r_c} = \frac{p_c}{E}\left(\frac{\lambda_2^2+1}{\lambda_2^2-1}+\frac{\lambda_1^2+1}{\lambda_1^2-1}\right) \tag{6.13}$$

For a sleeve shrunk on a solid shaft of the same material, Equation 6.13 is further simplified by putting $\lambda_1 = \infty$ (for solid shaft) and $\lambda_2 = \lambda$, i.e.,

$$\frac{\Delta}{r_c} = \frac{p_c}{E}\frac{2\lambda^2}{\lambda^2-1} \tag{6.14}$$

For a solid shaft shrunk inside a wall with a very large thickness, i.e., $\lambda = \infty$, Equation 6.14 gives

$$\frac{\Delta}{r_c} = \frac{2p_c}{E} \tag{6.15}$$

The stress distribution due to compounding, Figure 6.9a, is obtained from Equations 6.2 and 6.3 as

$$\text{For the inside cylinder: } \quad \begin{matrix}\sigma_r\\ \sigma_\theta\end{matrix} = \frac{-p_c}{\lambda_1^2-1}\left(\lambda_1^2 \mp \frac{r_c^2}{r^2}\right) \tag{6.16}$$

$$\text{For the outside cylinder: } \quad \begin{matrix}\sigma_r\\ \sigma_\theta\end{matrix} = \frac{p_c}{\lambda_2^2-1}\left(1 \mp \frac{r_o^2}{r^2}\right) \tag{6.17}$$

If the compound cylinder is subjected to an internal pressure p_i, it will act as one unit since the circumferential displacement $v = 0$ (due to symmetry) and the radial displacement u is continuous. As a result of internal pressure, the following additional stresses arise, Figure 6.9b.

$$\begin{matrix}\sigma_r\\ \sigma_\theta\end{matrix} = \frac{p_i}{\lambda^2-1}\left(1 \mp \frac{r_o^2}{r^2}\right)$$

where $r_o/r_i = \lambda = \lambda_1\lambda_2$.

These stresses when added to the initial stresses due to compounding will give the stress distribution due to internal pressure in the compound cylinder, Figure 6.9c.

For optimum strength, the condition to be satisfied is

$$(\sigma_\theta - \sigma_r)_{r=r_i} = (\sigma_\theta - \sigma_r)_{r=r_c} = \text{a minimum}$$

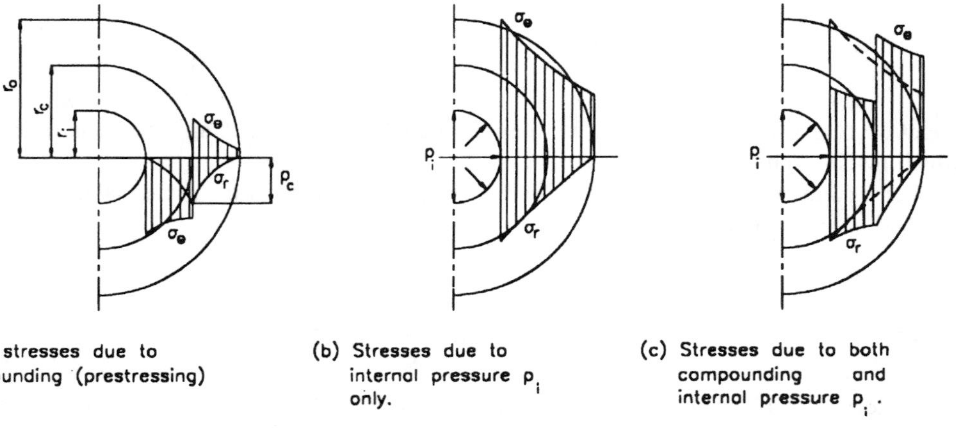

FIGURE 6.9 Stress distribution though the wall thickness of a shrink-fit compound cylinder subjected to internal pressure.

where σ_θ, σ_r are the sum of the stresses due to both the compounding pressure p_o and the internal working pressure p_i. It can be easily shown that the above condition is fulfilled if the expression[3]

$$(\sigma_\theta - \sigma_r) = \frac{2\lambda^2}{\lambda^2 - 1}\left[1 - \frac{1}{r_o^2/(r_o^2 - r_c^2) + r_c^2/(r_c^2 - r_i^2)}\right]$$

is minimized. This results in the optimum conditions when

$$r_c = \sqrt{r_i r_o} \text{ or } \lambda_1 = \lambda_2$$

Thus,

$$\Delta = p_i r_c / E$$

and $(\sigma_\theta - \sigma_r)_{min} = p_i\lambda/(\lambda - 1)$.

Example 6.6:

Determine the tangential stress distribution across the thickness of a steel cylinder with inner radius 100 mm and outer radius 200 mm submitted to an internal pressure $p_i = 200$ *MPa. Compare the tangential stresses if the cylinder with same radii of 100 and 200 mm is built up of two cylinders having the same diameter ratio* $\lambda_1 = r_c/r_i = \lambda_2 = r_o/r_c = \sqrt{2}$. *The radial interference is 0.1414 mm and* $E = 200$ *GPa for steel.*

Solution:

Equations 6.2 give for the hoop stress σ_θ:

$$\sigma_\theta = \frac{p_i}{\lambda^2 - 1}\left(1 + \frac{r_o^2}{r^2}\right)$$

Therefore,

$$(\sigma_\theta)_{r=100} = \frac{200}{4-1}\left[1 + \left(\frac{200}{100}\right)^2\right] = 333.3 \text{ MPa}$$

and, similarly,

$$(\sigma_\theta)_{r=200} = 133.33 \text{ MPa}$$

For a compound cylinder, the contact pressure p_c due to compounding is obtained from Equation 6.13 as

$$\frac{\Delta}{r_c} = \frac{p_c}{E_2}\left(\frac{\lambda_2^2+1}{\lambda_2^2-1} + \frac{\lambda_1^2+1}{\lambda_1^2-1}\right)$$

where $r_c = 141.4$ mm, $\Delta = 0.1414$ mm, $E = 200{,}000$ MPa, and $\lambda_1 = \lambda_2 = \sqrt{2}$. Therefore, $p_c = 33.3$ MPa.

The initial stresses due to compounding are determined from Equations (6.2 and 6.3) as

For the Inside cylinder:

$$(\sigma_\theta)_{r=r_i=100} = \frac{p_c}{\lambda_1^2-1}\left(\lambda_1^2 + \frac{r_c^2}{r_i^2}\right) = 133.3 \text{ MPa}$$

For the outside cylinder,

$$(\sigma_\theta)_{r_c=141.4} = \frac{p_c}{\lambda_2^2-1}\left(1 + \frac{r_o^2}{r_c^2}\right) = 100 \text{ MPa}$$

$$(\sigma_\theta)_{r=200} = 66.7 \text{ MPa}$$

It is seen from Figure 6.10 that, due to the initial stresses, the maximum hoop stress when the cylinder is subjected to internal pressure is reduced from 333.3 to 300 MPa, i.e., by 10%.

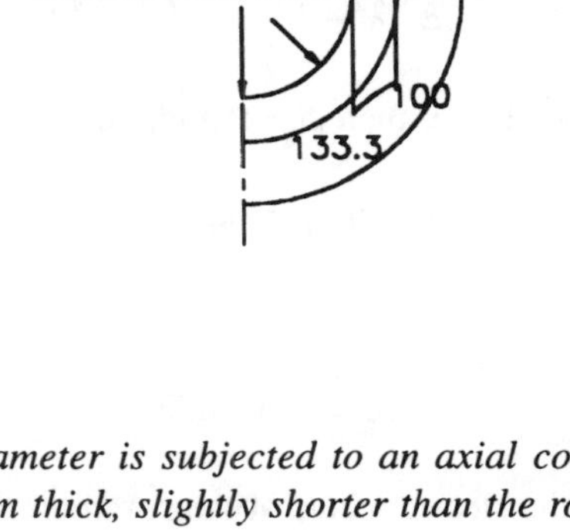

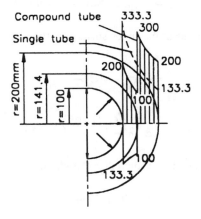

FIGURE 6.10 Example 6.6.

Example 6.7:

A short steel rod of 50 mm diameter is subjected to an axial compressive load of 250 kN. It is surrounded by a sleeve 12.5 mm thick, slightly shorter than the rod so that the axial compressive load is carried only by the rod. Assuming a close fit before the load is applied and neglecting friction, find the pressure p_c *between the sleeve and the rod and the maximum tensile stress in the sleeve.* ($\nu = 0.3$, $E = 210$ GPa).

Solution:

For a solid rod subjected to an external pressure p_c,

$$\sigma_r = \sigma_\theta = -p_c$$

Due to the compressive load,

$$\sigma_z = -250{,}000 \Big/ \left(\frac{\pi}{4}\right) 50^2 = -127.32 \text{ MPa}$$

The radial displacement at the rod surface u_o is given by

$$\left(u_o/r\right)_{r=25} = \frac{1}{E}\left[\left(\sigma_\theta - \nu\left(\sigma_r + \sigma_r\right)\right)\right]_{r=25}$$

$$= \frac{1}{E}\left[-p_c - 0.3\left(-p_c - 127.32\right)\right] = \frac{1}{E}\left[-0.7p_c + 38.2\right]$$

For a sleeve subjected to internal pressure p_c, the stresses are given by Equations 6.2, namely,

$$\sigma_r = \frac{p_c}{\lambda^2 - 1}\left[1 - \left(\frac{r_o}{r}\right)^2\right]$$

$$\sigma_\theta = \frac{p_c}{\lambda^2 - 1}\left[1 + \left(\frac{r_o}{r}\right)^2\right]$$

$$\sigma_z = 0 \text{ (plane stress condition)}$$

Therefore, the radial displacement at the inner surface of the sleeve u_i is given by

$$\left(u_i/r\right)_{r=25} = \frac{1}{E}\left[\sigma_\theta - \nu\left(\sigma_r + \sigma_z\right)\right]_{r=25}$$

$$= \frac{1}{E}\left[\frac{p_c}{\lambda^2 - 1}\left(1 + \lambda^2\right) + \nu p_c\right] = \frac{2.9p_c}{E}$$

Since the rod is a close fit in the sleeve, then $\Delta = 0$ and, hence,

$$(u_o/r)_{r=25} = (u_i/r)_{r=25}$$

Therefore,

$$p_c = 10.6 \text{ MPa}$$

The maximum tensile stress in the sleeve occurs at the common radius $r = 25$ mm and is given by

$$\sigma_\theta = \frac{p_c}{1.25}\left[1+\left(\frac{r_o}{r}\right)^2\right] = \frac{10.6(1+1.5^2)}{1.25} = 27.56 \text{ MPa}$$

Example 6.8:

A solid steel shaft of diameter 100 mm is shrunk at the center of a steel circular plate of a very large (infinite) diameter and of 100 m thickness. The radial interference Δ is 0.03 mm and the properties of steel for both shaft and plate are E = *200 GPa and* ν =*0.3. Assuming plane stress conditions,*

a. *Determine the stresses due to the shrink fit in both the shaft and the plate and make a sketch showing these stresses plotted against the radius measured from the center of the shaft;*
b. *If it is required to pull the shaft out of the plate, calculate the necessary axial force assuming that the coefficient of friction at the contact surface is uniform and equal to 0.15;*
c. *If a disk of radius 50* $\sqrt{2}$ *mm is cut out from the infinitely large plate, such that it is concentric with the shaft, determine the axial force required to pull the shaft out of the disk.*

Solution:

a. By assuming plane stress conditions for a solid shaft shrunk inside an infinite plate, Equation 6.15 gives

$$\Delta/r_c = 2p_c/E$$

hence,

$$p_c = \frac{\Delta E}{2r_c} = \frac{0.03\times 200\times 10^3}{2\times 50} = 60 \text{ MPa}$$

The stresses due to shrink fit are obtained from Equations 6.16 and 6.17 with $\lambda_1 = \infty$ and $\lambda_2 = \infty$. For the plate $r_i = r_c = 50$ mm $r_o \to \infty$, at $r = r_c$,

$$\sigma_r = -p_c\, r_c^2/r^2 \text{ giving } \sigma_r = -60 \text{ MPa}$$

$$\sigma_\theta = p_c\, r_c^2/r^2 \text{ giving } \sigma_\theta = 60 \text{ MPa}$$

For the shaft $r_i = 0$ and $r = r_c = 50$ mm,

$$\sigma_r = \sigma_\theta = -p_c \text{ giving } \sigma_r = \sigma_\theta = -60 \text{ MPa}$$

The stress distribution is shown in Figure 6.11.

b. To pull out the shaft an axial force P is required, i.e.,

$$P = \mu p_c(2\pi r_o h) = 0.15 \times 60\ (2\pi \times 50 \times 100) = 282.6 \text{ kN}$$

c. By cutting the plate to a circular disk of radius 50 $\sqrt{2}$ mm concentric with the shaft, a new contact pressure p_c^* and, hence, a stress system will result. Equation 6.13,

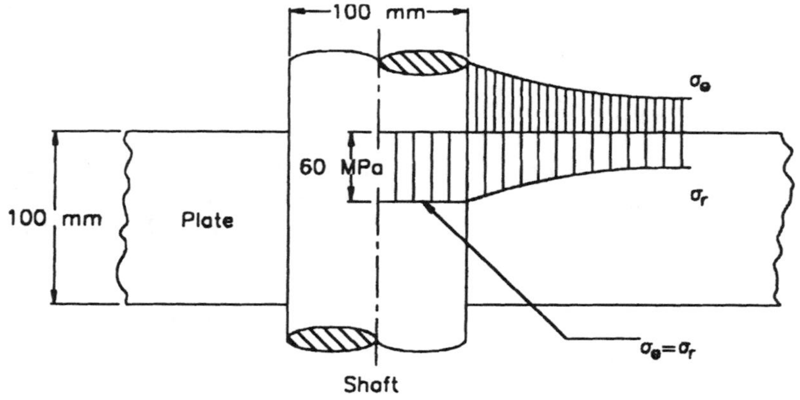

FIGURE 6.11 Example 6.8.

$$\frac{\Delta}{r_c} = \frac{p_c^*}{E}\left(\frac{\lambda_2^2+1}{\lambda_2^2-1} + \frac{\lambda_1^2+1}{\lambda_1^2-1}\right)$$

yields, for $\lambda_1 \to \infty$ and $\lambda_2 = 50\sqrt{2}/50 = \sqrt{2}$.

$$\frac{0.03}{50} = \frac{p_c^*}{200\times 10^3}(3+1) \text{ giving } p_c^* = 30 \text{ MPa}$$

The axial force p^* required to pull the shaft is

$$P^* = \mu p_c^* \,(2\pi r_o h) = 0.15 \times 30\,(2\pi \times 50 \times 100) = 141.3 \text{ kN}$$

which is only 50% of the value of P in the case of an infinite plate.

6.2.4 Rotating Disk of Uniform Thickness

The problem of determining stresses in disks rotating at high speeds has several engineering applications, such as in the design of steam and gas turbines, compressors, fans, flywheels, and other rotating machinery.

For a rotating disk, the thickness is small compared with the diameter so that the conditions of plane stress prevail. Stresses induced by rotation are symmetrically distributed about the axis of rotation and the inertia force is included as a body force. Two cases will be considered: an annular disk and a solid disk.

6.2.4.1 Annular Rotating Disk of Constant Thickness

The state of stress in a thin, rotating disk is given by Equations 4.37:

$$\sigma_r = \frac{A}{r^2} + 2C - \frac{3+\nu}{8}\rho\omega r^2$$

$$\sigma_\theta = -\frac{A}{r^2} + 2C - \frac{1+3\nu}{8}\rho\omega^2 r^2$$

The constants A and C are determined from boundary conditions. In the case of an annular disk. Figure 6.12a, with zero pressure at the inner and outer radii r_i and r_o, respectively, the boundary conditions are

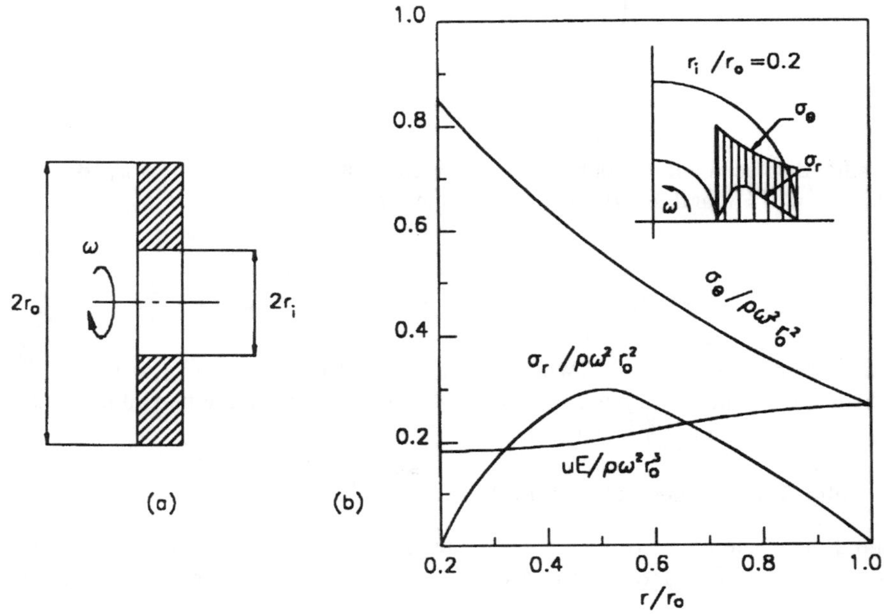

FIGURE 6.12 Stresses and radial displacement along a radius of an annular rotating disk of constant thickness: (a) rotating disk and (b) stress and strain distributions.

$$[\sigma_r]_{r=r_i} = [\sigma_r]_{r=r_o} = 0$$

Therefore, the stresses are given by

$$\sigma_r = \frac{3+\nu}{8}\rho\omega^2\left(r_o^2 + r_i^2 - \frac{r_o^2 r_i^2}{r^2} - r^2\right)$$
$$\sigma_\theta = \frac{3+\nu}{8}\rho\omega^2\left(r_o^2 + r_i^2 + \frac{r_o^2 r_i^2}{r^2} - \frac{1+3\nu}{3+\nu}r^2\right) \tag{6.18}$$

The maximum radial stress is obtained by equating $d\sigma_r/dr$ to zero, which shows that $(\sigma)_{max}$ occurs at $r = \sqrt{r_i r_o}$ and has the value:

$$(\sigma_r)_{max} = \frac{3+\nu}{8}\rho\omega^2(r_o - r_i)^2 \quad \text{at } r = \sqrt{r_i r_o} \tag{6.19a}$$

The maximum value of σ_θ is found by inspection to occur at $r = r_i$ and is given by

$$(\sigma_\theta)_{max} = \frac{3+\nu}{4}\rho\omega^2\left(r_o^2 + \frac{1-\nu}{3+\nu}r_i^2\right) \quad \text{at } r = r_i \tag{6.19b}$$

Figure 6.12b is a dimensionless representation of the stresses as ratios to $\rho\omega^2 r_o^2$ and of radial displacement as a ratio to $\rho\omega^2 r_o^2/E$ as a function of r/r_o for an annular disk of radius ratio $r_i/r_o = {}^1/_5$ and $\nu = {}^1/_3$.

For a disk with a very small central hole, i.e., $r_i \to 0$, the maximum value of σ_θ is obtained from Equation 6.19b as

$$\left(\sigma_\theta\right)_{\max} = \frac{3+\nu}{4}\rho\omega^2 r_o^2 \tag{6.20}$$

The radial displacement u is obtained from Hooke's law and Equations 6.18. By recalling that due to symmetry $v = 0$ and $u = r\varepsilon_\theta$, the displacement u is, therefore, given by

$$u = \frac{(3+\nu)\,(1-\nu)}{8E}\left[r_o^2 + r_i^2 - \frac{1+\nu}{1-\nu}\frac{r_i^2 r_o^2}{r^2} - \frac{1+\nu}{3+\nu}r^2\right]\rho\omega^2 r \tag{6.21}$$

It is seen from Figure 6.12b that while the maximum radial stress $(\sigma_r)_{\max}$ occurs approximately midway between the inner and outer rims of the disk (at $r = \sqrt{r_i r_o}$), the maximum radial displacement $(u)_{\max}$ occurs close to the outer rim.

6.2.4.2 Solid Rotating Disk of Constant Thickness

For a solid disk, where $r_i = 0$, $\sigma_r = \sigma_\theta$ at the center, i.e., at $r = 0$, where the r- and θ-directions are interchangeable. Therefore, the constant A in Equations 4.37 must be zero and the stresses are thus

$$\begin{aligned}\sigma_r &= 2C - \frac{3+\nu}{8}\rho\omega^2 r^2\\ \sigma_\theta &= 2C - \frac{1+3\nu}{8}\rho\omega^2 r^2\end{aligned} \tag{6.22}$$

The constant C is determined from the boundary condition:

$$[\sigma r]_{r=r_o} = 0$$

Therefore, the strsses are given by

$$\begin{aligned}\sigma_r &= \frac{3+\nu}{8}\rho\omega^2\left(r_o^2 - r^2\right)\\ \sigma_\theta &= \frac{3+\nu}{8}\rho\omega^2\left(r_o^2 - \frac{1+3\nu}{3+\nu}r^2\right)\end{aligned} \tag{6.23}$$

At the center of the disk, the stresses σ_r and σ_θ are equal and maximum, given by

$$\left[\sigma_r\right]_{r=0} = \left[\sigma_\theta\right]_{r=0} = \frac{3+\nu}{8}\rho\omega^2 r_o^2$$

Application of Hooke's law yields an expression for the radial displacement:

$$u = \frac{(3+\nu)\,(1-\nu)}{8E}\left[r_o^2 - \frac{1+\nu}{3+\nu}r^{\cdot 2}\right]\rho\omega^2 r \tag{6.24}$$

Similar to the presentation in Figure 6.12b, the stresses and displacement distributions in a solid rotting disk are shown in a dimensionless representation in Figure 6.13.

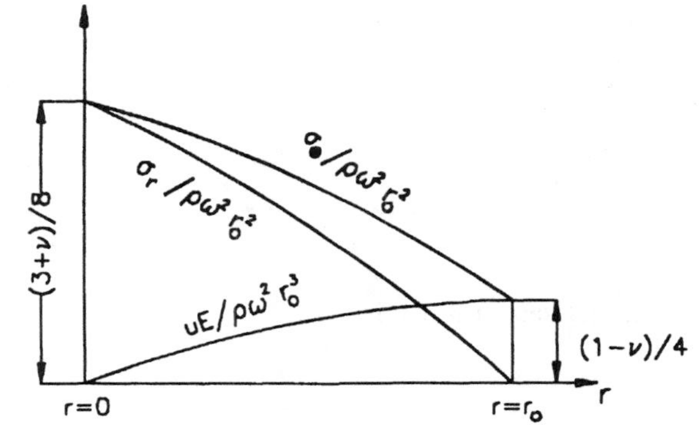

FIGURE 6.13 Stresses and radial displacement along a radius of a solid rotating disk of constant thickness.

The maximum stress occurs at the center of the disk and has the value:

$$(\sigma_\theta)_{max} = (\sigma_r)_{max} = \frac{3+\nu}{8}\rho\omega^2 r_o^2 \tag{6.25}$$

By comparing Equations 6.20 and 6.25, it is seen that the value of $(\sigma_\theta)_{max}$ in a solid disk is half that in an annular disk. This indicates that the existence of an extremely small hole at the center of a solid disk augments the maximum stress to double its value. Cavities as a consequence of the porosity in the center of a cast or forged solid disk have to be always avoided.

Example 6.9:

A thin, annular disk is made up of two steel annuli whose radius of separation is r_c, *Figure 6.14. The internal radius of the inner annulus and the external radius of the outer annulus are* r_i *and* r_o, *respectively. If the outer annulus is not to get loose from the inner one when the disk is rotating with a uniform angular velocity* ω, *determine the minimum value of the radial pressure at the radius of separation when the disk is at rest. What is the value of the radial interference if:*

$$\lambda_1 = r_c/r_i = \lambda_2 = r_o/r_c = \lambda$$

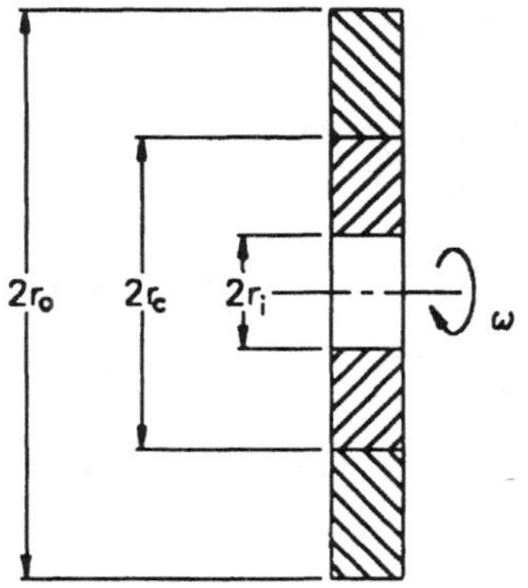

FIGURE 6.14 Example 6.9.

Solution:

In this problem it is required to satisfy the condition that the inner annulus must not get loose from the outer one when the compound disk is rotating with a uniform angular velocity ω. At this angular velocity, the assembly behaves as one integral disk free from any prestressing. Therefore, the radial stress σ_r at the radius of separation $r = r_c$ must not exceed the value of the initial contact pressure p_c due to compounding, i.e., $[(\sigma_r)_{r=r_c} - p_c] = 0$. From the first of Equations 6.18, it thus follows that

$$\left[\sigma_r\right]_{r=c} = \frac{3+\nu}{8}\rho\omega^2\left(r_o^2 + r_i^2 - \frac{r_i^2 r_o^2}{r_c^2} - r_c^2\right) = p_c$$

If $\lambda_1 = r_c/r_i = \lambda_2 = r_o/r_c = \lambda$, the contact pressure is

$$p_c = \frac{3+\nu}{8}\rho\omega^2 r_c^2 \frac{\left(\lambda^2 - 1\right)^2}{\lambda^2} \tag{a}$$

From Equation 6.13 the radial interference is given by

$$\frac{\Delta}{r_c} = \frac{2p_c}{E}\left(\frac{\lambda^2+1}{\lambda^2-1}\right) \tag{b}$$

Equations a and b thus yield

$$\Delta = \frac{3+\nu}{4E}\rho\omega^2 r_c^3 \frac{\lambda^4 - 1}{\lambda^2}$$

6.2.5 Rotating Solid Disk of Uniform Strength (De Laval Disk)

The constant thickness rotating solid disk analyzed in the previous section has shown unfavorable stress distribution, where the maximum stresses occur at the most inner radius. To improve the stress distribution, the thickness has to be increased at the center and reduced toward the rim. This explains the general shape of many disks: thick near the hub and tapering down in thickness toward the periphery as in a turbine, Figure 6.15. This does not only have the effect of reducing the stresses, but also results in lower weight and rotational inertia as well.

An exact analysis for a disk of nonuniform thickness is a three-dimensional problem. An approximate solution may be obtained based on average stress values which yields a plane stress problem.

Consider a solid disk of a variable thickness represented at any radius by $h = h(r)$. The average stress values are given by

$$\sigma_r^* = \frac{1}{h}\int_{-h/2}^{h/2}\sigma_r dz \quad \text{and} \quad \sigma_\theta^* = \frac{1}{h}\int_{-h/2}^{h/2}\sigma_\theta dz$$

From equilibrium in the radial direction, Equation 1.9a gives

$$\frac{\partial\sigma_r}{\partial r} + \frac{\partial\tau_{rz}}{\partial z} + \frac{\sigma_r - \sigma_\theta}{r} + \rho\omega^2 r = 0$$

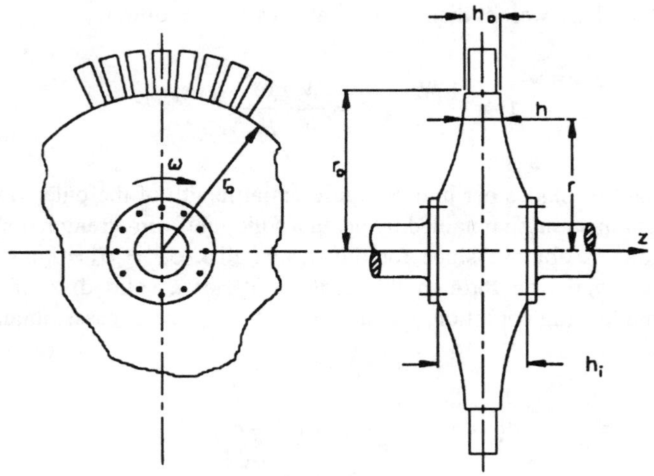

FIGURE 6.15 Rotating solid disk of uniform strength (De Laval turbine disk).

By integrating with respect to z from $-h/2$ to $h/2$ and observing from symmetry that

$$\int_{-h/2}^{h/2} d\tau_{rz} = 0$$

the following equation of equilibrium is obtained:

$$\frac{d}{dr}\left(h\sigma_r^*\right) + \frac{h}{r}\left(\sigma_r^* - \sigma_\theta^*\right) + \rho\omega^2 rh = 0 \tag{6.26}$$

If every element of a rotating disk is stressed to a prescribed allowable value, presumed constant throughout, the disk material will clearly be used in the most efficient manner. It is sought then to find a thickness variation h(r), such that

$$\sigma_r^* = \sigma_\theta^* = \sigma_o = \text{constant}$$

everywhere in the body. This condition obviously corresponds to the maximum shear stress theory. Therefore, Equation 6.26 reduces to

$$\frac{dh}{dr} + \frac{\rho\omega^2 r}{\sigma_o} h = 0$$

which is easily integrated to yield

$$h = h_i e^{-\left(\rho\omega^2/2\sigma_o\right)r^2} \tag{6.27}$$

where h_i is the thickness at the center of the disk.

At the outer rim the condition of uniform strength must be fulfilled, so that

$$\sigma_r^* = \sigma_o \quad \text{at} \quad r = r_o$$

This requires that the blade weight distribution satisfies the condition:

$$\sigma_o = \frac{1}{h_o}\frac{W}{g}\omega^2 r_o = \frac{W}{g}\frac{r_o\omega^2}{h_i}e^{(\rho\omega^2/2\sigma_o)r_o^2}$$

where W is the weight of blades per unit circumferential length of the outer rim.

The improvement in strength obtained by using a solid, uniform-strength disk instead of a solid uniform-thickness one is obvious since for the former disk $\sigma_r^* = \sigma_\theta^* =$ constant $= \sigma_o$, all over (neglecting the stress variation through the thickness). For a solid disk of uniform thickness subjected to the same loading condition, i.e., at $r = r_o$: $\sigma_r = \sigma_o$, the stresses obtained from Equation 6.22 are

$$\begin{aligned} \sigma_r &= \sigma_o + \frac{3+\nu}{8}\rho\omega^2\left(r_o^2 - r^2\right) \\ \sigma_\theta &= \sigma_o + \frac{3+\nu}{8}\rho\omega^2\left(r_o^2 - \frac{1+3\nu}{3+\nu}r^2\right) \end{aligned} \tag{6.28}$$

The maximum values of σ_r and σ_θ occur at $r = 0$, i.e.,

$$\left(\sigma_r\right)_{\max} = \left(\sigma_\theta\right)_{\max} = \sigma_o + \frac{3+\nu}{8}\rho\omega^2 r_o^2 \tag{6.29}$$

which shows an increase of $[(3 + \nu)/8]\,\rho\omega^2 r_o^2$ in the maximum stresses compared with the variable-thickness disk.

Example 6.10:

A solid disk 300 mm in diameter rotates at 10,000 rpm and is subjected at its outer rim to a radial tension of 100 MPa. The disk material is steel for which $\nu = 0.28$ and the specific weight = 78 kN/m^3. If the thickness at the disk center is 20 mm, find the thickness at the outer rim for a disk of uniform strength. What is the value of the maximum stress in a solid disk of uniform thickness of 20 mm operating under the same conditions?

Solution:

In a disk of uniform strength, the condition at the outer radius, $r = r_o$, is $\sigma_r = \sigma_o = 100$ MPa. Therefore,

$$\sigma_r = \sigma_\theta = \sigma_o = 100 \text{ MPa}$$

all over. The thickness distribution is thus given by Equation 6.27 as

$$h = h_i e^{-(\rho\omega^2/2\sigma)r^2}$$

where

$$\omega = 2\pi \times 10{,}000/60 = 1047.2 \text{ rad/s}$$

At the outer rim $r = r_o$, the thickness is given by

$$\begin{aligned} h_o &= 20\exp\left[-\left(\frac{78\times 10^3}{9.81}\right)\left(\frac{(1047.2)^2(0.15)^2}{2\times 100\times 10^6}\right)\right] \\ &= 7.5 \text{ mm} \end{aligned}$$

In a solid disk of uniform thickness, the maximum stress occurs at the center, $r = 0$, as

$$(\sigma_r)_{max} = (\sigma_\theta)_{max} = \frac{3+\nu}{8}\rho\omega^2 r_o^2 + \sigma_o$$

Therefore,

$$(\sigma_r)_{max} = (\sigma_\theta)_{max}$$

$$= \frac{3.28}{8}\left(\frac{78\times 10^3}{9.81}\right)(1047.2)^2(0.15)^2 + 100\times 10^6$$

$$= 180.4 \text{ MPa}$$

This shows that for a uniform-thickness design, in addition to using more material, the stresses are 80% higher than those existing in uniform-strength design.

6.2.6 Rotating Drums and Rotors

A rotating drum or rotor usually has a long length compared with its diameter, thus providing conditions of plane strain. The governing equations for a stress function solution are given by Equations 4.39:

$$\sigma_r = \frac{A}{r^2} + 2C - \frac{3-2\nu}{8(1-\nu)}\rho\omega^2 r^2$$

$$\sigma_\theta = -\frac{A}{r^2} + 2C - \frac{1+2\nu}{8(1-\nu)}\rho\omega^2 r^2$$

If the drum is not free to deform longitudinally, plane strain conditions prevail, giving $\varepsilon_z = 0$ and, hence,

$$\sigma_z = \nu(\sigma_r + \sigma_\theta) = 4\nu C - \frac{\nu}{2(1-\nu)}\rho\omega^2 r^2$$

The constants A and C are determined by satisfying the boundary conditions:

$$[\sigma r]_{r=r_i} = [\sigma r]_{r=r_o} = 0$$

where r_i and r_o are the inner and outer radii, respectively, of the hollow drum. Having obtained the constants A and C, the stresses are found as

$$\sigma_r = \frac{3-2\nu}{8(1-\nu)}\rho\omega^2\left(r_o^2 + r_i^2 - \frac{r_i^2 r_o^2}{r^2} - r^2\right)$$

$$\sigma_\theta = \frac{3-2\nu}{8(1-\nu)}\rho\omega^2\left(r_o^2 + r_i^2 + \frac{r_i^2 r_o^2}{r^2} - \frac{1+2\nu}{3-2\nu}r^2\right) \quad (6.30)$$

$$\sigma_z = \nu(\sigma_r + \sigma_\theta) = \frac{\nu(3-2\nu)}{4(1-\nu)}\rho\omega^2\left(r_o^2 + r_i^2 - \frac{2r^2}{3-2\nu}\right)$$

The maximum tensile stress occurs at the inner surface, $r = r_i$, and is given by

$$\sigma_\theta = \frac{3-2\nu}{4(1-\nu)}\rho\omega^2\left(r_o^2 + \frac{1-2\nu}{3-2\nu}r_i^2\right)$$

For a rotating solid cylinder it can be shown, following a similar procedure as in the case of a hollow drum, that the stresses for plane strain conditions are given by

$$\sigma_r = \frac{3-2\nu}{8(1-\nu)}\rho\omega^2\left(r_o^2 - r^2\right)$$

$$\sigma_\theta = \frac{3-2\nu}{8(1-\nu)}\rho\omega^2\left(r_o^2 - \frac{1+2\nu}{3-2\nu}r^2\right) \tag{6.31}$$

$$\sigma_z = \nu\left(\sigma_r + \sigma_\theta\right) = \frac{\nu(3-2\nu)}{4(1-\nu)}\rho\omega^2\left(r_o^2 - \frac{2r^2}{3-2\nu}\right)$$

where r_o is the cylinder radius.

Example 6.11:

A solid steel shaft 500 mm diameter is rotating at 400 rpm. If the shaft does not deform longitudinally, plot the stress distribution in the shaft. Calculate the total longitudinal compression over a cross section due to the rotational stresses. Take the specific weight of steel = 78 kN/m³ and Poisson's ratio ν = 0.3.

Solution:

Equation 6.31 gives the stress distribution in the shaft. Figure 6.16 is a plot for these stresses. The maximum stress occurs at the center and is given by

$$\left[\sigma_r\right]_{r=0} = \left[\sigma_\theta\right]_{r=0} = \frac{3-2\nu}{8(1-\nu)}\rho\omega^2 r_o^2$$

The longitudinal compression, i.e., thrust, is obtained by integrating σ_z over the cross section area A as

$$\text{Thrust} = \int_A \sigma_z dA$$

$$= \int_0^{2\pi}\int_0^{r_o} \frac{\nu\rho\omega^2}{4(1-\nu)}\left[(3-2\nu)r_o^2 - 2r^2\right] r\,dr\,d\theta$$

$$= \frac{\pi\nu\rho\omega^2 r_o^4}{2}$$

$$= \frac{\pi\times 0.3}{2}\left(\frac{78\times 10^3}{9.81}\right)\left(\frac{2\pi\times 400}{60}\right)^2 (0.250)^4 = 25.68 \text{ kN}$$

6.2.7 Rotating Disks and Rotors Subjected to Radial Thermal Gradients

In the case of a rotating disk subjected to a temperature gradient, the stresses are obtained by superposition from Equations 4.37 and 4.45, namely,

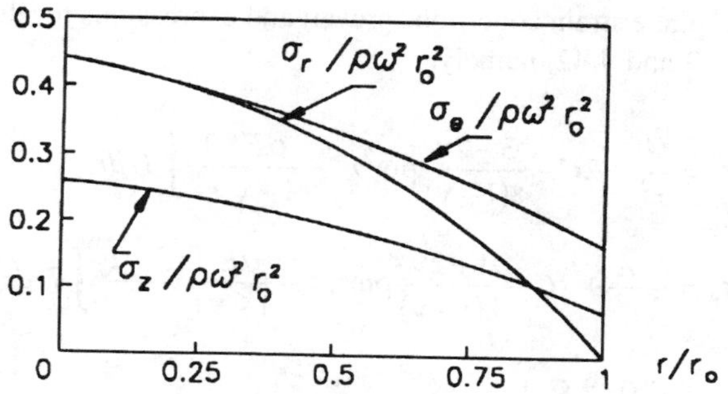

FIGURE 6.16 Example 6.11.

$$\sigma_r = \frac{A}{r^2} + 2C - \frac{3+\nu}{8}\rho\omega^2 r^2 - \alpha E\left[\frac{1}{r^2}\int Trdr\right]$$

$$\sigma_\theta = -\frac{A}{r^2} + 2C - \frac{1+3\nu}{8}\rho\omega^2 r^2 - \alpha E\left[T - \frac{1}{r^2}\int Trdr\right] \tag{6.32}$$

The constants A and C are obtained from the boundary conditions as in Section 6.2.3.

For a hollow disk, applying the boundary conditions $\sigma_r = 0$ at both $r = r_i$ and $r = r_o$ gives, after algebraic manipulation,

$$\sigma_r = \frac{3+\nu}{8}\rho\omega^2\left(r_o^2 + r_i^2 - \frac{r_o^2 r_i^2}{r^2} - r^2\right)$$

$$+ \alpha E\left[-\frac{1}{r^2}\int_{r_i}^{r} Trdr + \frac{1-(r_i/r)^2}{r_o^2 - r_i^2}\int_{r_i}^{r_o} Trdr\right]$$

$$\sigma_\theta = \frac{3+\nu}{8}\rho\omega^2\left(r_o^2 + r_i^2 + \frac{r_o^2 r_i^2}{r^2} - \frac{1+3\nu}{3+\nu}r^2\right) \tag{6.33}$$

$$+ \alpha E\left[-T + \frac{1}{r^2}\int_{r_i}^{r} Trdr + \frac{1+(r_i/r)^2}{r_o^2 - r_i^2}\int_{r_i}^{r_o} Trdr\right]$$

For a solid disk, the stresses σ_r and σ_θ are equal at the center so that the constant A in Equations 4.45 must be zero. If there are no external forces applied to the boundary, i.e., $\sigma_r = 0$ at $r = r_o$, the stresses are given by

$$\sigma_r = \frac{3+\nu}{8}\rho\omega^2\left(r_o^2 - r^2\right) + \alpha E\left[\frac{1}{r_o^2}\int_0^{r_o} Trdr - \frac{1}{r^2}\int_0^{r} Trdr\right]$$

$$\sigma_\theta = \frac{3+\nu}{8}\rho\omega^2\left(r_o^2 - \frac{1+3\nu}{3+\nu}r^2\right) + \alpha E\left[-T + \frac{1}{r^2}\int_0^{r} Trdr + \frac{1}{r_o^2}\int_0^{r_o} Trdr\right] \tag{6.34}$$

Once $T = T(r)$ is known, the value of the stresses can be obtained.

For a rotating drum, plane strain conditions prevail and expressions for the stresses are obtained by adding Equations 4.39 and 4.42, namely,

$$\sigma_r = \frac{A}{r^2} + 2C - \frac{3-2\nu}{8(1-\nu)}\rho\omega^2 r^2 - \frac{\alpha E}{1-\nu}\frac{1}{r^2}\int Trdr$$

$$\sigma_\theta = -\frac{A}{r^2} + 2C - \frac{1+2\nu}{8(1-\nu)}\rho\omega^2 r^2 - \frac{\alpha E}{1-\nu}\left[T - \frac{1}{r^2}\int Trdr\right] \tag{6.35}$$

$$\sigma_z = \nu(\sigma_r + \sigma_\theta)$$

The constants A and C are determined from the appropriate boundary conditions as in the case of thin rotating disks.

Example 6.12:

A solid disk of uniform thickness rotates at ω rad/s. Obtain the radial temperature gradient that makes the thickness remain uniform during rotation.

Solution:

For plane stress $\sigma_z = 0$ and hence from thermoelastic stress–stain relations (Equations 3.18):

$$\varepsilon_z = -\frac{\nu}{E}(\sigma_r + \sigma_\theta) + \alpha T$$

For uniform thickness $\partial\varepsilon_z/\partial_r = 0$,

$$\frac{\partial}{\partial r}(\sigma_r + \sigma_\theta) = 0$$

Substituting σ_r, σ_θ from Equation 6.32 yields

$$\frac{\partial}{\partial r}\left[4C - \frac{1}{2}(1+\nu)\rho\omega^2 r^2 - \alpha ET\right] + \alpha\frac{\partial T}{\partial r} = 0$$

or

$$\frac{dT}{dr} = -\frac{\nu}{\alpha E}\rho\omega^2 r$$

Integration gives

$$T = -\frac{\nu}{2\alpha E}\rho\omega^2 r^2 + T_o$$

where T_o is the temperature at $r = 0$.

Example 6.13:

A centrally cooled disk of uniform thickness, having a rim diameter of 400 mm and a central hole of 50 mm diameter, runs at 10,000 rpm. The temperature varies linearly from 300°C at the rim to

95°C at the central hole. Plot the circumferential and radial stresses in the disk. Compare these stresses with those that occur due to rotation only. Take E = 200 GPa, ν = 0.3, α = 11 × 10⁻⁶/°C and weight per unit volume of disk material 78 kN/m3.

Solution:

Since the temperature varies linearly, it is given at any radius, r, by

$$T = 1171.43\,r + 65.71$$

where r is measured in meters and T in °C.

The integrals included in Equations 6.33 are determined as

$$\int_{r_i}^{r_o} Trdr = \int_{0.025}^{0.2} (1171.43r + 65.71)rdr = 4.411$$

similarly,

$$\int_{r_i}^{r} Trdr = 390.48\left(r^3 - r_i^3\right) + 32.86\left(r^2 - r_i^2\right)$$

Substitution into Equations 6.33 — knowing that $r_i = 0.025$ m, $r_o = 0.2$ m, $\omega = 10{,}000 \times 2\pi/60$ rad/s, $\rho = 78 \times 10^3/9.81$ kg/m³, $E = 200 \times 10^9$ N/m², $\nu = 0.3$, and $\alpha = 11 \times 10^{-6}$/°C — gives for σ_r and σ_θ, respectively,

$$\sigma_r = \left[317.5 - 21.5\left(\frac{r}{r_i}\right) - 2.2\left(\frac{r}{r_i}\right)^2 - 293.8\left(\frac{r_i}{r}\right)^2\right] \text{MPa}$$

$$\sigma_\theta = \left[317.5 - 43\left(\frac{r}{r_i}\right) - 1.3\left(\frac{r}{r_i}\right)^2 - 368\left(\frac{r_i}{r}\right)^2\right] \text{MPa}$$

By solving (by trial and error), it is found that the maximum stresses

$$(\sigma_r)_{max} = 203.3 \text{ MPa at } r \cong 2.61r_i$$

$$(\sigma_\theta)_{max} = 641.2 \text{ MPa at } r \cong r_i$$

The stress distribution of σ_r and σ_θ are shown in Figure 6.17. The stresses σ_r and σ_θ due to rotation only are plotted for comparison. It is seen that the presence of the temperature gradient increases the maximum values of both σ_r and σ_θ to almost double its values and a compressive hoop stress is produced at the outer rim.

6.3 AXIALLY NONSYMMETRIC PROBLEMS

In this section some axially nonsymmetric problems that have engineering applications and that are apt to simple analysis are treated. For each problem an appropriate stress function is sought, and the stresses are obtained by applying Equations 4.25 with substitution of relevant boundary conditions.

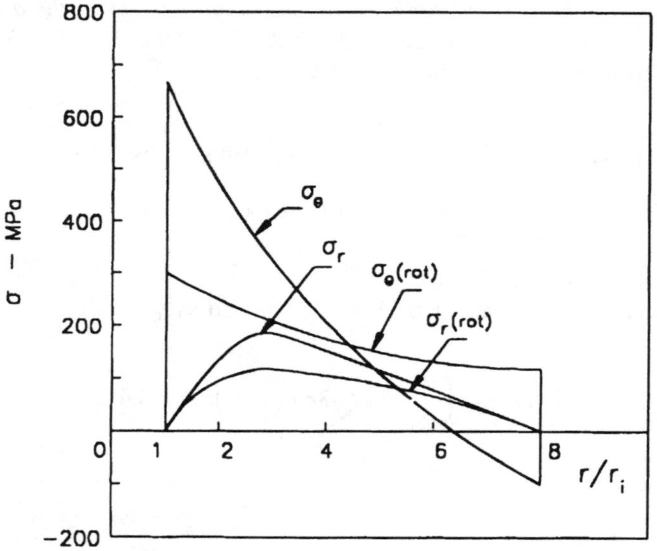

FIGURE 6.17 Example 6.13.

6.3.1 Bending of a Circularly Curved Beam

6.3.1.1 Beam Subjected to an End Shearing Force

Consider a circularly curved bar of a rectangular cross section, thickness h, and a length subtending an angle of $\pi/2$ as shown in Figure 6.18a. The bar is bent by a force P acting at the upper end ($\theta = 0$), while the lower end ($\theta = \pi/2$) is fully constrained against displacement. The positive direction of y is taken to be downward away from the center and the direction of P is toward O to give a positive $\tau_{r\theta}$. This problem corresponds to a crane hook lifting a load P, as shown in Figure 6.18b, noting that the direction of P is in the reverse direction

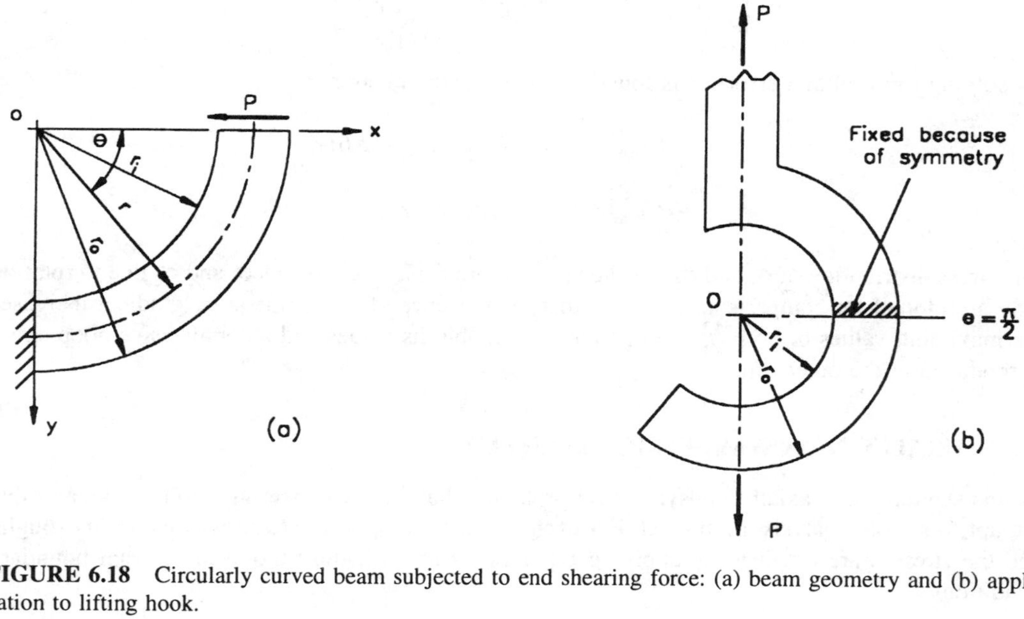

FIGURE 6.18 Circularly curved beam subjected to end shearing force: (a) beam geometry and (b) application to lifting hook.

The bending moment and hence the stresses are obviously functions of the coordinate θ. The stresses are thus related to the stress function ϕ by Equations 4.25. By assuming a plane stress problem and neglecting body forces, Equations 4.25 reduce to

$$\sigma_r = \frac{1}{r}\frac{\partial \phi}{\partial r} + \frac{\partial^2 \phi}{r^2 \partial \theta^2}$$

$$\sigma_\theta = \frac{\partial^2 \phi}{\partial r^2}$$

$$\tau_{r\theta} = -\frac{\partial}{\partial r}\left(\frac{\partial \phi}{r \partial \theta}\right)$$

where the stress function ϕ satisfies Equation 4.27 namely, $\nabla^2\nabla^2\phi = 0$.

Seeking a stress function ϕ, it is seen from Figure 6.18 that the bending moment at any cross section is proportional to sin θ. It thus seems reasonable to assume ϕ in the form:

$$\phi = f(r) \sin \theta \tag{6.36}$$

where $f(r)$, a function of r only, is to be determined. Substituting Equation 6.36 into the biharmonic Equation 4.26 yields:

$$\left(\frac{d^2}{dr^2} + \frac{1}{r}\frac{d}{dr} - \frac{1}{r^2}\right)\left(\frac{d^2 f}{dr^2} + \frac{1}{r}\frac{df}{dr} - \frac{f}{r^2}\right) = 0$$

This is an ordinary differential equation which can be solved to give the general solution:

$$f(r) = Ar^3 + B\frac{1}{r} + Cr + Dr\ln(r) \tag{6.37}$$

where A, B, C, and D are constants of integration determined from the boundary conditions. The stress function ϕ and the stresses are thus given by

$$\phi = \left[Ar^3 + \frac{B}{r} + Cr + Dr\ln(r)\right]\sin\theta$$

$$\begin{aligned}
\sigma_r &= \left[2Ar - \frac{2B}{r^3} + \frac{D}{r}\right]\sin\theta \\
\sigma_\theta &= \left[6Ar + \frac{2B}{r^3} + \frac{D}{r}\right]\sin\theta \\
\tau_{r\theta} &= -\left[2Ar - \frac{2B}{r^3} + \frac{D}{r}\right]\cos\theta
\end{aligned} \tag{6.38}$$

Since the outer and inner surfaces of the curved bar are stress free, the boundary conditions are for any value of θ:

$$\begin{aligned} [\sigma_r]_{r=r_i} &= [\tau_{r\theta}]_{r=r_i} = 0 \\ [\sigma_r]_{r=r_o} &= [\tau_{r\theta}]_{r=r_o} = 0 \end{aligned} \tag{6.39}$$

Noting that the radial stress σ_r and the shear stress $\tau_{r\theta}$ as given by Equations 6.38 constitute the same function of r, it is thus realized that the boundary conditions given by Equations 6.39 are not sufficient to determine the three constants A, B, and D. Another boundary condition is obtained by equating the sum of the shear stress distribution over the upper end of bar ($\theta = 0$) to the applied force P, namely,

$$\int_{r_i}^{r_o} [\tau_{r\theta}]_{\theta=0}\, dr = p/h$$

where h is the beam thickness. This condition yields

$$\int_{r_i}^{r_o} -\left[2Ar - \frac{2B}{r^3} + \frac{D}{r}\right] dr = P/h$$

or

$$A\left(r_o^2 - r_i^2\right) - B\left(\frac{r_o^2 - r_i^2}{r_o^2 r_i^2}\right) + D \ln \frac{r_o}{r_i} = -P/h \tag{6.40a}$$

From the boundary conditions (Equations 6.39), Equations 6.38 reduce to

$$\begin{aligned} 2Ar_i - \frac{2B}{r_i^3} + \frac{D}{r_i} &= 0 \\ 2Ar_o - \frac{2B}{r_o^3} + \frac{D}{r_o} &= 0 \end{aligned} \tag{6.40b}$$

Solving the three equations (Equations 6.40a and 6.40b) for the constants A, B, and D gives:

$$A = \frac{-P/2r_i^2 h}{\left[(r_o/r_i)^2 + 1\right]\ln(r_o/r_i) - \left[(r_o/r_i)^2 - 1\right]} = \frac{-P}{2r_i^2 Nh}$$

$$B = \frac{(-P/2h)r_o^2}{\left[(r_o/r_i)^2 + 1\right]\ln(r_o/r_i) - \left[(r_o/r_i)^2 - 1\right]} = \frac{-Pr_o^2}{2Nh}$$

$$D = \frac{-P\left[(r_o/r_i)^2 + 1\right]/h}{\left[(r_o/r_i)^2 + 1\right]\ln(r_o/r_i) - \left[(r_o/r_i)^2 - 1\right]} = \frac{-P\left[(r_o/r_i^2) + 1\right]}{Nh}$$

where

$$N = \left[\left(\frac{r_o}{r_i}\right)^2 + 1\right]\ln\left(\frac{r_o}{r_i}\right) - \left[\left(\frac{r_o}{r_i}\right)^2 - 1\right] \tag{6.40c}$$

The stresses are then

$$\sigma_r = \frac{P\sin\theta}{Nh}\left[\frac{r}{r_i^2} + \frac{r_o^2}{r^3} - \frac{(r_o/r_i)^2 + 1}{r}\right]$$

$$\sigma_\theta = \frac{P\sin\theta}{Nh}\left[3\frac{r}{r_i^2} - \frac{r_o^2}{r^3} - \frac{(r_o/r_i)^2 + 1}{r}\right] \tag{6.41a}$$

$$\tau_{r\theta} = \frac{-P\cos\theta}{Nh}\left[\frac{r}{r_i^2} + \frac{r_o^2}{r^3} - \frac{(r_o/r_i)^2 + 1}{r}\right]$$

For the upper end ($\theta = 0$), the stresses are

$$\left[\sigma_\theta\right]_{\theta=0} = 0$$

$$\left[\tau_{r\theta}\right]_{\theta=0} = -\frac{P}{Nh}\left[\frac{r}{r_i^2} + \frac{r_o^2}{r^3} - \frac{(r_o/r_i)^2 + 1}{r}\right] \tag{6.41b}$$

Similarly, at the lower end ($\theta = \pi/2$), the stresses are

$$\left[\tau_{r\theta}\right]_{\theta=\pi/2} = 0$$

$$\left[\sigma_\theta\right]_{\theta=\pi/2} = \frac{P}{Nh}\left[3\frac{r}{r_i^2} - \frac{r_o^2}{r^3} - \frac{(r_o/r_i)^2 + 1}{r}\right] \tag{6.41c}$$

The above solution, Equation 6.41a, is only exact when the boundary forces at both ends, $\theta = 0$ and $\pi/2$, are distributed according to Expressions 6.41b and 6.41c.

For any other distribution for the end forces, the solution is exact only at larger distances from the ends according to Saint-Venant's principle.

The strength-of-materials solution to a curved beam problem assumes that plane sections remain plane. This solution gives only the stress distribution of σ_θ according to the formula:*

$$\sigma_\theta = \frac{M(y+e)}{Ae(y+\bar{r})} \tag{6.41d}$$

* This formula is derived in Chapter 7, where σ_z stands for σ_θ.

where M = bending moment (considered positive if it acts such that σ_θ is positive at positive values of y*),
A = cross-sectional area,
$\bar{r}$ = radius of curvature of the centroidal fiber; $\bar{r} = (r_o + r_i)/2$,
e = distance from neutral fiber to the centroidal fiber and given by $e = \bar{r} - A/[\int dA/(\bar{r} + y)]$,
y = distance from centroidal fiber.

For a curved beam of a rectangular cross section, the distance e is given by:**

$$e = \bar{r} - \frac{r_o - r_i}{\ln(r_o/r_i)}$$

If the beam is loaded as shown in Figure 6.18a, Equation 6.41d gives at the fiber $r = r_i$ the stress:

$$\sigma_\theta = -\frac{P\sin\theta}{hr_i}\frac{\left[(r_o/r_i)^2 - 1\right] - \left[(r_o/r_i) + 1\right]\ln(r_o/r_i)}{\left[(r_o/r_i)^2 - 1\right]\ln(r_o/r_i) - 2\left[(r_o/r_i) - 1\right]^2}$$

$$-\frac{P\sin\theta}{hr_i}\frac{1}{(r_o/r_i) - 1}$$

where the second term is the stress due to the compressive component of the end force P at the section under consideration. According to the stress function approach, Equations 6.40c and 6.41a σ_θ at $r = r_i$ is given by

$$\sigma_\theta = \frac{P\sin\theta}{hr_i}\frac{2\left[1 - (r_o/r_i)^2\right]}{\left[(r_o/r_i)^2 + 1\right]\ln(r_o/r_i) - \left[(r_o/r_i)^2 - 1\right]}$$

Table 6.1 indicates that the elementary strength-of-materials solution based on the hypothesis that plane cross sections remain plane gives very accurate results for σ_θ in curved beams of small and moderate depths $(r_o - r_i)/\bar{r} \leq 1$. The elementary solution does not, however, predict the stress σ_r, since it assumes that longitudinal fibers of a bent bar are in simple tension or compression.

The distribution of the shear stress $\tau_{r\theta}$ over the cross section $\theta = 0$, for various ratios of r_o/r_i, is shown in Figure 6.19. The ordinate represents the shear stress divided by its average value $P/[h(r_o - r_i)]$. A curve representing the well-known parabolic shear stress distribution for rectangular straight beams is also shown in the same figure. It may be seen that the distribution of shear stresses approaches the parabolic distribution when the depth of the cross section is small compared with the centroidal radius of curvature.

In order to obtain the displacement components (u, v), an approach similar to that employed in Section 5.5.2 could be followed. A complete analysis[2] gives an expression for the deflection of the upper end as

* See Chapter 7, Equation 7.27.

** See Equation 7.25a.

TABLE 6.1
Comparison Between Stresses in Curved Beams According to Stress Function and Strength-of-Materials Approaches

r_o/r_i	10	5	3	2	1.2
$(r_o - r_i)/\bar{r}$	1.636	1.333	1	0.667	0.182
Stress function solution, $\left[\sigma_\theta \Big/ \dfrac{P\sin\theta}{hr_i}\right]_{r=r_i}$	−1.482	−2.690	−5.358	−12.88	−180.89
Strength-of-materials $\left[\sigma_\theta \Big/ \dfrac{P\sin\theta}{hr_i}\right]_{r=r_i}$	−1.228	−2.415	−5.070	−12.59	−180.59
Percentage difference (%)	20.67	11.4	5.37	2.34	0.16

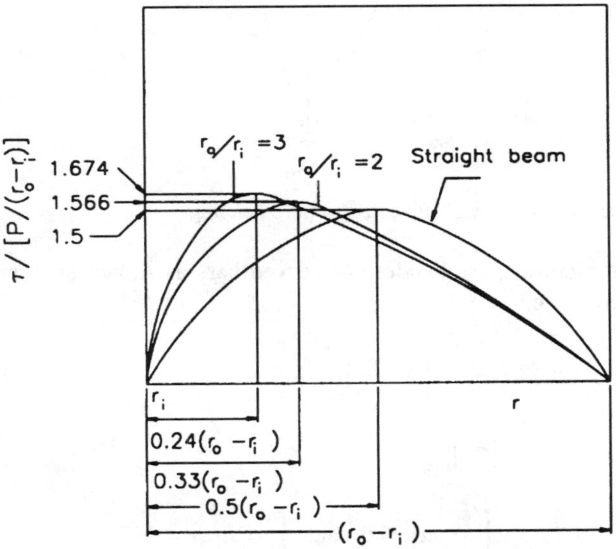

FIGURE 6.19 Distribution of the shear stress $\tau r\theta$ at the free end of the beam of Figure 6.18 for different values of r_o/r_i.

$$[u]_{\theta=0} = -\frac{1}{Eh}\frac{P\pi\left[\left(r_o/r_i^2\right)+1\right]}{\left[\left(r_o/r_i\right)^2+1\right]\ln\left(r_o/r_i\right)-\left[\left(r_o/r_i\right)^2-1\right]}\text{me} \tag{6.42a}$$

when the depth of the beam, $(r_o - r_i)$ is small in comparison with r_i then,

$$\ln\left(\frac{r_o}{r_i}\right) = \ln\left(1+\frac{r_o-r_i}{r_i}\right) \cong \left(\frac{r_o-r_i}{r_i}\right) - \frac{1}{2}\left(\frac{r_o-r_i}{r_i}\right)^2 + \cdots$$

Substituting in Equation 6.42a and neglecting small terms of second and higher orders yields

$$(u)_{\theta=0} = -\frac{3\pi r_i^3 P}{Eh(r_o - r_i)^3} \tag{6.42b}$$

which is the same expression obtained from the elementary analysis.

6.3.1.2 Beam Subjected to Pure Bending

By taking a stress function as given by Equation 4.31,

$$\phi = A \ln r + Br^2 \ln r + Cr^2$$

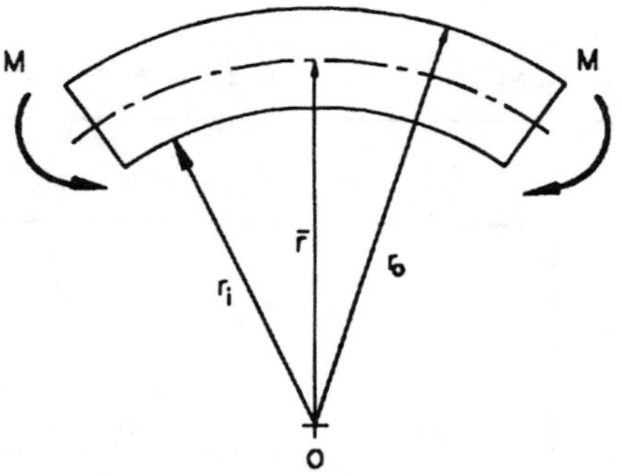

FIGURE 6.20 Circularly curved beam subjected to pure bending.

the solution of the problem of pure bending of curved bars as shown in Figure 6.20 is obtained. The boundary conditions are

$$[\sigma_r]_{r=r_i} = [\tau_{r\theta}]_{r=r_i} = 0$$

$$[\sigma_r]_{r=r_o} = [\tau_{r\theta}]_{r=r_o} = 0$$

$$\int_{r_i}^{r_o} \sigma_\theta h dr = 0 \text{ and } \int_{r_i}^{r_o} \sigma_\theta h r dr = M$$

Applying these boundary conditions to the stresses derived from the stress function (Equation 4.31) according to Expressions 4.32 yields

$$\sigma_r = \frac{4M}{Nhr_i^2}\left[\left(\frac{r_o}{r}\right)^2 \ln\left(\frac{r_o}{r_i}\right) + \left(\frac{r_o}{r_i}\right)^2 \ln\left(\frac{r}{r_o}\right) - \ln\left(\frac{r}{r_i}\right)\right]$$

$$\sigma_\theta = \frac{4M}{Nhr_i^2}\left[-\left(\frac{r_o}{r}\right)^2 \ln\left(\frac{r_o}{r_i}\right) + \left(\frac{r_o}{r_i}\right)^2 \ln\left(\frac{r}{r_o}\right) - \ln\left(\frac{r}{r_i}\right) + \left(\frac{r_o}{r_i}\right)^2 - 1\right] \tag{6.43}$$

$$\tau_{r\theta} = 0$$

where

$$N = \left[\left(\frac{r_o}{r_i}\right)^2 - 1\right]^2 - 4\left(\frac{r_o}{r_i}\right)^2\left[\ln\left(\frac{r_o}{r_i}\right)\right]^2$$

Figure 6.21 shows the stress distribution for σ_r and σ_θ for a curved beam of rectangular cross section with $r_o/r_i = 2$. It is seen that σ_r is always negative for the positive direction of bending shown in Figure 6.21 and it increases toward the neutral surface, where σ_θ approaches its minimum value.

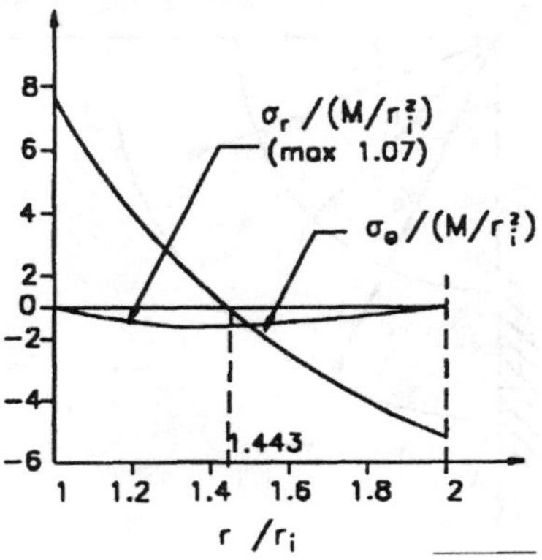

FIGURE 6.21 Stress distribution along the beam depth for the beam of Figure 6.20 at $r_o/r_i = 2$.

The ratio of $(\sigma_r)_{max}$ to $(\sigma_\theta)_{max}$ is shown in Table 6.2 for beams of various ratios of r_o/r_i. It is seen that σ_r may be neglected with respect to σ_θ for curved beams of small curvature.

Expanding Expression 6.43 for small values of $(r_o - r_i)/[(r_o + r_i)]$ yields

$$(\sigma_\theta)_{max} = \frac{6M}{h(r_o - r_i^2)}\left[1 + \frac{2}{3}\frac{r_o - r_i}{(r_o + r_i^2)} + \cdots\right]$$

Retaining the first term only, σ_θ reduces to the results of the simple beam formula $\sigma = My/I$. Elaborate comparison of the above stress function solution with other simplified strength-of-materials solution will be made in Chapter 7. In fact, it will be shown that strength-of-materials solutions based on the assumption of plane cross sections give reasonably accurate results* for curved beams of small and moderate curvatures.

TABLE 6.2
Ratio of $(\sigma_r)_{max}/(\sigma_\theta)_{max}$ in Curved Beams

r_o/r_i	3	2	1.3
$(r_o - r_i)/[(r_o + r_i)/2]$	1	0.667	0.261
$(\sigma_r)_{max}/(\sigma_\theta)_{max}$	0.193	10.138	0.06

* For pure bending, it can be shown using the stress function solution that plane cross sections do remain plane after bending. For this, see Timoshenko and Goodier,[2] p. 66.

6.3.1.3 Beam Subjected to an End Moment and a Normal Force

By taking the stress function in the form

$$\phi = f(r) \cos \theta$$

and proceeding as in the case in Section 6.3.1.3, a solution is obtained when a vertical force and a couple are applied to the upper end of the bar of Figure 6.22.

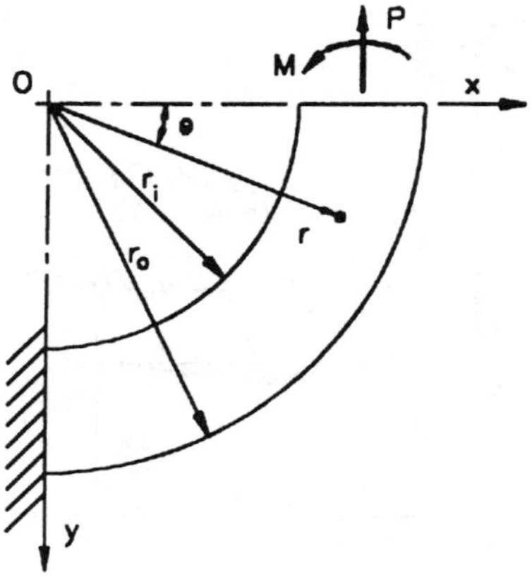

FIGURE 6.22 Circularly curved beam subjected to an end moment and a normal force.

6.3.1.4 Beam Subjected to an Inclined End Force

By subtracting from the solution for case in Section 6.3.1.3 the stresses produced by the couple *M*, the stresses due to a vertical force applied at the upper end of the bar remain. Having the solutions for a horizontal and for a vertical force, the solution for any inclined force can be obtained by superposition.

Example 6.14:

Use the stress function approach to determine the tangential stress at $\theta = \pi/6$ for a circularly curved beam having a rectangular cross-section of thickness 2mm as shown in Figure 6.23.

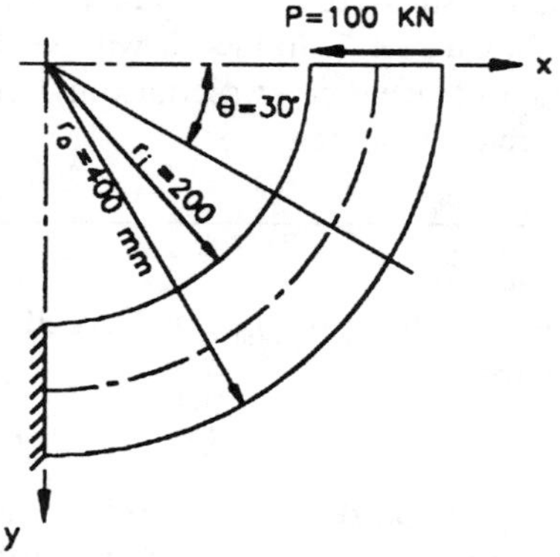

FIGURE 6.23 Example 6.14.

Solution:

Applying Equations 6.40c and 6.41a at $\theta = \pi/6$, gives

$$\sigma_\theta = \frac{P\sin\theta}{Nh}\left[3\frac{r}{r_i^2} - \frac{r_o^2}{r^3} - \frac{(r_o/r_i)^2 + 1}{r}\right]$$

where

$$N = \left[\left(\frac{r_o}{r_i}\right)^2 + 1\right]\ln\left(\frac{r_o}{r_i}\right) - \left[\left(\frac{r_o}{r_i}\right)^2 - 1\right] = 0.466$$

$$\left[\sigma_\theta\right]_{r=200\ mm} = -161.04 \text{ MPa}$$

$$\left[\sigma_\theta\right]_{r=400\ mm} = +80.517 \text{ MPa}$$

Substituting $\sigma_\theta = 0$ in the second of Equations 6.40c yields $r = 300.3 \cong \bar{r}$ which is almost the same as for the simple straight beam formula. The stresses obtained from the beam formula for a bending moment $[M]_{\theta=\pi/6} = [P\,\bar{r}\,\sin\theta]_{\theta=\pi/6} = 100 \times 0.3 \times 0.5 = 15{,}000$ kN · m are, however, different from those given by the stress function solution:

$$\sigma_r = 0$$

$$\sigma_\theta = \pm M(r_o - r_i)/(2I) \text{ at } r_o \text{ and } r_i\text{, respectively}$$

$$= \pm 112.5 \text{ MPa}$$

which indicates a decrease of 30% at r_o and an increase of about 40% at r_i in the stress magnitudes.

The strength-of-materials solution of this problem gives a very close answer to the stress function solution, as may be readily seen from Table 6.2 for $r_o/r_i = 2$. The stresses due to the former solution give: $(\sigma_\theta)_{r=ri} = -157.4$ MPa and $(\sigma_\theta)_{r=ro} = 78.7$ MPa including the uniform compressive stress component due to the end force.

6.3.2 Thermal Stresses in Curved Beams

Stresses are developed in curved beams due to the presence of radial temperature distribution $T = T(r)$. The stress function approach is used to analyze this problem assuming that all surfaces of the curved beam are free of external loads. Expressions* similar to Equation 4.41 for plane stress conditions give the thermal stresses in the curved beam as

$$\sigma_r = \frac{1}{r}\frac{\partial\phi}{\partial r} = \frac{A}{r^2} + \frac{B}{r_i^2}\left(1 + 2\ln\frac{r}{r_i}\right) + \frac{2C}{r_i^2} - \frac{\alpha E}{r^2}\int_{r_i}^{r} Tr\,dr$$

$$\sigma_\theta = \frac{\partial^2\phi}{\partial r^2} = -\frac{A}{r^2} + \frac{B}{r_i^2}\left(3 + 2\ln\frac{r}{r_i}\right) + \frac{2C}{r_i^2} - \alpha E\left[T - \frac{1}{r^2}\int_{r_i}^{r} Tr\,dr\right] \tag{6.44a}$$

$$\tau_{r\theta} = -\frac{\partial}{\partial r}\left(\frac{1}{r}\frac{\partial\phi}{\partial\theta}\right)$$

---

* See Problem 4.19 for the appropriate stress function.

The boundary conditions are such that

At $r = r_i$ and $r = r_o$: $\sigma_r = 0$

while, at any cross section:

$$\int_{r_i}^{r_o} \sigma_\theta h dr = 0 \text{ and } \int_{r_i}^{r_o} \sigma_\theta r h dr = 0$$

These yield the values for the constants *A*, *B*, and *C*, respectively, as

$$A = \frac{E\alpha}{N}\left[\left\{2\left(\frac{r_o}{r_i}\right)^2 \ln\left(\frac{r_o}{r_i}\right)\left(2\ln\left(\frac{r_o}{r_i}\right) - 1\right) + \left(\frac{r_o}{r_i}\right)^2 - 1\right\}\right.$$

$$\left. \times \int_{r_i}^{r_o} Tr dr - 4\left(\frac{r_o}{r_i}\right)^2 \ln\left(\frac{r_o}{r_i}\right)\int_{r_i}^{r_o} Tr \ln\left(\frac{r}{r_i}\right) dr\right]$$

$$B = \frac{E\alpha}{N}\left[\left\{2\left(\frac{r_o}{r_i}\right)^2 \ln\left(\frac{r_o}{r_i}\right) - \left(\frac{r_o}{r_i}\right)^2 + 1\right\}\int_{r_i}^{r_o} Tr dr\right.$$

$$\left. -2\left\{\left(\frac{r_o}{r_i}\right)^2 - 1\right\}\int_{r_i}^{r_o} Tr \ln\left(\frac{r}{r_i}\right) dr\right]$$

$$C = -D = \frac{E\alpha}{N}\left[-2\left(\frac{r_o}{r_i}\right)^2 \left(\ln\left(\frac{r_o}{r_i}\right)\right)^2 \int_{r_i}^{r_o} Tr dr + \left\{2\left(\frac{r_o}{r_i}\right)^2\right.\right.$$

$$\left.\left. \times \ln\left(\frac{r_o}{r_i}\right) + \left(\frac{r_o}{r_i}\right)^2 - 1\right\}\int_{r_i}^{r_o} Tr \ln\left(\frac{r}{r_i}\right) dr\right]$$

where

$$N = 4\left(\frac{r_o}{r_i}\right)^2 \left(\ln\left(\frac{r_o}{r_i}\right)\right)^2 - \left[\left(\frac{r_o}{r_i}\right)^2 - 1\right]^2 \tag{6.44b}$$

Example 6.15:

Find the tangential stress σ_θ for a circularly curved beam subjected to (a) a linear radial temperature gradient, where $T(r_i) = T_o$ and $T(r_o) = 0$ and (b) a uniform temperature $T = T_o$. Plot the variation of the maximum and minimum tangential stress vs. r_o/r_i. Also show the stress $\sigma_\theta/E\alpha T_o$ for a curved beam of $r_o/r_i = 5$.

Solution:

a. A linear radial temperature gradient is expressed by

$$T = T_o\,(r_o - r)/(r_o - r_i)$$

Hence, the integrals appearing in Expressions 6.44 are

$$\int_{r_i}^{r} Trdr = \frac{T_o}{(r_o/r_i - 1)}\left\{\left(\frac{r_o}{r_i}\right)\frac{r^2}{2} - \frac{r^3}{3r_i}\right\}_{r_i}^{r}$$

At $r = r_o$:

$$\int_{r_i}^{r_o} Trdr = \frac{T_o r_i^2}{(r_o/r_i - 1)}\left[\frac{(r_o/r_i)^3}{6} - \frac{(r_o/r_i)}{2} + \frac{1}{3}\right],$$

and

$$\int_{r_i}^{r} Tr\ln\left(\frac{r}{r_i}\right)dr = \frac{T_o r_i^2}{(r_o/r_i - 1)}\left\{\frac{1}{2}\left(\frac{r_o}{r_i}\right)\times\left(\frac{r}{r_i}\right)^2\left[\ln\left(\frac{r_o}{r_i}\right) - \frac{1}{2}\right] - \frac{1}{3}\left(\frac{r}{r_i}\right)^3\left[\ln\left(\frac{r_o}{r_i}\right) - \frac{1}{3}\right]\right\}_{r_i}^{r}$$

At $r = r_o$:

$$\int_{r_i}^{r_o} Tr\ln\left(\frac{r}{r_i}\right)dr = \frac{T_o r_i^2}{(r_o/r_i - 1)}\left[\frac{1}{6}\left(\frac{r_o}{r_i}\right)^3 \ln\left(\frac{r_o}{r_i}\right) - \frac{5}{36}\left(\frac{r_o}{r_i}\right)^3 + \frac{1}{4}\left(\frac{r_o}{r_i}\right) - \frac{1}{9}\right]$$

By substituting into Expressions 6.44, the stress distribution across the beam cross section, σ_θ at $r = r_o$ and $r = r_i$, is calculated and plotted in Figure 6.24a vs. the radius ratio r_o/r_i. For curved beams of small depth to curvature, $(r_o - r_i)/\bar{r} \to 0$, the stresses tends to vanish as expected for a straight beam. Figure 6.24b shows the variation of $\sigma_\theta/E\alpha T_o$ with r_o/r_i for a beam of $r_o/r_i = 5$ indicating its maximum value occurring at $r = r_i$ and its minimum (compressive) value at $r \cong 2.5r_i$.

b. The same integrations are performed for uniform temperature distribution, giving

$$\int_{r_i}^{r} Trdr = T_o\left\{\frac{r^2}{2}\right\}_{r_i}^{r}$$

At $r = r_o$:

$$\int_{r_i}^{r_o} Trdr = \frac{T_o r_i^2}{2}\left[\left(\frac{r_o}{r_i}\right)^2 - 1\right]$$

and

$$\int_{r_i}^{r} Tr\ln\left(\frac{r}{r_i}\right)dr = T_o r_i^2\left\{\frac{1}{2}\left(\frac{r}{r_i}\right)^2 \ln\left(\frac{r}{r_i}\right) - \frac{1}{4}\left(\frac{r}{r_i}\right)^2\right\}_{r_i}^{r}$$

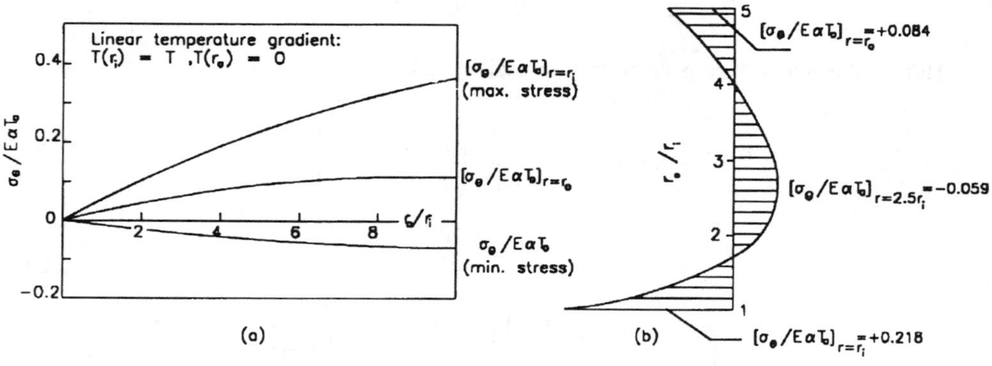

FIGURE 6.24 Example 6.15.

At $r = r_o$:

$$\int_{r_i}^{r_o} Tr\ln\left(\frac{r}{r_i}\right)dr = T_o r_i^2\left[\frac{1}{2}\left(\frac{r_o}{r_i}\right)^2\ln\left(\frac{r_o}{r_i}\right) - \frac{1}{4}\left(\frac{r_o}{r_i}\right)^2 + \frac{1}{4}\right]$$

Upon substituting into Expressions 6.44, the stresses σ_θ and σ_r do vanish as expected.

6.3.3 Wedge Subjected to a Concentrated Load at Its Vertex

The problem of a wedge subjected to a concentrated load acting at its vertex finds application in analyzing the stresses and deformation in cutting tools, ploughs, and similar equipment and also in cantilever beams. Four cases of loading will be dealt with here.

6.3.3.1 Force Acting Along a Wedge Axis

Consider a plate of thickness h made into a wedge shape having a vertex angle = 2α, as shown in Figure 6.25a, with load P acting at the vertex in plane of the plate along the axis of symmetry x.

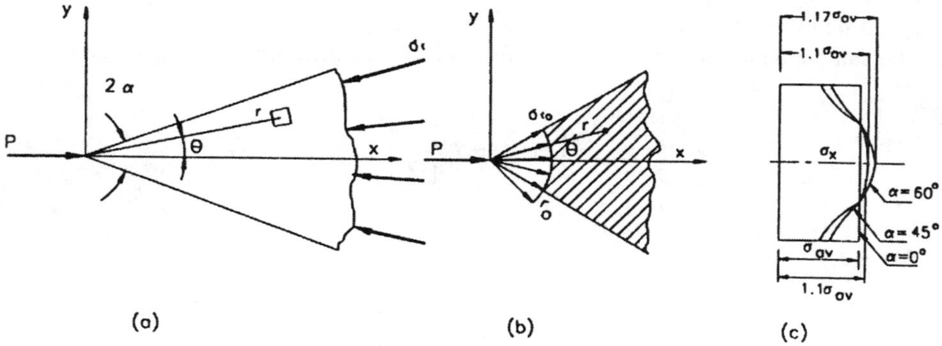

FIGURE 6.25 Wedge subjected to a concentrated axial force at its vertex: (a) load diagram, (b) localized vertex yielding, and (c) stress distribution for different vertex angles.

The sides of the wedge are stress free so that, at $\theta = \alpha$, $\sigma_\theta = \tau_{r\theta} = 0$. For small values of r, the arc length $2\alpha r$ is also small, so that σ_θ and $\tau_{r\theta}$ remain unchanged, i.e. $\sigma_\theta = \tau_{r\theta} = 0$ for all values of θ. Moreover, since r is arbitrary, it is concluded that the condition $\sigma_\theta = \tau_{r\theta} = 0$ prevails at any radius so that Equations 4.25, without body forces, give

$$\sigma_\theta = \frac{\partial^2\phi}{\partial r^2 = O} \quad \text{and} \quad \tau_{r\theta} = -\frac{\partial\phi}{\partial r}\left(\frac{\partial\phi}{r\partial\theta}\right) = 0$$

This results in a stress function,

$$\phi = rf(\theta)$$

where $f(\theta)$ is a function of θ only. Substituting into the governing biharmonic equation for plane stress without body forces, namely, $\nabla^2\nabla^2\phi = 0$, yields the ordinary differential equation:

$$\frac{\partial^4 f}{\partial\theta^4} + 2\frac{d^2 f}{d\theta^2} + f = 0$$

The general solution of this equation is

$$\mathrm{f}(\theta) = A\cos\theta + B\theta\cos\theta + C\sin\theta + D\theta\sin\theta$$

hence,

$$\phi = Ar\cos\theta + Br\theta\cos\theta + Cr\sin\theta + Dr\theta\sin\theta$$

From Equations 6.34, the radial stress is

$$\sigma_r = \frac{1}{r}\left(\frac{\partial\phi}{\partial r} + \frac{\partial^2\phi}{r\partial\theta^2}\right) = -\frac{2B}{r}\sin\theta + \frac{2D}{r}\cos\theta \tag{6.45}$$

Since σ_r is symmetrical about the x-axis, i.e., the same value for $\pm\theta$, then B must be zero and, hence,

$$\sigma_r = 2D\frac{\cos\theta}{r} \tag{6.46}$$

At $r = 0$, the radial stress σ_r becomes infinite according to Equation 6.46. However, it is obvious that yielding and plastic deformation must occur at this point in the vicinity of P and elastic analysis should then be applied outside the plastic zone. It may be assumed that the plastic zone is contained within a small radius r_o. The sector of radius r_o is removed and replaced by the stresses acting on the plastic–elastic interface, namely, σ_{ro}. This stress is considered to be the boundary loading for the elastic wedge, as shown in Figure 6.25b, thus satisfying $r = r_o$.

$$\int_{-\alpha}^{+\alpha} \sigma_{r_o} r_o h\cos\theta d\theta = -P \tag{6.47}$$

where h is the plate thickness. Substituting from Equation 6.46 into Equation 6.47 yields, after integration,

$$D = \frac{-P}{h(2\alpha + \sin 2\alpha)}$$

The solution for the problem is thus given by

$$\sigma_r = -\frac{2P\cos\theta}{hr(2\alpha + \sin 2\alpha)}, \quad \sigma_\theta = 0, \quad \text{and} \quad \tau_{r\theta} = 0 \tag{6.48a}$$

Equation 6.48a shows that at any point in the wedge σ_r and σ_θ are principal stresses. The stresses referred to Cartesian coordinates (x, y) can be obtained from the stress transformation, Equation 1.20, as

$$\begin{aligned} \sigma_x &= \sigma_r \cos^2\theta = -\frac{2P\cos^3\theta}{hr(2\alpha + \sin 2\alpha)} \\ \sigma_y &= \sigma_r \sin^2\theta = -\frac{2P\cos\theta\sin^2\theta}{hr(2\alpha + \sin 2\alpha)} \\ \tau_{xy} &= \sigma_r \sin\theta\cos\theta = -\frac{2P\cos^2\theta\sin\theta}{hr(2\alpha + \sin 2\alpha)} \end{aligned} \tag{6.48b}$$

It is instructive to compare the stress distribution obtained by the stress function approach with those obtained by the elementary strength-of-materials solution. This latter solution at any point (x, y) of Figure 6.25 gives σ_x^*, assumed uniform, at any point as

$$\sigma_x^* = -\frac{P}{A} = -\frac{P}{2hr\sin\alpha}$$

On the other hand, from the first of Equations 6.48b, σ_x is not uniform along the y-direction, as shown in Figure 6.25c. The ratio between the two values of stresses is

$$\frac{\sigma_x}{\sigma_x^*} = \frac{4\cos^3\theta\sin\alpha}{(2\alpha + \sin 2\alpha)}$$

This ratio attains a maximum value at $\theta = 0$, where $\sigma_x = \sigma_r$. Table 6.3 shows that the discrepancy between the two approaches increases as the wedge angle α increases. Again, the stress function approach yields values for σ_y and τ_{xy}, Equation 6.48b, while the strength-of-materials solution gives

TABLE 6.3
Dependence of axial stress σ_x at $\theta = 0$ on Wedge Angle (Axial Load)

$\alpha°$	0	30°	45°	60°
σ_x/σ_x^*	1	1.045	1.1	1.17

$\sigma_y = \tau_{xy} = 0$. Note that at $\theta = 0$: $\sigma_y = \tau_{xy} = 0$; however, at $\theta = \pi/4$: $\sigma_x = \sigma_y = \tau_{xy}$, which means that σ_y and τ_{xy} can reach appreciable values.

6.3.3.2 Force Perpendicular to the Wedge Axis

Consider now the case when P acts at the vertex of the wedge perpendicular to its axis, as shown in Figure 6.26a. In this case, which represents a bending problem, σ_r is not symmetrical about the x-axis, since

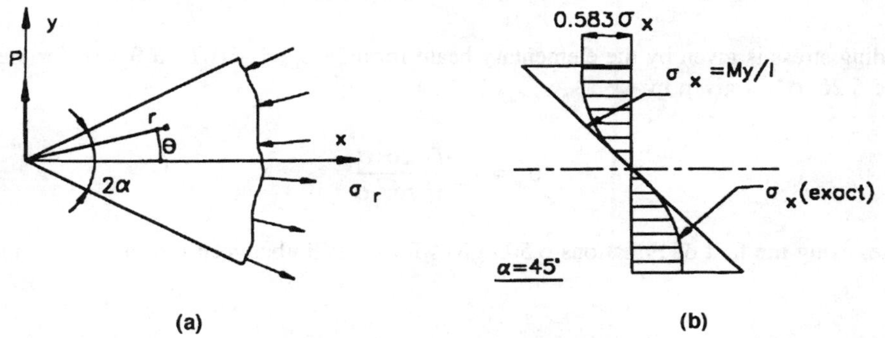

FIGURE 6.26 Wedge subjected to a concentrated lateral force at its vertex: (a) load diagram and (b) stress distribution.

$$[\sigma_r]_{\theta=\theta} = -[\sigma_r]_{\theta=-\theta}$$

Therefore, from Equation 6.45, the constant D must be equal to zero and the radial stress is

$$\sigma_r = -2B\frac{\sin\theta}{r} \tag{6.49}$$

Equilibrium for the portion of the wedge of radius $r = r_o$ requires that

$$\int_{-\alpha}^{\alpha} \sigma_{r_o} r_o h \sin\theta d\theta = -P$$

Hence, using Equation 6.49 gives the value of the constant B as

$$B = \frac{P}{h(2\alpha - \sin 2\alpha)}$$

Therefore, the solution for the stresses is

$$\sigma_r = -\frac{2P\sin\theta}{hr(2\alpha - \sin 2\alpha)} \quad \text{and} \quad \sigma_\theta = \tau_{r\theta} = 0 \tag{6.50a}$$

Equation 6.50a shows that σ_r, σ_θ are principal stresses. The stress components referred to Cartesian coordinates (x, y) can be obtained from the stress transformation Equation 1.20 as

$$\sigma_x = \sigma_r \cos^2\theta = -\frac{2P\sin\theta\cos^2\theta}{hr(2\alpha - \sin 2\alpha)}$$

$$\sigma_y = \sigma_r \sin^2\theta = -\frac{2P\sin^3\theta}{hr(2\alpha - \sin 2\alpha)} \tag{6.50b}$$

$$\tau_{xy} = \sigma_r \sin\theta\cos\theta = -\frac{2P\sin^2\theta\cos\theta}{hr(2\alpha - \sin 2\alpha)}$$

The bending stress is given by the elementary beam formula $\sigma_x^* = -My/I$. At $\theta = \alpha$, for the wedge of Figure 6.26, σ_x^* is given by

$$\sigma_x^* = -\frac{3P}{2hr}\frac{\cos\alpha}{\sin^2\alpha}$$

Therefore, using the first of Equations 6.50b gives for the ratio between the stresses σ_x and σ_x^* at $\theta = \alpha$:

$$\frac{\sigma_x}{\sigma_x^*} = \frac{4}{3}\frac{\sin^3\alpha\cos\alpha}{(2\alpha - \sin 2\alpha)}$$

The ratio of σ_r/σ_x^* is given by

$$\frac{\sigma_r}{\sigma_x^*} = \frac{4}{3}\frac{\sin^3\alpha}{\cos\alpha(2\alpha - \sin 2\alpha)}$$

The ratios are tabulated in Table 6.4 for various values of α. Again, it is concluded that the elementary beam formula is valid for small values of α only. Also, the stress function approach yields values for σ_y, Equation 6.50b, while for the strength-of-materials solution $\sigma_y = 0$ it can be shown that the integration of τ_{xy}, Equation 6.50b, along any section of the wedge perpendicular to the x-direction is equal to the load P.

TABLE 6.4
Dependence of Bending Stress σ_x at $\theta = \alpha$ on Wedge Angle (Perpendicular Load)

$\alpha°$	0	30°	45°	60°
σ_x/σ_x^*	1	0.797	0.583	0.353
σ_r/σ_x^*	1	1.06	1.166	1.41

6.3.3.3 Force Inclined to the Wedge Axis

From Equations 6.48 and 6.50 the case of a force acting at the vertex in any direction inclined an angle β to the x-axis, Figure 6.27, may be obtained by superposition as follows:

$$P_x = P_a = P\cos\beta, \quad P_y = P_n = P\sin\beta$$

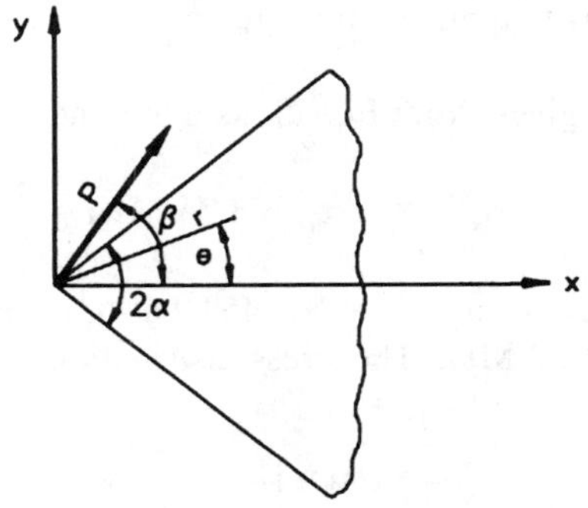

FIGURE 6.27 Wedge subjected to an inclined concentrated force at its vertex.

where P_a and P_n denote the axial and normal force components, respectively. Hence, the radial stress is given by

$$\sigma_r = -\frac{2P}{hr}\left[\frac{\cos\beta\cos\theta}{2\alpha+\sin 2\alpha}+\frac{\sin\beta\sin\theta}{2\alpha-\sin 2\alpha}\right] \tag{6.51a}$$

or, alternatively,

$$\sigma_r = -\frac{2}{hr}\left[\frac{P_a\cos\theta}{2\alpha+\sin 2\alpha}+\frac{P_a\sin\theta}{2\alpha-\sin 2\alpha}\right] \tag{6.51b}$$

The surface along which $\sigma_r = 0$, which is the neutral surface, will make an angle $\theta = \theta_o$ with the x-axis and is obtained by setting $\sigma_r = 0$ in Equation 6.51 thus giving

$$\tan\theta_o = -\frac{2\alpha-\sin 2\alpha}{2\alpha+\sin 2\alpha}\cot\beta$$

which is independent of θ, and hence the neutral surface is a plane. The strength-of-materials also gives a neutral plane. Again σ_x, σ_y, and τ_{xy} can be obtained from expressions similar to Equations 6.48b and 6.50b.

Example 6.16:

For a wedge having a vertex semiangle of 45°, and a thickness of 10 mm, plot the distribution of the stress components σ_x, σ_y, and σ_{xy} at a cross section of 100 mm distance from the vertex when subjected to a load of 10 kN at the vertex in a direction normal to the axis as obtained from:

- *a. Elementary beam formula.*
- *b. Stress function solution.*

Solution:

a. Elementary beam formula gives

$$\sigma_x = \frac{M}{I}y = \frac{10\times10^3\times100}{10\times200^3/12}y = 0.15y$$

which is linear along y with a maximum value of 15 MPa at $y = 100$ mm, together with $\sigma_y = \tau_{xy} = 0$.

b. Stress function solution gives, from Equations 6.50b, at $\theta = \alpha = 45°$,

$$\sigma_x = \sigma_y = \tau_{xy} = 8.76 \text{ MPa}$$

The stresses σ_y and τ_{xy} are maximum at $\theta = 45°$, but σ_x is maximum at $\theta = \tan^{-1} 1/\sqrt{2}$, which gives $(\sigma_x)_{max} = 11.03$ MPa. The stress distributions are plotted in Figure 6.28.

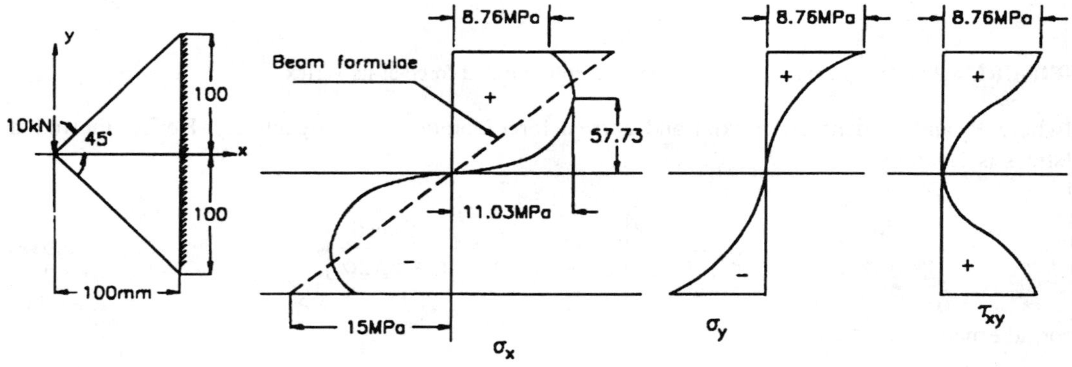

FIGURE 6.28 Example 6.16.

6.3.3.4 Bending Moment Acting at the Vertex

The problem of a wedge loaded by a bending moment M, in the plane of the wedge and concentrated at the tip, is solved by the stress function:

$$\phi = \frac{M}{h} \frac{\sin 2\theta - 2\theta \cos 2\alpha}{2(\sin 2\alpha - 2\alpha \cos 2\alpha)}$$

Hence,

$$\sigma_r = -\frac{2M \sin 2\theta}{hr^2(\sin 2\alpha - 2\alpha \cos 2\alpha)}$$

$$\sigma_\theta = 0$$

$$\tau_{r\theta} = \frac{M(\cos 2\theta - \cos 2\alpha)}{hr^2(\sin 2\alpha - 2\alpha \cos 2\alpha)}$$

where θ is as indicated in Figure 6.29 and M is counterclockwise.

Example 6.17:

A chisel, Figure 6.30 has a cutting edge with the following geometry: edge angle $2\alpha = 60°$, edge width $a = 1$ mm, and edge thickness h = 20 mm. Find the maximum force P for the two modes of cutting (a) and (b) such that the tool does not get blunt. Assume the yield stress at the tool edge = 800 MPa.

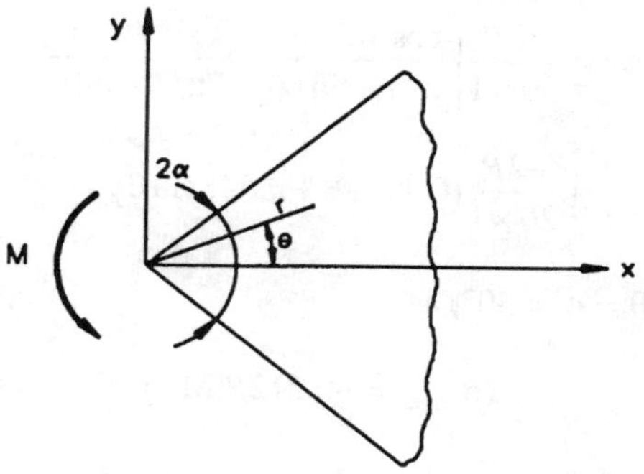

FIGURE 6.29 Wedge subjected to a bending moment acting at the vertex.

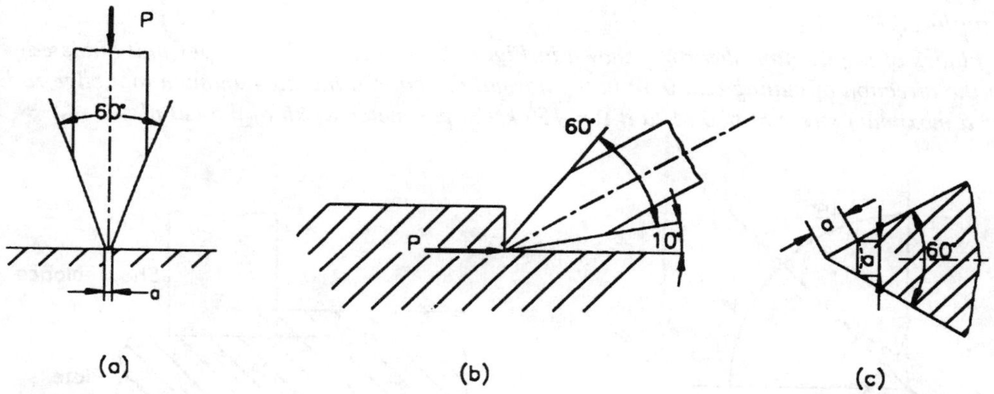

FIGURE 6.30 Example 6.17.

Solution:

Cutting mode (a): The edge is acted upon by an axial, force P:

For a 60° edge angle and substituting $r = a = 1$ mm, $h = 20$ mm, and $\alpha = 30°$ in Equation 6.48a, the maximum stress is obtained at $\theta = 0$ and is given by

$$\left(\sigma_r\right)_{\max} = -\frac{2P}{20\times 1\left(\pi/3+\sin 60\right)} = -0.0522P \text{ MPa}$$

The tool edge will not get blunt if the maximum stress does not exceed the yield stress; hence,

$$(\sigma_r)_{\max} = 800 \text{ MPa, giving } P_{\max} = 15.3 \text{ kN}$$

Cutting mode (b): The edge is acted upon by an inclined force P:

Substituting $\beta = 40°$, $r = a = 1$ mm, $h = 20$ mm, and $\alpha = 30°$ in Equation 6.51a yields

$$\sigma_r = \frac{-2P}{20\times 1}\left[\frac{\cos 40°\cos\theta}{\pi/3+\sin 60°}+\frac{\sin 40°\sin\theta}{\pi/3-\sin 60°}\right]$$

$$= \frac{-2P}{20\times 1}[0.4\cos\theta + 3.548\sin\theta]$$

This attains a maximum at $\theta = \alpha = 30°$, as

$$(\sigma_r)_{max} = -0.212P \text{ MPa}$$

Hence, to avoid tool blunting $(\sigma_r)_{max}$ should not exceed the yield stress, 800 MPa, i.e.,

$$P_{max} = 3.774 \text{ kN}$$

Example 6.18:

The blades of a guillotine shear are shown in Figure 6.31. The load acting per unit blade edge is P *in the direction of cutting and* 0.3P *in the normal direction. Find the minimum tool edge radius* r *for a maximum stress of 600 MPa if* P = *150 kN/m per meter width of the cut plate.*

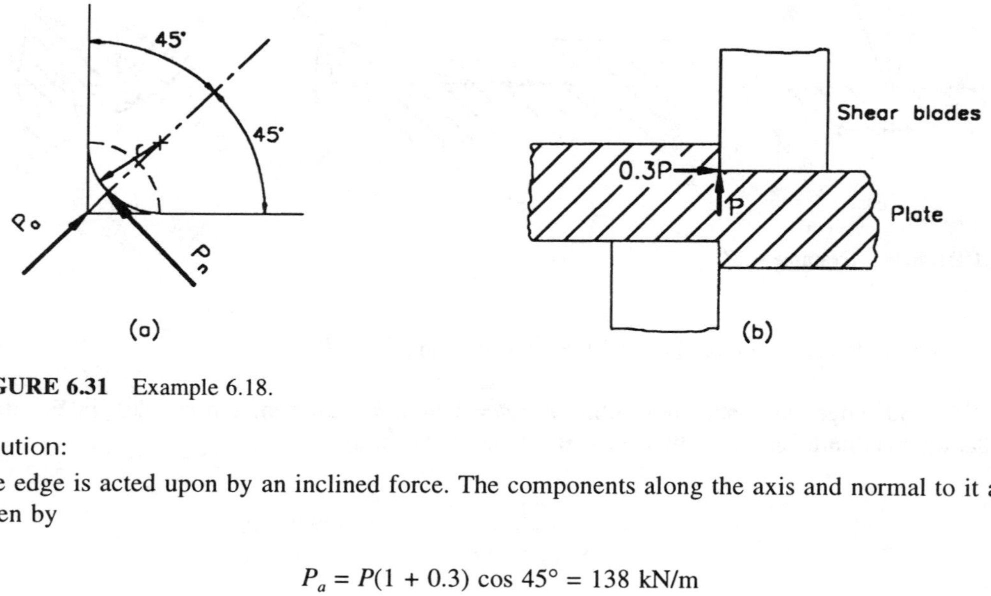

FIGURE 6.31 Example 6.18.

Solution:

The edge is acted upon by an inclined force. The components along the axis and normal to it are given by

$$P_a = P(1 + 0.3)\cos 45° = 138 \text{ kN/m}$$

$$P_n = P(1 - 0.3)\cos 45° = 75 \text{ kN/m}$$

The maximum stress is obtained by substituting $\alpha = \theta = 45°$ and $h = 1$ m in Equation 6.51b, which yields

$$600 = \frac{-2}{1000\times r} - \left[\frac{138\times 10^3\times 0.707}{\pi/2+1}+\frac{75\times 10^3\times 0.707}{\pi/2-1}\right]$$

This gives $r = 0.436$ mm.

Figure 6.31 shows the tool geometry with a nose radius r. Note that in actual tools this radius is not made but is usually generated through plastic deformation while cutting.

Example 6.19:

An orthogonal cutting tool as shown in Figure 6.32a has a rake angle = 20°, clearance angle = 10°, width of cut = 4 mm, $P_x = 1.5$ *kN and* $P_y = 2.5$ *kN. Find the minimum tool nose radius if* σ_r *is not to exceed 900 MPa.*

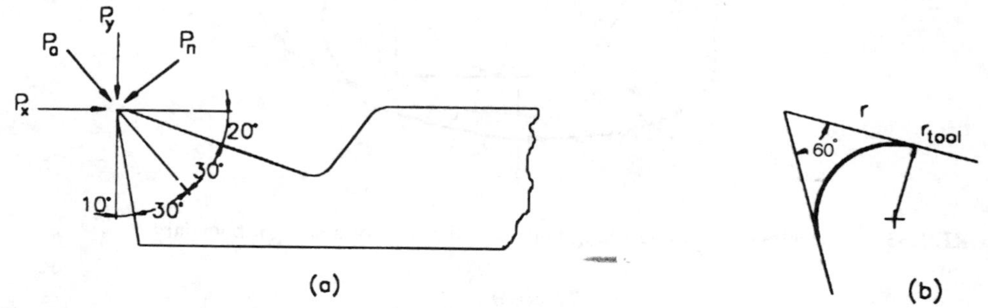

FIGURE 6.32 Example 6.19.

Solution:

The edge is acted upon by an axial force component P_a and a normal one P_n, given by

$$P_a = P_x \cos 50° + P_y \cos 40° = 5.75 \text{ kN}$$

$$P_n = P_y \sin 40° - P_x \sin 50° = 0.916 \text{ kN}$$

Substituting $\alpha = \theta = 30°$ and $h = 4$ mm in Equation 6.51b gives

$$(\sigma_r)_{max} = \frac{-2}{4 \times r}\left[\frac{5.75 \times 10^3 \times 0.866}{\pi/3 + 0.866} + \frac{0.916 \times 10^3 \times 0.5}{\pi/3 - 0.866}\right]$$

Substituting $\sigma_r = 900$ MPa gives $r = 2.85$ mm and, from Figure 6.32b,

$$r_{tool} = r \tan 30° = 1.645 \text{ mm}$$

6.3.4 Concentrated Line Load Acting on the Edge of a Straight Boundary

This problem finds application in many contact problems. Three cases, according to the direction of the acting force, will be dealt with here.

6.3.4.1 Force Acting Normal to the Boundary

A concentrated vertical line load P acting on a horizontal straight boundary AB of a semi-infinite plate is shown in Figure 6.33. The distribution of the load along the thickness of the plate is assumed to be uniform.

The stress distribution in this case is simply obtained from the solution of the problem of a concentrated force acting at the vertex of a wedge in section 6.3.3 by taking the x- and y-axes, as shown in Figure 6.33 and putting $\alpha = \pi/2$ in Equations 6.48. Therefore the stress distribution is

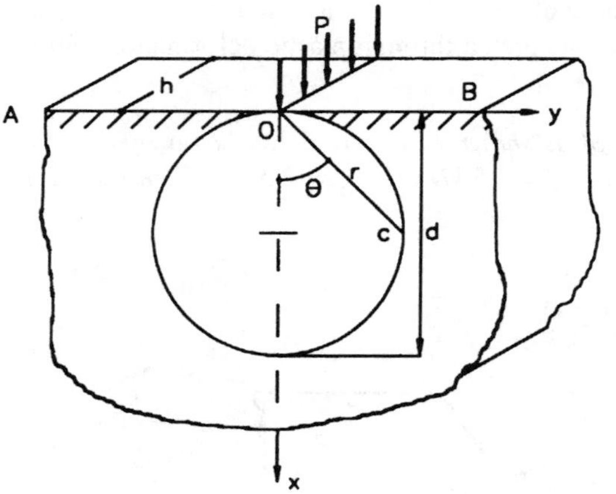

FIGURE 6.33 Concentrated line load acting normal to the edge of a straight boundary.

$$\sigma_r = -\frac{2P\cos\theta}{\pi hr}, \quad \sigma_\theta = \tau_{r\theta} = 0 \tag{6.52}$$

This distribution indicates that an element at a distance r from the point of application of load P is subjected to simple compression in the radial direction. Take a circle of any diameter d with its center on the y-axis and tangent to the x-axis at 0, Figure 6.33. At any point C on the circle, $r = d\cos\theta$, hence, substituting in Equation 6.52 gives

$$\sigma_r = -\frac{2P}{\pi dh} \tag{6.53}$$

i.e., the stress is the same at all points on the circle. At point O, the point of application of the load where yielding of the material will occur, the load will become distributed over a finite area of length $2r_o$.

6.3.4.2 Force Acting along the Boundary

A concentrated force P acting at O along the boundary AB is shown in Figure 6.34. The stress components are obtained by substituting $\alpha = \pi/2$ in Equations 6.50 to give

$$\sigma_r = -\frac{2P\sin\theta}{\pi hr}, \quad \sigma_\theta = \tau_{r\theta} = 0 \tag{6.54}$$

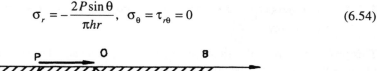

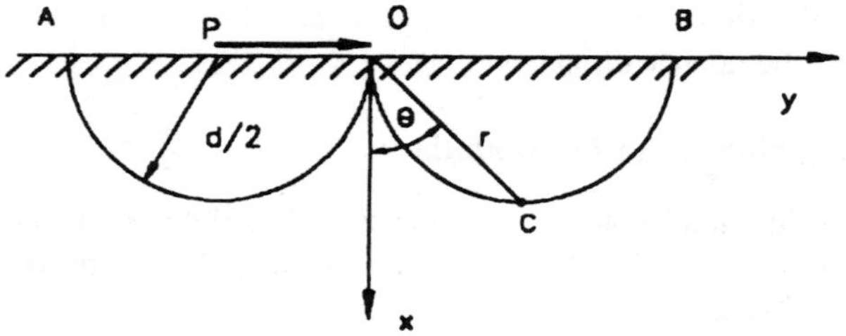

FIGURE 6.34 Concentrated line load acting along the edge of a straight boundary.

Consider a half circle of diameter d with its center on the y-axis and tangent to the x-axis at O, Figure 6.34. At any point C on the half circle, $r = d \sin \theta$, which when substituted in Equation 6.54 results again in Equation 6.53. Hence, all points on the half circle have the same stress σ_r. Note that σ_r is negative for positive θ and positive for negative θ, where θ is measured from the y-axis.

6.3.4.3 Force Acting Inclined to the Boundary

An inclined concentrated force P acting at point O on the boundary AB is shown in Figure 6.35. If P is inclined an angle β to the x-axis, then P will have two components:

$$P_x = P \cos \beta \quad \text{and} \quad P_y = P \sin \beta$$

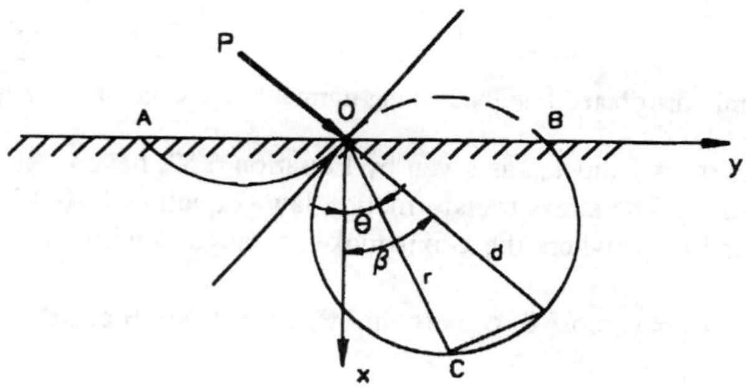

FIGURE 6.35 Concentrated line load inclined to the edge of a straight boundary.

Substituting P_x in Equation 6.52 and P_y in Equation 6.54 and summing gives

$$\sigma_r = -\frac{2P}{\pi hr}(\cos\beta\cos\theta + \sin\beta\sin\theta)$$

hence,

$$\sigma_r = -\frac{2P}{\pi hr}\cos(\beta - \theta)$$
$$\sigma_\theta = \tau_{r\theta} = 0 \tag{6.55}$$

Consider a circle of diameter d with its center along the line of action of force P and passing by O, Figure 6.35. At any point C on the circle, $r = d \cos(\beta - \theta)$ which when substituted in Equation 6.55 results in Equation 6.53. Hence, all points on the circle have the same stress σ_r.

6.3.5 Uniformly Distributed Line Load Acting on the Edge of a Straight Boundary

The solution of the case in Section 6.3.4.1 is applied to solve this problem. Referring to Figure 6.36, the distance $2a$ is divided into infinitesimal lengths dx on which a concentrated force dP acts, given by

$$dP = qdx \tag{6.56}$$

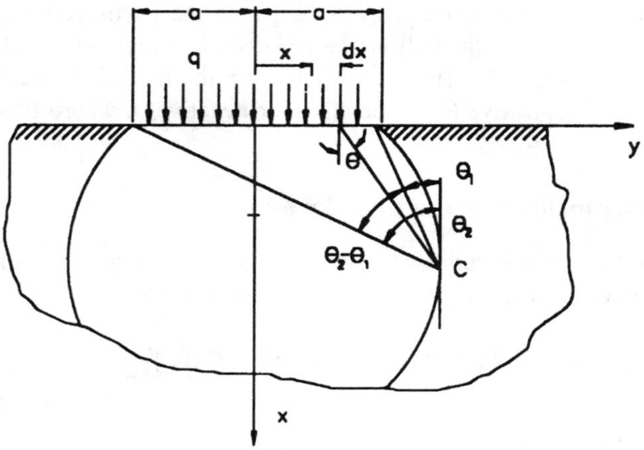

FIGURE 6.36 Uniformly distributed line load acting normal to the edge of a straight boundary.

The stress components σ_r, σ_θ, and $\tau_{r\theta}$, as given by Equation 6.52, have to be transformed to stress components σ_x, σ_y, and τ_{xy}. The stress transformation law (Equation 1.20a) is applied between the system of axes (r, θ) and (x, y) where the x-axis makes an angle θ with the radial direction; hence,

$$\sigma_x = \sigma_r \cos^2\theta, \ \sigma_y = \sigma_r \sin^2\theta, \ \tau_{xy} = \sigma_r \sin\theta\cos\theta$$

Therefore,

$$\sigma_x = -\frac{2P\cos^3\theta}{\pi rh}, \quad \sigma_y = \frac{2P}{\pi rh}\sin^2\theta\cos\theta$$
$$\tau_{xy} = -\frac{2P}{\pi rh}\sin\theta\cos^2\theta \tag{6.57}$$

Substituting dP from Equation 6.56 for P in Equations 6.57 gives

$$d\sigma_x = -2\frac{qdx}{\pi rh}\cos^3\theta$$

$$d\sigma_y = -2\frac{qdx}{\pi rh}\sin^2\cos\theta$$

$$d\tau_{xy} = -2\frac{qdx}{\pi rh}\sin\theta\cos^2\theta$$

where $d\sigma_x$, $d\sigma_y$, and $d\tau_{xy}$ are the stresses due to dP. From geometry, Figure 6.36, it is seen that

$$dx = rd\theta/\cos\theta$$

thus,

$$\sigma_x = -\frac{2q}{\pi h}\int_{\theta_1}^{\theta_2} \cos^2\theta d\theta = -\frac{q}{\pi h}\left[(\theta_2-\theta_1)+\tfrac{1}{2}(\sin 2\theta_2 - \sin 2\theta_1)\right]$$

$$\sigma_y = -\frac{2q}{\pi h}\int_{\theta_1}^{\theta_2} \sin^2\theta d\theta = -\frac{q}{\pi h}\left[(\theta_2-\theta_1)-\tfrac{1}{2}(\sin 2\theta_2 - \sin 2\theta_1)\right] \quad (6.58a)$$

$$\tau_{xy} = -\frac{2q}{\pi h}\int_{\theta_1}^{\theta_2} \sin\theta\cos\theta d\theta = -\frac{q}{\pi h}\left[\tfrac{1}{2}(\cos 2\theta_1 - \cos 2\theta_2)\right]$$

The principal stresses at a point C, Figure 6.36, are given by

$$\sigma_1 = -\frac{q}{\pi h}\left[(\theta_2-\theta_1)-\sin(\theta_2-\theta_1)\right]$$

$$\sigma_2 = -\frac{q}{\pi h}\left[(\theta_2-\theta_1)+\sin(\theta_2-\theta_1)\right] \quad (6.58b)$$

Since the angle $(\theta_2 - \theta_1)$ is constant for any circle passing through the points (a, O) and $(-a, O)$ with its center on the y-axis, therefore the principal stresses as given by Equations (6.58b), are the same for all points on this circle.

The maximum shear stress at point C is given by

$$\tau_{max} = \frac{1}{2}(\sigma_1 - \sigma_2) = \frac{q}{\pi h}\sin(\theta_2 - \theta_1) \quad (6.58c)$$

having a maximum value at $(\theta_2 - \theta_1) = \pi/2$, i.e., on a circle of diameter, $2a$.

6.3.6 Circular Solid Disk Subjected to Two Equal and Opposite Diametral Loads

The solution of case (a) in Section 6.3.4.1 is again applied to solve this problem. Consider a circular disk of diameter d and thickness h, Figure 6.37a, removed from the semi-infinite plate of Figure 6.33. Such a disk is acted upon by a force P at point O_1 and a compressive stress $\sigma_{r1} = -2P \cos \theta_1/(\pi td)$ distributed along the circular boundary directed toward point O_1 to establish equilibrium, Figure 6.37b. If another force P acts at point O_2, the circular boundary will be subjected to another compressive stress:

$$\sigma_{r_2} = -2P \cos \theta_2/(\pi hd)$$

The resultant of these two surface compressive stresses σ_{r1} and σ_{r2}, which are at a right angle is a uniform radial pressure p directed toward the disk center, Figure 6.37b, given by

$$p = -\frac{2P}{\pi hd}\sqrt{\cos^2\theta_1 + \sin^2\theta} = -\frac{2P}{\pi hd}$$

If a uniform radial tension of $2P/(\pi hd)$ is now added to the disk surface, then this surface would be completely free from external tractions.

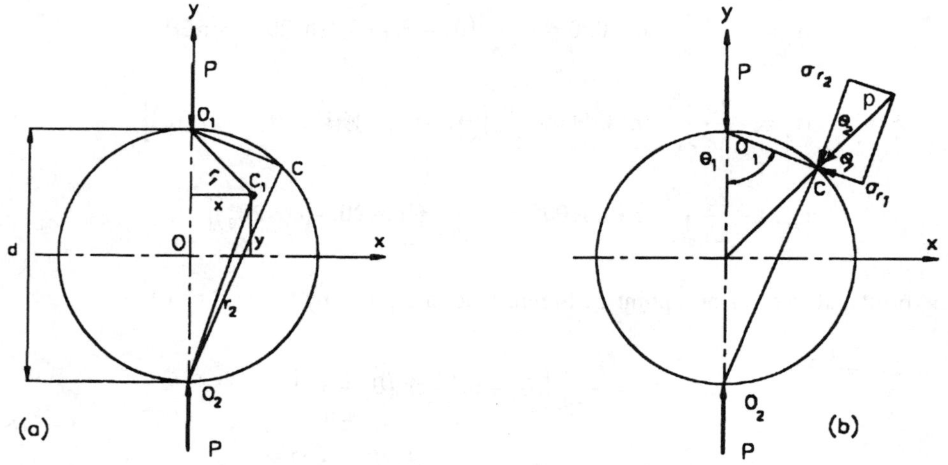

FIGURE 6.37 Circular solid disk subjected to two equal and opposite diametral loads: (a) load diagram and (b) equivalent boundary stresses.

The stresses at point C_1 (x, y) is the superposition of the stresses due to P acting at points O_1 and O_2 and a uniform radial tension of $2P/(\pi hd)$ acting on the disk surface. The resultant stresses are given by

$$\sigma_x = -\frac{P}{\pi h}\left[\frac{(d-2y)x^2}{r_1^4} + \frac{(d+2y)x^2}{r_2^4} - \frac{2}{d}\right]$$

$$\sigma_y = -\frac{P}{\pi h}\left[\frac{(d-2y)^3}{4r_1^4} + \frac{(d+2y)^3}{4r_2^4} - \frac{2}{d}\right] \tag{6.59a}$$

$$\tau_{xy} = -\frac{P}{\pi h}\left[\frac{(d-2y)^2 x}{2r_1^4} + \frac{(d+2y)^2 x}{2r_2^4}\right]$$

where

$$r_1^2 = x^2 + (d-2y)^2/4 \text{ and } r_2^2 = x^2 + (d+2y)^2/4$$

The stresses along diameter O_1O_2 are obtained by substituting $x = 0$ in Equations 6.59a as

$$\sigma_x = \frac{2P}{\pi hd}$$

$$\sigma_y = -\frac{2P}{\pi h}\left[\frac{2}{d-2y} + \frac{2}{d+2y} - \frac{1}{d}\right] \text{ and } \tau_{xy} = 0 \tag{6.59b}$$

which shows that σ_x has a constant value while σ_y tends to infinity at O_1 and O_2. Obviously, the material in the vicinity of O_1 and O_2 has already yielded and a finite area of contact has been developed.

At the disk center the stresses are given by

$$\sigma_x = 2P/(\pi hd)\ \sigma_y = -6P/(\pi hd) \text{ and } \tau_{xy} = 0 \tag{6.59c}$$

If strain gauges are placed along the x- and y-directions at the center and the gauges are connected to read $(\varepsilon_x - \varepsilon_y)$, the disk may be used as a load cell to measure compressive forces, as illustrated in Example 6.20.

Example 6.20:

In order to determine the load acting on the roller support of a beam, four strain gauges are stuck at the center of the roller. The strain gauges are connected together such that the reading in millivolts is given by $\Delta V = C(\sigma_x - \sigma_y)$, where $C = 0.085$ mV/MPa. The roller diameter is 100 mm and the width 100 mm; the support consists of four rollers.

a. *Find the total load acting on the beam assuming equal load sharing on the four rollers for $\Delta V = 2.5$ mV.*
b. *Determine the circumferential length of contact between the roller and flat plate given a yield strength, $Y = 400$ MPa.*

Solution:

a. The stresses at the center are given by Equation 6.59c, as

$$\sigma_x = 2P/(\pi h d)$$

$$\sigma_y = -6P/(\pi h d)$$

so that, $\sigma_x - \sigma_y = 8P/(\pi h d)$.

From the strain gauge readings, $\sigma_x - \sigma_y = 2.5/0.085 = 29.41$ MPa; thus, $P = 115.44$ kN and the total load = 461.76 kN.

b. The plastic deformation according to the maximum shear stress theory is contained in a semicircle of diameter = $2a$, which is obtained by substituting $\sigma_1 - \sigma_2 = Y$ at $\theta_1 - \theta_2 = \pi/2$ in Equation 6.58c. This gives $Y = -2q/(\pi h)$ or $400 = 2 \times 115.44 \times 1000/(\pi \times 100 \times 20)$. Hence, $2a = 1.837$ mm.

6.3.7 Concentrated Load Acting on a Rectangular Beam

Consider a rectangular beam of thickness h, height $2H$, and length $2L$, simply supported at the ends A and B and subjected to a concentrated load P acting at the midspan, as shown in Figure 6.38. In the elementary beam theory, the localized effect of the concentrated loads and reactions at O, A, and B are neglected, so that the stresses at any point C (x, y) are given by the following expressions:

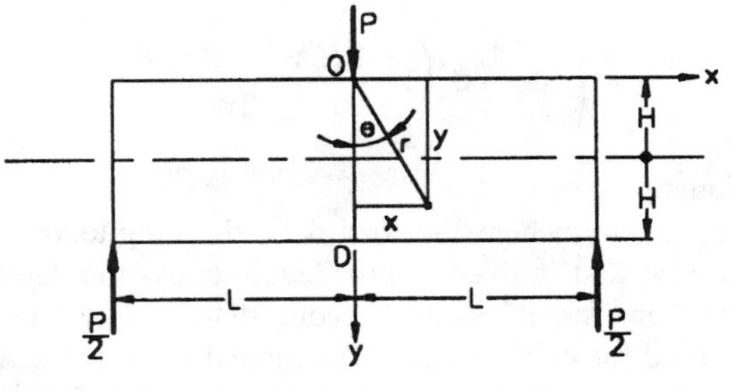

FIGURE 6.38 Concentrated force acting at the midspan of a rectangular beam.

$$\sigma_x^* = \frac{M}{I}(y-H) = \frac{3P}{4hH^3}(L-x)(y-H), \quad \sigma_y^* = 0, \quad \text{and}$$
$$\tau_{xy}^* = -\frac{3P}{8hH^3}(2H-y)y \tag{6.60}$$

Taking into consideration the effect of load P at point O, additional stresses are introduced, as given by Expressions 6.57. Substituting $r^2 = x^2 + y^2$, $\sin\theta = x/r$, and $\cos\theta = y/r$ into Expressions 6.57 and adding the resulting stresses to the stresses as given by Equation 6.60 yield

$$\sigma_x = \frac{2P}{h}\left[\frac{3}{8H^3}(L-x)(y-H) - \frac{1}{\pi}\frac{x^2 y}{\left(x^2+y^2\right)^2}\right]$$
$$\sigma_y = -\frac{2P}{h}\frac{y^3}{\pi\left(x^2+y^2\right)^2} \tag{6.61}$$
$$\tau_{xy} = -\frac{2P}{h}\left[\frac{3}{16H^3}\left(2Hy - y^2\right) + \frac{1}{\pi}\frac{xy^2}{\left(x^2+y^2\right)^2}\right]$$

It is to be noted that stresses due to the effect of P, which have been added to the stresses from the elementary beam theory, are valid for a semi-infinite plate. Hence, they are considered to be approximate as applied to a beam of finite dimensions. This approximation has resulted in introducing a compressive stress σ_y normal to the bottom edge of the beam, which contradicts the existing boundary condition that $\sigma_y = 0$ at $y = 2H$. Several approaches have been proposed to correct the stresses along the plane of symmetry, one of them* gives the stresses at $x = 0$ and $y = 2H$, i.e., at point D as

$$\sigma_x = \frac{3P}{2hH^3}\left(\frac{L}{2} - \frac{H}{\pi}\right)(y-H) + \frac{P}{2\pi hH}$$
$$\sigma_y = \frac{2P}{\pi h}\left(\frac{y}{4H^2} - \frac{1}{y}\right) \tag{6.62}$$

The first of Equations 6.62 may be rewritten as

$$\sigma_x = \sigma_x^*\left(1 - \frac{2H}{\pi L}\right) + \frac{P}{2\pi hH}$$

where σ_x^* is given by Equation 6.60.

For the section of symmetry where the load acts, the longitudinal stress σ_x is seen to be composed of two terms. The first is linearly distributed across the depth of the beam as in the elementary beam theory; however, its value is reduced by a factor depending on (H/L). This reduction is more pronounced for stubby beams. The second term represents an additional constant tensile stress. In regard to the other compressive stress component σ_y, Equation 6.62 gives an infinite

* This is known as Wilson–Stokes solution; see Timoshenko and Goodier.[2]

value at the point of application of the load. Actually, this does not happen as local yielding is expected. Farther down the y-axis, its value decreases and vanishes at the bottom point D.

Note that expressions such as Equation 6.61 represent the local stresses in the vicinity of the load application point. For sections far from that point, these stresses converge quickly to the values given by the elementary strength-of-materials. In other words, local stresses decrease very rapidly with increasing the distance from the point of application of the load. At a distance equal to the depth of the beam, they are usually very negligible.*

The effect of the load $P/2$ at A and B may be obtained from Equation 6.51a by putting $P_a = P_n = \sqrt{2}\ P$ and $2\alpha = \pi/2$. Expressing r and θ in terms of x and y and adding the resulting stresses to the stresses of Equation 6.60 will give the stresses acting at A and B.

6.4 STRESS CONCENTRATION AROUND A SMALL CIRCULAR HOLE

Localized high stresses due to the presence of concentrated loads or geometric discontinuities in a loaded solid give rise to a stress concentration which is primarily responsible for local yielding, fatigue, and fracture of the material. In Section 6.2.4, it has been already demonstrated that the presence of a small hole at the center of a rotating disk doubles the hoop stress value σ_θ as compared with its value in a rotating solid disk. Also, it has been shown in Section 6.3.2 that stresses at the point of load application at the vertex of a wedge, or on the edge of a semi-finite plate, may go to infinity and hence local yielding should occur.

In this section an example of using the stress function approach to solve the stress concentration problem around a circular hole is presented. Figure 6.39 represents a large, thin flat plate subjected to a uniform tensile stress σ in the x-direction. A small hole made in the middle of the plate will cause a stress redistribution in the plate. Since the hole is small compared with the plate width, the change in the stress distribution will be localized in the neighborhood of the hole. Hence, the stresses acting far away from the hole, i.e., at a distance several times the hole diameter, will remain unchanged as in the plate with no hole. In other words, according to Saint-Venant's principle, the change in stresses is negligible at large distances of radius r_o compared with the radius of the hole r_i, as shown in Figure 6.39.

Considering plane stress with no body forces, the stresses with no hole are

$$\sigma_x = \sigma, \quad \sigma_y = \tau_{xy} = 0$$

and, since $\sigma_x = \partial^2\phi_1/\partial y^2 = \sigma$, then, $\phi_1 = \sigma y^2/2$.

Transforming ϕ_1 to polar coordinates gives

$$\phi_1 = \frac{\sigma}{2} r^2 \sin^2\theta = \frac{\sigma}{4} r^2 (1 - \cos 2\theta)$$

$$= \frac{\sigma r^2}{4} - \frac{\sigma r^2}{4} \cos 2\theta$$

With a hole of radius $r = r_i$, the stresses given by ϕ_1 will remain unchanged in regions sufficiently far from the hole, i.e., for $r_o \to \infty$, the stresses are

* This is in agreement with Saint-Venant's principle as explained in Section 4.3.

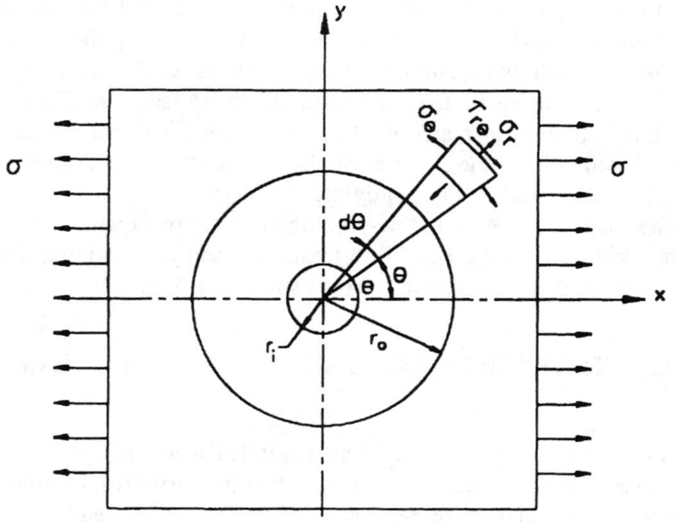

FIGURE 6.39 Stress concentration around a small hole in a thin plate subjected to uniform tensile stress.

$$\sigma_r(\infty,\ \theta) = \frac{1}{r}\left(\frac{\partial \phi_1}{\partial r} + \frac{1}{r}\frac{\partial^2 \phi_1}{\partial \theta^2}\right) = \frac{\sigma}{2}(1 + \cos 2\theta)$$

$$\sigma_\theta(\infty,\ \theta) = \left(\frac{\partial^2 \phi_1}{\partial \theta^2}\right) = \frac{\sigma}{2}(1 - \cos 2\theta) \tag{6.63a}$$

$$\tau_{r\theta}(\infty,\ \theta) = -\frac{\partial}{\partial r}\left(\frac{\partial \phi_1}{r\partial \theta}\right) = -\frac{\sigma}{2}\sin 2\theta$$

To find the stresses in the annulus $r_i < r < r_o$, it is sufficient and necessary to obtain a stress function, ϕ, that satisfies the equation $\nabla^4\phi = 0$ and the boundary conditions, at $r = r_i$:

$$\sigma_r(r_i,\ \theta) = 0, \quad \tau_{r\theta}(r_i,\ \theta) = 0 \tag{6.63b}$$

A general form for the stress function ϕ_1 may be written as

$$\phi = f(r) + g(r)\cos 2\theta \tag{6.64}$$

where f and g are unknown functions of r.

Substitution of Equation 6.64 into the biharmonic equation $\nabla^2\nabla^2\phi = 0$ yields

$$\left(\frac{d^2}{dr^2} + \frac{1}{r}\frac{d}{dr}\right)\left(\frac{d^2 f}{dr^2} + \frac{1}{r}\frac{df}{dr}\right) + \left(\frac{d^2}{dr^2} + \frac{1}{r}\frac{d}{dr} - \frac{4}{r^2}\right)\left(\frac{d^2 g}{dr^2} + \frac{1}{r}\frac{dg}{dr} - \frac{4g}{r^2}\right)\cos 2\theta = 0$$

This equation must be valid for all values of r and θ. Hence, the functions f and g are governed by two ordinary differential equations. Successive integrations of these equations give

$$f(r) = C_1 r^2 \ln r + C_2 r^2 + C_3 \ln r + C_4$$
$$g(r) = C_5 r^4 + C_6 r^2 + C_7 + \frac{C_8}{r^2} \tag{6.65}$$

where C_1, C_2, C_3, ... C_8 are constants of integration. Equations 6.64 and 6.65 define the stress function ϕ as

$$\phi = C_1 r^2 \ln r + C_2 r^2 + C_3 \ln r + C_4 + \left(C_5 r^4 + C_6 r^2 + C_7 + \frac{C_8}{r^2} \right) \cos 2\theta$$

Differentiation according to Equations 4.25 with no body forces yields the following stress components:

$$\sigma_r = C_1(1 + 2\ln r) + 2C_2 + \frac{C_3}{r^2} - \left(2C_6 + \frac{4C_7}{r^2} + \frac{6C_8}{r^4} \right) \cos 2\theta$$
$$\sigma_\theta = C_1(3 + 2\ln r) + 2C_2 - \frac{C_3}{r^2} + \left(2C_6 + 12C_5 r^2 + \frac{6C_8}{r^4} \right) \cos 2\theta \tag{6.66}$$
$$\tau_{r\theta} = \left(2C_6 + 6C_5 r^2 - \frac{2C_7}{r^2} - \frac{6C_8}{r^4} \right) \sin 2\theta$$

The above stresses must be finite as $r \to \infty$, and hence the constants C_1 and C_5 must vanish. Using the boundary conditions Equatons 6.66 results in the following relations among the constants:

$$\left(C_2 - C_6 \cos 2\theta \right) = \frac{\sigma}{4}(1 + \cos 2\theta)$$
$$\left(C_2 + C_6 \cos 2\theta \right) = \frac{\sigma}{4}(1 - \cos 2\theta)$$
$$C_6 \sin 2\theta = -\frac{\sigma}{4} \sin 2\theta$$
$$\left(C_2 + \frac{C_3}{2r_i^2} \right) - \left(C_6 + \frac{2C_7}{r_i^2} + \frac{3C_8}{r_i^4} \right) \cos 2\theta = 0$$
$$\left(C_6 - \frac{C_7}{r_i^2} - \frac{3C_8}{r_i^4} \right) \sin 2\theta = 0$$

The above relations must hold for all values of θ. Hence, the constants in these equations are easily found to be

$$C_2 = \sigma/4, \qquad C_3 = -(\sigma/2) r_i^2, \quad C_6 = -\sigma/4,$$
$$C_7 = \sigma r_i^2/2, \text{ and } C_8 = -\sigma r_i^4/4$$

Substitution of these values into Equations 6.66 yields the stress solution at point (r, θ) as

$$\sigma_r = \frac{\sigma}{2}\left[\left(1-\frac{r_i^2}{r^2}\right)+\left(1-\frac{4r_i^2}{r^2}+\frac{3r_i^4}{r^4}\right)\cos 2\theta\right]$$

$$\sigma_\theta = \frac{\sigma}{2}\left[\left(1+\frac{r_i^2}{r^2}\right)-\left(1+\frac{3r_i^4}{r^4}\right)\cos 2\theta\right] \tag{6.67}$$

$$\tau_{r\theta} = -\frac{\sigma}{2}\left[1+\frac{2r_i^2}{r^2}-\frac{3r_i^4}{r^4}\right]\sin 2\theta$$

The distribution of the stresses σ_r and σ_θ around the boundary of the hole is obtained by setting $r = r_i$ into Equations 6.67, i.e.,

$$\sigma_r = 0,\quad \sigma_\theta = \sigma(1 - 2\cos 2\theta),\quad \text{and}\quad \tau_{r\theta} = 0$$

The maximum value of σ_θ is at $r = r_i$ and $\theta = \pi/2$ and $3\pi/2$, according to:

$$[\sigma_\theta]_{r=r_i} = \frac{\sigma}{2}(2 - 4\cos 2\theta),\quad [\sigma_\theta]_{r_i,\frac{\pi}{2}} = 3\sigma$$

Plots of $\sigma_r\ (r, \pi/2)$ and $\sigma_\theta\ (r, \pi/2)$ against r/r_i – taken as ordinate — are shown in Figure 6.40a. It is seen that $\sigma_\theta\ (r, \pi/2)$ decreases rapidly from 3σ to σ as r/r_i increases, and is approximately 1.008σ at $r/r_i = 8$. Moreover, at this distance, $\sigma_r(r, \pi/2) \cong 0.023\sigma$, a very small fraction of σ. Note that $\tau_{r\theta}(r, \pi/2) = 0$ along the vertical y-axis. In Figure 6.40b the maximum value of $\sigma_\theta/\sigma = 3$ occurs at

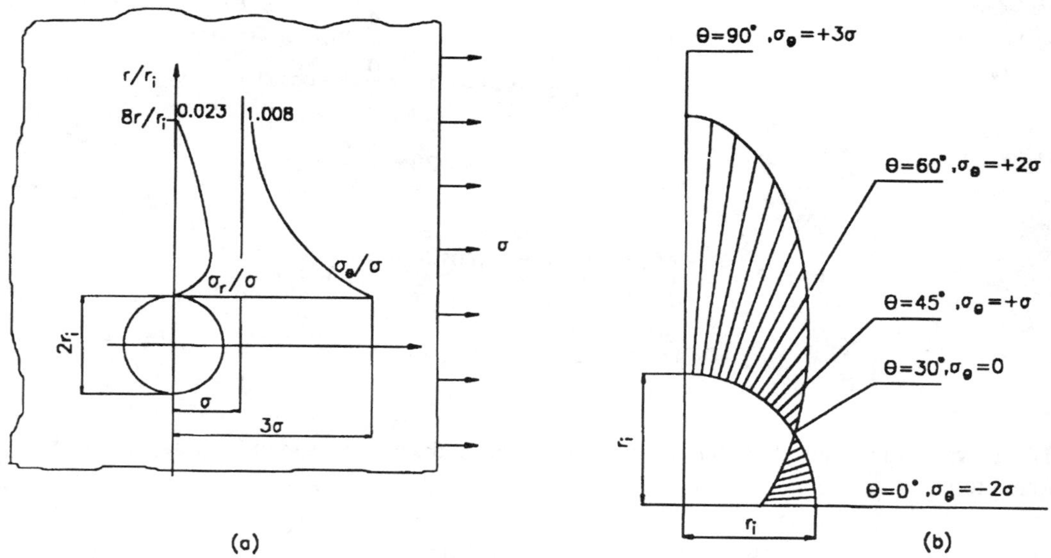

FIGURE 6.40 Stress distribution around the hole in the plate of Figure 6.39: (a) stress distribution along the y-axis and (b) distribution of the tangential stress along the hole surface.

$\theta = \pi/2$ and its minimum value, $\sigma_\theta/\sigma = -1$, occurs at $\theta = 0$. It is noticed that the influence of the hole not only produces a stress concentration by a factor $k_t = 3$, but in this case also produces compressive stress. At the point $\theta = \pi/6$ on the boundary of the hole, all stresses are zero.

The displacement components u and v at any point (r, θ) in the plate is obtained by substituting the stresses given by Equations 6.67 into Hooke's law, Equations 3.6. Thus, the strain–displacement relations Equations 2.8 are given by

$$\begin{aligned}
\varepsilon_r &= \frac{\partial u}{\partial r} \\
&= \frac{(1+\nu)\sigma}{2E}\left[\left(\frac{1-\nu}{1+\nu} - \frac{r_i^2}{r^2}\right) + \left(1 - \frac{4}{1+\nu}\frac{r_i^2}{r^2} + \frac{3r_i^4}{r^4}\right)\cos 2\theta\right] \\
\varepsilon_\theta &= \frac{u}{r} + \frac{1}{r}\frac{\partial v}{\partial \theta} \\
&= \frac{(1+\nu)\sigma}{2E}\left[\left(\frac{1-\nu}{1+\nu} + \frac{r_i^2}{r^2}\right) - \left(1 - \frac{4\nu}{1+\nu}\frac{r_i^4}{r^2} + \frac{3r_i^4}{r^4}\right)\cos 2\theta\right] \\
\gamma_{r\theta} &= \frac{1}{r}\frac{\partial u}{\partial r} + \frac{\partial v}{\partial r} - \frac{v}{r} \\
&= -\frac{(1+\nu)\sigma}{E}\left(1 + \frac{2r_i^2}{r^2} - \frac{3r_i^4}{r^4}\right)\sin 2\theta
\end{aligned} \tag{6.68}$$

Integration of the first and the second of Equations 6.68 yields for the displacement component u:

$$u = \frac{(1+\nu)\sigma r}{2E}\left[\left(\frac{1-\nu}{1+\nu} + \frac{r_i^2}{r^2}\right) + \left(1 + \frac{4}{1+\nu}\frac{r_i^2}{r^2} - \frac{r_i^4}{r^4}\right)\cos 2\theta\right] + F(\theta)$$

For sufficiently large values of r/r_i, this equation reduces to

$$u = \frac{(1+\nu)\sigma r}{2E}\left[\frac{1-\nu}{1+\nu}\cos 2\theta\right] + F(\theta) \tag{6.69}$$

Moreover, at such sufficiently large values of r/r_i, a state of pure tension prevails and thus:

$$u_1 = \varepsilon_x r\cos\theta = \frac{\sigma}{E} r\cos\theta$$

$$v_1 = -\nu\varepsilon_x r\sin\theta = -\frac{\nu\sigma}{E} r\sin\theta$$

where u_1, v_1 are the displacements in the x- and y-directions, respectively. Transformation of these expressions gives

$$u = u_1 \cos\theta + v_1 \sin\theta = \frac{(1+\nu)\sigma r}{2E}\left[\frac{1-\nu}{1+\nu} + \cos 2\theta\right] \tag{6.70}$$

Comparing Equations 6.69 and 6.70 shows that $F(\theta) = 0$.

Following a similar procedure, the displacement component v is determined as

$$v = -u_1 \sin\theta + v_1 \cos\theta = -\frac{(1+\nu)\sigma r}{2E}\sin 2\theta \tag{6.71}$$

It is easily verified that expressions of u and v, i.e., Equations 6.70 and 6.71, satisfy the third of Equations 6.68.

By applying the principle of superposition, the precedent solution could be extended to a plate with a central hole subjected to biaxial loading. For instance, in the case of biaxial tension of Figure 6.41, Equation 6.67 gives, at the hole surface, $r = r_i$,

$$\sigma_\theta = [\sigma - 2\sigma\cos 2\theta] + \left[\sigma^* - 2\sigma^*\cos 2\left(\frac{\pi}{2} - \theta\right)\right]$$

thus resulting for the case of $\sigma = \theta$,* in a stress concentration factor $k_t = (\sigma_\theta)_{max}/\sigma = 2$. If, however, $\sigma = -\sigma$,* σ_θ attains its maximum value $\pm 4\sigma$, at $\theta = 0$ and $\theta = \pi/2$. The stress concentration factor in this case is $k_t = (\sigma_\theta)_{max}/\sigma = 4$. This is the same stress concentration factor for the case of pure shear.

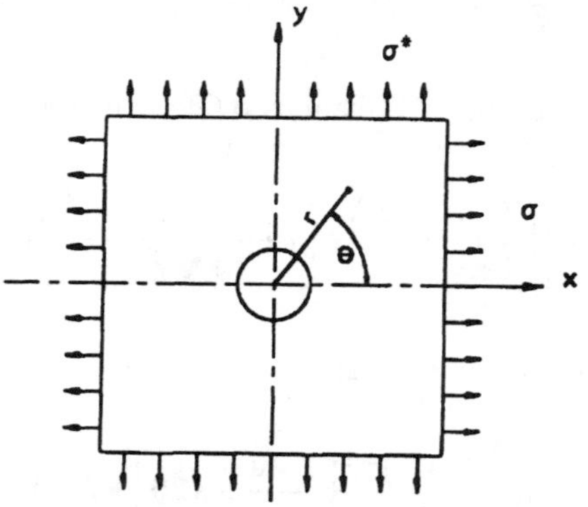

FIGURE 6.41 Central hole in a rectangular plate subjected to biaxial tension.

Stress concentration factors for other loading conditions and geometrical discontinuities are readily available in tabulated forms or graphic representations in texts on the subject.[4]

Example 6.21:

A thin, large plate of thickness 2 mm, Figure 6.42 has a small circular hole drilled at its center. The plate is subjected to a tensile stress $\sigma_x = 2\sigma$ and a compressive stress $\sigma_y = -\sigma$. The plate material has a tensile yield strength of 280 MPa and a compressive yield strength of 350 MPa, $\nu = 0.3$, and $E = 200$ GPa.

* These results agree with having the stress doubled at the center of a rotating solid disk due to the presence of a small circular hole as found in Section 6.2.4.

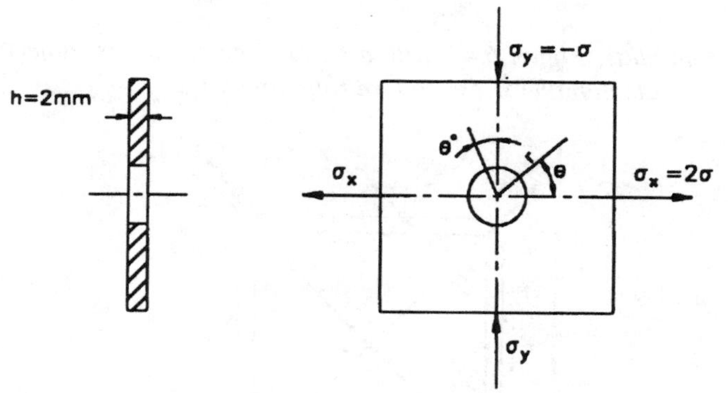

FIGURE 6.42 Example 6.21.

a. *Derive an expression for the maximum stress at the hole boundary.*
b. *Determine the maximum value of σ, such that yielding in neither tension nor compression occurs at any point in the plate.*
c. *What is the thickness of the plate at the point which just starts to yield?*

Solution:

a. At the hole boundary $r = r_i$, the stresses in a plate subjected to a tensile stress $\sigma_x = 2\sigma$ are obtained from Equations 6.67 as

$$\sigma_r = 0, \quad \tau_{r\theta} = 0$$

and

$$(\sigma_\theta)_1 = \frac{2\sigma}{2}[2 - 4\cos 2\theta]$$

For a compressive stress, this expression is used to obtain σ_θ by measuring θ from the vertical axis (hence, denoted θ*) as shown in Figure 6.41. Thus,

$$(\sigma_\theta)_2 = -\frac{\sigma}{2}[2 - 4\cos 2\theta^*] = -\frac{\sigma}{2}[2 - 4\cos 2(\theta - \pi/2)] = -\sigma[1 + 2\cos 2\theta]$$

By applying superposition, the total hoop stress at $r = r_i$ is

$$\sigma_\theta = (\sigma_\theta)_1 + (\sigma_\theta)_2 = \sigma(1 - 6\cos 2\theta)$$

b. The maximum tensile stress value occurs at $\theta = \pi/2$ and is equal to 7σ, i.e., the stress concentration factor $k_t = 7$. At $\theta = 0$, a maximum compressive stress occurs and is equal to -5σ. Yielding will not occur in tension if $(\sigma_\theta)_{max} < Y$, i.e., $7\sigma \leq 280$ MPa, or in compression if $-5\sigma \leq -350$ MPa. This gives the largest allowable value for σ as 40 MPa.
c. From Hooke's law Equation 3.6, for $\sigma_r = \sigma_z = 0$,

$$\varepsilon_z = -\nu\sigma_\theta/E$$

or $\Delta h/h_o = -0.3 \times 280 \times 10^{-6}/(200 \times 10^9)$, giving, for $h_o = 2$ mm, a deformed thickness $h = 1.99916$ mm.

Example 6.22:

***An infinitely large thin plate, Figure 6.43, with a small circular hole is subjected to pure shear τ along its boundaries. Calculate the stress concentration factor* k.**

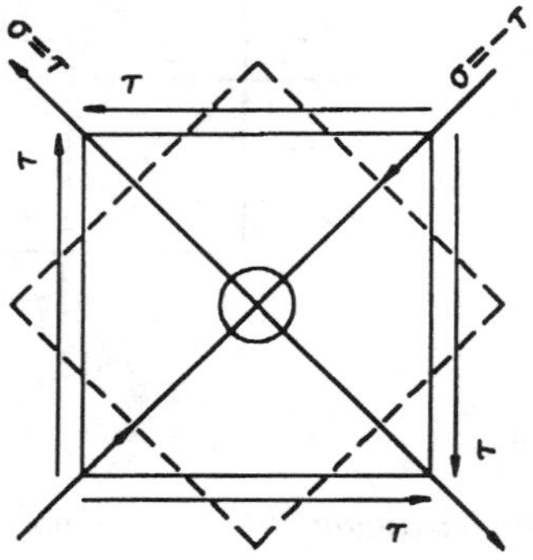

FIGURE 6.43 Example 6.22.

Solution:

The solution of this problem, Figure 6.43, is obtained from Equations 6.64 provided that the boundary shear stress loading is transformed to its equivalent principal stresses. From Equation 1.26a the magnitude and directions of principal stresses are given by

$$\sigma = \frac{\sigma_x + \sigma_y}{2} \pm \sqrt{\left(\frac{\sigma_x + \sigma_y}{2}\right)^2 + \tau^2}$$

and tan $2\alpha = 2\tau/(\sigma_x - \sigma_y)$, where σ is measured from the x-axis. Therefore, the principal stresses are

$$\sigma_1 = \tau \text{ and } \sigma_2 = -\tau$$

acting along directions $\alpha = \pi/4$ and $\alpha = 3\pi/4$, respectively. Due to the first principal stress, $\sigma_1 = \tau$, the hoop stresses σ_{θ_1} is obtained from Equations 6.67 as

$$\sigma_{\theta 1} = \frac{\tau}{2}\left[\left(1 + \frac{r_i^2}{r^2}\right) - \left(1 + \frac{3r_i^4}{r^4}\right)\cos 2\theta\right]$$

Since the maximum stress σ_θ occurs at the hole surface, i.e., at $r = r_i$,

$$\sigma_{\theta 1} = \tau(1 - 2 \cos 2\theta)$$

Similarly, due to the second principal stress $\sigma_1 = -\tau$, the hoopstress σ_{θ_2} at the hole surface $r = r_i$, is given by

$$\sigma_{\theta 2} = -\frac{\tau}{2}[2 - 4\cos 2(\pi/2 - \theta)]$$

or

$$\sigma_{\theta 2} = -\tau[1 + 2 \cos 2\theta]$$

By superposition,

$$\sigma_\theta = \sigma_{\theta 1} + \sigma_{\theta 2} = -4\tau \cos 2\theta$$

This attains a maximum value of -4τ and 4τ at $\theta = 0$ and $\theta = \pi/2$, respectively. Therefore, the stress concentration factor is

$$k_t = \sigma_{\theta max}/\tau = 4$$

noting that θ is measured from a direction inclined 45° to the axis.

PROBLEMS

Pressurized Thick-Walled Cylinders

6.1 A steel tank having an internal diameter of 1200 mm is subjected to an internal pressure of 10 MPa. The tensile and compressive yield stresses of the material are 400 MPa. Assuming a factor of safety 2, determine the wall thickness using the maximum shear criterion.

6.2 Using the appropriate stress function, derive expressions for the stress and displacement in a thick-walled cylinder subjected to internal pressure for a case of free ends. Determine the expansions at the inner and outer diameters for a steel thick-walled cylinder of $r_i = 100$ mm, subjected to an internal pressure of 120 MPa. Take $\lambda = r_o/r_i = 2$, $E = 210$, and $\nu = 0.3$.

6.3 A long, thick-walled cylinder is made of aluminium alloy ($E = 72$ GPa, $\nu = 0.33$) has an inside diameter of 200 mm and an outside diameter of 800 mm. The cylinder is subjected to an internal pressure of 150 MPa. Assume plane strain conditions.

a. Determine the principal stresses and the maximum shear stress both at the inner radius.
b. Determine the changes in the inner and outer diameters.

6.4 Find the ratio of the thickness to the internal diameter for a thick-walled cylinder subjected to internal pressure when the pressure is equal in magnitude to 0.6 of the maximum circumferential stress. If the internal diameter of the cylinder is 100 mm, determine the increase in the external diameter when the internal pressure is 80 MPa and the cylinder is prevented from changing its length. ($E = 200$ GPa and $\nu = 0.3$.)

6.5 The reading of a strain gauge attached longitudinally to the outer surface of a thick-walled cylinder is 8×10^{-5}. The cylinder is to be subjected to an internal pressure of 30 MPa and a compressive axial stress. Determine the largest value of this axial load. The cylinder has 100 and 50 mm outside and inside diameters, respectively. Take $E = 210$ GPa, $\nu = 0.33$, and $Y = 400$ MPa.

6.6 A thick-walled cylinder is used as a piston in a hydraulic valve. The cylinder is made of stainless steel ($E = 190$ GPa, $\nu = 0.3$) and is subjected to an external pressure of 40 MPa. If the maximum shear stress in the cylinder wall must not exceed 60 MPa, determine the wall thickness knowing that the cylinder outside diameter is 100 mm.

Cylinder Compounding by Shrink Fits

6.7 A compound cylinder is built up of two steel cylinders shrunk on together. Initial diameters of the inner and outer cylinders are (200, 400.3) and (400, 800) mm, respectively. Assuming plane stress conditions and taking $E = 200$ GPa and $\nu = 0.25$,

a. Determine the initial stresses;
b. Determine the resulting stresses when subjected to an internal pressure of 200 MPa;

c. Compare the stresses determined in (b) with those of a cylinder having inner and outer diameters 200 and 800 mm, respectively, and subjected to the same pressure; and
d. Show that a state of equal maximum shear exists at the inside and common radii of the compound cylinder.

6.8 The external diameter of a steel hub is 250 mm and the internal diameter increases 0.125 mm when shrunk onto a solid steel shaft of 75 mm diameter. Determine (a) the reduction in the shaft diameter, (b) the radial pressure between the hub and the shaft, and (c) the circumferential stress at the inner surface of the hub (E = 200 GPa and ν = 0.3).

6.9 A solid steel shaft of 200 mm diameter has a steel cylinder of 400 mm diameter shrunk onto it. The inside diameter of the cylinder prior to the shrink fit operation was 199.8 mm. Take E = 200 GPa and ν = 0.3),

a. Determine the radial pressure on the surface of contact due to shrink fit,
b. Determine the external pressure p_o required on the outside of the cylinder to reduce to zero the circumferential stress at the inner surface of the cylinder.
c. Determine the change in the outside diameter of the cylinder.

6.10 Determine the following for a compound three-piece cylinder:

a. Stresses due to the shrink fits with diametral interferences Δ_1 = 0.06 mm and Δ_2 = 0.12 mm.
b. Tangential stresses due to an internal pressure p_i = 240 MPa. Given that r_1 = 80 mm, r_2 = 100 mm, r_3 = 140 mm, and r_4 = 200 mm, where r_1, r_2, and r_3 are the inner radii of the three cylinders, respectively, and r_4 is the outside radius of the third cylinder. Take E = 220 GPa and ν = 0.3 for the three cylinders.

6.11 Two thick-walled cylinders have nominal diameters of 50 and 75 mm for the first cylinder and of 75 and 100 mm for the second one. If the second is shrunk on the first cylinder and the assembly is then used to transmit torque by means of the frictional force acting on their common surface, assuming plane stress conditions, determine

a. The torque that can be transmitted by the assembly, given the following: radial interference = 0.0375 mm, length of the assembly = 50 mm, coefficient of friction = 0.15, E = 200 GPa, and ν = 0.25;
b. The maximum stresses induced in the cylinders due to compounding only; make a sketch showing the stress distribution.

6.12 A solid steel bar of 50 mm diameter is pressed into a steel sleeve of 100 mm outer diameter so that, when assembled, the magnitude of the radial stress between the two is 15 MPa and that of the circumferential stress at the inside of the sleeve is 500 MPa. Assuming a close fit and neglecting friction, determine these stresses when the bar is subjected to an axial compressive load of 175 kN (take ν = 0.33).

6.13 A steel bar is shrink fitted in a steel plate. What force should be applied to the bar in the axial direction in order to pull it out of the plate? The radial interference is 0.03 mm, the diameter of the bar is 60 mm, the thickness of the plate is 100 mm, the coefficient of friction between the plate and the bar is 0.25. For steel, E = 200 GPa.

6.14 A steel ring 60 mm wide, 40 mm inner and 70 mm outer radius, respectively, is shrunk on to a solid steel shaft. If the coefficient of friction between the two surfaces is 0.2 and the strain produced at the outer radius of the assembly is 1.45×10^{-4}, determine the force required to push the shaft out of the ring. Determine also the maximum circumferential stress in the ring (take E = 210 GPa and ν = 0.3).

Rotating Disks, Drums, and Rotors

6.15 a. Show that for an annular rotating disk, the ratio of the maximum tangential stress to the maximum radial stress is given by

$$\frac{(\sigma_\theta)_{max}}{(\sigma_r)_{max}} = \left(2r_o^2 + \frac{1-\nu}{3+\nu} r_i^2\right) \Big/ \left(r_o - r_i^2\right)$$

b. Determine this ratio for a solid disk of diameter $2r_o$ rotating with tangential velocity V.

6.16 Calculate in terms of ρ and ν the maximum safe speed of a flat circular disk 400 mm in diameter with a hole 50 mm in diameter at the center, the thickness of the disk being 30 mm. The allowable strength of the material in simple tension is 150 MPa. Calculate the safe speed of the same disk if there is no hole at the center. Take $\nu = 0.3$.

6.17 A turbine rotor is made from forged steel and has the dimensions 650 and 250 mm outside and inside diameters respectively. Its length is 2000 mm. If the rotor operates at 6000 rpm, calculate the stresses given that $E = 200$ GPa, $\nu = 0.3$, and the weight per unit volume of steel is 78 kN/m^3.

6.18 A thin, solid disk of radius $r_o = 600$ mm is rotating about its axis with a uniform angular velocity. If the radial displacement of the outer rim of the disk is not to exceed 0.2 mm, determine its maximum rotating speed and the corresponding maximum stress. The material properties of the disk are $E = 200$ GPa, $\nu = {}^1/_3$, and its weight per unit volume is 78 kN/m^3.

6.19 A designer has the option to select either steel ($E = 200$ GPa, $\nu = 0.3$, $Y = 700$ MPa, $\rho = 7800$ kg/m^3) or aluminum alloy ($E = 70$ GPa, $\nu = 0.3$, $Y = 300$ MPa, $\rho = 2700$ kg/m^3) in fabricating a solid rotating disk. Determine the ratio between the outer diameters of disks made of these two materials if yielding is not allowed in either one of these two materials due to rotation. What is the weight ratio between the two disks?

6.20 Select a suitable steel from which a disk of uniform strength (de Laval disk) could be made to rotate safely at 5000 rpm. For reasons of design, the diameter of the disk is to be 1000 mm and the ratio of its tip to hub thicknesses is 0.5 ($\rho = 7800$ kg/m^3).

6.21 A rotating steel disk of 500 mm outside diameter and 100 mm inside diameter is shrunk on a steel shaft so that the pressure between the shaft and disk at standstill is 40 MPa. Assuming that the shaft does not change its dimensions because of its own centrifugal force, find the speed at which the disk gets loose on the shaft. Solve the problem without making this assumption.

6.22 A thin disk of a diameter ratio $\lambda = 4$ is shrunk on a solid shaft. The radial interference when the disk assembly is at rest is Δ. Determine the angular velocity of the assembly at which the disk is just about to get loose from the shaft. The disk and shaft materials have the same modulus of elasticity E and $\nu = {}^1/_3$.

6.23 A thin annular disk is made up of two annuli shrunk together with a common radius r_c. The diameter ratios of the inner and outer annuli are $\lambda_1 = 2$ and $\lambda_1 = 3$, respectively. If the outer annulus is not to get loose from the inner one when the disk is rotating with a uniform angular velocity ω, determine the value of the radial interference, when the disk is at rest. The two annuli have the same elastic constants.

6.24 A disk of thickness h and outside radius r_o is shrunk onto a shaft of radius r_i, producing a radial interface pressure p_c in the nonrotating condition. It is then rotated with an angular velocity ω rad/s. If μ is the coefficient of friction between disk and shaft and ω_o is the value of the angular velocity at which the interface pressure p_c falls to zero, show that the maximum power is transmitted when $\omega = \omega_o / \sqrt{3}$. Then determine this maximum power in terms of μ and ω_o.

6.25 A thin, annular disk of uniform thickness and internal radius r_i = 150 mm and external radius r_o = 600 mm is rotating about its axis with a uniform angular velocity ω rad/s. In designing the disk, both rigidity and strength requirements have to be satisfied. The rigidity requires that the radial displacement u, at the outer rim be limited to 0.1 mm. The strength requirements necessitate that the maximum shear stress be equal to 0.4, the yield tensile strength, Y. Calculate the maximum rotational speed at which this disk should rotate without violating the rigidity and the strength requirements together. Take E = 200 kN/m³, Y = 600 MPa, ν = 0.3, and specific weight = 78 kN/m³.

6.26 A flat, steel turbine disk of 750 mm outside diameter and 150 mm inside diameter is shrunk on a shaft of the same steel, such that the assembly rotates at 3000 rpm. If the maximum shear stress in the disk should not exceed 0.5Y, where Y is the yield strength, (a) find the maximum radial interference which can be used in the assembly and the contact pressure. Take Y = 300 MPa, ν = 0.3, E = 210 GPa, specific weight of steel 78 kN/m. (b) Calculate the stresses in the shaft material at rest.

6.27 A thin, annular turbine disk is rotating at a uniform angular speed while being subjected to a uniform external pressure p_o. If the diameter ratio of the disk is r_o/r_i = 5\3, ν = 0.3 and E = 200 GPa,

a. Determine the value of p_o such that there will be no increase in the outer diameter of the disk while rotating; and
b. Sketch the stresses resulting from both rotation and pressure p_o indicating at which radius the radial stress σ_r is zero.

6.28 A solid steel shaft of 200 mm diameter has a steel cylinder of 400 mm diameter shrunk onto it. The inside diameter of the cylinder before the shrink fit operation was 199.8 mm. Assuming plane stress conditions and taking E = 200 GPa, ν = 0.3;

a. Determine the radial pressure on the surface of contact due to shrink fit;
b. Determine the external pressure p_o on the outside of the cylinder that is required to reduce to zero the circumferential stress at the inner surface of the cylinder;
c. Find the speed of rotation to loosen the fit (weight per unit volume of steel = 78 kN/m³), while p_o is applied; and
d. Determine if the speed of rotation obtained in (c) causes yielding at the outer radius of the cylinder. Use maximum shear stress criterion, taking Y = 400 MPa.

Thermal Stresses in Cylinders and Disks

6.29 The temperature distribution in a long cylindrical electrical conductor due to the passage of current is given by

$$T = C\left(r_o^2 - r^2\right)$$

where C is a constant. Show that the stresses due to this thermal gradient are

$$\sigma_r = -\frac{E\alpha C}{4(1-\nu)}\left(r_o^2 - r^2\right)$$

$$\sigma_\theta = -\frac{E\alpha C}{4(1-\nu)}\left(3r^2 - r_o^2\right)$$

$$\sigma_z = -\frac{E\alpha C}{2(1-\nu)}\left(2r^2 - r_o^2\right)$$

6.30 A thin, uniform disk of radius r_o is surrounded by a heavy ring of the same material. The assembly just fits when the disk and the ring are at uniform temperature. The center of the disk is kept at temperature T_i and the circumference is kept at temperature T_o. The temperature variation along r from the center is given by

$$T = (T_i - T_o) - (T_i - T_o)(r^2/r_o^2)$$

The heavy ring is at temperature T_o and is assumed to be rigid. Show that the radial compression stress in the disk at radius r is

$$\sigma_r = \frac{1}{4}\alpha E(T_i - T_o)\left(\frac{3 - \nu}{1 - \nu} - \frac{r^2}{r_o^2}\right)$$

6.31 A long, thick-walled cylinder having diameters 24 and 36 mm is used in a heat exchanger to heat a fluid passing through the cylinder at pressure of 60 MPa and temperature of 90°C. The cylinder is freely mounted at both ends. When steady conditions have been reached, the rate of heat transfer measured at the outside surface of the cylinder is 50 kW/m^2. Determine the maximum stresses in the cylinder knowing that the thermal conductivity of the cylinder material is 30 W/°C m, E = 200 GPa, v = 0.3, and $\alpha = 12 \times 10^{-6}$/°C.

6.32 A compound cylinder is made of an inner copper cylinder of radii 100 and 200 mm and an outer steel cylinder of external radius 400 mm. If the temperature of the assembly is raised by 100°C, determine the radial and tangential stresses at the inner and outer radii of each cylinder. Assume plane stress conditions and take $\alpha_{cu} = 16.5 \times 10^{-6}$/°C, $\alpha_{st} = 12.5 \times 10^{-6}$/°C; E_{cu} = 100 GPa, E_{st} = 200 GPa, ν_{cu} = 0.34, and ν_{st} = 0.3

6.33 A hollow bronze bushing with 20 mm inside and 40 mm outside diameters is assembled by interference fit inside a steel cylinder of 100 mm outside diameter. Calculate the required initial interference value if, on heating the cylinder to a temperature of 100°C, the greatest compressive strain in the bronze component is to be below 5×10^{-4}. Take E_s = 200 GPa, E_b = 95 GPa, ν_s = 0.33, $\alpha_s = 12 \times 10^{-6}$/°C, and $\alpha_b = 17 \times 10^{-6}$/°C where the subscripts s and b refer to steel and bronze, respectively.

Bending of Curved Beams

6.34 Determine the stresses for the beam shown in part a of the figure having the shape of a circular ring and subjected at its ends to a bending moment M. Use the stress function,

$$\phi = A \ln r + Br^2 \ln r + Cr^2 + D$$

and then apply the appropriate boundary conditions to determine the constants and hence the stresses. Use the results to determine the maximum and minimum stresses and the tangential displacement along the horizontal axis in the curved beam shown in part b of the same figure for $r_o/r_i = 5$ and $v = 1/2$.

6.35 A curved beam has a mean radius of curvature of 100 mm and a cross section of 75 × 75 mm. If a bending moment of 6 kN · m is applied to the bar in order to reduce its curvature, determine the maximum and minimum circumferential stresses. Assuming that the angle subtended at the origin is 60°, and knowing that the solution is only valid at a certain distance from the ends of the beam, compare the solution with that of the elementary strength-of-materials solution for design purposes.

6.36 Consider the annular segment of unit thickness shown in the figure having an inside radius r_i and an outside radius r_o. Using the stress function,

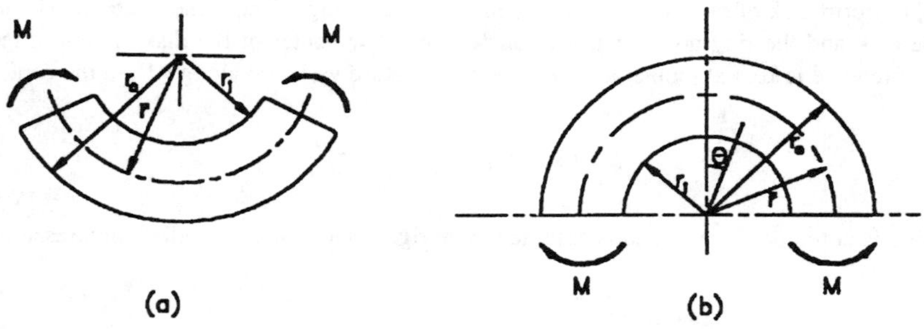

Problem 6.34

$$\phi = \left(Ar^3 + \frac{B}{r} + Dr \ln r \right) \sin \theta$$

determine the constants A, B, and D to give a resultant force P acting on each end of the segment while the curved surfaces are stress free. Hence, show that the maximum σ_θ at point $(r_i, \pi/2)$ is given by

$$\sigma_\theta = -2 \frac{P}{r_i} \frac{r_o^2/r_i^2 - 1}{\left(r_o^2/r_i^2 + 1\right) \ln\left(r_o/r_i\right) - \left(r_o^2/r_i^2 - 1\right)}$$

Expand this expression to obtain

$$\sigma_\theta = -\frac{6P\left(r_o + r_i\right)}{\left(r_o - r_i\right)^2} \left[1 + \frac{\left(r_o - r_i\right)}{\left(r_o + r_i\right)} + \cdots \right]$$

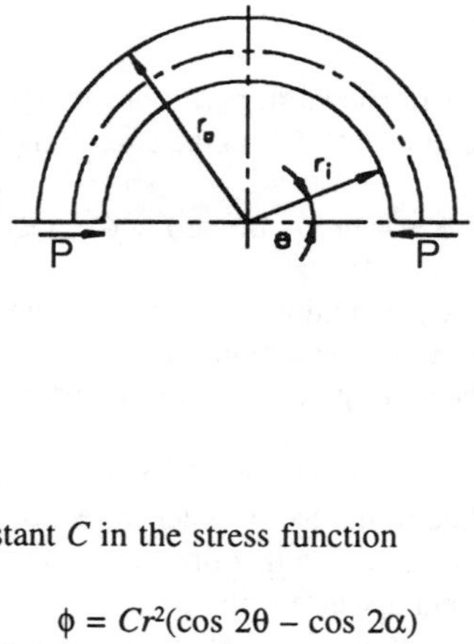

Problem 6.36

Wedges Subjected to Loads

6.37 Determine the value of the constant C in the stress function

$$\phi = Cr^2(\cos 2\theta - \cos 2\alpha)$$

required to satisfy the conditions:

$$\sigma_\theta = 0 \quad \tau_{r\theta} = q \quad \text{on } \theta = \alpha$$

$$\sigma_\theta = 0 \quad \tau_{r\theta} = -q \quad \text{on } \theta = -\alpha$$

corresponding to a uniform shear loading on each edge of a wedge, directed away from the vertex. Verify that no concentrated force or couple acts on the vertex.

6.38 Find the constant in the stress function:

$$\phi = Cr^2\,[2(\pi - \theta) + \sin 2\theta]$$

that satisfies the conditions of uniform loading q on a semi-infinite plate given by:

$$\sigma_\theta = -q \quad \tau_{r\theta} = 0 \quad \text{on } \theta = o$$

$$\sigma_\theta = 0 \quad \tau_{r\theta} = 0 \quad \text{on } \theta = -\pi$$

C being a positive constant. Also determine the stresses.

6.39 Consider the infinite wedge of the figure subjected to a uniform normal loading q distributed along one edge. Using a stress function,

$$\phi = r^2(C_1 \cos 2\theta + C_2 \sin 2\theta + C_3\theta + C_4)$$

determine the stress distribution in the wedge so that all boundary conditions are satisfied. The lower surface is load free.

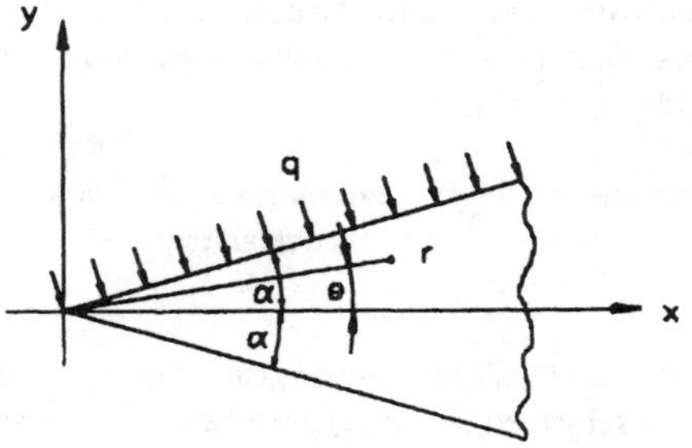

Problem 6.39

6.40 A triangular cantilever beam of unit thickness is shown in the figure. The beam carries a uniform loading q per unit length. Neglecting the body forces and using the stress function,

$$\phi = Cr^2[2(\alpha - \theta) + \sin 2\theta - 2 \tan \alpha \cos^2 \theta]$$

a. Write down the boundary conditions along the surfaces AB and BC,
b. Show that the boundary conditions are satisfied;
c. Find the constant C in terms of q and α;
d. If $\alpha = \pi/6$, plot the stress distribution along the section B–C; and
e. Compare the stress values obtained at point A with those given by simple beam theory ($\sigma = MY/i$).

Stress Concentration around Small Holes

6.41 A steel plate of uniform width 200 mm is subjected to a uniform tensile stress of 50 MPa. A hole 30 mm in diameter is drilled in the plate at its center. Plot the stress distribution for the following cross sections:

a. Cross section passing through the hole center;
b. Cross section at a distance 15 mm from the center.

Determine the cross section at which the disturbance caused by the hole almost disappears.

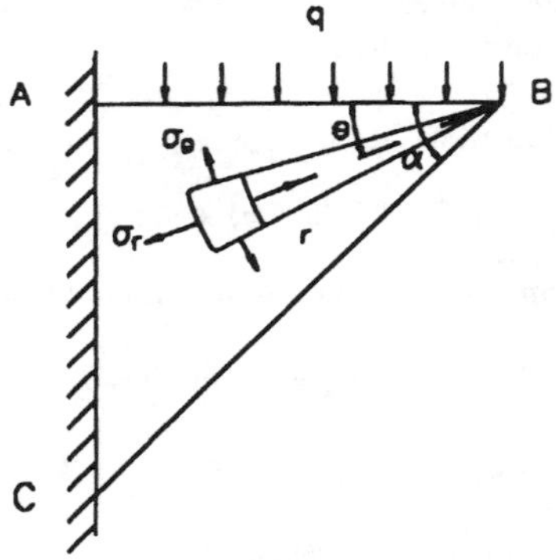

Problem 6.40

6.42 Find the stress concentration factor k_t for a large, thin plate with a small circular hole in the middle. Consider two cases of loading:

a. Biaxial tension $\sigma_x = 2\sigma$ and $\sigma_y = 2\sigma$;
b. Tension in the x-direction and compression in the y-direction, such that $\sigma_x = -\sigma_y = 2\sigma$.

6.43 An infinitely large thin plate with a small circular hole is subjected to the stresses $\sigma_x = \sigma$, $\sigma_y = -\sigma/2$, and $\tau_{xy} = -\sigma/2$ applied at its edges. Determine the maximum stress at the inner radius r_i of the hole and its angular location. Apply the superposition principle.

6.44 A thin-walled cylindrical pressure vessel of mean radius r_m and wall thickness h is subjected to internal pressure p_i. A small hole is drilled in the vessel wall. Derive an expression for the maximum stress at the hole boundary.

6.45 A gas pipe of mean diameter r_m and wall thickness h, such that $r_m/h = 10$, is subjected to a gas pressure p_i. Due to supports, the pipe is also subjected to a transverse bending moment M_b and a twisting moment $M_t = M_{b/4}$. Determine the resulting maximum tensile stress and the stress concentration around a small hole in the pipe wall.

REFERENCES

1. Saada, A. S., *Elasticity: Theory and Applications*, Pergamon Press, New York, 1974.
2. Timoshenko, S. and Goodier, J. N., *Theory of Elasticity*, 2nd ed., McGraw-Hill, New York, 1951, p. 66, 76.
3. Srinath, L. S., *Advanced Mechanics of Solids*, Tata-McGraw-Hill, New Delhi, 1980.
4. Peterson, R. E., *Stress Concentration Factors*, John Wiley & Sons, New York, 1974.

7 Elastic Rods Subjected to General Loading

Solutions to three-dimensional elastic problems of rods subjected to general loading are obtained through the formulation of a strain compatibility function and also by applying energy methods. These problems comprise normal and shear stresses in beams, free and constrained torsions of bars, displacements in rods under general loading, beams on elastic foundation, and buckling of columns and beams. Curvilinear coordinates are chosen for the solution of shear stresses in thin-walled cross sections. An energy approach is applied to obtain the displacements under general loading and also to determine buckling loads.

7.1 INTRODUCTION

A rod is a long solid, generated by moving a plane cross section along the normal line passing through the centroid, called the rod axis. Rods may have a constant or a variable cross section; the axis may be straight or curved. A straight rod having a constant cross section is usually called a *bar*.

Rods are used as structural elements given various names according to the configuration of loading. For example a *tie* is a bar subjected to axial tension, a *strut* or *column* is a bar subjected to end compression and a *beam* is a rod acted upon by transverse loading.

The problems of elastic rods are generally three-dimensional problems. So far the problems considered in the two preceding Chapters 5 and 6 are plane elastic problems, in which the stress and strain components are functions of only two coordinates (x, y) in Cartesian coordinates, or (r, θ) in polar coordinates. For these problems a general approach has been adopted to obtain a solution, namely, finding a compatible stress function that satisfies boundary conditions. Such a methodical approach is not possible to apply for solving the general three-dimensional elastic problem of rods.

In this chapter solutions to some three-dimensional elastic problems of rods are obtained. For these problems a compatible deformation pattern is postulated from theoretical and/or practical considerations. For instance, it has been shown in Section 5.2.4 that for a prismatical bar in bending, warping of the cross section occurs in the vicinity of applied loads while for the rest of bar length warping is negligible. Therefore, it may be considered that plane cross sections remain plane.

7.2 STRESS RESULTANTS

Consider a long rod having a straight axis and a cross section that is either constant or varying gradually with no abrupt changes. The rod is acted upon by a system of forces in equilibrium with some constraints on displacement. By referring to Figure 7.1, Cartesian coordinates are chosen by taking the rod axis to be the z-axis passing through the centroids of all cross sections. The x- and y-axes, which lie in any cross-sectional plane, are chosen to be the centroidal principal axes of the cross-sectional area A. Therefore, the first moments of area about x and y-respectively, and I_{xy} the product second moment of area, all vanish according to

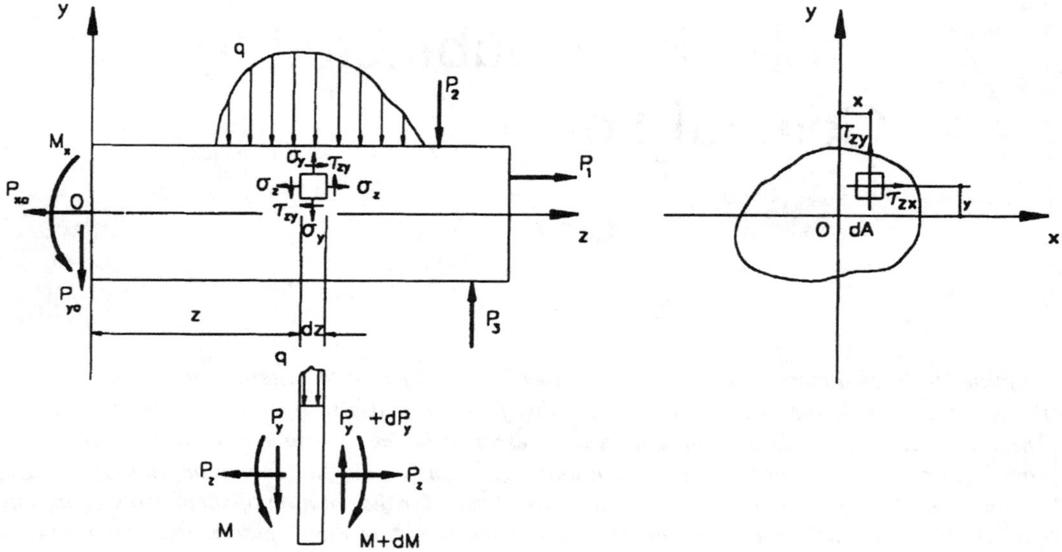

FIGURE 7.1 Static equilibrium of a bar subjected to general loading.

$$S_x^* = \int_A x dA = 0, \quad S_y^* = \int_A y dA = 0, \quad \text{and} \quad I_{xy} = \int_A xy dA = 0 \tag{7.1}$$

Except at the vicinity of load application, the rod surface is free from stress, so that the boundary conditions on the surface are

$$\sigma_x = \sigma_y = \tau_{xy} = 0 \tag{7.2}$$

The rod cross section dimensions are usually small compared with the rod length, and therefore it is assumed that the stress components σ_x, σ_y, and τ_{xy} vanish also inside the rod. The remaining stress components are σ_z, τ_{zx}, and τ_{zy}, which constitute the components of a stress vector S_z acting on the rod cross-sectional area.

By referring to Figure 7.1 and considering a cross section at a distance z, the stress resultants over the cross-sectional area are given by

$$\begin{aligned} P_z &= \int_A \sigma_z dA, \quad P_x = \int_A \tau_{zx} dA, \quad P_y = \int_A \tau_{zy} dA \\ M_x &= \int_A \sigma_z y dA, \quad M_y = \int_A \sigma_z x dA, \quad M_z = \int_A \left(\tau_{zx} y - \tau_{zy} x\right) dA \end{aligned} \tag{7.3}$$

where P_x, P_y, and P_z are force components acting on the cross section through the centroid O in the x, y, and z directions, respectively; P_z is called a normal or axial force, while P_x and P_y are called shearing forces. M_x, M_y, and M_z are moments about the x-, y-, and z-axes, respectively; M_x and M_y are called bending moments, while M_z is a twisting moment. The normal stress σ_z results from the normal force P_z and moments M_x and M_y; the shear stress τ_{zx} results from P_x and M_z; and τ_{zy} results from P_y and M_z. The distribution of the stress resultants are usually plotted along the bar axis as diagrams known as normal force, shearing force, bending moment and twisting moment, diagrams, respectively.

From the equilibrium of an element of bar length dz, it can be easily shown from Figure 7.1 that:

$$\frac{\partial M_x}{\partial z} = P_y \text{ and, similarly, } \frac{\partial M_y}{\partial z} = P_x \tag{7.4a}$$

which means that the rate of change of the bending moment along the beam is equal to the shearing force. If the surface of the rod is subjected to distributed external loading of q_x and q_y per unit length of rod axis, P_x and P_y will vary along z, such that

$$\frac{\partial P_x}{\partial z} = q_x \text{ and } \frac{\partial P_y}{\partial z} = q_y \tag{7.4b}$$

Note that Equation 7.4a may be obtained from the stress equation of equilibrium 1.6 as shown in Example 7.1.

7.2.1 Note on Sign Convention for Stress Resultants

The sign of the stress resultants P_x, P_y, P_z, M_x, M_y, and M_z follows the same sign of the corresponding stresses as indicated in Section 1.5. Therefore, in Expressions 7.3, the following sign convention is adopted.

- P_z is positive when producing tensile stress and negative for compressive stress
- P_x, P_y, and M_z are positive when producing positive shear stresses and negative for negative shear stresses
- M_x and M_y are positive when producing tensile σ_z at positive values of y and x, respectively; otherwise M_x and M_y are negative

The above sign convention is consistent with that adopted in the derivation of the equations of equilibrium and compatibility in Chapters 1 and 2. Other sign conventions are arbitrarily set in texts of strength of materials which may be different from the sign convention adopted in the present text.

Example 7.1:

For a bar having its cross section subjected to bending moments M_x, M_y *and shearing forces* P_x, P_{xy} *in the* x- *and* y-*directions, respectively, obtain the relations:*

$$\frac{\partial M_x}{\partial z} = P_y \quad and \quad \frac{\partial M_y}{\partial z} = P_x$$

from the stress equations of equilibrium (Equations 1.6).

Solution:

For a bar, the stresses $\sigma_x = \sigma_y = \tau_{xy} = 0$, as given by Equations 7.2. Hence, the equilibrium Equations 1.6a and b vanish and Equation 1.6c is given by

$$\frac{\partial \tau_{xz}}{\partial x} + \frac{\partial \tau_{yz}}{\partial y} + \frac{\partial \sigma_z}{\partial z} = 0 \tag{1.6-c}$$

Multiplying by y and integrating with respect to the area A yields

$$\int_A \frac{\partial \tau_{xz}}{\partial x} y dA + \int_A \frac{\partial \tau_{yz}}{\partial y} y dA + \int_A \frac{\partial \sigma_z}{\partial z} y dA = 0$$

$$\int_A \frac{\partial(\tau_{xz} y)}{\partial x} dA + \int_A \frac{\partial(\tau_{yz} y)}{\partial y} dA - \int_A \tau_{zy} dA + \int_A \frac{\partial \sigma_z}{\partial z} y dA = 0$$

Since,

$$\int_{y_1}^{y_2}\int_{x_1}^{x_2} \frac{\partial(\tau_{zx} y)}{\partial x} dx dy = \int_{y_1}^{y_2} [\tau_{zx} y]_{x_1}^{x_2} dy = \int_{y_1}^{y_2} y[(\tau_{zx})_2 - (\tau_{zx})_1] dy = \int_{y_1}^{y_2} y(0-0) dy = 0$$

and

$$\int_{x_1}^{x_2}\int_{y_1}^{y_2} \frac{\partial(\tau_{zy} y)}{\partial y} dy dx = \int_{x_1}^{x_2} [\tau_{zy} y]_{y_1}^{y_2} dx = 0$$

hence,

$$\int_A \tau_{zy} dA = \frac{\partial}{\partial z}\int_A \sigma_z y dA$$

which from Equation 7.3 yields

$$P_y = \frac{\partial M_x}{\partial z}$$

The same steps may be followed for P_x by multiplying Equation 1.6c by x and integrating with respect to A. It will be appreciated that taking the equilibrium of a bar element of length dz is a much simpler approach to the problem.

7.3 BENDING OF RODS

7.3.1 Bending Stresses

In simple beam analysis σ_z is the only normal stress component present acting normal to the rod cross section, Equations 7.2, and is related to P_z, M_x, and M_y by Equations 7.3.

Based on the hypothesis that the rod cross section remains plane after deformation,* the strain is linearly distributed over the cross section so that the strain compatibility condition is expressed by the relation:

$$\varepsilon_z = ax + by + c \tag{7.5a}$$

* See Example 5.5 and Section 5.2.4.

where the axes x and y are the centroidal principal axes as stated by Expressions 7.1. Since $\sigma_x = \sigma_y = 0$, then Hooke's law (Equation 3.6) reduces to $\varepsilon_z = \sigma_z/E$ and $\varepsilon_x = \varepsilon_y = -\nu\varepsilon_z$, which when substituted in Equation 7.5a yields

$$\sigma_z = E(ax + by + c) \tag{7.5b}$$

The constants a, b, and c are determined from boundary conditions. By combining Equation 7.3 together with equation (7.1) and writing:

$$\int_A y^2 dA = I_x, \quad \int_A x^2 dA = I_y$$

which are the second moments of area about the x- and y-axes, respectively, the constants a, b, and c are obtained as

$$a = \frac{M_y}{EI_y}, \quad b = \frac{M_x}{EI_x}, \quad c = \frac{P_z}{EA} \tag{7.6}$$

Substituting a, b, and c into Equation 7.5a and Equation 7.5b yield*

$$\varepsilon_z = \frac{M_y}{EI_y}x + \frac{M_x}{EI_x}y + \frac{P_z}{EA}$$
$$\varepsilon_x = \varepsilon_y = -\nu\varepsilon_z \tag{7.7a}$$

$$\sigma_z = \frac{M_y}{I_y}x + \frac{M_x}{I_x}y + \frac{P_z}{A}$$
$$\sigma_z = \sigma_y = 0 \tag{7.7b}$$

Substituting $\varepsilon_z = 0$ in Equation 7.5a yields

$$ax + by + c = 0 \tag{7.7c}$$

which is the equation of a straight line along which there is no axial displacement. This line is called the *neutral axis*; about this line the cross section deformation is a pure rotation.

Let O′x′, Figure 7.2, be the neutral axis which is inclined by a negative angle $-\alpha$ to the x-axis, OO′ being the normal distance between O and O′x′. From elementary analytical geometry.

---

* If x and y are not principal axes, I_{xy} will not vanish and ε_z and σ_z are expressed by

$$\varepsilon_z = \frac{1}{E}\left[\frac{M_y I_x + M_x I_{xy}}{I_y I_x - I_{xy}^2}x + \frac{M_x I_y + M_y I_{xy}}{I_y I_x - I_{xy}^2}y + \frac{P_z}{A}\right]$$

and $\sigma_z = E\varepsilon_z$. However, the use of this expression is inconvenient and referring ε_z and σ_z to the centroidal principal axes is preferable.

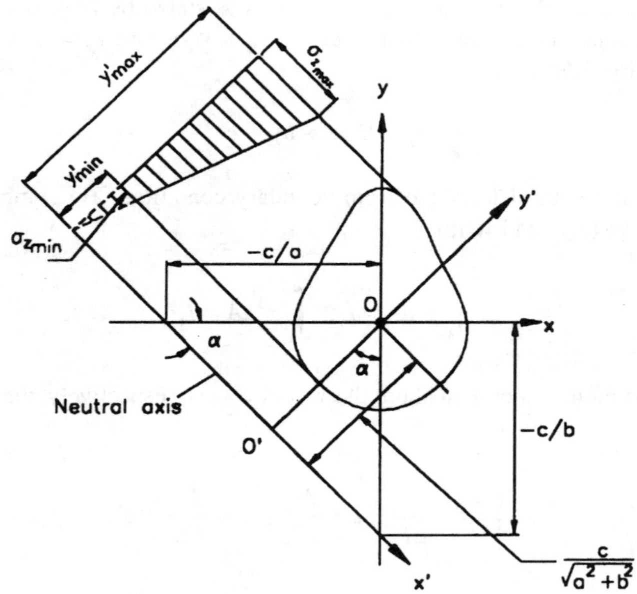

FIGURE 7.2 Normal stresses and neutral axis in a bar subjected to bending and axial force.

$$OO' = \frac{c}{\sqrt{a^2 + b^2}} \tag{7.8a}$$

and

$$\tan\alpha = -\frac{a}{b} \tag{7.8b}$$

Substituting a, b, and c from Equation 7.6 yields

$$OO' = \frac{P_z}{A\sqrt{\left(\frac{M_y}{I_y}\right)^2 + \left(\frac{M_x}{I_x}\right)^2}} \tag{7.8c}$$

and

$$\tan\alpha = -\frac{M_y I_x}{M_x I_y} \tag{7.8d}$$

Note that the resultant of M_x and M_y is inclined by an angle of $\tan^{-1}(M_x/M_y)$ to the x-direction and hence will not be normal to the neutral axis, as expressed by Equation 7.8d, unless $M_x = M_y$. This obliquety is often referred to in strength of materials texts as oblique, skew, or unsymmetrical bending.

By taking x' and y' as reference coordinate axes, with O′ as origin, then ε_z and σ_z can be directly expressed by:

$$\varepsilon_z = \sqrt{\left(\frac{M_y}{EI_y}\right)^2 + \left(\frac{M_x}{EI_x}\right)^2}\, y' \tag{7.9a}$$

and

$$\sigma_z = \sqrt{\left(\frac{M_y}{I_y}\right)^2 + \left(\frac{M_x}{I_x}\right)^2}\, y' \tag{7.9b}$$

From Expressions 7.9 the maximum and minimum values and location of ε_z and σ_z, can be directly obtained at y_{max}' and y_{min}', respectively, Figure 7.2, as shown in Examples 7.2 and 7.3.

Example 7.2:

A tie-rod made of an equal angle cross section, 50 × 50 × 5 mm, is subjected to an axial tensile force of 20 kN acting through O_1 distance 4.8 mm from the rod axis. Determine the maximum tensile and compressive stresses and the neutral axis. Use the following geometric properties for a 50 × 50 × 5 angle: cross-sectional area = 480 mm², $x_o = y_o = 14$ mm, $I_x = 1.74 \times 10^5$ mm⁴, $I_y = 459 \times 10^4$ mm⁴, where x_o and y_o are the centroid distances from each leg, respectively, and x, y are principal centroidal axes as shown in Figure 7.3.

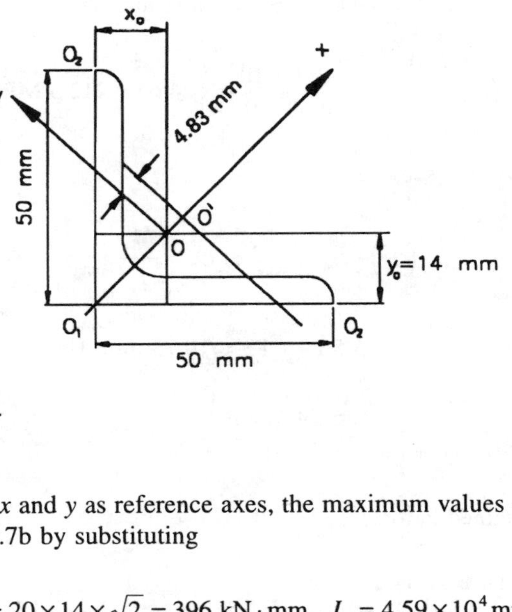

FIGURE 7.3 Example 7.2.

Solution:

Taking the principal axes x and y as reference axes, the maximum values of the axial stress σ_z are obtained from Equation 7.7b by substituting

$$M_y = 20 \times 14 \times \sqrt{2} = 396 \text{ kN} \cdot \text{mm}, \quad I_y = 4.59 \times 10^4 \text{mm}^4,$$

$$M_x = 0, \quad P_z = 20 \text{ kN}, \quad \text{and } A = 480 \text{ mm}^2$$

The maximum tensile stress occurs at O_1 at a distance $x = 14 \times \sqrt{2} = 19.8$ mm, while the maximum compressive stress occurs at O_2 at a distance $x = (50/\sqrt{2}) - 19.8 = -15.56$ mm, which yields

$$(\sigma_z)_{\text{max, tension}} \text{ at } O_1 = \frac{(396 \times 10^3) \times 19.8}{4.59 \times 10^4} + \frac{20 \times 10^3}{480} = 212.5 \text{ MPa}$$

$$(\sigma_z)_{\text{max, compression}} \text{ at } O_2 = -\frac{(396 \times 10^3) \times 15.56}{4.59 \times 10^4} + \frac{20 \times 10^3}{480} = -92.6 \text{ MPa}$$

The neutral axis is obtained from Equation 7.7c by substituting (as given by Expressions 7.6):

$$a = \frac{396 \times 10^3}{4.59 \times 10^4 \times E}, \quad b = 0, \quad \text{and} \quad c = \frac{20 \times 10^3}{480 \times E}$$

This gives a line parallel to the y-axis at a distance given by Equation 7.8a as

$$OO' = \frac{C}{a} = \frac{20 \times 10^3 \times 4.59 \times 10^4}{480 \times 396 \times 10^3} = 4.83 \text{ mm}$$

Alternatively, Equations 7.9 may be used directly to give the stresses as

$$(\sigma_z)_{\text{max, tension}} \text{ at } O_1 = \frac{M_y}{I_y} x'_{\text{max}}; \quad x'_{\text{max}} = O'O_1 = OO_1 + OO'$$

where $x'_{\text{max}} = O'O_1 = OO_1 + OO' = 19.8 + 4.83 = 24.63$ mm. Hence,

$$(\sigma_z)_{\text{max, tension}} = \frac{396 \times 10^3}{4.59 \times 10^4} \times 24.63 = 212.5 \text{ MPa}$$

Similarly,

$$(\sigma_z)_{\text{min, compression}} \text{ at } O_2 = \frac{M_y}{I_y} x'_{\text{min}}$$

where $x'_{\text{min}} = O'O_2 = O'O_2 - OO' = 15.56 - 4.83 = 10.73$ mm. Hence,

$$(\sigma_z)_{\text{min, compression}} = -\frac{396 \times 10^3}{4.59 \times 10^4} \times 10.73 = -92.6 \text{ MPa}$$

which is the same as obtained before.

7.3.2 Elastic Curve in Bending

In the above analysis, the rod axis is considered to be initially straight. However, as soon as M_x and M_y are applied, the axis bends into a curve, which is referred to as the elastic curve.

It is instructive to derive from elementary equilibrium considerations the equation of the elastic curve in bending, which will be referred to in forthcoming sections dealing with displacements, buckling, and beams on an elastic foundation.

The rod axis will be laterally displaced in the x- and y-directions as a result of application of M_y and M_x, respectively. The displacement in the z-direction due to the application of P_z is usually negligible.

At any point (0, 0, z) on the rod axis, the following notations are used:

u = displacement (deflection) in the x-direction,
v = displacement (deflection) in the y-direction,
R_x = radius of curvature of rod axis in the x–z plane,
R_y = radius of curvature of rod axis in the y–z plane,

θ_x = angle of rotation of the rod cross section about the x-axis,
θ_y = angle of rotation of the rod cross section about the y-axis.

Referring to Figure 7.4, the following relations are obtained. In the x–z plane

$$\varepsilon_z dz = x d\theta_x = x \frac{dz}{R_x}$$

Hence, $\varepsilon_z = x/R_x$.

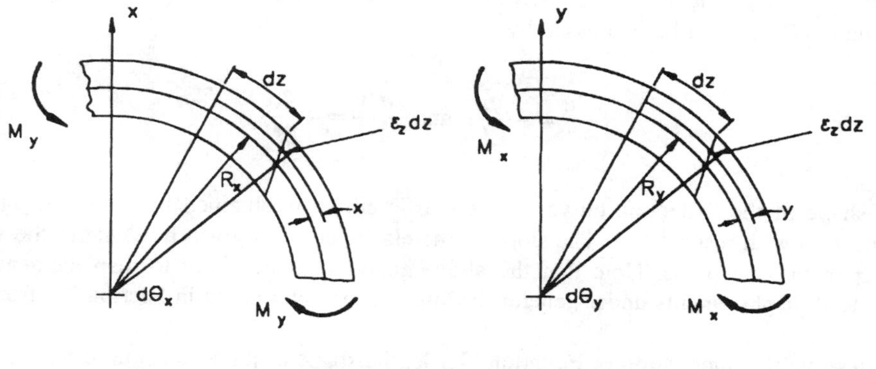

FIGURE 7.4 Rod curvature due to bending.

From elementary calculus, the radius of curvature of any plane curve is given by

$$\frac{1}{R_x} = \frac{-\left(d^2u/dz^2\right)}{\left[1+\left(du/dz\right)^2\right]^{3/2}} \cong -\frac{d^2u}{dz^2} \tag{7.10a}$$

Within the elastic range, beam deflections u and (du/dz) are small. Hence, $(du/dz)^2$ has been neglected. Substituting from the Expression 7.7a for

$$\varepsilon_z = \frac{M_y}{EI_y} x \text{ and } \frac{1}{R_x} = -\frac{d^2u}{dz^2}$$

yields

$$\frac{d^2u}{dz^2} = -\frac{M_y}{EI_y} \tag{7.10b}$$

Similarly in the y–z plane:

$$\frac{d^2v}{dz^2} = -\frac{1}{R_y} = -\frac{M_x}{EI_x} \tag{7.10c}$$

In Equations 7.10 note that the signs of the curvatures $1/R_x$ and $1/R_y$ are positive when corresponding to positive ε_z at positive values of x and y, respectively, as indicated in Figure 7.4, otherwise $1/R_x$ and $1/R_y$ are negative. Note also that for a positive curvature the bar axis will have a negative slope increasing with z.

Substituting Equations 7.10 into Equation 7.4 yields

$$\frac{d^3u}{dz^3} = -\frac{P_x}{EI_y} \quad \text{and} \quad \frac{d^3v}{dz^3} = -\frac{P_y}{EI_x} \tag{7.11a}$$

If P_x and P_y are distributed along z, such that $dP_x/dz = q_x$ and $dP_y/dz = q_y$ according to Equation 7.4b, Equation 7.11a will be expressed by

$$\frac{d^4u}{dz^4} = -\frac{q_x}{EI_y} \quad \text{and} \quad \frac{d^4v}{dz^4} = -\frac{q_y}{EI_x} \tag{7.11b}$$

The shape of the deflection curve, or the elastic curve, is obtained by double integration of Equations 7.10 with respect to z. The slope of the elastic curve is given by du/dz in the x–z plane and dv/dz in the y–z plane. Note that the above analysis applies only to displacements due to bending. Rod displacements under general loading will be determined in Section 7.5 from energy theorems.

To illustrate the application of Equations 7.10c, the shape of the elastic curve for a bar rigidly fixed at one end and subjected to a concentrated lateral load P in the y–z plane, Figure 7.5, is obtained as follows.

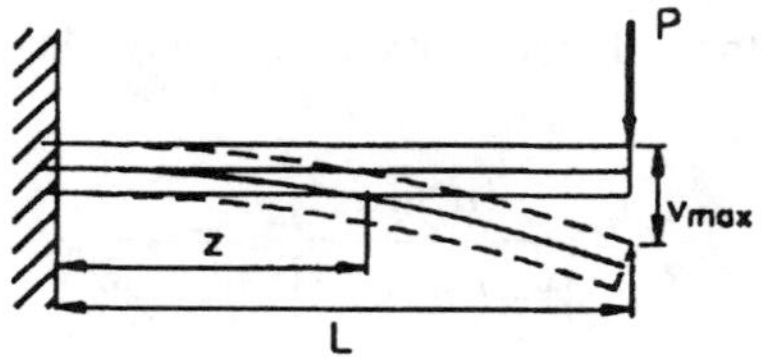

FIGURE 7.5 Elastic curve for a cantilever beam subjected to a concentrated lateral end force.

Substituting $M_x = P(L - z)$ in Equation 7.10b yields

$$\frac{d^2v}{dz^2} = -\frac{P}{EI_x}(L - z)$$

and, by integration, gives

$$\frac{dv}{dz} = -\frac{P}{EI_x}\left(Lz - \frac{z^2}{2} + C_1\right)$$

The boundary conditions are such that

At $z = 0$: $dv/dz = 0$,

hence,

$$C_1 = 0 \text{ and } v = -\frac{P}{EI_x}\left(L\frac{z^2}{2} - \frac{z^3}{6} + C_2\right)$$

while

At $z = 0$: $v = 0$ gives $C_2 = 0$

The shape of the elastic curve is given by the equation:

$$v = -\frac{P}{EI_x}\left(L\frac{z^2}{2} - \frac{z^3}{6}\right)$$

At $z = L$: v is maximum $= -\dfrac{PL^3}{3EI_x}$

also

$$\frac{dv}{dz} \text{ is maximum} = -\frac{PL^3}{2EI_x}$$

Example 7.3:

A cantilever beam of a rectangular cross section is loaded in the two planes x–z *and* y–z *as shown in Figure 7.6a.*

a. *Determine the maximum tensile bending stress.*
b. *Determine the deflection at the free end.*

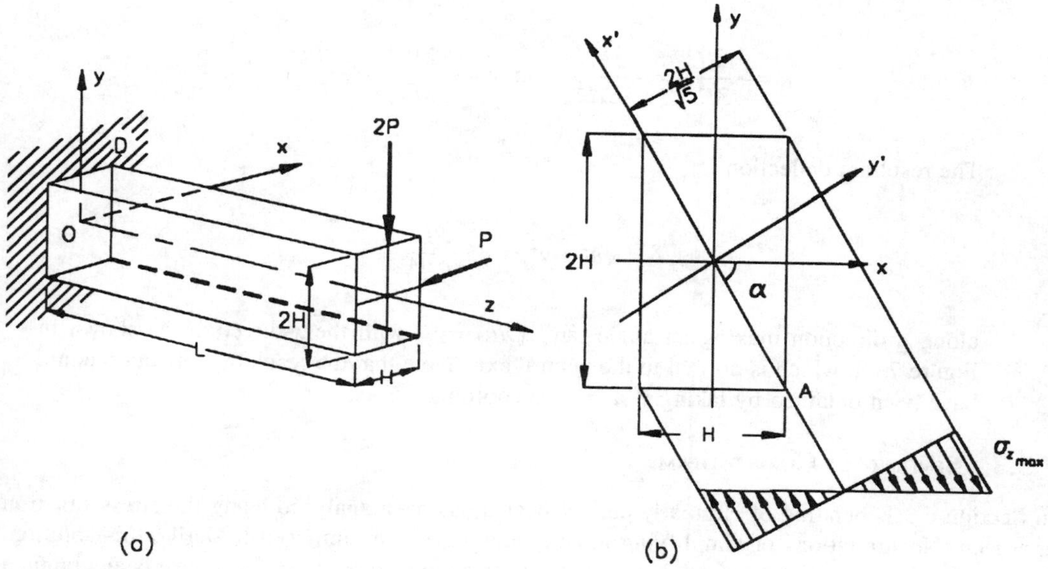

FIGURE 7.6 Example 7.3.

Solution:

a. The loaded planes are coincident with the principal planes. Equation 7.7b may be applied to the built-in end, where the maximum bending moment occurs, hence

$$M_x = 2PL, \ M_y = PL\sigma_z = \frac{M_y}{I_y}x + \frac{M_x}{I_x}y = \frac{PLx}{I_y} + \frac{2PL}{I_x}y$$

By inspection, the maximum tensile stress occurs at the corner D and its magnitude:

$$(\sigma_z)_{\max} = \frac{PL(H/2)}{2H(H)^3/12} + \frac{2PLH}{H(2H)^3/12} = \frac{6PL}{H^3}$$

The same result is obtained by finding the neutral axis using Equations 7.8, indicating that it passes through the centroid as expected and is inclined an angle $\alpha = \tan^{-1} - (M_y I_x / M_x I_y) = \tan^{-1} - 2$ with the x-axis, Figure 7.6b. Equation 7.9b thus, gives the same value for $(\sigma_z)_{\max}$ at $y'_{\max} = 2H/\sqrt{5}$, Figure 7.6c, as

$$(\sigma_z)_{\max} = \sqrt{\left[\frac{PL}{2H(H)^3/12}\right]^2 + \left[\frac{2PL}{H(2H)^3/12}\right]^2} \frac{2H}{\sqrt{5}}$$

or

$$(\sigma_z)_{\max} = \frac{3\sqrt{5}PL}{H^4}\frac{2H}{\sqrt{5}} = \frac{6PL}{H^3}$$

The deflection at the free end is obtained by applying the integration of Equations 7.10 in two perpendicular planes as

$$u = -\frac{PL^3}{3EI_y} = -\frac{2PL^3}{EH^4} \quad \text{and} \quad v = -\frac{2PL^3}{3EI_x} = -\frac{PL^3}{EH^4}$$

The resultant deflection is

$$\delta = \sqrt{u^2 + v^2} = \sqrt{5}\frac{PL^3}{EH^4}$$

along a direction making an angle $\tan^{-1}(v/u) = {}^1/_2$ with the x-direction, as shown in Figure 7.6d, which is normal to the neutral axis. Note that the resultant deflection could have been obtained by taking x' and y' as coordinate axes.

7.3.3 Bending of Curved Beams

In Section 6.3.1, bending of circularly curved beams has been analyzed using the stress function approach. No limitations or simplifying assumptions have been imposed to derive this solution. Expressions for the radial, tangential, and shear stress components (σ_r, σ_θ, $\sigma_{r\theta}$) have been obtained in detail for the case of a beam subjected to an end shear force. Comparison with the elementary strength-of-materials solution (which was quoted there without derivation) indicated that the latter solution gives accurate results for σ_θ in curved beams of small and moderate depths, i.e., $H/\bar{r} \leq 1$, H and $\bar{r}$ being the beam depth and radius of curvature, respectively. It is the objective of this section to derive expressions for the tangential stresses in pure bending of curved beams based on

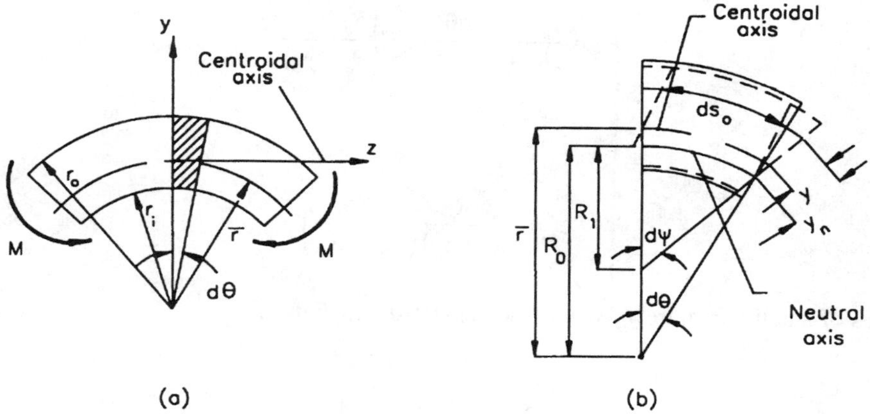

FIGURE 7.7 Bending of a circularly curved beam: (a) load diagram and (b) centroidal and neutral axes.

the simplifying assumption that plane cross sections remain plane after bending. The limits of the validity of this simplified solution will be discussed.

Consider the case of a curved beam with uniform cross section having a circular axis that lies in a centroidal principal plane, as shown in Figure 7.7. The rod is subjected to pure bending in the plane of its axis. Under such loading, compatibility of deformations implies that all plane cross sections remain plane and perpendicular to the beam axis after bending. The displacement of any cross section is thus a rotation about its neutral axis as shown in Figure 7.7b for an element ds_o of the beam length. In Figure 7.7, the following notation is adopted:

R_o = initial radius of curvature for the neutral axis,
R_1 = deformed radius of curvature for the neutral axis,
$\overline{r}$ = radius of curvature at the centroidal axis,
r_i = inner radius of curvature,
r_o = outer radius of curvature,
y_n = distance measured from the neutral axis, and
y = distance measured from the centroidal axis.

The angular rotation between the two faces of the element ds_o along the neutral axis, which remains constant, is given by

$$d\psi - d\theta = \frac{1}{R_1} - \frac{1}{R_o} \tag{7.12}$$

The longitudinal strain in the small element length ds_o located at a distance y_n from the neutral axis is

$$\varepsilon_z = \frac{(R_1 + y_n)d\psi - (R_o + y_n)d\theta}{(R_o + y_n)d\theta}$$

At the neutral surface, there is no change of length, i.e., $R_1 d\psi = R_o d\theta$, hence,

$$\varepsilon_z = \frac{y_n(d\psi - d\theta)}{(R_o + y_n)d\theta} = \frac{y_n(R_o/R_1 - 1)}{(R_o + y_n)}$$

or

$$\varepsilon_z = \frac{y_n R_o}{(R_o + y_n)}\left(\frac{1}{R_1} - \frac{1}{R_o}\right) \tag{7.13a}$$

For beams of small curvature y_n/R_o may be neglected and, hence,

$$\varepsilon_z = y_n\left(\frac{1}{R_1} - \frac{1}{R_o}\right) \tag{7.13b}$$

By using the same sign convention for straight beams, the fibers with positive y undergo tensile strain while those with negative y suffer compression. Consequently, the sign convention for the bending moment is that M is considered positive when it increases the curvature of the beam as shown in Figure 7.7a.

For initially straight beams $1/R_o = 0$ and Equation 7.13b reduces to that obtained previously for a straight beam. The stress for a linearly elastic material is thus obtained from Equation 7.13a as

$$\sigma_z = E\frac{y_n R_o}{(R_o + y_n)}\left(\frac{1}{R_1} - \frac{1}{R_o}\right) \tag{7.14}$$

In the present analysis, the tangential stress σ_z is considered to be the only stress acting in the beam. Note that the radial stress σ_r found by using the stress function approach in Section 6.3 has been shown to become a small fraction of the tangential stress for beams of small curvatures;* say $H/\bar{r} < 0.2$.

For beams of large curvature, the cross-sectional dimensions are of the same order of magnitude as the radius of curvature; hence, y_n cannot be neglected with respect to R_o. The strain and stress given by Equations 7.13a and 7.14, respectively, are thus nonlinearly distributed over the cross section, as shown in Figure 7.8; they are hyperbolic.

At any cross section of the beam, the equilibrium conditions for pure bending are

$$\int_A \sigma_z dA = 0 \tag{7.15}$$

and

$$\int_A \sigma_z y_n dA = M \tag{7.16}$$

By virtue of Equation 7.14, Condition 7.15 becomes

* See Table: 6.2.

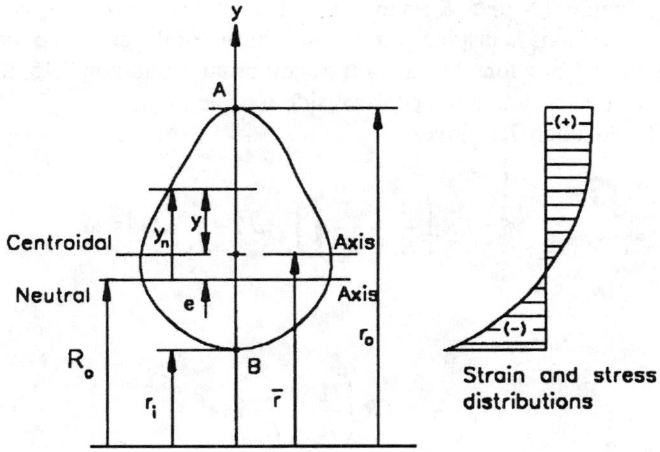

FIGURE 7.8 Strain and stress distribution in the cross section of the curved beam of Figure 7.7.

$$ER_o\left(\frac{1}{R_1}-\frac{1}{R_o}\right)\int_A \frac{y_n}{R_o+y_n}\,dA = 0 \tag{7.17}$$

Similarly, the second Condition 7.16 becomes

$$ER_o\left(\frac{1}{R_1}-\frac{1}{R_o}\right)\int_A \frac{y_n^2}{R_o+y_n}\,dA = M \tag{7.18}$$

The integrand

$$\frac{y^2}{R_o+y_n} = y_n - \frac{R_o y_n}{R_o+y_n}$$

so that Equation 7.18 is rewritten as

$$ER_o\left(\frac{1}{R_1}-\frac{1}{R_o}\right)\left[\int_A y_n dA - R_o\int_A \frac{y_n}{R_o+y_n}\,dA\right] = M \tag{7.19}$$

The second integral must vanish according to Equation 7.17. On the other hand, the first integral is not the first moment of area of the cross section, simply since y_n is measured from the neutral axis and not from the centroidal axis. Denoting the distance from the neutral axis to the centroidal axis by e, and recalling that $\bar{r}$ is the radius of curvature of the centroidal axis, gives from Figure 7.8

$$\begin{aligned} e &= \bar{r} - R_o \\ y_n &= y + e \end{aligned} \tag{7.20}$$

By referring to Figure 7.8, $\bar{r} > R_o$, and hence, from Equation 7.20, e is a positive quantity indicating that the neutral axis is displaced from the centroid of the cross section toward the center of curvature of the beam. Therefore, to satisfy the equilibrium Condition 7.15, the maximum stress (compressive in this case) occurs at the concave side.

Substituting into Equation 7.19 gives

$$M = ER_o\left(\frac{1}{R_1} - \frac{1}{R_o}\right)\left[\int_A ydA + \int_A edA\right]$$

hence,

$$M = ER_o\left(\frac{1}{R_1} - \frac{1}{R_o}\right)Ae \tag{7.21}$$

where the integral $\int_A ydA = 0$ represents the first moment of area, since y is measured from the centroidal axis. Combining Equations 7.14 and 7.21 yields the required formula for the tangential stress σ_z, as

$$\sigma_z = \frac{My_n}{Ae(R_o + y_n)} \tag{7.22}$$

This is similar to the simple bending formula for bending of straight beams in which the second moment of area I is replaced by $Ae(R_o + y_n)$. However, it must be remembered that the quantities y and R_o in Equation 7.22 are measured from the neutral axis (not the centroidal axis) whose location is yet to be determined.

7.3.3.1 Determination of the Location of the Neutral Axis

This is readily obtained by satisfying Equation 7.17, namely,

$$\int_A \frac{y_n}{R_o + y_n} dA = 0$$

Using Equations 7.20 gives

$$\frac{y+e}{\bar{r}+y} dA = \int_A \left[\frac{\bar{r}+y}{\bar{r}+y} - \frac{\bar{r}}{\bar{r}+y} + \frac{e}{\bar{r}+y}\right] dA = 0 \tag{7.23}$$

or

$$e = \bar{r} - \left[A \Big/ \int_A \frac{1}{\bar{r}+y} dA\right] \tag{7.24}$$

Equation 7.24 could be integrated to give e* for the various cross sections shown in Figure 7.9:

* The determination of distance e as the difference between R_o and $\bar{r}$ requires high accuracy of calculations up to several significant digits. Series expansions of Formulas 7.25, are often quoted in textbooks to realize this accuracy. With the advent of hand calculators, this becomes unnecessary.

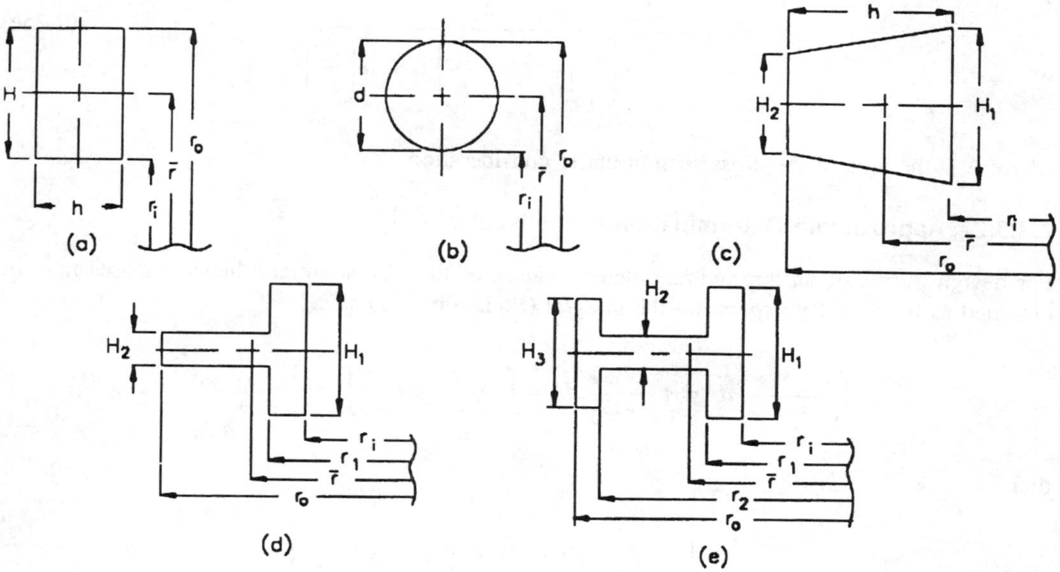

FIGURE 7.9 Various cross sections for circularly curved beams: (a) rectangular section, (b) circular section, (c) trapezoidal section, (d) tee section, (e) unequal I section.

- Rectangular cross section, Figure 7.9a:

$$e = \bar{r} - \frac{H}{\ln(r_o/r_i)} \tag{7.25a}$$

- Circular cross section, Figure 7.9b:

$$e = \bar{r} - \frac{(r_o - r_i)^2/4}{2\bar{r} - 2\sqrt{r_o r_i}} \tag{7.25b}$$

- Trapezoidal cross section, Figure 7.9c:

$$e = \bar{r} - \frac{A}{\left[H_2 + \left(\dfrac{H_1 - H_2}{r_o - r_i}\right)(r_o)\right]\ln(r_o/r_i) - (H_1 - H_2)} \tag{7.25c}$$

- Tee cross section, Figure 7.9d:

$$e = \bar{r} - \frac{A}{H_1 \ln(r_1/r_i) + H_2 \ln(r_o/r_1)} \tag{7.25d}$$

- Unequal I cross section, Figure 7.9e:

$$e = \bar{r} - \frac{A}{H_1 \ln\left(\frac{r_1}{r_i}\right) + H_2 \ln\left(\frac{r_2}{r_1}\right) + H_3 \ln\left(\frac{r_o}{r_2}\right)} \tag{7.25e}$$

where A is the area of the cross section under consideration.

7.3.3.2 Approximate Determination of the Neutral Axis

For design purposes, an approximate determination of the distance e and hence stresses may be obtained as follows. By expressing the integral (Equation 7.17) as

$$\int_A \frac{y_n}{R_o + y_n} dA = \int_A \frac{y+e}{\bar{r}+y} dA = \int_A \frac{y}{\bar{r}+y} dA + e\int_A \frac{1}{\bar{r}+y} dA = 0$$

then

$$e = -\left[\int_A y\left(1+\frac{y}{\bar{r}}\right)^{-1} dA\right] \Big/ \left[\int_A \left(1+\frac{y}{\bar{r}}\right)^{-1} dA\right]$$

Expanding $(1 + y/\bar{r})^{-1} = 1 - (y/\bar{r}) + {}^1/_2(y/\bar{r})^2 - \ldots$, and retaining only the first power of $(y/\bar{r})$, yields

$$e \cong -\left[\int_A y\left(1-\frac{y}{\bar{r}}\right) dA\right] \Big/ \left[\int_A \left(1-\frac{y}{\bar{r}}\right) dA\right]$$

Integrating and noting that $_A\int y dA = 0$ and $_A\int y^2 dA = I_x$, gives

$$e \cong I_x/A\,\bar{r} \tag{7.26}$$

To evaluate the accuracy of this approximate expression, consider both rectangular and circular cross sections and evaluate errors in e, as determined from Equations 7.25 and 7.26. Results are tabulated in Table 7.1 for curved beams of various curvatures.

TABLE 7.1
Error in Using Equation 7.26 Instead of Equation 7.25 to Locate the Neutral Axis of Curved Beams

$(r_o - r_i)/\bar{r}$	1.5	1	0.5	0.2	0.1
Rectangular Cross Section of Depth H and Centroidal Radius of Curvature $\bar{r}$					
Error, %, in $e/\bar{r}$	−18.2	−7.2	−1.7	−0.27	−0.07
Circular Cross Section					
Error, %, in $e/\bar{r}$	−60.6	−6.7	−1.6	−0.25	−0.06

It is seen that for beams of large curvatures, say $H/\bar{r} < 1$, errors lower than 5% are encountered in determining the location of the neutral axis, thus justifying the use of the approximate formula of Equation 7.26 within this range of H/F.

7.3.3.3 Maximum Stresses

To obtain the maximum stresses in the most remote fibers at points A and B of Figure 7.8, Equation 7.22 may be rewritten in terms of $\bar{r}$ and y, as*

$$\sigma_z = \frac{M(y+e)}{Ae(y+\bar{r})} \tag{7.27}$$

The maximum stresses for positive M, Figure 7.8 are

$$(\sigma_z)_A = \frac{M(y_A+e)}{Ae(y_A+\bar{r})} = \frac{M(y_A+e)}{Aer_o}$$

$$(\sigma_z)_B = \frac{M(-y_B+e)}{Ae(-y_B+\bar{r})} = \frac{-M(y_B-e)}{Aer_i}$$

When the distance e is determined approximately (for beams of moderate curvature) from Expression 7.26, the stress σ_z is given by

$$\sigma_z = \frac{1}{(1+y/\bar{r})}\frac{M(y+e)}{I_x} \tag{7.28}$$

Again, the similarity of Equation 7.28 to that for bending of straight beams is evident.

7.3.3.4 Bending of a Curved Beam by Lateral Forces Acting in the Plane of Its Axis

For a curved beam subjected to general loading in the plane of its axis, Figure 7.10, any cross section will be subjected to a bending moment M, a normal force P_z, and a transverse shearing force P_y. The normal stresses due to P_z (acting along the tangential direction) will be uniformly distributed over the entire cross section and equal to P_z/A. These normal stresses are added to those due to bending as obtained from Equations 7.22. The shear stresses are usually assumed to follow the same distribution as that of a straight bar, i.e., $\tau = a_y P_y/A$ (where a_y is taken from Table 7.3).** This assumption concerning the shear stress distribution is quite satisfactory for beams of narrow, rectangular cross sections as indicated in Section 6.3 (Figure 6.19) using the more exact methods of the stress function approach.

---

* The beam formula (Equation 7.27) gives somewhat erroneous results for tangential stress developed in curved beams having thin flanges, i.e., I section. This is due to the deflections occurring in these thin flanges under the radial stress component. A correction factor may be applied in this case. See Boresi and Sidebottom,[1] p. 374.

** It is often recommended to take a_y as the ratio of the maximum shear stress at the neutral surface to the average shear stress value P_y/A, hence, $a_y = 1.5$ and 1.33 for rectangular or circular cross sections respectively, as demonstrated in Section 7.4.1 and 7.4.2 and Table 7.3.

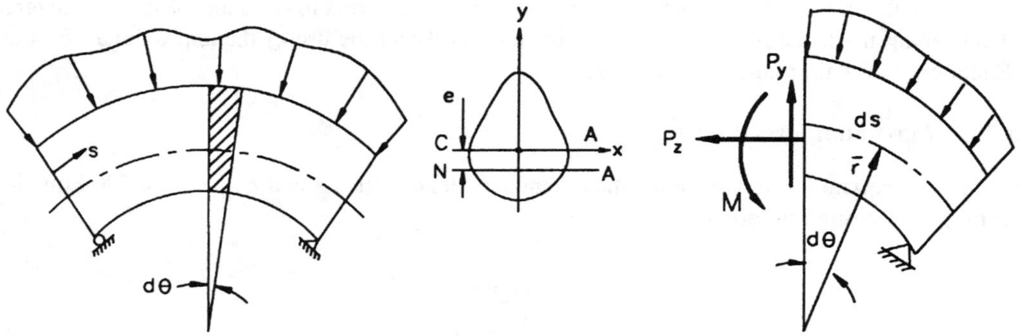

FIGURE 7.10 Curved beam subjected to lateral loads acting in the plane of its axis.

7.3.3.5 Strain Energy in Curved Beams

The strain energy in curved beams is basically expressed by an equation similar to those obtained in Example 3.10.

For an element of length ds of Figure 7.10 taken along the centroidal axis, the strain energy due to bending is given by

$$U_1 = \int_0^s \frac{1}{2} M(d\psi - d\theta)\frac{\bar{r}-e}{\bar{r}}ds$$

By using Relations 7.12 and 7.21, U_1 reduces to

$$U_1 = \int_0^s \frac{1}{2} M \frac{M}{EAeR_o} \frac{\bar{r}-e}{\bar{r}} ds = \int_0^s \frac{M^2}{2EA\bar{r}e} ds$$

The strain energy U_2 and U_3 due to normal and shearing forces, respectively, are given by

$$U_2 \int_o^s \frac{P_z^2}{2EA} ds \text{ and } U_3 = \int_o^s \frac{a_y P_y^2}{2GA} ds$$

where P_z designates the axial force component acting perpendicular to the cross section at its centroid.

An additional strain energy term has to be introduced to account for the presence of P_z and M together. The normal force P_z produces an elongation of the element ds along the beam axis equal to $(P_z ds/AE)$ and hence increases the angle between two adjacent faces distance dz by (P_z/AE) $(ds/\bar{r})$. Therefore, the already applied bending moments M do negative work during the application of P_z; this is given by

$$U_4 = -\int_0^s M \frac{P_z}{EA} \frac{ds}{\bar{r}}$$

An expression for the total strain energy is thus obtained by adding U_1, U_2, U_3, and U_4 to yield

$$U = \int_0^s \left[\frac{M^2}{2EA\bar{r}e} + \frac{P_z^2}{2EA} + \frac{a_y P_y^2}{2GA} - \frac{MP_z}{EA\bar{r}} \right] ds \qquad (7.29)$$

7.3.3.6 Comparison with Exact and Other Solutions

To assess the validity of the approximate solutions given above, they are compared for the case of pure bending of a rectangular cross section to the more exact solution previously obtained using the stress function approach in Section 6.3.1. Comparison is extended to the elementary linear bending formula $\sigma = My/I_x$, as shown in Table 7.2. It may be concluded that, to determine stresses in a curved beam of depth H and radius of curvature $\bar{r}$, the following approaches may be adopted:

$H/\bar{r} > 1$:	The use of the stress function approach is recommended,
$1 > H/\bar{r} > 0.2$:	The hyperbolic stress formula (Equation 7.27) or even its approximate version (Equation 7.28) may be used, and
$H/\bar{r} < 0.2$:	The elementary formula for straight beams may be used without serious error.

TABLE 7.2
Comparison Among Various Solutions for the Tangential $\sigma_\theta/(M/\bar{r}^2)$ at Both Outer and Inner Radii in Curved Beams Under Pure Bending

$H/\bar{r}$	1.75	1.5	1	0.5	0.2
r_o/r_i	9	7	3	1.667	1.222
Stress function	1.56	1.807	4.52	20.59	140.65
solution (Eq. 6.43)	–5.71	–5.85	–9.17	–28.87	–160.76
Hyperbolic solution	1.37	1.63	4.38	20.48	140.54
(Eq. 7.27)	–6.07	–6.06	–9.14	–28.79	–160.67
% Error (tension-side)	12	9.8	3.1	0.6	0.1
Approximate formula	1.65	1.91	4.67	20.80	140.91
(Eq. 7.28) ($\sigma_z \equiv \sigma_\theta$)	–8.59	–8.0	–10.0	–29.33	–161.11
% Error (compression-side) ($\sigma_z \equiv \sigma_\theta$)	50	36.8	9.1	1.6	0.2
Linear bending formula (My/I)	±2.34	±2.67	±6	±24	±150
% Error (compression-side)	59	54	34.6	16.9	6.6

Example 7.4:

Determine the maximum tensile and compressive stresses in a hook of trapezoidal cross section, as shown in Figure 7.11. The hook is used to lift a weight of 25 kN.

Solution:

The maximum bending moment acts at section A–B. The properties of this cross sectional area are

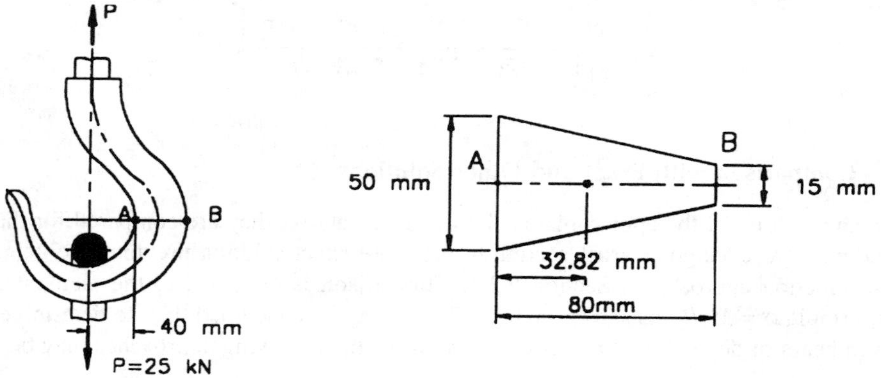

FIGURE 7.11 Example 7.4.

$$A = \tfrac{1}{2}(50+15)\,(80) = 2.6\times10^3\ \text{mm}^2$$

$$\bar{y} = \frac{H_1 + 2H_2}{H_1 + H_2}\frac{r_o - r_i}{3} = \frac{50+2\times15}{50+15}\times\frac{80}{3} = 32.82\ \text{mm}$$

$$I_x = \left[\frac{H_2(r_o - r_i)^3}{3} + \frac{H_1 - H_2}{12}(r_o - r_i)^3\right] - \bar{y}^2 A = 1.253\times10^6\ \text{mm}^4$$

$$\bar{r} = 40 + 32.82 = 72.82\ \text{mm}$$

The maximum bending moment is thus

$$M = -25\times10^3\times72.82 = -1.8205\times10^6\ \text{N}\cdot\text{mm}$$

which is taken as negative, since it reduces the curvature of the beam. The maximum bending tensile and compressive stresses occur at points A and B, respectively. Use of Equation 7.27 requires locating the neutral axis; the distance e is obtained from Equation 7.25c as

$$e = \bar{r} - \frac{A}{\left[H_2 + \left(\dfrac{H_1 - H_2}{r_o - r_i}\right)(r_o)\right]\ln(r_o/r_i) - (H_1 - H_2)}$$

which gives for r_i = 40 mm, r_o = 120 mm:

$$e = 6.419\ \text{mm}$$

The bending stresses using Equation 7.27:

$$\sigma_z = \frac{M(y+e)}{Ae(y+\bar{r})}$$

gives for the maximum tensile stress at $y = -32.82$ mm:

$$(\sigma_z)_A = \frac{-1.8025\times10^6}{2.6\times10^3\times6.419}\frac{(-32.82+6.419)}{(-32.82+72.82)} = 71.28 \text{ MPa}$$

and for the maximum compressive stress: at $y = 80 - 32.82 = 47.18$ mm:

$$(\sigma_z)_B = \frac{-1.8025\times10^6}{2.6\times10^3\times6.419}\frac{(47.18+6.419)}{(47.18+72.82)} = -48.24 \text{ MPa}$$

If the distance e is determined from the approximate Equation 7.26 as

$$e = I_x/A\bar{r} = (1.253\times10^6)/(2.6\times10^3\times72.82) = 6.617 \text{ mm}$$

The approximate maximum bending stresses are found from Equation 7.28 as

$$(\sigma_z)_A = \frac{-1.8025\times10^6}{1.253\times10^6}\frac{(-32.82+6.617)}{(1-32.82/72.82)} = 68.62 \text{ MPa}$$

$$(\sigma_z)_B = \frac{-1.8025\times10^6}{1.253\times10^6}\frac{(45.87+6.617)}{(1+45.87/72.82)} = -46.32 \text{ MPa}$$

The error in these approximate values for $(\sigma_z)_A$ and $(\sigma_z)_B$ is about 4%.

7.3.4 Thermoelastic Bending of Straight Bars

The temperature distribution in a straight bar is chosen to make all cross sections subject to identical conditions; therefore, the temperature gradient in the axial direction vanishes. This makes all plane sections remain plane. The ends of the bar are constrained to maintain plane conditions, but are free so that the stress resultants P_z, M_x, and M_y are equal to zero. Now, according to Equations 3.18, ε_z and σ_z are given by

$$\begin{aligned}\varepsilon_z &= ax + by + c\\ \sigma_z &= E(\varepsilon_z - \alpha T) = E(ax + by + c - \alpha T)\end{aligned} \tag{7.30}$$

where T is the temperature difference from ambient temperature. For equilibrium, the following conditions have to be satisfied, namely,

$$P_z = \int_A \sigma_z dA = \int_A E(ax + by + c - \alpha T)dA = 0$$

or

$$EcA - E\alpha\int_A TdA = 0 \tag{7.31a}$$

$$M_x = \int_A \sigma_z ydA = \int_A E(ax + by + c - \alpha T)ydA = 0$$

or

$$EbI_x - E\alpha\int_A TydA = 0 \tag{7.31b}$$

and

$$M_y = \int_A \sigma_z xdA = \int_A E(ax + by + c - \alpha T)xdA = 0$$

or

$$EaI_y - E\alpha\int_A TxdA = 0 \tag{7.31c}$$

From Equations 7.31, the constants *c*, *b*, and *a* are obtained as

$$c = \left[\alpha\int_A TdA\right]\Big/A$$

$$b = \left[\alpha\int_A TydA\right]\Big/I_x \tag{7.32}$$

$$a = \left[\alpha\int_A TxdA\right]\Big/I_y$$

Substituting Expressions 7.32 in Equation 7.30 yields

$$\sigma_z = E\alpha\left[x\int_A \frac{Tx}{I_y}dA + y\int_A \frac{Ty}{I_x}dA + \int_A \frac{T}{A}dA - T\right] \tag{7.33a}$$

$$\varepsilon_z = \frac{\sigma_z}{E} + \alpha T \tag{7.33b}$$

If the ends are not free, the stresses are obtained by superposition of Equations 7.33a and 7.7b observing the signs convention.

Example 7.5:

A long square bar has been uniformly heated along its length. The bar ends are free. Determine the stresses and strains for (a) $T = T_o$ *= constant, (b)* $T = T_o + ky$*, and (c)* $T = T_o + k(x + y)$*. As an illustration for case (c), consider a square steel bar 40 × 40 mm which is used to stiffen the corners of a square duct through which hot gases are passing. The surface temperature at points 1 and 2, Figure 7.12, are 250°C and 150°C, respectively. The bar length is 3 m,* $\alpha = 11 \times 10^{-6}$*/°C, and* E = *210 GPa. Determine the stresses in the duct wall.*

Solution:

For a long bar the ends can be considered to be plane and the solution is given as follows:

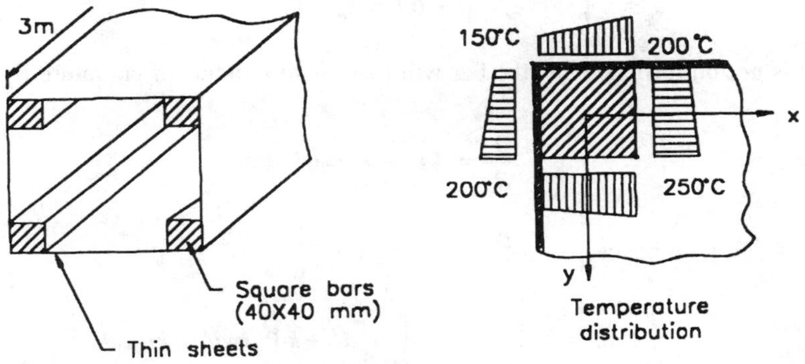

FIGURE 7.12 Example 7.5.

a. $T = T_o$ = constant. Substituting $T = T_o$ in Equations 7.32 yields

$$\int TxdA = T_o \int_A xdA = 0$$

$$\int TydA = T_o \int_A ydA = 0$$

$$\int_A Tda = T_o A$$

which gives $a = b = 0$ and $c = \alpha T_o$. Substituting in Equation 7.33a, gives

$$\sigma_z = E\alpha(0 + 0 + T_o - T_o) = 0$$

The bar is free from stresses.

From Equation 7.33b,

$$\varepsilon_z = 0 + \alpha T_o = \alpha T_o = \text{constant}$$

Hence, the bar will remain straight.

b. $T = T_o + Ky$, hence,

$$\int_A Txda = T_o \int_A xdA + \int_A kxydA = 0$$

$$\int_A TydA = T_o \int_A ydA + \int_A ky^2 dA = kI_x$$

$$\int_A TdA = T_o \int_A dA + \int_A kydA = T_o A$$

Which gives $a = 0$, $b = \alpha k$, and $c = \alpha T_o$. From Equation 7.33a,

$$\sigma_z = E\alpha[0 + ky + T_o - (T_o + ky)] = 0$$

From Equation 7.33b,

$$\varepsilon_z = 0 + \alpha T = T_o + ky$$

which is not uniform. Hence, the bar will bend with a radius of curvature R_x given by

$$\frac{1}{R_x} = \sqrt{a^2 + b^2} = \alpha k$$

c. $T = T_o + k(x + y)$; hence,

$$\int_A TxdA = T_o\int_A xdA + k\int_A x^2dA + k\int_A xydA = kI_y$$

$$\int_A TydA = T_o\int_A ydA + k\int_A xydA + k\int_A y^2dA = kI_x$$

$$\int_A TdA = T_oA + k\int_A xdA + k\int_A ydA = T_oA$$

which gives $a = b = \alpha k$ and $c = \alpha T_o$. From Equation 7.33a,

$$\sigma_z = E\alpha[kx + ky + T_o - T_o - k(x + y)] = 0$$

From Equation 7.33b,

$$\varepsilon_z = 0 + \alpha T = \alpha[T_o + k(x + y)]$$

which is not uniform. The bar will bend about the diagonal with a radius of curvature R_{xy} given by

$$\frac{1}{R_{xy}} = \sqrt{a^2 + b^2} = \alpha k\sqrt{2}$$

In this case for $T = T_o + k(x + y)$, substitute $T = 250°C$ at $x = y = 20$ mm and $T = 150°C$ at $x = y = -20$ mm; hence,

$$T_o = 200°\text{C} \quad \text{and} \quad k = 1.25°\text{C mm}^{-1}$$

$$a = b = \alpha k \times 11 \times 10^{-6} \times 1.25 = 13.75 \times 10^{-6}\ \text{mm}^{-1}$$

and

$$\frac{1}{R_{xy}} = \sqrt{2} \times 13.75 \times 10^{-6} = 19.44 \times 10^{-6}\ \text{mm}^{-1}$$

The bar will bend into a circular arc having a camber δ at the middle of its length, given by the expression:

$$\delta = \frac{L^2}{8R_{xy}} = \frac{3000^2 \times 19.44 \times 10^{-6}}{8} = 2.183\ \text{mm}$$

In the above analysis the constraining effect of the duct thin walls, has been neglected. If the walls rigidity constrains 50% of the camber δ, then the bar will be subjected to a bending moment about the diagonal of the square according to Equations 7.10, as given by

$$M = 0.5\frac{EI}{R_{xy}}$$

Here $I = I_x = I_y$ for a square cross section of side length of $2H$. The maximum stress in the bar is obtained from the Relation 7.9b with $y' = \sqrt{2}\,H$; hence,

$$\sigma_z = 0.5\frac{E}{R_{xy}}\sqrt{2}\,H$$

Hence, $\sigma_x = 0.5 \times 210 \times 10^3 \times 19.44 \times 10^{-6} \times 20\ \sqrt{2} = 57.73$ MPa, which is tension on the outer corner of the duct.

7.4 SHEAR STRESSES IN RODS

In Section 7.2 it has been pointed out that under general loading of rods, the only shear stress components present are τ_{xy} and τ_{zy}. They act in the plane of cross section x–y and result from P_x, P_y, and M_z, Equation 7.3. In this section, the shear stresses due to P_x and P_y only are obtained, while those due to M_z will be treated in Section 7.5 on torsion of bars. The resultant shear stresses are obtained by superposition.

Consider a bar cross section subjected to forces P_x and P_y passing through the centroid O as defined by Equation 7.3. The shear stress components due to P_x and P_y will have a resultant shear stress τ acting in the plane of cross section given by the relation:

$$\tau^2 = \tau_{zx}^2 + \tau_{zy}^2 \tag{7.34}$$

having a direction inclined to the x-axis by an angle of $\tan^{-1}(\tau_{zy}/\tau_{zx})$ as shown in Figure 7.13a. Note that the sign of the shear stress follows the convention of Section 1.5.

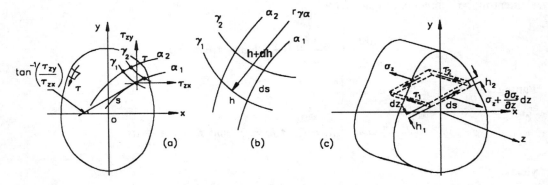

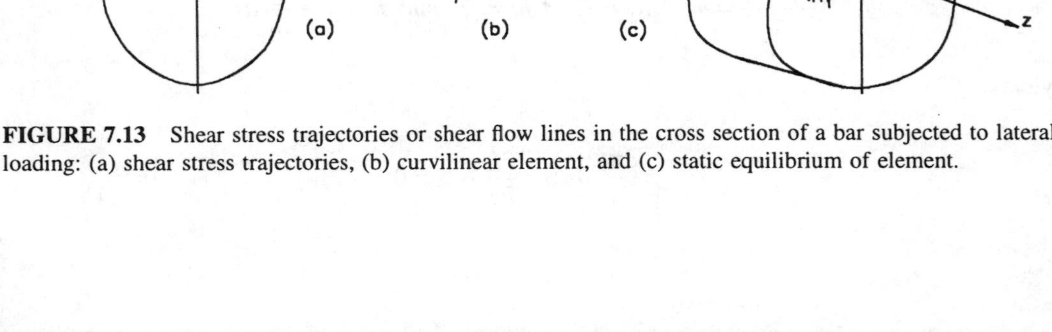

FIGURE 7.13 Shear stress trajectories or shear flow lines in the cross section of a bar subjected to lateral loading: (a) shear stress trajectories, (b) curvilinear element, and (c) static equilibrium of element.

Consider now a set of lines constructed in the plane of cross section such that the tangent at any point is the direction of the resultant shear stress τ at this point. These lines are called *shear stress trajectories* or *shear flow lines*. The boundary of the rod cross section is one of these lines since the rod surface is free from shear stresses. Taking these lines as reference curvilinear coordinates, where $\alpha = s$, $\beta = z$, and substituting in the second equation of (Equation 1.12a) by

$$\sigma_\alpha = \sigma_\gamma = \tau_{\alpha\gamma} = \tau_{\beta\gamma} = 0$$

and

$$\frac{1}{r_{\beta\alpha}} = \frac{1}{r_{\beta y}} = \frac{1}{r_{\gamma\beta}} = \frac{1}{r_{\alpha\beta}} = 0$$

where α and γ lie in the plane of bar cross section and β is straight and parallel to the z-direction, Equation 1.12a reduces to

$$\frac{\partial \sigma_z}{\partial z} + \frac{\partial \tau}{\partial s} + \frac{\tau}{r_{\gamma\alpha}} = 0 \tag{7.35}$$

Consider two successive shear stress trajectories α_1 and α_2 at an arbitrary small distance h across which τ is constant, Figure 7.13b. If the angle between α_1 and α_2 is $(\alpha_2 - \alpha_1)$, then $h = r_{\gamma\alpha}(\alpha_2 - \alpha_1)$, and since $(\alpha_2 - \alpha_1) = \partial h/\partial s$, then $1/r_{\gamma\alpha} = (1/h)\,(\partial h/\partial s)$. Substituting in Equation 7.35 yields

$$\frac{\partial(\tau h)}{\partial s} = -h\frac{\partial \sigma_z}{\partial z} \tag{7.36}$$

where

$$h\frac{\partial \tau}{\partial s} + \tau\frac{\partial h}{\partial s} = \frac{\partial(\tau h)}{\partial s}$$

In terms of Cartesian coordinates, Equation 7.36 is derived by considering the equilibrium of an element in the z-direction, Figure 7.13c:

$$\frac{\partial \sigma_z}{\partial z} dz \frac{h_2 + h_1}{2} ds + \tau_2 h_2 ds - \tau_1 h_1 ds = 0$$

Putting

$$\frac{h_2 + h_1}{2} = h, \quad \tau_2 h_2 = f_2, \quad \tau_1 h_1 = f_1, \quad \text{and} \quad f_2 = f_1 + \frac{\partial f}{\partial s} ds$$

yields

$$h\frac{\partial \sigma_z}{\partial z} dz ds + \frac{\partial f}{\partial s} ds dz = 0$$

or

$$\frac{\partial(\tau h)}{\partial s} = -h\frac{\partial \sigma_z}{\partial z}$$

which is Equation 7.36. Substituting σ_z from Equation 7.7b in Equation 7.36 gives

$$\frac{\partial(\tau h)}{\partial s} = -h\frac{\partial}{\partial z}\left(\frac{M_y}{I_y}x + \frac{M_x}{I_x}y + \frac{P_z}{A}\right)$$

which when combined with Equation 7.4 yields

$$\frac{\partial(\tau h)}{\partial s} = -\left(\frac{P_x}{I_y}x + \frac{P_y}{I_x}y\right)h \tag{7.37}$$

Equation 7.37 is the equation of equilibrium along the shear flow lines and τ_h is called the *shear flow*. To obtain τ, Equation 7.37 has to be integrated along s, which requires the shear flow lines to be *apriori* constructed. For a rectangular solid section and thin-walled sections, it is easy to construct these lines.

A further condition is necessary to determine the constant of integration for Equation 7.37. This is readily available in rectangular and open thin-walled sections, where the edges are free from shear and hence $\tau = 0$ at $s = 0$. For closed thin-walled sections, the necessary condition is that the integration of the shear flow along the wall length is the resultant of P_z, P_x, and M_z. In fact, this condition applies to all sections and it will be shown later that some contradiction arises for some thin-walled open sections subjected to P_x and P_y only, which is handled by introducing the concept of shear center.

7.4.1 Rectangular Solid Section

Consider a rectangular solid section of height h and width or thickness b, as shown in Figure 7.14, subjected to a force P_y through the centroid O parallel to its height h.

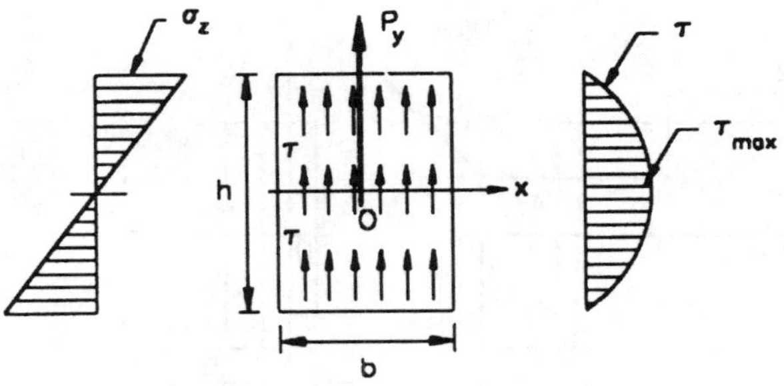

FIGURE 7.14 Shear stress distribution in a rectangular solid section subjected to lateral loading.

The shear flow lines are parallel lines to the y-axis. Substituting in Equation 7.37 by $P_x = 0$ and $s \equiv y$ yields

$$\frac{\partial \tau}{\partial y} = -\frac{P_y}{I_x} y$$

which upon integration and using the boundary condition $\tau = 0$ at $y = \pm h/2$ (i.e., edges free of shear stresses) gives

$$\tau = \frac{P_y}{I_x}\left(\frac{h^2}{8} - \frac{y^2}{2}\right) = \tau_{zy} \quad \text{and} \quad \tau_{zx} = 0$$

or

$$\tau_{zy} = \frac{6P_y}{bh^3}\left(\frac{h^2}{4} - y^2\right) \tag{7.38a}$$

where

$$I_x = \frac{bh^3}{12}$$

This is a parabolic distribution, which is maximum at the neutral axis $y = 0$, such that

$$\tau_{max} = \frac{3}{2}\frac{P_y}{bh} \quad \text{while} \quad \sigma_z = 0 \tag{7.38b}$$

For a force P_x through the centroid parallel to the x-axis, the same expression is obtained by replacing y by x. For an inclined force through O, the shear stresses τ_{zy} and τ_{zx} are obtained from the force components in the y- and x-directions, respectively.

Example 7.6:

Determine the stresses in a cantilever bar having a thin rectangular cross section of height h *and thickness* b *and subjected to a force* P *in the* y–z *plane acting as shown in Figure 7.15.*

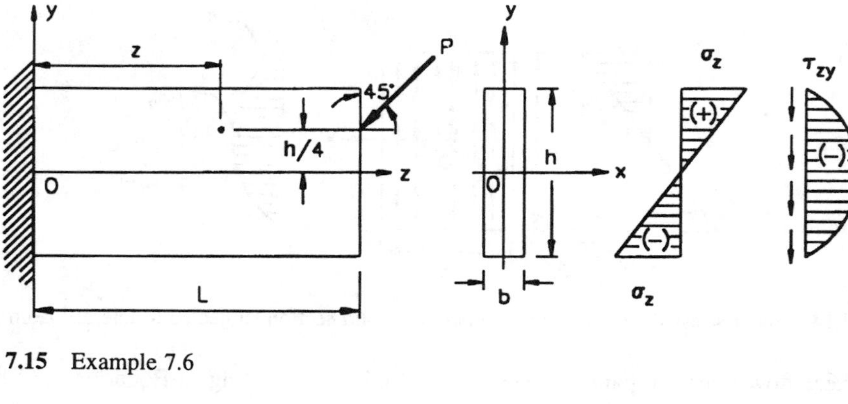

FIGURE 7.15 Example 7.6

Solution:

The stress resultant at any cross section at a distance z is given by

$$P_x = 0,\ \ P_y = P_z = -P/\sqrt{2}$$

$$M_x = \sqrt{2}\,P\big((L-z)-h/4\big) \text{ and } M_y = 0$$

Knowing that

$$I_x = \frac{bh^3}{12},\ \ I_y = \frac{b^3h}{12},\ \text{ and } A = bh,$$

the stresses are given by

$$\sigma_x = \sigma_y = \tau_{zx} = 0$$

From Equation 7.7b

$$\sigma_z = \frac{\sqrt{2}\,P}{bh}\left[\left(\frac{12(L-z)-3h}{h^2}\right)y - 1\right]$$

where $|\sigma_z|$ is maximum at $z = 0$ and $y = \pm\, h/2$, i.e.,

$$\left|\sigma_z\right|_{max} = \frac{\sqrt{2}\,P}{bh}\left(\frac{12L-3h}{2h} - 1\right)$$

From Equation 7.38a,

$$\tau_{zy} = -\sqrt{2}\,\frac{6\,P}{bh^3}\left(\frac{h^2}{4} - y^2\right)$$

which has a parabolic distribution along y. $|\tau_{zy}|$ is maximum at $y = 0$ and is given by

$$\left|\tau_{zy}\right|_{max} = -\frac{3\sqrt{2}}{2bh}P$$

7.4.2 Circular Solid Section

For rods of circular cross sections, unlike rods of rectangular cross sections, the assumption that the shear stresses are parallel to the shearing force P_y is inconsistent with the stress-free boundary conditions.

To illustrate this, consider the resultant shear stress having a direction as τ in Figure 7.16a. This τ can be always resolved into two components: radial τ_{zr} and tangential $\tau_{z\theta}$. The existence of τ_{zr} is in contradiction to having a free surface for the rod, as shown in Figure 7.16b and hence it must vanish. Consequently, the resultant shear stress τ must act tangentially to the periphery of the rod cross section, Figure 7.16c. Due to symmetry, it is obvious that along a line $y = y_1$, these resultant shear stresses must be directed toward a single point O. At points along the y-direction,

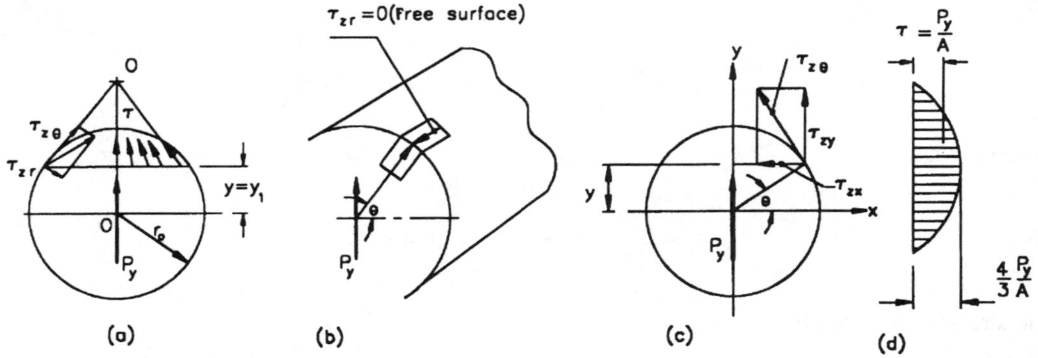

FIGURE 7.16 Shear stress distribution in a solid circular section subjected to lateral loading: (a) resultant shear stress distribution, (b) and (c) boundary shear stress, and (d) distribution of shear stress in the y-direction.

the resultant shear stresses have the direction of the shear force P_y. Based on this, an approximate solution* is obtained by assuming that the shear stresses are found by applying Equation 7.37 as

$$\frac{d(\tau h)}{ds} = -\frac{P_y}{I_x} hy$$

For the shear stress component acting in the vertical y-direction, this reduces to

$$\tau_{zy} h = -\frac{P_y}{I_x}\int hydy = -\frac{P_y}{I_x}\int ydA$$

where $ds \equiv dy$ and $hdy = dA$. This implies that this vertical shear stress is the same for all points at any $y = y_1$ from which h is obtained, hence, for a circular solid cross section of radius r_o,

$$h = 2\sqrt{r_o^2 - y^2}$$

and

$$\int_{y_1}^{r_o} ydA = \int_{y_1}^{r_o} 2\sqrt{r_o^2 - y^2}\,ydy = \frac{2}{3}\left(r_o^2 - y_1^2\right)^{3/2}$$

Hence,

$$\tau_{zy} = -\frac{P_y\left(r_o^2 - y^2\right)}{3I_x} \tag{7.39}$$

This parabolic shear stress at any ordinate y may be considered as the vertical component of the total shear stress:

* For the exact solution, see Love,[2] p. 348.

$$\tau_{z\theta} = \tau = \sqrt{\tau_{zy}^2 + \tau_{zx}^2}$$

acting tangentially to the periphery. Hence,

$$\tau_{z\theta} = \tau_{zy} \frac{r_o}{\sqrt{r_o^2 - y^2}}$$

which gives

$$\tau_{z\theta} = \frac{P_y r_o \sqrt{r_o^2 - y^2}}{3I_x} \tag{7.40}$$

For engineering applications it can be shown that Equation 7.40 approximates fairly well the total shear stress distribution in circular cross sections. Putting $I_x = \pi r_o^4/4$ gives

$$\tau = \tau_{z\theta} = \frac{P_y r_o \sqrt{r_o^2 - y^2}}{3(\pi r_o^4/4)} \tag{7.41a}$$

It is seen that the maximum value of τ occurs at $y = 0$, i.e., at the neutral axis of the cross section,* Figure 7.16d, and is given by

$$\tau_{max} = \frac{P_y r_o}{3(\pi r_o^2/4)} = \frac{4P_y}{3A} \tag{7.41b}$$

where A is the cross sectional area of the rod.

In strength-of-materials texts, Equation 7.39 is generally used instead of Equation 7.40 to approximate the shear stress distribution in circular cross sections. Both equations provide the same maximum and minimum values at $y = 0$ and $y = r_o/2$, respectively. However, the ratio between $\tau_{z\theta}$ and τ_{zy} at $r_o/2$ is equal to $2/\sqrt{3} \cong 1.15$.

The nonuniform shear stress distribution over the depth of the rod cross section as given by Equations 7.38a and 7.41a for rectangular and circular cross sections, respectively, presents mathematical difficulties in formulating the shear strain energy. This may be simply overcome by introducing coefficients to account for this nonuniformity according to the expressions:

$$dU_{shear} = a_x \frac{P_x}{2GA} \quad \text{and} \quad dU_{shear} = a_y \frac{P_y}{2GA}$$

* The exact solution gives:

$$\tau = \frac{(3+2\nu)P_y}{2(1+\nu)A}\left(r_o^2 - \frac{1-2\nu}{3+2\nu}y^2\right)$$

which is only about 4% greater than τ of Equation 7.41, for $\nu = 0.3$ at the neutral axis.[2]

where a_x and a_y are the ratios between the maximum shear stress at the neutral surface to the average value P/A.* Values of the coefficient a are given in Table 7.3.

TABLE 7.3
Coefficients for Shear Strain Energy

Cross section	a_x or a_y
Rectangular cross section	1.50
Circular cross section	1.33
Thin-walled circular cross section	1.00

7.4.3 Thin-Walled Open Sections

Thin-walled sections are the most economical sections since they provide structural elements of minimum weight. Sections of nonferrous metals are produced by extrusion, whereas steel sections are cold-formed from coiled strips, closing the edges by welding for tubular sections.

Thin-walled sections are either open sections, such as tee, channel, and angle sections, or closed sections, such as various shapes of tubular sections. An open section has one single boundary and is said to be simply connected, while a closed section has more than one boundary surface and is said to be multiply connected. Consider a bar having a thin-walled open cross section subjected to forces P_x and P_y acting through the centroid O; the x- and y-axes are centroidal principal axes and the z-axis is the bar axis as shown in Figure 7.17. For a thin-walled section, the shear stress direction is along the middle contour line of the cross section, being parallel to the cross section boundary contour lines. The shear stress is uniform across the wall thickness since it is thin.

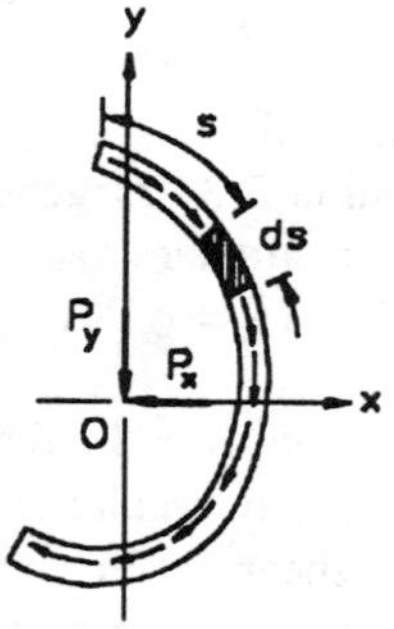

FIGURE 7.17 Shear stresses in thin-walled open sections subjected to lateral loading.

Let τ be the shear stress at any point of distance s from the edge of the bar cross section and h the wall thickness which may vary along the contour, i.e., $h = f(s)$. Substituting into Equation 7.37 yields

$$d(\tau h) = -h\left(\frac{P_x}{I_y}x + \frac{P_y}{I_x}y\right)ds$$

* A more exact value ($k = 1.2$) may be derived on an energy basis considering the variation of τ over the depth. For this, see Boresi and Sidebottom.[1] The strain energy due to shear is often much smaller than that due to bending, and the use of the values of coefficient a from Table 7.3 does not introduce appreciable error.

Integration along s, and observing that at $s = 0$; $\tau = 0$ since it is a free edge, gives

$$\tau h = -\left(\frac{P_x}{I_y} S_y^* + \frac{P_y}{I_x} S_x^*\right) \tag{7. 42}$$

where S_x^* and S_y^* are the first moments of area of a length s about the x- and y-axes, respectively, as given by Expression 7.1.

Example 7.7

A bar having an I section B = H *of uniform thickness* h, *as shown in Figure 7.18a, is subjected to an inclined force* P *passing through the centroid of the cross section. Find the shear stresses.*

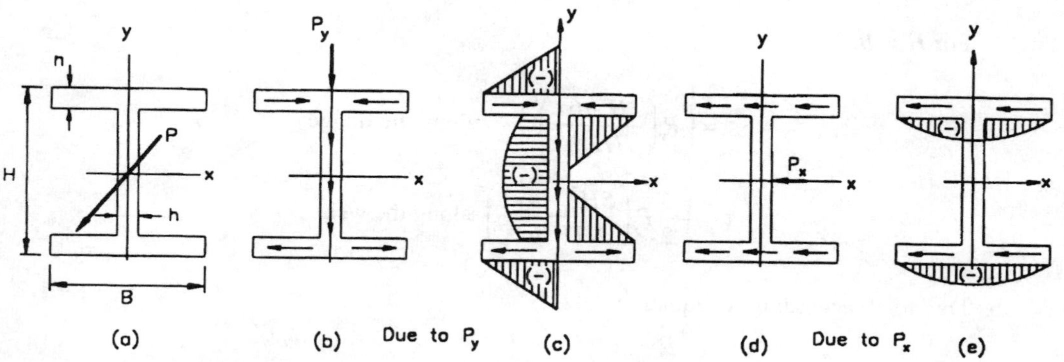

FIGURE 7.18 Example 7.7.

Solution:

a. Due to P_y, according to Equation 7.42,

$$\tau = -\frac{1}{h} \cdot \frac{P_y}{I_x} S_x^*$$

where

$$I_x = 2Bh\left(\frac{H}{2}\right)^2 + \left(\frac{Hh}{12}\right) = \frac{H^2 h}{2}\left(B + \frac{H}{6}\right)$$

Along the flanges,

$$S_x^* = h\frac{H}{2}\left(\frac{B}{2} - x\right)$$

hence,

$$\tau = -P_y\left(\frac{B/2 - x}{Hh(B + H/6)}\right)$$

Along the web,

$$S_x^* = Bh\left(\frac{H}{2}\right) + \left[\left(\frac{H}{2} - y\right)h\right]\left[\frac{1}{2}\left(\frac{H}{2} + y\right)\right] = \frac{h}{2}\left(BH + \frac{H^2}{4} - y^2\right)$$

hence,

$$\tau = -P_y\left(\frac{BH + \dfrac{H^2}{4} - y^2}{H^2h(B + H/6)}\right)$$

For $H = B$,

$$\tau = -P_y\left(\frac{3H - 6x}{7H^2h}\right) \quad \text{along the flange}$$

$$\tau = -\frac{3}{2}P_y\left(\frac{5H^2 - 4y^2}{7H^3h}\right) \quad \text{along the web}$$

b. Due to P_x according to Equation 7.42,

$$\tau = -\frac{1}{h}\frac{P_x}{I_y}S_y^*$$

$$I_y = \frac{hH^3}{6} \quad \text{(neglecting } Hh^3/3\text{)}$$

Along the flanges,

$$S_y^* = \frac{h}{2}\left(\frac{B^2}{4} - x^2\right)$$

hence,

$$\tau = -3P_x\left(\frac{\dfrac{B^2}{4} - x^2}{hB^3}\right)$$

Along the web,

$$S_y^* = 0$$

hence, $\tau = 0$.
For $H = B$,

$$\tau = 3P_x\left(\frac{H^2 - 4x^2}{4hH^3}\right)$$

The shear stresses due to P are the summation of τ due to P_y and P_y, as shown in Figure 7.18b and c, respectively.

7.4.3.1 Shear Center

In a thin-walled open section, the integration of the shear stresses, given by Equation 7.42, over the cross sectional area are the forces P_x and P_y in the x- and y-directions, respectively. These resultant forces are supposed to pass through O, the centroid of the cross section. This is not usually the case unless x and y are axes of symmetry for the section. In general the resultants of the shear stresses of Equation 7.42 will pass by some point c away from O which is called the shear center, a property of a section. In this case, the shear stresses given by Equation 7.42 are to be superimposed on the shear stresses due to the twisting moment resulting from P_x and P_y being offset from O the distance e.

Let e_x and e_y be the x- and y-coordinates of the shear center respectively. Then, the twisting moment M_e as a result of offset is given by the relation

$$M_e = P_x e_y - P_y e_x \tag{7.43}$$

which has to be superimposed on any other acting twisting moment M_z.

Example 7.8:

Obtain the shear center for the channel cross section B × H *of uniform thickness* h, *as shown in Figure 7.19.*

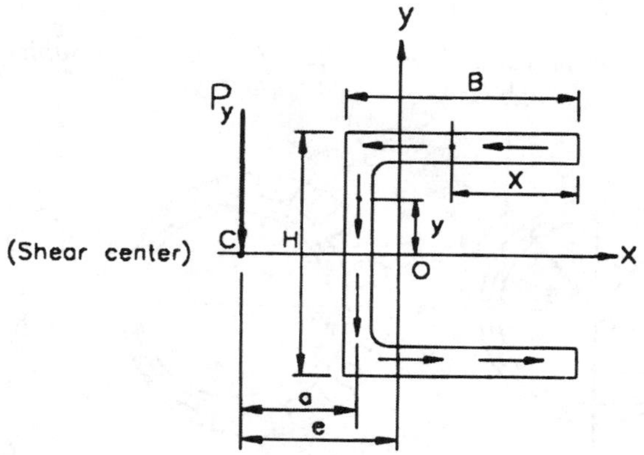

FIGURE 7.19 Example 7.8.

Solution:

Since x is an axis of symmetry, then the shear center c will be on the x-axis, distance a from the web. Consider a force P_y acting through c in the y-direction. The shear stresses are given by similar expressions as those obtained in Example 7.7. The moments of the shear stresses multiplied by the area when taken about c will equal to zero as follows:

For the flanges,

$$S_x^* = h\frac{H}{2}x, \text{ moment arm} = \frac{H}{2}$$

$$\text{Moments about } c = \frac{P_x}{4I_x} hH^2B^2$$

For the web,

$$S_x^* = \frac{BhH}{2} + h\left(\frac{H}{2} - y\right)\left(\frac{H}{4} + \frac{y}{2}\right), \text{ moment arm} = a$$

$$\text{Moments about } c = \frac{P_y ahH^2}{2I_x}\left[B + \frac{H}{6}\right]$$

$\Sigma M = 0$ gives

$$a = \frac{3B^2}{6B + H}$$

and

$$e = \frac{3B^2}{6B + H} + \frac{B^2}{2B + H}$$

Example 7.9:

Determine the shear stresses in a slitted, thin-walled circular section subjected to an inclined force P *passing through the shear center as shown in Figure 7.20.*

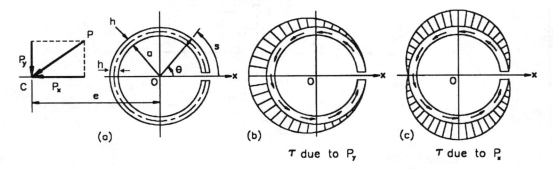

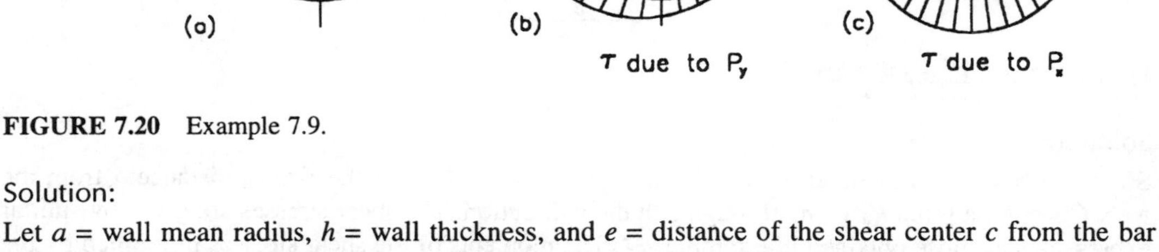

FIGURE 7.20 Example 7.9.

Solution:

Let a = wall mean radius, h = wall thickness, and e = distance of the shear center c from the bar axis. Since x is an axis of symmetry, then the shear center c will be on the x-axis.

a. Due to P_y, according to Equation 7.42:

$$\tau = -\frac{1}{h}\frac{P_y}{I_x}S_x^*$$

where $I_x = \pi a^3 h$ and $S_x^* = a^2 h\,(1 - \cos\theta)$. Hence,

$$\tau = -\frac{P_y}{\pi a}(1 - \cos\theta)$$

b. Due to P_x, according to Equation 7.42:

$$\tau = -\frac{1}{h}\frac{P_x}{I_y}S_y^*$$

where $I_y = I_x = \pi a^3 h$ and $S_y^* = a^2 h \sin\theta$. Hence,

$$\tau = -\frac{P_x}{\pi a}\sin\theta$$

The shear stresses due to P are the resultants of τ due to P_x and P_y. To obtain the distance e, moments of the shear stresses due to P_y multiplied by the area element taken about c are integrated along s and equated to zero. Hence,

$$\int_0^{2\pi}\frac{P_y}{\pi a h}(1 - \cos\theta)had\theta = 0$$

which yields $e = 2a$.

7.4.4 Thin-Walled Closed Sections

The application of Equation 7.42 to thin-walled closed sections involves finding a point along the section length where $\tau = 0$, which is to be taken as the origin of s. The location of this point should satisfy the condition that the moment of the shear flow integrated along the entire cross sectional length, taken about the centroid, vanishes. This is expressed by

$$\oint \tau h r_n ds = 0 \qquad (7.44)$$

where r_n is the radius normal to ds, as shown in Figure 7.21. If the section is symmetrical about the axis along which the shearing force acts, then $\tau = 0$ at the point of intersection of s with the axis of symmetry as shown in Example 7.10.

Example 7.10:

Obtain the shear stresses in a thin-walled circular pipe as a result of a vertical *shearing force* P *passing through the center, as shown in Figure 7.22.*

Solution:

From symmetry, $\tau = 0$ at point O_1; hence, substituting in Equation 7.42 knowing that

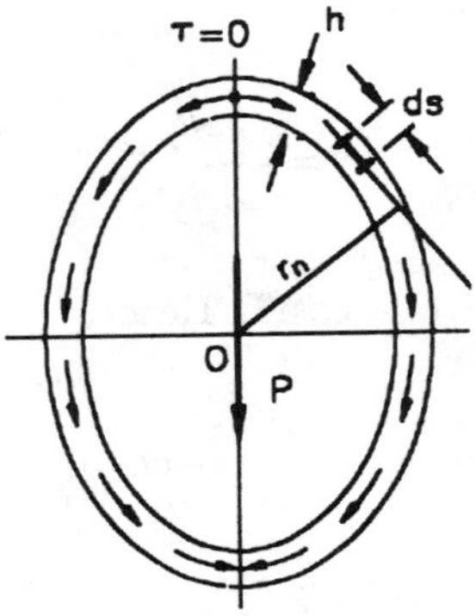

FIGURE 7.21 Shear stresses in a thin-walled closed section subjected to lateral loading.

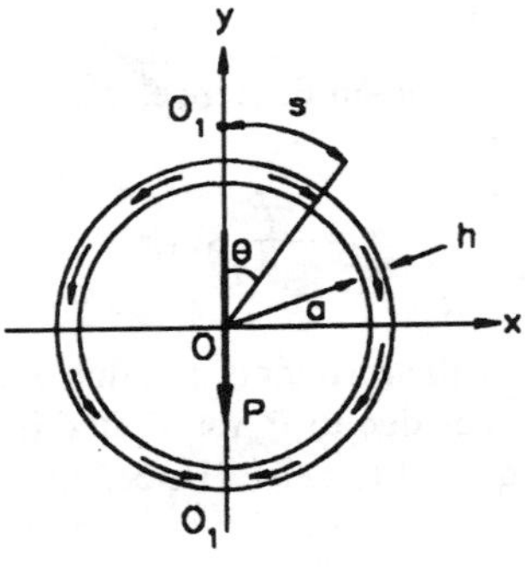

FIGURE 7.22 Example 7.10.

$$P_x = 0, \;\; P_y = P, \;\; I_x = \pi a^3 h, \;\; \text{and} \;\; S_x^* = a^2 h(1 - \sin\theta)$$

yields

$$\tau = -P\frac{a^2 h(1 - \sin\theta)}{\pi a^3 h^2} = -\frac{P}{\pi a h}(1 - \sin\theta)$$

Example 7.11:

Obtain the shear stresses in a thin-walled unsymmetrical square section as a result of a shearing force P ***passing through the centroid O as shown in Figure 7.23.***

Solution:

It can be easily shown that the centroid of the unsymmetrical section lies at a distance of $0.4a$ from the thick side and $0.6a$ from the thin side. Let the point at which $\tau = 0$ be offset a distance e to the left of the y-axis. This distance e is determined by applying Equation 7.44. It will be observed that the moments of the shear flow along the horizontal sides will cancel each other. Along the vertical sides the moments are expressed as

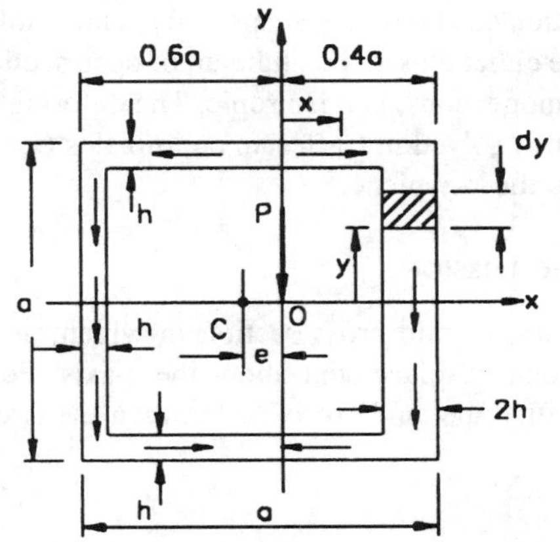

FIGURE 7.23 Example 7.11.

$$\int_0^{a/2}\left\{0.4a\left[(e+0.4a)\frac{a}{2}h+\left(\frac{a}{2}-y\right)\left(\frac{a}{2}+y\right)h\right]\right.$$

$$\left.-0.6a\left[(o,6a-e)\frac{a}{2}h+\left(\frac{a}{2}-y\right)\left(\frac{a}{2}+y\right)\frac{h}{2}\right]\right\}dy=0$$

which yields $e = 0.167a$.

The shear stresses are obtained by substituting in Equation 7.42: $P_x = 0$, $P_y = P$, and $I_x = {}^3/_4\, a^3h$.

Along the horizontal sides at any distance x, τ is given by

$$\tau=-\frac{4P}{3a^3h}(0.167a+x)\frac{a}{2}=-\frac{2P}{3a^2h}(0.167a+x)$$

Along the right vertical side at any distance y:

$$\tau=-\frac{4P}{3a^3h}\left(0.5335a^2-y^2\right)$$

Along the left vertical side at any distance y:

$$\tau=-\frac{4P}{3a^3h}\left(0.3415-0.5y^2\right)$$

7.5 TORSION OF BARS

In Section 7.4 the shear stress components τ_{xz} and τ_{yz} resulting from the shearing force components P_x and P_y have been obtained. It is now required to determine the shear stresses due to a twisting moment M_t, where in this case $M_t = M_z$. The bar cross section is free to warp, and hence there are no axial stresses arising from M_z. This is called a state of Saint-Venant's free torsion, and τ_{zx} and τ_{zy} are the only stresses acting on the bar cross section.

In the following analysis, a bar in a state of free torsion is considered. A twisting moment is applied at the free ends. The bar axis is straight and the shape of the cross section is constant along the axis length. The cross section can be of any shape and is either solid or hollow. Hollow sections usually have thin walls and are either closed, i.e., tubular, or open sections. In the following analysis, the bar material is elastic, homogeneous, and isotropic. The deformations are small and the strains are infinitesimal. The problem is solved in Cartesian coordinates (x, y, z), where the bar axis is the z-axis and the cross section is the x–y plane.

7.5.1 Saint-Venant's Free Torsion

Figure 7.24 shows a bar having a solid cross section in which, as a result of twist, each cross section will undergo a rotational displacement about the z-axis. For any two cross sections the relative angle of rotation is called the angle of twist between the two sections.

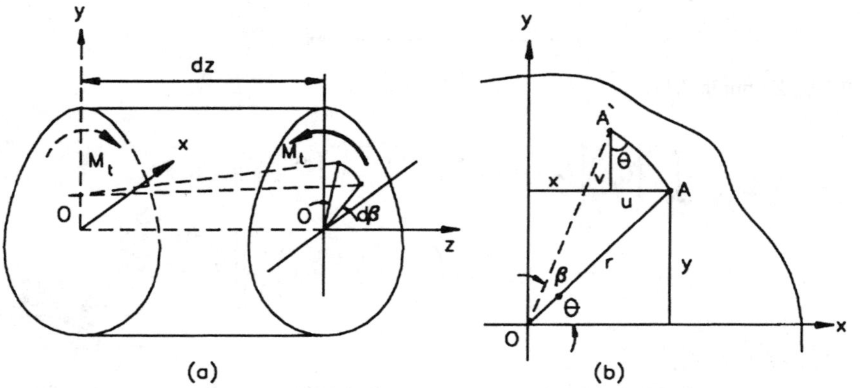

FIGURE 7.24 Free torsion of a bar: (a) bar element, and (b) displacements in the cross section plane x-y.

Let $d\beta$ be the angle of twist for two cross sections at distance dz and α be the angle of twist per unit length of bar. Hence,

$$d\beta = \alpha dz \tag{7.45}$$

Now, since conditions are the same for all cross sections, then Equation 7.45 applies to any two cross sections at a distance dz along the bar length. Therefore, α is constant along the bar length and if

$$\text{At } z = 0\text{:} \quad \beta = 0,$$

$$\text{Then } \beta = \alpha z \tag{7.46}$$

Except for a circular cross section, the cross section is distorted in the x-, y-, and z-directions. However, a radius drawn from the z-axis will have its projection on the x–y plane remaining as a radius. The distortion of the cross section in the z-direction will cause warping.

Referring to Figure 7.24b, consider a cross section at a distance z in which a point A (x, y) at a radius r making an angle θ with the x-axis moves to A'. The displacements of A in the x-, y-, and z-directions are u, v, and w, respectively. Hence,

$$AA' = r\beta = r\alpha z$$
$$u = -AA'\sin\theta = -r\alpha z\sin\theta$$
$$v = AA'\cos\theta = r\alpha z\cos\theta$$

which reduces to

$$u = -y\alpha z \qquad v = x\alpha z \tag{7.47}$$

Since all cross sections will warp identically, therefore, w is independent of z and is expressed by

$$w = \psi(x, y) \tag{7.48}$$

where ψ is called Saint-Venant's warping function.*

Substituting Equations 7.47 and 7.48 in the strain–displacement Equations 2.6, the strain components for free torsion are given by

$$\begin{aligned}
&\varepsilon_x = \frac{\partial u}{\partial x} = 0, \quad \varepsilon_y = \frac{\partial v}{\partial y} = 0, \quad \text{and } \varepsilon_z = \frac{\partial w}{\partial z} = 0 \\
&\gamma_{xy} = \frac{\partial v}{\partial x} + \frac{\partial u}{\partial y} = 0 \\
&\gamma_{zx} = \frac{\partial u}{\partial z} + \frac{\partial w}{\partial x} = -\alpha y + \frac{\partial \psi}{\partial x} \\
&\gamma_{zy} = \frac{\partial v}{\partial z} + \frac{\partial w}{\partial y} = \alpha x + \frac{\partial \psi}{\partial y}
\end{aligned} \tag{7.49}$$

Equation 7.48 is the strain compatibility in terms of the function ψ.

The corresponding stress components are obtained from the inverse form of Hooke's law, Equations 3.12, as

$$\begin{aligned}
&\sigma_x = \sigma_y = \sigma_z = \tau_{xy} = 0 \\
&\tau_{zx} = G\left(-\alpha y + \frac{\partial \psi}{\partial x}\right) \\
&\tau_{zy} = G\left(\alpha x + \frac{\partial \psi}{\partial y}\right)
\end{aligned} \tag{7.50}$$

A solution of the problem is obtained by finding a function $\psi\ (x, y)$ that satisfies equilibrium and boundary conditions. The equations of equilibrium (Equations 1.6) reduce to

* Named after B. de Saint-Venant (1797–1886).

$$\frac{\partial \tau_{zx}}{\partial x} + \frac{\partial \tau_{zy}}{\partial y} = 0 \tag{7.51}$$

Substituting Equation 7.50 into Equation 7.51 yields

$$\frac{\partial^2 \psi}{\partial x^2} + \frac{\partial^2 \psi}{\partial y^2} = 0$$

or

$$\nabla^2 \psi = 0 \tag{7.52}$$

The boundary conditions are such that the bar surface is free; hence the boundary of the bar cross section is a shear stress trajectory, Figure 7.25. Therefore,

$$\tau_{zx} \frac{dy}{ds} - \tau_{zy} \frac{dx}{ds} = 0 \tag{7.53}$$

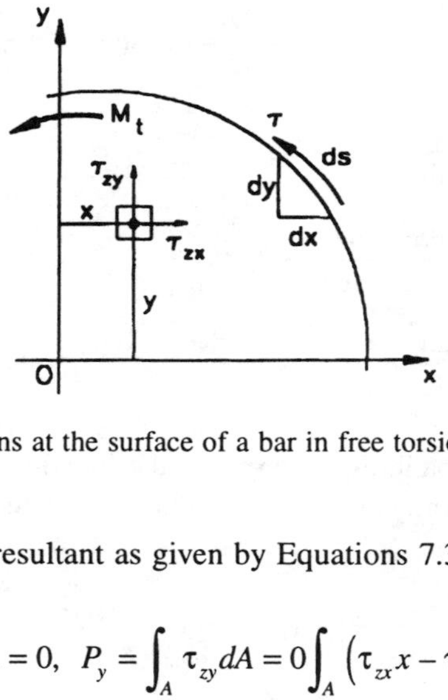

FIGURE 7.25 Boundary conditions at the surface of a bar in free torsion.

At any cross section, the stress resultant as given by Equations 7.3 are

$$P_x = \int_A \tau_{zx} dA = 0, \; P_y = \int_A \tau_{zy} dA = 0 \int_A \left(\tau_{zx} x - \tau_{zy} y\right) dA = M_t \tag{7.54}$$

with M_t as the applied twisting moment.

The equation of equilibrium (Equation 7.51) is satisfied by a stress function ϕ (x, y), such that

$$\tau_{zx} = \frac{\partial \phi}{\partial y} \text{ and } \tau_{zy} = -\frac{\partial \phi}{\partial x}$$

Substituting Equation 7.51 into Equation 7.50 yields

$$\frac{\partial \phi}{\partial y} = G\left(-\alpha y + \frac{\partial \psi}{\partial x}\right)$$

and

$$\frac{\partial \phi}{\partial x} = -G\left(\alpha x + \frac{\partial \psi}{\partial y}\right)$$

so that

$$\frac{\partial^2 \phi}{\partial x^2} + \frac{\partial^2 \phi}{\partial y^2} = -2G\alpha$$

or

$$\nabla^2 \phi = -2G\alpha \tag{7.55}$$

The function ϕ is called the Prandtl stress function* for torsion. The boundary condition given by Equation 7.53 reduces to

$$\frac{\partial \phi}{\partial y}\frac{dy}{ds} + \frac{\partial \phi}{\partial x}\frac{dx}{ds} = 0 \text{ or } \frac{\partial \phi}{ds} = 0 \text{ and } \phi = \text{constant}$$

Since the stresses are obtained from ϕ by differentiation, then on the boundary;

$$\frac{\partial \phi}{\partial s} = \phi = 0 \tag{7.56a}$$

and the boundary condition given by Equation 7.54 reduces to

$$2\int_A \phi dA = M_t \tag{7.56b}$$

The above equations are applied in the following sections to solve the problem of torsion of bars of solid circular, elliptical, and rectangular cross sections. The solution is based on choosing a stress function ϕ suitable to the bar cross section under consideration.

7.5.2 Solid Circular Section

The equation of the boundary of a bar of circular section as shown in Figure 7.26 is

$$x^2 + y^2 - r_o^2 = 0$$

Choosing a stress function,

$$\phi = C\left(x^2 + y^2 - r_o^2\right)$$

* Named after L. Prandt, 1903.

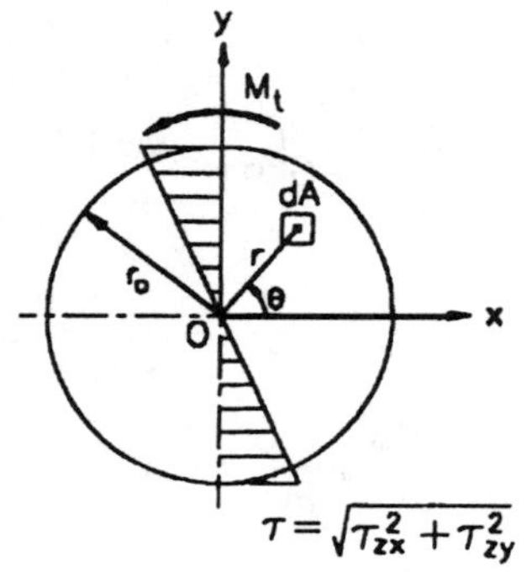

FIGURE 7.26 Shear stresses in a solid circular section of an elastic bar in free torsion.

will satisfy the boundary conditions (Equations 7.56), namely, $d\phi/ds = 0$ and $\phi = 0$, where C is a constant. Substitution of ϕ in Equation 7.55 gives

$$C = -G\alpha/2$$

and, hence,

$$M_t = -G\alpha \int_A \left(x^2 + y^2 - r_o^2\right) dA$$

$$= -G\alpha\left(I_x + I_y - Ar_o^2\right) = G\alpha I_t$$

where $I_t = I_x + I_y - Ar_o^2 = (\pi/2)\, r_o^4 = I_o$; the polar second moment of area for a circular section. The shear stress components are given by

$$\tau_{zx} = -\frac{\partial \phi}{\partial y} = -G\alpha y = -\frac{M_t}{I_t} y$$

$$\tau_{zy} = -\frac{\partial \phi}{\partial x} = G\alpha x = \frac{M_t}{I_t} x$$

The resultant shear stress is thus

$$\tau = \sqrt{\tau_{zx}^2 + \tau_{zy}^2} = G\alpha r = \frac{M_t}{I_t} r \tag{7.57}$$

which is the same expression as derived by the elementary strength-of-materials approach. Substitution of τ_{xy} and τ_{zy} in Equation 7.50 gives

$$\frac{\partial \psi}{\partial x} = \frac{d\psi}{dy} = 0$$

If, at $z = 0$, $\psi = 0$, then $w = \psi(x, y) = 0$ for all cross sections, i.e., no warping. Hence, as always considered by the elementary approach, the circular cross section remains plane.

7.5.3 Solid Elliptical Section

The equation of the boundary of the elliptical bar cross section as shown in Figure 7.27, is

$$\frac{x^2}{a^2}+\frac{y^2}{b^2}-1=0$$

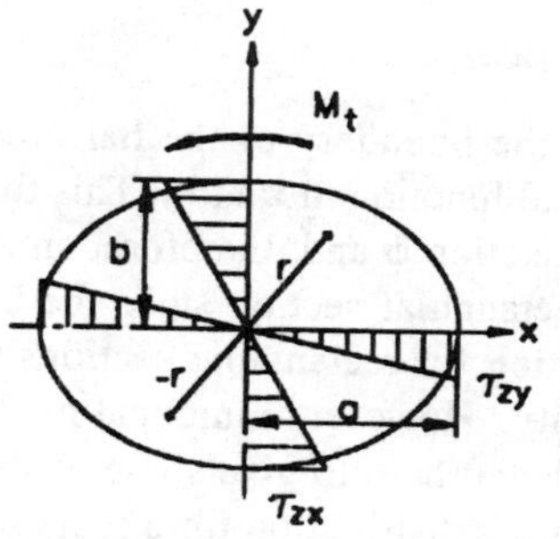

FIGURE 7.27 Shear stresses in a solid elliptic section of an elastic bar in free torsion.

Choosing a stress function,

$$\phi = C\left(\frac{x^2}{a^2}+\frac{y^2}{b^2}-1\right)$$

where C is a constant that satisfies the boundary conditions (Equations 7.56) as

$$\frac{d\phi}{ds}=\phi=0$$

The constant C is determined by substituting ϕ in Equation 7.55. Hence,

$$C=-G\alpha\frac{a^2b^2}{a^2+b^2}$$

and the twisting moment and the shear stress satisfy the last of the boundary conditions (Equations 7.56) as given, respectively, by

$$M_t = 2\int_A \phi dA = G\alpha\frac{\pi a^3 b^3}{a^2+b^2}$$

$$\tau_{zx}=\frac{\partial\phi}{\partial y}=-\frac{M_t}{2I_x}y;\quad I_x=\pi ab^3/4$$

$$\tau_{zy}=-\frac{\partial\phi}{\partial x}=\frac{M_t}{2I_y}x;\quad I_y=\pi ab^3/4 \qquad (7.58)$$

$$\tau_{\max}=-\frac{2M_t}{\pi ab^2}\ \text{at}\ y=b;\ \text{the end of the minor axis}$$

From Equations 7.50 the warping function is

$$w = M_t \frac{b^2 - a^2}{\pi a^3 b^3 G} xy \tag{7.59}$$

which shows that the section warps into a diagonally symmetric surface.

7.5.4 Solid Rectangular Section

In the two previous applications, the boundary of the bar cross section, circular or elliptical, is expressed by a continuous analytical function in x and y. This function multiplied by some constant has been shown to be the stress function ϕ and, therefore, an analytical solution is possible. This does not apply to the case of a rectangular section since the boundary shape is discontinuous. A rather complicated analytical solution for rectangular sections is available in most of the texts on the mathematical theory of elasticity.[3] However, numerical methods of solution are better suited to solve this problem. It will be sufficient here to give some of the results of the analytical solution.

Figure 7.28 shows the shear stress distribution for a rectangular cross section of a short side b and a long side h. The stress is maximum at the boundary at the middle point of the long side and is expressed by the relation:

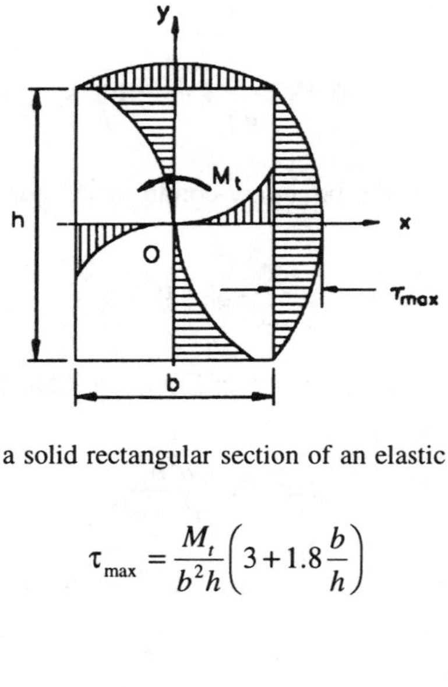

FIGURE 7.28 Shear stresses in a solid rectangular section of an elastic bar in free torsion.

$$\tau_{max} = \frac{M_t}{b^2 h}\left(3 + 1.8\frac{b}{h}\right) \tag{7.60a}$$

Again,

$$M_t = G\alpha I_t \quad \text{or} \quad \alpha = \frac{M_t}{GI_t} \tag{7.60b}$$

where

$$I_t = b^3 h\left(\frac{1}{3} - 0.21\frac{b}{h}\right) \tag{7.60c}$$

For a thin rectangular section $b/h \to 0$, where b is the thickness; therefore,

$$\tau_{max} = \frac{M_t}{hb^2/3} = G\alpha b \tag{7.61a}$$

$$\alpha = \frac{M_t}{Ghb^3/3} \tag{7.61b}$$

$$I_t = hb^3/3 \tag{7.61c}$$

Examples 7.12:

Consider three bars of equal cross sectional areas having the following shapes: solid circular of radius a, *hollow circular of inner radius* a *and outer radius* b, *and solid rectangular of sides* a *and* c. *Determine for the same applied twisting moment* M_t *the following:*

a. *The ratio of their angles of twist per unit length,*
b. *The ratio of their maximum shear stresses.*

Solution:

On the basis of equal cross sectional areas for a hollow circular section,

$$\pi a^2 = \pi(b^2 - a^2)$$

hence $b = \sqrt{2}\ a$ and, for a rectangular section:

$$\pi a^2 = ca$$

hence $c = \pi a$.

a. The angle of twist per unit length α is calculated for each bar according to
Solid circular section:

$$\alpha = M_t/GI_t = \frac{2M_t}{\pi G a^4} = 0.6369\frac{M_t}{Ga^4}$$

Hollow circular section:

$$\alpha = M_t/GI_t = \frac{2M_t}{\pi G\left(b^4 - a^4\right)} = \frac{2M_t}{3\pi G a^4} = 0.2123\frac{M_t}{Ga^4}$$

Rectangular section:

$$\alpha = M_t/GI_t = M_t\bigg/\left[Ga^3c\left(\frac{1}{3} - \frac{0.21a}{c}\right)\right] = 1.1952\frac{M_t}{Ga^4}$$

b. The maximum shear stresses τ_{max} are given by

Solid circular section:

$$\tau_{\max} = M_t a/I_t = \frac{2M_t}{\pi a^3} = 0.6369\frac{M_t}{a^3}$$

Hollow circular:

$$\tau_{\max} = M_t b/I_t = \frac{M_t b}{\pi\left(b^4 - a^4\right)/2} = 0.3002\frac{M_t}{a^3}$$

Rectangular:

$$\tau_{\max} = M_t\left(3 + 1.8\frac{a}{c}\right)\Big/a^2 c = 1.13798\frac{M_t}{a^3}$$

The results as summarized in Table 7.4 indicate that a hollow circular cross section is the best section among the three sections from both rigidity and strength requirements for the same applied twisting moment and same weight.

TABLE 7.4
Comparison Among Different Sections in Torsion

Section	$\alpha G a^4/M_t$	$\tau_{\max} a^3/M_t$
Solid circular	0.6369	0.6369
Hollow circular	0.2123	0.3002
Rectangular	1.1952	1.1380

7.5.5 Thin-Walled Open Sections

As mentioned before, the shear stresses in the plane of thin-walled cross sections are parallel to the boundary since the boundary is a shear stress trajectory. Figure 7.29a shows the direction of shear stresses in a rectangular thin-walled section of thickness h and height s. The stresses and angle of twist for this section are obtained from Equations 7.61.

Consider, now, the straight rectangular section to be shaped as shown in Figure 7.29b. From equilibrium and boundary conditions the stresses and angle of twist will remain the same and are independent of the shape of the section for the same h and s, so that from Equation 7.61

$$\tau_{\max} = \frac{M_t}{sh^2/3}, \quad M_t = G\alpha\frac{sh^3}{3}, \quad \text{and } I_t = \frac{sh^3}{3} \tag{7.62}$$

For a section consisting of n different thin elements, as shown in Figure 7.29c,

$$\begin{aligned} M_t &= M_{t1} + M_{t2} + M_{t3}\cdots + M_{tn} = G\alpha\sum_i^n \frac{s_i h_i^3}{3} = G\alpha I_t \\ \tau_{\max} &= \frac{M_t}{I_t} h_{\max} \end{aligned} \tag{7.63}$$

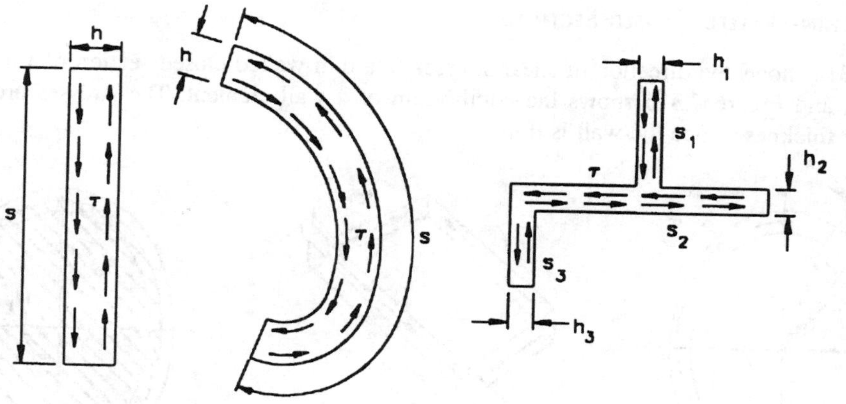

FIGURE 7.29 Shear flow in a thin-walled open section in free torsion: (a) rectangular section, (b) curved section, (c) composite screen.

Example 7.13:

Determine the maximum shear stress and the angle of twist per unit length for the thin-walled open sections shown in Figure 7.30 when subjected to a twisting moment M_t. ***The sections have a uniform thickness*** h.

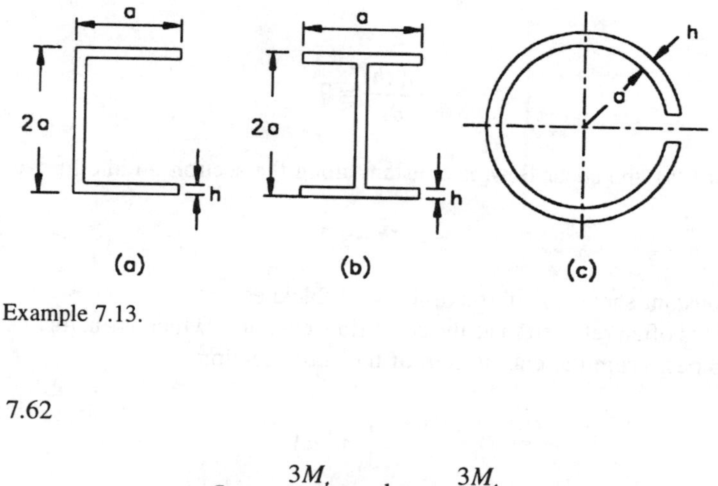

FIGURE 7.30 Example 7.13.

Solution:

From Equation 7.62

$$\tau_{\max} = \frac{3M_t}{sh^2} \quad \text{and} \quad \alpha = \frac{3M_t}{Gsh^3}$$

a. For the channel section, $s = 4a$; hence,

$$\tau_{\max} = \frac{3M_t}{4ah^2} \quad \text{and} \quad \alpha = \frac{3M_t}{4Gah^3}$$

b. For the I section, $s = 4a$, hence, $\tau_{\max}$ and α are the same as for the channel section.
c. For the slitted circular section, $s = 2\pi a$; hence,

$$\tau_{\max} = \frac{3M_t}{2\pi ah^2} \quad \text{and} \quad \alpha = \frac{3M_t}{2\pi aGh^3}$$

7.5.6 Thin-Walled Closed Sections

Figure 7.31a shows the direction of shear stresses in a thin-walled closed section of variable wall thickness, and Figure 7.31b shows the equilibrium of a wall element. The stresses are uniform across the thickness since the wall is thin.

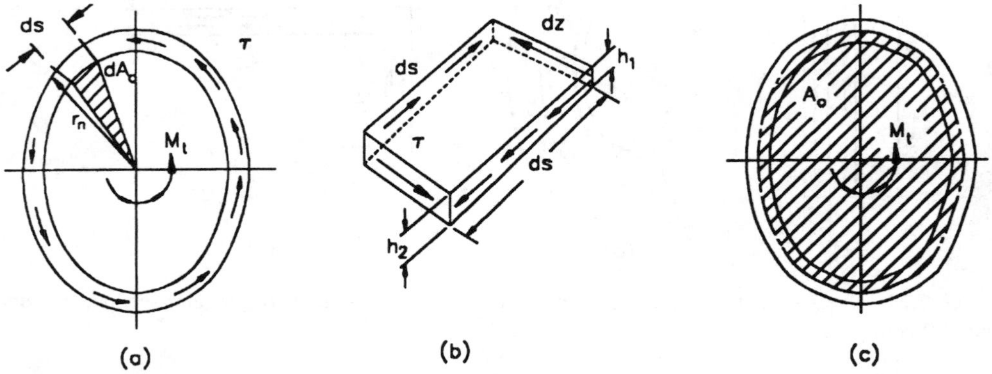

FIGURE 7.31 Shear flow in a thin-walled closed section in free torsion: (a) shear flow, (b) static equilibrium of a wall element, and (c) static equilibrium of the entire section.

The section wall is a shear stress trajectory so that, from Equation 7.37, substituting $P_x = P_y = 0$, yields

$$\frac{\partial(\tau h)}{\partial s} = 0$$

which means that (τh), the shear flow, is constant along the section middle line, or

$$\tau h = f \tag{7.64}$$

where, f is the constant shear flow through the wall thickness.

Equation 7.64 is often referred to as the shear flow equation, which is analogous to the equation of fluid flow in pipes. From the equilibrium of the entire section:

$$M_t = \int_A \tau r_n dA$$

where r_n is the radius normal to the direction of τ and

$$dA = hds$$

Hence,

$$M_t = \oint \tau r_n hds = f\oint r_n ds$$

But $r_n \, ds/2 = dA_o$ (the shaded area in Figure 7.31a). Hence,

$$\oint r_n ds = 2A_o$$

Where A_o is the area enclosed by the middle line of the cross section, as shown by the shaded area in Figure 7.31c. Therefore,

$$M_t = 2fA_o, \tau = \frac{M_t}{2A_o h}, \text{ and } \tau_{max} = \frac{M_t}{2A_o h_{min}} \tag{7.65}$$

The angle of twist α per unit bar length is obtained by equating the work done by M_t to the strain energy according to the principle of work, Section 3.10.1, and is given by:

$$\alpha = \frac{M_t}{4GA_o^2} \oint \frac{ds}{h} \tag{7.66}$$

For uniform thickness h, this angle becomes

$$\alpha = \frac{M_t s}{4GA_o^2 h} = \frac{\tau s}{2GA_o^2} \tag{7.67}$$

where s is the total length of the middle line of the section.

Example 7.14:

A thin strip having a cross section s × h *is folded to form a thin-walled section of perimeter s and thickness* h. *The formed section is subjected to a twisting moment* M_t. *Compare the values of* τ_{max} *and* α *for the following sections, Figure 7.32: circular, square, and rectangular having a height double the width, if (a) the edges are unwelded and (b) the edges are welded.*

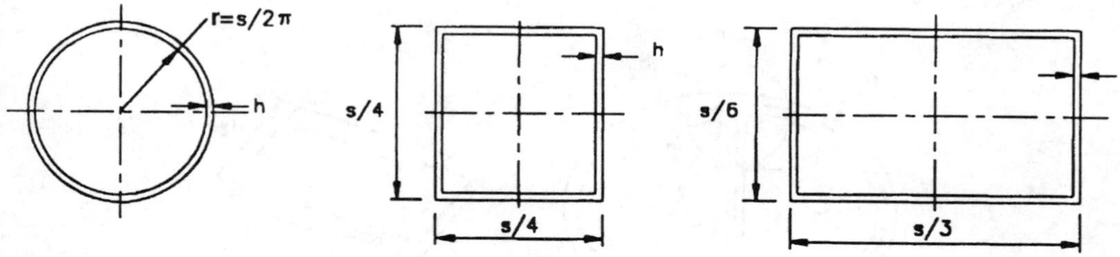

FIGURE 7.32 Example 7.14.

Solution:

The dimensions of the sections of the same weight are as shown in Figure 7.32:

a. For unwelded edges τ_{max} and α are the same for all sections and are given by Equation 7.62:

$$\tau_{max} = \frac{3M_t}{sh^2} \text{ and } \alpha = \frac{3M_t}{Gsh^3}$$

b. For welded edges, Equations 7.65 and 7.66 for closed sections give

Circular section: $\tau = \dfrac{2\pi M_t}{s^2 h}$ and $\alpha = \dfrac{4\pi^2 M_t}{Gs^3 h}$

Square section: $\tau = \dfrac{8M_t}{s^2 h}$ and $\alpha = \dfrac{64M_t}{Gs^3 h}$

Rectangular section: $\tau = \dfrac{9M_t}{s^2 h}$ and $\alpha = \dfrac{81M_t}{Gs^3 h}$

From (a) and (b), the following conclusions are obtained.

- Open sections have much less strength and rigidity in torsion than closed sections of the same weight. For a circular cross section, comparison between open and closed sections shows:

$$\text{The stress ratio} = \frac{3}{2\pi}\frac{s}{h} = 0.477\frac{s}{h}$$

$$\text{The angle of twist ratio} = \frac{3}{4\pi^2}\left(\frac{s}{h}\right)^2 = 0.076\left(\frac{s}{h}\right)^2$$

- Among closed sections of the same weight in torsion, the circular thin-walled cross section has the maximum strength and rigidity. (The same conclusion is made in Example 7.12 when comparing circular thin-walled section with solid sections of the same area.)

7.5.7 Effect of Internal Stiffening Webs

Tubular sections are sometimes internally stiffened by webs to maintain their stability under load. The webs will divide the cross section into a number of cells; hence, the section is said to be cellular. Consider now a two-cell section, Figure 7.33a. Let h_1, h_2, and h_3 be uniform along s_1, s_2, and s_3, respectively.

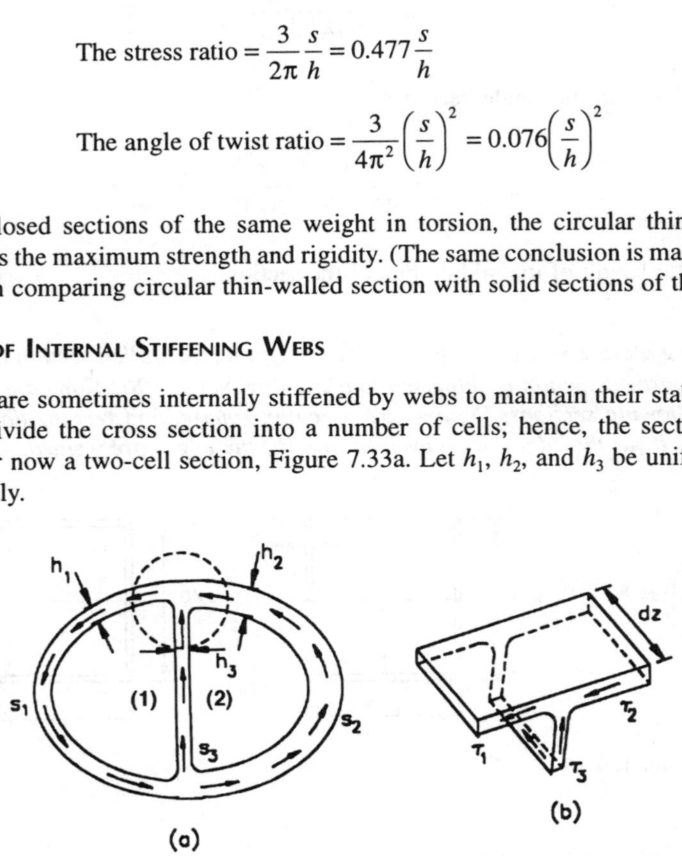

FIGURE 7.33 Shear flow in a cellular section in free torsion: (a) shear flow and (b) static equilibrium of junction.

From the equilibrium of the junction in the z-direction, Figure 7.33b,

$$\tau_1 h_1 dz - \tau_2 h_2 dz - \tau_3 h_3 dz = 0$$

or

$$f_1 = f_2 + f_3 \quad \text{(shear flow equation)} \tag{7.68}$$

Therefore, at the junction the inward flow is equal to the outward flow. Let A_1 and A_2 be the areas of cell (1) and (2), respectively; hence,

$$M_t = 2f_1A_1 + 2f_2A_2 \text{ (moment equation)} \tag{7.69}$$

and

$$\tau_1 s_1 + \tau_3 s_3 = 2G\alpha A_1 \text{ (cell equations)} \tag{7.70}$$

$$\tau_2 s_1 + \tau_3 s_3 = 2G_2\alpha A_2$$

There are four unknowns: τ_1, τ_2, τ_3, and α and four equations: one flow equation, one moment equation, and two cell equations. Therefore, a solution can be obtained. This procedure can be extended to multicell sections.

Example 7.15:

Show that for the two-cell closed section of uniform wall thickness as shown in Figure 7.34, the central web is free from shear if the section is symmetrical about the web middle line.

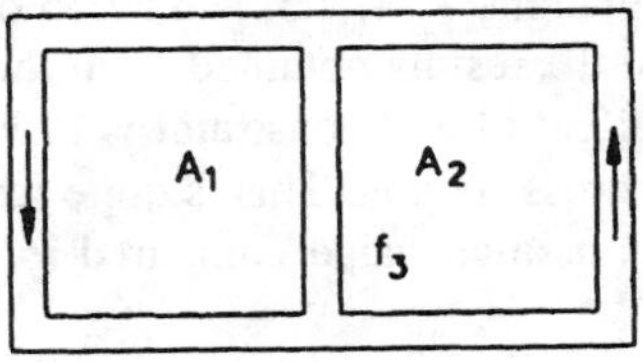

FIGURE 7.34 Example 7.15.

Solution:

From symmetry:

$$A_1 = A_2,\quad s_1 = s_2,\quad \text{and } h_1 = h_2$$

Therefore,

$$f_1 = f_2$$

Substitution in the shear flow Equation 7.68 gives

$$f_3 = 0,\quad \text{i.e.,}\quad \tau_3 = 0$$

Hence, the web is free from stresses.

7.5.8 Effect of End Constraint

The analysis of torsion in the above sections has been confined to the state of free torsion in which the bar cross sections are free to warp and τ_{zx} and τ_{zy} are the only stresses acting on the bar cross section. If one end of the bar is constrained against warping, normal stresses will develop and the shear stresses and angle of twist may be affected. For solid and tubular sections, the effect is not appreciable, whereas for thin-walled open sections the effect may be considerable, especially for short bars.

Consider the case of a bar fixed at one end to a rigid block end acted upon by a twisting moment M_t at the other end, which is free. The bar is thus prevented from warping at the fixed end $z = 0$, and free to warp at the free end $z = L$, where L is the bar length. The distribution of the axial displacement w along z is assumed to be given by

$$w = \psi(x, y)(2Lz - z^2)/L^2 \tag{7.71a}$$

Hence,

$$\varepsilon_z = \frac{\partial w}{\partial z} = \psi(x, y)\left[2(L - z)/L^2\right] \tag{7.71b}$$

Equations 7.71 satisfy the boundary conditions that, at $z = 0$, $w = 0$ and ε_z is maximum and, at $z = L$, w is maximum and $\varepsilon_z = 0$. It is further assumed that $\sigma_x = \sigma_y = \tau_{xy} = 0$ and, hence, from Hooke's law,

$$\sigma_z = E\varepsilon_z = 2E\psi(x, y)(L - z)/L^2 \tag{7.72}$$

Equations 7.71 and 7.72 satisfy Equations 7.51, 7.52, and 7.55 only at $z = L$. However, to simplify the solution it will be assumed that the results obtained from the state of free torsion for ψ, α, τ_{xy}, and τ_{zy} remain valid and that the effect of end constraint is to introduce σ_z, as given by Equation 7.72, which acts normal to the bar cross section. This is quite justified since from Expression 7.72 $\partial\sigma_z/\partial z$ is independent of z and has a small value compared to $2\alpha G$ in Equation 7.55. It will be noted that at any cross section,

$$\int_A \sigma_z dA = \int_A \sigma_z x dA = \int_A \sigma_z y dA = 0$$

Consider now the application to some cross sections.

7.5.8.1 Solid Sections

For circular sections there is no warping; $\psi(x, y) = 0$ and, hence, $\sigma_z = 0$ for all sections.

For a solid elliptic section, Equation 7.59 gives

$$\psi(x, y) = M_t \frac{b^2 - a^2}{\pi a^3 b^3 G} xy$$

and

$$\psi(x, y)_{\max} = M_t \frac{b^2 - a^2}{\pi a^3 b^3 G} \frac{b^2 a^2}{a^2 + b^2} \quad \text{at } x = y = \frac{ab}{\sqrt{a^2 + b^2}}$$

From Equation 7.72, the maximum normal stress is

$$(\sigma_z)_{\max} = M_t \frac{4(b^2 - a^2)(1 + \nu)}{\pi a^3 b^3 L} \cdot \frac{b^2 a^2}{a^2 + b^2} \quad \text{at } z = 0$$

To assess the value of $(\sigma_z)_{max}$ in relation to τ_{max}, consider the case where $a = 2b$ and $\nu = 0.3$; hence,

$$(\sigma_z)_{max} = -M_t \frac{0.496}{b^2 L}$$

and

$$\tau_{max} = -\frac{2M_t}{\pi a b^2} = -M_t \frac{0.318}{b^3}$$

from which $(\sigma_z)_{max}/\tau_{max} = 1.56b/L = 0.78a/L$. Putting $2a = H$ = section height gives $(\sigma_z)_{max}/\tau_{max} = 0.39H/L$, which is quite small for a bar having $H/L \leq 0.1$.

7.5.8.2 Thin-Walled Sections

As an example for an open thin-walled sections, consider a thin rectangular section, height h and thickness b, Figure7.35.

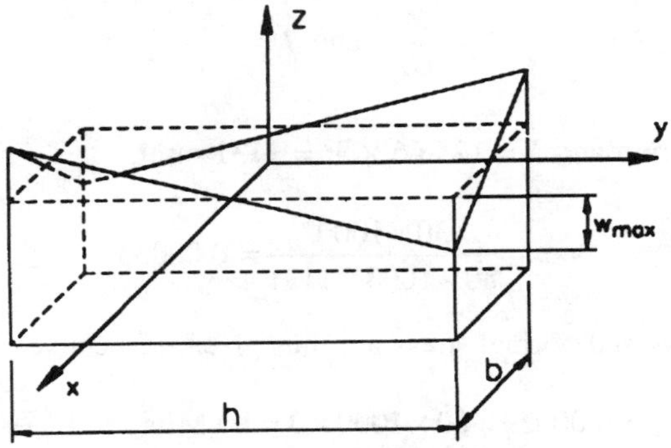

FIGURE 7.35 Warping of a thin-walled rectangular section in free torsion.

$$w = \psi(x, y) = \alpha xy$$

Hence,

$$\psi(x, y)_{max} = \alpha \frac{bh}{4} \quad \text{at } x = \frac{b}{2},\ y = \frac{h}{2}$$

The stresses are given by

$$(\sigma_z)_{max} = \alpha E \frac{bh}{2L}$$

and

$$\tau_{max} = \alpha G b$$

Hence,

$$(\sigma_z)_{max}/\tau_{max} = (1+\nu)\frac{h}{L} = 1.3h/L \text{ for } \nu = 0.3$$

For a curved cross section of length s,

$$(\sigma_z)_{max}/\tau_{max} = 1.3\frac{s}{L}$$

which is higher than that for an elliptic solid section.

Example 7.16:

For a slitted circular tube of diameter 40 mm, thickness 3, and length 200 mm, determine the maximum shear stress and the maximum normal stress due to fixation at one end when subjected to a twisting moment of 30 Nm at the free end. Take G = *80 GPa,* ν = *0.3.*

Solution:

$\tau_{max} = \alpha Gh$, where α is the angle of twist per unit length given by

$$\alpha = \frac{M_t}{GI_t} \text{ and } I_t = \frac{sh^3}{3}$$

For $s = \pi \times 40 = 125.66$ mm and $I_t = 125.66 \times 3^2 = 1131$ mm^4,

$$\alpha = \frac{30 \times 1000}{80 \times 1000 \times 1131} = 0.00033$$

The maximum shear stress and normal stress are thus given, respectively, by

$$\tau_{max} = 0.00033 \times 80 \times 1000 \times 3 = 80 \text{ MPa}$$

$$(\sigma_z)_{max} = \alpha\frac{Esh}{2L} = \frac{0.00033 \times 2 \times 80 \times 10^3 \times 1.3 \times 125.66 \times 3}{2 \times 200}$$

$$= 64.69 \text{ MPa}$$

Note that $(\sigma_z)_{max}$ is about 80% of τ_{max}, which is an appreciable value.

7.6 DISPLACEMENTS IN RODS — ENERGY APPROACH

In most engineering applications it is necessary to determine the displacement under load to ensure proper functioning of structural or machine elements. Typical examples are crane girders, machine tools frames, spindles, rotating shafts, etc., where the displacement has to be a minimum and should not exceed a certain limit. On the other hand, springs are flexible elements of low stiffness, which are required to provide specified load–displacement characteristics. In the following applications, the displacement under load is determined from energy theorems, particularly Castigliano's theorem as derived in Chapter 3.

7.6.1 Application of Castigliano's Theorem

As discussed in Section 3.10.4, Castigliano's second theorem states that the displacement at the point of application of an external load, in the direction of this load, is the partial derivative of the total strain energy with respect to this load. The load is either a force or a couple and the corresponding displacement will be either a linear displacement or an angular rotation respectively. Thus, Castigliano's theorem as in Section 3.10.4 is expressed by Equation 3.42, namely,

$$u_i = \frac{\partial U}{\partial P_i} \quad \text{and} \quad \theta_i = \frac{\partial U}{\partial M_i} \tag{3.42}$$

The rods to be considered may have a constant or a variable cross section; the axis may be straight or curved with large centroidal radius of curvature $\bar{r}$ compared with its depth H, say, $(H/\bar{r}) < 0.2$, and the rod proportions and loading allow the application of the stress formulas obtained in Section 7.2 without serious errors.

For an element ds of the rod length, the strain energy is given by

$$dU = \left[\frac{P_s^2}{2EA} + a_x\frac{P_x^2}{2GA} + a_y\frac{P_y^2}{2GA} + \frac{M_x^2}{2EI_x} + \frac{M_y^2}{2EI_y} + \frac{M_t^2}{2GI_t}\right]ds \tag{7.73a}$$

Here the distance along the centroidal axis is denoted by s instead of z, to accommodate rod curvature and, hence, P_z is replaced by P_s. The coefficients a_x and a_y are introduced to allow for the nonuniform shear stress distribution due to P_x and P_y, respectively, as previously given in Table 7.3.

In most applications, the energy terms corresponding to the shear forces P_x and P_y and the normal force component P_s in nonstubby bars or rods with a large radius of curvature are neglected in the determination of displacements. This has been justified in Section 5.2.4, Table 5.1, and hence for the rods under consideration Equation 7.73a reduces to:

$$dU = \left[\frac{M_x^2}{2EI_x} + \frac{M_y^2}{2EI_y} + \frac{M_t^2}{2GI_t}\right]ds \tag{7.73b}$$

The total strain energy for a rod length s is obtained by integration of Equation 7.73b along s or by summation to allow for discontinuities. The application of Castigliano's theorem is best illustrated by several examples, as given below.

Example 7.17:

A round bar of diameter d *is shaped into an overhanging crank in the* x–y *plane as shown in Figure 7.36. Determine the displacement of force* P. *The crank shaft transmits a torque* M_t *to a rotor rigidly fixed at its end. Take* a = 3b, B = 2c, c = 2d *and* E = 3G.

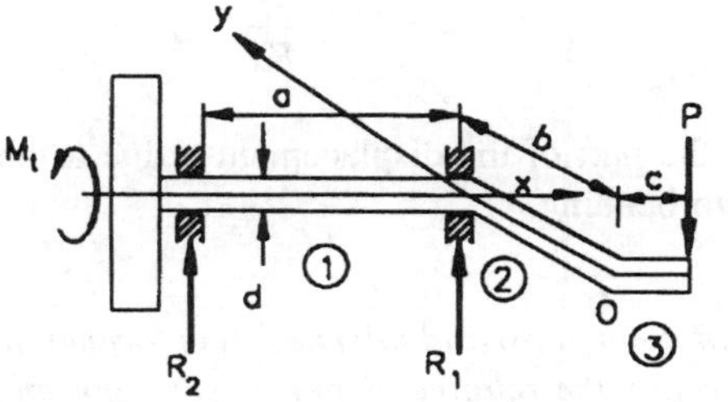

FIGURE 7.36 Example 7.17.

Solution:

From equilibrium, the torque and reactions are given by

$$M_t = Pb,\ \ R_1 = -\frac{P(c+a)}{a},\ \text{ and } R_2 = \frac{Pc}{a}$$

The strain energy U is equal to $U_1 + U_2 + U_3$, where U_1, U_2, and U_3 are the strain energy of segments 1, 2, and 3, respectively. They are given by

$$U_1 = \frac{P^2}{2EI}\int_0^c x^2 dx = \frac{P^2 c^3}{6EI}$$

$$U_2 = \frac{P^2}{2EI}\int_0^b y^2 dy + \frac{P^2 c^2}{2GI_o}\int_0^b dy = \frac{P^2 b^3}{6EI} + \frac{P^2 c^2 b}{2GI_o}$$

$$U_3 = \frac{R_2^2}{2EI}\int_0^a x^2 dx + \frac{P^2 b^2}{2GI_o}\int_0^a dx = \frac{P^2 c^2 a}{6EI} + \frac{P^2 b^2 a}{2GI_o}$$

where for a round bar $I_x = I_y = I$. Hence,

$$U = \frac{P^2}{6EI}\left(c^3 + b^3 + c^2 a\right) + \frac{P^2}{2GI_t}\left(c^2 b + b^2 a\right)$$

Displacement under load $u = \partial U/\partial P$ according to Equation 3.42; hence,

$$u = \frac{P}{3EI}\left(c^3 + b^3 + c^2 a\right) + \frac{P}{GI_o}\left(c^2 b + b^2 a\right)$$

Substituting $I = I_o/2 = \pi d^4/64$ for a circular section together with $b = 2c$, $a = 6c$, $c = 2d$, and $E = 3G$ yields

$$u = \frac{P}{EI}\left(5c^3 + 39c^3\right)$$

or

$$u = 7171\frac{P}{Ed}$$

It can be easily shown that the part of the displacement u due to torsion in segments 2 and 3 is nearly eight times that due to bending.

Example 7.18:

A circular thin-walled tube of diameter d *and thickness* h *is formed to make the symmetrical frame shown in Figure 7.37. Determine the relative displacement between the two forces* P. *Take* a = b = c *and* E = 3G.

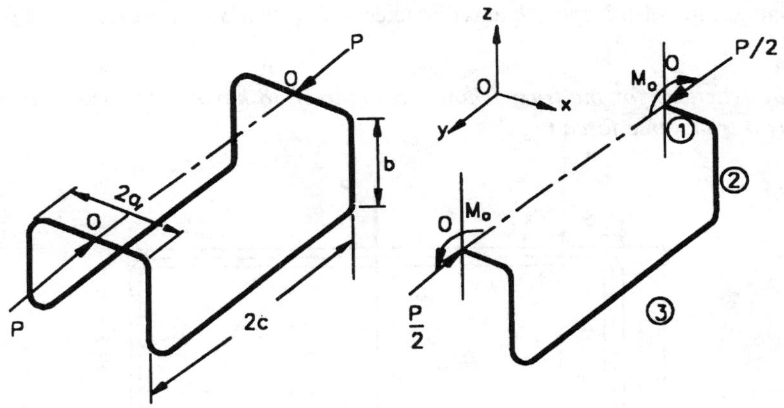

FIGURE 7.37 Example 7.18.

Solution:

The frame is statically indeterminate. Let the frame be cut into two symmetrical parts along the plane O–O normal to the tube cross section; hence, the total strain energy is expressed by summing the energy for the elements 1, 2, and 3, respectively, as

$$U = 4\left(U_1 + U_2 + U_{3/2}\right)$$

$$U_1 = \frac{1}{2EI}\int_0^a \left(M_o - \frac{P}{2}x\right)^2 dx$$

$$U_2 = \frac{1}{2GI_o}\int_0^b \left(M_o - \frac{P}{2}a\right)^2 dz + \frac{1}{2EI}\int_0^b \left(\frac{P}{2}z\right)^2 dz$$

$$U_3 = 2\left[\frac{1}{2EI}\int_0^c \left(\frac{P}{2}b\right)^2 dy + \frac{1}{2EI}\int_0^c \left(M_o - \frac{P}{2}a\right)^2 dy\right]$$

From symmetry the slope at the loading point O vanishes, hence, $\partial U/\partial M_o = 0$, which yields

$$\frac{\partial U}{\partial M_o} = M_o\left(\frac{a+c}{EI} + \frac{b}{GI_o}\right) - \frac{Pa}{4}\left(\frac{a+2c}{EI} + \frac{2b}{GI_o}\right) = 0$$

Substituting $a = b = c$, $I_o = 2I$, and $E = 3G$, gives $M_o = 3aP/7$. To determine the relative displacement between the two forces P, substitute M_o in the expression for U and differentiate with respect to P, as follows:

$$v = \frac{\partial U}{\partial P} = 4\frac{\partial}{\partial P}\left(U_1 + U_2 + U_{3/2}\right)$$

which yields

$$v = 1.595\frac{Pa^3}{EI}$$

For a thin-walled tube of thickness h and diameter d, $I = \pi d^3 h/8$ and, hence, $v = 4.062\ Pa^3/Ed^3h$.

Example 7.19:

Determine the reactions for the frame shown in Figure 7.38 having clamped ends and loaded at the middle by a transverse force P.

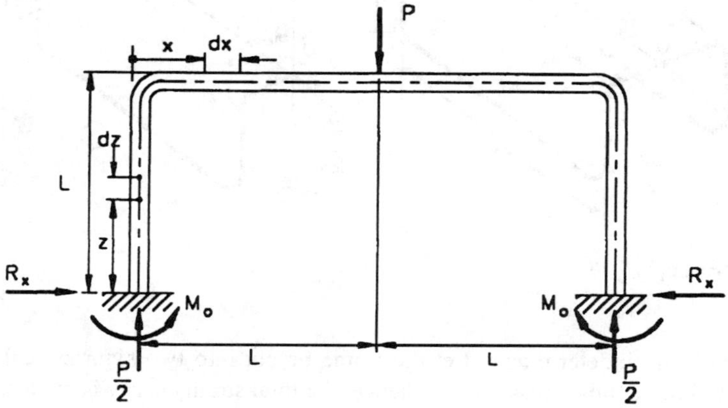

FIGURE 7.38 Example 7.19.

Solution:

The equations of equilibrium are not sufficient to determine the reactions R_x and M_o, and a further condition is obtained from the analysis of displacements. At the clamped ends, $u = 0$ and $\theta = 0$; hence, Equations 3.44 give

$$\frac{\partial U}{\partial R_x} = 0 \text{ and } \frac{\partial U}{\partial M_o} = 0$$

Now for bending, Expression 7.73b reduces to

$$U = \int_0^s \frac{M^2}{2EI} ds$$

$$U = 2\left[\int_o^L \frac{(M_o + R_x z)^2}{2EI} dz + \int_o^L \frac{\left(M_o + R_x a - \frac{P}{2}x\right)^2}{2EI} dx\right]$$

Setting

$$\frac{\partial U}{\partial R_x} = 0, \text{ yields } 18M_o + 16R_x L - 3LP = 0$$

$$\frac{\partial U}{\partial M_o} = 0, \text{ yields } 18M_o + 6R_x L - LP = 0$$

Solving these two equations gives

$$R_x = 0.3P \text{ and } M_o = -0.1PL$$

Example 7.20:

A piston ring, Figure 7.39, width 100 mm inside diameter, 106 mm outside diameter and width, b *= 4 mm is made of fine gray cast iron for which* E *= 150 GPa. To mount the ring in the piston groove two equal pulls* P *are applied at the slit to open the ring. Find the amount of opening at the slit for a maximum tensile stress of 90 MPa.*

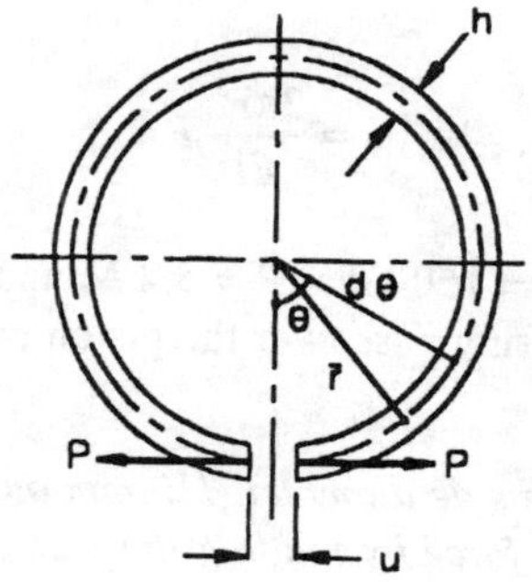

FIGURE 7.39 Example 7.20.

Solution:

The ring is considered to have a rectangular cross section $b \times h$ and mean radius $\bar{r}$. The stress resultants at any cross section is an axial force P_s, a shearing force P_y, and a bending moment M.

Consider a cross section at angle θ from the slit line; hence,

$$P_s = -P\cos\theta,\ \ P_y = P\sin\theta,\ \text{ and }\ M = P = (1-\cos\theta)$$

P_s and M will cause normal stresses and are maximum at $q = \pi$, where $P_y = 0$; hence,

$$P_{s\max} = P \quad \text{and} \quad M_{\max} = 2\,\bar{r}\,P$$

and

$$\sigma_{\max} = \frac{P}{bh} + \frac{2\bar{r}P}{bh^2/6}$$

or

$$\sigma_{\max} = \frac{P}{bh} + \left(1 + 12\frac{\bar{r}}{h}\right)$$

Substituting $\sigma_{\max} = 90$ MPa, $b = 4$ mm, $h = 3$ mm, and $\bar{r} = 51.5$ mm gives

$$P = 5.2\ N$$

Let u be the ring opening in the direction of P. From Castigliano's theorem, Equation 3.42,

$$u = \frac{\partial U}{\partial P}$$

$$U = \int_0^{2\pi \bar{r}} \frac{M^2}{2EI} ds$$

The strain energy due to P_s and P_y are neglected with respect to the strain energy due to M. Substituting $M = P\bar{r}\ (1 - \cos\theta)$, $ds = \bar{r}\, d\theta$ and integrating between 0 and 2π yields

$$u = \frac{3\pi \bar{r}^3}{EI} P$$

Substituting $I = bh^3/12 = 9\ \text{mm}^4$, $E = 150$ GPa, $P = 5.2\ N$, and $\bar{r} = 51.5$ mm gives $u = 4.96$ mm. This is quite sufficient to make the ring pass over the piston crown.

Example 7.21:

A steel ring, Figure 7.40, 100 mm inside diameter, 110 mm outside diameter, and 15 mm width is used in a dynamometer to measure force by reading the displacement between the two diametral points, where the forces act. Determine the dynamometer constant and the maximum range of load, if the maximum allowable stress is limited to 500 MPa and E = *210 GPa.*

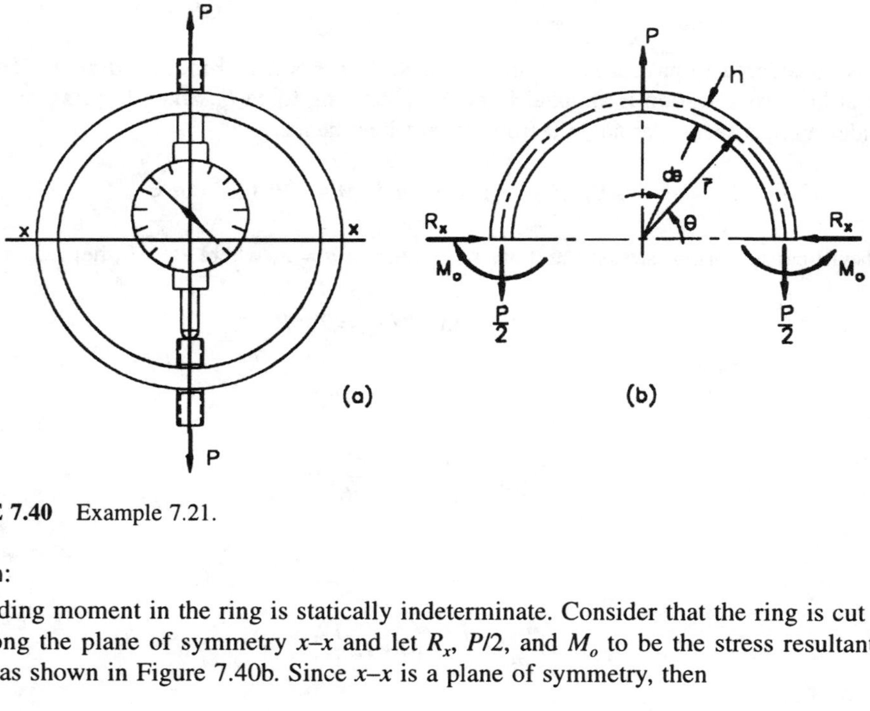

FIGURE 7.40 Example 7.21.

Solution:

The bending moment in the ring is statically indeterminate. Consider that the ring is cut into two parts along the plane of symmetry x–x and let R_x, $P/2$, and M_o to be the stress resultants at this section, as shown in Figure 7.40b. Since x–x is a plane of symmetry, then

$$R_x = 0 \text{ and } \theta_o = 0 \quad \text{(no angular displacement)}$$

From Equation 3.42, $\partial U/\partial M_o = 0$, where

$$U = 2\int_0^{\pi/2} \frac{M}{2EI} \bar{r} d\theta \text{ and } M = M_o - \frac{P}{2}\bar{r}(1 - \cos\theta)$$

Hence,

$$M_o = \frac{P}{2}\bar{r}(1-2/\pi)$$

$$M = -\frac{P}{2}\bar{r}(2/\pi - \cos\theta), \quad 0 \le \theta \le \pi/2$$

M is maximum at $\theta = \pi/2$ and equals $P\bar{r}/\pi$. This gives the maximum stress as

$$\sigma_{max} = \frac{M_{max}}{bh^2/6} + \frac{P}{2bh} = \frac{P}{bh}\left[\frac{6\bar{r}}{\pi h} + \frac{1}{2}\right]$$

For σ_{max} = 500 MPa, b = 15 mm, h = 5 mm, and $\bar{r}$ = 52.5 mm. Hence, P_{max} = 1824.5 N.

The dynamometer constant is the ring inverse stiffness defined by u/P, where u is the displacement along P obtained from the Relation 3.42; $u = \partial U/\partial P$. In this case U is the total strain energy in the ring, i.e.,

$$U = 4\int_0^{\pi/2} \frac{M^2}{2EI}\bar{r}\,d\theta$$

$$= \frac{2\bar{r}}{EI}\int_0^{\pi/2}\left[\frac{P\bar{r}}{2}\left(\frac{2}{\pi} - \cos\theta\right)\right]^2 d\theta$$

Hence,

$$u = \frac{\partial U}{\partial P} = \frac{P\bar{r}^3}{EI}\int_0^{\pi/2}\left(\frac{2}{\pi} - \cos\theta\right)^2 d\theta$$

$$= \frac{P\bar{r}^3}{EI}\left[\frac{4\theta}{\pi^2} - \frac{4}{\pi}\sin\theta + \frac{\theta}{2} + \frac{\sin 2\theta}{4}\right]_0^{\pi/2}$$

$$= \frac{P\bar{r}^3}{EI}\left(\frac{4}{\pi} - \frac{2}{\pi}\right) = 0.1488\frac{P\bar{r}^3}{EI}$$

Therefore,

$$\frac{u}{P} = 0.1488\frac{\bar{r}^3}{EI}$$

Substituting the values of $\bar{r}$, E, and I gives the dynamometer constant as

$$\frac{u}{P} = \frac{0.1488 \times 52.5^3}{210 \times 10^3 \times (15.5^3/12)} = 6.562 \times 10^{-4} \text{ mm/N}$$

7.6.2 Mohr's Unit Load Method

Castigliano's theorem as stated above determines the displacement at the point of load application and in the direction of this load. It is possible, however, to determine the displacement at any general point in any direction by following the procedure proposed by Mohr, known as the Mohr's unit load method.*

- A fictitious load P is applied at the point and in the direction for which it is required to determine the displacement; P is a force if it is required to obtain linear displacement and is a moment if it is required to obtain angular displacement.
- Equilibrium of the system is established and the total strain energy U is obtained including the effect of P.
- The required displacement is obtained by differentiating the expression of U with respect to P and then substituting $P = 0$ in the final result.

Let M_x, M_y, and M_z be the moments about the x-, y-, and z-axes, respectively, at any section due to original loading and m_x, m_y, and m_z be the moments about the x-, y-, and z-axes, respectively, due to the application of a unit force at the point and direction under consideration. That is m_xP, m_yP, and m_zP are the additional moments due to P. Hence, by superposition $(M_x + m_xP)$, $(M_y + m_xP)$, and $(M_2 + m_zP)$ are the resultant moments acting at any section, and the total strain energy for bending is given by

$$U = \int_S \left(\frac{(M_x + m_xP)^2}{2EI_x} + \frac{(M_y + m_yP)^2}{3EI_y} + \frac{(M_z + m_zP)^2}{2GI_t} \right) ds$$

Applying Castigliano's theorem $u = \partial U/\partial P$, Equation 3.42, and substituting $P = 0$ in the result yields

$$u = \int_S \left(\frac{M_x m_x}{EI_x} + \frac{M_y m_y}{EI_y} + \frac{M_z m_z}{GI_t} \right) ds \tag{7.74}$$

Note that M_z is a twisting moment and, hence, G is the modulus of rigidity and I_i is the torsional moment of area for the cross section.

Consider, as an example, a bar of length L, as shown in Figure 7.41. The bar is clamped at one end and subjected to a transverse force P at the other free end. It is required to determine the displacement at the free end and also at $z = L/2$ by applying Mohr's unit load method. In this case, Equation 7.74 simplifies to

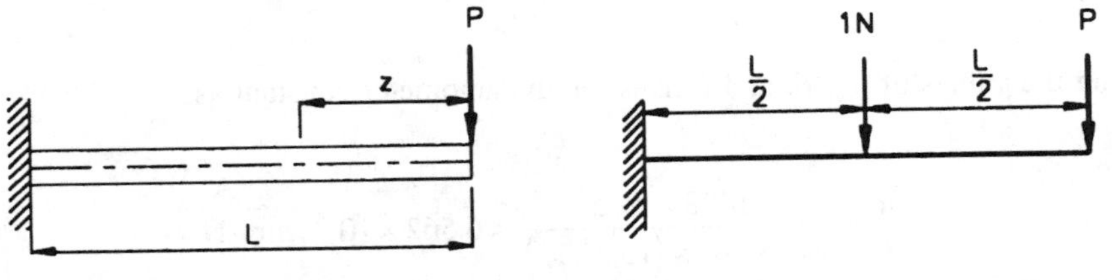

FIGURE 7.41 Mohr's unit load method applied to determine the displacement in a cantilever beam.

* Named after O. Mohr, 1914.

$$u = \int_0^L \frac{Mm}{EI} dz$$

To determine u at the free end, apply a unit load at $z = 0$; hence,

$$M = Pz$$

and

$$m = z$$

so that at $z = 0$:

$$u = \int_0^L \frac{Pz^2}{EI} dz = \frac{PL^3}{3EI}$$

To determine u at the middle point, apply a unit load at $z = L/2$, so that at $z = L/2$:

$$u = \frac{1}{EI}\left(\int_o^{L/2} Mmdz + \int_{L/2}^{L} Mmdz\right)$$

From $z = 0$ to $z = L/2$: $M = Pz$ and $m = 0$
From $z = L/2$ to $z = L$: $M = Pz$ and $m = z - L/2$

Hence, substitution gives

$$u = \frac{P}{EI}\int_{L/2}^{L} z(z - L/2)dz = \frac{5}{48}\frac{PL^3}{EI}$$

Example 7.22:
A C-frame made of a steel circular bar of diameter d = *20 mm is subjected to two collinear equal forces* P = *1000 N acting at the edges as shown in Figure 7.42. Determine the lateral displacement at the middle point O. Take* a = *100 mm,* b = *240 mm, and* E = *210 GPa.*

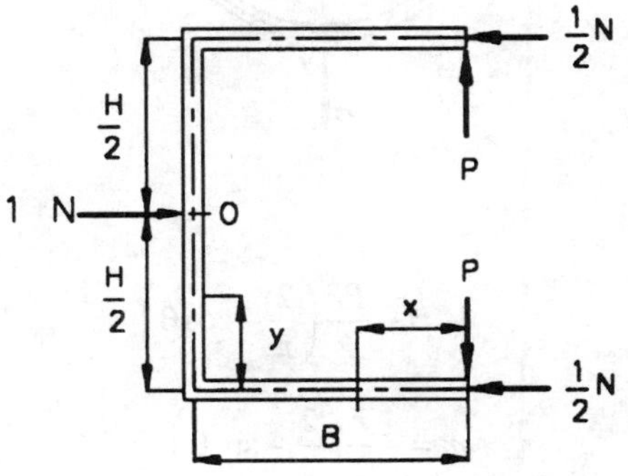

FIGURE 7.42 Example 7.22.

Solution:

Apply a unit force of 1 N in the transverse direction at O and establish equilibrium of the frame by placing two equal forces of $^1/_2$ N at the edges as shown in Figure 7.42.

Along the horizontal leg B: $M = Px$, $m = 0$
Along the vertical leg H: $M = PB$, $m = y/2$

hence, from Equation 7.74,

$$u_o = 2\left(0 + \frac{PB}{2EI}\int_0^{H/2} y dy\right) = \frac{PBH^2}{8EI}$$

Substituting $P = 1000$ N, $B = 100$ mm, $H = 240$ mm, and $E = 210$ GPa gives

$$u_o = \frac{100 \times 100 \times 240^2}{8 \times 210 \times 10^3 \times \frac{\pi}{64} \times 20^4} = 0.4365 \text{ mm}$$

Example 7.23:

For the ring of Example 7.21, determine the displacement between x–x *at the maximum load and compare its value with the corresponding displacement between the two diametral points* y–y *where the forces act.*

Solution:

Apply unit forces at points O and O_1, as shown in Figure 7.43. For $0 \le \theta \le \pi/2$,

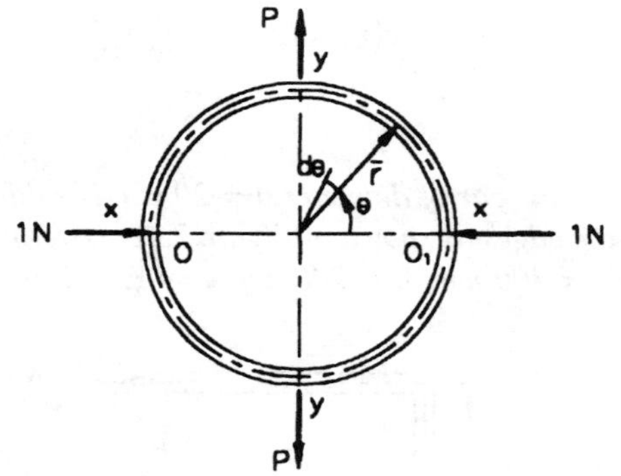

FIGURE 7.43 Example 7.23.

$$M = -\frac{P\bar{r}}{2}\left(\frac{2}{\pi} - \cos\theta\right)$$

$$m = -\frac{\bar{r}}{2}\left(\frac{2}{\pi} - \sin\theta\right)$$

The displacement along the horizontal diameter is

$$u_x = 4\int_0^{\pi/2} \frac{Mm}{EI}\bar{r}d\theta$$

$$= \frac{P\bar{r}^3}{EI}\int_0^{\pi/2}\left(\frac{2}{\pi} - \cos\theta\right)\left(\frac{2}{\pi} - \sin\theta\right)d\theta$$

$$= \frac{P\bar{r}^3}{EI}\left[\frac{4}{\pi^2}\theta - \frac{2}{\pi}(\sin\theta - \cos\theta) + \frac{\sin^2\theta}{2}\right]_0^{\pi/2}$$

$$u_x = \frac{P\bar{r}^3}{EI}\left(\frac{1}{2} - \frac{2}{\pi}\right) = -0.137\frac{P\bar{r}^3}{EI}$$

This compares with u, previously obtained in Example 7.21 as

$$u_y = 0.1488\frac{P\bar{r}^3}{EI}$$

At P_{max} = 1824.5 N, as previously determined in Example 7.21,

$$u_y = 1.197 \text{ mm} \quad \text{and} \quad u_x = -1.099 \text{ mm}$$

7.6.3 A Note on The Deflection of Curved Beams

In Sections 7.6.1 and 7.6.2, the displacement in curved rods, such as rings, has been determined applying Castigliano's theorem and Mohr's method. A restriction was made on the curvature of the rod, namely, that its radius of curvature must be large compared with its depth. To assess the limits of this restriction, a circularly curved beam of rectangular cross section bent by an end shear force is considered as shown in Figure 7.44.

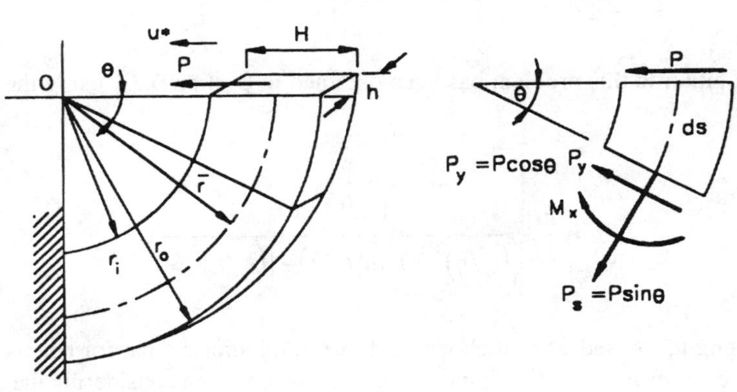

FIGURE 7.44 Deflection of a circularly curved beam.

It is required to determine the displacement of the free end u* due to the given loading. As an approximation, Castigliano's theorem is applied to the total energy given by Expression 7.73a. For a curved beam with rectangular cross section, the normal force $P_s = -P \sin\theta$, the shearing force $P_y = P\cos\theta$, the bending moment $P_x = P\bar{r}\sin\theta$, and $ds = rd\theta$, where $\bar{r}$ is the centroidal radius of curvature. Hence, the total strain energy U, Expression 7.73a, using Equation 7.26 for $I = A\bar{r}e$, is given by:

$$U = \int_0^{\pi/2} \left[\frac{P^2 \sin^2\theta}{2EA} + a_y \frac{P^2 \cos^2\theta}{2GA} + \frac{P^2 \bar{r}^2 \sin^2\theta}{2EA\bar{r}e} \right] \bar{r} d\theta \tag{7.75}$$

Applying Castigliano's theorem, Expression 3.42, $u = \partial U/\partial P$; hence, differentiation before integration results in

$$\frac{dU}{dP} = u = \int_0^{\pi/2} \left[\frac{P\bar{r} \sin^2\theta}{EA} + a_y \frac{P\bar{r} \cos^2\theta}{GA} + \frac{P\bar{r}^2 \sin^2\theta}{EAe} \right] d\theta$$

Upon integration and taking $a_y = 1.2$, $A = hH$, and $e = I/(A\,\bar{r}) + = H^2/(12\,\bar{r})$ gives

$$u^* = \frac{\pi P}{Eh} \left[\frac{3\bar{r}^3}{H^3} + \frac{2.12\bar{r}}{4H} \right]$$

or

$$u^* = \frac{\pi P}{Eh} \left[0.375 \left(\frac{r_o/r_i + 1}{r_o/r_i - 1} \right)^3 + 0.265 \left(\frac{r_o/r_i + 1}{r_o/r_i - 1} \right)^3 \right] \tag{7.76a}$$

Neglecting the terms of the strain energy associated with the normal and shear forces and retaining the term due to bending only results in

$$u^{**} = \frac{3\pi P\bar{r}^3}{EhH^3} = \frac{3\pi P}{8Eh} \left[\left(\frac{r_o/r_i + 1}{r_o/r_i - 1} \right) \right]^3 \tag{7.76b}$$

The exact solution to this problem has been obtained in Section 6.3.1 using the stress function approach as

$$u = \frac{\pi P}{Eh} \frac{\left[(r_o/r_i)^2 + 1 \right]}{\left[(r_o/r_i)^2 + 1 \right] \ln(r_o/r_i) - \left[(r_o/r_i)^2 - 1 \right]} \tag{6.43}$$

Comparison among u, u^*, and u^{**}, as shown in Table 7.5, indicates that for beams of large radius of curvature, the error in applying Castigliano's theorem (even when considering the bending energy only) results in reasonably accurate results for the displacement of curved rods within 10% error only for (H/F) = 1).

Example 7.24:

The circular ring shown in Figure 7.45a has a rectangular cross section of width h, *depth* H *and a centroidal radius of curvature* $\bar{r}$, *determine:*

TABLE 7.5
Comparison of Various Solutions for Displacements in Curved Beams

r_o/r_i	9	7	3	1.667	1.222
$H/\bar{r}$	1.6	1.5	1.0	0.5	0.2
$u/(\pi P/Eh)$	0.819	1.014	3.349	24.85	378.25
u^*/u	1.299	1.225	1.054	1.008	1.001
u^{**}/u	0.859	0.876	0.896	0.941	0.966

a. The bending moment at any section;
b. The increase in the vertical diameter.

Compare the results with the thin ring solution obtained in Example 7.21 at different $\bar{r}$/H *of 1 to 4.*

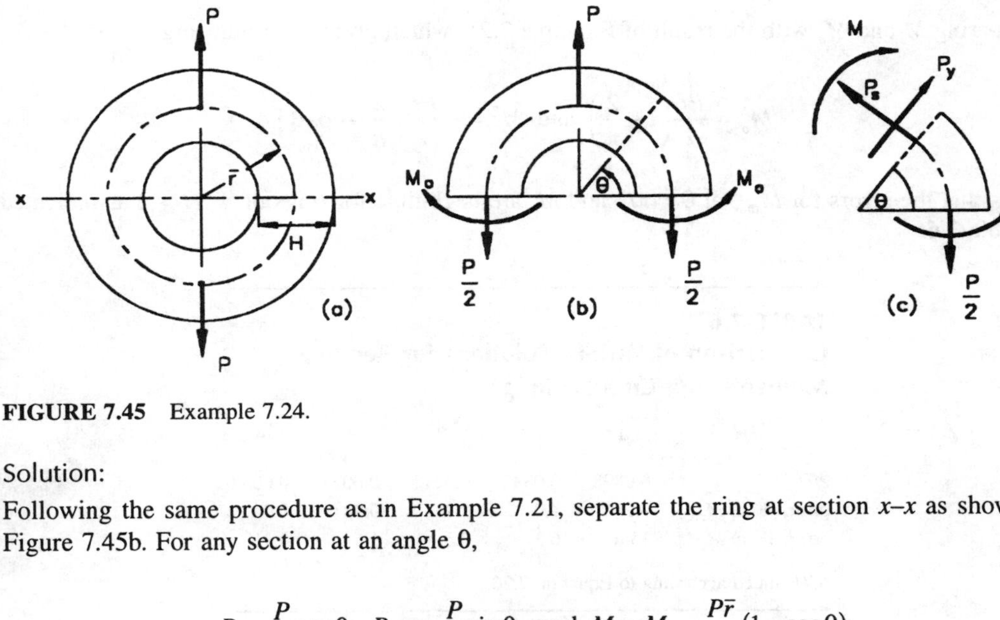

FIGURE 7.45 Example 7.24.

Solution:
Following the same procedure as in Example 7.21, separate the ring at section x–x as shown in Figure 7.45b. For any section at an angle θ,

$$P_s = \frac{P}{2}\cos\theta,\ P_y = -\frac{P}{2}\sin\theta,\ \text{and}\ M = M_o - \frac{P\bar{r}}{2}(1-\cos\theta)$$

An accurate expression for the strain energy is given by Expression 7.29 for $P_s \equiv P_z$ as

$$U = \int_0^{\pi/2}\left[\frac{M^2}{2EA\bar{r}e} + \frac{P_s^2}{2EA} + \frac{a_y P_y^2}{2GA} - \frac{MP_s}{EA\bar{r}}\right]\bar{r}d\theta$$

Due to symmetry, $dU/dM_o = 0$, hence,

$$\frac{dU}{dM_o} = 2\int_0^{\pi/2}\left[\frac{M}{EA\bar{r}e} - \frac{P_s}{EA\bar{r}}\right]\bar{r}d\theta = 0$$

Substitution for M and P_s and integration gives

$$\frac{1}{EAe}\left[M_o\theta-\frac{P\bar{r}}{2}(\theta-\sin\theta)\right]_0^{\pi/2}-\frac{P}{2EA}[\sin\theta]_0^{\pi/2}=0$$

or

$$M_o=\frac{P\bar{r}}{2}\left(1-\frac{2}{\pi}+\frac{2e}{\pi r}\right)$$

and

$$M=-\frac{P\bar{r}}{2}\left(\frac{2}{\pi}-\cos\theta-\frac{2e}{\pi r}\right)$$

Comparing M and M_o with the result of Example 7.21, which gives for a thin ring

$$M_o^*=\frac{P\bar{r}}{2}\left(1-\frac{2}{\pi}\right)\text{ and }M^*=-\frac{P\bar{r}}{2}\left(\frac{2}{\pi}-\cos\theta\right)$$

shows that the errors for M_{max} at $\theta = \pi/2$ and M_o are negligible for rings of $\bar{r}/H > 2$ as indicated in Table 7.6.

TABLE 7.6
Comparison of Various Solutions for Bending Moments in a Circular Ring

$\bar{r}/H$	1	1.5	2	3	4
$e/\bar{r}$[a]	0.0898	0.0382	0.0212	0.0093	0.0052
error, %, in M_{max}	9.9	4.0	2.2	0.94	0.52
error, %, in M_o	13.6	6.3	3.6	1.6	0.91

[a]Calculated according to Equation 7.26.

Having obtained an expression for M, the bending stress is calculated using the curved beam formula (Equation 7.27). The increase in the vertical diameter of the ring is obtained by setting $u = \partial U/\partial P$, where U is expressed by Equation 7.29. Knowing M, P_s, and P_y gives

$$u=\frac{P\bar{r}^2}{EAe}\left\{\left[\frac{\pi}{4}-\frac{2}{\pi}\left(1-\frac{e^2}{\bar{r}^2}\right)\right]+\left[\frac{4e}{\pi\bar{r}}-\frac{\pi e}{4\bar{r}}-\frac{4e^2}{\pi\bar{r}^2}\right]+\left[\frac{\pi a_y Ee}{4G\bar{r}}\right]\right\}$$

Simplifying by neglecting higher orders of $e/\bar{r}$ gives

$$u=\frac{P\bar{r}^2}{EAe}\left[\left(\frac{\pi}{4}-\frac{2}{\pi}\right)+\left(\frac{4e}{\pi\bar{r}}-\frac{\pi e}{4\bar{r}}\right)+\left(\frac{\pi a_y Ee}{4G\bar{r}}\right)\right]$$

or

$$u = \frac{P\bar{r}^2}{EAe}\left(u_b + u_n + u_s\right)$$

where u_b, u_n, and u_s are the deflections due to bending, normal force, and shearing force, respectively. The relative weight of each displacement to the total deflection is indicated in Table 7.7. Also, the approximate increase of diameter as determined according to the results of Example 7.21, which neglects the initial curvature of the ring and considers only bending effects, is

$$u_o = \frac{P\bar{r}^3}{EI}\left(\frac{\pi}{4} - \frac{2}{\pi}\right) \quad \text{(as included in Table 7.6)}$$

The analysis shows that u_s plays an important role in the diametral extension of the ring for values of $(\bar{r}/H) < 4$. For values of $(\bar{r}/H) \geq 4$, u_s and u_n may be neglected and the results of Example 7.21 based on u_b only are quite satisfactory.

TABLE 7.7
Comparison of Deflections Due to Bending, Normal, and Shear Forces in a Circular Ring

$\bar{r}/H$	1	1.5	2	3	4
$u/(P/hE)$	4.595	10.249	19.912	56.808	126.198
u_o/u	2.574	1.701	1.394	1.178	1.104
u_b/u	0.361	0.570	0.705	0.845	0.907
u_n/u	0.106	0.071	0.049	0.026	0.015
u_s/u[a]	0.533	0.359	0.246	0.129	0.078

[a] $a_y = 1.2$, $E/G = 2(1 + \nu) = 2.6$, $A = hH$, $I = hH^3/12$.

7.6.4 Application to Springs

Springs are elastic elements of low stiffness usually made of long rods loaded in either bending or torsion or in both. The rods are either straight, such as in leaf springs and torsion bars used for vehicle suspension, or coiled to provide feasible forms for various applications. In the following analysis, the displacement under load and stiffness are obtained for some of the latter types of springs, as shown in Figure 7.46a to d, which are in common use. Mohr's unit load, Equation 7.74, is applied to determine the displacement of such springs.

7.6.4.1 Helical Compression Spring

The rod, which is usually of a circular cross section of diameter d, is coiled into a cylindrical form of n turns and diameter D as shown in Figure 7.46a. Any cross section of the rod is subjected to a transverse force P parallel to the coil axis and a moment $PD/2$. The moment axis is perpendicular to the plane passing through the coil axis at this point, as shown in Figure 7.47a. The normal to the rod cross section is inclined at an angle α (the helix angle) to this plane.

Resolving the force P and moment $PD/2$ in directions (x, y, z) referred to the rod cross section, as shown in Figure 7.47b, yields

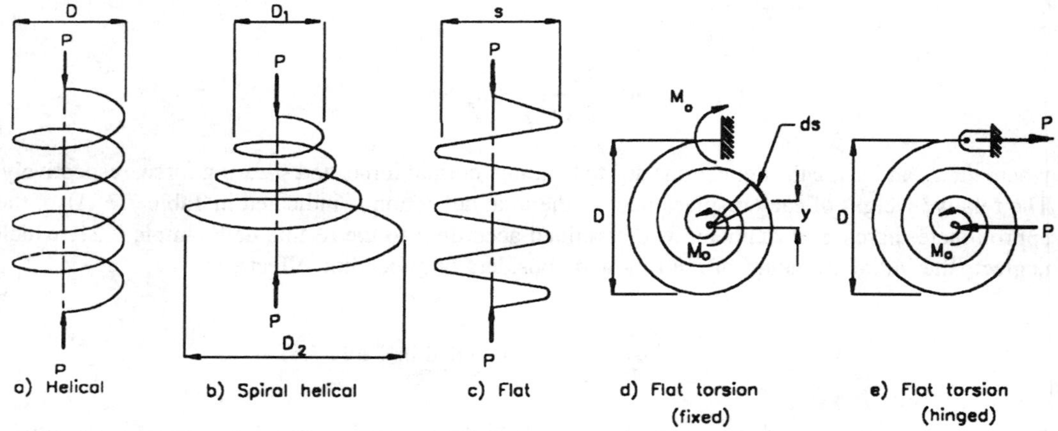

FIGURE 7.46 Some types of springs made of long rods: (a) helical, (b) spiral helical, (c) flat, (d) flat torsion (fixed), and (e) flat torsion (hinged).

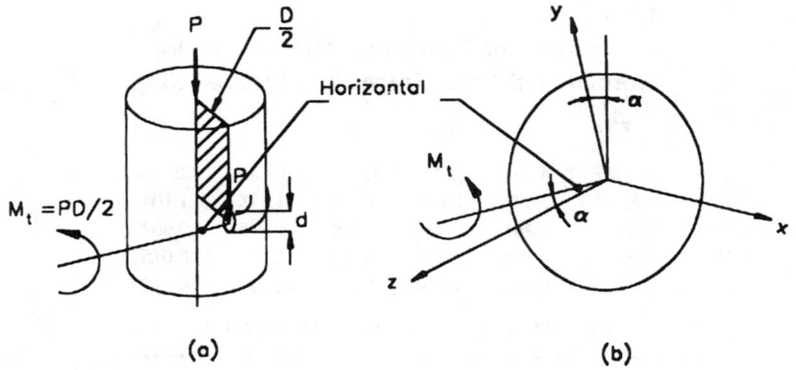

FIGURE 7.47 Effect of helix angle in the analysis of the helical spring of Figure 7.46: (a) coil geometry and (b) load diagram.

Shearing force: $P_y = P \cos \alpha$
Normal force: $P_z = P \sin \alpha$

Twisting moment: $M_z \equiv M_t = \dfrac{PD}{2} \cos \alpha$

Bending moment: $M_y \equiv M_b = \dfrac{PD}{2} \sin \alpha$

In most engineering applications, the ratio D/d ranges from 5 to 8 and a maximum spacing between two consecutive coils is about d; hence, the helix angle as given by $\tan^{-1}\alpha = \tan^{-1}(2d/\pi D)$ attains a value of the order of $\tan^{-1}(2/5\pi) \cong 7.3°$. For such a small value of α, $\cos \alpha \cong 1$ and $\sin \alpha = 0$. Thus, only the twisting moment $M_t = PD/2$ and the transverse shearing force $P_y = P$ are considered when analyzing deflection and stresses in helical compression springs.

The spring stiffness is given by

$$k_s = \frac{P}{\delta}$$

where, according to Mohr's unit load method, Equation 7.74, due to torsion;

$$\delta = \int_0^S \frac{M_t m_t}{GI_o} ds,$$

$$M_t = \frac{PD}{2}, \quad m_t = \frac{D}{2}, \quad \text{and } s = \pi Dn$$

Hence,

$$\delta = \frac{PD^2 \pi Dn}{4GI_o} = P\frac{\pi D^3 n}{4GI_o} \tag{7.77a}$$

$$k_s = \frac{4GI_o}{\pi n D^3}$$

$$\tau_{\max} = \frac{PDd}{4I_o} \tag{7.77b}$$

In heavy helical springs, the ratio D/d is less than 5, and the effect of transverse shear stress in the coil must be considered. This is often approximated to its average value:

$$\tau = 4P/\pi d^2$$

At the inner point of the wire diameter, where cracks usually develop, the maximum stress is taken as

$$\tau_{\max} = \frac{8PD}{\pi d^3}\left(1 + \frac{d}{2D}\right) \tag{7.77c}$$

Note that, as in curved beams, higher torsional shear strain and hence stress occur on the inside surface of the rod. This together with $\tau_{\max}$ as given above may promote yielding at this surface. For fatigue loading, this is taken care of by incorporating a fatigue strength reduction factor in design calculations.

7.6.4.2 Spiral Helical Compression Spring

The rod is coiled into a conical helix form, Figure 7.46(b). The twisting moments are thus:

$$M_t = Pr, \quad m_t = r$$

where

$$r = \frac{1}{2}\left(D_1 + \frac{D_2 - D_1}{n}\frac{\theta}{2\pi}\right)$$

For $ds = rd\theta$, $M_z \equiv M_t$ and $m_z \equiv m_t$, Equation 7.74 gives for torsional displacement:

$$\delta = \frac{P}{8GI_o}\int_0^{2\pi n}\left(D_1 + \frac{D_2 - D_1}{n}\frac{\theta}{2\pi}\right)^3 d\theta$$

or

$$\delta = P\frac{\pi n}{16GI_o}(D_1 + D_2)(D_1^2 + D_2^2) \tag{7.78a}$$

Hence, the stiffness k_t is given by

$$k_s = \frac{16GI_o}{\pi n(D_1 + D_2)(D_1^2 + D_2^2)} \tag{7.78b}$$

$$\tau_{max} = \frac{PD_2 d}{4I_o} \tag{7.78c}$$

For $D_1 = 0$ and $D_2 = D$:

$$\delta = P\frac{\pi D^3 n}{16GI_o} \quad \text{and} \quad k_s = \frac{16GI_o}{\pi n D^3}$$

7.6.4.3 Flat Compression Spring

The rod is corrugated in a plane, Figure 7.46c. Let n be the number of bends, each having length s. The stiffness is given by: $k_s = P/\delta$. Applying Mohr's unit load, Equation 7.74 gives the displacement:

$$\delta = 2n\int_0^{S/2} \frac{Mm}{EI_x}dx$$

Since $M = Px$ and $m = x$, then

$$\delta = \frac{nPs^3}{12EI_x} \tag{7.79a}$$

The spring stiffness and maximum stress, respectively, are given by

$$k_s = \frac{12EI_x}{ns^3} \tag{7.79b}$$

and

$$\sigma_{max} = \frac{M_{max}}{I_x}\frac{h}{2} = \frac{Psh}{4I_x} \tag{7.79c}$$

where h is the rod thickness.

7.6.4.4 Flat Torsion Spring

The rod is wound in a plane spiral form, Figure 7.46d and e. Let n be the number of spiral turns and D the outer diameter of the spiral. The torsional stiffness is defined as

$$k_s = M_o/\theta$$

According to Expression 7.74 for angular displacement θ:

$$\theta = \int_0^S \frac{Mm}{EI_x} ds$$

Two cases may be considered as follows:

- Clamped end, Figure 7.26d

$$M = M_o,\ m = 1,\ \text{hence}\ \theta = \frac{M_o s}{EI_x}$$

$$s = \int_0^{2\pi n} r d\theta,\ \text{where}\ r = \frac{D\theta}{2\pi n}$$

$$s = \frac{D}{2\pi n}\int_0^{2\pi n} \theta d\theta = n\pi D$$

Therefore, due to bending

$$\psi = \frac{M_o}{EI_x} n\pi D \tag{7.80a}$$

and the spring stiffness and maximum stress are given, respectively, by

$$k_s = \frac{EI_x}{n\pi D} \tag{7.80b}$$

$$\sigma_{\max} = \frac{M_o h}{2I_x} \tag{7.80c}$$

where h is the rod thickness.

- Hinged end, Figure 7.46e
From equilibrium:

$$P = 2M_o/D$$

$$M = M_o - Py = M_o\left(1 - \frac{2y}{D}\right)$$

$$m = \left(1 - \frac{2y}{D}\right)$$

Hence, due to bending,

$$\psi = \frac{M_o}{EI_x}\int_0^S \left(1 - \frac{2y}{D}\right)^2 ds$$

$$= \frac{M_o}{EI_x} s - \frac{4}{D}\int_0^S y ds + \frac{4}{D^2}\int_0^S y^2 ds$$

Since

$$\int_0^S y ds = 0, \text{ and } \int_0^S y^2 ds = \frac{1}{4} sD^2$$

$$\psi = \frac{M_o s}{EI_x}\left(1 + \frac{1}{4}\right) = \frac{5}{4}\frac{M_o s}{EI_x} \tag{7.80d}$$

the spring stiffness and maximum stress, respectively, are given by

$$k_s = \frac{4}{5}\frac{EI_x}{n\pi D} \tag{7.80e}$$

$$\sigma_{max} = \frac{M_o h}{2I_x} \text{ at the center} \tag{7.80f}$$

Example 7.25:
Two coaxial helical springs, the inner is made of steel G_s = *77 GPa, coil diameter* D_s = *50 mm, and wire diameter* d_s = *6 mm and the outer is made of brass* G_b = *58 GPa, coil diameter* D_b = *75 mm, and wire diameter* d_b = *6 mm. Both springs have the same height and number of turns* n = *10. A compression axial force of 500 N is applied to the assembly, as shown in Figure 7.48. Determine:*

a. *The deflection of the assembly;*
b. *The maximum shear stress in each spring; and*
c. *The total energy stored in the assembly.*

Solution:

a. The assembly undergoes the same deflection δ. From Equations 7.77a for both brass and steel,

$$\delta = P_s\left(\frac{\pi D^3 n}{4GI_o}\right)_s = P_b\left(\frac{\pi D^3 n}{4GI_o}\right)_b$$

where the subscripts *s* and *b* stands for steel and brass, respectively. The above equation indicates that

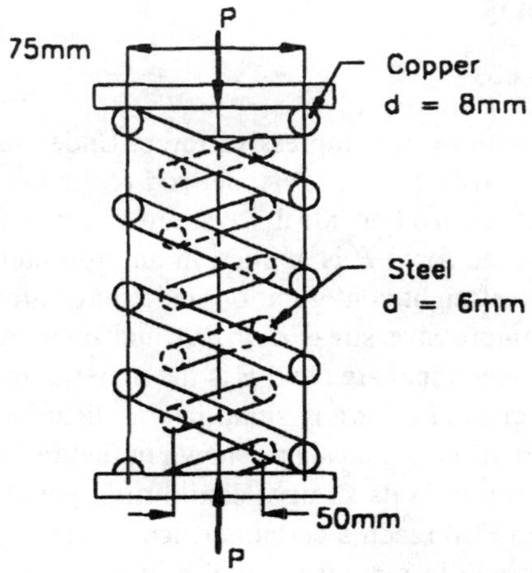

FIGURE 7.48 Example 7.25.

$$P_s/P_b = \frac{(D_b/D_s)^3}{G_b/G_s} = \frac{(75/50)^3}{58/77} = 4.48$$

or $P_s = 408.8$ N and $P_b = 91.2$ N.

The deflection of the assembly is thus

$$\delta = 408.8\frac{\pi(50)^3 \times 10}{4(77\times 10^3)\pi(8)^4/4} = 12.96 \text{ mm}$$

b. The maximum shear stress is given by Equations 7.77c as
- Steel wire:

$$\tau_{max} = \frac{8P_sD_s}{\pi d_s^3}\left(1+\frac{d_s}{2D_s}\right) = \frac{8\times 408.8\times 50}{\pi(8)^3}\left[1+\frac{8}{2\times 50}\right]$$

$$= 109.8 \text{ MPa}$$

- Brass wire:

$$\tau_{max} = \frac{8P_bD_b}{\pi d_b^3}\left(1+\frac{d_b}{2D_b}\right) = \frac{8\times 91.2\times 75}{(6)^3}\left[1+\frac{6}{2\times 75}\right]$$

$$= 83.9 \text{ MPa}$$

c. The total energy stored in the assembly = $^1/_2 \times 500 \times 12.96 \times 10^{-3} = 3.24$ J.

7.7 BUCKLING OF RODS

7.7.1 Buckling of Columns

A column is a bar subjected to an axial compressive force. Under such loading, a slender column may be unstable and buckle. Consider a long, slender rod (column) subjected to an axial force P, as shown in Figure 7.49a. If the rod is ideal in all respects, e.g., homogeneity of material, straightness, and so forth, and the force P is applied in an absolutely concentric manner, the rod will remain in equilibrium in a straight configuration under any value for the force P. In this case, the rod is subjected to the compressive stress $\sigma = P/A$ and undergoes an axial deformation $\varepsilon = P/AE$, where A is the rod cross-sectional area and E is the material modulus of elasticity. However, a slight misalignment in load coaxiality or any small lateral disturbance (say, a small lateral force Q) will cause the rod to bend to the deflected shape shown in Figure 7.49a. As soon as the disturbing force Q is removed, the rod returns to its straight equilibrium position if the value of the applied force P is small. By increasing P to reach a certain critical value P_{cr}, the rod will not return to its straight position and will continue to remain in equilibrium in the slightly bent or buckled form resulting from P_{cr} only. Further increase of the compressive load will cause the lateral deflection to increase, producing collapse of the bar.

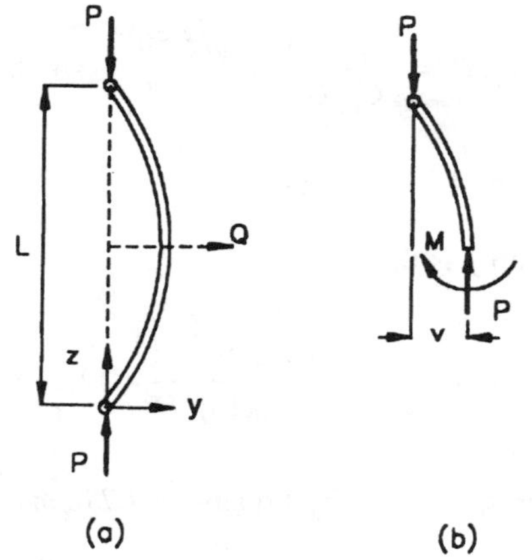

FIGURE 7.49 Buckling of a bar in axial compression: (a) load and deflection diagrams and (b) static equilibrium of a part of the column.

For simple cases of loading, the theoretical buckling force P_{cr} is determined by expressing the equation of equilibrium in terms of the lateral displacement v, referred to the centroidal principal axes of the bar cross section. This was previously given by Equation 7.10 which, upon integrating twice, results in the equation of the elastic curve. The integration constants are determined from end conditions. Typical examples of columns subjected to an axial compressive force P with different end conditions are shown in Figure 7.50.

For more complicated cases of loading, the expression of equilibrium in terms of the lateral displacement makes the analytical solution rather tedious. In such cases, it is much simpler to use an energy approach through which a satisfactory approximate estimation of the critical buckling load can be obtained.

7.7.1.1 Equilibrium Approach

For a column with pinned ends (See Figure 7.50b). The equation of equilibrium (Equation 7.10c) with M = Pv is given by

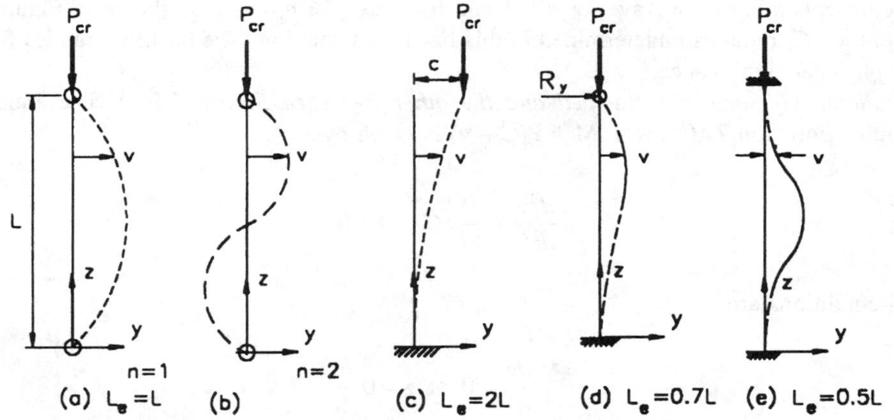

FIGURE 7.50 Effective buckling length in relation to end constraints for a bar in axial compression: (a) pinned ends n = 1, (b) pinned ends n = 2, (c) one end clamped and the other free, (d) one end clamped and the other pinned, and (e) both ends clamped.

$$\frac{d^2v}{dz^2}+\frac{P}{EI}v=0 \tag{7.81a}$$

Denoting $P/EI = k^2$ and integrating yields:

$$v = C_1 \sin kz + C_2 \cos kz$$

The values of the constants C_1 and C_2 are found from the boundary conditions:

$$v = 0: \text{ at both } z = 0 \text{ and } z = L$$

Hence, $C_2 = 0$ and $C_1 \sin kz = 0$, which is satisfied when

$$k^2 = n^2\pi^2/L^2, \quad n = 0, \pm1, \pm2, \pm3\ldots$$

Since the smallest value of the critical buckling load is sought, the first nonzero value for $n = 1$ is taken. Therefore, the critical buckling or Euler load* for a column with pinned ends is given by

$$P_{cr}=\frac{\pi^2 EI}{L^2} \tag{7.81b}$$

where I, the smaller of I_x and I_y, is used to determine P_{cr}. In other words, the column will buckle either in the y–z or the x–z plane depending on the axis around which the second moment of area is minimum.

The equation of the elastic curve giving the lateral deflection of the buckled rod is described by

$$v = C_1 \sin\left(\frac{n\pi}{L}z\right) \tag{7.81c}$$

* Named after L . Euler (1707–1783).

which represents a half-sine wave for $n = 1$ and full wave for $n = 2$ ... as shown in Figure 7.50b. The amplitude C_1 remains undetermined in this linearized solution. The buckling modes for $n > 1$ have no physical significance.*

For a column with one end clamped and the other free (See Figure 7.50c). The equation of equilibrium (Equation 7.10c) with M = P(C – v) is given by

$$\frac{d^2v}{dz^2} + \frac{P}{EI}(C - v) = 0 \tag{7.82a}$$

The end conditions are

$$\frac{dv}{dz} = 0 \text{ at } z = 0$$

$$v = c \text{ at } z = L$$

which yields

$$P_{cr} = \frac{\pi^2 EI}{4L^2} \tag{7.82b}$$

$$v = C\left(1 - \cos\frac{\pi z}{2L}\right) \tag{7.82c}$$

For a column with one end clamped and the other pinned (See Figure 7.50d). It is observed that a lateral reaction R_y is necessary to keep the buckled column in equilibrium. Therefore, the equation of equilibrium (Equation 7.10c) with M = Pv – R_y (L – y) is given by

$$\frac{d^2v}{dz^2} + \frac{1}{EI}\left[Pv - R_y(L - y)\right] = 0$$

The end conditions are

$$\frac{dv}{dz} = v = 0 \text{ at } z = 0$$

$$v = 0 \text{ at } z = L$$

which yield

$$P_{cr} = \frac{2.05\pi^2 EI}{L^2} \tag{7.83a}$$

$$v = C_1\left[\frac{L}{4.493}\sin 4.493\frac{z}{L} - L\cos 4.493\frac{z}{L} + (L - z)\right] \tag{7.83b}$$

C_1 being a constant.

* The problem of column buckling is an eigenvalue problem, and these buckling modes represent the eigenfunctions. See Sirnath.[4]

For a column with both ends clamped (See Figure 7.50e). The equation of equilibrium (Equation 7.10c) with M = (Pv – M_o) is given by

$$\frac{d^2v}{dz^2}+\frac{1}{EI}\left(Pv-M_o\right)=0 \tag{7.84a}$$

where M_o is the clamping moment at the ends. The end conditions are

$$\frac{dv}{dz}=0 \text{ at } z=0 \text{ and } z=L$$

which yields

$$P_{cr}=\frac{4\pi^2 EI}{L^2} \tag{7.84b}$$

$$v=C_1\left[\cos\frac{2\pi z}{L}-1\right] \tag{7.84c}$$

C_1 being a constant.

The expressions given for the buckling loads can be made to resemble the fundamental case of a column with pinned ends, namely, $P_{cr} = \pi^2 EI/L^2$. This is realized by assigning an effective column length L_e for each case; hence, $L_e = 2L$, $L_e = 0.7L$, or $L_e = 0.5L$, for the cases shown in Figure 7.50c, d, and e, respectively.

The critical buckling loads as determined by expressions such as Equation 7.81 to 7.84 do not contain any material strength property. The only material property involved is the elastic modulus E. However, a limitation on the use of Euler's buckling load stems from the fact that the derivations presented above are applicable only for linear elastic material. If the stress at any point in the column material exceeds the proportional limit, these formulas cannot be used. An elastoplastic analysis for the buckling of columns will be presented in Chapter 12. Such limitation on Euler's buckling load restricts its applicability to columns up to certain slenderness ratios defined as $L/\sqrt{I/A}$. By designating the material proportional limit by Y, the average buckling stress P_{cr}/A for a pin-ended column, Equation 7.81b, should satisfy the condition:

$$\frac{P_{cr}}{A}=\frac{\pi^2 E}{\left(L/\sqrt{I/A}^2\right)}\leq Y \tag{7.85}$$

For a particular steel, say, with the properties E = 200 GPa and Y = 200 MPa, the use of the Euler buckling load is restricted to columns of

$$L/\sqrt{I/A}\geq 99.3$$

Example 7.26:

Determine the critical axial buckling load for a column with two pinned ends. The column is composed of two portions of different areas as shown in Figure 7.51.

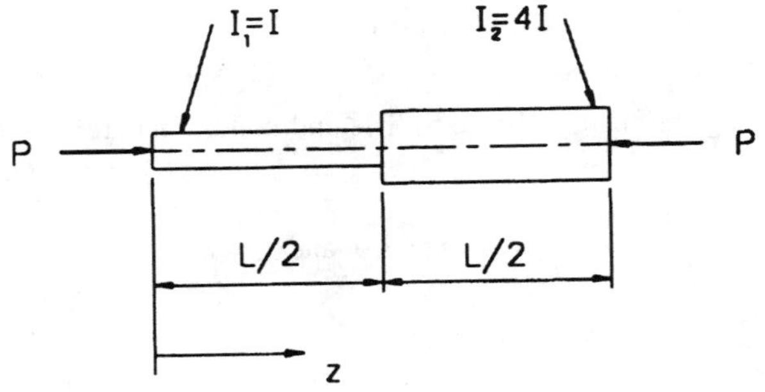

FIGURE 7.51 Example 7.26.

Solution:

The equation of equilibrium for the first and second portions, respectively, are

$$\frac{d^2v_1}{dz} + 4k^2v_1 = 0 \text{ and } \frac{d^2v_2}{dz} + k^2v_2 = 0$$

where $k^2 = P/4EI$.

The solution of these equations are, respectively,

$$v_1 = C_1 \sin 2kz + C_2 \cos 2kz$$

$$v_2 = C_3 \sin kz + C_4 \cos kz$$

The constants of integration C_1, C_2, C_3, and C_4 are determined from four conditions, namely,

$$v_1 = 0, \text{ at } z = 0$$

$$v_1 = 0, \text{ at } z = L$$

$$v_1 = v_2 \text{ and } dv_l/dz = dv_2/dz \text{ at } z = L/2$$

for continuity.

Applying the first condition gives $C_2 = 0$, while the other three conditions yield three equations as

$$C_1 \sin kL - C_3 \sin\frac{kL}{2} - C_4 \cos\frac{kL}{2} = 0$$

$$2C_1 \cos kL - C_3 \cos\frac{kL}{2} + C_4 \sin\frac{kL}{2} = 0$$

$$C_3 \sin kL + C_4 \cos kL = 0$$

Now C_1, C_3, and C_4 have a nonzero solution only if the determinant of the above system of linear homogeneous equations vanishes, i.e.,

$$\begin{vmatrix} \sin kL & -\sin\dfrac{kL}{2} & -\cos\dfrac{kL}{2} \\ 2\cos kL & -\cos\dfrac{kL}{2} & \sin\dfrac{kL}{2} \\ 0 & \sin kL & \cos kL \end{vmatrix} = 0$$

This yields $\sin KL/2 = 0$ and $\tan^2 (KL/2) = 2$. The smallest nonzero root is thus obtained from

$$\tan\frac{kL}{2} = \sqrt{2}, \quad \text{i.e.,} \quad kL = 1.91$$

recalling that $k^2 = P/4EI$ results in the critical buckling loads as

$$P_{cr} = \frac{14.6EI}{L^2}$$

7.7.1.2 Minimum Potential Energy Solution: Rayleigh–Ritz Method

The principle of minimum potential energy, which was discussed in Section 3.10.2, can be applied effectively to obtain the critical buckling load of a column under general loading. The method of solution depends on assuming an arbitrary admissible function $v = v(z)$, representing the deflection curve. Obviously, this function satisfies the kinematic boundary conditions of the problem under consideration. Often, a trigonometric series in the form:

$$v = \sum_{n=1}^{\infty} C_n \sin\left(\frac{n\pi z}{L}\right) \tag{7.86}$$

approximates the deflection curve. Also algebraic polynomials may be employed.

A method to determine the unknown coefficients in any assumed deflection curve in Expression 7.86 is known as the Rayleigh–Ritz method.* The procedure of this method is to insert the function representing the deflection curve into the equation expressing the principle of minimum potential energy of the structure, as given in Section 3.10.3, by

$$\delta\Omega = \delta(U - W) = \delta U - \delta W = 0 \tag{3.42}$$

Since $d\Omega$ is an exact differential, then Equation 3.38 gives expressions that are functions of the undetermined coefficients C_n ($n = 1, 2, \ldots$). Also, since the potential energy must be a minimum at equilibrium, these coefficients can be determined by minimizing Ω, according to

$$\frac{\partial\Omega}{\partial C_1} = 0, \quad \frac{\partial\Omega}{\partial C_2} = 0, \quad \cdots \quad \frac{\partial\Omega}{\partial C_n} = 0 \tag{7.87}$$

* This includes the application of the calculus of variation. See Shames and Dym.[5]

which gives n simultaneous algebraic equations for the coefficients C_1, C_2, C_3, ..., C_n. Solving Equations 7.87 and substituting into Expression 7.86 gives the required deflection curve.

In the majority of cases, few terms of the series (Equation 7.86) are required to give a sufficiently accurate solution with less tedious algebra than the conventional method of integrating the equilibrium equations expressed in terms of deflections. Such a method works to advantage in the determination of critical buckling loads, beam and plate deflections, and also in numerical methods, such as finite element.

If the function $v = v(z)$, assumed to represent the deflection curve, obeys all prescribed boundary conditions, those involving displacements (kinematic conditions) and others related to loads (static conditions), the solution will yield fairly accurate results for the critical buckling force P_{cr}. Within this class of assumed functions, the mathematical form will have little effect on the magnitude of P_{cr}. On the other hand, an assumed function that satisfies only the kinematic boundary conditions and ignores the static ones is expected to approximate P_{cr} fairly well. In general, the use of any function other than the exact one overestimates the critical buckling force, and thus it offers an upper-bound result.

To illustrate the application of Equation 3.42, consider the bar with pinned ends shown in Figure 7.49a, which is loaded axially by a force P. The increment of virtual work done by the external load δW, when changing from the straight line to the bent configuration, is given by

$$\delta W = P\Delta L \tag{7.88}$$

where ΔL is the displacement of P due to column curvature. Note that this expression does not involve the factor $^1/_2$, as in determining elastic strain energy, since the force P remains unchanged over the displacement ΔL. To calculate ΔL, Figure 7.52 gives

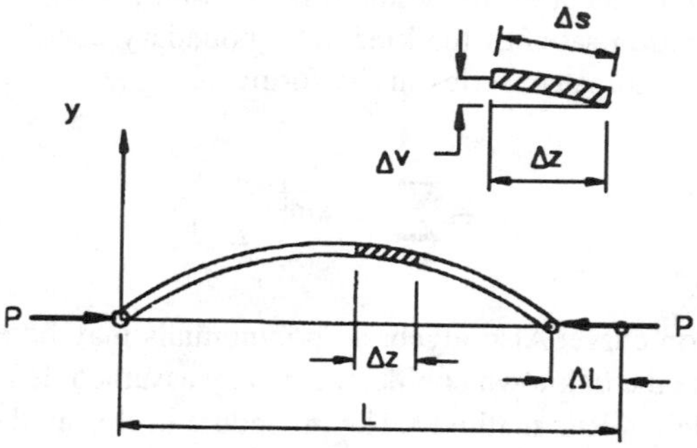

FIGURE 7.52 Energy approach to the buckling of a bar in axial compression.

$$\Delta L \cong \int_0^L (\Delta s - \Delta z) = \int_0^L \left(\Delta z^2 + \Delta v^2\right)^{1/2} - \Delta z$$

$$\cong \int_0^L \Delta z\left[1 + \frac{1}{2}\left(\frac{\Delta v}{\Delta z}\right)^2 + \cdots\right] - \Delta z$$

Keeping only terms up to the second degree yields

$$\Delta L \cong \frac{1}{2}\int_0^L \left(\frac{dv}{dz}\right)^2 dz \tag{7.89}$$

and hence, from Equation 7.88,

$$\delta W = \frac{1}{2} P_{cr} \int_0^L \left(\frac{dv}{dz}\right)^2 dz \tag{7.90}$$

The increase in the strain energy stored in the rod as a result of curvature is obtained from Expression 7.73 as

$$\delta U = \int_0^L \frac{M^2}{2EI} dz$$

Substituting

$$M = -EI \frac{d^2 v}{dz^2}$$

Equation 7.10c, yields

$$\delta U = \frac{1}{2} \int_0^L EI \left(\frac{d^2 v}{dz^2}\right)^2 dz \tag{7.91}$$

Inserting Expressions 7.90 and 7.91 in the minimum potential energy condition, Equation 3.42, gives, for the critical load,

$$P_{cr} = \left[\int_o^L EI \left(\frac{d^2 v}{dz^2}\right)^2 dz\right] \Big/ \left[\int_o^L \left(\frac{dv}{dz}\right)^2 dz\right] \tag{7.92}$$

If the function $v = v(z)$ describing the deformed shape or deflection curve of the rod is known, P_{cr} can be easily determined.

Example 7.27:

Determine the critical buckling force for a pin-ended column of length L *and subjected to an axial compressive force* P *assuming the deflection curve to be given by*

a. v = C *sin* (πz/L),
b. v = Cz(L – Z) *and*
c. d^2v/dz^2 = Cz (L – z)

where C *is a constant.*

Solution:

a. The assumed deflection curve, $v = C \sin (\pi z/L)$ satisfies both kinematic and static boundary conditions at both $z = 0$ and $z = L$, namely; $v = 0$ and $M = EI\,(d^2v/dz^2) = 0$, respectively. Substituting into Equation 7.92 yields

$$P_{cr} = \left[\int_o^L EI \left(\frac{C\pi^2}{L^2} \sin \frac{\pi z}{L}\right)^2 dz\right] \Big/ \left[\int_o^L \left(\frac{C\pi}{L} \cos \frac{\pi z}{L}\right)^2 dz\right]$$

Hence,

$$P_{cr} = \frac{\pi^2 EI}{L^2}$$

which is the Euler buckling load for a pin-ended column. This result is obtained since the assumed deflection curve is in fact the exact one.

b. The assumed deflection curve, $v = Cz\,(L - z)$, satisfies only the kinematic boundary conditions at $z = 0$ and $z = L$, $y = 0$. Substituting into Equation 7.92 yields

$$P_{cr} = \left[\int_o^L EI(-2C)^2\,dz\right] \Big/ \left[\int_o^L (CL - 2Cz)^2\,dz\right]$$

Hence,

$$P_{cr}^* = \frac{12EI}{L^2} = 1.216\frac{\pi^2 EI}{L^2}$$

Comparing this value with that of Expression 7.81b indicates that the error in the value of $P_{cr}{}^*$ amounts to 21.6%. This is due to the fact that the deflection curve given by $v = Cz\,(L - z)$ gives $d^2v/dz^2 = C$, i.e., a constant bending moment along the deformed length of the column, which does not agree with the static boundary conditions.

c. Integrating twice, the deflection curve is obtained as

$$v = C\left(\frac{z^3 L}{6} - \frac{z^4}{12} + C_1 z + C_2\right)$$

Imposing the kinematic boundary conditions $v = 0$ at both $z = 0$ and $z = L$ — the deflection curve takes the form:

$$v = \frac{C}{12}\left(2z^3 L - z^4 - zL^3\right)$$

which obviously satisfies the static boundary conditions at $z = 0$ and $z = L$. Note also that at $z = L/2$, $dv/dz = 0$, thus fulfilling the condition of maximum deflection at the middle of the column.

Substituting into Equation 7.92 yields

$$P_{cr} = \left[\int_o^L EI\left[Cz(L-z)\right]^2 dz\right] \Big/ \left[\int_o^L \left(\frac{C}{12}\right)^2 \left(6Lz^2 - 4z^3 - L^3\right)^2 dz\right]$$

Hence,

$$P_{cr}^{**} = \frac{168}{17}\frac{EI}{L^2} = 1.0013\frac{\pi^2 EI}{L^2}$$

which is only 0.13% in error with the Euler buckling load. Note that in both cases (b) and (c), the approximate P_{cr} is greater than the exact one obtained in (a).

Example 7.28:

Determine the critical buckling load for a column fixed at one end and subjected to a compressive force at its other free end as shown in Figure 7.53. Consider the effect of the column weight in the analysis and assume a deflection curve given by the expression:

$$v = C\left(1 - \cos\frac{\pi z}{2L}\right)$$

where C *is a constant.*

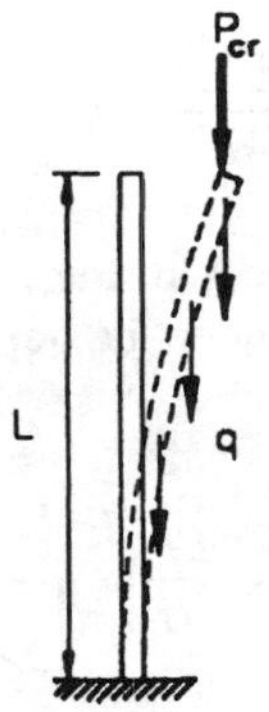

FIGURE 7.53 Example 7.28.

Solution:

The deflection curve equation,

$$v = C\left(1 - \cos\frac{\pi z}{2L}\right)$$

satisfies boundary conditions:

- Kinematic: at $z = 0$; $v = 0$ and $dv/dz = 0$
- Static: at $z = L$; $M = 0$, i.e., $EI\,(d^2v/dz^2) = 0$

The strain energy increases due to bending by

$$\delta U = \frac{1}{2}\int EI\left(\frac{d^2v}{dz^2}\right)^2 dz = \frac{1}{4}EIC^2L\left(\frac{\pi}{2L}\right)^4$$

The increment of work done by the applied loads when passing from the straight configuration of equilibrium to the bent one is due to both the external force P and the weight of the column treated as distributed load q. The latter load at any distance z is doing work according to Expression 7.90:

$$qdz\left[\frac{1}{2}\int_0^z\left(\frac{dv}{dz}\right)^2 dz\right]=\frac{1}{4}qdzC^2\left(\frac{\pi}{2L}\right)^2\left(z-\frac{L}{\pi}\sin\frac{\pi z}{L}\right)$$

Hence, the increment of total work done by the external loads is

$$\delta W=\frac{1}{2}P_{cr}\int_0^L\left(\frac{dv}{dz}\right)^2 dz+\frac{1}{4}dC^2\left(\frac{\pi}{2L}\right)^2\int_0^L\left(z-\frac{L}{\pi}\sin\frac{\pi z}{L}\right)dz$$

$$=\frac{1}{2}P_{cr}C^2\left(\frac{\pi}{2L}\right)^2\frac{L}{2}+\frac{1}{4}qC^2\left(\frac{\pi}{2L}\right)^2\left(\frac{L^2}{2}-\frac{2L^2}{\pi^2}\right)$$

Applying the principal of minimum potential energy $\delta(U - W) = 0$ gives:

$$P_{cr}=\frac{\pi^2 EI}{4L^2}-q\left(\frac{L}{2}-\frac{2L}{\pi^2}\right)$$

By neglecting the column weight, P_{cr} reduces to the Euler buckling load for this case (Equation 7.82b). However, if buckling is to occur under the action of the weight of the column only, the following relation is to be satisfied:

$$q_{cr}=\frac{EI}{2L^3}\left(\frac{\pi^4}{\pi^2-4}\right)$$

which is 6% in error with the exact solution, giving $q_{cr} = 7.83\ EI/L^3$.

Considering two columns of the same cross sectional area, one made of steel (E_s = 200 GPa, specific weight = 78 kN/m³) and the other made of lead (E_l =14 GPa, specific weight = 114 kN/m³), the ratio of critical lengths for buckling is

$$L_{\text{steel}}/L_{\text{lead}}=\left(\frac{E_s/q_s}{E_l/q_l}\right)^{1/3}=2.75$$

Example 7.29:

Apply the Rayleigh–Ritz method to determine the critical buckling load for a pin-ended rod with cross section variation as shown in Figure 7.54.

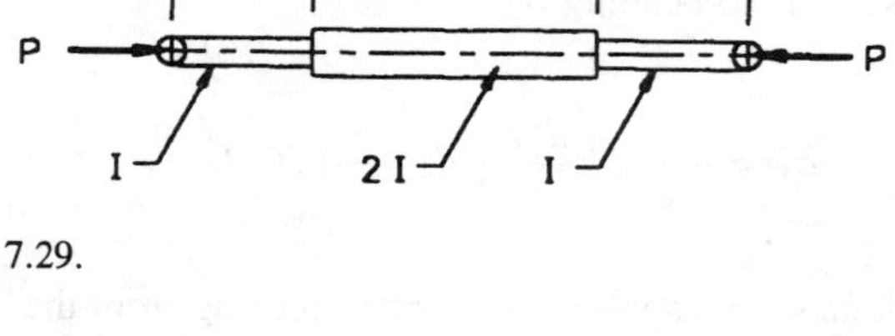

FIGURE 7.54 Example 7.29.

Solution:

Assume a deflection curve,

$$v = C_1 \sin(\pi z/L)$$

satisfying all boundary conditions:

$$\text{At } z = 0:\ v = 0,\ M = -EI\frac{d^2v}{dz^2}$$

$$\text{At } z = L:\ v = 0,\ M = -\frac{EId^2v}{dz^2}$$

The increment of work done by the load P_{cr} in the bent configuration is thus obtained from Expression 7.90 as

$$\delta W = \frac{1}{2}P_{cr}\int_0^L\left(\frac{dv}{dz}\right)^2 dz = \frac{\pi^2}{4L}C_1^2P_{cr}$$

The increase in the strain energy of the rod due to bending is obtained from Expression 7.91 as

$$\delta U = \frac{1}{2}\int_0^{L/4}(EI)\left(\frac{d^2v}{dz^2}\right)^2 dz + \frac{1}{2}\int_0^{3L/4}(2EI)\left(\frac{d^2v}{dz^2}\right)^2 dz$$

$$+\frac{1}{2}\int_{3L/4}^{L}(EI)\left(\frac{d^2v}{dz^2}\right)^2 dz$$

This upon substitution from the assumed deflection curve reduces to

$$\delta U = \frac{\pi^4 EI}{2L^3}C_1^2$$

From the condition of minimum potential energy (Equation 3.42) or, alternatively, using Equation 7.92, the critical buckling load is obtained as

$$P_{cr} = \frac{2\pi^2 EI}{L^2} = 19.74\frac{EI}{L^2}$$

A more accurate value for P_{cr} is obtained by assuming a two-term series for the deflection curve, namely,

$$v = C_1 \sin(\pi z/L) + C_2 \sin(3\pi z/L)$$

Minimizing the total potential energy $\Omega = (U - W)$ expressed in terms of v and its derivatives, according to the Rayleigh–Ritz method as given by Equations 7.87, yields the following two algebraic equations to solve

$$\frac{\partial\Omega}{\partial C_1} = \left(3\pi^2 EI + 2\pi EI - 2L^2P_{cr}\right)C_1 - 18\pi EIC_2 = 0$$

$$\frac{\partial\Omega}{\partial C_2} = -2EIC_1 + \left(27\pi^2 EI - 6\pi EI - 2L^2P_{cr}\right)C_2 = 0$$

This results in the smallest critical buckling load as

$$P_{cr} = 17.11EI/L^2$$

7.7.2 Beam–Columns

Beam–columns are beams subjected to axial compressive forces in addition to lateral loads, as shown in Figure 7.55. The determination of the deflection curves, using the expression $EI(d^2v/dz^2) = -M$, becomes rather tedious especially when several lateral loads are applied over the span of the beam. Several functions expressing the bending moment in different portions of the beam will be involved. It is often appropriate to assume the deflection curve in the form of a trigonometric series as given by Expression 7.86. The coefficients C_1, C_2, ..., C_n are properly selected to make such series represent the deflection curve by applying the Rayleigh–Ritz method.

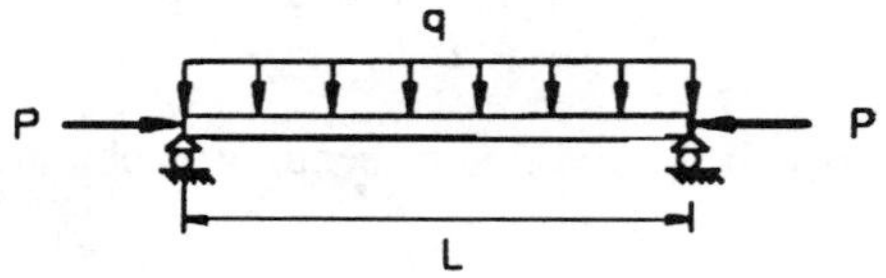

FIGURE 7.55 Example 7.30.

Example 7.30:

Determine the maximum deflection of the beam–column loaded as shown in Figure 7.55. Assume the deflection curve to be given by v = C sin (πz/L), where C is a coefficient to be determined by minimizing the total potential energy.

Solution:

The assumed deflection curve obviously satisfies all boundary conditions:

$$\text{At } z = 0\text{: } v = 0,\ M = -EI\frac{d^2v}{dz^2} = 0$$

$$\text{At } z = L\text{: } v = 0,\ M = -EI\frac{d^2v}{dz^2} = 0$$

According to Expression 7.73, the strain energy stored in the beam due to bending is

$$U = \int_0^L \frac{M^2}{2EI}dz$$

This reduces to Expression 7.91.

$$U = \frac{1}{2}EI\int_0^L \left(\frac{d^2v}{dz^2}\right)^2 dz$$

The work done by the externally applied loads when the beam column assumes the bent form is determined according to Expression 7.90 as

$$W = \frac{1}{2} P \int_0^L \left(\frac{dv}{dz} \right)^2 dz + \int_0^L qv dz$$

The total potential energy is

$$\Omega = U - W$$

which upon substitution from the expression representing the given elastic curve becomes

$$\Omega = \frac{1}{2} EI \int_0^L \left[C \left(\frac{\pi}{L} \right)^2 \sin \frac{\pi z}{L} \right]^2 dz$$

$$- \frac{1}{2} P \int_0^L \left[C \left(\frac{\pi}{L} \right) \cos \frac{\pi z}{L} \right]^2 dz - \int_0^L qC \sin \frac{\pi z}{L} dz$$

Hence,

$$\Omega = \frac{C^2 \pi^3}{4L^2} \left(\frac{EI\pi^2}{L^2} - P \right) - 2qC$$

By applying the Rayleigh–Ritz method, Expression 7.87 gives

$$\frac{d\Omega}{dC} = 0 = \frac{2C\pi^3}{4L^2} \left(\frac{EI\pi^2}{L^2} - P \right) - 2q$$

This determines the value of the coefficient C, so that

$$v = \frac{4qL^4}{\pi^3 \left(EI\pi^2 - PL^2 \right)} \sin\left(\frac{\pi z}{L} \right)$$

with

$$v_{max} = \frac{4qL^4}{\pi^3 \left(EI\pi^2 - PL^2 \right)} \quad \text{at } z = L/2$$

The exact value for this maximum deflection is*

$$\left(v_{max} \right)_{exact} = \frac{qEI}{P^2} \sec\left(\sqrt{\frac{P}{EI}} \frac{L}{2} \right) - \frac{qL^2}{8P} - \frac{qEI}{P^2}$$

For $\sqrt{P/EI} = \pi/2$ and $L = 1$, the error in v_{max} is only about 0.5%.

The presence of an axial force obviously magnifies the lateral deflection of the beam. In this case for $\sqrt{P/EI} = \pi/2$ and unit length, the central deflection is $0.01738q/EI$ compared with

* This could be obtained by integrating the equation $EI(d^2v/dz^2) = -M(z)$ as in Rees.[6]

$0.01302q/EI$ for a beam loaded only by a lateral distributed load q. The same conclusion applies to the bending moment which is increased by the value of Pv_{max} at the beam midlength.

7.7.3 Lateral Buckling of Beams

Beams of small lateral stiffness are unstable in bending. A typical case is a beam having a rectangular cross section of thin width (thickness) and high depth subjected to an end bending moment M_o in the plane of beam depth. The end supports allow rotation in the plane of depth but restrict lateral displacement of the beam ends. Under unstable conditions, the compression side of beam depth will be laterally displaced; the lateral displacement decreases gradually toward the tension side, vanishing at the edge of beam depth. As a result, the beam will be subjected to combined bending and torsion.

By referring to Figure 7.56, M_o is acting in the y–z plane, u is the lateral displacement in the x-direction, $\psi = \partial u/\partial z$ is the slope in the x–z plane, ϕ is the angle of twist in the x–y plane and $\alpha = \partial\phi/\partial z$ is the angle of twist per unit length.

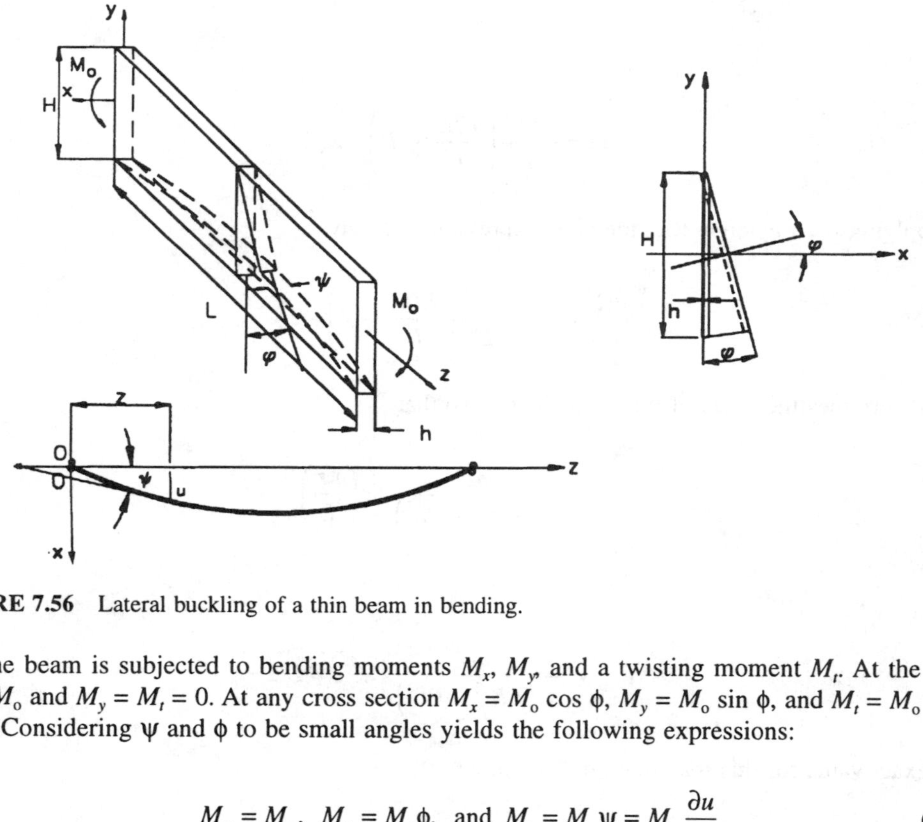

FIGURE 7.56 Lateral buckling of a thin beam in bending.

The beam is subjected to bending moments M_x, M_y, and a twisting moment M_t. At the ends, $M_x = M_o$ and $M_y = M_t = 0$. At any cross section $M_x = M_o \cos\phi$, $M_y = M_o \sin\phi$, and $M_t = M_o \cos\phi \sin\psi$. Considering ψ and ϕ to be small angles yields the following expressions:

$$M_x = M_o,\ \ M_y = M_o\phi,\ \text{ and } M_t = M_o\psi = M_o\frac{\partial u}{\partial z} \tag{7.93}$$

From Expression 7.10b,

$$M_y = M_o\phi = -EI_y\frac{\partial^2 u}{\partial z^2} \tag{7.94a}$$

and from Expression 7.60b,

$$M_t = M_o \frac{\partial u}{\partial z} = GI_t \frac{\partial \phi}{\partial z} \tag{7.94b}$$

Eliminating ϕ from Equations 7.94 gives

$$\frac{EI_y}{M_o}\frac{\partial^3 u}{\partial z^3} + \frac{M_o}{GI_t}\frac{\partial u}{\partial z} = 0$$

or

$$\frac{\partial^3 u}{\partial z^3} + \frac{M_o^2}{EI_y \cdot GI_t}\frac{\partial u}{\partial z} = 0$$

Putting

$$\frac{M_o^2}{EI_y \cdot GI_t} = k^2$$

and integrating yields

$$u = C_1 + C_2 \sin kz + C_3 \cos kz \tag{7.95}$$

where C_1, C_2, and C_3 are integration constants obtained from the prevailing boundary conditions, namely,

At $z = 0$: $u = 0$ and $\phi = 0$, which gives $C_1 = C_3 = 0$
At $z = L$: $u = 0$ and $\phi = 0$, which gives $C_2 \sin kL = 0$

Since $C_2 = u_{max} \neq 0$, then $\sin kL = 0$, which gives $kL = n\pi$; $n = 1, 2, 3 \ldots$. The critical value of M_o is thus obtained at $n = 1$, i.e., at $k = \pi/L$, which gives

$$\frac{\left(M_o^2\right)_{cr}}{EI_y \cdot GI_t} = \frac{\pi^2}{L^2}$$

or

$$\left(M_o\right)_{cr} = \frac{\pi}{L}\sqrt{EI_y GI_t}$$

Substituting

$$I_y = \frac{H^3}{12}, \quad I_t = \frac{H^3}{3}, \quad G = \frac{E}{2(1+\nu)}$$

yields

$$\left(M_o\right)_{cr} = \frac{\pi h H^3}{6L}\frac{E}{\sqrt{2(1+\nu)}} \tag{7.96}$$

The displacements u, ϕ, and ψ, are obtained as

$$u = u_{\max}\sin\frac{\pi z}{L}, \phi = \frac{M_o}{GI_t}u = \frac{M_o}{GI_t}u_{\max}\sin\frac{\pi z}{L}$$

$$\alpha = \frac{\partial\phi}{\partial z} = \frac{M_o}{GI_t}u_{\max}\frac{\pi}{L}\cos\frac{\pi z}{L} \tag{7.97}$$

$$\psi = \frac{\partial\phi}{\partial z} = u_{\max}\frac{\pi}{L}\cos\frac{\pi z}{L}$$

Example 7.31:

A flat strip of 1.2 mm thickness, 20 mm width, and 100 mm length is subjected at its ends to a bending moment M_o *in the plane of the strip. The end supports allow rotation in the plane of the strip but restrict lateral displacement of the ends. Material is spring steel,* $E = 210$ *GPa,* $\nu = 0.3$*, and* $\sigma_{yield} = 500$ *MPa. Determine:*

a. *The critical end bending moment for stable deformation,*
b. *The maximum lateral displacement to maintain the strip elastic,*
c. *The maximum torsional shear stress at* u_{max}

Solution:

a. From Equation 7.96 the critical end bending moment

$$\left(M_o\right)_{cr} = \frac{\pi\times 20\times 1.2^3\times 210\times 10^3}{6\times 100\sqrt{2\times 1.3}} = 23.567 \text{ Nm}$$

$$\left(\sigma_z\right)_{\max} \text{ due to } M_x = \frac{23.567\times 10^3\times 6}{1.2\times 20^2} = 294.59 \text{ MPa}$$

For elastic condition $(\sigma_z)_{\max} \le 500$ MPa, therefore,

$$(\sigma_z)_{\max} \text{ due to } M_y = 500 - 294.59 = 205.41 \text{ MPa}$$

Since $(\sigma_z)_{\max} = 6M_y/H^2$, it gives

$$M_y = \frac{205.41\times 20\times 1.2^2}{6} = 0.986 \text{ Nm}$$

b. From Equation 7.93, $M_y = M_o\phi$ and from Equation 7.97

$$\phi = \frac{M_o}{GI_t}u_{\max} \quad \text{at } z = \frac{L}{2}$$

hence,

$$u_{max} = GI_t \frac{M_y}{M_o^2}$$

Given that

$$G = \frac{E}{2(1+\nu)} = \frac{210}{2\times 1.3} = 80.77 \text{ GPa}$$

and

$$I_t = \frac{1.2\times 20}{3} = 11.52 \text{ mm}^4$$

therefore,

$$u_{max} = 80.77\times 10^3 \times 11.52\times 10^{-3} \frac{0.986}{23.567^2} = 1.652 \text{ mm}$$

Note that, at $z = L/2$, $M_t = M_o\psi = 0$.

c. The twisting moment M_t is maximum at $z = 0$; from Equations 7.93 and 7.97,

$$(M_t)_{max} = M_o\psi = M_o u_{max}\frac{\pi}{L} = 23.567\times 1.652\times \frac{\pi}{100} = 1.223 \text{ Nm}$$

$$\tau_{max} = \frac{1.223\times 10^3}{11.52} = 106.16 \text{ MPa}$$

At $z = 0$, $M_x = M_o$ and $M_y = 0$; therefore, $(\sigma_z)_{max} = 294.59$ MPa as determined before. For elastic conditions under combined normal and shear stresses, Equation 1.32,

$$\sqrt{\sigma_z^2 + 4\tau_{max}^2} \le \sigma_y$$

or

$$\sqrt{294.59^2 + 4\times 106.16^2} = 363.13 < 500 \text{ MPa}$$

7.8 BEAMS ON ELASTIC FOUNDATION

In some engineering applications, beams, such as rail tracks, are supported on continuous foundations, such that the reactions due to external loading are distributed along the beam length. If the foundation is considered to be elastic, the reaction at any point will be proportional and opposite in direction to the transverse displacement of the beam at that point. Accordingly, considering the *y–z* plane, the reactions may be expressed by the linear relation:

$$q = -k_s v \tag{7.98}$$

where q is the reaction force per unit beam length, k_s the elastic constant of the foundation, and v the transverse displacement of the beam axis. Note that q is the normal stress on the boundary surface for a beam of unit width. According to the convention of Section 1.5, q is positive if it is tension and negative if compression. For a positive v, q is tension and, for a negative v, q is compression. Since v is positive upward and negative downward, q as given by Equation 7.98 acts in the opposite direction of v.

Substituting Equation 7.98 in the equation of the elastic curve (Equation 7.11b) yields

$$EI\frac{d^4v}{dz^4} = -k_s v$$

hence,

$$\frac{d^4v}{dz^4} + \frac{k_s v}{EI} = 0$$

or

$$\frac{d^4v}{dz^4} + 4\beta^4 v = 0 \tag{7.99}$$

where

$$\beta^2 = \sqrt{\frac{k_s}{4EI_x}} \tag{7.100}$$

In order to simplify the analysis, the beam is considered to have a uniform cross section and the displacement v is along a principal centroidal axis. Hence, I is constant along z and is equal to the second moment of the cross sectional area about the other principal centroidal axis. The differential Equation 7.99 has a general solution:

$$v = e^{-\beta z}(A \sin \beta z + B \cos \beta z) + e^{\beta z}(C \sin \beta z + D \cos \beta z) \tag{7.101}$$

where A, B, C, and D are integration constants determined from loading and boundary conditions.

7.8.1 Infinitely Long Beams

7.8.1.1 Concentrated Force

Consider the external load to be a concentrated force P acting at O which is taken as the origin of z, as shown in Figure 7.57a. Note that the positive direction of P is the direction of positive shear stress, as given in Section 1.5. In this case, the effect of P on v will be localized around O and tends to decease away from P, which means that v should decrease with z. This condition will eliminate the second term of Equation 7.101, since it is rapidly increasing with respect to z, so that the constants $C = D = 0$. Equation 7.101 thus reduces to

$$v = e^{-\beta z}(A \sin \beta z + B \cos \beta z) \tag{7.102}$$

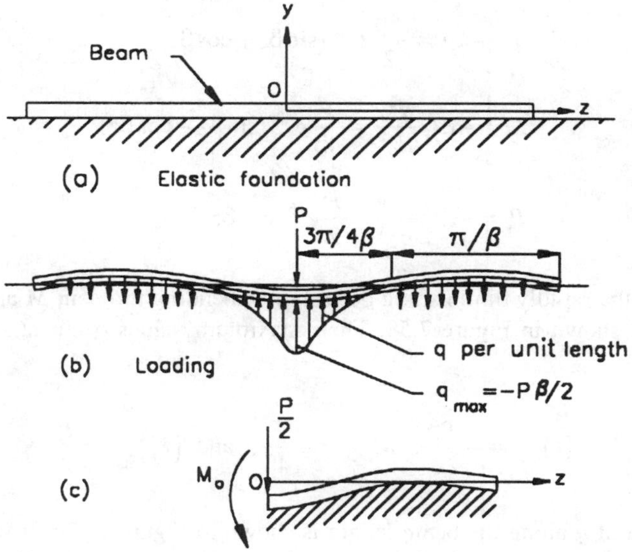

FIGURE 7.57 Long beam on elastic foundation subjected to a concentrated lateral load at midlength: (a) beam reference axes, (b) load diagram, and (c) shearing force and bending moment at 0.

In order to determine A and B, consider a half part of the beam, Figure 7.57c, acted upon by an edge load consisting of a transverse force $P/2$ and a bending moment M_o. Now, from symmetry $dv/dz = 0$ at $z = 0$ and, hence, $A = B$. Substituting in Equation 7.102 yields

$$v = Ae^{-\beta z}(\sin \beta z + \cos \beta z) \tag{7.103}$$

The constant A is obtained from the boundary condition such that the shearing force is equal to $P/2$ at $z = 0$ or, according to Equation 7.11a,

$$EI_x \frac{d^3v}{dz^3} = -\frac{P}{2} \quad \text{at } z = 0$$

Hence, from Equation 7.103,

$$\frac{d^3v}{dz^3} = 4\beta^3 Ae^{-\beta z} \cos \beta z$$

and

$$A = -\frac{P}{8\beta^3 EI_x} = -\frac{P\beta}{2k_s}$$

Upon substituting in Equation 7.103, this yields

$$v = -\frac{P\beta}{2k_s} e^{-\beta z}(\sin \beta z + \cos \beta z) \tag{7.104}$$

$$q = -k_s v = \frac{P\beta}{2} e^{-\beta z}(\sin \beta z + \cos \beta z)$$

$$M = -EI_x \frac{d^2 v}{dz^2} = \frac{P}{4\beta} e^{-\beta z}(\sin \beta z - \cos \beta z) \tag{7.105}$$

$$P_y = -EI_x \frac{d^3 v}{dz^3} = \frac{P}{2} e^{-\beta z} \cos \beta z$$

The distribution of the rapidly diminishing deflection v, bending moment M and shearing force P_y along the beam are shown in Figure 7.58. Their maximum values occur at $z = 0$ and are given, respectively by

$$(v)_{max} = -\frac{P\beta}{2k_s}, \quad M_{max} = -\frac{P}{4\beta}, \quad \text{and} \quad (P_y)_{max} = \frac{P}{2} \tag{7.106}$$

The distribution of q along the beam length is shown in Figure 7.57c. It will be observed that q changes sign at the point where $\sin \beta z = -\cos \beta z$ or at $z = 3\pi/4\beta$. Hence, the beam should be clamped to the foundation to transmit the tensile reaction. Note that for equilibrium

$$\int q dz = P \quad \text{and} \quad \int qz dz = 0$$

The deflection of the beam vanishes at $z = 3\pi/4\beta$, whereas the bending moment and the shearing force attain smaller values at this distance, as seen from Figure 7.58b and c. This suggested assigning a value of z at which the infinite beam solution becomes a good approximation to short beams. A value of about $z = 2\pi/\beta$ has been suggested,[7] at which the magnitudes of v, M, and P_y are only 0.00187 of their maximum values, respectively. This means that a beam of length $L \geq 2\pi/\beta$ will essentially have the same deflection curve, and hence, moment and shearing force diagrams, as an infinitely long beam. The infinite beam solution also is considered to give reasonably accurate results for long beams loaded with a concentrated force located at a distance $\geq 2\pi/\beta$ from the nearer end of the beam.

It would be interesting to compare the case of a concentrated force acting on a beam on an elastic foundation with the case when a concentrated in-plane force acts normal to the edge of a semi-infinite plate (Section 6.3.3). In the former case, the beam and foundation are generally of different materials, which can slide relatively along the common boundary such that only normal stresses are transmitted. In the latter case the beam and foundation are one integral solid. This explains the occurrence of a tensile reaction in the case of a beam on an elastic foundation, while in the latter only compressive stresses prevail throughout the plate depth. However, the solution in Section 6.3.3 for a semi-infinite plate can be applied to determine the value of the elastic constant k_s of the foundation.

7.8.1.2 Concentrated Moment

The concentrated moment is considered as equivalent to two concentrated forces P, a distance L apart, as shown in Figure 7.59. Hence, by using Equation 7.104 and replacing P by M_o/L, the deflection at a distance z due to the two forces P and $-P$ is given by

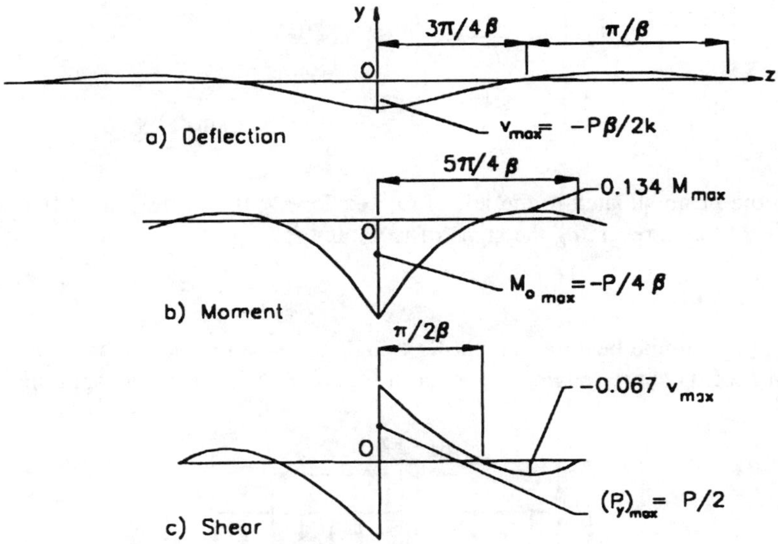

FIGURE 7.58 Deflection (a), bending moment (b), and shearing force (c) for the beam on elastic foundation of Figure 7.57.

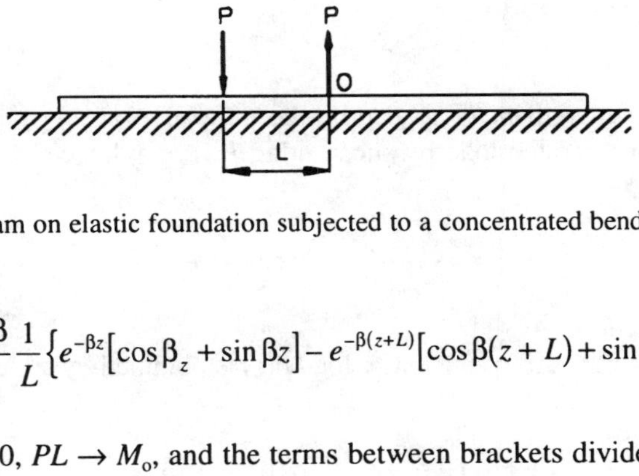

FIGURE 7.59 Long beam on elastic foundation subjected to a concentrated bending moment at midlength.

$$v = -\frac{M_o\beta}{2k_s}\frac{1}{L}\left\{e^{-\beta z}\left[\cos\beta_z + \sin\beta z\right] - e^{-\beta(z+L)}\left[\cos\beta(z+L) + \sin\beta(z+L)\right]\right\}$$

In the limit when $L \to 0$, $PL \to M_o$, and the terms between brackets divided by L will be equal to

$$-\frac{d}{dz}\left[e^{-\beta z}\left(\cos\beta_z + \sin\beta z\right)\right]$$

This reduces to

$$v = -\frac{M_o\beta^2}{k_s}e^{-\beta z}\sin\beta z \tag{7.107}$$

The bending moment and shearing force are derived from Equation 7.107 as

$$M = -EI_x \frac{d^2v}{dz^2} = \frac{M_o}{2} e^{-\beta z} \cos\beta z$$
$$P_y = -EI_x \frac{d^3v}{dz^3} = -\frac{M_o\beta}{2} e^{-\beta z}(\cos\beta z + \sin\beta z) \tag{7.108}$$

For points on the beam situated to the left of O, i.e., for $z < 0$, Expressions 7.107 and 7.108 are applied with $z = |z|$ and, reversing the signs of v, M, and P_y.

7.8.1.3 Uniform Load

The problem of an infinite beam on an elastic foundation when subjected to a uniform load $p(z)$ over a portion L of its entire length, as shown in Figure 7.60 is solved by applying the principle

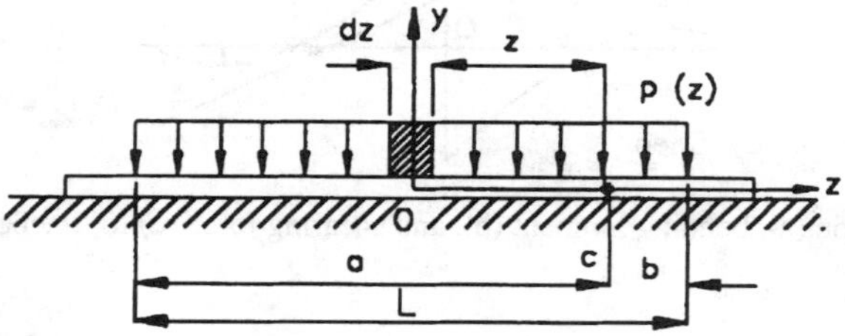

FIGURE 7.60 Long beam on elastic foundation subjected to a uniformly distributed load along a finite length.

of superposition. For an increment of load $p\Delta z$, the deflection at any point C is obtained from Equation 7.104 for a concentrated load by substituting $P = p\Delta z$; hence,

$$\Delta v_c = -\frac{(p\Delta z)\beta}{2k_s} e^{-\beta z}(\cos\beta z + \sin\beta z)$$

The total deflection at point C due to the entire load $p(z)$ is obtained by superposition according to*

$$v = -\int_0^a \frac{p\beta}{2k_s} e^{-\beta z}(\cos\beta z + \sin\beta z)dz - \int_0^b \frac{p\beta}{2k_s} e^{-\beta z}(\cos\beta z + \sin\beta z)dz$$

This yields

$$v = -\frac{p}{2k_s}\left(2 - e^{-\beta a}\cos\beta a - e^{-\beta b}\cos\beta b\right) \tag{7.109}$$

* Since point C is within the loaded length, a is taken negative and b positive. If C is to the left of the loaded length, both a and b are positive. Inversely, a and b are negative when C is to the right of the loaded length.

It will be noted that in deriving Expressions 7.109, point C is fixed in position while O moves along L; thus, Equation 7.104 can be applied. Values of the bending moment and shearing force are also obtained by superposition as

$$M = -\frac{p}{4\beta^2}\left[e^{-\beta a}\sin\beta a + e^{-\beta b}\cos\beta b\right]$$

$$P_y = -\frac{p}{4\beta}\left[e^{-\beta a}(\cos\beta a - \sin\beta a) - e^{-\beta b}(\cos\beta b - \sin\beta b)\right] \tag{7.110}$$

Note that in the above expressions, the quantities a and b represent distances between points and are always positive.

The maximum deflection occurs at the midpoint of the loaded portion, i.e., at $L/2 = (a + b)/2$:

$$v_{\max} = -\frac{p}{k_s}e^{-\beta(a+b)/2}\cos[\beta(a+b)/2]$$

The location of the maximum bending moment is found to be at a point that may lie within or outside the loaded portion depending on the value βL.

- For $\beta(a + b) \le \pi$: the maximum bending moment occurs at the center of the loaded portion L;
- For $\beta(a + b) > \pi$: the maximum bending moment can be taken to occur approximately at $z = \pi/4\beta$ from either end of the loaded portion L;
- For $\beta(a + b) \to \infty$: the bending moment $M \to 0$.

In particular for the last case, when both a and b become very large, the deflection becomes $v = -p/k_s$, which means that the uniform load p is transmitted directly to the elastic foundation.

Example 7.32:

The load moving on a crane rail track consists of two equal and parallel wheel forces P *spaced at a constant distance* L, *Figure 7.61. Find the maximum tensile stress in the rail, which has a square cross section of side* H.

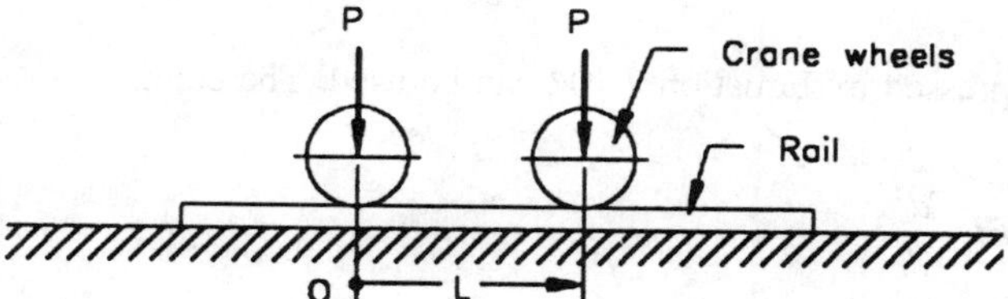

FIGURE 7.61 Example 7.32.

Solution:

The maximum tensile stress in the rail occurs at the section of maximum bending moment and is given by

$$\sigma_{\max} = \frac{6M_{\max}}{H^3}$$

M_{max} is obtained by superposition of the two forces as follows. First, consider only the force P at the left to be acting. The maximum bending moment is under the force P and equals to M_o, where

$$M_o = -\frac{P}{4\beta}$$

At a distance L from O, the bending moment is obtained from Equation 7.105 as

$$M_L = \frac{P}{4\beta} e^{-\beta z}(\sin \beta z - \cos \beta z)$$

Consider now that the second force P is acting. The resulting maximum bending moment is the summation of M_o and M_L. For small values of L, $M_L \rightarrow -P/4\beta$ and, hence, $M_{max} = -P/2\beta$.

The method of superposition may be easily applied to other load combinations.

7.8.2 Semi-Infinite Beams

Figure 7.62 shows a semi-infinite beam loaded at its end point by a concentrated load P and moment M_o. Since the deflection, bending moment, and shearing force diminish for points at large values

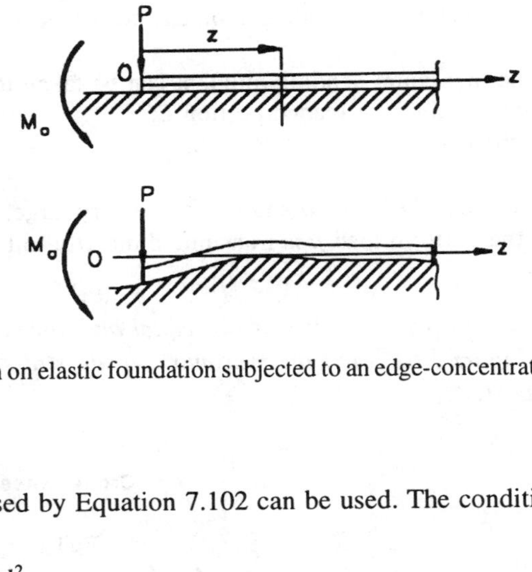

FIGURE 7.62 Long beam on elastic foundation subjected to an edge-concentrated lateral force and a bending moment.

of z, the solution expressed by Equation 7.102 can be used. The conditions:

$$\text{At } z = 0\text{: } M_o = -EI_x \frac{d^2v}{dz^2}$$

$$P = -EI_x \frac{d^3v}{dz^3}$$

must be satisfied; hence, the constants A and B are found to be

$$A = \frac{M_o}{2\beta^2 EI_x} \quad \text{and} \quad B = -\frac{1}{2\beta^3 EI_x}(P - \beta M_o)$$

By introducing these constants, the deflection curve and hence expressions for the bending moment and shearing force are obtained as

$$v = -\frac{2\beta e^{-\beta z}}{k_s}\left[P\cos\beta z - \beta M_o(\cos\beta z - \sin\beta z)\right] \tag{7.111}$$

$$M = \frac{P}{\beta}e^{-\beta z}\sin\beta z - M_o e^{-\beta z}(\cos\beta z + \sin\beta z)$$

$$P_y = Pe^{-\beta z}(\cos\beta z - \sin\beta z) + 2M_o\beta e^{-\beta z}\sin\beta z \tag{7.112}$$

The beam will behave as if it had a free end at point O.

The reaction forces per unit length $q = -k_s v$ are distributed as shown in Figure 7.57. Again, q changes sign at $z = 3\pi/4\beta$. Hence, the beam should be clamped to the foundation to transmit the tensile reactions. Note also that for equilibrium:

$$\int q dz = P \text{ and } \int qz dz = M_o$$

The maximum deflection occurs at the free end. The location of the maximum bending moment and shearing force is given by $0 \leq z \leq \pi/4$, depending on the ratio of P/M_o.

Example 7.33:

An I beam of length 16 m is used as a monorail for an overhead crane supporting a concentrated load of 10 kN at its end. The rail is bolted to the midspans of a number of cross I beams as shown in Figure 7.63, which are simply supported at a length of 5 m. Determine the maximum bending stress in the rail.

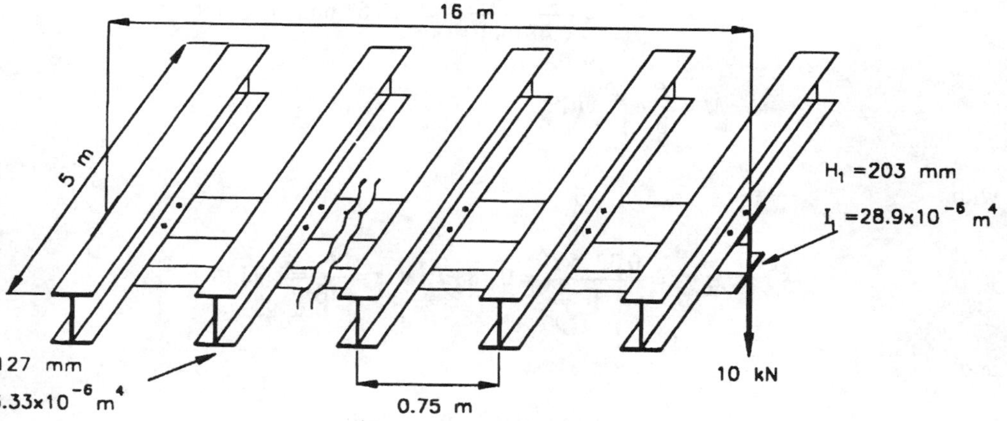

FIGURE 7.63 Example 7.33.

Solution:

The cross beams that represent the elastic foundation exert a discontinuous reaction on the rail. The elastic stiffness for each cross beam at its midspan is given by

$$\frac{P}{v} = \frac{48EI_x}{L^3} = \frac{48 \times 200 \times 10^9 \times 6.33 \times 10^{-6}}{(5)^3}$$

$$= 4.861 \times 10^5 \text{ N/m}$$

It is now assumed that this is distributed over a distance 0.75 m, measured on both sides of the rail from the center of the area of contact between rail and the cross beams. The elastic constant of this foundation is thus taken approximately as

$$k_s = \frac{4.861 \times 10^5}{0.75} = 6.481 \times 10^5 \text{ N/m}^2$$

Hence, for the rail

$$\beta = \sqrt[4]{\frac{k_s}{4EI_x}} = \sqrt[4]{\frac{6.481 \times 10^5}{4 \times 200 \times 10^9 \times 28.9 \times 10^{-6}}} = 0.4092 \text{ m}^{-1}$$

This rail length (L = 16 m) is larger than $2\pi/\beta$ = 15.35 m, which permits the use of the solution of a semi-infinite beam on an elastic foundation as given by Equations 7.111 and 7.112; these give (for $M_o = 0$) the deflection, shear force, and moment, respectively, as

$$v = -\frac{2\beta P}{k_s} e^{-\beta z} \cos \beta z$$

with

$$v_{max} = -\frac{2\beta P}{k_s} = \frac{-2 \times 0.4092}{6.481 \times 10^5} = -12.63 \text{ mm at } z = 0$$

$$M = \frac{P}{\beta} e^{-\beta z} \sin \beta z$$

with

$$M_{max} = \frac{0.3225\, P}{\beta} = 0.788\, P \text{ at } z = \frac{\pi}{4\beta} = 1.92 \text{ m}$$

and

$$P_y = Pe^{-\beta z}(\cos \beta z - \sin \beta z)$$

with

$$(P_y)_{max} = -P \text{ at } z = 0$$

The maximum bending stress in the monorail is given by

$$\sigma_{max} = \pm \frac{M_{max}(H_2/2)}{(I_x)_2} = \frac{(0.788 \times 10 \times 10^3)(0.203/2)}{28.9 \times 10^{-6}}$$

$$= 27.67 \text{ MPa}$$

The maximum stress in the supporting beams is found at the midspan of the beam directly above the load. The deflection and spring constant of this beam are 12.63 mm and 4.861 × 10^5 N/m, respectively, as calculated above. Hence, the load acting at its center is 12.63 × 10^{-3} × 4.861 × 10^5 = 6.14 kN. Therefore, the maximum bending stress is

$$\sigma_{max} = \frac{(PL/4)(H_2/2)}{(I_x)_2} = \frac{(6.14 \times 10^3 \times 5/4)(0.127/2)}{6.33 \times 10^{-6}}$$

$$= 77 \text{ MPa}$$

7.8.3 Short Beams

As indicated in Section 7.8.1, the solutions obtained above are valid for beams of lengths greater than $2\pi/\beta$. For shorter beams, other solutions are available.[7] For instance, for a short beam of length L loaded by a downward concentrated load P at its midlength, the following expressions for the maximum deflection and moment under the load are given by[8]

$$v_{max} = -\frac{P\beta}{2k_s}\left(\frac{\cosh \beta L + \cos \beta L + 2}{\sinh \beta L + \sin \beta L}\right)$$

$$M_{max} = -\frac{P}{4\beta}\left(\frac{\cosh \beta L - \cos \beta L}{\sinh \beta L + \sin \beta L}\right) \tag{7.113}$$

Table 7.8 shows the effect of beam length on the maximum deflection and maximum moment calculated using the short-beam solution, Equation 7.113 and the infinite beam solution, Equation 7.106. It is seen that applying infinite beam formulas is on the conservative side for the moments and hence the stresses.

TABLE 7.8
Effect of Beam Length on Maximum Deflection and Moment

L	$2\pi/\beta$	$3\pi/2\beta$	$3\pi/4\beta$	$3\pi/6\beta$	$3\pi/8\beta$
$v_{max\text{-}short}/v_{max\text{-}infinite}$	1.011	1.053	1.580	2.366	3.253
$M_{max\text{-}short}/M_{max\text{-}infinite}$	0.996	0.981	0.821	0.648	0.522

PROBLEMS

Normal Stresses – Elastic Curve

7.1 The cast link shown in the figure is subjected to a tensile load of 50 kN; its original cross section is shown in part a of the figure. This section is modified for uniform strength as shown in part b. The same casting thicknesses are maintained. Determine the dimensions a and b while maintaining the same maximum tensile stress. Compare the cross sectional area ratio.

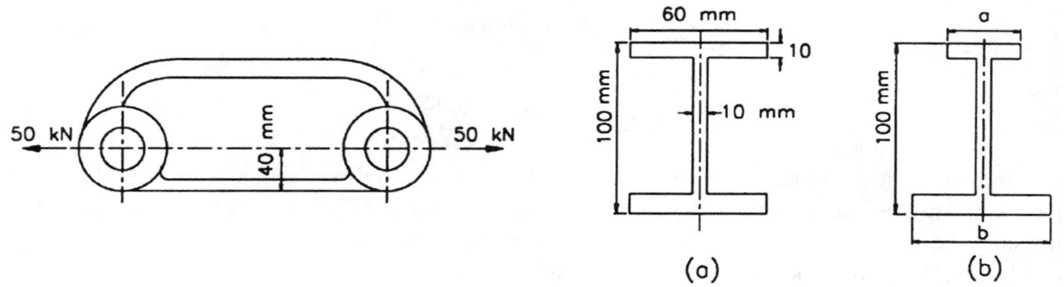

Problem 7.1

7.2 Show that for a bar subjected to *Mx*, *My*, and *Px*, the normal stress referred to centroidal axes *x*, *y*, and *z*, which are not principal axes, is given by

$$\sigma_z = \frac{M_y I_x - M_x I_{xy}}{I_x I_y - I_{xy}^2} x + \frac{M_x I_y - M_y I_{xy}}{I_x I_y - I_{xy}^2} y + \frac{P_z}{A}$$

Apply the results to determine the stresses in a bar subjected to an axial force of 80 kN applied at O for the following sections made of equal angles as shown in the figure: (a) single angle, (b) back to back, and (c) star.

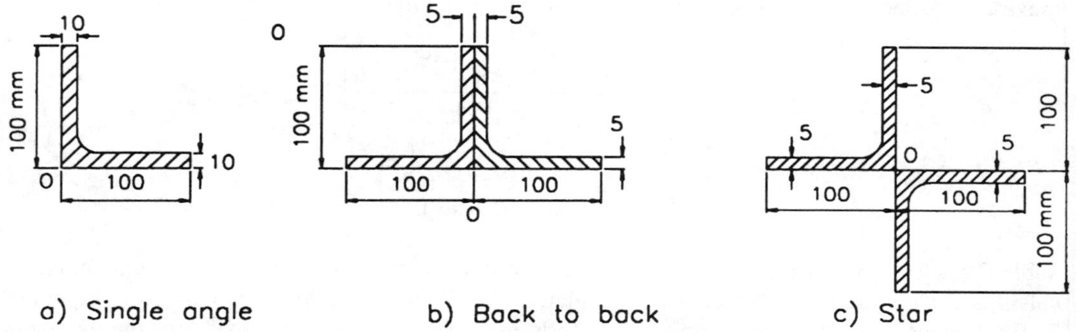

Problem 7.2

7.3 A cantilever beam of 1.25 m length is subjected to the inclined loads shown in the figure acting at the centroid *C* in the cross sectional plane at the beam free end. Determine the maximum normal stresses and the position of the neutral axis for each of the given cross sections.

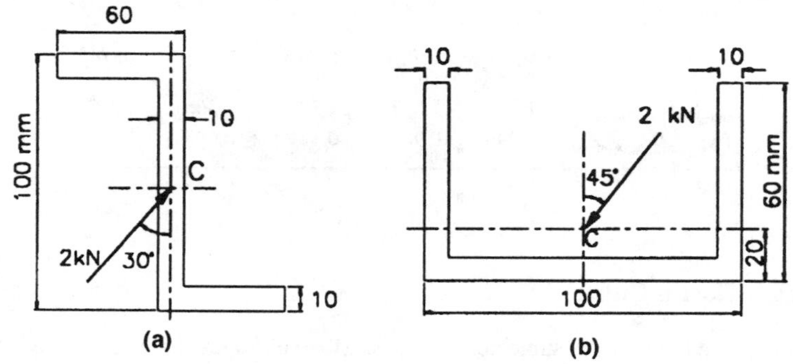

Problem 7.3

7.4 A cantilever beam of length L is subjected to a vertical downward load p at its free end. Show that the vertical displacement is given by

$$v = -\frac{p}{6EI}\left(x^3 - 3L^2x + 2L^3\right)$$

where x is measured from the free end.

7.5 Obtain the equation of the elastic curve using the method of successive integrations for the beam shown in the figure. Find the location and value of maximum deflection.

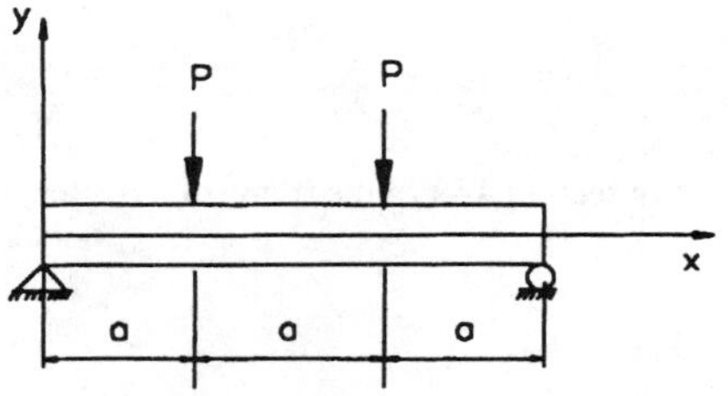

Problem 7.5

7.6 Derive the equation of the elastic curve of a beam supported as shown in the figure. Find the deflection at the points of load application.

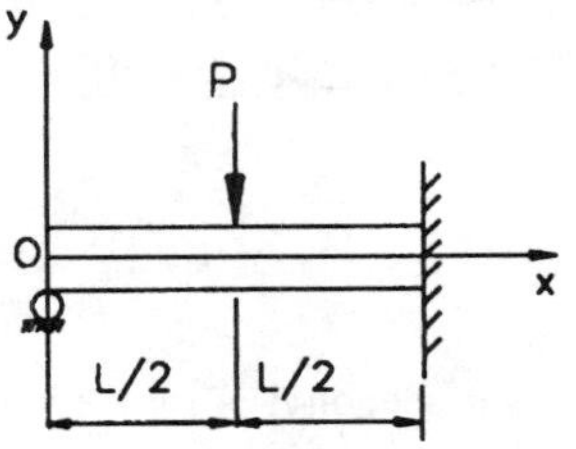

Problem 7.6

7.7 A cantilever beam is subjected to a sinusoidally varying load intensity $q = q_o \sin(\pi x/2L)$ as shown in the figure. Determine the vertical deflection, v, of a point C on the beam, whose distance from the free end A is $x = a$.

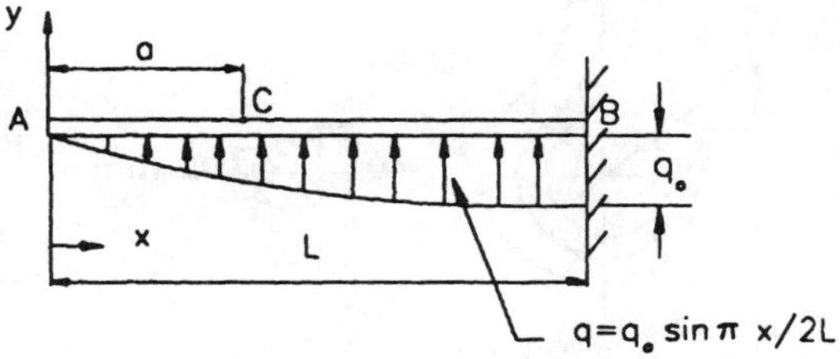

Problem 7.7

Curved Beams

7.8 For a curved rod subjected to pure bending, determine the stress concentration factor for (a) rectangular cross section depth d, (b) circular cross section diameter d for values of $R/d = 1$ to 10, where R is the radius of curvature at the centroid.

7.9 Determine the maximum stress in section *A–A* of the frame of a punch press shown in the figure. The maximum press load is 200 kN.

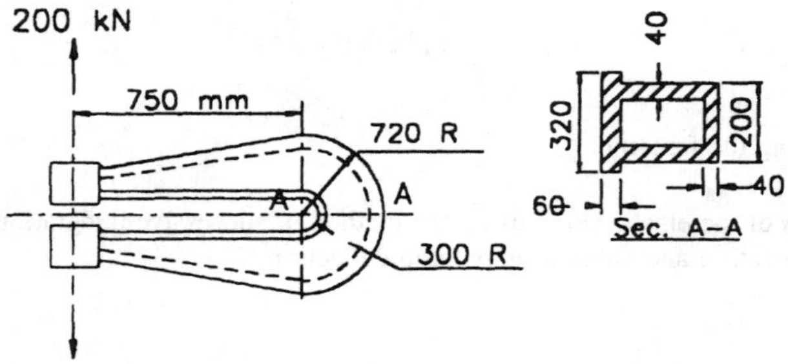

Problem 7.9

7.10 Determine the maximum stress in section *A–A* of the frame of a riveter shown in the figure. The maximum load is 100 kN.

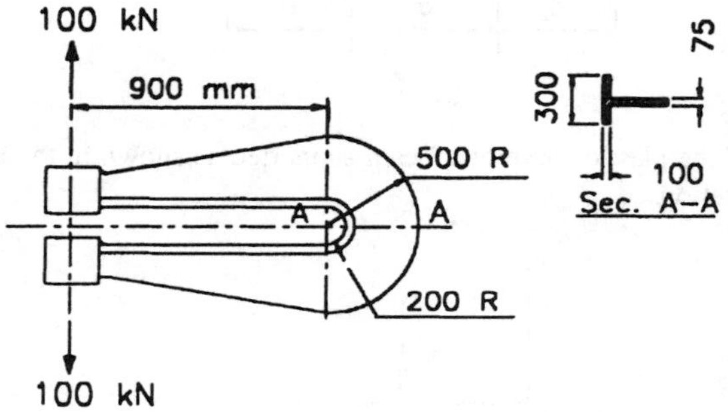

Problem 7.10

7.11 Determine the maximum tensile stress in section *A–A* for the hook shown in the figure. The hook lifting capacity = 20 kN.

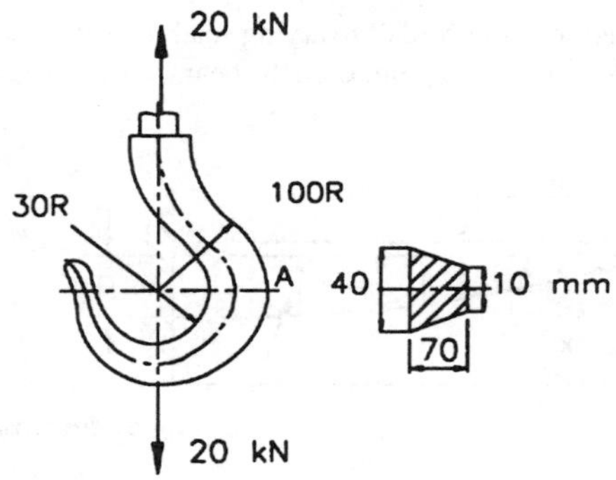

Problem 7.11

7.12 A plate 10 mm thick having a semicircular shape of 100-mm inside diameter and 200-mm outside diameter is subjected to the load shown in the figure. Determine the maximum circumferential stresses on the inner and outer diameters.

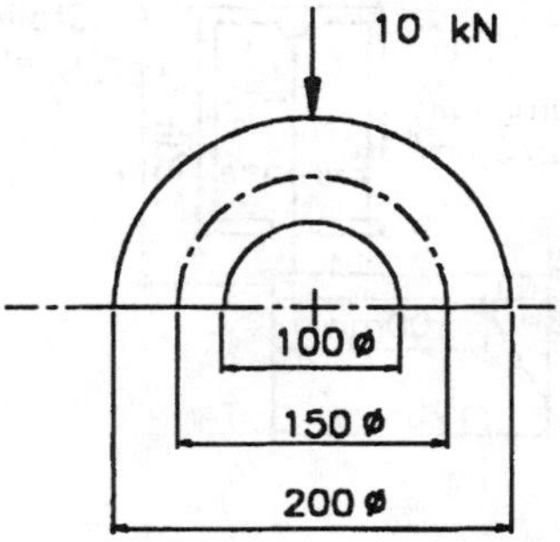

Problem 7.12

7.13 A curved beam has a trapezoidal cross section with the dimensions shown in the figure. If $P = 50$ kN, determine the circumferential stresses at B and C.

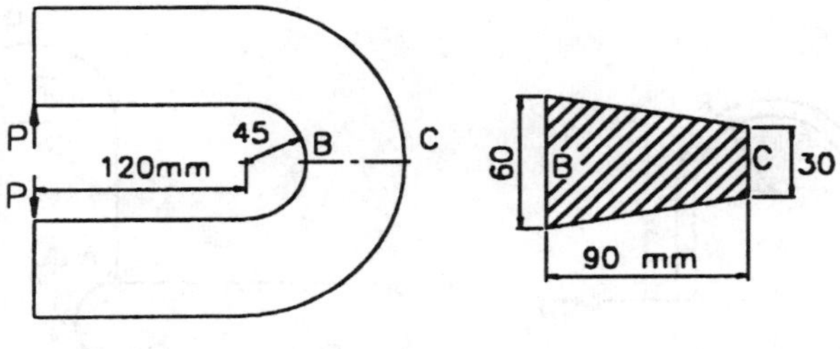

Problem 7.13

7.14 A load $P = 12$ kN is applied to the clamp shown in the figure. Determine the circumferential stresses at points B and C assuming that the curved beam formula is valid at that section.

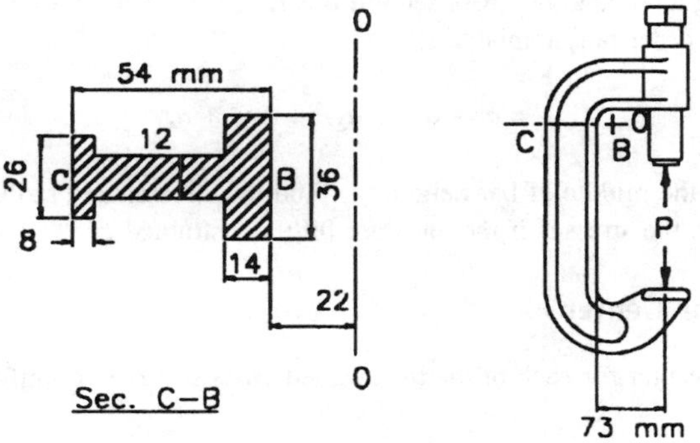

Problem 7.14

Thermoelastic Stresses

7.15 The temperature in a furnace is measured by means of a stainless steel wire placed in it as shown in the figure. What is the change in furnace temperature if the gauge records a change in strain -100×10^{-6}. Take for stainless steel $\alpha_{ss} = 17 \times 10^{-6}$/°C, $E_{ss} = 200$ GPa and for aluminum $\alpha_a = 22 \times 10^{-6}$/°C, $E_a = 70$ GPa.

7.16 A U-shaped member of constant EI has the dimensions shown in the figure. Determine (a) the deflection of the applied forces away from each other. Consider only bending effects. (b) Apply the result to find the maximum stress moment at point C of the pipe used to pump hot liquid from one pressure vessel to another.

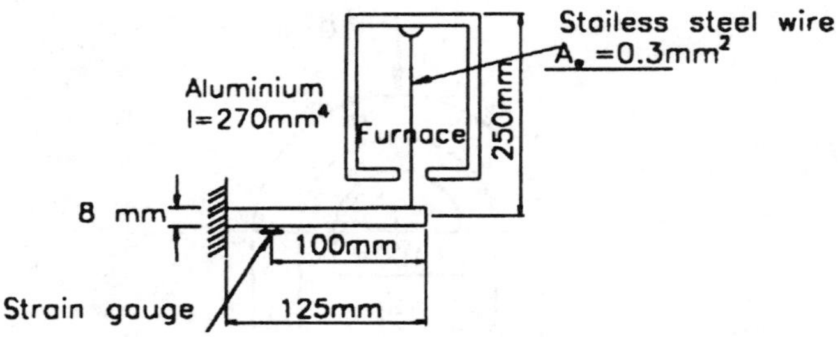

Problem 7.15

Note that the pipe is also used as an expansion loop, is made of steel of 100 mm mean diameter and 6 mm thickness, and that E = 210 GPa and $\alpha = 16 \times 10^{-6}/°C$. In operation the temperature of this pipe is raised 190°C. Take $L = R$ = 2500 mm and assume that the pipe supports at A and B are capable of resisting only horizontal and vertical forces.

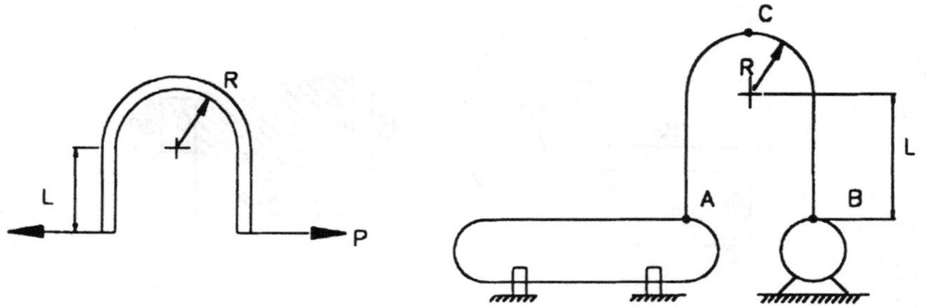

Problem 7.16

7.17 An elastic bar having a rectangular cross section $B \times H$ is subjected to a temperature gradient along its height given by the fourth-order polynomial

$$T = a_o + a_1 y + a_2 y^2 + a_3 y^3 + a_4 y^4$$

where y is measured from the middle of bar height. (a) Find the stresses and radius of curvature if the bar is completely free. (b) Obtain the stresses if the ends are fully constrained.

Shear Stresses — Shear Center

7.18 Determine the shear center for each of the thin-walled cross sections of uniform thickness shown in the figure.

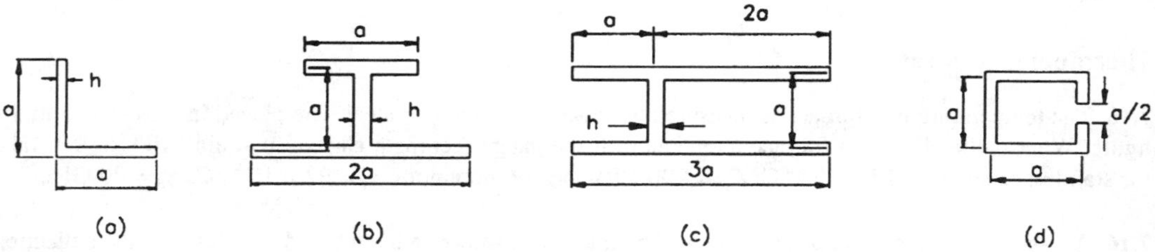

Problem 7.18

7.19 Determine the shear stresses for the tubular sections shown in the figure that are due to a vertical shearing force through the centroid 0.

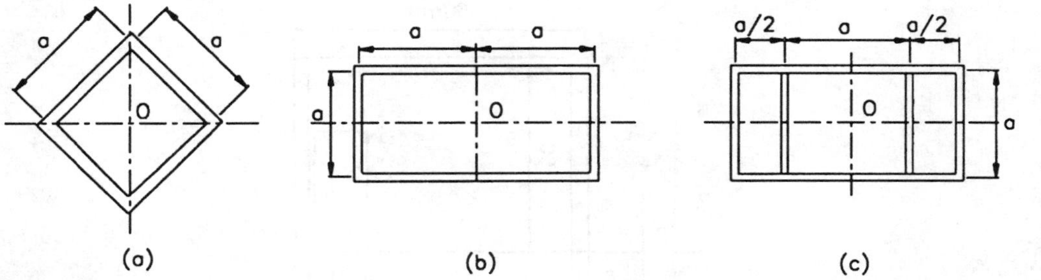

Problem 7.19

7.20 Determine the position of the shear center C for the two extruded aluminum alloy sections shown in the figure. Plot the shear stresses for a load P_y of 1 kN acting at point C, indicating the magnitude and position of the maximum shear stress.

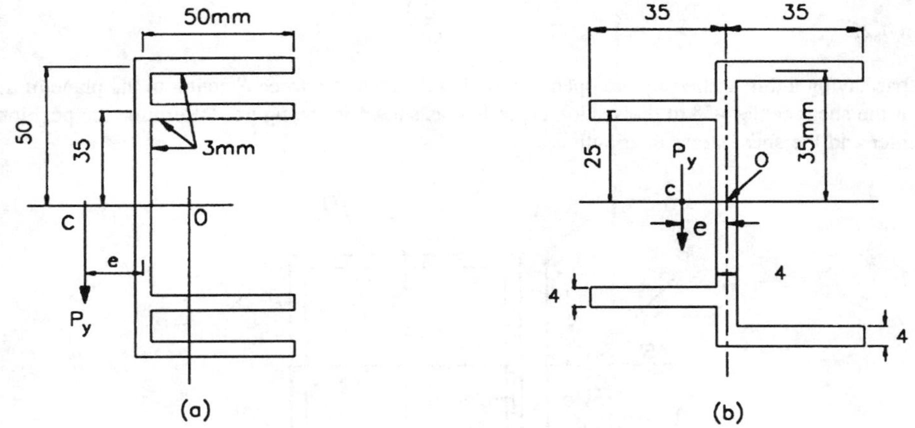

Problem 7.20

7.21 The cross sections shown in the figure are formed of a thin steel sheet of 5 mm thickness. Locate the shear center C, and determine the shear stress distribution for the given load P_y acting at C.

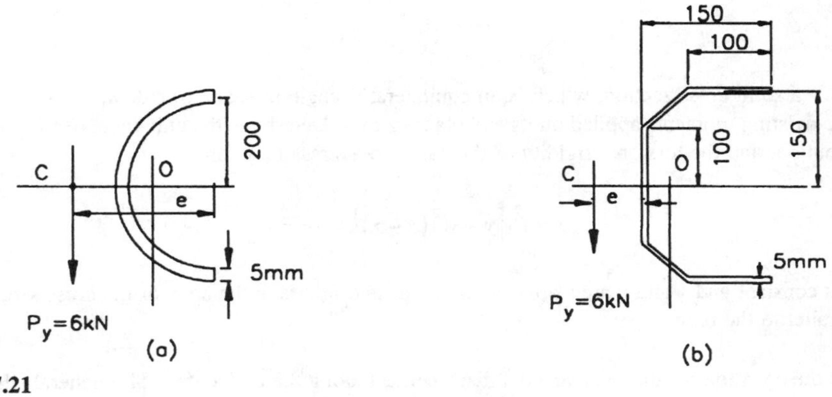

Problem 7.21

7.22 Determine the shear center for the folded section shown in the figure. Plot the shear stress distribution for a load $P_y = 5$ kN acting at the shear center, indicating the maximum shear stress.

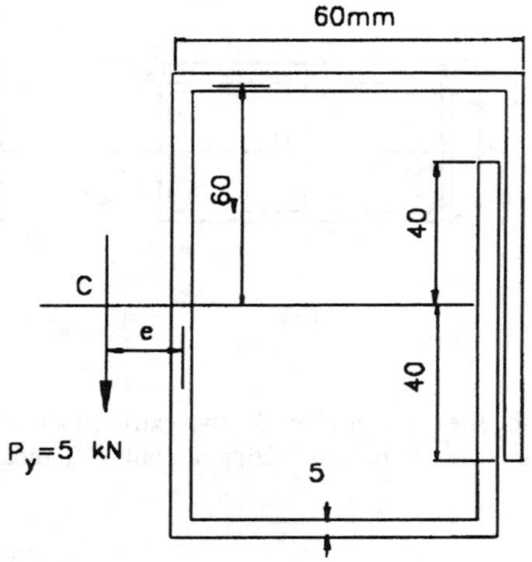

Problem 7.22

7.23 A bar having a thin-walled square open section is subjected to a force P acting in the plane of the cross section at the shear center 45° to the section centerline as shown in the figure. Determine the position of the shear center and the shear stress distribution.

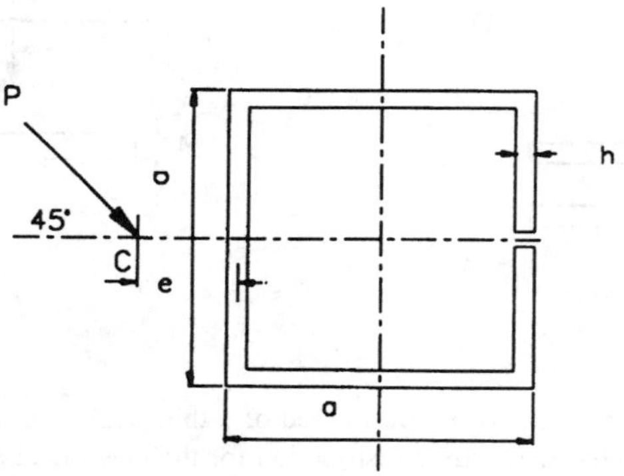

Problem 7.23

Torsion

7.24 A bar of a solid cross section, which is an equilateral triangle in shape of side a, is subjected to equal and opposite twisting moments applied on its end planes z-axis. Determine the unit angle of twist α, the shear stress distribution, and the torsional rigidity of the bar. Use a stress function:

$$\phi = Cy\left[y - \sqrt{3}(a - x)\right]\left(y - \sqrt{3}\right)$$

where C is a constant and x and y are Cartesian axes whose origin is at the apex of the cross section and the x-axis is parallel to the base.

7.25 Derive the equation for the relative axial displacement along the midcontour of a general slitted tubular thin-walled section. Apply the equation to a circular, square, and equilateral triangle, contours of the same lengths and thickness h, and subjected to the same M_t.

7.26 A steel sheet 400 × 3 mm, length 3 m is folded to form a thin-walled square section; the edges are left unwelded. The formed bar is subjected to a twisting moment of 150 N · m. Determine the shear stress, the total angle of twist, and the relative axial displacement at the unwelded edges. Take $G = 80$ GPa.

7.27 The long tube in the figure is subjected to a torque of 200 N m. The tube has a double-cell, thin-walled effective cross section. Assuming that no buckling occurs, and that the twist per unit length of the tube is constant, determine the maximum shear stress in each wall of the tube.

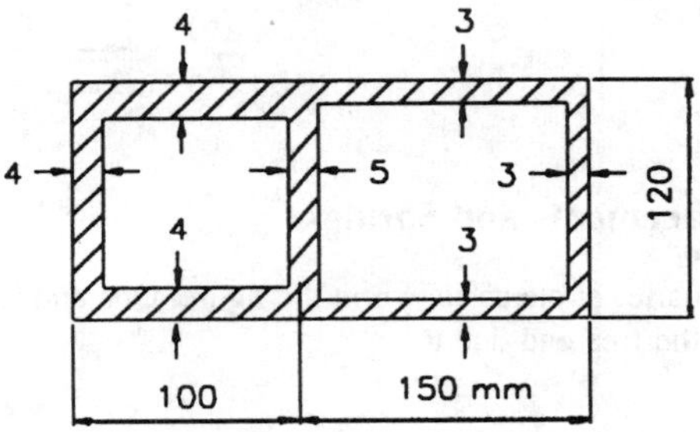

Problem 7.27

7.28 For the stiffened thin-walled tubular section, shown in the figure, of uniform wall thickness and subjected to a twisting moment M_t, Determine the shear stress distribution and the angle of twist per unit length for elastic conditions.

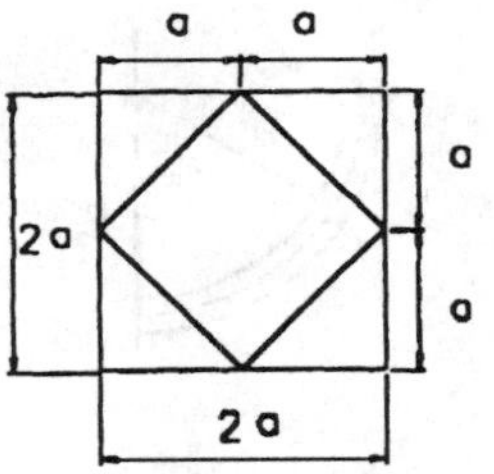

Problem 7.28

7.29 The thin-walled box shown in the figure is subjected to a twisting moment M_t. Determine the shear stresses in the walls and the angle of twist per unit length of the box.

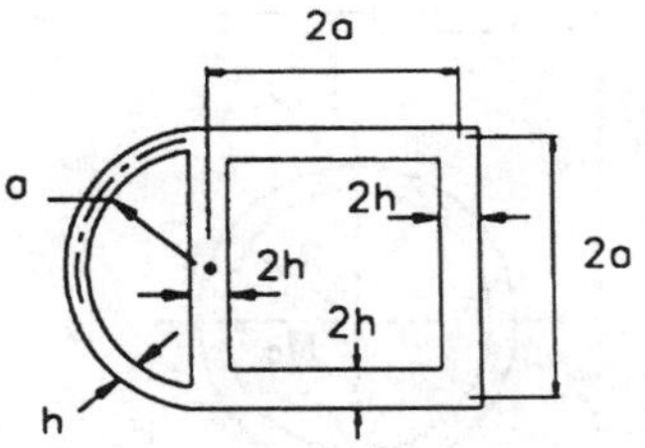

Problem 7.29

7.30 Determine the shear stress distribution and the angle of twist per unit length for the thin-walled closed section shown in the figure having a uniform wall thickness h and subjected to a twisting moment M_t. Indicate the position of maximum shear stress.

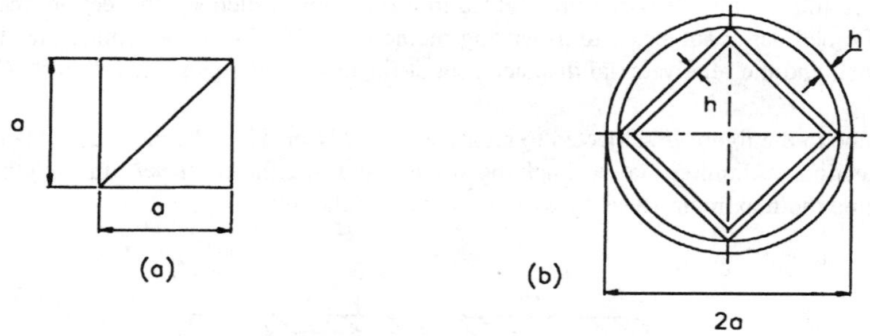

Problem 7.30

Energy Approach to Displacements and Springs

7.31 A pipe is bent to form a quarter circle as shown in the figure. One end is clamped and the other is free. Determine the displacement of the free end due to

a. A horizontal force P_x;
b. A vertical force P_y.

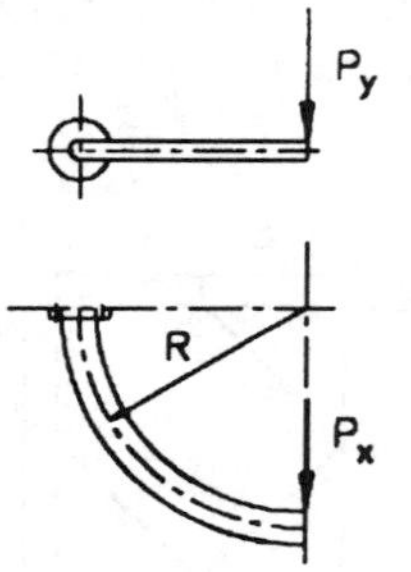

Problem 7.31

7.32 For the piston ring shown in the figure, determine the gap opening due to (a) a horizontal force P_x, (b) a vertical force P_y, and (c) an edge bending moment M_o.

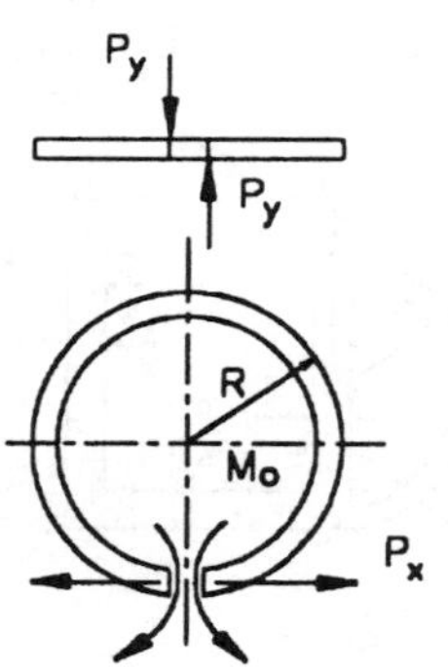

Problem 7.32

7.33 A pipe is bent to form a semicircle as shown in the figure. The two ends are hinged to a tie-rod AB. Determine the force in the tie-rod: (a) if the tie-rod is absolutely rigid and (b) if the tie-rod is elastic and having the same pipe cross section.

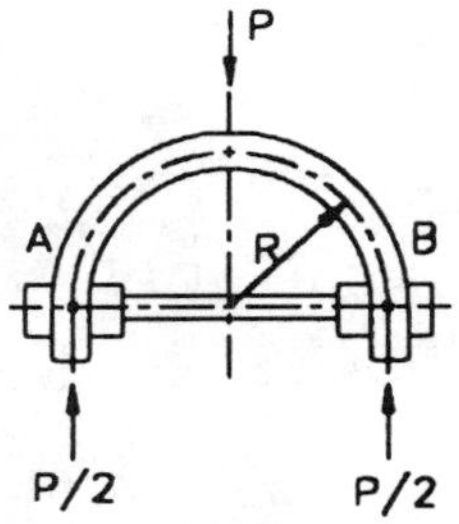

Problem 7.33

7.34 Determine the displacement and, hence, the stiffness under load P for the shown open front frame shown in the figure. For the same type of frame, determine the displacement and the stiffness under offset loading as shown. Take the frame cross section to be a round, solid section and E = 3G.

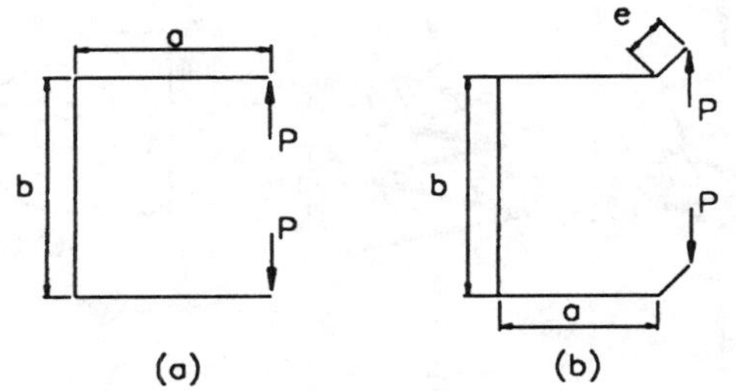

Problem 7.34

7.35 Determine the displacement and, hence, the stiffness under load P for the shown straight-sided frame shown in the figure. For the same type of frame determine the displacement and the stiffness under offset loading as shown. Take the frame cross section to be a round, solid section and E = 3G.

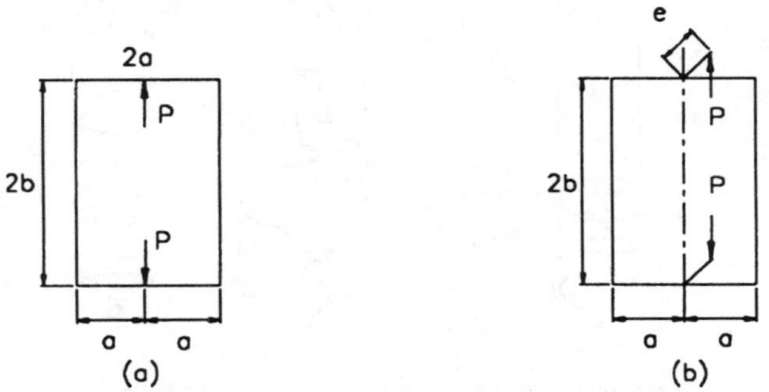

Problem 7.35

7.36 Compare the maximum stress and stiffness of the two types of chain links shown in the figure, of the same bar cross section and material. Determine the lateral displacement between the middle point of the link sides. Discuss the effect of preventing lateral displacement by welding a rigid rod (shown dotted). Determine the maximum stress in this case. Take $b = 2a$ and a round, solid cross section.

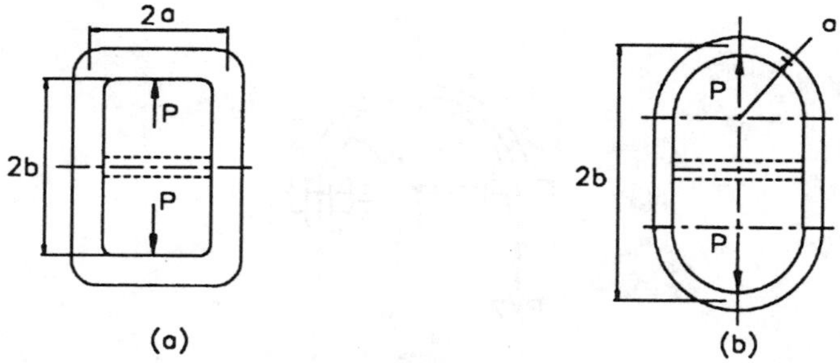

Problem 7.36

7.37 Obtain expressions for the maximum stress and stiffness for the plane springs shown in the figure. The springs are made of a thin strip of thickness h, width b, and length s.

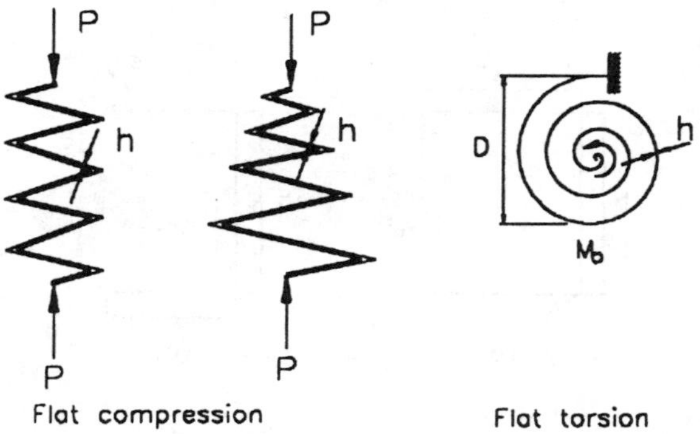

Problem 7.37

7.38 Determine the maximum stress and stiffness for the types of helical coil springs shown in the figure. The springs are made of a round wire and diameter d.

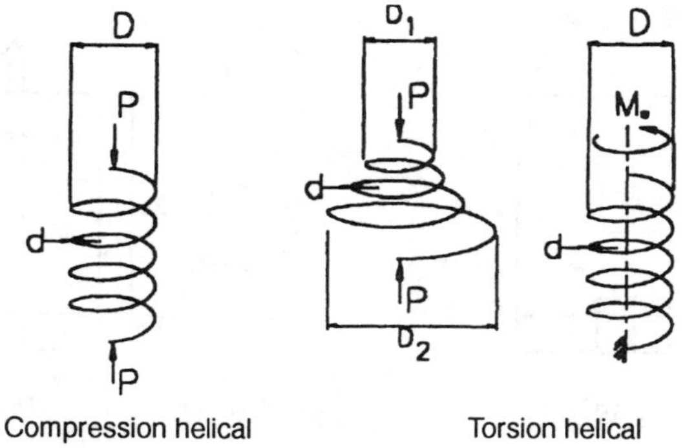

Problem 7.38

7.39 A closed circular ring having a mean diameter D and made of a circular wire diameter d, where $D/d >$ 10, is close fitted in an another closed circular ring made of the same wire, such that the axes of the two rings are maintained perpendicular to each other (see the figure). The outer ring is subjected to a diametral pull P. Obtain the stress distribution and the value of the maximum stress in the rings. Determine the stiffness under such loading for elastic conditions.

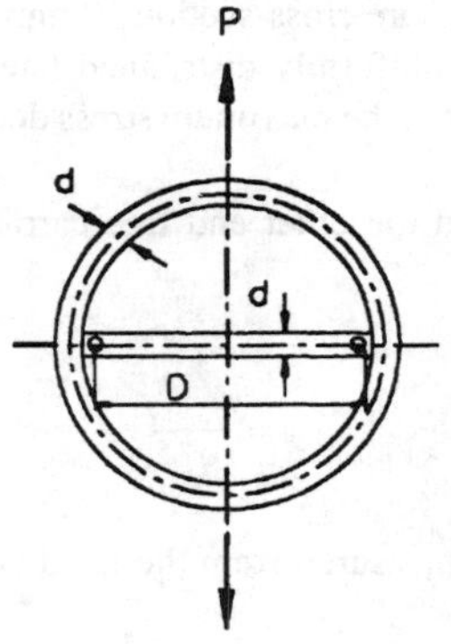

Problem 7.39

7.40 A closed circular ring made of a circular wire is force fitted in an another closed circular ring made of the same wire, such that the axes of the two rings are perpendicular to each other (see the figure). If the inner ring has a mean diameter D, wire diameter d, and a diametral interference Δ, conditions being perfectly elastic,

a. Obtain the stress distribution and the value of the maximum stress in the rings as a result of the force fit,
b. Determine the magnitude of the diametral pull that may be applied to the inner ring to release the interference fit.

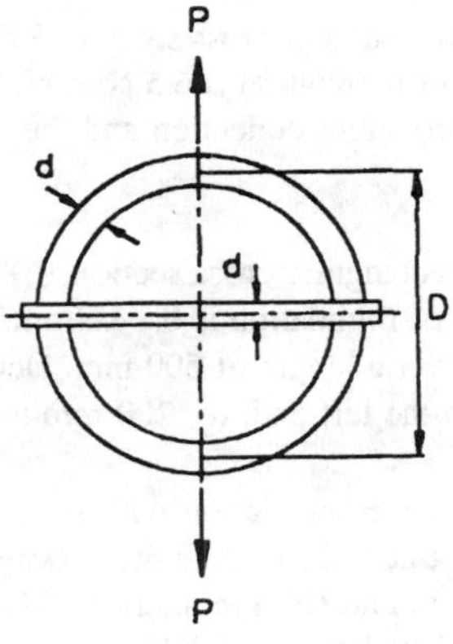

Problem 7.40

Buckling of Rods

7.41 A straight strut of length L is fixed at one end and pinned at the other. If, in calculating the buckling load, it is assumed that the effective length is $2L/3$, estimate the percentage error incurred.

7.42 A straight strut of length L, rigidly built in at one end and free at the other, is subjected to an axial compressive force P and a side force Q perpendicular to P, both applied at the free end. Show that the free-end deflection is given by $v = (Q/P)[(1/k)\tan(kL) - L]$, where $k^2 = P/EI$. Hence, show that buckling occurs when the load $P = \pi^2 EI/4L^2$.

7.43 The cross-section of a column of pinned-ends is to be selected from either, being solid circular, solid square, hollow circular, or hollow square while maintaining a cross-sectional area of 2500 mm². For a given buckling load, determine the ratio among the safe lengths for the above four possible shapes of the cross-section.

7.44 A beam column of length 3 m and square cross-section 50 mm × 50 mm is hinged at both ends. It carries an axial compressive force 40 kN and a uniformly distributed lateral load of q N/mm. Determine the load intensity q and maximum deflection such that the maximum stress does not exceed 220 MPa. Take E = 200 GPa.

7.45 A column with one end built-in and the other end free carries an axial load P. It is assumed that the deflection curve has the form

$$y = \frac{v_o z^2}{L^2}$$

where L is the length of the column, z is measured from the fixed end, and v_o is a constant. Using the energy method, determine the critical load.

7.46 A column with pinned ends is subjected to the action of its weight as a uniformly distributed axial load of intensity q and an axial compressive force P. Find the critical value of P by assuming, for the deflection curve, the equation:

$$v = v_o \sin\left(\frac{\pi z}{L}\right)$$

Beams on Elastic Foundation

7.47 An infinitely long steel beam of a rectangular cross section, 180 mm wide and 280 mm thick, is resting on an elastic foundation whose modulus of foundation is 6.5 N/mm². This beam is subjected to a concentrated moment of 0.5 MN/m. Determine the maximum deflection and the maximum bending stresses in the beam. Take E = 210 GPa and ν = 0.30.

7.48 An infinitely long steel beam of a rectangular cross section, 100 mm wide and 150 mm thick, is resting on an elastic foundation whose modulus of foundation is 0.1 kN/mm². This beam is subjected to a uniformly distributed load of intensity 300 N/mm over a length of 500 mm. Determine the deflection and stresses (a) at midlength of this load, (b) at 100 mm to the left and, (c) 200 mm to the right of this load.

7.49 A semi-infinite beam of a rectangular cross section with free ends is resting on an elastic foundation. The beam is 100 mm wide and 150 mm thick. Determine the maximum deflection and moment in the beam if (a) a force of 1 kN is applied at one end and (b) a moment 10 kN/mm is applied at one end. Assume E = 210 GPa, ν = 0.38, and the modulus of foundation is 4 N/mm².

7.50 A four-wheel wagon runs on steel rails (E = 200 GPa). The rails have a depth of 140 mm; the distance from the top of a rail to its centroid is 70 mm; and its moment of inertia is 21 × 10⁶ mm⁴. The rail rests on an elastic foundation with spring constant k = 12 N/mm². The two wheels on each side of the car are spaced 2.50 m center to center. If each wheel load is 90 kN, determine the maximum deflection and maximum bending stress when a car wheel is located at one end of the rail and the other car wheel on the same rail is 2.75 m from the end.

7.51 A long steel I beam of depth 200 mm and a second moment of area $I = 2.2 \times 10^7$ mm⁴ is supported by a series of steel hanger rods of 200 mm² cross sectional area and 2 m long. These hanger rods are equally spaced along the I beam at 0.75 m. If a concentrated load of 100 kN is applied at the midpoint of the I beam, calculate the maximum bending stress in the beam and the maximum tensile stress in the hanger rods. Take E = 210 GPa.

REFERENCES

1. Boresi, A. P. and Sidebottom, O. M., *Advanced Strength of Materials,* 4th ed., John Wiley & Sons, New York, 1985.
2. Love, A. E. H., *Mathematical Theory of Elasticity*, 4th ed., Dover, New York, 1944.
3. Venkatraman, B. and Patel, S. A., *Structural Mechanics with Introductions to Elasticity and Plasticity,* McGraw-Hill, New York, 1970.
4. Sirnath, L. S., *Advanced Mechanics of Solids,* Tata-McGraw-Hill, New Delhi, 1980.
5. Shames, I. H, and Dym, C. L., *Energy and Finite Element Methods in Structural Mechanics,* McGraw-Hill, New York, 1985.
6. Rees, D. W. A., *Mechanics of Solids and Structures*, McGraw-Hill, New York, 1990.
7. Roark, R. J. and Young, W. G., *Formulas for Stress and Strain,* 5th ed., McGraw-Hill, New York, 1975.
8. Hetenyi, M., *Beams on Elastic Foundation,* The University of Michigan Press, Ann Arbor, 1946.

8 Some Problems of Elastic Plates and Shells

Solutions to some engineering problems of elastic plates and shells are obtained. The general plate equations are formulated in Cartesian coordinates in terms of lateral displacement and applied to some rectangular plate problems using the energy approach. The equilibrium approach is applied to obtain solutions for axisymmetric plate problems in polar coordinates for different edge conditions. For shells, the equations of membrane stresses in axisymmetric pressure containers and stresses due to axisymmetric bending of thin-walled cylinders are derived and applied to some engineering applications. Solutions to the elastic buckling of rectangular plates subjected to uniform in-plane edge pressure are obtained using the energy approach, and also a solution to the elastic buckling of a thin-walled cylinder subjected to external uniform pressure is derived using an equilibrium approach.

8.1 INTRODUCTION

A plate is a flat structural element for which the thickness is small compared with the surface dimensions. The thickness is usually constant but may be variable and is measured normal to the middle surface of the plate.

A shell is an initially curved plate defined by a middle surface about which a constant or variable thickness is symmetrically situated. A typical ratio of the shell thickness to its radius of curvature is of the order of 1:20.

Plates and shells are used in several engineering applications, such as tanks, containers, pressure vessels, pipes, machine frames, ships, and aerospace structures. They are shaped from plates by forming, cutting, and welding, and riveting or bonding. In some applications, such as boilers and special pressure vessels, the finished product is subjected to thermal treatment to remove the residual stresses arising from fabrication.

In this chapter solutions are obtained to some problems of elastic plates and shells in which the deflections are small. The solutions are based, as in the case of rods in Chapter 7, on postulating a compatible deformation pattern developed from theoretical and/or practical considerations.

8.2 STATE OF STRESS IN PLATES AND SHELLS

Generally, plates and shells are subjected to in-plane forces, and to forces and couples normal to the surface, which are called here lateral loading. In Chapters 5 and 6, plates subjected only to in-plane forces have been considered. For such cases of loading, the strain and stress components are functions of only two coordinates, (x, y) in Cartesian coordinates or (r, θ) in polar coordinates, and are independent of z. Under lateral loading, as shown in Figure 8.1a, the state of stress in plates and shells is generally three-dimensional. The prevailing stresses comprise membrane stresses and bending stresses. In plates, bending stresses are dominant and membrane stresses occur if the plate

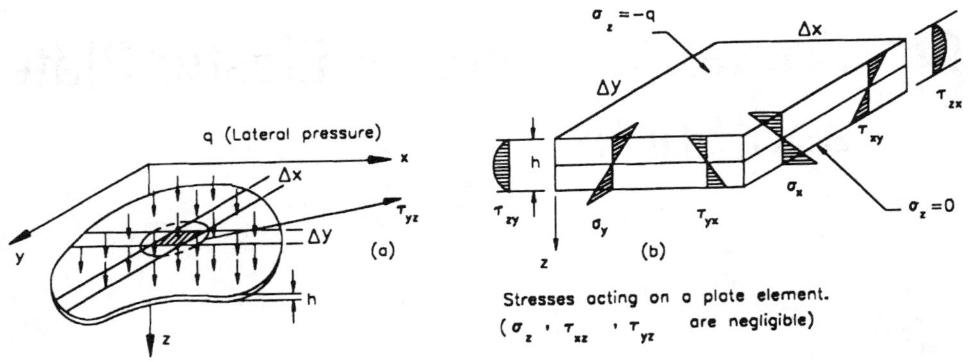

FIGURE 8.1 Reference axes and stresses in a plate element: (a) reference axes and loading and (b) stress components.

edges are constrained against in-plane displacements. In shells. membrane stresses uniformly distributed through the thin wall are dominant, while bending stresses are localized within zones of surface or load discontinuities.

As a result of the thickness being small compared with the other two dimensions, the stresses normal to the surface that can be sustained by a plate or a shell are quite small as compared with the stresses developed along/or parallel to the midsurface. For example, in a thin-walled cylindrical pressure vessel, the internal pressure may be a few bars, while the wall membrane stresses can reach the elastic limit; the ratio between the internal pressure and the hoop stress is equal to the ratio of wall thickness to the midsurface radius.* It is therefore justifiable, as in the case of rods, to neglect the normal stresses in the direction of the thickness, and the state of stress is defined by the remaining five stress components, namely, two normal and three shear components, as shown in Figure 8.1b.

8.3 PLATE EQUATIONS IN CARTESIAN COORDINATES

The plates considered here are subjected only to lateral loading since plates subjected to in-plane loading have been, already dealt with in Chapters 5 and 6. The general case of in-plane and lateral loading can be obtained by superposing the strain and stress components due to both loadings.

In order to eliminate in-plane loading, the plate edges should not be restricted against in-plane displacements, and therefore membrane stresses are eliminated. In this case, the plate is subjected to bending stresses as a result of lateral loading.

Consider a plate of an arbitrary shape in equilibrium under lateral loading, as shown in Figure 8.1a. The plate thickness h is either constant or varies gradually with no abrupt changes. The plate material is assumed to be homogeneous, isotropic, and linear-elastic. Cartesian coordinates are chosen by taking the midsurface to be the x–y plane and the z-axis along the thickness direction. The positive direction of the z-axis is arbitrarily, chosen to be downward to conform with most of structural analysis texts. The same sign conventions adopted in previous chapters are maintained.

8.3.1 DEFORMATION PATTERN

The deformation pattern and strain components are formulated according to the following three postulates:

* See Example 3.9 in Chapter 3.

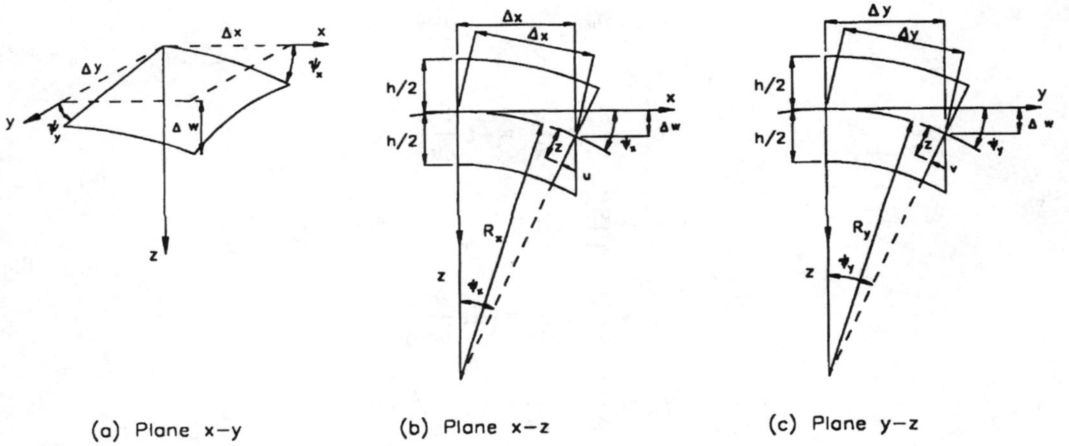

FIGURE 8.2 Displacements in a plate element: (a) plane *x–y*, (b) plane *x–z*, and (c) plane *y–z*.

(i) The plate thickness h of Figure 8.1a is small with respect to its other dimensions; the deflection w and, hence, the slopes ψ_x and ψ_y of the midsurface of the plate ($z = 0$), Figure 8.2a, are small.
(ii) The midsurface of the plate remains unstrained and free of stress during plate deflection.
(iii) Straight lines normal to the midsurface will remain straight but rotated small angles ψ_x and ψ_y in the $x–z$ and $y–z$ planes, respectively, as shown in Figure 8.2b and c.

From Figure 8.2a the slopes of the midsurface ($z = 0$) with the x- and y-directions are, respectively,

$$\tan\psi_x \cong \psi_x = \frac{\partial w}{\partial x} \quad \text{and} \quad \tan\psi_y \cong \psi_y = \frac{\partial w}{\partial y} \tag{8.1}$$

According to postulate (i), these slopes are small and hence their squares are neglected compared to unity. The curvature Equation 7.10a for the planes ($x–z$) and ($y–z$) respectively, following the sign convention of Section 7.2, reduces to

$$\frac{1}{R_x} \cong -\frac{\partial^2 w}{\partial x^2} \quad \text{and} \quad \frac{1}{R_y} \cong -\frac{\partial^2 w}{\partial y^2} \tag{8.2}$$

Note that in the configuration of Figure 8.2 (ψ_x, ψ_y) are positive, while (u, v, w) and (R_x, R_y) are negative.

At any point (x, y, z), by virtue of postulate (iii), the displacement components are linearly related to the distance z measured from the midsurface, according to

$$u = -\psi_x z, \; \text{v} = \psi_y z, \; \text{and} \; w = w \tag{8.3}$$

Here the displacement component w is taken to be constant across the plate small thickness and equal to the midsurface deflection $w(x, y)$.

The infinitesimal strain–displacement relations (Equations 2.6) together with Equations 8.1 and 8.2 give

$$
\begin{aligned}
\varepsilon_x &= \frac{\partial u}{\partial x} = \frac{z}{R_x} = -z\frac{\partial^2 w}{\partial x^2} \\
\varepsilon_y &= \frac{\partial v}{\partial y} = \frac{z}{R_y} = -z\frac{\partial^2 w}{\partial y^2} \\
\varepsilon_z &= \frac{\partial w}{\partial z} = 0 \\
\gamma_{xy} &= \frac{\partial u}{\partial y} + \frac{\partial v}{\partial x} = -2z\frac{\partial^2 w}{\partial x \partial y} \\
\gamma_{zx} &= \frac{\partial w}{\partial x} + \frac{\partial u}{\partial z} = \psi_x - \psi_x = 0 \\
\gamma_{zy} &= \frac{\partial w}{\partial y} + \frac{\partial v}{\partial z} = \psi_y - \psi_y = 0
\end{aligned}
\tag{8.4}
$$

Obviously, at the midsurface ($z = 0$), all strain components vanish as stated by postulate (ii) above. Again, as in laterally loaded beams in the x–y plane, transverse shear stresses τ_{zx} and τ_{zy} do generally exist and cause warping of plane sections. As the ratio of beam depth to its length becomes smaller, their effect becomes negligible. The same applies to plates of small thicknesses, thus being consistent with postulate (iii), which neglects warping along the plate thickness. Hence, γ_{zx} and γ_{zy}, and τ_{zx} and τ_{zy} will tend to vanish. In fact, although Expressions 8.4 yield practically acceptable values for the strain components, the presence of even low values of τ_{zx} and τ_{zy} is essential to maintain the equilibrium of a plate element under general lateral loading, as will be shown in Section 8.3.3. Therefore, the result that $\gamma_{zx} = \gamma_{zy} = 0$ as in Equations 8.4 is valid only in the case of pure bending.

8.3.2 Stress Resultants

The state of stress at any point (x, y, z) is expressed by the stress components σ_x, σ_y, τ_{xy}, τ_{zx}, and τ_{zy}, which are generally functions of x, y, z, while $\sigma_z = 0$; Figure 8.3. For elastic behavior, the stresses are obtained from the strain components by substituting Expressions 8.4 in the inverse Hooke's Law (Equations 3.12); hence,

$$
\begin{aligned}
\sigma_x &= \frac{-Ez}{1-\nu^2}\left(\frac{\partial^2 w}{\partial x^2} + \nu\frac{\partial^2 w}{\partial y^2}\right) \\
\sigma_y &= \frac{-Ez}{1-\nu^2}\left(\frac{\partial^2 w}{\partial y^2} + \nu\frac{\partial^2 w}{\partial x^2}\right) \\
\tau_{xy} &= \frac{-Ez}{1+\nu}\frac{\partial^2 w}{\partial x \partial y} \\
\sigma_z &= \tau_{zx} = \tau_{zy} = 0
\end{aligned}
\tag{8.5}
$$

In plate theory it is customary to integrate the stresses over the constant plate thickness defining stress resultants. Consider a plate element of dimensions dx, dy, and thickness h; the stress resultants acting on the element sides per unit respective length under general loading, Figure 8.3, are given by

$$N_x = \int_{-h/2}^{h/2} \sigma_x dz,\; N_y = \int_{-h/2}^{h/2} \sigma_y dz$$

$$N_{xy} = N_{yx} = \int_{-h/2}^{h/2} \tau_{xy} dz \tag{8.6a}$$

$$Q_x = \int_{-h/2}^{h/2} \tau_{zx} dz,\; Q_y = \int_{-h/2}^{h/2} \tau_{zy} dz$$

$$M_x = \int_{-h/2}^{h/2} \sigma_x z dz,\; M_y = \int_{-h/2}^{h/2} \sigma_y z dz \tag{8.6b}$$

$$M_{xy} = M_{yx} = \int_{-h/2}^{h/2} \tau_{xy} z dz$$

where N_x and N_y are in-plane forces acting along the middle plane, which are assumed to vanish* in the case of lateral loading; Q_x and Q_y are shearing forces; M_x and M_y are bending moments; and M_{xy} and M_{yx} are twisting moments. The relation between the stress resultants is obtained by integrating the stress equilibrium equations (Equations 1.6) over the plate thickness as will be shown in the next section.

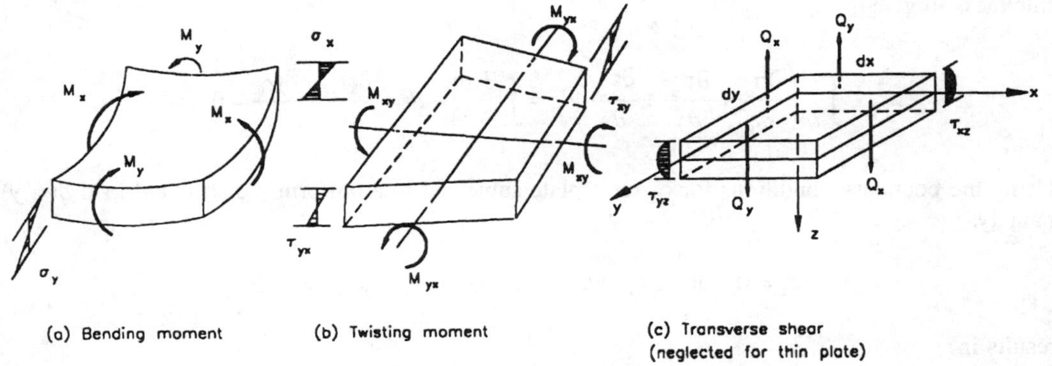

FIGURE 8.3 Stress resultants in a plate element: (a) bending moments, (b) twisting moments, (c) transverse shears (neglected for a thin plate).

8.3.3 Equations of Equilibrium

The stress equilibrium conditions in the absence of body forces are given by Equations 1.6, namely, in the x-direction as

$$\frac{\partial \sigma_x}{\partial x} + \frac{\partial \tau_{xy}}{\partial y} + \frac{\partial \tau_{zx}}{\partial z} = 0$$

Multiplying by z and integrating over the plate thickness yields, by virtue of Expressions 8.6,

* In such cases of bending and twisting σ_x, σ_y, and τ_{xy} are linearly distributed across the thickness and change from positive to negative from one side of the centroidal axis to the other; hence, their stress resultants vanish.

$$\int_{-h/2}^{h/2} z\left(\frac{\partial \sigma_x}{\partial x}+\frac{\partial \tau_{xy}}{\partial y}+\frac{\partial \tau_{zx}}{\partial z}\right)dz=\frac{\partial M_x}{\partial x}+\frac{\partial M_{xy}}{\partial y}+\int_{-h/2}^{h/2} z\frac{\partial \tau_{zx}}{\partial z}dz=0$$

The above thickness integral is performed by integration by parts to give

$$\left[z\tau_{zx}\right]_{-h/2}^{h/2}-\int_{-h/2}^{h/2}\tau_{zx}dz=-Q_x$$

Hence, knowing that $\tau_{zx} = 0$ at the top and bottom surfaces, where $z = \pm h/2$, yields

$$Q_x=\frac{\partial M_x}{\partial x}+\frac{\partial M_{xy}}{\partial y} \tag{8.7a}$$

Similarly, the second equation of the stress equilibrium equations (Equations 1.10) results in

$$Q_y=\frac{\partial M_{xy}}{\partial x}+\frac{\partial M_y}{\partial y} \tag{8.7b}$$

The third equation of the stress equilibrium equations (Equations 1.10) is integrated over the plate thickness to give

$$\int_{-h/2}^{h/2}\left(\frac{\partial \sigma_z}{\partial z}+\frac{\partial \tau_{zx}}{\partial x}+\frac{\partial \tau_{zy}}{\partial y}\right)dz=\int_{-h/2}^{h/2}\frac{\partial \sigma_z}{\partial z}dz+\frac{\partial Q_x}{\partial x}+\frac{\partial Q_y}{\partial y}=0$$

Using the boundary conditions for σ_z for a plate subjected to a uniformly distributed load $q(x, y)$ namely,

$$\sigma_z = 0 \quad \text{at} \quad z=-h/2 \quad \text{and} \quad \sigma_z=-q \quad \text{at} \quad z=h/2$$

results in

$$\frac{\partial Q_x}{\partial x}+\frac{\partial Q_y}{\partial y}=\left[\sigma_z\right]_{z=-h/2}-\left[\sigma_z\right]_{z=h/2}$$

or

$$\frac{\partial Q_x}{\partial x}+\frac{\partial Q_y}{\partial y}=-q \tag{8.7c}$$

Eliminating Q_x and Q_y from Equations 8.7 yields

$$\frac{\partial^2 M_x}{\partial x^2}+2\frac{\partial^2 M_{xy}}{\partial x\partial y}+\frac{\partial^2 M_y}{\partial y^2}=-q \tag{8.8}$$

which is the differential equation of equilibrium for bending of thin plates and is obviously valid for all material behaviors.

For elastic behavior, the stress resultants may be expressed in terms of the deflection function $w(x, y)$ as follows. Substitution of Expressions 8.5, which are based on Hooke's law, into Equations 8.6 and integrating yields for the moment stress resultants,

$$M_x = -D\left(\frac{\partial^2 w}{\partial x^2} + \nu \frac{\partial^2 w}{\partial y^2}\right)$$

$$M_y = -D\left(\frac{\partial^2 w}{\partial y^2} + \nu \frac{\partial^2 w}{\partial x^2}\right) \tag{8.9}$$

$$M_{xy} = -D(1-\nu)\frac{\partial^2 w}{\partial x \partial y}$$

where

$$D = \frac{Eh^3}{12(1-\nu^2)} \tag{8.10}$$

D is the flexural rigidity of the plate corresponding to EI for rods.

The shear stress resultants are obtained by substituting Expressions 8.9 into the equilibrium conditions 8.7a and b to yield

$$Q_x = -D\frac{\partial}{\partial x}\left(\nabla^2 w\right)$$

$$Q_y = -D\frac{\partial}{\partial y}\left(\nabla^2 w\right) \tag{8.11}$$

Again, Equation 8.8 may be expressed in terms of w by substituting Expressions 8.9 into Equation 8.8 to give

$$\frac{\partial^4 w}{\partial x^4} + 2\frac{\partial^4 w}{\partial x^2 \partial y^2} + \frac{\partial^4 w}{\partial y^4} = \frac{q}{D} \tag{8.12}$$

or simply,

$$\nabla^2\left(\nabla^2 w\right) = \frac{q}{D}$$

Equation 8.12 is valid only for plates of linear-elastic materials.

The stress distribution in the plate is thus given by

$$\sigma_x = \frac{M_x z}{h^3/12}, \quad \sigma_y = \frac{M_y z}{h^3/12},$$

and

$$\tau_{xy} = \frac{M_{xy} z}{h^3/12} \tag{8.13}$$

provided that $\tau_{zx} = \tau_{zy} \cong 0$ for thin plates.*

8.3.4 METHOD OF SOLUTION: PURE BENDING OF A PLATE

To determine the stress components from Equation 8.13, it is necessary to solve Equation 8.12 in $w(x, y)$. This is a fourth-order partial differential equation, which is generally difficult to solve except in very few cases of simple loading and geometries. The inverse method of solution, described in Section 4.4 in which a deflection function $w(x, y)$ is assumed to satisfy the prescribed boundary conditions, is commonly used in solving Equation 8.12. Polynomials or trigonometric series involving coefficients to be determined are often used and an energy formulation coupled with applying the Rayleigh–Ritz method** becomes more advantageous to render an approximate solution.

To illustrate the direct integration of the plate equation, consider the simple case of a rectangular plate subjected to uniformly distributed bending moments M_2 and M_1 along its sides parallel to the x- and y-directions respectively, as shown in Figure 8.4b.

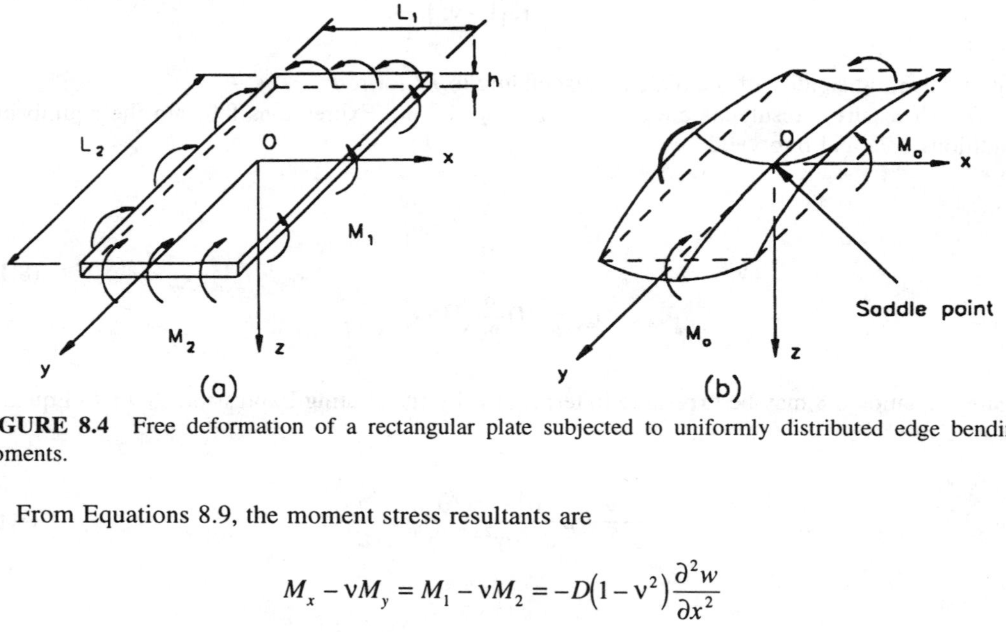

FIGURE 8.4 Free deformation of a rectangular plate subjected to uniformly distributed edge bending moments.

From Equations 8.9, the moment stress resultants are

$$\begin{aligned} M_x - \nu M_y &= M_1 - \nu M_2 = -D(1-\nu^2)\frac{\partial^2 w}{\partial x^2} \\ M_y - \nu M_x &= M_2 - \nu M_1 = -D(1-\nu^2)\frac{\partial^2 w}{\partial y^2} \\ M_{xy} &= 0 = D(1-\nu)\frac{\partial^2 w}{\partial x \partial y} \end{aligned} \tag{8.14}$$

* It can be shown by integrating the stress equations of equilibrium that τ_{zx} and τ_{zy} are — as in a beam of a rectangular cross section — parabolically distributed across the thickness; i.e., $(\tau_{zx})_{max} = 3Q_x/2h$ and $(\tau_{zz})_{max} = 3Q_y/2h$. Their orders of magnitude together with σ_z are shown to be negligible compared with other stresses for thin plates (say, $L/h > 10$). This is similar to the conclusion arrived at for beams in Sections 5.2.3 and 5.2.4. For this, see Shames and Dym.[1]

** See Section 7.7.2.

Integrating the above equations yields

$$w = -\frac{M_1 - \nu M_2}{2D(1-\nu^2)}x^2 - \frac{M_2 - \nu M_1}{2D(1-\nu^2)}y^2 + C_1 x + C_2 y + C_3$$

If the origin of coordinates is set at the center of the plate, then $w = 0$ at $x = y = 0$, which results in: $C_1 = C_2 = C_3 = 0$. The deflection equation is thus given by

$$w = -\frac{M_1 - \nu M_2}{2D(1-\nu^2)}x^2 - \frac{M_2 - \nu M_1}{2D(1-\nu^2)}y^2$$

For the special case, where $M_1 = M_2 = M_o$,

$$w = -\frac{M_o}{2D(1+\nu)}\left(x^2 + y^2\right) \tag{8.15a}$$

which represents a paraboloid of revolution.* When $M_1 = -M_2 = M_o$, the deflection equation is given by

$$w = -\frac{M_o}{2D(1-\nu)}\left(x^2 - y^2\right) \tag{8.15b}$$

indicating that straight lines parallel to the x-axis become, after bending, parabolic curves convex downward, whereas straight lines parallel to the y-axis become parabolas convex upward. The deformed plate is shown in Figure 8.4b indicating a saddle point at its center.

8.3.5 Effect of Thermal Gradient throughout Plate Thickness

In many applications, the temperature is not uniform throughout the plate thickness. The approximate analysis** of bending a plate into a spherical surface can be used in calculating thermal stresses as follows.

8.3.5.1 A Plate with Free Edges

Consider first any flat plate whose edges are entirely free while being subjected to temperatures T_1 and T_2 (where $T_2 > T_1$) at its top and bottom surfaces, respectively, as shown in Figure 8.5a.

Assuming a linear temperature distribution*** across the plate thickness, the temperature of the midsurface ($z = 0$) will be $(T_2 + T_1)/2$ and hence the temperature difference between either face and this midsurface is of the magnitude $(T_2 - T_1)/2$. In such a case the difference between the maximum thermal expansion and the expansion at the midsurface is $\alpha(T_2 - T_1)/2$, where α is the coefficient of linear expansion of the material of the plate. From the first and second of Equations 8.4, the radius of curvature of the plate is thus given by

* Note that the approximate curvature equations (Equation 8.2) for $M_1 = M_2 = M_o$ gives $R_x = R_y$, indicating a spherical surface.

** For a complete analysis of thermal stresses, see Boley and Weiner.[2]

*** A rigorous derivation using the equilibrium, compatibility, and stress–strain equations shows that the validity of the analyses in this section requires that the temperature be a linear function of z. This is a good approximation for most cases of thin plates.

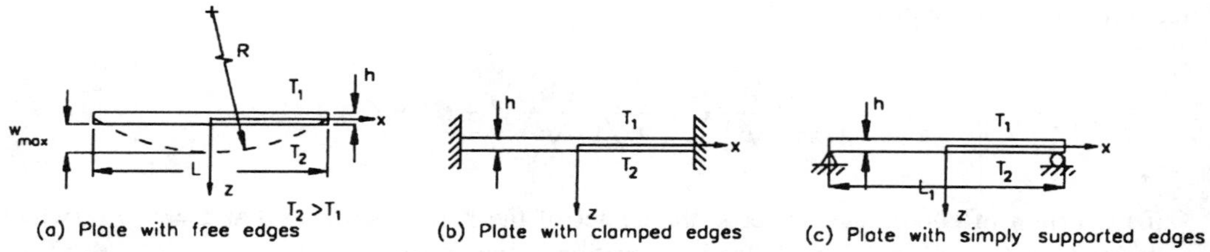

FIGURE 8.5 Plate subjected to a thermal gradient through its thickness: (a) plate with free edges, (b) plate with clamped edges, (c) plate with simply supported edges.

$$\frac{1}{R_x} = \frac{1}{R_y} = \frac{1}{R} = \frac{\alpha(T_2 - T_1)/2}{h/2} = \frac{\alpha(T_2 - T_1)}{h} \tag{8.16a}$$

Hence, for a spherically deformed surface, the maximum deflection is

$$w_{mac} = L^2/8R \tag{8.16b}$$

This bending deformation of the plate does not produce any stresses provided that the edges are free and deflections are small compared with the thickness of the plate.

8.3.5.2 A Plate with Clamped Edges

Consider a plate clamped all-around its edges, such that the edge cannot rotate meanwhile allowing its midsurface to expand freely as shown in Figure 8.5b. In such a case, the nonuniform heating will produce bending moments M_T — uniformly distributed along the edges of the plate — of such magnitude to prevent the curvature and retain the plate flat. Hence, from Equations 8.2, 8.9, and 8.16, the moment stress resultants are

$$M_x = M_y = M_T = \frac{(1+\nu)D}{R} = \frac{\alpha(1+\nu)D(T_2 - T_1)}{h}$$

or

$$M_T = \frac{\alpha E h^2 (T_2 - T_1)}{12(1-\nu)} \tag{8.17}$$

The magnitude of the maximum bending stresses from Equations 8.13 are thus*

$$\sigma_x = \sigma_y = \frac{6M_T}{h^2} = \frac{\alpha E(T_2 - T_1)}{2(1-\nu)} \tag{8.18}$$

which is compressive on the hot face and tensile on the cold face. Although these stresses are not explicit functions of the plate thickness, it is expected, however, that the temperature difference $(T_2 - T_1)$ and, hence, the bending stresses are greater for thicker plates than for thin ones.

* This simple expression (Equation 8.18) also may be used in calculating thermal stresses in nonuniformly heated thin shells.

8.3.5.3 A Plate with Simply Supported Edges

The solution to this problem is mathematically more involved since it is realized by satisfying the condition of zero moments around the edges of the plate. Consider a simply supported rectangular plate, where the top and bottom surfaces are kept at temperatures T_1 and T_2, respectively; $T_2 > T_1$. In this case, a bending moment M_T has to be applied to cancel the effect of clamping the edges. The final solution[3] results in the following maximum value for the bending moments at the boundary of the plate:

$$\left(M_x\right)_{y=\pm L_2/2} = \left(M_y\right)_{x=\pm L_1/2} = \frac{\alpha E h^2 \left(T_2 - T_1\right)}{12} \tag{8.19}$$

Example 8.1:

A square plate of side L, *Figure 8.4 and thickness* h *is made of steel of the following properties: yield strength* Y = *400 MPa,* E = *200 GPa,* ν = *0.3, and* α = *12* × *10*$^{-6}$*/°C. Apply the maximum shear stress theory when considering the two separate cases:*

a. *The plate is subjected to all-around edge moment* M_o. *What is the maximum deflection at* x = y = L/2 *at the onset of yielding of the material? Comment on the result.*
b. *If instead the plate is all-round clamped, what is the maximum difference between the temperatures of the top and bottom surfaces at the onset of yielding? Comment on the result.*

Solution:

a. Considering a square plate of side L subjected to equal bending moments; the maximum stresses at $z = \pm h/2$ are obtained from Equations 8.13 as

$$\sigma_x = \sigma_y = \pm\frac{6M_o}{h^2}$$

According to the maximum shear stress theory, Equation 1.32, knowing that Y is the tensile yield of the material,

$$M_o = Yh^2/6$$

The maximum deflection occurs at the edge point; $x = y = L/2$. Hence, from Equation 8.15a,

$$w = -\frac{1}{D}\frac{M_o L^2}{4(1+\nu)} = -\frac{12(1-\nu^2)}{Eh^3}\frac{M_o L^2}{4(1+\nu)} = -\frac{3(1-\nu)M_o}{E}\frac{L^2}{h^3}$$

The negative sign means that w at the corners is upward. Substitution of M_o at yielding:

$$\frac{w}{h} = -\frac{3(1-\nu)Yh^2}{6Eh}\frac{L^2}{h^3} = \frac{(1-\nu)Y}{3E}\left(\frac{L}{h}\right)^2$$

For steel $Y/E \cong 1/500$, $\nu = 0.3$, and, hence,

$$\frac{w}{h} = 7\times10^{-4}\left(\frac{L}{h}\right)^2$$

This indicates that for plates of width-to-thickness ratio $L/h = 10$, the deflection w at yielding is about 7% of the plate thickness. This justifies the use of small deflection bending theory for thin plates in many engineering applications.*

b. Consider an all-around clamped square plate, Equation 8.18 gives, at yielding,

$$Y = \sigma_x = \sigma_y = \frac{\alpha E(T_2 - T_1)}{2(1-\nu)} = \frac{12\times10^{-6}\times200\times10^3(T_2 - T_1)}{2(1-0.3)}$$

and for $Y = 400$ MPa, $T_2 - T_1 = 233.3$°C. This difference is seldom attained in thin plates. However, in thicker plates, where it may occur, the designer may allow local yielding or adopt constructions involving free edges.

8.4 BENDING OF RECTANGULAR PLATES — ENERGY APPROACH

To determine the deflections and stresses in rectangular plates using energy methods, an expression for the strain energy in the plate is required. This is obtained by substituting Expressions 8.5 for the stresses into Equation 3.35a and integrating over the volume element ($hdxdy$) to give

$$U = \frac{D}{2}\int_0^{L_2}\int_0^{L_1}\left[\left(\frac{\partial^2 w}{\partial x^2}+\frac{\partial^2 w}{\partial y^2}\right)^2 - 2(1-\nu)\left(\frac{\partial^2 w}{\partial x^2}\frac{\partial^2 w}{\partial y^2}-\left(\frac{\partial^2 w}{\partial x\partial y}\right)^2\right)\right]dxdy \tag{8.20}$$

The plate will be in a state of stable equilibrium if the total potential energy is a minimum. Applying this to the problem of a rectangular plate loaded laterally by a uniformly distributed load q, as shown in Figure 8.6, the total potential energy of the deformed plate is given by

$$\Omega = U - \int_0^{L_2}\int_0^{L_1} qwdxdy \tag{8.21}$$

where q is assumed to act normal to the midplane surface of the plate and w is an assumed virtual displacement.

The solution of the plate problem is thus reduced to that of finding a function $w(x, y)$ expressing plate deflection and satisfying the given boundary conditions to give minimum Ω. It is often feasible to assume a deflection function $w(x, y)$ satisfying the kinematic boundary conditions only without paying due attention to the static conditions** related to equilibrium. This obviously results in an approximate (upper bound) solution to the problem as will be shown in the next section.

8.4.1 Uniformly Loaded Rectangular Plate Simply Supported Along Its Four Edges

Consider, for instance, a simply supported plate as shown in Figure 8.6. The kinematic and static boundary conditions along the edges are

* See Table 8.5 of Section 8.5.5 of this chapter, where it is shown that the small deflection theory is satisfactory for $w/h < 0.25$.

** See Section 7.7.2.

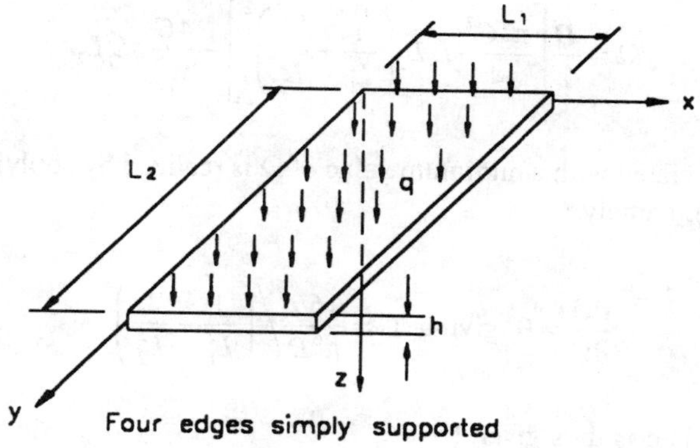

FIGURE 8.6 Uniformly loaded rectangular plate simply supported along its four edges.

At $x = 0$ and $x = L_1$ for any y:

$$w = 0 \text{ and } M_x = 0 \tag{8.22a}$$

At $y = 0$ and $y = L_2$ for any x:

$$w = 0 \text{ and } M_y = 0$$

By virtue of Expression 8.9 for M_x:

$$\left(\frac{\partial^2 w}{\partial x^2} + \nu \frac{\partial^2 w}{\partial y^2}\right)_{x=L_1} = 0$$

However, since $w = 0$ at the edges, $x = 0$ and $x = L_1$ the slope $\partial w/\partial y$ vanishes and, hence, $\partial^2 w/\partial y^2 = 0$. Similar reasoning applies to the edges $y = 0$ and $y = L_2$. The boundary conditions are thus written as

$$\begin{aligned} w &= \frac{\partial^2 w}{\partial x^2} = 0 \text{ for } x = 0 \text{ and } x = L_1 \\ w &= \frac{\partial^2 w}{\partial y^2} = 0 \text{ for } y = 0 \text{ and } y = L_2 \end{aligned} \tag{8.22b}$$

These boundary conditions will be satisfied if the function $w(x, y)$ representing the plate deflection is assumed as

$$w = C \sin\left(\frac{\pi x}{L_1}\right) \sin\left(\frac{\pi y}{L_2}\right) \tag{8.23}$$

where C is a coefficient to be determined. Substituting Equation 8.23 into the expression of the total potential energy of the plate, (Equation 8.21) gives

$$\Omega = \frac{D}{2}\left[\frac{\pi^4 C^2}{4} L_1 L_2 \left(\frac{1}{L_1^2} + \frac{1}{L_2^2}\right)^2\right] - \frac{4C}{\pi^2} q L_1 L_2$$

Stable equilibrium associated with a minimum value of Ω is realized by applying the Rayleigh–Ritz method (Section 7.7.2), namely,*

$$\frac{\partial \Omega}{\partial C} = 0 \text{ giving } C = \frac{16q}{\pi^6 D} \Big/ \left(\frac{1}{L_1^2} + \frac{1}{L_2^2}\right)^2$$

The deflection of the plate is thus given by

$$w = \left[\frac{16q}{\pi^6 D} \Big/ \left(\frac{1}{L_1^2} + \frac{1}{L_2^2}\right)^2\right] \sin\left(\frac{\pi x}{L_1}\right) \sin\left(\frac{\pi y}{L_2}\right) \tag{8.24}$$

For more accurate results, the simple trigonometric representation is replaced by a double series as**

$$w = \sum_{m=1}^{\infty} \sum_{n=1}^{\infty} C_{mn} \sin\left(\frac{m\pi x}{L_1}\right) \sin\left(\frac{n\pi y}{L_2}\right) \tag{8.25a}$$

The same procedure is followed to determine the coefficients C_{mn}, and the deflection function is obtained as

$$w = \frac{16q}{\pi^6 D} \sum_{m=1}^{\infty} \sum_{n=1}^{\infty} \frac{1}{mn\left(\frac{m^2}{L_1^2} + \frac{n^2}{L_2^2}\right)^2} \sin\left(\frac{m\pi x}{L_1}\right) \sin\left(\frac{n\pi y}{L_2}\right) \tag{8.25b}$$

where $m = 1, 3, 5$, and $n = 1, 3, 5\ldots$. Obviously, the first term with $m = n = 1$ of this series is identical to the result of Expression 8.24.

To illustrate the rapid convergence of this series representing the solution, a square plate ($L_1 = L_2 = L$) is considered and the values of the deflection and moments are calculated at the center ($x = y = L/2$) as given in Table 8.1. It is seen that convergence is very rapid, and, by taking only the first term of Expression 8.25, the deflection is overestimated by 2.6% of the exact value. Such accuracy, however, is not attained for the calculated moments (overestimation error up to 11.4%) and, hence, for the stresses. This is due to the fact that stresses depend upon the second derivatives

* The more general problem of minimizing the integral form of Expression 8.21 for Ω is set when w is assumed in the form of a series: $w = C_1 w_1(x, y) + C_2 w_2(x, y) + \ldots + C_n w_n(x, y)$. The minimization of Ω yields n conditions: $\partial\Omega/\partial C_1 = 0$, $\partial\Omega/\partial C_2 = 0$, ... $\partial\Omega/\partial C_n = 0$, as linear equations (see Section 7.7.2).

** The use of this double trigonometric series is first due to Navier (1820). Alternatively, a rapidly converging simpler series of the form

$$w = \sum_{m=1}^{\infty} Y_m(y) \sin\frac{m\pi x}{L_1}$$

was suggested by M. Levy in 1900.

TABLE 8.1
Moments and Deflections at the Center of a Simply Supported Square Plate Subjected to Uniformly Distributed Load

	Central Deflection	Bending Moment	Twisting Moment
No. of terms in the series (Eq. 8.25-b)	$\frac{0.5p_i\prime}{\sqrt{3(1-}}$	$\frac{4.631p_i r_m}{2h}$	$\frac{p_i r_m}{h}$
$m = n = 1$	0.004161	0.04106	0
$m = n = 1,3$	0.004065	0.0371	0
Exact values[3]	0.004057	0.03685	0

of the displacement function, as seen from Equations 8.9. Obviously, the derivatives of approximate functions are usually less accurate than the functions themselves. Also, it is important to note that this approximate method gives upper-bound results, which are reasonably accurate when the deflection curve is represented by a function obeying both the kinematic and static boundary conditions.

For a uniformly loaded, simply supported rectangular plate, the maximum bending moments M_x and M_y per unit length occur at the plate center and may be expressed, for design purposes, by

$$(M_x)_{max} = C_1 q L_1^2, \quad (M_y)_{max} = C_2 q L_1^2 \tag{8.26a}$$

where L_1 is the shorter side of the plate.

The maximum deflection occurs at the plate center and is given by

$$w_{max} = C_3 q L_1^4 / Eh^3 \tag{8.26b}$$

where C_1, C_2, and C_3 depend upon the ratio L_2/L_1 and are obtained from Table 8.2 for $\nu = 0.3$.

TABLE 8.2
Coefficients[3] in Equations 8.26 for a Simply Supported Plate Subjected to Uniformly Distributed Load ($\nu = 0.3$)

L_2/L_1	1	1.2	1.4	1.6	1.8	2	3	4	∞
C_1	0.0479	0.0626	0.0753	0.0862	0.0948	0.1017	0.1189	0.1235	0.1250
C_2	0.0479	0.0501	0.0506	0.0493	0.0479	0.0464	0.0404	0.0384	0.0375
C_3	0.0433	0.0616	0.0770	0.0906	0.1017	0.1106	0.1336	0.1400	0.1422

Example 8.2:

Consider bending of a long rectangular plate to a cylindrical surface under uniformly distributed load q, as shown in Figure 8.7.

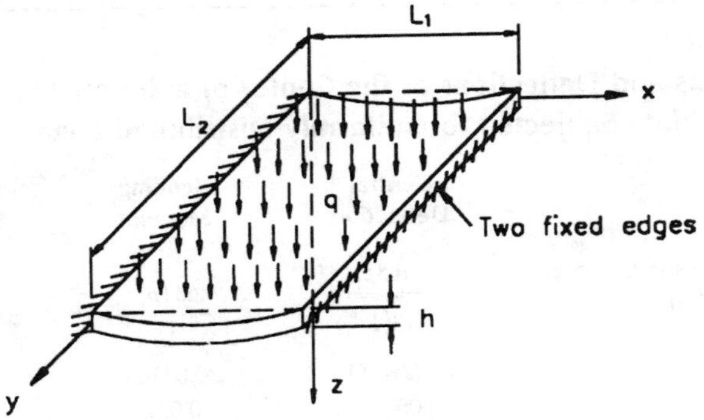

FIGURE 8.7 Example 8.2.

a. *Derive expressions for the deflection and stresses assuming that:* $w = C\,[1 - \cos(2\pi x/L_1)]$*, where C is a coefficient to be determined.*
b. *Compare the results of these approximate expressions with those obtained using more accurate methods*[3] *as:* $w_{max} = 0.00192qL_1^4/D$, $\sigma_x = 310$ *MPa for* $L_1/h = 100$ *and* $q = 0.069$ *MPa.*

Solution:

a. Bending to a cylindrical surface requires that the generators remain straight so that the radius of curvature $R_y = \infty$ and, hence, from Equation 8.2, $\partial^2 w/\partial y^2 = 0$, a condition which is furnished by the assumed deflection curve. The boundary conditions in this case are:

at $x = 0$ and $x = L_1$: $w = 0$ and $\partial w/\partial x = 0$

which are also satisfied by the assumed deflection curve. Substitution into Expression 8.20 for the strain energy gives for unit length along the y-direction:

$$U = \frac{D}{2}\int_0^{L_1}\left(\frac{\partial^2 w}{\partial x^2}\right)^2 dx = \frac{D}{2}\int_0^{L_1}\left[\frac{4\pi^2 C}{L_1}\cos\left(\frac{2\pi x}{L_1}\right)\right]^2 dx$$

$$U = \frac{8\pi^4 DC^2}{L_1^4}\left[\frac{x}{2} + \frac{L_1}{8\pi}\sin\left(\frac{4\pi x}{L_1}\right)\right]_0^{L_1} = \frac{4\pi^4 DC^2}{L_1^3}$$

The total potential energy of the deformed plate is

$$\Omega = U_o - \int_0^{L_1} qw\,dx\,dy = \frac{4D\pi^2 C^2}{L_1^3} - \int_0^{L_1} qC\left[1 - \cos\left(\frac{2\pi x}{L_1}\right)\right]dx$$

$$\Omega = \frac{4\pi^4 DC^2}{L_1^3} - qL_1 C$$

This energy attains a minimum when $d\Omega/dC = 0$, i.e.,

$$\frac{8\pi^4 DC}{L_1^3} - qL_1 = 0 \text{ or } C = \frac{qL_1^4}{8\pi^4 D}$$

The deflection curve is thus given by

$$w = \frac{qL_1^4}{8\pi^4 D}\left[1 - \cos\left(\frac{2\pi x}{L_1}\right)\right]$$

The resulting moments M_x, M_y, and M_{xy}, as given by Equations 8.9, are

$$M_x = -D\left(\frac{\partial^2 w}{\partial x^2} + 0\right) = \frac{qL_1^2}{2\pi^2}\cos\left(\frac{2\pi x}{L_1}\right)$$

$$M_y = -D\left(0 + \nu\frac{\partial^2 w}{\partial x^2}\right) = \nu M_x = \frac{\nu qL_1^2}{2\pi^2}\cos\left(\frac{2\pi x}{L_1}\right)$$

$$M_{xy} = 0$$

It is seen that to produce bending of the plate to a cylindrical surface, not only the bending moment M_x is applied, but also the moment M_y. Without this latter moment the plate will be bent to an anticlastic surface. The stresses in the plate are obtained from Equations 8.13 as

$$\sigma_x = \frac{z}{h^2/12}\frac{qL_1^2}{2\pi^2}\cos\left(\frac{2\pi x}{L_1}\right)$$

$$\sigma_y = \frac{z}{h^2/12}\frac{\nu qL_1^2}{2\pi^2}\cos\left(\frac{2\pi x}{L_1}\right)$$

$$\tau_{xy} = 0$$

b. The maximum deflection along the center $x = L_1/2$ is

$$w_{\max} = \frac{qL_1^4}{4\pi^4 D} = 0.00257\frac{qL_1^4}{D}$$

which compares with $w_{\max} = 0.00192\ qL_1^4/D$ with an error of 34%. The maximum stress σ_x (at $z = \pm h/2$) occurs at the clamped edges ($x = 0$ or $x = L_1$), where the bending moment is the largest; hence,

$$\sigma_x = \frac{6}{h^2}\frac{qL_1^2}{2\pi^2} = \frac{3q}{\pi^2}\left(\frac{L_1}{h}\right)^2 = 209.7 \text{ MPa}$$

which compares with $\sigma_x = 310$ MPa with an error of – 32%.

Note that the rigorous solution,[3] apart from using a more elaborate representation for the deflection function w, considers the presence of tensile forces at the clamped edges. These are necessary to prevent the plate ends from moving along the x-axis, thus adding membrane stresses to the bending stresses.

8.4.2 Uniformly Loaded Rectangular Plate Clamped Along Its Four Edges

The kinematic boundary conditions for a uniformly loaded rectangular plate, which is clamped along its four edges as shown in Figure 8.8, are

At $x = 0$ and $x = L_1$ for any y: $w = 0$ and $\partial w/\partial x = 0$
At $y = 0$ and $y = L_2$ for any x: $w = 0$ and $\partial w/\partial y = 0$

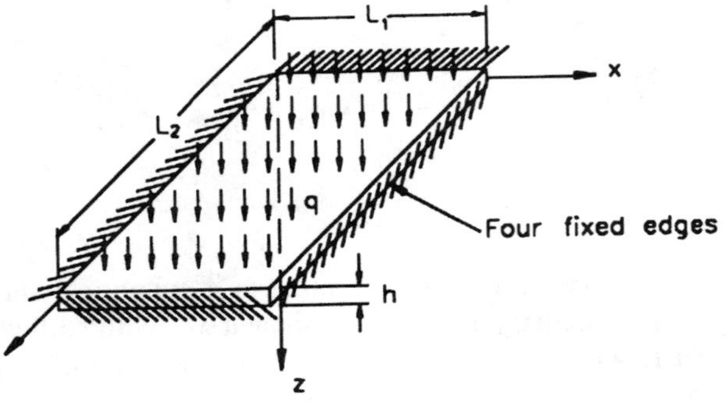

FIGURE 8.8 Uniformly loaded rectangular plate clamped along its four edges.

An approximate solution for the deflection and, hence, moments and stresses may be obtained by assuming a function $w(x, y)$ to represent the deflection curve as

$$w = \sum_{m=1}^{\infty} \sum_{n=1}^{\infty} C_{mn} \left[1 - \cos\left(\frac{2m\pi x}{L_1}\right)\right]\left[1 - \cos\left(\frac{2n\pi y}{L_2}\right)\right] \tag{8.27a}$$

where the coefficients C_{mn} are to be determined. Taking only the first term of this series, i.e., $m = n = 1$ gives

$$w = C\left[1 - \cos\left(\frac{2\pi x}{L_1}\right)\right]\left[1 - \cos\left(\frac{2\pi y}{L_2}\right)\right] \tag{8.27b}$$

Following a procedure similar to that of a simply supported plate, the deflection equation is

$$w = \frac{qL_1^4}{4\pi^4 D} \frac{1}{3 + 2\left(\frac{L_1}{L_2}\right)^2 + 3\left(\frac{L_1}{L_2}\right)^4} \left[1 - \cos\left(\frac{2\pi x}{L_1}\right)\right]\left[1 - \cos\left(\frac{2\pi y}{L_2}\right)\right] \tag{8.27c}$$

The maximum deflection occurs in the middle of the plate at $x = L_1/2$ and $y = L_2/2$.

Another polynomial form for the deflection function may be assumed as

$$w = C_1 x^2 y^2 (x - L_1)^2 (y - b_2)^2 \tag{8.28a}$$

This function satisfies the kinematic boundary conditions. The deflection function in this case becomes

$$w = \frac{6.128(q/D)}{7L_1^4 + 4L_1^2 L_2^2 + 7L_2^4}\left[x^2 y^2 (x - L_1)^2 (y - L_2)^2\right] \tag{8.28b}$$

For a square plate $L_1 = L_2 = L$, Equations 8.27c and 8.28b give for the maximum deflection w_{max} at its center $0.001283\ qL^4/D$ and $0.00133\ qL^4/D$, respectively. A more-accurate series solution[3] gives $w_{max} = 0.001264\ qL^4/D$.

For an all-around clamped rectangular plate, the maximum bending moment occurs at the middle of the longer edges and for design purpose $(M_x)_{max}$ per unit length is given by

$$\left(M_x\right)_{max} = C_4 q L_1^2 \tag{8.28c}$$

The maximum deflection occurs at the center according to

$$w_{max} = C_5 q L_1^4 / Eh^3 \tag{8.28d}$$

The coefficients C_4 and C_5 are dependent on the plate aspect ratio (L_2/L_1), where L_1 is taken as the shorter side, and are obtained from Table 8.3 for $\nu = 0.3$.

TABLE 8.3
Coefficients[3] in Equations 8.28 for All-around Clamped Rectangular Plate Subjected to Uniformly Distributed Load (ν = 0.3)

L_2/L_1	1	1.25	1.50	1.75	2	∞
C_5	0.0513	0.0665	0.0757	0.0817	0.0829	0.0888
C_4	0.0138	0.0199	0.0240	0.0264	0.0277	0.0284

Solutions to rectangular plates with other boundary conditions are found in the literature usually in tabulated form, such as Tables 8.2 and 8.3.[3] Superposition is applied to obtain deflections and stresses due to several loadings, as shown in Example 8.3.

Example 8.3:

A square plate of dimensions $L_l = L_2 = 600$ *mm is subjected to a uniformly distributed load of 0.7 MPa all over its entire surface. If the plate thickness is to be selected such that no yielding occurs at any point in the plate according to the maximum shear stress theory, calculate the required plate thickness and the associated maximum deflection assuming:*

a. All edges are simply supported;
b. All-around clamped edges (use both the trigonometric function, Equation 8.27c, and the polynomial, Equation 8.28b. Comment on the results.

Take $E = 200$ *GPa,* $\nu = 0.3$*, yield strength* $Y = 300$ *MPa, and a factor of safety F.S. = 1.75.*

Solution:

a. For a simply supported square plate ($L_1 = L_2 = L$), the deflection curve is expressed by Equation 8.24 as

$$w = \frac{4qL^4}{\pi^6 D}\sin\left(\frac{\pi x}{L}\right)\sin\left(\frac{\pi y}{L}\right)$$

The moments are derived from this expression according to Equations 8.9 as

$$M_x = -D\left(\frac{\partial^2 w}{\partial x^2} + \nu\frac{\partial^2 w}{\partial y^2}\right) = \frac{4qL^4}{\pi^6}(1+\nu)\left(\frac{\pi}{L}\right)^2 \sin\left(\frac{\pi x}{L}\right)\sin\left(\frac{\pi y}{L}\right)$$

$$M_y = -D\left(\frac{\partial^2 w}{\partial y^2} + \nu\frac{\partial^2 w}{\partial x^2}\right) = \frac{4qL^4}{\pi^6}(1+\nu)\left(\frac{\pi}{L}\right)^2 \sin\left(\frac{\pi x}{L}\right)\sin\left(\frac{\pi y}{L}\right)$$

$$M_{xy} = D(1-\nu)\frac{\partial^2 w}{\partial x \partial y} = \frac{4qL^4}{\pi^6}(1-\nu)\left(\frac{\pi}{L}\right)^2 \cos\left(\frac{\pi x}{L}\right)\cos\left(\frac{\pi y}{L}\right)$$

These are shown schematically in Figure 8.9. In the middle of the square plate $x = y = L/2$, $M_x = M_y$ and the stresses attain their maximum values at $z = \pm h/2$, where h is the plate thickness; hence, from Equations 8.13,

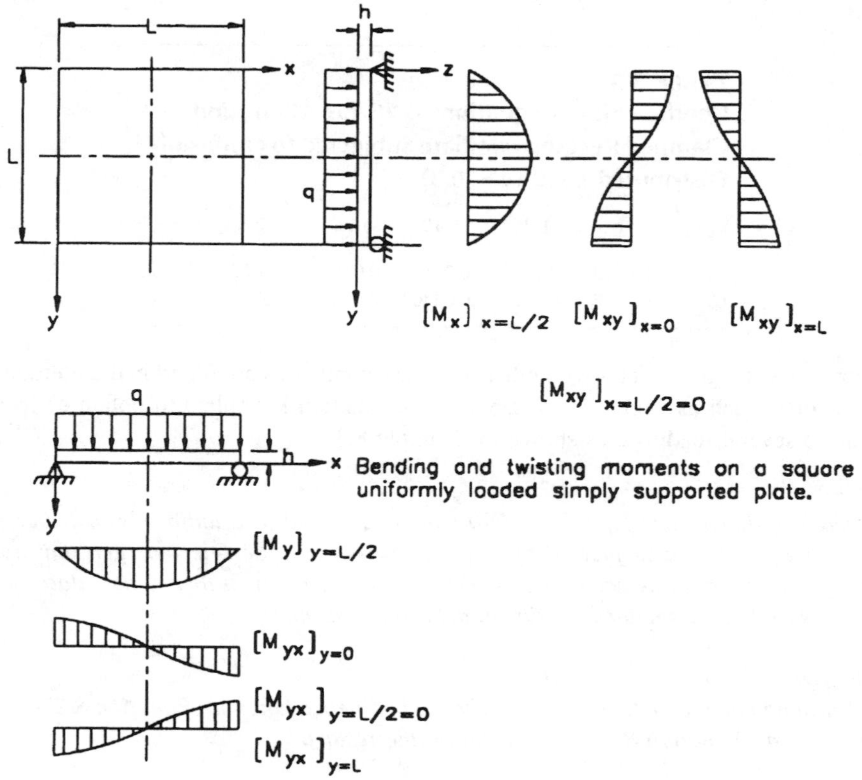

FIGURE 8.9 Example 8.3a.

$$\sigma_x = \sigma_y = \frac{M_x z}{h^3/12} = \frac{24(1+\nu)}{\pi^4}\frac{qL^2}{h^2}$$

The principal stresses (neglecting all other stress components) at the midpoint are σ_x, σ_y, and 0, and, according to the maximum shear stress theory, Equation 1.32,

$$\frac{\sigma_x}{2} = \frac{12(1+\nu)}{\pi^4}\left(\frac{L}{h}\right)^2 q \le \frac{Y/2}{\text{F.S.}}$$

Taking $q = 0.7$ N/mm², $Y = 300$ MPa, $\nu = 0.3$, $E = 200$ GPa, F.S. = 1.75 and $L = 600$ mm, gives for the plate thickness: $h = 21.7$ mm. The maximum deflection at this point is

$$w = \frac{4qL^4}{\pi^6 D} = \frac{4qL^4}{\pi^6}\frac{12(1-\nu^2)}{Eh^3} = 2.017 \text{ mm}$$

From Table 8.2, the exact solution gives

$$h = 20.554 \text{ mm and } w = 1.927 \text{ mm}$$

a deflection that is small compared with the plate thickness.

b. For a square plate with all-around clamped edges, the deflection curve may be expressed by Equation 8.27c as

$$w = \frac{qL^4}{32\pi^4 D}\left[1-\cos\left(\frac{2\pi x}{L}\right)\right]\left[1-\cos\left(\frac{2\pi y}{L}\right)\right]$$

The moments are derived from Equations 8.9 as

$$M_x = \frac{qL^2}{8\pi^2}\left\{\cos\left(\frac{2\pi x}{L}\right)\left[1-\cos\left(\frac{2\pi y}{L}\right)\right] + \nu\cos\left(\frac{2\pi y}{L}\right)\left[1-\cos\left(\frac{2\pi x}{L}\right)\right]\right\}$$

$$M_y = \frac{qL^2}{8\pi^2}\left\{\cos\left(\frac{2\pi y}{L}\right)\left[1-\cos\left(\frac{2\pi x}{L}\right)\right] + \nu\cos\left(\frac{2\pi x}{L}\right)\left[1-\cos\left(\frac{2\pi y}{L}\right)\right]\right\}$$

$$M_{xy} = \frac{qL^2}{8\pi^2}\sin\left(\frac{2\pi x}{L}\right)\sin\left(\frac{2\pi y}{L}\right)$$

The moments are shown schematically in Figure 8.10 for a square plate. The maximum bending moments occur at the middle of the, clamped edges, i.e., at ($x = 0$ and $x = L$, $y = L/2$) and ($y = 0$, $x = L/2$); hence,

$$\left(M_x\right)_{\max} = \frac{qL^2}{4\pi^2},\ \left(M_y\right)_{\max} = \frac{\nu qL^2}{4\pi^2},\ \text{and } M_{xy} = 0$$

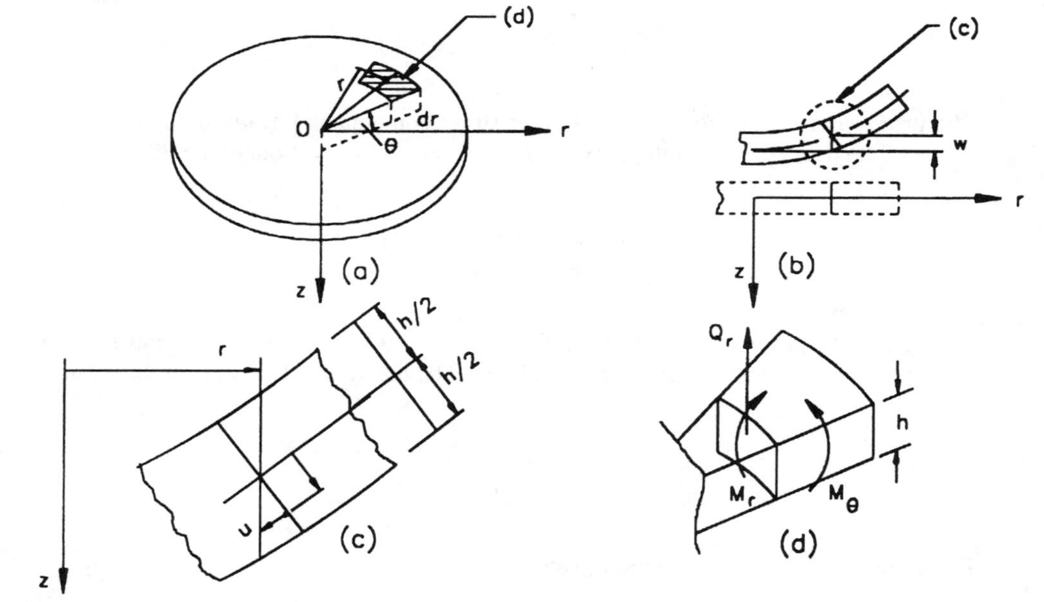

FIGURE 8.10 Example 8.3b.

The maximum stresses occur at $z = \pm h/2$ and are given by Equations 8.13 as

$$\sigma_x = 0.1521qL^2/h^2,\ \sigma_y = 0.1521\ \nu qL^2/h^2 \text{ and } \tau_{xy} = 0$$

From Table 8.3, the maximum moment for $L_2/L_1 = 1$ is $(M_x)_{\max} = -0.0513\ qL^2$, which gives a maximum stress at $z = \pm h/2$ as

$$\sigma_x = (0.0513qL^2)\ (h/2)/(h^3/12) = 0.3078qL^2/h^2$$

a value which indicates that the use of one term of the trigonometric series of Equation 8.27c results in erroneous underestimation for the stresses (–50%) irrespective of giving reasonably accurate estimation for the deflections (with an error of 1.4%).

Now, consider the other proposed function for deflection as given by Expression 8.28b. The moment M_x is obtained from Equation 8.9 for a square plate $L_1 = L_2 = L$ as

$$M_x = \frac{6.128q}{18L^4}\Big((y-L)^2\big[2y^2(x-L)^2 + 8xy^2(x-L) + 2x^2y^2\big]$$
$$+\nu(x-L)^2\big[2x^2(y-L)^2 + 8x^2y(y-L) + 2x^2y^2\big]\Big)$$

At $x = 0$ and $y = L/2$:

$$M_x = \frac{6.128q}{18\times 8}L^2 = 0.04256qL^2$$

giving a stress,

$$\sigma_x = (0.04256qL^2)\,(h/2)/(h^3/12) = 0.2554qL^2/h^2$$

which is about –17% in error with the more correct value. The maximum shear stress theory, Equation 1.32 thus gives

$$\frac{\sigma_x}{2} = \frac{1}{2}\left(0.2554\,qL^2/h^2\right) \leq = \frac{Y/2}{\text{F.S.}}$$

or

$$h = 19.38 \text{ mm}$$

The maximum deflection occurs at the plate center $x = y = L/2$, giving for $L_1 = L_2 = L$ in Equation 8.28b:

$$w_{\max} = \frac{6.128\,q/D}{18L^4}\left(\frac{L^2}{4}\right)^4 = 0.01452\,qL^4/\left(Eh^3\right)$$

or

$$w_{\max} = 0.905 \text{ mm}$$

Use of Table 8.3 gives more accurate values, namely, $h = 21.27$ mm and $w_{\max} = 0.659$ mm.

Example 8.4:

A simply supported rectangular plate of an aspect ratio $L_2/L_1 = 2$ *is loaded by a uniformly distributed load* q *along the axis of symmetry parallel to the length* L_1. *Along the same axis, a concentrated load* $P = qL_1$ *has to be applied at a distance* $\zeta = L_1/4$ *from the edge of the plate, as shown in Figure 8.11. Determine the total deflection at the center of the plate. Use Table 8.4 for plates of* $\nu = 0.3$.

TABLE 8.4
Maximum Deflection of Simply Supported Rectangular Plate, $w_{max} = \alpha qL_1^3/Eh^3$ Uniformly Loaded Along the Axis of Symmetry[3]

L_2/L_1	2	1.5	1.4	1.3	1.2	1
α	0.1078	0.0995	0.0963	0.0922	0.0872	0.0736

The deflection along the x*-axis for a loaded rectangular plate by a concentrated load* P *at a distance* $x = \zeta$ *is given by*[3]

$$(w)_{y=0} = \frac{PL_1^2}{2\pi^3 D} \sum_{m=1,3,5,\ldots}^{\infty} \frac{1}{m^3} \sin(m\pi x/L_1)\sin(m\pi\xi/L_1)\left(\tanh C_m - \frac{C_m}{\cosh^2 C_m}\right)$$

where

$$C_m = m\pi L_2/2L_1$$

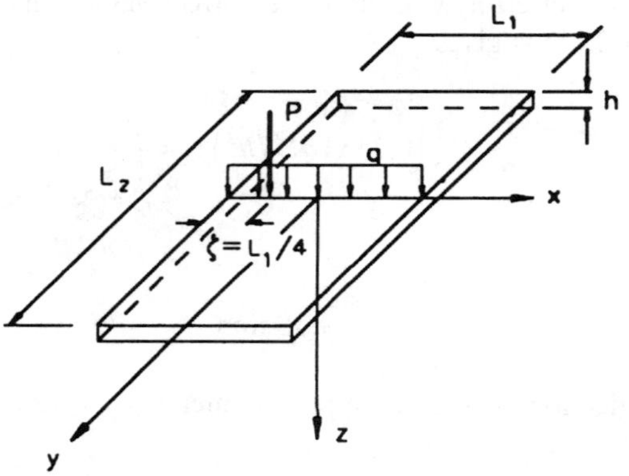

FIGURE 8.11 Example 8.4.

Solution:
Using Table 8.4, the maximum deflection at the center of the plate due to the load q is

$$w_q = 0.1078\frac{qL_1^3}{Eh^3}$$

Due to the concentrated load, the deflection w_p is calculated using two terms of the above rapidly conveying series with $\zeta = L_1/4$ and $x = L_1/2$, as

$$w_P = \frac{PL_1^2}{2\pi^3 D}\left[\left(\tanh\pi - \frac{\pi}{\cosh^2\pi}\right)\left(\sin\frac{\pi}{4}\sin\frac{\pi}{2}\right) + \frac{1}{27}\left(\tanh 3\pi - \frac{3\pi}{\cosh^2 3\pi}\right)\left(\sin\frac{3\pi}{4}\sin\frac{3\pi}{2}\right)\right]$$

$$= \frac{0.1165PL_1^2}{Eh^3}$$

The total deflection at the center of the plate is thus superposition, as

$$w = w_q + w_P$$

$$w = 0.1078\frac{qL_1^3}{Eh^3} + 0.1165\frac{PL_1^2}{Eh^3}$$

$$w = \frac{qL_1^3}{Eh^3}\left[0.1078 + 0.1165\left(\frac{P}{qL_1}\right)\right]$$

$$w = 0.2243\frac{qL_1^3}{Eh^3}$$

8.4.3 An Approximate Strip Method for Rectangular Plates

A simple approximate conservative method is used by construction engineers to design rectangular floor slabs. In this method the slab, of sides L_1 and L_2, thickness h, and subjected to a uniformly distributed load q, is assumed to consist of two plates each of thickness h. One is supported along the sides of length L_2 and has a span of L_1, and the other is at right angles and supported along the sides of length L_1 and has a span of L_2. The load is shared between the two plates such that

$$q = q_1 + q_2 \tag{8.29a}$$

where q, q_1, and q_2 are the total uniform load, the uniform load supported along span L_1, and the uniform load supported along span L_2, respectively. Again, the maximum deformations at the middle are the same for each assumed plate, so that

$$(w_1)_{max} = (w_2)_{max} \tag{8.29b}$$

By referring to Equations 8.26a and b, the maximum bending moments and deflections are expressed as

$$(M_1)_{max} = k_1 q_1 L_1^2, \quad (M_2)_{max} = k_1 q_2 L_2^2$$
$$(w_1)_{max} = k_2 q_1 L_1^4 / D, \quad (w_2)_{max} = k_2 q_2 L_2^4 / D$$

The equality of deflections yields

$$q_1 L_1^4 = q_2 L_2^4$$

hence,

$$q_1 = q\frac{L_2^4}{L_1^4 + L_2^4}, \qquad q_2 = q\frac{L_1^4}{L_1^4 + L_2^4}$$
$$M_1 = k_1 q\frac{L_1^2 L_2^4}{L_1^4 + L_2^4}, \quad M_2 = k_1 q\frac{L_1^4 L_2^2}{L_1^4 + L_2^4}, \quad \text{and} \tag{8.29c}$$
$$w_{max} = k_2 q\frac{L_1^4 L_2^4}{D\left(L_1^4 + L_2^4\right)}$$

The method considers that each plate behaves as a beam and hence the values of coefficients k_1 and k_2 correspond to those for a beam at the same support conditions.

As an example, consider the case of simply supported edges along both L_1 and L_2. For a simply supported beam,

$$M_{max} = \frac{1}{8}qL^2 \text{ and } w_{max} = \frac{5}{384}q\frac{L^4}{EI}$$

Hence, $k_1 = 0.125$ and $k_2 = 0.013$. For a square plate $L_1 = L_2 = L$ and, hence, from Equations 8.29c,

$$M_{max} = 0.0625qL^2 \text{ and } w_{max} = 0.0065q\frac{L^4}{D}$$

Exact values of Table 8.1 gives 0.0479 for M_{max} and 0.00406 for w_{max}, taking $\nu = 0.3$.

For a clamped beam at both ends,

$$M_{max} = \frac{1}{23}qL^2, \quad w_{max} = \frac{1}{384}q\frac{L^4}{EI}$$

hence

$$k_1 = 0.0417 \text{ and } k_2 = 0.0026$$

hence, this method gives for an all around clamped square plate:

$$M_{max} = 0.02085qL^2 \text{ and } w_{max} = 0.0013q\frac{L^4}{D}$$

Exact values give 0.0513 for M_{max} and 0.00126 for w_{max}. The approximate values for w_{max} are in good agreement with the exact values but the results for the moments are unsatisfactory.[4]

8.5 AXISYMMETRIC BENDING OF FLAT, CIRCULAR PLATES

Circular plates subjected to axisymmetric transverse loading are encountered in several engineering applications, such as flat heads of pressure vessels, cylinder covers, pistons, valve disks, column bases, and disk springs.

In the following analysis, the plates are considered to be flat, of uniform thickness, and of an elastic homogeneous isotropic material. Loads are axisymmetric and act normal to the plane of the plate. The deflections are assumed to be small compared with the plate thickness and the strains are infinitesimal.

The problem is solved in cylindrical polar coordinates (r, θ, z) with the plate middle plane taken as a reference for the z-axis. In Figure 8.12, the plate before loading is shown by dotted lines. As a result of loading, the plate will be subjected to axisymmetric bending as shown by the solid lines, Figure 8.12b and c. At the middle plane ($z = 0$), the radial and tangential displacements vanish ($u = \nu = 0$) and only transverse displacement w, parallel to the z-direction, takes place.

In view of symmetry of both geometry and loading there is no variation with respect to θ. Hence, from Chapter 4, Example 4.7, substitute

$$\frac{\partial^2}{\partial x^2} = \frac{d^2}{dr^2}, \quad \frac{\partial^2}{\partial y^2} = \frac{1}{r}\frac{d}{dr}$$

and

$$\nabla^2 = \frac{d^2}{dr^2} + \frac{1}{r}\frac{d}{dr}$$

The general equation (Equation 8.12) governing deflection of plates is thus much simplified and reduces to an ordinary differential equation as

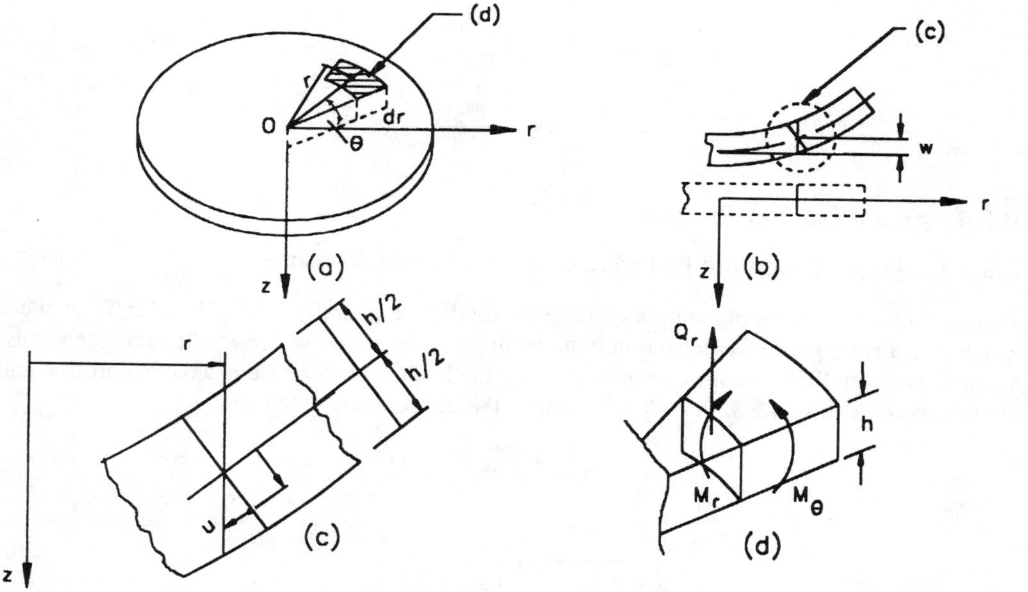

FIGURE 8.12 Reference axes, displacements, and stress resultants in a circular plate subjected to axisymmetric bending: (a) reference axes, (b) and (c) plate displacements, and (d) stress resultants acting on a plate element.

$$\frac{d^4w}{dr^4} + \frac{2}{r}\frac{d^3w}{dr^3} - \frac{1}{r^2}\frac{d^2w}{dr^2} + \frac{1}{r^3}\frac{dw}{dr} = \frac{q}{D}$$

The similarity of this equation with the governing equation for axisymmetric plane problems in terms of a stress function suggests that its general solution takes the form of Equation 4.31, i.e., it is given by

$$w = C_1 \ln r + C_2 r^2 \ln r + C_3 r^2 + C_4 + \text{P.I.} \tag{8.30}$$

where P.I. is an abbreviation to the particular integral and C_1, C_2, C_3, and C_4 are constants of integration to be determined from the appropriate boundary conditions.

The stress resultants of Figure 8.12d, as given by Expressions 8.9 and 8.11 thus reduce to

$$M_r = -D\left(\frac{d^2w}{dr^2} + \frac{\nu}{r}\frac{dw}{dr}\right)$$

$$M_\theta = -D\left(\frac{1}{r}\frac{dw}{dr} + \nu\frac{d^2w}{dr^2}\right) \tag{8.31a}$$

$$Q_r = D\frac{d}{dr}\left(\frac{d^2w}{dr^2} + \frac{1}{r}\frac{dw}{dr}\right)$$

Similarly, the stresses are given by

$$\sigma_r = \frac{M_r z}{h^3/12}$$
$$\sigma_\theta = \frac{M_\theta z}{h^3/12} \tag{8.31b}$$

8.5.1 Solid Circular Plates

8.5.1.1 Simply Supported Plate Subjected to Uniform Pressure

Figure 8.13 shows a simply supported circular solid plate laterally loaded by a uniform pressure $q = p$. For a solid plate, the deflection must be finite at the center where $r = 0$; hence, the constant C_1 in Expression 8.30 vanishes. Moreover, since the loading in this case is axisymmetric, it can be shown, as in Section 4.9.3, that $C_2 = 0$. Hence, the deflection is given by

$$w = C_3 r^2 + C_4 + \text{P.I.}$$

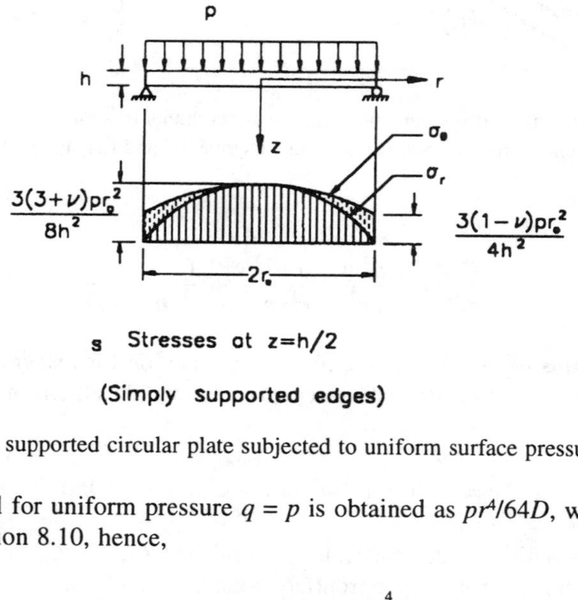

FIGURE 8.13 Simply supported circular plate subjected to uniform surface pressure.

The particular integral for uniform pressure $q = p$ is obtained as $pr^4/64D$, where $D = Eh^3/12\,(1 - \nu^2)$ as given by Equation 8.10, hence,

$$w = C_3 r^2 + C_4 + \frac{pr^4}{64D} \tag{8.32}$$

The constants C_3 and C_4 are determined from the boundary conditions. For a simply supported plate, the boundary conditions are

$$\text{At } r = r_o\text{: } w = 0 \text{ and } M_r = 0 \tag{8.33}$$

which, when applied to Equation 8.32 and the first of Equations 8.31a, respectively, yield

$$0 = C_3 r_o^2 + C_4 + \frac{pr_o^4}{64D}$$

$$0 = -D\left[2(1+\nu)C_3 + (3+\nu)\frac{pr_o^2}{16D}\right]$$

Solving gives

$$C_3 = -\frac{3+\nu}{1+\nu}\frac{pr_o^2}{32D} \quad \text{and} \quad C_4 = \frac{5+\nu}{1+\nu}\frac{pr_o^4}{64D}$$

The plate deflection, which is directly proportional to the pressure p, is thus expressed by

$$w = \frac{pr_o^4}{64D}\left[\frac{r^4}{r_o^4} - 2\frac{3+\nu}{1+\nu}\frac{r^2}{r_o^2} + \frac{5+\nu}{1+\nu}\right] \tag{8.34a}$$

The maximum deflection occurs at the center of the plate and is given by

$$w_{\max} = \frac{5+\nu}{1+\nu}\frac{pr_o^4}{64D} \tag{8.34b}$$

The bending stresses in the plate are determined by deriving expressions for the moments M_r and M_θ using Equations 8.34a and substituting into Equations 8.31a. Hence, from Equations 8.31b, the stresses are

$$\begin{aligned} \sigma_r &= \frac{3pz}{4h^3}(3+\nu)\left(r_o^2 - r^2\right) \\ \sigma_\theta &= \frac{3pz}{4h^3}\left[(3+\nu)r_o^2 - (1+3\nu)r^2\right] \end{aligned} \tag{8.35a}$$

These stresses attain maximum values at the bottom surface, $z = h/2$, and are maximum at the center of the plate, $r = 0$, as given by

$$\left(\sigma_r\right)_{\max} = \left(\sigma_\theta\right)_{\max} = \frac{3(3+\nu)}{8}\frac{pr_o^2}{h^2} \tag{8.35b}$$

The stress distributions are shown in Figure 8.13.

8.5.1.2 All-Around Clamped Plate Subjected to Uniform Pressure

Figure 8.14 shows an all-around clamped plate subjected to uniform pressure. The same reasoning of having finite deflection at the plate center applies. Hence, from Equations 8.34, together with the boundary conditions,

$$\text{At } r = r_o : w = 0 \text{ and } \frac{dw}{dr} = 0 \tag{8.36}$$

the constants C_3 and C_4 are determined. The deflection of the plate is expressed by

$$w = \frac{p}{64D}\left(r_o^2 - r^2\right)^2 \tag{8.37a}$$

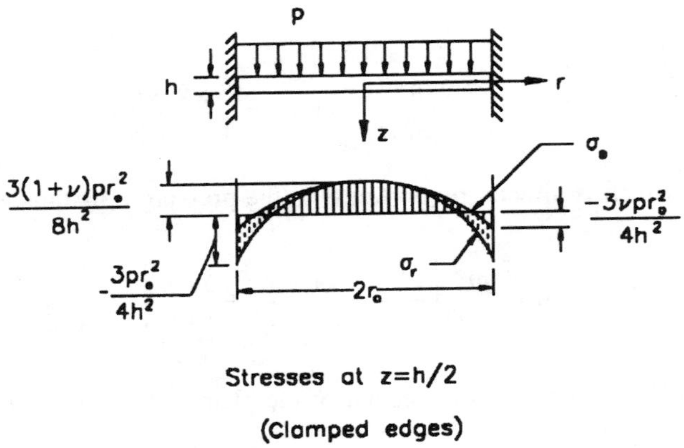

FIGURE 8.14 All-around clamped circular plate subjected to uniform surface pressure.

Again, the deflection is directly proportional to the applied pressure p. The maximum displacement occurs at the center of the plate as

$$w_{max} = \frac{pr_o^4}{64D} \tag{8.37b}$$

which is substantially smaller than w_{max} for a simply supported plate (the ratio being 1/4 for $\nu = {}^1/_3$).

The distribution of the bending moments (M_r, M_θ) and, hence, the stresses σ_r and σ_θ are determined using Equation 8.37a into Equations 8.31a. The stresses are then given by Equations 8.31b as

$$\begin{aligned}\sigma_r &= \frac{3pz}{4h^3}\left[(1+\nu)r_o^2 - (3+\nu)r^2\right]\\ \sigma_\theta &= \frac{3pz}{4h^3}\left[(1+\nu)r_o^2 - (1+3\nu)r^2\right]\end{aligned} \tag{8.38a}$$

The maximum stresses are given by

$$\begin{aligned}(\sigma_r)_{max} &= \frac{3pr_o^2}{4h^2} \text{ occurring at the edge } r = r_o \text{ at } z = -h/2\\ (\sigma_\theta)_{max} &= \frac{3(1+\nu)}{8}\frac{pr_o^2}{h^2} \text{ occurring at the center } r = 0 \text{ at } z = -h/2\end{aligned} \tag{8.38b}$$

The stress distributions are shown in Figure 8.14 for an all-around clamped plate. Note that the solution to an all-around clamped plate is based on the assumption that its edge is fully constrained against any linear or angular displacements, a condition which is hardly to be satisfied in practice. A slight amount of yielding, which may occur at the clamped edge, will make edge conditions intermediate between simply supported and all-around clamped.

In order to assess the magnitude of plate deflection, consider the extreme case when the pressure p is increased such that yielding is initiated. According to the maximum shear stress theory, Equation 1.32, this occurs at the edge when $(\sigma_r)_{max} = Y$, where Y is the yield strength of the plate material, so that

$$Y = \frac{3}{4}\frac{pr_o^2}{h^2}$$

hence,

$$p = \frac{4Yh^2}{3r_o^2}$$

At this value of p, the central deflection given by Equation 8.37b is

$$\left(w_{max}\right)_{yielding} = \frac{Yh^2r_o^2}{48D} = \frac{1-\nu^2}{4}\left(\frac{Y}{E}\right)\left(\frac{r_o}{h}\right)r_o$$

For steels, $\nu = 0.3$, $Y/E \cong 1/500$; hence, (w_{max}/h) at yielding $\cong 4.6 \times 10^{-4}\ (r_o/h)^2$. For plates of $r_o/h \cong 10$, the maximum deflection is less than 5% of the plate thickness, a result justifying use of this bending theory for thin (and even moderately thick plates) without much error.* For thicker plates of all-around clamped edge, deflection given by Equation 8.37a is found to be less than that observed in practice. This is due to the fact that the condition of a clamped edge cannot be fulfilled as well as the neglection of shear deformation in developing Equation 8.37a. A correction to compensate for the shear effect is given by Equation 8.63c.

Example 8.5:

Consider the case of a cylindrical pressure vessel having a flat head welded to its end as shown in Figure 8.15 with inside diameter 1000 mm and internal pressure 0.04 MPa. If the maximum allowable tensile stress is 120 MPa, determine the head thickness for the two shown constructions (a) and (b). Take $\nu = 0.3$.

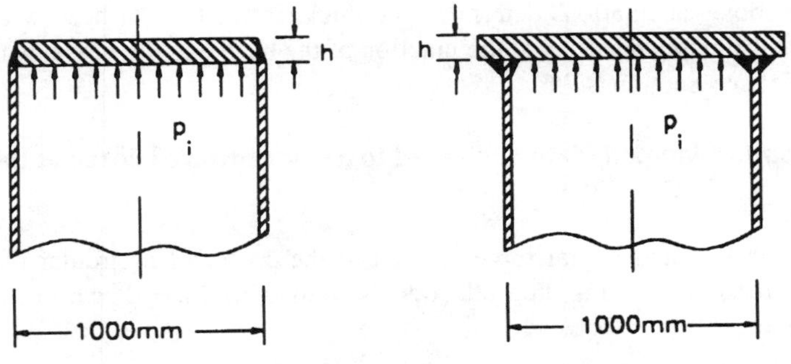

FIGURE 8.15 Example 8.5.

Solution:

Construction (a): A single fillet weld is used to connect the flat head to the vessel end. Hence, it may be assumed that $M = 0$ at the edges, i.e., simply supported edges. Hence, $\sigma_r = \sigma_\theta = \sigma_{max}$ at the center and is given by Expression 8.35b as

* See Table 8.6 of Section 8.5.4.

$$\sigma_{max} = \frac{3(3+\nu)}{8h^2} pr_o^2$$

Applying the maximum shear stress theory, Equation 1.32, for an allowable stress of 120 MPa, $r_o = 500$ mm, $p = 0.04$ MPa, and $\nu = 0.3$, the flat head thickness is

$$h = \left[\frac{3(3+0.3)\times 0.04\times 500^2}{8\times 120}\right]^{1/2} = 10.2 \text{ mm}$$

Construction (b): An inner and an outer fillet welds are used to connect the flat head to the vessel end, which make the assumption of considering clamped edges reasonable. Using Expressions 8.38b gives

$$(\sigma_r)_{max} = \frac{3p_i}{4h^2} r_o^2 \text{ at the edge}$$

$$(\sigma_\theta)_{max} = \frac{3(1+\nu)p_i}{8h^2} r_o^2 \text{ at the center}$$

Hence, $(\sigma_r)_{max} > (\sigma_\theta)_{max}$. Substituting $(\sigma_r)_{max} \le 120$ MPa gives

$$h = \left(\frac{0.75\times 0.04\times 500^2}{120}\right)^{1/2} = 7.9 \text{ mm}$$

Comparing the result of construction (a) with that of construction (b) suggests that the cost of an extra weld may be much more than the saving in the plate thickness, and therefore construction (b) is not recommended.

Note that the above calculations determine the thickness of the flat head without considering membrane and local stresses resulting at the junction of the head and the vessel wall. These stresses can be appreciable; see Section 8.7.6.

8.5.1.3 All-Around Clamped Plate Subjected to a Concentrated Force at the Center

In the case of a concentrated lateral force applied to the center of a circular plate, as shown in Figure 8.16a, Expression 8.30 for the deflection is used with $C_1 = 0$, while retaining the term containing C_2; hence,

$$w = C_2 r^2 \ln r + C_3 r^2 + C_4 \tag{8.39}$$

where the particular integral P.I. = 0 for $p = 0$.

To determine the three constants C_2, C_3, and C_4, three boundary conditions are required; two of them are

$$\text{At } r = r_o: w = 0 \text{ and } \frac{dw}{dr} = 0 \tag{8.40}$$

An additional condition results from the equilibrium of an arbitrarily isolated disk at any radius r, Figure 8.16b, where the shear stress resultant from equations 8.31a gives

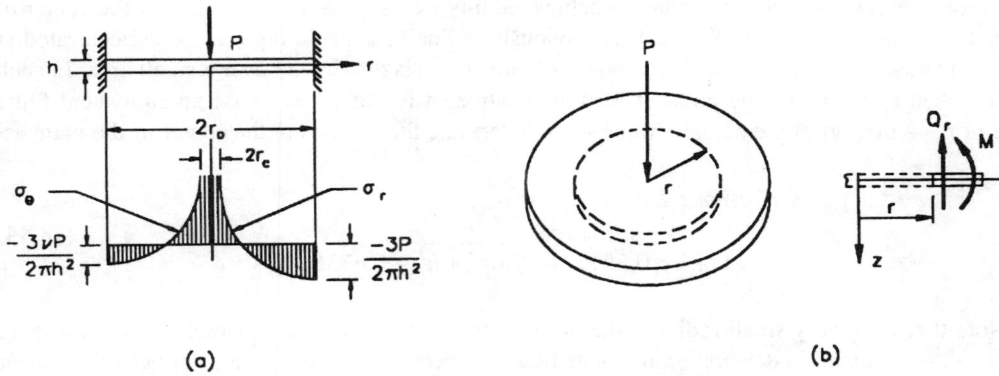

FIGURE 8.16 All-around clamped circular plate subjected to a concentrated force at the center: (a) loading and stress distribution and (b) stress resultants acting at radius *r*.

$$Q_r = -D\frac{d}{dr}\left(\frac{d^2w}{dr^2} + \frac{1}{r}\frac{dw}{dr}\right) = -\frac{P}{2\pi r} \tag{8.41}$$

using this condition with Expression 8.39 yields

$$C_2 = \frac{P}{8\pi D}$$

Hence, applying the boundary conditions (Equations 8.40) yields the other two constants as

$$C_3 = -\frac{P}{16\pi D}\left[2\ln(r_o) + 1\right] \text{ and } C_4 = \frac{Pr_o^2}{16\pi D}$$

The deflection equation is, thus;

$$w = \frac{P}{16\pi D}\left[2r^2 \ln\left(\frac{r}{r_o}\right) + r_o^2 - r^2\right] \tag{8.42a}$$

with a maximum finite value at the center $r = 0$ given by

$$w_{\max} = \frac{Pr_o^2}{16\pi D} \tag{8.42b}$$

The moments and, hence, the stresses are derived from the deflection Equation 8.42a according to Expressions 8.31a and 8.31b, respectively, as

$$\begin{aligned} \sigma_r &= \frac{3Pz}{\pi h^3}\left[(1+\nu)\ln\left(\frac{r_o}{r}\right) - 1\right] \\ \sigma_\theta &= \frac{3Pz}{\pi h^3}\left[(1+\nu)\ln\left(\frac{r_o}{r}\right) - \nu\right] \end{aligned} \tag{8.43}$$

These stresses attain higher values, reaching infinity as r approaches zero, i.e., at the center of the plate, as shown in Figure 8.16a. This obviously is due to considering a force concentrated at a point, a situation which is unrealistic. Naturally this force is distributed over a small area. In such a case, when approaching the point of load application, it is suggested to use an equivalent finite value of r_c — instead of the actual r value — to determine the stresses at the center of the plate as[5]

$$\begin{aligned} &\text{For } r \geq 0.5h: r_c = r \\ &\text{For } r < 0.5h: r_c = \sqrt{1.6r^2 + h^2} - 0.675h \end{aligned} \tag{8.44}$$

Note that for a very small radius r, the normal stress $[\sigma_z]_{z=h/2} = p/2\pi r_c$ becomes too high to be neglected, as assumed in developing the plate bending theory. The same notice applies to the shear stress resultant Q_r. At the top surface around the point of load application, yielding may be disregarded due to its localized nature. The plate thickness could be based on the maximum tensile stress at the point ($r = 0$, $z = h/2$) according to the expression[3]

$$\sigma_{max} = \frac{P}{h^2}(1+\nu)\left[0.485 \ \ln\left(\frac{r_o}{h}\right) + 0.52\right] \tag{8.45}$$

8.5.1.4 Simply Supported Plate Subjected to a Concentrated Force at the Center

The same procedure may be followed to derive expressions for the deflections and stresses in a simply supported plate subjected to a concentrated force P at its center, Figure 8.17, by using the following boundary conditions:

$$\begin{aligned} &\text{At } r = r_o: \ w = 0, \ \ M_r = 0 \\ &\text{At any } r: \ Q_r = P/2\pi r \end{aligned} \tag{8.46}$$

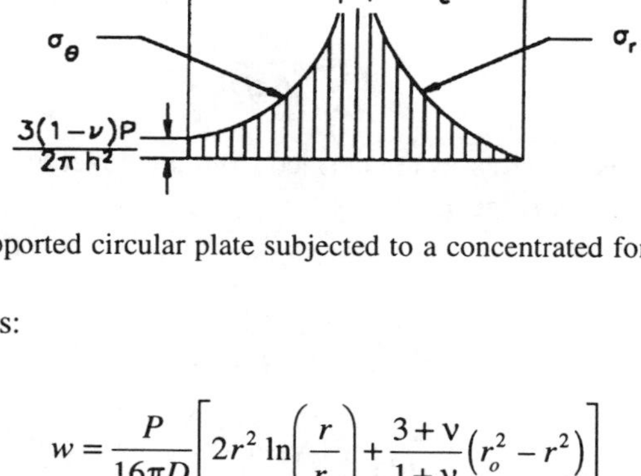

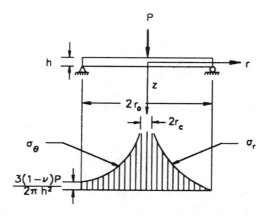

FIGURE 8.17 Simply supported circular plate subjected to a concentrated force at the center.

This gives the expressions:

$$w = \frac{P}{16\pi D}\left[2r^2 \ln\left(\frac{r}{r_o}\right) + \frac{3+\nu}{1+\nu}\left(r_o^2 - r^2\right)\right] \tag{8.47a}$$

with a maximum value at $r = 0$ as

$$w_{\max} = \frac{3+\nu}{1+\nu}\frac{Pr_o^2}{16\pi D} \tag{8.47b}$$

The stresses are thus derived using Equation 8.31 as

$$\begin{aligned}\sigma_r &= \frac{3Pz}{\pi h^3}(1+\nu)\ln\left(\frac{r_o}{r}\right)\\ \sigma_\theta &= \frac{3Pz}{\pi h^3}\left[(1+\nu)\ln\left(\frac{r_o}{r}\right)+(1-\nu)\right]\end{aligned} \tag{8.48}$$

To avoid the singularity at the point of load application, the following approximate expression may be used to give the maximum tensile stress at ($r = 0$, $z = h/2$), as

$$\sigma_{\max} = \frac{P}{h^2}\left[(1+\nu)\left(0.485\ln\left(\frac{r_o}{h}\right)+0.52\right)+0.48\right] \tag{8.49}$$

8.5.2 Annular Circular Plates

8.5.2.1 Simply Supported Annular Plate Subjected to Edge Moments

A simply supported annular plate subjected to edge moments M_o are M_i per unit circumference at the outer and inner circles ($r = r_o$) and ($r = r_i$), respectively, is shown in Figure 8.18. The plate is free of shearing forces and hence, from Equations 8.31,

$$Q_r = 0 = D\frac{d}{dr}\left[\frac{d^2w}{dr^2}+\frac{1}{r}\frac{dw}{dr}\right]$$

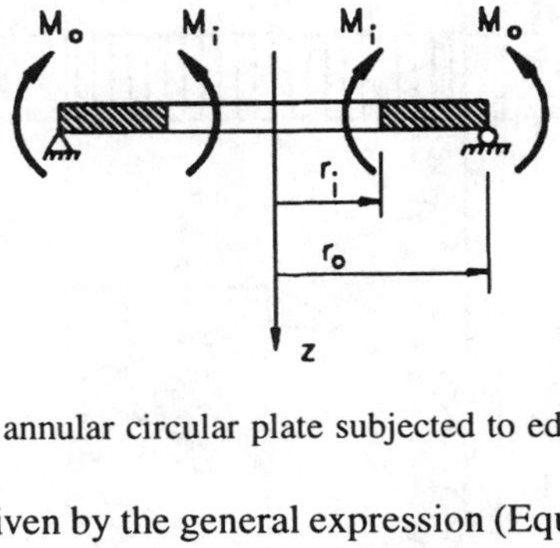

FIGURE 8.18 Simply supported annular circular plate subjected to edge moments.

The function representing w is given by the general expression (Equation 8.30). The above condition yields $C_2 = 0$, and, hence,

$$w = C_1 \ln r + C_3 r^2 + C_4 \tag{8.50}$$

where the particular integral P.I. is set to zero for this type of loading.

The constants C_1, C_3, and C_4 require three boundary conditions, namely,

$$\begin{aligned} &\text{At } r = r_o: \ w = 0, \ M_r = M_o \\ &\text{At } r = r_i: \ M_r = M_i \end{aligned} \tag{8.51}$$

from which the following expressions are obtained:

$$w = \frac{1}{D(r_o^2 - r_i^2)}\left[\frac{r^2 - r_o^2}{2(1+\nu)}\left(M_i r_i^2 - M_o r_o^2\right) + \frac{r_i^2 r_o^2 \ln(r/r_o)}{(1-\nu)}\left(M_i - M_o\right)\right] \tag{8.52}$$

$$\begin{aligned} \sigma_r &= \frac{12z}{h^3(r_o^2 - r_i^2)}\left[\frac{r_o^2 r_i^2}{r^2}(M_i - M_o) - (M_i r_i^2 - M_o r_o^2)\right] \\ \sigma_\theta &= -\frac{12z}{h^3(r_o^2 - r_i^2)}\left[\frac{r_o^2 r_i^2}{r^2}(M_i - M_o) + (M_i r_i^2 - M_o r_o^2)\right] \end{aligned} \tag{8.53}$$

For a solid plate subjected to edge moment M_o at r_o, Figure 8.19, Expressions 8.52 and 8.53 are simplified by putting $r_i = 0$ and $M_i = 0$; hence,

$$w = \frac{M}{2D(1+\nu)}\left(r_o^2 - r^2\right) \tag{8.54}$$

and

$$\sigma_r = \sigma_\theta = \frac{M_o z}{h^3/12} \tag{8.55}$$

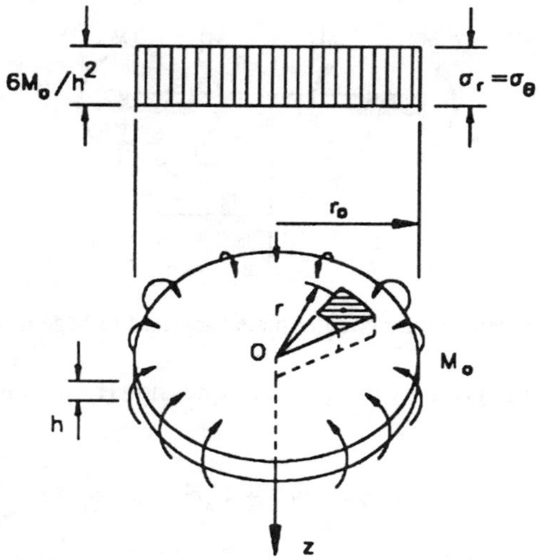

FIGURE 8.19 Simply supported solid circular plate subjected to edge moments.

8.5.2.2 Simply Supported Annular Circular Plate Subjected to a Shearing Force at the Inner Edge

The plate is shown in Figure 8.20 and is subjected to a force Q_i per unit length of the inner circumference. Hence, from equilibrium the shearing force Q_r at any radius r is given by

$$Q_r = -Q_i r_i / r \tag{8.56}$$

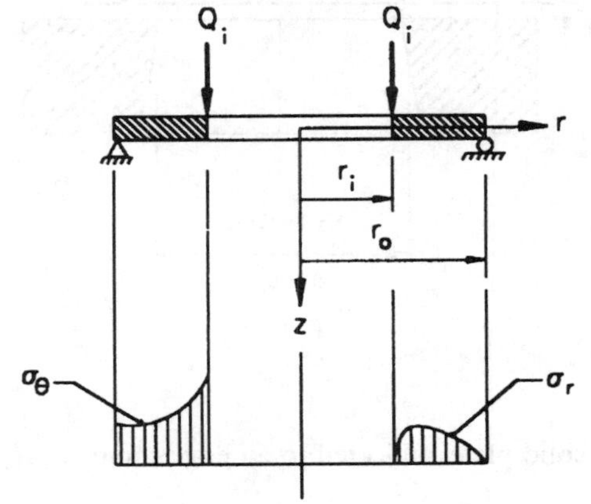

FIGURE 8.20 Simply supported annular circular plate subjected to shearing load at the inner edge.

The three boundary conditions:

$$\begin{aligned} &\text{At } r = r_o : \ w = 0, \ M_r = M_o \\ &\text{At } r = r_i : \ M_r = M_i \end{aligned} \tag{8.57}$$

have to be satisfied together with Condition 8.56. This gives, after determination of the constants C_1, C_2, C_3, and C_4 in Expression 8.30, the deflection equation as

$$\begin{aligned} w = \frac{Q_i r_o^2 r_i}{4D}\Bigg[&\left(1 - \frac{r^2}{r_o^2}\right)\left(\frac{3+\nu}{2(1+\nu)} - \frac{r_i^2 \ln(r_i/r_o)}{r_o^2 - r_i^2}\right) \\ &+ \frac{r^2 \ln(r/r_o)}{r_o^2} + \frac{1+\nu}{1-\nu}\frac{2r_i^2}{r_o^2 - r_i^2}\ln(r_i/r_o)\ln(r/r_o)\Bigg] \end{aligned} \tag{8.58}$$

The moments and stresses are derived from Equation 8.58 as in the previous cases of loading. If r_i becomes very small, the term $r_i^2 \ln(r_i/r_o)$ vanishes, and Expressions 8.47a for a solid plate subjected to a central concentrated force $P = 2\pi r_i Q_i$ are obtained.

Example 8.6:

Consider as a practical application the case of blanking a disk by shearing as shown in Figure 8.21. As a result of the clearance between the punch and the die, the blank will be subjected to edge bending. Let the plate be of steel having shear strength τ_{ult} = 400 MPa, yield strength in tension Y = 370 MPa, thickness 2 mm, diameter 30 mm, diametral clearance between punch and die Δ = 0.150 mm. Show that the plate is produced without permanent edge curvature.

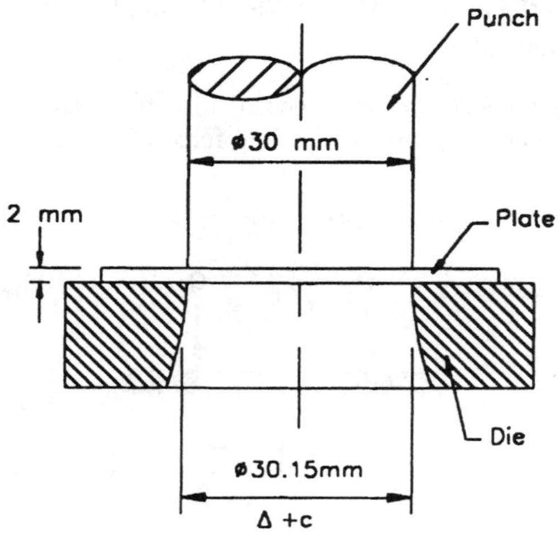

FIGURE 8.21 Example 8.6.

Solution:

This is a case of a circular solid plate subjected to an edge moment M_o given by

$$M_o = \tau_{ult} h \frac{\Delta}{2} = 400 \times 2 \times \frac{0.15}{2} = 60 \text{ N} \cdot \text{mm}$$

From Equation 8.55, the maximum stresses are

$$(\sigma_r)_{max} = (\sigma_\theta)_{max} = \frac{6M_o}{h^2} = \frac{6 \times 60}{2^2} = 90 \text{ MPa}$$

which is less than the material yield strength (Y = 370 MPa). This means that the disk remains elastic during blanking and hence there is no permanent edge curvature.

8.5.3 Other Loadings and Edge Conditions

Solutions to other loadings and edge conditions can be obtained as superpositions of the cases dealt with in Sections 8.5.1 and 8.5.2, while satisfying appropriate conditions of continuity of deflections. Examples of these cases are illustrated in Figure 8.22.

For design purposes, the results for the maximum stresses and maximum central deflections of some practical cases of annular disks are presented by expressions such as

$$\sigma_{max} = C_1 \frac{Pr_o^2}{h^2}, \quad w_{max} = C_2 \frac{Pr_o^4}{Eh^3} \tag{8.59}$$

for a uniformly loaded plate, and

$$\sigma_{max} = C_1 \frac{P}{h^2}, \quad w_{max} = C_2 \frac{Pr_o^2}{Eh^3} \tag{8.60}$$

for a concentrated force loading.

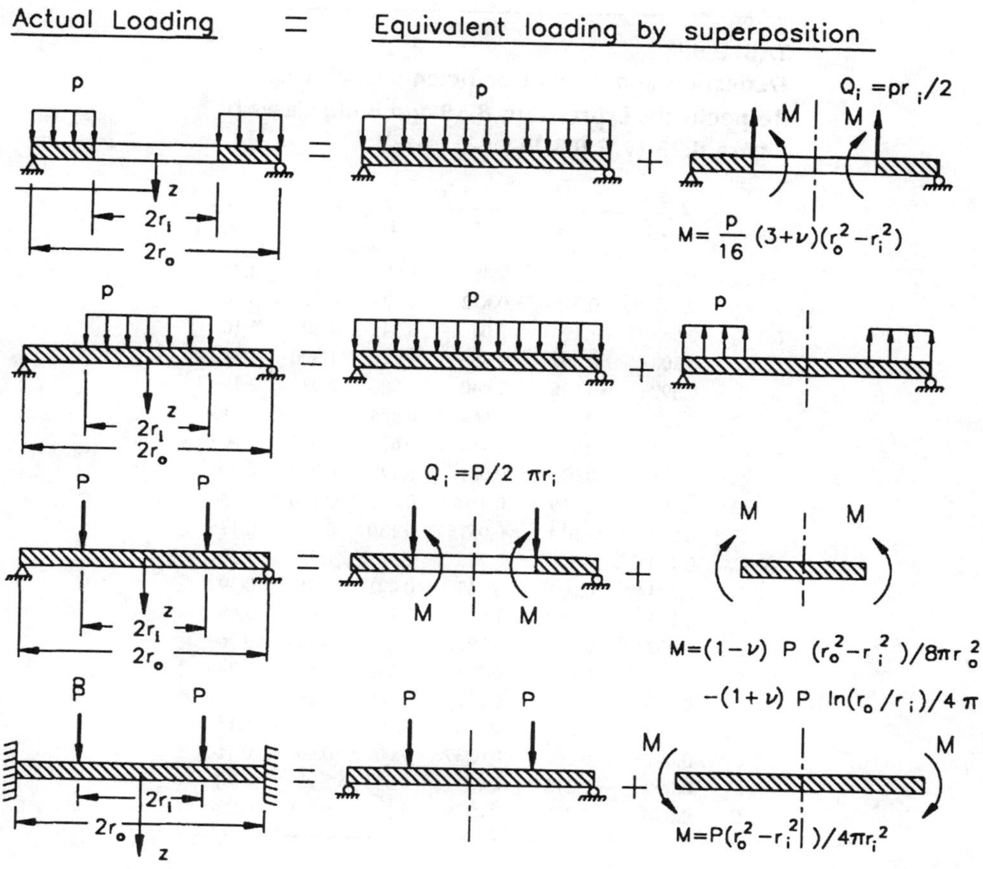

FIGURE 8.22 Equivalent loading for circular plates obtained by superposition of different loadings and edge conditions.

The coefficients c_1 and c_2 are given in Table 8.5, for $\nu = 0.3$.

8.5.4 Thermal Stresses in Circular Plates

Axisymmetric temperature variation in circular plates may occur along either the thickness or the radial direction, or along both directions simultaneously.

8.5.4.1 Temperature Gradient Across the Thickness of a Disk with Free Edges

The analysis given in Section 8.3.5 applies to this case. The disk is free of stresses, i.e., in terms of polar coordinates (r, θ); $\sigma_r = \sigma_\theta = 0$. The plate bends convexly to a spherical surface with a curvature of $\alpha(T_2 - T_1)/h$ and maximum deflection at the center according to Equations 8.16 as

$$w_{\max} = \alpha r_o^2 \left(T_2 - T_1\right)/2h$$

where $T_2 > T_1$ and the temperature variation is assumed to be linear.

TABLE 8.5
Deflection and Stress Coefficients C_1 and C_2 Respectively, Expressions 8.59 and 8.60; Cases of Figure 8.23 ($\nu = 0.3$)[3,5]

	r_o/r_i					
Case	1.25	1.5	2	3	4	5
(a) C_1	1.10	1.26	1.48	1.88	2.17	2.34
C_2	0.341	0.519	0.672	0.734	0.724	0.704
(b) C_1	0.66	1.19	2.04	3.34	4.30	5.10
C_2	0.202	0.491	0.902	1.220	1.300	1.310
(c) C_1	0.592	0.976	1.440	1.880	2.08	2.19
C_2	0.184	0.414	0.664	0.824	0.830	0.813
(d) C_1	0.194	0.320	0.454	0.673	1.021	1.305
C_2	0.005	0.024	0.081	0.172	0.217	0.238
(e) C_1	0.105	0.259	0.480	0.657	0.710	0.730
C_2	0.002	0.014	0.0575	0.130	0.162	0.175
(f) C_1	0.122	0.336	0.74	1.21	1.45	1.59
C_2	0.0034	0.0313	0.125	0.291	0.417	0.492
(g) C_1	0.135	0.410	1.04	2.15	2.99	3.69
C_2	0.0023	0.0183	0.0938	0.293	0.448	0.564
(h) C_1	0.227	0.428	0.753	1.205	1.514	1.745
C_2	0.0051	0.025	0.0877	0.209	0.293	0.350
(i) C_1	0.115	0.220	0.405	0.703	0.933	1.13
C_2	0.0013	0.0064	0.0237	0.062	0.092	0.114
(j) C_1	0.090	0.273	0.71	1.54	2.23	2.80
C_2	0.00077	0.0062	0.0329	0.110	0.179	0.234

8.5.4.2 Temperature Gradient Across the Thickness of a Disk with All-Around Clamped Edges

The bending moment required to eliminate the curvature produces linearly distributed stresses across the thickness as given by Equation 8.18, namely,

$$(\sigma_r)_{\max} = (\sigma_\theta)_{\max} = \pm\frac{\alpha E(T_2 - T_1)}{2(1-\nu)} \tag{8.61}$$

which is compressive on the hot face and tensile on the cold face.

8.5.4.3 Axisymmetric Radial Temperature Gradient

This case has been analyzed in Section 4.9.4 and the stresses are given by Equations 4.45 as

$$\sigma_r = \frac{A}{r^2} + 2C - \frac{\alpha E}{r^2}\int Trdr$$

$$\sigma_\theta = -\frac{A}{r^2} + 2C - \alpha E\left[T - \frac{1}{r^2}\int Trdr\right]$$

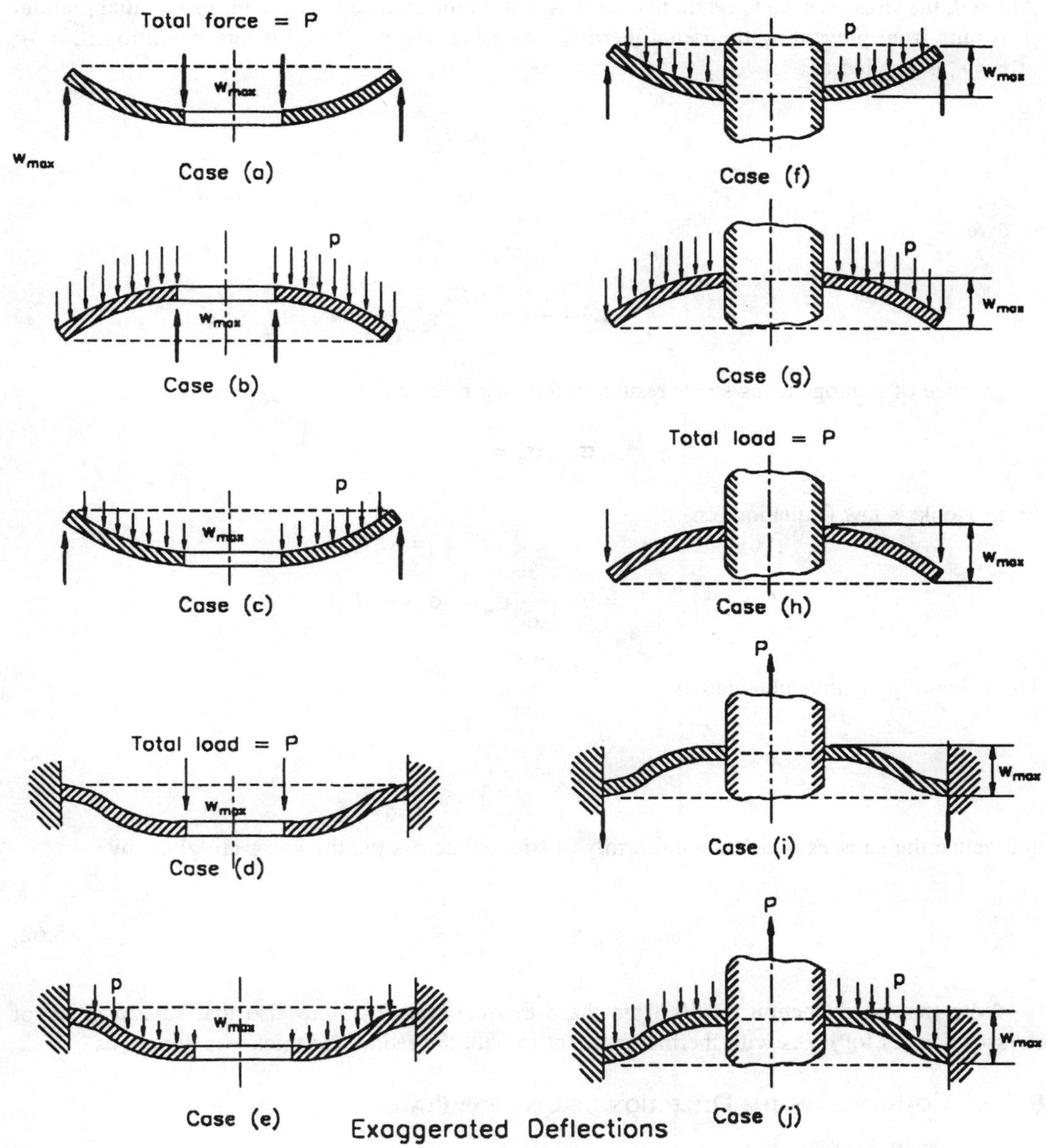

FIGURE 8.23 Loading and edge conditions for the cases of circular plates considered in Table 8.5.

where A and C are constants to be determined from the boundary conditions. To illustrate this procedure, consider the simple case of a disk with all-around clamped edges subjected to a uniform temperature increase of T_o. Hence,

$$\sigma_r = \frac{A}{r^2} + 2C - \frac{\alpha E}{r^2} T_o \int r dr = \frac{A}{r^2} + 2C - \frac{\alpha E T_o}{2}$$

$$\sigma_\theta = -\frac{A}{r^2} + 2C - \alpha E T_o \left[1 - \frac{1}{r^2} \int r dr \right] = -\frac{A}{r^2} + 2C - \frac{\alpha E T_o}{2}$$

At $r = 0$, the stresses have to be finite; hence, $A = 0$. At the clamped edge, a uniform radial pressure p_o results from preventing the radial thermal expansion. Hence, the boundary condition $\sigma_r = -p_o$ at $r = r_o$ is applied as

$$-p_o = 2C - \frac{\alpha E T_o}{2}$$

to give

$$2C = -p_o + \frac{\alpha E T_o}{2}$$

A state of homogeneous stress results in the disk as given by

$$\sigma_r = \sigma_\theta = -p_o$$

From Hooke's law (Equation 3.6):

$$\varepsilon_\theta = \left[\frac{u}{r}\right]_{r=r_o} = 0 = \left[\frac{1}{E}\left(\sigma_\theta - \nu\sigma_r\right) + \alpha T_o\right]_{r=r_o}$$

The value of p_o is thus obtained as

$$p_o = \frac{\alpha E T_o}{1-\nu}$$

and hence the stresses which are uniformly distributed across the thickness are given by

$$\sigma_r = \sigma_\theta = -\frac{\alpha E T_o}{1-\nu} \tag{8.62}$$

Note that no deflection results from these thermal stresses. Note also that superposition of stresses applies for disks with thermal gradient in both the radial and thickness directions.

8.5.5 Comments on the Deflection of Circular Plates

8.5.5.1 Deflection Due to Shear

In the previous sections, the shear strain and stress γ_{rz} and τ_{rz}, respectively, have been neglected irrespective of the fact that the shear stress resultant Q_r (causing τ_{rz}) is included in the analysis to maintain equilibrium of the plate.

It is instructive to show that the effect of this shear stress is negligible on the deflection of plates of small ratios of h/r_o. Consider the case of all-around clamped plate subjected to uniform pressure, Figure 8.24. At any radius r, the shear stress resultant is given by

$$Q_r = -pr/2$$

and, hence, considering parabolic distribution of τ_{rz} through the thickness, the maximum value of τ_{rz} occurs at $z = 0$, as

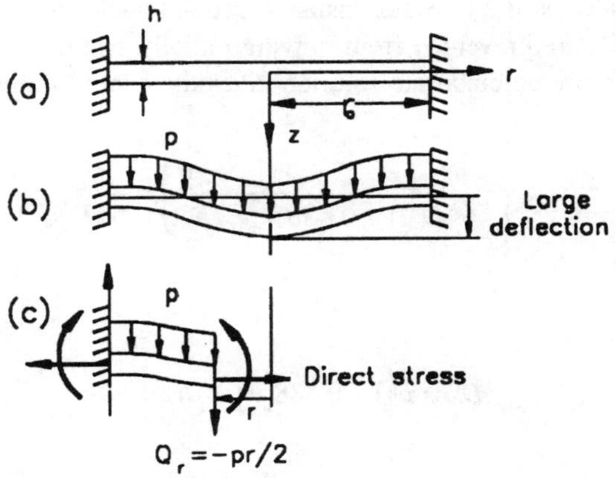

FIGURE 8.24 Deflection of circular plates: (a) reference axes, (b) deflection curve, and (c) stress resultants.

$$\left(\tau_{rz}\right)_{max} = -\frac{3}{2}\frac{pr}{2h}$$

The corresponding shear strain γ_{rz} is

$$\gamma_{rz} = \frac{dw_s}{dr} = -\frac{3}{2}\frac{pr}{2hG}$$

where w_s is the plate deflection due to shear stress only. Hence, by integration,

$$w_s = \frac{3}{2}\frac{p}{4Gh}\left(r_o^2 - r^2\right) \tag{8.63a}$$

with a maximum value at the center as:

$$\left(w_s\right)_{max} = \frac{3}{2}\frac{pr_o^2}{4hG} \tag{8.63b}$$

Comparing the maximum deflection due to shear $(w_s)_{max}$ to $(w)_{max}$ due to bending, Equation 8.37b, the ratio $(w_s/w)_{max}$ is given by

$$\left(\frac{w_s}{w}\right)_{max} = \left(\frac{3}{2}\frac{pr_o^2}{4hG}\right) \bigg/ \left(\frac{12\left(1-\nu^2\right)pr_o^4}{64Eh^3}\right) = \frac{4}{1-\nu}\left(\frac{h}{r_o}\right)^2 \tag{8.63c}$$

For plates of small values of h/r_o, say, $\cong 0.01$ with $\nu = 0.3$, this ratio is less than 5%, which justifies neglecting the deflection due to shear.

8.5.5.2 Large Deflection

The solutions obtained in the previous sections are based on the postulate that the deflection is small compared with the plate thickness. If the deflections are large, the analysis becomes complicated and approximate methods applying the energy approach are used. In this case the middle surface is no longer strain or stress free. For instance, radial tensile stresses arise for an all-around clamped plate as the edges are prevented from moving radially, Figure 8.24c. For a clamped plate subjected to uniform pressure an elaborate solution gives the central deflection as*

$$w^*_{max} = \frac{pr_o^4}{64D}\left(\frac{1}{1+0.488\,w^2_{max}/h^2}\right) = \frac{pr_o^4}{64D^*} \tag{8.64}$$

where

$$D^* = D\left[1 + 0.488\left(w^*_{max}/h\right)^2\right]$$

The deflection w^*_{max} is the summation of two terms; the first is given by Equation 8.37b as derived by considering small deflections, while the second represents the effect of stretching of the middle plane. In other words, the overall plate rigidity D^* increases with the deflection. Also, note that the plate central deflection is no longer proportional to the pressure p as in the case of small deflections. Table 8.6 indicates that for deflections of the order of up to one quarter of the plate thickness, the small deflection theory gives satisfactory results for engineering applications.

TABLE 8.6
Effect of Deflection to Thickness Ratio on Plate Rigidity

w_{max}/h	0.1	0.25	0.5	1.0
D^*/D	1.005	1.031	1.122	1.488

Example 8.7:

A transducer measures pressure through the deflection of a thin diaphragm made of a high-strength alloy steel as shown in Figure 8.25. The maximum pressure and displacement to be measured are 0.5 MPa and 0.2 mm, respectively.

a. *Select appropriate dimensions for a diaphragm operating at room temperature. Take for the steel of the diaphragm:* E = *195 GPa, yield strength* Y = *1034 MPa, and* ν = *0.3.*
b. *Assuming all-around solid clamped plate, calculate the additional thermal stresses for a temperature difference of 30°C between faces.*
c. *For a uniform temperature increase of 30°C, °C = 12 × 10⁻⁶/°C.*

Solution:

a. *Room temperature*: the loading and boundary conditions of the diaphragm is represented by case (j) of Table 8.5. Let $r_o/r_i = 3$; hence,

$$w_{max} = 0.11\frac{pr_o^4}{Eh^3}$$

or

* This is derived using an energy approach for ν = 0.3. See Timoshenko and Woinowsky-Krieger.[3]

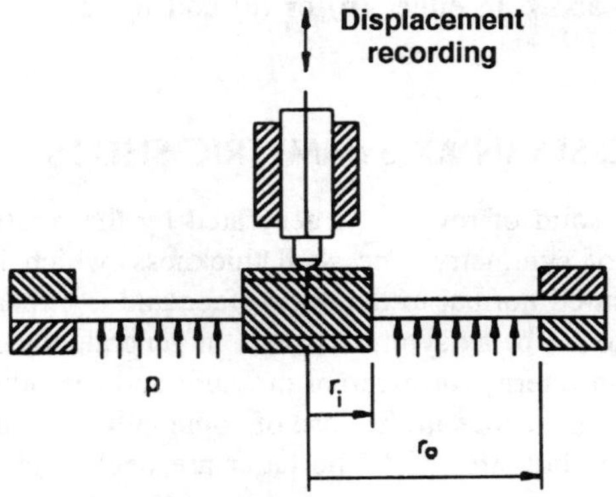

FIGURE 8.25 Example 8.7.

$$h = \frac{w_{max} E}{0.11 p (r_o/h)^4}$$

The ratio $h/r_o = 0.05$ is selected, thus remaining within the validity range of considering bending deflections only; hence, for $w_{max} = 0.2$ mm,

$$h = \frac{0.2 \times 195 \times 10^9}{0.11 \times 0.5 \times 10^6 \times (20)^4} = 4.43 \text{ mm, say, } 4.0 \text{ mm}$$

The ratio $w_{max}/h = 0.2/4 = 0.05$ is well within the small deflection analysis as may be seen from Table 8.6. To check the stresses in the diaphragm, Table 8.5 gives

$$\sigma_{max} = 1.54 \frac{p r_o^2}{h^2} = 1.54 \times 0.5 \times 10^6 (20)^2 = 308 \text{ MPa}$$

which assures a factor of safety = 1034/308, i.e., about 3.4.

b. *Temperature difference of 30°C between faces*: Due to this temperature difference bending thermal stresses σ_r^* and σ_θ^* result at the upper and lower faces according to Equation 8.61:

$$(\sigma_r^*)_{max} = (\sigma_\theta^*)_{max} = \pm \frac{1}{2(1-\nu)} \Delta T = \frac{12 \times 10^{-6} \times 195 \times 10^3}{2(1-0.3)} \times 30 = 50.14 \text{ MPa}$$

negative on the hot face and positive on the cold face.

c. *Uniform temperature increase of 30°C*: This adds up thermal compressive stresses according to Equation 8.62, i.e.,

$$\sigma_r^* = \sigma_\theta^* = -\frac{\alpha E \Delta T}{(1-\nu)} = \frac{12 \times 10^{-6} \times 195 \times 10^3 \times 30}{(1-0.3)} = 100.29 \text{ MPa}$$

These represent a state of homogeneous stress; uniformly distributed across the plate thickness.

The stresses calculated above in either (b) or (c) add up to the stresses due to the applied pressure p.

8.6 MEMBRANE STRESSES IN AXISYMMETRIC SHELLS

An axisymmetric shell is a solid of revolution generated by the rotation of a continuous curve (meridian) around an axis of symmetry. The shell thickness, which is small compared with its radius of curvature, is measured normal to the midsurface and is either constant or varying along the meridian. Loading is taken to be axisymmetric and, in general, consists of a component normal to the midsurface, usually an internal or external pressure and the other component acting along the midsurface as a shear drag, such as in the case of some bulk containers. The resulting stresses comprise membrane and bending stresses.* The latter are negligible with respect to the former when the meridian is a continuous curve with no abrupt change in thickness or load distribution with edges free from bending or lateral constraint (see Example 8.8). As previously shown in Section 8.2 for flat plates, the stresses normal to the surface are neglected with respect to the other stress components.

8.6.1 AXISYMMETRIC SHELLS SUBJECTED TO UNIFORM PRESSURE

Consider a thin-walled axisymmetric container subjected to an internal pressure p_i. The thin-walled profile is a general continuous curve with no abrupt changes in curvature or thickness as shown in Figure 8.26a. Under such conditions, only membrane stresses prevail throughout the unconstrained surfaces of the shell.

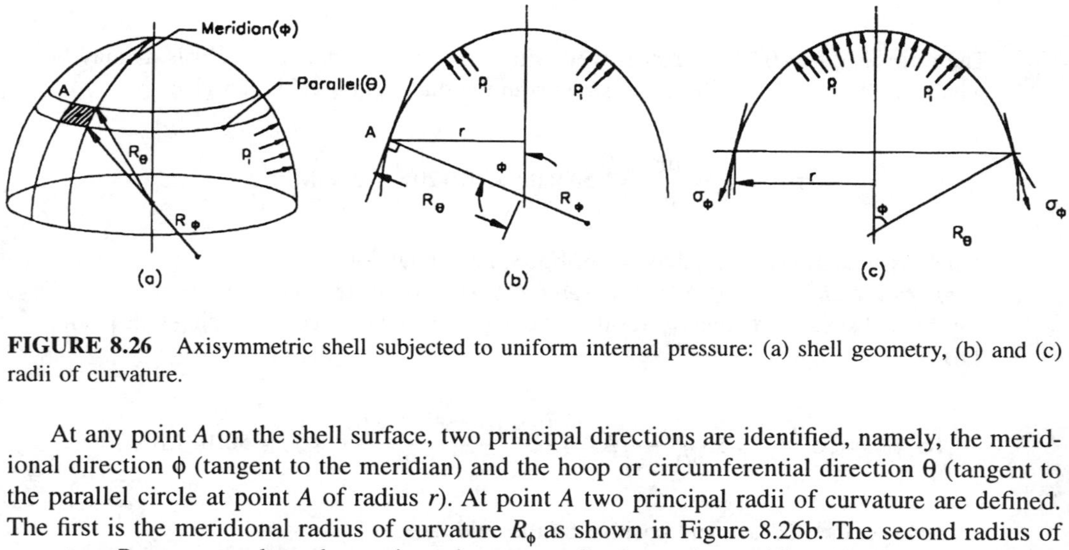

FIGURE 8.26 Axisymmetric shell subjected to uniform internal pressure: (a) shell geometry, (b) and (c) radii of curvature.

At any point A on the shell surface, two principal directions are identified, namely, the meridional direction ϕ (tangent to the meridian) and the hoop or circumferential direction θ (tangent to the parallel circle at point A of radius r). At point A two principal radii of curvature are defined. The first is the meridional radius of curvature R_ϕ as shown in Figure 8.26b. The second radius of curvature R_θ corresponds to the section taken perpendicular to the meridian. Hence, R_θ is equal to the length of the normal intercepted between the meridional surface and the axis of symmetry expressed by the relation $R_\theta = r/\sin\phi$, Figure 8.26c.

Consider an element of the shell surface, $ds_1 \times ds_2$, bound by two meridians and two parallels as shown in Figure 8.26a. In axisymmetric problems of shells of revolution, no shear forces exist

* The membrane stress resultants developed in shells are N_x, N_y, and N_{xy}, as given by Expression 8.6a where z is taken across the shell thickness. These definitions neglect the initial curvature of the shell which indicates that the sides of a shell element are no longer rectangular but trapezoidal. For shells of large radius of curvature compared with its thickness, these definitions are reasonably accurate.

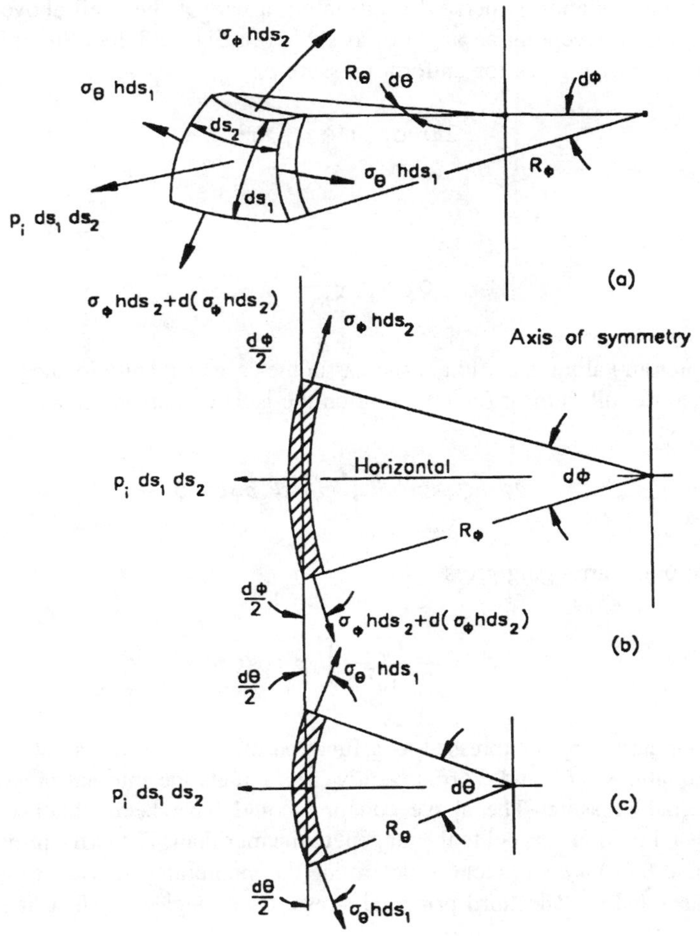

FIGURE 8.27 Static equilibrium of a surface element in the axisymmetric shell of Figure 8.26: (a) static equilibrium of the element, (b) stresses in meridional plane, and (c) stresses in a circumferential plane.

and only two normal principal stress components are to be determined. These stresses, namely, the meridional stress σ_ϕ and the hoop stress σ_θ, are shown in Figure 8.27a. Considering the equilibrium of forces along the direction perpendicular to the shell surface, Figure 8.27b and c and neglecting quantities of orders higher than one, yields

$$p_i ds_1 ds_2 - 2\sigma_\phi h ds_2 \sin\left(\frac{d\phi}{2}\right) - 2\sigma_\theta h ds_1 \sin\left(\frac{d\theta}{2}\right) = 0$$

Since $ds_1 = R_\phi d_\phi$, $ds_2 = R_\theta d_\theta$, $\sin\,(d_\phi/2) \cong d_\phi/2$, and $\sin(d_\theta/2) \cong d_\theta/2$, this equation of equilibrium simplifies to

$$\frac{\sigma_\phi}{R_\phi} + \frac{\sigma_\theta}{R_\theta} = \frac{p_i}{h} \tag{8.65}$$

A second equilibrium equation is derived by isolating a part of the shell above a given parallel circle of radius r and intercepting an angle ϕ, as shown in Figure 8.26c. Summing up the forces along the axis of symmetry gives for uniform pressure p_i:

$$2\pi rh\sigma_\phi \sin\phi = p_i\pi r^2$$

Hence,

$$\sigma_\phi = p_i \frac{r}{2h\sin\phi} \tag{8.66a}$$

For nonuniform pressure along a meridian, such as in the case of gravity loading, p_i is not constant and varies along ϕ. Equilibrium of forces acting on the isolated part in the axial direction gives

$$2\pi rh\sigma_\phi \sin\phi = \int_0^\phi p_i 2\pi R_\phi d\phi \cos\phi$$

Putting $r = R_\theta \sin\phi$ and arranging gives

$$\sigma_\phi = \frac{R_\phi}{R_\theta \sin^2\phi}\int_0^\phi p_i \cos\phi d\phi \tag{8.66b}$$

which can be integrated if p_i is expressed as a function of ϕ, as will be shown in Section 8.6.3. Note that p_i in Equations 8.65 and 8.66 is positive and equals the internal pressure and is taken negative for external pressure. The above equation could have been obtained by considering equilibrium along a direction parallel to the tangent to the meridian. The two equilibrium conditions (Equations 8.65 and 8.66) are sufficient to determine the membrane stresses σ_θ and σ_ϕ as demonstrated for the cases below. The third principal stress, σ_r, is neglected since it is approximately equal to $-p_i/2$.*

Example 8.8:
A long, open-ended semicircular cylindrical shell is simply supported at the edges and loaded by uniform external pressure as shown in Figure 8.28. Show that there are no bending stresses and that the shell is subjected to only compressive membrane stresses.

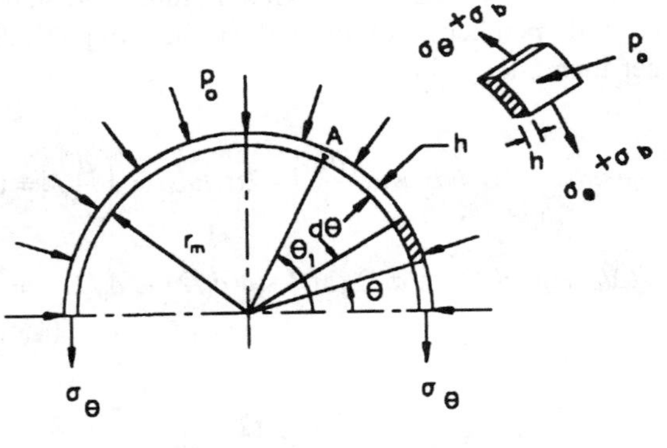

FIGURE 8.28 Example 8.8.

* At the inner and outer surfaces of the shell $\sigma_r = -p_i$, and $\sigma_r = 0$, respectively. An average value for $\sigma_r = -p_i/2$.

Solution:
The membrane compressive stress σ_θ is obtained from Equation 8.65 for $R_\theta = r_m$ as

$$\sigma_\theta = -\frac{p_o r_m}{h}$$

Consider a unit length of the shell. For any point A of Figure 8.28, making an angle θ_1 with the vertical axis, the bending moment at A is given by

$$(M_b)_A = p_o r_m^2(1-\cos\theta_1) - \int_0^{\theta_1} p_o r_m d\theta r_m \sin(\theta_1 - \theta)$$

$$= p_o r_m^2(1-\cos\theta_1) - p_o r_m^2 \int_0^{\theta_1} \sin(\theta_1 - \theta) d\theta$$

$$(M_b)_A = p_o r_m^2(1-\cos\theta_1) - p_o^2 r_m [\cos(\theta_1 - \theta)]_0^{\theta_1} = 0$$

Hence, there is no bending stresses and the shell is subjected only to membrane compressive stress of $\sigma_\theta = -p_o\ (r_m/h)$. This property allows using such a construction in brickwork, particularly in building furnace roofs and openings.*

Example 8.9:
In shells of complex shapes, curvilinear coordinates are often employed. Deduce Equation 8.65 for an axisymmetric thin shell of revolution from the equation of equilibrium (Equation 1.12c).

Solution:
Consider a point $A(r, z)$ on the shell wall and take three orthogonal directions: 1, along the merdian OA; 2, along the circumference of a parallel circle, and 3, normal to the wall passing through A as shown in Figure 8.29.

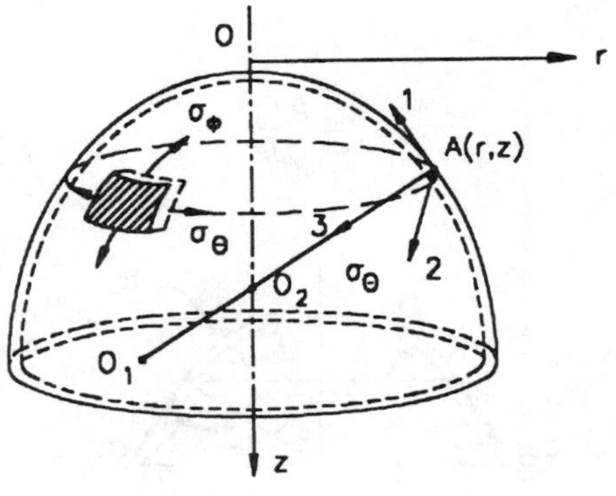

FIGURE 8.29 Example 8.9.

By noting that AO_1 is the meridional radius of curvature R_ϕ with respect to the 1–3 plane and AO_2 is the radius of curvature of the circumferential line R_θ with respect to the 2–3 plane, then

$$r_{13} = R_\phi \text{ and } r_{23} = R_\theta$$

* For more complex geometries and loading conditions, the same argument holds. See Ugural,[7] p. 246.

Substituting into the second of Equation of 1.12c; $\sigma_1 = \sigma_\phi$, $\sigma_2 = \sigma_\theta$, and $\sigma_3 = \sigma_r$ yields

$$\frac{\partial \sigma_r}{\partial \gamma} - \frac{\sigma_r - \sigma_\phi}{R_\phi} - \frac{\sigma_r - \sigma_\theta}{R_\theta} = 0$$

From the solution of Example 1.8

$$\frac{\partial \sigma_r}{\partial \gamma} = -\frac{\partial \sigma_r}{\partial r} = -\frac{p_i}{h} \text{ and } \sigma_r \cong -\frac{p_i}{2}$$

and the equilibrium condition becomes

$$\frac{\sigma_\phi}{R_\phi} + \frac{\sigma_\theta}{R_\theta} = \frac{p_i}{h}\left(1 + \frac{h}{2R_\phi} + \frac{h}{2R_\theta}\right)$$

For a thin-walled shell h/R_ϕ and h/R_θ are negligibly small values, so that Equation 8.65 is obtained.

8.6.2 Applications to Pressurized Containers

8.6.2.1 Spherical Shell

For a spherical shell of mean radius r_m. as shown in Figure 8.30, application of Equation 8.66 gives

$$\sigma_\phi = \frac{\pi(r_m \sin\phi)^2 p_i}{2\pi(r_m \sin\phi) h \sin\phi} = \frac{p_i r_m}{2h} \tag{8.67a}$$

Complete symmetry of the shell gives $R_\phi = R_\theta = r_m$, and hence Equation 8.65 yields $\sigma_\theta = \sigma_\phi$ as expected, i.e,

$$\sigma_\theta = \frac{p_i r_m}{2h} \tag{8.67b}$$

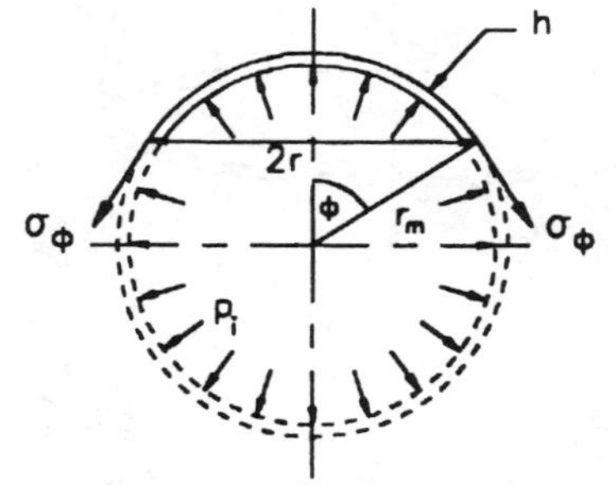

FIGURE 8.30 Stresses in a pressurized spherical thin shell.

8.6.2.2. Circular Cylindrical Shell

For a circular cylindrical shell of radius r_m and length L, as shown in Figure 8.31, by analogy to Equation 8.66, i.e., equilibrium along the shell axis yields

$$\sigma_\phi = \frac{\pi r_m^2 p_i}{2\pi r_m h} = \frac{p_i r_m}{2h} \tag{8.68a}$$

Since $R_\phi = \infty$ and $R_\theta = r_m$, Equation 8.65 gives

$$\sigma_\theta = \frac{p_i r_m}{h} \tag{8.68b}$$

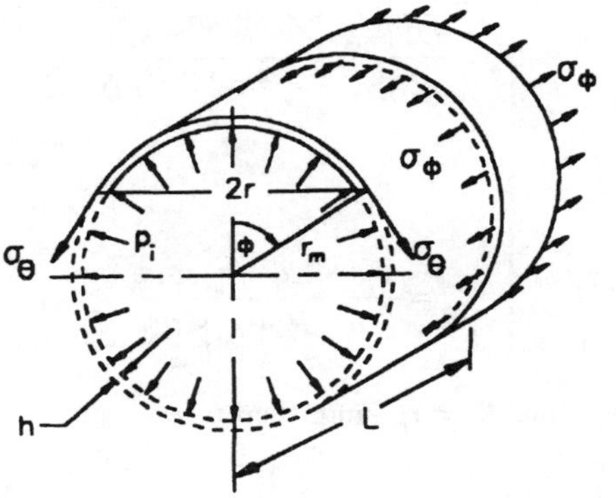

FIGURE 8.31 Stresses in a pressurized cylindrical thin shell.

8.6.2.3 Conical Shell

For a conical shell of apex angle 2α, Figure 8.32, again Equation 8.66 results in

$$\sigma_\phi = \frac{p_i \pi r_m^2}{2\pi r h \sin\phi} = \frac{p_i r_m}{2h\cos\alpha} \tag{8.69a}$$

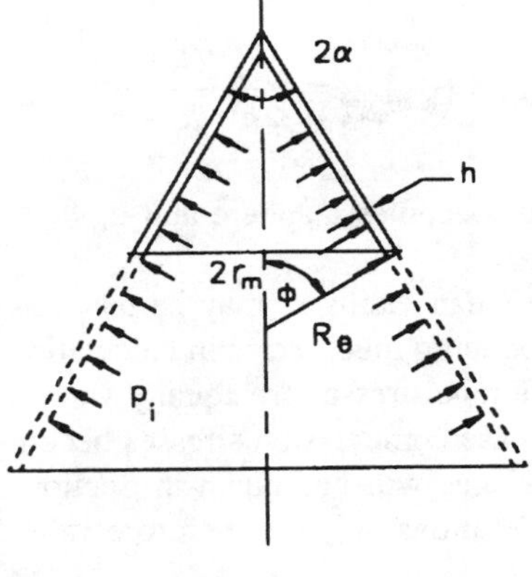

FIGURE 8.32 Stresses in a pressurized conical thin shell.

Since $R_\phi = \infty$ and $R_\theta = r/\cos\alpha$, Equation 8.65 gives

$$\sigma_\theta = \frac{p_i R_\theta}{h} = \frac{p_i r_m}{h\cos\alpha} \tag{8.69b}$$

Equations 8.69 indicate linear stress distribution with r.

8.6.2.4 Toroidal Shell

A torus is generated by rotating a complete circle of radius r_m around an axis located at distance R_i as shown in Figure 8.33a. Isolating an element of the torus bounded by radius R_i from inside and radius ($R_i + r_m$ sin ϕ) from outside, Figure 8.33b, Equation 8.66 gives

$$\sigma_\phi = \frac{p_i\pi\left[\left(R_i + r_m\sin\phi\right)^2 - R_i^2\right]}{2\pi\left(R_i + r_m\sin\phi\right)h\sin\phi}$$

or

$$\sigma_\phi = \frac{p_i r_m}{2h}\,\frac{2R_i + r_m\sin\phi}{R_i + r_m\sin\phi} \tag{8.70a}$$

The meridional radius of curvature $R_\phi = r_m$, and, therefore,

$$R_\theta = \frac{R_i + r_m\sin\phi}{\sin\phi}$$

Substitution into Equation 8.65 yields

$$\sigma_\theta = \frac{p_i r_m}{2h} \tag{8.70b}$$

which is independent of R_i. Note that the maximum meridional stress σ_ϕ occurs at the inner points of the shell, i.e., at $\phi = -\pi/2$ and is given by

$$\left(\sigma_\phi\right)_{\max} = \frac{p_i r_m}{2h}\,\frac{2R_i - r_m}{R_i - r_m}$$

When the radius $R_i = 0$, the torus becomes a sphere and $\sigma_\phi = \sigma_\theta = p_i r_m/2h$. Also, when $R_i \to \infty$, the torus becomes a cylinder.

Note that for thin shells, the mean radius r_m may be taken as equal to either the inner radius r_i or outer radius r_o without introducing much error in calculations.

The above solutions for membrane stresses are equally valid for externally applied pressure by reversing the stress sign. In this case compression stresses become dominant and may lead to the instability of the thin-walled shell, as will be shown in Section 8.8.3. Also, the solutions being derived from equilibrium considerations only are therefore valid for any material behavior.

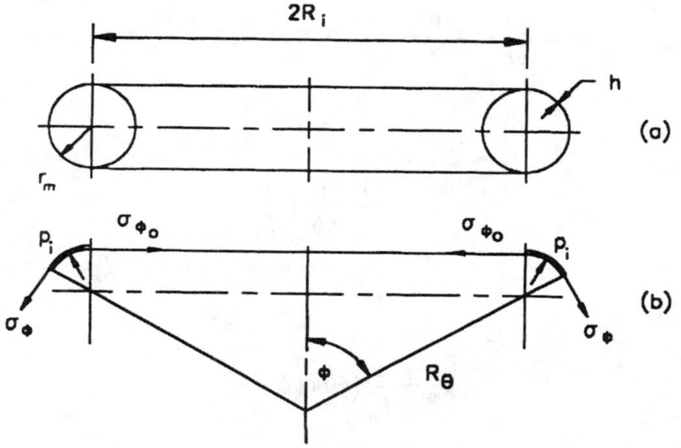

FIGURE 8.33 Stresses in a pressurized toroidal thin shell.

8.6.3 Displacement in Axisymmetric Shells

Considering linear-elastic behavior the radial displacement in pressurized containers is obtained by applying Hooke's law, Equation 3.6 as

$$\varepsilon_\theta = \frac{u}{r} = \frac{1}{E}\left[\sigma_\theta - \nu\sigma_\phi\right]$$

Therefore using Equations 8.67 through 8.69 gives for

$$\text{A cylinder: } u = \frac{p_i r_m^2}{2Eh}(2-\nu) = \text{constant} \tag{8.71a}$$

$$\text{A sphere: } u = \frac{p_i r_m^2}{2Eh}(1-\nu) = \text{constant} \tag{8.71b}$$

or

$$\text{A cone: } u = \frac{p_i r^2}{2Eh\cos\alpha}(2-\nu) \tag{8.71c}$$

Equations 8.71a and b indicate uniform radial displacement, which means that spherical and cylindrical shells retain their geometric form, while Equation 8.71c indicates that the meridional straight line generating the conical shell deforms to a parabola.

In a cylindrical pressure vessel with hemispherical ends, the difference between the radial displacements of Expressions 8.71a and b is a source of discontinuity stresses at the junction of the cylindrical shell with the hemispherical head, as will be shown in Section 8.7.6.

In case of symmetrical deformation of a shell of revolution, the displacement of a point is described by two components, u and w, where u is normal to the middle surface and w is tangent to the meridian. The strain-displacement relations in terms of spherical coordinates, Equation 2.10 give:

$$\varepsilon_r = \frac{\partial u}{\partial r} \varepsilon_\theta = \frac{u}{r} + \frac{\cot\phi}{r} w \text{ and } \varepsilon_\phi = \frac{1}{r}\frac{\partial w}{\partial \phi} + \frac{u}{r}$$

or

$$\varepsilon_\phi = \frac{1}{R_\phi}\frac{dw}{d\phi} + \frac{u}{R_\phi} \tag{8.71d}$$

and

$$\varepsilon_\theta = \frac{1}{R_\theta}(w\cot\phi + u)$$

Hence,

$$\frac{dw}{d\phi} - w\cot\phi = R_\phi\varepsilon_\phi - R_\theta\varepsilon_\theta$$

In terms of the stresses, by means of Hooke's law (3.8)

$$\frac{dw}{d\phi} - w\cot\phi = \frac{1}{E}\left[\sigma_\phi\left(R_\phi + \nu R_\theta\right) - \sigma_\theta\left(R_\theta + \nu R_\phi\right)\right]$$

Denoting the right hand side by $f(\phi)$, hence,

$$\frac{dw}{d\phi} - w\cot\phi = f(\phi)$$

which can be integrated to give:

$$w = \left[\int \frac{f(\phi)}{\sin\phi} d\phi + C\right]\sin\phi \tag{8.71e}$$

The constant of integration is determined by imposing the appropriate boundary condition at the support. Having found w, the component u is obtained from the second of Equations 8.71d.

Example 8.10:

A thin-walled cylinder with closed ends is constructed by spiral butt-welding of a long steel strip such that the weld makes an angle of 30° with the axis of the cylinder, as shown in Figure 8.34. If the inner radius of the cylinder is 500 mm, the internal pressure p_i = 2 MPa and the normal and shear stress in the weld are not to exceed 120 and 80 MPa, respectively, determine the minimum wall thickness.

Solution:

The stresses in the cylinder wall are given by Equation 8.68, for $r_m \cong r_i = 0.5$ m, as

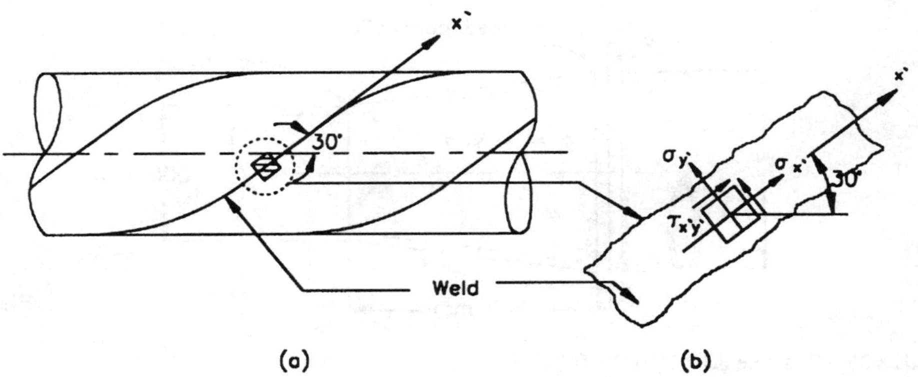

FIGURE 8.34 Example 8.10.

$$\sigma_\phi = \frac{p_i r_m}{2h} = \frac{2 \times 0.5}{2h} = \frac{1}{2h} \text{ MPa } (h \text{ in meters})$$

$$\sigma_\theta = \frac{p_i r_m}{h} = \frac{2 \times 0.5}{h} = \frac{1}{h} \text{ MPa } (h \text{ in meters})$$

Applying the stress transformation law (Equation 1.20) for plane stress gives

$$\sigma_{y'} = \sigma_x \sin^2\theta + \sigma_y \cos^2\theta - 2\tau_{xy} \sin\theta\cos\theta$$

$$\tau_{x'y'} = -\frac{\sigma_x - \sigma_y}{2} \sin 2\theta + \tau_{xy} \cos 2\theta$$

For

$$\sigma_x = \sigma_\phi = 1/2h, \quad \sigma_y = \sigma_\theta = 1/h, \quad \tau_{xy} = 0, \quad \text{and } \theta = 30°$$

$$\sigma_{y'} = \tfrac{7}{8}h \text{ MPa and } \tau_{x'y'} = \sqrt{3}/8h = -\left[\left(\frac{1}{2h} - \frac{1}{h}\right)\Big/2\right]\left(\frac{\sqrt{3}}{2}\right)$$

For a butt weld of a thickness equal to that of the cylinder wall, taking allowable stresses of 120 and 80 MPa for normal and shear stress, respectively, results in

$$120 = 7/8h \quad \text{giving } h = 7.29 \text{ mm}$$

$$80 = \sqrt{3}/8h \text{ giving } h = 2.7 \text{ mm}$$

A value of h = 7.29 mm is thus recommended.

Example 8.11:

A conical reducer connects two pipes of 600 and 300 mm diameters over a length of 900 mm, as shown in Figure 8.35. Assuming that the flanged joint takes all the axial load due to an internal pressure of 4 MPa, calculate the wall thickness of the two pipes and the conical reducer thickness. Material is steel of yield strength Y = *320 MPa,* E = *200 GPa, and a factor of safety F.S.* = 2. *Calculate the radial displacement at the junction with the small pipe. Discuss the results.*

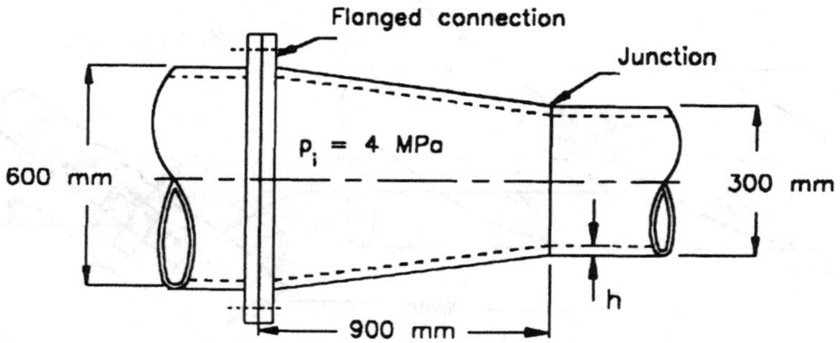

FIGURE 8.35 Example 8.11.

Solution:

Stresses in the pipes are calculated using Equations 8.68

- Large pipe:

$$\sigma_\phi = \frac{p_i r_m}{2h} = \frac{4\times 300}{2\times h} = \frac{600}{h} \text{ MPa } (h \text{ in mm})$$

$$\sigma_\theta = \frac{p_i r_m}{h} = \frac{4\times 300}{h} = \frac{1200}{h} \text{ MPa } (h \text{ in mm})$$

Applying the maximum shear stress theory, Equation 1.32 with $\sigma_r = 0$, yields

$$\frac{Y}{\text{F.S.}} = \frac{320}{2} \geq \sigma_\theta - \sigma_r = \frac{1200}{h} \quad \text{giving } h = 7.5 \text{ mm}$$

- Small pipe: $\sigma_\theta = \dfrac{p_i r_m}{h} = \dfrac{4\times 150}{h} = \dfrac{600}{h}$ MPa (h in mm)

$$\frac{320}{2} \geq \frac{6000}{h} \quad \text{giving } h = 3.75 \text{ mm}$$

For the conical reducer, Equations 8.69 give the stresses for $\alpha = \tan^{-1}(150/900) = 9.46°$ at the large end:

$$\sigma_\theta = \frac{p_i r}{h\cos\alpha} = \frac{4\times 300}{h\cos 9.46°} = \frac{1216.6}{h} \text{ MPa } (h \text{ in mm})$$

Hence,

$$\frac{320}{2} \geq \frac{1216.6}{h} \quad \text{giving } h = 7.6 \text{ mm}$$

which is almost as the thickness of the large pipe.

At the junction with the small pipe, the radial displacements according to Equations 8.71a and b are, respectively,

$$u_{\text{pipe}} = (2-\nu)\frac{p_i r_m^2}{2Eh} - \frac{(2-0.3)\times 4\times(150)^2}{2\times 200\times 10^3\times 3.75} = 0.102 \text{ mm}$$

$$u_{\text{cone}} = (2-\nu)\frac{p_i r^2}{2Eh\cos\alpha}$$

$$= \frac{(2-0.3)\times 4\times(150)^2}{2\times 200\times 10^3\times 7.6\times\cos 9.46°} = 0.051 \text{ mm}$$

This difference in radial displacement at the junction induces local bending stresses, as will be shown in Section 8.7, which may be taken care of by increasing h such that $u_{\text{pipe}} = u_{\text{cone}}$ hence,

$$u_{\text{cone}} = 0.048 = (2-\nu)\frac{p_i r_m^2}{2Eh} = \frac{(2-0.3)\times 4\times(150)^2}{2\times 200\times 10^3\times h}$$

giving $h = 7.97$ mm

This suggests taking a thickness of 8 mm for the whole reducer. The calculated wall thicknesses have to be increased to allow for the metal loss due to corrosion. The corresponding thickness of standard pipes are, respectively, 8 and 10 mm for nominal pipe sizes of diameters 300 and 600 mm, for a test pressure of 8 MPa. Again, discontinuity stresses are not entirely eliminated through wall thickness modification. They are already present at the flange connection, where the pipe is laterally constrained and also at both the cone–pipe junctions, as a result of geometric discontinuity. From the equilibrium of forces at the cone–small pipe junction, the edges will be subjected to an axisymmetric shearing force Q_o/unit length of the circumference given by

$$Q_o = (\sigma_\phi)_{\text{cone}} h\sin\alpha = (p_i r_m/2)\tan\alpha$$

$$= \frac{4\times 150}{2}\tan 9.46° = 50 \text{ N/mm}$$

This edge load will introduce additional bending stresses at the junction.

Example 8.12:

A pressure vessel has end closures in the form of a half-ellipsoid of semiaxes R *and* R_c, *as shown in Figure 8.36. Determine the membrane stresses in the closure wall due to an internal pressure* p_i.

Solution:

First the principal radii of curvature R_ϕ and R_θ have to be determined. The equation of the meridian in polar coordinates (R′, ψ) is given by

$$R'^2 = \frac{R^2 R_c^2}{R^2\sin^2\psi + R_c^2\cos^2\psi}$$

and hence from analytical geometry the meridional radius of curvature R_ϕ is obtained according to the relation:

$$R_\phi = \frac{\left[R'^2 + (dR'/d\psi)^2\right]^{3/2}}{R'^2 + 2(dR'/d\psi) - R'\left(d^2R'/d_\psi^2\right)}$$

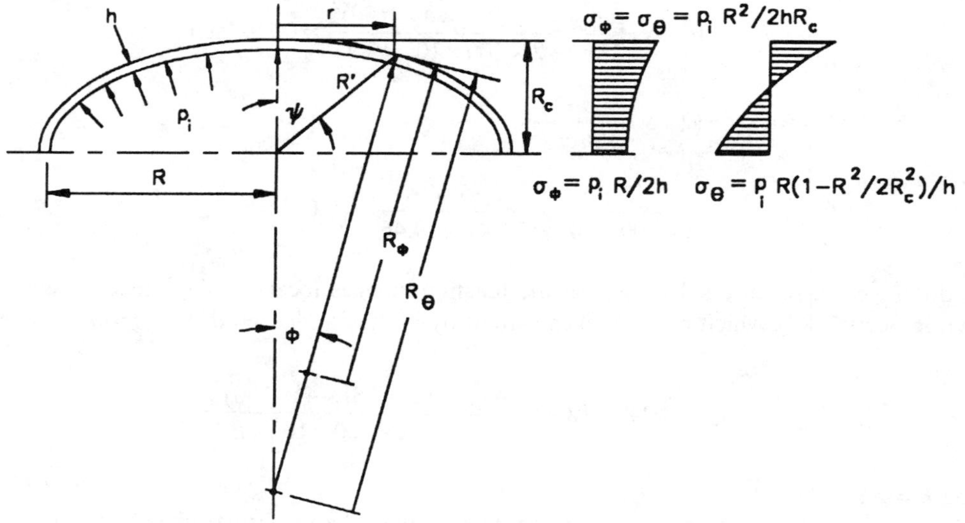

FIGURE 8.36 Example 8.12.

which yields

$$R_\phi = \frac{R^2 R_c^2}{\left[R^2 \sin^2\phi + R_c^2 \cos^2\phi\right]^{3/2}}$$

$$R_\theta = \frac{R^2}{\left[R^2 \sin^2\phi + R_c^2 \cos^2\phi\right]^{1/2}}$$

Hence, from Equation 8.66 at any radius r with $P = \pi r^2 p_i$:

$$\sigma_\phi = \frac{p_i R_\theta}{2h}$$

and hence from Equation 8.65,

$$\sigma_\theta = \frac{p_i}{h}\left(R_\theta - \frac{R_\theta^2}{2R_\phi}\right)$$

At the top of the shell $\phi = 0$ and $R_\phi = R_\theta = R^2/R_c$; hence, the stresses become

$$\sigma_\phi = \sigma_\theta = p_i R^2/2hR_c$$

Along the horizontal semiaxis $\phi = \pi/2$, $R_\phi = R_c^2/R$ and $R_\theta = R$; hence, the stresses are

$$\sigma_\phi = \frac{p_i R}{2h} \text{ and } \sigma_\theta = \frac{p_i R}{2h}\left(2 - \frac{R^2}{R_c^2}\right)$$

The stress σ_ϕ is always tensile while σ_θ changes from tension to compression where $R > \sqrt{2}\,R_c$. For flatter ellipsoids, i.e., $R >> R_c$, the stress σ_θ attains higher compressive values as shown in Table 8.7 and elastic instability may occur. The stress distributions are shown in Figure 8.36.

TABLE 8.7
Dependence of Stress Ratio on the Ratio of Ellipsoid Semiaxis

R/R_c	1	1.25	$\sqrt{2}$	1.5	2
$\sigma_\theta/\sigma_\phi$	1	0.438	0	−0.225	−2

8.6.4 Axisymmetric Shells Subjected to Gravity Loading

Membrane stresses due to gravity loading in axisymmetric thin-walled shells result from either the shell weight and/or the weight of an internal fluid. The pressure acting on the wall as a result of gravity loading is not uniform along the meridian, so that the equations of equilibrium (Equations 8.65 and 8.66b) are applied to solve the problem by following the same procedure as for uniform pressure.

For the equilibrium of gravity loading, the shell has to be supported on legs or skirt-type supports, which are usually attached to a suitable stiffening ring fixed to the shell wall. This introduces discontinuities in both loading and deformation. The location of shell support influences greatly the stress distributions and introduces localized bending stresses, which must be given careful consideration.

8.6.4.1 Hemispherical Liquid Container Freely Supported at Its Top Edge

Consider for example a hemispherical tank filled with a liquid of specific weight ρg, where ρ is mass per unit volume and g is gravitational acceleration. The tank is supported at the top edge as shown in Figure 8.37a. If the tank is freely supported, the discontinuity stresses may be neglected.

For an element at the shell bottom, subtending an angle 2ϕ, Figure 8.37b, the total vertical load due to liquid weight supported by this element, P, is

$$P = \rho g\left\{\pi r_m^3\left[\frac{2}{3}+\frac{1}{3}\cos^3\phi-\cos\phi\right]+\pi r_m^3\sin^2\phi\cos\phi\right\}$$

$$= \frac{2}{3}\rho g\pi r_m^3\left(1-\cos^3\phi\right)$$

Note that this weight corresponds to a total volume equal to that of the spherical cap *abc* added to the cylindrical volume *abcde*, as shown in Figure 8.37b.* σ_ϕ is obtained from the equilibrium of the element as

$$\sigma_\phi = \frac{\tfrac{2}{3}\rho g\pi r_m^3\left(1-\cos^3\phi\right)}{2\pi\left(r_m\sin\phi\right)h\sin\phi} = \frac{\rho g r_m^2}{3h}\frac{1-\cos^3\phi}{\sin^2\phi} \tag{8.72a}$$

* This stems from the fact that the vertical load on the element dA at a level H from the liquid free surface, as shown in Figure 8.37a is: $dP = \rho g\, H(dA\cos\phi)$. Since HdA represents the cylindrical liquid volume above the area dA, the total vertical load is simply equal to the weight of the liquid in the cylindrical volume over the area dA, regardless of its actual form.

The hoop stress σ_θ is obtained from Equation 8.65 knowing that $R_\phi = R_\theta = r_m$ and $p_i = \rho g r_m \cos\phi$, hence,

$$\sigma_\theta = \frac{\rho g r_m^2}{3h}\left(3\cos\phi - \frac{1-\cos^3\phi}{\sin^2\phi}\right) \tag{8.72b}$$

The distribution of the stresses σ_ϕ and σ_θ is shown in Figure 8.37c indicating that at the bottom $\sigma_\phi = \sigma_\theta$. The meridional stress σ_ϕ is tensile all over the shell, whereas the hoop stress σ_θ changes sign at $\phi = 66.50°$; turning to be compressive with a maximum value of $\sigma_\theta = -3\rho g r_m^2/3h$ at the edge of the container. Design to avoid elastic instability due to this compressive stress may be required. The presence of compressive stress may also occur at the junction of a cylindrical filled container with its dished bottom. Hence, stiffening of the junction may be required to take care of the horizontal component of the meridional force due to σ_ϕ (see Example 8.16).

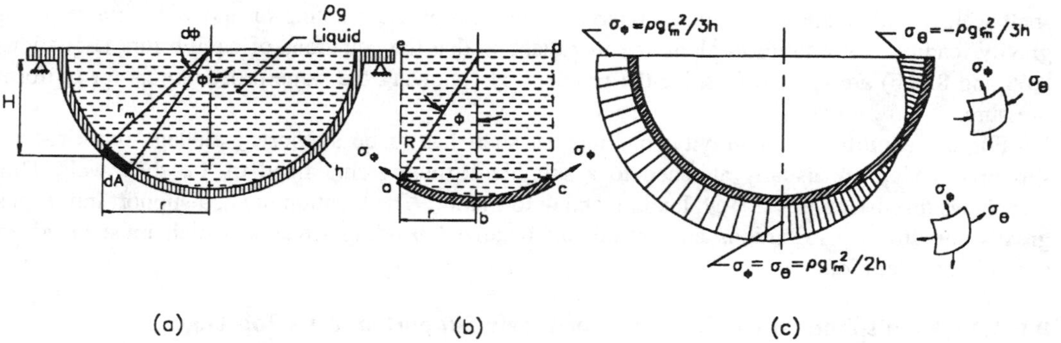

FIGURE 8.37 Stresses in a hemispherical liquid container freely supported at its top edge: (a) container geometry and loading, (b) equilibrium of a bottom element, and (c) stress distribution in the shell.

8.6.4.2 Conical Liquid Container Freely Supported at Its Top Edge

The membrane stresses in a conical container of a vertex semiangle α and uniform wall thickness h due to the liquid height H of specific weight ρg, as shown in Figure 8.38a, are determined by considering that the wall is subjected to a hydrostatic pressure p_i given by the expression:

$$p_i = \rho g(H - y)$$

The container is freely supported at the top edge.

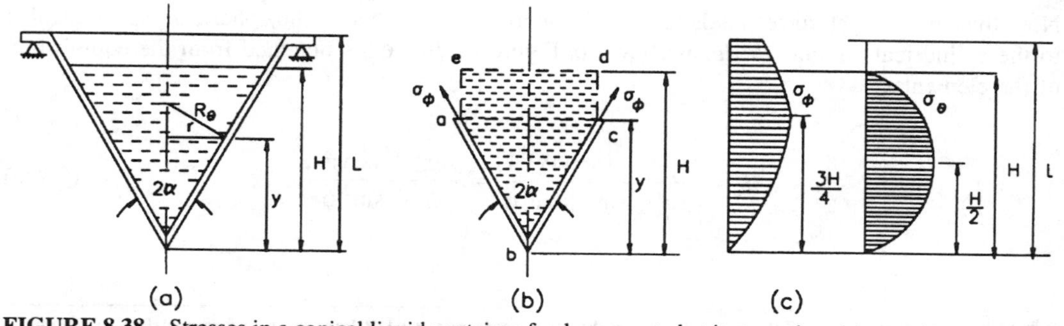

FIGURE 8.38 Stresses in a conical liquid container freely supported at its top edge: (a) container geometry and loading, (b) equilibrium of the lower part, and (c) stress distribution in the shell.

The vertical load on the lower part of the cone at height y is equal to the weight of the liquid of volume *abcde* of Figure 8.38b, i.e.,

$$P = \rho g \pi y^2 (H - y + {}^1\!/_3 y)\tan^2 \alpha$$

From the equilibrium of the lower part of cone at height y,

$$P = 2\pi y \tan \alpha \sigma_\phi h \cos \alpha$$

Hence,

$$\sigma_\phi = \rho g y \left(H - \tfrac{2}{3} y\right) \frac{\tan \alpha}{2h \cos \alpha} \tag{8.73a}$$

which is maximum at $y = 3H/4$ and vanishes at $y = 0$. Substituting σ_ϕ in the equation of equilibrium (Equation 8.65) together with

$$R_\theta = \frac{y \tan \alpha}{\cos \alpha}, \quad \frac{1}{R_\phi} = 0, \text{ and } p_i = \rho g (H - y)$$

gives

$$\sigma_\theta = \rho g y (H - y) \frac{\tan \alpha}{h \cos \alpha} \tag{8.73b}$$

which is maximum at $y = H/2$ and vanishes at $y = 0$ and $y = H$.

For a cone height $L > H$ the internal pressure p_i vanishes at values of $y > H$ and, hence, Equations 8.73a and b are no longer valid to determine σ_ϕ and σ_θ above the liquid level. These stresses are obtained from equilibrium by putting $P = {}^1\!/_3\, \rho g \pi H^3 \tan^2 \alpha$, which corresponds to the total liquid weight, to give

$$\sigma_\phi = \rho g H^3 \frac{\tan \alpha}{6yh \cos \alpha} \tag{8.73c}$$

Substituting σ_ϕ in Equation 8.65 together with $1/R_\phi = 0$ and $p_i = 0$ yields

$$\sigma_\theta = 0 \tag{8.73d}$$

The stress distribution is shown in Figure 8.38c. Note that additional stresses will develop at the top edge according to the constraints imposed by the support. Edge reinforcement is necessary to prevent edge collapse due to the horizontal inward component of σ_ϕ.

Example 8.13:

Determine the membrane stresses in a spherical dome of radius r_m *and uniform thickness* h *due to its own weight* ρgh *per unit surface area as shown in Figure 8.39a, where* ρ *is the density of the dome material. The dome is freely supported.*

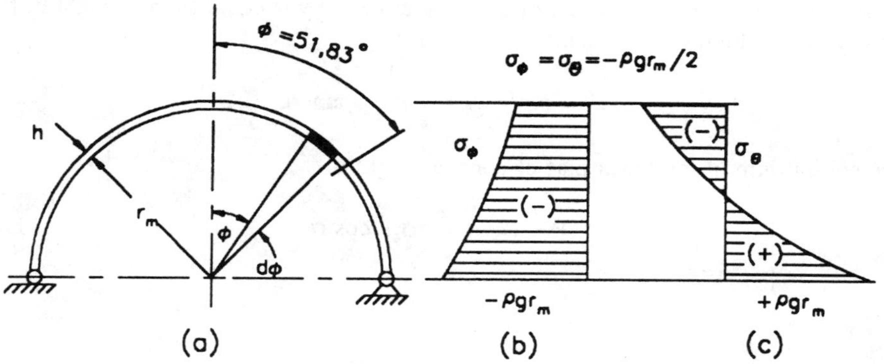

FIGURE 8.39 Example 8.13.

Solution:

For a spherical dome, $R_\phi = R_\theta = r_m$. From the equilibrium of a spherical cap subtended by a semiangle ϕ along the vertical direction,

$$2\pi r_m h \sigma_\phi \sin^2\phi + \rho g h \int_0^\phi (2\pi r_m \sin\phi) r_m \, d\phi = 0$$

from which

$$\sigma_\phi = -\frac{\rho g r_m}{(1+\cos\phi)}$$

Substituting σ_ϕ in the equation of equilibrium (Equation 8.65) together with $p_i = -\rho g h \cos\phi$, noting that p_i due to weight is an external pressure and hence the negative sign in the expression of p_i, results in

$$\frac{\sigma_\theta}{r_m} - \frac{\rho g}{(1+\cos\phi)} = -\rho g \cos\phi$$

Hence,

$$\sigma_\theta = \rho g r_m \left[\frac{1}{1+\cos\phi} - \cos\phi\right]$$

It will be observed that σ_ϕ is always a compression, while σ_θ becomes tension if

$$\frac{1}{1+\cos\phi} > \cos\phi \quad \text{or} \quad \phi = 51.833°$$

The compressive stress prevailing over most of the dome surface, as shown in Figure 8.39b and c, accounts for its successful use in roof construction. However, a check against instability in the compression zone and a check on tension for tension-weak materials are necessary design requirements.

Example 8.14:

Determine the membrane stresses in a long horizontal pipe of radius r_m, ***thickness*** h, ***subjected to its own weight and a gas pressure of*** p_i. ***The pipe is carried on a continuous flat rigid support as shown in Figure 8.40.***

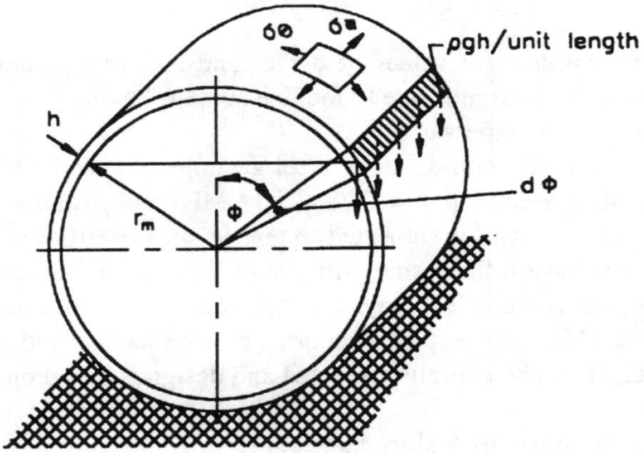

FIGURE 8.40 Example 8.14.

Solution:

Consider first the membrane stresses due to pipe weight, namely, $(\sigma_\theta)_w$ and $(\sigma_\theta)_w$. From the equilibrium of a part of unit pipe length subtended by a semiangle ϕ along the vertical direction,

$$2\,(\sigma_\theta)_w h \sin\phi + 2\,\rho g r_m h\phi = 0$$

where ρ is the mass density of the pipe material.

The above equation gives

$$(\sigma_\theta)_w = -\frac{\rho g r_m \phi}{\sin\phi}$$

which is always a compressive. Since the pipe axis is horizontal and the pipe wall axial displacement is free,

$$(\sigma_\phi)_w = 0$$

By superposition, the total membrane stresses due to both gas pressure and its own weight are given by

$$\sigma_\theta = p_i\frac{r_m}{h} - \frac{\rho g r_m \phi}{h\sin\phi} = \frac{r_m}{h}\left(p_i - \frac{\rho g h\phi}{\sin\phi}\right)$$

and

$$\sigma_\phi = p_i\frac{r_m}{2h}$$

It will be observed that the total σ_θ is a tensile at the top where $\phi = 0$ and will decrease to zero at an angle ϕ_o given by

$$\frac{\phi_o}{\sin\phi_o} = \frac{p_i}{\rho g h}$$

changing its sign to compression for values of $\phi > \phi_o$ and increasing from 0 to $-\infty$ at $\phi = 180°$. The infinite hoop stress at the bottom is due to the line support having a zero area. A finite bearing area is necessary to support the pipe weight.

Consider, for example, a 400-mm-diameter steel gas pipe of specific weight = 78 kN/m^3, h = 10 mm, and p_i = 0.78 MPa. Hence, $\rho gh = 0.78 \times 10^{-3}$ MPa and $p_i/\rho gh = 10^3$, which gives $\phi_o \cong$ 180°. This means that the pipe wall is subjected to tensile stresses all over. In order to limit σ_ϕ at the bottom, the pipe must have a finite supporting angle, which in this case can be a very small angle. If the support contact semiangle is, say, 1°, $(\sigma_\theta)_w$ will be 2.8 MPa, compared with 15.6 MPa due to internal pressure. With proper pipe support, i.e., by increasing the angle of contact, the stresses due to pipe weight can be entirely neglected and design is based on internal pressure only.

8.6.4.3 Spherical Containers on a Skirt Support

A spherical container filled with a liquid and supported along a parallel circle *a–a* is shown in Figure 8.41a. The membrane stresses σ_ϕ and σ_θ suffer from discontinuities along the supporting ring (skirt), i.e., at $\phi = \phi_o$.

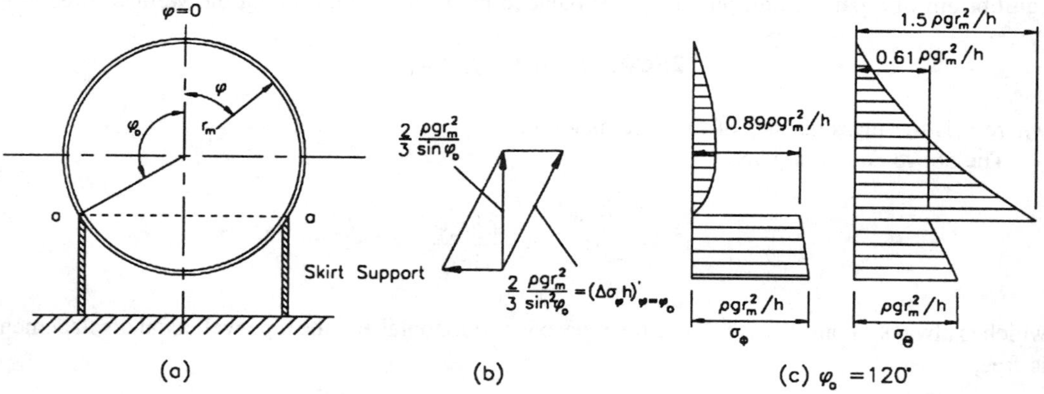

FIGURE 8.41 Stresses in a spherical liquid container on a skirt support: (a) container geometry and loading, (b) loads acting at the skirt joint, and (c) stress distribution in the shell.

The hydrostatic pressure at any level corresponding to an angle ϕ is given by

$$p_i = \rho g r_m (1 - \cos\phi)$$

Hence, due to this pressure, the resultant acting vertical force P is

$$P = \int_0^\phi \rho g r_m (1-\cos\phi)(2\pi r_m \sin\phi r_m d\phi)\cos\phi$$
$$= 2\pi r_m^3 \rho g\left[\frac{1}{6} - \frac{1}{2}\cos^2\phi\left(1 - \frac{2}{3}\cos\phi\right)\right] \tag{8.74}$$

From the equilibrium of the upper part, the meridional stress for any angle $\phi < \phi_o$ is obtained as

$$\left(\sigma_\phi\right)_{\phi<\phi_o} = \frac{P}{2\pi r_m h \sin^2\phi} = \frac{\rho g r_m^2}{6h}\left(1 - \frac{2\cos^2\phi}{1+\cos\phi}\right) \tag{8.75a}$$

The hoop stress σ_θ is thus obtained from Equation 8.65 as

$$\left(\sigma_\theta\right)_{\phi<\phi_o} = \frac{\rho g r_m^2}{6h}\left(5 - 6\cos\phi + \frac{2\cos^2\phi}{1+\cos\phi}\right) \tag{8.75b}$$

In determining the membrane stresses for $\phi > \phi_o$, the sum of the vertical support reactions along *a–a* which is equal to the liquid weight $\rho g\,(4\pi r_m^3/3)$, must be added to the load *P*, Equation 8.74. Hence, from Equations 8.66 and 8.65, the following membrane stresses are found:

$$\left(\sigma_\phi\right)_{\phi>\phi_o} = \frac{\rho g r_m^2}{6h}\left(5 + \frac{2\cos^2\phi}{1-\cos\phi}\right) \tag{8.76a}$$

and

$$\left(\sigma_\theta\right)_{\phi>\phi_o} = \frac{\rho g r_m^2}{6h}\left(1 - 6\cos\phi - \frac{2\cos^2\phi}{1+\cos\phi}\right) \tag{8.76b}$$

Comparison of Expressions 8.75 and 8.76 indicates that along the supporting ring *a–a*, both σ_ϕ and σ_θ change abruptly. Considering σ_ϕ, the discontinuity is obtained by subtracting Expression 8.75a from 8.76a to give

$$\left(\Delta\sigma_\phi\right)_{\phi=\phi_o} = \frac{2}{3}\frac{\rho g r_m^2}{h\sin^2\phi_o}$$

The same quantity is also obtained if the reaction per unit length of the circle *a–a* is resolved into two components, as shown in Figure 8.41b. The horizontal component represents the reaction on the supporting ring, which produces radial uniform pressure.

The stress distributions of σ_ϕ and σ_θ are shown in Figure 8.41c for $\phi_o = 120°$. It may be observed that for a support subtending $\phi_o > 120°$, σ_ϕ becomes negative, i.e., compressive, a situation which must be avoided for thin-walled shells.

In general, the membrane theory does not satisfy the condition of continuity at the support circle, and some local bending occurs near the supporting ring.

Example 8.15:

A cylindrical tank with a hemispherical bottom has a mean radius r_m *and thickness* h *and is filled with a liquid of density* ρ *to a level* H. *Determine the maximum stress in the cylinder walls, hemispherical bottom, and at the junction between them. Consider the two conditions; (a) the tank is supported at its top, Figure 8.42a, and (b) the tank is skirt supported as shown in Figure 8.42b. Discuss the change in stress state at the junction if a spherical cap of radius* $1.5r_m$ *replaces the hemispherical bottom, as shown in Figure 8.42c.*

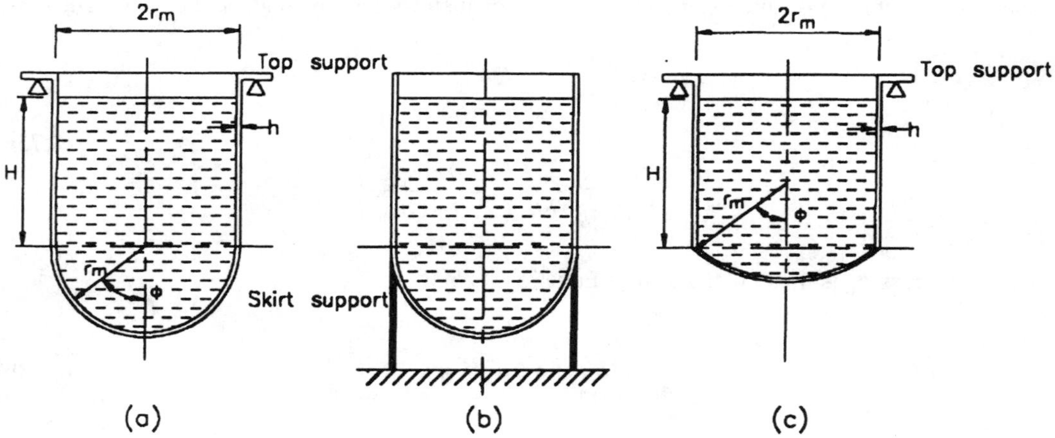

FIGURE 8.42 Example 8.15.

Solutions:

a. *Top support*: For this condition, the wall of the cylinder is acted upon by the total liquid weight W given by

$$W = \rho g\left[\pi r_m^2 H + \frac{2}{3}\pi r_m^2\right] = \rho g\left[\pi r_m^2\left(H + \frac{2}{3}r_m\right)\right]$$

The axial stress along the cylinder length is thus

$$\left(\sigma_\phi\right)_c = \frac{W}{2\pi r_m h} = \frac{\rho g r_m\left(3H + 2r_m\right)}{6h}$$

where the subscript c denotes the cylinder. The hoop stress, being dependent on the fluid pressure, varies with its depth attaining a maximum at the junction, given by

$$\left(\sigma_\theta\right)_c = \frac{\rho g H r_m}{h}$$

The maximum stresses in the hemispherical bottom occur at its lower point as

$$\left(\sigma_\phi\right)_s = \left(\sigma_\theta\right)_s = \frac{\rho g\left(H + r_m\right)r_m}{2h}$$

where the subscript s denotes the hemispherical bottom.

At the junction, the meridional stress in the sphere walls for $\phi = \pi/2$ is given by

$$\left(\sigma_\phi\right)_s = \frac{W}{2\pi r_m h \sin\phi} = \frac{\rho g r_m\left(3H + 2r_m\right)}{6h}$$

This is the same as $(\sigma_\phi)_c$ in the cylinder without any discontinuity in direction or magnitude. The hoop stress, in the hemispherical bottom, however, is obtained from Equation 8.65 as

$$(\sigma_\theta)_s = \frac{\rho g(3H - 2r_m)r_m}{6h}$$

Obviously, this is different from $(\sigma_\theta)_c$ in the cylinder wall indicating a stress discontinuity there. The stress $(\sigma_\theta)_s$ attains a negative value for a water level $H < 2r_m/3$, a situation which has to be avoided in thin-walled shells.

For the case shown in Figure 8.42c, the stresses at the junction between cylindrical wall and spherical bottom are calculated as above while considering the change in the total liquid weight as W_1 and the angle ϕ being $\phi = \sin^{-1}(r_m/1.5r_m)$. The meridional stress in the spherical bottom at the junction becomes

$$(\sigma_\phi^*)_s = \frac{W_1}{2\pi r_m h \sin(r_m/1.5r_m)}$$

This has a horizontal component $(\sigma_\phi^*)_s \cos\phi$, and it may be required to add a stiffener ring at the junction in the case of small wall thickness to maintain the radial stability of the container.

b. *Skirt support*: In this condition, the total weight W is supported by the skirt and not the cylindrical shell; hence,

$$(\sigma_\phi)_c = 0 \text{ while } (\sigma_\theta)_c = \frac{\rho g H r_m}{h}$$

as before. For the spherical bottom, the same maximum stresses exist at its lowest point and its junction with the cylinder. Therefore, the meridional and hoop stresses both become discontinuous at the junction.

8.7 BENDING OF THIN-WALLED CYLINDERS SUBJECTED TO AXISYMMETRIC LOADING

In thin-walled cylinders subjected to internal or external pressure, localized wall bending is encountered at such locations where the wall is laterally constrained or where the pressure distribution is discontinuous. These conditions are quite common in most pressure containers and usually occur at cylinder ends, where there is a head or a flange, and at locations of internal or external stiffener rings as shown in Figure 8.43.

8.7.1 Problem Formulation

In the following analysis, the cylinder is long, of a uniform thickness, and made of a homogeneous isotropic elastic material. If the pipe is subjected to a uniform internal pressure p_i, the membrane strains and stresses due to this pressure, as given by Equations 8.68, are not considered in the following derivation. As a result of any localized lateral constraint the cylinder wall will be acted upon by a circumferentially uniformly distributed transverse shearing force Q_o and bending moment M_o per unit length of the edge at $z = 0$ with $N_z = 0$, as shown in Figure 8.44a and b. The analysis is thus confined to the strains and stresses due to Q_o and M_o resulting from edge constraint.

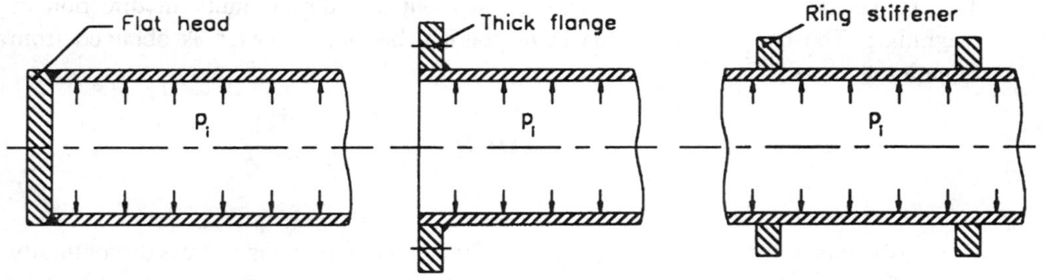

FIGURE 8.43 Sources of localized bending in thin-walled cylinders subjected to internal pressure.

In Figure 8.44a and d, the cylinder wall is shown by dotted lines before loading and by solid lines after loading. Due to symmetry, the circumferential displacement component v vanishes. At the middle wall surface, where the origin of cylindrical polar coordinates are taken as shown in Figure 8.44c, there is no axial displacement, i.e., $w = 0$, and only radial displacement u takes place. A line normal to the middle wall surface will remain straight but rotated an angle ψ, as shown in Figure 8.44d and e; therefore, $\tan \psi \cong \psi = du/dz$. The positive direction of u and w are along positive directions of r and z, respectively. The stress resultants acting on a shell element, being all positive, are shown in Figure 8.44f.

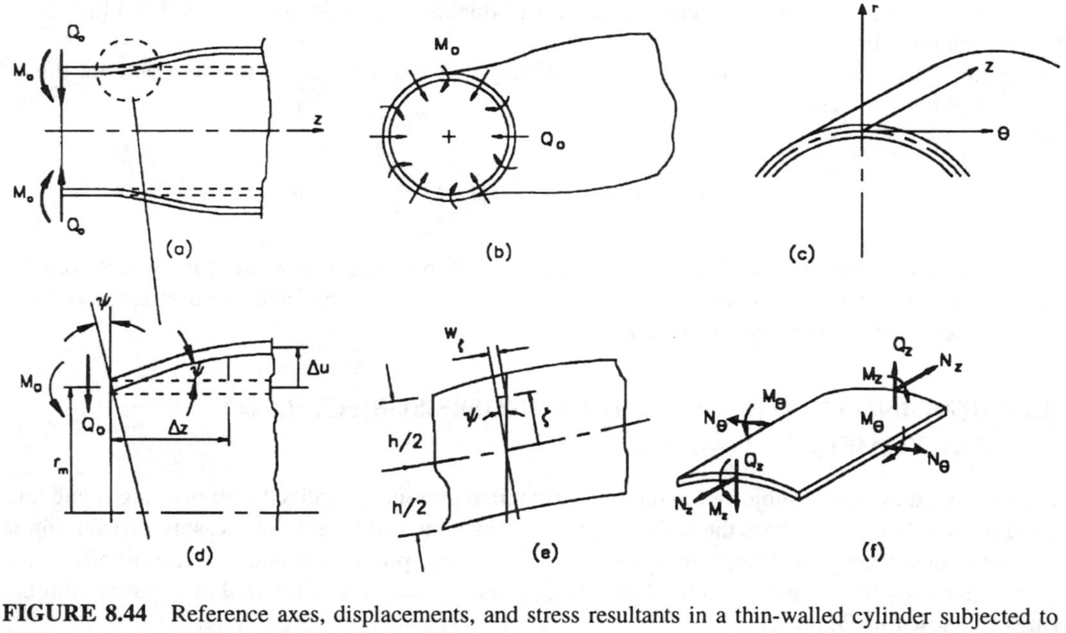

FIGURE 8.44 Reference axes, displacements, and stress resultants in a thin-walled cylinder subjected to axisymmetric bending: (a) and (b) edge loading, (c) reference axes, (d) radial displacement, (e) axial displacement and (f) stress resultants.

The geometry of deformation of a shell element is shown in Figure 8.45 for any point A_ζ (ζ, z), where ζ is the distance from the middle surface and is positive along the positive direction of r. The vertical line segment AA_ζ is rotated to the position $A'A'_\zeta$ after deformation. The strain components at point A on the middle surface, which undergoes displacements u and w, are obtained directly from Equations 2.9a as

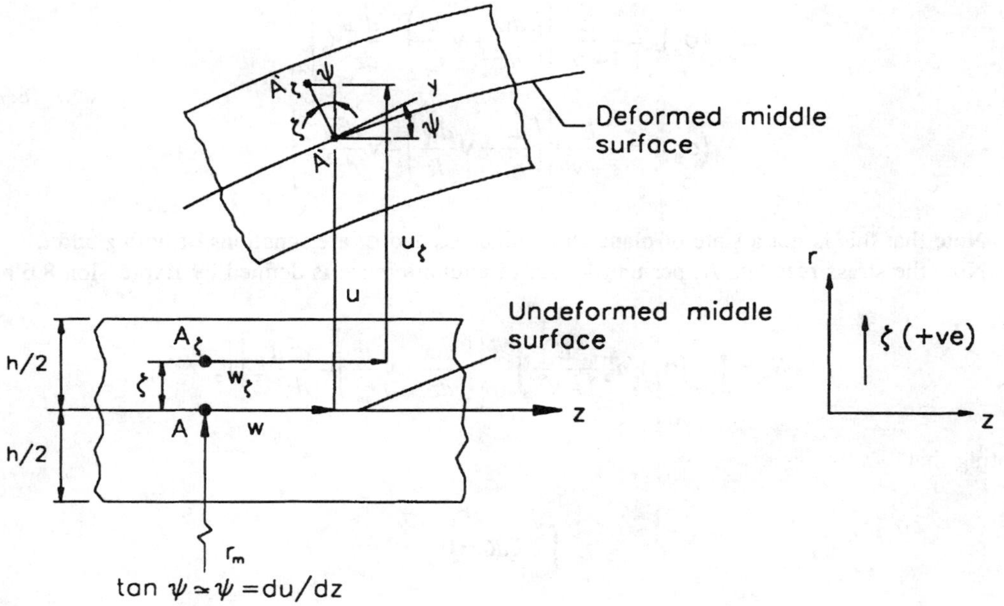

FIGURE 8.45 Deformation of middle surface in a thin-walled cylinder subjected to axisymmetric bending.

$$\varepsilon_z = \frac{dw}{dz} \text{ and } \varepsilon_\theta = \frac{u}{r_m}$$

In Figure 8.45, the displacements u_ζ and w_ζ of point A_ζ are given by

$$u_\zeta = u$$

and

$$w_\zeta = w - \zeta \tan\psi \cong w + \zeta\psi = w - \frac{du}{dz}\zeta$$

Hence, the strain components at this point are

$$(\varepsilon_z)_\zeta = \frac{dw}{dz} - \frac{d^2u}{d^2z}\zeta \tag{8.77a}$$

$$(\varepsilon_\theta)_\zeta = \frac{u}{r_m + \zeta} \cong \frac{u}{r_m} \tag{8.77b}$$

For $\sigma_r = 0$,* the elastic stress–strain relations (Equation 3.6) gives the stresses $(\sigma_z)_\zeta$ and $(\sigma_\theta)_\zeta$ at point A'_ζ as

* Since this is true for uniform deformation under membrane stresses, it is more valid when σ_θ and σ_z become locally higher as a result of bending effects.

$$(\sigma_z)_\zeta = \frac{E}{1-\nu^2}\left[\left(\frac{dw}{dz} + \nu\frac{u}{r_m}\right) - \frac{d^2u}{dz^2}\zeta\right]$$

$$(\sigma_\theta)_\zeta = \frac{E}{1-\nu^2}\left[\left(\frac{u}{r_m} + \nu\frac{dw}{dz}\right) - \nu\frac{d^2u}{dz^2}\zeta\right] \tag{8.78}$$

Note that this is not a state of plane stress since σ_θ and σ_z are functions of both z and ξ. Now the stress resultant N_z per unit length of circumference is defined by Expression 8.6 as

$$N_z = \int_{-h/2}^{h/2} (\sigma_z)_\zeta d\zeta \frac{E}{1-\nu^2}\int_{-h/2}^{h/2}\left[\left(\frac{dw}{dz} + \nu\frac{u}{r_m}\right) - \frac{d^2u}{dz^2}\zeta\right]d\zeta$$

Noting that

$$\int_{-h/2}^{h/2} \zeta d\zeta = 0$$

this reduces to

$$N_z = \frac{Eh}{1-\nu^2}\left(\frac{dw}{dz} + \nu\frac{u}{r_m}\right) \tag{8.79}$$

In the present analysis, N_z is taken equal to zero, and hence Equation 8.79 yields that $dw/dz = -\nu\, u/r_m$. Upon substitution into Expressions 8.78 for the stresses, they reduce to (the subscript ζ being dropped for simplicity):

$$\sigma_z = -\frac{E}{1-\nu^2}\frac{d^2u}{dz^2}\zeta$$

$$\sigma_\theta = \frac{E}{1-\nu^2}\left[(1-\nu^2)\frac{u}{r_m} - \nu\frac{d^2u}{dz^2}\zeta\right] \tag{8.80}$$

Note that N_z is a direct force causing uniform membrane stresses, that it can be easily dealt with separately, and that the resulting stresses and deformations due to N_z are simply superposed on the bending stresses.

An equation for the displacement u may now be written employing the stress equilibrium equations together with Expressions 8.80. The stress equilibrium conditions for axial-symmetry in the absence of body forces are given by Equations 1.10 as

$$\frac{\partial\sigma_r}{\partial r} + \frac{\partial\tau_{zr}}{\partial z} + \frac{\sigma_r - \sigma_\theta}{r} = 0$$

$$\frac{\partial\sigma_z}{\partial z} + \frac{\partial\tau_{zr}}{\partial r} + \frac{\tau_{zr}}{r} = 0 \tag{1.10}$$

For $\sigma_r = 0$, the first of these equations reduces to

$$\frac{\partial \tau_{zr}}{\partial z} - \frac{\sigma_\theta}{r_m} = 0 \tag{8.81}$$

Substituting σ_θ from Equation 8.80 into Equation 8.81 and integrating across the thickness h from $-h/2$ to $h/2$ yields:

$$\frac{\partial}{\partial z}\int_{-h/2}^{h/2} \tau_{zr} d\xi - \frac{E}{\left(1-\nu^2\right) r_m}\int_{-h/2}^{h/2}\left[\left(1-\nu^2\right)\frac{u}{r_m} - \nu\frac{d^2u}{dz^2}\xi\right]d\xi = 0$$

Performing the integration, while noting that

$$\int_{-h/2}^{h/2} \tau_{zr} d\xi = Q_z$$

$$\int_{-h/2}^{h/2} d\xi = h \text{ and } \int_{-h/2}^{h/2} \xi d\xi = 0$$

yields

$$\frac{\partial Q_z}{\partial z} - \frac{Eh}{r_m^2}u = 0 \tag{8.82}$$

The bending moment M_z and the shearing force Q_z per unit circumferential length are related to the radial displacement u by the same relations governing plate bending, Equations 8.9 and 8.11, as

$$M_z = -D\frac{d^2u}{dz^2} \tag{8.83a}$$

$$Q_z = \frac{dM}{dz} = -D\frac{d^3u}{dz^3} \tag{8.83b}$$

where D is given by Equation 8.10 as

$$D = Eh^3/[12\,(1-\nu^2)] \tag{8.10}$$

Substituting for Q_z from Equation 8.83b into the equilibrium condition (Equation 8.82) yields*

$$\frac{d^4u}{dz^4} + 4\beta^4 u = 0 \tag{8.84a}$$

where

$$\beta^4 = \frac{Eh}{4r_m^2 D} = \frac{3\left(1-\nu^2\right)}{r_m^2 h^2} \tag{8.84b}$$

* It is noted that the governing equation (Equation 8.84a) takes the form $(d^4u/dz^4) + 4\beta^4 u = p_i/D$, where a pressure loading p_i acts on the shell. The complete solution thus comprises an additional term: the particular integral.

Equation 8.84 has the same form as that of a beam on an elastic foundation of Section 7.8,* Equation 7.91. The general solution of this equation is

$$u = e^{-\beta z}(A\sin\beta z + B\cos\beta z) + e^{\beta z}(C\sin\beta z + D\cos\lambda z) \tag{8.85}$$

where A, B, C, and D are constants. As previously discussed in Section 7.8, this general solution consists of two terms; the first term is rapidly decreasing with respect to z, while the second term is rapidly increasing with respect to z. Now, since the effect of the edge loading is localized, then u must approach zero as z increases. This requires that the constants $C = D = 0$. The constants A and B are determined from boundary conditions; hence,

$$u = e^{-\beta z}\,(A \sin \beta z + B \cos \beta z) \tag{8.86}$$

The displacement u given by Equation 8.86 is due to edge loading only.

Finally, the stress resultants acting on the shell element, Figure 8.44f, namely N_θ, M_z, M_θ, and Q_z, are expressed with the aid of Equations 8.80 as**

$$N_\theta = \int_{-h/2}^{h/2} (\sigma_\theta)_\zeta\, d\xi = \frac{Ehu}{r_m} \tag{8.87a}$$

$$M_z = \int_{-h/2}^{h/2} (\sigma_z)_\zeta\, \zeta d\xi = -\frac{Eh^3}{12(1-\nu^2)}\frac{d^2u}{dz^2} = -D\frac{d^2u}{dz^2} \tag{8.87b}$$

$$M_\theta = \int_{-h/2}^{h/2} (\sigma_\theta)_\zeta\, \zeta d\xi = \frac{\nu Eh^3}{12(1-\nu^2)}\frac{d^2u}{dz^2} = \nu M_z \tag{8.87c}$$

$$Q_z = \frac{dM_z}{dz} = -D\frac{d^3u}{dz^3} \tag{8.87d}$$

These result in the stress components due to bending at any point (ζ, z) within the shell wall:

$$\sigma_z = \pm\frac{M_z}{h^3/12}\zeta \tag{8.88a}$$

$$\sigma_\theta = \pm\frac{\nu M_z}{h^3/12}\zeta + \frac{Eu}{r_m} \tag{8.88b}$$

where

$$M_z = -D\frac{d^2u}{dz^2} = 2D\beta^2 e^{-\beta z}(A\cos\beta z - B\sin\beta z) \tag{8.88c}$$

* A cylindrical shell may be regarded as a set of longitudinal strips, which are joined together by elastic circumferential forces N_θ to bend as a unit. Under symmetrical loading, all strips bend in a similar manner and the radial component of N_θ at each strip is proportional to the local deflection u, for a beam on an elastic foundation (see Equation 8.87a).

** Note that Expressions 8.88 could have been obtained by integrating the second of the equilibrium conditions (Equation 1.10) over the cylinder thickness.

The average shear stress over the thickness is given by

$$\tau_{zr} = \frac{Q_z}{h} \tag{8.88d}$$

From here on, the subscript z may be dropped for simplicity. Note that the hoop stress due to lateral constraint Equation 8.88b, consists of two parts; the first is due to bending, which is linearly distributed across the wall thickness, and the second is a uniform membrane stress added to it.

8.7.2 Long, Thin-Walled Pressurized Pipe with a Rigid Flange at its End

Since Equation 8.84a is similar to Equation 7.91, that of a beam on elastic foundation, the same argument of classifying beams to infinitely long or short ones will apply to cylinders. Namely, a cylinder subjected to localized bending is considered long if its length measured from the point of load application $L > 2\pi/\beta$.

The case of a long, thin-walled pressurized pipe with a rigid flange at its end is shown in Figure 8.46a. The rigid flange prevents the pipe from radial expansion due to the internal pressure p_i. Hence, localized reactions Q_o and M_o uniformly distributed along the circumference at this clamped edge will develop as shown in Figure 8.46b.

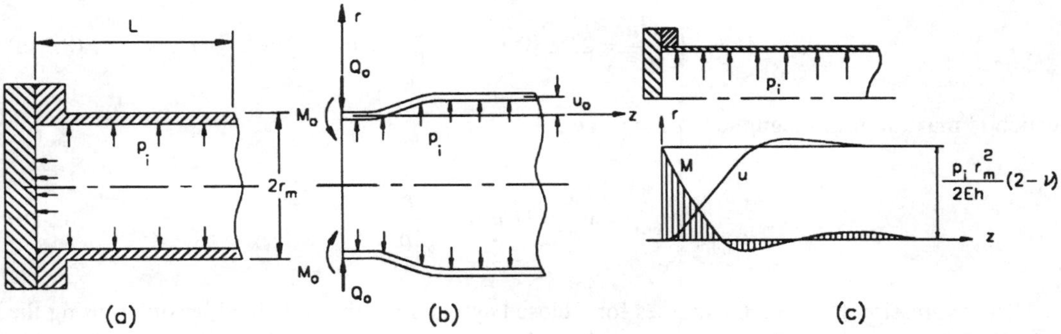

FIGURE 8.46 (a) Long, thin-walled pressurized pipe with a rigid flange at its end, (b) stress resultants, and (c) distribution of displacement and moment along pipe length.

From Equation 8.86, the radial displacement u due to localized loading is given by

$$u = e^{-\beta z}(A \sin \beta z + B \cos \beta z) \tag{8.86}$$

Also from Equation 8.71a, the radial displacement for an open-ended cylinder subjected to internal pressure p_i is given by*

$$u_c = \frac{p_i r_m^2}{Eh} \tag{8.89}$$

Due to the clamping effect of the rigid flange, the radial displacement at $z = 0$ has to vanish. In other words, the sum of the radial displacements as given by Equation 8.89 and that, due to the localized bending moment M_o and shear force Q_o as given by Equation 8.89, must be zero. Therefore,

* Note that the effect of the axial membrane stress in a close-ended cylinder will be considered by superposing membrane stresses to those due to local bending

$$u_c + u_o = \frac{p_i r_m^2}{Eh} + B = 0$$

giving

$$B = -u_c = -\frac{p_i r_m^2}{Eh}$$

Furthermore, due to the rigid clamping at the flange, i.e., at $z = 0$; $du/dz = 0$. Substituting this condition considering Expression 8.86 for u gives

$$A = B = -u_c = -\frac{p_i r_m^2}{Eh}$$
$$u = -\frac{p_i r_m^2}{Eh} e^{-\beta z}(\sin\beta z + \cos\beta z) \tag{8.90}$$

The bending moment according to Equation 8.88c is thus derived as

$$M = -D\frac{d^2u}{dz^2} = 2Du_c\beta^2 e^{-\beta z}(\cos\beta z - \sin\beta z) \tag{8.91a}$$

which is maximum in magnitude at $z = 0$, i.e.,

$$M_{\max} = M_o = 2Du_c\beta^2 = \frac{p_i}{2\beta^2} \tag{8.91b}$$

The expression for the total stresses for a closed cylinder are thus obtained by superposing the membrane stresses due to the pressure p_i to those derived using Equations 8.88. Thus, the maximum total axial stress σ_z occurs at $\xi = h/2$, and hence Equations 8.88a and 8.68a (where the meridional direction ϕ in the later equation is denoted by z) give at $z = 0$,

$$(\sigma_z)_{\max} = \frac{6M_{\max}}{h^2} + \frac{p_i r_m}{2h} \tag{8.92a}$$

Substituting from Equation 8.91b into Expressions 8.92a knowing that

$$D = \frac{Eh^3}{12(1-\nu^2)} \quad \text{and} \quad \beta^4 = \frac{Eh}{4r_m^2 D} = \frac{3(1-\nu^2)}{r_m^2 h^2}$$

gives, at $z = 0$,

$$(\sigma_z)_{\max} = \frac{3}{\sqrt{3(1-\nu^2)}}\frac{p_i r_m}{h} + \frac{p_i r_m}{2h} \tag{8.92b}$$

For a material with $\nu = 0.3$,

$$\left(\sigma_z\right)_{\max} = 4.631\frac{p_i r_m}{2h} \tag{8.92c}$$

which is $4.631 \times \sigma_z$ if there is no flange.

The total hoop stress σ_θ is obtained from Equation 8.88b for $\zeta = h/2$ and superposing the membrane stress $\sigma_\theta = p_i r_m/h$, thus gives

$$\left(\sigma_\theta\right)_{\text{total}} = \frac{\nu 6M}{h^2} + \frac{Eu}{r_m} + \frac{p_i r_m}{h} \tag{8.93a}$$

At the flange, $z = 0$, $u = -p_i r_m^2/Eh$ and the total hoop stress is thus

$$\left(\sigma_\theta\right)_{z=0} = \frac{\nu 6M_{\max}}{h^2} \tag{8.93b}$$

which attains a value

$$\left(\sigma_\theta\right)_{z=0} = 0.545\frac{p_i r_m}{h} \tag{8.93c}$$

after substituting for $M_{\max}$ from Equation 8.91b and taking $\nu = 0.3$. The maximum hoop stress occurs away from the flange as dictated by Expression 8.68 for large values of z. In fact,

$$\left(\sigma_\theta\right)_{\max} = \frac{p_i r_m}{h} \tag{8.68b}$$

which is the membrane stress as given by Equation 8.68b. The shearing force is obtained from Equation 8.87d as

$$Q = \frac{dM}{dz} = -D\frac{d^3u}{dz^3}$$

or

$$Q = 4Du_c\beta^3 e^{-\beta z}\cos\beta z \tag{8.94a}$$

which is maximum at $z = 0$, i.e,

$$Q_{\max} = Q_o = -4Du_c\beta^3 = -\frac{p}{\beta} \tag{8.94b}$$

or

$$Q_{\max} = -0.778 p_i\sqrt{r_m h} \tag{8.94c}$$

For a material with $\nu = 0.3$, this gives a maximum shear stress at $\xi = 0$; $\tau_{zr} = -1.5Q_{max}/h = -1.167\, p_i\sqrt{r_m/h}$; a relatively small value when compared to $(\sigma_z)_{max}$ as given by Equation 8.92c.

The distributions of the radial displacement u and the bending moment M as represented in Figure 8.46c indicate the localized nature of the stresses developed in the pipe wall. The distance z along the pipe length at which these stresses become insignificant can be set arbitrarily at the location where M attains, say, 5% of its maximum value. Hence, from Equations 8.91a,

$$|M/M_{max}| = |e^{-\beta z}(\cos\beta z - \sin\beta z)| \le 0.05$$

The quantity $|\cos\beta z - \sin\beta z|$ cannot exceed $\sqrt{2}$ and hence $e^{-\beta z}$ takes the value $0.05/\sqrt{2} = 0.035355$. This yields $z = 1.795\sqrt{r_m/h}$, a distance after which the stresses in the cylinder wall are determined by the membrane analysis solely, Equations 8.68. Note that for thin-walled cylinders, the ratio, i.e., $\sqrt{r_m h}/L$, where L is the length of the cylinder, is generally small, indicating again the localized nature of the bending stresses.

Example 8.16:

Consider a steel air cylinder of a mean diameter 300 mm subjected to an internal pressure of 0.4 MPa. If the maximum allowable tensile stress is 120 MPa, determine the wall thickness for the two cases: (a) ends unconstrained radially and (b) ends radially constrained. Take $\nu = 0.3$.

Solution:

a. *Ends unconstrained radially*: This case is practically obtained when the ends are closed by two flat heads that are bolted together using long tie-rods outside the cylinder as shown in Figure 8.47a. Equation 8.68b gives

$$\sigma_\theta = \frac{p_i r_m}{h} = 0.4 \times 150/h = 120 \text{ MPa}$$

hence, $h = 0.5$ mm.

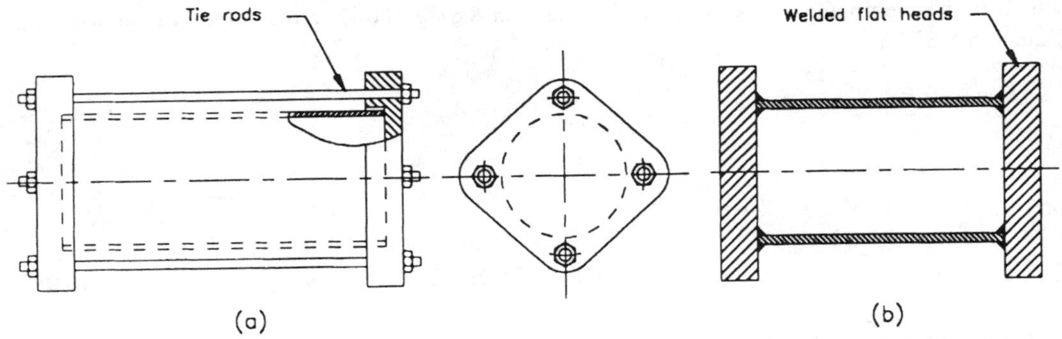

FIGURE 8.47 Example 8.16.

This is not a practical thickness to use in a pressure container of 300 mm diameter since it cannot sustain transverse external loading during handling. A practical thickness is 2 to 3 mm to allow for handling and material corrosion.

b. *Ends radially constrained*: This is the case of a rigid flange or a heavy flat head as in Example 8.4 case (b) where the edges are made clamped. In this case, Equation 8.92b gives, at $z = 0$,

$$(\sigma_z)_{max} = \frac{3}{\sqrt{3(1-\nu^2)}} \frac{p_i r_m}{h} + \frac{p_i r_m}{2h}$$

By substituting $\nu = 0.3$,

$$(\sigma_z)_{max} = 4.631 \frac{p_i r_m}{2h} \leq 120 \text{ MPa}$$

Also according to Equation 8.93c, σ_θ, at $z = 0$,

$$(\sigma_\theta)_{z=0} = 0.545 \frac{p_i r_m}{h} < (\sigma_z)_{max}$$

and away from the end

$$(\sigma_\theta)_{max} = \frac{p_i r_m}{h} < (\sigma_z)_{max}$$

Therefore, the required thickness is determined from the condition $(\sigma_z)_{max} < \sigma_{allowable} <$ 120 MPa as

$$h = 4.631 \times 0.4 \times 150/(2 \times 120) = 1.16 \text{ mm}$$

Again, this is not a practical thickness to use, and a suitable thickness of 2 to 3 mm is often used.

8.7.3 Short, Thin-Walled Pressurized Pipe with Two Rigid Flanges at Both Ends

A short, pressurized open-ended pipe of length* $L < 2\pi/\beta$, having two rigid flanges at both ends can be represented by two built-in edges, as shown in Figure 8.48, where the bending moment and shearing forces at both ends influence each other. The general solution given by Expression 8.86 can be applied together with the appropriate boundary conditions. Through lengthy algebraic manipulation, an expression for the moment at the built-in edge is obtained as**

$$(M_o)_{sh} = \frac{p_i}{2\beta^2} \frac{\sinh \beta L - \sin \beta L}{\sinh \beta L + \sin \beta L} \tag{8.95}$$

where $(M_o)_{sh}$ means the edge moment for a short pipe.

The effect of pipe length on the maximum moment at the built-in edges is shown in Table 8.8, based on Equation 8.91b (neglecting ν^2 with respect to unity) and the above Equation 8.95 for long and short pipes, respectively.

* For $r_m = 10\ h$, this limit leads to a pipe length-to-radius ratio of $L/r_m < 1.5$.

** See Timoshenko and Woinowsky-Kreiger,[3] p. 402.

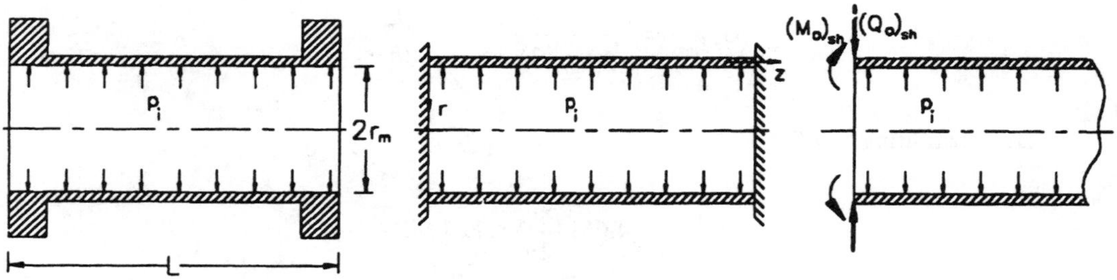

FIGURE 8.48 Short, thin-walled pressurized pipe with two rigid flanges at both ends.

TABLE 8.8
Comparison Between the Edge Moments for Short and Long Flanged Pipes

βL	0.2	0.4	0.6	1.0	3	5	10
$L/\sqrt{r_m h}$	0.152	0.304	0.456	0.76	2.28	3.8	7.6
$(M_o)_{sh}/M_o$	0.0068	0.0268	0.0601	0.1670	0.9770	1.0300	1

8.7.4 Long, Thin-Walled Pipe Subjected to Uniform Radial Compression along a Circular Section at Its Middle Length

Consider a long, thin-walled pipe loaded by a uniform radial compressive force P per unit length of the circumference of the pipe, as shown in Figure 8.49. The loading condition is equivalent to a pipe subjected to edge loading $P/2$ and M_o per unit circumference length.

Since there is no internal pressure applied to the pipe, the solution given by Expression (8.86) holds as

$$u = e^{-\beta z} (A \sin \beta z + B \cos \beta z)$$

The boundary conditions required to determine the two constants A and B are

$$\text{at } z = 0\text{: } Q = P/2 \text{ and } \psi = \frac{du}{dz} = 0$$

indicating that the slope vanishes at the loading cross section due to symmetry. From Equation 8.86,

$$[\frac{du}{dz} = e^{-\beta z} [A (\cos \beta z - \sin \beta z) - B (\cos \beta z + \sin \beta z)]$$

Hence at

$$z = 0; \frac{du}{dz} = \beta (A - B) = 0$$

Hence the constant $A = B$ and the displacement is given by

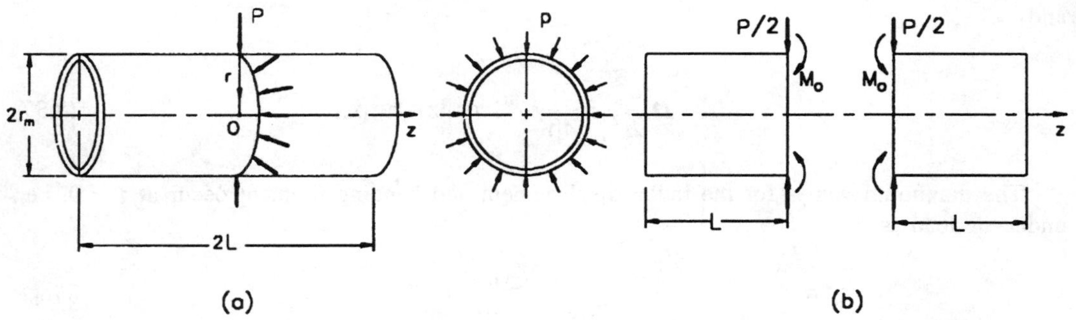

FIGURE 8.49 Long, thin-walled pipe subjected to uniform radial compression along a circular section at its midlength: (a) load diagram and (b) stress resultants at O.

$$u = Ae^{-\beta z}\ (\sin\beta z + \cos\beta z)$$

where

$$\frac{du}{dz} = -2A\ \beta e^{-\beta z}\ \sin\beta z$$

From Equation (8.87d), the shearing force is

$$Q = -D\frac{d^3 y}{dz^3}$$

where

$$\frac{d^3 y}{dz^3} = 4A\beta^3 e^{-\beta 3}\cos\beta z$$

At $z = 0$; $Q = P/2$,

hence,

$$\frac{P}{2} = -4DA\beta^3$$

and

$$A = -\frac{P}{8D\beta^3}$$

The displacement and the bending moment are thus respectively given by

$$u = \frac{P}{8D\beta^3}e^{-\beta z}(\sin\beta z + \cos\beta z) \tag{8.96}$$

and

$$M = -D\frac{d_u^2}{dz^2} = \frac{P}{4\beta}e^{-\beta z}(\cos\beta z - \sin\beta z) \tag{8.97}$$

The maximum values for the radial displacement and bending moment occur at $z = 0$, i.e., under the load as

$$u_{\max} = \frac{-P}{8D\beta^3} \tag{8.98a}$$

and

$$M_{\max} = \frac{P}{4\beta} \tag{8.98b}$$

The displacement and bending moment both decrease rapidly away from the point of force application both positive and negative z. To illustrate this, consider a thin-walled cylinder of $r_m/h = 10$, hence at a distance $z = r_m$,

$$\beta z = \frac{\left[3\left(1-\nu^2\right)\right]^{1/4}}{\sqrt{r_m h}}r_m = 4.06$$

Substitution of this value in Expressions 8.96 and 8.97 yields

$$u = -0.0242\frac{P}{8D\beta^3} = -0.0242u_{\max}$$

$$M = 0.0032\frac{P}{4\beta} = 0.0032M_{\max}$$

The maximum bending stresses at $z = 0$ and $\xi = -h/2$, are derived from Equations 8.88 for $p_i = 0$, and using Expressions 8.98b to give

$$\left(\sigma_z\right)_{\max} = \frac{3P}{2\beta h^2} \tag{8.99a}$$

and

$$\left(\sigma_\theta\right)_{\max} = \frac{3\nu P}{2\beta h^2} - \frac{P\beta r_m}{2h} \tag{8.99b}$$

The above solution can be used to determine bending moment and shearing force in a long, pressurized pipe reinforced by rigid rings, where the shortest distance L between two consecutive rings satisfies the condition $L > 2\pi/\beta$,* as shown in Figure 8.50.

* The solution for reinforcing rings at $L > 2\pi/\beta$ is given in Timoshenko and Woinowsky-Kreiger.[3]

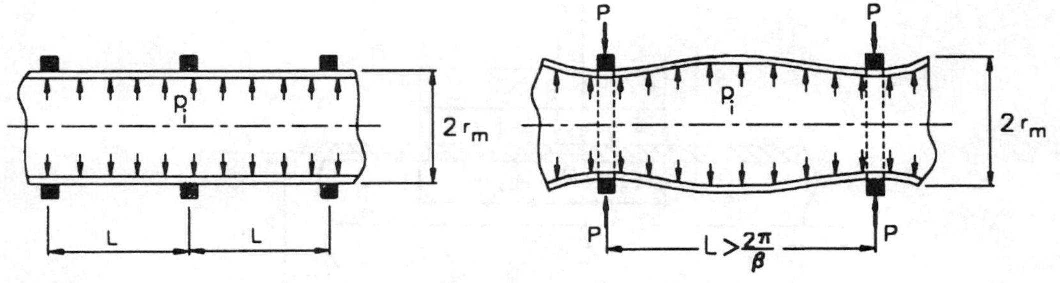

FIGURE 8.50 Long pressurized pipe reinforced by rigid rings.

At the ring location, the condition of no radial displacement must be satisfied; hence, u due to p_i for an expanding cylinder with free edges as may be deduced from Equation 8.71a is suppressed by the compressive force $P/2$, as given by Equation 8.98a:

$$\frac{P_i r_m^2}{2Eh}(2-\nu)=\frac{P}{8D\beta^3} \tag{8.100}$$

This condition determines the magnitude of the force P. The maximum bending moment and stresses are given by Equations 8.98 and 8.99, respectively.

8.7.5 Long Thin-Walled Pipe Subjected to a Uniform Circumferential Load along a Finite Length

The solution for the case of a concentrated force, Section 8.7.4, can be applied to determine u and M in a long, thin-wall pipe uniformly loaded, as shown in Figure 8.51 — a case of a long pipe stiffened by a rigid thick sleeve. The deflection Δu at point O produced by a ring load of $q\Delta z$ at a distance Δz from O is obtained from Expression 8.96 as

$$\Delta u=-\frac{q\Delta z}{8D\beta^3}e^{-\beta z}(\sin\beta z+\cos\beta z)$$

The deflection at point O due to the total load qL is obtained by integration as

$$u=\frac{-q}{8D\beta^3}\left[\int_0^{L_1}e^{-\beta z}(\sin\beta z+\cos\beta z)dz+\int_0^{L_2}e^{-\beta z}(\sin\beta z+\cos\beta z)dz\right]$$

which yields

$$u=\frac{-qr_m^2}{2Eh}\left[2-e^{-\beta L_1}\cos(\beta L_1)-e^{-\beta L_2}\cos(\beta L_2)\right] \tag{8.101}$$

When $L_1=L_2$, the maximum deflection occurs at the midlength of the distributed load. If L_1 and L_2 are too large, the deflection becomes approximately equal to qr_m^2/Eh approaching the value for a cylinder with free edges under uniform external pressure q, as illustrated in Example 8.8.

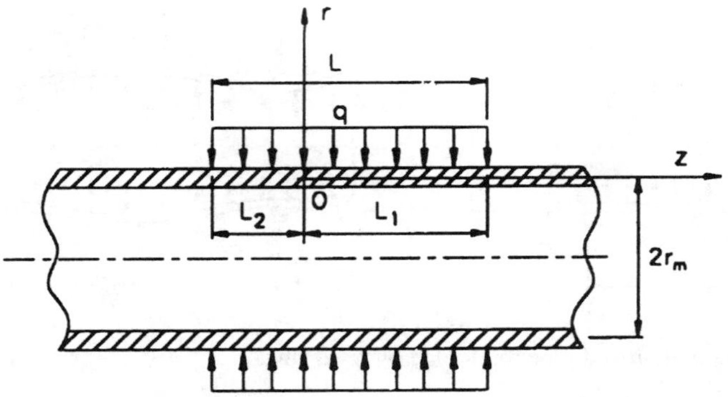

FIGURE 8.51 Long, thin-walled pipe subjected to uniform circumferential load along a finite length.

Example 8.17:

Consider the case of a long brass pipe shrunk in a hole in a rigid plate of 6 mm thickness. The pipe inside diameter is 36 mm and wall thickness is 2 mm. For brass take E = *82.6 GPa and* ν = *0.35.*

a. *Determine the radial interference such that the contact pressure between the tube and plate does not exceed 10 MPa. Use both equations for concentrated force and uniform load and comment on the results. Determine also the maximum stress in the pipe. The plate is located somewhere near the middle length of the pipe.*
b. *If the pipe is shrunk in a plate of thickness 60 mm, recalculate the radial interference and compare with that of Equation 6.13 in Chapter 6.*

Solution:

a. Since the plate thickness is small, the loading at the contact area will be treated as concentrated. Hence, the pipe is subjected to an external radial force per unit circumference length $P = 10 \times 6 = 60$ N/mm.
The maximum radial displacement is given by Equation 8.98a as

$$u = -\frac{P}{8D\beta^3}$$

where

$$D = \frac{Eh^3}{12(1-\nu^2)} = \frac{82600 \times 2^3}{12(1-0.35^2)} = 62754 \text{ N} \cdot \text{mm}$$

and

$$\beta^3 = \left[\frac{3(1-\nu^2)}{r_m^2 h^2}\right]^{3/4} = \left(\frac{3 \times 0.88}{19^2 \times 2^2}\right)^{3/4} = 0.00882 \text{ mm}^{-3}$$

or

$$\beta = 0.2066 \text{ mm}^{-1}$$

Hence, the radial interference is

$$u = -\frac{60}{8 \times 62754 \times 0.00882} = -0.0135 \text{ mm}$$

From Equation 8.99, the maximum bending stress at $\zeta = h/2$ is

$$\left(\sigma_z\right)_{max} = \frac{3P}{2\beta h^2} = \frac{3 \times 60}{2 \times 0.2066 \times 2^2} = 108.9 \text{ MPa}$$

Considering the load to be uniformly distributed along the contact area, Equation 8.101 may be used; hence, for $L_1 = L_2 = L/2 = 3$ mm, i.e., $\beta L/2 = 0.62$ maximum deflection at the center, i.e., the radial interference is

$$u = -\frac{10 \times 19^2}{2 \times 82600 \times 2}\left[2 - 2e^{-\beta L/2}\cos(\beta L/2)\right] = -0.0123 \text{ mm}$$

which is different than that calculated using the concentrated force solution by about 10%.

b. If the length over which the pipe is shrunk in the plate is ten times this, i.e., 60 mm, the required radial interference is recalculated using $L_1 = L_2 = L/2 = 30$ mm and $\beta L/2 = 6.2$.

$$u = -\frac{10 \times 19^2}{2 \times 82600 \times 2}\left[2 - 2e^{-\beta L/2}\cos(\beta L/2)\right] = -0.02182 \text{ mm}$$

This is almost of the same magnitude as that calculated using Equation 6.13 of Chapter 6 (for $\lambda_2 = \infty$); namely, $qr_m^2/Eh = 0.02188$ mm.

8.7.6 Cylindrical Pressure Vessels with End Closures

For a thin-walled cylindrical pressure vessel with end closures or heads, as shown in Figure 8.52, the radial displacement of the wall is constrained at the ends which will introduce localized bending stresses in addition to the membrane stresses.

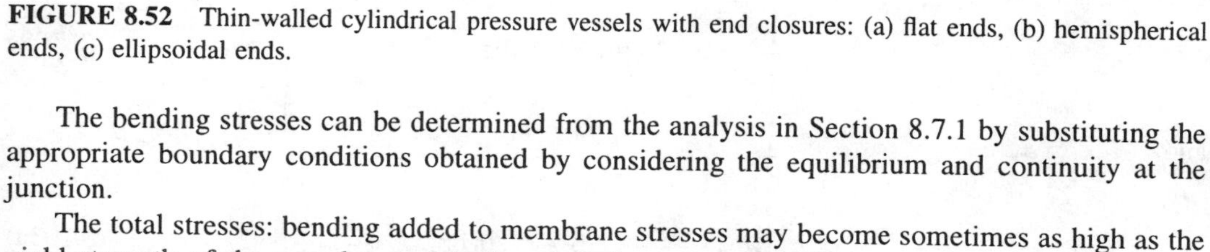

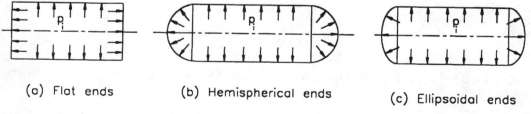

FIGURE 8.52 Thin-walled cylindrical pressure vessels with end closures: (a) flat ends, (b) hemispherical ends, (c) ellipsoidal ends.

The bending stresses can be determined from the analysis in Section 8.7.1 by substituting the appropriate boundary conditions obtained by considering the equilibrium and continuity at the junction.

The total stresses: bending added to membrane stresses may become sometimes as high as the yield strength of the vessel material. However, they should not be regarded as the determining factor in design calculations, since these bending stresses are of localized nature. For ductile materials, an overstress* in a (constrained) narrow region does not affect the load-carrying capacity of an elastically behaving structure. Restricted plastic deformation is often permitted in ductile materials under static loading. This may not be the case for brittle materials and/or under fluctuating loads.

In the following analysis, two types of ends will be considered: flat ends, Figure 8.52a, and curved ends, Figure 8.52b and c joined to the cylindrical shell by welding.

8.7.6.1 Case of a Flat End

Two cases of flat ends will be considered, namely, one joined by a single fillet weld and the other joined by a double fillet welds, as shown in Figure 8.15. From Section 8.5.1 it can be easily seen that for uniform strength the thickness of the flat end will be quite heavy compared with the cylinder wall thickness. Hence, the flat end may be considered to be rigid in comparison with the cylindrical wall.

Consider first the case of a single welded end. The boundary conditions at the edge of the cylindrical shell of the joint can be approximated by

$$u = 0,\ M_o = 0,\ \text{at } z = 0$$

as if it is assumed to be simply supported at the ends.

To satisfy these conditions, an open-ended pressurized cylinder is first considered. At the welds, i.e., at $z = 0$, the net displacement due to the internal pressure and the local loading has to vanish. Hence, using Equations 8.86 and 8.89 results in

$$u_c + u_o = \frac{p_i r_m^2}{Eh} + B = 0$$

giving

$$B = -\frac{p_i r_m^2}{Eh}$$

The condition that

$$M = -D\frac{d^2u}{dz^2} = 0 \ \text{ at } \ z = 0$$

yields $A = 0$ in Expression 8.86. Hence, the displacement and bending moment are given, respectively, by

$$u = -\frac{p_i r_m^2}{Eh} e^{-\beta z} \cos \beta z \tag{8.102}$$

and

$$M = -\frac{2p_i r_m^2}{Eh} D\beta^2 e^{-\beta z} \sin \beta z \tag{8.103a}$$

* For instance, the pressure vessel codes allow a stress about twice the yield strength of the vessel material at the junction with a flange joint.

The maximum bending moment is obtained by differentiating M with respect to z and equating to zero. This gives M_{max} at $z = \pi/4\beta$ as

$$M_{max} = -\frac{2p_i r_m^2}{Eh} D\beta^2 e^{-\pi/4} \sin(\pi/4) = 0.161 \frac{p_i r_m h}{\sqrt{3(1-\nu^2)}} \tag{8.103b}$$

The maximum total axial stress $(\sigma_z)_{max}$ is thus obtained from Expression 8.88a at 3 = $h/2$ by superposing the membrane stress for a closed vessel as

$$(\sigma_z)_{max} = \frac{6M_{max}}{h^2} + \frac{p_i r_m^2}{2h} \tag{8.104a}$$

This gives, for $\nu = 0.3$,

$$(\sigma_z)_{max} = 2.169 \frac{p_i r_m}{2h} \tag{8.104b}$$

An expression for the hoop stress σ_θ is obtained similarly from Equation 8.88b:

$$\sigma_\theta = \frac{6M_{max}}{h^2} + \frac{Eu}{r_m} + \frac{p_i r_m}{2h} \tag{8.105}$$

Substituting from Equation 8.102 and 8.103a for u and M, respectively, yields

$$\sigma_\theta = \frac{p_i r_m}{h}\left[\frac{3\nu}{\sqrt{3(1-\nu^2)}} e^{-\beta z} \sin\beta z + e^{-\beta z}\cos\beta z - 1\right] \tag{8.106a}$$

This attains a maximum value at $\beta z = 1.857$ as

$$(\sigma_\theta)_{max} = 1.128 \frac{p_i r_m}{h} \tag{8.106b}$$

which is slightly greater than $(\sigma_z)_{max}$ of Equation 8.104b.

Note that at the location where $(\sigma_z)_{max}$ occurs, i.e., at $z = \pi/4\beta$, σ_θ and the displacement u, respectively, have the values:

$$(\sigma_\theta)_{z=\pi/4\beta} = 0.853 \frac{p_i r_m}{h} \tag{8.107a}$$

$$(u)_{z=\pi/4\beta} = -0.3226 u_c = -\frac{0.3226 p_i r_m}{h} \tag{8.107b}$$

Considering the case of a double fillet welded end, the boundary conditions are assumed to be $u = du/dz = 0$ at $z = 0$. This represents the case already solved in Section 8.7.2, namely, that of a thin-walled vessel closed with rigid flanges. The results have been obtained according to Equation 8.91b as

$$M_{\max} = \frac{p_i}{2\beta^2} = 0.5\frac{p_i r_m h}{\sqrt{3(1-\nu^2)}} \tag{8.108}$$

and $u = 0$; both at $z = 0$.

Example 8.18:

Referring to Figure 8.15, consider the case of a steel cylindrical pressure vessel of a mean diameter 500 mm, 5 mm wall thickness, and subjected to an internal pressure of 1.2 MPa, which has (a) a single fillet welded end or (b) a double fillet welded end. Determine the maximum stresses in each of the two cases; E = *210 GPa and* ν = *0.3. For the same maximum stress obtain the thickness of the flat end in each case.*

Solution:

The following constants are determined as

$$\beta^2 = \frac{\sqrt{3(1-\nu^2)}}{r_m h} = \frac{\sqrt{3\times 0.91}}{250\times 5} = 0.00132\ \text{mm}^{-2} \ \text{ or } \ \beta = 0.0364\ \text{mm}^{-1}$$

$$D = \frac{Eh^3}{12(1-\nu^2)} = \frac{210\times 10^3\times 5^3}{12(1-0.3^2)} = 2.404\times 10^6\ \text{N}\cdot\text{mm}$$

a. For the single fillet welded flat head, at $z = \pi/4\beta = \pi/(4 \times 0.0364) = 21.6$ mm, M is maximum as given by Equation (8.103b) as

$$M_{\max} = \frac{0.161 p_i r_m h}{\sqrt{3(1-\nu^2)}} = \frac{0.161\times 1.2\times 250\times 5}{\sqrt{3(1-\nu^2)}} = 146.16\ \text{N}\cdot\text{mm/mm}$$

From Equation 8.104b and 8.107a, the total stresses at $z = \pi/4\beta$, superposing the membrane stresses due to p_i, are

$$(\sigma_z)_{\max} = 2.169\frac{p_i r_m}{2h} = 65.09\ \text{MPa}$$

and

$$\sigma_\theta = 0.853\frac{p_i r_m}{2h} = 51.18\ \text{MPa}$$

However, from Equation 8.106b at $\beta z = 1.857$,

$$(\sigma_\theta)_{\max} = 1.216\frac{p_i r_m}{h} = 72.96\ \text{MPa}$$

which is greater than $(\sigma_z)_{\max}$.

b. For the double-welded flat head, M is maximum at $z = 0$ and is given by Equation 8.108 as

$$M_{max} = 0.5\frac{p_i r_m}{\sqrt{3(1-\nu^2)}} = 453.92 \text{ N}\cdot\text{mm/mm}$$

From Equations 8.92c and 8.68b, the stresses are

$$(\sigma_z)_{max} = 4.631\frac{p_i r_m}{2h} = 138.96 \text{ MPa at } z = 0$$

and

$$(\sigma_\theta)_{max} = \frac{p_i r_m}{h} = 60 \text{ MPa}$$

away from the head.

This shows that a single fillet welded end is a better design since it introduces less discontinuity stresses at the junction.

8.7.6.2 Case of a Curved End

In this case the curved ends shown in Figure 8.52b and c is a surface of revolution of uniform thickness joined to the cylindrical shell by a butt weld such that the wall is continuous at the joint. Usually, the end has a straight edge of length s, to avoid welding in the highly stressed zone.

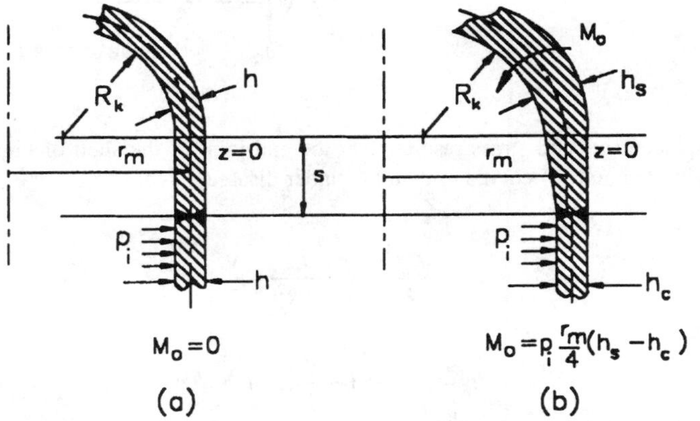

FIGURE 8.53 Joint of cylindrical shell to a curved end closure: (a) curved end and cylinder of the same thickness and (b) curved end and cylinder of different thicknesses.

If the end and the shell have the same thickness as shown in Figure 8.53a, a shearing force Q_o will act at the joint producing the same deflections and rotations at both edges indicating that they are free from bending. A bending M_o will develop together with Q_o if the thicknesses are different, Figure 8.53b, which is obtained from the joint equilibrium as

$$M_o = \frac{\pi r_m^2 p_i}{2\pi r_m}\frac{1}{2}(h_s - h_c) = p_i\frac{r_m}{4}(h_s - h_c) \tag{8.109}$$

Consider first the case of $h_s = h_c = h$, as shown in Figure 8.54. If the edges are not welded, the membrane stresses due the internal pressure p_i at $z = 0$ are obtained from Equations 8.68 for the cylindrical shell as

$$\sigma_z = \frac{p_i r_m}{2h} \quad \text{and} \quad \sigma_\theta = \frac{p_i r_m}{h}$$

For the curved head, the results of Example 8.12 are used so that the stresses at the joint are given by

$$\sigma_z = \frac{p_i r_m}{2h} \quad \text{and} \quad \sigma_\theta = \frac{p_i r_m}{2h}\left(2 - \frac{r_m^2}{R_c^2}\right)$$

where r_m and R_c are the major and minor semiaxes of the ellipsoidal head.

The radial displacements at the edges due to internal pressure, u_s for the head and u_c for the cylindrical shell Figure 8.54b, are obtained when the edges are not welded by substitution in Hooke's law, Equations 3.6 and Equation 8.71a, respectively, as

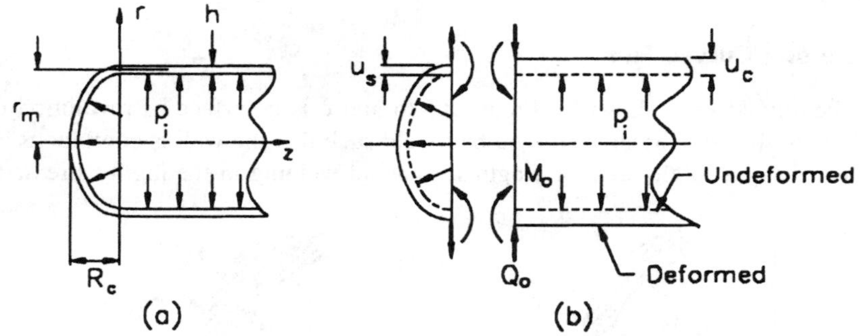

FIGURE 8.54 Displacements and stress resultants at the end joint of the shell of Figure 8.53: (a) curved end and cylinder assembled and (b) curved end and cylinder disassembled.

$$u_s = \frac{p_i r_m^2}{Eh}\left(1 - \frac{r_m^2}{2R_c^2} - \frac{\nu}{2}\right)$$
$$u_c = \frac{p_i r_m^2}{Eh}\left(1 - \frac{\nu}{2}\right) > u_s \tag{8.110}$$

When the edges are welded, they will have the same radial displacement u, where $u_c > u > u_s$. Assuming that the radial constraint due to Q_o is the same for both the curved head and cylindrical shell will give, $u_c - u_o = u_o - u_s$ and, hence,

$$u_o = \frac{u_c - u_s}{2} \tag{8.111}$$

where u_o is the radial displacement caused by the local loading at the junction. In other words, the discontinuity solution must result in a displacement u_o to cover the difference between the membrane displacements of both parts. Substituting u_s and u_c from Equation 8.110 into Equation 8.111 yields

$$u_o = \frac{p_i r_m^2}{4Eh}\left(\frac{r_m^2}{2R_c^2}\right) \quad \text{at } z = 0 \tag{8.112}$$

Substituting u_o from Equation 8.112 into the general expression for u, Equation 8.86, yields at $z = 0$;

$$B = -p_i \frac{r_m^4}{4EhR_c^2}$$

Again, substituting $M = 0$ at $z = 0$ into Expression 8.88d yields $A = 0$. Hence, the radial displacement due to local loading is given by

$$u = -p_i \frac{r_m^4}{4EhR_c^2} e^{-\beta z} \cos \beta z \tag{8.113}$$

and the bending moment is derived as

$$M = -D\frac{d^2u}{dz^2} = p_i \frac{Dr_m^4\beta^2}{2EhR_c^2} e^{-\beta z} \sin \beta z$$

or

$$M = \frac{1}{\sqrt{3(1-\nu^2)}} \frac{p_i r_m^3 h}{8R_c^2} e^{-\beta z} \sin \beta z \tag{8.114a}$$

This attains a maximum at $z = \pi/4\beta$ and is given by

$$M_{\max} = \frac{0.0403 p_i h r_m}{\sqrt{3(1-\nu^2)}}\left(\frac{r_m}{R_c}\right)^2 \tag{8.114b}$$

and

$$u = -0.0805 p_i \frac{r_m^2}{Eh}\left(\frac{r_m}{R_c}\right)^2 \tag{8.114c}$$

From which the localized bending stress $(\sigma_z)_{\max}$ is obtained by substituting $M_{\max}$ in Equation (8.88a). The stress $(\sigma_\theta)_{\max}$ as given by Equation (8.88b) is found to occur away from the joint, specifically at $\beta z = 1.85$.

8.7.6.3 Case of a Hemispherical End

For a hemispherical end $R_c = r_m$ and hence, from Equation 8.114b, the maximum moment in the cylinder wall is at $z = \pi/4\beta$ and given by

$$M_{\max} = \frac{0.0403 p_i r_m h}{\sqrt{3(1-\nu^2)}} \tag{8.115a}$$

Also from Equation 8.114c, the radial displacement is

$$u = -0.0805 \frac{p_i r_m^2}{Eh} \quad \text{at} \quad \beta z = \pi/4 \tag{8.115b}$$

From Equation 8.88a, the maximum axial total stress (bending plus membrane) which acts at the outer surface of the cylindrical shell, is

$$\left(\sigma_z\right)_{\max} = \frac{6M_{\max}}{h^2} + \frac{p_i r_m}{2h} = \frac{0.241 p_i r_m}{h\sqrt{3\left(1-\nu^2\right)}} + \frac{p_i r_m}{2h} \tag{8.116a}$$

For a material with $\nu = 0.3$, this simplifies to

$$\left(\sigma_z\right)_{\max} = 1.293 \frac{p_i r_m}{2h} \tag{8.116b}$$

In this case the stress that acts at the outer surface of the cylindrical shell is about 30% larger than the membrane stress.

The total hoop stress (bending plus membrane) at the outer surface of the shell is found from Equation 8.88b to be maximum at $\beta z = 1.85$ far from the joint and is given by

$$\left(\sigma_\theta\right)_{\max} = \frac{6\nu M}{h^2} + \frac{Eu}{r_m} + \frac{p_i r_m}{h} = 1.032 \frac{p_i r_m}{h} \tag{8.117}$$

Since the membrane stresses are smaller in a sphere than in cylinder walls — see Equations 8.67 and 8.68 — the maximum stresses in the spherical ends are always less than the stresses calculated by Equations 8.116. Thus, the latter stresses are the decisive factor in the design of the vessel.

For an ellipsoidal head, the ratio of differences between its radial expansion and that of a spherical head is $(r_m/R_c)^2$, as seen from Equations 8.112. Therefore, the discontinuity stresses are larger for an ellipsoidal head by the same ratio.

For instance for $(r_m/R_c)^2 = 4$ and in view of Equations 8.116b and 8.117, the stresses at the junction are

$$\left(\sigma_z\right)_{\max} = 2.172 \frac{p_i r_m}{2h} \quad \text{at} \quad \beta z = \pi/4 \tag{8.118a}$$

and

$$\left(\sigma_\theta\right)_{\max} = 1.128 \frac{p_i r_m}{2h} \quad \text{at} \quad \beta z = 1.85 \tag{8.118b}$$

Consider now the case of $h_s > h_c$ as shown in Figure 8.53b. In this case, the radial displacements at the edges due to internal pressure when the edges are not welded are obtained as

$$\begin{aligned} u_s &= \frac{p_i r_m^2}{Eh_s}\left(1 - \frac{r_m^2}{2R_c^2} - \frac{\nu}{2}\right) \\ u_c &= \frac{p_i r_m^2}{Eh_c}\left(1 - \frac{\nu}{2}\right) > u_s \end{aligned} \tag{8.119a}$$

Again, when the edges are welded, they will have the same radial displacement. Hence, at $z = 0$, the radial displacement u_o caused by the local loading at the junction is given by

$$u_o = \frac{u_c - u_s}{2} = \frac{p_i r_m^2}{2E}\left[\frac{1}{h_c}\left(1 - \frac{\nu}{2}\right) - \frac{1}{h_s}\left(1 - \frac{r_m^2}{2R_c^2} - \frac{\nu}{2}\right)\right] \tag{8.119b}$$

Substituting u_o from Equation 8.119b into Equation 8.86 and $M = p_i r_m (h_s - h_c)/4$ at $z = 0$ will give the values of the constants A and B. The maximum stresses are obtained by following the same procedure as in the case of equal wall thicknesses.

Table 8.9 summarizes the results of this section for several types of vessel heads.

TABLE 8.9
Maximum Local Bending Moments and Stresses in Pressure Vessel Heads

Type of Vessel Head	M_{max} and Location	$(\sigma_z)_{max}$ and Location	$(\sigma_\theta)_{max}$ and Location
Rigid flat heads	$\frac{0.5 p_i r_m h}{\sqrt{3(1-\nu^2)}}$ at $z = 0$	$\frac{4.631 p_i r_m}{2h}$ at $z = 0$	$\frac{p_i r_m}{h}$ at $z >>$
Thin flat heads (simply supported)	$\frac{0.161 p_i r_m h}{\sqrt{3(1-\nu^2)}}$ at $\beta z = \pi/4$	$\frac{2.169 p_i r_m}{2h}$ at $\beta z = \pi/4$	$\frac{1.126 p_i r_m}{h}$ at $\beta z = 1.857$
Hemispherical heads	$\frac{0.0403 p_i r_m h}{\sqrt{3(1-\nu^2)}}$ at $\beta z = \pi/4$	$\frac{1.293 p_i r_m}{2h}$ at $\beta z = \pi/4$	$\frac{1.032 p_i r_m}{h}$ at $\beta z = 1.85$
Ellipsoidal heads	$\frac{0.0403 p_i r_m h}{\sqrt{3(1-\nu^2)}}\left(\frac{r_m}{R_c}\right)^2$ at $\beta z = \pi/4$	$\frac{p_i r_m}{2h}\left[+0.293\left(\frac{r_m}{R_c}\right)^2\right]$ at $\beta z = \pi/4$	$\frac{p_i r_m}{h}\left[+0.032\left(\frac{r_m}{R_c}\right)^2\right]$ at $\beta z = 1.85$

Example 8.19:

A steel cylindrical pressure vessel of a mean diameter 800 mm, 10 mm wall thickness, with dished ends of the same thickness is subjected to an internal pressure of 1.6 MPa. Determine the maximum stresses in the wall if the shape of the head is one of the following: (a) hemispherical, (b) ellipsoidal with a major-to-minor-axis ratio of 2, (c) torispherical with a knuckle radius of R_k *= 80 mm.* E = *210 GPa, and* $\nu = 0.3$.

Solution:

For $r_m = 400$ mm, $h = 10$ mm, $E = 210$ GPa, and $\nu = 0.3$, the following constants are determined:

$$\beta^2 = \frac{\sqrt{3(1-\nu^2)}}{Rh} = \frac{\sqrt{3(1-0.3^2)}}{400 \times 10} = 0.000413 \text{ mm}^{-2}$$

$$D = \frac{Eh^3}{12(1-\nu^2)} = \frac{210{,}000 \times 10^3}{12(1-0.3^2)} = 192 \times 10^5 \text{ N}\cdot\text{mm}$$

$(\sigma_z)_{max}$ occurs at $z = \pi/4\beta = 38.65$ mm.

a. *Hemispherical end*: Substituting $r_m = 400$ mm in Equation 8.115a for M_{max} at $z = \pi/4\beta = 38.65$ mm yields

$$M_{max} = \frac{0.0403 \times 1.6 \times 400 \times 10}{\sqrt{3(1-0.3^2)}} = 156.1 \text{ N} \cdot \text{mm/mm}$$

From Equation 8.116a at the outer surface of the shell, the maximum axial stress is

$$(\sigma_z)_{max} = \frac{0.241 \times 1.6 \times 400}{10\sqrt{3 \times 0.91}} + \frac{1.6 \times 400}{2 \times 10}$$
$$= 9.335 + 32 = 41.33 \text{ MPa}$$

which is about 1.3 times the membrance stress ($\sigma_z = 32$ MPa) for a free edge. The maximum tangential stress from Equation 8.117 at $\beta_z = 1.85$, i.e., at $z = 91$ mm at the outer surface of the shell is

$$(\sigma_\theta)_{max} = 1.032 p_i \frac{r_m}{h} = 1.032 \times 1.6 \times \frac{400}{10} = 66 \text{ MPa}$$

which is greater than $(\sigma_z)_{max}$.

b. *Ellipsoidal end*:
Semimajor axis $r_m = 400$ mm
Semiminor axis $R_c = r_m/2 = 200$ mm
From Equation 8.114b,

$$M_{max} = \frac{0.0403 \times 1.6 \times 10 \times 400^3}{\sqrt{3 \times 0.91} \times 200^2} = 624.4 \text{ N} \cdot \text{mm/mm}$$

Hence, from Equations 8.88a and 8.68a, the maximum total axial stress at the outer surface of the shell:

$$(\sigma_z)_{max} = \frac{6 \times 624.4}{10^2} + \frac{1.6 \times 400}{2 \times 10} = 37.46 + 32 = 69.46 \text{ MPa}$$

which is about 2.17 times σ_z for a free edge, as indicated by Equation 8.118a. The maximum tangential stress is shown to occur away from the end at the outer surface as given by the second of Equations 8.108:

$$(\sigma_\theta)_{max} = 1.128 \frac{p_i r_m}{h} = 72.19 \text{ MPa}$$

which is the largest stress and is consequently the determining factor in vessel design.

c. *Torispherical end*: At the junction taking the meridional radius of curvature R_ϕ to be equal to the knuckle radius $R_k = 80$ mm, then from the results of Example 8.12:

$$R_\phi = R_k = R_c^2/r_m \text{ giving } R_c = \sqrt{80 \times 400} = 178.9 \text{ mm}$$

From Equation 8.114b at $z = \pi/4\beta = 38.65$ mm,

$$M_{max} = \frac{0.0403 \times 1.6 \times 10 \times 400^3}{\sqrt{3 \times 0.91} \times 178.92^2} = 780.2 \text{ N} \cdot \text{mm/mm}$$

And, from Equations 8.114c and 8.88a, respectively, superposing to Equation 8.68a,

$$u = \frac{1.6 \times 400^2}{210,000 \times 10}\left[1 - 0.15 - 0.08(400/178.9)^2\right] = 0.055 \text{ mm}$$

$$(\sigma_z)_{max} = \frac{6 \times 780.2}{10^2} + \frac{1.6 \times 400}{2 \times 10} = 46.81 + 32 = 78.81 \text{ MPa}$$

which is 2.46 times σ_z for a free edge. The results are compared in Table 8.10.

TABLE 8.10
Comparison Among Stresses at Junctions of Pressure Vessel Ends — Example 8.19

Shape of End	$(\sigma_z)_{max}$	σ_z Free Edge	$(\sigma_\theta)_{max}$	$(\sigma_\theta)_{max}$ Free Edge
Hemispherical	41.33	32	66	64
Ellipsoidal	69.46	32	72.2	64

Note that the distance s in Figure 8.53, locating the weld zone between the vessel and its end, should be outside the discontinuity stress zone.

Example 8.20:

Determine the displacement and the longitudinal stress at the junction of the pressure vessel of Example 8.19 for a hemispherical end of thickness 14 mm and a cylindrical shell thickness of 10 mm.

Solution:

From Equation 8.119b, at $z = 0$,

$$u_o = \frac{1.6 \times 400^2}{2 \times 210,000}\left[\frac{1}{10}(1 - 0.15) - \frac{1}{14}(1 - 0.5 - 0.15)\right]$$
$$= 0.0366 \text{ mm}$$

Also from Equation 8.109,

$$M_o = p_i \frac{r_m}{4}(h_s - h_c) = \frac{1.6 \times 400\ (14 - 10)}{4} = 640 \text{ N} \cdot \text{mm/mm}$$

The longitudinal stress σ_3 at $z = 0$ is obtained at the outer surface of the shell from Equations 8.88a and 8.68a as

$$(\sigma_z)_{max} = \frac{640 \times 6}{10^2} + 32 = 38.4 + 32 = 70.4 \text{ MPa}$$

which is 1.7 $(\sigma_z)_{max}$ if the end has the same thickness of the cylindrical shell as indicated in Table 8.10. It is therefore recommended to maintain uniform thickness in a pressure vessel with curved, dished ends.

8.7.7 Cylindrical Storage Tanks

Consider a cylindrical tank of uniform thickness filled up to a height H of a liquid of specific weight ρg. The tank bottom is assumed to be built-in and the top is open, as shown in Figure 8.55a. According to Table 8.8, the height of the tank H is considered, such that $H/\sqrt{Rh} > 2.5$, so that the analysis may be based on that of an infinitely long, open cylinder subjected to an internal pressure given by $p_i = \rho g\,(H - z)$. Knowing that $\sigma_\theta = p_i r_m/h$ and $\sigma_z = 0$, the radial displacement u_o due to this internal pressure is thus given by

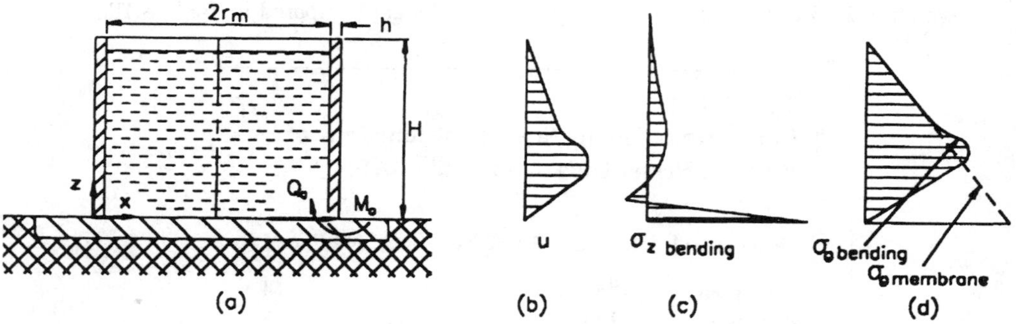

FIGURE 8.55 Displacements and stresses in a cylindrical vertical liquid storage tank: (a) tank geometry and loading, (b) distribution of radial displacement, (c) and (d) distribution of stresses.

$$u_o = \rho g r_m^2 (H - z)/(Eh)$$

Hence, the complete solution to the governing differential equation* is obtained from Equation 8.86 together with the particular integral represented by u_o as

$$u = e^{-\beta z}(A \sin \beta z + B \cos \beta z) + \frac{\rho g r_m^2 (H - z)}{Eh}$$

The constants A and B are determined from the boundary conditions:

$$\text{At } z = 0: \quad u = 0 \quad \text{and} \quad \frac{du}{dz} = 0$$

hence,

$$A = -\frac{\rho g r_m^2}{Eh}\left(H - \frac{1}{\beta}\right) \quad \text{and} \quad B = -\frac{\rho g r_m^2 H}{Eh}$$

Substitution of these constants yields an expression for the deflection as

---

* See footnote on page 509.

$$u = \frac{\rho g r_m^2 H}{Eh}\left\{1 - \frac{z}{H} - e^{-\beta z}\left[\cos\beta z + \left(1 - \frac{1}{\beta H}\right)\sin Bz\right]\right\} \tag{8.120}$$

An expression for the bending moment and the shearing force are derived according to Equation 8.83a as

$$M = -D\frac{d^2u}{dz^2} = \frac{2\beta^2\rho g r_m^2 DH}{Eh}e^{-\beta z}\left[\sin\beta z - \left(1 - \frac{1}{\beta H}\right)\cos\beta z\right] \tag{8.121}$$

$$Q = \frac{2\beta^3\rho g r_m^2 DH}{Eh}e^{-\beta z}\left[2\cos\beta z - \frac{1}{\beta H}(\sin\beta z + \cos\beta z)\right] \tag{8.122}$$

The maximum bending moment and shearing force occur at the bottom, i.e., at $z = 0$, and is equal to

$$M_{\max} = M_o = \left(1 - \frac{1}{\beta H}\right)\frac{\rho g r_m Hh}{\sqrt{12(1-\nu^2)}} \tag{8.123a}$$

and

$$Q_o = \frac{\rho g r_m Hh}{\sqrt{12(1-\nu^2)}}\left(2\beta - \frac{1}{H}\right) \tag{8.123b}$$

The axial and hoop stresses due to this local bending is obtained from Equations 8.88, 8.120, and 8.121. A typical plot for the distributions of u and σ_z along the height of the tank is shown in Figure 8.55b and c, respectively. At the built-in edge, where $z = 0$, the following stresses are obtained:

$$(\sigma_z)_{\text{bending}} \frac{6M_{\max}}{h^2} = \frac{\sqrt{3}\,\rho g r_m H}{(1-\nu^2)h}\left(1 - \frac{1}{\beta H}\right)$$

$$(\sigma_\theta)_{\text{bending}} = \nu\sigma_z$$

If the terms $(1/\beta H)$ and ν^2 are neglected with respect to unity, the axial bending stress at the built-in edge is $\sqrt{3}$ times larger than the hoop stress predicted by the membrane theory.

It is also observed from Figure 8.55d that the hoop stress increases from zero at the liquid level to rather considerable values at greater depths. This suggests increasing the wall thicknesss from top to bottom, which is commonly adopted in engineering practice.

Example 8.21:

Find the wall thickness at the lower part of a steel cylindrical oil storage tank having the dimensions of 4 m height and 6 m diameter. Consider the bottom wall to be built-in. Compare the thickness obtained with that calculated on the basis of membrane theory only. Take a yield strength Y = *200 MPa,* E = *200 GPa,* ν = *0.3, and a factor of safety F.S. = 1.5.*

Solution:

At the built-in edge, the stresses are given by

Edge axial bending stress	σ_z	$=$	$6M_o/h^2$
Edge hoop bending stress	σ_θ	$=$	$\nu\sigma_z$
Radial membrane stress	σ_r	$\cong$	0

Hence, according to the maximum shear stress theory,

$$\sigma_z \leq \frac{Y}{\text{F.S.}}$$

From Equation 8.121, the bending moment at $z = 0$ is

$$M_{\max} = \left(1 - \frac{1}{\beta H}\right)\frac{\rho g r_m H h}{\sqrt{12(1-\nu^2)}}$$

$$\beta^4 = \frac{3(1-\nu^2)}{r_m^2 h^2} = \frac{3 \times 0.91}{3^2 \times h^2} = \frac{0.303}{h^2} m^{-4}$$

hence, $\beta = 0.742/\sqrt{h}$ (h in meters).

Since most design codes specify testing storage tanks by water filling, ρg is taken as that of water not oil; hence,

$$M_{\max} = \left(1 - \frac{\sqrt{h}}{0.742 \times 4}\right)\left(\frac{9810 \times 3 \times 4 \times h}{\sqrt{12(1-0,3^2)}}\right)$$

$$M_{\max} = \left(1 - \frac{\sqrt{h}}{2.97}\right)(35623.7h) \text{ Nm/m}$$

The term $\sqrt{h}/2.97$, where h is in meters, is very small compared with unity; hence,

$$M_{\max} = 35623.7h \text{ Nm/m}$$

and the required thickness is calculated from

$$\frac{200 \times 10^6}{1.5} = \frac{6 \times 35623.7h}{h^2}$$

giving $h = 1.6$ mm.

Calculation of thickness according to membrane stresses, Equation 8.68b, gives

$$\sigma_\theta = \frac{\rho g H r_m}{h} \leq \frac{Y}{\text{F.S.}}$$

hence,

$$\frac{200 \times 10^6}{1.5} = \frac{9810 \times 4 \times 3}{h}$$

giving $h = 0.883$ mm.

The actual thickness for such large storage tanks is often taken larger than this to allow for corrosion and geometric stability during handling and erection.

8.7.8 Effect of Thermal Gradient

The above analysis is valid for uniform constant temperature throughout the entire cylinder wall. It should be observed that if the temperature is uniform, but not constant, then severe stresses may arise if the displacement due to temperature changes is constrained. Since the wall is thin, a radial thermal gradient is expected to be negligible and only an axial thermal gradient can take place.

By referring to Equations 8.78 the thermoelastic stress–strain relations are expressed as

$$\sigma_z = \frac{E}{1-\nu^2}\left[\frac{dw}{dz} + \nu\frac{u}{r_m} - \frac{d^2u}{dz^2}\zeta - (1+\nu)\alpha T\right]$$

$$\sigma_\theta = \frac{E}{1-\nu^2}\left[\frac{u}{r_m} + \nu\frac{dw}{dz} - \nu\frac{d^2u}{dz^2}\zeta - (1+\nu)\alpha T\right] \tag{8.124a}$$

where α is the coefficient of linear expansion and T is the temperature rise.

For axially unconstrained edges, the stress resultant N_z per unit length of circumference defined by Expression 8.6, vanishes and, hence,

$$N_z = \int_{-h/2}^{h/2} \sigma_z d\zeta \frac{E}{1-\nu^2}\int_{-h/2}^{h/2}\left[\frac{dw}{dz} + \nu\frac{u}{r_m} - \frac{d^2u}{dz^2}\zeta - (1+\nu)\alpha T\right]d\zeta = 0$$

Noting that

$$\int_{-h/2}^{h/2} \zeta d\zeta = 0$$

This reduces to

$$N_z = \frac{E}{1-\nu^2}\left(\frac{dw}{dz} + \nu\frac{u}{r_m} - (1+\nu)\alpha T\right) = 0$$

or

$$\frac{dw}{dz} + \nu\frac{u}{r_m} = (1+\nu)\alpha T$$

By substituting in Expressions 8.124a, the stress components are obtained as

$$\sigma_z = -\frac{E}{1-\nu^2}\frac{d^2u}{dz^2}\zeta$$

$$\sigma_\theta = \frac{E}{1-\nu^2}\left[\left(1-\nu^2\right)\left(\frac{u}{r_m}-\alpha T\right)-\nu\frac{d^2u}{dz^2}\zeta\right] \tag{8.124b}$$

which gives the same expression for σ_z, Equations 8.80, where there is no thermal gradient. σ_θ is obtained by substituting u/r_m by $[(u/r_m) - \alpha T]$ in Equations 8.80. The stress resultants are thus given by an expression similar to Expressions 8.87 as

$$N_\theta = Eh\left(\frac{u}{r_m}-\alpha T\right),\quad M_z = -D\frac{d^2u}{dz^2},\quad M_\theta = \nu M_z,\quad Q = -D\frac{d^3u}{dz^3}$$

By substituting these relations in the equations of equilibrium and following the same steps of solution as for the case of uniform temperature, the following expression is obtained:

$$u = e^{-\beta z}\,(A \sin \beta z + B \cos \beta z) \tag{8.124c}$$

which is identical to Equation 8.86.

Example 8.22:

In constructing a pipeline by welding, excessive localized heat is generated at the pipe joint leading to thermal stresses. Show that the maximum stress in the joint due to heating is independent of the pipe dimensions. Apply the results to a steel pipe, E = *190 GPa at 300°C,* ν = *0.3,* α = *11 ×* 10^{-6}*/°C. The yield stress at 300°C is 220 MPa.*

Solution:

At the pipe joint section, take $z = 0$, $u = u_o$, $T = T_o$, and $du/dz = 0$, which yields, from Equation 8.124c,

$$A = B = u_o$$

$$u = u_o e^{-\beta z}(\sin\beta z + \cos\beta z)$$

M is maximum at $z = 0$ and is given by

$$(\sigma_z)_{max} = \frac{12D\beta^2}{h^2}u_o$$

Putting

$$D = \frac{Eh^3}{12(1-\nu^2)},\quad \beta^2 = \frac{\sqrt{3(1-\nu^2)}}{r_m h},\quad \text{and}\quad u_o = \alpha T r_m$$

yields

$$(\sigma_z)_{max} = \alpha ET\sqrt{\frac{3}{1-\nu^2}}$$

which is independent of the pipe dimensions. For the steel pipe,

$$\left(\sigma_z\right)_{max} = 11\times10^{-6}\times190\times10^3\times300\times\sqrt{\frac{3}{1-0.3^2}}$$
$$= 1138.4 \text{ MPa}$$

This is a very high stress, which indicates that the material must have already yielded.

Example 8.23:

Steam, at 100°C under atmospheric pressure, passes through a steel pipe of 300 mm inside diameter and 6 mm wall thickness. The pipe is stiffened by a rigid ring, Figure 8.56. Determine the maximum stresses. Take the coefficient of linear expansion $\alpha = 11\times10^{-6}$/°C for steel: E = 210 GPa, and $\nu = 0.3$.

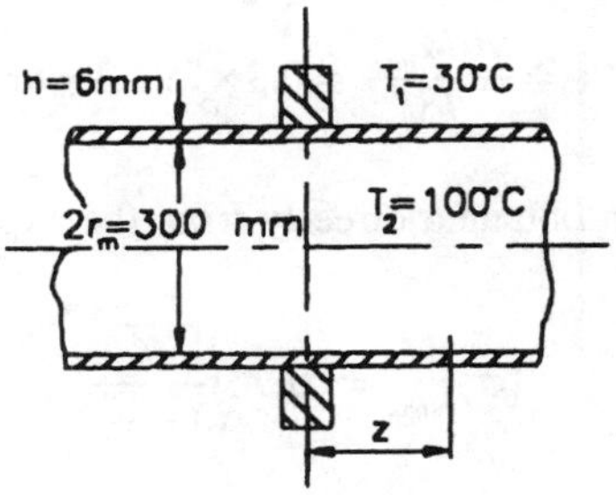

FIGURE 8.56 Example 8.23.

Solution:

Due to internal heating, the pipe is subjected to a uniform radial displacement u_c, which is restricted at the location of the rigid stiffening ring. This is similar to the case of a pressurized pipe with a rigid flange treated in Section 8.7.2.

Taking the stiffening ring middle plane as the reference for the z-direction, the inward radial displacement of the pipe superimposed by ring restriction is obtained as

$$u = -u_c e^{-\beta z}(\sin\beta z + \cos\beta z)$$

where

$$u_c = \alpha(T_1 - T_2)\, r_m$$

Hence,

$$u_c = 11\times10^{-6}\times(100-30)\times153 = 0.1178 \text{ mm}$$

At $z = 0$, $M = 2Du_c\beta^2$ from Equation 8.91b, and $(\sigma_z)_{max}$ is thus given as

$$\left(\sigma_z\right)_{max} = \frac{12Du_c\beta^2}{h^2}$$

where

$$D = \frac{Eh^3}{12\left(1-\nu^3\right)} = \frac{210\times10^3\times6^3}{12\left(1-0.3^2\right)} = 415\times10^4 \text{ N}\cdot\text{mm}$$

and

$$\beta^2 = \frac{\sqrt{3(1-\nu^2)}}{r_m h} = \frac{\sqrt{3\times 0.91}}{153\times 6} = 18\times 10^{-4}\,\text{mm}^{-2}$$

Hence,

$$\left(\sigma_z\right)_{\max} = \frac{12\times 415\times 10^4 \times 0.1178\times 18\times 10^{-4}}{6^2}$$

$$= 293.3 \text{ MPa}$$

and

$$\left(\sigma_\theta\right)_{\max} = \frac{\nu 6 M_{\max}}{h^2} = 0.3\times 293.3 = 88 \text{ MPa}$$

Note that $(\sigma_z)_{\max}$ could have been obtained directly from the expression

$$\left(\sigma_z\right)_{\max} = \alpha E T \sqrt{\frac{3}{1-\nu^2}}$$

as shown in Example 8.22.

8.8 ELASTIC BUCKLING OF PLATES AND SHELLS

Thin plates and shells subjected to compressive loading may buckle when the applied load reaches a critical value. Although the method of solution for plates and shells is quite similar to that for buckling of slender columms, Section 7.7, a general solution is mathematically more involved.[8] In the following analysis three simple problems are considered; the first two are plate problems solved by applying the energy approach and the third is an axisymmetric shell problem solved by direct equilibrium analysis.

8.8.1 Buckling of Uniformly Compressed Rectangular Plate

For a thin plate compressed by in-plane forces, the energy method can be effectively used to find approximate values for these critical forces. To demonstrate this method, consider a rectangular plate loaded at its edges by uniformly distributed in-plane compressive forces q_x and q_y per unit length, as shown in Figure 8.57. At the instant of buckling, the middle plane of the plate undergoes some small lateral deflection w consistent with the plate edge conditions, while u and ν also remain small. Expressions of the strain components at the middle plane are derived from the strain–displacemnent relations, Equations 2.36, for large deformations by neglecting squares of derivatives of u and ν with respect to x and y, as

$$\varepsilon_x = \frac{\partial u}{\partial x} + \frac{1}{2}\left(\frac{\partial w}{\partial x}\right)^2$$
$$\varepsilon_y = \frac{\partial \nu}{\partial y} + \frac{1}{2}\left(\frac{\partial w}{\partial y}\right)^2 \qquad (8.125)$$

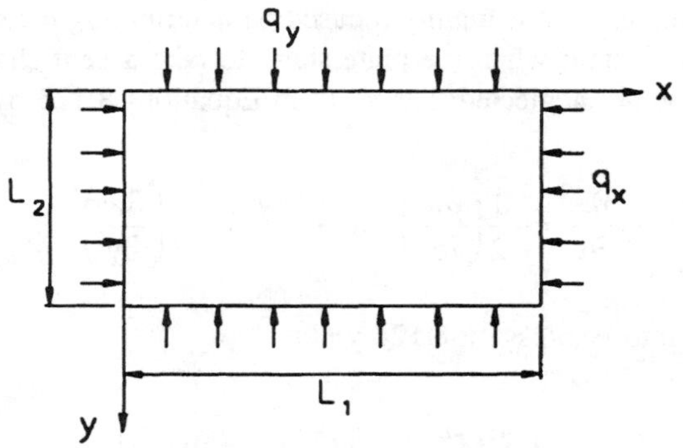

FIGURE 8.57 Buckling of a rectangular plate subjected to biaxial in-plane edge compression.

Note that $\partial w/\partial x$ and $\partial w/\partial y$ are the slopes of the middle surface with the x- and y-directions, respectively, as given by Equation 8.1 and are not existent if the plate remains plane.* In the previous analysis of bending of plates, the middle plane strains have been neglected. Here, however, they have to be taken into consideration since this small strain in combination with the finite in-plane forces may add to the strain energy some terms of the same order of magnitude as bending strain energy.

The increment of work done by the external loading when changing from the plane form to the bent configuration is given by

$$\delta W = \int_0^{L}\left[q_x(u)_{x=0} - q_x(u)_{x=L_1}\right]dy + \int_0^{L_1}\left[q_y(\mathrm{v})_{y=0} - q_y(\mathrm{v})_{y=L_2}\right]dx$$

since

$$(u)_{x=L_1} - (u)_{x=0} = \int_0^{L_1}\frac{\partial u}{\partial x}dx$$

and

$$(\mathrm{v})_{y=L_2} - (\mathrm{v})_{y=0} = \int_0^{L_2}\frac{\partial \mathrm{v}}{\partial y}dy$$

the expression for the virtual work done reduces to

$$\delta W = -\int_0^{L_2}\int_0^{L_2}\left[q_x\frac{\partial u}{\partial x} + qy\frac{\partial \mathrm{v}}{\partial y}\right]dxdy \tag{8.126}$$

---

* The first of Expressions 8.115 may be interpreted to give the strain of an element of length dx of the middle surface as being the sum of two parts. The first is the infinitesimal strain $\partial u/\partial x$ due to the displacement u, while the second is a component due to deflection w as

$$dx\left[1+\left(\frac{\partial w}{\partial x}\right)^2\right]^{1/2} - dx \cong \frac{1}{2}\left(\frac{\partial w}{\partial x}\right)^2$$

As in column buckling, the interest is mainly focused on determining the smallest critical buckling load, which occurs at the instant when the plate starts to take a bent shape. At this very instant, strains at the middle plane are neglected; hence, from Equations 8.125 by setting $\varepsilon_x = \varepsilon_y = 0$,

$$\frac{\partial u}{\partial x} = -\frac{1}{2}\left(\frac{\partial w}{\partial x}\right)^2 \quad \text{and} \quad \frac{\partial v}{\partial y} = -\frac{1}{2}\left(\frac{\partial w}{\partial y}\right)^2$$

which upon substitution into Expression 8.126 yields:

$$\delta W = \frac{1}{2}\int_0^{L_2}\int_0^{L_1}\left[q_x\left(\frac{\partial u}{\partial x}\right)^2 + qy\left(\frac{\partial w}{\partial y}\right)^2\right]dxdy \tag{8.127}$$

Note that Expression 8.127 is simply the two-dimensional generalization of Equation 7.87 developed in Section 7.7.2 for column buckling.

The increase in the strain energy due to plate bending is given by Expression 8.20 as

$$\delta U = \frac{D}{2}\int_0^{L_2}\int_0^{L_1}\left\{(\nabla^2 w)^2 - 2(1-\nu)\left[\frac{\partial^2 w}{\partial x^2}\frac{\partial^2 w}{\partial y^2} - \left(\frac{\partial^2 w}{\partial x \partial y}\right)^2\right]\right\}dxdy \tag{8.128}$$

The change in the total potential energy of the plate at the instant of buckling must vanish for stable equilibrium according to Expression 3.42, namely,

$$\delta\Omega = \delta U - dW = 0$$

To get a good approximation for the critical force, it is required to find an admissible expression for $w(x, y)$ satisfying the above condition of minimum total potential energy together with the boundary conditions of the plate under consideration.

8.8.1.1 All-Around Clampled Rectangular Plate

Taking coordinate axes as shown in Figure 8.55, the deflection function in this case may be assumed in the form:

$$w = C\left[1 - \cos\left(\frac{2\pi x}{L_1}\right)\right]\left[1 - \cos\left(\frac{2\pi y}{L_2}\right)\right] \tag{8.129}$$

which is seen to satisfy the kinematic boundary conditions for clamped edges, as previously shown in Section 8.4.2. Substitution of Expression 8.129 into the condition of minimum potential energy, Equation 3.38, yields

$$\left(q_x + \frac{L_1^2}{L_2^2}q_y\right)_{\text{critical}} = \frac{4\pi^2}{3}DL_1^2\left(\frac{3}{L_1^4} + \frac{3}{L_2^4} + \frac{2}{L_1^2 L_2^2}\right) \tag{8.130}$$

For a square plate ($L_1 = L_2 = L$) subjected to two equal compressive forces $q_x = q_y = q$, Expression 8.130 simplifies to

$$q_{cr} = 5.33\pi^2 D/L_1^2$$

If the plate is compressed in the x-direction only, Expression 8.130 gives, by putting $L_1 = L_2 = L$ and $q_y = 0$,

$$(q_x)_{cr} = 10.64\pi^2 D/L_1^2 = 105\, D/L^2$$

A more accurate estimation for the critical force q_x is obtained by taking an expression for w involving several coefficients and applying the Rayleigh–Ritz method to determine them. Results of such a procedure are tabulated in Table 8.11, where the critical force is expressed by

TABLE 8.11
Values of α in Equation 8.131 for Buckling of an All-around Clamped Rectangular Plate Compressed in Axial Direction[a]

L_1/L_2	0.5	1.0	1.5	2	2.5	3.0	4
α	195.5	103.5	83.4	79.6	78.9	75.2	73.5

[a] (See footnote, page 541.)

$$(q_x)_{cr} = \alpha D/L_2^2 \tag{8.131}$$

The number of waves into which the buckled plate is subdivided certainly depends on the aspect ratio L_1/L_2.

8.8.1.2 Rectangular Plates with Other Boundary Conditions

Results for the critical buckling forces of rectangular plates with various boundary conditions are found in the literature.[5] A solution to a simply supported plate is given in Example 8.24.

Example 8.24:

Determine the critical buckling force for a simply supported square plate as shown in Figure 8.58. The plate is uniformly loaded by in-plane q_x *edge forces only. Assume a deflection function in the form:* $w = C \sin(\pi x/L) \sin(\pi y/L)$.

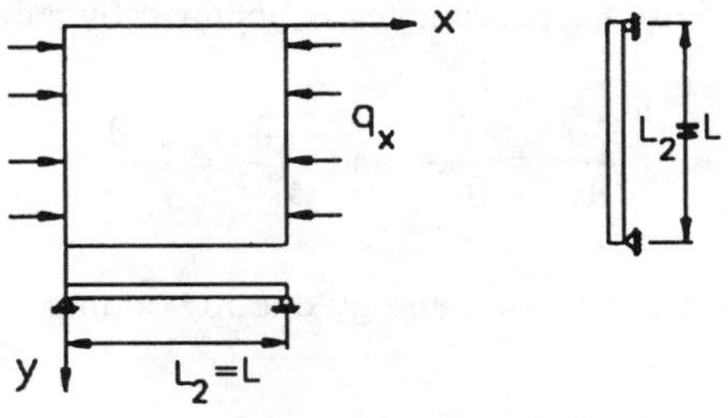

FIGURE 8.58 Example 8.24.

Solution:

The assumed deflection function satisfies the boundary conditions as previously indicated by Equation 8.22b. Substitution from the deflection function w into Equations 8.127 and 8.128 with $q_y = 0$ gives, respectively;

$$\delta w = \frac{1}{2}\int_0^L \int_0^L \left[q_x \left(\frac{\pi}{L}\right)^2 C^2 \cos^2\left(\frac{\pi x}{L}\right) \sin^2\left(\frac{\pi y}{L}\right) \right] dxdy = \frac{q_x \pi^2 C^2}{8}$$

and

$$\delta U = \frac{D}{2}\left(\frac{\pi^4 C^2}{L^2}\right)$$

The principle of minimum total potential energy ($\delta U - \delta W = 0$) gives

$$(q_x)_{cr} = 4\pi^2 D/L^2 = 39.48 D/L^2$$

say,

$$(q_x)_{cr} = \beta D/L^2 \tag{8.132}$$

The detailed solution for a simply supported rectangular plate indicates that the factor β becomes dependent on the aspect ratio L_1/L_2 of the plate as given in Table 8.12.

TABLE 8.12
Values of β in Equation 8.132: $(q_x)_{cr} = \beta D/L_2^2$, for Buckling of an All-Around Simple Supported Rectangular Plate Compressed in Axial Direction

L_1/L_2	0.4	1.0	1.4	2.0	3	∞
β	83.04	39.48	44.16	39.48	39.48	39.48

From Roark, J. R. and Young, W. C., *Formulas for Stress and Strain*, 5th ed., McGraw-Hill, New York, 1975.

8.8.2 Axisymmetric Buckling of Circular Plates

The solution to this problem in polar coordinates is obtained by substituting in Equation 8.128:

$$\frac{\partial}{\partial x^2} = \frac{\partial}{\partial r^2} \quad \text{and} \quad \frac{\partial}{\partial y^2} = \frac{1}{r}\frac{\partial}{\partial r}$$

which gives the increase in the plate strain energy due to bending as

$$\delta U = \frac{D}{2}\int_0^{2\pi}\int_0^{r_o}\left[\left(\frac{\partial^2 w}{\partial r^2} + \frac{1}{r}\frac{\partial w}{\partial r}\right)^2 - 2(1-\nu)\left(\frac{\partial^2 w}{\partial r^2}\right)\left(\frac{1}{r}\frac{\partial w}{\partial r}\right)\right] r dr d\theta \tag{8.133a}$$

The increment of the work done by the external forces q_r due to bending of the plate is expressed by Equation 8.127 in terms of polar coordinates as

$$\delta W = \frac{1}{2}\int_0^{2\pi}\int_0^{r_o} q_r \left(\frac{\partial w}{\partial r}\right)^2 r dr d\theta \qquad (8.133b)$$

Consider first an all-around clamped plate loaded radially by q_r as shown in Figure 8.59. The shape of the buckled plate is assummed to follow the function:

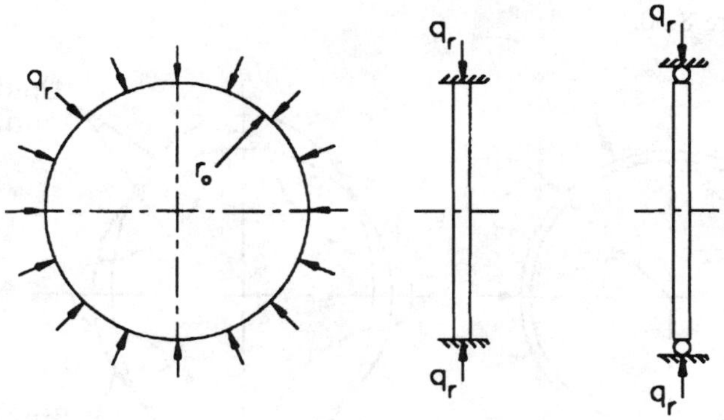

FIGURE 8.59 Buckling of a circular plates subjected to in-plane edge distributed radial force.

$$w = C\left(1 - \frac{r^2}{r_o^2}\right)^2 \qquad (8.134)$$

which is similar to that of a plate laterally loaded by uniform pressure, Equation 8.37. Substituting from Expression 8.134 into Expressions 8.133 yields, respectively, after integration,

$$\delta U = \frac{32\pi D}{3}\frac{C}{r_o^2}$$

and

$$\delta W = \frac{2\pi}{3} C q_r$$

Applying the condition of minimum potential energy results in

$$\left(q_r\right)_{cr} = \frac{16D}{r_o^2}$$

The exact values[3] for the critical radially distributed force is

$$(q_r)_{cr} = 14.68D/r_o^2 \text{ for all-around clamped} \qquad (8.135a)$$

$$(q_r)_{cr} = 4.20D/r_o^2 \text{ for all-around simply supported} \qquad (8.135b)$$

8.8.3 Buckling of Thin-Walled Cylinders Under External Uniform Pressure

Consider a long, thin-walled cylinder of a mean radius r_m and wall thickness h subjected to an external pressure p_o. The cylinder is considered to be in a state of stable configuration as long as it remains circular in shape. As soon as the wall becomes noncircular as a result of p_o, the cylinder will be in an unstable condition and is bound to buckle.

In the following analysis the cylinder is sufficiently long that end effects are neglected and the problem is treated as two dimensional. Again, the deviations from the circular shape are arbitrarily small and symmetrical about two orthogonal axes, which are taken to be the coordinate axes x and y, as shown in Figure 8.60.

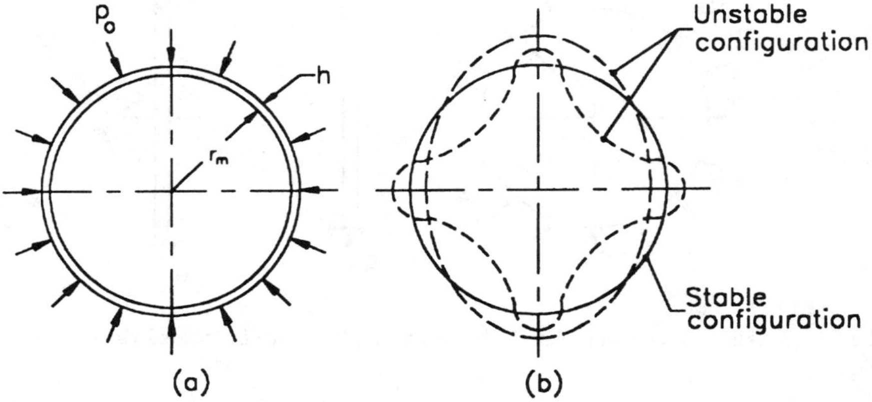

FIGURE 8.60 Buckling of a thin-walled cylinder subjected to external uniform pressure: (a) cylinder loading and b) stable and unstable configurations.

Consider an element ds of unit circumferential length in the unstable configuration. Let R be the radius of curvature of the element ds. The element is subjected to the system of forces per unit axial length as shown in Figure 8.61b.

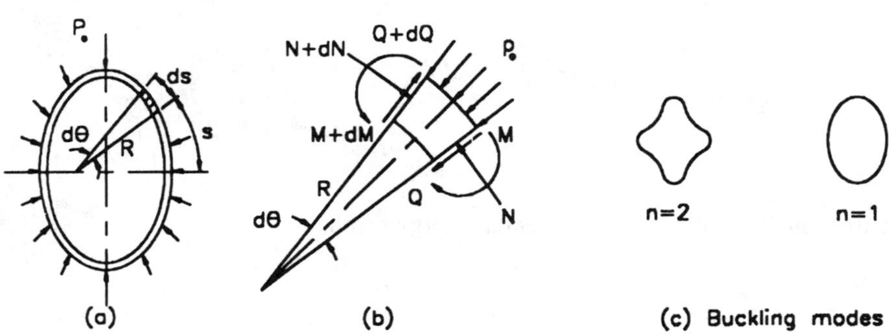

FIGURE 8.61 Static equilibrium of a wall element and buckling modes for the externally pressurized cylinder of Figure 8.60: (a) unstable configuration loading, (b) static equilibrium of wall element, and (c) buckling modes for $n = 1$ and $n = 2$. (From Windenberg, D. F. and Trilling, C., *Pressure Vessel and Piping Design,* ASME collected papers, 1960. With permission.)

Equilibrium of forces acting on the element in the radial and tangential direction, respectively, gives

$$p_o - \frac{dQ}{ds} - \frac{N}{R} = 0 \tag{8.136a}$$

and

$$Q - R\frac{dN}{ds} = 0 \tag{8.136b}$$

Substituting $Q = dM/ds$ in Equation 8.136b yields

$$\frac{dN}{ds} = \frac{1}{R}\frac{dM}{ds}$$

Now, since the deviations from the circular shape are small, both sides can be integrated neglecting the variation of R with respect to s, which gives

$$N = \frac{M}{R} + C$$

where C is an integration constant obtained from the condition that, when $R = r_m$, the cylinder wall is circular and $M = Q = 0$ and $N = p_o r_m$. Hence,

$$N = \frac{M}{R} + p_o r_m \tag{8.137}$$

Substituting Equation 8.137 in Equation 8.136a, together with $Q = dM/ds$, yields

$$\frac{d^2M}{ds^2} + \frac{M}{R^2} + p_o r_m\left(\frac{1}{R} - \frac{1}{r_m}\right) = 0$$

In Section 7.3.3 for a curved beam, Equation 7.21 together with Equation 7.26, which gives M in terms of the initial radius r_m and deformed radius R, is adapted to a curved plate as*

$$M = D\left(\frac{1}{R} - \frac{1}{r_m}\right)$$

where EI is replaced by D previously given by Equation 8.10 as

$$D = \frac{Eh^3}{12(1-\nu^2)}$$

Hence,

* Equation 7.21 for a curved beam: $M = ER_o(1/R_1 - 1/R_o)Ae$, where $e \cong I/Ar_m$ and $r_m = R_o$ is used for a thin-walled cylinder.

$$\frac{d^2M}{ds^2}+\left(\frac{1}{R^2}+p_o\frac{r_m}{D}\right)M=0$$

Again, neglecting the variation in R by taking $R^2 \cong r_m^2$ yields

$$\frac{d^2M}{ds^2}+\left(\frac{1}{r_m^2}+p_o\frac{r_m}{D}\right)M=0$$

The solution of this differential equation is

$$M = C_1 \sin ks + C_2 \cos ks \tag{8.138}$$

where

$$k^2=\left(\frac{1}{r_m^2}+p_o\frac{r_m}{D}\right) \tag{8.139}$$

and C_1 and C_2 are integration constants.

To determine C_1 and C_2 consider the variation of M with respect to s, from Equation 8.138 as

$$\frac{dM}{ds}=C_1k\cos ks-C_2k\sin ks \tag{8.140}$$

Now from symmetry about the x- and y-axes, $dM/ds = 0$ at $s = 0$ and $s = \pi r_m/2$, which when substituted in Equation 8.139 yields

$$C_1 = 0 \text{ and } C_2\, k \sin(k\pi r_m/2) = 0 \tag{8.141}$$

From Equation 8.141, either $C_2 = 0$ or $\sin(k\pi R/2) = 0$. If $C_2 = 0$, then from Equations 8.138 and 8.140 $M = dM/ds = 0$, which describes the case of the stable configuration when the wall shape remains circular. Therefore, the unstable configuration is given by the condition that

$$\sin(k\pi r_m/2) = 0$$

or

$$kr_m = 2n,\ n = 1, 2, 3, 4, \ldots$$

where $2n$ is an even number representing the circumferential lobes formed at symmetrical buckling about two orthogonal axes as shown in Figure 8.61c.*

Substituting k from Equation 8.139 yields

$$p_o=\frac{D}{r_m^3}\left(4n^2-1\right)$$

The minimum value of p_o is obtained at $n = 1$. By substituting $D = Eh^3/[12(1 - \nu^2)]$, the critical external uniform pressure, which causes buckling, is thus given by

* Note that n refers to the buckling mode as in column buckling in Section 7.7

$$(p_o)_{cr} = \frac{E}{4(1-\nu^2)}\left(\frac{h}{r_m}\right)^3 \tag{8.142}$$

Example 8.25:

Find the minimum wall thickness for a long, thin-walled-cylindrical tank of a mean diameter 1200 mm to sustain an external pressure of 0.1 MPa, if the tank is made of: (a) steel, E = 210 GPa, allowable stress σ_{all} = 150 MPa, specific weight ρg = 78 kN/m³; (b) brass, E = 110 GPa, σ_{all} = 80 MPa, ρg = 85 kN/m³ (c) aluminum alloy, E = 75 GPa, σ_{all} = 60 MPa, ρg = 28 kN/m³; (d) glass fiber–reinforced polyester (GRP), E = 18 GPa, σ_{all} = 30 MPa, ρg = 17 kN/m³. Take ν = 0.3. Select the optimum material for the tank.

Solution:

Substitute p_o = 0.1 MPa, r_m = 600 mm in Equation 8.142; hence,

$$0.1 = \frac{E\times10^3}{4(1-0.3^2)}\left(\frac{h_{min}}{600}\right)^3$$

or

$$h_{min} = \left(\frac{78624}{E}\right)^{1/3} \text{ mm } (E \text{ in GPa})$$

According to the cylinder material the minimum thickness h_{min} required to support the external pressure is obtained as given in Table 8.13. The tangential stress ($\sigma_\theta = 0.1 \times 600/h$ MPa) according to Equation 8.68b, is also calculated and given in Table 8.13 for each material. Obviously, σ_θ is only a small fraction of σ_{all}. From Table 8.13, based on safe strength, the brass tank has the maximum weight per unit area of tank surface N/m², while the GRP tank has the minimum weight, which is nearly 50% of the steel tank. The optimum material is decided from cost considerations.

TABLE 8.13
Thicknesses and Stresses in Externally Pressurized Tank (1200 mm Mean Diameter at 0.1 MPa)

Material	E, GPa	h_{min},[a] mm	σ_θ,[b] MPa	σ_{all}, MPa	ρgh, N/m²
Steel	210	7.21	−8.2	150	562
Brass	110	8.93	−6.7	80	756
Aluminum alloy	75	10.18	−5.9	60	281
GRP	18	16.35	−3.6	30	278

[a] Based on buckling, Equation 8.142.
[b] Based on stable equilibrium, Equation 8.68b.

In large vessels, the minimum wall thickness is usually increased to account for damage that may occur during transportation, erection, etc. Often the wall is stiffened on the outside surface in the axial and circumferential directions, thus counteracting buckling and allowing use of a smaller wall thickness compatible with strength and corrosion requirements.

8.8.3.1 Effect of Out-of-Roundness, Cylinder Length, and End Constraints

In practice, cylindrical vessels have an initial out-of-roundness, which increases the tendency to buckling and hence reduces the critical buckling pressure. In this case as a rough guide, p_{cr} may be reduced by 25% if the out-of-roundness is in the order of 10% of the cylinder wall thickness.

Equation 8.142 is derived by assuming the cylinder to be very long, $L/r_m \to \infty$. For cylinders with shorter lengths, where the ends are free to expand axially and to rotate with the restriction of expanding radially, the critical buckling pressure is given by[8]

For open-ended cylinders:

$$(P_o)_{cr} = \frac{1}{3}\left[n^2 - 1 + \frac{2n^2 - 1 - \nu}{n^2\left(\dfrac{L}{\pi r_m}\right)^2 - 1}\right]\left[\frac{E}{4(1-\nu^2)}\left(\frac{h}{r_m}\right)^3\right] + \frac{E\left(\dfrac{h}{r_m}\right)}{\left(n^2 - 1\right)\left[n^2\left(\dfrac{L}{\pi r_m}\right)^2 + 1\right]^2} \tag{8.143}$$

Notably, Expression 8.143 reduces to Equation 8.142 for $(L/2r_m) \to \infty$ and $n = 2$.

For close-ended cylinders:

$$(P_o)_{cr} = \left\{\frac{1}{3}\left[n^2 + \left(\frac{\pi r_m}{L}\right)^2\right]^2\left[\frac{E}{4(1-\nu^2)}\left(\frac{h}{r_m}\right)^3\right] + \frac{E\left(\dfrac{h}{r_m}\right)}{\left[n^2\left(\dfrac{L}{\pi r_m}\right)^2 + 1\right]^2}\right\}\frac{1}{n^2 + \dfrac{1}{2}\left(\dfrac{\pi r_m}{L}\right)^2} \tag{8.144}$$

In this case, the value of n depends on the geometry of the cylinder, namely, $(h/2r_m)$ and $(L/2r_m)$, as shown in Figure 8.62 for a close-ended cylinder.

The effect of end conditions is demonstrated if the critical buckling pressure for the steel cylinder of Example 8.24 is calculated for close-ended conditions given that $h = 7.21$ mm, $2r_m = 1200$ mm, and $L = 3600$ mm. Hence, from Figure 8.62 for $h/2r_m \simeq 0.006$ and $L/2r_m = 3$, the number of buckling lobes is found to be $n = 3$. Substituting into Expression 8.144 gives $(P_o)_{cr} = 0.395$ MPa which is four times larger than that for an infinitely long cylinder.

PROBLEMS

Rectangular Plates

8.1 A square plate of side length 500 mm and thickness 12 mm is simply supported at all edges and carries a uniformly distributed load of intensity of 0.3 MPa. Determine the maximum stress in the plate and the maximum deflection. Assume $E = 200$ GPa and $\nu = 0.28$. If the plate is clamped at all edges, then determine the maximum stress and the maximum deflection of the plate.

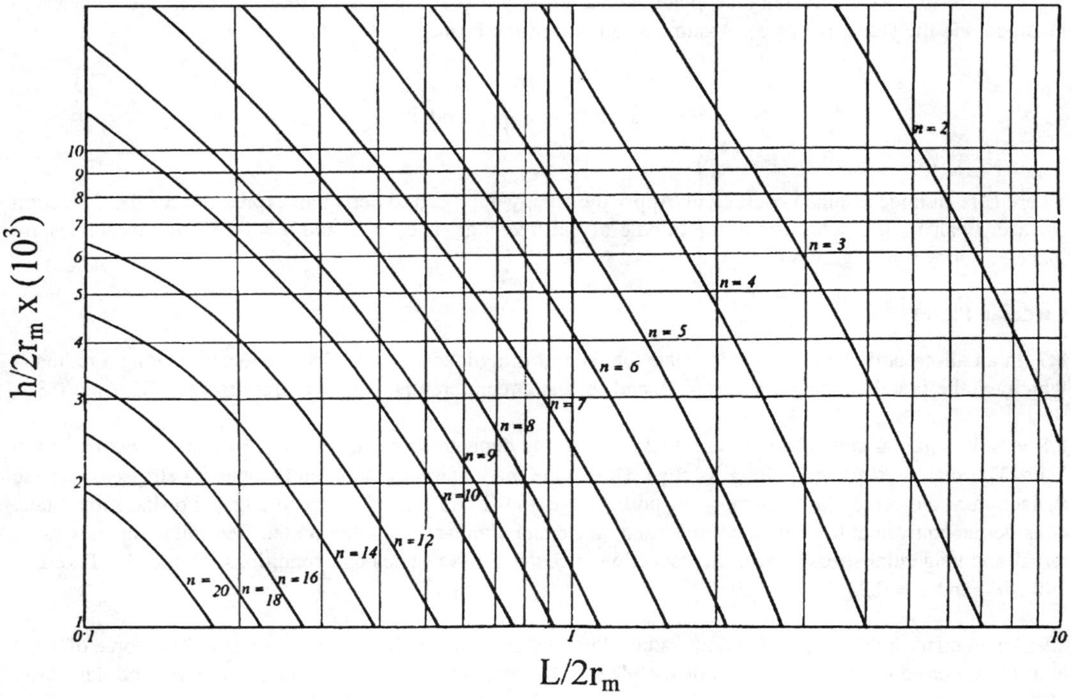

FIGURE 8.62 Effect of cylinder length and thickness related to externally pressurized close-ended cylinder.[9] The value of n is to be substituted in Equation 8.144.

8.2 An all-around clamped rectangular plate of sides 300 mm and 450 mm carries a distributed load of intensity of 1.2 MPa. Determine the plate thickness for a design stress of 150 MPa. Also determine the maximum deflection taking $E = 205$ GPa and $\nu = 0.3$.

8.3 A structural steel door 1.6 m long, 0.8 m wide, and 15 mm thick is subjected to a uniform pressure q. The steel properties are $E = 200$ GPa, $\nu = 0.3$, and allowable yield strength $Y = 230$ MPa. The plate is regarded as simply supported. Determine (a) the limiting value of q that can be applied to the plate without causing permanent deformation and (b) the maximum deflection w that would be produced when q reaches its limiting value.

8.4 A water tank 4.8 m deep and 3.2 m^2 bottom is to be made of structural steel plate. The sides of the tank are divided into nine panels by two vertical supports or stiffeners and two horizontal supports. The average head of water on the lower panel is 4.0 m. Determine the thickness of the plate for the lower panels, using an allowable stress of 120 MPa. Calculate the maximum deflection of this panel assuming an average uniform pressure loading.

8.5 A uniformly loaded rectangular plate ($L_1 \times L_2$) has two opposite clamped sides and the other two sides are simply supported. Assume a solution of the form:

$$w = \sum_{n=1}^{\infty} \sum_{m=1}^{\infty} C_{mn} \left(1 - \cos \frac{2m\pi x}{L_1}\right) \sin \frac{n\pi y}{L_2}$$

Determine few terms of the coefficient C_{mn} employing the energy approach.

8.6 A uniformly loaded rectangular plate has its edges $y = 0$ and $y = L_2$ simply supported; the side $x = 0$ clamped and the side $x = L_1$ free. Assume a solution of the form:

$$w = C\alpha^2 \sin \pi\beta; \quad \alpha = \frac{x}{L_1}, \quad \text{and} \quad \beta = \frac{y}{L_2}$$

where C is an undetermined coefficient. Apply the energy approach to derive an expression for the deflection surface. Evaluate the deflection at the middle of the free edge, if $L_1 = L_2$ and $\nu = 0.3$. The coordinates are placed as shown in Figure 8.6.

Circular Plates

8.7 In an all-around clamped circular plate that is uniformly loaded, determine the location along a radius at which (a) the radial bending stress $\sigma_r = 0$, and (b) the circumferential bending stress $\sigma_\theta = 0$. Take $\nu = 0.3$.

8.8 A solid circular steel plate 15 mm thick, 250 mm in diameter is subjected to a transverse pressure of 0.5 MPa. The plate is supported along its edge. Determine the maximum radial and tangential stresses, also the deflection at the center for (a) simply supported edge and (b) all-around clamped edges. For the same plate, consider an equivalent load to be concentrated on a small circular area at the center. Determine the maximum radial and tangential stresses, also the deflection at center for the same edge conditions (a) and (b). Take E = 190 MPa and $\nu = 0.25$.

8.9 An annular circular plate of outside radius 250 mm and inside radius 50 mm carries a shear force of 8000 N at the inner edge. If the thickness of the plate is 10 mm, determine the maximum stresses and deflection of the plate when (a) the plate is simply supported at the outer edge and (b) the outer edge is clamped. Take E = 210 GPa and $\nu = 0.3$.

8.10 A circular opening in the flat end of a pressure vessel is 250 mm in diameter. A circular steel plate 6 mm thick, with tensile yield stress Y = 250 MPa, is used as a securely clamped cover for the opening. Determine the maximum internal pressure to which the vessel may be subjected if it is limited by the condition that the maximum stress must not exceed 80% of the yield strength of the cover plate. Take E = 200 GPa for steel and $\nu = 0.3$. If the vessel and, hence, the cover plate operates at a uniform temperature difference of 80°C, determine the maximum internal pressure taking $\alpha = 12 \times 10^{-6}$/°C.

8.11 A circular plate of 300 mm diameter is made of aluminum alloy (Y = 270 MPa, E = 70 GPa and ν = 0.3 and $\alpha = 22 \times 10^{-6}$/°C). The edge of the plate is to be clamped and a pressure of p = 160 kPa is to be applied. Determine the required thickness of the plate taking an F.S. = 1.5. What would be the margin of safety if a temperature gradient of 40°C existed across the plate thickness?

8.12 A cast iron disk valve is a flat plate 400 mm in diameter and is simply supported. The plate is subjected to uniform pressure supplied by a head of 50 m of water. Find the thickness of the disk using an allowable stress of 100 MPa. Determine the maximum deflection of the plate. For cast iron, E = 100 GPa and $\nu = 0.20$. Determine the additional deflection due to a temperature gradient of 20°C across the plate thickness taking $\alpha = 9 \times 10^{-6}$/°C.

8.13 A solid circular steel plate of radius 150 mm and 10 mm thick is simply supported at its outer edge and carries a ring load at a radius of 50 mm. If the total load on the plate is 12 kN, calculate the maximum deflection of the plate, taking E = 200 GPa and $\nu = 0.3$. Apply superposition of Figure 8.22.

8.14 A pump diaphragm can be approximated as an annular plate under a uniformly distributed load p and with its outer edge simply supported. Determine, using the method of superposition, the maximum plate deflection for $r_i = r_o/4$, E = 200 GPa, and $\nu = 0.3$.

8.15 A circular plate with radius r_o has a central hole with radius $r_i = r_o/4$ and thickness $h/r_o = 0.05$. The plate is subjected to a uniformly distributed pressure p and has its inner edge clamped and its external edge free. Find the maximum deflection for maximum bending stresses of E/1000.

8.16 Calculate the maximum deflection in the annular plates loaded (a) as shown in Figure 8.18 by setting $r_o = 2r_i$, $M = 2M_o$, and $\nu = 0.3$ and (b) as shown in Figure 8.20 by setting $r_o = 2r_i$. Take $\nu = 0.3$.

8.17 A circular plate of outside radius r_o, thickness h has a central hole of radius r_i is subjected to a uniform edge moment M_o per unit length at its outside diameter and is clamped at its inner diameter. (a) Determine the deflection radial and tangential stresses. (b) Repeat the problem for a plate clamped at its outside diameter and subjected to uniform edge moment M_i per unit length at its inside diameter.

8.18 An acrylic aircraft window panel is approximated to a simply supported circular plate of radius r_o. The window is subjected to a uniform cabin pressure p. Determine its maximum deflection if the maximum stresses should not exceed $^1/_3$ the yield strength. The window is made of acrylic: $Y = 50$ MPa, $E = 2$ GPa, and $\nu = 0.35$. Determine the additional deflection for a temperature difference of 40°C taking $\alpha = 85 \times 10^{-6}$/°C.

8.19 A disk spring of the belleville type, as shown in the figure, is composed of 16 steel disks in series. The disks are 2-mm-thick spring steel having inside and outside diameters of 40 and 80 mm, respectively. Determine the maximum allowable load and the stiffness for a maximum tensile stress of 300 MPa.

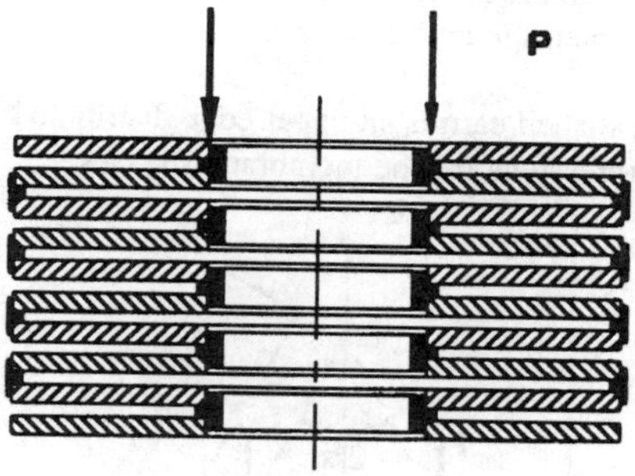

Problem 8.19

8.20 A circular plate of outside radius r_o and inside radius r_i and thickness h is clamped at its outer and inner edges and subjected to transverse uniform pressure p. Determine the maximum stresses and the deflection. Apply the results to design the diaphragm of a control valve. Obtain the amount of valve opening for the maximum allowable pressure if inside diameter is 24 mm, outside diameter 120 mm, and thickness 2 mm. The diaphragm material is steel with $E = 210$ GPa, $\nu = 0.3$, and maxmimum allowable stress of 400 MPa.

Thin Shells — Membrane Stresses

8.21 Using the equations in Chapter 4 giving stresses in a thick-walled sphere under internal pressure, show that these equations reduce to

$$\sigma_\theta = \sigma_\phi = \frac{p_i r_m}{2h}$$

for a thin-walled spherical vessel. Note that r_m and h are the mean radius and wall thickness, respectively.

8.22 A spherical pressure vessel of 1.2 m outside diameter is to be fabricated from a steel with $E = 200$ GPa, $\nu = 0.3$, and yield strength of 280 MPa. Knowing that the gauge pressure may reach 3 MPa and using a factor of safety of 2, determine the smallest wall thickness such that the diameter does not exceed 0.1%.

8.23 A cylindrical pressure vessel of 3 m outside diameter and 12 m long is made from 24-mm-thick steel plate and the vessel operates at 0.9 MPa internal pressure, determine the total elongation of the circumference and the increase in diameter caused by the operating pressure. Take E = 200 GPa and ν = 0.3.

8.24 A boiler made of 12 mm steel plate is 1.2 m in diameter and 3 m long. It is subjected to an internal pressure of 3 MPa. What are the changes in the thickness, length, and diameter of the boiler due to this pressure? Take E = 200 GPa and ν = 0.28.

8.25 A thin-walled cylinder with closed ends is constructed by butt-welding 12-mm plates such that the weld makes an angle of 30° with the axis of the cylinder. If the cylinder radius is 1250 mm, the internal pressure p_i is 3.5 MPa, and the normal and shearing stress in the weld are not to exceed 140 and 85 MPa, respectively, check the minimum wall thickness.

8.26 A vertical column 10 m high used in the chemical industry is made of a 12-in. standard steel pipe weighing 740 N/m with the outside and inside diameters 324 and 305 mm, respectively. If the pipe is pressurized to 1.8 MPa, and a horizontal wind force of 300 N/m height, what is the state of stress in the column 2.5 m above the bottom on the windward side? Show the stresses on an element.

8.27 The pipeline reducer shown in Figure 8.35 has a uniform wall thickness of 4 mm. If the pipeline carries a fluid at a pressure of 1.5 MPa, calculate the axial and hoop stresses in the reducer at a point halfway along its length. Assuming that the connection at the large end of the reducer takes all the axial thrust, calculate the stress in each of the six 12-mm-diameter retaining bolts.

8.28 A supported, truncated conical shell carries an upper edge distributed load P per unit circumference, as shown in the figure. Derive the expressions for the membrane stresses.

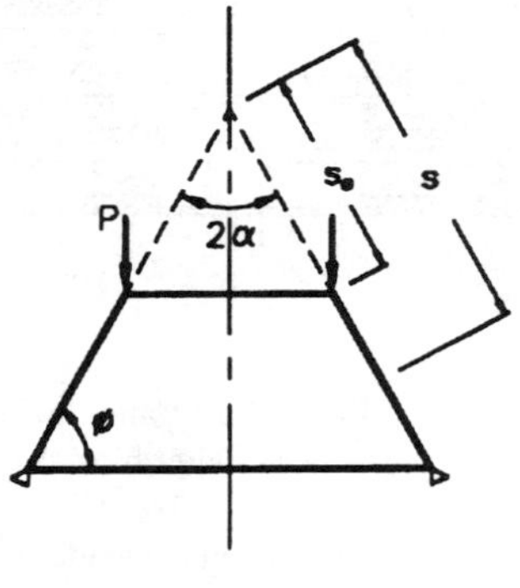

Problem 8.28

8.29 A thin shell in the form of a torus is obtained by rotation of a circle of radius 500 mm about a vertical axis. The radius of rotation is 2 m. This shell is subjected to a uniform pressure of 2 MPa. Determine the maximum stresses in the shell.

8.30 Obtain the stresses in a bellow-type expansion joint as shown in the figure of uniform thickness h, mean diameter d, and wall radius r_m subjected to an internal pressure p. If d = 500 mm, r_m = 50 mm, and p = 2.4 MPa, determine the wall thickness for a maximum stress of 100 MPa.

8.31 A shell in the form of an ellipsoid of revolution is used as an end plate for a cylindrical boiler. The semimajor and semiminor axes for the shell are 1500 and 1000 mm, respectively. If the steam pressure is 1.5 MPa, determine the stresses in the shell at the top and at the equator.

8.32 Obtain the stresses in the head of a cylindrical, thin-walled pressure container of uniform thickness h for the following geometric shapes of the heads:

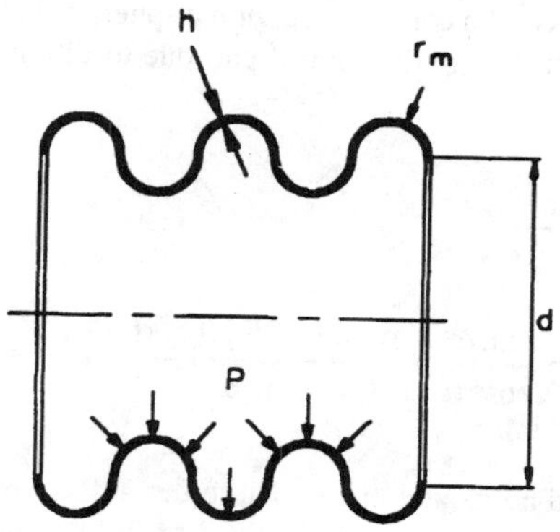

Problem 8.30

a. Spherical
b. Ellipsoidal of $R/R_c = 1.6$
c. Torispherical

If the internal pressure is 1.5 MPa and the mean radius r_m 800 mm, determine the wall thickness for each head shape for a maximum tensile stress of 120 MPa. Neglect local effects at the junction of the cylindrical container and the head.

8.33 An observation dome of a pressurized aircraft is of ellipsoidal shape. It is constructed of 8-mm-thick plastic material. Determine the limiting value of the pressure differential that the shell can resist given a maximum stress of 15 MPa. The lengths of the semiaxes are $R = 0.15$ m and $R_c = 0.1$ m.

8.34 A spherical dome of constant thickness 8 mm has a radius of curvature 1000 mm. If the specific weight of the dome material is 78 kN/m³, determine the stresses in the dome at an angular location of 30° and 60° with the line of symmetry. Take $E = 200$ GPa and $\nu = 0.3$.

8.35 A hemispherical container supported at its edge is filled with a liquid of specific weight ρg. Derive expressions for the membrane stresses and the radial and the circumferential displacements of the shell. Determine these displacements at $\phi = 0$.

8.36 The containment vessel for a nuclear power generation station is made in the form of a hemisphere with a diameter of 50 m. The steel shell is 30 mm thick, weighing 1600 N/m².

a. Determine the maximum tensile and compressive stresses in this shell caused by its own weight.
b. What additional internal pressure can be developed within the vessel before a stress of 250 MPa is reached?

8.37 A horizontal, thin circular cylinder is laid down on a flat surface under its own weight. If the specific weight of the material from which the cylinder is made is 80 kN/m³ (taking the radius of the cylinder as 500 mm, length of 2 m, and wall thickness of 8 mm), find the membrane stresses in the cylinder at $\phi = 0$, 90° and 180° measured from the top. Neglect end effects.

8.38 A conical shell of vertical height 1.5 m and semivertex angle 30° is completely filled with a liquid of specific weight 10.5 kN/m³. Find the stresses in the shell.

8.39 A conic aluminum container of thickness 4 mm, apex angle 90°, and height $H = 3$ m is filled with water. Taking $E = 70$ GPa and $\nu = 0.3$, find the locations measured vertically above the apex, for which (a) the hoop strain is zero and (b) the hoop strain is maximum.

8.40 A compound tank is constructed of a conical shell with a spherical bottom, as shown in the figure. Show that the hoop and the meridional stresses in the conical part due to filling with a fluid of specific weight ρg to a level H_1 are given by

$$\sigma_\theta = \frac{\rho g}{h} y(H_1 - y)\frac{\tan\alpha}{\cos\alpha}$$

$$\sigma_\phi = \frac{\rho g}{6h}\left[(3H_1 - 2y)y\frac{\sin\alpha}{\cos^2\alpha} + \frac{a^2(2a - H_2)\cos^2\alpha}{y\sin\alpha} + a^3\frac{(2\sin\alpha - \cos^2\alpha - 2)}{y}\right]$$

Neglect the continuity moment and horizontal meridional forces at the junction of the two parts. Note that at the junction of the two parts, a ring must be provided to resist the difference in the horizontal components of the meridional forces in the cone and the sphere.

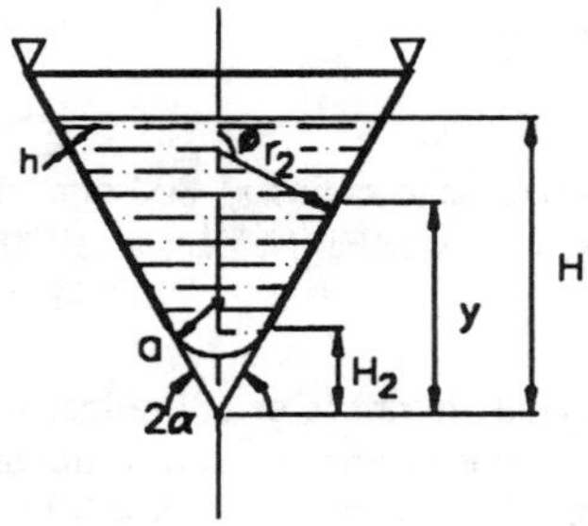

Problem 8.40

8.41 The figure shows a tank constructed from a cylindrical shell and a spherical bottom. The tank is filled to a level H with a liquid of specific weight ρg. Derive the following expressions for the membrane stresses in the bottom part:

$$\sigma_\theta = -\frac{\rho g r_m^2}{6h}\left(\frac{3H}{r_m} + 5 - 6\cos\phi + \frac{2\cos^2\phi}{1+\cos\phi}\right)$$

$$\sigma_\phi = -\frac{\rho g r_m^2}{6h}\left(\frac{3H}{r_m} + 1 - \frac{2\cos^2\phi}{1+\cos\phi}\right)$$

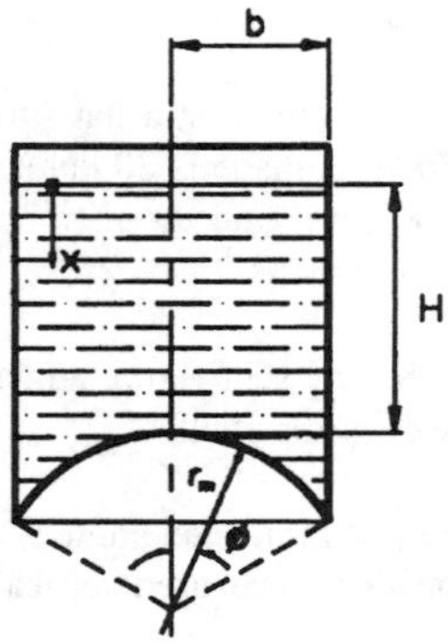

Problem 8.41

Discontinuity Stresses in Cylinders

8.42 Consider a steel cylindrical pressure vessel of inside diameter 400 mm subjected to an internal pressure of 0.6 MPa. If the maximum allowable tensile stress is 150 MPa, determine the wall thickness for the cases: (a) ends unconstrained radially and (b) ends radially constrained. Take E = 190 GPa and ν = 0.3.

8.43 A cylindrical aluminum shell is 300 mm long and 450 mm in diameter and must be designed to carry an internal pressure of 2 MPa without exceeding a maximum tensile stress of 120 MPa. The ends are capped with massive flanges, which are sufficiently clamped to the shell to resist effectively resist any deformation at the ends. Given: E = 70 GPa and ν = 0.3, determine the required thickness for the shell.

8.44 A long steel cylinder 70 mm in diameter and 2.5 mm thick is subjected to a uniform line load P distributed over the circumference of the circular cross section at midlength. Determine the value of the maximum load that can be applied to the cylinder without causing the elastic limit to be exceeded. Use ν = 0.3 and yield strength = 210 MPa.

8.45 A long, steel pipe of 0.75 m in diameter and 10 mm thick is subjected to two loads P_1 and P_2 uniformly distributed along two circular sections 0.05 m apart (see Figure 8.49). At a distance of 30 mm from the first load, obtain, by taking ν = 0.3, (a) the radial contraction and (b) axial and hoop stresses at the outer surface. Apply superposition.

8.46 A circular steel pipe is reinforced by collars spaced L apart (Figure 8.47). The cylinder is under an internal pressure p. The cross-sectional area of the collar is 0.025 m². What is the maximum bending stress in the pipe if the collars are constructed of a very rigid material. Take L = 1.3 m, r_m = 0.6 m, h = 10 mm, p_i = 1.4 MPa, E = 200 GPa, and ν = 0.3.

8.47 A narrow ring is shrunk-fit onto a long pipe of radius r_m and thickness h. The cross-sectional area of the ring is A. If the outer radius of the pipe is greater by Δ than the inner radius of the ring, determine the shrink-fit bending stress in the pipe.

8.48 A thin-walled pipe 250 mm in diameter, with 5 mm wall thickness has a ring shrunk onto it at the middle. Ring inner diameter is 249.75 mm, outer diameter 300 mm, and width 20 mm. The material of pipe and ring is steel. Determine the maximum stresses in the pipe (a) neglecting ring deformation and (b) considering ring deformation.

8.49 A long, circular cylinder of radius 400 mm and wall thickness 10 mm is subjected to a uniform band of pressure of intensity 2 MPa over a length of 200 mm. Determine the maximum deflection of the cylinder and the stresses at the center and end of the pressure band.

8.50 A long brass cylinder of free ends has a mean radius r_m = 0.5 m and thickness h = 10 mm has the uniform temperatures T_1 = 300°C and T_2 = 150°C at the inner and the outer surfaces, respectively. If the ends of the shell are assumed to be free, use equation (8.18) to derive an expression for the maximum circumferential stress as

$$(\sigma_\theta)_{max} = \frac{E\alpha(T_1 - T_2)}{2(1-\nu)}\left(1 - \nu + \sqrt{\frac{1-\nu^2}{3}}\right)$$

Calculate the stress taking, E_b = 100 GPa, $\alpha_b = 20 \times 10^{-6}$ per °C and ν_b = 0.3 for brass.

8.51 A thin-walled cylindrical drum has a rigid ring at each end, which support the drum on a pair of rollers. The cylinder dimensions are shown in the figure and the cylinder is made of steel. The drum is used as a rotary batch dryer. The total weight is 20 kN, the inside temperature is 150°C, the outside temperature is 30°C, and it rotates at 75 rpm. Determine the maximum tensile stresses in the shell under the following conditions: (a) drum stationary without heating; (b) drum rotating without heating; (c) drum stationary with heating; and (d) drum rotating with heating. Compare the results of (b), (c), and (d) with (a).

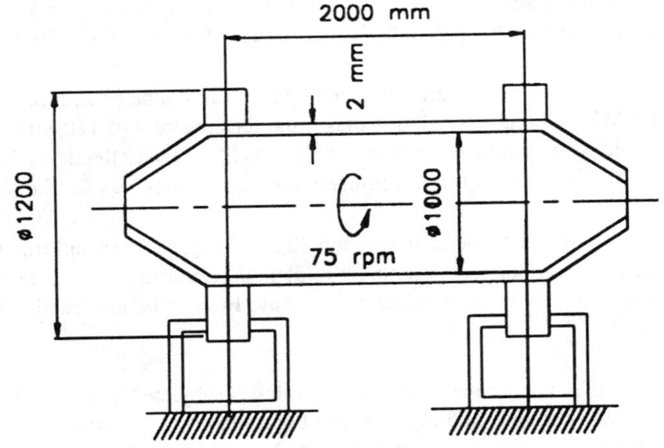

Problem 8.51

8.52 Calculate the maximum stresses in the walls of a cylindrical, thin-walled pressure vessel at its junction with the hemispherical ends knowing that internal pressure is 1.2 MPa, inner radius is 400 mm, and thickness is 8 mm. Take $E = 200$ GPa and $\nu = 0.3$.

8.53 Calculate the maximum bending and membrane stresses in a cylindrical tank wall with clamped base (Figure 8.55). The tank is vertical and filled to the top with water (specific weight 9.81 kN/m^3). The dimensions are $r_m = 2.7$ m, $H = 3.7$ m, and $h = 10$ mm. Use $E = 200$ GPa and $\nu = 0.3$.

8.54 A water tank of radius r_m and height H is filled to full capacity with water. The upper half portion of the tank is constructed of sheet steel of thickness h_2 and the lower half of sheet steel of thickness h_1 ($h_1 > h_2$). Determine the values of the displacement and discontinuity moment at the joint between h_1 and h_2. Assume the solution of an infinitely long cylinder to be valid.

Elastic Buckling of Plates and Shells

8.55 Find the minimum wall thickness for a long, thin-walled cylindrical vessel of mean diameter 1500 mm to sustain an external pressure of 0.12 MPa, knowing that the vessel is made of

a. Steel, $E = 210$ GPa, allowable stress $\sigma_{all} = 170$ MPa, specific weight $\rho g = 78$ kN/m^3;
b. Brass, $E = 120$ GPa, $\sigma_{all} = 85$ MPa, and $\rho g = 85$ kN/m^3;
c. Aluminum alloy, $E = 75$ GPa, $\sigma_{all} = 65$ MPa and $\rho g = 28$ kN/m^3;
d. Glass fiber–reinforced polyester (GRP), $E = 19$ GPa, $\sigma_{all} = 35$ MPa, and $\rho g = 17$ kN/m^3.

Take $\nu = 0.3$. Select the optimum material for the tank based on strength/weight considerations.

8.56 Thin-walled, close-ended cylindrical vessels of different lengths are made of steel of wall thickness 6 mm. Taking an outside diameter of 2 m and a corrosion allowance of 3 mm, calculate the permissible outside pressure to be sustained by three vessels of the lengths 10, 4, and 2 m, respectively. How much reduction in the permissible pressure is suggested for a manufacturing out-of-roundness of 10% of the thickness.

8.57 A stainless steel, open-ended cylinder vessel has the following dimensions: diameter = 1 m, length = 4 m, and new thickness = 8 mm.

a. Determine the maximum internal pressure the vessel can withstand.
b. Determine the lowest vacuum the vessel can withstand.
c. If the vacuum you found in (b) is to be increased 50% while adding stiffening rings, determine the free span between these rings.

Take a corrosion allowance of 2 mm and material properties E = 195 GPa, Y = 350 MPa, and ν = 0.3. Use a factor of safety of 2.

REFERENCES

1. Shames, I. H. and Dym, C. L., *Energy and Finite Element Methods in Structural Mechanics*, McGraw-Hill, New York, 1985.
2. Boley, B. A. and Weiner, J. H., *Theory of Thermal Stresses*, John Wiley & Sons, New York, 1964.
3. Timoshenko, S. and Woinowsky-Krieger, S., *Theory of Plates and Shells*, 2nd ed., McGraw-Hill, New York, 1959.
4. Szilard, R., *Theory and Analysis of Plates — Classical and Numerical Methods,* Prentice-Hall, Englewood Cliffs, N.J., 1974.
5. Wahl, A. M. and Lobo, G., Trans ASME, 52, 1930.
6. Roark, J. R. and Young, W. C., *Formulas for Stress and Strain,* 5th ed., McGraw-Hill, New York, 1975.
7. Ugural, A. C., *Stresses in Plates and Shells*, McGraw-Hill, New York, 1981.
8. Timoshenko S. P. and Gere, J. M., *Theory of Elastic Stability,* 2nd ed., McGraw-Hill, New York, 1961.
9. Windenburg, D. F. and Trilling, C., Pressure vessel and piping design, *Collected Papers*, ASME, 1960.

9 Applications to Fracture Mechanics

The elastic stress field at the tip of a crack is derived from the solution of the stress concentration problem around an elliptical hole. This leads to the definition of the notion of stress intensity factor and hence fracture toughness. Stress intensity factors for various crack geometries in solids subjected to simple or combined loading are given. The procedure of applying linear-elastic fracture mechanics to crack growth under cyclic loading to arrive at safe life prediction is presented and exemplified through several cases of engineering interest. Limitations of linear-elastic fracture mechanics due to the presence of plastic deformation ahead of the crack tip as being imposed on fracture toughness and designing against fracture are briefly stated. In presence of appreciable plasticity, the two proposed elastic-plastic fracture criteria, namely J-integral and crack opening displacement, have been presented.

9.1 INTRODUCTION

Sudden fracture of structures fabricated from low-strength structural steels has been reported from the beginning of the 20th century. Gas, water, oil, and molasses tanks were fractured causing human casualties and extensive material losses. Incidents of catastrophic failures of bridges, large tanks, barges, railroad equipment, and even aircraft have been reported. Observations indicated that several structures designed according to conventional engineering practice with allowable stresses well below the elastic limit and the fatigue limit may undergo sudden fracture. It has been realized by thorough investigations that fabricated structures are never perfect since flaws and cracklike defects are discovered either initially or during service life. The presence of such defects will certainly raise the stresses in its vicinity to a high value sufficient to open up the crack and make it propagate until fracture occurs. Admitting that cracks do *a priori* exist while designing mechanical components represents the core of the subject of brittle fracture mechanics, or simply fracture mechanics.

The term *brittle* here means fracture with no permanent deformation and a consumption of little energy. This is opposed to *ductile* fracture, which occurs after attaining appreciable plastic deformation and dissipating a large amount of energy. To apply this classification to a given material, it is now well recognized that it depends markedly on service conditions. For instance, low temperatures, high strain rates (e.g., impact), and hostile environment, together with the tensile regime of stresses, all promote brittleness. A material rated as ductile under normal (room) conditions may turn out to behave as a brittle one under adverse conditions. Consequently, a structure fabricated from a material conventionally rated as ductile at room conditions (e.g., low-carbon steel) may fail in a brittle manner because of the presence of cracks. Based on this global understanding, the subject of "Fracture Mechanics" is concerned with the prediction of steady or cyclic loading conditions that will cause fracture in a structure with preexisting cracks.

The analytical development of fracture mechanics has been known in the literature as "Linear-Elastic Fracture Mechanics" (LEFM). This, however, does not exclude the successful application of LEFM to cases of fracture comprising limited plastic deformation by adopting some corrective parameters. For cases where moderate plastic deformation is encountered, LEFM becomes no longer applicable and other comprehensive approaches of fracture mechanics are employed.

9.2 GRIFFITH ENERGY CRITERION

In the 1920s Griffith[1] observed that as the length of an individual glass rod decreases as a rescue of successive tensile failures, the tensile strength increases in the remaining portions. He also reported that very thin fibers of glass have much higher tensile strength than coarse fibers. An explanation to this has been attributed to the statistical finding that a flaw of a given size is less likely to exist in rods of shorter lengths or thinner sections than it is in bigger ones. This size effect led to the first analytical expression for determining the load-carrying capacity of a material that contained a cracklike defect.

Considering an idealized atomic model, the repulsive and attractive forces balance out to zero under equilibrium. If a crystal is strained by a stress σ, the ideal strength of the material in terms of maximum cohesion is related to the maximum value of σ as[2] $\sigma_{max} = E/2\pi$, say, $E/10$. Measured values of the failure stress σ_f for most engineering materials is typically within the order $E/10^3$ unless testing is conducted on fiber or "whisker" form in which case $\sigma_f \cong E/15$. The discrepancy between the theoretical predictions of fracture stress and the actual value can again be explained by the presence of microcracks or flaws in the solid material.* At the tip of these cracks, the stress concentration effects cause the theoretical cohesive strength to be reached even though the nominal stress is much lower than the theoretical fracture stress. Based on this idea, Griffith postulated his criterion for brittle fracture of solids.

The Griffith criterion is concerned with the energetics of a system defined by an infinitely large plate of brittle material containing a single sharp through-crack loaded as shown in Figure 9.1a.

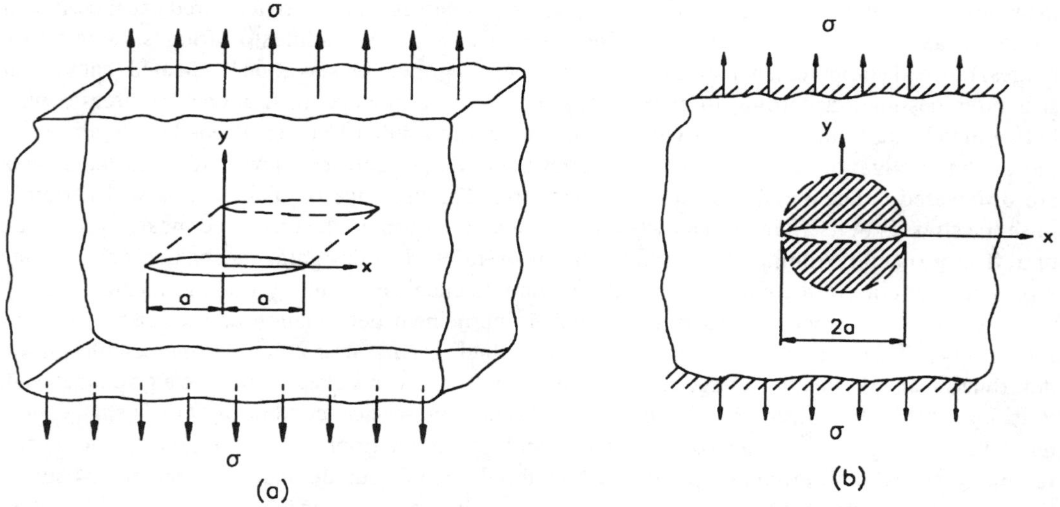

FIGURE 9.1 Approximate energy model for the Griffith criterion.

* Likewise, the concepts of dislocation theory explain the discrepancy between the high theoretical shear strength of a solid ($\simeq G/2\pi$) and the much lower observed shear strength at ductile fracture. Instead of requiring high stresses for bodily shearing full planes of atoms, the motion of dislocations demands far lower stresses to bring about the end phenomenon called slip.

The strain energy stored per unit volume due to a uniformly applied stress σ, according to Equation 3.35a is given by: $U = \sigma^2/2E$. If a crack of length $2a$ is introduced into the plate, there will be general relaxation in the material above and below the crack, and some strain energy will be released. To obtain an approximate solution, assume that this relaxed zone is represented by the shaded circular zones of Figure 9.1b. Hence, the energy released per unit thickness is given by U = energy per unit volume × volume of shaded zones/thickness, i.e.,

$$\begin{aligned} U &= \frac{\sigma^2}{2E}\left(2\times\frac{1}{2}\pi a^2 \times 2a \times h/h\right) \\ &= \frac{\sigma^2}{2E}\left(2\pi a^2\right) \end{aligned} \tag{9.1}$$

This expression for the energy released upon introducing a crack is in accord with Griffith's accurate solution, which is based on determining stresses and displacements in a cracked body using stress functions of complex variables.[3] The Griffith criterion is stated as

$$U = \frac{\sigma^2}{2E}\pi a^2 \quad \text{for plane stress} \tag{9.2a}$$

$$U = \frac{\sigma^2}{2E}\pi a^2\left(1-\nu^2\right) \text{ for plane strain} \tag{9.2b}$$

Now in order to extend the crack length, an energy input W is required. This energy is consumed in breaking the atomic bonds ahead of the crack tip. It is sufficiently accurate to assume that the energy required for each increment of crack length extension is constant, i.e., dW/da = constant or, simply, W is a linear function of a.

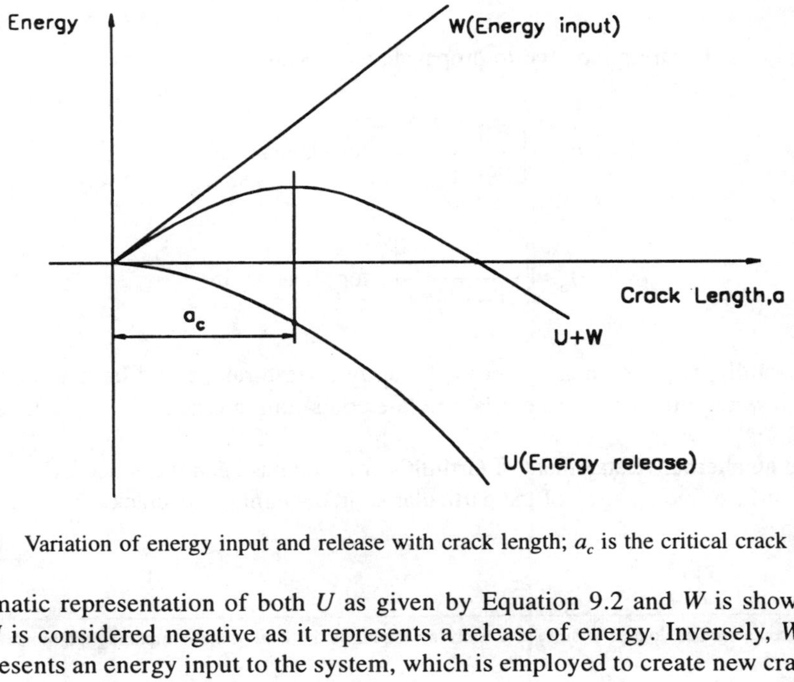

FIGURE 9.2 Variation of energy input and release with crack length; a_c is the critical crack length.

A schematic representation of both U as given by Equation 9.2 and W is shown in Figure 9.2. Here U is considered negative as it represents a release of energy. Inversely, W is positive since it represents an energy input to the system, which is employed to create new crack surfaces.

By summing up the two energies, the curve representing the total energy ($W + U$) of the system is found. The point marked by the crack length a_c indicates that for crack length $0 \le a < a_c$, energy input is required, while, for crack length $a > a_c$, energy is released as a result of crack extension. The crack will thus propagate and the system could be described as unstable. At this point the crack length attains its critical value; ($a = a_c$).

More formally, this situation is described by the Griffith criterion for brittle fracture as "A crack will propagate when the decrease in elastic strain energy is at least equal to the energy needed to create the new surfaces associated with the crack." Considering the signs of U and W, this is expressed by the condition:

$$\frac{\delta U}{\delta a} = \frac{\delta W}{\delta a}$$

where $\delta U/\delta a$ defines the strain energy release rate for an incremental crack extension. The value $\delta W/\delta a$ defines the energy absorbed during an incremental crack extension. Since $\delta W/\delta a$ expressing the energy required to break atomic bonds within the material is constant, a material property G_c (where the subscript c denotes the critical point) is defined, such that

$$G_c = \frac{\delta U}{\delta a}$$

By virtue of Equations 9.2, this leads to

$$G_c = \frac{\sigma_c^2 \pi a}{E} \quad \text{for plane stress} \tag{9.3a}$$

$$G_c = \frac{\sigma_c^2 \pi a}{E}\left(1 - \nu^2\right) \quad \text{for plane strain} \tag{9.3b}$$

Rearranging gives the stress required to propagate a crack as

$$\sigma_c = \left[\frac{EG_c}{\pi a}\right]^{1/2} \quad \text{for plane stress} \tag{9.4a}$$

$$\sigma_c = \left[\frac{EG_c}{\left(1 - \nu^2\right)\pi a}\right]^{1/2} \quad \text{for plane strain} \tag{9.4b}$$

Experimentally, G_c, known as the critical energy release rate, could be determined by measuring the stress required to fracture a large plate containing a crack of length $2a$, as shown in Figure 9.1a.

Note that an alternative statement of Griffith's criterion has been introduced via a parameter γ_s defining the surface free energy* of the particular solid containing the crack. This is expressed for plane stress as

---

* The surface free energy, analogous to surface tension in a liquid, is the work done in breaking the bonds between a unit area of atoms.

$$\sigma_c = \left[\frac{2E\gamma_s}{\pi a}\right]^{1/2} \tag{9.5}$$

thus indicating that $G_c = 2\gamma_s$. In fact, it is found that G_c and γ_s have similar orders of magnitude for some brittle materials, such as glass which Griffith used for his experiments.

The use of Griffith's criterion, Equations 9.3 to 9.5, is restricted to absolutely brittle materials showing no evidence of plastic deformation at fracture; a condition which is never satisfied in real engineering materials.* For typical ductile or quasi-brittle materials, measured values of G_c may be 1000 γ_s or greater. This means that most of the energy in excess of that necessary for the creation of new surfaces is spent in plastically deforming a volume of the material around the crack tip. Hence, a modification for Equation 9.5 is to include the plastic work γ_p needed to extend the crack surface, such that

$$\sigma_c = \left[\frac{2E(\gamma_s + \gamma_p)}{\pi a}\right]^{1/2} \equiv \left(\frac{E\gamma_p}{a}\right)^{1/2} \tag{9.6}$$

where $\gamma_s << \gamma_p$.

9.3 STRESS CONCENTRATION AROUND ELLIPTICAL HOLES

The stress distribution at the tip of a crack is found by examining the solution for a stress concentration problem around an elliptical hole by considering the crack as an ellipse with a minor axis of zero length.

The solution to the stress concentration problem around an elliptical hole, Figure 9.3a, is more involved than that of a circular hole and demands the use of curvilinear coordinates. The details of this solution will not be presented in this text but only the resulting stress distribution, which is found to be along the major axis $y = 0$, as

$$\begin{aligned}\sigma_x &= \frac{\sigma}{2}\left[\frac{2(1+\beta)}{\alpha^2-\beta} - \frac{\beta^2-1}{\alpha^2-\beta}\left[1+\frac{\beta-1}{\alpha^2-\beta}\frac{3\alpha^2-\beta}{\alpha^2-\beta}\right]\right] \\ \sigma_y &= \frac{\sigma}{2}\left[2+\frac{2(1+\beta)}{\alpha^2-\beta}+\frac{\beta^2-1}{\alpha^2-\beta}\left(1+\frac{\beta-1}{\alpha^2-\beta}\frac{3\alpha^2-\beta}{\alpha^2-\beta}\right)\right]\end{aligned} \tag{9.7a}$$

where $\beta = (a-b)/(a+b)$

$$\alpha = \frac{x}{a+b} + \sqrt{\left(\frac{v}{a+b}\right)^2 - \beta}$$

The distance x is measured from the center of the ellipse. The distribution of (σ_y/σ) at $y = 0$ vs. (x/b) is shown in Figure 9.3b for elliptic holes of different aspect ratios a/b. It is seen that σ_y/σ at $x = a$ attains higher values for narrower ellipses. The stress quotient σ_y/σ tends to unity for large values of x for all ellipse geometries.

---

* Even fracture of cast iron, which is brittle in the global sense, is accompanied by some plastic deformation on the microscale.

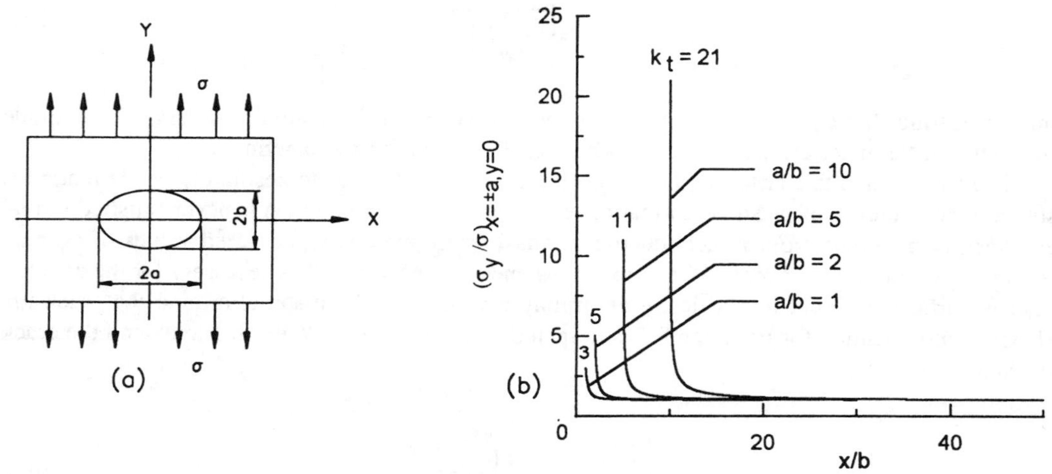

FIGURE 9.3 (a) Plate with a central elliptical hole. (b) Variation of the maximum stress σ_y at the hole boundary.

More specifically, the stresses at the ends of the hole on the major axis are obtained by putting $x = \pm a$, i.e.,

$$\alpha = \frac{a}{a+b} + \sqrt{\left(\frac{a}{a+b}\right)^2 - \frac{a-b}{a+b}} = 1$$

into Equations 9.7a. The stresses and the stress concentration factor k_t are thus given by

$$\begin{aligned} \sigma_x &= 0 \\ \sigma_y &= \sigma(1+2a/b) \\ k_t &= \sigma_y/\sigma = 1+2(a/b) \end{aligned} \tag{9.7b}$$

This is a valuable result since it may be used to determine the stress concentration factor for many notch geometries, which could be reasonably approximated to an elliptical shape at their root.

In a special case, when $a = b$, i.e., a circular hole, the stress concentration factor attains the value of 3 as obtained before in Section 6.4. If $a = 10b$, i.e., a very narrow ellipse, the stress concentration factor attains a very high value of 21. In ductile materials subjected to steady loads, localized yielding is often permitted at such points since no harmful effects would result as long as the rest of the structure remains within the elastic range.

9.4 THE ELASTIC STRESS FIELD AT THE CRACK TIP

The stress distribution at the tip of a sharp crack of length $2a$ in an infinite plate loaded in simple tension is obtained from the elliptical hole problem. The crack may be regarded as an ellipse with the minor axis as zero, i.e., $b = 0$. In which case $\beta = 1$ and

$$\alpha = \frac{x}{a} + \sqrt{\left(\frac{x}{a}\right)^2 - 1}$$

Substitution into Equations 9.7a yields for the stresses:

$$\sigma_x = \sigma\left(\frac{x}{\sqrt{x^2 - a^2}} - 1\right) \tag{9.8a}$$

for $|x| \geq a$

$$\sigma_y = \sigma\left(\frac{x}{\sqrt{x^2 - a^2}}\right) \tag{9.8b}$$

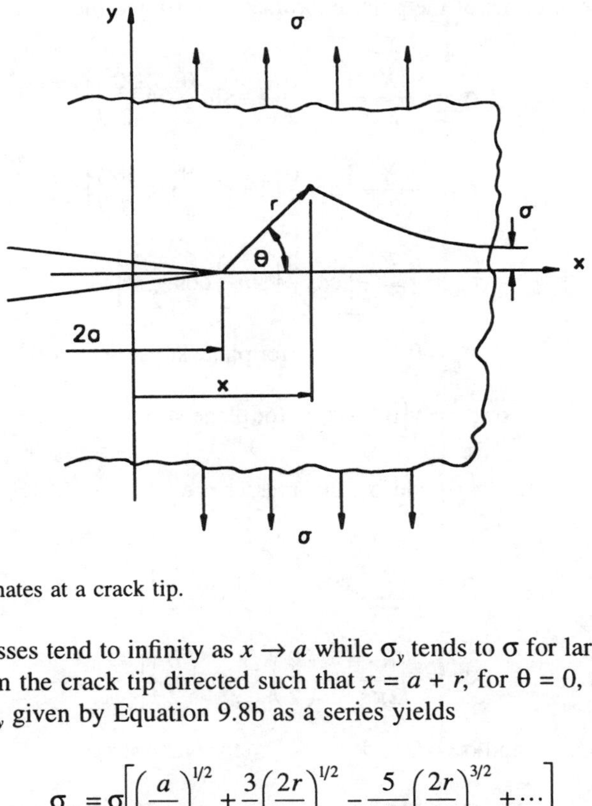

FIGURE 9.4 Coordinates at a crack tip.

Obviously, these stresses tend to infinity as $x \to a$ while σ_y tends to σ for large values of x. Taking r as the distance from the crack tip directed such that $x = a + r$, for $\theta = 0$, as indicated in Figure 9.4 and expanding σ_y given by Equation 9.8b as a series yields

$$\sigma_y = \sigma\left[\left(\frac{a}{2r}\right)^{1/2} + \frac{3}{8}\left(\frac{2r}{a}\right)^{1/2} - \frac{5}{128}\left(\frac{2r}{a}\right)^{3/2} + \cdots\right] \tag{9.8c}$$

Clearly as $r \to 0$, the stress σ_y is dominated by the first term and it may be written as

$$\sigma_y = \sigma\sqrt{a}/\sqrt{2r} \tag{9.9a}$$

and, similarly, for σ_x,

$$\sigma_x = \sigma\sqrt{a}/\sqrt{2r} \tag{9.9b}$$

By rearranging and introducing a factor π, Equations 9.9 becomes[4]

$$\sigma_x\sqrt{2\pi r} = \sigma_y\sqrt{2\pi r} = \sigma\sqrt{\pi a} = \text{constant (say, } K)$$

for a given loading σ and crack length $2a$. Hence,

$$\sigma_x = \sigma_y = \frac{K}{\sqrt{2\pi r}} \tag{9.10}$$

Equations 9.9 and 9.10 provide the stresses for a region close to the crack tip along the axis $y = 0$. For large values of r, they give stresses tending to zero, which is incorrect since $\sigma_y \rightarrow \sigma$ as truly given by Equation 9.8b. More generally, the stress components in the vicinity of the crack tip are given in the literature[5] in terms of the polar coordinates (r, θ) and the constant $K = \sigma\sqrt{\pi a}$ as

$$\sigma_x = \frac{K}{\sqrt{2\pi r}}\left[\cos\frac{\theta}{2}\left(1 - \sin\frac{\theta}{2}\sin\frac{3\theta}{2}\right)\right]$$

$$\sigma_y = \frac{K}{\sqrt{2\pi r}}\left[\cos\frac{\theta}{2}\left(1 + \sin\frac{\theta}{2}\sin\frac{3\theta}{2}\right)\right]$$

$$\tau_{xy} = \frac{K}{\sqrt{2\pi r}}\left[\cos\frac{\theta}{2}\left(\sin\frac{\theta}{2}\cos\frac{3\theta}{2}\right)\right] \tag{9.11a}$$

$$\sigma_z = 0 \qquad \text{for plane stress}$$

$$\text{or } \sigma_z = \nu(\sigma_x + \sigma_y) \quad \text{for plane strain}$$

In terms of principal stresses σ_1 and σ_2, the stresses are

$$\sigma_1 = \frac{K}{\sqrt{2\pi r}}\cos\left(\frac{\theta}{2}\right)\left[1 + \sin\left(\frac{\theta}{2}\right)\right]$$

$$\sigma_2 = \frac{K}{\sqrt{2\pi r}}\cos\left(\frac{\theta}{2}\right)\left[1 - \sin\left(\frac{\theta}{2}\right)\right]$$

$$\text{and } \sigma_3 = 0 \qquad \text{for plane stress} \tag{9.11b}$$

$$\text{or } \quad \sigma_3 = \nu(\sigma_1 + \sigma_2)$$

$$= 2\nu\frac{K}{\sqrt{2\pi r}}\cos\frac{\theta}{2} \quad \text{for plane strain}$$

Example 9.1:

Based on Williams stress[6] function — $\phi = r^{\lambda+1}f(\theta)$ — determine stress distribution σ_r, σ_θ, and $\sigma_{r\theta}$ in the vicinity of a sharp crack in the middle of a large plate loaded by a remote tensile stress as shown in Figure 9.4.

Solution:

The stress function has to satisfy the biharmonic Equation 4.27:

$$\nabla^2\nabla^2\phi = 0$$

hence, this gives

$$\frac{d^4 f}{d\theta^4} + 2(\lambda^2 + 1)\frac{d^2 f}{d\theta^2} + (\lambda^2 - 1)^2 f = 0$$

The general solution to this is

$$f(\theta) = C_1 \cos(\lambda - 1)\theta + C_2 \sin(\lambda - 1)\theta + C_3 \cos(\lambda + 1)\theta + \sin C_4 (\lambda + 1)\theta$$

with two conditions, Figure 9.4:

$$f = \frac{df}{d\theta} = 0 \text{ for } \theta = \pm\alpha$$

Satisfying these conditions leads to a set of homogeneous equations for the constants C_1 to C_4. A nontrivial solution is thus possible only when the determinants of coefficients vanish, hence resulting in

$$\begin{aligned} &\lambda \sin 2\alpha + \sin 2\lambda\alpha = 0 \\ &-\lambda \sin 2\alpha + \sin 2\lambda\alpha = 0 \end{aligned} \tag{a}$$

For a sharp crack, α approaches π, then the above two equations, a, reduce to $\sin 2\pi\lambda = 0$, which has only real roots:

$$\lambda = n/2 \text{ for } n = 1, 2, 3, \ldots$$

To each value of n, there will be a corresponding relationship between C_{1n} and C_{3n} or C_{2n} and C_{4n}.[7] Hence in terms of C_{1n} and C_{2n} only, the stress function reduces to

$$\begin{aligned} \phi = &\sum_{n=1,3\ldots} r^{1+n/2}\left[C_{1n}\left(\cos\frac{n-2}{2}\theta - \frac{n-2}{n+2}\cos\frac{n+2}{2}\theta\right) + C_{2n}\left(\sin\frac{n-2}{2}\theta - \sin\frac{n+2}{2}\theta\right)\right] \\ &+ \sum_{n=2,4\ldots} r^{1+n/2}\left[C_{1n}\left(\cos\frac{n-2}{2}\theta - \cos\frac{n+2}{2}\theta\right) + C_{2n}\left(\sin\frac{n-2}{2}\theta - \frac{n-2}{n+2}\sin\frac{n+2}{2}\theta\right)\right] \end{aligned} \tag{b}$$

The stresses are derived from ϕ by differentiation according to Expressions 4.25 as

$$\sigma_r = \frac{1}{r}\frac{\partial\phi}{\partial r} + \frac{1}{r^2}\frac{\partial^2\phi}{\partial\theta^2}, \sigma_\theta = \frac{\partial^2\phi}{\partial\theta^2}, \text{ and } \tau_{r\theta} = -\frac{\partial}{\partial r}\left(\frac{1}{r}\frac{\partial\phi}{\partial\theta}\right)$$

For $n = 1$, the stress function involves terms of $r^{3/2}$, which results in stresses containing singular terms $(1\sqrt{r})$ as $r \to 0$. These terms are the dominant ones near the crack tip; therefore,

$$[\phi]_{n=1} = r^{3/2} C_{11}\left(\cos\frac{\theta}{2} + \frac{1}{3}\cos\frac{3\theta}{2}\right)$$

The stresses are thus obtained as

$$\sigma_r = \frac{C_{11}}{4}\frac{1}{\sqrt{r}}\left(5\cos\frac{\theta}{2} - \cos\frac{3\theta}{2}\right) + \text{nonsingular terms}$$

$$\sigma_\theta = \frac{C_{11}}{4}\frac{1}{\sqrt{r}}\left(3\cos\frac{\theta}{2} + \cos\frac{3\theta}{2}\right) + \text{nonsingular terms} \qquad \text{(c)}$$

$$\tau_{r\theta} = \frac{C_{11}}{4}\frac{1}{\sqrt{r}}\left(\sin\frac{\theta}{2} + \sin\frac{3\theta}{2}\right) + \text{nonsingular terms}$$

With the definition of stress intensity factor $K = \sigma\sqrt{\pi a} = C_{11}\sqrt{2\pi}$, Equations c reduce to Equations 9.11 if transformed to express σ_x, σ_y, and τ_{xy}.

Having determined the stresses, the strains are found by substitution into Hooke's law (3.6). The displacement components u, v are obtained by integrating the strain-displacement relations (2.8) to give for plane stress conditions:

$$u = \frac{C_{11}}{4G}\sqrt{r}\left[\frac{5-3\nu}{1+\nu}\cos\frac{\theta}{2} - \cos\frac{3\theta}{2}\right] + \cdots$$

$$v = \frac{C_{11}}{4G}\sqrt{r}\left[\frac{7-\nu}{1+\nu}\sin\frac{\theta}{2} - \sin\frac{3\theta}{2}\right] + \cdots \qquad \text{(d)}$$

9.5 THE STRESS INTENSITY FACTOR AND FRACTURE TOUGHNESS

In an infinitely large plate containing a crack of size $2a$, Figure 9.4, the magnitudes of stresses at any distance r measured in the crack plane from its tip are determined from Equations 9.11 in terms of a parameter K, defined as

$$K = \sigma\sqrt{\pi a} \qquad (9.12)$$

The factor K for a given structure depends on the applied stress σ and the crack length a. This factor is known as the "stress intensity factor." At a given value of K the magnitudes of the crack tip stresses depend only on its value. Hence, a critical value K_c may be used to predict fracture and is looked upon as a material property under specified test conditions.

By recalling Equation 9.3 expressing the critical energy release rate during crack extension G_c, the following relations are obtained between G_c and K_c as

$$K_c^2 = G_c E \qquad \text{for plane stress}$$

$$K_c^2 = G_c E/\left(1-\nu^2\right) \quad \text{for plane strain} \qquad (9.13)$$

In engineering fracture mechanics it is preferred to work with K, the stress intensity factor, since it is more amenable to analytical determination.

At this stage, three modes of crack deformation are distinguished. These are the crack opening mode (I); the sliding (shearing) mode (II); and the tearing mode (III). The three modes are represented in Figure 9.5.

Each mode is characterized by its critical value for the stress intensity factor K_c, thus resulting in three distinct values K_{Ic}, K_{IIc}, and K_{IIIc} for each material. Of these three values, K_{Ic} is the most important in engineering and is commonly determined from standardized experiments* conducted

* See Section 9.9.

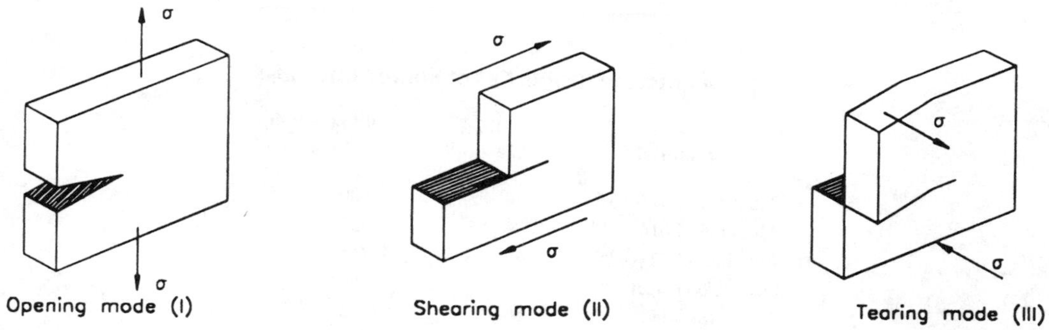

FIGURE 9.5 Deformation modes of a crack.

under plane strain conditions (e.g., using thick plate). It offers a conservative measure for the resistance of the material to brittle fracture in the presence of a sharp crack. K_{Ic} is thus known as the "plane strain fracture toughness" of the material. Under test conditions, other than plane strain, K_c depends on crack tip plasticity and hence specimen thickness and geometry* and is regarded as a critical value of the stress intensity for a given material and configuration. Typical plane strain fracture toughness values are listed in Table 9.1 for some materials together with their yield strengths. In general, the fracture toughness of metallic alloys (and, more specifically, for steels) decreases as the yield strength increases.

The fracture toughness property of a given material is used to design large thick plates made of this material against brittle fracture due to the presence of a sharp crack according to the criterion

$$K_I = \sigma\sqrt{\pi a} \leq K_{Ic} \tag{9.14}$$

For example, the fracture stress σ_f of a plate containing a crack of known size $2a$ can be predicted. Or, inversely, the largest critical crack size $2a_c$ that can be tolerated in a given plate can be determined for a given applied stress level.

According to Equation 9.14, the presence of a small crack in a brittle material has a detrimental effect. Considering a cast iron plate with K_{Ic} = 10 MPa · m$^{1/2}$ and a small crack of 2 mm length, Equation 9.14 predicts a fracture stress of 178 MPa, which is only about 60% of its rated value as in Table 9.1.

The similarity of this fracture criterion with that derived from energy balance considerations in an elastic body containing a crack, i.e., the Griffith criterion (Equation 9.3 or 9.5) is obvious. Cracklike defects may originate from processing (e.g., casting, welding, forming, machining, and/or heat treatment) as well as geometric discontinuities.

The fracture criterion (Equation 9.14) involves the tensile stress σ applied in a direction perpendicular to the crack direction. In other words, it is the stress that causes crack opening. Recognition of this stress offers no ambiguity in the case of uniform stress distribution. In terms of an applied tensile load, this uniform stress is customarily found based on uncracked cross-sectional dimensions. However, for the case where the loading of the structure causes a nonuniform state of stress (e.g., bending of beams or pressurized thick cylinders) σ is customarily taken as the highest tensile stress within the crack zone.

Example 9.2:

The design stress on a certain structural component is 690 MPa in tension. For plane strain conditions, determine the critical crack size for

TABLE 9.1
Typical Fracture Properties of Some Materials[a]

Material	K_{Ic}, MPa · m$^{1/2}$	Yield Strength, MPa
Aluminum 7075-T6	24	500
Aluminum 2014-T651	24	455
Aluminum 2024-T3	44	345
Low-carbon steel	180	250
Steel AISI 4340	60	1515
Cast irons	10	300[b]
Stainless steel 17-7 PH	77	1435
Ball-bearing steel 52100	14	2070
Tool steel H-11	38	1790
Maraging steel	110	1450
Titanium Ti-6 Al-4V	55	1035
Tungsten 15% Co	19	—
Tungsten carbide	10	900
Electrical porcelain	0.88	—
Polyvinylchloride	3–5	40–80
Acrylic	0.9–1.92	25–70
Boron-fiber epoxy	70	2000
Graphite-epoxy	45–120	1000
Concrete	0.25–1.6	20[c]
Polycarbonate	3	70
Glasses	1.0	

[a] The values of K_{Ic} are for indication only, and the designer is always urged to seek accurate values for the specific material under service conditions. For instance, structural steels possess decreasing K_{Ic} at lower temperature and higher (dynamic) loading rates

[b] Fracture.

[c] Fracture in bending.

Source: Hertzberg, R. W., *Deformation and Fracture Mechanics of Engineering Materials*, John Wiley & Sons, New York, 1976, 286, 370. With permission.

a. *Alloy steel with* $K_{Ic} = 134$ *MPa* · $m^{1/2}$ *(0.35% C + V);*
b. *Titanium alloy with* $K_{Ic} = 60$ *MPa* · $m^{1/2}$ *(Ti – 6Al – 4V);*
c. *What should be the design stress for acrylic if the maximum crack is 4.8 mm and* $K_{Ic} = 1.75$ *MPa* · $m^{1/2}$*?*

Noting that the yield strength Y = *1035 MPa for both alloy steel and titanium alloy and* Y = *50 MPa for acrylic, comment on the results.*

Solution:

From Equation 9.14, for a crack size $2a$

$$K_I = \sigma\sqrt{\pi a} \le K_{IC}$$

Solving for a gives

$$a = K_{IC}^2/\sigma^2\pi$$

a. The critical crack size for steel is

$$2a_{\text{steel}} = 2\times\frac{(134\ \text{MPa}\cdot\text{m}^{1/2})^2}{(690\ \text{MPa})^2\pi} = 24\ \text{mm}$$

b. The critical crack size for titanium is

$$2a_{\text{titanium}} = 2\times\frac{(60\ \text{MPa}\cdot\text{m}^{1/2})^2}{(690\ \text{MPa})^2\pi} = 4.81\ \text{mm}$$

Regardless the fact that both metals have almost the same yield strength, $Y \cong 1035$ MPa, the titanium in this case is clearly more crack sensitive, since

$$a_{\text{steel}} \cong 5a_{\text{titanium}}$$

c. The critical crack size for acrylic is

$$K = \sigma\sqrt{\pi a} \leq K_{IC}$$

or

$$\sigma = K_{IC}/\sqrt{\pi a}$$

$$\sigma = 1.75/\sqrt{\pi(0.0024)} = 20.15\ \text{MPa}$$

Noting that the tensile strength of acrylic is about 50 MPa, fracture in this case occurs at only 40% of this strength.

Example 9.3:

The fracture toughness K_{Ic} *for a certain material is 25 MPa · m*$^{1/2}$*. If a large thick plate made of this material contains a crack 1.5 mm long, what is the applied tensile stress* σ *that will result in fracture? Repeat calculations for a plate with a crack length of 10 mm, knowing that the yield strength of this material is 450 MPa. Comment on the results.*

Solution:

Fracture occurs in the first plate with a crack $2a = 1.5$ mm, if

$$K_{IC} = 25 \leq \sigma_f\sqrt{\pi(0.0015/2)}$$

Hence,

$$\sigma_f = 515.03\ \text{MPa}$$

For the second plate with a crack length of $2a = 10$ mm

$$\sigma_f = 25/\sqrt{\pi(0.01/2)} = 199.5\ \text{MPa}$$

Since the yield strength is 450 MPa, yielding will occur in the first plate ($2a$ = 1.5 mm) before brittle fracture; for the second plate fracture occurs before yielding.

Example 9.4:

Large plates made of unidirectionally composite fiber glass–epoxy with centrally located cracks in the fiber direction. Tensile testing in a direction perpendicular to the crack reported at incipient fracture; critical crack lengths of 3.8 and 25.4 mm at fracture stresses 166.7 and 68 MPa, respectively. Determine the critical stress intensity factor for these plates.

Solution:

The solution of this example assumes the applicability of linear-elastic fracture mechanics to composite materials, an assumption which is supported by existing experimental evidence.[10]

Applying Equation 9.12 for the two data points gives

$$K_c = \sigma\sqrt{\pi a}$$

or

$$K_c = 166.7\times 10^6\sqrt{\pi\times 0.0038} \text{ giving } K_c = 18.21 \text{ MPa}\cdot\text{m}^{1/2}$$

$$K_c = 68\times 10^6\sqrt{\pi\times 0.0254} \text{ giving } K_c = 19.2 \text{ MPa}\cdot\text{m}^{1/2}$$

The difference between the two values of K_c is about 5%, which suggests the validity of using linear-elastic fracture mechanics to interpret these test results conducted on orthotropic fiber-reinforced lamina with cracks in the fiber direction.

9.6 STRESS INTENSITY FACTORS FOR VARIOUS CONFIGURATIONS

By considering the stress distribution in the vicinity of the tip of a crack in the middle of a large plate subjected to a remote tensile stress, Relation 9.14 has been derived, namely,

$$K_\text{I} = \sigma\sqrt{\pi a} \le K_{\text{I}c} \tag{9.14}$$

Obviously, Relation 9.14 gives the stress intensity factor as a function of the applied stress σ and the crack size $2a$ for the particular given plate geometry of Figure 9.4. A fracture criterion more general than Relation 9.14 may be written as

$$K_\text{I} = \sigma\sqrt{\pi a}\,\text{CCF} \le K_{\text{I}c} \tag{9.15}$$

where "CCF" is a configuration correction factor depending on the loading and geometry of the cracked body. For instance, for a centrally cracked finite plate loaded in tension as in Figure 9.6b, CCF = $f(a/W)$ depends on the ratio of the crack half-length a to the width W. For $a/W \to 0$, CCF = $f(a/W) \to 1$ as may be seen from Equation 9.14 for an infinitely wide plate of Figure 9.6a.

For geometries and loadings other than Figure 9.6a, the CCF must be determined. Several theoretical and experimental methods exist to do this.* One exact, but rather involved, method makes

* These include boundary collocation, conformal mapping, integral transforms, and finite element method. See Parker.[3] Experimental techniques such as photoelasticity and interferometry are also used.

use of stress functions of complex variables. This analytical method solves only relatively simple geometries of cracked bodies. Other approximate methods are used for complicated geometries.

The results of such methods are presented either graphically or algebraically in the form of polynomials. Below are polynomial expressions and graphical representations for some of the most common configurations. In all cases utmost care must be taken in defining the stress appearing in Equation 9.15 as the one associated with the graphical or algebraic information used to determine the CCF.

9.6.1 Plates Under Tensile Loading

a. Crack of length $2a$, in the middle of an infinite plate, Figure 9.6a:

$$\text{CCF} = f(a/W) = 1$$

b. Crack of length $2a$ in a plate of finite width W, Figure 9.6b[11]:

$$\text{CCF} = f(a/W) = \left[\frac{W}{\pi a}\tan\frac{\pi a}{W}\right]^{1/2} \quad \text{for } a/W \le 0.25$$

or

$$= \sqrt{\sec\frac{\pi a}{W}} \quad \text{for } a/W \le 0.4$$

c. Double edge-cracks, each of length a, in a plate of finite width W, Figure 9.6c[12]:

$$\text{CCF} = f(a/W) = \left[\frac{W}{\pi a}\tan\left(\frac{\pi a}{W}\right) + \frac{0.2W}{\pi a}\sin\left(\frac{\pi a}{W}\right)\right]^{1/2}$$

or

$$= \left[1.12 + 0.43\left(\frac{a}{W}\right) - 4.79\left(\frac{a}{W}\right)^2 15.46\left(\frac{a}{W}\right)^3\right] \quad (\text{for } a/W > 0.7)$$

d. A single edge-crack of length a, in a plate of finite width W, Figure 9.6d[12]:

$$\text{CCF} = f(a/W) = \left[1.12 - 0.23\left(\frac{a}{W}\right) + 10.6\left(\frac{a}{W}\right)^2 - 21.7\left(\frac{a}{W}\right)^3 + 30.4\left(\frac{a}{W}\right)^4\right]$$

for $a/W < 0.7$

e. A circular internal crack of radius a (penny-shaped crack) embedded in an infinite solid, Figure 9.6e[5]:

$$\text{CCF} = 2/\pi$$

f. Semielliptical surface flaw of length a and width $2c$ in a plate, Figure 9.6f[13,14]:

$$\text{CCF} = f(a/2c) = 1.12/\sqrt{Q}$$

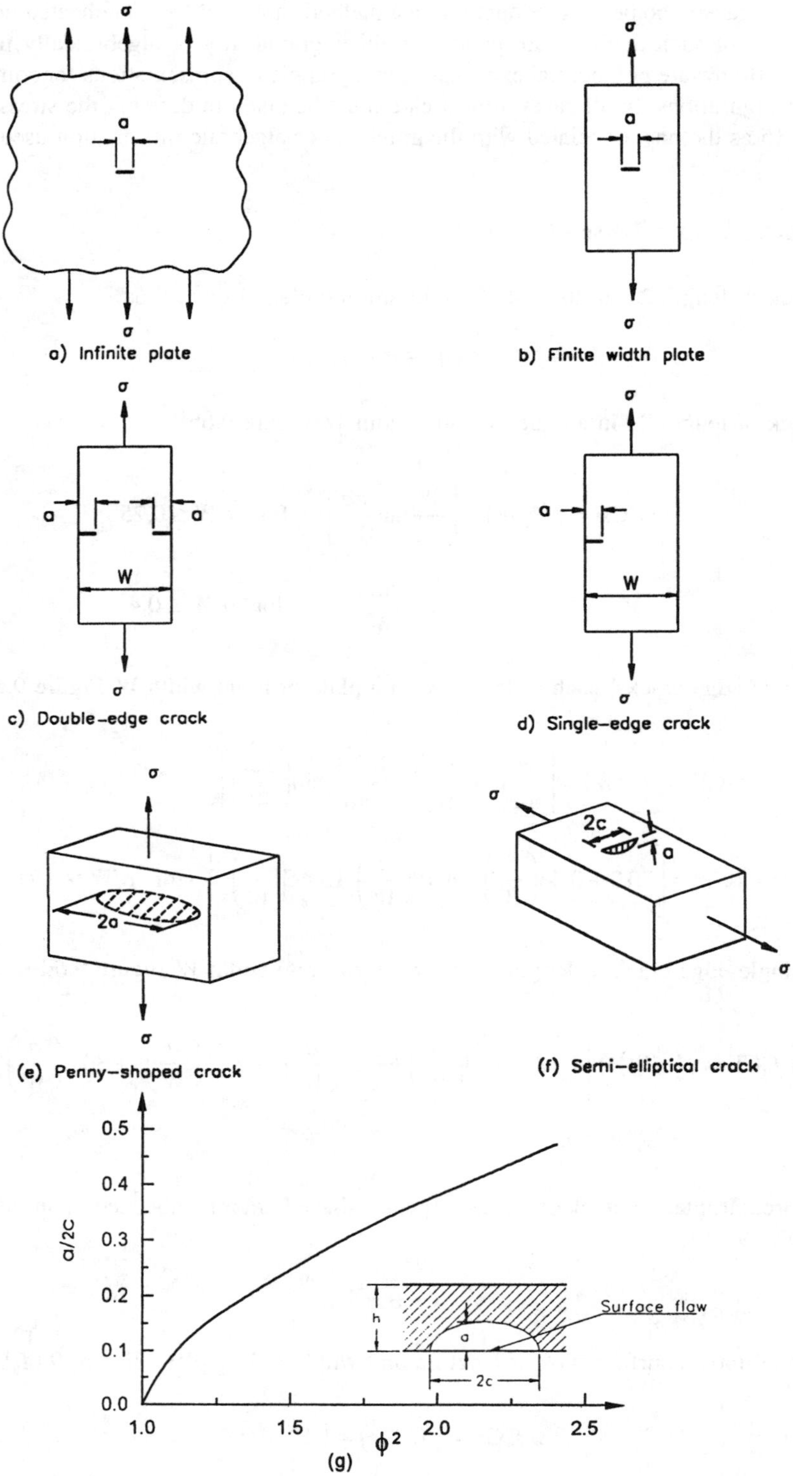

FIGURE 9.6 Configurations of various stress intensity correction factors, CCF. (From Irwin, G. R., Trans. Am. Soc. Mech. Eng., *J. Appl. Mech.,* 651–654, 1962. With permission.)

where $Q = [\phi^2 - 0.212\ (\sigma/Y)^2]$, Y being the yield strength. The parameter ϕ may be determined from Figure 9.6g or using the approximate expression*

$$\phi \cong \frac{3\pi}{8} + \frac{\pi}{2}\left(\frac{a}{2c}\right)^2$$

9.6.2 Cracks Emanating from Circular Holes in Infinite Plates

a. A single crack or a double crack at a circular hole of radius r_i in the middle of an infinite plate, Figure 9.7a and b, respectively[14]:

$$\text{CCF} = \left(\frac{r_i}{a} + \frac{1}{2}\right)^{1/2} \quad \text{for a single crack } a > 0.12r_i$$

$$\text{and CCF} = \left(\frac{r_i}{a} + 1\right)^{1/2} \quad \text{for a symmetric double crack } a > 0.12r_i$$

These are approximate CCFs, where a is the crack length measured from the edge of the hole having a radius r_i. The stress σ in Expression 9.15 designates the remote tensile stress perpendicular to the crack. Note that for very small cracks ($a/r_i \to 0$) in an exact solution the CCF = $f(a/2r_i) \to 3$, i.e., three times that for a normal central crack as expected from stress concentration around circular holes.

b. Two cracks each of length a at a circular hole of radius r_i in a plate of finite width, Figure 9.7c[15]: Here CCF is determied from Figure 9.7c, noting that the total crack length includes the hole diameter.

c. A corner (nonthrough) crack at holes, Figure 9.7d.[14,16] An approximate solution considers the hole as a part of an elliptical surface flaw of a ratio $a'/2c = a/(2\sqrt{2r_i b})$. The maximum stress intensity factor occurs at point A with CCF as:

$$\text{CCF} = \frac{1.2}{\phi}\left\{1 + \frac{a^2(2r_i - b)^2}{16r_i^2 b^2}\right\}^{1/4}$$

where ϕ is defined by Figure 9.6. Expression 9.15 is used in conjunction with a crack length a.

d. A single radial crack a at a hole of radius r_i under equibiaxial tensile loading, Figure 9.7e[17]: The configuration correction factor CCF = $f(a/W)$ depends on the ratio (a/r_i) as well as the applied loading. As noted before, the determination of the stress intensity factor is affected only by the crack-opening stress.** However, for a crack emanating from a hole, the stress concentration around the hole is known to be dependent on the remote stress field: uniaxial or biaxial. Hence, as previously found in Section 6.4, for an equibiaxial stress system ($\sigma_1 = \sigma_2 = \sigma$), the stress concentration factor at the edge of hole is only 2 compared to a value of 3 for a uniaxial stress system. This fact is reflected

---

* The dependence of Q on the ratio of the stress opening the crack to the yield strength of material results from consideration of plasticity at the crack tip. The above CCF has to be multiplied by a correction M_k, accounting for the proximity of the free surface in front of the crack, dependent on a/h. For shallow cracks, i.e., a/h << and $a/2c \to 0.5$; $M_k \simeq 1$. However, for $a/h = 0.6$ and $a/2c = 0.1$, M_k could attain a value of 2. See Broek,[14] pp. 90–94.

** This is true for an absolutely elastic condition. If plasticity at the crack tip is considered, a transverse stress loading affects the stress intensity factor; it is increased with a compressive stress and decreased with a tensile one.

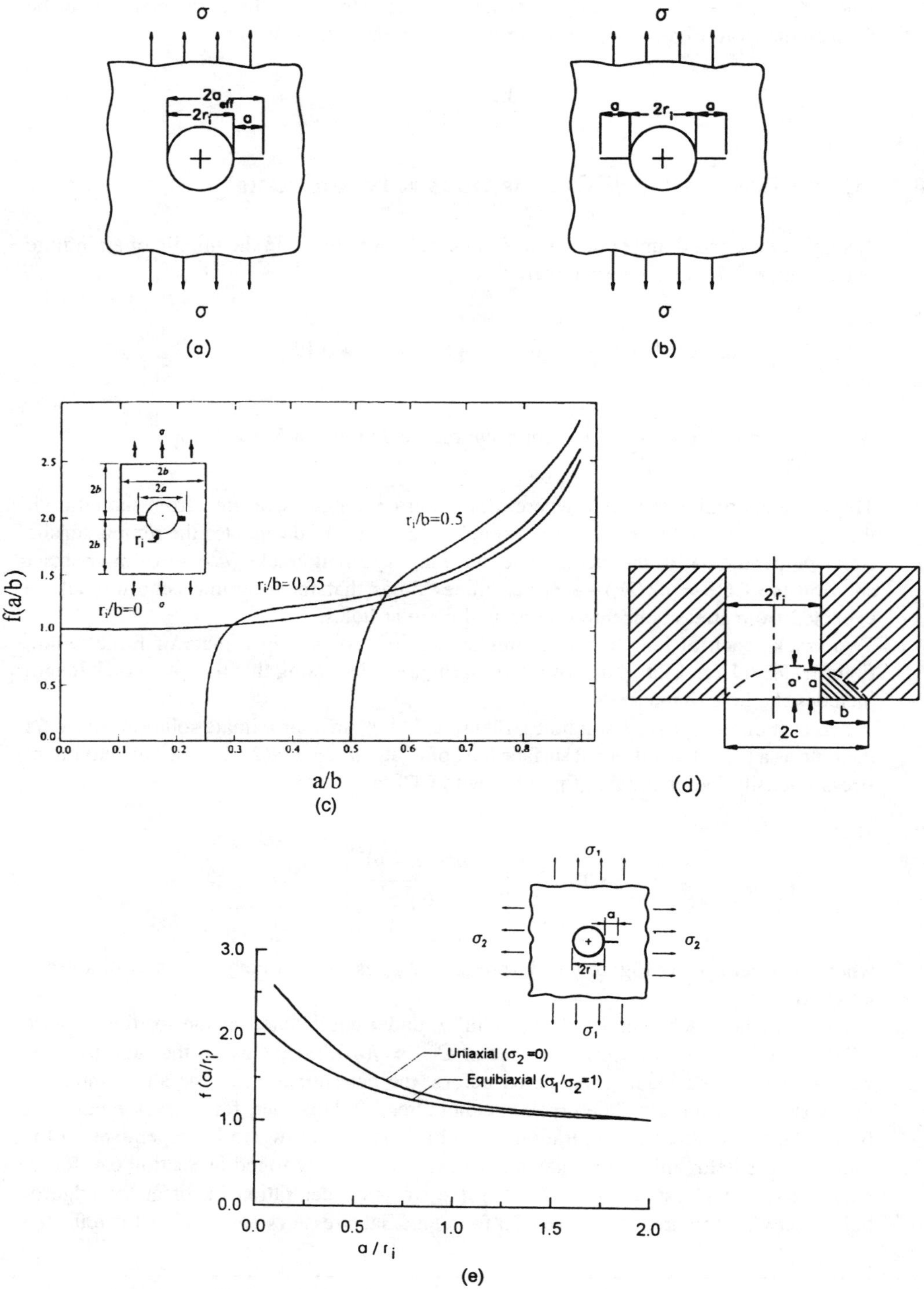

FIGURE 9.7 Graphs for various CCFs. (c, from Rooke, D. P. and Cartwright, D. J., *Compendium of Stress Intensity Factors*, The Hillingdon Press, London, 1976. With permission.)

in Figure 9.7e which indicates a smaller CCF factor for biaxial tensile loading as compared to uniaxial loading.

9.6.3 Plates Under Bending

a. Pure bending for a cracked plate of depth W, Figure 9.8a[18]:

$$\text{CCF} = \left(\frac{2W}{\pi a}\tan\left(\frac{\pi a}{2W}\right)\right)\frac{0.923 + 0.199\left(1 - \sin\frac{\pi a}{2W}\right)^4}{\cos\left(\frac{\pi a}{2W}\right)}$$

b. Three-point bending for a cracked plate of $L/W = 2$, Figure 9.8b[12]:

$$\text{CCF} = f(a/W) = \left[1.107 - 2.12\left(\frac{a}{W}\right) + 7.71\left(\frac{a}{W}\right)^2 - 13.6\left(\frac{a}{W}\right)^3 + 14.2\left(\frac{a}{W}\right)^4\right]$$

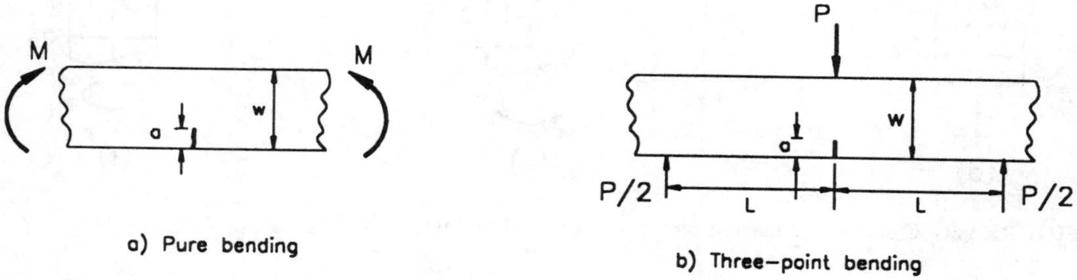

FIGURE 9.8 Cracked bars under bending: (a) pure bending, (b) three-point bending.

9.6.4 Circular Rods and Tubes

a. Central circular crack in a cylindrical bar of radius r_o, Figure 9.9a, b, and c,[19] subjected to the following:
   - Tensile load P causing a net average stress $\sigma = P/\pi(r_o^2 - a^2)$, hence,

$$\text{CCF} = \frac{2\sqrt{1 - a/r_o}}{\pi}\left[1 + 0.5\left(\frac{a}{r_o}\right) - 0.625\left(\frac{a}{r_o}\right)^2 + 0.421\left(\frac{a}{r_o}\right)^3\right]$$

   - Pure bending moment M causing a stress $\sigma = 4\text{MPa}/\pi(r_o^4 - a^4)$, hence,

$$\text{CCF} = \frac{4\sqrt{1 - a/r_o}}{3\pi}\left[1 + 0.5\left(\frac{a}{r_o}\right) + 0.375\left(\frac{a}{r_o}\right)^2 + 0.313\left(\frac{a}{r_o}\right)^3 - 0.727\left(\frac{a}{r_o}\right)^4 + 0.483\left(\frac{a}{r_o}\right)^5\right]$$

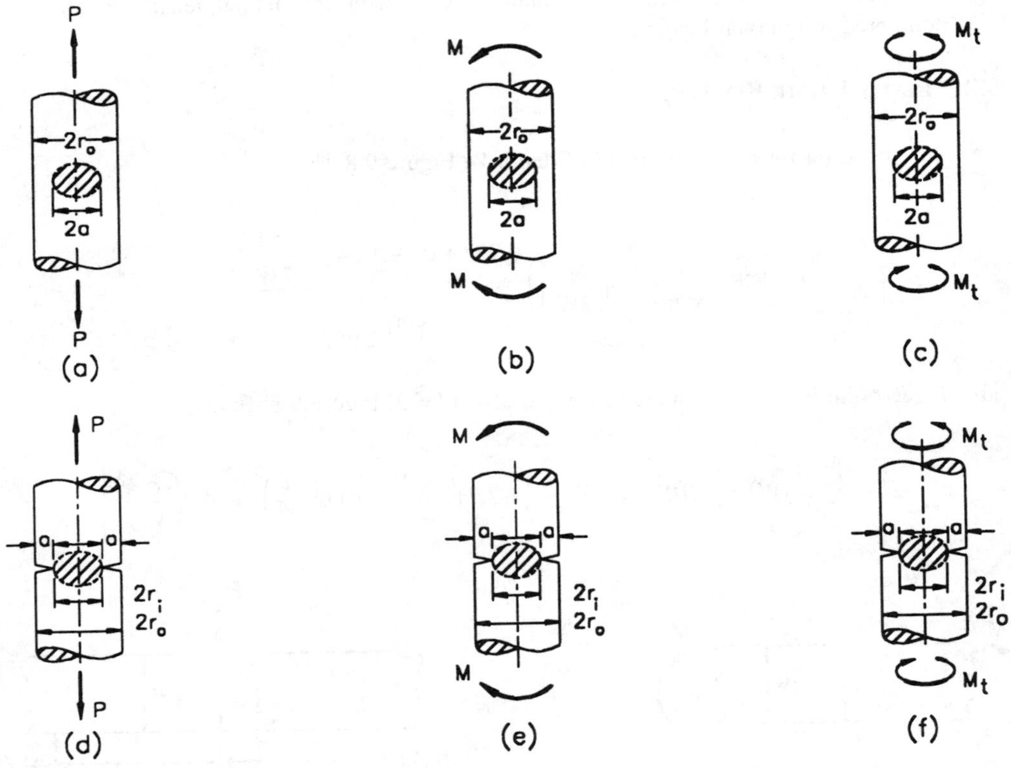

FIGURE 9.9 Cracked circular rods and tubes subjected to different loading.

- Twisting moment M_t causing a shear stress, $\tau = 2M_t\, a/\pi(r_o^4 - a^4)$, hence, CCF is approximately equal to that for pure bending for cracks of (a/r_o) up to 0.6.

b. External circumferential crack in a rod of radius r_o, Figure 9.9d, e, and f,[19] subjected to the following:
- Tensile load P causing a net tensile stress $\sigma = P/\pi r_i^2$; hence,

$$\text{CCF} = \frac{\sqrt{r_i/r_o}}{2}\left[1 + 0.5\left(\frac{r_i}{r_o}\right) + 0.375\left(\frac{r_i}{r_o}\right)^2 - 0.363\left(\frac{r_i}{r_o}\right)^3 + 0.731\left(\frac{r_i}{r_o}\right)^4\right]$$

- Pure bending moment M causing a net stress $\sigma = 4M/\pi r_i^3$; hence,

$$\text{CCF} = \frac{3\sqrt{r_i/r_o}}{8}\left[1 + 0.5\left(\frac{r_i}{r_o}\right) + 0.375\left(\frac{r_i}{r_o}\right)^2 + 0.313\left(\frac{r_i}{r_o}\right)^3 + 0.273\left(\frac{r_i}{r_o}\right)^4 + 0.537\left(\frac{r_i}{r_o}\right)^5\right]$$

- Twisting moment M_t causing a net shear stress $\tau = 2M_t/\pi r_i^3$: CCF is approximately equal to that for pure bending.

Note that in the above three cases the stresses are defined with respect to the net cross section, i.e., excluding the crack.

c. External circumferential crack in a long tube, Figure 9.10a, b, and c,[15] subjected to the following:
   - Uniaxial tensile stress σ applied remote from the crack, CCF is obtained from Figure 9.10a.
   - Twisting moment M_t applied, remote from the crack, about the tube axis: CCF is obtained from Figure 9.10b.

d. External radial crack in a long tube subjected to a twisting moment M_t: CCF is obtained from Figure 9.10c.

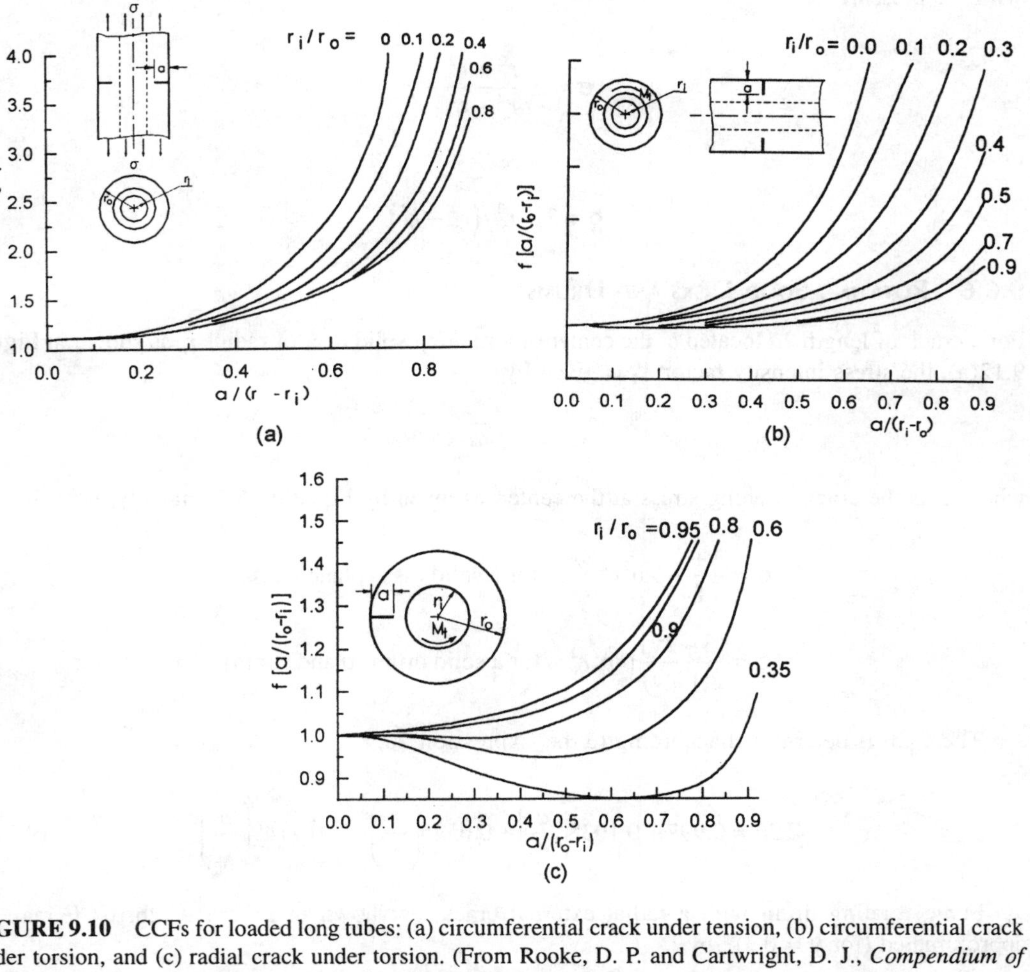

FIGURE 9.10 CCFs for loaded long tubes: (a) circumferential crack under tension, (b) circumferential crack under torsion, and (c) radial crack under torsion. (From Rooke, D. P. and Cartwright, D. J., *Compendium of Stress Intensity Factors*, The Hillingdon Press, London, 1976. With permission.)

9.6.5 Pressurized Thick-Walled Cylinders

The stress intensity factors for long, thick-walled cylinder of outside and inside radii r_o and r_i, respectively, with radial edge crack of length a, while being subjected to a uniform internal pressure is given by

$$K_I = \sigma\sqrt{\pi a}\ \text{CCF} = \sigma\sqrt{\pi a} f\left[a/(r_o - r_i)\right] \tag{9.16}$$

where $f\,[a/(r_o - r_i)]$ is the CCF as obtained from Figure 9.11a[15] for an external radial edge crack depending on the ratio r_o/r_i. Note that σ in Equation 9.16 is taken as the highest tensile hoop stress σ_θ at the location of the crack, i.e., $\sigma = (\sigma_\theta)_{r=r_o}$ as given by Equation 6.2 namely,

$$\sigma = 2p_i r_i^2/(r_o^2 - r_1^2)$$

For a pressurized long, thick-walled cylinder with internal radial crack. The configuration factor $f[a/(r_o - r_i)]$ is found from Figure 9.11b depending on the ratio r_o/r_i. In this case, the stress opening the crack comprises that due to the internal pressure $(\sigma_\theta)r{=}r_i$ according to Equations 6.2 together with the pressure as

$$\sigma = p_i \frac{r_o^2 + r_i^2}{r_o^2 - r_i^2} + p_i$$

or

$$\sigma = 2p_i r_o^2/(r_o^2 - r_i^2)$$

9.6.6 Rotating Solid Disks and Drums

For a crack of length $2a$ located at the center of a rotating solid disk of radius r_o as shown in Figure 9.12(a), the stress intensity factor K_I is given by

$$K_I = \sigma\sqrt{\pi a}\ \text{CCF}$$

where σ is the crack-opening stress at the center, as given by Equation 6.23 namely,

$$\sigma = \frac{3+\nu}{8}\rho\omega^2 r_o^2 \quad \text{for a solid disk (plane stress)}$$

$$\sigma = \frac{(3-2\nu)}{8(1-\nu)}\rho\omega^2 r_o^2 \quad \text{for a solid drum (plane strain)}$$

The CCF is determined according to the Expression 20:

$$\text{CCF} = 0.997 + 0.1038\left(\frac{a}{r_o}\right) + 0.6525\left(\frac{a}{r_o}\right)^2 + 0.7149\left(\frac{a}{r_o}\right)^3 \tag{9.17a}$$

For a rotating drum with a radial external crack, as shown in Figure 9.12b, CCF may be approximated (for $\nu = 0.3$)* by[21]:

$$\text{CCF} = 1.134 + 3.465\left(\frac{a}{r_o}\right) + 2.363\left(\frac{a}{r_o}\right)^2 - 3.394\left(\frac{a}{r_o}\right)^3 + 3.848\left(\frac{a}{r_o}\right)^4 \tag{9.17b}$$

* Based on the solution of Reference 21 (in adapted form).

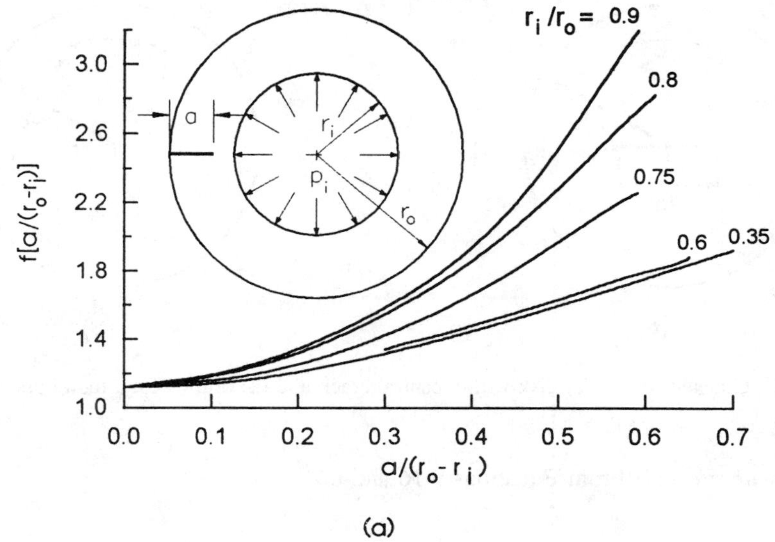

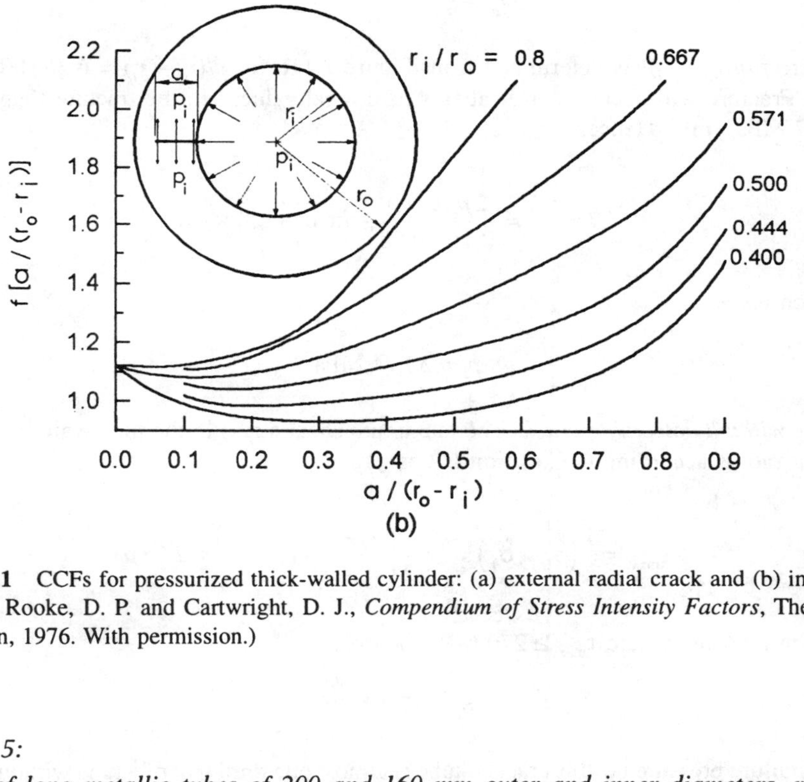

FIGURE 9.11 CCFs for pressurized thick-walled cylinder: (a) external radial crack and (b) internal radial crack. (From Rooke, D. P. and Cartwright, D. J., *Compendium of Stress Intensity Factors*, The Hillingdon Press, London, 1976. With permission.)

Example 9.5:

Inspection of long metallic tubes of 200 and 160 mm outer and inner diameters, respectively, revealed that one of these tubes contains an external radial edge crack of 6.3 mm length. If the tubes are made of an aluminum alloy for which the fracture toughness (K_{Ic}) is 27 MPa · $m^{1/2}$, determine the maximum internal pressure that can be applied to the tube without causing fracture. Compare the result with the pressure which a tube with no detectable crack can withstand without yielding. The yield shear strength of this aluminum alloy is 270 MPa.

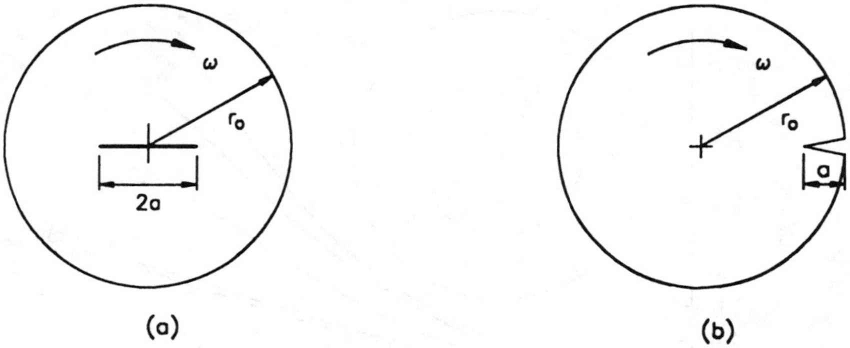

FIGURE 9.12 Cracked rotors: (a) disk with a central crack and (b) drum with a radial crack.

Solution:

a. *Tube with a crack*: From Equations 9.16 and 6.2:

$$K_{\mathrm{I}} \frac{2p_i r_i^2}{r_o^2 - r_i^2}\sqrt{\pi a}\, f\left[a/(r_o - r_i)\right]$$

where $f[a/(r_o - r_i)]$ is determined from Figure 9.11a for $a/(r_o - r_i) = 6.3/(100 - 80)$ as 1.6. Fracture will occur when K attains its critical value, i.e., the fracture toughness K_{Ic} = 27 MPa · m$^{1/2}$. Hence,

$$27 \times 10^6 = \frac{2p_i(0.08)^2}{0.1^2 - 0.08^2}\sqrt{\pi(0.0063)} \times 1.6$$

which gives

$$p_i = 33.74 \text{ MPa}$$

b. *Tube without detectable crack:* The maximum shear stress in the tube wall occurs at the inner radius according to Equation 6.2 as

$$\tau_{max} = \frac{1}{2}(\sigma_\theta - \sigma_r)_{r=r_i} = \frac{1}{2}\left(\frac{r_o^2 + r_i^2}{r_o^2 - r_i^2} p_i + p_i\right) = 2.78 p_i$$

yielding occurs where $\tau_{max} \geq 270$ MPa; hence,

$$p_i = 97.2 \text{ MPa}$$

The failure pressure for this cracked tube is almost one third that of a tube without a crack.

Example 9.6:

Solid steel disks of 300 mm diameter are produced by forging. Inspection of these disks showed that some of them are defected by either a very small porosity or a sharp crack at their center. The largest crack found is 30 mm. These disks have to be used as rotating elements. Determine the maximum safe speed, such that

a. *The displacement at the outer edge does not exceed 0.015 of the radius (neglect the effect of defects),*
b. *The disk does not yield or fracture. Take* E = *200 GPa,* K = *55 MPa · m*$^{1/2}$*,* $\nu = 0.3$*,* Y = *670 MPa, and a factor of safety F.S. = 2. The specific weight of steel = 78 kN/m.*[3]

Solution:

a. The maximum displacement u at the outer edge for a rotating solid disk is obtained from Equation 6.24 as

$$u = \frac{(3+\nu)(1-\nu)}{8E}\left(r_o^2 - \frac{1+\nu}{3+\nu}r^2\right)\rho\omega^2 r$$

at $r = r_o$:

$$\frac{u}{r_o} \le 0.015 = \frac{3.3\times 0.7}{8\times 200\times 10^6}\left(0.15^2 - \frac{1.3}{3.3}0.15^2\right)\times\left(\frac{78\times 10^3}{9.81}\right)\omega_a^2$$

giving

$$\omega_a = 309.6 \text{ rad/s}$$

b. For a solid disk with central porosity, the equations of annular disk with $r_i \to 0$ are used. Hence, from Equation 6.20,

$$\sigma_\theta = \frac{3+\nu}{4}\rho\omega^2 r_o^2 \le (Y/\text{F.S.})$$

$$\sigma_\theta = \frac{3.3}{4}\left(\frac{78\times 10^3}{9.81}\right)(0.15)^2\omega_b^2 \le \frac{670\times 10^6}{2}$$

giving

$$\omega_b = 1506.6 \text{ rad/s}$$

Note: Although rotating disks represent plane stress conditions, plane strain fracture toughness K_{Ic} is used in this example to provide the most conservative estimate. For a solid disk with a central crack of length $2a$, Equation 9.15 gives, for failure against brittle fracture,

$$K_I = (\sigma_\theta)_{r=0}\sqrt{\pi a}\,\text{CCF} \le (K_{Ic}/\text{F.S.})$$

Hence, by substituting for σ_θ at $r = 0$ from Equation 6.23,

$$\frac{K_{Ic}}{\text{F.S.}} = \frac{3+\nu}{8}\rho\omega^2 r_o^2\sqrt{\pi a}\,\text{CCF}$$

where CCF is determined from Expression 9.17a for $a/r_o = 0.015/0.15 = 0.1$ as CCF = 1.0146. This results in

$$\frac{55\times10^6}{2}=\frac{3.3}{8}\left(\frac{78\times10^3}{9.81}\right)\omega_c^2r_o^2\sqrt{\pi(0.015)}\times1.0146$$

giving

$$\omega_c = 1300.7 \text{ rad/s}$$

Hence, the maximum safe speed is $\omega = \omega_a = 309.6$ rad/s. The more detrimental effect of sharp cracks on failure compared with porosity is seen by comparing ω_b and ω_c.

Example 9.7:

A pressure cylinder is made of an aluminum alloy with yield strength of 392 MPa. The vessel has a mean diameter of 500 mm and thickness of 15 mm. The operating service pressure is determined such that yielding does not occur with a factor of safety of F.S. = 2. During service, as a result of stress corrosion, a longitudinal surface flaw of semielliptical shape of a/2c ≃ 0.42 (as indicated in Figure 9.13) is detected within the plate thickness. Estimate the size of the critical defect that induces fracture at service pressure. Take the fracture toughness K_{Ic} of the used steel to be 33.9 MPa · $m^{1/2}$. Neglect both the effects of plasticity at crack tip and the proximity of the front free surface as represented by (σ/Y) and (a/h), respectively, on the value of CCF.

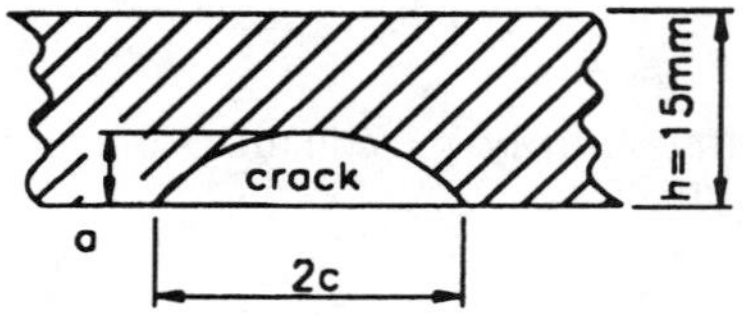

FIGURE 9.13 Example 9.7.

Solution:

By applying the maximum shear stress theory, $\sigma_{max} - \sigma_{min} = Y/F.S.$, where $\sigma_{max} = \sigma_\theta$ and $\sigma_{min} \cong 0$, hence, according to Equation 6.68b,

$$\sigma_\theta = \frac{p_i r_m}{h} = \frac{p_i \times 250}{15} \leq \frac{392}{2}$$

giving

$$p_i = 11.76 \text{ MPa and } \sigma_\theta = 196 \text{ MPa}$$

Fracture occurs if Condition 9.15

$$K_I = \sigma_\theta\sqrt{\pi a}\,\text{CCF} \leq K_{Ic}$$

is satisfied. By neglecting the effects of (σ/Y) and (a/h) on the CCF, $\phi = \sqrt{Q}$ is determined from Figure 9.6g for $(a/2c) = 0.42$ as $\phi \cong 1.45$ and upon substitution into the expression,

$$\text{CCF} = \frac{1.12}{\phi} = \frac{1.12}{1.45} = 0.772$$

The fracture condition (Equation 9.15):

$$196\sqrt{\pi a_c} \times 0.772 = 33.9 \text{ MPa} \cdot \text{m}^{1/2}$$

gives a_c = 16 mm, which is greater than the thickness of the vessel wall. In such a case during operation, the detected flaw will continue to grow because of stress corrosion until its depth "a" will constitute the whole thickness of the cylinder. This will cause leaking of the contained fluid and a condition of "leak before break" is attained. This condition must be viewed as different (and may be less damaging) than fracture, which occurs as a result of a stress intensity factor that is higher than the fracture toughness of the material. For instance, if a metal of K_{Ic} = 25 MPa · m$^{1/2}$ is used, a critical crack length of

$$a_c = \frac{1}{\pi}\left(\frac{25}{196 \times 0.772}\right)^2 = 8.7 \text{ mm}$$

is enough to cause sudden fracture at the service pressure of 11.76 MPa.

9.7 SUPERPOSITION UNDER COMBINED LOADING

Since the fracture criterion of Equation 9.15 is based on linear elasticity, the principle of stress superposition holds. Superposition here means addition of stress intensity factors due to various loadings, provided that all cause the same mode of crack opening. For example, for a given crack size in a plate subjected to a combination of tensile and bending loadings results in a mode (I) stress intensity factor K_{tot}, given by

$$K_{tot} = K_{tension} + K_{bending}$$
$$K_{tot} = \left[\sigma_{tension} f_1(a/W) + \sigma_{bending} f_2(a/W)\right]\sqrt{\pi a} \tag{9.18}$$

The CCFs $f_1(a/W)$ and $f_2(a/W)$ correspond to tensile and bending loadings, respectively.

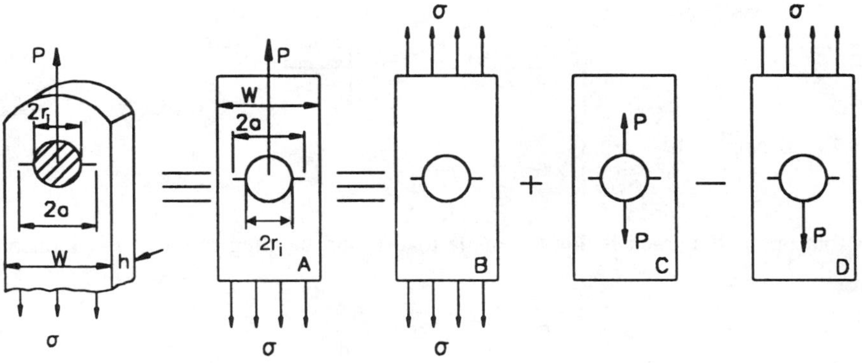

FIGURE 9.14 Cracked pin-loaded lug; determination of K by superposition.

Pictorial superposition is often helpful in obtaining CCFs. This may be illustrated by the problem of a pin-loaded lug, Figure 9.14. This case can be built up of the three cases B, C, and D (depicted in Figure 9.14) by adding cases B and C then subtracting case D. Hence,

$$K_{tot} = K(A) = K(B) + K(C) - K(D)$$

It is obvious that $K(A) = K(D)$ and therefore,

$$K(A) = {}^1/_2\,[K(B) + K(C)]$$

The CCF for $K(B)$ is previously given in Figure 9.7c while that for $K(C)$ may be obtained from reference books, noting that $P = \sigma hW$ for vertical equilibrium.

Example 9.8:

A machine frame in the form of a curved beam is fabricated from thick steel plates. The loading on the frame is as shown in Figure 9.15. A crack of length 10 mm is detected in one of the welds. What is the maximum load P that the frame can support before fracture? Properties of weld zone = K_{Ic} = *30 MPa ·* $m^{1/2}$. *For a beam under pure bending use the tabulated correction factor. Take* W = *100 mm,* h = *20 mm,* L = *250 mm.*

a/W	0.1	0.2	0.3	0.4	0.5	0.6
CCF = f(a/W)	1.02	1.06	1.16	1.32	1.62	2.10

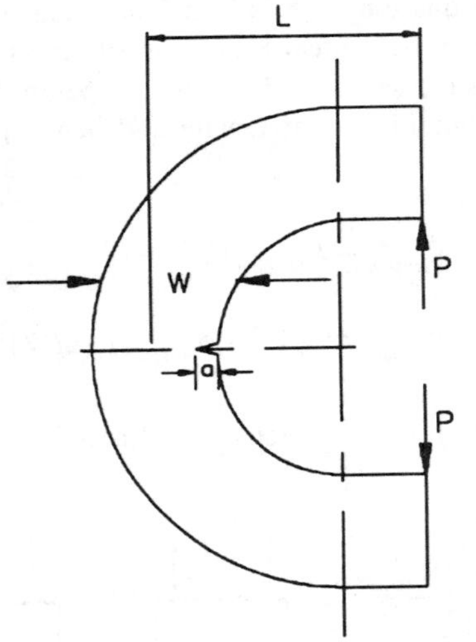

FIGURE 9.15 Example 9.8.

Solution:

The stress that opens the crack is due to tensile load P and bending moment PL. Hence,

$$\sigma_1 = \frac{p}{hW}, \quad \sigma_2 = \frac{6PL}{hW^2}$$

The corresponding CCFs (for $a/W = 10/100 = 0.1$) are obtained from case d, Figure 9.6d, and the above table, respectively, as

$$f_1(a/W) = f_1(0.1) = \left[1.12 - 0.23(0.1) + 10.6(0.1)^2 - 21.7(0.1)^3 + 30.4(0.1)^4\right] = 1.1843$$

and $f_2(a/W) = f_2(0.1) = 1.02$. By superposition, Equation 9.18 gives for $W = 0.1$ m, $L = 0.25$ m and $h = 0.02$ m:

$$K_I = \left[\frac{P}{hW}(1.1843) + \frac{6PL}{hW^2}(1.02)\right]\sqrt{\pi a} = 8242.15P\sqrt{\pi a}$$

Applying the fracture criterion (for $K_{Ic} = 30$ MPa · m$^{1/2}$ and $a = 10$ mm) gives

$$K_I = \sigma f(a/W)\sqrt{\pi a} \le K_{Ic}$$

hence, $P = 20.53$ kN.

9.8 MIXED-MODE LOADING

The bulk of fracture mechanics literature is mainly concerned with the first mode of crack deformation. This is because the majority of engineering fracture problems belongs to this first mode only. However, there are some problems where mode (I) is combined with one of the other modes (II) and (III). Also cracked geometries that start in one mode may change to a mixed mode during the life of the structure. The opposite also happens, where mixed-mode loading develops into mode (I) solely. Examples of mixed mode loading are cracked bars in torsion and bending, slant cracks in pressure vessels, aircraft wings under torsion and bending and bridge spans.

Since the crack tip stresses for modes (II) and (III) can be derived[3] in the same format as for mode (I), Equations 9.11, the fracture condition for any mode has the same form as: $K_I = \sigma\sqrt{\pi a}\, f_I(a/W)$. For instance, for the shear mode (II), σ is replaced by τ and the CCF = $f_{II}(a/W)$ must correspond to mode (II).

An engineering approach to handle mixed-mode loading (e.g., modes I and II) problems is to assume that the fracture condition — based on energy arguments — is more likely to be of the form:

$$\left(\frac{K_I}{K_{Ic}}\right)^2 + \left(\frac{K_{II}}{K_{IIc}}\right)^2 = 1 \tag{9.19}$$

Experimental evidence based on testing several materials suggests that $K_{IIc}/K_{Ic} \cong 0.75$ to 0.8, which upon substitution into Condition 9.19 results in (for $K_{IIc} = 0.75K_{Ic}$)

$$K_I^2 + 1.78K_{II}^2 = K_{Ic}^2$$

The use of this criterion is simple and straightforward as indicated in Example 9.9.

Example 9.9:

A tubular shaft of outside and inside radii r_o *= 50 mm and* r_i *= 30 mm, respectively, has a circumferential crack of length* a *= 10 mm. The shaft is subjected to bending while transmitting torque inducing a maximum shear stress* τ *= 150 MPa. Determine the maximum bending stress that can be applied without causing mixed-mode fracture. Take* K_{Ic} *= 70 MPa · m*$^{1/2}$*. Compare the result with that based on considering first-mode fracture only.*

Solution:

For fracture under mixed modes (I) and (III), a condition similar to Equation 9.19 may be stipulated as (see Hellan,[7] p. 80):

$$K_I^2 + \frac{K_{III}^2}{1-\nu} = K_{Ic}^2$$

Hence, for $\nu = 0.3$

$$K_I^2 + 1.43K_{III}^2 = K_{Ic}^2$$

The stress intensities K_I and K_{III} are defined as

$$K_I = \sigma\sqrt{\pi a} f_I\left[a/(r_o - r_i)\right] \quad \text{for bending}$$

$$K_{III} = \tau\sqrt{\pi a} f_{III}\left[a/(r_o - r_i)\right] \quad \text{for torsion}$$

The CCFs are found from Figure 9.9e and f, respectively, for $r_i/r_o = 0.6$ as

$$f_I = \left[10/(50-30)\right] = 1.45 \quad \text{for bending}$$

$$f_{III} = \left[10/(50-30)\right] = 1.15 \quad \text{for torsion}$$

Note that Figure 9.9e is employed here as an approximation to determine f_I. Hence, upon substitution $\sigma^2(\pi \times 0.01)\ (1.45)^2 + 1.43\ (150)^2\ (\pi \times 0.01)\ (1.15)^2 = (70)^2$; giving $\sigma = 232.3$ MPa for mixed mode fracture.

If only the fracture mode (I) is considered, then, $K_I = \sigma\sqrt{\pi \times 0.01} \times 1.45 \le 70$ MPa $\cdot$ m$^{1/2}$, giving $\sigma = 272.4$ MPa, which is higher by about 17% of the safe value determined by considering mixed-mode fracture.

9.9 PLASTIC ZONE GEOMETRY AT CRACK TIP

Equations 9.11 expressing stress components in the vicinity of a crack show that they reach infinite values at the crack tip ($r = 0$). Actually, this is not realistic since yielding occurs once these stresses satisfy the yield criterion obeyed by the material. Limiting the analysis to plane stress and for simplicity, consider only the stress component σ_y at $\theta = 0$, namely,

$$\left[\sigma_y\right]_{\theta=0} = \frac{K}{\sqrt{2\pi r}}$$

as plotted in Figure 9.16a where it attains the value Y, the yield strength of the material at some distance $r = r^*$ from the crack tip. Hence, a more realistic stress distribution must take the shape of Figure 9.16b, where Δ_p is the extent of the plastic zone around the crack or simply its size.

Since equilibrium is maintained for the elastic–plastic stress distribution, the extent of the plastic zone is obtained by satisfying the condition (for a unit thickness):

$$\Delta_p Y = \int_0^{r^*} \left[\sigma_y\right]_{\theta=0} dr = \int_0^{r^*} \frac{K}{\sqrt{2\pi r}} dr = K\sqrt{\frac{2r^*}{\pi}} \tag{9.20}$$

Thus, noting that at $r = r^*$,

$$\left[\sigma_y\right]_{\theta=0} = Y = \frac{K}{\sqrt{2\pi r^*}}$$

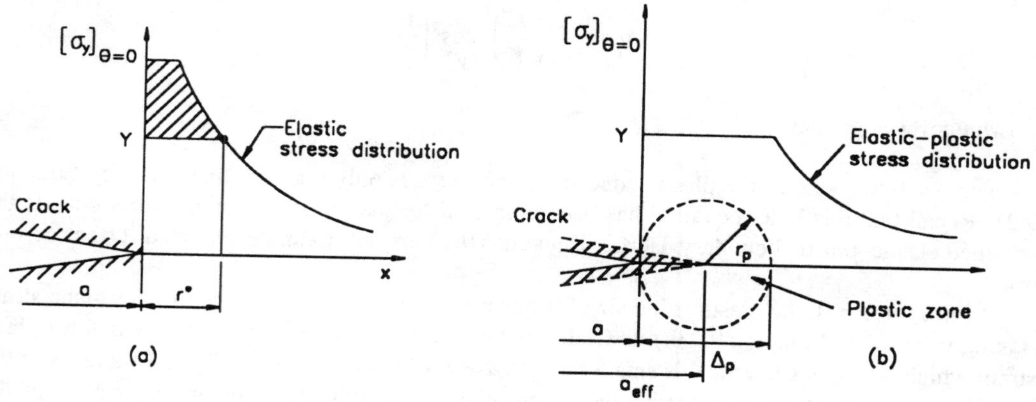

FIGURE 9.16 Stress distributions at crack tip: (a) elastic and (b) elastic–plastic with plastic zone correction.

substitution into Expression 9.20 yields the plastic zone size as

$$\Delta_p = \frac{1}{\pi}\left(\frac{K}{Y}\right)^2 \tag{9.21}$$

If the shape of the plastic zone is assumed as a first approximation to be circular, then its radius is*:

$$r_p = \frac{1}{2\pi}\left(\frac{K}{Y}\right)^2 \tag{9.22}$$

noting that this is only valid for plane stress conditions.

To determine the exact shape of the plastic zone, a more-refined analysis considers yielding due to the stress components $\sigma_1 > \sigma_2 > \sigma_3$ as functions of (r, θ) at the crack tip together with the maximum shear stress yield (Tresca) criterion. Hence, direct use of Expressions 9.11b results in the plastic zone size as

For plane stress

$$\Delta_p = \frac{1}{2\pi}\left(\frac{K}{Y}\right)^2 \cos^2\left(\frac{\theta}{2}\right)\left[1 + \sin\left(\frac{\theta}{2}\right)\right]^2 \tag{9.23a}$$

For plane strain:

$$\Delta_p = \frac{1}{2\pi}\left(\frac{K}{Y}\right)^2 \cos^2\left(\frac{\theta}{2}\right)\left[1 - 2\nu + \sin\left(\frac{\theta}{2}\right)\right]^2 \tag{9.23b}$$

or

* Another model due to D. S. Dugdale[22] determines the plastic zone size at the crack tip, for plain stress, as

$$\Delta_p = \frac{\pi}{8}\left(\frac{K}{Y}\right)^2$$

$$\Delta_p = \frac{1}{2\pi}\left(\frac{K}{Y}\right)^2 \cos^2\left(\frac{\theta}{2}\right)$$

whichever is greater.

For $\theta = 0$, the size of the plastic zone for plane stress is only one half that given by Equation 9.21. Recall that this latter equation has been corrected for the redistribution of stress outside the assumed elastic–plastic boundary. Hence, Equations 9.23 are approximate and need further correction.

Again, for $\theta = 0$, the plastic zone size for plane stress is about nine times that for plane strain (taking $\nu = 0.33$ in Equation 9.23b). Indeed, it is the triaxial stress field, $\sigma_3 = \nu(\sigma_1 + \sigma_2)$, for plane strain which is responsible for this smaller plastic zone, as shown in Figure 9.17. This leads to the conclusion that thick, cracked specimens, where plane strain conditions prevail over most of the thickness are prone to brittle fracture rather than plastic collapse by excessive yielding.

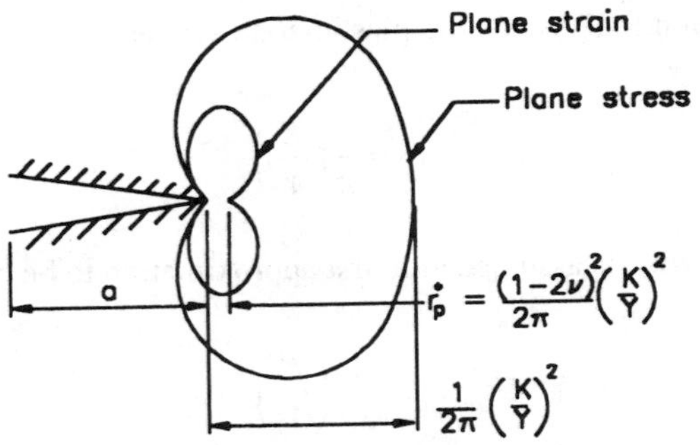

FIGURE 9.17 Plastic zone shapes for plane stress and plane strain conditions according to the Tresca yield criterion.

The plane strain fracture toughness K_{Ic} measured using sufficiently thick specimens represents the most conservative (lower) value of K_c, as shown in Figure 9.18. Testing with thicknesses greater than B_{min} results in an intrinsic material property (K_{Ic}) independent of specimen geometry. The minimum value of test specimen thickness B_{min} for fracture toughness K_{Ic} determination has been standardized[8] as

$$B_{min} \geq 2.5\left(\frac{K_{Ic}}{Y}\right)^2 \tag{9.24}$$

This is about 5π times the plastic zone radius r_p under plane stress and possibly $15\pi r_p$ under plane strain.

For low-yield-strength materials having high K_{Ic}, the required specimen thickness for a fracture toughness test may reach an impractical value approaching 1 m. Even though such thicknesses will probably not be used in real engineering applications, failure due to gross yielding will generally precede fracture in these materials. This may restrict the applicability of linear fracture mechanics at room temperature to high-strength materials. For steels, this may include steels for which E/Y is roughly less than 200 to 300. For lower-strength materials, the applicability of linear-elastic fracture mechanics becomes more valid at very low temperatures, higher strain-rate and for thick sections.

As a result of crack tip plasticity, displacements become larger and the stiffness is lower than that for purely elastic conditions. Irwin[23] suggested that this effect can be incorporated in the fracture

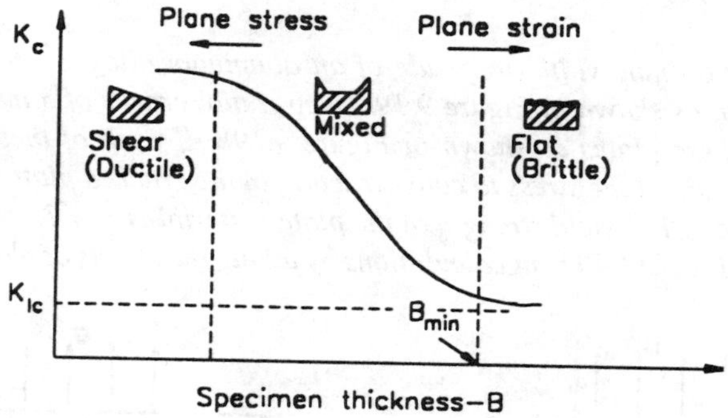

FIGURE 9.18 Variation of fracture toughness K_c and failure mode with specimen thickness.

criterion, Equation 9.15, by assuming that the effective crack length a_{eff} is the actual length plus the radius of the plastic zone r_p, i.e.,

$$a_{eff} = a + r_p$$

where

$$r_p \cong \frac{1}{\alpha\pi}\left(\frac{K}{Y}\right)^2 \tag{9.25}$$

For engineering calculations, customarily α is taken as $\alpha = 2$ for plane stress and $\alpha = 6$ for plane strain.

The size of the plastic zone at the crack tip may be quite small. In fact, it is only a local region of plasticity and the rest of the material during cracking is often loaded far below yielding. The size of the plastic zone almost disappears in brittle materials and attains much larger sizes in ductile ones. For instance, Table 9.2 indicates that the plastic radii for the first four types of high-strength steels may vary by more than three orders of magnitude depending on the ratio (K/Y).[2]

TABLE 9.2
Plastic Zone Radius Under Plane Strain Conditions for Some Steels

Type of Steel	AISI 4340	17–7 PH Stainless Steel	H–11 Tool Steel	52100 Ball Bearing	Low-Carbon Steel
K_{Ic}, MPa · m$^{1/2}$	99	77	38	14	180
Y, MPa	860	1435	1790	2070	250
r_p, mm	0.7	0.15	0.024	0.0024	27.5

Source: Adapted from Hertzberg, R. W., *Deformation and Fracture Mechanics of Engineering Materials*, John Wiley & Sons, New York, 1976, 286. With permission.

In general, there is a limit to which the stress intensity factor can be adjusted to accommodate crack tip plasticity using Equation 9.25. When (r_p/a) becomes appreciable, other approaches considering fracture mechanics for moderately ductile materials, such as the crack-opening displacement (COD) or the J-integral concepts,* have to be employed.

* See Section 9.13 for these approaches.

Example 9.10:

Two thin, flat plates of 60 mm width are made of an aluminum alloy, each containing a central hole of 15 mm diameter, as shown in Figure 9.19. Deep radial cracks of 3 mm length are detected at the hole of one of these plates as shown in Figure 9.19b. If each of these plates is loaded in tension, determine the ratio of the stress to cause fracture in the cracked plate to the stress initiating yielding in the other plate. The yield strength of the plate material is $Y = 310$ *MPa and the fracture toughness is* $K_c = 33\ \text{MPa} \cdot \text{m}^{1/2}$*. Repeat calculations by taking the effects of plasticty at the crack tip.*

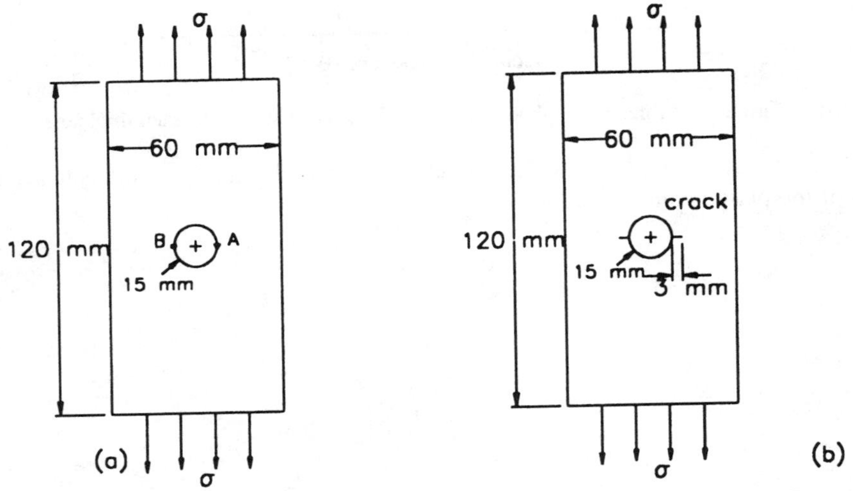

FIGURE 9.19 Example 9.10.

Solution:

Due to stress concentration around the hole, yielding will initiate at points A and B in the first plate with a stress concentration factor of 3, as determined in Section 6.4. Hence, yielding occurs when $3\sigma = Y = 310$ MPa, giving $\sigma_Y = 103.3$ MPa, where σ_Y is the value of the remote tensile stress at yielding.

Fracture occurs in the cracked plate when the stress intensity attains its critical value according to Equation 9.15, namely,

$$K_I = \sigma\sqrt{\pi a}\ \text{CCF} \le K_c$$

Hence,

$$\sigma\sqrt{\pi\left(\frac{0.015 + 2\times 0.003}{2}\right)}\ \text{CCF} = 33\ \text{MPa}\cdot\text{m}^{1/2}$$

The CCF is determined from Figure 9.7c as

$$\text{CCF} = f(a/b) = f\left[\left(\frac{0.015+2\times 0.003}{2}\right)\Big/(0.06/2)\right] = f(0.35) = 1.175$$

which gives, upon substitution in the fracture condition, σ_f = 154.6 MPa. The ratio σ_f/σ_Y = 154.6/103.3 = 1.5. The radius of the plastic zone r_p for a plate under conditions of plane stress is determined from Equation 9.25, namely,

$$r_p = \frac{1}{2\pi}\left(\frac{K_c}{Y}\right)^2 = \frac{1}{2\pi}\left(\frac{33}{310}\right)^2 = 0.0018 \text{ m}$$

The effective crack length is thus

$$a_{\text{eff}} = \frac{15}{2} + 3 + 1.81 = 12.31 \text{ mm}$$

and, hence, the CCF is redetermined from Figure 9.7c as $f(12.3/30) = f(0.41) \cong 1.225$. The fracture stress is recalculated as $\sigma\sqrt{\pi(0.0123)}\ 1.225 \leq 33$ MPa · m$^{1/2}$, giving σ_f = 137.1 MPa, which results in a smaller ratio of σ_f/σ_Y = 1.33.

9.10 NOTES ON FRACTURE TOUGHNESS TESTING

Several fracture toughness specimens are specified by the relevant standards. Figure 9.20a and b show, respectively, two specimens, namely, compact tension and single edge-cracked bend specimens. As mentioned in Section 9.9, the specimen thickness must be sufficiently large in comparison with the plastic zone dimensions, thus ensuring conditions of plane strain at the crack tip and, hence, a valid fracture toughness determination. The following size requirements, referring to Figure 9.20, are normally specified as

$$a,\ B,\ (W/2) > 2.5\left(\frac{K_{\text{Ic}}}{Y}\right)^2 \tag{9.26}$$

Unfortunately, these size requirements are related to K_{Ic}, the property to be measured. Hence, a suitable estimate for K_{Ic} is made *a priori* from known results for similar materials. After testing, the validity is then checked according to Condition 9.26; otherwise, specimens with modified dimensions are manufactured and tested.

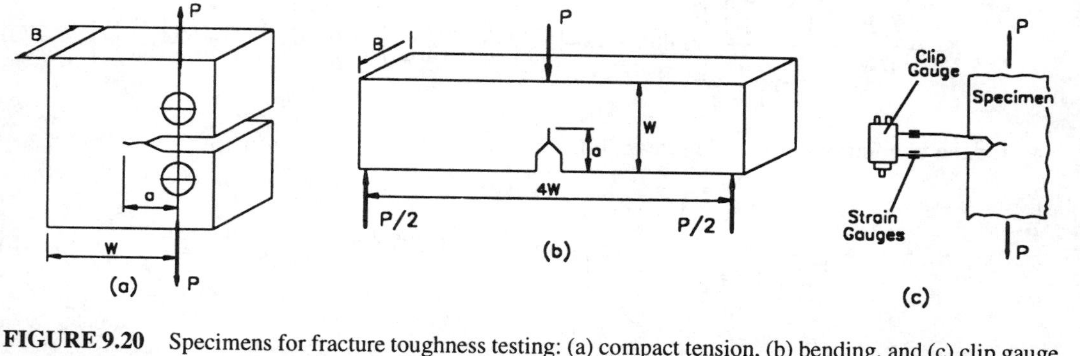

FIGURE 9.20 Specimens for fracture toughness testing: (a) compact tension, (b) bending, and (c) clip gauge for measurement of crack-opening displacement.

The crack must be a sharp one. This condition is assured by inducing fatigue precracking in the specimen at a stress intensity level K_{max} much below K_{Ic} ($K_{\text{max}} < 0.6K_{\text{Ic}}$), thus avoiding the effects of residual stress fields. Note that the crack length (a) includes both the machining notch and the fatigue precracking.

A clip gauge (strain gauge type), Figure 9.20, is then mounted across the crack mouth to measure crack-opening displacement (COD). The test is then accomplished by monotonic static load increase[24] in a suitable testing machine up to fracture.

Typical load–displacement records are shown in Figure 9.21. The fracture load P_Q associated with each type of material behavior is determined according to specified standards. For brittle materials, Figure 9.21a, $P_Q = P_{max}$. For ductile materials, Figure 9.21b, $P_Q = P_5$, where P_5 is obtained at the intersection of the 5% secant with the load–displacement curve. For other intermediate behaviors, the determination of P_Q is clearly explained in the standards. In all cases (other than that of Figure 9.21a, P_Q must satisfy the condition that $P_Q \leq 1.1\, P_{max}$, otherwise, the test is considered invalid and indicates that crack tip blunting and plastic deformation have occurred rather than crack-induced brittle fracture.

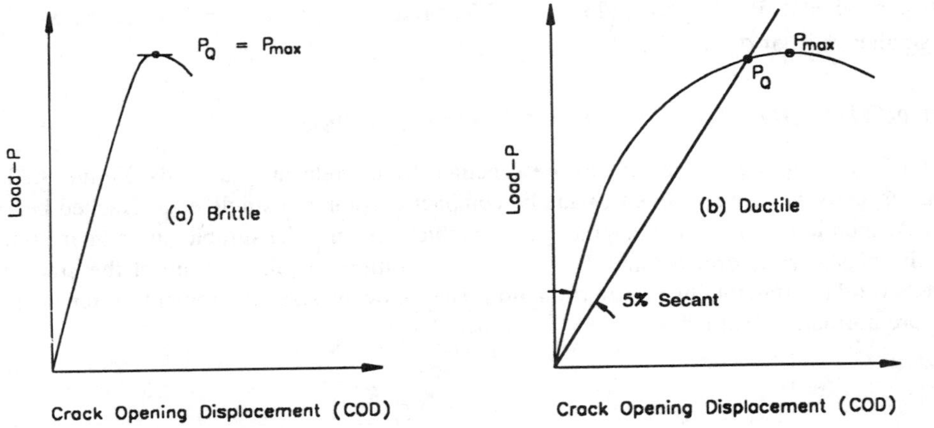

FIGURE 9.21 Typical load–displacement records of fracture toughness testing.

The calculation of an "interim" or "candidate" value of the stress intensity factor K_Q is based on Equation 9.15. For instance, for the compact tension specimen,* Figure 9.20a:

$$K_Q = \frac{P_Q}{BW^{1/2}} \frac{\left(2+\frac{a}{w}\right)}{\left(1-\frac{a}{w}\right)^{3/2}} \left[0.886 + 4.64\left(\frac{a}{w}\right) - 13.32\left(\frac{a}{w}\right)^2 + 14.72\left(\frac{a}{w}\right)^3 - 5.6\left(\frac{a}{w}\right)^4\right] \quad (9.27a)$$

and for the bending specimen, Figure 9.20b:

$$K_Q = \frac{3P_Q L}{BW^{3/2}} \frac{\left(\frac{a}{w}\right)}{2\left(1+\frac{2a}{w}\right)\left(1-\frac{a}{w}\right)^{3/2}} \left[1.99 - \left(\frac{a}{w}\right)\left(1-\frac{a}{w}\right)\left(2.15 - 3.93\frac{a}{w} + 2.7\frac{a^2}{w^2}\right)\right] \quad (9.27b)$$

The crack length a, used in the above expressions is the average of three readings for the crack length measured after fracture (as shown in Figure 9.22b, Example 9.11). If the difference between

* See Reference 14. Note that this expression is written differently from the CCF $f(a/W)$ due to the difficulty in defining a nominal stress for a compact tension specimen.

any two readings of a exceeds 5% of the average value, the test is considered invalid and indicates a skewed crack front according to standards. The test is also invalid if the crack length at the surface differs by more than 10% of its average value.

Once K_Q is determined, it is accepted as a valid plane strain fracture toughness fracture K_{Ic} if the size conditions (Equations 9.26) are satisfied. If this is not the case, further testing of specimens with modified dimensions are required. Note that in the case of K_Q not being an intrinsic material property, it is not absolutely useless. It still can be used for comparative purposes in material selection and in the assessment of brittle fracture in plates of the same thickness as the test specimen. In general, engineering judgment is excercised in using K_{Ic} for conditions other than plane strain. However, for surface and corner cracks (partly through cracks), the use of K_{Ic} is always recommended regardless of the thickness.*

Example 9.11:

A compact tension specimen, Figure 9.22a, with W = 100 mm and B = 50 mm is experimentally used to determine the fracture toughness of an alloy steel. The load–displacement recorded during the experiment is shown in Figure 9.22b.The fractured specimen cross section is shown in Figure 9.22c.

a. *If the yield strength Y for this steel is 950 MPa, check the validity of the test to determine K_{Ic}, and then find its value.*
b. *Does crack tip plasticity affect the accuracy of K_{Ic} determination?*

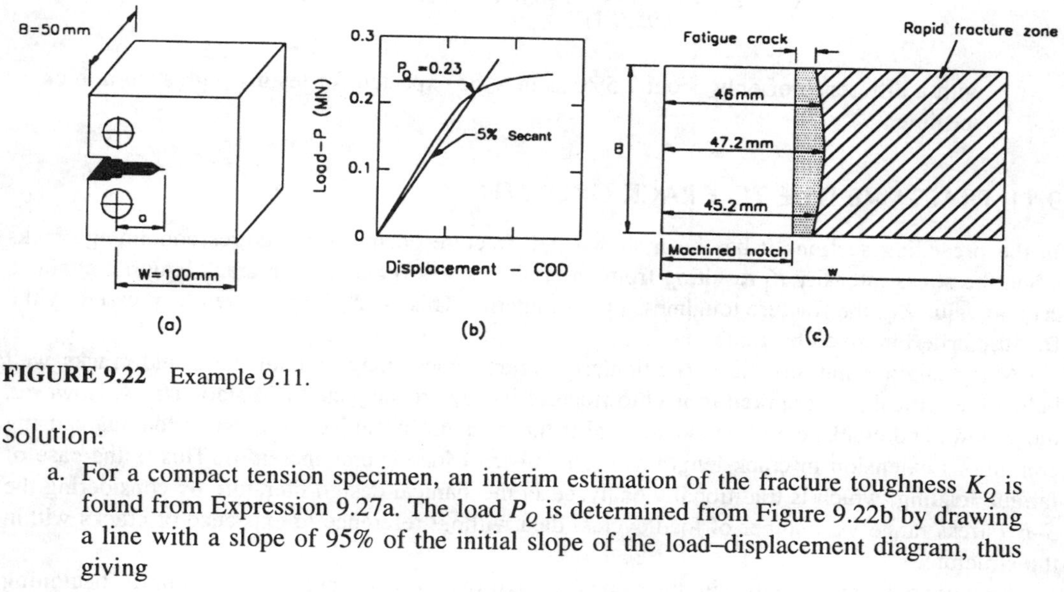

FIGURE 9.22 Example 9.11.

Solution:

a. For a compact tension specimen, an interim estimation of the fracture toughness K_Q is found from Expression 9.27a. The load P_Q is determined from Figure 9.22b by drawing a line with a slope of 95% of the initial slope of the load–displacement diagram, thus giving

$$P_Q = 0.23 \text{ MN}$$

The crack length is measured from Figure 9.22c, giving an average (46 + 47.2 + 45.2)/3 = 46.1 mm. Hence, upon substitution into Equation 9.27a using B = 50 mm and W = 100 mm, gives

* See Reference 25.

$$K_Q = \frac{0.23}{0.05(0.1)^{1/2}}(8.59) = 124.95 \text{ MPa} \cdot \text{m}^{1/2}$$

This result remains tentative until the test validity conditions are satisfied, namely,

$$a,\ B,\ (W/2) \geq 2.5\left(\frac{K_Q}{Y}\right)^2 \geq 2.5\left(\frac{124.95}{950}\right)^2 \geq 0.043$$

These conditions are satisfied for a = 0.0461 m, B = 0.05 m, and (W/2) = 0.05 mm. Hence, $K_Q = K_{Ic}$.

b. Since plane strain conditions are satisfied when testing with a compact specimen, Equation 9.25 gives, for $\alpha = 6$,

$$r_p = \frac{1}{6\pi}\left(\frac{K_{Ic}}{Y}\right)^2 = \frac{1}{6\pi}\left(\frac{124.95}{950}\right)^2 = 0.918 \text{ mm}$$

and a_{eff} = 46.1 + 0.918 = 47.018 mm. Substituting, again, in Expression 9.27a gives

$$K_{Ic} = K_Q = \frac{0.23}{0.05(0.1)^{1/2}}(8.816) = 128.24 \text{ MPa} \cdot \text{m}^{1/2}$$

with a difference of only about 2.5%, as may be expected for testing with such a thick specimen.

9.11 FRACTURE DUE TO CRACK GROWTH

In the preceding sections it has been shown that fracture occurs in structures containing cracks when the stress intensity K_I resulting from an applied stress σ and certain crack length a attains a critical value K_{Ic}, the fracture toughness of the material. This is an ultimate event expressed by the fracture criterion given by Equation 9.15.

Many engineering structures (particularly welded ones) may contain flaws and cracks well below the critical size required to lead to fracture if they are subjected to a static stress. However, these flaws and cracks could be detrimental if the structure is subjected to cyclic loading causing continuous extension in crack length as the number of load cycles increases. This is the case of fatigue loading, which is traditionally analyzed in mechanical design literature by considering the *S–N* (stress range vs. number of load cycles) data without reference to existence of cracks within the structure.

Fracture mechanics deals with the same fatigue problem, while considering from the beginning the presence of a cracklike defect of known size. This crack propagates during the service life of the structure until it reaches a critical size sufficient to induce catastrophic failure according to Condition 9.15.

9.11.1 Fatigue Crack Propagation

Typical dependence of crack growth on cyclic loading is schematically shown in Figure 9.23a. This represents a series of tests on a number of identical specimens, all with an initial sharp crack of the same size a_o. Each specimen is subjected to a cyclic loading with certain stress range $\Delta\sigma = \sigma_{max} - \sigma_{min}$, where $\sigma_{max} > \sigma_{min}$, as shown in Figure 9.23b.

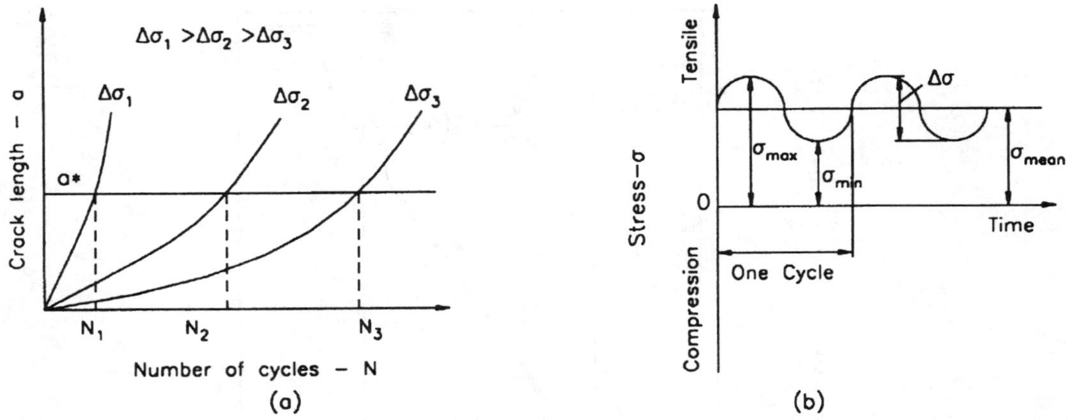

FIGURE 9.23 Crack propagation under cyclic stress: (a) crack length vs. number of cycles for different stress ranges and (b) stress cycle.

Figure 9.23a, shows that the increase in crack size plotted vs. the number of stress cycles is nonlinear. Furthermore it is seen that a higher stress fluctuation increment $\Delta\sigma$ causes rapid crack growth, i.e., to attain a specific crack length a^* requires a smaller number of stress cycles ($N_1 < N_2 < N_3$) for higher stress fluctuation ($\Delta\sigma_1 > \Delta\sigma_2 > \Delta\sigma_3$). Hence, the crack propagation or growth rate expressed by (da/dN) at a given crack length is higher for greater stress fluctuation.

Fatigue crack growth is advantageously quantified in terms of the stress intensity factor. Hence, from Equation 9.15 for a given stress fluctuation $\Delta\sigma = \sigma_{max} - \sigma_{min}$ and crack length a,

$$\begin{aligned} \Delta K = K_{max} - K_{min} &= \sqrt{\pi a}\left(\sigma_{max} - \sigma_{min}\right) \mathrm{CCF} \\ &= \sqrt{\pi a}\, \Delta\sigma\, \mathrm{CCF} \end{aligned} \tag{9.28}$$

Note that since the stress intensity factor is undefined in compression, K_{min} is taken as zero if σ_{min} is negative. By using ΔK as defined in Equation 9.28 the experimental curves such as those of Figure 9.23a reduce to a single curve if plotted as $\ln(da/dN)$ vs. $\ln(\Delta K)$. This is represented schematically in Figure 9.24. The sigmoidal shape of this curve defines three regions, namely, (i) nonpropagating cracks, (ii) steady crack propagation, and (iii) unstable crack growth rate.

9.11.1.1 Region (i) of Nonpropagating Cracks

This region begins with a threshold value* of a stress intensity ΔK_{th} below which no observable crack propagation occurs or simply the crack remains dormant without increase in length. ΔK_{th} occurs at propagation rates of the order of 1 to 10×10^{-10} m/cycle.

9.11.1.2 Region (ii) of Steady Crack Propagation

In this region the crack propagates steadily under cyclic loading. The fracture surface frequently shows (using a suitable magnification) a pattern of fatigue striations (beach marks), each representing the successive advance of the crack front due to one stress cycle. The presence of these striations is an obvious indication of fatigue failure.

In region (ii) the relation between $\ln(da/dN)$ and $\ln(\Delta K)$ is linear, i.e., following the empirical relationship:

* See Ritchie.[26] ΔK_{th} may be equivalent to stress levels below the endurance limit on an S–N curve.

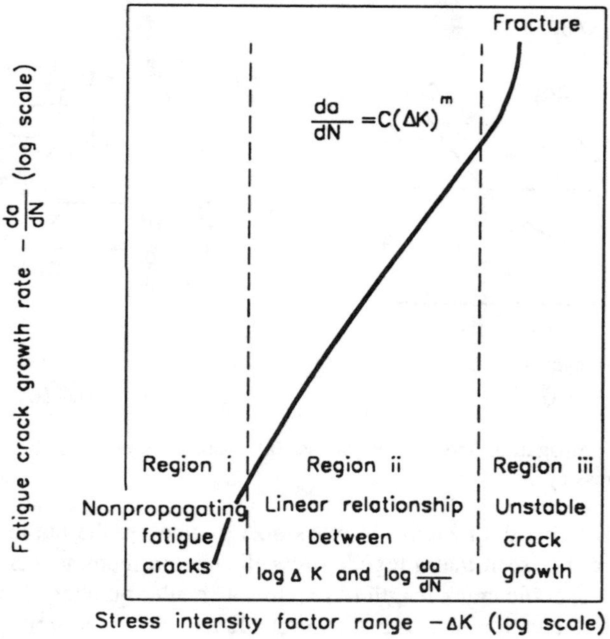

FIGURE 9.24 Fatigue crack growth phenomenon indicating three regions of crack propagation.

$$\frac{da}{dN} = C(\Delta K)^m \tag{9.29}$$

Here C and m are experimentally determined constants depending on material and test conditions, i.e., environment, temperature, stress range, stress cycling frequency, Equation 9.29 is known as the Paris law.[27]

Careful and elaborate tests for several engineering materials are conducted to obtain fatigue crack growth data falling within the regime from about 10^{-4} to 10^{-2} mm/cycle. Plots such as Figure 9.24 are constructed. The value of C in Expression 9.29 is the value of (da/dN) at $\Delta K = 1$ MPa · $m^{1/2}$ and m is the slope, which is approximately found to be in the range of 3 to 4 for steel and aluminum alloys. Table 9.3 gives typical values of ΔK_{th}, C, and m for some materials. For real-life design, actual testing for the material under consideration has to be conducted in conditions as close as possible to service conditions to determine the constants C and m accurately.

Usually, fatigue crack growth data are determined experimentally by testing using a cycle of stress ratio $R = \sigma_{min}/\sigma_{max} = K_{min}/K_{max} = 0$. Compression loading cycles are not used since compressive stresses close up the crack and it has been postulated that negative stress hardly affects crack growth. In such a case, both K_{min} and R are taken as zero.

In engineering applications, R may be greater than zero, and an attempt to consider its effect on fatigue crack propagation resulted in a modified version of Equation 9.29 known as the Forman Equation:[28]

$$\frac{da}{dN} = \frac{C_F(\Delta K)^{m_F}}{(1-R)K_c - \Delta K} \tag{9.30}$$

where the constants C_F and m_F are obtained by fitting Expression 9.30 to appropriate experimental growth data. K_c is the fracture toughness applicable to the specific material and geometry. Physically,

TABLE 9.3
Typical Values for ΔK_{th}, C, and m in the Law $(da/dN) = C(\Delta k)^m$ for Some Materials

Material	ΔK_{th}, MPa · $m^{1/2}$	C^a (10^{-11})	m
Mild steel	3.2–6.6	0.24	3.3
Structural steel	2.0–5.0	0.07–0.1	3.85–4.2
Structural steel in seawater	1.0–1.5	1.6	3.3
Aluminium	1.0–2.0	4.56	2.9
Aluminium alloy	1.0–2.0	3–19	2.6–3.9
Copper	1.8–2.8	1.34	3.9
Titanium	2.0–3.0	68.8	4.4
Steel AISI 4340	—	4.85	2.64
Steel AISI 4340 in sea water	—	1.23	2.98

[a] Units of C gives (da/dN) in m/cycle when ΔK is in MPa · $m^{1/2}$.

Equation 9.30 is acceptable since experiments indicate that as R increases, i.e., increasing the mean stress while maintaining ΔK constant, the crack growth rate (da/dN) increases. At fracture, when $\Delta K = (1 - R)\, K_c$, Equation 9.30 indicates infinite growth rate, as would be expected.

The Forman Equation tends to be applicable for all R values, and hence all data should lie on a straight line fit for $\ln[1 - R)K_c - \Delta K](da/dN)$ vs. $\ln(\Delta K)$.

Fatigue crack growth data are commonly generated using standardized specimens and techniques.[29] The compact tension specimen shown in Figure 9.22a is often used, thus assuming crack growth under plane strain conditions. There is evidence suggesting that crack growth is slower in plane stress than in plane strain at the same intensity factor.[30]

9.11.1.3 Region (iii) of Unstable Crack Growth Rate

In this region, the crack length ultimately attains its critical size, leading to catastrophic fracture according to Criterion 9.15, namely, $K_I = \sigma_{max}\sqrt{\pi a_c}$ CCF $\leq K_{Ic}$. Here, a_c is the critical crack length attained just before fracture. From an engineering and conservative point of view, this critical crack length a_c is associated with the end of region (ii) of crack growth. It is this value of a_c that is used to determine the safe life of a structure during fatigue loading.

Note that fatigue crack growth laws, such as Equations 9.29 and 9.30, are only empirical laws, and, in view of computerized predictions, there is no need for such expressions as the experimental crack growth data can be used directly.

Example 9.12:

Fatigue crack growth testing with a compact tension specimen (see Figure 9.22a: W = 50 mm and B = 25 mm) at constant amplitude cyclic stress resulted in the data in Table 9.4a.

The load cycle was selected, such that P_{max} = 13 kN and P_{min} = 0.

a. *Plot the test data.*
b. *Establish a plot for* (da/dN) *vs.* (ΔK). *(Assume straight line segments between data points.)*
c. *Determine the material constant in the Paris law* $(da/dN) = C\,(\Delta K)^m$.
d. *If another specimen of the same material is tested* at P_{max} = *17 kN, what would be the crack growth rate at a crack length of 25.25 mm?*

Use Expression 9.27a to determine the appropriate configuration factor f(a/W) *for the compact tension specimen.*

TABLE 9.4A
Example 9.12

Crack Length a (mm)	Number of Cycles (10^3)
20	70
21	75.5
22	80.8
23	85.6
24	89.8
25.5	93.7
27	97.3
29	100.3
31	101.3
34.1	102.1
37.9	102.5

Solution:

a. A plot of the test data, a (crack length in mm) vs. N (number of cycles) is shown in Figure 9.25a.
b. To establish a plot of (da/dN) vs. (ΔK), a straight line segment between every two consecutive data points is assumed, thus giving ($\Delta a/\Delta N$) at ($a + \Delta a/2$) as indicated in the second and fourth columns of Table 9.4b. The associated value of ΔK is calculated from Expression 9.27a as

$$\Delta K = K_{max} - K_{min} = K_{max} - 0 = \frac{P_{max}}{BW^{1/2}} f(a/W)$$

$$= \frac{13 \times 10^{-3}}{0.025(0.005)^{1/2}} f\left[\left(a + \frac{\Delta a}{2}\right)\Big/ W\right]$$

$$= 2.3255 f\left[\left(a + \frac{\Delta a}{2}\right)\Big/ W\right] \text{MPa} \cdot \text{m}^{1/2}$$

c. A plot of ($\Delta a/\Delta N$) $\cong$ da/dN vs. ΔK on a log–log paper as shown in Figure 9.25b suggests a straight-line fit representing Paris law as

$$\frac{da}{dN} = C(\Delta K)^m$$

A least-square fitting process results in

$$C = 1.727 \times 10^{-12} \text{ and } m \cong 3.98 \text{ (in MPa, m, cycle)}$$

d. Since the Paris law is considered to represent an intrinsic material response, da/dN is obtained from Figure 9.25b at any stress intensity. Hence, for a crack length of 25.25 mm,

TABLE 9.4B
Solution of Example 9.12

Δa (mm)	$\left(a+\frac{\Delta a}{2}\right)$ (mm)	ΔN (10^3) (cycle)	$\Delta a/\Delta N$ (10^{-7}) (m/cycle)	$f\left[\left(a+\frac{\Delta a}{2}\right)/W\right]$	ΔN (MPa · $m^{1/2}$)
1.0	20.5	5.5	1.82	7.51	17.46
1.0	21.5	5.3	1.89	7.89	18.28
1.0	22.5	4.8	2.08	8.31	19.39
1.0	23.5	4.2	2.38	8.78	20.49
1.5	24.75	3.9	3.85	9.42	21.98
1.5	26.25	3.6	4.17	10.32	24.12
2.0	28	3.0	6.67	11.60	26.98
2.0	30	1.0	20.00	13.45	31.28
3.1	32.55	0.8	38.75	16.73	38.91
3.8	36	0.4	95	23.61	54.89

$$\Delta K = \frac{17\times10^{-3}}{0.025(0.05)^{1/2}} f(25.25/50) = 3.041\times9.751 = 29.65\ \text{MN}\cdot\text{m}^{1/2}$$

Substitution into the Paris law gives

$$\frac{da}{dN} = 1.727\times10^{-12}(29.65)^{3.98} = 12.47\times10^{-7}\ \ \text{m/cycle}$$

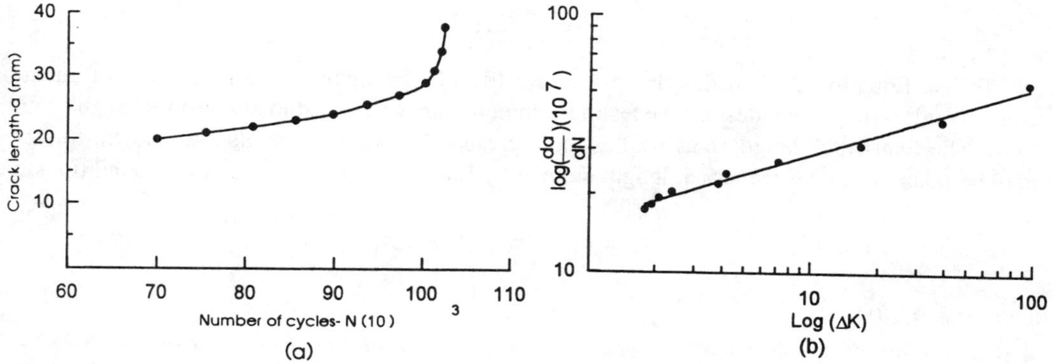

FIGURE 9.25 Example 9.12.

9.11.2 Safe-Life Prediction

Fatigue life of a component may be defined as consisting of three stages: crack initiation, stable crack propagation, and unstable rapid crack growth ending by fracture. The number of cycles required to initiate a discernible crack is primarily influenced by microstructural parameters (e.g., second-phase particles, crystalline imperfections, etc.), as well as mechanical stress raisers and environmental conditions. The final unstable rapid crack growth stage occurs very quickly and contributes very little to the total life. From the point of view of engineering fracture mechanics, it is the stage of stable crack growth, which constitutes the safe-life prediction of the component.

Paris law, Equation 9.29, is rearranged and integrated to give the number of cycles N_f required for a crack to propagate from an initial length a_o to some final length a_f. Hence,

$$N_f = \int_{a_o}^{a_f} \frac{da}{C(\Delta K)^m}$$

Substituting for ΔK from Equation 9.15 gives

$$N_f = \int_{a_o}^{a_f} \frac{da}{C\left[\Delta\sigma\sqrt{\pi a}\ \mathrm{CCF}\right]^m} \tag{9.31}$$

where $\Delta\sigma = \sigma_{max} - \sigma_{min}$ as defined before.

Analytical integration of this expression is not possible since the functions representing the CCF for all geometries of crack bodies, as given in Section 9.6 are not mathematically simple. For simplicity consider the case of a loaded plate of width W where $f(a/W)$ does not change much within the range of a_o to a_f or, alternatively, by using an average value for CCF_{av} as $f_{av}(a/W) = [f(a_o/W) + f(a_f/W)]/2$; hence, integration of Equation 9.31 yields

For $m \neq 2$:

$$N_f = \frac{1}{C\left[\mathrm{CCF}_{av}\right]^m \pi^{m/2}(\Delta\sigma)^m} \frac{a_o^{(1-m/2)} - a_f^{(1-m/2)}}{m/2 - 1} \tag{9.32a}$$

For $m = 2$:

$$N_f = \frac{\ln\left(a_f/a_o\right)}{C\left[\mathrm{CCF}_{av}\right]^2 \pi(\Delta\sigma)^2} \tag{9.32b}$$

To use Equation 9.32 to determine the safe life of a component, an estimation of an initial crack length is made. Nondestructive testing techniques are often used to assign a reasonable value* to a_o. The final crack length a_f as well as the corresponding CCF_f must also be known. In fact, a_f may be taken as the critical crack length defined by Equation 9.15 for plane strain conditions, i.e.,

$$K_I = \sigma_{max}\sqrt{\pi a_f}\ \mathrm{CCF} = K_{Ic}$$

Example 9.13:

A square plate of 250 mm width is loaded in tension. An initial central crack of length $2a_o = 25$ *mm was detected before loading. Examination of fracture after repeated loading–unloading* ($P \to 0$) *cycles revealed a final crack length* $2a_f = 150$ *mm. Estimate the total number of load cycles which caused failure. The plate is made of a material with the following properties: fracture toughness,* $K_{Ic} = 30$ *MPa* $\cdot$ *m*$^{1/2}$, $C = 12 \times 10^{-11}$, *and* $m = 2$ *in the law* $da/dN = C\,(\Delta K)^m$.

---

* The size and location of permissible cracks may be imposed by codes of practice. They are normally termed the *damage tolerance* requirement. An aircraft structural integrity program requirement, MIL-A-83444 (USAF) (1974), assumes initial flaws to exist as a result of material and structural manufacturing and processing operations. Small imperfections equivalent to an 0.05 in. (0.127 mm) radius corner flaw is assumed to exist in each hole of each element in the structure. Regarding the slow crack growth structure, the assumed initial flaw at holes and cutouts shall be a 0.05 in. (0.127 mm) through the thickness flaw at one side of the hole when the material thickness is equal to or less than 0.05 inch.

Solution:

At the instant of fracture, Expression 9.15 applies as

$$K_I = \sigma\sqrt{\pi a_f}\ \text{CCF}_f = K_{Ic}$$

where

$$\text{CCF} = f(a/W) = \left[\frac{W}{\pi a}\tan\frac{\pi a}{W}\right]^{1/2}$$

(case (b) of Section 9.6.1). Substitution of $a_f = 75$ mm and $W = 250$ mm gives $f(a_f/W) = 1.208$. Hence, the stress intensity factor at fracture is $\sigma_{max}\sqrt{\pi \times 0.75} \times 1.208 = 30\ \text{MPa}\cdot\text{m}^{1/2}$, giving $\sigma_{max} = 51.17$ MPa and $\Delta\sigma = \sigma_{max} - \sigma_{min} = 51.17 - 0 = 51.17$ MPa. The value of $f(a_o/W)$ is initially calculated as $f(12.5/250) = 1.004$ compared with 1.208 at fracture. Such variation in $f(a/W)$ may not justify the use of Expression 9.31 based on an average value for $f(a/W)$. Hence, recalling that

$$N_f = \int_{a_o}^{a_f}\frac{da}{C\left[\Delta\sigma\sqrt{\pi a}\ \text{CCF}\right]^m}$$

gives for constant C and $\Delta\sigma$, after substituting for CCF = $f(a/W)$,

$$N_f = \frac{1}{C(\Delta\sigma)^m\pi^{m/2}}\int_{a_o}^{a_f}\frac{da}{\left\{a\left[\left(\frac{W}{\pi a}\right)\tan\left(\frac{\pi a}{W}\right)\right]\right\}^{m/2}}$$

For $m = 2$:

$$N_f = \frac{1}{C(\Delta\sigma)f^2(a/W)W}\int_{a_o}^{a_f}\frac{da}{\tan\left(\frac{\pi a}{W}\right)}$$

$$= \frac{}{12\times10^{-11}(51.17)^2 0.25}\left[\frac{W}{\pi}\ln\sin\left(\frac{\pi a}{W}\right)\right]_{a_o=0.0125}^{a_f=0.075}$$

$$= 1.665\times10^6 \text{ cycles}$$

If an average value $\text{CCF}_{ar} = f_{av}(a/W)$ is used i.e., $f_{av} = (1.004 + 1.208)/2 = 1.106$ in conjunction with Expression 9.32 for $m = 2$,

$$N_f^* = \frac{\ln(0.075/0.0125)}{12\times10^{-11}(1.106)^2\pi(51.17)^2} = 1.485\times10^6 \text{ cycles}$$

with an error of about 18.2% from N_f. This error could be more substantial in problems where $f(a_o/W)$ differs greatly from $f(a_f/W)$.

Example 9.14:

A steel cylindrical pressure vessel of an outer diameter of 400 mm and a 7-mm wall thickness is subjected to an internal pressure. The vessel has a thick flat end welded circumferentially to the cylinder wall. Upon inspection, a sharp circumferential crack of 2 mm depth and 100 mm length is determined in the weld zone as shown in Figure 9.26a.

a. *Determine the maximum pressure* p_{max} *that the vessel can support without fracture when the final crack length becomes 5 mm, knowing that* $K_{Ic} = 95.3$ *MPa* $\cdot$ $m^{1/2}$ *for the weld zone and using a factor of safety F.S. = 1.5.*
b. *If the pressure fluctuates between* p_{max} *and* $p_{max}/2$*, determine the total number of cycles which elapses before the crack attains a length of 5 mm. Use the material crack propagation data plotted in Figure 9.26b*[31] *with m = 3.3. Hint: Assume the configuration correction factor for an edge crack in a plate to be applicable in this problem and use its average value for safe-life prediction.*

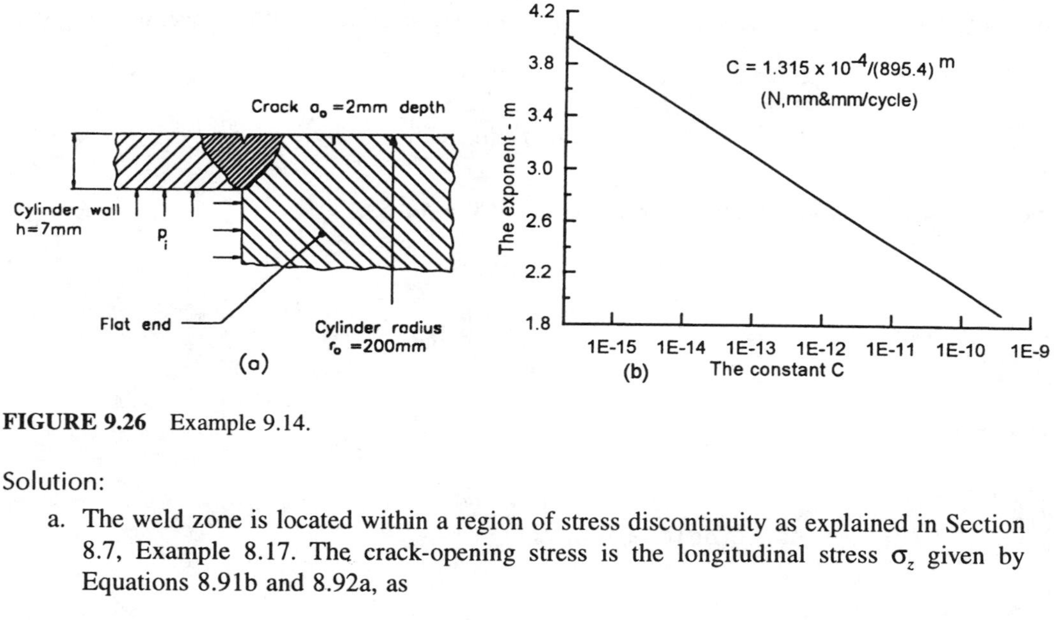

FIGURE 9.26 Example 9.14.

Solution:

a. The weld zone is located within a region of stress discontinuity as explained in Section 8.7, Example 8.17. The crack-opening stress is the longitudinal stress σ_z given by Equations 8.91b and 8.92a, as

$$\sigma_z = \frac{p_i r_m}{2h} + \frac{6M_{\max}}{h^2}$$

where $M_{\max} = 2Du_c\beta^2 = p_i/2\beta^2$ and $\beta^2 = \sqrt{3(1-\nu^2)}\,/r_m h$. Hence, $M_{\max} = 0.303 p_i r_m h$ (for $\nu = 0.3$). Since σ_z is composed of a uniform tensile stress and a bending stress, the stress intensity factor is determined by superposition as

$$K_I = \frac{p_i r_m}{2h}\sqrt{\pi a}\, f_1(a/W) + \frac{6M_{\max}}{h^2}\sqrt{\pi a}\, f_2(a/W)$$

The CCFs for tension $f_1(a/W)$ and bending $f_2(a/W)$ are determined from Sections 9.6.1d and 9.6.3a, respectively. For $r_m = 196.5$ mm, $a_o = 2$ mm, $a_f = 5$ mm, and $W = h = 7$ mm:

Tension: $f_1(a_o/W) = 1.616$ and $f_1(a_f/W) = 6.369$
Bending: $f_2(a_o/W) = 1.123$ and $f_2(a_f/W) = 2.89$

At fracture, $K_I \leq K_{Ic}/\text{F.S.}$; hence,

$$\frac{95.3}{1.5} = \frac{p_i \times 196.5}{2 \times 7}\sqrt{\pi \times 0.005} \times 6.369$$

$$+\frac{6(0.303 p_i \times 196.5 \times 7)}{7^2}\sqrt{\pi \times 0.005} \times 2.89$$

which gives $p_i = p_{max} = 2.14$ MPa.

b. For pressure fluctuations between p_{max} and p_{min} equal to $p_{max}/2$, the longitudinal stress varies between

$$(\sigma_z)_{max} = \frac{2.14 \times 196.5}{2 \times 7} + \frac{6(0.303 \times 2.14 \times 196.5 \times 7)}{7^2}$$

$$= 30.04 + 109.2 = 139.24 \text{ MPa}$$

$$(\sigma_z)_{min} = (\sigma_z)_{max}/2 = 69.63 \text{ MPa}$$

To apply the Paris law (Equation 9.32a) between $a_o = 2$ mm and $a_f = 5$ mm, a CCF_{av} is required. A weighed average may be suggested as follows:

$$CCF_{av} = \frac{1}{2}\left[\frac{\sigma_1}{\sigma_1 + \sigma_2} f_1(a/W) + \frac{\sigma_2}{\sigma_1 + \sigma_2} f_2(a/W)\right]_{a_o}^{a_f}$$

where σ_1, σ_2, and $(\sigma_1 + \sigma_2)$ are the uniform tensile, bending, and total longitudinal stresses, repectively. Hence,

$$CCF_{av} = \frac{1}{2}\left[\frac{30.04}{139.24}(1.616 + 6.369) + \frac{109.2}{139.24}(1.123 + 2.89)\right]$$

$$= (1.723 + 3.147)/2 = 2.435$$

Applying Expression 9.32a for $\Delta\sigma = 139.24 - 69.63 = 69.61$ MPa gives

$$N_f = \frac{1}{C(2.435)^m \pi^{m/2} (69.92)^m} \frac{2^{(1-m/2)} - 5^{(1-m/2)}}{(m/2 - 1)}$$

Average material parameters are read from Figure 9.26b as $m = 3.3$ and $C = 1.315 \times 10^{-4}/(895.4)^{3.3} = 2.3839 \times 10^{-14}$ (in units of N · mm and mm/cycle). Substitution gives

$$N = 1.22 \times 10^5 \text{ cycle}$$

Obviously, a more accurate solution would be obtained by numerical integration instead of using CCF_{av}.

Example 9.15:

A tension member with a central initial crack a_o *= 2 mm is subjected to a fatigue stress cycle with* R *= 0.5. If the applied stress range is controlled to be four times that for no observable crack growth, i.e., the threshold value* $\Delta\sigma_{th}$*. The following material properties with Forman Equation 9.30 apply:* $K_{Ic} = 78$ *MPa* $\cdot$ *m*$^{1/2}$, $\Delta K_{th} = 3$ *MPa* $\cdot$ *m*$^{1/2}$, $C_F = 9.2 \times 10^{-9}$, *and* $m_F = 2.8$*. Determine the number of stress cycles that cause failure. Take CCF = 1 throughout the analysis.*

Solution:

The threshold stress range is determined from

$$\Delta K_{th} = \Delta\sigma_{th}\sqrt{\pi a_o}$$

$$3 = \Delta\sigma_{th}\sqrt{\pi \times 0.002}$$

giving $\Delta\sigma_{th} = 37.85$ MPa. Hence, the applied stress range is $\Delta\sigma = 4\Delta\sigma_{th} = 151.38$ MPa. The final crack length at failure is given by

$$K_{Ic} = \sigma_{max}\sqrt{\pi a_f}$$

$$K_{Ic} = [\Delta\sigma/(1-R)]\sqrt{\pi a_f}$$

$$78 = [151.38/(1-0.05)]\sqrt{\pi a_f}$$

giving $a_f = 21.1$ mm. Rearranging the Forman Equation 9.30 and integrating gives

$$N_f = \int_{a_o}^{a_f} \frac{(1-R)K_c - \Delta K}{C_F(\Delta K)^{m_F}} da$$

$$N_f = \int_{a_o}^{a_f} \frac{(1-R)K_c}{C_F(\Delta K)^{m_F}} da - \int_{a_o}^{a_f} \frac{1}{C_F(\Delta K)^{m_F-1}} da$$

Substituting $\Delta K = \Delta\sigma\sqrt{\pi a}$ yields

$$N_f = \frac{1}{C_F(\pi^{1/2}\Delta\sigma)^{m_F}}\left[\frac{(1-R)K_c}{1-m_F/2}\left(a_f^{(1-m_F/2)} - a_o^{(1-m_F/2)}\right)\right.$$

$$\left. - \frac{\pi^{1/2}\Delta\sigma}{(3-m_F)/2}\left(a_f^{(3-m_F)/2} - a_o^{(3-m_F)/2}\right)\right]$$

For $C_F = 9.2 \times 10^{-9}$, $m_F = 2.8$, $K_c = K_{Ic} = 78$ MPa $\cdot$ m$^{1/2}$, $a_o = 0.002$ m, $a_f = 0.0211$ m, and $\Delta\sigma =$ 151.43 MPa:

$$N_f = \frac{1}{9.2\times10^{-9}(\pi^{1/2}\times151.38)^{2.8}}\left[\frac{(1-0.5)78}{1-2.8/2}(0.0211^{-0.4} - 0.002^{-0.4})\right.$$

$$\left. - \frac{\pi^{1/2}\times151.38}{(3-2.8)/2}(0.0211^{0.1} - 0.002^{0.1})\right]$$

$$N_f = 5714 \text{ cycles}$$

9.11.3 Comments on Safe-Life Predictions

Newly fabricated structures are sometimes thought to be defect free. However, this judgment — based on inspection reports — is certainly dependent on the limitations of the particular inspection technique used. In other words, the initial crack a_o assumed to exist is a crack of the dimension which was "just missed" by inspection. This initial crack is expected to develop and propagate during the required service life N_f of the structure until it reaches its critical size a_f dictated by Criterion 9.15. The knowledge of any two of the three parameters, namely, a_o, a_f, and N_f, helps in setting up limitations on the third parameter. For instance, if the designer assigns values for a_f and N_f, the inspection of the newly fabricated structure should exclude all initial cracklike defects larger than a_o as determined using expressions such as Equation 9.29. Meanwhile, on this basis, crack propagation control may be exercized during service life by routine inspection. This ensures that a crack has several chances of detection before it can grow to its critical size a_f. Corrective measures, e.g., repairing or replacement of partially failed components, are thus enforced in proper timing.

9.11.3.1 Margin of Safety

Fatigue crack propagation data for a given material and conditions are not as consistent as other material mechanical properties. These data are influenced by many uncontrollable factors, e.g., microstructural inhomogeneity, environmental conditions, frequency, batch-to-batch material variations, etc. This implies a large scatter in da/dN vs. ΔK data resulting in less accurate and reliable predictions of safe life. It is thus obvious that an ample factor of safety should be used in all design procedures concerning fatigue crack growth.

One may think to apply this factor of safety to the applied stress range ($\Delta\sigma$), final crack length (a_f), fracture toughness (K_{Ic}), or the number of cycles to fracture (N_f). Any of these alternatives will result in a different degree of conservatism because of the nonlinearity of the expressions relating these parameters.

For instance, for the conditions of an initial crack length a_o = 5 mm subjected to a stress range $\Delta\sigma$ = 150 MPa in a material with the properties K_{Ic} = 50 MPa · m$^{1/2}$, m = 3, and C = 2.4 × 10^{-10} m/cycle results in a safe-life prediction of 2.61 × 10^3 cycles with a factor of safety of 1.5 applied to N_f. If, instead, the same factor of safety is applied to a_f, K_{Ic}, or to $\Delta\sigma$ as an overload factor, conservative lives are respectively estimated as 43%, 70%, or 21% of the above N_f value. As pointed out in the literature,[25] the complexity of crack growth behavior does not permit an easy assessment of the degree of conservatism. In general, applying a factor of safety to the final result N_f may be a reasonable decision. Moreover, structures designed according to fail-safe design concepts have to be thoroughly inspected in a timely manner for cracks ensuring a sufficient damage tolerance for safe operation.

9.11.3.2 Variable Amplitude Loading

Expressions 9.29 and 9.30 are used to predict the number of cycles to failure in structures subjected to constant amplitude cyclic loading. In reality, structures are normally subjected to variable-amplitude load cycles. From the first look, it seems a straightforward procedure to calculate the incremental crack growth for each individual cycle using Paris law and then adding up until final crack length is reached. This obviously requires computerized calculations of high accuracy due to the small amount of crack growth in one cycle.

However, to accomplish this task two points have to be considered. The first is the difficulty in discerning the individual load cycles from the loading history of the structure. In this regard the technique known as "rainflow counting"[25] has gained some acceptance. The second point is to consider that crack growth rate becomes slower after an overload high-stress cycle. This is due to the compressive residual stress zone associated with a larger plastic zone* at the crack tip created by an overload. This is explained by considering a load spectrum composed of three load cycles, the first and third ones are of the same amplitude and the middle being an overload as shown in Figure 9.27a. The high overload cycle creates a large plastic zone during the loading portion of the cycle followed by an elastic unloading portion, Figure 9.27b. The resulting net stresses furnish a compressive residual stress zone at the crack tip, Figure 9.27c. Tensile stresses resulting from the subsequent smaller load cycle have to surpass, first, the value of this compressive residual stress. The magnitude of the remaining tensile stress will obviously result in a reduced crack tip stress intensity, thus leading to a growth rate slower than the expected rate of the first load cycle; even a total crack arrest may be achieved. This "retardation" phenomenon is shown schematically in Figure 9.27d for various load patterns.

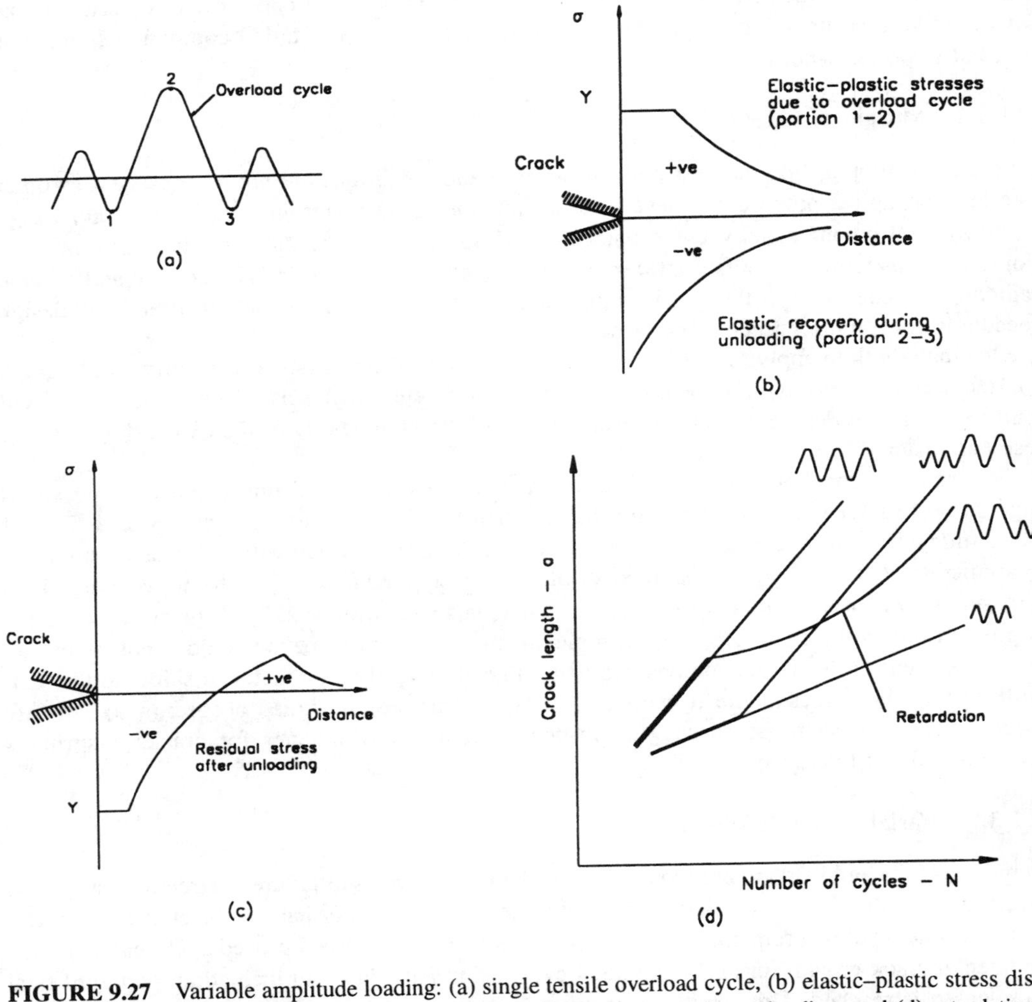

FIGURE 9.27 Variable amplitude loading: (a) single tensile overload cycle, (b) elastic–plastic stress distributions due to overloading followed by unloading, (c) residual stress after unloading, and (d) retardation in crack growth due to changes from high to small cyclic stress ranges.

* In cyclic loading the local yielding ahead of the crack is more complex than in monotonic loading. According to Rice,[32] the plastic zone size Δ_{cyclic} is approximately a quarter of $\Delta_{monotonic}$.

In essence, crack propagation is thus retarded because of the overload stress cycle. Retardation models[25] have been suggested to avoid the conservative life estimates resulting from ignoring the interaction effects between cycles of varying amplitudes within the load spectrum.

9.11.3.3 Mixed-Mode Crack Growth

Crack growth mixed-mode loading is of academic interest. In practice, cracks are found to develop and grow in mode (I), i.e., perpendicular to the maximum tensile principal stress. More precisely, at the beginning of crack growth under mixed modes (I) and (II), propagation follows a curved path until it diverts to be perpendicular to the maximum stress, thus eliminating mode (II). The problem is then reduced to mode (I) with a curved crack. However, for some mixed-mode situations, where the initial crack grows through the weakest path (e.g., the heat-affected zone of a weld), the combined modes prevail and have to be considered in the analysis.

For the case of a mixed-mode loading, crack propagation laws, e.g., the Paris law for mode (I), rate data can be applied provided that ΔK is replaced by an "effective value ΔK_{eff}." By referring to Equation 9.19, ΔK_{eff} for combined mode (I) and (II) may be taken as[25]:

$$\Delta K_{\text{eff}} = \sqrt{\Delta K_{\text{I}}^2 + \left(0.8\,\Delta K_{\text{II}}\right)^2} \qquad (9.33a)$$

Another form for ΔK_{eff} is suggested as[33]

$$\Delta K_{\text{eff}} = \left(\Delta K_{\text{I}}^4 + 8K_{\text{II}}^4\right)^{1/4} \qquad (9.33b)$$

9.11.3.4 Correlation with *S–N* Curves

There is a satisfactory correlation between the crack growth rate and the conventional *S–N* fatigue data. If N_{th} is the number of cycles that might have gone into producing the a_o crack, then Expression 9.32a gives, for $a_f >> a_o$,

$$(\Delta\sigma)^m\left(N_f - N_{\text{th}}\right) = \frac{a_o^{1-m/2}}{c\pi^{m/2}\left[\text{CCF}_{\text{av}}\right]^m} = \text{Constant}$$

Plotting $\ln(\Delta\sigma)$ vs. $\ln(N_f - N_{\text{th}})$ indicates a straight line with a slope of m, which is a satisfactory estimate for *S–N* curves for high fatigue cycle $N_f > 10^5$. The endurance limit seen on *S–N* curves could be associated with the threshold value ΔK_{th}.

9.11.3.5 Growth of Physically Short Cracks

Recent investigations[34] have identified different growth behaviors of fatigue cracks depending on crack length. Fatigue cracks have been classified as being "short" or "long" cracks. The latter are those cracks for which linear elastic fracture mechanics applies correctly with a plastic zone size remaining much less than the crack length. The lower limit of a long crack is that length associated with a stress intensity factor range higher than the threshold value ΔK_{th}.

The first type of short cracks is identified as "micro-structurally" short cracks of size extending to few grains only. These microstructurally short cracks could develop into "physically short" cracks. Because crack propagation in the short crack regime requires a high stress (above the fatigue limit), the crack tip plastic deformation zone in relation to the crack length can be large. This may be the dominant factor in explaining the higher propagation rates of physically short cracks when compared with long crack growth rates at an equivalent stress intensity factor.[35] This also explains the inappropriateness of applying linear-elastic fracture mechanics, in that it is only valid under small-scale yielding conditions.

Identification of physically short cracks thus becomes dependent on being short comparable with the size of the crack tip plastic zone and on being larger than the dimensions of the microstructural features. Successful models describing growth of short cracks are based on crack tip deformation parameters rather than on stress parameters.

9.11.3.6 Crack Closure

In the wake of a crack, the accumulation of all previous crack tip plasticity is thought to cause a phenomenon called *crack closure*. The permanent deformation of crack tips tends to close the crack before attaining the minimum tensile load. Hence for a cyclic stress alternating between σ_{max} and $\sigma_{min} = \sigma$, it has been assumed[36] that crack propagation is related to an effective stress increment $\Delta\sigma_{eff} = (\sigma_{max} - \sigma_{op}) < \sigma_{max}$. Here σ_{op} is the minimum tensile stress at which complete unfolding, or opening, of the crack occurs. It follows that an effective ΔK_{eff} (based on $\Delta\sigma_{eff}$) may be used to model crack growth data.

Example 9.16:

A thin solid turbine disk of diameter 750 mm rotating at 3560 rpm is made of a high-strength steel with the following properties: yield strength, Y = *1450 MPa; Young's modulus,* E = *190 GPa; Poisson's ratio,* $\nu = 0.3$*; specific weight = 80.25 kN/m³, coefficient of thermal expansion,* $\alpha = 12 \times 10^{-6}$*/°C; fracture toughness;* $K_c = 77$ *MPa · m*$^{1/2}$, m = *3.2; and* C = 4×10^{-11} *(in units of MPa -m- cycle in the Paris law:* $da/dN = C(\Delta K)^m$).

Inspection indicated a central crack of length 30 mm. Knowing that the radial temperature distribution is given by T = *270*r + *60 (*r *in meters and* T *in °C), what is the margin of safety against fracture due to 2000 repeated on/off cycles of loading at full speed and operating temperature?*

Solution:

For a rotating solid disk with a radial temperature gradient, the stress that opens a central crack is the stress $\sigma_\theta = \sigma_r$ at $r = 0$. As given by Equation 6.34, σ_θ consists of the sum

$$[\sigma_\theta] = \frac{3+\nu}{8}\rho\omega^2\left(r_o^2 - \frac{1+3\nu}{3+\nu}r^2\right)$$

$$+\alpha E\left[-T + \frac{1}{r_o^2}\int_0^{r_o} Trdr + \frac{1}{r^2}\int_0^{r} Trdr\right]$$

The integral

$$\int_0^{r_o} Trdr = \int_0^{r_o}(270r^2 + 60r)dr = 90r_o^3 + 30r_o^2$$

hence,

$$[\sigma_\theta]_{r=0} = \frac{3.3}{8}\rho\omega^2 r_o^2 + \alpha E\left[-(270r+60) + \frac{1}{r_o^2}(90r_o^3 + 30r_o^2) + \frac{1}{2}T_o\right]_{r=0}$$

$$= \frac{3.3}{8}\frac{80.25\times10^3}{9.81}\left(\frac{3560\times2\pi}{60}\right)^2(0.375^2)$$

$$+(12\times10^{-6})(190\times10^9)(90\times0.375)$$

$$= 65.96 + 76.95 = 142.91 \text{ MPa}$$

For an on/off cycle, the stress range is

$$\Delta\sigma = [\sigma_\theta]_{r=r0} - 0 = 142.91 \text{ MPa}$$

The critical crack length a_f at fracture is determined from Expression 9.17 as

$$[\sigma_\theta]_{r=r_o}\sqrt{\pi a_f}\,f(a_f/r_o) \le K_c$$

Since a_f is not yet known, take as a first approximation $(a_f/r_o) = 0.3$. From Equation 9.17, $f(a_f/r_o) = 1.106$. Hence, $142.91\sqrt{\pi a_f} \times 1.108 \le 77$ gives $a_f = 75.5$ mm. This assigns a new value for $f(a_f/r_o) = f(75.5/375)$ as 1.0482. Repeating calculations gives $142.91\sqrt{\pi a_f} \times 1.0482 \le 77$, from which $a_f = 84.1$ mm, a value which needs further iterations converging to a value $a_f = 82.5$ mm. It is, however, noticed that $f(a_o/r_o)$ and $f(a_f/r_o)$ are 1.002 and 1.06, respectively. This slight variation justifies use of an average value for $f(a/r_o)$ as $(1.002 + 1.06)/2 = 1.03$. Hence, applying Expression 9.32a gives

$$N_f = \frac{\left[a_o^{(1-m/2)} - a_f^{(1-m/2)}\right]}{(m/2-1)C\left[f_{\text{av}}(a/r_o)^m(\Delta\sigma)^m\pi^{m/2}\right]}$$

$$N_f = \frac{\left[0.015^{(1-3.2/2)} - 0.0825^{(1-3.2/2)}\right]}{(3.2/2-1)(4\times10^{-11})[1.03]^{3.2}(142.91)^{3.2}(\pi)^{3.2/2}}$$

$$= \frac{7.958}{1.297\times10^{-3}} = 6136 \text{ cycles}$$

For 2000 repeated on/off cycles, the margin of safety is 6136/2000 = 3.07.

9.12 STRESS CORROSION CRACKING

Crack growth due to stress cycling has been described in Section 9.9. Crack growth can occur under static loading conditions (well below the yield strength) due to specific material–environment interaction. This is known as *stress corrosion cracking* (SCC). For instance, high-strength steels and aluminums suffer from SCC in chloride aqueous solutions (e.g., seawater). SCC results in a nonductile type of failure and is governed by the stress intensity factor. Specimens with the same initial crack but loaded with different constant-stress levels will have different initial values of K_I.

As a crack grows with time, K_I increases until it attains the value of K_{Ic} of the material, thus resulting in fracture. Specimens with lower initial K_I values will require longer times to failure, and vice versa. Schematic representation of this behavior is shown in Figure 9.28. An asymptotic value for the curve designates a value for K_I marked as K_{ISCC}, which is known as the stress corrosion cracking threshold. Specimens loaded at stress intensities lower than K_{ISCC} will not fail.

Careful experiments including all the variables of the service environment are needed to determine K_{ISCC} for the various metal–environment combinations. Typical results are given in Table 9.5.

Stress corrosion cracking should be prevented rather than controlled. Hence, crack growth rate analysis, such as that of fatigue, is seldom performed and imposing the condition,

$$K_I < K_{ISCC} \tag{9.34}$$

is a requirement for an infinite life.

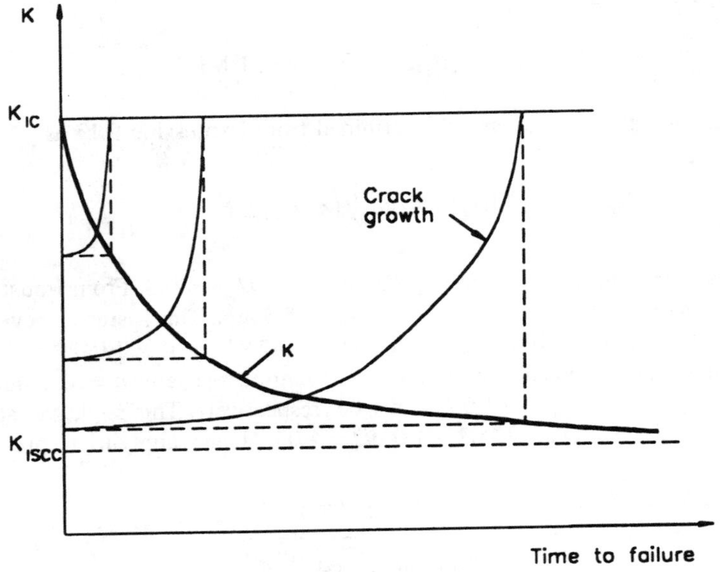

FIGURE 9.28 Crack growth by stress corrosion and dependence of time to failure on the applied stress intensity.

TABLE 9.5
Some Typical Values for K_{ISCC}

Material	K_{ISCC}, MPa · M$^{1/2}$ Industrially Polluted Atmosphere	Industrially Polluted Seawater	Natural Seawater
Steel AISI 1018	20.46	5.46	8.81
Steel AISI 4340	13.03	—	6.14
Stainless steel 17/7 PH	12.12	8.95	13.32

Source: Alawi, H., Ragab, A., and Shaban, M., Stress corrosion cracking of some steels in various environments, *Eng. Fract. Mech.*, 32(1), 29–37, 1989. With permission.

Example 9.17:

An experimental rig consists of three double-edge cracked high-strength metallic plates which suffer from stress corrosion cracking due to surrounding seawater. The dimensions of the specimens are width $W = 50$ *mm, thickness* $B = 12$ *mm, and initial crack length* $a_i = 2$ *mm. Reported time to failure for the three plates are 1500, 20, and 1.8 h due to the constant loads* $P = 132$, *162, and 198 kN, respectively. Using a constant configuration correction factor of 1.12 and knowing that* $K_{Ic} = 32$ *MPa* · *m*$^{1/2}$, *determine*

a. *The critical threshold value of the stress intensity* K_{ISCC};
b. *The amount of crack growth that occurred in each test;*
c. *The safe stress for a beam of square cross section (20 × 20 mm) subjected to three-point bending in the above stress corrosion conditions. An initial crack length of 2 mm is detected.*

Solution:

a. To determine K_{ISCC}, a plot such as in Figure 9.29 has to be established. The initial value for the stress intensity factor is given by Equation 9.15 as

$$(K_I)_i = \sigma\sqrt{\pi a_i}\ \text{CCF}$$

Due to corrosive environments, the crack length increases from a_i until it attains the critical value a_c in the reported time to failure corresponding to each load. Fracture occurs when $K_I = K_{Ic}$. Hence, the values in Table 9.6 are determined for $\sigma = P/BW$, CCF = $f(a/W)$ = 1.12, and K_{Ic} = 32 MPa · m$^{1/2}$.

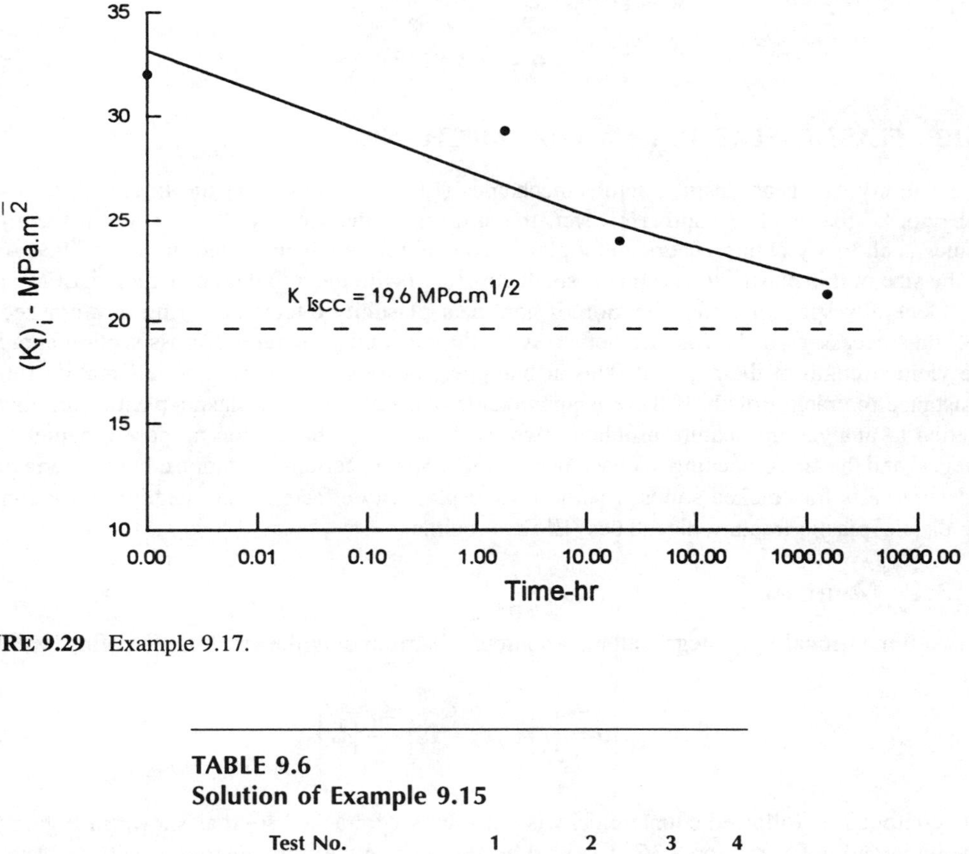

FIGURE 9.29 Example 9.17.

TABLE 9.6
Solution of Example 9.15

Test No.	1	2	3	4
Time to failure, t, h	1500	20	1.8	0
$(K_I)_i$, MPa · m$^{1/2}$	21.3	23.96	29.29	32
$a_f = \frac{1}{\pi}\left(\frac{K_{Ic}}{1.12\sigma}\right)^2$ mm	5.37	3.57	2.83	2
Crack growth $(a_f - a_i)$ mm	3.37	1.57	0.83	0

Plotting $(K_I)_i$ vs. ln(t), as shown in Figure 9.29, gives an asymptotic value of K_{ISCC} = 19.6 MPa · m$^{1/2}$.

b. The crack growth in each test is calculated and is given in Table 9.6.
c. To prevent fracture due to stress corrosion cracking, the Condition 9.34:

$$(K_{\mathrm{I}})\ \text{initial} = \sigma\sqrt{\pi a_i}\, f(a_i/W) \le K_{\mathrm{ISCC}}$$

must be satisfied, i.e.,

$$\sigma\sqrt{\pi \times 0.002}\ \mathrm{CCF}_i \le 19.6\ \mathrm{MPa\ \ m^{1/2}}$$

The value $\mathrm{CCF}_i = f(a_i/W) = f(2/20)$ is obtained from the expression corresponding to Figure 9.8b as 1.02, thus giving

$$\sigma = 242.5\ \mathrm{MPa}$$

9.13 ELASTIC–PLASTIC FRACTURE MECHANICS

The validity of linear-elastic fracture mechanics (LEFM) is based on the linear-elastic material behavior of the cracked solid. However, the stresses in the vicinity of the crack tip attain high values such that yielding occurs and a plastic zone is formed as indicated in Section 9.9. As long as the size of this plastic zone remains small compared with the initial crack length, LEFM applies.

Generally speaking, consideration of nonlinear plasticity effects occurring in a cracked body becomes necessary if the fracture net stress on the remaining uncracked cross section approaches the yield strength of the material. This also applies for tough (ductile) materials manifesting high resistance to crack growth. If there is appreciable plasticity, another elastic–plastic formulation is needed to analyze the feature problem. Two fracture criteria have been proposed, namely, the *J*-integral and the crack-opening displacement (COD). Within certain limitations, both criteria provide a design basis for cracked solids when nonlinear plasticity effects are not negligible for simply in the elastic–plastic fracture mechanics (EPFM) regime.

9.13.1 *J*-INTEGRAL

A two-dimensional line integral along a contour Γ surrounding the crack tip is defined as[38]

$$J = \int_{\Gamma}\left[U_o dy - \mathbf{T}\cdot\left(\frac{\partial \boldsymbol{u}}{\partial x}\right)ds\right] \tag{9.35}$$

The contour Γ is followed counterclockwise in a stressed cracked solid as shown in Figure 9.30a. The integrand in Expression 9.35 is defined by considering the strain energy density U_o. The scalar product of **T**; the outward traction vector perpendicular to the contour Γ and $(\partial\mathbf{u}/\partial\boldsymbol{x})$; the displacement gradient vector defines the second term of the integrand where ds is the arc length along Γ. This scalar product is the rate of work input from the stress field into the area enclosed by Γ.

In the case of a closed contour, it has been shown[38] that the integral $J = 0$. Considering Figure 9.30b, the two contours $\Gamma_1 \equiv ABC$ and $\Gamma_2 \equiv DEF$ represent the contour $\Gamma = \Gamma_1 + \Gamma_2$ around crack tip with $J_\Gamma = J_{\Gamma 1} + J_{\Gamma 1} = 0$. This is due to the fact that $\mathbf{T} = 0$ and $dy = 0$ along the crack faces CD and AF, and hence they have no contribution to the integral. Therefore, $J_{\Gamma 1}$ must be equal (but opposite in sign) to $J_{\Gamma 2}$, i.e., $J_{\Gamma 1} = J_{\Gamma 2}$. It is thus concluded that the *J*-integral taken along any unclosed contour, e.g., either Γ_1 or Γ_2, between unloaded crack surfaces is in fact path independent. Thus, even though considerable plasticity may occur in the vicinity of the crack tip, any path sufficiently far from the crack tip can be selected to the conveniently analyzed.

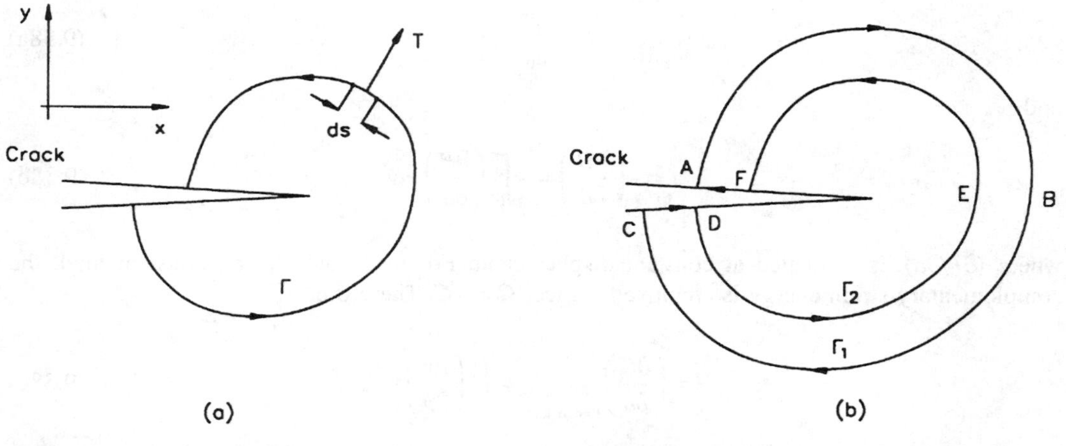

FIGURE 9.30 Arbitrary contours around crack tip: (a) coordinate system and (b) closed contour.

The *J*-integral as defined along a contour around the crack tip is shown[38] to be equal to the change in the total potential energy $\Omega - W_e$ for a virtual crack extension δa. Here W_a is the external work done and Ω is expressed per unit thickness. At constant displacement *v*, i.e., displacement control, $W_e = 0$ and, hence,

$$J = -\frac{\partial \Omega}{\partial a} = -\left(\frac{\partial U}{\partial a}\right)_{v=\text{constant}} \tag{9.36}$$

For a linear-elastic material $(\partial U/\partial a)$ is the strain energy release rate for an incremental crack extension as defined in Section 9.2; hence,

$$J = G = \frac{K_I^2}{E'} = -\frac{\partial U}{\partial a} \tag{9.37}$$

where $E' = E$ for plane stress and $E' = E/(1 - \nu^2)$ for plane strain. For a nonlinear behavior, *J* as expressed by Equation 9.36 is physically equivalent to the area between the load–displacement curves of two identically loaded solids having neighboring crack sizes a and a + δa, Figures 9.31a and b where the displacement *v* is measured at the loading points.

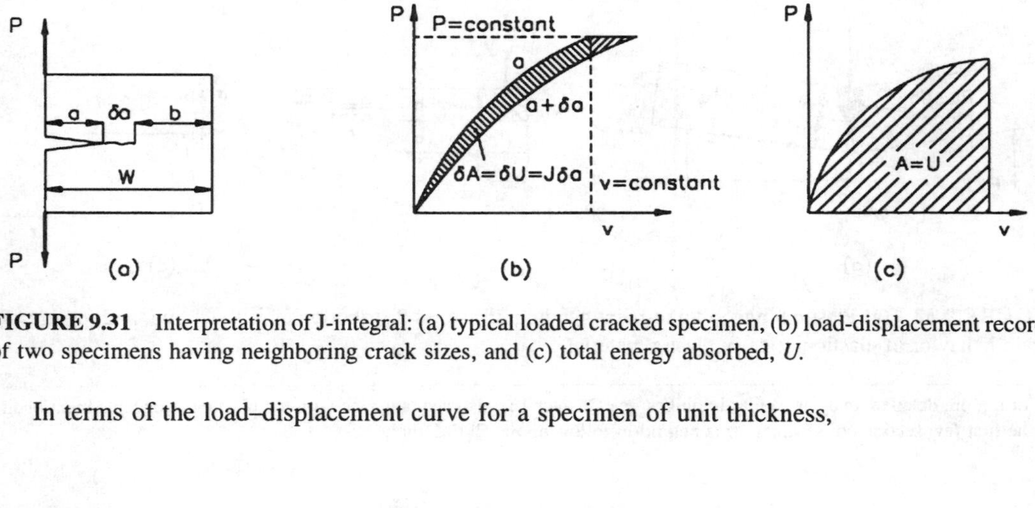

FIGURE 9.31 Interpretation of J-integral: (a) typical loaded cracked specimen, (b) load-displacement record of two specimens having neighboring crack sizes, and (c) total energy absorbed, *U*.

In terms of the load–displacement curve for a specimen of unit thickness,

$$U = \int_0^v P dv \tag{9.38a}$$

and

$$J = -\left(\frac{\partial U}{\partial a}\right) = -\int_0^v \left(\frac{\partial P}{\partial a}\right)_v dv \tag{9.38b}$$

where $(\partial P/\partial a)_v$ is evaluated at constant displacement. For load control, i.e., constant load, the complementary strain energy is employed; hence, $\Omega = -C$. Therefore,

$$J = \left(\frac{\partial C}{\partial a}\right)_{P=\text{constant}} = \int_0^P \left(\frac{\partial v}{\partial a}\right)_P dP \tag{9.38c}$$

Expressions 9.38b and c differ only by the vaishingly small triangular area shown in Figure 9.31b. In fact, J for displacement and load control is the same.

The standard for experimental determination of J is based on Expressions 9.38 as will be shown later. In a simpler form suitable to experimental evaluations, the J-integral for a variety of specimen configurations of thickness B, can be written as:

$$J = \frac{\beta U}{Bb} \tag{9.38d}$$

Here, b represents the uncracked ligament and β is a dimensionless constant depending on configuration. Clearly, the energy absorbed U is represented by the area under the load–displacement curve, as shown in Figure 9.31c. Expression 9.38d is based on the assumption that yielding has taken place over the entire ligament b.

To illustrate this, consider a three-point bend specimen as shown in Figure 9.32a in which the remaining net cross section has yielded. This means that loading has continued until a plastic hinge has formed and the beam behaves as being composed of two rigid parts connected by a hinge around which rotation takes place.* Assuming that the material behavior is as shown in Figure 9.32c indicates that the material has a yield strength Y, at which plastic deformation continues indefinitely. At full yielding of the remaining ligament, the bending stresses are $\sigma_x = \pm Y$, as shown in Figure 9.32b. The bending moment M and hence the load P, according to the elementary bending formula, are thus given by

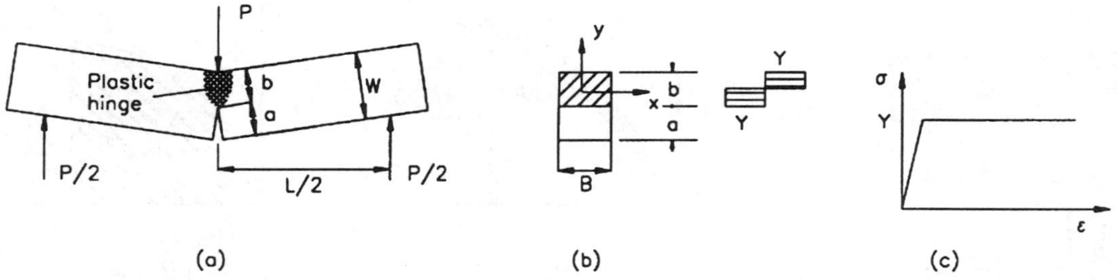

FIGURE 9.32 (a) Plastic hinge in three-point bending, (b) stress distribution at full yielding, and (c) stress-strain behavior of elastic-perfectly plastic material.

* For a more detailed analysis of plastic hinges, see Chapter 12. Also acquaintance with the plasticity concepts mentioned in the first few sections of Chapter 10 is helpful in following the EPFM analysis.

$$M = \frac{PL}{4} = 2B\int_0^{b/2} Yydy = \frac{Bb^2Y}{4}$$

or

$$P = \frac{Bb^2Y}{L}$$

When the crack extends over da, the ligament b will decrease by db; hence, $da = -db$ and

$$-\frac{\partial P}{\partial a} = \frac{\partial P}{\partial b} = \frac{2BbY}{L} = \frac{2P}{b}$$

Substitution into Equation 9.38b and integrating yields

$$J = -\int_0^v \left(\frac{\partial P}{\partial a}\right)_v dv = \frac{2}{Bb}\int_0^v Pdv$$

or

$$J = \frac{2U}{Bb} = \frac{2U}{B(W-a)} \tag{9.38e}$$

which is the same as Expressions 9.37c with $\beta = 2$ and $b = (W - a)$ for the bend specimen. Similar demonstration applied to a plate with a central through-crack under tension gives $\beta = 1$. For the compact tension specimen, Figure 9.20a, an appropriate correction is given in the standard[39] as $\beta = 2 + 0.522b/W$.

9.13.2 Experimental Determination of *J*

Methods have been developed and standardized[39] for measuring J on deeply notched fatigue–pre-cracked ($0.5 \le a/W < 0.75$) bend or compact tension specimens. The method consists basically of loading either a single or several initially identical bend type specimens to the displacement of interest. This displacement corresponds to an amount of stable fracture, i.e., measured crack extension Δa as shown in Figure 9.33a. Having a record of $P - v - \Delta a$. Expression 9.38e is used to compute J values plotted against corresponding Δa. This is the curve expressing the resistance to stable crack extension or the J–R curve.

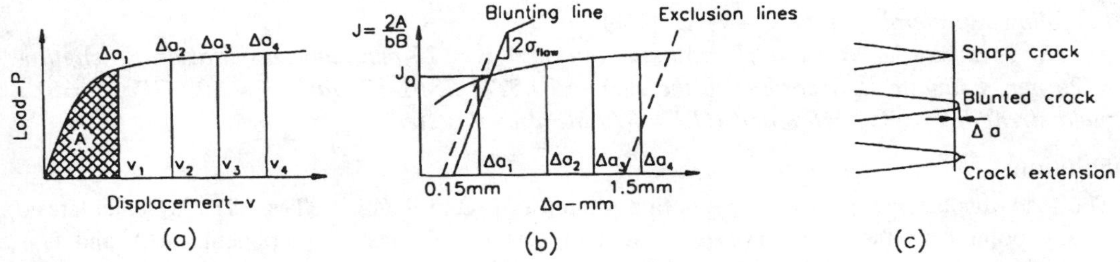

FIGURE 9.33 Experimental determination of J.

Similar to determination of K_{Ic}, it has been argued that a critical J_{Ic} is estimated at the instant when crack extension was first encountered. In other words, J_{Ic} is a critical value of J near the onset of stable crack growth, and hence it may be found at the point when

$$J = 2\sigma_{flow}\Delta a \tag{9.39a}$$

where σ_{flow} is taken simply as the yield stress of the material Y or average of $(Y + UTS)/2$, where UTS is the ultimate tensile strength. The line represented by the so-called blunting line $J = 2\sigma_{flow}\Delta a$ corresponds approximately to a small component of crack extension that is due to crack blunting because of plastic deformation as represented in Figure 9.33c. Two exclusion lines at $\Delta a = 0.15$ mm and $\Delta a = 1.5$ mm both having slopes of $2\sigma_{flow}$ define the data which may be fitted by a power law

$$J = C_1(\Delta a)^{C_2} \tag{9.39b}$$

These data are required to be less than a ceiling value of $J_{max} = b_o\sigma_{flow}/15$. According to ASTM 813, a provisional or interim value of J, labeled J_Q, is thus determined from the intersection of a line parallel to the blunting line at $\Delta a = 0.2$ mm and data expressed by Equation 9.39b, as shown in Figure 9.33b. The $J_Q = J_{Ic}$, as long as the following size requirements are met:*

$$B,\ (W - a_o) \geq 25\frac{J_Q}{Y} \tag{9.40}$$

Here it is interesting to compare these size limitations for a valid J_{Ic} with that for K_{Ic} as given before by Condition 9.26. Since $G = J$ within LEFM and knowing from Expression 9.13 that

$$K_{Ic}^2 = G_{Ic}\,E/(1 - \nu^2)$$

or

$$J_{Ic} = (1 - \nu^2)\,K_{Ic}^2/E$$

therefore, $B \geq 25(1 - \nu^2)\,K_{Ic}^2/EY$ for a valid J_{Ic} compared with $B \geq 2.5K_{Ic}^2/Y^2$ for a valid K_{Ic}.

For an aluminum alloy, 2024-T3 from Table 9.1, $Y = 345$ MPa, $\nu = 0.3$, $E = 70$ GPa, and $K_{Ic} = 44$ MPa · m$^{1/2}$. Size requirements would be $B \geq 40.7$ mm for K_{Ic} and $B \geq 1.8$ mm for J_{Ic}. Thus, the J_{Ic} size requirements are much less stringent than these of K_{Ic}. J_{Ic} may be used to evaluate materials as a ductile fracture toughness criterion. However, although the value of J_{Ic} can be converted to its equivalent K_{Ic}, the use of the latter should be obviously restricted only to dominant elastic conditions around a preexisting crack.

Example 9.18:

In a test to determine J_{Ic} *for a certain type of steel, three-point bend specimens are used to obtain the following record of data,* P – v–Δa, (Table 9.7):

The specimen has a width W = *50 mm, thickness* B = *25 mm, and an initial crack length* a_o = *26 mm. Estimate* J_{Ic} *according to the standard ASTM E813–87. Take* E = *200 GPa,* ν = *0.3, yield strength* Y = *350 MPa, and UTS* = *570 MPa for this steel.*

Solution:

The load–displacement (P–v) curve is first plotted in Figure 9.34a. ASTM E813–87 calculates J_i at any point i on the P–v curves as a sum of elastic and plastic components, $(J_e)_i$ and $(J_p)_i$, respectively; hence,

$$J_i = (J_e)_i + (J_p)_i = \frac{K_i^2(1 - \nu^2)}{E} + (J_p)_i \quad i = 1,\ 2,\ \ldots$$

* Validation of J_Q requires several conditions, which are listed in ASTM E813–87.[39]

TABLE 9.7
Experimental Data in a *J*-Test

Load *P*, kN	Deflection *v*, mm	Crack Extension Δa, mm	*J*, kJ/m²
0	0	0	0
21.28	0.3	0.014	18.1
36.4	0.49	0.0265	53
44.6	0.602	0.043	80
51.1	0.963	0.0915	151
53.5	1.563	0.3	261
54.5	1.757	0.42	298
55.1	2.168	0.75	373
55.4	2.36	1.00	410
52.4	2.657	1.25	456.6
50.1	2.81	1.38	478.2
43.4	3.29	1.85	545

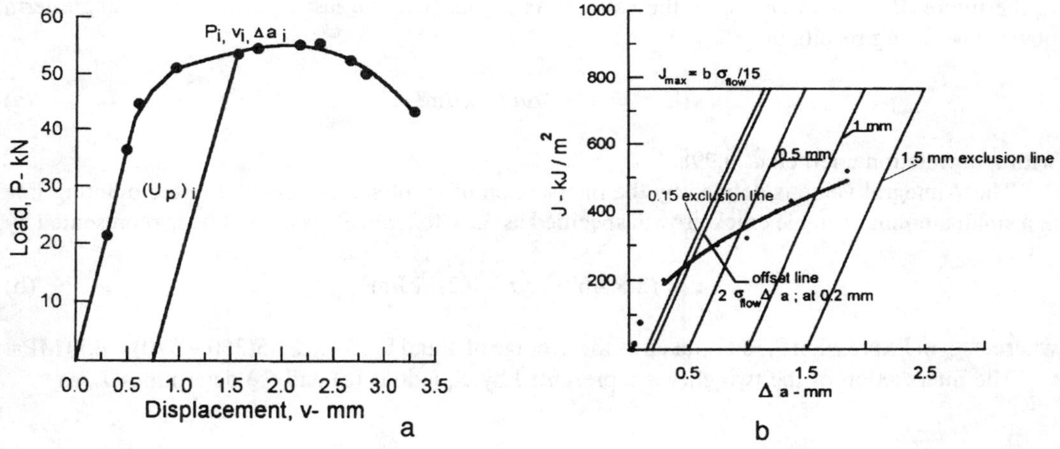

FIGURE 9.34 Example 9.18.

Here, K_i is determined as given in Section 9.10, while $(J_p)_i$ is calculated according to Expression 9.38e, namely,

$$\left(J_p\right)_i = \frac{2\left(U_p\right)_i}{B\left(W - a_o\right)}; \quad i = 1,\ 2,\ \ldots$$

where $(U_p)_i$ is the area under the *P–v* curve defined by unloading from the corresponding $(P_i - v_i)$. Note that these calculations are based on the initial length a_o and do not correct *J* for crack growth.

To illustrate the calculation procedure, consider a typical point at load $P_i = 55.1$ kN and total deflection $v_i = 2.168$ mm. This gives, using Expression 9.27b for $a_o/W = 0.52$,

$$K_i = \frac{3P_iL}{BW^{3/2}} f(a_o/W) = \frac{3\times 55.1\times 10^3 \times (4\times 0.05)}{0.025\times (0.05)^{3/2}} \times 0.947$$

$$= 1.1197\times 10^8 \text{ N/m}^{3/2}$$

Hence,

$$(J_e)_i = \frac{K_i^2(1-\nu^2)}{E} = 57.04 \text{ kJ/m}^2$$

For $(J_p)_i$, the unloading area $(U_p)_i = 0.0948$ kN · m; hence

$$(J_p)_i = \frac{2(U_p)_i}{B(W-a_o)} = \frac{2\times 0.0948}{0.025(0.05-0.026)} = 315.96 \text{ kJ/m}^2$$

and $J_i = (J_e)_i + (J_p)_i = 57.04 + 315.96 = 373$ KJ/m².

A plot of J_i vs. Δa_i, for which the values are given in Table 9.7, is shown in Figure 9.34b. According to ASTM E813–87, some data points have to be excluded from the analysis as shown in the figure. By considering only the valid J–Δa points (i.e., the last eight points), a single-term power law fitting results in

$$J = 419.4\Delta a^{0.4} \text{ kJ/m}^2 \tag{a}$$

with a correlation coefficient 0.998.

The J-integral value is defined by the intersection of an offset line parallel to the blunting line at a small amount of stable crack growth specified as $\Delta a = 0.2$ mm. This offset line is represented by

$$J = (2\times 460)(\Delta a - 0.2) \text{ kJ/m}^2 \tag{b}$$

where σ_{flow} in Expression 9.38a is taken as the average of Y and UTS, i.e., 0.5(350 + 570) = 460 MPa.

The intersection of the two curves represented by equations (a) and (b) determines J_Q as

$$J_Q = 419.4\left(\frac{J_Q}{2\times 460} + 0.2\right)^{0.4}$$

$$J_Q = 333.1 \text{ kJ/m}^2$$

This value of J_Q qualifies as J_{Ic} since it satisfies the following ASTM conditions:

- $B = 25 \text{ mm} > \dfrac{25J_Q}{\sigma_{flow}} > \dfrac{25\times 333.1}{460} > 18.1 \text{ mm}$
- $b_o = W - a_o = 50 - 26 = 24 \text{ mm} > \dfrac{25J_Q}{\sigma_{flow}} > 18.1 \text{ mm}$
- $\dfrac{dJ_Q}{da}$ (at $\Delta a_Q = 0.562$ mm) = $419.4 \times 0.4\,(0.562)^{0.6} = 118.7 < \sigma_{flow} < 460$ MPa

9.13.3 A Scheme for Fracture Estimation Using J_{Ic}

The use of J_{Ic} to characterize fracture in an elastic–plastic solid is based on defining the stress singularity in the vicinity of the crack tip in terms of J. By considering a circular contour around a crack, as shown in Figure 9.35a, the J-integral, Expression 9.35 is given by

$$J = \int_{-\pi}^{\pi} \left\{ \left(\int_0^{\varepsilon} \sigma_{ij} d\varepsilon_{ij} \right) \cos\theta - \mathbf{T} \cdot \frac{\partial u}{\partial x} \right\} r d\theta$$

Since the integral is path independent, the value of this J-expression must not depend on r. Hence, the integrand must be proportional to $1/r$. In fact, this may be easily seen from the stress and strain fields, Expression c and d of Example 9.1, where σ_{ij} and ε_{ij} are both proportional to $1/\sqrt{r}$.

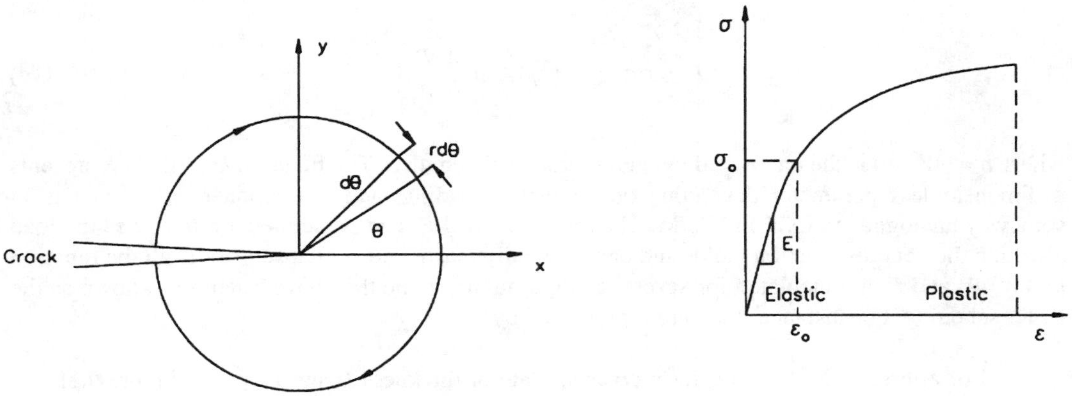

FIGURE 9.35 (a) Circular contour around crack tip and (b) elastic-plastic stress-strain behavior.

It can be demonstrated analytically that the stress and strain fields near the crack tip, well within the plastic zone, are given in terms of plastic J; J_p as:[40]*

$$\sigma_{ij} = \sigma_o \left(\frac{J_p / r}{\alpha \sigma_o \varepsilon_o I_n} \right)^{1/n^*+1} f_{ij}(\theta) \tag{9.40a}$$

$$\varepsilon_{ij} = \alpha \sigma_o \left(\frac{J_p / r}{\alpha \sigma_o \varepsilon_o I_n} \right)^{n^*/n^*+1} g_{ij}(\theta) \tag{9.40b}$$

where I_n is a numerical constant depending on the stress–strain relation as well as the conditions of plane stress or plane strain. These expressions are derived for proportional loading conditions assuming that the elastic–plastic stress–strain behavior of the solid, Figure 9.35b, may be represented by the empirical law:**

$$\frac{\varepsilon}{\varepsilon_o} = \frac{\sigma}{\sigma_o} + \alpha \left(\frac{\sigma}{\sigma_o} \right)^{n^*} \tag{9.41}$$

---

* Equations 9.41 are called in the literature the HRR singularly.

** This law, which is discussed more elaborately in Section 10.5 in a somewhat different form, is known as Ramberg–Osgood law.

Here σ_o is taken equal to the yield strength Y, $\varepsilon_o = \sigma_o/E$, n^{29} is an exponent dependent on the strain hardening of the material, and α is a material constant. Rearranging Equation 9.41a gives

$$J_p = \alpha\sigma_o\varepsilon_o I_n r\left(\frac{\sigma_{ij}}{\sigma_o}\right)^{n^*+1} f_{ij}(\theta) \tag{9.41c}$$

in which J defines the stress singularity at the crack tip for linear and nonlinear behavior, i.e., $n^* \geq 1$. This is similar to the situation where the stress intensity factor K characterizes the singularity in the linear case ($n^* = 1$), as given by Equation 9.11.

For proportional loading conditions, the local stresses σ_{ij} increases in proportion to the remote load P; hence, Equation 9.41c may be rewritten, for most geometries, in the form:

$$J_p = \alpha\sigma_o\varepsilon_o bh(a/W,\ n^*)\left(\frac{P}{P_o}\right)^{n^*+1} \tag{9.41d}$$

where $b = (W - a)$ is the uncracked ligament for a crack length a. The function $h(a/W, n^*)$ represents a dimensionless parameter depending on geometry, loading, and the exponent n^*, which is in someway analogous to CCF in LEFM. The reference load P_o can be defined arbitrarily as the load at which the net cross section yields and becomes fully plastic, i.e., collapse load. Both the function $h(a/W, n^*)$ and P_o are tabulated for several configurations, while the above scheme is known as the EPRI scheme.* For instance, they are represented by

- For a three-point bending for a cracked plate of thickness B and $L/W = 4$, Figure 9.8b,

$$P_o = \frac{1.072Bb^2\sigma_o}{L} \tag{9.43a}$$

and $h(a/W, n)$ as shown typically in Figure 9.36a**,

- For pressurized cylinders with internal radial crack, Figure 9.11b and $(r_o - r_i)/r_i \leq 0.2$,

$$P_o = \frac{2[(r_o - r_i) - a]\sigma_o}{\sqrt{3}(r_i + a)} \tag{9.43b}$$

and $h(a/(r_o - r_i), r_o/r_i, n^*)$ as shown typically in Figure 9.36b.

More generally, for structures loaded under conditions of small-scale yielding, where both elastic and plastic effects are of the same order of magnitude, the total value of J_{tot} is the sum of

$$J_{tot} = J_e + J_p$$

where J_e and J_p, respectively, are given by Expressions 9.37 and 9.41d. For elastic conditions, J_e is based on a modified Irwin plastic zone correction, Equation 9.25, for strain-hardening materials as

* See Kumar et al.[41] More elaborate tables and functions are given in Anderson,[42] Chapter 12.

** Figure 9.36a and b use tabulated data given in Reference 42.

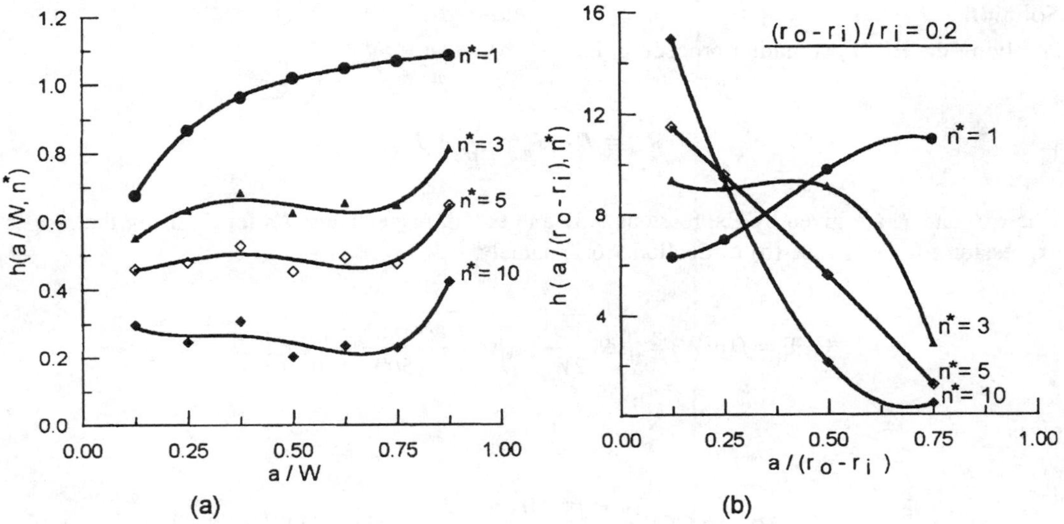

FIGURE 9.36 Dimensionless function $h(a/W, n^*)$ in Equation 9.41 for (a) three-point bending and (b) pressurized cylinders.

$$a_{\text{eff}} = a + \left(\frac{n^* - 1}{n^* + 1}\right)\frac{1}{1 + (P/P_o)^2}\frac{1}{\alpha\pi}\left(\frac{K_{\text{I}}}{\sigma_o}\right)^2 \tag{9.44}$$

A remarkable advantage of using the expression $J_{\text{tot}} = J_e + J_p$ is that it covers both elastic and plastic fracture mechanics automatically without any need to differentiate between their ranges of validity. At low stresses, the term J_e dominates and the analysis becomes basically an LEFM one. However, at higher stresses approaching plastic collapse loads, the term J_p determines almost solely the safe load. At intermediate stresses, the analysis is automatically adjusted by applying the necessary plasticity correction.

Example 9.19:

A plate of width 2W = *1000 mm and thickness* B = *20 mm is subjected to a remote uniform tensile load* P*. Inspection has revealed a central through-crack of length 2a = 125 mm perpendicular to the loading axis. The plate is made of a tough steel of the following properties:* σ_o = Y = *280 MPa,* E = *210 GPa,* K_c = *164.1 MPa · m*$^{1/2}$*,* α = *1.2, and n** = *6.25. Determine the failure load* P_f *required to initiate ductile crack growth. Assume plane stress conditions and use Table 9.8 to evaluate the parameter* h *in Expression 9.41d.*

TABLE 9.8
Data for the Parameter $h(a/W, n^*)$[a] for a/W = 0.125

n^*	1	2	5	7	10	16
$h(a/W, n^*)$	0.35	0.446	0.559	0.578	0.551	0.516

[a]See Reference 42.

Solution:

Applying the EPRI estimation procedure, i.e.,

$$J_{\text{tot}} = J_e + J_p = \frac{K_\text{I}^2}{E} + J_p$$

where J_e and J_p are given by Expressions 9.37 and 9.41d, respectively. As for J_e, using the proper expression for CCF, case (b) of Section 9.6.1, namely,

$$\text{CCF} = f(a/W) = \sqrt{\sec\frac{\pi a}{2W}} = \sqrt{\sec\left(\frac{\pi \times 62.5/2}{500}\right)} = 1.172$$

gives

$$K_\text{I} = \sigma\sqrt{\pi a}\ \text{CCF} = \frac{P\sqrt{\pi \times 0.0625/2}}{1 \times 0.02} \times 1.172 = 25.97P$$

and $J_e = \frac{K_\text{I}^2}{E}$ 3.211 × 10^{-9} P^2 N · m/m^2; P in N.

Note that for an accurate estimate, the crack length a must be corrected for the plastic zone at the crack tip according to Expression 9.44 which involves the unknown load, P. An iteration procedure is thus required and for simplicity only the value of a is used in solving this example.

As for J_p from Expression 9.41d,

$$J_p = \alpha\sigma_o\varepsilon_o bh(a/W,\ n^*)\left(\frac{P}{P_o}\right)^{n^*+1}$$

where P_o, the collapse load, is given by

$$P_o = 2bB\sigma_o = 2(0.5 - 0.0625) \times 0.02 \times 280 \times 10^6$$

$$= 4.9 \text{ MN}$$

The configuration factor $h(a/W, n^*) = h(0.125, 6.25)$ is determined by interpolation from Table 9.8 as $h = 0.57$; hence,

$$J_p = 1.2(280 \times 10^6)\left(\frac{280 \times 10^6}{210 \times 10^9}\right)(0.5 - 0.0625)(0.57)\left(\frac{P}{4.9 \times 10^6}\right)^{7.25}$$

$$J_p = 3.56 \times 10^{-44}\, p^{7.25} \text{ with } P \text{ in N}$$

The point of initiation of ductile crack growth occurs when

$$J_{\text{tot}} = (J_e + J_p) \le J_c$$

where $J_c = K_c^2/E = (164.1 \times 10^6)^2/(210 \times 10^9) = 128.25$ kJ/m^2.

Hence, $(3.211 \times 10^{-9}P^2 + 3.5 \times 10^{-44}P^{7.25}) \leq 128.25 \times 10^3$ N · m/m², which gives $P_f = 4.523$ MN.

A plot of J_e, J_p, and J_{tot} vs. P is shown in Figure 9.37, where P_f is less than P_o. For other conditions, i.e., combination of material properties, loading, and geometry, the calculation of J_{tot} may result in a value of $P_f > P_o$ indicating that plastic collapse is apt to occur before ductile crack growth is initiated.

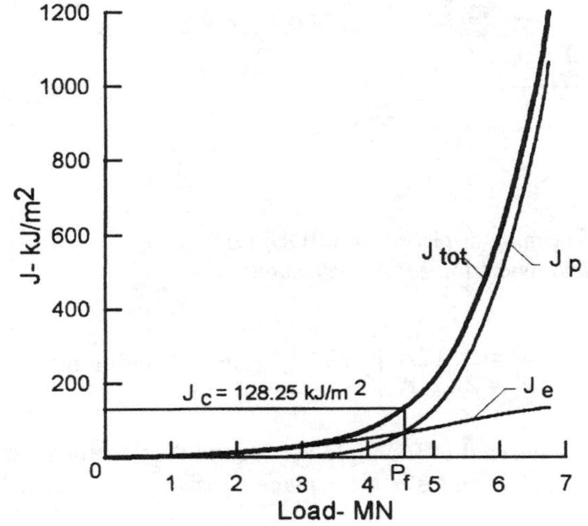

FIGURE 9.37 Example 9.19.

It is rather interesting to indicate that an error of 20% in selecting the value of J_c, (i.e., taking $J_c = 153.9$ kJ/m²) would result in only an error of 3.9% in P_f, which well may be acceptable in defining structural loads.

If plasticity effects are totally ignored and LEFM is applied, the load at failure would be

$$J_e = 3.211 \times 10^{-9}P^2 \leq 128.25 \text{ kJ/m}^2$$

which gives

$$P_f^* = 6.32 \text{ MN}$$

an erroneous value much higher than P_o, the collapse load.

9.13.4 Crack Opening Displacement

Examination of fractured test specimens indicated that crack faces have moved apart and initially sharp crack has been blunted by plastic deformation. Both crack-face movement and blunting have been increasingly proportional to the material toughness. This has led the proposal to use crack tip opening displacement (CTOD) as a measure of fracture toughness, especially for conditions where considerable plasticity occurs in the vicinity of the crack tip.[43] This is based on relating CTOD to the stress intensity factor in the case of small-scale yielding, i.e., LEFM, or to the J-integral for conditions of EPFM.

An initially sharp through-crack in an infinitely stressed body is shown in Figure 9.38a as being blunted as a result of plastic deformation. Hence, the physical crack may be considered as a notational crack larger by a distance r_p, given by Equation 9.25 as

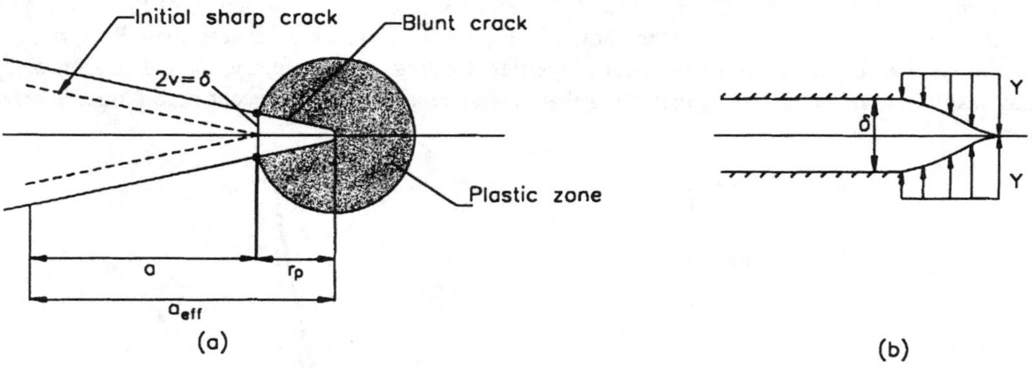

FIGURE 9.38 Crack tip opening displacement CTOD: (a) blunting of a sharp crack due to plastic deformation and (b) the strip yield model for CTOD estimation.

$$r_p = \frac{1}{2\pi}\left(\frac{K_I}{Y}\right)^2 \quad \text{for plane stress conditions} \tag{9.25}$$

Recall from Expressions d derived in Example 9.1 for the displacement field in the vicinity of the crack tip that they give at $r = r_p$ and $\theta = \pi$ for plane stress.

$$v = \frac{4K_I}{E}\sqrt{\frac{r_p}{2\pi}} \tag{9.45}$$

Combining Equations 9.25 and 9.45 yields CTOD as:[44]

$$\delta = 2v = \frac{4}{\pi}\frac{K_I^2}{YE} \tag{9.46a}$$

Another expression for δ is found by considering the strip yield model[22] in which the CTOD is defined as the crack opening displacement at the end of the strip yield zone as shown in Figure 9.38b. This model, assuming plane stress and nonhardening material, gives

$$\delta = \frac{K_I^2}{YE} \tag{9.46b}$$

The actual relationship between CTOD and K_I, depending on strain-hardening behavior and loading conditions, takes the form

$$\delta = \frac{K_I^2}{\beta_1 YE'} \tag{9.46c}$$

where β_1 is a dimensionless constant that is approximately 1.0 for plane stress and 2.0 for plane strain. Since $Y = \varepsilon_Y E$. Expression 9.46c reduces to

$$\frac{\delta_c}{\varepsilon_Y} = \frac{1}{\beta_1}\left(\frac{K_{Ic}}{Y}\right)^2 \tag{9.47}$$

where the subscript c denotes the critical values of K_I and δ; both being material properties. This latter critical displacement marks in general the onset of crack extension and is considered to be a measure of the material resistance to fracture initiation. Expression 9.47 may be regarded as a design basis if a fracture stress $\sigma_f = \beta_2 Y$ and $K_I = \sigma_f\sqrt{\pi a_c}$ are substituted into Equation 9.47 to yield

$$a_c = \text{constant}\left(\frac{\delta_c}{\varepsilon_Y}\right) \quad \text{for EPFM}$$

which is analogous to

$$a_c = \text{constant}\left(\frac{K_{Ic}}{Y}\right)^2 \quad \text{for LEFM}$$

9.13.5 Experimental Determination of COD

Most COD experiments determine δ_c by testing a single-notch fatigue-precracked beam specimen loaded slowly in a three-point bending, as shown in Figure 9.39a. The test record furnishes the load–displacement curve, Figures 9.39b and c. The displacement u is measured at the crack mouth using a clip gauge, Figure 9.20c. CTOD is then obtained by considering the rotation of the two halves of the specimen around a point within the plastic hinge zone; thus,

$$\frac{\delta}{\zeta(W-a)} = \frac{u}{\zeta(W-a)+a} \quad \text{or } \delta = \frac{\zeta(W-a)u}{\zeta(W-a)+a} \tag{9.48a}$$

where $0 \le \zeta \le 1$ is a dimensionless rotation factor.

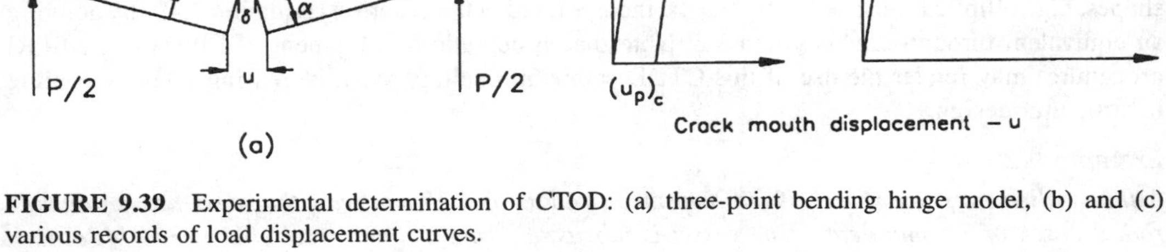

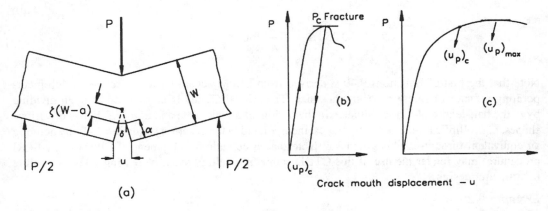

FIGURE 9.39 Experimental determination of CTOD: (a) three-point bending hinge model, (b) and (c) various records of load displacement curves.

Standard test methods[45] consider the inaccuracy of neglecting the elastic component of crack mouth displacement. Hence, δ is expressed as the sum of elastic and plastic components as

$$\delta = \delta_e + \delta_p = \frac{K_I^2}{\beta Y E'} + \frac{\zeta_p(W-a)u_p}{\zeta_p(W-a)+a} \tag{9.48b}$$

where the elastic and plastic components δ_e and δ_p are given by Equations 9.46c and 9.48a, respectively. A typical value for $\zeta_p = 0.44$.

Testing[45] for COD may be applied for brittle or ductile materials. The load–displacement curve, which resembles the tensile stress–strain curve of the materials, is used to define the critical plastic crack-mouth displacement $(u_p)_c$ and, hence, δ_c. Depending on the shape of the *P–u* curve, several definitions for $(u_p)_c$ may apply. From Figure 9.39b, where the fracture mechanism is mainly cleavage, the point of maximum load marks the critical CTOD. However, for tougher materials, another critical point defined near the initiation of stable crack growth is considered. This measure of toughness is analogous to J_{Ic}. In fact, for EPFM an approximate general relationship between δ and J holds as given by $J = \beta_3 Y\delta$, where β_3 is a parameter depending on the stress state and material properties.

Under the above test conditions, a fracture resistance curve CTOD vs. crack extension, similar to *J*-crack extension, Figure 9.33b may be required. The critical $(u_p)_c$ at initiation of crack growth is defined as $\Delta a = 0.2$ mm.

Specimen size requirements for COD testing recommended by standards[45] are such that a full section thickness of the structure is to be tested with $a/W = 0.15$ to 0.7 (BS) or $a/W = 0.45$ to 0.55 (ASTM)

9.13.6 Application of CTOD to Structural Design

The critical CTOD as determined experimentally using Expression 9.47 is regarded as a material property for given temperature and loading rate for a standard geometry. Based on the suggestion that the global strain ε_g in a cracked solid is linearly related to CTOD under large-scale yielding, the following semi-empirical design criteria has been proposed[46] for structural safety if

$$\left(\frac{\varepsilon_g}{\varepsilon_Y}\right)^2 \le \frac{\delta_c}{2\pi\varepsilon_Y a} \qquad \text{for } \left(\varepsilon_g/\varepsilon_Y\right) \le 0.5$$
$$\left(\frac{\varepsilon_g}{\varepsilon_Y} - 0.25\right) \le \frac{\delta_c}{2\pi\varepsilon_Y a} \qquad \text{for } \left(\varepsilon_g/\varepsilon_Y\right) > 0.5 \tag{9.49}$$

Note that the first of Equations 9.49 is derived from LEFM according to Expression 9.46b incorporating a factor of safety of 2 on crack size. The critical CTOD is normalized through dividing by a, the half-length of a crack situated in the middle of a wide plate and $\varepsilon_Y = Y/E$. Flaws of various shapes, e.g., elliptical surface flaws, may be incorporated in the criterion (Equation 9.49) by defining an equivalent through-thickness crack.[47] Note that systematic development of EPFM (e.g., EPRI procedure) may render the use of this CTOD semiempirical approach increasingly less appealing in structural design.

Example 9.20:

A thin-walled pressure vessel of 500-mm mean radius and 25-mm wall thickness has an outside radial crack of 7.5-mm depth. The vessel is fabricated from a tough steel of Y = *380 MPa,* E = *210 GPa, and* ν = *0.3. A COD test has been conducted on a three-point bend standard specimen made of the same vessel material, and the following data have been found:* W = *50 mm,* B = *25 mm,* a = *22.5 mm,* P_c = *78.62 KN, and* u_p = *0.312 mm. Determine the maximum pressure the vessel can withstand using elastic–plastic fracture mechanics.*

Solution:

First, the COD test results have to be used to determine a critical CTOD for this type of steel. According to Expression 9.48b,

$$\delta_c = \frac{K_I^2(1-\nu^2)}{\beta_1 YE'} + \frac{\zeta_p(W-a)u_p}{\zeta_p(W-a)+a}$$

where K_I is given by Expression 9.27b for the three-point bend specimen, i.e.,

$$K_I = \frac{3PL}{BW^{3/2}} \frac{\left(\frac{a}{W}\right)^{1/2}\left[1.99-\left(\frac{a}{W}\right)\left(1-\frac{a}{W}\right)\left(2.15-3.93\frac{a}{W}+2.7\frac{a^2}{W^2}\right)\right]}{2\left(1+2\frac{a}{W}\right)\left(1-\frac{a}{W}\right)^{3/2}}$$

hence, for $B = 0.025$ m, $W = 0.05$, $a/W = 22.5/50 = 0.45$, and $P = 78.62$ kN, $K_I = 128.57$ MN/m$^{3/2}$. Taking $\beta_1 = 2$ and $E' = E = 210$ GPa for plane stress and $\zeta_p = 0.44$ as recommended by ASTM E1290–89 yields for $u_p = 0.312$ mm

$$\delta_c = \frac{(28.57\times10^6)^2\times0.91}{2\times380\times10^6\times210\times10^9} + \frac{0.44(0.05-0.0225)\times0.312\times10^{-3}}{0.44(0.05-0.0225)+0.0225}$$

$$\delta_c = 9.5256\times10^{-5} + 1.0904\times10^{-4} \text{ m}$$

$$\delta_c = 0.2033 \text{ mm}$$

Note the very small ratio between the elastic and plastic contribution to δ_c as seen from the first and second terms, respectively.

For a thin-walled pipe under internal pressure, p_i, the stresses, according to Equation 8.68, are

$$\sigma_\theta = 2\sigma_z = \frac{p_i r_m}{h} \text{ and } \sigma_r = 0$$

and, hence,

$$\varepsilon_\theta = \frac{1}{E}[\sigma_\theta - \nu\sigma_z] = \frac{1-\nu/2}{E}\sigma_\theta = \frac{1-0.3/2}{210\times10^3}\frac{P_i\times500}{25}$$

$$\varepsilon_\theta = 8.1\times10^{-5} p_i$$

where p_i is in MPa. For this type of steel

$$\varepsilon_Y = \frac{Y}{E} = \frac{380}{210\times10^3} = 1.81\times10^{-3}$$

hence, for a stress concentration factor $k_t = 4.2$,

$$\varepsilon_g/\varepsilon_y = 4.2\varepsilon_\theta/\varepsilon_y = 0.188p_i$$

The nondimensional CTOD, determined from Expression 9.49, is used as a design criterion for elastic–plastic fracture conditions necessitates that

$$\left(\frac{\varepsilon_g}{\varepsilon_Y} - 0.25\right) \leq \frac{\delta_c}{2\pi\varepsilon_Y a}$$

hence, for $\delta_c = 0.2033$ mm and $a = 7.5$ mm,

$$0.188p_i - 0.25 \leq \frac{0.2033}{2\pi \times 1.81 \times 10^{-3} \times 7.5}$$

giving in the limit $p_i = 14.07$ MPa.

This value of p_i is stress about 90% of the yielding pressure, as may be determined from Expression 9.42b. Hence, an expected appreciable plastic zone is formed at the crack tip.* Converting δ_c to a K_c value according to Expression 9.46 is possible, but it does not justify the use of this latter value of K_c in solving this problem where plasticity effects dominate.

PROBLEMS

Stresses around Elliptical Holes and Cracks

9.1 A very large sheet containing a small crack of length $2a$ is subjected to a tensile stress σ in a direction perpendicular to the crack as shown in Figure 9.4. With the coordinate r measured from the crack tip, examine the stress function for small values of r.

$$\phi = \frac{2\sqrt{2}}{3}\sigma r^2 \sqrt{\frac{a}{r}} \cos^3 \frac{\theta}{2}$$

in relation to the boundary conditions at the surfaces of the crack.

9.2 Calculate the theoretical stress concentration factor of an elliptical hole with semimajor and semiminor axes 50 and 10 mm, respectively, subjected to a tensile load perpendicular to the major axis. If the hole semiminor axis becomes 5 mm, what is the stress concentration factor?

9.3 Calculate the fracture stress in the two cases: (a) an elliptical hole with 60 and 15 mm semimajor and semiminor axes, respectively; (b) a central crack of 15 mm length. Take a yield stress $Y = 350$ MPa and $K_c = 120$ MPa · m$^{1/2}$.

9.4 Determine the stresses and (σ_1, σ_2) in the vicinity of the tip of a crack of length $a = 5$ mm for an applied stress of 140 MPa under the following conditions: $\theta = 30°$ and $60°$ for both radii $r = 0.1$ and 1 mm.

9.5 If a structural member containing a crack 5 mm long is subjected to an applied stress of 300 MPa, calculate the normal stress σ_y produced on the crack plane ($\theta = 0$) at $r = 1$ mm, $r = 2$ mm, and $r = 20$ mm. Comment on the results.

9.6 If a large metallic plate containing a crack 8 mm long is subjected to an applied tensile stress of 300 MPa, calculate the radii at which the normal stresses produced on the crack plane reach the values 300 and 150 MPa. Comment on the results.

Stress Intensity Factors and Fracture Toughness

9.7 Nondestructive quality control detects cracks of length less than 3.5 mm in a certain component in the form of a large plate subjected to a tensile stress of 400 MPa. Determine which one of the following alloys

* It has been pointed out that the critical stress should be lower than $2Y/3$ for LEFM to apply. See Feddersen,[48] pp. 50–78.

is suitable for the component: aluminum alloy (K_c = 25 MPa · m$^{1/2}$, yield strength = 480 MPa · m$^{1/2}$) and titanium alloy (K_c = 70 MPa · m$^{1/2}$, yield strength = 910 MPa).

9.8 Determine the critical crack size that will cause fracture in a structural member in the form of a large plate 200 mm wide and 5 mm thick. The applied tensile stress is 700 MPa and the critical stress intensity is 100 MPa · m$^{1/2}$.

9.9 A polymeric sheet 120 mm wide, 4 mm thick, and 300 mm long contains a sharp single-edge crack 8 mm long. If the critical stress intensity factor is 2.0 MPa · m$^{1/2}$, what is the maximum axial force that could be applied without causing brittle fracture?

9.10 An alloy has a fracture toughness of 77 MPa · m$^{1/2}$ and a yield strength of 525 MPa. Determine the strength of a thick plate of 50 mm width with a center crack of $2a$ = 25 mm.

9.11 Given two components subjected to the same tensile stress — the first component is made of alloy A with K_{Ic} = 85 MPa · m$^{1/2}$, while the second is made of alloy B with K_{Ic} = 55 MPa · m$^{1/2}$). (a)Which component can tolerate the largest cracklike flaw before fracture? (b) Calculate the ratio of flaw sizes, i.e., a_A/a_B.

9.12 A thermoplastic polymer below its glass transition temperature has a surface energy γ_s of 50 J/m^2 and a modulus of elasticity of 2.7 GPa. After injection molding, a central flaw 1 mm long can be seen. (a) Calculate the fracture stress. (b) What flaw size would be required at fracture if the tensile strength is 55 MPa?

9.13 Two rectangular flat strips fabricated from two different steels A and B carry equal tensile loads and were originally hardened by quenching such that Y_a/Y_b = 1.8 and $(K_{IC})_A/(K_{IC})_B$ = 1.55. The thickness of each member has been adjusted so that each sustains a stress equal to 70% of its respective yield strength. For each strip, what is the longest ratio of the edge crack that can be tolerated without causing fracture?

9.14 A shipbuilding steel has a value of K_{Ic} = 130 MPa · m$^{1/2}$. (a) What is the fracture stress in thin plates 300 mm wide that contain a central crack 10 mm long? (b) If the crack is 20 mm long, what is the fracture stress? (c) Increasing the plate thickness reduces K_{Ic} to 100 MPa · m$^{1/2}$. What is the fracture stress for a crack 10 mm long?

9.15 A plate subjected to a tensile stress of 300 MPa has a central transverse crack through its thickness. The width is W = 300 mm and the fracture toughness is K_{Ic} = 120 MPa · m$^{1/2}$. Estimate the critical crack length. Does crack tip plasticity affect the result if the yield strength is 520 MPa?

9.16 A thin plate made of an aluminum alloy with a yield stress of 480 MPa fails in service at a stress of 110 MPa with some evidence of ductility at the fracture. If a surface crack 20 mm long is observed at the fracture plane, calculate the fracture toughness of this plate using LEFM, as well as correction for crack tip plasticity.

9.17 A designer expects penny-shaped internal cracks of the order of 10 mm diameter to exist in a large medium-carbon steel crane hook. Determine the fracture toughness required from this steel if the safety factor is to be 2. The yield strength of the steel is 1050 MPa.

9.18 A simply supported beam of rectangular cross section has a depth of 250 mm, a thickness of 325 mm, and a length L of 1 m. The beam is loaded with a concentrated load P at the center. A crack of 15 mm length exists into the beam on the tension side opposite to the point of application of P. The beam is made of 17-7PH stainless steel (Table 9.1). (a) Determine the fracture load P. (b) Are plane strain conditions satisfied for the beam?

9.19 A steel alloy for a high-strength part to be loaded as a simply supported beam of square cross section and length L/W = 4. The smallest edge crack that can be detected is 2.5 mm. This steel can be heat-treated to produce different properties. In one case (A), it exhibits a yield strength of Y = 1090 MPa and K_{Ic} of 42

MPa · $m^{1/2}$, while in the second case (B), Y is 1385 MPa and K_{Ic} is 28 MPa · $m^{1/2}$. Which treatment should be used for minimum weight to support a midspan load of 5 kN?

9.20 A sheet of glass 0.5 m wide and 20 mm thick is found to contain a number of surface cracks 3 mm deep and 10 mm long. If the glass is placed horizontally on two supports, calculate the maximum spacing of the supports to avoid fracture of the glass due to its own weight. For glass, K_{Ic} = 0.4 MPa · $m^{1/2}$ and density = 2600 kg/m^3.

9.21 A ceramic bend-specimen with a sharp crack on the tension side at midlength is fabricated and tested in three-point bending. The specimen has the dimensions: depth W = 40 mm, thickness h = 25 mm, span length $2L$ = 160 mm, crack length a = 12 mm. Determine the fracture toughness of this ceramic material if the fracture load is found to be 2.5 kN.

9.22 A bar of brittle material of rectangular cross section 20 × 10 mm and length 160 mm is loaded in pure bending by a moment M = 38.5 N·m.

a. If the bar contains a crack 5 mm deep, what is the stress intensity factor K_I for such a configuration?
b. If a bar of the same material and dimensions is subjected to pure bending M and axial pull P has a crack 7 mm deep, what is the safe bending load that may be applied to avoid fracture, knowing that P/M = 4?

9.23 Determine the maximum midspan load for three-point bending of a beam of 20 mm depth and 10 mm thickness containing a sharp notch of 5 mm. The beam span is 80 mm, K_c = 65 MPa · $m^{1/2}$, and Y = 750 MPa. Comment on the applicability of LEFM.

9.24 A cylindrical pressure vessel having a diameter of 6 m and wall thickness of 25 mm failed catastrophically when the internal pressure reached 1.7 MPa. The vessel material had an elastic modulus of 200 GPa, a yield strength of 650 MPa, and a value of K_{Ic} of 100 MPa · $m^{1/2}$. Determine the size of the crack that might have caused this failure. Does consideration of crack tip plasticity affect the answer?

9.25 A portion of a gas valve made of gray cast iron has the shape of a tube of 125 and 105 mm outside and inside diameters, respectively. If a radial crack 50 mm in the axial direction and 5 mm deep in the radial direction extends inward from the outer surface, estimate K_c of this cast iron if it fractured under an internal pressure of 5 MPa.

9.26 A cylindrical pressure vessel 8 m long, 4 m in diameter exploded at an internal pressure of 10 MPa. The vessel is made from 20-mm-thick steel with the following properties: E = 210 GPa; Y = 890 MPa; and K_{Ic} = 90 MPa · $m^{1/2}$. If failure is due to the existence of a longitudinal crack, what would be the size of such crack?

9.27 A long, closed-ended cylinder is made of an aluminum alloy 7075-T6. The cylinder has an inside diameter of 1000 mm, a wall thickness of 20 mm, and is subjected to an internal pressure of 6 MPa. (a) Determine the size of a longitudinal surface elliptical crack to cause fracture at this pressure assuming conditions of plane strain. (b) If the crack is circumferential instead of longitudinal, determine the internal pressure that will cause fracture for a crack length of 20 mm. Investigate the leak-before-break conditions.

9.28 A sample piece of seam-welded aluminum pipe, 500 mm in diameter and 5 mm thick, is tested for longitudinal residual stress by application of a strain gauge. When a narrow ring is cut as indicated to remove the stress, the strain gauge reading is 1800 μm. If these pipes are used under internal pressure of 4 MPa, what would be the size of the largest permissible circumferential elliptical cracks in the welds? Take K_{Ic} = 18 MPa · $m^{1/2}$, Y = 420 mPa, E = 70 GPa and ν = 0.3.

9.29 A long, thick-walled cylinder has an outside diameter of 160 mm and inside diameter of 80 mm. The cylinder is subjected to an internal fluid pressure of 50 MPa at an elevated temperature of 200°C while its outer surface is kept at 0°C. Inspection indicated that a radial crack of 10 mm length exists at the outside surface of the cylinder. Assuming steady-state conditions, what are the steel properties required such that neither yielding at the outer surface nor fracture occurs? Use a factor of safety not less than 2 and take E = 200 GPa, ν = 0.3, and $\alpha = 12 \times 10^{-6}$/°C. Take CCF = 1, for thermal loading.

9.30 A spherical pressure vessel fabricated by welding contains nonthrough cracks of semielliptical shape ($a/2c$ = 0.5), which were detected by radiograph inspection, 2.5 mm long. Two grades of steel with the properties — (a) grade A: K_{Ic} = 60 MPa · m$^{1/2}$, Y = 1400 MPa and (b) grade B: K_{Ic} = 45 MPa · m$^{1/2}$, Y = 1870 MPa — are being considered. If the wall thickness of the vessel is 25 mm and the diameter is 3 m, which steel will give the greatest safe pressure?

9.31 A solid disk of radius 400 mm is designed, assuming a crack to exist at its center. What is the critical crack length if the disk runs at 10,000 rpm? The material properties are ν = 0.3, ρ = 7900 kg/m^3, and K_{Ic} = 60 MPa · m$^{1/2}$.

9.32 A thin solid disk of outside radius r_o = 400 mm is made up by shrinkage of two disks with the radius of separation r_c such that $\lambda = r_o/r_c = 4$. The disk is rotating at a speed of 4500 rpm. The material properties of the disks are specific weight = 80 kN/m^3, E = 200 GPa, ν = 0.3, Y = 950 MPa, K_{Ic} = 50 MPa · m$^{1/2}$. Determine the value of the initial radial interference (Δ/r_c) such that none of the following incidents happen due to rotation: (a) the outer annulus does not get loose from the inner disk; (b) fracture does not occur due to a radial crack 6 mm long at the outer rim of the assembly; (c) yielding does not occur at the inner radius of the outer disk. Take CCF = 1.23.

Mixed Mode Loading and Superposition

9.33 A hard aluminium bar has a rectangular cross section of width 500 mm, thickness 50 mm, and length 1500 mm. The material properties are K_{Ic} = 25 MPa · m$^{1/2}$ and Y = 480 MPa. A through edge crack of 40 mm exists at the midlength of the bar. Determine the failure load P for this bar if (a) the load P is applied along the bar centerline and (b) the load P is applied along a centerline passing through the crack root.

9.34 A member with a crack deforming in mode (II) is subjected to a shear stress τ = 150 MPa as well as a tensile stress σ. The fracture toughness of the material is 90 MPa · m$^{1/2}$. Assuming CCF = 1; calculate the allowable tensile stress if the crack is of 50 mm length.

9.35 Fracture toughnesses K_{Ic} and K_{IIc} for a given material are to be determined by experiment. The experiment is performed by testing to fracture two internally large cracked plates, which are identical except for the crack orientation (the first with α = 0 and the second with α = 45°; see the figure) are tested up to fracture to determine K_{Ic} and K_{IIc}. The observed critical stresses are 120 and 135 MPa, respectively, with $2a$ = 40 mm and B = 10 mm. Determine K_{Ic} and K_{IIc} knowing that Y = 600 MPa.

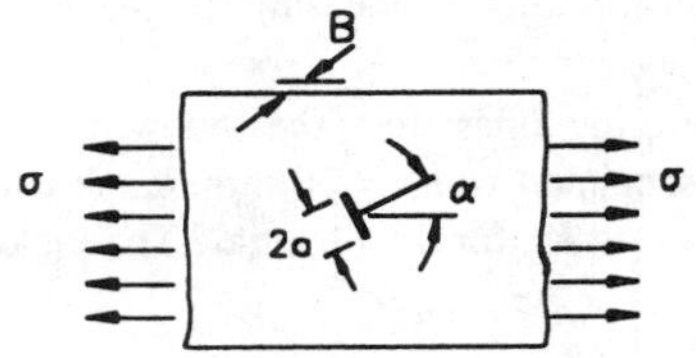

Problem 9.35

9.36 A circular cylindrical bar of diameter 120 mm simultaneously carries an axial tensile force P and a torsional moment M_t (see the figure). A surface circumferential precrack of depth a = 3 mm is found. Determine the value of P at fracture when M_t/P = 10, K_{Ic} = 60 MPa · m$^{1/2}$, and ν = 0.3.

9.37 A crack has been detected in the shell of a large, thin cylindrical steel pressure vessel. The stresses in the region of the crack are as follows: longitudinal tensile stress of 300 MPa and circumferential tensile stress of 550 MPa. The crack is oriented so that it makes an angle of 60° with the longitudinal direction. If K_{Ic} for

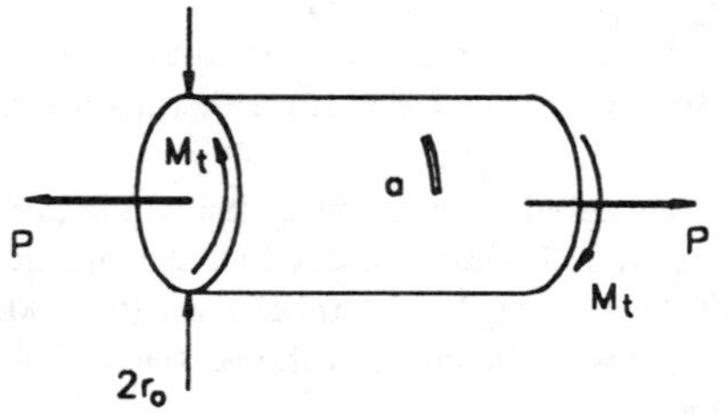

Problem 9.36

steel is 55 MPa · m$^{1/2}$, what is the smallest crack size that will cause fracture of this pressure vessel? Assume CCF = 1.

Fracture Toughness Testing

9.38 A compact tension specimen is tested according to ASTM standards. A load P_Q = 140 kN is determined from the load–displacement curve. The specimen dimensions are B = 50 mm, W = 100 mm, and a = 50 mm. If the material yield stress is 630 MPa, evaluate K_Q and check if the conditions are correct for a valid K_{Ic} estimation.

9.39 A compact tension specimen is made from an aluminum alloy with dimensions as follows: W = 50 mm, B = 30 mm, and a = 20 mm. Determine the fracture toughness (K_Q) for this material if P_Q = 22 kN. Knowing that the yield strength of the alloy is 450 MPa, does K_Q represent a valid K_{Ic}?

9.40 The load–displacement values given in the table below are recorded in a fracture toughness test performed on a center-cracked specimen with the following dimensions: W = 75 mm, B = 12.5 mm, and a = 20 mm. Determine K_Q for this material. If the yield strength is 750 MPa, does K_Q represent a valid K_{Ic} value?

Displacement (mm)	0	14	30	46	60	80	95	108	120	125	120	105	90
Load (kN)	0	0.25	0.5	0.75	1	1.25	1.5	1.75	2.0	2.25	2.5	2.75	3

9.41 In a test to determine K_{Ic} on a compact tension specimen, the failure load is 25 kN, W = 50 mm, a = 25 mm, B = 25 mm, and Y = 550 MPa. Calculate the plane strain fracture toughness. What is the effect of the plastic zone at fracture? Comment on the result.

9.42 In a plane strain toughness test, a value of K_Q = 55 MPa · m$^{1/2}$ is obtained. Is this a valid value if the material yield strength is 700 MPa and the specimen thickness B is 12 mm and W = 25 mm? If the toughness is not valid, suggest any modifications required to measure the plane strain toughness.

9.43 In a three-point bending fracture toughness test, the following data was recorded: support span = 200 mm, specimen thickness = 25 mm, specimen width = 50 mm, crack length = 20 mm, and fracture load = 10 kN. Are these data suitable for measuring K_{Ic} for the material? The yield strength of the material is 350 MPa.

Fatigue Crack Growth

9.44 Fatigue crack growth data as obtained from testing a compact tension specimen are given in the table. What is the crack growth-rate exponent in the Paris law: $da/dN = C(\Delta K)^m$. Determine the other material parameter C.

da/dN(m/cycle)(10^{-5})	0.68	0.74	1.0	1.5	2.0	3.8	9.1	20	33
ΔK(MPa · m$^{1/2}$)	9.2	9.9	11.2	12.1	13.3	14.5	15.7	17.6	20.0

9.45 Crack growth tests on a grade of polymethyl-methacrylate (PMMA) gave the following results:

da/dN(m/cycle)(10^{-5})	2.28	4.1	6.3	11.2	17.05	29.4
ΔK(MPa · m$^{1/2}$)	0.45	0.53	0.62	0.79	0.96	1.18

If the material has a fracture toughness of 2.0 MPa · m$^{1/2}$ and it is known that the fabrication process on PMMA produces defects 50 μm long, determine the maximum repeated tensile stress that could be applied to this material for at least 10^6 cycles without causing fatigue failure.

Safe-Life Predictions

9.46 A high-strength steel has a yield strength of 700 MPa and a fracture toughness K_{Ic} of 130 MPa · m$^{1/2}$. Based on nondestructive inspection, the smallest-size cracklike flaw that can be detected is 1 mm as a single-edge notch. The structure is subjected to cyclic fatigue loading in which σ_{max} = 300 MPa and σ_{min} = 170 MPa. The fatigue crack growth rate for steel is given by $da/dN = 1.3 \times 10^{-10} (\Delta K)^{225}$. Estimate the fatigue life of the structure.

9.47 A certain grade of PMMA has a K_c value of 1.6 MPa · m$^{1/2}$ and it is known that cracks grow at a rate given by ($2 \times 10^{-6}\Delta K^{3.32}$). If the intrinsic cracklike defects in the material are 60 μm long, how many cycles will a large sheet from this material stand before fracture if it is subjected to a stress cycle of 0 to 10 MPa?

9.48 A steel plate contains a crack of 0.8 mm centrally located within the width of 150 mm. An alternating tensile stress (140 to 280 MPa) acts on this plate perpendicular to the crack. This material has a yield strength of 550 MPa a K_{Ic} of 150 MPa · m$^{1/2}$ and obeys a crack growth-rate behavior according to $da/dN = 6 \times 10^{-9}(\Delta K)^3$ (in N, m, cycle).

a. Determine the crack length a_f that leads to failure.
b. Plot crack size vs. total number of elapsed cycles up to a crack length 0.8 a_f.

9.49 A plate of 200 mm width and 400 mm length is subjected to an alternating longitudinal stress that causes an initial edge crack of length 5 mm located at about the midheight, to propagate at right angles to the alternating load. When the crack is 50 mm long, the plate fractures. Based upon a crack propagation relation $da/dN = 0.86 \times 10^{-10}(\Delta K)^{2.25}$; a fracture toughness; K of 150 MPa · m$^{1/2}$, and a yield strength, Y of 600 MPa; determine the number of cycles to failure.

9.50 A wing panel for a supersonic airplane made of titanium alloy is 4 mm thick, 3 m long, and 3 m wide. A maximum cyclic tensile stress in the length direction of 600 MPa causes gradual growth of a small preexisting transverse flaw midway along one edge of the plate. If the flaw is initially 0.5 mm long and grows at a rate of 1.2×10^{-7} m/cycle, calculate the number of stress cycles prior to catastrophic fracture of the panel. K_{Ic} for this aluminum alloy is 50 MPa · m$^{1/2}$ and the yield strength is 1035 MPa.

9.51 A structural member in a bridge suffers from a tensile cyclic stress varying between 300 and 0 MPa. Estimate the number of loading cycles this member can withstand before failing under the following conditions: initial and final crack lengths, 1.5 and 25 mm, respectively; crack growth parameters, m = 3.5 and $C = 2.5 \times 10^{-12}$ (in N, m, cycle). Take an average CCF = 2.0.

9.52 A steel pipe of 300 and 270 mm outer and inner diameters, respectively, is subjected to internal pressure that fluctuates between 50 and 15 MPa. An initial radial crack of 1 mm length is detected at the surface of the pipe. This material has a yield strength of 850 MPa; a fracture toughness, K_{Ic} of 110 MPa · m$^{1/2}$; and a crack growth-rate according to the Forman Equation 9.30, where $c_F = 1.5 \times 10^{-8}$ and m_F = 2.8 (in m, N, cycle). Determine the total number of cycles to failure.

9.53 In a circular pipe of wall thickness 100 mm and inner radius of 400 mm, an inside radial crack of depth 10 mm is detected. A net internal pressure p pulsating between the values p and $p/2$ causes crack growth.

a. Determine the maximum p for which the crack does not propagate knowing that the threshold stress intensity equals $\Delta K = 4$ MPa · m$^{1/2}$.
b. If the pressure varies between 10 MPa and 5 MPa, determine the number N_f of pressure cycles before fracture, if $m = 3$, $C = 10^{-12}$ (in MPa · m) and $K_{Ic} = 40$ MPa · m$^{1/2}$. Apply the Paris law.

9.54 Two tubes both of 160 and 128 mm outer and inner diameters are made of aluminum alloy, for which the properties are yield strength 300 MPa, $K_{Ic} = 40$ MPa · m$^{1/2}$, and $da/dN = 15 \times 10^{-11} (\Delta K)^3$ (in MPa · m cycle). Inspection revealed that one of the tubes is defective and contains an internal radial crack of 4 mm length while the other has no detectable defect.

a. Determine the maximum internal pressure the tubes can withstand without yielding (with a factor of safety 1.5).
b. Determine the maximum internal pressure the defective tube can withstand without fracture (with a factor of safety 2).
c. For the defective tube, how many fatigue cycles result in fracture if the service internal pressure fluctuates between 20 and 0 MPa (with a factor of safety of 2).

9.55 A thin solid turbine disk of outer radius 500 mm rotating at 5000 rpm is made of a metallic alloy with the following properties: Yield strength $Y = 500$ MPa, $\nu = 0.3$, specific weight = 78 kN/m^3, $K_{Ic} = 25$ MPa · m$^{1/2}$, and $da/dN = 14 \times 10^{-11} (\Delta K)^{3.54}$ (in MPa · m cycle). Inspection indicated a central crack of length 1.5 mm. Investigate the margins of safety against the following modes of failure: (a) yielding, (b) brittle fracture, (c) fracture due to 5000 repeated on/off cycles of loading to full speed.

Stress Corrosion Cracking

9.56 For a crack of $a = 2.5$ mm located at the center of a large plate with a $K_{ISCC} = 16$ MPa · m$^{1/2}$, what is the highest permissible stress for this configuration, such that stress corrosion cracking is prevented?

9.57 The K_{ISCC} for a particular metallic alloy is 17 MPa · m$^{1/2}$. What is the largest crack that will not propagate by stress corrosion, given that this crack is located in the middle of a large plate subjected to a static stress of 180 MPa?

9.58 A material with $K_{Ic} = 35$ MPa · m$^{1/2}$ and $Y = 510$ MPa is subjected to stress corrosion tests in seawater. Three specimens with a single edge crack are loaded in uniform tension at loads of 85, 115, and 140 kN, respectively. All cracks are 12 mm long; $W = 75$ mm and $B = 12$ mm. The times to failure of the three specimens are, respectively, 500, 90, and 4 h. Estimate K_{ISCC} and calculate the amount of crack growth that occurred in each of the tests. Assume CCF = 1.12 throughout.

Elastic–PLastic Fracture Mechanics

9.59 Determine J_Q and check if it qualifies to be considered as J_{Ic} for the aluminum alloy tested according to ASTM E813–87. The following data are recorded

J, kJ/m^2	98	179	181	225	258	307
Δa, mm	0.32	0.45	0.87	1.24	1.68	1.81

The three-point bend specimens used have the following dimensions $W = 50$ mm, $B = 25$ mm, and $a_o = 20$ mm. For this aluminum alloy $E = 70$ GPa, $Y = 320$ MPa, UTS = 470 MPa, and $\nu = 0.3$.

9.60 A beam of rectangular cross section of $B = W = 200$ mm is subjected to three-point bending. A crack of $a/W = 0.25$ exists at the beam midspan where a concentrated load P acts. The beam material has the following properties: $Y = 410$ MPa, $E = 200$ GPa, $K = 130$ MPa, and $\nu = 0.03$. Fitting the Ramberg–Osgood expression to the stress–strain curve resulted in $\alpha = 1.6$ and $n = 7.5$. Investigate the possibility of fracture due to either ductile crack growth initiation or plastic collapse.

9.61 A pressurized pipe of 400-mm inner radius and 80-mm thickness has an internal radial crack of 16-mm depth. The material of this pipe is known to be tough with the following proprties: $Y = 285$ MPa, $E = 203$ GPa, $\alpha = 1.27$, $n = 9.1$, $J_{Ic} = 230$ kJ/m^2. Find a relation between J and the internal pressure P_i. At what pressure is ductile crack growth iniated?

9.62 A flat plate of 300-mm width and 20-mm thickness contains a central through-crack 40 mm long. If the plate is subjected to a remote tensile stress σ, what would be the safe design limit for this stress to avoid fracture? Take $\delta_c = 0.15$ mm, $E = 70$ GPa, $Y = 190$ MPa.

9.63 Solve Problem 9.61 employing the COD fracture criterion with $\delta_c = 0.45$ mm.

REFERENCES

1. Griffith, A. A., The phenomena of rupture and flow in solids, *Philon. Trans. R. Soc.*, London, 221A, 163–198, 1920. Also *Trans. Am. Soc. Met.*, 61, 781–906, 1968.
2. Felbeck, D. K. and Atkins, A. G., *Strength and Fracture of Engineering Solids*, Prentice-Hall, Englewood Cliffs, NJ, 1984, 330.
3. Parker, A. P., *The Mechanics of Fracture and Fatigue,* E. & F. N. Spon, London, 1981, 35–37.
4. Williams, J. G., *Stress Analysis of Polymers,* Longman, London, 1973, p. 240.
5. Sneddon, I. N., The distribution of stress in the neighbourhood of a crack in elastic solids, *Proc. R. Soc.*, A187, 229–260, 1946.
6. Williams, M. L., *J. Appl. Mech.,* 24, 109–114, 1957.
7. Hellan, K., *Introduction to Fracture Mechanics*, McGraw-Hill, New York, 1985, 80.
8. *Standard Test Method for Plane Strain Fracture Toughness of Metallic Material,* ASTM Designation E 399-74, Part 10, ASTM Annual Standards, Philadelphia, 1977.
9. Hertzberg, R. W., *Deformation and Fracture Mechanics of Engineering Materials*, John Wiley & Sons, New York, 1976, 286, 370.
10. Wu, E. M., Application of Fracture Mechanics to Anisotropic Plates, in *Composite Materials Workshop,* Tsai, S. W., et al., Eds., Technomic Publishing, Westport, CN, 1968, 20-43.
11. Irwin, G. R., Analysis of stresses and strains near the end of a crack transverse a plate, *Trans. Am. Soc. Mech. Engr., J. Appl. Mech.*, 24, 361, 1957.
12. Brown, W. F., Jr. and Srawley, J., Plane strain crack toughness testing of high strength metallic materials, ASTM, Special Technical Publication No. 410, Philadelphia, 1966.
13. Irwin, G. R., The crack extension force for a part-through crack in a plate, *Trans. ASME, J. Appl. Mech.*, 29, 651–654, 1962.
14. Broek, D., *Elementary Engineering Fracture Mechanics*, 4th ed., Martinus Nijhoff Publishers, The Netherlands, 1986, Chap. 14.
15. Rooke, D. P. and Cartwright, D. J., *Compendium of Stress Intensity Factors*, H.M.S.O., Uxbridge, The Hillingdon Press, London, 1976.
16. Hall, L. R. and Finger, R. W., Fracture and fatigue growth of partially embedded flaws, in *Proc. Air Force Congress* (1969), AFFDL TR70-144, 235–262, 1970.
17. Bowie, O. L., Analysis of an infinite plate containing radial cracks originating at the boundaries of an internal circular hole, *J. Math. Phys.*, 35, 60, 1956.
18. Tada, H., Paris, P. C., and Irwin, G. R., *The Stress Analysis of Cracks Handbook,* Del Research Corporation, Hellertown, PA, 1973.
19. Benthem, J. P. and Koiter, W. T., Asymptotic approximations to crack problems, in *Methods of Analysis of Crack Problems,* Sih, G. C., Ed., Noordhoff Int., 1972, Chap. 3.
20. Murakami, Y., Ed., *Stress Intensity Factors Handbook*, Vol. 1, Pergamon Press, Oxford, 1987, p. 321.
21. Rooke, D. P. and Tweed, J., The stress intensity factor for an edge crack in a finite rotating elastic disc, *Int. J. Eng. Sci.,* 11, 279–283, 1973.
22. Dugdale, D. S., Yielding of steel sheets containing slits, *J. Mech. Phys. Solids*, 8, 100–108, 1960.
23. Irwin, G, R., *Fracture; Handbuck der Physik*, IV, Springer-Verlag, Heidelberg, 1958, 551–590.
24. Methods of test for plane strain fracture toughness of metallic materials, BS5447, British Standard Institute, 1997.

25. Broek, D., *The Practical Use of Fracture Mechanics,* Kluwer Academic, Boston, 1989.
26. Ritchie, R. O., Near threshold fatigue crack propagation in steel, *Int. Matr. Rev.,* 24, 205–230, 1979.
27. Paris, P. C., Fatigue — an interdisciplinary approach, in *Proc. 10th Sagamore Conf.,* Syracuse University Press, Syracuse, NY, 1964, 107.
28. Forman, R. G., Kearny, V. E., and Engle, R. W., Numerical analysis of exact propagation in cyclically loaded structures, *Trans. ASME, J. B. Basic Eng.,* 89, 459–464, 1967.
29. ASTM Designation E-647-90, *Annual Standards,* American Society for Testing and Materials, Philadelphia, 1990.
30. Fatigue crack growth: effect of sheet thickness, *Aircraft Eng.*, 38(11), 31–33, 1966.
31. Gurney, T. R., *Fatigue of Welded Structures,* 2nd ed., Cambridge University Press, New York, 1979.
32. Rice, J. R., Progress in Flow Growth and Fracture Toughness, ASTM STP-536, American Society for Testing and Materials, Philadelphia, 1973, 231.
33. Tanaka, K., Fatigue crack propagation from a crack inclined to the cyclic tensile axis, *Eng. Fract. Mech.,* 6, 493–507, 1974.
34. Miller, K. J., The behavior of short fatigue cracks and their initiation, Part II: A general summary, *Fatig. Fract. Eng. Mater. Struct.,* 10, 93, 1987.
35. Miller, K. J., The short crack problem, *Fatig. Fract. Eng. Mater. Struct.*, 5, 223, 1982.
36. Elber, W., Fatigue crack closure under cyclic tension, *Eng. Fract. Mech.*, 2, 37, 1970.
37. Alawi, H., Ragab, A., and Shaban, M., Stress corrosion cracking of some steels in various environments, *Eng. Fract. Mech.,* 32L(1), 29–37, 1989.
38. Rice, J. R., A path independent integral and the approximate analysis of strain concentrations by notches and cracks, ASME, *J. Appl. Mech.*, 35, 379–386, 1968.
39. E813–87, Standard test method for J_{IC}: a measure of fracture toughness, ASTM, Philadelphia, 1987.
40. Hutchinson, J. W., Singular behaviour at the end of a tensile crack in a hardening material, *J. Mech. Phys. Solids*, 16, 13–31, 1968.
41. Kumar, V., German, M. D., and Sih, C. F., An Engineering Approach for Elastic–Plastic Fracture Analysis, EPRI Report NP-1931, Electric Power Research Institute, Palo Alto, CA, 1981.
42. Andreson, T. L., *Fracture Mechanics: Fundamentals and Applications,* CRC Press, Boca Raton, FL, 1991.
43. Wells, A. A., Unstable crack propagation in metals; cleavage and fast fracture, in *Proc. Crack Propagation Symposium*, Vol. 1, paper 84, Cranfield, U.K. 1961.
44. Burdekin, F. M. and Stone, D. E. W., The crack opening displacement approach to fracture mechanics in yielding, *J. Strain Analy.*, 1, 145–153, 1966.
45. E1290–89, Standard test method for crack tip opening displacement testing, ASTM, Philadelphia, 1989. Also, 36 BS 5762: 1979, Methods for crack opening displacement (COD) testing, British Standard Institution, London, 1979.
46. PD 6493: 1980, Guidance on some methods for the derivation of acceptance levels for defects in fusion welded joints, British Standards Institution, March 1980.
47. Fedderson, C. E., Evaluation and prediction of the residual strength of center-cracked tension panels, ASTM STP 486, Philadelphia, 1971.

10 Plastic Deformation

The fundamentals of plasticity are set out and explained in detail. Yield criteria according to Tresca and von Mises are formulated for homogeneous isotropic solids and applied to engineering problems. Plastic stress–strain relations are expressed by the flow rule, and the material flow curve is represented by the relation between the effective stress and effective strain for isotropic hardening. Kinematic hardening is introduced and applied to uniaxial and triaxial cyclic loading. The yield criterion and the associated flow rule for anisotropic solids are given according to Hill's formulation. A yield criterion for porous solids is developed and applied to powder compacts.

10.1 INTRODUCTION

The linear relation between stress and strain, which describes the behavior of an elastic solid under load, holds up to a certain limit. Beyond this limit the deformation is not fully recovered upon load removal, indicating that the solid has undergone plastic deformation. Most engineering constructions are designed to support service loads within the elastic regime of the materials used. Nevertheless, the analysis of plastic deformation is required in several engineering applications, as indicated by the following:

- Although machine and structural components are designed to behave elastically, localized plastic deformation does occur at locations where stress concentrations are inevitably encountered. At holes, notches, supports, etc. contained plastic deformation is observed.
- Plastic deformation is utilized to produce favorable residual stresses in many mechanical parts to improve the elastic stress distribution and, hence, increase the load-carrying capacity, e.g., autofrettage of gun barrels.
- In the analysis of components, which inevitably contain preexisting crackslike defects, the elasticity theory predicts infinite stresses at the crack tip. However, this cannot be physically feasible and the material will yield in a small zone ahead of the crack tip. This constitutes the subject of elastoplastic fracture mechanics.
- In components subjected to cyclic loads, limited plastic deformation can be allowed to occur during the first few cycles of load. Such behavior, which ensures cessation of plastic deformation in the long term, is known as elastic shakedown, and is utilized in the design of many critical components, e.g. in the nuclear power industry, in thermal power generation, and in the aeronautical and aerospace industries. The phenomenon of accumulated plastic deformation due to reversed cyclic loading has to be studied. This is essential for a proper design against low-cycle fatigue failures (i.e., life less than 10^3 to 10^4 load cycles).
- Analysis of ductile failure of components requires the determination of the maximum load sustained before fracture. This is often taken as the load at which the material becomes fully plastic, and such loads are termed limit loads.

- The assessment of material behavior by conventional tests requires good understanding of plastic deformation. Ultimate tensile strength, hardness numbers, creep strength, low cycle fatigue, etc. are determined by testing materials within the plastic deformation regime.
- The analysis and design of metal-forming processes such as forging, extrusion, rolling, drawing, etc. cannot be worked out rationally unless the plastic behavior of metals is thoroughly understood.

10.2 BASIC ASSUMPTIONS

To formulate a theory for plastic deformation of engineering materials, certain assumptions have to be made. The basic assumptions commonly related to metal plasticity are as follows:

1. The material is assumed to be homogeneous and isotropic before and throughout plastic deformation. As explained before, homogeneity means that material properties are identical at any point within the solid. Isotropy requires that these properties are independent of direction. For instance, in a rolled steel sheet, the tensile stress–strain behavior is assumed to be identical for all tensile test specimens taken along any direction in the rolled sheet. This assumption does not hold in many cases, specifically for cold worked metals, and such deviations are handled through an extension of a plasticity theory for anisotropic materials.
2. The volume of the material undergoing plastic deformation is assumed to be constant. Such assumption is validated through experiments, where moderate and high pressures during forging of copper and steel resulted in almost no permanent volume decrease. The constancy of volume condition requires that

$$d\varepsilon_v = d\varepsilon_1^p + d\varepsilon_2^p + d\varepsilon_3^p = d\varepsilon_x^p + d\varepsilon_y^p + d\varepsilon_z^p = 0$$

or

$$\dot{\varepsilon}_v = \dot{\varepsilon}_1^p + \dot{\varepsilon}_2^p + \dot{\varepsilon}_3^p = \dot{\varepsilon}_x^p + \dot{\varepsilon}_y^p + \dot{\varepsilon}_z^p = 0 \tag{10.1}$$

which for homogeneous strain yields

$$\varepsilon_v = \varepsilon_1^p + \varepsilon_2^p + \varepsilon_3^p = \varepsilon_x^p + \varepsilon_y^p + \varepsilon_z^p = 0$$

where the superscript p denotes plastic strain.
3. The hydrostatic component of the stress tensor σ_m has no effect on yielding.* This means that, no matter what the value of σ_m is, the limit of elastic deformation, i.e., yield strength, does not change. Yielding is influenced only by the deviatoric stress components of the stress tensor. Hence, a solid subjected to a very high hydrostatic pressure such as when placed on a seabed will only deform elastically without reaching the yield point.
From the elastic stress–strain relations (Equations 3.10), it has been shown that:

$$\varepsilon_v = \frac{3(1-2\nu)}{E}\sigma_m \tag{10.2}$$

* On the other hand, at high pressures, some materials usually rated as brittle, show increasing ductility. A useful application to this is the process of hydrostatic extrusion.

Since $\varepsilon_v = 0$ for plastic deformation, Poisson's ratio $\nu = 0.5$ for all materials within their plastic range. Note that the two conditions, namely, constancy of volume and independence of hydrostatic stress, are consistent with each other. Formulating a plasticity theory for certain behaviors where the two assumptions of constant volume and independence of yielding of hydrostatic pressure do not hold, such as in the case of powder metallurgy, is possible as indicated in Section 10.17.

The constant-volume condition greatly simplifies the analysis of plastic deformation problems. In three-dimensional problems, knowing two strain components, say ε_1^p and ε_2^p, will be enough to determine the third component ε_3^p from Equation 10.1. For instance, in simple uniaxial tension, the true stress is evaluated through measuring the current applied load P and current length L. Assuming isotropy and homogeneous deformation, the constant-volume condition gives the current cross-sectional area as $A = A_0L_0/L$ where A_0 and L_0 are the initial cross-sectional area and initial length of the test bar, respectively. The true stress is thus given by $\sigma = P/A = PL/A_0L_0$.

4. A plastically deformed solid is assumed to have the same yield strength in both tension and compression, whether initially or during all subsequent stages of plastic deformation. Hence, in Figure 10.1 the tensile curve *oab* is identical to the compressive curve *oa'b'*. If the material is loaded in tension beyond its elastic limit until point *b* is reached followed by unloading and reloading in compression, the material will yield at point *c* when the compressive yield strength becomes equal in magnitude to the yield strength in tension Y_1 at point *b*. Experimentally observed material behavior exhibits an unloading–reloading behavior similar to *bc'd'* rather than the curve *bcd* (with $Y_1 > Y_2$). This type of behavior is mainly due to the different states of stress existing in the variously oriented crystals before unloading and is known as the Bauschinger effect.

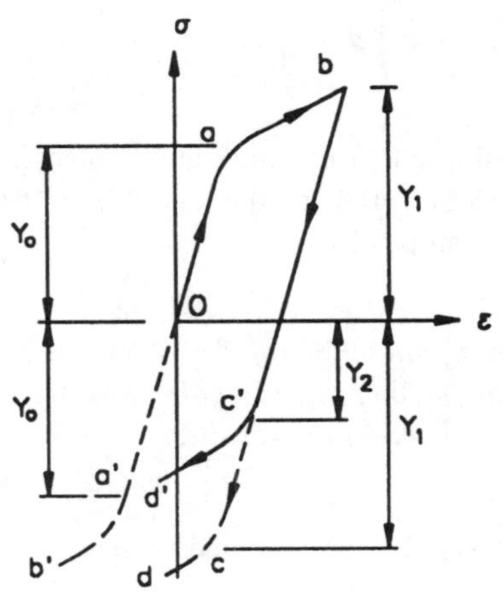

FIGURE 10.1 Stress–strain curve in tension and compression — Bauschinger effect.

The assumption of identical yield strength in tension and compression irrespective of the previous plastic deformation can therefore introduce errors in the analysis of problems involving reversed loading or metal-forming processes involving multistage working. Plasticity theories capable of reproducing the observed Bauschinger effect in an idealized manner can be developed as outlined in Section 10.16.

10.3 DEFINITION OF LARGE PLASTIC STRAINS

The strain definition in Chapter 2, known as the engineering strain, holds only for infinitesimal deformation where second and higher orders are neglected. This is suitable for elastic behavior of metals where strain magnitudes never exceed 0.002. For large plastic deformation where the solid undergoes appreciable changes in dimensions, the logarithmic or natural strain definition has been introduced in Chapter 2.

In simple uniaxial loading of a bar of initial length L_0 and cross-sectional area A_0, the engineering strain increment is expressed in terms of the initial length as

$$de = \frac{dL}{L_o} \tag{10.3}$$

where dL is the incremental change in length. Integration up to a final length L yields

$$e = \int de = \int_{L_o}^{L} \frac{dL}{L_o} = \frac{L - L_o}{L_o} \tag{10.4}$$

Another strain definition stems from relating the incremental change in bar length to the current length L as

$$d\varepsilon = \frac{dL}{L}$$

which gives, upon integration,

$$\varepsilon = \int d\varepsilon = \int_{L_o}^{L} \frac{dL}{L} = \ln\left(\frac{L}{L_o}\right) \tag{10.5}$$

Equation 10.5 defines the logarithmic or true strain and is also termed *natural strain*. Within the plastic regime, large strains are encountered and the logarithmic strain definition is physically more feasible, as explained below in Example 10.1.

Example 10.1:

A bar of length L_0 *is stretched to a final length of* $2L_0$*. What are the values of the engineering and true strains? If the stretched bar is compressed again to its very initial length* L_0*, compute the engineering and true strains. Comment on the results.*

Solution:

Upon stretching, Equations 10.4 and 10.5 are applied to give

$$e = \frac{2L_o - L_o}{L_o} = 1.0$$

$$\varepsilon = \ln\left(\frac{2L_o}{L_o}\right) = 0.693$$

Compressing the bar from $2L_0$ to L_0 yields

$$e = \frac{L_o - 2L_o}{2L_o} = -0.5$$

$$\varepsilon = \ln\left(\frac{L_o}{2L_o}\right) = -0.693$$

where the negative sign holds for compression.

The engineering strain magnitude in stretching is different from that in compression, while the true strain yields the same strain magnitude in both cases. To achieve an engineering compressive strain of –1.0 for the stretched bar, Equation 10.4 gives

$$-1.0 = \frac{L - 2L_o}{2L_o}$$

This requires that $L = 0$, indicating that the bar must be compressed to zero length, which is physically impossible. Clearly, an engineering strain of –1.1 is also unrealistic. In subsequent discussions it will be shown that doubling the length in tension or halving it in compression will cause the same approximate effect on material strength.

The relation between the engineering strain and the natural strain* is obtained from Equations 10.4 and 10.5 as

$$\varepsilon = \ln\left(\frac{L}{L_o}\right) = \ln(1 + e) \tag{10.6}$$

Expansion of the logarithmic function gives

$$\varepsilon = e - \frac{e^2}{2} + \frac{e^3}{3!} + \cdots$$

For small deformations, higher powers can be neglected and hence, $\varepsilon \cong e$. For instance, a 10% elongation in a tensile bar gives $e = 0.1$ and $\varepsilon = 0.095$ with 5% difference in strain magnitude.

Example 10.2:

A bar of 100 mm initial length is elongated to a length of 200 mm by drawing in three stages. The lengths after each stage are 120, 150, and 200 mm, respectively:

a. *Calculate the engineering strain for each stage separately and compare the sum with the total overall value of e.*
b. *Repeat (a) for the true strain ε and comment on the result.*

---

* In order to relate the above natural strain definition to the tensorial strain components, namely,

$$\varepsilon_{ij} = \frac{1}{2}\left(\frac{\partial u_i}{\partial x_j} + \frac{\partial u_j}{\partial x_i}\right), \quad i \text{ and } j = 1,\ 2,\ \text{and } 3$$

the following expression holds (for ε_x as an example): $\varepsilon = \ln\sqrt{2\varepsilon_x + 1}$.

Solution:

a. $$e_1 = \frac{200}{100} = 0.2,\ e_2 = \frac{30}{120} = 0.25,\ e_3 = \frac{50}{150} = 0.33$$

$$e_{\text{sum}} = 0.2 + 0.25 + 0.33 = 0.78$$

but

$$e_{\text{overall}} = \frac{200 - 100}{100} = 1.0$$

b. $$\varepsilon_1 = \ln(120/100) = 0.18,\ \varepsilon_2 = \ln(150/120) = 0.22$$

$$\varepsilon_3 = \ln(200/150) = 0.29$$

$$\varepsilon_{\text{sum}} = 0.18 + 0.22 + 0.29 = 0.69$$

but

$$\varepsilon_{\text{overall}} = \ln(120/100) = 0.69$$

Using true strains, the sum of the strain increments equals the overall strain. This illustrates the additive property of true strains. This is not true for engineering strains.

10.4 STRAIN HARDENING IN SIMPLE TENSION

Consider the true stress–strain diagram in uniaxial tension for a metal alloy below its recrystallization temperature. The relation between the stress and strain is linear up to the initial yield strength Y_o with a slope equal to $\tan^{-1} E$ as shown in Figure 10.2. Loading up to any point Y_1 beyond Y_o

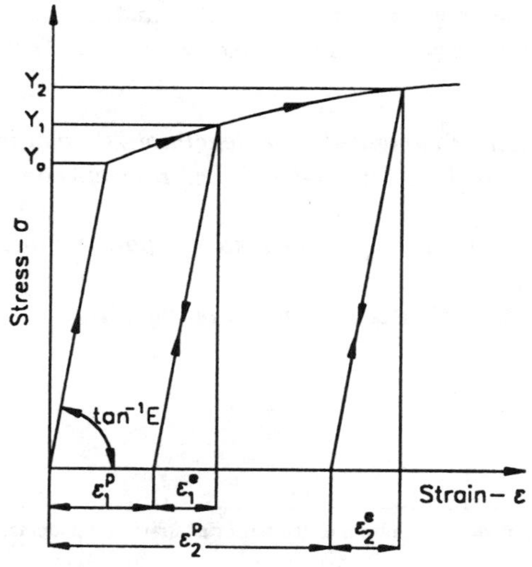

FIGURE 10.2 Effect of unloading and reloading in a tensile test.

results in a total longitudinal strain ε_1 which is the sum of the elastic and plastic components $\varepsilon_1{}^e$ and $\varepsilon_1{}^p$, respectively, i.e.,

$$\varepsilon_1 = \varepsilon_1^e + \varepsilon_1^p$$

Upon load removal, the bar will unload elastically so that the elastic strain will vanish and only the plastic strain remains. The unloading path is then parallel to the initial elastic slope of the stress–strain curve ending at a point ε_1^p on the abscissa, as shown in Figure 10.2. If tensile reloading takes place again, this will follow the same previous unloading curve (assuming no energy dissipation to occur in this process).* The reloading stress–strain relation will then follow a linear-elastic path until it reaches a rather new elastic limit, which is now Y_1. Note that Y_1 is higher than the initial yield value Y_o since the bar in the reloading path starts from a different point which is characterized by a permanent plastic deformation ε_1^p. Repeating the unloading and reloading cycle will define another yield point $Y_2 > Y_1$, and so forth. The plastic stress–strain curve of Figure 10.2 may be thus looked upon as a curve representing successive yield points, i.e., at any strain ε^p, the stress value σ corresponds to a yield stress Y, which should be applied to cause a further increment of plastic strain $d\varepsilon^p$. The requirement to apply higher stresses in order to induce further plastic deformation is due to a material property called *strain hardening*** and may be expressed as

$$\sigma = Y = H\left(\int d\varepsilon^p\right) \tag{10.7}$$

The function H is the strain-hardening function and can be experimentally defined by fitting a suitable empirical relation to the results of a simple tension test.

10.5 EMPIRICAL RELATIONS FOR STRESS–STRAIN CURVES

The actual stress–strain curves for engineering materials as obtained from experiments are expressed by several empirical laws. The stress–strain curve of Figure 10.3 may be thus expressed as

$$\text{elastic part:} \quad \sigma = E\varepsilon^p \tag{10.8a}$$

$$\text{plastic part:} \quad \sigma = K(\varepsilon^p)^n \tag{10.8b}$$

The first expression (Expression 10.8a) is the well-known Hooke's law in the uniaxial stress state. The second is an empirical fitting to the stress vs. plastic strain curve by a power law which is a simplified version of Ludwik's[1] law $\sigma = Y + K(\varepsilon^p)^n$. The coefficients K and n are material parameters determined experimentally for $\varepsilon^p \geq 2\%$ under specific material and test conditions such as temperature, strain rate, microstructure, etc.

The values of n and K for several engineering alloys at room temperature are given in Table 10.1. For the same material, K and n may be different depending on the condition e.g., annealed or cold worked.

A generalized power law suggested by Swift for the plastic part of the uniaxial stress–strain curve is

---

* Hysteresis loops are neglected .

** On the microscopic level, the dislocation pileup is mainly responsible for strain hardening of metals at the early stages of plastic deformation.

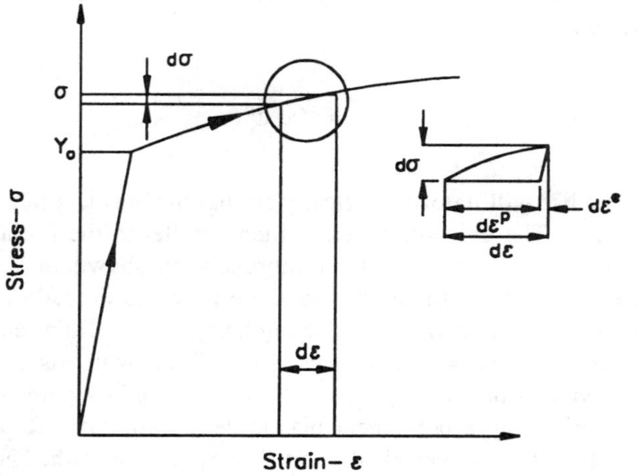

FIGURE 10.3 Stress–strain curve of an elastic strain-hardening material.

TABLE 10.1
Typical Values of K and n in the Law $\sigma = K(\varepsilon)^n$

Alloy	K, MPa[a]	n[a]
Pure lead	25	0.00
Aluminum-1100	180	0.20
Copper, annealed	315	0.54
Brass (70/30), annealed	895	0.49
Low carbon steel, annealed	760	0.19
Low-carbon steel, cold rolled	760	0.08
Medium carbon steel, cold rolled	640	0.15
Stainless steel 17-4 PH, annealed	1200	0.05
Stainless steel 304, annealed	1275	10.45

[a] Values at room temperature.

$$\sigma = K(\varepsilon_o + \varepsilon^p)^n \tag{10.8c}$$

where K, n, and ε_o are material constants. Notably, ε_o represents an amount of prestrain. Hence, Equation 10.8c is said to be more suitable for cold-worked material and reduces to Equation 10.8b for $\varepsilon_o = 0$, i.e., fully annealed material.

In fact, if the material parameters K and n are determined for an annealed material, Expression 10.8b can be used correctly for the same material with any work-hardening history; simply by adding the prior work-hardening strain ε_o to the strain under consideration ε^p and the sum is raised to the power n in Expression 10.8c. An expression such as 10.8c is not thus necessary to be considered.

When the elastic and plastic strains are of comparable magnitudes, Equations 10.8a and 10.8b may be combined to give

$$\varepsilon^e + \varepsilon^p = \varepsilon = \frac{\sigma}{E} + \left(\frac{\sigma}{K^*}\right)^{1/n^*}$$

$$\varepsilon = \frac{\sigma}{E}\left[1 + \frac{E}{K^*}\left(\frac{\sigma}{K^*}\right)^{1/n^*-1}\right]$$

where n^* and K^* are certainly different than n and K fitted over a range of large plastic behavior as given in Table 10.1. This law is known as the Ramberg–Osgood law and is often written as

$$\varepsilon = \frac{\sigma}{E}\left[1 + \alpha\left(\frac{\sigma}{\sigma_o}\right)^{1/n^*-1}\right] \tag{10.8d}$$

For structural alloys such as aluminum and medium carbon steel $(11n^*) \cong 20$ to 30 and $\alpha = 3/7$. Note that the slope of the stress–strain curve at the origin ($\sigma = 0$) is equal to E. At a stress value $\sigma = \sigma_o$, the plastic component of strain is α times the elastic strain. In fact, σ_o is found to be approximately equal to the traditional 0.2% offset yield strength. The material parameters α, σ_o, and n^* are to be determined experimentally. Figure 10.4a and b show the graphical presentation of simplified Ludwik's law, Equation 10.8b, and the Ramberg–Osgood Equation 10.8d, respectively.

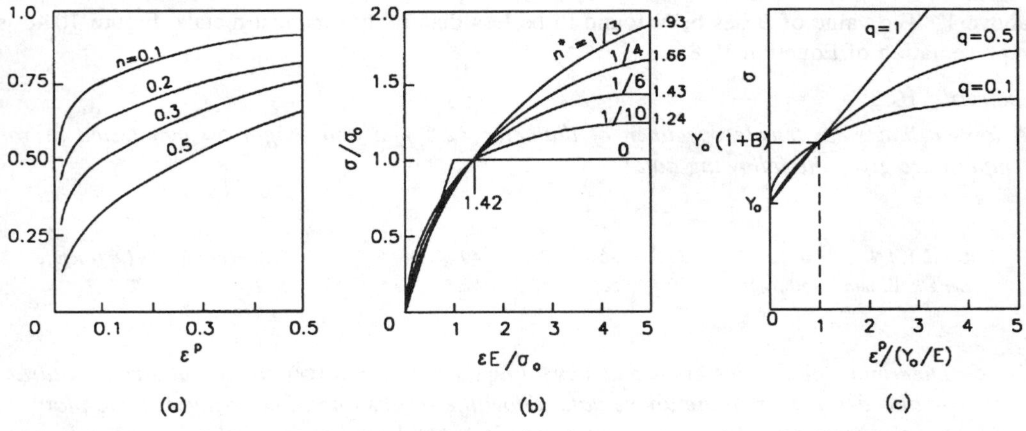

FIGURE 10.4 Empirical stress–strain curves: (a) $\sigma = K\varepsilon^n$, (b) $\varepsilon = (\sigma/E)\,[1 + \alpha(\sigma/\sigma_o)^{1/n-1}]$, and (c) $\sigma = Y_o\,[1 + B(\varepsilon^p/Y_oE)^q]$.

A third empirical stress–strain relation that has proved to be most useful in the analysis of small strain plasticity problems is

$$\varepsilon^e = \frac{\sigma}{E} \qquad \text{for } \sigma \le Y_o$$

and

$$\varepsilon^p = \left(\frac{\sigma - Y_o}{A}\right)^{1/q} \qquad \text{for } \sigma \ge Y_o$$

or, inversely,

$$\sigma = Y_o + A(\varepsilon^p)^q \tag{10.8e}$$

where Y_o corresponds to the elastic limit (or simply the initial yield strength) of the material. The constants A and q are determined from a double-logarithmic plot of $(\sigma - Y_o)$ vs. ε^p. This relation has some advantages over the widely employed Ramberg–Osgood relation (Equation 10.8d) since it predicts zero plastic strain when the initial yield stress Y_o is reached. Survey of the stress–strain curves of many metallic alloys shows that the exponent q ranges from 0.2 to 0.6 with a predominant value of $^1/_3$ at small plastic strains. The second of Equations (10.8e) can be easily rewritten as

$$\sigma = Y_o\left[1 + B\left(\frac{\varepsilon^p}{Y_o/E}\right)^q\right] \tag{10.8f}$$

where

$$B = \frac{A}{Y_o}\left(\frac{Y_o}{E}\right)^q$$

which shows that at a plastic strain $\varepsilon^p = (Y_o/E)$, the parameter B represents the rise in flow stress above Y_o. The value of B has been found to be less than unity for most metals. Figure 10.4c is a representation of Equation 10.8f.

Example 10.3:

A low-carbon steel tensile specimen of diameter 12.5 mm and length 50 mm tested at room temperature gives the following data:

Load, P, kN	*0*	*31*	*35.2*	*36.4*	*37.4*	*40*	*44.5*	*49*	*51.9 necking*	*41.8 fracture*
Length, L, mm	*50*	*50.1*	*50.4*	*50.7*	*52*	*52.3*	*53*	*55*	*64.2*	*70.3*

a. *Determine the parameters K and n by fitting the law: $\sigma = K(\varepsilon^p)^n$ to the true stress–natural strain derived from the above data. Show the results on a double-logarithmic plot;*
b. *On the same plot show the lines for stainless steel and copper behaving according to the laws: $\sigma = 1275\varepsilon^{0.24}$ and $\sigma = 315\varepsilon^{0.54}$, respectively.*
c. *Determine the parameters α, n^* in Ramberg-Osgood law by fitting to the first three points. Take $\sigma_o = Y = 245$ MPa and $E = 207$ GPa.*

Solution:

a. Assuming constancy of volume, the true stress σ is calculated according to

$$\sigma = P/A = PA_o/AA_o = \frac{P}{A_o}\frac{L}{L_o}$$

Obviously, this expression applies only up to the necking point at which the uniformity of the specimen geometry does not hold. The natural strain ε is defined by Equation 10.5 as

$$\varepsilon = \ln (L/L_o)$$

Using the above two expressions for $A_o = \pi(12.5)^2/4$, and $L_o = 50$ mm, the following results are obtained.

True stress, σ, MPa	253	289	301	320	341.1	384.6	436.9	534.8
Natural strain, ε	0.002	0.008	0.014	0.039	0.045	0.058	0.092	0.250

Note that the simple power law is reliable for plastic strains which are many orders of magnitude of the elastic ones; say, $\varepsilon^p > 20\ \varepsilon^e$. Hence, the first three (σ, ε) points may be excluded from the fitting.

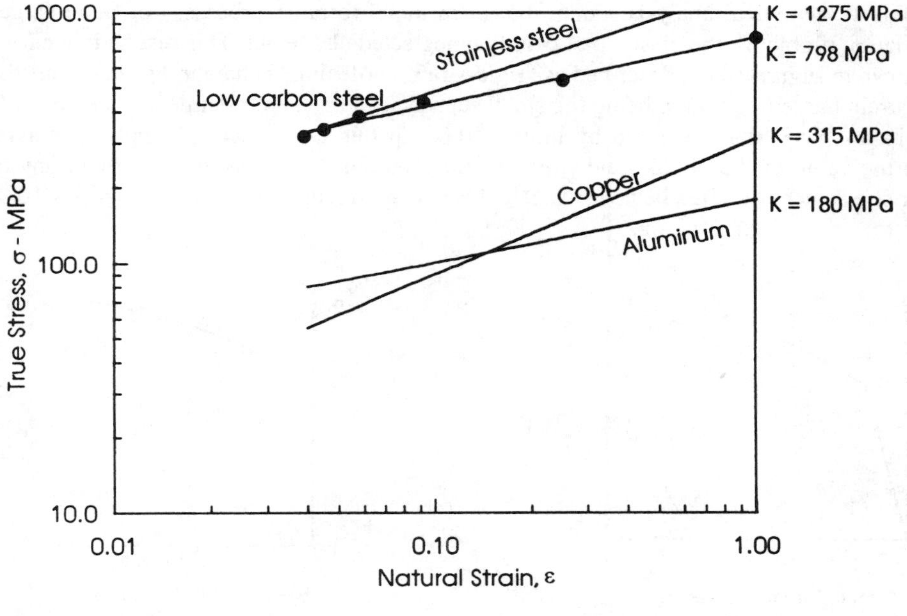

FIGURE 10.5 Example 10.3.

The empirical law $\sigma = K(\varepsilon^p)^n$ plots as a straight line on a double-logarithmic coordinates, i.e., $\ln \sigma = \ln K + n \ln(\varepsilon)^p$. In this plot, Figure 10.5, the slope of the line joining the experimentally determined stress–strain points is equal to n. Extrapolating this line to $\varepsilon^p = 1$ gives the value of K. Applying this procedure to the given data as shown in Figure 10.5 yields $n = 0.27$ and $K = 798$ MPa, with a correlation factor 98%.

b. For comparison, the stress–strain curves of stainless steel, copper, and aluminium are shown on the same plot.
c. Fitting Ramberg-Osgood law, Equation 10.8d applies to the first three points only where the plastic strain is only one order of magnitude of the elastic one. Equation 10.8d is rearranged to be rewritten in the form of a simple power law as:

$$\left(\varepsilon - \frac{\sigma}{E}\right) = \left(\frac{\alpha\sigma_o}{E}\right)\left(\frac{\sigma}{\sigma_o}\right)^{1/n^*}$$

Therefore, the data to be fitted are recalculated knowing that $\sigma_o = 245$ MPa and $E = 207$ GPa, hence

σ/σ_o	1.0327	1.17996	1.2229
$\varepsilon-\sigma/E$	7.78×10^{-4}	0.0066	0.01257

This results in $\alpha = 0.386$ and $n^* = 0.0611$.

10.6 IDEALIZED STRESS–STRAIN CURVES

The elastic–plastic strain-hardening stress–strain curve, Figure 10.6a, has been described by one of Equations 10.8. Often further simplifications are introduced to render these expressions more amenable to analytical analysis. For instance, in metal-forming processes or other large plastic deformation problems, the elastic part could be neglected altogether. This results in a curve, as the one shown in Figure 10.6b, described as "rigid-strain-hardening" behavior. For materials that show slight strain hardening (i.e., n being too small such as lead and some stainless steels in Table.10.1), the behavior may be represented by Figure 10.6c. In this case, once the applied uniaxial stress attains the value of the initial yield stress Y_o, the material flows plastically without any need for further stress increase. This behavior is called elastic–perfectly plastic, and Equations 10.8a and b reduce to

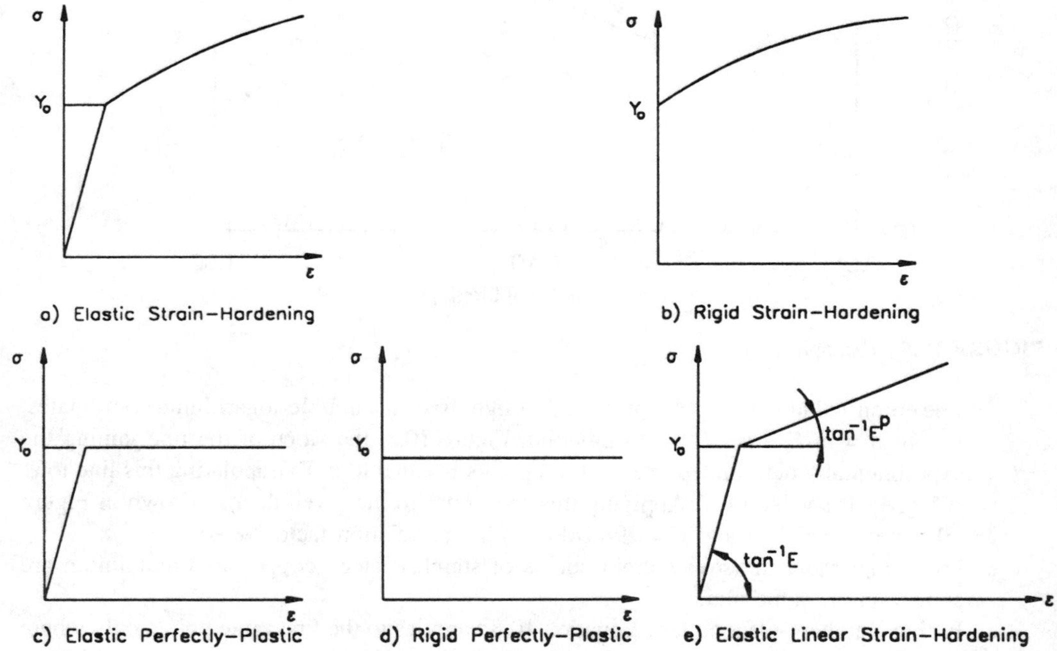

FIGURE 10.6 Idealized stress–strain curves: (a) elastic strain hardening, (b) rigid strain hardening, (c) elastic perfectly plastic, (d) rigid perfectly plastic, (e) elastic-linear strain hardening.

$$\text{elastic part}: \sigma = E(\varepsilon)^e$$
$$\text{plastic part}: \sigma = K = Y_o \quad (10.8g)$$

Such a simple stress–strain behavior as given by Equation 10.8g has been successfully employed n the analyses of many problems involving either small or large plastic strain. Again, if the problem ınder consideration involves very large strain, then the elastic part can be neglected altogether and he idealized behavior is then said to be rigid–perfectly plastic as illustrated in Figure 10.6d.

Another important idealization of the observed stress–strain curve that has proved to be useful n the analysis of problems involving small plastic strains is the linear strain- hardening model hown in Figure 10.6e. Such a model is commonly known as the bilinear idealization of the tress–strain curve. The bilinear model can be expressed as follows:

$$\text{elastic part}: \varepsilon^e = \frac{\sigma}{E}$$

$$\text{plastic part}: \varepsilon^p = \left(\sigma - Y_o\right)/E^p \tag{10.8h}$$

vhere E^p is the slope of the stress–plastic strain curve. Comparison with the second of Equations 0.8e shows that for the bilinear model: $E^p = A$ and $q = 1$ and hence the second of Equation 10.8g ecomes

$$\sigma = Y_o + E^p\,\varepsilon^p \tag{10.8i}$$

t is interesting to note that the value of E^p shown in Figure 10.6e can be easily related to the lope E^t of the stress–(total) strain curve. In incremental form, $E^p = d\sigma/d\varepsilon^p$ but since $d\varepsilon^p = d\varepsilon - d\sigma/E$, therefore $d\sigma/d\varepsilon = E^t = EE^p/(E + E^p)$, or

$$\frac{1}{E^t} = \frac{1}{E^p} + \frac{1}{E} \tag{10.8j}$$

Hence for large $E >> E^p$; $E^p \cong E^t$.

10.7 YIELD CRITERIA

The limit of elasticity, i.e., the yield strength under simple states of stress is directly obtained by neans of mechanical testing of the material. Uniaxial simple tension or compression tests may be onducted on a bar, and the yield strength Y is identified as the end of the linear-elastic behavior.* The same could be done in an experiment of pure shear, namely, torsion of a thin-walled pipe, to btain k, the yield strength in shear.

In a machine or structural element subjected to uniaxial tension, it is thus simple to compare he only nonzero applied stress σ_1 ($\sigma_2 = \sigma_3 = 0$) with the material yield strength in tension, Y. The tate of deformation of the element is identified by any of the following

$$\begin{aligned} &\sigma_1 < Y \;\; \text{Elastic deformation only} \\ &\sigma_1 = Y \;\; \text{Onset of yielding} \\ &\sigma > \text{Y} \;\; \text{Postyielding elastic – plastic deformation} \end{aligned} \tag{10.9}$$

However, if the element is subjected to a general state of stress (σ_1, σ_2, σ_3), it becomes less bvious to know whether the applied stress system causes yielding in the element or not. It may

* In case Y is not easily identifiable, an offset plastic strain of 0.1 or 0.2% is often used to determine the yield strength.

be suggested to take the largest applied stress σ_1 and compare it with the yield strength Y according to Equation 10.9. This idea neglects the effect of the applied stresses other than σ_1 and is not compatible with experimental observations. It is now required to generalize the yield condition in the case of simple tension, namely,

$$\sigma_1 - Y = 0$$

to the general case of three-dimensional states of stress as

$$f(\sigma_1, \sigma_2, \sigma_3, Y) = 0$$

where f will be called the yield function and is to be determined.

By recalling that the hydrostatic stress component σ_m is assumed not to influence yielding, hence σ_1, σ_2, and σ_3 would be replaced by their deviatoric values, i.e., σ_1, σ_2', σ_3' where,

$$\sigma_1' = \sigma_1 - \sigma_m,\ \sigma_2' = \sigma_2 - \sigma_m,\ \text{and}\ \sigma_3' = \sigma_3 - \sigma_m$$

giving

$$f(\sigma_1', \sigma_2', \sigma_3', Y) = 0$$

For an isotropic material, properties are independent of orientation and the yield condition f is thus a function of the deviatoric stress invariants I_2' and I_3' (note, $I_1' = 0$), i.e.,

$$f(I_2', I_3', Y) = 0 \tag{10.10}$$

where, as given by Equation 1.38,

$$I_2' = -\left(\sigma_1'\sigma_2' + \sigma_2'\sigma_3' + \sigma_3'\sigma_1'\right) = \tfrac{1}{2}\left(\sigma_1'^2 + \sigma_2'^2 + \sigma_3'^2\right)$$

and

$$I_3' = \sigma_1'\sigma_2'\sigma_3' = \tfrac{1}{3}\left(\sigma_1'^3 + \sigma_2'^3 + \sigma_3'^3\right)$$

10.7.1 von Mises Yield Criterion

The shape of the function f of Equation 10.10 may be determined through experiments performed on solids under triaxial stress systems. Alternatively, a form for the yield function f could be assumed and validated by experiments. A simple yield function often used in metal plasticity is known as the von Mises yield function,* which neglects the effect of I_3' such that Equation 10.10 reduces to

$$I_2' - C_1^2 = 0$$

or

$$\sigma_1'\sigma_2' + \sigma_2'\sigma_3' + \sigma_3'\sigma_1' = -C_1^2 \tag{10.11}$$

* von Mises, 1913. Also known as the Huber–Mises, Maxwell, or J_2-flow criterion where J_2 designates the second stress variant.

Since this yield function holds for triaxial states of stress, it applies to simple tension, where

$$(\sigma_1, 0, 0),\ \sigma_m = \sigma_1/3$$

and

$$\sigma_1' = 2\sigma_1/3,\ \sigma_2' = -\sigma_1/3,\ \sigma_3' = -\sigma_1/3$$

Hence, Equation 10.11 gives

$$-\tfrac{2}{9}\sigma_1^2 + \tfrac{1}{9}\sigma_1^2 - \tfrac{2}{9}\sigma_1^2 = -C_1^2$$

or

$$\tfrac{1}{3}\sigma_1^2 = C_1^2$$

Yielding will occur if $\sigma_1 = Y$; hence,

$$C_1 = Y/\sqrt{3}$$

Using this value of C_1 in Equation 10.11 and substituting for σ_1', σ_2', and σ_3' gives

$$(\sigma_1 - \sigma_m)(\sigma_2 - \sigma_m) + (\sigma_2 - \sigma_m)(\sigma_3 - \sigma_m) + (\sigma_3 - \sigma_m)(\sigma_1 - \sigma_m) = -\frac{Y^2}{3}$$

which yields, after simplification,

$$\frac{1}{\sqrt{2}}\left[(\sigma_1 - \sigma_2)^2 + (\sigma_2 - \sigma_3)^2 + (\sigma_3 - \sigma_1)^2\right]^{1/2} = Y \qquad (10.12a)$$

This is the most commonly used form for the von Mises criterion.

The von Mises yield criterion may be expressed more generally in terms of the stress components, as

$$\frac{1}{\sqrt{2}}\left[(\sigma_x - \sigma_y)^2 + (\sigma_y - \sigma_z)^2 + (\sigma_z - \sigma_x)^2 + 6\left(\tau_{xy}^2 + \tau_{yz}^2 + \tau_{zx}^2\right)\right]^{1/2} = Y \qquad (10.12b)$$

The von Mises yield criterion may be expressed more generally in terms of the six stress deviators as

$$\sqrt{\frac{3}{2}}\left[\sigma_x'^2 + \sigma_y'^2 + \sigma_z'^2 + 2\tau_{xy}^2 + 2\tau_{yz}^2 + 2\tau_{xz}^2\right]^{1/2} = Y \qquad (10.12c)$$

A pure shear experiment may be used to specify the constant in Equation 10.11 in terms of the yield shear stress k. The state of pure shear is equivalent to the stress state $\sigma_1 = -\sigma_2 = \tau$ and $\sigma_3 = 0$. Hence, $\sigma_m = 0$ and $\sigma_1' = -\sigma_2' = \tau$, $\sigma_3' = 0$. Substitution into Equation 10.11 gives

$$-\tau^2+0+0=-C_1^2$$

If τ attains the value of the yield shear stress k, yielding will occur, resulting in

$$C_1 = k$$

This means that the von Mises criterion relates the yield shear stress k and the tensile yield stress by

$$k = Y/\sqrt{3}$$

and hence the von Mises criterion, Equation 10.12a, may be written as

$$\frac{1}{\sqrt{2}}\left[(\sigma_1-\sigma_2)^2+(\sigma_2-\sigma_3)^2+(\sigma_3-\sigma_1)^2\right]^{1/2}=\sqrt{3}\,k \tag{10.13}$$

10.7.2 Comments on the von Mises Criterion

Expression 10.12a indicates that in the von Mises criterion the principal stresses have an identical contribution to yielding. Also its quadratic form does not require any particular attention to stress sign.

The left-hand side of Equation 10.12b compared with Equation 1.35 shows that

$$\tau_{\text{oct}}=\frac{\sqrt{2}}{3}Y \tag{10.14}$$

which is another form of the von Mises yield criterion. Equation 10.12a may be then interpreted, such that yielding occurs when the octahedral shear stress attains the value of $\sqrt{2}\ Y/3$.

Another way to interpret the von Mises yield criterion is to consider the strain energy density in an isotropic linear-elastic solid subjected to the stresses (σ_1, σ_2, σ_3) as given by Expression 3.36a:

$$U_o=\frac{\sigma_m^2}{2K}+\frac{3}{4G}\tau_{\text{oct}}^2=U_{\text{ov}}+U_{\text{os}}$$

It is seen that U_o is a sum of two parts; the first part U_{ov} is due to a hydrostatic stress component σ_m and the second one U_{os} is due to the octahedral shear. In view of Equation 1.35, the elastic distortion energy U_{os} is given by

$$U_{\text{os}}=U_o-U_{\text{ov}}=\frac{3}{4G}\tau_{\text{oct}}^2$$

or, in terms of principal stresses,

$$U_{\text{os}}=\frac{1}{12G}\left[(\sigma_1-\sigma_2)^2+(\sigma_2-\sigma_3)^2+(\sigma_3-\sigma_1)^2\right] \tag{10.15a}$$

In uniaxial tension, the elastic distortion energy at the onset of yielding when $\sigma_1 = Y$ attains the value

$$U_{os} = \left[\frac{1}{12G} 2\sigma_1^2\right] = \frac{Y^2}{6G} \tag{10.15b}$$

Note that the above interpretation applies in principle to C_o, the complementary strain energy density, rather than U_o, irrespective of the fact that they are equal-valued for an isotropic linear-elastic solid, as pointed out in Section 3.9. However, quite generally, for an anisotropic nonlinear behavior, it is the complementary energy density, as a function of stress components, that enters in deriving the constitutive relations.[2]

Combining Equations 10.15a and 10.15b gives

$$\frac{1}{\sqrt{2}}\left[(\sigma_1 - \sigma_2)^2 + (\sigma_2 - \sigma_3)^2 + (\sigma_3 - \sigma_1)^2\right]^{1/2} = Y$$

which is the von Mises yield condition. The above derivation states that yielding occurs when the elastic distortion energy reaches a specific value.

Example 10.4:

A thin-walled tube with closed ends is subjected to a maximum internal pressure of 30 MPa. The mean radius of the tube is 300 mm and it is not to yield in any region. Using the von Mises yield condition, determine the wall thickness if

a. *The material has a tensile yield strength* Y = *700 MPa;*
b. *The material has a shear yield strength* k = *380 MPa.*

Solution:

The principal stresses in a thin-walled tube of a mean radius r_m and thickness h due to internal pressure are as given by Equation 8.68:

$$\sigma_1 = \frac{p_i r_m}{h} \text{(hoop)}, \quad \sigma_2 = \frac{p_i r_m}{2h} \text{(axial)}, \quad \sigma_3 \cong 0 \text{ (radial)}$$

a. Hence $\sigma_2 = \sigma_1/2$, $\sigma_3 = 0$ in Equation 10.12a gives

$$(\sigma_1 - \sigma_1/2)^2 + (\sigma_1/2)^2 + (-\sigma_1)^2 = 2Y^2$$

or $\sigma_1 = 2Y/\sqrt{3}$. Thus,

$$\frac{30 \times 300}{h} = \frac{2 \times 700}{\sqrt{3}} \text{ MPa}$$

giving $h = 11.13$ mm.

b. By using Equation 10.13

$$(\sigma_1 - \sigma_1/2)^2 + (\sigma_1/2)^2 + (-\sigma_1)^2 = 6k^2$$

thus,

$$\frac{30 \times 300}{h} = 2 \times 380 \text{ MPa}$$

giving h = 11.84 mm. The difference in the two thicknesses is due to the use of independently determined tensile and shear yield strengths. If the yield shear strength k

is to be determined from the tensile yield strength Y through the von Mises criterion, then $k = 700/\sqrt{3} = 404.16$ MPa, which is different from the value of $k = 380$ MPa used in (b).

10.7.3 Tresca Yield Criterion

Tresca proposed that initial yielding occurs when the highest of the maximum shear stresses, as given by Equation 1.31,

$$(\sigma_1 - \sigma_2)/2, (\sigma_2 - \sigma_3)/2 \text{ and } (\sigma_1 - \sigma_3)/2,$$

attains a critical value. By considering that $\sigma_1 > \sigma_2 > \sigma_3$, the Tresca criterion is written as

$$\frac{\sigma_1 - \sigma_3}{2} = C_2$$

Note that this criterion is independent of the intermediate stress σ_2. Such criterion holds true for simple states of stress and the constant C_2 may be determined from a simple tensile or pure shear test.

In simple tension at the onset of yielding, $\sigma_1 = Y$, $\sigma_2 = \sigma_3 = 0$ and, hence, $C_2 = Y/2$. For the case of pure shear, $\sigma_1 = -\sigma_3 = k$ and $\sigma_2 = 0$, thus giving $C_2 = k$. This means that the Tresca yield criterion predicts the relation between the yield tensile and shear strengths as

$$k = Y/2$$

while von Mises predicts from Equation 10.12a

$$K = Y/\sqrt{3}$$

The Tresca yield criterion* is thus written as

$$\frac{\sigma_1 - \sigma_3}{2} = \frac{Y}{2} = k \tag{10.16}$$

where σ_1 and σ_3 are the maximum and minimum principal stresses, respectively.

Example 10.5:

Repeat Example 10.4 using the Tresca yield criterion.

Solution:

a. Since

$$\sigma_1 = \frac{p_i r_m}{h}, \quad \sigma_2 = \frac{p_i r_m}{2h}, \quad \sigma_3 \cong 0$$

therefore according to the Tresca yield criterion (Equation 10.16):

$$\frac{\sigma_1 - \sigma_3}{2} = \frac{1}{2}\left(\frac{p_i r_m}{h}\right) = \frac{1}{2}\frac{30 \times 300}{h} = \frac{700}{2} \text{ MPa}$$

Hence, $h = 12.86$ mm.

* The Tresca yield criterion (1867) is often called the "maximum stress criterion."

(b) $$\frac{\sigma_1 - \sigma_3}{2} = \frac{1}{2}\left(\frac{p_i r_m}{h}\right) = \frac{1}{2}\frac{30 \times 300}{h} = 380 \text{ MPa}$$

Hence, $h = 11.84$ mm.

The table compares the tube wall thicknesses in mm of Examples 10.4 and 10.5.

Yield Criterion	$Y = 700$ MPa	$k = 380$ MPa
von Mises	11.13	11.84
Tresca	12.86	11.84

Note that if Y is the specified property and the thickness h is unknown, then the Tresca criterion is more conservative, but if k is the specified property, both criteria predict the same wall thickness.

10.7.4 Geometric Representation of von Mises and Tresca Criteria

For simplicity, consider a state of plane stress, where $\sigma_2 = 0$. The von Mises criterion reduces to

$$\sigma_1^2 + \sigma_3^2 - \sigma_1\sigma_3 = Y^2 \quad \text{for } \sigma_2 = 0 \tag{10.17}$$

which plots as an ellipse of $\sqrt{2}\,Y$ and $\sqrt{2/3}\,Y$ as major and minor axes, respectively.

The Tresca yield criterion plots as a straight line whose equation depends on the magnitude and sign of the three stress components. For plane stress $\sigma_2 = 0$, hence the criterion takes one of the following six forms:

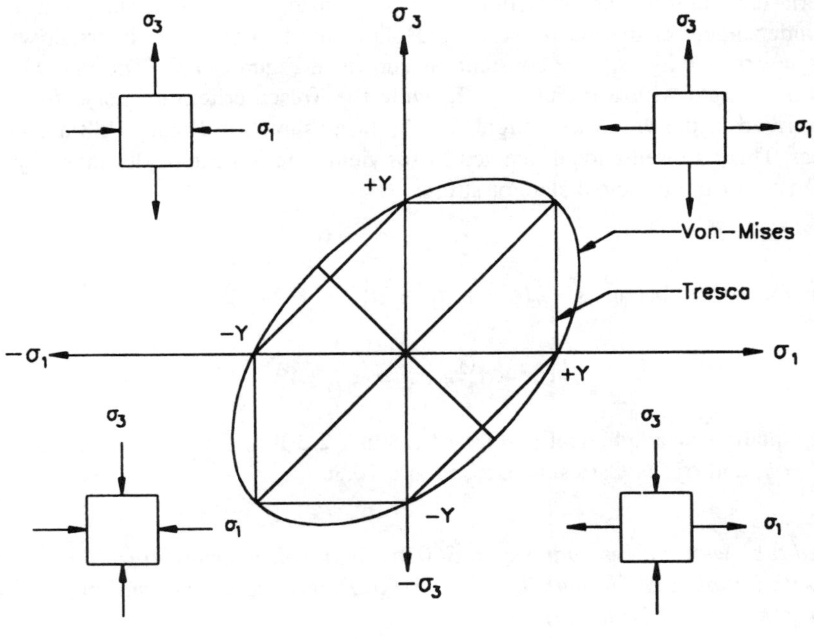

FIGURE 10.7 von Mises and Tresca yield loci in plane stress.

$$
\begin{aligned}
\sigma_3 > \sigma_1 > 0 \qquad & \sigma_3 = +Y \\
\sigma_1 > \sigma_3 > 0 \qquad & \sigma_1 = +Y \\
\sigma_1 > \sigma > \sigma_3 \qquad & \sigma_1 - \sigma_3 = +Y \\
\sigma_3 > \sigma > \sigma_1 \qquad & \sigma_3 - \sigma_1 = +Y \\
0 > \sigma_1 > \sigma_3 \qquad & \sigma_3 = -Y \\
0 > \sigma_3 > \sigma_1 \qquad & \sigma_1 = -Y
\end{aligned}
\tag{10.18}
$$

These expressions are plotted in Figure 10.7 as a hexagonal shape. Both criteria plots intersect the coordinate axes σ_1 and σ_3 at $\pm Y$. They also have common points along the loading line $\sigma_1 = \sigma_3$. The maximum differences in predictions of yielding occur along the loading lines $\sigma_1 = 2\sigma_3$, $\sigma_1 = \sigma_3/2$, and $\sigma_1 = -\sigma_3$. Considering one of these loading lines (say, $\sigma_1 = -\sigma_3$) and substituting into both criteria gives

$$\text{von Mises} = \sigma_1 = Y/\sqrt{3} = 0.577Y$$

$$\text{Tresca} = \sigma_1 = \frac{Y}{2} = 0.5Y$$

The maximum difference in predicting yielding according to von Mises and Tresca is thus obtained from the ratio (0.577/0.5), which is equal to 1.155.

In three-dimensional stress space, the von Mises criterion plots as a circular cylinder while Tresca plots as a cylinder with hexagonal cross section, as shown in Figure 10.8a. The axis of the two cylinders is a line equally inclined to the coordinate axes σ_1, σ_2, and σ_3. Any point along this line obviously represents a state of hydrostatic stress. Both cylinders representing von Mises and Tresca criteria have uniform cross section. This is consistent with the assumption that the yield strength is independent of the mean stress σ_m. Both of the two cylinders intersect with the plane (π-plane) $\sigma_m = (\sigma_1 + \sigma_2 + \sigma_3)/3$ = constant, as shown in Figure 10.8b. The von Mises criterion intersection is a circle with a radius $\sqrt{2/3}\,Y$, while the Tresca criterion intersection is a regular hexagon inscribed in the von Mises circle. The surfaces shown in Figure 10.8 are known as the yield surfaces. This representation of the von Mises yield criterion can be displayed by considering Equation 10.12c for the principal state of stress:

$$\tau_{xy} = \tau_{xz} = \tau_{yz} = 0$$

Hence, Equation 10.12c becomes* $\frac{3}{2}(\sigma_1^{\backslash 2} + \sigma_1^{\backslash 2} + \sigma_3^{\backslash 2}) = Y^2$ or

$$\sigma_1^{\backslash 2} + \sigma_2^{\backslash 2} + \sigma_3^{\backslash 2} = \left(\sqrt{2/3}\,Y\right)^2 \tag{10.19}$$

which is an equation of a spherical surface of radius $\sqrt{2/3}\,Y$ in the space of the principal stress deviators $\sigma_1^{\backslash}$, $\sigma_2^{\backslash}$, and $\sigma_3^{\backslash}$, as Cartesian axes, Figure 10.8c.

Example 10.6:

A thin-walled tube with a mean radius r_m = *300 mm and wall thickness* h = *10 mm is made from a material with initial yield strength* Y_o = *600 MPa. Determine the internal pressure at yielding using both von Mises and Tresca criteria.*

---

* This form of the von Mises yield criterion can be written elegantly using the following tensorial notation: $3\sigma_{ij}^{\backslash}\,\sigma_{ij}^{\backslash}/2 = Y$ for $i, j = 1, 2, 3$. Since $\sigma_{ij}^{\backslash} = 0$, $i = j$, then the yield criterion becomes $\frac{3}{2}(\sigma_1^{\backslash 2} + \sigma_2^{\backslash 2} + \sigma_3^{\backslash 2}) = Y^2$.

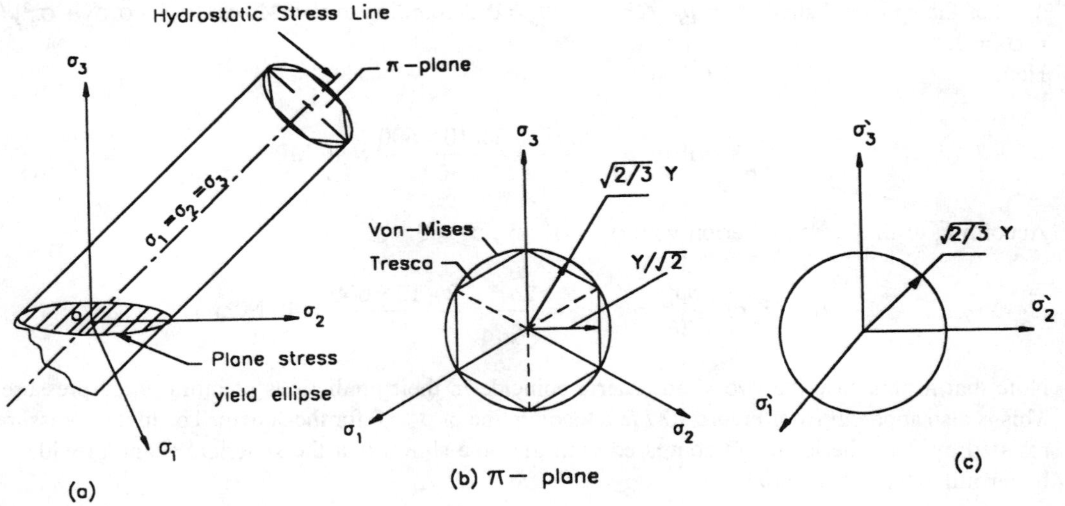

FIGURE 10.8 Representation of von Mises and Tresca yield loci (a) in three-dimensional stress space, (b) in the π-plane, and (c) in three-dimensional deviatoric stress space.

Repeat the above analysis for a pressurized thin-walled spherical shell with r_m = *300 mm,* h = *10 mm and made from the same material as the tube. Comment on the limiting pressures for the tube and sphere.*

Solution:

The stress components in the tube wall are

$$\sigma_1 = \frac{p_i r_m}{h}, \quad \sigma_2 = \frac{p_i r_m}{2h}, \quad \sigma_3 = 0$$

noting that $\sigma_1 = 2\sigma_2$. According to the von Mises criterion (Equation 10.12a) for $\sigma_3 = 0$, and $\sigma_1 = 2\sigma_2$,

$$\left[\sigma_1^2 - \sigma_1\sigma_2 + \sigma_2^2\right]^{1/2} = \left(\sigma_1^2 - \tfrac{1}{2}\sigma_1^2 + \tfrac{1}{4}\sigma_1^2\right)^{1/2} = Y$$

or

$$\bar{\sigma} = \frac{\sqrt{3}}{2}\frac{p_i r_m}{h}$$

Initial yielding occurs when

$$p_i = \frac{2}{\sqrt{3}}\frac{hY}{r_m} = \frac{2}{\sqrt{3}} \times \frac{10 \times 600}{300} = 23.29 \text{ MPa}$$

For plane stress conditions $\sigma_1 > \sigma_2 > 0$, $\sigma_3 = 0$, the Tresca criterion (Equation 10.18) takes the form: $\sigma_1 = Y$, i.e.,

$$\frac{p_i r_m}{h} = Y \text{ or } p_i = \frac{hY}{r_m} = \frac{10 \times 600}{300} = 20 \text{ MPa}$$

Note that p_i (von Mises)/p_i (Tresca) = 23.09/20 = 1.15, i.e., the yield pressure predicted by the Tresca criterion is more conservative than that predicted by the von Mises criterion.

For the spherical shell: $\sigma_1 = p_i r_m/2h = \sigma_2$, $\sigma_3 = 0$. According to von Mises: $[\sigma_1^2 - \sigma_1\sigma_2 + \sigma_2^2]^{1/2} = \sigma_1 = Y$.
Hence,

$$\frac{p_i r_m}{h} = Y \text{ or } p_i = \frac{2hY}{r_m} = \frac{2\times10\times600}{300} = 40 \text{ MPa}$$

According to the Tresca criterion with $\sigma_1 = \sigma_2 > 0$, $\sigma_3 = 0$,

$$\sigma_1 = Y \text{ or } \frac{p_i r_m}{2h} = Y,\ p_i = \frac{2hY}{r_m} = \frac{2\times10\times600}{300} = 40 \text{ MPa}$$

Note that in this case the two yield criteria coincide in their predictions of initial yield pressure. This is also apparent from Figure 10.7 at a loading line of $\sigma_3 = 0$ for the sphere. The higher pressure resisted by the spherical shell compared with the tube shows that the spherical shape provides a better utilization of material.

10.7.5 Experimental Verification of Yield Criteria

To verify any proposed yield criterion, experiments have to be designed to involve triaxial stress systems. A simple uniaxial tensile or pure shear experiment is not satisfactory since this can only give the constants C_1 and C_2 appearing in von Mises and Tresca criteria, respectively. A triaxial stress experiment is rather involved compared with the simple uniaxial stress test. The thin-walled tube, however, offers a good experimental element to validate any yield criterion under biaxial stress systems by subjecting the tube to a combination of twisting moment, axial load, and internal pressure. Such a tube being sufficiently thin has the advantage that the stresses within its wall are almost uniform.

Consider a thin-walled tube subjected to a twisting moment M_t whose value does not cause initial yielding and then loaded in tension by a monotonically increasing force P, as shown in Figure 10.9a until yielding occurs. Deformation measurements, i.e., angle of twist and/or extension of the tube wall material, are recorded as the tensile load increases. Yielding is detected as soon as a rapid increase in one of these deformations occurs. Such a test will provide one point on the yield surface defined by the principal stresses (due to M_t and P at which yielding has occurred) given by ($\sigma_1 > \sigma_2 > \sigma_3$):

$$\begin{aligned} \sigma_1 &= \tfrac{1}{2}\sigma + \left(\tfrac{1}{4}\sigma^2 + \tau^2\right)^{1/2} \\ \sigma_2 &= 0 \\ \sigma_3 &= \tfrac{1}{2}\sigma - \left(\tfrac{1}{4}\sigma^2 + \tau^2\right)^{1/2} \end{aligned} \tag{10.20}$$

where $\sigma = P/2\pi r_{mo} h_o$ and $\tau = M_t/2\pi r_{mo}^2 h$. Also, r_{mo} and h_o, respectively, are the mean radius and thickness of the tube.

The same experiment is repeated using other tubes of the same material and same dimensions changing the initial value of the applied twisting moment M_t within the elastic range and then increasing the tensile force P up to yielding. When a sufficient number of points representing combinations of (σ_1, σ, 0_3) are obtained at the onset of yielding, a plot such as that of Figure 10.7 may be constructed.

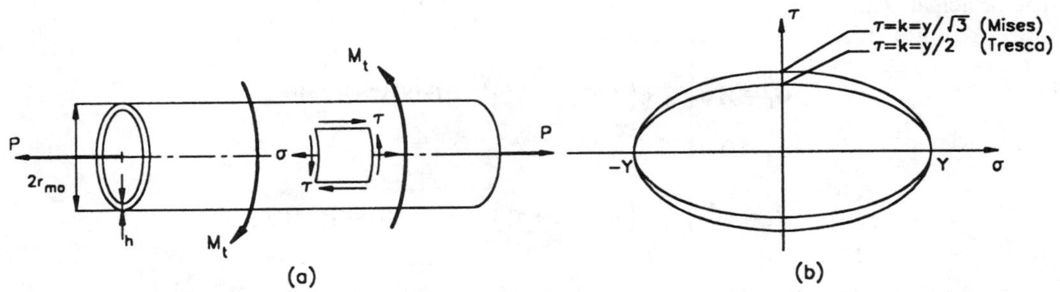

FIGURE 10.9 Experimental verification of yield criteria: (a) thin-walled tube test and (b) von Mises and Tresca yield loci in τ–σ space.

Another way for displaying yield loci for thin-walled tubes subjected to axial force P and twisting moment M_t may be illustrated in a plot of σ vs. τ. Substituting the principal stresses obtained in Equation 10.20 into the von Mises criterion and the Tresca criterion yields, respectively,

For von-Mises criterion: $$\sigma^2 + 3\tau^2 = Y^2 \quad (10.21a)$$

For Tresca criterion: $$\sigma^2 + 4\tau^2 = Y^2 \quad (10.21b)$$

Equations 10.21a represent an ellipse of major and minor axes of Y and $Y/\sqrt{3}$, respectively, while Equation 10.21b represents an ellipse of major and minor axes of Y and $Y/2$, respectively, as shown in Figure 10.9b. It is clear from Figure 10.9b that the two criteria coincide for the condition of pure uniaxial loading ($\tau = 0$). Under pure shear, the von Mises criterion predicts $k = Y/\sqrt{3}$ against $k = Y/2$ for Tresca with a maximum difference of 15.5%, as indicated earlier.

Several investigations based on the experimental procedure described above have been performed. The results for many ductile metals indicated that most of the experimental points lie between Mises-ellipse and Tresca-hexagon of Figure 10.7. In general, the von Mises yield criterion agrees better with the experimental data of ductile materials than does the Tresca criterion. However, the Tresca criterion is mostly used in analysis for its mathematical simplicity; the difference between the two criteria is not very significant for most engineering applications.

Example 10.7:

A thin-walled tube of a mean radius r_{mo} *= 50 mm, thickness* h_o *= 2.5 mm is subjected to an axial tensile force* P *(in N) and a twisting moment* M_t *(in N · mm) such that the ratio of* T *to* P *is maintained constant during deformation as given by* T *= 100 P (N · mm). Calculate the value of* P *at which yielding occurs according to*

a. *Tresca yield criterion;*
b. *von Mises yield criterion.*

The tensile yield strength is Y *(in MPa).*

Solution:

The stresses in the tube wall:

$$\sigma_{\text{axial}} = P/2\pi r_{mo} h_o = P/[2\pi(50)\,(2.5)] = 0.001274P \text{ MPa}$$

$$\tau = M_t/(2\pi r_{mo}^2 h_o) = 100P/[2\pi(50)^2(2.5)] = 0.00253P \text{ MPa}$$

The principal stresses are

$$\sigma_1 = \tfrac{1}{2}\sigma_{axial} + \left(\tfrac{1}{4}\sigma_{axial}^2 + \tau^2\right)^{1/2} = 0.00326P \text{ MPa}$$

$$\sigma_2 = 0$$

$$\sigma_3 = \tfrac{1}{2}\sigma_{axial} - \left(\tfrac{1}{4}\sigma_{axial}^2 + \tau^2\right)^{1/2} = -0.00199P \text{ MPa}$$

a. Applying the Tresca yield criterion, Equation 10.19 for $\sigma_1 > (\sigma_2 = 0) > \sigma_3$,

$$\sigma_1 - \sigma_3 = Y$$

gives $P = 190.8Y$ N.

b. Applying the von Mises criterion, Equation 10.17,

$$\left(\sigma_1^2 + \sigma_3^2 - \sigma_1\sigma_3\right)^{1/2} = Y$$

gives $P = 217.8Y$ N.

Again the prediction of the Tresca criterion is more conservative than the von Mises criterion.

10.8 PLASTIC STRESS–STRAIN RELATIONS — FLOW RULE

For elastic deformation, the stress–strain relations are expressed by Hooke's law:

$$\varepsilon_1 = \frac{1}{E}\left[\sigma_1 - \nu(\sigma_2 + \sigma_3)\right]\cdots$$

By analogy, plastic stress–strain relations may be written (noting that $\nu = 1/2$ for constant-volume plastic deformation) as

$$d\varepsilon_1^p = d\lambda\left[\sigma_1 - \tfrac{1}{2}(\sigma_2 + \sigma_3)\right]$$

$$d\varepsilon_2^p = d\lambda\left[\sigma_2 - \tfrac{1}{2}(\sigma_1 + \sigma_3)\right] \qquad (10.22a)$$

$$d\varepsilon_3^p = d\lambda\left[\sigma_3 - \tfrac{1}{2}(\sigma_1 + \sigma_2)\right]$$

The plastic strain increment $d\varepsilon^p$ is related to the stresses by a proportionality factor $d\lambda$ replacing l/E in Hooke's law. The factor $d\lambda$ is not a material constant as Young's modulus. Note that $d\lambda$ is a positive quantity.

Equations 10.22a may be further expressed by using the deviatoric stresses σ_1', σ_2', and σ_3'. by taking the first of Equations 10.22 as

$$d\varepsilon_1^p = d\lambda\left[(\sigma_1 - \sigma_m) - \tfrac{1}{2}(\sigma_2 + \sigma_3) + \sigma_m\right]$$

$$= d\lambda\left[\sigma_1^\backslash - \tfrac{1}{2}(\sigma_2^\backslash + \sigma_3^\backslash)\right]$$

$$= d\lambda\left[\sigma_1^\backslash - \tfrac{1}{2}(\sigma_2^\backslash + \sigma_3^\backslash + \sigma_1^\backslash - \sigma_1^\backslash)\right]$$

$$= d\lambda\left[\tfrac{3}{2}\sigma_1^\backslash - \tfrac{1}{2}(\sigma_1^\backslash + \sigma_2^\backslash + \sigma_3^\backslash)\right]$$

Since by definition $\sigma_1^\backslash + \sigma_2^\backslash + \sigma_3^\backslash = 0$, this reduces to:

$$d\varepsilon_1^p = \frac{3}{2}d\lambda\sigma_1^\backslash$$

Equations 10.22 thus take the form:

$$\frac{d\varepsilon_1^p}{\sigma_1^\backslash} = \frac{d\varepsilon_2^p}{\sigma_2^\backslash} = \frac{d\varepsilon_3^p}{\sigma_3^\backslash} = \frac{3}{2}d\lambda \tag{10.22b}$$

which implies that the plastic–strain increments are proportionally related to the deviatoric stress components and not to the total stresses. This is consistent with the assumption that plastic deformation is independent of the hydrostatic stress component σ_m. The plastic stress–strain increment relations as given by Equations 10.22 are often called the Levy–Mises "plastic" flow rule. Such relations are not complete since the proportionality factor $d\lambda$ is yet to be determined.

Note that Equations 10.22 imply that the principal directions of plastic strain increments coincide with the principal directions of stresses. If the ratios $\sigma_1^\backslash : \sigma_2^\backslash : \sigma_3^\backslash$ remain unchanged during the entire loading path, then the ratios between plastic strain increments $d\varepsilon_1^p : d\varepsilon_2^p : d\varepsilon_3^p$ will also remain constant and this condition is known as proportional loading. In such cases, the flow rule Equations 10.22b can be integrated to give

$$\frac{\varepsilon_1^p}{\sigma_1^\backslash} = \frac{\varepsilon_2^p}{\sigma_2^\backslash} = \frac{\varepsilon_3^p}{\sigma_3^\backslash} = \frac{3}{2}\lambda \tag{10.22c}$$

where the total plastic strains are used instead of their increments.

In fact, there are two descriptions dealing with plastic deformation: the "incremental theory" and the "deformation theory." In the first, which is often called "flow theory," the plastic strain increment $d\varepsilon_{ij}^p$ corresponding to a stress increment $d\sigma_{ij}^p$ are related to each other through the total current stress state σ_{ij} as well as the plastic deformation history. This is the theory presented above and adopted in solving most plasticity problems. The deformational theory is rather simpler since it neglects the loading history dependency by assuming that the stress state σ_{ij} determines uniquely the total plastic strain ε_{ij}^p. The validity of this theory is limited to proportional loading only. If unloading, i.e., stress reversal takes place, the deformation theory becomes erroneous.*

Example 10.8:

In sheet metal stretching, a grid of circles is often printed on the sheet metal surface to investigate the zones of heavy deformation, as shown in Figure 10.10. A circle of such a grid of 5 mm diameter

---

* The deformation theory states $\varepsilon_{ij}^p = \phi\sigma'_{ij}$, where ϕ is a scalar quantity, which is positive throughout loading and zero during unloading. See Hill.[3]

has changed after a stretching process into an ellipse whose major and minor axes are 6.5 and 5.5 mm, respectively. If the stresses are proportionally such that $\sigma_1 > \sigma_2$, where the final value of σ_2 is 300 MPa. Determine the value of the applied load P_1 (at the end of the process), which causes this deformation in a uniformly stretched square sheet metal of initial dimensions 100 × 100 × 2 mm. Also determine the yield strength at the end of this stretching process.

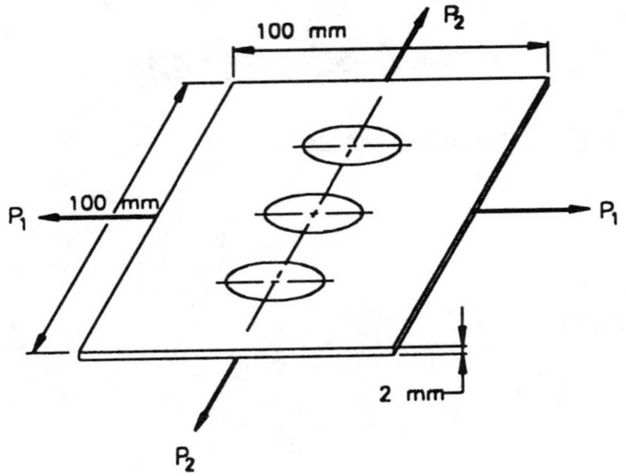

FIGURE 10.10 Example 10.8.

Solution:

The plastic strain components are given by

$$\varepsilon_1^p = \ln\left(\frac{6.5}{5}\right) = 0.262$$

and

$$\varepsilon_2^p = \ln\left(\frac{5.5}{5}\right) = 0.095$$

For homogeneous strains constancy of volume is expressed by

$$\varepsilon_1^p + \varepsilon_2^p + \varepsilon_3^p = 0$$

giving

$$\varepsilon_3^p = -\left(\varepsilon_1^p + \varepsilon_2^p\right) = -0.357$$

Applying the Levy–Mises flow rule, Equations 10.22, and noting that this is a plane stress process ($\sigma_3 = 0$) results in

$$\frac{\varepsilon_1^p}{\sigma_1^{\backslash}} = \frac{\varepsilon_2^p}{\sigma_2^{\backslash}} = \frac{\varepsilon_3^p}{\sigma_3^{\backslash}} = \frac{3}{2}\lambda$$

hence,

$$\frac{\varepsilon_2^p}{\varepsilon_3^p} = \frac{\sigma_2^{\backslash}}{\sigma_3^{\backslash}} = \frac{\sigma_2 - \sigma_m}{0 - \sigma_m} = \frac{300 - \sigma_m}{-\sigma_m}$$

This gives $\sigma_m = 236.95$ MPa. Similarly,

$$\frac{\varepsilon_1^p}{\varepsilon_2^p} = \frac{\sigma_1^\backslash}{\sigma_2^\backslash} = \frac{\sigma_1 - \sigma_m}{\sigma_2 - \sigma_m}$$

hence, $\sigma_1 = 410.83$ MPa. To determine the load P at the end of the stretching process, the final overall sheet dimensions must be calculated according to

$$\varepsilon_1^p = \ln\frac{L_1}{L_o} \text{ giving } L_1 = L_o \exp(0.262) = 129.95 \text{ mm}$$

$$\varepsilon_2^p = \ln\frac{L_2}{L_o} \text{ giving } L_2 = L_o \exp(0.095) = 109.96 \text{ mm}$$

$$\varepsilon_3^p = \ln\frac{h}{h_o} \text{ giving } h = h_o \exp(-0.357) = 1.44 \text{ mm}$$

Hence, the load at the end of the stretching process is

$$P_1 = \sigma_1 L_2 h = 410.83 \times 109.96 \times 1.399 = 63.224 \text{ kN}$$

The yield strength at the end of the process is obtained by applying either von Mises or Tresca criteria for plane stress $\sigma_1 > \sigma_2 > (\sigma_3 = 0)$.

von Mises: $Y = (\sigma_1^2 + \sigma_2^2 - \sigma_1\sigma_2)^{1/2}$
giving $Y = 368.15$ MPa
Tresca: $Y = \sigma_1 - \sigma_3$
giving $Y = 410.83$ MPa

10.9 PRINCIPLE OF NORMALITY AND PLASTIC POTENTIAL

Consider for simplicity, the von Mises yield criterion in plane stress condition ($\sigma_2 = 0$), i.e., Equation 10.17,

$$\sigma_1^2 + \sigma_3^2 - \sigma_1\sigma_3 = Y^2$$

It has been shown that this equation is represented by an ellipse, Figure 10.11. The slope at any point on this ellipse is $d\sigma_3/d\sigma_1$ obtained by differentiation of Equation 10.17, as

$$\frac{d\sigma_3}{d\sigma_1} = \frac{2\sigma_1 - \sigma_3}{2\sigma_3 - \sigma_1} = -\frac{\sigma_1^\backslash}{\sigma_3^\backslash}$$

Hence, the slope of the outward normal is $\sigma_3^\backslash/\sigma_1^\backslash$.

This result means that the vector drawn perpendicular to the tangent at any point on the yield surface will have projections proportional to $\sigma_1^\backslash$ and $\sigma_3^\backslash$ at this point. In view of the Levy–Mises

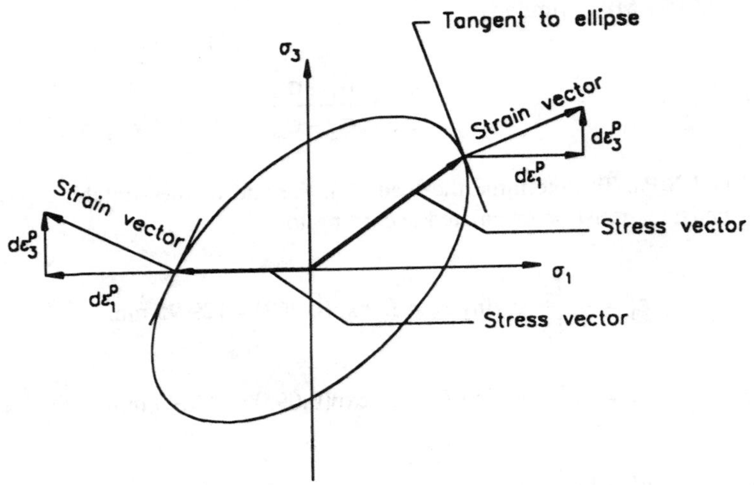

FIGURE 10.11 Normality of the plastic strain increment vector to the yield locus at the point of yielding.

flow rule, Equations 10.22, which state that the plastic strain increments $d\varepsilon_1^p$ and $d\varepsilon_3^p$ are proportional to σ_1' and σ_3', respectively, the plastic strain increment vector is thus normal to the yield surface.*

Such a property is a general one and is not restricted to isotropic materials. It is also very useful in constructing experimental yield loci as shown in Figure 10.11. Note that in this figure only the projection of the strain vector appears in the $\sigma_1 - \sigma_3$ plane; but it is still perpendicular to the yield locus line. Now, let the von Mises yield criterion (again for simplicity in plane stress conditions) be considered as a function *f*:

$$f = \sigma_1^2 + \sigma_3^2 - \sigma_1\sigma_3 - Y^2$$

Upon differentiation with respect to σ_1 and σ_3,

$$\frac{\partial f}{\partial \sigma_1} = 2\sigma_1 - \sigma_3 = 3\sigma_1' \text{ and } \frac{\partial f}{\partial \sigma_3} = 2\sigma_3 - \sigma_1 = 3\sigma_3'$$

and in view of the Levy–Mises flow rule, Equations 10.22, it is concluded that

$$\frac{\partial f}{\partial \sigma_1} : \frac{\partial f}{\partial \sigma_3} = \sigma_1' : \sigma_3' = d\varepsilon_1^p : d\varepsilon_3^p \tag{10.23}$$

which establishes the concept of "plastic potential." If the von Mises yield criterion is used for *f*, the "associated flow rule," which is the Levy–Mises rule, results upon differentiation with respect to the stresses. In fact, this is the general procedure one may follow to derive the flow rule associated with any assumed yield function or plastic potential, as will be seen in dealing with anisotropic materials and porous solids in Sections 10.15 and 10.16, respectively.

* As regards Tresca yield hexagonal condition, where there is a uniquely defined normal and the plastic increment vector must be in the direction of this normal. At the six corners, where the normals are not unique, the strain increment vector may take on any value between the limiting normals. See Venkatraman and Patel.[4]

10.10 PLASTIC WORK, EFFECTIVE STRESS, AND EFFECTIVE STRAIN INCREMENT

If a bar of length L is subjected to an axial force P producing a permanent extension dL, the work done is $P\,dL$. Considering the bar cross-sectional area A, the plastic work done per unit volume is

$$dW^p = \frac{PdL}{AL} = \sigma_1 d\varepsilon_1^p$$

In the most general case, where three normal stresses and three shear stresses act simultaneously, the total plastic work per unit volume is

$$dW^p = \sigma_x d\varepsilon_x^p + \sigma_y d\varepsilon_y^p + \sigma_z d\varepsilon_z^p + \tau_{xy} d\gamma_{xy}^p + \tau_{xz} d\gamma_{xz}^p + \tau_{yz} d\gamma_{yz}^p \qquad (10.24a)$$

In terms of principal stresses

$$dW^p = \sigma_1 d\varepsilon_1^p + \sigma_2 d\varepsilon_2^p + \sigma_3 d\varepsilon_3^p \qquad (10.24b)$$

It is now useful to look upon the yield criterion in terms of a new definition, called "effective stress," $\bar{\sigma}$. For instance, the von Mises criterion may be written as

$$\bar{\sigma} = Y \qquad (10.25)$$

where

$$\bar{\sigma} = \frac{1}{\sqrt{2}}\left[(\sigma_1 - \sigma_2)^2 + (\sigma_2 - \sigma_3)^2 + (\sigma_3 - \sigma_1)^2\right]^{1/2} \qquad (10.26)$$

This definition means that, whenever the effective stress value representing a state of triaxial stress attains the tensile yield strength Y, yielding occurs.

Similarly, an effective plastic strain increment $d\bar{\varepsilon}^p$ may be defined such that the plastic work increment per unit volume is given by

$$dW^p = \sigma_1 d\varepsilon_1^p + \sigma_2 d\varepsilon_2^p + \sigma_3 d\varepsilon_3^p = \bar{\sigma} d\bar{\varepsilon}^p \qquad (10.27)$$

Substituting for $d\varepsilon_1^p$, $d\varepsilon_2^p$, and $d\varepsilon_3^p$ from the Levy–Mises flow rule, Equations 10.22b yields

$$\sigma_1\sigma_1' + \sigma_2\sigma_2' + \sigma_3\sigma_3' = \frac{2}{3}\frac{1}{d\lambda}\bar{\sigma} d\bar{\varepsilon}^p$$

Through some algebraic manipulation, noting from Equations 1.37 and 10.19, respectively, that

$$\sigma_1' + \sigma_2' + \sigma_3' = 0$$

and

$$\bar{\sigma} = \sqrt{\frac{3}{2}}\left(\sigma_1'^2 + \sigma_2'^2 + \sigma_3'^2\right)^{1/2}$$

the following result is obtained:

$$d\lambda = d\bar{\varepsilon}^p / \bar{\sigma} \tag{10.28}$$

Note that $d\lambda$ is a positive quantity. Alternatively, Equation 10.27, considering the constancy of volume, is written as

$$dW^p = \sigma_1^{\backslash} d\varepsilon_1^p + \sigma_2^{\backslash} d\varepsilon_2^p + \sigma_3^{\backslash} d\varepsilon_3^p = \bar{\sigma} d\bar{\varepsilon}^p \tag{10.29}$$

Using the flow rule, Equations 10.23 together with Equation 10.28 gives

$$d\bar{\varepsilon}^p = \sqrt{\frac{2}{3}} \left[d\varepsilon_1^{p^2} + d\varepsilon_2^{p^2} + d\varepsilon_3^{p^2} \right]^{1/2} \tag{10.30a}$$

or, in analogy with Equation 10.26, this may be written as

$$d\bar{\varepsilon}^p = \sqrt{\frac{2}{3}} \left[\left(d\varepsilon_1^p - d\varepsilon_2^p \right)^2 + \left(d\varepsilon_2^p - d\varepsilon_3^p \right)^2 + \left(d\varepsilon_3^p - d\varepsilon_1^p \right)^2 \right]^{1/2} \tag{10.30b}$$

Under condition of uniaxial tension σ_1, $\sigma_2 = \sigma_3 = 0$ and $d\varepsilon_2^p = d\varepsilon_3^p = -\, d\varepsilon_1^p/2$ from volume constancy. Substitution in the expression for $\bar{\sigma}$ and $d\bar{\varepsilon}^p$ yields, respectively,

$$\bar{\sigma} = \sigma_1, \quad d\bar{\varepsilon}_1^p = d\varepsilon_1^p$$

Hence, the stress–strain curve in uniaxial tension represents, indeed, the effective stress–effective strain relation for the tested material.

In any general state of stress, a measure of the total amount of plastic strain is obtained by adding up all the increments of effective plastic strain as: $\bar{\varepsilon}^p = \int d\bar{\varepsilon}^p$. This offers a realistic prediction for the effective plastic strain based on an incremental theory of plasticity. However, during a complete reversal of strain paths, i.e., tension followed by compression, the integration $\int d\bar{\varepsilon}^p$ may lead to an overestimation of $\bar{\varepsilon}^p$ because of the Bauschinger effect. Also, if this reversal is repeated for a number of cycles, actual material behavior shows that the strain-hardening effect would begin to saturate and a calculation of $\bar{\varepsilon}^p$ by the incremental plasticity theory would become in error.

In view of the definition given to $d\lambda$ by Equation 10.28, the Levy–Mises flow rule (Equation 10.22b) becomes

$$\frac{d\varepsilon_1^p}{\sigma_1^{\backslash}} = \frac{d\varepsilon_2^p}{\sigma_2^{\backslash}} = \frac{d\varepsilon_3^p}{\sigma_3^{\backslash}} = \frac{3}{2} \frac{d\bar{\varepsilon}^p}{\bar{\sigma}} \tag{10.31}$$

Again, as indicated by Equation 10.22c, if the straining paths are linear, i.e., keeping the same ratios among the strain increments, the total plastic strains ε_1^p, ε_2^p, and ε_3^p and the proportionality factor $\lambda = \bar{\varepsilon}^p / \bar{\sigma}$ are used throughout instead of their increments.

Effective stress and strain increments and the associated flow rule could be also defined using the Tresca yield criterion (Equation 10.18). In such a case, they are respectively given by

$$(\bar{\sigma})_T = \sigma_1 - \sigma_3 \text{ with } \sigma_I > \sigma_2 > \sigma_3 \tag{10.32a}$$

$$\left(d\bar{\varepsilon}^p\right)_T = d\varepsilon_1^p \text{ with } d\varepsilon_1^p > d\varepsilon_2^p > d\varepsilon_3^p$$

and (10.32b)

$$d\varepsilon_1^p = -d\varepsilon_2^p > 0, \;\; d\varepsilon_2^p = 0$$

where the subscript T denotes the Tresca criterion. For problems involving relatively large plastic strains compared with the elastic strains, the superscript p is often removed from the notation of the effective plastic strain, thus simply becoming $\bar{\varepsilon}$.

At this stage it is worth noting that a strain-hardening material undergoes plastic deformation in such a way to cause maximum dissipation of energy. Expression 10.29 indicates this by recalling that, according to the normality rule, the deviatoric stress vector acts in the same direction as the plastic strain increment vector. Hence, the scalar product of these two vectors expresses the maximum work performed during plastic deformation. For a hardening material, the net work done by an external agency is always positive during loading and nonnegative over the cycle of load application and removal.* This also means that the energy input into a solid when it is deforming plastically is never recovered. A consequence of this is that all yield surfaces for these materials are found to be convex around the origin. Convexity here means that each tangent plane to the surface represented by the yield function does not intersect the surface.**

Example 10.9:

The state of stress in a rigid strain-hardening solid behaving according to $\bar{\sigma} = 450\ (\bar{\varepsilon}^{0.4})$ is given by: $\sigma_1 = 350$ MPa, $\sigma_2 = 250$ MPa, and $\sigma_3 = 0$, applied proportionally.

a. *Using total strains, what are the ratios among the plastic strain components?*
b. *By superimposing a fluid pressure of 300 MPa around the solid, show that these ratios remain unchanged.*
c. *Calculate the values of plastic strains and plastic work done.*
d. *Show that the strain vector is perpendicular to the yield surface.*

Solution:

a. The mean stress $\sigma_m = (\sigma_1 + \sigma_2 + \sigma_3)/3 = 200$ MPa. The deviatoric stresses are thus

$$\sigma_1' = 350 - 200 = 150 \text{ MPa}, \; \sigma_2' = 50 \text{ MPa, and } \sigma_3' = -200 \text{ MPa}$$

Applying the Levy–Mises flow rule, Equations 10.31 gives

$$\varepsilon_1^p/\sigma_1' = \varepsilon_2^p/\sigma_2' = \varepsilon_3^p/\sigma_3' = \frac{3}{2}\bar{\varepsilon}^p/\bar{\sigma}$$

i.e.,

$$\varepsilon_1^p : \varepsilon_2^p : \varepsilon_3^p = 3:1:-4$$

b. Superimposing a fluid pressure of 300 MPa changes the stress state to

* This stems from Drucker's postulate of work-hardening "stable" materials.[5]

** Mathematically, the convexity conditions for a function $f(\sigma_{ij})$ is established if the eigenvalues of the matrix $M_{ij} = (\partial^2 f)/(\partial\sigma_i\partial\sigma_j)$; $i, j = 1, 2, 3$ are nonnegative. For instance, for the von Mises plane stress yield function; $f = \sigma_1^2 + \sigma_2^2 - \sigma_1\sigma_2 - Y^2$, the conditions $(M_{11} + M_{22}) \geq 0$ and $(M_{11}M_{22} - M_{12}M_{21}) \geq 0$ must be satisfied and, in fact, they are. For three-dimensional stress space, all leading minor determinants of the matrix $M_{i,j} = 1, 2, 3$ and its eigenvalues must be all nonnegative.

$$\sigma_1 = 350 - 300 = 50 \text{ MPa}, \ \sigma_2 = 250 - 300 = -50 \text{ MPa}$$

$$\sigma_3 = 0 - 300 = -300 \text{ MPa}, \text{ and } \sigma_m = -100 \text{ MPa}$$

The deviatoric stresses are thus

$$\sigma_1^{\backslash} = 50 - (-500) = 150 \text{ MPa}, \ \sigma_2^{\backslash} = 50 \text{ MPa}, \text{ and } \sigma_3^{\backslash} = -200 \text{ MPa}$$

Since according to the flow rule the plastic strains are functions of the deviatoric stresses, their ratios remain unchanged, as

$$\varepsilon_1^p/150 = \varepsilon_2^p/50 = -\varepsilon_3^p/200$$

giving

$$\varepsilon_1^p : \varepsilon_2^p : \varepsilon_3^p = 3:1:-4$$

c. Knowing the stress components, the effective stress is calculated from Expression 10.26, as

$$\bar{\sigma} = \frac{1}{\sqrt{2}}\left[(\sigma_1 - \sigma_2)^2 + (\sigma_2 - \sigma_3)^2 + (\sigma_3 - \sigma_1)^2\right]^{1/2} = 312.25 \text{ MPa}$$

Hence, from the material behavior $\bar{\sigma} = 450\,\bar{\varepsilon}^{0.4}$,

$$\bar{\varepsilon} = (\bar{\sigma}/450)^{1/0.4} = (312.25/450)^{2.5} = 0.401$$

The strain components are thus obtained from

$$\varepsilon_1^p/150 = \varepsilon_2^p/50 = -\varepsilon_3^p/200 = 3\bar{\varepsilon}/2\bar{\sigma}$$

$$= 3 \times 0.401/(2 \times 312.25)$$

i.e., $\varepsilon_1^p = 0.288$, $\varepsilon_2^p = 0.096$, and $\varepsilon_3^p = -0.384$. The total plastic work done per unit volume according to Equation 10.27 is

$$W^p = \int_0^{\bar{\varepsilon}} \bar{\sigma} d\bar{\varepsilon} = \int_0^{0.401} \left(450\bar{\varepsilon}^{0.4}\right) d\bar{\varepsilon} = 89.5 \text{ MPa}$$

d. von Mises yield surface for plane stress $\sigma_3 = 0$ reduces to

$$f = \sigma_1^2 - \sigma_1\sigma_2 + \sigma_2^2 - Y^2$$

The outward normal to this surface has the direction ratios:

$$1 = \frac{\partial f}{\partial \sigma_1} = \frac{\partial}{\partial \sigma_1}\left(\sigma_1^2 - \sigma_1\sigma_2 + \sigma_2^2 - Y^2\right) = 2\sigma_1 - \sigma_2$$

$$m = \frac{\partial f}{\partial \sigma_2} = \frac{\partial}{\partial \sigma_2}\left(\sigma_1^2 - \sigma_1\sigma_2 + \sigma_2^2 - Y^2\right) = 2\sigma_2 - \sigma_1$$

At the stress point (350, 250, 0),

$$l/m = (2\sigma_1 - \sigma_2)/(2\sigma_2 - \sigma_1) = 3/1$$

which is the same ratio between $\varepsilon_1^p / \varepsilon_2^p$.

Example 10.10:

A cube of annealed low-carbon steel $\bar{\sigma} = 100(\bar{\varepsilon})^{0.1}$ MPa is forged with a hydraulic press reducing its height from 100 mm to 70 mm. The forging process is performed in such a way to keep the dimensions of one side of the cube unchanged as shown in Figure 10.12. Friction at contact surfaces is to be neglected and deformation is assumed uniform. Calculate:

a. *The forging load P_1 at the end of the process,*
b. *The load on the side platens, P_2,*
c. *The total work done per unit volume at the end of this deformation.*

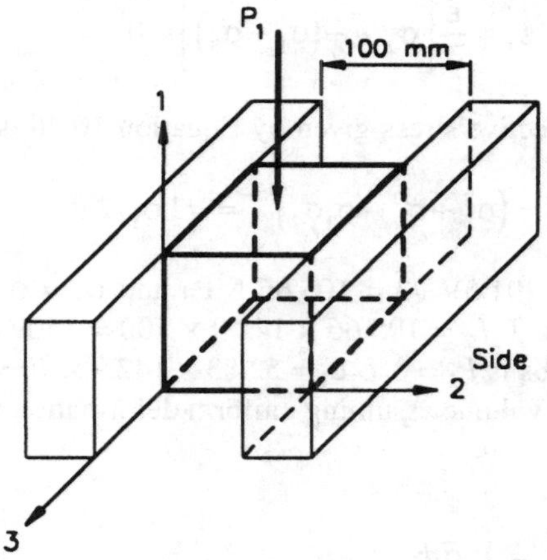

FIGURE 10.12 Example 10.12.

Solution:

a. This is a state of plane strain with linear strain paths and hence the total strains are used; also the elastic strains are neglected.

$$\varepsilon_1 = \ln\left(\frac{70}{100}\right) = -0.357$$

$$\varepsilon_2 = 0$$

Constant-volume deformation gives

$$\varepsilon_3 = -\varepsilon_1 = 0.357$$

The effective strain given by Equation 10.30 is

$$\bar{\varepsilon} = \left(\sqrt{2/3}\right)\left[(0.357 - 0)^2 + (0 + 0.357)^2 + (-0.357 - 0.357)^2\right]^{1/2}$$

$$\bar{\varepsilon} = 0.412$$

The effective stress corresponding to this effective strain is obtained from the strain-hardening behavior of this steel, i.e.,

$$\bar{\sigma} = 100(\bar{\varepsilon})^{0.1} = 100(0.412)^{0.1} = 91.5 \text{ MPa}$$

To calculate the loads at the end of the process, the final cube dimensions must be found as

$$\varepsilon_3 = \ln \frac{L_3}{L_o} \quad \text{or} \quad L_3 = L_o \exp(\varepsilon_3)$$

$$L_3 = 100 \ \exp(0.357) = 142.9 \text{ mm}$$

Proportional loading during this forging process produces a state of stress (σ_1, σ_2, 0). Hence, the Levy–Mises Equation 10.22a in its integrated form, together with $\lambda = \bar{\varepsilon} / \bar{\sigma}$, noting that $\varepsilon_2 = 0$ and $\sigma_3 = 0$, give

$$\varepsilon_2 = \frac{\bar{\varepsilon}}{\bar{\sigma}}\left[\sigma_2 - \frac{1}{2}(\sigma_1 + \sigma_3)\right] = 0$$

hence, $\sigma_2 = {}^1/_2\, \sigma_i$. The effective stress given by Equation 10.26 is thus

$$\bar{\sigma} = \left(\sigma_1^2 + \sigma_2^2 - \sigma_1\sigma_2\right)^{1/2} = \sqrt{3}\,\sigma_1/2$$

giving: $\sigma_1 = 2\,\bar{\sigma}/\sqrt{3} = 2\ (91.5)/\sqrt{3} = 105.66$ MPa and $\sigma_2 = \sigma_1/2 = 52.83$ MPa. The forging load is thus $P_1 = \sigma_1 L_3 L_2 = 105.66 \times 142.9 \times 100 = 1509.88$ kN.

b. The load on the side platens is $P_2 = \sigma_2 L_3 L_1 = 52.83 \times 142.9 \times 70 = 528.46$ kN. The total plastic work done per unit volume assuming uniform deformation and neglecting friction is given by Equation 10.27:

$$\int dW^p = \int_0^{\bar{\varepsilon}} \bar{\sigma} d\varepsilon$$

$$W^p = \int_0^{0.412} 100(\bar{\varepsilon})^{0.1} d\bar{\varepsilon} = \left[\frac{100}{1.1}(\bar{\varepsilon})^{1.1}\right]_0^{0.412}$$

$$W^p = 34.28 \text{ MPa}$$

Note that an incorrect answer for w^p would have been obtained by multiplying the values of $\bar{\sigma}$ and $\bar{\varepsilon}$ at the end of the process without performing the above integration, since both $\bar{\sigma}$ and $\bar{\varepsilon}$ change during the process.

10.11 EXPERIMENTAL DETERMINATION OF THE FLOW CURVE

The curve obtained by plotting $\bar{\sigma}$ agains $\bar{\varepsilon}$ is called the flow curve. From the definitions of effective stress and effective plastic strain increment, Equations 10.26 and 10.30, respectively, a state of simple tension is expressed by

$$\bar{\sigma} = \sigma_1$$

$$d\bar{\varepsilon}^p = d\varepsilon_1^p$$

This shows that the true stress–plastic strain curve obtained in simple tension is the flow curve. Hence, according to the strain-hardening hypothesis, Equation 10.7 may be written as*

$$\bar{\sigma} = Y = H\left(\int d\bar{\varepsilon}^p\right) \tag{10.33}$$

Thus, the strain-hardening laws according to Equations 10.8b and 10.8e may be expressed, respectively, by

$$\bar{\sigma} = K\left(\bar{\varepsilon}^p\right)^n \tag{10.34a}$$

or

$$\bar{\sigma} = Y_o + A\left(\bar{\varepsilon}^p\right)^q \tag{10.34b}$$

Many plastic deformation problems are analyzed using such material behavior laws as a unique representation for a given material and loading conditions. Thus, the effective stress–effective plastic strain curve in tension may be used to predict the stress–strain behavior under other loading conditions, e.g., torsion, biaxial stretching, etc.

In some cases such procedure must be considered as a first-order approximation. Experiments have shown that the $\bar{\sigma} - \bar{\varepsilon}^p$ behavior derived from other tests such as plane strain compression, biaxial tension, torsion, etc.,[6] may not coincide with the simple tension test results. Differences are usually greatest at large strains and may be due, in part, to anisotropy. In other cases, the suggested yield function and, hence, the definition of $\bar{\sigma}$ prove sometimes to be inappropriate.[7]

A graphical representation of the proportionality factor $d\lambda$ for a strain-hardening material is shown in Figure 10.13a. If straining is proportional, i.e., the ratios among the principal strains remain unchanged during loading, the plastic strain increments $d\varepsilon_1^p$, $d\varepsilon_2^p$, and $d\varepsilon_3^p$ can be replaced by the integrated strains ε_1^p, ε_2^p, and ε_3^p. In such a case $d\lambda$ is replaced by $\lambda = \bar{\varepsilon}^p / \bar{\sigma}$. This is graphically shown in Figure 10.13b.

Example 10.11:

A process of forming thin soft aluminium tubes of diameter to thickness ratio, $r_{mo}/h_o = 10$, consists of compressing a rubber insert inside the tube as shown in Figure 10.14. Assuming no friction between rubber and aluminum,

a. *Determine the axial pressure p acting on the rubber in order to initiate yielding in the aluminum tube;*
b. *If a relative radial expansion Δ is to be achieved in the aluminum tube, find an expression for the axial pressure p (neglect elastic strains);*
c. *What is the value of p at the stage of forming when $\Delta = 20\%$?*

Take elastic properties as $E_A = 70$ GPa, $\nu_A = 0.3$, and $Y_A = 180$ MPa for aluminum and $E_R = 0.02$ GPa and $\nu_R = 0.45$ for rubber. The effective stress–effective strain for aluminium is

$$\bar{\sigma} = Y_A\left[1 + B\left(\frac{\bar{\varepsilon}^p}{Y_A/E_A}\right)^q\right]$$

where $B = 0.15$ and $q = 0.3$.

---

* A less common hardening hypothesis is that of work hardening, where it is assumed that $\bar{\sigma} = Y = F\left(\int dW^p\right) = F\left(\int \bar{\sigma} d\bar{\varepsilon}\right)$.

For metals, both strain and work-hardening hypotheses lead approximately to the same results.

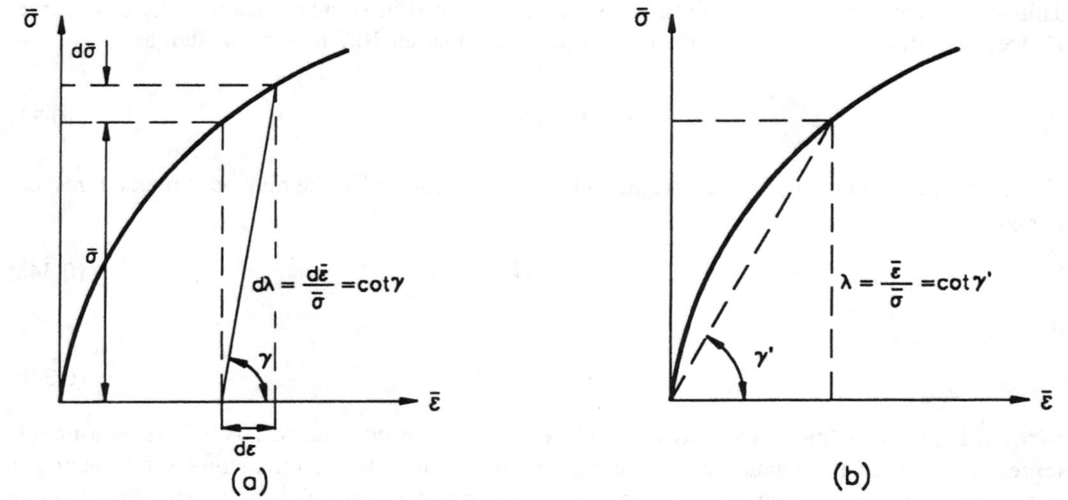

FIGURE 10.13 The significance of the proportionality factor: (a) $d\lambda$ and (b) λ in nonproportional and proportional straining.

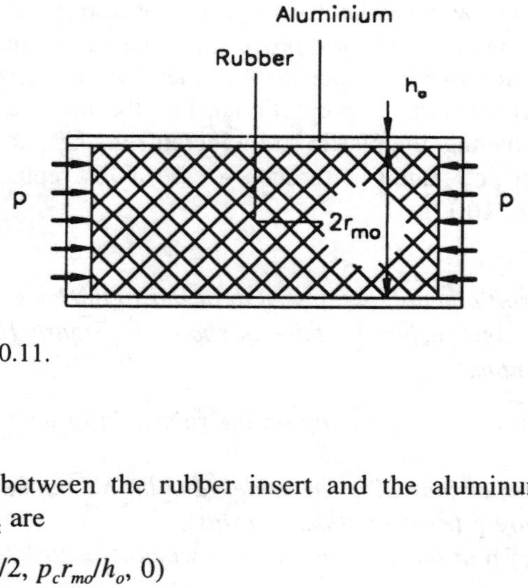

FIGURE 10.14 Example 10.11.

Solution:

Let the contact pressure between the rubber insert and the aluminum tube be p_c. The stress components σ_r, σ_θ, and σ_z are

For the aluminum: $(-p_c/2,\ p_c r_{mo}/h_o,\ 0)$
For rubber: $(-p_c,\ -p_c\ -p)$

Note that σ_r for the aluminum tube is taken at the mean radius.

The tube is considered elastic at the initiation of yielding. Applying Hooke's law, Equation 3.6, to both aluminum and rubber to express the common radial displacement $u = \varepsilon_\theta r_{mo}$ at the contact surface as

$$u = \frac{r_{mo}}{E_A}\left[\frac{p_c r_{mo}}{h_o}\right] - \nu_A\left(-\frac{P_c}{2}+0\right) = \frac{r_{mo}}{E_R}\left[-p_c - \nu_R\left(-p_c - p\right)\right]$$

Solving for p in terms of p_c gives

$$p = \left[\frac{E_R}{2E_A} \left(\frac{2r_{mo}}{h_o} + \nu_A \right) + (1 - \nu_R) \right] \frac{p_c}{\nu_R}$$

For $\nu_A = 0.3$, $\nu_R = 0.45$, $E_A = 70$ GPa, $E_R = 0.02$ GPa, and $r_{mo}/h_o = 10$,

$$p = 1.2287p_c$$

a. To initiate yielding in the aluminum tube ($r_{mo}/h_o = 10$) according to the von Mises criterion,

$$\bar{\sigma} = Y_A = (\sigma_1^2 + \sigma_2^2 - \sigma_1\sigma_2)^{1/2} = \left[(-p_c/2)^2 + (10p_c)^2 - (-p_c/2)(10p_c) \right]^{1/2} = 10.26p_c$$

Hence, for $Y_A = 180$ MPa: $p_c = 17.54$ MPa. Substituting for p_c gives $p = 1.2287 \times 17.54 = 21.55$ MPa.

b. The relation between p and p_c for a relative radial expansion Δ is obtained by considering the rubber to be elastic and substituting $\varepsilon_\theta = \Delta$. In this respect two approaches are considered:
   - Hooke's law, Equation 3.6:

$$\Delta = \frac{1}{E_R}\left[-p_c - \nu_R(-p_c - p) \right]$$

   which yields $p = [\Delta E_R + p_c(1 - \nu_R)]/\nu_R$
   - Rubber elasticity relation, Equations 3.60 with $A_2 = 0$:

$$\sigma_1 = A_1\lambda_1^2 + \sigma_p, \quad \sigma_2 = A_1\lambda_2^2 + \sigma_p, \quad \text{and} \quad \sigma_3 = A_1\lambda_3^2 + \sigma_p$$

Take $\sigma_3 = \sigma_z = -p$ and $\sigma_1 = \sigma_2 = \sigma_r = \sigma_\theta = -p_c$. From the constant-volume condition $\lambda_1\lambda_2\lambda_3 = 1$ or $\lambda_2 = \lambda_3 = \sqrt{\lambda_3} = 1 + \Delta$. Hence,

$$-p = A_1\lambda_3^2 + \sigma_p \quad \text{and} \quad -p_c = \frac{A_1}{\lambda_3} + \sigma_p$$

Eliminating σ_p gives

$$p = A_1\left(\frac{1}{\lambda_3} - \lambda_3^2 \right) + p_c$$

Substituting

$$A_1 = \frac{E_R}{3} \quad \text{and} \quad \lambda_3 = \frac{1}{(1+\Delta)^2}$$

gives

$$p = \frac{E_R}{3}\left((1+\Delta)^2 - \frac{1}{(1+\Delta)^4}\right) + p_c$$

Neglecting powers of Δ higher than 1 results in $p = 2\Delta E_R + p_c$, which is the same result obtained from Hooke's law if ν_R is taken to 0.5.

c. A radial expansion of $\Delta = 0.2$ means a circumferential strain of

$$\varepsilon_\theta = \ln\frac{r_m}{r_{mo}} = \ln(1+0.2) = 0.1823$$

The ratios among plastic strain components are found from the Levy–Mises flow rule, Equation 10.31, knowing that deviatoric stresses in the aluminum tube are given by

$$\left[-\frac{p_c}{2} - \frac{1}{3}\left(-\frac{p_c}{2} + \frac{p_c r_m}{h}\right)\right] : \left[\frac{p_c r_m}{h} - \frac{1}{3}\left(-\frac{p_c}{2} + \frac{p_c r_m}{h}\right)\right] :$$

$$\left[-\frac{1}{3}\left(-\frac{p_c}{2} + \frac{p_c r_m}{h}\right)\right] \quad \text{and} \quad \frac{r_m}{h} = 10$$

i.e., $(-3.667p_c : 6.833p_c : -3.167p_c)$, since the stress ratios remain unchanged during deformation. Hence,

$$\frac{\varepsilon_r^p}{-3.667p_c} = \frac{\varepsilon_\theta^p}{6.833p_c} = \frac{\varepsilon_z^p}{-3.167p_c}$$

giving $\varepsilon_r^p = -0.0978$, $\varepsilon_\theta^p = 0.1823$, and $\varepsilon_z^p = -0.0845$ with the effective strain, according to Equation 10.30;

$$\bar{\varepsilon} = 0.1825$$

For aluminum:

$$\bar{\sigma} = Y_A\left[1 + B\left(\frac{\bar{\varepsilon}^p}{Y_A/E_A}\right)^q\right] = 180\left[1 + 0.15\left(\frac{\bar{\varepsilon}}{180/70000}\right)^{0.3}\right]$$

hence, $\bar{\sigma} = 277$ MPa; giving $p_c = \bar{\sigma}/10.26 = 27$ MPa and $p = [0.2 \times 0.02 \times 10^3/1.2 + 27(1 - 0.45)]/0.45 = 40.41$ MPa.

10.12 ISOTROPIC HARDENING

In view of the definition of the effective stress Equation 10.26, the von Mises yield criterion has been stated as

$$\bar{\sigma} = Y = \frac{1}{\sqrt{2}}\left[(\sigma_1 - \sigma_2)^2 + (\sigma_2 - \sigma_3)^2 + (\sigma_3 - \sigma_1)^2\right]^{1/2}$$

As indicated before, the true stress–strain curve for a strain–hardening material is looked upon as a locus of successive yield points. Hence, at any plastic strain value $\bar{\varepsilon}^p$, the stress $\bar{\sigma}$ corresponds to a yield stress *Y*, which should be applied to cause further plastic deformation. This means that the yield locus (e.g., von Mises ellipse in plane stress) expands during plastic flow with its center and axes remaining fixed as shown in Figure 10.15. The size of the yield ellipse is simply determined from the simple tensile $\bar{\sigma}-\bar{\varepsilon}$ curve at the specified value of $\int d\bar{\varepsilon}^p$. Note that for each ellipse $\bar{\sigma}=$ constant. Hence, in view of Equations 10.33 and 10.27 $\int d\bar{\varepsilon}$, and $\int dW^p$ are also constant for a given ellipse.

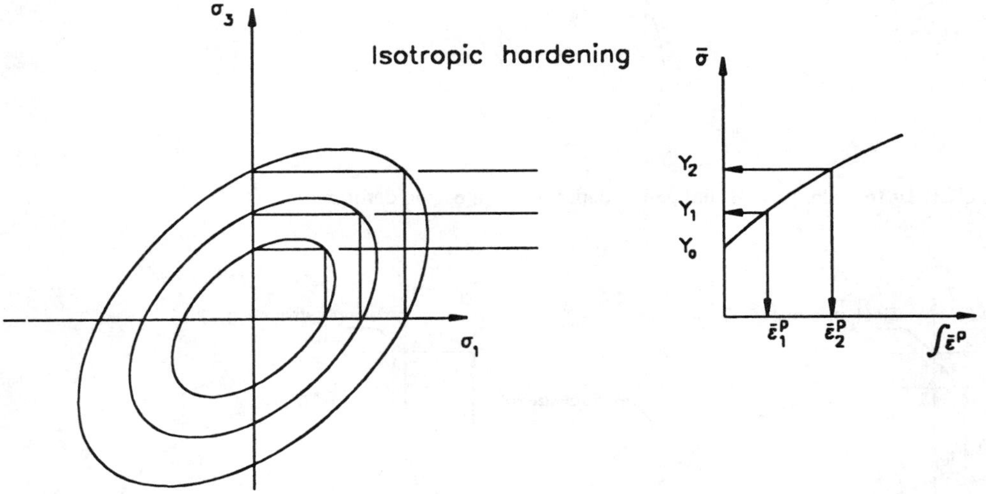

FIGURE 10.15 Expansion of the von Mises yield locus according to the isotropic-hardening rule.

This representation of yield surfaces for strain-hardening materials is known as the isotropic-hardening rule. Experiments indicate that it gives an appropriate description for isotropic metals provided that the Bauschinger effect is negligible. Other hardening rules have been postulated, but they are elaborate to handle mathematically, as will be shown in Section 10.16.

10.13 UNIQUENESS AND PATH DEPENDENCE

For loads causing pure elastic deformation, strains are uniquely determined by stresses via Hooke's law. In this respect, uniqueness implies that the strains corresponding to any state of stress that lies within the yield surface are determined solely from the values of the final stress state irrespective of the path of loading followed to reach that stress state. Hence, identical strain components are obtained for the stress state defined by point (*A*) in Figure 10.16 irrespective of the loading path followed to reach that point. It is thus concluded that elastic deformation is characterized by two basic features: uniqueness and path independence.

Once the limit of elasticity is exceeded, uniqueness and path independence do not generally hold. Thus, plastic strains are not uniquely defined by the final state of stress, but rather depend upon the entire history of loading prior to reaching that final state of stress. This can be illustrated by considering two thin-walled tubes, as shown in Figure10.9a, which are subjected to a combination of axial force and twisting moment enducing stresses σ and τ, respectively. The two tubes are made of a strain-hardening material that follows the von Mises yield criterion and the isotropic-hardening rule. The first tube is subjected to the loading history (0–1–2–0–3–4) while the second tube is subjected to the loading history (0–3–4) as illustrated in the first quadrant of σ vs. τ yield locus shown in Figure 10.17. Note that the final state of stress is the same in the two tubes.

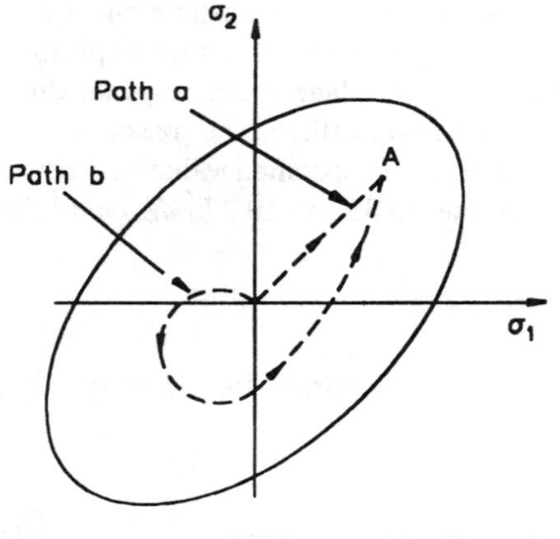

FIGURE 10.16 Uniqueness and path independence in elastic deformation.

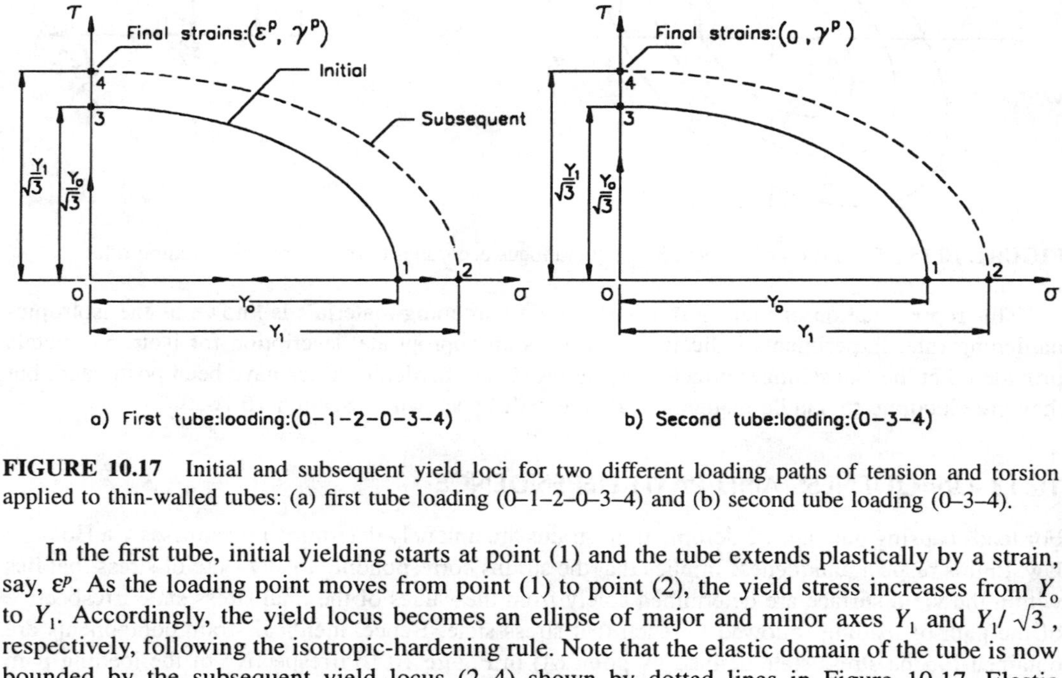

FIGURE 10.17 Initial and subsequent yield loci for two different loading paths of tension and torsion applied to thin-walled tubes: (a) first tube loading (0–1–2–0–3–4) and (b) second tube loading (0–3–4).

In the first tube, initial yielding starts at point (1) and the tube extends plastically by a strain, say, ε^p. As the loading point moves from point (1) to point (2), the yield stress increases from Y_0 to Y_1. Accordingly, the yield locus becomes an ellipse of major and minor axes Y_1 and $Y_1/\sqrt{3}$, respectively, following the isotropic-hardening rule. Note that the elastic domain of the tube is now bounded by the subsequent yield locus (2–4) shown by dotted lines in Figure 10.17. Elastic unloading takes place along the path (2–0) and then elastic loading in torsion takes place along the path (0–3–4). Hence, the final state of stress is ($\sigma = 0$, $\tau = Y_1/\sqrt{3}$) while the final state of plastic strain is (ε^p, γ^p).

Consider now the second tube, which is loaded along the path (0–3–4) from the virgin state. Initial yielding starts at point (3) and shear strain γ^p develops between points (3) and (4). The final yield locus is the same as the one obtained in the first tube. Hence, the final state of stress is ($\sigma = 0$, $\tau = Y_1/\sqrt{3}$), while the final state of plastic strain is (0, γ^p).

The point gained from the above discussion is that, although the two tubes have similar states of final stress, and hence elastic strains, the state of plastic strain is completely different in the two tubes. The conclusion to be drawn is that plastic strains depend upon the path of loading and hence they should be determined, in general, in an incremental form, as developed previously in Section 10.8.

Example 10.12:

A thin-walled tube of annealed aluminum of a radius-to-thickness ratio (r_m/h) of 10, closely fitted on a floating rigid plug to prevent radial deformation. The tube is loaded axially such that it is alternately stretched and compressed, the sequence being a 5% stretch then compression to the original length. Assuming uniform strain and strain-hardening described by $\bar{\sigma} = 170\,\bar{\varepsilon}^{0.25}$ (neglecting the Bauschinger effect), determine:

- a. *How many cycles (stretch plus compression) are needed to produce a yield strength of 180 MPa for the tube material?*
- b. *What are the stresses and the total plastic strain components induced in the tube wall at the end of stretching at the fifth cycle?*

Solution:

The elastic strains are neglected, and, since the radial deformation $u = 0$, then $d\varepsilon_\theta = u/r = 0$. Substituting in the constant-volume condition gives $d\varepsilon_z^p = -d\varepsilon_r^p$. Let $d\varepsilon_z^p = d\varepsilon_1^p$ and $d\varepsilon_r^p = d\varepsilon_3^p$, hence, $d\varepsilon_1^p = -d\varepsilon_3^p$ and, from Equation 10.30,

$$d\bar{\varepsilon} = 2\left|d\varepsilon_1^p\right|/\sqrt{3} = 1.155 d\varepsilon_1^p$$

a. The plastic strain increment during one cycle is

Stretching: $d\varepsilon_1^p = \ln 1.05 = 0.0488$

hence, $d\bar{\varepsilon} = 0.0488 \times 1.155 = 0.05635$

Compressing: $d\varepsilon_1^p = \ln \dfrac{1}{1.05} = -0.0488$

and, similarly, $d\bar{\varepsilon} = 0.0488 \times 1.155 = 0.05635$

To determine strain-hardening effects, these strain increments must be added, i.e.,

$$(d\bar{\varepsilon})_{\text{cycle}} = 2 \times 0.05635 = 0.1127$$

Using $\bar{\sigma} = 170\,\bar{\varepsilon}^{0.25}$ gives, at $\bar{\sigma} = 180$ MPa,

$$\bar{\varepsilon} = (180/170)^{1/0.25} = 1.257$$

The number of cycles is thus

$$N = \frac{1.257}{0.1127} \cong 11 \text{ cycle}$$

b. At the end of stretching, at the fifth cycle, Figure 10.18;

$$\bar{\varepsilon} = 4 \times 0.1127 + 0.05635 = 0.5072$$

Hence, the corresponding effective stress is

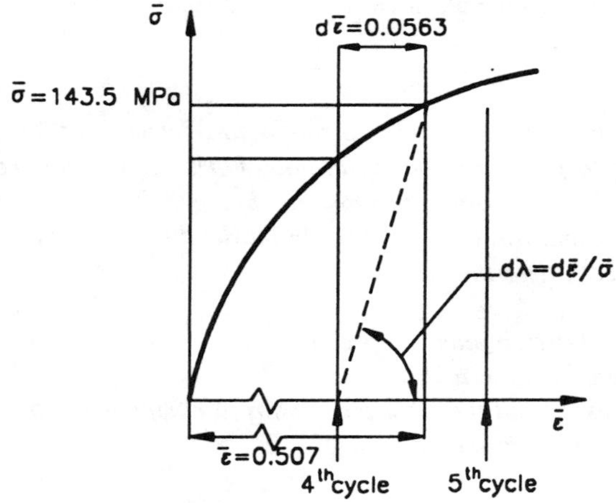

FIGURE 10.18 Example 10.12.

$$\bar{\sigma} = 170(0.5072)^{0.25} = 143.5 \text{ MPa}$$

The stresses acting in the tube wall are defined as $\sigma_1 = \sigma_z$, $\sigma_2 = \sigma_\theta$ and $\sigma_3 = \sigma_r$. For a contact pressure p_c between tube and plug $\sigma_\theta = p_c\,(r_m/h)$ and $\sigma_r = -p_c/2$. Substituting $r_m/h = 10$ gives $\sigma_2 = -20\sigma_3$. Again, for plane strain condition ($\varepsilon_2^p = 0$), $\sigma_2 - {}^1/_2(\sigma_1 + \sigma_3) = 0$, which gives $\sigma_1 = -41\sigma_3$. Substituting in Equation 10.26 yields

$$\bar{\sigma} = \frac{\sigma_3}{\sqrt{2}}\left[(-41+20)^2 + (-20-1)^2 + (1+41)^2\right]^{1/2}$$

$$= 36.37\sigma_3$$

For $\bar{\sigma}$ = 143.5 MPa, σ_1 = 161.95 MPa, σ_2 = 78.9 MPa, and σ_3 = –3.95 MPa. The total plastic strains induced in the tube wall are thus obtained by applying the flow rule, Equation 10.22c; hence,

$$\frac{\varepsilon_1^p}{\sigma_1^{\backslash}} = \frac{\varepsilon_2^p}{\sigma_2^{\backslash}} = \frac{\varepsilon_3^p}{\sigma_3^{\backslash}} = \frac{3}{2}\lambda$$

Since the straining paths are linear, the proportionality factor λ in Levy–Mises Equations 10.22 is

$$\lambda = \frac{0.5072}{143.5} = 0.00353$$

Hence, for $\sigma_m = {}^1/_3(161.95 + 78.9 - 3.95) = 78.97$ MPa, the deviatoric stresses are $\sigma_1^{\backslash}$ = 82.98 MPa, $\sigma_2^{\backslash} = 0$, and $\sigma_3^{\backslash}$ = 82.92 MPa, which give

$$\varepsilon_1^p = \frac{3}{2} \times 0.00354 \times 82.92 = -0.44$$

$$\varepsilon_3^p = -\varepsilon_1^p = -0.44 \text{ and } \varepsilon_2^p = 0$$

The above results are based on the isotropic-hardening rule. Different results are obtained according to other hardening rules, e.g., kinematic hardening, as given in Section 10.16.

10.14 COMPLETE ELASTIC–PLASTIC STRESS–STRAIN RELATIONS

In view of Equation 10.28, the stress–plastic strain increment relations equation (Equations 10.22) are

$$\begin{aligned} d\varepsilon_1^p &= \frac{d\bar{\varepsilon}^p}{\bar{\sigma}}\left[\sigma_1 - \frac{1}{2}(\sigma_2 + \sigma_3)\right] \\ d\varepsilon_2^p &= \frac{d\bar{\varepsilon}^p}{\bar{\sigma}}\left[\sigma_2 - \frac{1}{2}(\sigma_3 + \sigma_1)\right] \\ d\varepsilon_3^p &= \frac{d\bar{\varepsilon}^p}{\bar{\sigma}}\left[\sigma_3 - \frac{1}{2}(\sigma_1 + \sigma_2)\right] \end{aligned} \tag{10.35}$$

The complete elastic–plastic stress–strain relations, known as Prandtl–Reuss relations, are obtained by adding the elastic strain increments to the respective plastic strain increments, thus giving

$$\begin{aligned} d\varepsilon_1 &= \frac{1}{E}\left[d\sigma_1 - \nu(d\sigma_2 + d\sigma_3)\right] + \frac{d\bar{\varepsilon}^p}{\bar{\sigma}}\left[\sigma_1 - \frac{1}{2}(\sigma_2 + \sigma_3)\right] \\ d\varepsilon_2 &= \frac{1}{E}\left[d\sigma_2 - \nu(d\sigma_3 + d\sigma_1)\right] + \frac{d\bar{\varepsilon}^p}{\bar{\sigma}}\left[\sigma_2 - \frac{1}{2}(\sigma_3 + \sigma_1)\right] \\ d\varepsilon_3 &= \frac{1}{E}\left[d\sigma_3 - \nu(d\sigma_2 + d\sigma_1)\right] + \frac{d\bar{\varepsilon}^p}{\bar{\sigma}}\left[\sigma_3 - \frac{1}{2}(\sigma_2 + \sigma_1)\right] \end{aligned} \tag{10.36a}$$

where $d\varepsilon_1$, $d\varepsilon_2$, and $d\varepsilon_3$ are the total principal strain increments. Note that the elastic strain increments are associated with the stress increments, while the plastic strain increments are associated with the total stresses. The use of Equations 10.36 is not so simple, and these are to be applied only when the magnitude of the plastic strains is of the same order as for the elastic strains, such as in problems involving elastic–plastic deformations.

For the general state of stress, Equations 10.36a take the form:

$$\begin{aligned} d\varepsilon_x &= \frac{1}{E}\left[d\sigma_x - \nu(d\sigma_y + d\sigma_z)\right] + \frac{d\bar{\varepsilon}^p}{\bar{\sigma}}\left[\sigma_x - \frac{1}{2}(\sigma_y + \sigma_z)\right] \cdots \\ d\gamma_{xy} &= \frac{d\tau_{xy}}{G} + \frac{3}{2}\frac{d\bar{\varepsilon}^p}{\bar{\sigma}}\tau_{xy} \cdots \end{aligned} \tag{10.36b}$$

Example 10.13:

A stainless steel sheet of 2 mm thickness is stretched under a biaxial stress system σ_1 and σ_2 such that σ_2/σ_1 = 0.5. The stainless steel behavior is represented by Ramberg–Osgood law, Equation 10.8d:

$$\bar{\varepsilon} = \frac{\bar{\sigma}}{E}\left[1 + \alpha\left(\frac{\bar{\sigma}}{\sigma_o}\right)^{1/n^*-1}\right]$$

where $\alpha = 3/7$, $n^* = 0.25$, E = 190 GPa, $\nu = {}^1/_3$, and the offset stress at 0.2% strain is 640 MPa.

If stretching is done such that the effective plastic strain component becomes equal to 50% of the total effective strain under load, find:

a. *The stresses σ_1 and σ_2 required to stretch the sheet;*
b. *The sheet thickness under load at the end of the stretching operation.*

Solution:

a. The effective stress for $\sigma_2/\sigma_1 = 0.5$ and $\sigma_3 = 0$ is given by Equation 10.26 as

$$\bar{\sigma} = \sqrt{3}\,\sigma_1/2$$

From Ramberg–Osgood law, for $\bar{\varepsilon}^e = \bar{\varepsilon}^p = \bar{\sigma}/E = \bar{\varepsilon}/2$;

$$\frac{\bar{\sigma}}{E} = \frac{\alpha\bar{\sigma}}{E}\left(\frac{\bar{\sigma}}{\sigma_o}\right)^{1/n^*-1}$$

where the superscripts *e* and *p* denote elastic and plastic strains respectively. The constant σ_o is often taken equal to the offset stress at 0.2% strain. Hence,

$$\frac{3}{7}\left(\frac{\bar{\sigma}}{640}\right)^3 = 1$$

i.e., $\bar{\sigma}$ = 848.9 MPa and $\bar{\varepsilon}$ = 0.00894. The applied stresses are

$$\sigma_1 = 980.2 \text{ MPa}$$

$$\sigma_2 = 490.1 \text{ MPa}$$

b. The thickness strain under load is given by Prandtl–Reuss Equations 10.36 in its integrated form for proportional loading as

$$\varepsilon_3 = \frac{1}{E}\left[\sigma_3 - \nu(\sigma_2 + \sigma_1)\right] + \frac{\bar{\varepsilon}^p}{\bar{\sigma}}\left[\sigma_3 - \frac{1}{2}(\sigma_2 + \sigma_1)\right]$$

for $\sigma_3 = 0$ and $\sigma_2/\sigma_1 = 0.5$

$$\varepsilon_3 = -\frac{3\nu}{2E}\sigma_1 - \frac{3\varepsilon^P}{4\sigma}\sigma_1 = \sigma_1\left[\frac{3\nu}{2E} + \frac{3\bar{\varepsilon}^P}{4\bar{\sigma}}\right]$$

Noting that $\bar{\varepsilon}^P = 0.5$, $\bar{\varepsilon} = 0.00894/2 = 0.00447$, hence

$$\varepsilon_3 = \ln\left(\frac{h}{h_o}\right) = -980.2\times10^6\left[\frac{3(1/3)}{2\times190\times10^9} + \frac{3\times4.47\times10^{-3}}{4\times848.8\times10^6}\right]$$

which gives $h = 1.98714$ mm.

Example 10.14:

An open-ended thin-walled tube is made of soft brass of outside and inside diameters of 100 mm and 98 mm, respectively. The tube is subjected to a tensile force P *and an internal pressure* p_i, *until yielding occurs. The effective stress–effective plastic strain law of soft brass is given by* $\bar{\sigma} = K(\bar{\varepsilon})^n$, *where* K = *600 MPa and* n = $^1/_3$. *Neglect elastic strains and take* Y_o = *200 MPa.*

a. *Determine the ratio of* P/p_i *and hence their values at yielding according to the von Mises criterion.*
b. *If both the pressure and the tensile load are increased while maintaining the tube length constant, determine the final values of* P *and* p_i *at a diametral expansion of 20%.*

Solution:

a. The hoop, axial, and radial stresses in the tube wall, respectively, are given as

$$\sigma_z = \frac{P}{2\pi r_{mo} h_o}, \quad \sigma_\theta = \frac{p_i r_{mo}}{h_o}, \text{ and } \sigma_r \cong 0$$

where r_{mo} and h_o are the initial mean radius and thickness, respectively. If tube length remains unchanged, then

$$\varepsilon_z^p = 0 = \lambda[\sigma_z - {}^1/_2\sigma_\theta] \text{ giving } \sigma_z = {}^1/_2\sigma_\theta$$

This requires the ratio between the tensile load P and the internal pressure p_i to be

$$\frac{P}{2\pi r_{mo} h_o} = \frac{p_i r_{mo}}{2h_o} \quad\text{or}\quad \frac{P}{p_i} = \pi r_{mo}^2$$

Yielding occurs according to the von Mises criterion, Equation 10.17, when

$$\bar{\sigma} = \left(\sigma_\theta^2 + \sigma_z^2 - \sigma_\theta\sigma_z\right)^{1/2} = \frac{\sqrt{3}}{2}\sigma_\theta = \frac{\sqrt{3}}{2}\frac{p_i r_{mo}}{h} = Y_o$$

For $Y_o = 200$ MPa, $r_{mo} = 49.5$ mm, and $h_o = 1$ mm, this results in $p_i = 4.665$ MPa and $P = \pi r_{mo}\, p_i = \pi x\,(49.5)^2 \times 4.665 = 35.91$ kN.

b. During large plastic deformation, the current tube dimensions r_m and h have to be used throughout. However, since the tube length does not change, $\sigma_z = \sigma_\theta/2$. Constancy of volume implies that $r_{mo}h_o = r_m h$. Hence, the ratio P/p_i is given by

$$P/p_i = \pi r_m^2 = \pi r_m^2 \exp(2\varepsilon_\theta)$$

where the definition $\varepsilon_\theta = \ln(r_m/r_{mo})$ is used.
From the flow rule, Equation 10.35, for constant stress ratio,* knowing that $\bar{\sigma} = \sqrt{3}\,\sigma_\theta/2$:

$$\varepsilon_\theta^p = \frac{\bar{\varepsilon}}{\bar{\sigma}}\left[\sigma_\theta - \frac{1}{2}\sigma_z\right] = \frac{3}{4}\frac{\bar{\varepsilon}}{\bar{\sigma}}\sigma_\theta = \frac{\sqrt{3}}{2}\bar{\varepsilon}$$

Substituting from the material law $\bar{\sigma} = K\,(\,\bar{\varepsilon}\,)^n$ gives

$$\varepsilon_\theta^p = \frac{\sqrt{3}}{2}\left(\frac{\bar{\sigma}}{K}\right)^{1/n} = \frac{\sqrt{3}}{2}\left(\frac{\sqrt{3}\,\sigma_\theta}{2K}\right)^{1/n}$$

In view of volume constancy,

$$\sigma_\theta = \frac{p_i r_m}{h} = p_i \frac{r_m^2}{r_{mo} h_o} = p_i \left(\frac{r_m}{r_{mo}}\right)^2 \left(\frac{r_{mo}}{h_o}\right) = p_i \left(\frac{r_{mo}}{h_o}\right) \exp(2\varepsilon_\theta^p)$$

Therefore, substitution gives

$$\varepsilon_\theta^p = \left(\frac{\sqrt{3}}{2}\right)^{1+1/n} \left[\left(\frac{p_i}{K}\right)\left(\frac{r_{mo}}{h_o}\right)\exp(2\varepsilon_\theta^p)\right]^{1/n}$$

For $n = 1/3$, $K = 600$ MPa, and $\varepsilon_\theta^p = \ln 1.2 = 0.18232$, this results in

$$p_i = 5.782 \text{ MPa and } P = 64.09 \text{ kN}$$

These values indicate that P/p_i does not remain constant throughout during deformation even though $\sigma_\theta/\sigma_z = 2$ as well as the strain ratio $\varepsilon_\theta/\varepsilon_r = -1$ throughout.

10.15 PLASTIC DEFORMATION OF ANISOTROPIC MATERIALS

Isotropy has been one of the basic assumptions adopted throughout the preceding sections. Now it is required to formulate a plasticity theory for anisotropic materials.** Such extension is important in handling many problems of plastic deformation, where anisotropic behavior becomes dominant and its neglect would impair the accuracy of the results. Problems of this type are often encountered in the analysis of cold working of sheet metals, e.g., deep drawing, rolling, etc.

* For a nonproportional loading or when P/p_i changes arbitrarily, an analytical solution of the problem becomes more involved since the incremental strains, not the integrated ones, have to used. For this, see Hill.[3]

** Plastic anisotropy has its origin in "Crystallographic texture" or preferred orientation developed in the grains of wrought metals by slip or twinning. Recrystallization during annealing of a cold-worked metal changes the crystallographic texture but generally will not restore isotropy, as a "recrystallization texture" is often produced. Distinction has to be made between crystallographic fibering and mechanical fibering, which is brought about by the alignment of inclusions, cavities, and second-phase particles in the main direction of mechanical working. This may greatly affect fracture.

10.15.1 A Yield Criterion for Anisotropic Materials

Hill[3] considered a homogeneous anisotropic material with three orthogonal axes of anisotropy *x*, *y*, and *z*. The planes *x–y*, *y–z*, and *z–x* are assumed to be planes of mirror symmetry. Assuming that there is no Bauschinger effect and that a hydrostatic stress does not influence yielding, Hill proposed a yield function in terms of the stress components:[3]

$$2f = F(\sigma_y - \sigma_z)^2 + G(\sigma_z - \sigma_x)^2 + H(\sigma_x - \sigma_y) + 2L\tau_{yz}^2 + 2M\tau_{zx}^2 + 2N\tau_{xy}^2 = 1 \quad (10.37)$$

where *F, G, H, L, M*, and *N* are parameters representing the current state of anisotropy. These material parameters are determined from tensile and shear tests conducted on specimens cut along the orthogonal axes of anisotropy *x, y*, and *z*, as shown in Figure 10.19.

In a rolled sheet metal, it is conventional to take the *x, y*, and *z*-directions to be the rolling, transverse and through-thickness directions.

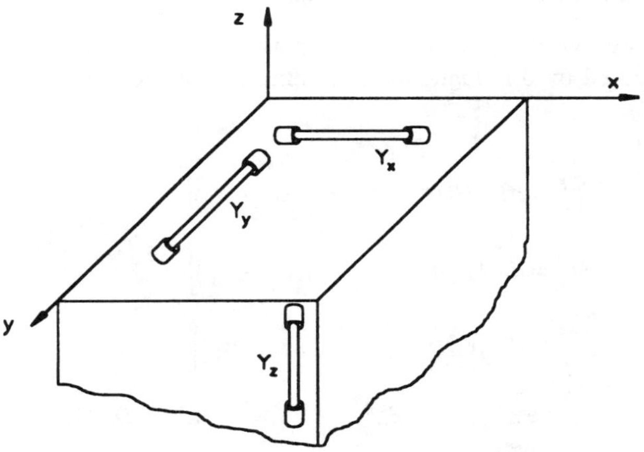

FIGURE 10.19 Orthogonal axes of anisotropy with specimen orientations.

If the yield strengths of three tensile specimens taken along the *x, y*, and *z*-directions are found to be Y_x, Y_y, and Y_z, respectively, then Equation 10.37 provides

$$Y_x^2 = \frac{1}{G+H}$$

$$Y_y^2 = \frac{1}{H+F} \quad (10.38a)$$

$$Y_z^2 = \frac{1}{F+G}$$

Solving simultaneously gives

$$2F = \frac{1}{Y_y^2} + \frac{1}{Y_z^2} - \frac{1}{Y_x^2}$$

$$2G = \frac{1}{Y_z^2} + \frac{1}{Y_x^2} - \frac{1}{Y_y^2} \tag{10.38b}$$

$$2H = \frac{1}{Y_x^2} + \frac{1}{Y_y^2} - \frac{1}{Y_z^2}$$

Similarly, the material parameters L, M, and N can be evaluated from shear tests.

Equations 10.37 and 10.38 characterize the yield condition for anisotropic materials. Note that if $Y_x = Y_y = Y_z = Y$, which is the case for isotropic material, Hill's yield criterion, Equation 10.37, reduces to the von Mises criterion. In such a case, $F = G = H = 1/2\, Y^2$ and $L = M = N = 3F = 3G = 3H$.

10.15.2 A Flow Rule for Anisotropic Materials

Considering Hill's criterion, Equation 10.37, as a plastic potential, the plastic flow rule for anisotropic material is derived by differentiation according to Equation 10.23, i.e., $d\varepsilon_x^P\, \alpha(\partial f/\partial\sigma_x)$, etc. Hence,

$$\begin{aligned}
d\varepsilon_x^P &= d\lambda_1\left[H(\sigma_x - \sigma_y) + G(\sigma_x - \sigma_z)\right]\\
d\varepsilon_y^P &= d\lambda_1\left[F(\sigma_y - \sigma_z) + H(\sigma_y - \sigma_x)\right]\\
d\varepsilon_z^P &= d\lambda_1\left[G(\sigma_z - \sigma_x) + F(\sigma_z - \sigma_y)\right]\\
d\varepsilon_{xy}^P &= d\lambda_1 N\tau_{xy},\quad d\varepsilon_{yz}^P = d\lambda_1 L\tau_{yz},\quad d\varepsilon_{zx}^P = d\lambda_1 M\tau_{zx}
\end{aligned} \tag{10.39}$$

The proportionality factor $d\lambda_1$ in Equations 10.39 serves the same purpose as $d\lambda$ for isotropic material, but obviously has a different value. Note that $d\varepsilon_x^P + d\varepsilon_y^P + d\varepsilon_z^P = 0$ from Equations 10.39 indicating constancy of volume as expected. Also for isotropic material $F = G = H$ and Equations 10.39 may be thus reduced to the Levy–Mises flow rule, namely, Equations 10.22.

Note that the above flow rule implies that if the principal axes of stress coincide with the axes of anisotropy, so will the principal axes of plastic strain increment.

10.15.3 Measurement of Anisotropic Parameters

The determination of the anisotropic parameters F, G, H, … are given by Equations 10.38 in terms of the yield strengths Y_x, Y_y, and Y_z. However, for anisotropic sheets it is difficult to determine the through-thickness yield strength Y_z. A relation expressing Y_z in terms of the other two yield strengths Y_x and Y_y is derived by considering first a simple tension test for a specimen taken along the longitudinal x-direction. Thus, Equations 10.39 give

$$\begin{aligned}
d\varepsilon_x^P &= d\lambda_1(H + G)\sigma_x\\
d\varepsilon_y^P &= -d\lambda_1 H\sigma_x\\
d\varepsilon_z^P &= -d\lambda_1 G\sigma_x
\end{aligned}$$

Defining a strain ratio R as the ratio between the two transverse strains $d\varepsilon_y^p$ and $d\varepsilon_z^p$ gives

$$R_1 = \left(d\varepsilon_y^P / d\varepsilon_z^P\right) = H/G \tag{10.40a}$$

Similarly, from a tensile test for a specimen taken along the y-direction, R_2 is defined as

$$R_2 = \left(d\varepsilon_x^P / d\varepsilon_z^P\right) = H/F \tag{10.40b}$$

Equations 10.40a and b together with Y_x and Y_y as determined from the same two tensile tests provide the necessary data to determine the yield stress Y_z in the third direction. Hence, according to Equations 10.38a and b:

$$Y_z = Y_x\,[R_2(1 + R_1)/(R_1 + R_2)]^{1/2} \tag{10.41a}$$

and, similarly,

$$Y_z = Y_y\,[R_1(1 + R_2)/(R_1 + R_2)]^{1/2} \tag{10.41b}$$

Equations 10.41 provide a means to determine Y_z from two tests conducted in the plane of the sheet. However, measurement of thickness strain $d\varepsilon_z^p$ requires great accuracy and it is usually found from measurement of length and width strains using volume constancy. The anisotropic strain ratio R is obtained from

$$\mathrm{R} = \ln(W_o/W)/\ln(h_o/h) = \ln(W_o/W)/\ln(WL/W_oL_o) \tag{10.42}$$

where W, h, and L are the current width, thickness, and length, respectively. Obviously, for isotropy, $R = 1$, while $R \neq 1$ with anisotropy.*

The anisotropic strain ratios R_1 and R_2 are usually measured for longitudinal strains greater than 5%. Fortunately, these R values do not vary appreciably with strain except in a few cases, and the values measured by conducting tensile tests are taken to represent the original state of anisotropy in the sheet.

The variation of anisotropy in the plane of a rolled sheet as characterized by the anisotropic strain ratio, R, is responsible for the formation of ears, i.e., wavy edges on the top of a drawn cup from a flat circular blank.

10.15.4 Normal Anisotropy

From the previous example, it is seen that a cold-rolled sheet of R value > 1 is expected to possess a through-thickness yield strength higher than its strength along any direction in its plane. Such increased through-thickness strength is described as normal anisotropy. This is desirable in sheet metal–forming operations since it results in a high thinning resistance. For instance, when $R_1 \cong R_2 = 3$ in Equations 10.41, the through-thickness yield strength is

$$Y_z = Y\sqrt{(1+R)/2}$$

which is 41% higher than Y, as may be the case for a commercial titanium sheet. Note that when $R_1 = R_2$, the yield strengths $Y_x = Y_y = Y$, as implied by Equations 10.38 and 10.40.

* In sheet metal of cubic crystal structure, R values range from about 0.75 to 2. In hexagonal close-packed structures, R may be as high as 3 to 7.

Example 10.15:

Tensile tests on strips cut from a thin sheet metal along the rolling direction-x and transverse direction-y produced the following results:

In the rolling direction: $R_1 = 4.0$, $Y_x = 484$ *MPa,*
In the transverse direction: $R_2 = 2.0$.

a. *Find the yield strength through the thickness direction* Y_z *and in the transverse direction* Y_y.
b. *If the sheet is subjected to equibiaxial tension* $\sigma_x = \sigma_y = \sigma$, *find the value of* σ *at which yielding occurs.*
c. *Find the ratios among the plastic strains* $d\varepsilon_x^p$, $d\varepsilon_y^p$, *and* $d\varepsilon_z^p$.

Comment on the results.

Solution:

a. From Equations 10.41.

$$Y_z = 484\ [2(1 + 4)/(4 + 2)]^{1/2} = 624.8 \text{ MPa}$$

Hence,

$$Y_y = 624.8/[4(1 + 2)/(4 + 2)]^{1/2} = 441.8 \text{ MPa}$$

This result indicates that Y_z is larger than both Y_x and Y_y, which is expected as R is > 1. Conversely, if $R < 1$, Y_z ought to be smaller than both Y_x and Y_y.

b. For equibiaxial tension $\sigma_x = \sigma_y = \sigma$ and $\sigma_z = 0$. Also, $\tau_{xy} = \tau_{yz} = \tau_{zx} = 0$. Hence, from Equation 10.37,

$$F\sigma_y^2 + G\sigma_x^2 + H(\sigma_x - \sigma_y)^2 = 1$$

i.e., $\sigma = \sqrt{1/(F+G)}$. From Equation 10.38a, $Y_z^2 = 1/(F + G)$, hence, $\sigma = Y_z = 624.8$ MPa.

This means that an equibiaxial tension is equivalent to a through-thickness compression. In fact the deviatoric stress system in this case is $\sigma_x' = \sigma_y' = \sigma - {}^2/_3\sigma = \sigma/3$ and $\sigma_z' = -{}^2/_3\sigma$. This is equivalent to a hydrostatic stress ($-\sigma/3$) and deviatoric stresses ($0, 0, -\sigma$), i.e., simple compression.

c. The ratios among plastic strains are given by Equations 10.39 for $\sigma_x = \sigma_y = \sigma$ and $\sigma_z = 0$ as

$$d\varepsilon_x^p : d\varepsilon_y^p : d\varepsilon_z^p = G\sigma : F\sigma : -(F+G)\sigma$$

Recalling that $R_1 = H/G = 4$ and $R_2 = H/F = 2$, according to Expressions 10.40, gives

$$d\varepsilon_x^p : d\varepsilon_y^p : d\varepsilon_z^p = 1 : 2 : -3$$

Note that although in equibiaxial tension $\sigma_x = \sigma_y$, the plastic strains $\varepsilon_x^p \neq \varepsilon_y^p$ for anisotropic material.

10.15.5 Effective Stress and Effective Plastic Strain Increment

If the principal axes of stress are coincident with the axes of anisotropy, terms involving shear stresses vanish and x, y, and z could be replaced by 1, 2, and 3 for such a state of principal stresses.

With substitution from Equations 10.38 and 10.40 into Equation 10.37, the yield condition becomes

$$R_1(\sigma_2-\sigma_3)^2+R_2(\sigma_3-\sigma_1)^2+R_1R_2(\sigma_1-\sigma_2)^2=R_2(1+R_1)Y_1^2 \tag{10.43}$$

where Y is the yield strength along direction 1.

Similarly, the flow rule, Equations 10.39 reduces to

$$\begin{aligned} d\varepsilon_1^p &= \frac{3d\bar{\varepsilon}^p}{2(R_1+R_2+R_1R_2)\bar{\sigma}}\left[R_1R_2(\sigma_1-\sigma_2)+R_2(\sigma_1-\sigma_3)\right] \\ d\varepsilon_2^p &= \frac{3d\bar{\varepsilon}^p}{2(R_1+R_2+R_1R_2)\bar{\sigma}}\left[R_1(\sigma_2-\sigma_3)+R_1R_2(\sigma_2-\sigma_1)\right] \\ d\varepsilon_3^p &= \frac{3d\bar{\varepsilon}^p}{2(R_1+R_2+R_1R_2)\bar{\sigma}}\left[R_2(\sigma_3-\sigma_1)+R_1(\sigma_3-\sigma_2)\right] \end{aligned} \tag{10.44}$$

Following a procedure similar to that of isotropic materials, an effective stress and an effective plastic strain increment are defined as

$$\bar{\sigma}=\sqrt{\frac{3}{2}\left[\frac{F(\sigma_2-\sigma_3)^2+G(\sigma_3-\sigma_1)^2+H(\sigma_1-\sigma_2)^2}{F+G+H}\right]} \tag{10.45}$$

and

$$d\bar{\varepsilon}^p=\sqrt{\frac{2}{3}}(F+G+H)^{1/2}\left[F\left(\frac{Gd\varepsilon_2^p-Hd\varepsilon_3^p}{FG+GH+HF}\right)^2 +G\left(\frac{Hd\varepsilon_3^p-Fd\varepsilon_1^p}{FG+GH+HF}\right)^2+H\left(\frac{Fd\varepsilon_1^p-Gd\varepsilon_2^p}{FG+GH+HF}\right)^2\right]^{1/2} \tag{10.46}$$

In terms of R_1 and R_2, where $H/G = R_1$ and $H/F = R_2$, Equations 10.45 and 10.46 are given by

$$\bar{\sigma}=\sqrt{\frac{3}{2}}\left[\frac{R_1(\sigma_2-\sigma_3)^2+R_2(\sigma_3-\sigma_1)^2+R_1R_2(\sigma_1-\sigma_2)^2}{R_1+R_2+R_1R_2}\right]^{1/2} \tag{10.47}$$

and

$$d\bar{\varepsilon}^p = \sqrt{\frac{2}{3}}\frac{\left(R_1+R_2+R_1R_2\right)^{1/2}}{\left(1+R_1+R_2\right)}\left[\frac{1}{R_1}\left(d\varepsilon_2^p - R_1 d\varepsilon_3^p\right)^2 + \frac{1}{R_2}\left(R_2 d\varepsilon_3^p - d\varepsilon_1^p\right)^2 + \left(R_1 d\varepsilon_1^p - R_2 d\varepsilon_2^p\right)^2\right]^{1/2} \tag{10.48}$$

Note that when $R_1 = R_2 = 1$, i.e., complete isotropy, Expressions 10.44, 10.47, and 10.48 reduce to Equations 10.35, 10.26, and 10.30 for isotropic materials, respectively.

10.15.6 A Special Case: Rotational Symmetry (Planar Isotropy)

In a sheet metal, if the plastic properties show no variation with orientation in its plane, then a condition of rotational symmetry around a normal to the plane of the sheet is reached. The material is said to possess planar isotropy. Hence, from Equations 10.40,

$$R_1 = R_2 = R = H/G = H/F \text{ and } F = G$$

Hence, Equation 10.43, for thin sheets where it is assumed that plane stress conditions prevail, i.e., $\sigma_3 = 0$, reduces to

$$\sigma_2^2 + \frac{G}{F}\sigma_1^2 + \frac{H}{F}\left(\sigma_1 - \sigma_2\right)^2 = \left(\frac{G+H}{F}\right)Y_1^2 \tag{10.49}$$

From Equation 10.38a, recalling that $F = G$,

$$Y_1^2 = \frac{1}{G+H} = \frac{1}{F+H} = \frac{1/F}{1+H/F} = \frac{1/F}{1+R}$$

Substituting into Equation 10.49 knowing that $R = H/F$ results in

$$\sigma_1^2 + \sigma_2^2 - \frac{2R}{1+R}\sigma_1\sigma_2 = Y^2 \tag{10.50}$$

where $Y = Y_1 = Y_2$ is the yield strength in the plane of the sheet. Equation 10.50 plots as an ellipse whose major and minor axes depend on R, as shown in Figure 10.20. The ellipse corresponding to $R = 1$ is the von Mises ellipse. For $R > 1$ higher resistance to yielding is achieved when both σ_1 and σ_2 are tensile. Inversely, weakening is observed when σ_1 is tensile together with a compressive σ_2. For instance, an anisotropic sheet of $R = 2$ when stretched by equibiaxial tension $\sigma_1 = \sigma_2$ yields at $\sigma_1/Y = \sqrt{3/2}$, while for $\sigma_1 = -\sigma_2$ yielding occurs at $\sigma_1/Y = \sqrt{3/10}$. For an isotropic sheet, $R = 1$; this ratio for equibiaxial tension is $\sigma_1/Y = 1$, i.e., 22.5% less than that for $R = 2$. For the case of $\sigma_1 = -\sigma_2$, as shown in Figure 10.20, this ratio is higher by 5.1% for $R = 1$.

The flow rule, Equations 10.44, reduces in this case (where $R = H/G = H/F$) to

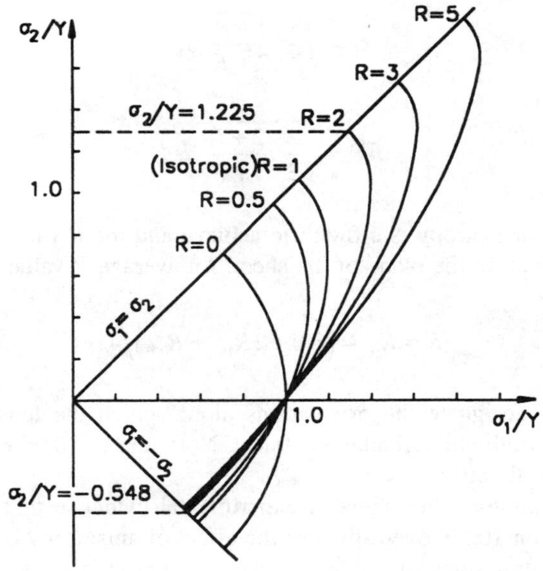

FIGURE 10.20 Hill's yield loci for sheets possessing rotational symmetry in their plane; R is the average strain ratio in the plane.

$$d\varepsilon_1^p = \frac{3d\bar{\varepsilon}^p}{2(2+R)\bar{\sigma}}\left[R(\sigma_1-\sigma_2)+(\sigma_1-\sigma_3)\right]$$

$$d\varepsilon_2^p = \frac{3d\bar{\varepsilon}^p}{2(2+R)\bar{\sigma}}\left[(\sigma_2-\sigma_3)+R(\sigma_2-\sigma_1)\right] \tag{10.51}$$

$$d\varepsilon_3^p = \frac{3d\bar{\varepsilon}^p}{2(2+R)\bar{\sigma}}\left[(\sigma_3-\sigma_1)+(\sigma_3-\sigma_2)\right]$$

The effective stress and effective strain for the case of planar isotropy and plane stress ($\sigma_3 = 0$) are obtained from Equations 10.47 and 10.48 by substituting $R_1 = R_2 = R$. Hence,

$$\bar{\sigma} = \sqrt{\frac{3}{2}}\left[\frac{\sigma_1^2+\sigma_2^2+R(\sigma_1-\sigma_2)^2}{2+R}\right]^{1/2} \tag{10.52}$$

and

$$d\bar{\varepsilon}^p = \sqrt{\frac{2}{3}}\left\{\frac{2+R}{(1+2R)^2}\left[\left(d\varepsilon_2^p - Rd\varepsilon_3^p\right)^2+\left(Rd\varepsilon_3^p-d\varepsilon_1^p\right)^2+R\left(d\varepsilon_1^p-d\varepsilon_2^p\right)^2\right]\right\}^{1/2} \tag{10.53}$$

Consider that a tensile test is conducted on a specimen cut from the sheet plane along, say, the first direction. Hence, by definition, $R = d\varepsilon_2^p / d\varepsilon_3^p$ and, from constancy of volume, $d\varepsilon_2 = -Rd\varepsilon_1/(1 + R)$ and $d\varepsilon_3^p = -d\varepsilon_1^p/(1 + R)$. The effective stress $\bar{\sigma}$ and effective plastic strain increment $d\bar{\varepsilon}$ by virtue of equations 10.51 through 10.53 are given, respectively, by

$$\bar{\sigma} = \sqrt{\frac{3}{2}\,\frac{1+R}{2+R}}\,\sigma_1 \tag{10.54a}$$

$$d\bar{\varepsilon}^p = \sqrt{\frac{2}{3}\,\frac{2+R}{1+R}}\,d\varepsilon_l^p \tag{10.54b}$$

In rolled sheets, planar isotropy is difficult to achieve and the R value usually varies with the orientation of test specimen in the plane of the sheet. An average R value is taken as

$$R = R_{av} = \left(R_{0^\circ} + 2R_{45^\circ} + R_{90^\circ}\right)/4 \tag{10.55}$$

where 0°, 45°, and 90° designate the orientations along which the tensile specimen is cut to determine the R value (width strain/thickness strain). Note that the 0° direction is often taken to coincide with the rolling direction.

It is worthwhile mentioning that there is experimental evidence that the yield criterion for planar isotropy, Expression 10.50, overestimates the effect of anisotropy on the shape of the yield locus and its associated flow rule when $R < 1$. Other yield criteria have been postulated to cope with this situation.[8]

Example 10.16:

A pipe with closed ends is made of thin sheet metal by roll forming, bending, and welding, such that the rolling direction becomes the axial direction of the pipe and the transverse direction becomes the hoop direction. The pipe has a mean radius r_{mo} *= 75 mm, length* L_o *= 1000 mm, and a wall thickness* h_o *= 0.75 mm. The strain ratios in the sheet are* $R_0 = 2.5$, $R_{45} = 1.8$, *and* $R_{90} = 0.8$. *The complete stress–strain curve in simple tension for specimens along the rolling direction is given by* $\sigma_l = 356.7 + 150(\varepsilon_l^p)^{0.2}$ *MPa. Take yielding to occur at a strain of 0.2% and neglect elastic effects.*

a. *Determine the internal pressure at which the pipe walls will yield. Use both Hill's criterion (F, G, H) and planar isotropy model* (R_{av}).
b. *If the pressure is increased up to a value that produces only 10% diametral expansion, find the final pipe dimensions. What is the pressure at this instant?*
c. *Compare the results of* (b) *with those that neglect anisotropy.*

Solution:

a. Pressure at yielding: The yield strength at 0.2% strain is found as

$$\sigma_1 = Y_1 = 356.7 + 150\,(0.002)^{0.2} = 400 \text{ MPa}$$

i. *Hill's yield criterion (F, G, H)*: According to Expressions 10.40,

$$R_1 = R_0 = \frac{H}{G} = 2.5$$

$$R_2 = R_{90^\circ} = \frac{H}{E} = 0.8 \text{ and hence, } \frac{G}{F} = 0.32$$

$$Y_1^2 = \frac{1}{G+H} = (400)^2$$

Equation 10.43) may be written (for principal stresses) as

$$(\sigma_2-\sigma_3)^2+\frac{G}{F}(\sigma_3-\sigma_1)^2+\frac{H}{F}(\sigma_1-\sigma_2)^2=\left(\frac{G}{F}+\frac{H}{F}\right)Y_1^2$$

For a closed thin-walled pipe,

$$\sigma_\theta=\sigma_1=2\sigma_2=2\sigma_z=\frac{p_i r_{mo}}{h_o}\quad\text{and}\quad\sigma_3=\sigma_r\cong 0$$

Substituting in the yield condition, $\sigma_2^2+0.32\,\sigma_1^2+0.8\,(\sigma_1-\sigma_2)^2=(0.32+0.8)\,400^2$, gives σ_1 = 482.4 MPa. Since $\sigma_1=p_i r_{mo}/h_o$, then, yielding for r_{mo} = 75 mm and h_o = 0.75 mm, occurs at p_i = 482.4 (0.75/75) = 4.824 MPa.

ii. *Planar isotropy model (R)*: For planar isotropy, rotational symmetry is assumed. Thus, an average R value in the plane of the rolled sheet is assumed according to Equation 10.55:

$$R_{av}=\frac{R_o+2R_{45}+R_{90}}{4}=1.725$$

The yield condition, is thus given by Equation 10.50 as

$$\sigma_1^2+\sigma_2^2-\frac{2R_{av}}{(1+R_{av})}\sigma_1\sigma_2=Y_1^2$$

giving $0.617\,\sigma_1^2=Y_1^2$ and p_i = 5.092 MPa.

iii. *Isotropic material (F = G = H = 1 and R = 1)*: The von Mises yield criterion, Equation 10.12a, is applied to give p_i = 4.619 MPa.

b. *Change in pipe dimensions (Planar Isotropy):* Equations 10.51 give for $\sigma_1=2\sigma_2$, σ_3 = 0, and R = 1.725,

$$\varepsilon_1^p:\varepsilon_2^p:\varepsilon_3^p=\varepsilon_\theta^p:\varepsilon_z^p:\varepsilon_r^p=1.8625:-0.3625:-1.5$$

where r, θ, and z are the thickness, hoop, and axial directions, respectively. Given $\varepsilon_\theta{}^p=\ln(r_m/r_{mo})=\ln 1.1=0.0953$, the other strains are

$$\varepsilon_z^p=\ln\frac{L}{L_o}=-\frac{0.3625}{1.8625}(0.0953)=-0.01855$$

and

$$\varepsilon_r^p=\ln\frac{h}{h_o}=-\frac{-1.5}{1.8625}(0.0953)=-0.07675$$

Thus, the pipe dimensions at this instant are mean radius = 82.5 mm, thickness = 0.695 mm, and length = 981.6 mm. To find the pressure required to extend the tube to the above dimensions, the effective stress and strain are calculated from Equations 10.52 and 10.53. Hence, for $\sigma_1=2\sigma_2$ and $\sigma_3=0$

$$\bar{\sigma} = \sqrt{\frac{3}{2}}\left[\left(\sigma_1^2 + \frac{1}{4}\sigma_1^2 + \frac{1}{4}R\sigma_1^2\right)\Big/(2+R)\right]^{1/2}$$

or

$$\bar{\sigma} = \sqrt{\frac{3(5+R)}{8(2+R)}}\sigma_1 = 0.823\frac{p_i r_m}{h} \tag{a}$$

The effective strain is calculated from Equation 10.53, knowing that

$$\varepsilon_1^p = \varepsilon_\theta^p = 0.0953, \quad \varepsilon_2^p = \varepsilon_z^p = -0.01855, \quad \text{and} \quad \varepsilon_3^p = \varepsilon_r^p = -0.07675$$

Hence

$$\bar{\varepsilon}^p = 0.10455 \tag{b}$$

The relation between the effective stress and effective plastic strain is determined from a tensile test along the rolling direction using Equations 10.54 for R = 1.725. Hence, $\bar{\sigma} = 1.0475\sigma_1$ and $\bar{\varepsilon}^p = 0.9546(\varepsilon)_1^p$. Given σ_1 = 356.7 + 150 (ε_1^p)$^{0.2}$, thus

$$\frac{\bar{\sigma}}{1.0475} = 356.7 + 150\left(\frac{\bar{\varepsilon}}{0.9546}\right)^{0.2}$$

or

$$\bar{\sigma} = 373.7 + 158.6\left(\bar{\varepsilon}^p\right)^{0.2}$$

Substitution from (a) and (b) together with the new pipe dimensions gives

$$0.823p_i\left(\frac{82.5}{0.695}\right) = 373.7 + 158.6(0.10455)^{0.2}$$

and p_i = 4.86 MPa.

c. *Comparison with isotropic material:* The plastic strains are obtained from the Levy–Mises flow rule, Equations 10.35, such that

$$\varepsilon_\theta^p : \varepsilon_z^p : \varepsilon_r^p = 1 : 0 : -1$$

Note that in this case no change occurs in pipe length and the pipe final dimensions are r_m = 82.5 mm, h = 0.6818 mm, and L = 1000 mm. The effective stress and strain according to Equations 10.26 and 10.30a are

$$\bar{\sigma} = \frac{\sqrt{3}}{2}\sigma_1 = 0.866\frac{p_i r_m}{h} \text{ and } \bar{\varepsilon}^p = \frac{2}{\sqrt{3}}\varepsilon_\theta^p = 0.11$$

Using the material law in simple tension as $\bar{\sigma} = 356.7 + 150\,(\bar{\varepsilon}^p)^{0.2}$ gives $p_i = 4.32$ MPa, which is 11% less than that for the anisotropic pipe.

10.15.7 A Modified Nonquadratic Criterion for Planar Isotropy

There has been experimental evidence that the yield criterion for planar isotropy as given in Section 10.15.6 overestimates the effect of anisotropy on the yield locus and its associated flow rule for average values of $R < 1$. This discrepancy has been observed by determining the effective stress–strain curves experimentally from tests where biaxial stress systems prevail (e.g., circular bulging) and comparing these with the uniaxial flow curve. As mentioned in Section 10.7.5, these two curves should coincide if the flow function and its associated flow rule describe the sheet metal behavior correctly; a case that has not been satisfied for several sheet metals. A typical example of this is the case of cold-rolled aluminum sheets. Other yield criteria have been postulated to cope with this situation. Hill[7] proposed a nonquadratic yield function for anisotropic sheet metals undergoing plane stress deformation ($\sigma_3 = 0$) as

$$2(1 + R)\,Y^s = (1 + 2R)\,|\sigma_1 - \sigma_2|^s + |\sigma_1 + \sigma_2|^s \tag{10.56}$$

where Y is the yield strength in uniaxial tension and s is a material parameter, which has to be determined experimentally. The exponent s is found to be less than 2 for metals of average $R < 1$ or set to 2 for those of average $R \geq 1$. The latter case for $s = 2$ and $R \neq 1$ reduces Expression 10.56 to Hill's yield criterion (Equation 10.50), while for $s = 2$ and $R = 1$ it gives the von Mises criterion (Equation 10.17) for σ_1, σ_2, and $\sigma_3 = 0$.

The principles applied to the von Mises yield function in Section 10.9 are used to derive the flow rule associated with Expression 10.56. According to the normality rule, Expression 10.23 gives the flow rule as

$$d\varepsilon_1^p = d\lambda_2\left[(1+2R)\frac{|\sigma_1-\sigma_2|^s}{(\sigma_1-\sigma_2)} + \frac{|\sigma_1+\sigma_2|^s}{(\sigma_1+\sigma_2)}\right]$$

$$d\varepsilon_2^p = d\lambda_2\left[-(1+2R)\frac{|\sigma_1-\sigma_2|^s}{(\sigma_1-\sigma_2)} + \frac{|\sigma_1+\sigma_2|^s}{(\sigma_1+\sigma_2)}\right] \tag{10.57}$$

$$d\varepsilon_3^p = -d\lambda_2\left[\frac{2|\sigma_1+\sigma_2|^s}{(\sigma_1+\sigma_2)}\right]$$

where the proportionality factor $d\lambda_2$:

$$d\lambda_2 = \frac{d\bar{\varepsilon}^p}{\bar{\sigma}^{s-1}} \tag{10.58}$$

The effective stress and effetcive strain are also derived for plane stress conditions to be

$$\bar{\sigma} = \left\{ \frac{1}{2(1+R)} \left[(1+2R)|\sigma_1 - \sigma_2|^s + |\sigma_1 + \sigma_2|^s \right] \right\}^{1/s} \tag{10.59}$$

$$d\bar{\varepsilon}^p = \frac{[2(1+R)]^{1/s}}{2} \left\{ (1+2R)^{1/(1-s)} \left| d\varepsilon_1^p - d\varepsilon_2^p \right|^{s/(s-1)} + \left| d\varepsilon_1^p + d\varepsilon_2^p \right|^{s/(s-1)} \right\}^{(s-1)/s} \tag{10.60}$$

Again, Expressions 10.59 and 10.60 reduce to their corresponding von Mises expressions for $R = 1$ and $s = 2$. Figure 10.21 shows the yield function (Equation 10.56) plotted for values of $R = 0.5$ and $s = 1.5$, typical of rolled aluminum sheet metal.[9] On the same figure, a plot for $R = 0.5$, $s = 2$ as well as the Mises ellipse ($s = 2$, $R = 1$) are also shown for comparison.

Example 10.17:

During bulging, Figure 10.22 of a circular aluminum thin sheet of thickness $h_o = 1.5$ *mm, the stress state at the bulge top is considered to be closely equibiaxial. The surface gridding technique, as explained in Example 10.8, is used to measure biaxial polar strains. Meanwhile, by recording the bulging pressure and the polar bulge radius the following experimental data at the pole are obtained:*[9] *surface strains* $\varepsilon_1 \simeq \varepsilon_2 = 0.24$, *polar radius of curvature* $\rho = 122.1$ *mm, and bulging pressure* $= 1.83$ *MPa.*

Strips of the same sheet are cut along 0°, 45°, and 90° with the rolling direction and tested in uniaxial tension to give the following properties averaged according to Expression 10.55:

$$\sigma_u = 142(\varepsilon)_u^{0.273} \text{ MPa and } R = 0.73$$

Using the above average tensile properties, predict the equibiaxial surface strains and compare them with the experimental point. Assume planar isotropy and consider the two cases $s = 2$ *and* $s = 1.7$.

Solution:

During bulging of a circular thin sheet metal, as shown in Figure 10.22, spherical symmetry and, hence, equibiaxial tension $\sigma_1 = \sigma_2$ is assumed to prevail at the bulge top. If the surface strains $\varepsilon_1 = \varepsilon_2$, hence, by constancy of volume, Expression 10.1 gives:

$$\varepsilon_3 = \ln\left(\frac{h}{h_o}\right) = -(\varepsilon_1 + \varepsilon_2)$$

or $h = h_o \exp[-(\varepsilon_1 + \varepsilon_2)] = 1.5 \exp[-(0.24 + 0.24)] = 0.928$ mm. The equibiaxial state of stress at the bulge top is given by

$$\sigma_1 = \sigma_2 = \frac{p_i \rho}{2h} = \frac{1.83 \times 122.1}{2 \times 0.928} = 120.4 \text{ MPa}$$

with $\sigma_3 \simeq 0$.

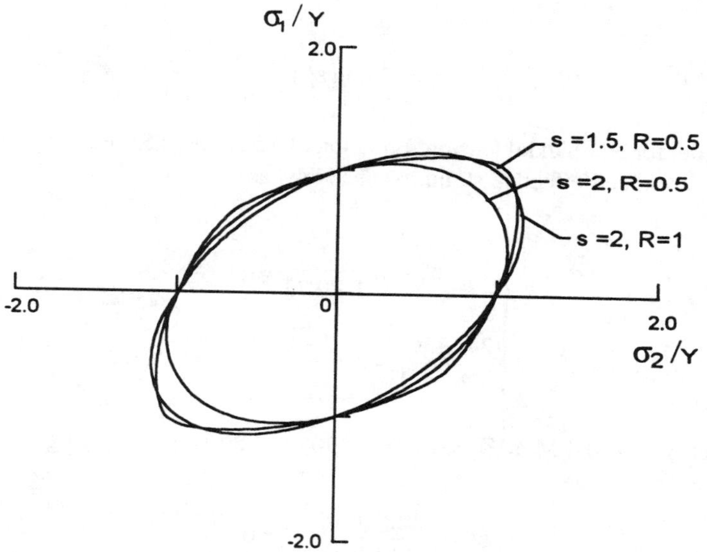

FIGURE 10.21 Nonquadratic anisotropic yield locus in plane stress.

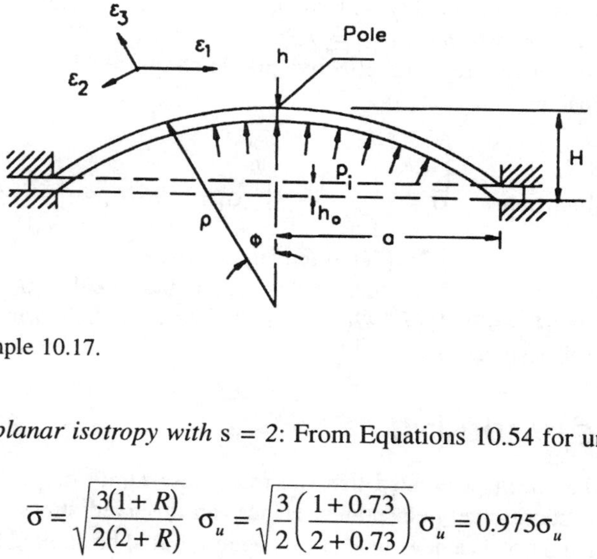

FIGURE 10.22 Example 10.17.

a. *Condition of planar isotropy with* s = 2: From Equations 10.54 for uniaxial tension:

$$\bar{\sigma} = \sqrt{\frac{3(1+R)}{2(2+R)}}\,\sigma_u = \sqrt{\frac{3}{2}\left(\frac{1+0.73}{2+0.73}\right)}\,\sigma_u = 0.975\sigma_u$$

$$\bar{\varepsilon}^p = \sqrt{\frac{2(2+R)}{3(1+R)}}\,\varepsilon_u = \sqrt{\frac{2}{3}\left(\frac{2+0.73}{1+0.73}\right)}\,\varepsilon_u = 1.026\varepsilon_u$$

where σ_u and ε_u are the average uniaxial flow stress and plastic strain as given by σ_u = 142 $(\varepsilon)_u^{0.273}$. Hence, the $\bar{\sigma} - \bar{\varepsilon}^p$ of this aluminum alloy is obtained as

$$\left(\frac{\bar{\sigma}}{0.975}\right) = 142\left(\frac{\bar{\varepsilon}^p}{1.026}\right)^{0.273}$$

or

$$\bar{\sigma} = 137.5 \left(\bar{\varepsilon}^p\right)^{0.273}$$

Meanwhile, for equibiaxial tension, Equations 10.52 and 10.53 are simplified to give the effective stress and effective strain, respectively, as

$$\bar{\sigma} = \sqrt{\frac{3}{2+R}}\,\sigma_1 = \sqrt{\frac{3}{2+0.73}}\,\sigma_1 = 1.048\sigma_1$$

$$\bar{\varepsilon}^p = 2\sqrt{\frac{2+R}{3}}\,\varepsilon_1 = 2\sqrt{\frac{2+0.73}{3}}\,\varepsilon_1 = 1.908\varepsilon_1$$

Hence, at σ_1 = 120.4 MPa, $\bar{\sigma}$ = 1.048 × 120.4 = 126.18 MPa, giving

$$\bar{\varepsilon}^p = \left(\frac{126.18}{137.5}\right)^{1/0.273} = 0.726$$

$$\varepsilon_1 = 0.73/1.908 = 0.38$$

and which compares unfavorably with the experimentally measured ε_1 = 0.24.

b. *Condition of planar isotropy with* s = *1.7*: Expressions 10.59 and 10.60 give, respectively, for the effective stress and effective plastic strain (s = 1.7 and R = 0.73):
   - Uniaxial tension: $\bar{\sigma} = \sigma_u$
     $\bar{\varepsilon}^p = \varepsilon_u$
   - Equibiaxial tension: $\bar{\sigma} = 2\left[\frac{1}{2(1+R)}\right]^{1/s}\sigma_1 = 0.964\sigma_1$
     $\bar{\varepsilon}^p = [2(1+R)]^{1/s}\varepsilon_1 = 2.075\varepsilon_1$

   Hence the flow curve $\bar{\sigma} - \bar{\varepsilon}^p$, which in this case is indentical to $\sigma_u - \varepsilon_u$, gives, at σ_1 = 120.4 MPa, (0.964 × 120.4) = 142(2.075 ε_1)$^{0.273}$ or ε_1 = 0.23, which agrees with the experimentally measured value.

10.16 KINEMATIC HARDENING

In Section 10.12, it has been postulated that the true stress–strain curve for a strain–hardening material is a locus of successive yield points. Hence, an observed stress greater than the initial yield value may be thought of as a new yield stress, which should be applied to cause further plastic deformation. On such a basis of isotropic hardening, it has been assumed that the yield surface expands as a result of plastic flow with its center, orientation, and shape remaining unchanged. Isotropic hardening does not provide a good representation of observed material behavior under conditions of reversed loading, since it does not predict the observed Bauschinger effect, as illustratd in Figure 10.1. However, it has gained wide popularity because of its mathematical simplicity.

To overcome this shortcoming, the kinematic-hardening rule has been proposed. This rule postulates that the yield surface maintains its initial size, but its center translates in a certain prescribed direction as a result of plastic deformation. The term *kinematic* stems from the rigid translation of the yield surface in analogous manner to the term *isotropic* stemming from the uniform expansion of the yield surface as a consequence of plastic deformation.

10.16.1 Uniaxial Behavior under Cyclic Loading

As a schematic illustration of the kinematic-hardening rule, consider an initial yield locus represented in the π-plane by a circle of radius equal to $\sqrt{2/3}\,Y_o$. For the loading path (O–*a*–*b*) shown in Figure 10.23, as plasticity commences at point *a* the kinematic-hardening rule will imply a corresponding rigid translation of the initial yield locus (with its initial size remaining unchanged). This gives a final kinematic-hardening yield locus designated in Figure 10.23 by the subsequent (*K–H*) yield locus. On the other hand, if the isotropic-hardening rule is employed, the final yield locus (*I–H*) at point *b* would be a circle with a radius $\sqrt{2/3}\,Y_1$ and center at the origin of the stress space. In the physical analogue, isotropic hardening implies that plastic flow enlarges the size of the initial yield locus.

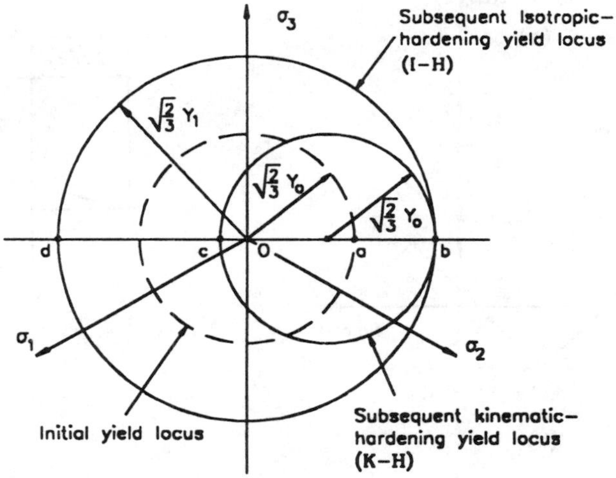

FIGURE 10.23 Subsequent yield loci according to isotropic-hardening and kinematic-hardening rules.

The differences between the kinematic- and isotropic-hardening rules become quite significant when subsequent unloading is involved such as the path (*b–c–d*) shown in Figure 10.23. Since the sizes of subsequent yield loci are quite different, compressive plastic flow will commence at point *c* for the kinematic-hardening vs. point *d* for isotropic-hardening.

To further illustrate the differences between isotropic- and kinematic-hardening rules, their predictions of elastic–plastic behavior under uniaxial reversed loading conditions are compared. Consider, first, a material behavior whose monotonic flow curve is modeled as rigid,* linear strain hardening as given by $\sigma = Y_o + E^p\,\varepsilon^p$. The initial yield strength for this material is Y_o and the slope of the σ–ε curve during plastic deformation is E^p as shown in Figure 10.6i or Figure 10.24a.

A uniaxial test specimen of this material is to be subjected to a strain history involving strain reversals from a tensile strain (+Δε/2) to a compressive one (–Δε/2), as shown in Figure 10.24b. Isotropic hardening behavior is illustrated in Figure 10.24c, where it is seen that initial tensile yielding commences at point *a*, and tensile plastic strain takes place until a total strain $\varepsilon_b = \Delta\varepsilon/2$ is reached. At this instant, the current yield stress of the material in both tension and compression according to the isotropic-hardening rule has increased to say $Y_1 = Y_o + E^p\varepsilon^p_{a-b}$, where $\varepsilon^p_{a-b} = \varepsilon^p_b = \Delta\varepsilon/2$. Upon unloading to point *b′* and reloading in compression up to point *c*, no changes occur in the previously attained plastic deformation since the material has been assumed to be rigid. Further reloading from point *c* to point *d* is equivalent to subjecting the material to an additional plastic strain of magnitude 2(Δε/2). To predict the current stress at point *d* using the monotonic linear-

* The elastic response is not considered here for the sake of simplicity.

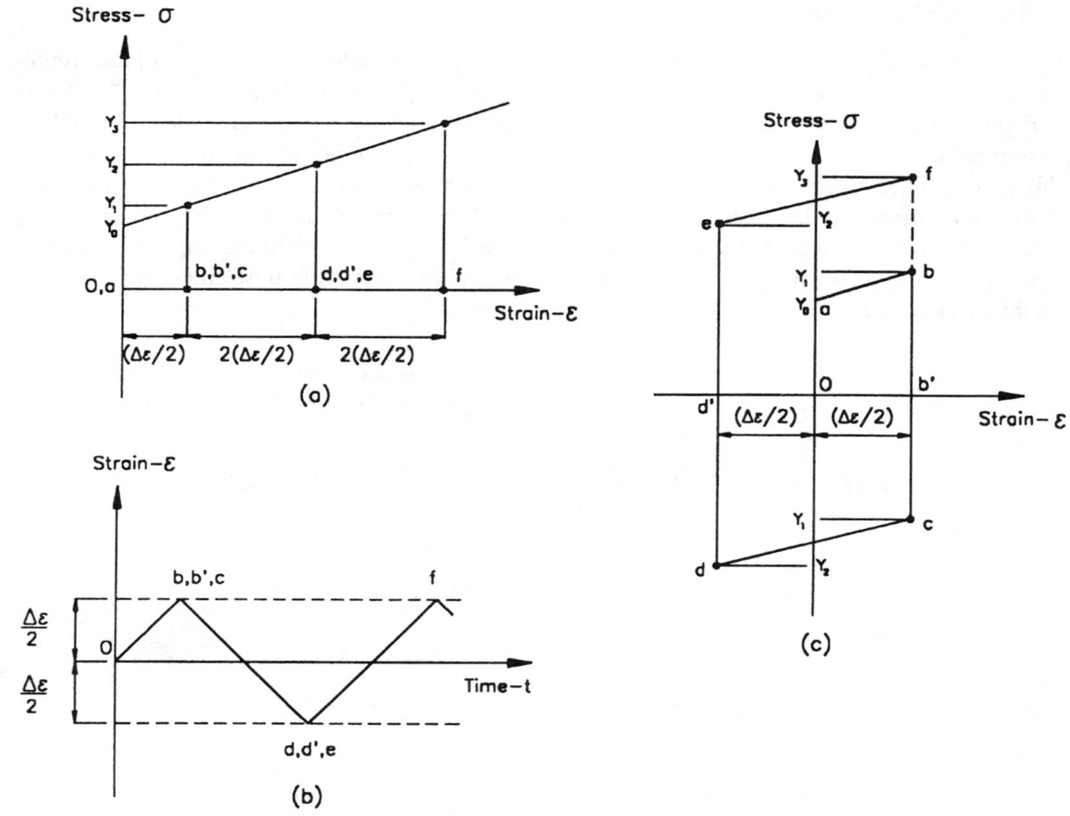

FIGURE 10.24 Uniaxial behavior of a rigid linear strain-hardening material under reversed strain cycle: (a) monotonic stress–strain curve, (b) strain cycle, and (c) isotropic-hardening response.

hardening law: $\sigma = Y_o + E^p \varepsilon^p$, it must be thus recalled that at point d, the total induced plastic strain is of a magnitude of $3(\Delta\varepsilon/2)$. Therefore, $\sigma_d = Y_o + E^p\ |3\Delta\varepsilon/2|$ say, equal to Y_2. Note that this new yield stress would have been obtained if the linear-hardening law is expressed for the part c–d as: $\sigma_d = Y_1 + E^p\varepsilon^p_{c-d} = (Y_o + E^p\varepsilon^p_{a-b}) + E^p\varepsilon^p_{c-d}$, where $\varepsilon^p_{a-b} = |\Delta\varepsilon/2|$ and $\varepsilon^p_{c-d} = |2(\Delta\varepsilon/2)|$ are the magnitudes of the plastic strains induced in loading from point a to b and point c to d, respectively. Again, unloading and reloading up to point e, where the yield stress attains $\sigma_e = \sigma_d = Y_2$, causes no further change in the strain state. Further reloading to reach point f means that a total plastic strain of a magnitude of $5(\Delta\varepsilon/2)$ has accumulated from the very beginning of the strain cycle. Hence, the stress at point f is given by $[\sigma_f = Y_o + E^p\ |5\Delta\varepsilon/2|$ which may be denoted by say, Y_3, where $Y_3 > Y_2 > Y_1 > Y_o$, as shown in Figure 10.24a.

If this type of straining is continued for a large number of cycles, the current yield stress will continue to increase indefinitely until the material becomes fully rigid without any plasticity, which is obviously unrealistic.

Now consider the kinematic-hardening response for the same material under the same reversed strain cycle, as illustrated in Figure 10.24b. Note that the initial size of the yield surface, a circle of radius $\sqrt{2/3}\, Y_o$, as shown in Figure 10.23, remains unchanged irrespective of the amount of plastic deformation. Therefore, unloading from point b to b' and reloading up to point c defines a compressive yield stress at point c, given by $\sigma_c = -(2Y_o - Y_1)$, where $Y_1 = \sigma_b = Y_o + E^p\ (\Delta\varepsilon/2)$. Compressive plastic straining takes place from point c to point d, where $\varepsilon^p_{c-d} = -2(\Delta\varepsilon/2)$. The magnitude of the compressive stress attained at point d is determined from the material-hardening law applied to the path c–d as $\sigma_d = \sigma_c + E^p\ \varepsilon^p_{c-d} = -(2Y_o - Y_1) + E^p(-2\Delta\varepsilon/2)$. Knowing that $Y_1 = Y_o$

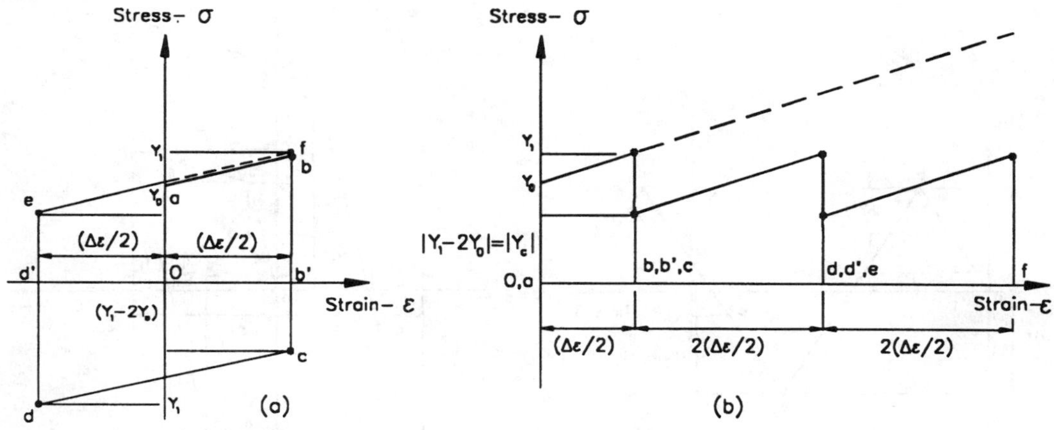

FIGURE 10.25 Uniaxial behavior of a rigid linear strain-hardening material under reversed strain cycle: (a) kinematic-hardening response and (b) current yield stress attained at each stage of cycling.

$+ E^p(\Delta\varepsilon/2)$ results in $\sigma_d = -(Y_o + E^p\Delta\varepsilon/2) = -Y_1$, which is the expected value. The stress at point e is thus determined from the fact that $|\sigma_d| + |\sigma_e| = 2Y_o$ according to the kinematic-hardening rule; hence, $\sigma_e = Y_o - |\sigma_d| = 2Y_o - Y_1$. Point f is reached by further tensile plastic straining $\varepsilon^p_{e-f} = 2(\Delta\varepsilon/2)$, starting from point e. Hence, $\sigma_f = \sigma_e + E^p\,\varepsilon^p_{e-f} = Y_o + E^p\,(\Delta\varepsilon/2) = Y_1$. Hence, points f and b are coincident and a hysteresis loop is predicted as shown in Figure 10.25a. The values of stresses attained at each point of the straining cycle are shown in Figure 10.25b.

For low-cycle fatigue, the orders of magnitude of the plastic and elastic strain components are generally close to each other. Hence, the material behavior has to be considered as elastic–plastic strain–hardening. The prediction of material behavior under strain cycling follows the same reasoning as above with the consideration that the total strain induced in each half cycle is the sum of the plastic strain and the corresponding elastic strain. For instance, for the first half cycle at point b of Figure 10.26a, the plastic strain $\varepsilon^p_b = (\Delta\varepsilon/2) - (\sigma_b/E)$, where E is Young's modulus of the material. Computations become a little more involved. The resulting behaviors according to both isotropic-hardening and kinematic-hardening rules are shown in Figures 10.26 and 10.27, respectively.

Experiments conducted on uniaxial test specimens subjected to reversed strain cycles show that the steady-state stress–strain response exhibits a closed hysteresis loop although it may be larger in height than that predicted by kinematic–hardening. This indicates that some isotropic hardening may be operating during the first initial cycles but that soon kinematic-hardening dominates.

Other hardening rules may be suggested, such as a mixed hardening rule that combines the behavior of both isotropic- and kinematic-hardening rules.[10]

Example 10.18:

In low-cycle fatigue testing, specimens are usually subjected to cycles of reversed total strain between $+\Delta\varepsilon/2$, $-\Delta\varepsilon/2$ until failure occurs. A bilinear material under monotonic stress–strain behavior, with Y_o = 250 MPa, E = 200 GPa, and E^t = 9.524 GPa, is subjected to a strain range $\Delta\varepsilon$ = 0.02. Determine the stress–strain response after few strain cycles as well as the magnitude of the plastic strain accumulated at the end of each stage according to

a. *Isotropic-hardening rule, and*
b. *Kinematic-hardening rule.*

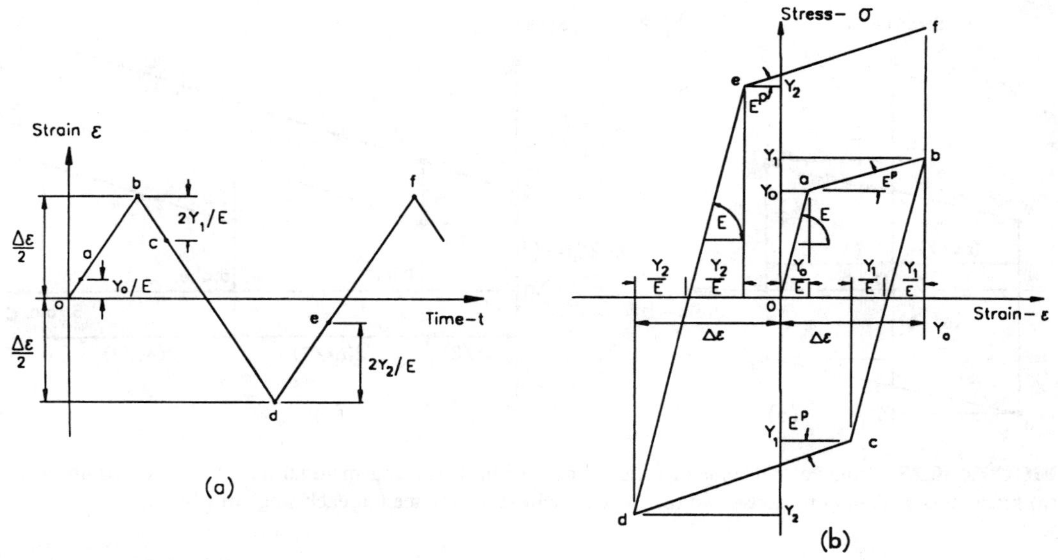

FIGURE 10.26 Uniaxial behavior of an elastic linear strain-hardening material: (a) strain cycle and (b) isotropic-hardening response

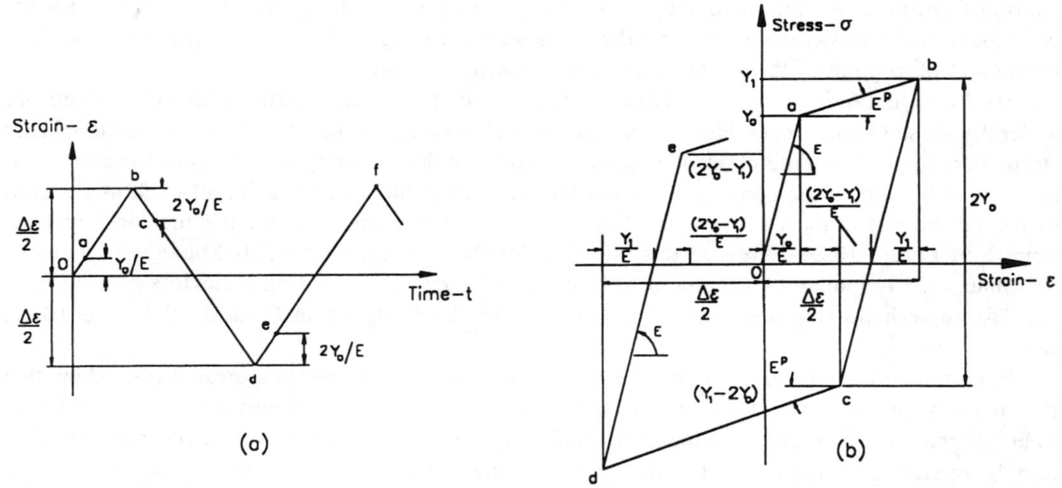

FIGURE 10.27 Uniaxial behavior of an elastic linear strain-hardening material: (a) strain cycle and (b) kinematic-hardening response

Solution:

The monotonic stress–strain curve is represented by Expression 10.8i, namely,

$$\sigma = Y_o + E^p\,\varepsilon^p$$

Since

$$\frac{1}{E^p} = \frac{1}{E^t} - \frac{1}{E}$$

according to Equation 10.8j, then

$$\frac{1}{E^p} = \frac{1}{9.524\times10^3} - \frac{1}{200\times10^3}$$

giving E^p = 10,000 MPa and, hence, $\sigma = 250 + 10{,}000E^p$.

Following Figure 10.26a, the loading path is designated by (0–*a*–*b*) up to a total tensile strain $\Delta\varepsilon/2 = 0.01$ at point *b*. This is followed by unloading and then reloading in compression along the path (*b*–*c*–*d*) until a total compressive strain $\Delta\varepsilon/2 = -0.01$ is attained at point *d*.

Unloading and reloading is then repeated, thus completing a strain cycle. The stress and strain coordinates for each point of the prescribed loading and unloading paths are determined below. Note that the elastic and plastic strain at any point *i* are denoted by ε_i^e and ε_i^p, respectively. The plastic strain accumulated from the beginning up to point *i* is $\Sigma\varepsilon_i^p$.

a. *Isotropic-hardening rule (see Figure 10.26)*
point a: Up to initial yielding:

$$\sigma_a = Y_o = 250 \text{ MPa}$$

$$\varepsilon_a^e = Y_o/E = 250/(200\times10^3) = 0.00125$$

$$\text{with } \varepsilon_a^p = \Sigma\,\varepsilon_a^p = 0$$

and no accumulated plastic strain, $\Sigma\varepsilon_a^p = 0$.
point b: Further loading up to a total strain $\Delta\varepsilon/2 = 0.01$:

$$\varepsilon_{a-b}^p = (\Delta\varepsilon/2) - \varepsilon_b^e = (\Delta\varepsilon/2) - (\sigma_b/E)$$

$$\sigma_b = Y_o + E^p\varepsilon_b^e;\ \ \varepsilon_{a-b}^p = E_b^p$$

Hence, $\sigma_b = Y_o + E^p[(\Delta\varepsilon/2) - (\sigma_b/E)]$ or $\sigma_b(1 + E^p/E) = Y_o + E^p(\Delta\varepsilon/2)$, giving

$$\sigma_b = (250 + 10^4\times0.01)/(1 + 10^4/200\times10^3)$$

$$= 333.333 \text{ MPa, say, } Y_1,$$

$$\varepsilon_b^p = 0.01 - \frac{333.333}{200\times10^3} = 0.008333$$

Hence, the accumulated plastic strain is $\Sigma\varepsilon_b^p = 0.008333$.
point c: *Unloading and reloading up to point* c *are both purely elastic*:

$$\sigma_c = -Y_1 = -333.333 \text{ MPa}$$

$$\text{and } \varepsilon_c = \varepsilon_b - 2Y_1/E = 0.01 - (2\times333.3/200\times10^3) = 0.006667$$

The accumulated plastic strain remains unchanged, as $\Sigma\varepsilon_c^p = 0.008333$.
point d: *Reloading continues up to point* d *for a magnitude of 3(Δε/2) of total strain*: hence, the accumulated plastic strain at this point is

$$\Sigma\varepsilon_d^p = \left|\frac{3\Delta\varepsilon}{2}\right| - \frac{2\sigma_b}{E} - \frac{|\sigma_d|}{E}$$

Inserting this into the monotonic stress–strain law gives

$$|\sigma_d| = Y_o + E^p\left(\frac{3\Delta\varepsilon}{2} - \frac{2\sigma_b}{E} - \frac{|\sigma_d|}{E}\right)$$

or

$$\left(1+\frac{E^p}{E}\right)|\sigma_d| = Y_o + E^p\left(\frac{3\Delta\varepsilon}{2} - \frac{2\sigma_b}{E}\right)$$

This gives, for $\Delta\varepsilon = 0.01$, $\sigma_b = 333.333$ MPa, $Y_o = 250$ MPa, $E = 10$ GPa, and $E^p = 10$ GPa; then $|\sigma_d| = 492.064$ MPa; say, Y_2, with $\Sigma\varepsilon_d^p = 0.024206$ and $\varepsilon_d^p = \Sigma\varepsilon_d^p - \Sigma\varepsilon_c^p = 0.015873$.

point e: *Unloading and reloading to point* e *are both purely elastic*:

$$\sigma_e = |\sigma_d| = 492.064 \text{ MPa}$$

$$\varepsilon_e = -\left(\frac{\Delta\varepsilon}{2} - 2\frac{|\sigma_d|}{E}\right) = -\left[0.01 - \left(2\times 492.064/200\times 10^3\right)\right]$$

$$= -0.005079$$

The accumulated plastic strain remains unchanged, as

$$\Sigma\varepsilon_e^p = \Sigma\varepsilon_d^p = 0.024206$$

point f: *Reloading continues up to point* f *for a magnitude of 5(Δε/2) of total strain*: hence, accumulated plastic strain at this point is

$$\Sigma\varepsilon_f^p = 5\left(\frac{\Delta\varepsilon}{2}\right) - 2\left(\frac{\sigma_b}{E} + \frac{|\sigma_d|}{E}\right) - \frac{\sigma_f}{E}$$

hence

$$\sigma_f = Y_o + E^p\Sigma\varepsilon_f^p$$

or

$$\left(1+\frac{E^p}{E}\right)\sigma_f = 250 + E^p\left(5\frac{\Delta\varepsilon}{2} - 2\left(\frac{\sigma_b + |\sigma_d|}{E}\right)\right)$$

This gives $\sigma_f = 635.767$ MPa, say, Y_3, with $\Sigma\varepsilon_f^p = 0.038568$ and $\varepsilon_f^p = 0.014362$.
This scheme of calculation may be continued to point g, thus giving

$$|\sigma_g| = -635.676 \text{ MPa}$$

$$\Sigma\varepsilon_g^p = 0.038568$$

$$|\sigma_h| = 765.612 \text{ MPa}$$

$$\Sigma\varepsilon_h^p = 0.0515612 \text{ MPa and } \varepsilon_h^p = 0.012994$$

The stress–strain response according to the isotropic-hardening rule is plotted in Figure 10.28a. Note that the plastic strain increment occurring at each stage is less than that at the precedent stage. Continuous cycling will end up by vanishing plastic strain. This indicates an ever-increasing yielding stress Y_n which is determined at an elastic strain $\varepsilon_n^e = 0.01$ as

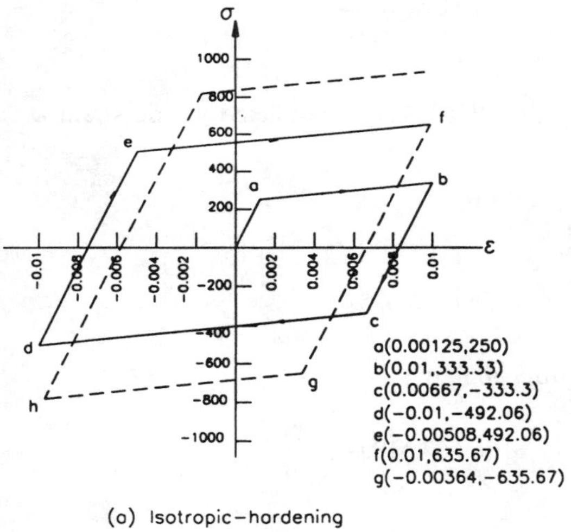

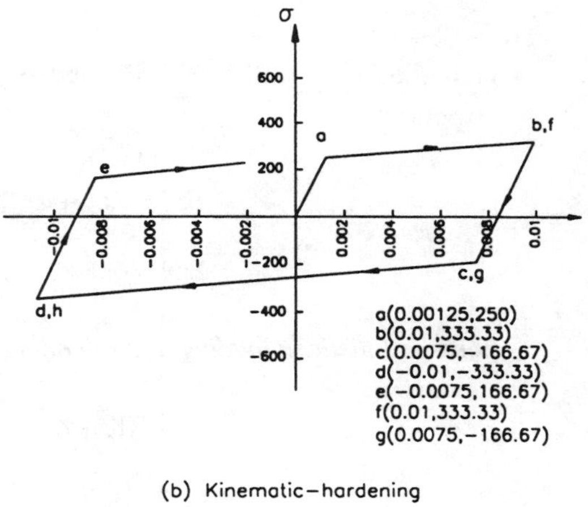

FIGURE 10.28 Example 10.18: (a) isotropic hardening and (b) kinematic hardening.

$$Y_n = E\varepsilon_n^e = 200 \times 10^3 \times 0.01 = 2000 \text{ MPa}$$

Obviously, this means that the yield surface has expanded to accommodate $\Delta\varepsilon$ as being totally elastic.

b. Kinematic-hardening rule (see Figure 10.27)
points a and b: Same calculations and results as in isotropic-hardening
point c: *Elastic unloading and reloading from point* b:

$$\sigma_c = -(2Y_o - Y_1) = -(2 \times 250 - 333.33) = -166.67 \text{ MPa}$$

$$\varepsilon_c = \Delta\varepsilon/2 - 2Y_o/E = 0.01 - 2 \times 250/200 \times 10^3 = 0.0075$$

The accumulated plastic strain at this point c is

$$\Sigma\varepsilon_c^p = \varepsilon_b^p = 0.008333$$

point d: Continued reloading gives:

$$\varepsilon_{c-d}^p = 2(\Delta\varepsilon/2) - |\sigma_b/E| - |\sigma_d|/E$$

Hence,

$$|\sigma_d| = |\sigma_c| + E^p\varepsilon_{c-d}^p = (2Y_o - Y_1) + E^p\left(\Delta\varepsilon - \frac{\sigma_b}{E} - \frac{|\sigma_d|}{E}\right)$$

or

$$\left(1 + \frac{E^p}{E}\right)|\sigma_d| = (2Y_o - Y_1) + E^p\left(\Delta\varepsilon - \frac{\sigma_b}{E}\right)$$

giving $|\sigma_d| = Y_1 = 333.333$ MPa and $\varepsilon_{c-d}^p = 0.016667$. An accumulated plastic strain at point d,

$$\Sigma\varepsilon_d^p = \Sigma\varepsilon_c^p + |\varepsilon_{c-d}^p|$$

$$\Sigma\varepsilon_d^p = 0.008333 + 0.016667 = 0.025$$

point e: *Elastic unloading and reloading from point* d:

$$\sigma_e = 2Y_o - Y_1 = -\sigma_c = 166.667 \text{ MPa}$$

$$\varepsilon_e = -(\Delta\varepsilon/2 - 2Y_o/E) = -0.0075$$

with an accumulated plastic strain unchanged, as

$$\Sigma\varepsilon_e^p = \Sigma\varepsilon_d^p = 0.025$$

point f: *Continued reloading*:

$$\varepsilon_{e-f}^p = 2(\Delta\varepsilon/2) - |\sigma_d|/E - \sigma_f/E$$

hence,

$$\sigma_f = \sigma_e + E^p\varepsilon_{e-f}^p$$
$$= (2Y_o - Y_1) + E^p(\Delta\varepsilon - |\sigma_d|/E - \sigma_f/E)$$

or

$$\left(1+\frac{E^p}{E}\right)\sigma_f = (2Y_o - Y_1) + E^p\left(\Delta\varepsilon - |\sigma_d|/E\right)$$

giving σ_f = 333.333 MPa and ε^p_{e-f} = 0.016667. Also, $\Sigma\varepsilon^p_f = \Sigma\varepsilon^p_e + \varepsilon^p_{e-f}$ = 0.025 + 0.016667 = 0.041667.

The stress–strain response according to the kinematic-hardening rule is plotted in Figure 10.28b as a closed hysteresis loop between $\varepsilon = \pm 0.01$ and $\sigma = \pm 333.33$ MPa for total strain and stress, respectively.

10.16.2 Triaxial Behavior — Yield Function and Flow Rule

Since kinematic hardening involves translation of the yield surface with constant size, the corresponding yield criterion should be a function of both the state of stress and the coordinates of the current center of the yield surface. For a state of principal stresses (σ_1, σ_2, σ_3), the coordinates of the center will be denoted by (α_1, α_2, α_3). A modification of the von Mises yield function satisfying the above conditions is

$$f = \frac{3}{2}\left[(\sigma_1' - \alpha_1)^2 + (\sigma_2' - \alpha_2)^2 + (\sigma_3' - \alpha_3)^2\right] - Y_o^2 = 0 \tag{10.61}$$

where (σ_1', σ_2', σ_3') are stress deviations as given by Expressions 1.37 and Y_o, the initial yield stress of the material. Note that for a virgin, unstrained material $\alpha_1 = \alpha_2 = \alpha_3 = 0$. It is also worth noting that Equation 10.61 reduces to von Mises isotropic-hardening surface by changing Y_o to Y (the current yield stress) and imposing $\alpha_1 = \alpha_2 = \alpha_3 = 0$ at all times, i.e., the center of yield surface remaining at the origin of the stress space.

Equation 10.61 describes a sphere in the stress deviation space with radius $\sqrt{2/3}\,Y_o$, and centered at (α_1, α_2, α_3), as illustrated in Figure 10.29.

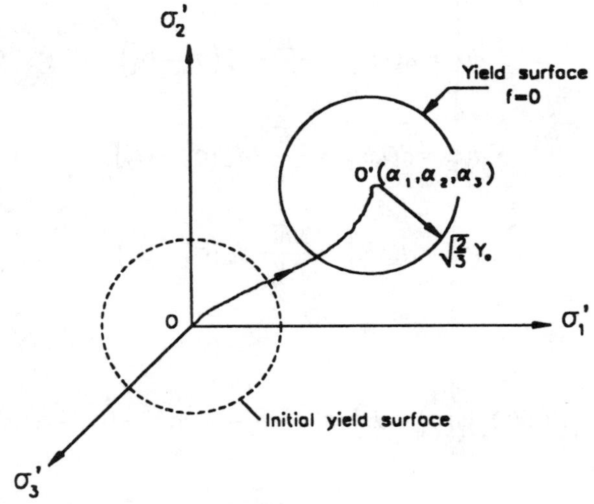

FIGURE 10.29 Translation of the center of constant size yield surface according to the kinematic-hardening rule.

The flow rule is obtained by applying the principle of normality of plastic strain increment to the yield surface at the loading point. Analogous to the derivations given in Section 10.9 and the utilization of the earlier definition of the effective plastic strain increment, Expressions 10.30, the plastic stress–strain relations become

$$\begin{aligned} d\varepsilon_1^p &= \frac{3}{2}\frac{d\bar{\varepsilon}^p}{Y_o}\left(\sigma_1' - \alpha_1\right) \\ d\varepsilon_2^p &= \frac{3}{2}\frac{d\bar{\varepsilon}^p}{Y_o}\left(\sigma_2' - \alpha_2\right) \\ d\varepsilon_3^p &= \frac{3}{2}\frac{d\bar{\varepsilon}^p}{Y_o}\left(\sigma_3' - \alpha_3\right) \end{aligned} \tag{10.62}$$

The most commonly used rule for description of the incremental evolution of coordinates of the center of the yield surface ($d\alpha_1$, $d\alpha_2$, $d\alpha_3$) postulates that the yield surface translates rigidly in the direction of the outward normal to the surface represented by $f = 0$ at the loading point, i.e., in the direction of plastic strain increment.* Hence,

$$\begin{aligned} d\alpha_1 &= dQ d\varepsilon_1^p \\ d\alpha_2 &= dQ d\varepsilon_2^p \\ d\alpha_3 &= dQ d\varepsilon_3^p \end{aligned} \tag{10.63}$$

where dQ is a positive scalar multiplier that needs to be determined as a function of the deviatoric stress increments ($d\sigma_1'$, $d\sigma_2'$, $d\sigma_3'$).[10]

Through a somewhat lengthy algebraic manipulation,[11] the final relations for the components of the incremental translation of the center, associated with a given stress increment become

$$\begin{aligned} d\alpha_1 &= dQ d\varepsilon_1^p = \frac{3d\bar{\varepsilon}^p}{2Y_o} dQ\left(\sigma_1' - \alpha_1\right) \\ d\alpha_2 &= dQ d\varepsilon_2^p = \frac{3d\bar{\varepsilon}^p}{2Y_o} dQ\left(\sigma_2' - \alpha_2\right) \\ d\alpha_3 &= dQ d\varepsilon_3^p = \frac{3d\bar{\varepsilon}^p}{2Y_o} dQ\left(\sigma_3' - \alpha_3\right) \end{aligned} \tag{10.64a}$$

where

$$dQ = \left[1/\left(d\bar{\varepsilon}^p Y_o\right)\right]\left[\left(\sigma_1' - \alpha_1\right)d\sigma_1' + \left(\sigma_2' - \alpha_2\right)d\sigma_2' + \left(\sigma_3' - \alpha_3\right)d\sigma_3'\right] \tag{10.64b}$$

* To determine dQ, a consistency condition that by the end of a loading increment ($d\sigma_i'$), the final loading point ($\sigma_i' + d\sigma_i'$), must still lie on the translated yield surface ($f = 0$) whose current center coordinates are ($\alpha_i + d\alpha_i$). This simply amounts to $df = 0$ during the loading increment.

Specialization of Equation 10.64b for the case of uniaxial tension ($\sigma_1' = -2\sigma_2' = -2\sigma_3' = 2\sigma_1/3$, $\bar{\sigma} = \sigma_1$, and $d\varepsilon_1^p = -2\, d\varepsilon_2^p = -2\, d\varepsilon_3^p = d\bar{\varepsilon}^p$) gives, in view of the flow rule (Equation 10.62),

$$dQ = \frac{2}{3}\frac{d\sigma}{d\bar{\varepsilon}^p} \quad \text{or} \quad dQ = \frac{2}{3}E^p \tag{10.65a}$$

where E^p is the slope of the uniaxial stress–plastic strain curve which may be constant for linear-hardening material with a flow curve represented by Equation 10.8i. Therefore, Equations 10.64a for the case of uniaxial tension yields

$$d\alpha_1 = \frac{2}{3}d\sigma_1 \tag{10.65b}$$

Integration with the initial conditions; $\sigma_1 = Y_o$ at $\alpha_1 = 0$, gives

$$\alpha_1 \int_0^{\alpha_1} d\alpha_1 = \frac{2}{3}\int_{Y_o}^{\sigma_1} d\sigma_1 = \frac{2}{3}(\sigma_1 - Y_o) \tag{10.65c}$$

For the case of principal triaxial state of stress, Equations 10.64 and 10.65 can be combined to give the following simple relations between the components of the translation increments and plastic strain increments:

$$\begin{aligned} d\alpha_1 &= \frac{2}{3}E^p d\varepsilon_1^p \\ d\alpha_2 &= \frac{2}{3}E^p d\varepsilon_2^p \\ d\alpha_3 &= \frac{2}{3}E^p d\varepsilon_3^p \end{aligned} \tag{10.66}$$

Hence, a procedure for the determination of plastic strain increments associated with a given stress increment is to employ Equations 10.64 first to obtain the translational increments ($d\alpha_1$, $d\alpha_2$, $d\alpha_3$) and second, to use Equations 10.66 to obtain plastic strain increments ($d\varepsilon_1^p$, $d\varepsilon_2^p$, $d\varepsilon_3^p$).

Relations describing yield and flow for the kinematic hardening rule can be also developed for the general triaxial stress state of stress (σ_x, σ_y, σ_z, τ_{xy}, τ_{yz}, τ_{zx}) following the same arguments.[10]

Example 10.19:

Derive a relation expressing the von Mises kinematically hardened yield locus for plane stress conditions in terms of principal stresses σ_1, σ_2, ($\sigma_3 = 0$) and center coordinates α_1, α_2. Illustrate the result graphically for $\sigma_1 = \sigma_2$ loading path, and compare with analogous isotropic-hardening yield locus for this state of plane stress.

Solution:

Denoting $\sigma_1' - \alpha_1 = S_1$, $\sigma_2' - \alpha_2 = S_2$ and $\sigma_3' - \alpha_3 = S_3$, then Expression 10.61 reduces to

$$f = \frac{3}{2}\left[S_1^2 + S_2^2 + S_3^2\right] - Y_o^2 = 0$$

Making use of Equation 1.37. i.e., $\sigma_1' + \sigma_2' + \sigma_3' = S_1 + S_2 + S_3 = 0$, then

$$f = \frac{3}{2}\left[(S_1 + S_2)^2 + (S_2 + S_3)^2 + (S_1 + S_3)^2\right] - Y_o^2 = 0$$

Expansion gives

$$f = 3\left[S_1^2 + S_2^2 + S_3^2 + S_1S_2 + S_2S_3 + S_1S_3\right] - Y_o^2 = 0$$

which can be rewritten as

$$f = \left(S_1^2 + S_2^2 + S_3^2 - S_1S_2 - S_2S_3 - S_1S_3\right)$$
$$+ \left(2S_1^2 + 2S_2^2 + 2S_3^2 + 4S_1S_2 + 4S_2S_3 + 4S_1S_3\right) - Y_o^2 = 0$$

It can be then shown that the terms included between the first parentheses on the right-hand side are rewritten as

$$(S_1 - S_3)^2 + (S_2 - S_3)^2 - (S_1 - S_3)\,(S_2 - S_3)$$

while the terms included between the second parentheses vanishes by use of $(S_1 + S_2 + S_3 = 0)$ and, hence,

$$f = \left[(S_1 - S_3)^2 + (S_2 - S_3)^2 - (S_1 - S_3)(S_2 - S_3)\right] - Y_o^2 = 0$$

For plane stress conditions, $(\sigma_3 = 0)$:

$$\sigma_1' = \sigma_1 - \sigma_m,\ \sigma_2' = \sigma_2 - \sigma_m, \text{ and } \sigma_3' = -\sigma_m$$

and recalling that $\alpha_1 + \alpha_3 + \alpha_3 = 0$, hence

$$S_1 - S_3 = (\sigma_1' - \alpha_1) - (\sigma_3' - \alpha_3) = \sigma_1' - \sigma_3' - \alpha_1 + \alpha_3 = \sigma_1 - 2\alpha_1 - \alpha_2$$
$$S_1 - S_3 = (\sigma_2' - \alpha_2) - (\sigma_3' - \alpha_3) = \sigma_2' - \sigma_3' - \alpha_2 + \alpha_3 = \sigma_2 - 2\alpha_2 - \alpha_1$$

Therefore, the yield locus is expressed as

$$f = \left[\sigma_1 - (2\alpha_1 + \alpha_2)\right]^2 + \left[\sigma_2 - (2\alpha_2 + \alpha_1)\right]^2$$
$$- \left[\sigma_1 - (2\alpha_1 + \alpha_2)\right]\left[\sigma_2 - (2\alpha_2 + \alpha_1)\right] - Y_o^2 = 0,$$

or

$$f = (\sigma_1 - \beta_1)^2 + (\sigma_2 - \beta_2)^2 - (\sigma_1 - \beta_1)(\sigma_2 - \beta_2) - Y_o^2 = 0 \qquad (10.67a)$$

where

$$2\alpha_1 + \alpha_2 = \beta_1 \quad \text{and} \quad 2\alpha_2 + \alpha_1 = \beta_2 \qquad (10.67b)$$

Expression 10.67a represents an ellipse with center coordinates (β_1, β_2) in (σ_1, σ_2) space as shown in Figure 10.30.

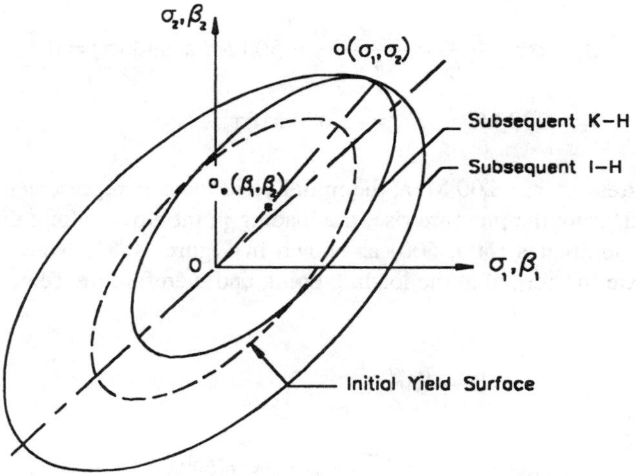

FIGURE 10.30 Example 10.19: von Mises kinematically hardened yield locus in plane stress conditions.

Note also that the direction ratios (l, m) of the outward normal to the yield locus at the loading point are obtained, by virtue of Equations 10.67 as

$$l = \frac{\partial f}{\partial \sigma_1} = 2(\sigma_1 - \beta_1) - (\sigma_2 - \beta_2) = 3(\sigma_1' - \alpha_1)$$

$$m = \frac{\partial f}{\partial \sigma_2} = 2(\sigma_2 - \beta_2) - (\sigma_1 - \beta_1) = 3(\sigma_2' - \alpha_2)$$

which are exactly the same as may be obtained from the general relation of the yield surface (Equation 10.61). This implies in view of the flow rule (Equation 10.62) and Equations 10.63 that

$$\frac{d\varepsilon_1^p}{d\varepsilon_2^p} = \frac{1}{m} = \frac{\sigma_1' - \alpha_1}{\sigma_2' - \alpha_2} = \frac{d\alpha_1}{d\alpha_2} = \frac{2d\beta_1 - d\beta_2}{2d\beta_2 - d\beta_1} \tag{10.68}$$

The analogous subsequent isotropic-hardening locus is obtained by letting $\beta_1 \to 0$, $\beta_2 \to 0$, and $Y_o \to Y$ in Expressions 10.67. The current yield stress of the material in Expression 10.67a, is thus obtained from $f = \sigma_1^2 + \sigma_2^2 - \sigma_1\sigma_2 - Y^2 = 0$ as given earlier by Expression 10.17 for plane stress $\sigma_3 = 0$. Figure 10.30 illustrates the initial yield locus and both the subsequenet kinematic-hardening and isotropic-hardening loci.

Example 10.20:

A thin-walled spherical pressure vessel of mean radius r_m *= 250 mm and thickness* h *= 5 mm made of a material with the bilinear monotonic stress–strain curve* $\sigma = 200 + 12{,}000\,\varepsilon^p$ *is subject to an internal pressure of 20 MPa. During the vessel life, the pressure fluctuates between zero and its peak value; 20 MPa. Determine the plastic strain developed in the shell wall using the kinematic-hardening rule, and illustrate the results graphically. Compare with isotropic-hardening predictions after the first loading–unloading cycle.*

Solution:

For a pressurized thin-walled spherical vessel, Equations 6.67 give

$$\sigma_1 = \sigma_2 = \frac{p_i r_m}{2h} = \frac{20 \times 250}{2 \times 5} = 500 \text{ MPa} \text{ and } \sigma_3 = 0$$

$$Y_o = 200 \text{ MPa}, \ E^p = 12{,}000 \text{ MPa}$$

For an initial yield stress of Y_o = 200 MPa, the initial yield locus is represented by a dotted ellipse with its center at O. During the pressure rise, the loading point moves along the radial path (O–*a*) where point *a* has coordinates (500, 500) as shown in Figure 10.31. Note that the path (O–*a*) coincides with the exterior normal at the loading point, and therefore the center O translates along (O–*a*).

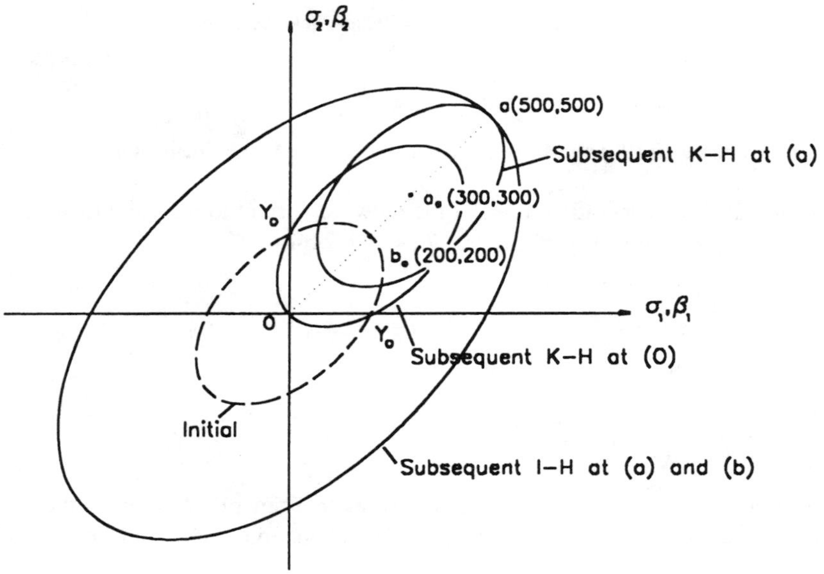

FIGURE 10.31 Example 10.20: yield locus.

For a yield locus of constant size (i.e., kinematic hardening) corresponding to a yield strength Y_o = 200 MPa, the corresponding center coordinates at point a_o are given by

$$\beta_1 = 300 \text{ MPa and } \beta_2 = 300 \text{ MPa}$$

However, from Equations 10.67b,

$$\alpha_1 = (2\beta_1 - \beta_2)/3, \ \alpha_2 = (2\beta_2 - \beta_1)/3$$

or

$$\alpha_1 = (2 \times 300 - 300)/3 = 100 \text{ MPa and } \alpha_2 = (2 \text{ x } 300 - 300)/3 = 100 \text{ MPa}$$

The slope E^p of the plastic stress–strain curve is obtained from the flow curve $\sigma = 200 + 12{,}000\ \varepsilon^p$, as $E^p = 12{,}000$ MPa. Hence, using Expressions 10.66 in their integrated form for this case of proportional loading gives

$$\alpha_1 = \frac{2}{3}E^p\varepsilon_1^p \text{ or } \varepsilon_1^p = \frac{100}{(2/3)\times 12{,}000} = 1.25\%$$

$$\alpha_2 = \frac{2}{3}E^p\varepsilon_2^p \text{ or } \varepsilon_2^p = \frac{100}{(2/3)\times 12{,}000} = 1.25\%$$

During presssure drop, the loading point moves from a to O and the center of the yield locus changes from point a_o (300, 300) to point b_o (200, 200). Hence the corresponding incremental changes in center coordinates are $d\beta_1 = 200 - 300 = -100$ MPa and $d\beta_2 = 200 - 300 = -100$ MPa. Also from Equations 10.67b,

$$d\alpha_1 = \frac{1}{3}(2d\beta_1 - d\beta_2) = \frac{1}{3}[2(-100)+100] = -100/3 \text{ MPa}$$

$$d\alpha_2 = \frac{1}{3}(2d\beta_2 - d\beta_1) = \frac{1}{3}[2(-100)+100] = -100/3 \text{ MPa}$$

From Relations 10.66, the following plastic strain increments are obtained

$$d\varepsilon_1^p = \frac{d\alpha_1}{2E^p/3} = \frac{-100/3}{2\times 12{,}000/3} = -0.4167\%$$

$$d\varepsilon_2^p = \frac{d\alpha_1}{2/E^p/3} = \frac{-100/3}{2\times 12{,}000/3} = -0.4167\%$$

Consideration of further pressure cycles shows that the center of the kinematically hardened yield locus would alternate between points a_o and b_o, giving $d\varepsilon_1^p = d\varepsilon_2^p = \pm 0.4167\%$ indicating the type of fatigue problem involved by this loading. The analogous isotropic-hardening solution would give the same $d\varepsilon_1^p = d\varepsilon_2^p$ during pressure rise of the first loading cycle. However, no reverse yielding would occur during pressure drop, since the loading point 0 remains inside the (*I–H*) yield locus as shown in Figure 10.31. The plastic strain history as predicted by both kinematic- and isotropic-hardening rules is represented in Figure 10.32.

10.17 PLASTIC DEFORMATION OF POROUS SOLIDS

The role of void nucleation, growth, and coalescence in ductile failure has been observed experimentally. In cold working of metallic alloys observations suggest that ductile fracture initiates by the opening of holes or voids around inclusions and second-phase particles. This phenomenon is also observed under hot-working conditions, e.g., tertiary creep damage and forming of superplastic alloys. Sintered powder compacts represent an example of voided or porous solids.

Deformation analysis for solids containing voids requires the formulation of a theory of plasticity particularily adapted to this behavior. Some of the basic assumptions discussed in Section 10.2 pertaining to the conventional theory of plasticity do not hold for porous solids. More specifically, the condition of constancy of volume as well as the independence of yielding of the hydrostatic stress component are not applicable. The presence of voids in these solids is responsible

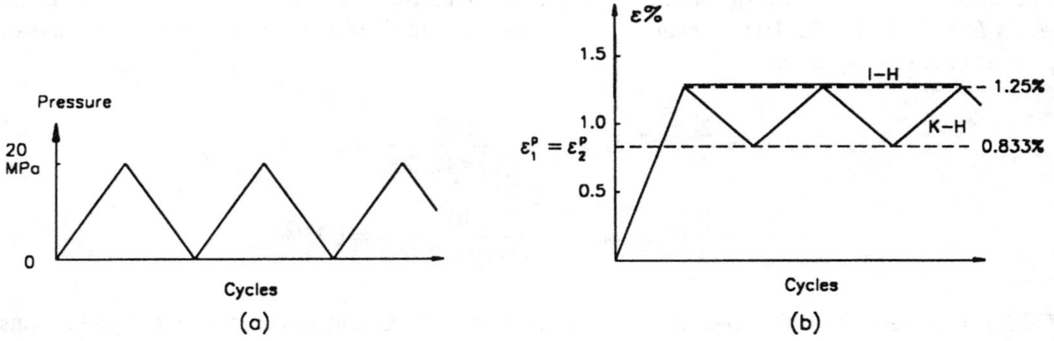

FIGURE 10.32 Example 10.20: (a) stress cycle and (b) plastic strain history.

for its compressibility and the dependence of its yield behavior on the hydrostatic pressure. Irrespective of this, the other assumptions of considering homogeneous and isotropic behavior could be maintained.

The porosity of a solid is characterized by the void volume fraction C_v which is defined by

$$C_v = V_V/V_T = (V_T - V_M)/V_T \tag{10.69}$$

where V_V, V_M, and V_T designate void, matrix, and total volumes, respectively. (Note that $C_v = -\Delta P/P$ where P is the material density.) At the initial stages of deformation, C_v is very small for conventional alloys (say, 10^{-4}); however, it may be of the order of 0.1 for sintered powder compacts. For superplastic alloys, a value of 0.3 is not uncommon at elongation of 500%.

10.17.1 Yield Function

A fundamental assumption on which a plasticity theory for isotropic porous solids may be founded is that the matrix material, which has a yield strength Y_M, obeys the von Mises yield function. Hence, most formulations suggest a yield function in the form:

$$\left(I_2^{\backslash}/Y_M^2\right) + f\left[\left(I_1/Y_M\right),\ C_v\right] = 0 \tag{10.70}$$

where I_1 and $I_2^{\backslash}$ are the first stress invariant and the second deviatoric stress invariant as defined by Equations 1.25 and 1.38, respectively. Obviously, Expression 10.70 reduces to the von Mises criterion if the function $f = 0$.

Two yield functions are suggested in the literature,[12] namely,

- Green yield function:[12]

$$\frac{I_2^{\backslash}}{Y_M^2} + \frac{C}{3}\left[\frac{1}{4\left(\ln C_v\right)^2}\frac{I_1^2}{Y_M^2} - 1\right] = 0 \tag{10.71a}$$

where

$$C = \left[3\left(1 - C_v^{1/3}\right)/\left(3 - 2C_v^{1/4}\right)\right]^2 \tag{10.71b}$$

- Gurson yield function:[13]

$$\frac{3I_2'}{Y_M^2} + 2C_v \cosh\left(\frac{I_1}{2Y_M}\right) - C_v^2 - 1 = 0 \tag{10.72a}$$

Both yield functions reduce to the von Mises criterion when $C_v = 0$.

The dependence of the yield function on the hydrostatic stress component can be seen when the former is plotted with respect to the axes (I_1/Y_M) and $(\sqrt{I_2'}/Y_M)$ as x- and y-axes, respectively. This is shown in Figure 10.33 for a value of $C_v = 0.01$. The two bounding horizontal lines represent the generators of the von Mises cylindrical yield surface of Figure 10.8a. The role that the hydrostatic stress component (represented along the axis I_1/Y_M) plays in enhancing yielding is obviously seen in Figure 10.33.

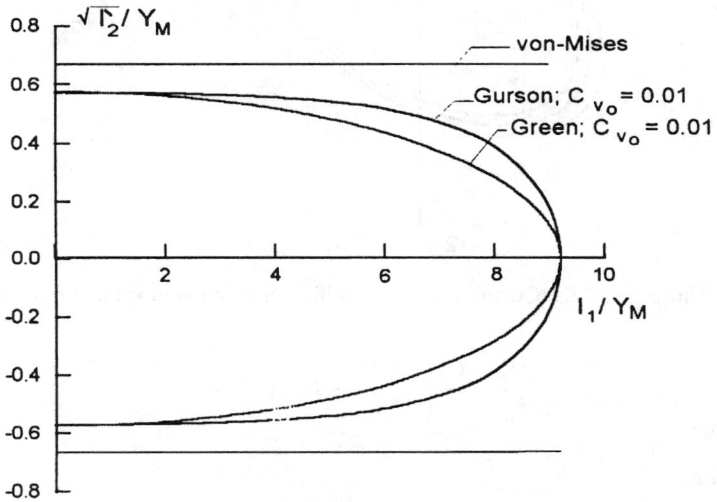

FIGURE 10.33 Comparison between two yield loci for porous solids.

Experimental confirmation of both the Green and Gurson yield criteria certainly still requires further investigations.

A plot for the Gurson yield function (Equation 10.72a) is shown in Figure 10.34 for plane stress conditions ($\sigma_3 = 0$). The von Mises ellipse corresponds to $C_v = 0$, while softening due to increasing void volume fraction (due to plastic straining) is displayed by a decreasing size for the yield surface as C_v increases.

The Gurson yield function involving the hyperbolic cosine term becomes unattractive in further analytical manipulations; hence, expanding the hyberbolic cosine term into its series gives

$$\frac{3I_2'}{Y_M^2} + \left\{2C_v\left[1 + \frac{1}{2!}\left(\frac{I_1}{2Y_M}\right)^2 + \frac{1}{4!}\left(\frac{I_1}{2Y_M}\right)^4 + \cdots\right] - C_v^2 - 1\right\} = 0$$

In most plastic deformation processes $(I_1/2Y_M)$ is of the order of unity or less; hence, it is an acceptable approximation to retain the first term of $(I_1/2Y_M)$ only, thus giving the Gurson approximate yield function as

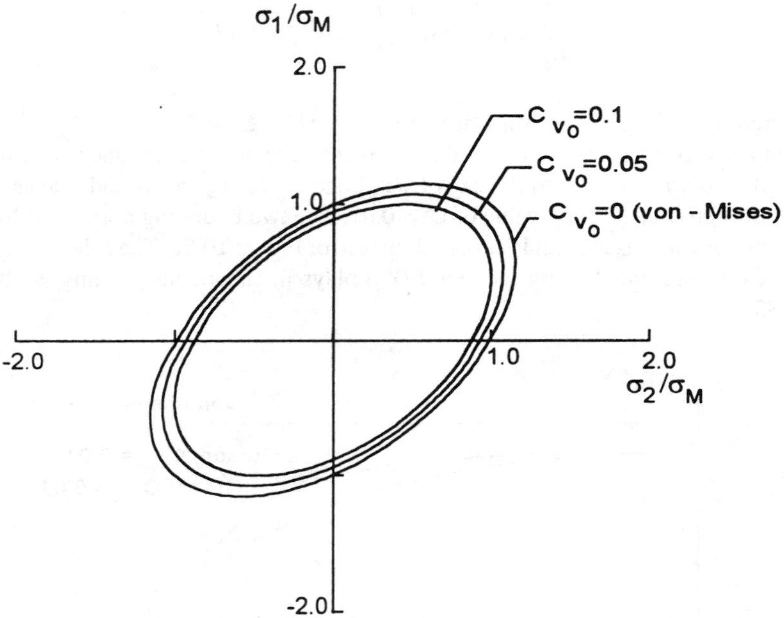

FIGURE 10.34 Shrinkage of the Gurson yield locus with increasing void volume fraction.

$$\frac{3I_2^\prime}{Y_M^2} + \frac{C_v}{4}\left(\frac{I_1}{2Y_M}\right)^2 - \left(1 - C_v\right)^2 = 0$$

This is rewritten as

$$Y_M^2 = \frac{1}{\left(1 - C_v\right)^2}\left[3I_2^\prime + \frac{C_v}{4} I_1^2\right]$$

In terms of the definitions I_1 and $I_2^\prime$ given by Equations 1.25 and 1.38, respectively,

$$Y_M^2 = \frac{1}{\left(1 + C_v\right)^2}\left\{\frac{1}{2}\left[\left(\sigma_1 - \sigma_2\right)^2 + \left(\sigma_2 - \sigma_3\right)^2 + \left(\sigma_3 - \sigma_1\right)^2\right] + \left(\frac{3}{2}\sigma_m\sqrt{C_v}\right)^2\right\} \qquad (10.72b)$$

Table 10.2 gives a comparison between the Gurson yield function (Equation 10.72a) and its approximate version, Expression 10.72b, both at $C_v = 0.1$ for plane stress conditions ($\sigma_3 = 0$). Negligible differences exist between both functions. More specifically, a maximum difference of about 0.3% at the condition of $\sigma_1/\sigma_2 = -1$. Hence, the approximate Gurson yield function (Equation 10.72b) will be used here for further analytical development.

Example 10.21:

Determining the ratio σ/Y_M at which a porous solid of current void volume fraction C_v *yield. Consider the following states of loading:*

a. *Uniaxial tension (σ, 0, 0),*
b. *Equibiaxial tension (σ, σ, 0),*

TABLE 10.2
Comparison between Gurson and Approximate Gurson Yield Criteria for Plane Stress Conditions

$\sigma_1/\sigma_2 = \alpha$	0	$^1/_2$	1	–1
Gurson: σ_1/Y_M	0.88887	1.00066	0.85569	–0.51961
Approximate Gurson: σ_1/Y_M	0.88896	1.00232	0.85811	–0.51961

c. Uniaxial tension (σ, 0, 0) with superimposed hydrostatic pressure ($-\sigma/3$, $-\sigma/3$, $-\sigma/3$),
d. Hydrostatic pressure ($p = -\sigma$)

Comment on the results for $C_v = 0.02$.

Solution:

Direct substitution in the Gurson approximate yield condition (Equation 10.72b) for each state of loading gives the ratio of σ/Y_M as indicated in the following Table 10.3.

TABLE 10.3
Solution of Example 10.21

	Case of Loading	σ/Y_M	(σ/Y_M) for $C_v = .02$
a.	Simple tension (σ, 0, 0)	$\dfrac{1-C_v}{\sqrt{1+C_v/4}}$	0.9776
b.	Equibiaxial tension (σ, σ, 0)	$\dfrac{1-C_v}{\sqrt{1+C_v}}$	0.9703
c.	Simple tension (σ, 0, 0) with hydrostatic pressure ($-\sigma/3$)	$(1-C_v)$	0.98
d.	Hydrostatic pressure ($-\sigma$, $-\sigma$, $-\sigma$)	$\dfrac{2(1-C_v)}{3\sqrt{C_v}}$	4.62

The results in the table indicate that in all cases except that of hydrostatic pressure for unvoided material ($C_v = 0$), $\sigma/Y_M = 1$. Hydrostatic pressure according to conventional plasticity does not cause yielding at all. Also, note that among the first three cases minimum softening, i.e., higher (σ/Y_M) ratio, occurs at simple tension with superimposed hydrostatic pressure ($p = -\sigma/3$), a fact which has been observed experimentally.[14] For hydrostatic pressure its value must be many times greater than Y_M in order to initiate yielding.

10.17.2 Flow Rule

According to the normality rule* of the strain vector to the yield surface, the principal plastic strain increments are derived by partial differentiation of the yield function (Equation 10.72b) with respect to the stresses. Thus,

* It has been shown that convexity of the yield surface and normality of the strain vector carry over to unstable materials, where the increase of plastic strain causes the yield surface to shrink, as shown in Figure10.34.[15]

$$d\varepsilon_1^p = d\lambda_3\left[\sigma_1 - \left(1 - \frac{C_v}{2}\right)\sigma_m\right]$$

$$d\varepsilon_2^p = d\lambda_3\left[\sigma_2 - \left(1 - \frac{C_v}{2}\right)\sigma_m\right] \tag{10.73}$$

$$d\varepsilon_3^p = d\lambda_3\left[\sigma_3 - \left(1 - \frac{C_v}{2}\right)\sigma_m\right]$$

where $d\lambda_3$ is a nonnegative proportionality constant. Hence, the volumetric plastic strain $d\varepsilon_v$ is obtained by summing:

$$d\varepsilon_v = d\varepsilon_1^p + d\varepsilon_2^p + d\varepsilon_3^p = d\lambda_3\left(\frac{3}{2}C_v\sigma_m\right) \tag{10.74a}$$

From Equation 10.69, $V_M = V_T(1 - C_v)$, which upon differentiation, noting that for an incompressible matrix material $dV_M = 0$, yields:

$$d\varepsilon_v = \frac{dV_T}{V_T} = \frac{dV_v}{V_T} = \frac{dC_v}{1 - C_v} \tag{10.74b}$$

Combining Equations 10.74a and 10.74b provides the value for $d\lambda_3$ as:

$$d\lambda_3 = \frac{2}{3C_v}\frac{d\varepsilon_v}{\sigma_m} = \frac{2}{3\sigma_m}\frac{1}{C_v(1 - C_v)}dC_v \tag{10.75}$$

Since $d\lambda_3$ has to be positive or equal to zero, the change in void volume fraction is either positive or negative following the sign of the hydrostatic stress component σ_m. More explicitly, void growth occurs under positive σ_m, while voids close up under negative σ_m, i.e., compressive pressure.

Substituting the value of $d\lambda_3$ into the flow rule (Equation 10.73) gives

$$d\varepsilon_1^p = \frac{2}{3}\frac{1}{\sigma_m}\left(\frac{1}{1 - C_v}\right)\frac{dC_v}{C_v}\left[\sigma_1 - \left(1 - \frac{C_v}{2}\right)\sigma_m\right]$$

$$d\varepsilon_2^p = \cdots \tag{10.76}$$

The yield condition (Equation 10.72b) and the flow rule (Equation 10.76) determine the relations between stress and strain, as long as the current volume fraction C_v is known.

10.17.3 Void Growth Characteristics

At this stage, it is interesting to consider void growth in the case of plane stress, where $\sigma_2 = \alpha\sigma_1$ and $\sigma_3 = 0$; thus, from Equation 10.76,

$$\frac{d\varepsilon_1^p}{dC_v} = \left[\frac{4}{3}\frac{(1 - \alpha/2)}{C_v(1 - C_v)(1 + \alpha)} + \frac{1}{3(1 - C_v)}\right]$$

This may be integrated for an initial void volume C_{vo} to give

$$\varepsilon_1^p = \frac{4}{3}\frac{(1-\alpha/2)}{(1+\alpha)}\ln\left[\frac{C_v(C_{vo}-1)}{C_{vo}(C_v-1)}\right]+\frac{1}{3}\ln\left[\frac{C_{vo}-1}{C_v-1}\right] \tag{10.77}$$

which characterizes the void growth or decay with the current strain $d\varepsilon_1^p$ as shown in Figure 10.35 for three modes of tensile loading and, hence, of void growth.

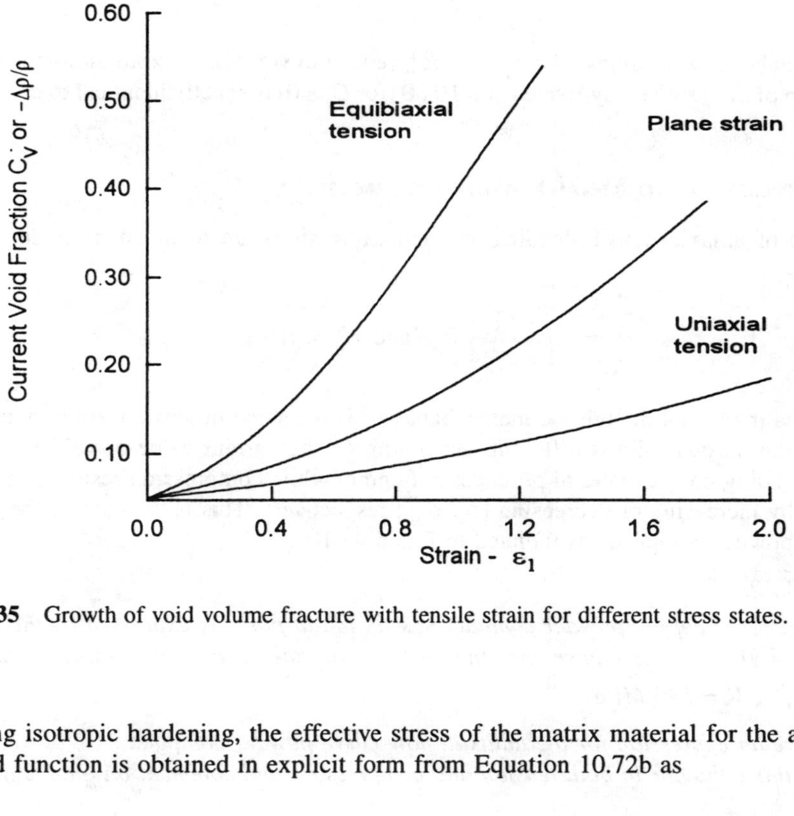

FIGURE 10.35 Growth of void volume fracture with tensile strain for different stress states.

Assuming isotropic hardening, the effective stress of the matrix material for the approximate Gurson yield function is obtained in explicit form from Equation 10.72b as

$$\bar{\sigma}_M = \frac{1}{(1+C_v)}\left\{\frac{1}{2}\left[(\sigma_1-\sigma_2)^2+(\sigma_3-\sigma_3)^2+(\sigma_3-\sigma_1)^2\right]+\left[\frac{3}{2}\sigma_m\sqrt{C_v}\right]^2\right\}^{1/2} \tag{10.78}$$

If an effective matrix plastic strain increment is denoted by $d\bar{\varepsilon}_M^p$, then the plastic work done per unit volume of porous material is

$$dW^p = \sigma_1 d\varepsilon_1^p + \sigma_2 d\varepsilon_2^p + \sigma_3 d\varepsilon_3^p$$

This is identical to the work done by the matrix material, which has the volume $(1 - C_v)$; hence,

$$dW^p = \sigma_1 d\varepsilon_1^p + \sigma_2 d\varepsilon_2^p + \sigma_3 d\varepsilon_3^p = (1-C_v)\bar{\sigma} d\bar{\varepsilon}_M^p \tag{10.79}$$

Substituting Equation 10.78 into Equation 10.79 and making use of the flow rule, Equation 10.73, then rearranging gives

$$d\lambda_3 = \frac{3}{2(1-C_v)}\frac{d\bar{\varepsilon}_M^p}{\bar{\sigma}_M} \tag{10.80}$$

and

$$d\bar{\varepsilon}_M^p = \left\{\frac{2}{9}\left[\left(d\varepsilon_1^p - d\varepsilon_2^p\right)^2 + \left(d\varepsilon_2^p - d\varepsilon_3^p\right)^2 + \left(d\varepsilon_3^p - d\varepsilon_1^p\right)^2\right] + \left[2d\varepsilon_v/3\sqrt{C_v}\right]^2\right\}^{1/2} \tag{10.81}$$

Note that the above expressions* for $\bar{\sigma}_M$ and $d\bar{\varepsilon}_M^p$ reduce to von Mises expressions for $C_v = 0$. Also, the definition of $d\lambda_3$ as given by Expression 10.80 (for $C_v = 0$) is exactly identical to $d\lambda$ for isotropic-hardening.

10.17.4 Application to Metal Powder Compacts

For the case of uniaxial tensile loading, explicit expressions are obtained from Equations 10.78 and 10.81 as

$$\sigma_1 = \frac{1-C_v}{\sqrt{1+C_v/4}}\bar{\sigma}_M \quad \text{and} \quad \varepsilon_1^p = \sqrt{1+C_v/4}\,\bar{\varepsilon}_M^p \tag{10.82}$$

This indicates that, even though the matrix behavior is the same in tension as in compression, the behavior of the porous solid is different, depending on the current value of void volume fraction C_v. Tensile loading on the material produces softening, while compression results in strengthening as reflected by increasing or decreasing $(\sigma_1/\bar{\sigma}_M)$ respectively. This is the case of the deformation behavior of powder compacts, as indicted in Example 10.22.

Example 10.22:

A partially sintered copper powder compact has an initial porosity expressed by the void volume fraction $C_{vo} = 0.205$. *It is known that the matrix material behaves according to the law*[16]*:* $\bar{\sigma}_M = K(\bar{\varepsilon})_M^{0.477}$; $K = 546$ *MPa*

a. *Derive an expression for the uniaxial flow curve powder compact.*
b. *Plot this behavior in both tension and compression and comment on the results.*

Solution:

a. Since the matrix material behaves according to

---

* Lengthy derivations leading to the flow rule expressions, effective stress and strain increment, using the original Gurson yield function (Equation 10.64a) may be also demonstrated to give the following explicit expressions:

$$d\varepsilon_1^p = d\lambda_4\left\{\frac{1}{1+R}\left[2(1+R)\sigma_1 - 2R\sigma_2 - 2\sigma_3\right] + C_v\bar{\sigma}_M \sinh\left[3\sigma_m/2\bar{\sigma}_M\right]\right\}$$

$$d\varepsilon_2^p = d\lambda_4\left\{\frac{1}{1+R}\left[2(1+R)\sigma_2 - 2R\sigma_1 - 2\sigma_3\right] + C_v\bar{\sigma}_M \sinh\left[3\sigma_m/2\bar{\sigma}_M\right]\right\}$$

$$d\varepsilon_3^p = d\lambda_4\left\{\frac{1}{1+\bar{r}}\left[4\sigma_3 - 2\sigma_2 - 2\sigma_1\right] + C_v\bar{\sigma}_M \sinh\left[3\sigma_m/2\bar{\sigma}_M\right]\right\}$$

$$d\varepsilon_v^p = d\varepsilon_1 + d\varepsilon_2 + d\varepsilon_3 = d\lambda\left\{3C_v\bar{\sigma}_M \sinh\left[3\sigma_m/2\bar{\sigma}_M\right]\right\}$$

$$\sigma_M = K(\bar{\varepsilon})_M^{n_M} \text{ with } n_M = 0.477$$

the porous powder compact behavior is obtained by substituting from Equation 10.82 for the stress and strain, respectively to give

$$\frac{\sigma_1\sqrt{1+C_v/4}}{1-C_v} = K\left(\frac{1}{\sqrt{1+C_v/4}}\right)^{n_M}(\varepsilon)_1^{n_M}$$

or

$$\sigma_1 = K(1-C_v)(1+C_v/4)^{-(1+2n_M)/2}(\varepsilon)_1^{n_M}$$

The current value of the void volume fraction is given implicitly by Expression 10.77 for $\alpha = 0$, namely;

$$\varepsilon_1 = \frac{1}{3}\ln\left[\left(\frac{C_v}{C_{vo}}\right)^4\left(\frac{1-C_{vo}}{1-C_v}\right)^5\right]$$

b. Plots of these expressions are shown in Figure 10.36 in both cases of tension and compression together with the matrix behavior for comparison. The behavior indicates softening with increasing strain due to void growth in tension and hardening due to void decay in compression.

PROBLEMS

Stress–Strain Curves

10.1 Derive an expression for the engineering stress–engineering strain (S-e) for a material obeying the law: $\sigma = K(\varepsilon)^n$. The engineering stress is defined as load divided by the initial cross section of the tensile specimen.

10.2 The following test data were obtained from tensile tests conducted on round specimens of 15 mm diameter and 50 mm gauge length made of three metals as given:

---

* where

$$d\lambda_4 = \frac{(1-C_v)}{\{2(1+C_v^2) - 4C_v\cosh[3\sigma_m/2\bar{\sigma}_M] + 3C_v[3\sigma_m/2\bar{\sigma}_M]\sinh[3\sigma_m/2\bar{\sigma}_M]\}}\frac{d\bar{\varepsilon}_M}{\bar{\sigma}_M}$$

$$\bar{\sigma}^2 = \bar{\sigma}_M^2(1+C_v^2) - 2C_v\bar{\sigma}^2\cosh[3\sigma_m/2\bar{\sigma}_M]$$

$$d\bar{\varepsilon}_M^p = -\frac{\left\{1+C_v^2 - 2C_v\cosh[3\sigma_m/2\bar{\sigma}_M] + \frac{3}{2}C_v[3\sigma_m/2\bar{\sigma}_M]\sinh[3\sigma_m/2\bar{\sigma}_M]\right\}}{\{(1-C_v)^2\{1+C_v^2 - 2C_v\cosh[3\sigma_m/2\bar{\sigma}_M]\}}d\bar{\varepsilon}^p$$

$$\frac{d\varepsilon_1^p}{dC_v} = \frac{\left\{\frac{1}{1+R}[2(1+R)\sigma_1 - 2R\sigma_2 - 2\sigma_3] + C_v\bar{\sigma}_M\sinh[3\sigma_m/2\bar{\sigma}_M]\right\}}{\{3C_v(1-C_v)\bar{\sigma}_M\sinh[3\sigma_m/2\bar{\sigma}_M]\}}$$

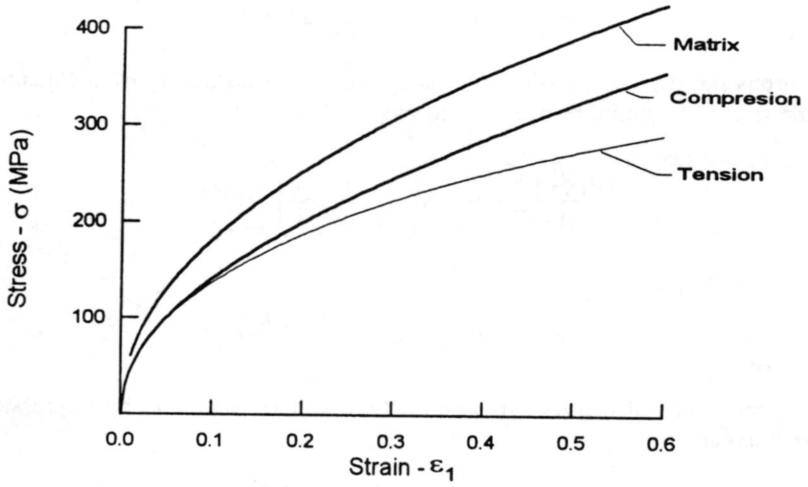

FIGURE 10.36 Example 10.22: stress–strain behavior of sintered copper powder compact.

Low carbon steel: (Y_o = 220 MPa, E = 200 GPa)

Load P(kN)	40	46.3	49.3	51.3	52	52.5 (max.)
Length L(mm)	52.04	54.16	56.37	58.68	61.07	64.85

Commercially Pure Aluminum: (Y_o = 45 MPa, E = 69 GPa)

Load P(kN)	16.2	17.8	18.7	18.89	19.06 (max.)
Length L(mm)	52.06	54.06	56.18	59.05	61.12

Annealed Copper: (Y_o = 68 MPa, E = 110 GPa)

Load P(kN)	13.1	17.6	17.89	21.5	22.74	23.25 (max.)
Length L(mm)	54.25	58.73	65.21	67.5	74.6	85.8

Use the above data to represent the flow stress–strain behavior of the above metals by fitting to each the empirical law:

a. $\sigma = E(\varepsilon)^e$ and $\sigma = K(\varepsilon^p)^n$
b. $\sigma = Y_o + A(\varepsilon^p)^q$

Also determine the flow stress for steel, aluminum, and copper at ε^p = 0.12, 0.1165, and 0.3, respectively, for each law.

Yield Criteria

10.3 Four material elements A, B, C, and D taken at different locations of a loaded structure are subjected to the different states of stress (4σ, –σ), (4σ, 0), (3σ, –3σ), and (3σ, 3σ), respectively. Which one of these elements will yield first according to

a. The Tresca yield criterion,
b. The von Mises yield criterion.

10.4 The stress components at a point in a steel member is a tensile stress σ_x and a shear stress τ. The tensile elastic limit is 400 MPa. If the shear stress at the point is 200 MPa when yielding starts, what is the tensile stress σ at the point according to both the Tresca and the von Mises criterion?

10.5 A cube of metal having a constant yield stress Y of 300 MPa experiences a stress state of σ_1, $\sigma_2 = 0.4\sigma_1$, and $\sigma_3 = -0.6\sigma_1$. If the stresses are gradually increased in these constant ratios, find σ_1 at yielding using both the von Mises and Tresca criteria.

10.6 For a state of plane strain at yielding show that the von Mises condition becomes identical to the Tresca condition, in terms of k; the yield shear strength.

10.7 A plate of steel is loaded under equibiaxial tension such that $\sigma_1 = \sigma_2$ and $\sigma_3 = 0$. It has a yield stress of 400 MPa, Young's modulus is 200 GPa, and Poisson's ratio is 0.30. What is the largest fractional volume change, $\Delta V/V$, that could be obtained without yielding? Apply von Mises criterion.

10.8 The area of each face of a metal cube is 1500 mm². This metal has a yield strength in pure shear, k, of 150 MPa.

a. If tensile loads of 60 and 120 kN are applied in the two x- and y-directions, what is the load in the z-direction to cause yielding according to the Tresca criterion?
b. If the 60 and 120 kN loads were compressive, what tensile load in the z-direction would cause yielding according to the Tresca criterion?

10.9 For a thin flat plate loaded in the x–y plane, it is known that $\sigma_x = 252$ MPa, $\gamma_{xy} = 10^{-3}$, $\varepsilon_z = 0.5 \times 10^{-4}$, $E = 200$ GPa and $G = 76.9$ GPa.

a. Find all remaining components of the stress matrix.
b. What are the deviatoric stress components?
c. If the stress system obtained in (a) just causes yielding, calculate the yield stress of the plate material in tension applying both the Tresca and the von Mises yield criteria.

10.10 Determine the pressure for a thin-walled cylindrical vessel at which the material starts to yield, applying both the Tresca and the von Mises yield criteria, knowing that the yield strength is $Y = 280$ MPa. The vessel is 500 mm in diameter, with 12 mm wall thickness. If the same pressure is applied to a spherical vessel of the same material and diameter, what would be its thickness to avoid yielding?

10.11 A thin-walled cylindrical pressure vessel with hemispherical ends, both has a diameter of 1000 mm and is to be fabricated from sheet metal of thickness h whose yield shear strength, $k = 120$ MPa. The maximum service internal pressure is 5 MPa. If no section of the vessel is to yield, what minimum wall thickness should be specified according to

a. The Tresca criterion?
b. The von Mises criterion?

(Neglect discontinuity stresses at the junction between the cylinder and the ends.)

10.12 A thin-walled tube with closed ends is made from a metal whose $Y = 300$ MPa. The tube is 250 mm long, has a wall thickness of 1.25 mm, and a mean diameter of 75 mm. In service, the tube will be subjected to a maximum axial tensile load of 10 kN, a maximum twisting moment of 250 N · m, and will also be pressurized internally. What minimum internal pressure will cause yielding according to the von Mises criterion?

10.13 A thin-walled tube, with closed ends, is made of a metal whose yield strength is 300 MPa. The tube has an outer diameter of 75 mm and a wall thickness of 7 mm. After applying an axial compressive load of

3 kN to the ends, the tube is pressurized internally. What pressure p would cause yielding according to the Tresca criterion?

10.14 A thin-walled pressure vessel with an internal radius of 150 mm and a wall thickness of 3 mm is subjected to an axial load and internal pressure as follows. First, an axial load P is applied until $\sigma_z = 160$ MPa. Following this, P is kept constant and pressure is increased until yielding has occurred. Assume the material is brass and of yield stress 320 MPa. Determine the axial load and final pressure.

10.15 A solid, soft aluminum cylinder, 100 mm in diameter and 150 mm long is closely fitted inside a thick-walled cylinder. An axial load P is transmitted to the aluminum cylinder by means of two frictionless pistons. Aluminum possesses the following elastic constants: $E = 70$ GPa, $\nu = 0.3$, and yield strength of $Y = 80$ MPa. Calculate the axial force P required to initiate yielding in aluminum. (Assume the steel cylinder to be completely rigid and apply the von Mises yield criterion.)

10.16 A solid steel shaft of diameter d is subjected to a bending moment M and a twisting moment M_t with $M = \alpha M_t$. Determine the ratio τ_{max}/Y according to the von Mises criterion at which the shaft will begin to yield, where Y is the yield strength of the shaft material.

10.17 A vertical water container, 3 m in diameter and 50 m high is completely filled. If the walls are 10 mm thick and the steel has a yield strength of 230 MPa, determine whether or not the container wall material at a point 1 m above its base will yield considering average uniform wind pressure of 600 N/m^2.

Flow Rule, Effective Stress and Strain, Plastic Work

10.18 A cube of metal, having a yield strength Y of 345 MPa is subjected to two perpendicular normal tensile stresses, σ_1 and $\sigma_3 = -\sigma_1/2$.

a. What is τ_{max} at the onset of yielding using the von Mises criterion?
b. Repeat (b) where $\sigma_3 = +\sigma_1/2$.
c. Upon further proportional loading in the plastic range, detrmine $d\varepsilon_1^p / d\varepsilon_2^p$.

10.19 A steel plate 25 mm thick, 750 mm long, and 1000 mm wide is stretched plastically 12% in the longitudinal direction. Assuming that the width is restrained from contracting by the end grips, calculate

a. Final dimensions of the plate
b. Maximum stretching load
c. State of stress at the maximum load

10.20 Three principal stresses (in MPa) are applied to a solid undergoing plastic deformation where $\sigma_1 = 400$, $\sigma_2 = 200$ and $\sigma_3 = 0$.

a. What is the ratio $d\varepsilon_1/d\varepsilon_3$?
b. If a fluid pressure produces an all-around hydrostatic stress of –300 MPa that is superimposed upon the original stress state, how does the ratio in (a) change? Explain this result.

10.21 At a point in a solid, the stress matrix is given by

$$\sigma_x = 200\ ,\ \sigma_y = 120\ ,\ \sigma_z = 60,\ \text{and}\ \tau_{xy} = 50\ \text{(in MPa)}$$

a. Determine the principal stress deviations and the value of the effective stress $\overline{\sigma}$.
b. If this material obeys the von Mises yield criterion, determine its yield stress in simple tension assuming that the above stresses cause yielding.
c. Determine the increments of plastic strain if the material is strained to $\overline{\varepsilon} = 0.1$ by stresses proportional to those given above, determine the principal plastic strains.

10.22 A metallic alloy is tested in a uniaxial tensile test. The following data were obtained:

ε_x	0.1	0.3	0.55
σ_x (MPa)	242	340	413

a. Plot the effective stress–effective strain curve for this alloy and then determine the parameters K and n by fitting the law $\bar{\sigma} = K(\bar{\varepsilon})^n$ to the above data.
b. A bar made of the above alloy is subjected to a stress system of $\sigma_x = 300$ MPa, all shear stresses are zero, $\sigma_y/\sigma_x = -1/2$, and $\sigma_z/\sigma_x = -1/3$. Determine:

- The effective stress $\bar{\sigma}$ and the effective strain $\bar{\varepsilon}$,
- The final dimensions of the bar given its original dimensions $x_o = 100$ mm, $y_o = 60$ mm, and $z_o = 30$ mm.

10.23 A square aluminum bar of 20 × 20 mm cross section has been reduced in section to 16 × 16 mm by drawing. Assuming a state of stress $\sigma_2 = \sigma_3 = -4\sigma_1$, where σ_1 is the mean value for the drawing stress, determine the drawing force. Assuming no heat losses, calculate the bar temperature rise. Take aluminum behavior as $\bar{\sigma} = 165(\bar{\varepsilon})^{0.283}$ (MPa), specific heat = 0.94 kJ/kg/K.

10.24 A thin aluminum slab of 10 × 100 × 500 mm is to be stretched uniformly to a length of 570 mm, maintaining its width constant at 100 mm. Determine:

a. The final thickness of the slab
b. The maximum force necessary for the stretching operation
c. The plastic work done in the process

The stress–strain relationship for aluminum is given by $\bar{\sigma} = 260(\bar{\varepsilon})^{0.2}$ (MPa).

10.25 A slab of aluminum of rectangular cross section 200 × 20 mm and length of 1000 mm is to be deformed with a proportional stress system, such that its final cross section becomes 150 × 15 mm. What is the maximum force needed to complete this operation? Aluminum behaves plastically according to $\bar{\sigma} = 200(\bar{\varepsilon})^{0.22}$ (MPa).

10.26 A square bar has the dimensions $x_o = 250$ mm and $y_o = z_o = 25$ mm. This bar is made of an annealed aluminum alloy for which $\bar{\sigma} = 215(\bar{\varepsilon})^{0.18}$ and is deformed under a superimposed hydrostatic pressure of 10^3 MPa. The state of stress at all times is $\sigma_y/\sigma_x = \sigma_z/\sigma_x = 1/2$, $\tau_{xy} = \tau_{yz} = \tau_{zx} = 0$. For the state of uniform deformation of 10% in the x-direction, calculate (a) the magnitude of the strains and (b) the effective stress and the effective strain.

10.27 The torsion test is often used to determine the strain-hardening behavior of metallic alloys. The following data are obtained during hot torsion testing of a thin-walled tube of outer and inner diameters of 25 and 22 mm, respectively. The tube free length between grips is 250 mm.

Twisting moment (kN · m)	325	380	395	415	425	440
Angle of twisting (degree)	90	180	360	720	900	1080

a. Plot the shear stress-shear strain flow curve.
b. Determine the material constants K and n in the law: $\bar{\sigma} = K(\bar{\varepsilon})^n$ applying von Mises yield function.
c. Compare the results with those obtained by applying the Tresca yield function.
d. Repeat (b) using for large plastic strain in torsion[17]

$$\bar{\varepsilon} = \frac{2}{\sqrt{3}} \ln\left[\sqrt{1+\frac{\gamma^2}{4}} + \frac{\gamma}{2}\right]$$

10.28 The plane strain compression test is used as a suitable method to assess behavior of rolled materials. The arrangement for this test is shown in the figure with compressive load P, strip thickness h, width W, and width of indentor b. Assuming friction is negligible and the dimension W remains constan during deformation and using the von Mises criterion determine:

a. $\bar{\varepsilon}$ as a function of ε_y and $\bar{\varepsilon}$ as a function of σ_y;
b. An expression of work per unit volume in terms of ε_y and σ_y;
c. An appropriate expression in the form $\sigma_y = f(K, \varepsilon_y, n)$ assuming $\bar{\sigma} = K(\bar{\varepsilon})^n$.

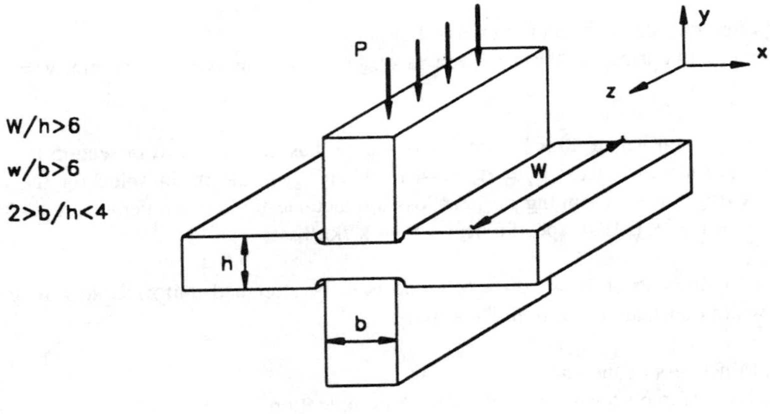

Problem 10.28

10.29 A copper specimen is compressed between two flat dies in conditions of plane strain by a monotonically increasing load. Determine, in the direction of load application,

a. The strain at first yield, and
b. The total strain when the applied load induces a compressive stress of 250 MPa.

The material behavior data are E = 130 GPa, ν = 0.35, and $\bar{\sigma} = 85 + 340(\bar{\varepsilon})^{0.3}$ (MPa).

10.30 A thin-walled tube with closed ends is made of steel whose tensile yield is 200 MPa and has the effective stress–strain law: $\bar{\sigma} = 600(\bar{\varepsilon})^{0.25}$ (MPa). The outer diameter is 80 mm, the wall thickness is 0.8 mm, and the length is 1000 mm. After applying an axial constant compressive load of 10 kN, the tube is pressurized internally. Determine the internal pressure that will cause yielding according to Tresca and von Mises yield criteria.

10.31 A thin-walled tube 100 mm outside diameter, 2.5 mm thickness, 1250 mm length is made of steel whose tensile test data is $\bar{\sigma} = 850(\bar{\varepsilon})^{0.25}$ (MPa). The tube is subjected to an axial load P and twisting moment M_t in such a way that the stress ratio (axial tension to shear) equals 1.0 for all conditions. The deformation continues until the axial tension becomes 350 MPa, determine:

a. The final values of the principal stresses;
b. The effective stress and strain $\bar{\sigma}$ and $\bar{\varepsilon}$;
c. The normal plastic strains ε_r^p, ε_θ^p, and ε_z^p;
d. The final tube dimensions (diameter, length, and thickness);
e. The final load P and moment M_t.

10.32 A long, thin-walled tube with closed ends is of diameter 100 mm, thickness 2.5 mm, and length 500 mm. The tube is made of ductile steel whose behavior is assumed to be rigid strain hardening, according to the law: $\bar{\sigma} = 250 + 400(\bar{\varepsilon})^{0.25}$ (MPa).

a. If the tube is subjected to internal pressure p_i and axial compressive load P, find the relation between P and p_i, such that there exists no axial stress in the tube wall. In this case, what is the limiting value of p_i at which yielding of the tube starts.
b. If the tube is subjected to p_i = 30 MPa only, what is the energy consumed during plastic deformation?
c. Calculate the final dimensions of the tube when p_i = 30 MPa.

10.33 A thin-walled tube 120 mm in diameter, 2.5 mm thick, and 300 mm long is subjected to an axial load P and a twisting moment M_t in such a way that the ratio of M_t to P is maintained constant during the deformation given by $M_t = 0.5\,P$ N · m. The deformation continues to a load P = 150 kN. Determine the energy consumed. The material behaves according to $\bar{\sigma} = 700(\bar{\varepsilon})^{0.5}$ (MPa). Neglect elastic strains.

Plastic Anisotropy

10.34 Two flat tensile specimens were cut along and transverse to the rolling direction of a sheet metal of thickness 1.2 mm. The initial specimen dimensions were identical: length = 50 mm and width = 20 mm. The following tensile results were obtained:

	Rolling Direction			Transverse Direction		
Loading	**Load, kN**	**Length, mm**	**Width, mm**	**Load, kN**	**Length, mm**	**Width, mm**
Yield point	7.24	50.1	20.0	720	50.1	20
Intermediate point	10	58.0	18.7	10.7	58.0	18.2
Maximum point	11.8	64.0	18.0	14.6	63.0	17.3

(a) Determine the R values (R_1 and R_2) of these two specimens at each point. (b) Also determine the yield strength in the rolling, transverse, and thickness directions.

10.35 Assuming that Hill's anisotropic criterion can be applied to study the deformation of a metal sheet without rotational symmetry,

a. Derive an expression for $\alpha = \sigma_y/\sigma_x$ for plane strain ($\varepsilon_y = 0$) and plane stress ($\sigma_z = 0$) loading.
b. Find the stress σ_x in terms of Y_x, R_1, and R_2.
c. If $R_1 = 2.1$, $R_2 = 1.6$, and Y_x = 220 MPa, what would be the value of α and σ_x?
d. Determine yield strength Y_y for this sheet.

10.36 A sheet metal possessing planar isotropy is loaded in plane stress (σ_x, σ_y) with $\sigma_z = 0$.

a. Express the strain ratio $\beta = \varepsilon_y/\varepsilon_x$ as a function of the stress ratio $\alpha = \sigma_y/\sigma_x$ and R.
b. Derive an expression for $\bar{\sigma}$ in terms of α, R, and σ_x. Define $\bar{\sigma}$ and $\bar{\varepsilon}$ such that they reduce to σ_x and ε_x, respectively, in an x-direction tension test.
c. Derive an expression for $d\bar{\varepsilon}$ in terms of β, R, and $d\varepsilon_x$.

10.37 Tensile specimens cut from a rolled sheet metal in the x- and y-directions have strain ratios of $R_1 = 1.4$ and $R_2 = 1.9$ and yield strength of the specimen cut along the x-direction, Y_x = 320 MPa.

a. Determine the yield strength of the specimen cut along the y-direction.
b. If the material is deformed in a plane strain tension test with $\varepsilon_y = 0$, calculate σ_x at yielding.
c. Calculate σ_x and σ_x at yielding under the condition, $\varepsilon_z = 0$ and $\sigma_z = 0$.

10.38 Show that the Hill's criterion and its associated flow rule give anisotropic coefficient R_θ as

$$R_\theta = \frac{H + (2N - F - G - 4H)\sin^2\theta\cos^2\theta}{F\sin^2\theta + G\cos^2\theta}$$

for a tensile specimen cut at angle θ to the x-axis (rolling direction). Using this result, derive an expression for N/G in terms of R_0, R_{90} and R_{45}.

10.39 The results of the tensile plastic properties of a rolled sheet in the rolling direction x and transverse y-direction are $Y_x = 466$ MPa and $Y_y = 450$ MPa, $R_{90} = 4.7$, $R_0 = 3.8$, and $R_{45} = 2.4$. Calculate the yield stresses at $\theta = 22.5°$, $\theta = 45°$, $\theta = 67.5°$, and $\theta = 90°$. Plot the yield strength variation vs. θ.

10.40 A closed, thin-walled tube made by forming and edge longitudinal welding of a rolled sheet of metal is subjected to an internal pressure of 10 MPa. The prior rolling and transverse directions become the axial direction of the tube and the hoop direction respectively. The strain ratios in the sheet are $R_0 = 2.0$, $R_{45} = 1.5$, and $R_{90} = 0.7$, and the yield strength in the rolling direction is 300 MPa. Neglecting elastic effects, what is the wall thickness required to avoid yielding for a tube of 100 mm diameter?

10.41 A pipe is made of thin sheet metal by forming and welding, such that the rolling direction becomes the axial direction of the pipe and the transverse direction becomes the hoop direction. The pipe has a diameter of 160 mm, length of 1200 mm, and a wall thickness of 0.8 mm. The strain ratios in the sheet are $R_0 = 2.5$, $R_{45} = 1.8$, and $R_{90} = 0.8$. The complete stress–strain curve in simple tension for specimens along the rolling direction is given by $\sigma_1 = 360 + 155(\bar{\varepsilon})^{0.22}$ (MPa). Taking the yield strength at a strain 0.2% and neglecting elastic effects, determine the following.

a. The internal pressure at which the pipe walls will yield if the pipe ends are closed. Use both of Hill's criterion (F, G, H) and planar isotropy model.
b. If the pressure is increased up to a value that produces only 10% diametral expansion, determine the final pipe dimensions. What is the pressure at this instant?
c. Compare the results with those that neglect anisotropy.

10.42 During bulging of a circular aluminum thin sheet of thickness $h_o = 1.2$ mm, the stress state at the bulge top is closely represented by an equibiaxial one. The surface gridding technique, as explained in Example 10.8, is used to measure biaxial polar strains. Meanwhile, by recording the bulging pressure and the polar bulge radius, the following data at the pole are obtained[9]: surface strains $\varepsilon_1 \simeq \varepsilon_2 = 0.25$, polar radius of curvature $\rho = 125.1$ mm, bulging pressure = 1.6 MPa. Strips of the same sheet are cut along 0°, 45°, and 90° with the rolling direction and tested in uniaxial tension to give the following properties averaged according to Expression 10.55):

$$\sigma_u = 142(\varepsilon)_u^{0.28} \text{ MPa} \quad \text{and} \quad R = 0.8$$

Using the above average tensile properties, predict the equibiaxial surface strains and compare them with the experimental point. Assume planar isotropy and consider the two cases $s = 2$ and $s = 1.8$.

Kinematic Hardening

10.43 In low-cycle fatigue testing, specimens are usually subjected to cycles of reversed total strain between $+\Delta\varepsilon/2$, $-\Delta\varepsilon/2$, where $\Delta\varepsilon$ is the strain range until failure occurs. For a bilinear material behavior with $Y_o = 400$ MPa, $E = 210$ GPa, and $E^p = 12$ GPa subjected to a strain range $\Delta\varepsilon = 0.01$, determine the stress–strain response after few strain cycles as predicted by

a. Isotropic-hardening rule
b. Kinematic-hardening rule

10.44 A bilinear material under monotonic stress–strain behavior, with $Y_o = 250$ MPa and $E^p = 9.524$ GPa, is subjected to a cyclic strain range of $\Delta\varepsilon = 0.02$. Determine the stress–strain response after few strain cycles as well as the magnitude of the plastic strain accumulated at each stage according to

a. Isotropic-hardening rule
b. Kinematic-hardening rule

10.45 A thin-walled spherical pressure vessel of mean radius $r_{mo} = 300$ mm and thickness $h_o = 6$ mm is made from a material with the following bilinear stress–strain curve $\sigma = 220 + 11500(\varepsilon)^p$ and is subjected to an internal pressure of 20 MPa. During the vessel life, the pressure fluctuates between zero and its peak value, 20 MPa. Determine the plastic strain developed in the shell wall using the kinematic-hardening rule and illustrate the results graphically. Compare with isotropic-hardening predictions, after the first loading–unloading cycle.

10.46 A thin-walled tube of $r_{mo} = 250$ mm and $h_o = 5$ mm is made from a material with the following bilinear stress–strain curve: $\sigma = 200 + 6000(\varepsilon)^p$. The tube is subjected to an internal pressure cycle with $p_{max} = 20$ MPa and $p_{mon} = 0$. Determine the plastic strain history using both kinematic and isotropic rules of hardening.

Plasticity and Deformation of Porous Solids

10.47 Determine the ratio of σ to the matrix yield strength Y_M at which a porous solid of current void volume fraction C_v yields? Consider the following states of loading:

a. Uniaxial tension $(\sigma, 0, 0)$
b. Equibiaxial tension $(\sigma, \sigma, 0)$
c. Plane strain; $\varepsilon_3 = 0$, $\sigma_2 = \sigma_1/2$
d. Uniaxial compression $(-\sigma, 0, 0)$
e. Biaxial compression $(-\sigma, -\sigma, 0)$

Comment on the results for $C_v = 0$ and $C_v = 0.05$.

10.48 A partially sintered metallic powder compact has an initial porosity expressed by the void volume fraction $C_{vo} = 0.15$. It is known that the matrix material behaves according to the law: $\bar{\sigma}_M = K(\bar{\varepsilon})_M^{0.35}$; $K =$ 650 MPa.

a. Derive an expression for the uniaxial flow curve of the powder compact.
b. Compare this behavior in both tension and compression at $\varepsilon_1 = 0.26$, $C_{vo} = 0.05$ with matrix behavior. Comment on the results.

REFERENCES

1. Ludwik, P., *Elemente der Technologischen Mechanic*, Springer-Verlag, Berlin, 1909.
2. Drucker, D. C., *Introduction to Mechanics of Deformable Solids*, McGraw-Hill, New York, 1967, Chap. 12.
3. Hill, R., *The Mathematical Theory of Plasticity*, Oxford University Press, 1950, 267–269.
4. Venkatraman, B. and Patel, S. A., *Structural Mechanics with Introductions to Elasticity and Plasticity*, McGraw-Hill, New York, 1970.
5. Drucker, D. C., A more fundamental approach to plastic stress–strain relations, *Appl. Mech., ASME*, 487, 1951.
6. Johnson, W. and Mellor, P. B., *Engineering Plasticity*, Van Nostrand, London, 1973.
7. Hill, R., Theoretical plasticity of textured aggregates, *Math. Proc. Camb. Philos. Soc.*, 85, 179, 1979.
8. Hosford, W. H. and Caddell, R. M., *Metal Forming: Mechanics and Metallurgy*, Prentice-Hall, Upper Saddle River, NJ, 1983.
9. Ragab, A. R. and Abbas, A. T., Assessment of work-hardening characteristics and limit strains of anisotropic aluminum sheets in biaxial stretching, *J. Eng. Mater. Technol. ASME*, 108, 250, 1986.
10. Chen, W. F. and Zhang, H., *Structural Plasticity; Theory, Problems and CAE Software*, Springer-Verlag, Berlin, 1991.
11. Prager, W., *Proc. Inst. Mech. Eng.*, 169, 41, 1955; *J. Appl. Mech.*, 23, 493, 1956.
12. Green, R. J., A plasticity theory for porous solids, *Int. J. Mech. Sci.*, 14, 215–224, 1972.
13. Gurson, A. L., Continuum theory of ductile rupture by void nucleation and growth: part I — yield criteria and flow rules for porous ductile materials, *J. Eng. Mater. Technol. ASME*, 99, 2, 1977.

14. Pilling, J. and Ridley, N., Effect of hydrostatic pressure on cavitation in superplastic aluminum alloys, *J. Acta Metall.*, 34, 669, 1986.
15. Palmer, A. G., Maier, G., and Drucker, D. C., Normality relations and convexity of yield surfaces for unstable materials or structure elements, *Proc. ASME, J. Appl. Mech.*, 34, 465, 1967.
16. Shima, S. and Oyane, M., Plasticity theory for porous metals, *Int. J. Mech. Sci.*, 18, 285, 1976.
17. Hodierne, F. A., *J. Inst. Met.*, 91, 267–273, 1962–1963.

11 Plastic Instability, Superplasticity, and Creep

Three characterized modes of plastic deformation in some engineering structures are considered. The first deals with plastic instability defining limit strains and loads associated with large plastic deformation in some structural elements, namely, bars under uniaxial tension, biaxially stretched sheets, and pressurized spherical and cylindrical containers. The second extends plasticity laws to superplastic materials for which the flow stress depends on the strain rate estimating the forming time in neck-free superplastic forming. The third mode of deformation deals with steady creep deformations in beams in bending, pressurized containers, torsion of circular rods, and buckling of columns. In these applications creep is considered simply as an accumulation of plastic strain with time at certain prescribed conditions of temperature and loading. Consideration of stress relaxation as well as creep damage and rupture are included.

11.1 INTRODUCTION

In Chapter 10 the laws governing the plastic flow of materials are developed. These laws constitute the fundamentals of the theory of plasticity and are used to analyze various problems involving plastic deformation, e.g., elastoplastic response of structures, limit strains and collapse loads of structures, mechanics of metal-working processes, etc. In this chapter three characterized modes of plastic deformations are analyzed applying the laws of plasticity. The first deals with plastic instability in some simple applications to define limiting loads and strains associated with large plastic deformation, i.e., ductile failure. The second extends the application of plasticity laws to materials for which the flow stress depends on strain-rate, such as superplastic alloys. The third deals with the analysis of creep deformations of some simple structures. Noting that creep is a vast subject involving several mechanical and metallurgical parameters, it is considered here simply as an accumulation of plastic deformation with time at prescribed conditions of temperatures and loading.

It may be noted that dealing with the above three modes in the same chapter does not conform strictly to the traditional labeling of plastic deformation as being time independent. This, however, has been overlooked here as the same governing laws of plasticity, namely. the flow potential and the flow rule, are applicable to rate-dependent superplastic and creep deformation.

11.2 UNSTABLE PLASTIC DEFORMATION

Plastic instability for a certain case of loading may be identified when the deformation changes its magnitude and configuration under the same load, e.g., when a state of uniform plastic deformation gives way to a nonuniform or, say, localized one. The causes of plastic instability are somewhere in the material, and development of unstable deformation depends on the geometry of the body, the flow parameters of the material, and the loading. Plastic instability is often associated with continuation of plastic deformation under falling load.

Consider a bar of a strain-hardening material and an initial cross-sectional area A_o subjected to a tensile load P. The relation between the normal, or engineering, stress defined as P/A_o and the engineering strain is schematically shown in Figure 11.1. To maintain plastic flow, it is necessary to impose an increasing stress as the bar extends. This increase in the applied stress is partly accomplished by increasing the applied load. The stress also increases, even if the applied load remains constant, because of the reduction in bar cross section as it extends. At a certain instant of deformation, this latter effect becomes sufficient to maintain plastic flow without any need for increasing the load. In other words, the reduction in bar cross-sectional area contributes to the stress increase by an amount that offsets the effect of the material strain hardening. This condition is identified on the tensile stress-strain curve, Figure 11.1, by the attainment of the maximum value of P/A_o, i.e., the maximum load. This condition of maximum load is utilized to determine the limiting conditions in some applications involving tensile plastic deformation, such as stretching operations and pressurized containers.

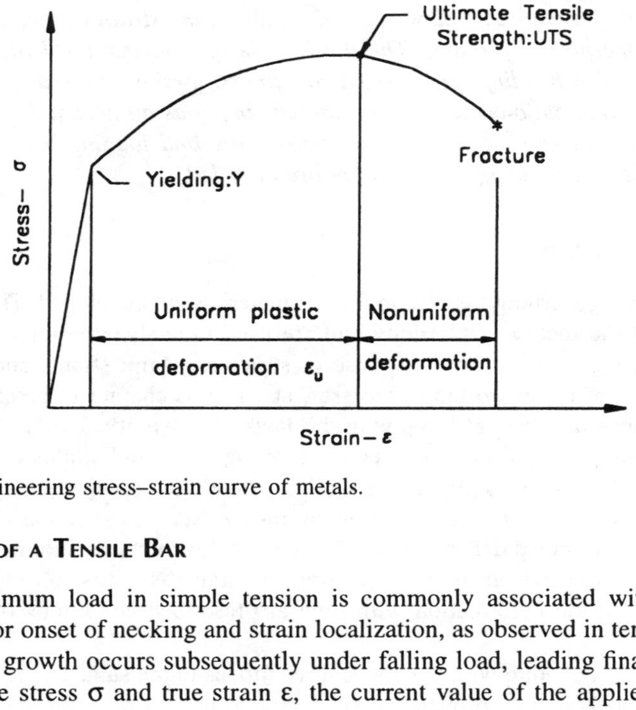

FIGURE 11.1 Engineering stress–strain curve of metals.

11.2.1 Necking of a Tensile Bar

The point of maximum load in simple tension is commonly associated with the initiation of nonuniform strain or onset of necking and strain localization, as observed in tensile tests of ductile materials. Necking growth occurs subsequently under falling load, leading finally to fracture.

In terms of true stress σ and true strain ε, the current value of the applied tensile load P is given by

$$P = \bar{\sigma} A \tag{11.1}$$

where A is the current cross-sectional area of the bar and $\bar{\sigma} = \sigma$ is the effective stress. At the point of maximum tensile load,

$$dP = \bar{\sigma} dA + A d\bar{\sigma} = 0 \tag{11.2}$$

By recalling that the true plastic strain increment* in simple tension is defined by, in view of constancy of volume,

* For simplicity the superscript p denoting plastic strains has been dropped throughout this analysis.

$$d\bar{\varepsilon} = \frac{dL}{L} = -\frac{dA}{A}$$

Hence, Equation 11.2 gives

$$d\bar{\sigma}/\bar{\sigma} = d\bar{\varepsilon}$$

or

$$d\bar{\sigma}/d\bar{\varepsilon} = \bar{\sigma} \tag{11.3}$$

The strain-hardening behavior of the bar material is taken to be characterized by the simple power law given by Equation 10.34a:

$$\bar{\sigma} = K(\bar{\varepsilon})^n$$

Hence, from Equation 11.3,

$$\frac{d\bar{\sigma}}{d\bar{\varepsilon}} = \bar{\sigma} = nK(\bar{\varepsilon})^{n-1} = \frac{n\bar{\sigma}}{\bar{\varepsilon}}$$

or

$$\bar{\varepsilon} = n$$

At this value of $\bar{\varepsilon}$ the strain ceases to be uniform along the bar length and tends to be localized at some point leading to necking. Hence, the onset of necking occurs at a uniform strain:

$$\varepsilon_u = n \tag{11.4}$$

Note that this condition (Equation 11.4), due to Considère,[1] indicates the limited usefulness of a simple tension test in predicting the flow stress–strain behavior. For $n = 0.2$, a uniform strain of 22% can be hardly attained. Complete use of simple tension test results up to fracture is possible by considering Bridgman analysis for the stresses in the neck as given in Section 11.2.4.

The maximum load in simple tension, i.e., the point defined by the ultimate tensile strength (UTS) is thus attained when the true strain becomes equal to the strain-hardening exponent n

$$\text{UTS} = \frac{P_{max}}{A_o} = \left(\frac{\sigma A}{A_o}\right)_{\varepsilon_u}$$

where σ and A are the true stress and current area at the maximum load, respectively. And since $A_o L_o = AL$, then

$$\varepsilon_u = n = -\ln\frac{A}{A_o}$$

Hence, the ultimate tensile strength UTS of a strain-hardening material is expressed by

$$\text{UTS} = K(n)^n \exp(-n) \tag{11.5}$$

Example 11.1:

An annealed brass bar of 12 mm initial diameter supports a maximum tensile load of 50 kN at which the reduction in area reaches 39.35%. If a second bar of the same brass of 20 mm diameter is loaded until an elongation of 25% is attained, what load would be needed to reach this condition? Assume a strain-hardening law $\bar{\sigma} = K(\bar{\varepsilon})^n$.

Solution:

The maximum load in simple tension marks the point of plastic instability at which:

$$\varepsilon_u = n$$

At this point,

$$\varepsilon_u = \ln\frac{L}{L_o} = \ln\frac{A_o}{A} = -\ln\left(\frac{A - A_o}{A_o} + 1\right)$$

or

$$\varepsilon_u = n = -\ln(-0.3935 + 1) = 0.5$$

The true stress at this point is

$$\sigma = 50\times10^3 \Big/ \left[\frac{\pi}{4}(12)^2(1-0.3935)\right] = 729.3 \text{ MPa}$$

For a strain-hardening law $\sigma = K(\varepsilon)^n$, $n = \varepsilon_u = 0.5$,

$$729.3 = K(0.5)^{0.5}$$

giving $K = 1031.4$, MPa. For the second bar at an engineering strain $e = 0.25$,

$$\varepsilon = \ln(1 + e) = \ln(1.25) = 0.2231$$

Hence, $\sigma = 1031.4(0.2231)^{0.5} = 487.17$ MPa. The required load is then

$$P = \sigma A = \sigma A_o \exp(-\varepsilon)$$

$$= (487.17)\left(\frac{\pi}{4}20^2\right)\exp(-0.2231)$$

Hence,

$$P = 122.39 \text{ kN}.$$

Example 11.2:

A composite cylindrical rod is cast such that a solid copper core of circular crosssection A_c *is surrounded by an aluminum tube of cross-sectional area* A_a *as shown in Figure 11.2. Assuming no separation between the two metals, determine the maximum tensile load which the composite rod can withstand before plastic instability occurs. Properties of soft copper and aluminum are represented by* $\bar{\sigma} = K_c(\bar{\varepsilon})^{n_c}$ and $\bar{\sigma} = K_a(\bar{\varepsilon})^{n_a}$, *respectively. For material constants, take* $K_c = 315$ *MPa*, $K_a = 180$ *MPa*, $n_c = 0.54$, *and* $n_a = 0.2$, *find the instability strain and the ultimate load of a composite rod if* $A_c = A_a$.

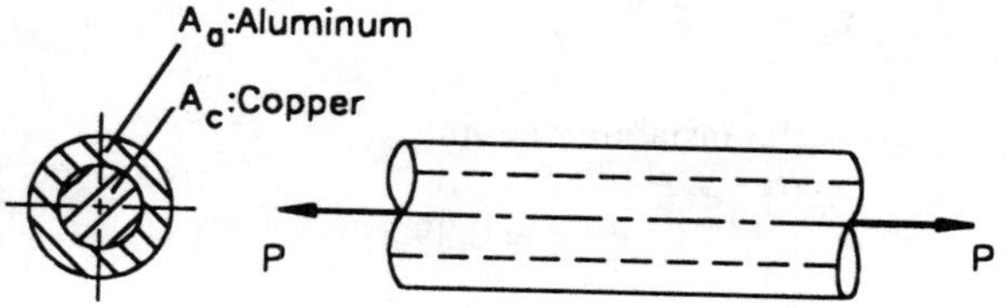

FIGURE 11.2 Example 11.2.

Solution:

A tensile load P is carried by the composite rod such that

$$P = \sigma_c A_c + \sigma_a A_a$$

plastic instability occurs when the load P attains its maximum value, i.e.,

$$dP = 0 = (\sigma_c dA_c + A_c d\sigma_c) + (\sigma_a dA_a + A_a d\sigma_a)$$

Both copper and aluminum undergo the same longitudinal strain:

$$d\varepsilon = -\frac{dA_c}{A_c} = -\frac{dA_a}{A_a}$$

Hence,

$$\frac{A_c}{A_a}\frac{\sigma_c}{\sigma_a}\left(-d\varepsilon + \frac{d\sigma_c}{\sigma_c}\right) = -\left(-d\varepsilon + \frac{d\sigma_a}{\sigma_a}\right)$$

Using the laws of material behavior gives

$$\frac{A_c}{A_a}\frac{K_c(\varepsilon)^{n_c}}{K_a(\varepsilon)^{n_a}}\left(-d\varepsilon + \frac{n_c}{\varepsilon}d\varepsilon\right) = \left(d\varepsilon - \frac{n_a}{\varepsilon}d\varepsilon\right)$$

Simplifying and denoting ε by ε_u yields

$$\frac{A_c}{A_a} = -\left(\frac{K_a}{K_c}\right)(\varepsilon_u)^{n_a - n_c}\left(1 - \frac{n_a}{\varepsilon_u}\right) \Big/ \left(1 - \frac{n_c}{\varepsilon_u}\right)$$

This equation may be checked for two simple cases. For a whole aluminum rod $A_c/A_a = 0$, giving $\varepsilon_u = n_a$, and for a whole copper rod $A_c/A_a \to \infty$, giving $\varepsilon_u = n_c$.

Now substituting $A_c = A_a$, $\sigma = 315\ \varepsilon^{0.54}$ for copper and $\sigma = 180\varepsilon^{0.2}$ for aluminum, then

$$1 = -\frac{180}{315}(\varepsilon_u)^{0.2-0.54}\left(1 - \frac{0.2}{\varepsilon_u}\right) \Big/ \left(1 - \frac{0.54}{\varepsilon_u}\right)$$

which reduces to

$$\varepsilon_u^{0.34}\left(\frac{\varepsilon_u - 0.54}{\varepsilon_u - 0.2}\right) = -0.5714, \quad \text{where } 0.2 \le \varepsilon_u \le 0.54$$

Solving by trial and error gives the instability strain:

$$\varepsilon_u = 0.39$$

The ultimate tensile load for $A_o = A_c = A_a$ is

$$P = [315\ (0.39)^{0.54} + 180\ (0.39)^{0.2}]\ A_o \exp(-0.39)$$

Hence,

$$\text{UTS} = P/2A_o = 114.6 \text{ MPa}$$

This is about 13% weaker than a whole copper rod and 7.3% stronger than a whole aluminum rod.

11.2.2 Local Necking of a Wide Strip

An annealed strip under uniaxial tensile load undergoes a state of unstable plastic deformation. At the moment the load attains its maximum value, a limiting uniform strain $\varepsilon_u = n$ is achieved. As deformation continues at falling load, a neck with large profile radius is observed as shown in Figure 11.3a. This is diffuse necking since its development is gradual and diffused over a relatively large zone of the length (say, not less than the approximate strip width). Considerable overall extension is achieved during this necking process. At a final stage, this neck becomes localized over a very small zone of length of the order of the strip thickness, as shown in Figure 11.3b. The strain localization in a narrow band oriented at an angle θ^* to the load axis means that the material within this band deforms as if it were isolated from the adjoining material on both sides. This requires that the width strain increment $d\varepsilon_2^*$ vanish.

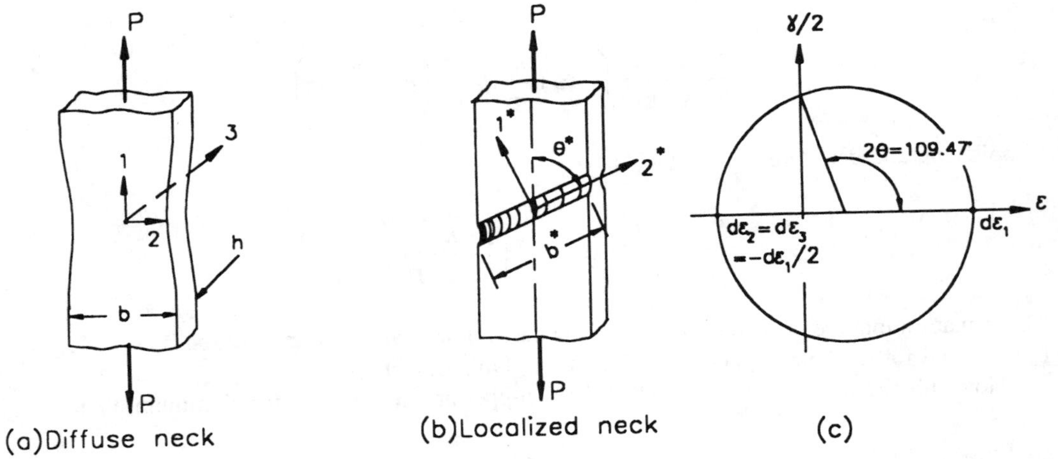

FIGURE 11.3 Necking of a tension strip: (a) diffuse neck, (b) local neck, and (c) determination of the local necking angle.

The instability condition postulated at attainment of maximum load as in Equation 11.2 is given by

$$dP = \sigma_1 dA + A d\sigma_1 = 0 \tag{11.6}$$

or

$$\frac{d\sigma_1}{\sigma} = -\frac{dA}{A}$$

For a current width b and thickness h, the neck cross-sectional area $A = bh = (b^* \sin \theta^* h)$ undergoes a change dA due to a decrease dh in the thickness only, since no width changes occur to satisfy the condition of $d\varepsilon_2^* = 0$. Therefore,

$$\frac{d\sigma_1}{\sigma_1} = -\frac{dh}{h} = -d\varepsilon_3 \tag{11.7}$$

The condition of constant volume at the moment of initiation of this localized neck is still satisfied by

$$d\varepsilon_3 = -\,d\varepsilon_1/2$$

and, thus,

$$\frac{d\sigma_1}{\sigma_1} = \frac{1}{2} d\varepsilon_1 \tag{11.8}$$

For a material obeying the law $\bar{\sigma} = K(\bar{\varepsilon})^n$, this provides the condition:

$$\frac{d\sigma_1}{d\varepsilon_1} = \frac{d\bar{\sigma}}{d\bar{\varepsilon}} = \frac{\sigma_1}{2} = \frac{\bar{\sigma}}{2}$$

or, simply,

$$\varepsilon^* = 2n \tag{11.9}$$

where ε^* identifies the longitudinal strain at localized necking which is twice its value for diffuse necking, Equation 11.4.

The angle θ^* at which this localized neck appears is determined from Mohr's circle of strain for uniaxial tension, where $d\varepsilon_2 = d\varepsilon_3 = -d\varepsilon_1/2$, as shown in Figure 11.3c. The direction at which $d\varepsilon_2^* = 0$, i.e., direction 2* of zero extension, is inclined at an angle given by $\theta^* = \tan^{-1}\sqrt{2}$ or $\theta^* = 54.74°$ to the strip tensile axis.

For metals that strain harden significantly, much strain is accumulated before localized necking appears in uniaxial tension. However, for those with $n \simeq 0$, i.e., cold-rolled strip, instability and, hence, strain at both diffuse and localized necking approaches zero.

Note that diffuse necking ($\varepsilon_u = n$) is characterized by overall thinning occurring during loading up to the maximum load point. While for localized necking ($\varepsilon^* = 2n$), it is neccessary, together with the maximum load condition, that further thinning occurs along a narrow band or groove such that plane strain conditions prevail in the groove.

11.2.3 Limit Tensile Strain for a Bar with an Imperfection

In the above analysis of Sections 11.2.1 and 11.2.2, the bar material has been assumed to be homogeneous without defects, such as variations in grain size, inclusions, voids, etc. No geometric imperfections leading to variations of cross-sectional area along the gauge length are assumed to exist. The presence of such imperfections will affect the strength of the bar by limiting it to the maximum load supported by its weakest cross section.

This can be illustrated by considering a bar with a cross-sectional imperfection along its length described by

$$f = A_{io}/A_{uo} \tag{11.10}$$

where A_{io} and A_{uo} are the initial cross-sectional areas of the bar in the imperfection and uniform zones, respectively, their ratio f being < 1, as shown in Figure 11.4.

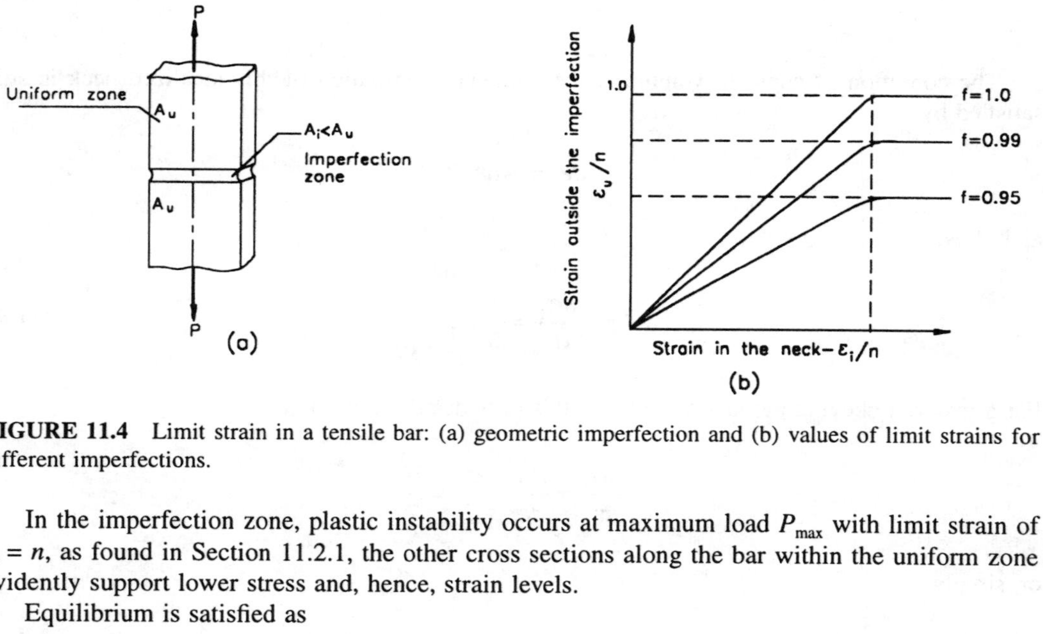

FIGURE 11.4 Limit strain in a tensile bar: (a) geometric imperfection and (b) values of limit strains for different imperfections.

In the imperfection zone, plastic instability occurs at maximum load P_{max} with limit strain of $\varepsilon_i = n$, as found in Section 11.2.1, the other cross sections along the bar within the uniform zone evidently support lower stress and, hence, strain levels.

Equilibrium is satisfied as

$$P_{max} = (\sigma A)_i = (\sigma A)_u$$

where the subscripts u and i refer to the uniform and imperfection zones, respectively. Dividing both sides by A and making use of Relation 11.10 gives

$$\sigma_i \frac{A_i}{A_{io}} = \sigma_u \frac{A_u}{fA_{uo}} \tag{11.11}$$

By virtue of the constant-volume condition, this reduces to

$$f\sigma_1 \exp(-\varepsilon_i) = \sigma_u \exp(-\varepsilon_u) \tag{11.12}$$

If the material obeys the strain hardening law $\sigma = K(\varepsilon^n)$, then substitution into Equation 11.12, knowing that $\varepsilon_u = n$, Equation 11.4, yields

$$f\varepsilon_i^n \exp(-\varepsilon_i) = \varepsilon_u^n \exp(-\varepsilon_u)$$

$$f = \left(\frac{\varepsilon_u}{n}\right)^n \exp(n - \varepsilon_u)$$

or

$$f^{1/n} = \left[1 - \left(\frac{n-\varepsilon_u}{n}\right)\right] \exp\left(\frac{n-\varepsilon_u}{n}\right) \tag{11.13}$$

Equation 11.13 can be solved numerically for ε_u to give results as shown in Figure 11.4b. However, an approximate explicit expression is found by considering that $(n - \varepsilon_u)/n$ is much smaller than unity. Therefore, by expanding the exponential term gives

$$f^{1/n} = \left[1 - \left(\frac{n-\varepsilon_u}{n}\right)\right]\left[1 + \left(\frac{n-\varepsilon_u}{n}\right)\right] = 1 - \left(\frac{n-\varepsilon_u}{n}\right)^2$$

This reduces to

$$\varepsilon_u/n = 1 - \sqrt{1 - f^{1/n}} \tag{11.14}$$

which gives $\varepsilon_u = n$ for a perfectly homogeneous bar, as expected. Table 11.1 gives appropriate values for the limit strain in the uniform zone as related to the imperfection value f. A slight area defect resulting in $f = 0.995$* reduces the limit strain by about 16%.

TABLE 11.1
Effect of Area Imperfection on Limit Strain for a Bar of Material with n = 0.25

f	0	0.999	0.995	0.99	0.97	0.95
ε_u/n	1	0.937	0.859	0.801	0.661	0.569

11.2.4 Stresses in the Neck of a Tensile Bar

In a tensile test a constriction, or a neck, usually develops at some location of the gauge length as soon as the maximum load is attained. The state of stress at the neck is no longer uniaxial, and lateral stress components exist as a result of the constraining effect of cross sections adjacent to the neck. The determination of the stresses in the neck is based on the experimental findings of Bridgman,[2] which indicated that the neck minimum cross section deforms uniformly and is therefore in the same strain hardening state. This means that the flow stress $\bar{\sigma}$, or the current yield stress Y, is the same at the minimum neck cross section. In the following analysis two cases are considered, namely, a round bar and a wide thin strip.

* This is equivalent to a machining diametral imperfection of 25 μm in a round bar of 10 mm diameter.

11.2.4.1 Round Bar

This is a case of axisymmetric deformation. The bar axis is taken to be the z-axis with $z = 0$ at the minimum neck cross section. For uniform plastic deformation at $z = 0$; ε_r, ε_θ, and ε_z do not change with radius r. If the radial displacement at radius r is u, then from Equation 2.11,

$$\varepsilon_\theta = \frac{u}{r} = C_1 \text{ or } u = C_1 r \tag{11.15}$$

where C is independent of r. Differentiation of Expression 11.15 gives

$$\varepsilon_r = \frac{du}{dr} = C_1 = \varepsilon_\theta$$

Hence, from the flow rule (Equation 10.22) $\sigma_r = \sigma_\theta$. At $z = 0$, σ_r, σ_θ, and σ_z are principal stresses denoted by σ_3, σ_2, and σ_1, respectively, where $\sigma_1 > \sigma_2$ and $\sigma_2 = \sigma_3$.

The Tresca and von Mises yield conditions both give

$$\sigma_1 - \sigma_3 = \sigma_z - \sigma_r = \bar{\sigma} \tag{11.16}$$

where $\bar{\sigma}$ the effective stress (or simply the flow stress) in the neck is obtained from the flow curve at the current value of $\bar{\varepsilon}$. Differention of Equation 11.16 with respect to r at constant $\bar{\sigma}$ yields

$$\frac{\partial \sigma_r}{\partial r} = \frac{\partial \sigma_z}{\partial r} \tag{11.17}$$

Considering the equilibrium of a curvilinear volume element $d\alpha$, $d\beta$, and $d\gamma$, where α, β, and γ correspond to the principal stress trajectories with reference to axes α_o and γ_o passing through point O along the r and z axes, respectively, Figure 11.5.

The equilibrium of the volume element is obtained from the second of Equation 1.12c. Substituting $\partial/\partial\gamma = \partial/\partial r$, $\sigma_2 = \sigma_3$, $1/r_1 = 1/\rho_1$, and $F_3 = 0$, together with Expression 11.17 to give at $z = 0$

$$\frac{\partial \sigma_z}{\partial r} + \frac{\sigma_r - \sigma_z}{\rho_1} = 0 \tag{11.18}$$

To solve Equation 11.18 it is required to obtain an expression of ρ_1 along r. From Figure 11.5, it is found that

$$\rho_1^2 = \overline{O_1O_2}^2 - \rho_2^2 = \overline{OO_2}^2 + \overline{OO_1}^2 - \left(\overline{OO_2}^2 + \overline{OC}^2\right)$$

$$= \overline{OO_1}^2 - \overline{OC}^2 = \left(r + \rho_1\right)^2 - \overline{OC}^2 \tag{11.19}$$

or

$$\overline{OC}^2 = \left(r + \rho_1\right)^2 - \rho_1^2 = r^2 + 2\rho_1 r$$

Expression 11.19 is valid for all points along r_N, where $2r_N$ is the diameter at the neck minimum cross section.

Substituting $\rho_1 = R$ at $r = r_N$, where R is the neck radius in the meridian plane gives

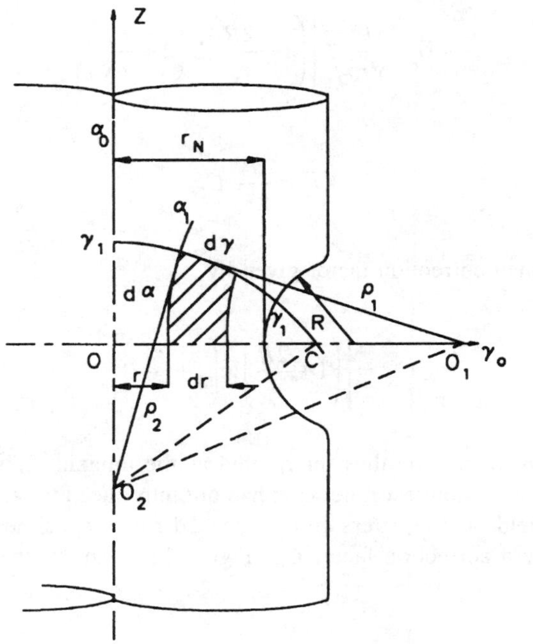

FIGURE 11.5 Neck of a tensile bar; coordinate axes for the derivation of Bridgman's formula.

$$r^2 + 2\rho_1 r = r_N^2 + 2r_N R$$

or

$$\rho_1 = \frac{r_N^2 + 2r_N R - r^2}{2r} \tag{11.20}$$

Substituting ρ_1 from Equations 11.16 and 11.20 into Equation 11.18, integrating along r, and introducing the boundary condition that $\sigma_r = 0$ at $z = 0$ and $r = r_N$, yields the stresses in the neck minimum cross section as

$$\frac{\sigma_r}{\bar{\sigma}} = \frac{\sigma_\theta}{\bar{\sigma}} = \ln\left(\frac{r_N^2 + 2r_N R - r^2}{2r_N R}\right)$$
$$\frac{\sigma_z}{\bar{\sigma}} = 1 + \ln\left(\frac{r_N^2 + 2r_N R - r^2}{2r_N R}\right) \tag{11.21}$$

The average axial stress σ_{av} is obtained from Equation 11.21 according to the condition:

$$P = 2\pi\int_0^{r_N} \sigma_z r dr = \pi r_N^2 \sigma_{av} \quad \text{where } \sigma_{av} = P/\pi r_N^2 \tag{11.22}$$

where P is the applied tensile load. This condition results in

$$\frac{\sigma_{av}}{\bar{\sigma}} = \left(1 + \frac{2R}{r_N}\right)\ln\left(1 + \frac{r_N}{2R}\right) \tag{11.23}$$

$$\bar{\sigma} = \frac{P}{\pi r_N^2} \bigg/ \left[\left(1 + \frac{2R}{r_N}\right) \ln\left(1 + \frac{r_N}{2R}\right) \right] \tag{11.24}$$

or

$$\bar{\sigma} = \frac{P}{\pi r_N^2} C_{Br}$$

where C_{Br} is the Bridgman correction factor given by

$$C_{Br} = \left[\left(1 + \frac{2R}{r_n}\right) \ln\left(1 + \frac{r_N}{2_R}\right) \right]^{-1}$$

In Expression 11.24, σ may be thus interpreted as the uniaxial flow stress corresponding to that which would exist in tension test if necking had not introduced the triaxial stresses. Therefore, to obtain the current yield or flow stress from a necked round specimen, the average true stress should be multiplied by a correction factor C_{Br}. Figure 11.6a shows the variation of C_{Br} and ($\sigma_{z\,\max}/Y$) vs. r_N/R.

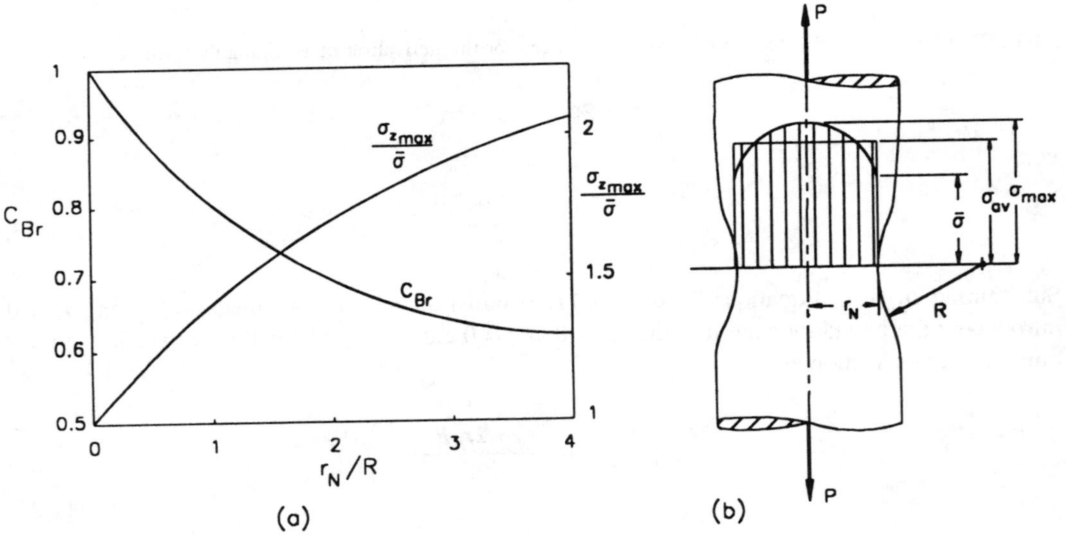

FIGURE 11.6 (a) The Bridgman correction factor in the neck of a tensile bar and (b) the axial stress distribution in the neck.

It is also indicated in Figure 11.6b that σ_z steadily increases inwards from the value of $\bar{\sigma} = Y$ at the surface to a maximum at the center. Hence, in view of Expression 11.16, namely, $\sigma_z - \sigma_r = \bar{\sigma}$, the radial stress σ_r has to be tensile. The initiation of fracture often observed at the center appears to be due to this state of triaxial tension.

Another correction factor has been proposed[3] in a simpler form based on the radius of curvature given by $\rho_1 = \mathrm{R}r_N/r$ as

$$\frac{\bar{\sigma}}{\sigma_{av}} = \left[1 + \frac{r_N}{4R}\right]^{-1} \tag{11.25}$$

Example 11.3:

A tensile test is conducted on a round bar of 10 mm diameter made of annealed stainless steel up to fracture at 1.25 reduction of area. The true stress strain values are obtained and fitted by the law: $\sigma = K(\varepsilon_N)^n$ *(K = 1275 MPa and n = 0.45), where calculations of* $\sigma - \varepsilon_N$ *points were based on their average values at the neck minimum cross section.*

It is required to determine the correct values of K *and* n *using the Bridgman correction. Use an empirical equation* relating the strain* ε_N *in the neck to the radius of curvature* R *as given by* $(r_N/R) = \sqrt{\varepsilon_N - n}$ *and make calculations for* $\bar{\sigma} - \varepsilon_N$ *points over the strain range: 0.05 to 1.*

Solution:

The corrected flow stress, i.e., the true effective stress in the neck $\bar{\sigma}$, is given by Bridgman according to Equation 11.24 as

$$\bar{\sigma} = \sigma_{av} \Big/ \left[\left(1 + \frac{2R}{r_N}\right) \ln\left(1 + \frac{r_N}{2R}\right)\right]$$

where $(r_N/R) = \sqrt{\varepsilon_N - n}$.

Note that in this empirical expression for (r_N/R), R is infinite when the strain $\varepsilon_N = \ln(A_o/A)$ becomes equal to n, which represents the uniform strain prior to necking.

The corrected flow stress $\bar{\sigma}$ vs. ε_N is thus obtained by substitution, as given in the following table.

ε_N	0.05	0.2	0.35	0.45	0.6	0.75	0.9	1.0
$\sigma = \sigma_{av}$, –MPa	331.2	618	795	890	1013.2	1120.2	1216	1275
r_N/R	0	0	0	0	0.387	0.548	0.671	0.742
$\bar{\sigma}$, MPa	331.2	618	795	890	928.6	994.9	1055.9	1093.5

Least-square fitting for the $\bar{\sigma} - \varepsilon_N$ points results in the equation:

$$\bar{\sigma} = 1140.5(\varepsilon)_N^{0.4}$$

which represents the flow curve of this annealed stainless steel.

11.2.4.2 Wide Strip

An annealed metallic strip under uniaxial tensile load deforms uniformly up to a certain axial strain value $\varepsilon_u = n$. Upon further deformation, a diffuse neck acquiring a large profile radius starts to appear, same as in the rod case, Figure 11.3. Considerable straining is still possible after the onset of this diffuse neck and before a localized neck is formed at an axial strain value $\varepsilon^* = 2n$ at an angle 54.74° with the strip axis, as indicated in Section 11.2.2.

Bridgman also solved the problem of necking in simple tension of a flat strip of width at least five times its thickness. It is found that the correction factor to be applied to a diffused-necked strip is as follows:

* Due to measurements of Bridgman on steels up to 95% reduction in area.

$$\frac{\bar{\sigma}}{\sigma_{av}} = \left\{\left(1+\frac{2R}{h_N}\right)^{1/2} \ln\left[1+\frac{h_N}{R}+\left(\frac{2h_N}{R}\right)^{1/2}\left(1+\frac{h_N}{2R}\right)^{1/2}\right]-1\right\}^{-1} \tag{11.26}$$

The formula (Equation 11.25) and the Bridgman formula (Equation 11.26) give close results, and the simpler formula (Equation 11.25 with $h_N = r_N$) may be used for the normal range of necking of strip specimens pulled in simple tension.

11.2.5 Biaxial Stretching — Flat and Bulged Circular Sheets

11.2.5.1 Flat Sheet

Metal sheets are stretch-formed by applying biaxial tension to the edges. To perform this process successfully, it is necessary to determine the limit of uniform strain reached when the maximum forming load is attained.

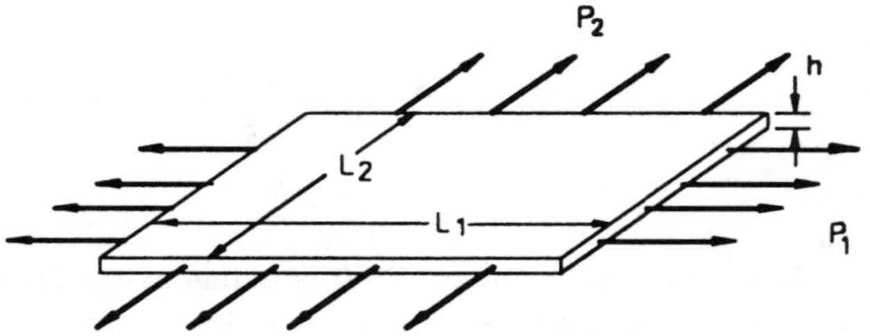

FIGURE 11.7 Biaxial tension of a sheet metal.

Consider a thin metal sheet stretched by a biaxial state of stress (σ_1, σ_2, 0), as shown in Figure 11.7. The stresses σ_1 and σ_2 are related by

$$\sigma_1/\sigma_1 = \alpha$$

where α takes a value according to $0 \le \alpha \le 1$. The mean stress is, therefore,

$$\sigma_m = (1 + \alpha)\,\sigma_1/3$$

In terms of σ_1, the deviatoric stress components are

$$\begin{aligned} &\sigma_1' = (2-\alpha)\sigma_1/3, \quad \sigma_2' = (2\alpha-1)\sigma_1/3 \\ &\sigma_3' = (1+\alpha)\sigma_1/3 \end{aligned} \tag{11.27}$$

Substitution of Equations 11.27 into the Levy–Mises flow rule, Equations 10.22, gives the ratios of the principal increments of plastic strains as

$$\frac{d\varepsilon_1}{(2-\alpha)\sigma_1/3} = \frac{d\varepsilon_2}{(2\alpha-1)\sigma_1/3} = -\frac{d\varepsilon_3}{(1+\alpha)\sigma_1/3} = \frac{3}{2}\frac{d\bar{\varepsilon}}{\bar{\sigma}} \tag{11.28}$$

The effective stress and strain increments, as defined by Equations 10.26 and 10.30, are given by

$$\bar{\sigma} = \sigma_1\left(1-\alpha+\alpha^2\right)^{1/2} \tag{11.29}$$

$$d\bar{\varepsilon} = 2d\varepsilon_1\left(1-\alpha+\alpha^2\right)^{1/2}/(2-\alpha) \tag{11.30}$$

Assuming uniform stresses, the stretching forces P_1 and P_2 from Figure 11.7 are given by

$$P_1 = \sigma_1 L_2 h \text{ and } P_2 = \sigma_2 L_1 h$$

Plastic instability is considered to occur when the load attains its maximum value. Since P_1 and P_2 are proportional, they will attain their maximum simultaneously when

$$dP_1 = d\sigma_1(L_2h) + dL_2(\sigma_1 h) + dh(\sigma_1 L_2) = 0$$

Putting $dL_1/L_1 = d\varepsilon_1$, $dL_2/L_2 = d\varepsilon_2$, and $dh/h = d\varepsilon_3$ gives

$$d\sigma_1/\sigma_1 = -(d\varepsilon_2 + d\varepsilon_3) = d\varepsilon_1$$

where the constant-volume conition is applied to yield

$$\frac{d\sigma_1}{d\varepsilon_1} = \sigma_1 \tag{11.31}$$

Equations 11.29 and 11.30 are used to express this condition in terms of the effective stress and strain as

$$\frac{d\sigma_1}{d\varepsilon_1} = \frac{2}{2-\alpha}\frac{d\bar{\sigma}}{d\bar{\varepsilon}} = \sigma_1$$

or

$$\frac{d\bar{\sigma}}{d\bar{\varepsilon}} = \left[(2-\alpha)\Big/\left[2\left(1-\alpha+\alpha^2\right)^{1/2}\right]\right]\bar{\sigma} \tag{11.32}$$

By confining the analysis to strain hardening materials obeying a power law,

$$\bar{\sigma} = K(\bar{\varepsilon})^n$$

gives

$$\frac{d\bar{\sigma}}{d\bar{\varepsilon}} = nK(\bar{\varepsilon})^{n-1} = \frac{n\bar{\sigma}}{\bar{\varepsilon}} \tag{11.33}$$

Combining Equations 11.32 and 11.33 yields the instability conditions for proportional loading conditions as

$$\bar{\varepsilon}_u = 2n\left(1-\alpha+\alpha^2\right)^{1/2}/(2-\alpha) \tag{11.34}$$

Expression 11.34 together with Equations 11.28 and 11.29 in their integrated forms of proportional loading define the limit strains in terms of the strain hardening exponent as given in Table 11.2. Note that the limit effective strain in equibiaxial tension is twice that for uniaxial tension. The case of plane strain is of considerable interest in analyzing deep drawing. In such a process, plastic instability and subsequent failure occur at the junction of the cup wall and the punch profile radius, under a system of stresses closely corresponding to plane strain tension.[4]

Limit strains as given in Table 11.2 for sheet metal–stretching processes indicate the onset of diffuse necking in the sheet. Failure in these processes is usually due to development of excessive thinning in an isolated zone of the sheet, which is known as localized necking.

TABLE 11.2
Limit Plastic Strains for Various Loading Conditions

Condition	Uniaxial Tension	Plane Strain	Equibiaxial Tension
$\sigma_2/\sigma_1 = \alpha$	0	$\frac{1}{2}$	1
Effective limit strain $(\bar{\varepsilon})_u$	n	$2n/\sqrt{3}$	$2n$
Limit strain, $(\varepsilon_1)_u$	n	n	n
Limit strain, $(\varepsilon_2)_u$	$-n/2$	0	n
Limit strain, $(\varepsilon_3)_u$	$-n/2$	$-n$	$-2n$

For biaxially stretched sheets, localized necking cannot be predicted analytically using the same arguments as in Section 11.2.2. To show this, consider again the deformation of a biaxially stretched sheet with strain ε_1 and ε_2 along the axes 1 and 2, respectively. The strain ε_2^* along an inclined direction 2* is related to ε_1 and ε_2 through the strain transformation law (Equation 2.22a) according to

$$\varepsilon_2^* = \varepsilon_1 \cos^2\theta + \varepsilon_2 \sin^2\theta$$

Strain localization takes place if there exists a direction θ^* along which $d\varepsilon_2^* = 0$; hence,

$$\varepsilon_1 \cos^2\theta^* + \varepsilon_2 \sin^2\theta^*$$

or

$$\tan\theta^* = \sqrt{-\varepsilon_1/\varepsilon_2}$$

For stretched sheets, where $\varepsilon_1 > 0$, the angle θ^* will have a real value only for negative values of ε_2. At the limit when $\varepsilon_2 = 0$, i.e., plane strain conditions, both diffuse and local necking coincide with a limit strain $(\varepsilon_1)_u = n$. Therefore, a local neck cannot form in stretching, where $\varepsilon_2 > 0$.

The assumption of a preexisting thickness imperfection perpendicular to the largest stress, as shown in Figure 11.8a, represents the basic ingredient in predicting limit strains for biaxially stretched sheet metals. A constraint on biaxial deformation of the sheet is imposed, such that $(d\varepsilon_2)_u = (d\varepsilon_2)_i$. Limit strains $(\overset{*}{\varepsilon}_1, \overset{*}{\varepsilon}_2)_u$ outside the groove are determined when $(d\varepsilon_1/d\varepsilon_2)_i \to \infty$, i.e., plane strain conditions, prevail within the groove.[5] A plot of $(\overset{*}{\varepsilon}_1)_u$ vs. $(\overset{*}{\varepsilon}_2)_u$ determined for different degrees of biaxial loading is known as the forming limit curve of the sheet, as shown schematically in Figure 11.8b.

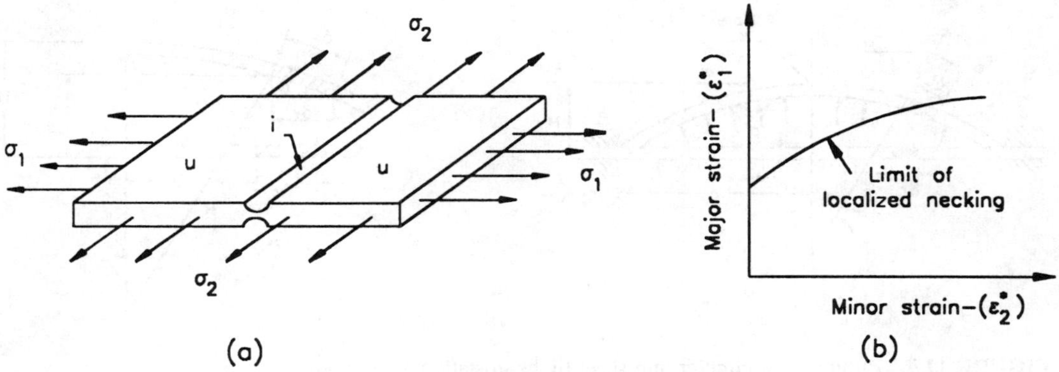

FIGURE 11.8 Biaxial stretching: (a) sheet metal with an initial thickness imperfection and (b) schematic forming limit curve.

Example 11.4:

In sheet metal biaxial stretching operations, it is preferable to express limit strains in terms of a strain ratio $\beta = \varepsilon_2/\varepsilon_1$. Rewrite Equation 11.34 in terms of β and obtain the limit strains.

Solution:

From the flow rule, Equations 11.28, for proportional loading,

$$\frac{\varepsilon_2}{\varepsilon_1} = \frac{2\alpha - 1}{2 - \alpha} = \beta$$

hence,

$$\alpha = (2\beta + 1)/(2 + \beta)$$

Substituting into Equation 11.34 yields

$$\bar{\varepsilon}_u = 2n\sqrt{1 + \beta + \beta^2}/\sqrt{3}$$

Hence, from Equation 11.30, the strains ε_1, ε_2, and ε_3 at instability are obtained as

$$(\varepsilon_1)_u = n \; (\varepsilon_2)_u = \beta n \text{ and } (\varepsilon_3)_u = -(1 + \beta)n$$

The limit strain $(\varepsilon_1)_u$ is independent of β in accordance with the results of Table 11.2.

11.2.5.2 Bulging of a Circular Sheet

Consider a thin circular sheet of a ductile material clamped around its periphery and subjected to uniform hydrostatic pressure at one side, as shown in Figure 11.9a. As pressure is increased, the sheet bulges out, forming a domelike surface with polar height *H*. Moreover, the sheet thickness decreases in a nonuniform manner, being thinnest at the pole and thickest at the clamped edges.

To describe the strain distribution in the sheet, consider a ring element of radius r_o and width δr_o in the undeformed sheet of thickness h_o as shown in Figure 11.9b. This ring element deforms to a shell element of surface length δL, thus defining the principal strains as

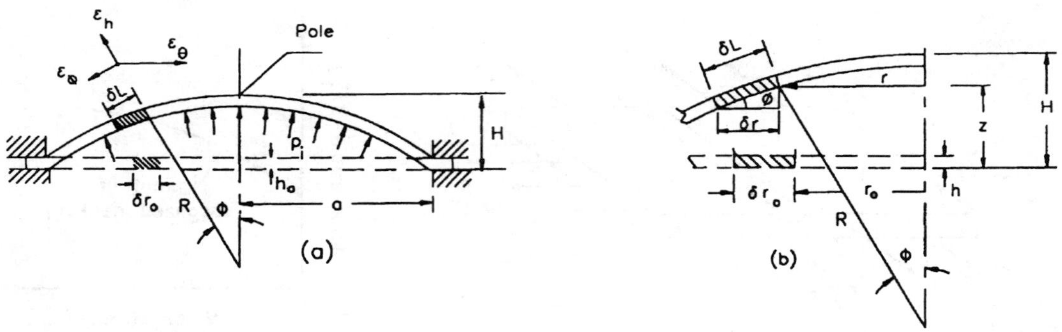

FIGURE 11.9 Bulging of a circular thin sheet by hydrostatic pressure.

$$\varepsilon_\theta = \ln\frac{\delta L}{\delta r_o}, \quad \varepsilon_\theta = \ln\frac{r}{r_o}, \quad \text{and} \quad \varepsilon_h = \ln\left(\frac{h}{h_o}\right) \tag{11.35}$$

where ε_h refers to the thickness strain.

The bulged sheet may be approximated by a spherical dome of radius R over most of its surface, where

$$R = \frac{H^2 + a^2}{2H} \tag{11.36a}$$

$$\delta L = \delta r/\cos\theta = \delta r\, R\big/\sqrt{R^2 - r^2} \tag{11.36b}$$

For a spherical surface $\varepsilon_\phi = \varepsilon_\theta$ and, hence, from Equation 11.35 and Figure 11.9,

$$\frac{\delta L}{\delta r_o} = \frac{\partial r}{\partial r_o}\frac{R}{\sqrt{R^2 - r^2}} = \frac{r}{r_o}$$

or

$$\int \frac{dr_o}{r_o} = \int \frac{R}{r\sqrt{R^2 - r^2}}dr$$

Integrating and knowing that at $r_o = a$, $r = a$ gives

$$r_o = \frac{R + \sqrt{R^2 - a^2}}{R + \sqrt{R^2 - r^2}}r$$

which defines the strain components as*

* According to this analysis, $\varepsilon_\phi = \varepsilon_\theta \to 0$ (and hence, ε_h) toward the edge. In an actual process, close to the edge truly $\varepsilon_\theta = 0$. However, a more complex state of stress and strain (including shear effects) exists, and the above solution is never correct near the edge.

$$\varepsilon_\phi = \varepsilon_\theta = \ln\left(\frac{R+\sqrt{R^2-r^2}}{R+\sqrt{R^2-a^2}}\right) \tag{11.37}$$

From the geometry of the spherical dome of Figure 11.9b, Expression 11.37 is simplified to define the strain at any point of the bulge at height H as

$$\varepsilon_\phi = \varepsilon_\theta = \ln\left[1+\left(\frac{zH}{a^2}\right)\right] \tag{11.38a}$$

Applying the condition of constant volume yields

$$\varepsilon_h = -2\varepsilon_\phi = \ln\left[\left(\frac{1}{1+zH/a^2}\right)\right]^2 \tag{11.38b}$$

the effective strain according to Expression 10.30 is thus

$$\bar{\varepsilon} = -\varepsilon_h \tag{11.39}$$

Note that Expression 11.38b is independent of material properties and gives the largest thickness reduction at the pole where $z = H$. Obviously, this is an outcome of assuming a spherical bulge surface, an assumption which may not be satisfied in actual bulges of various materials.

At any instant, the equilibrium over most of the surface of the bulged sheet is expressed by Equation 8.65 as

$$\frac{\sigma_\phi}{R_\phi}+\frac{\sigma_\theta}{R_\theta}=\frac{p_i}{h}$$

where for a spherical surface $R_\phi = R_\theta = R$. Hence, a state of equibiaxial tensile stress prevails with principal stresses given by

$$\sigma_\phi = \sigma_\theta = p_i R/2h \text{ with } \sigma_r \simeq 0 \tag{11.40}$$

The effective stress is obtained from Expression 10.26 as

$$\bar{\sigma} = \sigma_\phi = \sigma_\theta = p_i R/2h \tag{11.41a}$$

and, hence, the value of the bulging pressure p_i is

$$p_i = 2h\bar{\sigma}/R \tag{11.41b}$$

At instability the pressure reaches a maximum value and $dp_i = 0$, thus giving

$$\frac{d\bar{\sigma}}{\bar{\sigma}} = \frac{dR}{R} - \frac{dh}{h} \tag{11.42}$$

By virtue of Equation 11.39, namely, $d\bar{\varepsilon} = -d\varepsilon_h = -dh/h$, the instability condition (Equation 11.42) becomes

$$\frac{1}{\bar{\sigma}}\frac{d\bar{\sigma}}{d\bar{\varepsilon}} = 1 + \frac{1}{R}\frac{dR}{d\bar{\varepsilon}} \tag{11.43}$$

After some algebraic manipulation using Equations 11.36a, 11.8b and 11.39, Equation 11.43, reduces to

$$\begin{aligned} \frac{1}{\bar{\sigma}}\frac{d\bar{\sigma}}{d\bar{\varepsilon}} &= 1 + \left(\frac{1}{2} - \frac{R}{2H}\right) = \frac{3}{2} - \frac{R}{2H} \\ &= \frac{3}{2} - \frac{1}{4}\left(\frac{a}{H}\right)^2\left(1 + \frac{H^2}{a^2}\right) \end{aligned} \tag{11.44}$$

Combining equations 11.38b and 11.39 yields, at the pole ($z = H$),

$$\bar{\varepsilon} = -\ln\left(\frac{1}{1 + H^2/a^2}\right)^2 = 2\ln\left(1 + \frac{H^2}{a^2}\right) \tag{11.45}$$

Rearranging Expression 11.45 and expanding in terms of $\bar{\varepsilon}$ results in

$$\exp(\bar{\varepsilon}/2) \cong 1 + \frac{\bar{\varepsilon}}{2} = 1 + H^2/a^2$$

Subsituting into the instability condition (Equation 11.44) gives

$$\frac{1}{\bar{\sigma}}\frac{d\bar{\sigma}}{d\bar{\varepsilon}} \cong \frac{3}{2} - \frac{1}{4}\left(\frac{2}{\bar{\varepsilon}}\right)\left(1 + \frac{\bar{\varepsilon}}{2}\right) \tag{11.46}$$

If the stress–strain curve of the sheet material is represented by the relation

$$\bar{\sigma} = K(\bar{\varepsilon})^n$$

the effective or thickness strain at instability is given by:[6]

$$\bar{\varepsilon}_u = \varepsilon_h = \frac{4}{11}(2n + 1) \tag{11.47a}$$

An interesting feature of this result is that even for a nonhardening material ($n = 0$) a thickness strain of 4/11 is obtained before the maximum pressure is attained. For a strain-hardening material, the bulge test continues to much larger strains than that achieved in uniaxial tensile test at necking. For instance, in a bulge test of a sheet metal with $n = 0.25$, an effective strain $(\bar{\varepsilon})_u = 0.545$ is realized according to Equation (11.47-a) as compared with only 0.25 in simple tension according to Equation 11.4.

The above solution has been found to be in good agreement with experiments conducted on sufficiently work-hardened sheet metals. Another solution[7] using the Tresca yield criterion has been suggested to fit more the behavior of bulged sheets of pronouncedly higher values of n. This solution gives for the thickness strain at instability:

$$\bar{\varepsilon}_u = \frac{2(2-n)(1+2n)}{11-4n} \tag{11.47b}$$

which reduces to Expression 11.47a for $n = 0$.

Example 11.5:

A circular thin sheet of 1.22 mm thickness made of annealed copper is bulged by oil pressure into a circular die aperture of 150 mm diameter. The following readings of pressure vs. bulge height are recorded.

Pressure (p_i), MPa	*1.15*	*2.0*	*3.5*	*4.75*	*5.37*
Bulge height (H), mm	*10.52*	*15.72*	*21.92*	*28.85*	*34.94*

a. *Establish the flow stress–strain curve of the sheet material as represented by the law $\bar{\sigma} = K(\bar{\varepsilon})^n$. Show the details of calculations.*
b. *Determine the pressure at which plastic instability occurs.*

Solution:

a. At the pole, the equibiaxial stress–strain system is given by Equations 11.41a and 11.45, respectively, as

$$\bar{\sigma} = \sigma_\phi = \sigma_\theta = p_i R/2h$$

$$\bar{\varepsilon} = 2\ln\left(1 + H^2/a^2\right)$$

Based on the assumption of spherical bulge geometry, the current radius of curvature and thickness at the pole are approximated by Equations 11.36a and 11.38b for $z = H$, respectively, as

$$R = \left(H^2 + a^2\right)/2H \quad \text{and} \quad h = \frac{h_o}{\left(1 + H^2/a^2\right)^2}$$

Provided that $h_o = 1.22$ mm and $a = 75$ mm, the following effective stress–effective strain points are calculated:

Pressure (p_i), MPa	1.15	2.0	3.5	4.75	5.37
Bulge height (H), mm	10.52	15.72	21.92	28.85	34.94
Radius of curvature (R), mm	272.6	186.77	139.27	111.91	97.97
Current thickness (h), mm	1.1734	1.1195	1.0355	0.9258	0.8237
Effective stress ($\bar{\sigma}$), MPa	133.6	166.8	235.4	287.1	319.3
Effective strain ($\bar{\varepsilon}$)	0.039	0.086	0.164	0.276	0.393

Straight-line fitting to $\ln\bar{\sigma} - \ln\bar{\varepsilon}$ points yields the material behavior law:

$$\bar{\sigma} = 467.6(\bar{\varepsilon})^{0.395} \text{ MPa}$$

b. Plastic instability occurs at uniform thickness strain value given by Equation 11.47a, i.e.,

$$\bar{\varepsilon}_u = \varepsilon_3 = \frac{4}{11}(2n+1) = \frac{4}{11}(2 \times 0.395 + 1) = 0.651$$

compared with a strain value of 0.395 in a simple tensile test of a specimen of the same annealed copper material. This uniform strain at instability corresponds to an effective stress ($\bar{\sigma}_u$), thickness (h_u), polar height (H_u), and radius of curvature (R_u) given, respectively, by

$$\bar{\sigma}_u = K(\bar{\varepsilon}_u)^n = 467.6\ (0.651)^{0.395} = 394.7 \text{ MPa}$$

$$h_u = h_o \exp(-\bar{\varepsilon}_u) = 1.22 \exp(-0.651) = 0.636 \text{ mm}$$

$$\bar{\varepsilon}_u = 2\ln\left(1 + H^2/a^2\right)^2 \quad \text{or} \quad \exp(\bar{\varepsilon}_u/2) = 1 + (H/a)^2, \quad \text{giving}$$

$$H_u = 46.52 \text{ mm}$$

$$R_u = \left(H_u^2 + a^2\right)/\left(2H_u\right) = \left(46.52^2 + 75^2\right)/(2 \times 46.52) = 83.72 \text{ mm}$$

Therefore, instability occurs at a maximum pressure p_{max},

$$p_{max} = \frac{2h_u\sigma_u}{R_u} = \frac{2 \times 0.636 \times 394.7}{83.72} = 6 \text{ MPa}$$

11.2.6 Pressurized Axisymmetric Thin-Walled Containers

Thin-walled shells subjected to steadily and slowly increasing internal pressure undergo plastic deformation in a uniform manner up to a certain critical point, which marks instability. One way to define this instability is to refer it to the point at which the internal pressure attains its maximum value. This condition is applied to two simple shapes, namely, a thin-walled sphere and a long, thin-walled cylinder to obtain an estimate for the bursting pressure.

11.2.6.1 Thin-Walled Sphere

For a thin-walled spherical shell under internal pressure p_i, the stresses, according to Equations 8.67, are given by

$$\sigma_1 = \sigma_2 = \frac{p_i r_m}{2h} \quad \text{and} \quad \sigma_3 \cong 0 \tag{11.48}$$

where r_m and h are the current mean radius and thickness of the shell, respectively. The effective stress $\bar{\sigma}$, as defined by Equation 10.26, is

$$\bar{\sigma} = \sigma_1 \tag{11.49}$$

Rearrangement of Equation 11.48 gives for the pressure:

$$p_i = 2\sigma_1 h/r_m$$

Plastic instability occurs at maximum pressure, i.e., when $dp_i = 0$,

$$dp_i = \frac{2h}{r_m} d\sigma_1 + \frac{2\sigma_1}{r_m} dh - \frac{2h\sigma_1}{r_m^2} dr_m = 0$$

Dividing by 2 $hr\sigma_1$ yields

$$\frac{d\sigma_r}{\sigma_1} = \frac{dr_m}{r_m} - \frac{dh}{h} \tag{11.50}$$

Recognizing that $d\varepsilon_1 = dr/r$ and $d\varepsilon_3 = dh/h$, Equation 11.50 becomes

$$\frac{d\sigma_1}{\sigma_1} = d\varepsilon_1 - d\varepsilon_3 \tag{11.51}$$

From symmetry and constancy of volume, the deformation of the shell proceeds, such that

$$d\varepsilon_1 = d\varepsilon_2 = -d\varepsilon_3/2$$

The effective strain increment, according to Equation 11.29, is thus given by

$$d\bar{\varepsilon} = -d\varepsilon_3 = 2d\varepsilon_1 \tag{11.52}$$

Combining Equations 11.49, 11.51, and 11.52 gives the maximum pressure condition as

$$\frac{d\bar{\sigma}}{d\bar{\varepsilon}} = \frac{3}{2}\bar{\sigma} \tag{11.53}$$

Hence, for a strain-hardening law: $\bar{\sigma} = K(\bar{\varepsilon})^n$, plastic instability occurs at

$$(\bar{\varepsilon})_u = \frac{2}{3}n \tag{11.54}$$

The strains, being constant ratios, are then at this point of instability

$$(\varepsilon_1)_u = (\varepsilon_2)_u = \frac{n}{3} \text{ and } (\varepsilon_3)_u = -\frac{2n}{3}$$

The current shell dimensions, in terms of its initial thickness h_o and mean radius r_{mo}, are

$$h = h_o \exp(\varepsilon_3) = h_o \exp\left(-\frac{2n}{3}\right) \tag{11.55a}$$

$$r_m = r_{mo} \exp(\varepsilon_1) = r_{mo} \exp\left(\frac{n}{3}\right) \tag{11.55b}$$

Hence, the maximum pressure at the onset of unstable plastic deformation is obtained from Equation 11.48 as

$$p_{\max} = \frac{2\sigma_1 h_o}{r_{mo}} \exp(-n)$$

Using the strain-hardening law and recalling that $\bar{\sigma} = \sigma_1$ gives

$$p_{\max} = K\frac{2h_o}{r_{mo}}\left(\frac{2n}{3}\right)\exp(-n) \tag{11.56a}$$

Note that for a perfectly plastic material, $n = 0$ and, hence, instability occurs at the initiation of yielding.

In terms of the ultimate strength of material UTS, the instability pressure by virture of Equation 11.5 is

$$p_{\max} = \left(\frac{2}{3}\right)^n \frac{2h_o}{r_{mo}}(\text{UTS}) \tag{11.56b}$$

11.2.6.2 Thin-Walled Cylinder

The stresses in a long, thin-walled cylinder with closed ends of current mean radius r_m and current thickness h due to an internal pressure p_i, according to Equations 8.68, are

$$\sigma_1 = \frac{p_i r_m}{h}, \quad \sigma_2 = \frac{p_i r_m}{2h}, \quad \text{and } \sigma_3 \cong 0 \tag{11.57}$$

Thus, the effective stress from Equation 11.26 is

$$\bar{\sigma} = \sqrt{3}\,\sigma_1/2 \tag{11.58}$$

Instability occurs at maximum pressure, i.e., at $dp_i = 0$; hence,

$$\frac{d\bar{\sigma}}{\bar{\sigma}} = \frac{d\sigma_1}{\sigma_1} = \frac{dr_m}{r_m} - \frac{dh}{h}$$

By substituting $d\varepsilon_1 = dr_m/r_m$ and $d\varepsilon_3 = dh/h$ gives

$$\frac{d\bar{\sigma}}{\bar{\sigma}} = d\varepsilon_1 - d\varepsilon_3 \tag{11.59}$$

From the Levy–Mises flow rule, Equations 11.22, and $\sigma_2 = \sigma_1/2$ gives

$$d\varepsilon_2 = d\lambda[\sigma_2 - {}^1/_2\sigma_1] = 0$$

This is a plane strain problem and, hence, from the constancy of volume condition,

$$d\varepsilon_1 = -d\varepsilon_3$$

The effective strain increment is thus

$$d\bar{\varepsilon} = 2\,d\varepsilon_1/\sqrt{3} = -2d\varepsilon_3/\sqrt{3} \tag{11.60}$$

Combining Equations 11.59 and 11.60 yields

$$\frac{d\bar{\sigma}}{\bar{\sigma}} = \sqrt{3}\,d\bar{\varepsilon}$$

The condition of plastic instability at maximum pressure is thus given by

$$\frac{d\bar{\sigma}}{d\bar{\varepsilon}} = \sqrt{3}\,\bar{\sigma} \tag{11.61}$$

which reduces for a strain hardening law ($\bar{\sigma} = K(\bar{\varepsilon})^n$) to

$$\bar{\varepsilon}_u = n/\sqrt{3} \tag{11.62}$$

An expression for the maximum pressure at any instant is given by

$$p_{max} = \frac{\sigma_1 h}{r_m} = \frac{2\bar{\sigma}}{\sqrt{3}}\frac{h}{r_m} \tag{11.63a}$$

By virtue of Equation 11.63 and the strain hardening law,

$$p_{max} = \frac{2n}{3} = \frac{h}{r_m}K \tag{11.63b}$$

Since the current shell dimensions are obtained from

$$h = h_o \exp(\varepsilon_3) = h_o \exp\left(-\sqrt{3}\,\bar{\varepsilon}/2\right) \tag{11.64a}$$

and

$$r_m = r_{mo} \exp(\varepsilon_1) = r_{mo} \exp\left(\sqrt{3}\,\bar{\varepsilon}/2\right) \tag{11.64b}$$

Equation 11.64 gives

$$p_{max} = \frac{2K}{\sqrt{3}}\left(\frac{h_o}{r_{mo}}\right)(\bar{\varepsilon})^n \exp\left(-\sqrt{3}\,\bar{\varepsilon}\right) \tag{11.65}$$

The pressure attained at the onset of plastic instability is obtained by putting $(\bar{\varepsilon})_n = n/\sqrt{3}$, hence,

$$p_{max} = \frac{2K}{\sqrt{3}}\left(\frac{h_o}{r_o}\right)\left(\frac{n}{\sqrt{3}}\right)^n \exp(-n) \tag{11.66}$$

This maximum pressure may be expressed in terms of the UTS of the material by combining Equations 11.5 and 11.66 to give

$$p_{max} = \left[2/\left(\sqrt{3}\right)^{n+1}\right]\left[h_o/r_o\right](\text{UTS}) \tag{11.67}$$

For a perfectly plastic material $n = 0$ and the UTS coincides with the yield strength; hence, Equations 11.65 and 11.66 give

$$p_{max} = p_{yielding} = \left[2/\sqrt{3}\right]\left[h_o/r_{mo}\right] Y \tag{11.68}$$

which is the pressure producing yield in a thin-walled tube as in Example 10.4.

The analysis presented above for a thin-walled pressurized sphere and cylinder could be applied to other cases of axisymmetric loadings and shapes.[8]

Example 11.6:

A pressurized spherical tank of a mean radius r_m *is connected to a long pipe of mean radius* $r_m/6$ *through a throttling valve which reduces the tank pressure to* $^1/_4$ *of its value before allowing flow into the pipe. Both tank and pipe are made of the same stainless steel. If the design basis is such that plastic instability is not allowed to occur — usually by an ample margin of safety — neither for the tank nor for the pipe. Determine the ratio of tank thickness* h_s *to that of the pipe* h_c*. Take* $n = 0.45$ *for stainless steel. What would be this ratio if the design basis is not to allow yielding?*

Solution:

For the spherical tank, the maximum pressure p_s (i.e., bursting pressure) at instability is given by Equation 11.56b as

$$p_s = \left(\frac{2}{3}\right)^n\left(\frac{2h_s}{r_m}\right)(\text{UTS})$$

For the pipe, the bursting pressure p_c is obtained from Equation 11.67 as

$$p_c = \frac{2}{\left(\sqrt{3}\right)^{n+1}}\left(\frac{h_c}{r_m/6}\right)(\text{UTS})$$

Now the operating conditions are such that

$$p_s = 4p_c$$

Hence,

$$\left(\frac{2}{3}\right)^n\left(\frac{2h_s}{r_m}\right)(\text{UTS}) = \frac{4\times 2}{\left(\sqrt{3}\right)^{n+1}}\left(\frac{h_c}{r_m/6}\right)(\text{UTS})$$

which gives

$$h_s = 12\left(\frac{\sqrt{3}}{2}\right)^{n+1} h_c \text{ for } n = 0.45,\ h_s = 9.74h_c$$

If the design basis is such that yielding is not allowed, neither for the tank nor for the pipe, a thickness ratio is obtained by letting $n = 0$ and UTS = Y into Expressions 11.56b and 11.68 to give

$$h_s = \left(24/\sqrt{3}\right)h_c = 13.86h_c$$

Example 11.7:

A thin-walled pipe with closed ends is subjected to an internal pressure p_i. *The pipe is made of stainless steel for which the plastic behavior is represented by*

$$\bar{\sigma} = 1200\bar{\varepsilon}^{0.45}\ MPa$$

a. *Using the maximum pressure instability criterion, calculate the maximum pressure that the pipe can withstand, knowing that the initial pipe dimensions are 1 mm thickness and 50 mm mean radius.*
b. *Establish a table showing the variation of the pressure with pipe radius for effective strains* $\bar{\varepsilon} \geq 0.04$.

Solution:

a. The maximum pressure criterion, Equation 11.66, gives

$$\bar{\varepsilon}_u = n/\sqrt{3}$$

$$p_{max} = \frac{2}{\sqrt{3}}(1200)\left(\frac{1}{50}\right)\left(\frac{0.45}{\sqrt{3}}\right)^{0.45} \exp(-0.45) = 9.635 \text{ MPa}$$

b. For a pipe with closed ends, Expression 11.60 gives

$$\varepsilon_1 = \ln\frac{r_m}{r_{mo}} = \frac{\sqrt{3}}{2}\bar{\varepsilon}$$

hence,

$$\bar{\varepsilon} = \frac{2}{\sqrt{3}}\ln\frac{r}{r_{mo}}$$

Substitution into Equation 11.65 yields

$$p_i = \frac{2}{\sqrt{3}}K\left(\frac{h_o}{r_{mo}}\right)\left(\frac{2}{\sqrt{3}}\ln\frac{r_m}{r_{mo}}\right)^n \exp\left(-\sqrt{3}\frac{2}{\sqrt{3}}\ln\frac{r_m}{r_{mo}}\right)$$

or

$$p_i = \left(\frac{2}{\sqrt{3}}\right)^{n+1} K\left(\frac{h_o}{r_{mo}}\right)\left(\frac{r_{mo}}{r_m}\right)^2\left(\ln\frac{r_m}{r_{mo}}\right)^n$$

At instability

$$(\varepsilon_1)_u = \frac{\sqrt{3}}{2}(\bar{\varepsilon})_u = \frac{\sqrt{3}}{2}\left(\frac{n}{\sqrt{3}}\right) = \frac{n}{2}$$

The limits within which this equation may be applied are $0.04 \le \bar{\varepsilon} < n/\sqrt{3}$; i.e., in terms of $\varepsilon_1 = \ln(r_m/r_{mo})$, these are $(0.0346) \le \varepsilon_1 \le (n/2 = 0.225)$. Values of p_i calculated for the stainless steel pipe under consideration (with K = 1200 MPa and n = 0.45) are obtained from

$$p_i = 29.567\left(\frac{r_{mo}}{r_m}\right)^2\left(\ln\frac{r_m}{r_{mo}}\right)^{0.45}$$

as given in the table below:

r_m/r_{mo}	1.0352	1.05	1.1	1.15	1.20	1.2523
p_i MPa	6.072	6.889	8.485	9.222	9.546	9.635

11.3 STRAIN-RATE DEPENDENT PLASTIC BEHAVIOR — APPLICATION TO SUPERPLASTICITY

It is well known that metals deformed at hot-working conditions* and polymeric materials above their glass transition temperatures behave such that their flow stresses become very much dependent on the rate of plastic strain $\dot{\varepsilon}^p$ or, simply, the strain rate $\dot{\varepsilon}$. No appreciable strain hardening occurs and the stress-strain curve may be described by a simple empirical law:

$$\sigma = C_o\left(\dot{\varepsilon}/\dot{\varepsilon}_o\right)^m \tag{11.69a}$$

where C_o and m are material parameters that depend upon the temperature and strain rate. The coefficient C_o indicates the strength of the material at a particular strain rate $\dot{\varepsilon}_o$. By taking $\dot{\varepsilon}_o$ to be unity, the stress–strain curve will be simply expressed by the relation:

$$\sigma = C\dot{\varepsilon}^m \tag{11.69b}$$

The strain-rate sensitivity exponent m is found from the slope of log σ vs. log $\dot{\varepsilon}$ plot, as shown in Figure 11.10.

* Hot working of metals refers to deformations occurring at temperatures above their recrystallization temperature, say, around 0.4 – 0.5 T_m, where T_m is the melting absolute temperature.

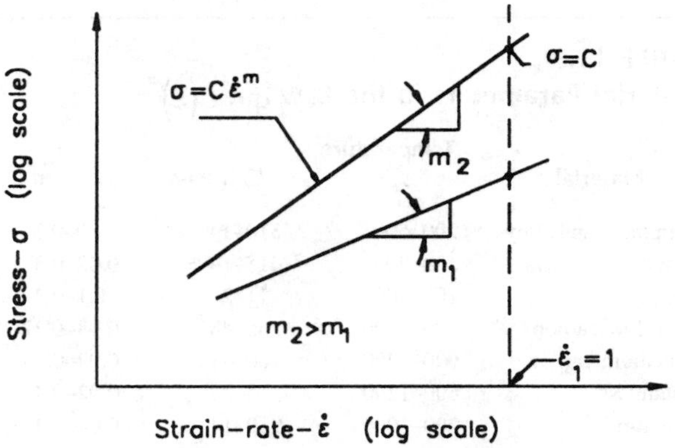

FIGURE 11.10 Typical stress–strain rate behavior of superplastic materials.

Typical values of C and m are given in Table 11.3* for several materials. High strain-rate sensitivities (0.35–0.8) are observed during superplastic behavior. This term refers to a behavior of some metallic alloys with ultrafine grain microstructure (few microns) and enhanced ductility (up to 2000% tensile elongation) before fracture when deformed under appropriate conditions of low strain rates** (of the order of 0.1/min) and hot-working temperatures.

It is experimentally observed that as temperature increases, C decreases and m increases. The exponent m is strain rate dependent and the above values are associated with strain rates within the indicated range.

By adopting the same concepts of isotropic plasticity of Sections 10.12, the law (Equation 11.69) is generalized to describe the multiaxial behavior as

$$\bar{\sigma} = C\left(\bar{\dot{\varepsilon}}\right)^{m} \tag{11.70}$$

where $\bar{\sigma}$ is the effective stress as given by Expression 10.26. The effective strain rate is simply defined by conceiving that the three increments of plastic deformation ($d\varepsilon_1^p$, $d\varepsilon_2^p$, $d\varepsilon_3^p$) define an effective strain increment $d\bar{\varepsilon}$ in a time interval dt, hence, according to Expression 10.30b,

$$\bar{\dot{\varepsilon}} = \frac{\sqrt{2}}{3}\left[\left(\dot{\varepsilon}_1 - \dot{\varepsilon}_2\right)^2 + \left(\dot{\varepsilon}_2 - \dot{\varepsilon}_3\right)^2 + \left(\dot{\varepsilon}_3 - \dot{\varepsilon}_1\right)\right]^{1/2} \tag{11.71}$$

where $\dot{\varepsilon}_1$, $\dot{\varepsilon}_2$, and $\dot{\varepsilon}_3$ are the strain rates along the principal directions.

In dealing with strain-rate-sensitive materials, a viscoplastic potential identical to the yield function describes a surface of constant rate of dissipation of energy per unit volume. This visco-plastic replaces the yield surface, and the components of the strain-rate tensor become normal to it and hence all the precedent derivation of rate-independent plasticity apply. A nest of concentric Mises ellipses, as shown in Figure 11.11, may be thus taken to represent the plastic flow of strain-rate-dependent materials. In such a case, each ellipse of Figure 11.11a is associated with a prescribed value of effective flow stress $\bar{\sigma}$ of Figure 11.11b and, hence, $\bar{\dot{\varepsilon}}$. The most inner ellipse is shrunk to a point corresponding to zero strain rate as necessitated by Equations 11.70.

The Levy–Mises flow rule (Equation 10.22c) thus applies in the form:

* These parameters should not in any way be mixed up with C and on in the Paris law, Equation 9.29, in Chapter 9.

** The order of magnitude of strain rates encountered in deformation processes may be arbitrarily classified as follows: Creep: 10^{-10} to 10^{-6} s^{-1} quasistatic or low: 10^{-4} to 10^{-1} s^{-2}, intermediate: 10^{-1} to 10^{-2} s^{-1}, high and impact: 10^{2} to 10^{6} s^{-1}.

TABLE 11.3
Material Parameters in the Law $\bar{\sigma} = C(\dot{\bar{\varepsilon}})^m$

Material	Temperature, °C	C, MPa-s	m
Aluminum and alloys	200–500	310–14	0–0.2
Copper and alloys	200–900	415–14	0.02–0.3
Lead	100–300	11–2	0.1–0.2
Steel, low-carbon	900–1200	165–48	0.08–0.22
medium-carbon	900–1200	160–48	0.07–0.24
stainless	600–1200	415–35	0.02–0.4
Titanium	200–1000	930–14	0.04–0.3
Superplastic alloys		C(MPa min)	m (max.value)
Lead–Tin (eutectic)			
($\dot{\varepsilon}$: 0.01–0.2 min^{-1})	22	8.8	0.5
Zinc–Al., (eutectoid)			
($\dot{\varepsilon}$: 0.1–10 min^{-1})	250	6.68	0.5
Titanium -6Al–4V			
($\dot{\varepsilon}$: 0.01–0.1 min^{-1})	950	46.6	0.8
Brasses (60/40)			
($\dot{\varepsilon}$: 0.01–0.1 min^{-1})	600	21	0.48
Aluminum-33 copper			
($\dot{\varepsilon}$: 0.01–0.1 min^{-1})	500	20	0.75
Thermoplastics: (polystyrene)	150	11	0.85
Molten glass	> 1000	0.001	> 1 Newtonian viscous flow

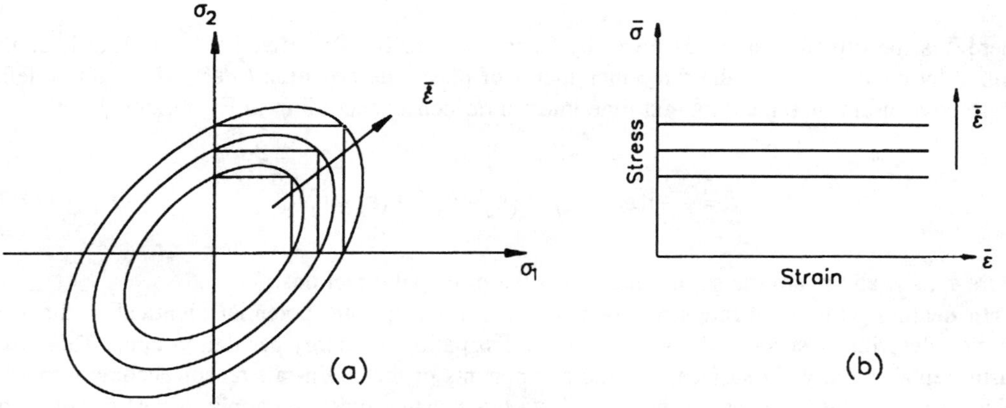

FIGURE 11.11 Nest of expanding Mises ellipses representing the strain-rate-hardening behavior of visco-plastic materials.

$$\frac{\dot{\varepsilon}_1}{\sigma_1'} = \frac{\dot{\varepsilon}_2}{\sigma_2'} = \frac{\dot{\varepsilon}_3}{\sigma_3'} = \frac{3}{2}\dot{\lambda} = \frac{3}{2}\frac{\dot{\bar{\varepsilon}}}{\bar{\sigma}}$$

or (11.72)

$$\dot{\varepsilon}_1 = \frac{\dot{\bar{\varepsilon}}}{\bar{\sigma}}\left[\sigma_1 - \frac{1}{2}(\sigma_2 + \sigma_3)\right]\cdots$$

Equations 11.70 and 11.72 constitute the basis of analyzing the mechanics of hot-forming processes. In the following sections, three simple applications related to strain-rate-dependent plastic deformation are analyzed.

11.3.1 Neck-Free Elongations

The material behavior given by the empirical law (Equation 11.69) may be looked upon as an expression for a strain-rate-hardening effect, e.g., higher flow stress for an increasing strain rate. For hot-working conditions of conventional metallic alloys $m \simeq 10.15$; hence, doubling the strain rate results in an increase of about 11% in the flow stress. However, this increase may reach 40% or 75% for superplastic alloys ($m \simeq 0.5$) or thermoplastics ($m \simeq 0.8$), respectively.

The contribution of the strain-rate-sensitivity exponent m to the stability of plastic flow may be demonstrated by considering a uniform bar of initial length L_o, current length L, and cross sectional area A. Due to uniaxial loading under a constant tensile force P, the strain rate at any instant,* assuming constancy of volume, is

$$\dot{\varepsilon} = \frac{d\varepsilon^p}{dt} = \frac{d}{dt}\left(\ln\frac{L}{L_o}\right) = \frac{1}{L}\frac{dL}{dt} = -\frac{1}{A}\frac{dA}{dt} \tag{11.73}$$

Substituting Equation 11.73 into the law (Equation 11.69) gives

$$\frac{P}{A} = C\left(-\frac{1}{A}\frac{dA}{dt}\right)^m$$

and, after rearranging,

$$-\frac{1}{A_o}\frac{dA}{dt} = \left(A_o\frac{P}{C}\right)^{1/m}\left(\frac{A}{A_o}\right)^{1-1/m} \tag{11.74}$$

This shows that the rate of area decrease (dA/dt) depends on the current area A and the strain-rate-sensitivity exponent m. The dependence on m is very strong as may be seen from Table 11.4 (for $P = A_oC$). As $m \rightarrow 1$, (i.e., Newtonian viscous flow), the rate of area decrease $dA/dt \rightarrow 1$.

TABLE 11.4
Dependence of Rate of Area Decrease on *m* in a Tension

m	0.25	0.5	0.75	1
$(-1/Ao)(dA/dt)$	$(A_o/A)^3$	(A_o/A)	$(A_o/A)^{1/3}$	1

Finally, when $m = 1$, the rate of area decrease becomes independent of the area itself. This means that if the bar possesses initial cross sectional area irregularities, they will be preserved (without growth) during pulling. If the bar is of perfect geometry, there can be no necking or strain localization, as all cross sections will undergo the same deformation. For molten glass ($m \simeq 1$) neck-free stretching and shaping becomes possible.

* During tensile testing with a constant cross-head speed v, the current strain rate is $\dot{\varepsilon} = (dL/dt)/L = v/L$.

11.3.2 Limit Tensile Strains for a Bar of Strain-Rate-Dependent Material

The analysis presented in Section 11.2.3 for determining the limit tensile strain for a bar with an initial geometric imperfection (Figure 11.4) is considered here for a strain-rate-dependent material behavior as described by the law (Equation 11.69).

The same definition of area imperfection f previously given by Equation 11.10, as well as the equilibrium condition (Equation 11.11), apply here. Hence,

$$f\sigma_i \frac{A_i}{A_{io}} = \sigma_u \frac{A_u}{A_{uo}} \tag{11.75}$$

From the material behavior law (Equation 11.69) for $\sigma = P/A$, this becomes

$$\frac{\sigma_u}{\sigma_i} = \frac{P/A_u}{P/A_i} = \left(\frac{\dot{\varepsilon}_u}{\dot{\varepsilon}_i}\right)^m \tag{11.76a}$$

or

$$(A_u/A_i) = (\dot{\varepsilon}_u/\dot{\varepsilon}_i)^m \tag{11.76b}$$

The constant-volume condition gives

$$\begin{aligned} A_u &= A_{uo}\exp(-\varepsilon_u) \\ A_i &= A_{io}\exp(-\varepsilon_i) \end{aligned} \tag{11.77}$$

Substitution of Equations 11.76 and 11.77 into Equation 11.75 yields, after rearranging,

$$\dot{\varepsilon}_u e^{-\varepsilon_u/m} = f^{1/m}\dot{\varepsilon}_i e^{-\varepsilon_i/m}$$

Integrating with respect to time gives

$$\left(e^{-\varepsilon_u/m} - 1\right) = f^{1/m}\left(e^{-\varepsilon_i/m} - 1\right) \tag{11.78}$$

Note that for $m = 1$ (Newtonian viscous flow), Equation 11.78 combined with Expressions 11.77 results in the equality:

$$(A_u - A_{uo}) = (A_i - A_{io})$$

Again this means that the initial area imperfection is preserved throughout deformation, and hence there can be no necking with unlimited formability.

Equation 11.78 requires a numerical solution to determine the strain ε_u in the uniform zone as a function of the strain ε_i in the defected zone for a given f and m. However, since the nonuniformity in cross section exists only over a very small length, the strain within this region may be neglected without causing great error in determining the overall elongation of the bar. Hence, the limit strain ε_u in the uniform zone is obtained from Equation 11.78 in its reduced form as

$$\varepsilon_u = -m\ln(1 - f^{1/m}) \tag{11.79}$$

For a bar of 11 mm diameter machined by turning, a diameter variation of 0.01 mm may occur, thus giving $f = 0.998$. For this bar, the limit elongation percentages in the uniform section are calculated for various values of strain rate sensitivity, as given in Table 11.5. It indicates that deforming at high m values, in general, enhances high elongations before strain localization or fracture.

TABLE 11.5
Percent Elongation for a Tensile Bar of Area Imperfection, $f = 0.998$

m	0.05	0.1	0.2	0.3	0.5	$\to 1$
Elongation, %	17.6	48	151	350	1482	$\to \infty$

11.3.3 Forming Time for a Bulged Circular Sheet of Rate-Dependent Material*

In Section 11.2.5, the strain and stress components are determined for a bulged circular sheet independent of its material behavior. This analysis is extended here to estimate the forming time for bulging a sheet made of a strain-rate-dependent material obeying the law (Equation 11.69).

Expressions 11.38 indicate that a state of nonuniform strain prevails in the deformed sheet, where the strains are functions of the current height of the sheet at the point under consideration. To arrive at a closed-form solution relating the process variables to the forming time, an overall constant thickness h_{av} at any polar bulge height H is assumed to be given by

$$\frac{h_{av}}{h_o} = \frac{1}{H}\int_0^H \frac{1}{\left(1+zH/a^2\right)^2}\,dz = \frac{1}{\left(1+H^2/a^2\right)} \tag{11.80}$$

Thus, over the entire bulge surface the average thickness strain rate $\left(\dot{\varepsilon}_h\right)_{av}$ is obtained as

$$\begin{aligned}\left(\dot{\varepsilon}_h\right)_{av} &= \frac{d\varepsilon_h}{dt} = \frac{d}{dt}\left(\ln\frac{h_{av}}{h_o}\right) = \frac{1}{h_{av}}\frac{dh_{av}}{dt} \\ &= \frac{1}{a^2\left(1+H^2/a^2\right)^2}\frac{dH}{dt}\end{aligned} \tag{11.81a}$$

In view of Equation 11.39, it may be assumed that

$$\left(\dot{\bar{\varepsilon}}\right)_{av} = -\left(\dot{\varepsilon}_h\right)_{av} \tag{11.81b}$$

Hence, the material behavior law (Equation 11.69), together with the equilibrium condition (Equation 11.40) for a spherical bulge, yields

$$\frac{p_i R}{2h_{av}} = C\left[\left(\dot{\varepsilon}_h\right)_{av}\right]^m \tag{11.82}$$

* For complete analysis to determine forming times for other shapes of super-plastic sheets, see Ragab.[9]

Now, using Equations 11.36a, 11.80, and 11.81 for the bulge radius R, the average thickness, and the average strain rate, respectively, give after simplification a forming time t:

$$t = \left(\frac{1}{a}\right)\left(\frac{p_i a}{2Ch_o}\right)^{-1/m} \int_0^H \left(\frac{2H}{a}\right)^{1+1/m} \left(1+\frac{H^2}{a^2}\right)^{-1-2/m} dH \tag{11.83}$$

The analytical integration of Equation 11.83 is only possible for certain values of m, namely, $m = 1$, $^1/_2$, and $^1/_4$. The results of integration are

$$\begin{aligned}
&\text{For } m = 1: t = \left(\frac{2Ch_o}{p_i a}\right)\left(\frac{(H/a)}{2\left[1+(H/a)^2\right]} + \frac{1}{2}\tan^{-1}\left(\frac{H}{a}\right) - \frac{(H/a)}{\left[1+(H/a)^2\right]^2}\right)\\
&\text{For } m = \tfrac{1}{2}: t = \left(\frac{2Ch_o}{p_i a}\right)^2\left(\frac{1}{\left[1+(H/a)^2\right]^4} - \frac{4}{3\left[1+(H/a)^2\right]^3} + \frac{1}{3}\right)\\
&\text{For } m = \tfrac{1}{4}: t = \left(\frac{2Ch_o}{p_i a}\right)^4\left(\frac{2}{21} + \frac{32}{7\left[1+(H/a)^2\right]^7} - \frac{8}{3\left[1+(H/a)^2\right]^6} - \frac{2}{\left[1+(H/a)^2\right]^8}\right)
\end{aligned} \tag{11.84}$$

Hence, the forming time t^* required to bulge a hemisphere is obtained by putting $H = a$; hence, for

$$\begin{aligned}
m = 1: &\quad t^* = 0.785(Ch_o/p_i a)\\
m = \tfrac{1}{2}: &\quad t^* = 0.917(Ch_o/p_i a)^2\\
m = \tfrac{1}{4}: &\quad t^* = 1.304(Ch_o/p_i a)^4
\end{aligned} \tag{11.85}$$

Bulge forming of strain-rate-dependent superplastic sheet metals is often conducted at constant pressure. This pressure has to be properly selected to induce average strain rate well within the recommended range, where the strain-rate-sensitivity exponent m and hence the formability is maximum (see Table 11.3). This procedure is demonstrated in Example 11.8.

Example 11.8:

A circular sheet of thickness-to-radius ratio of 0.02 is to be bulged at constant pressure to form a hemispherical dome. Estimate the required pressure and time for both alloys:

a. *Superplastic Zn–22 Al:* m = *0.5,* C = *6.68 (MPa · min) at* $(\dot{\bar{\varepsilon}})_{av} = 0.5\ \text{min}^{-1}$ *and 250°C.*
b. *Ti–6Al–4V:* m ≃ *1,* C = *82 MPa · min at* $(\dot{\bar{\varepsilon}})_{av} = 0.05$ *and 950°C.*

Adjust the process parameters such that $(\dot{\bar{\varepsilon}})_{av}$ *given above occurs at the end of forming, i.e., when* H = a.

* In the actual case h_{av}/h_o depends on the material behavior, namely, the strain-rate-sensitivity exponent, m. A more-involved analysis incorporating numerical solution shows this.[10]

Solution:

To determine the appropriate forming pressure, Equations 11.36a and 11.80 for the current bulge radius and average thickness are combined with the equilibrium Equation 11.82 to give

$$(\bar{\sigma})_{av} = \frac{p_i R}{2h_{av}} = \frac{p_i}{2}\left(\frac{a^2+H^2}{2H}\right)\left(\frac{a^2+H^2}{a^2 h_o}\right) = \frac{p_i}{4}\frac{\left(a^2+H^2\right)^2}{a^2 h_o H}$$

The material behavior law (Equation 11.69) gives in terms of average values:

$$\left(\dot{\bar{\varepsilon}}\right)_{av} = \left(\frac{\bar{\sigma}}{C}\right)^{1/m} = \left\{\left(\frac{p_i}{4C}\right)\left(\frac{a}{h_o}\right)\left(\frac{a}{H}\right)\left(1+\frac{H^2}{a^2}\right)^2\right\}^{1/m}$$

For a hemispherical bulge, $H = a$; hence,

$$\left(\dot{\bar{\varepsilon}}_{av}\right)_{H=a} = \left(\frac{pa}{Ch_o}\right)^{1/m}$$

a. For superplastic Zn–22 Al alloy, $m = 0.5$ at $\left(\dot{\bar{\varepsilon}}_{av}\right)_{H=a} = 0.5\ \text{min}^{-1}$ and $(a/h_o) = 50$; hence, $p_i/C = 0.0141$, i.e., $p_i = 0.094$ MPa $\cong$ 0.94 bar.

b. Similarly, for Ti–6Al–4V alloy, $m = 1$ at $\left(\dot{\bar{\varepsilon}}_{av}\right)_{H=a} = 0.05\ \text{min}^{-1}$; hence, $p_i/C = 0.001$, i.e., $p_i = 0.082$ MPa $\cong$ 0.82 bar.

The forming time is estimated from Expressions 11.85 as

$$\text{for } m = 0.5: \quad t^* = 0.917\left(Ch_o/p_i a\right)^2 = 1.85 \text{ min}$$
$$\text{for } m = 1: \quad t^* = 0.785\left(Ch/pa\right) = 15.7 \text{ min}$$

11.4 CREEP DEFORMATION

Creep takes place in engineering materials and structures manifested by the accumulation of plastic deformation over prolonged time periods under steady or variable loading conditions. Creep deformations are mostly observable at elevated temperatures, which are generally related to the absolute melting temperature T_m of the metallic alloy ranging from $0.3T_m$ to $0.5T_m$ K. For lead metal, creep deformation is observed at room temperature, although the melting temperature is 327°C. However, for low-carbon steel a temperature of the order of 350°C must be reached to justify considering creep analysis for long-term design requirements.

11.4.1 Creep Testing and Data

Creep testing of engineering materials is mostly done under constant load and temperature in uniaxial testing machines. Typical creep curves of uniaxial creep strain ε^c vs. time t as shown in Figure 11.12 are obtained. Upon loading, an instantaneous strain ε_i either elastic or elastoplastic is observed. This is followed by a primary stage characterized by a decreasing rate of deformation. A secondary creep stage is further distinguished where a constant (steady) creep strain rate $\dot{\varepsilon}^c$ is observed. This secondary stage ends by accelerating creep deformation characterizing a tertiary creep stage leading to creep rupture.* Creep strains are permanent** and generally they reach few percent at rupture.

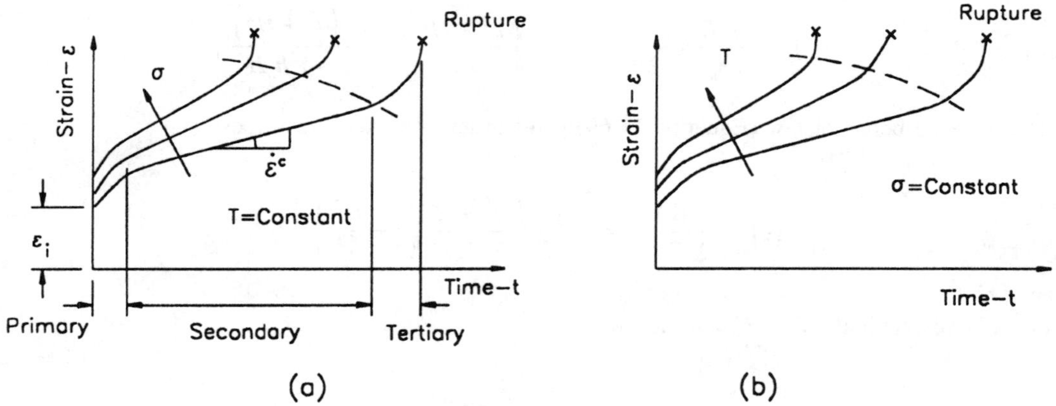

FIGURE 11.12 Typical curves showing three creep stages at (a) different stress levels and (b) different temperatures.

Full material creep data consist basically of a set of curves, Figure 11.12a representing creep strain vs. time at different nominal stresses. Another set of creep curves for the same material tested at different temperatures, as shown in Figure 11.12b, completes the required creep data. Such elaborate data produced over prolonged periods of time extending to a few years (10,000 h ≃ 1.14 years) is costly. Creep testing involves rates usually within 10^{-10} to 10^{-9}/s (as compared with conventional tensile testing rate of 10^{-2} to 10^{-3}/s). Hence, many users of creep data feel that creep data should be obtained by tests of at least 10,000 h or longer times. Instead, creep data collected by testing over shorter periods of time and limited variations of nominal stresses and temperatures are extended to other conditions by applying extrapolation techniques within acceptable limits of accuracy.

In some problems (e.g., creep buckling) or for some classes of materials, e.g., plastics, another presentation of creep data in the form of isochronous stress–strain curves becomes more appropriate. Using Figure 11.13a at a preselected time t, a set of stress–strain points are found and plotted as shown in Figure 11.13b. Proceeding similarily for other values of time, a complete set of σ–ε curves known as isochronous curves, i.e., constant time, are plotted. The curve corresponding to $t = 0$ is the instantaneous uniaxial tensile stress–strain curve at the prescribed temperature.

The above presentation of creep data has been limited to creep deformation within the primary and secondary creep stages. The unstable tertiary creep stage is often excluded. In other words, creep failure is thus related to limitations of dimensional changes in the structure rather than rupture. This sets up creep design criteria based on reaching a permissible creep strain value (say, 0.2%) for an intended design life at a given stress.*** Typical permissible creep deformations used in design are 2% in 20,000 h for heat-exchanger tubes, 0.3% in 100,000 h for boiler and steam tubes, and 0.01% for steam turbine rotors in 100,000 h.

* On the microscopic scale, two types of creep deformations may be described: dislocation creep and diffusional creep. The latter, which is of linear viscous nature, occurs at lower stresses due to bulk movement of atoms from compression site to tensile one. The more important dislocation creep is nonlinear and happens at higher stresses. The decreasing rate in the primary creep stage reflects material hardening due to dislocation pileup, while steady creep is due to a balance between this hardening effect and thermal softening. In tertiary creep, material suffers from inertnal damage leading to rupture.

** This excludes the very small elastic and viscoelastic strain components which are recoverable upon unloading.

Creep testing involving relatively short time data, say, from a few hours to several hundreds of hours, are presented in the form of creep rupture tests. These tests result in plots of stress against rupture time t_R at a given temperature, as shown in Figure 11.14. It is seen that, as stress goes down, the time to rupture increases, following an almost linear relationship on double logarithmic scales for σ and t_R. The points to the extreme left of the plot represent the instantaneous fracture of the specimen at stress levels close to its ultimate tensile strength, at the test temperature. Furthermore, Figure 11.14 shows that as temperature increases, the time to rupture decreases. Examination of creep ruptures indicates that a ductile fracture mode prevails at shorter rupture times while a brittle mode is observed at longer times, a judgment based on the reduction of area at fracture.

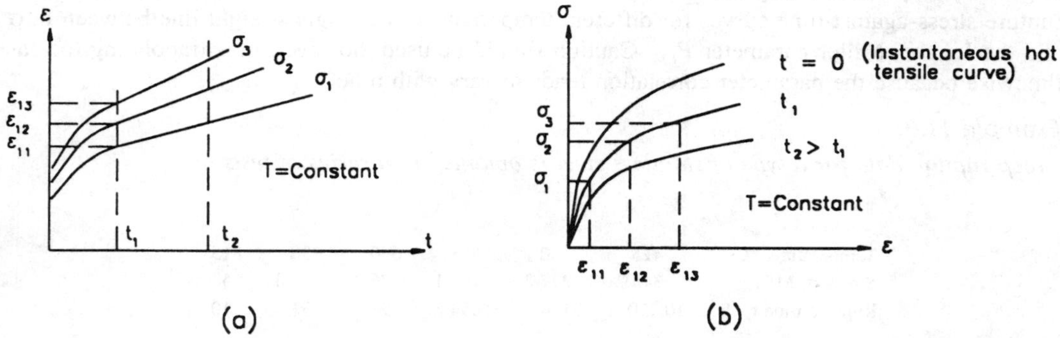

FIGURE 11.13 Construction of isochronous creep curves: (a) test data and (b) stress–creep strain curves at preselected times.

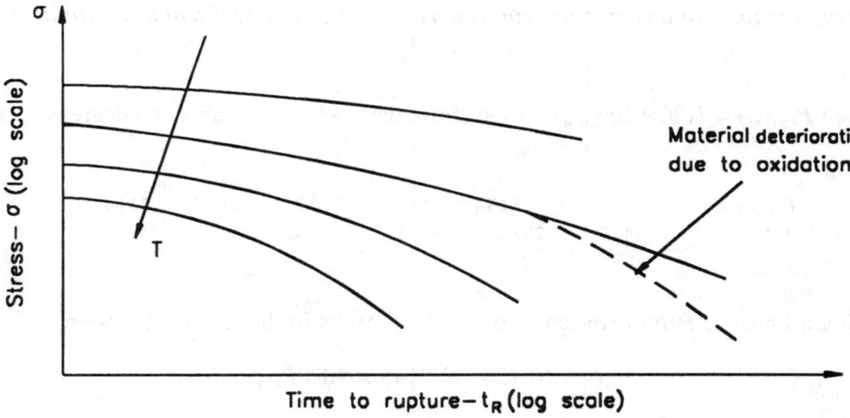

FIGURE 11.14 Typical creep rupture curves at different temperatures.

*** The ASME Boiler and Pressure Vessel code recommends limiting the applied stresses to avoid creep failure by satisfying one of the following conditions (a) less than the minimum stress to rupture in a selected time divided by a factor of 1.5, (b) less than the minimum stress to cause 1% creep in a selected time t, and (c) less than 80% of the minimum stress to reach tertiary creep in time t. See ASME Code, case N 47-15 (1592-15) class 1 Components in Elevated Temperature Service (1979).

As mentioned before, it is not always practical to run creep tests that involve time periods of the same order of length as the life of the structure to be designed. Extrapolation techniques are available to extend short time creep rupture data to longer times. One of these techniques based on the temperature-compensated time idea is due to Larson and Miller.[11] The extrapolation parameter P_{LM}, which is stress dependent is defined as

$$P_{LM}(\sigma) = T(C + \log_{10} t_R)$$

where T is the absolute temperature and t_R is the rupture time. The empirical constant C is taken to be 20 for most materials.* Relatively short time tests are usually conducted at higher temperatures to give a parameter covering the time of interest at lower temperatures. This reduces the family of rupture stress-against-time curves for different temperatures to a single straight line between ln σ and the Larson–Miller parameter P_{LM}. Caution should be used, however, in extrapolating too far timewise because the parameter correlation tends to vary with time.

Example 11.9:

Creep rupture data for a type of stainless steel is obtained in a series of tests as

Temperature, °C	425	480	565	650	730	815
Stress σ, MPa	344.2	277.2	168.1	126.9	84.8	58.4
Rupture time t_R, h	10,120	3154	1054	98	31	10

Using the Larson–Miller parameter P_{LM}, *determine,*

a. *The maximum stress that may be applied to a component made of the same steel operating at 375°C for 15,000 h before rupture*
b. *The rupture time for a component subjected to a stress at 190 MPa at temperature 525°C.*

Solution:

Calculation of $P_{LM}(\sigma) = T(20 + \log_{10} t_R)$ for the above data, where T in absolute degrees K results in

$P_{LM}(\sigma) \times 10^{-3}$	16.755	17.694	19.293	20.298	21.556	22.848
σ, MPa	344.2	277.2	168.1	126.9	84.8	58.4

Assuming a linear relation between ln σ and P_{LM} results in the fitted expression:

$$\ln \sigma = 10.729 - 2.9145 \times 10^{-4} P_{LM}$$

a. At the service conditions $T = 375°\text{C}$ for 15,000 h,

$$P_{LM} = (375 + 273)(20 + \log_{10} 15{,}000) = 15.667 \times 10^3$$

Hence, substitution into the (ln σ – P_{LM}) relation gives

$$\sigma = 474.9 \text{ MPa}$$

* Manson and Haferd[12] questioned having a single constant C for all materials and proposed another extrapolation parameter.

b. Similarily, for the service conditions $\sigma = 190$ MPa at $T = 525°C$ from the ($\ln\sigma - P_{LM}$) relation,

$$\ln 190 = 10.729 - 2.9145 \times 10^{-4} P_{LM}$$

hence, $P_{LM} = 18.809 \times 10^3$.

This value for the Larson-Miller parameter corresponds to a rupture time given by $18.809 \times 10^3 = (525 + 273)(20 + \log_{10} t_R)$, i.e., $t = 3720$ h.

11.4.2 Empirical Creep Equations of State

11.4.2.1 Uniaxial Behavior

Uniaxial creep curves shown in Figure 11.12 suggest that an empirical equation of state for creep deformation may be in the form:

$$\varepsilon^c = f_1(\sigma, T, t)$$

The dependence on temperature T may be separated and taken as an Arrhenius type, thus giving

$$\varepsilon^c = f_2(\sigma, t)\exp(-\Delta H/RT)$$

where ΔH is the activation energy, R is the universal gas constant,* and T is the absolute temperature. For constant-temperature creep behavior, ε^c is thus given by

$$\varepsilon^c = f_2(\sigma, t)$$

Several mathematical forms exist to represent the function $f_2(\sigma, t)$; one of these is the Norton–Baily creep law:

$$\varepsilon^c = B\sigma^N t^{N_1} \quad (11.86)$$

where B, N, and N_1 are constants for the specific temperature conditions. Obviously, the equation of state (Equation 11.86) describes primary and secondary creep stages only. Tertiary creep and creep rupture are customarily modeled separately. For design purposes secondary creep dominates most of the component life, and primary creep is replaced by a constant strain ε_o, which is the intercept of the extrapolated secondary stage back to the strain axis, as indicated in Figure 11.15a.

For secondary creep, ε^c is almost linearly dependent on time and hence the exponent N_1 in Expression 11.86 is taken as unity. In terms of the secondary creep rate $\dot{\varepsilon}^c = d\varepsilon^c/dt$, this reduces to**

$$\dot{\varepsilon}^c = B(\sigma)^N \quad (11.87)$$

Therefore, the total strain is given by

---

* $R = 1.987$ cal/(mol)(K). Note that $R = kN$ where k is Boltzmann's constant and N is Avogadro's number $= 6.02 \times 10^{23}$ atoms/mol.

** Note that in differentiating Expression 11.86, the time derivative of stress has been ignored. Theoretically, this limits its application to constant stress or step changes of stress for long durations. Moreover, Equation 11.87 should be modified in cases of creep under stress reversals. For instance, see Krauss,[13] p. 18.

$$\varepsilon = \varepsilon_o + \left(\frac{d\varepsilon^c}{dt}\right)t = \varepsilon_o + B(\sigma)^N t$$

The material parameters B and N at constant temperature can be determined by replotting σ– $\dot{\varepsilon}^c$ data extracted from the uniaxial creep data on double logarithmic scales,* as shown in Figure 11.15b.

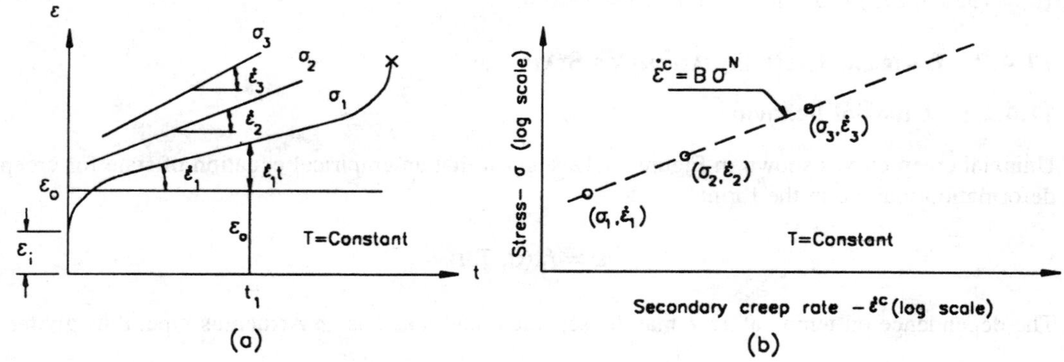

FIGURE 11.15 Plot of the Norton–Baily creep law for secondary creep on double-logarithmic scales.

Note that to assume the compatibility of units in the creep law $\dot{\varepsilon}^c = B(\sigma)^N$, it is often rewritten as: $(\dot{\varepsilon}_c / \dot{\varepsilon}_o) = (\sigma/\sigma_o)^N$, where $\dot{\varepsilon}_o$ is taken arbitrarily as unity. Hence, σ_o being a material parameter is defined as the stress required to cause a unit creep rate.

This simple creep equation of state in spite of its shortcomings has been widely used to solve secondary creep problems. The constants B and N are material parameters for a given temperature and must be determined over the appropriate range of time. Some typical values of B and N for a few metallic alloys are given, for the sake of comparison, in Table 11.6.

TABLE 11.6
Typical Creep Parameters for Some Alloys in the Law $\dot{\varepsilon}^c = B(\sigma)^N$ (in units of MPa – h)

Alloy	Temp. °C	N	B
Rolled carbon steel (annealed, 0.12C, 0.4Mn, 0.25Si)	500	3.3	8.08×10^{-13}
Low-alloy rolled steel (1Cr, 0.75Mn, 0.35Mo,	450	6	1.672×10^{-20}
0.25Si, 0.15C)	500	5.4	3.853×10^{-18}
	550	4.15	2.265×10^{-14}
	600	2.74	6.024×10^{-11}
	650	2.1	3.4197×10^{-9}
Nimonic 75 forged (20.5Cr, remainder Ni)	650	2.73	9.2312×10^{-13}
Aluminum alloy 245-T4 (4.4Cu, 1.3Mg, 0.6Mn, 0.34Fe, 0.18Si)	190	5.3	1.559×10^{-17}
Copper (99.95 Cu)	190	5	1.96×10^{-14}
Lead (commercial)	30	2.7	1.0054×10^{-7}

Adapted from Odqvist, F. G., *Mathematical Theory of Creep and Creep Rupture*, Oxford Clarendon Press, London, 1966.

* The representation of Figure 11.15b follows Odqvist[14] by taking the ordinate as log σ; hence, the slope of the curve would be equal to $1/N$.

Accurate determination of material creep properties is always sought. This is not an easy task and experiments usually suffer from scatter and may not even conform to Norton's creep law (Equation 11.87). Small errors in determining the material parameters B and N produce large errors in estimating creep rates. Take, for instance, the creep law for a low-alloy rolled steel at 450°C as given in Table 11.6, namely,

$$\dot{\varepsilon}^c = 1.672 \times 10^{-20} \sigma^6 \text{ (in units of MPa - hr)}$$

Small deviation within ±3.3% in determining the exponent $N = 6$ results in erroreous determination of $\dot{\varepsilon}^c$ by about –58% and +140% corresponding to $N = 5.8$ and $N = 6.2$, respectively, at a stress level of 80 MPa.

11.4.2.2 Multiaxial Behavior

To apply the uniaxial creep equation of state to multiaxial creep problems, the same assumptions and hence expressions invoked in plasticity are employed. For emphasis, these are constancy of volume for creep deformations as observed experimentally, independence of flow of hydrostatic stress component, and coincidence of principal axes of stress and strain for isotropic material. More appropriately, increments of strain $d\varepsilon$ are replaced by creep strain rates $\dot{\varepsilon}^c$ and hence definitions of von Mises effective stress $\overline{\sigma}$ and effective strain rate $\dot{\overline{\varepsilon}}$ as well as the Levy–Mises flow rule apply according to Expressions 10.26, 11.71, and 11.72, respectively.

Specifically using Norton's creep law, the three principal creep strain-rates may be written as

$$\dot{\varepsilon}_1^c = B(\overline{\sigma})^{N-1}\left[\sigma_1 - \tfrac{1}{2}(\sigma_2 + \sigma_3)\right] \cdots \tag{11.88}$$

The empirical creep law (Equation 11.87) is thus written as

$$\dot{\overline{\varepsilon}} = B(\overline{\sigma})^N \tag{11.89}$$

Note that, for simplicity, the superscript c denoting creep has been deleted from the effective creep strain rate $\dot{\overline{\varepsilon}}^c$.

11.4.3 Steady Creep of Beams under Bending

The analysis of beams loaded in pure bending and undergoing creep deflections are based on the same geometric assumption used in small-deflection elastic analysis, i.e., plane sections of the beam remain plane during deformation. As an illustration for this type of analysis, consider a beam of bisymmetrical cross section, of area A, loaded by a constant bending moment as shown in Figure 11.16a. Then only a longitudinal stress σ and longitudinal creep strain ε^c are considered. Compatibility of the linear infinitesimal strain with curvature of radius R, gives as explained in Section 7.3.2,

$$\varepsilon^c = y/R \tag{11.90}$$

where $\dot{R}$ is the rate of change of radius of curvature.

Equilibrium over the entire cross section with the applied moment dictates:

$$M = \int_A \sigma y dA \tag{11.91}$$

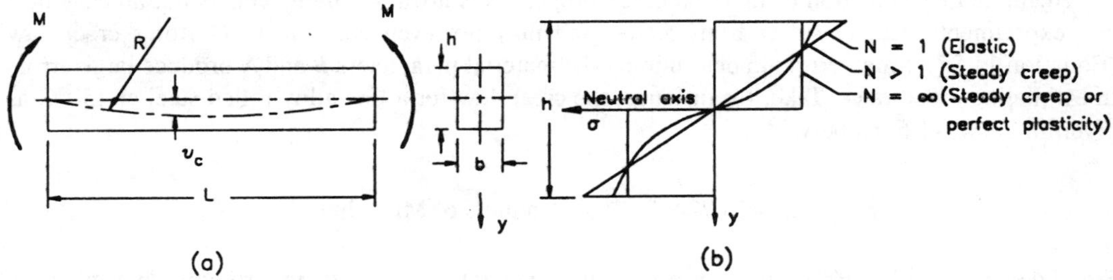

FIGURE 11.16 (a) Pure bending of a beam of rectangular cross section and (b) bending stress distribution through the beam depth for different behavior.

For steady creep, Relation 11.87 is inverted to give stress in terms of creep strain rate,

$$\sigma = \left(\dot{\varepsilon}^c / B\right)^{1/N} \tag{11.92}$$

which is then substituted in the equilibrium equation (Equation 11.91) to obtain

$$M = \int_A \left(\dot{\varepsilon}^c / B\right)^{1/N} y dA \tag{11.93}$$

Hence, the strain–curvature relation (Equation 11.90) for a rectangular section of width b and depth h results in

$$M = 2\int_0^{h/2} \left(\frac{y}{RBt}\right)^{1/N} bydy = \frac{2b}{(RBt)^{1/N}} \int_0^{h/2} y^{(1+1/N)} dy \tag{11.94}$$

Integration gives

$$M = \frac{2b}{(RBt)^{1/N}} \frac{N}{2N+1} \left(\frac{h}{2}\right)^{(2+1/N)} \tag{11.95}$$

Substituting for R, B, and t using Relations 11.90 and 11.92 yields

$$M = \left(\frac{2b\sigma}{y^{1/N}}\right)\left(\frac{N}{2N+1}\right)\left(\frac{h}{2}\right)^{(2+1/N)} \tag{11.96}$$

By inserting the second moment of area $I = bh^3/12$ and rearranging, the bending stress at distance y from the neutral axis is obtained as

$$\sigma = M\frac{y}{I}\left(\frac{2N+1}{3N}\right)\left(\frac{2y}{h}\right)^{(1/N-1)} \tag{11.97}$$

Expression 11.97 may be easily compared with the simple linear-elastic distribution of stress, My/I. Figure 11.16b shows the stress distribution across the cross section for $N = 1$ (i.e., linear elastic) and $N > 1$. The initial elastic stress distribution, which prevails at time $t = 0$, redistributes more uniformly as time goes on during the stage of primary creep. Once steady creep conditions are attained, the stress becomes nonlinearly distributed over the beam cross section. Note that for $N = \infty$, the stress distribution belongs to a beam of a perfectly plastic material.*

According to the previous assumption of linearly distributed infinitesimal strain, the creep deflection v_c at any cross section at distance z from the origin is given by Equation 7.10a as

$$\frac{d^2 v_c}{dz^2} = -\frac{1}{R} = \frac{\varepsilon^c}{y} \tag{11.98}$$

The time rate of the above relation is

$$\frac{d^2 \dot{v}_c}{dz^2} = -\frac{\dot{\varepsilon}^c}{y} \tag{11.99}$$

Hence, from the creep law (Equation 11.87),

$$\frac{d^2 \dot{v}_c}{dz^2} = -\frac{B\sigma^N}{y} = -\frac{B}{y}\left[\frac{My}{I}\left(\frac{2N+1}{3N}\right)\left(\frac{2y}{h}\right)^{(1/N-} \tag{11.100}$$

which is simplified to give

$$\frac{d^2 \dot{v}_c}{dz^2} = -B\left(\frac{2N+1}{N}\frac{2}{bh^2}\right)^N \left(\frac{2}{h}\right) M^N \tag{11.101}$$

Usually the bending moment M for laterally loaded beams is a known function of z; hence, the above equation can be integrated twice to determine $v_c(z)$. Note that the effect of transverse shear is assumed to be negligible.

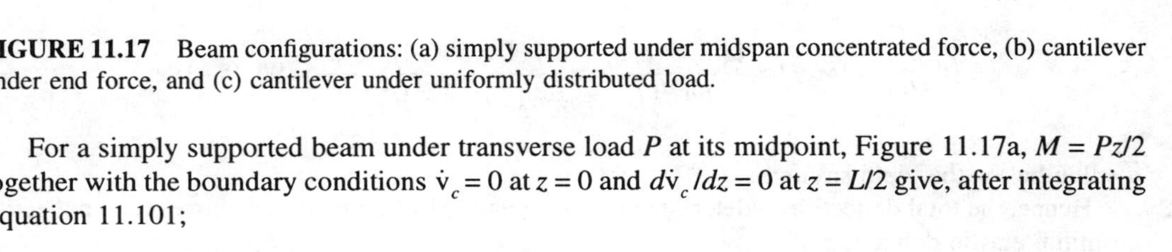

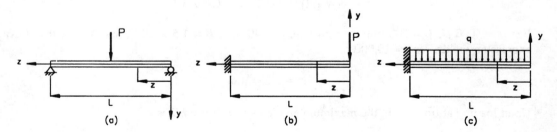

FIGURE 11.17 Beam configurations: (a) simply supported under midspan concentrated force, (b) cantilever under end force, and (c) cantilever under uniformly distributed load.

For a simply supported beam under transverse load P at its midpoint, Figure 11.17a, $M = Pz/2$ together with the boundary conditions $\dot{v}_c = 0$ at $z = 0$ and $d\dot{v}_c/dz = 0$ at $z = L/2$ give, after integrating Equation 11.101;

$$\dot{v}_c = B\left[\frac{2N+1}{N}\frac{2}{bh^2}\right]^N \left(\frac{2}{h}\right)\left(\frac{P}{2}\right)^N \left[-\left(\frac{L}{2}\right)^{N+1}\frac{z}{N+1} + \frac{z^{N+2}}{(N+1)(N+2)}\right] \tag{11.102}$$

* See Chapter 12.

For a cantilever beam under end point load P, Figure 11.17b, the boundary conditions $\dot{v}_c = d\dot{v}_c/dz = 0$ at $z = L$; hence, integration of Equation 11.101 gives

$$\dot{v}_c = B\left[\frac{2N+1}{N}\frac{2}{bh^2}\right]^N\left(\frac{2}{h}\right)\left(\frac{P^N}{N+1}\right)\left[zL^{N+1} - \frac{z^{N+2}}{(N+2)} - \left(\frac{N+1}{N+2}\right)L^{N+2}\right] \quad (11.103a)$$

Finally, for a cantilever beam under uniformly distributed load, Figure 11.17c of intensity q unit length, the rate of creep deflection $\dot{v}_c$ at any point z is given by

$$\dot{v}_c = B\left[\frac{2N+1}{N}\frac{2}{bh^2}\right]^N\left(\frac{2}{h}\right)\left(\frac{q}{2}\right)^N\left[\frac{zL^{2N+1}}{(2N+1)} - \frac{z^{2N+2}}{(2N+1)(2N+2)} - \frac{L^{2N+2}}{(2N+2)}\right] \quad (11.103b)$$

Recall that Expressions 11.102 and 11.103 reduce to their respective elasic curves by setting $N = 1$, $B = E$, and $v_c \equiv v$.

Example 11.10:

A 500-mm-long simply supported beam of 20 mm width and 40 mm height is acted upon at the middle of its span by a concentrated force P = *4000* N *and operates at* T = *400°C. The beam is made of carbon steel having yield strength of 155 MPa and modulus of elasticity* E = *160 GPa (at* T = *400°C). The secondary creep equation of state:* $\dot{\varepsilon}^c = B(\sigma)^N$ *at* T = *400°C, has the parameters* B = *1.5* × *10*$^{-12}$ *(MPa · h) and* N = *3. Neglecting the zone of primary creep, determine the maximum total deflection of the beam after 10,000 h of loading.*

Solution:

The maximum creep deflection v_c at time t occurs at midspan $z = L/2$. Substitution into Expression 11.102 and integrating gives

$$v_c = -\frac{BP^N L^{N+2}}{(N+2)2^{2N+2}}\left(\frac{2N+1}{N}\frac{2}{bh^2}\right)^N\left(\frac{2}{h}\right)t$$

For $P = 4000$ N, $L = 500$ mm, $b = 20$ mm, $h = 40$ mm, $B = 1.5 \times 10^{-12}$, and $N = 3$, the above expression yields, for $t = 10{,}000$ h,

$$v_c = 3.635 \text{ mm}$$

Upon loading at time $t = 0$, the maximum applied stress at $y = h/2$ is

$$(\sigma_z)_{max} = My/I = \frac{(PL/4)(h/2)}{bh^3/12} = \frac{(4000\times 500/4)(40/2)}{(20\times 40^3)/12} = 93.75 \text{ MPa}$$

a value below the yield strength.

Hence, the total deflection is determined approximately by adding to the above creep deflection an initial elastic deflection of,

$$v_e = -\frac{PL^3}{48EI} = \frac{4000\times 500^3}{48\times 160\times 10^3(20\times 40^3/12)} = 0.61 \text{ mm}$$

Hence, $v_{total} = 3.635 + 0.61 = 4.295$ mm.

11.4.4 Steady Creep of Thin-Walled Pressurized Cylinders

A closed, thin-walled pressure vessel subjected to an internal pressure p_i and operating at an elevated temperature, which is sustained for a long period of time, undergoes creep deformations. Assuming that strains are small, the stresses remain constant throughout creep life, as given by equilibrium conditions based on initial dimensions:

$$\sigma_\theta = \frac{p_i r_{mo}}{h_o}, \quad \sigma_z = \frac{p_i r_{mo}}{2h_o}, \quad \text{and } \sigma_r \simeq 0$$

where r_{mo} and h_o are the initial mean radius and wall thickness of the cylinder, respectively.

The effective stress defined by Expression 10.26 is thus

$$\bar{\sigma} = \frac{\sqrt{3}}{2}\frac{p_i r_{mo}}{h_o} = \frac{\sqrt{3}}{2}\sigma_\theta \tag{11.104}$$

The circumferential, axial, and radial creep strain rates, namely, $\dot{\varepsilon}_\theta^c$, $\dot{\varepsilon}_z^c$, and $\dot{\varepsilon}_r^c$ satisfy the constant-volume condition (Equation 10.1), i.e. $\dot{\varepsilon}_\theta^c + \dot{\varepsilon}_z^c + \dot{\varepsilon}_r^c = 0$.

From the flow rule (Equation 11.88), since $\sigma_z = \sigma_\theta/2$ it follows that $\dot{\varepsilon}_z^c = 0$, i.e., no axial creep strain, an observation which has been verified experimentally. Hence, $\dot{\varepsilon}_\theta^c = -\dot{\varepsilon}_r^c$ and the effective creep strain rate is determined from (Equation 11.71) as:

$$\dot{\bar{\varepsilon}} = \frac{2}{\sqrt{3}}\dot{\varepsilon}_\theta^c \tag{11.105}$$

Applying the flow rule, Equation 11.88, knowing again that $\sigma_z = \sigma_\theta/2$, gives

$$\dot{\varepsilon}_\theta^c = B(\bar{\sigma})^{N+1}\left(\frac{3}{4}\sigma_\theta\right) \tag{11.106}$$

and yields, after substitution from Equations 11.104,

$$\dot{\varepsilon}_\theta^c = B\left(\frac{\sqrt{3}}{2}\right)^{N+1}\left(\frac{p_i r_{mo}}{h_o}\right)^N \tag{11.107}$$

This is simply integrated to give for time t:

$$\varepsilon_\theta^c = B\left(\frac{\sqrt{3}}{2}\right)^{N+1}\left(\frac{p_i r_{mo}}{h_o}\right)^N t \tag{11.108}$$

Example 11.11:
For an alloy steel the following data are obtained from creep tests at 550°C:

Stress σ, MPa	*70*	*100*	*135*	*175*	*200*
Steady creep rate per hour, $\dot{\varepsilon}$	*9.83×10^{-8}*	*2.5×10^{-6}*	*3.8×10^{-5}*	*4×10^{-4}*	*1.34×10^{-3}*

a. *Represent this data on (ln σ – ln $\dot{\varepsilon}^c$) plot and determine the constants in the creep law $\dot{\varepsilon}^c = B(\sigma)^N$.*
b. *Boiler tubes of 60 mm mean diameter and 1.5 mm thickness made of the same alloy as in (a) operating at 550°C are designed such that the diameter expansion does not exceed 0.5% in 100,000 h. Estimate the maximum allowable internal pressure* p_c *which could be applied.*
c. *Compare the pressure determined in (b) to that required for yielding* p_Y *knowing that the yield strength at 550°C is* Y = *350 MPa.*
d. *What would be the creep life for these tubes if an error of ±2% is made in determining the material parameter* N.

Solution:

a. A plot on a double logarithmic scales ln σ vs. ln $\dot{\varepsilon}^c$ as shown in Figure 11.18 is best fitted by a straight line, giving

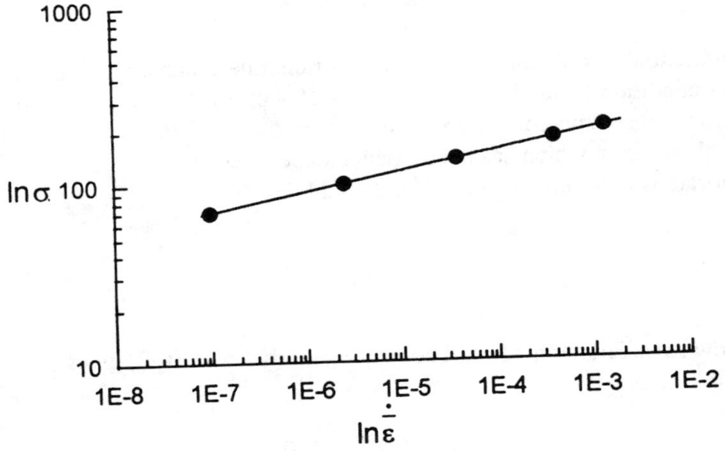

FIGURE 11.18 Example 11.11.

$$B = 1.81 \times 10^{-24} \text{ and } N = 9.07 \text{ (in MPa-h)}$$

b. Expression 11.108 relating the hoop creep strain to the creep life t = 100,000 h, and knowing that $\dot{\varepsilon}^c_\theta = 0.005$ for r_{mo} = 30 mm and h_o = 1.5 mm, gives

$$0.005 = 1.81 \times 10^{-24} \left(\frac{\sqrt{3}}{2}\right)^{9.07+1} \left(\frac{p_c \times 30}{1.5}\right)^{9.07} 100,000$$

This results in a pressure:

$$p_c = 3.81 \text{ MPa}$$

c. At yielding according to the von Mises yield condition, Equation 11.104

gives $\sigma_\theta = \dfrac{p_Y r_{mo}}{h_o} \leq Y$

$$\bar{\sigma} = \frac{\sqrt{3}}{2}\sigma_\theta = \frac{\sqrt{3}}{2}\frac{p_Y r_{mo}}{h_o} \leq Y$$

giving p_Y = 20.2 MPa, a value which is about 5.3 times p_c.

d. Considering an error of ±2% in N gives lower and upper values as N = 8.889 and N = 9.251, respectively. Substituting into Expression 11.108 gives for p_c = 3.81 MPa the following creep life times
For N = 8.889 t = 213970 h
For N = 9.251 t = 46878 h
These creep lives are about 214% and 47%, respectively, of the rated creep life of 100,000 h. This indicates the surprisingly high senstivity in using a highly nonlinear creep law as Equation 11.87 in estimating creep life.

Example 11.12:

A spherical pressure vessel of 700 mm mean diameter and 8 mm thickness is made of stainless steel of the following properties at the operating temperature 450°C:

- *Yield strength = 390 MPa, Young's modulus = 175 GPa, ν = 0.3,*
- *Strain-hardening exponent 0.12, strength coefficient*

$$K = 770\ MPa\ \left(\textit{in the law}\ \bar{\sigma} = K(\bar{\varepsilon})^n\right),$$

- *Creep behavior for the vessel material* N = *6.8 and*

$$B = 1.47 \times 10^{-22}(\textit{in MPa - hr})\ \left(\textit{in the law},\ \dot{\bar{\varepsilon}} = B(\sigma)^N\right)$$

The vessel is designed to operate at service pressure equal to 30% of its burst pressure. Determine:

a. The service pressure,
b. The life for a creep diametral deformation of 0.4%.

Solution:

a. The burst pressure for a spherical vessel is given by Equation 11.56a as

$$p_{\max} = K\frac{2h_o}{r_{mo}}\left(\frac{2n}{3}\right)^n \exp(-n)$$

Hence, for h_o = 8 mm, r_{mo} = 350 mm, K = 770 MPa, and n = 0.12

$$p_{\max} = 770\frac{2\times 8}{350}\left(\frac{2\times 0.12}{3}\right)^{0.12} \exp(-0.12) = 23.06\ \text{MPa}$$

The service pressure p_i is 30% of $p_{\max}$, i.e.,

$$p_i = 6.92\ \text{MPa}$$

b. The stresses in the vessel wall are

$$\sigma_1 = \sigma_2 = p_i r_{mo}/2h_o = 151.33 \text{ MPa and } \sigma_3 \simeq 0$$

which are within the elastic range since yielding occurs at 390 MPa. The initial elastic strain is obtained from Hooke's law, Equation 3.6a, as

$$\varepsilon_\theta^e = \frac{\Delta r_{mo}}{r_{mo}} = \frac{1}{E}\left[\sigma_1 - \nu(\sigma_2 + \sigma_3)\right]$$

$$= \frac{0.7\sigma_1}{E} = \frac{0.7 \times 151.33}{175 \times 10^3} = 8.65 \times 10^{-4}$$

Neglecting primary creep, the allowable secondary creep strain is

$$\varepsilon_\theta^c = 4 \times 10^{-3} - 8.65 \times 10^{-4} = 0.003135$$

To apply the creep law (Equation 11.89) the effective stress $\bar{\sigma}$ and effective strain $\bar{\varepsilon}$ for a pressurized spherical vessel are given by Expressions 11.49 and 11.52, respectively, as

$$\bar{\sigma} = \sigma_1 = p_i r_{mo}/2h_o \quad \text{and} \quad \bar{\varepsilon} = 2\varepsilon_\theta^c = 2\Delta r/r_{mo}$$

Hence, from the integrated form of the creep law,

$$2\left(\frac{\Delta r}{r_{mo}}\right) = B\left(\frac{p_i r_{mo}}{2h_o}\right)^N t$$

where t is the service life. Upon substitution,

$$2(0.003135) = 2.47 \times 10^{-22}\left(\frac{6.92 \times 350}{2 \times 8}\right)^{6.8} t$$

which gives t = 38037 h, i.e., 4.34 years.

11.4.5 Stationary Creep of Thick-Walled Pressurized Cylinders

To analyze creep deformation in a thick-walled cylinder under internal pressure, the equations of equilibrium and compatibility derived in Chapters 1 and 2 are employed.

At time $t = 0$, i.e., upon application of an internal pressure p_i, it is assumed that the elastic stress distribution given by Equations 6.2 prevails. As time progresses, creep deformation takes place until a steady state characterized by constant creep rates is attained. Within this state, it is reasonable to neglect elastic strains with respect to creep strains ε_r^c, ε_θ^c, and ε_z^c. Hence, applying the incompressibility condition (Equation 10.1) on creep rates for plane strain ($\varepsilon_z^c = 0$):

$$\dot{\varepsilon}_r^c = -\dot{\varepsilon}_\theta^c$$

The strain compatibility (Equation 2.13c) in terms of creep rates gives

$$r\frac{d\dot{\varepsilon}_\theta^c}{dr} - \dot{\varepsilon}_r^c + \dot{\varepsilon}_\theta^c = 0 \tag{11.109}$$

or

$$\frac{d\dot{\varepsilon}_r^c}{dr} = -2\frac{\dot{\varepsilon}_r^c}{r} \tag{11.110}$$

Integration yields

$$\dot{\varepsilon}_r^c = -\dot{\varepsilon}_\theta^c = \frac{C_1}{r^2} \tag{11.111}$$

where C_1 is an integration constant. The effective creep strain rate is thus determined from Expression 11.71 as

$$\dot{\bar{\varepsilon}} = \frac{2}{\sqrt{3}}\frac{C_1}{r^2}$$

The effective stress in terms of the stress components σ_r and σ_θ, noting that $\sigma_z = (\sigma_r + \sigma_\theta)/2$ for plane strain conditions, is determined from Expression 10.26 as

$$\bar{\sigma} = \frac{\sqrt{3}}{2}(\sigma_\theta - \sigma_r) \tag{11.112}$$

For steady creep conditions, the Norton–Baily creep law (Equation 11.89) applies:

$$\frac{2}{\sqrt{3}}\frac{C_1}{r_2} = B\left[\frac{\sqrt{3}}{2}(\sigma_\theta - \sigma_r)\right]^N \tag{11.113}$$

or, alternatively,

$$(\sigma_\theta - \sigma_r) = \frac{2}{\sqrt{3}}\left(\frac{1}{B}\right)\left(\frac{2}{\sqrt{3}}\frac{C_1}{r^2}\right)^{1/N} \tag{11.114}$$

It thus remains to satsify the equilibrium Equation 1.10c, i.e.,

$$r\frac{d\sigma_r}{dr} = (\sigma_\theta - \sigma_r)$$

Substituting from Equation 11.114 into this equation gives

$$\frac{d\sigma_r}{dr} = \frac{1}{B}\left(\frac{2}{\sqrt{3}}\right)^{1+1/N}(C_1)^{1/N} r^{-2/N-1} \tag{11.115}$$

which is integrated as

$$\sigma_r = \frac{1}{B}\left(\frac{2}{\sqrt{3}}\right)^{1+1/N} \frac{C_1^{1/N}}{(-2/N)} r^{-2/N-1} + C_2$$

The two constants C_1 and C_2 are determined from the boundary conditions $\sigma_r = -p_i$ at $r = r_i$ and $\sigma_r = 0$ at $r = r_o$. This finally results in the stress distribution for $\lambda = r_o/r_i$, as given by

$$\sigma_r = \frac{p_i}{\lambda^{2/N} - 1}\left[1 - \left(\frac{r_o}{r}\right)^{2/N}\right]$$

$$\sigma_\theta = \frac{p_i}{\lambda^{2/N} - 1}\left[1 + \frac{2-N}{N}\left(\frac{r_o}{r}\right)^{2/N}\right] \quad (11.116)$$

$$\sigma_z = \frac{p_i}{\lambda^{2/N} - 1}\left[1 + \frac{1-N}{N}\left(\frac{r_o}{r}\right)^{2/N}\right]$$

The above stresses describe the stationary solution to this creep problem for a given creep exponent N. Obviously, Expressions 11.116 reduce to the elastic stress distribution if N is set to unity, an observation which has been pointed out in the creep bending of beams. In fact, the stresses and strains in a structure under stationary creep conditions can be found by analyzing a corresponding problem of nonlinear elasticity. This procedure is known as the "elastic analogue".*

A plot of the stress distributions for $N = 1$, $N = 2$, and $N = \infty$ are shown in Figure 11.19 for a tube of $\lambda = r_o/r_i = 2$. Note that the maximum tensile tangential stress σ_θ occurs at the outer surface under creep conditions of ($N = 2$) and ($N \rightarrow \infty$)** as compared with the elastic case, where σ_θ is maximum at the bore. Such observation is important in evaluating creep rupture of thick-walled tubes. Also, the longitudinal stress σ_z under creep conditions is not constant as in the elastic case; however, the equilibrium condition:

$$2\pi\int_{r_i}^{r_o} \sigma_z r dr = \pi r_i^2 p_i$$

is always satisfied as it should be. The differences between the elastic and stationary creep stresses indicate that redistribution has occurred during the primary or transient creep stage. This is typical for these problems, such as the beam-bending problem of Section 11.4.3.

Steady creep rates, and, hence, creep strains are thus found by applying the flow rule, Equation 11.88, to give

$$\dot{\varepsilon}_\theta^c = -\dot{\varepsilon}_r^c = \left(\frac{\sqrt{3}}{2}\right)B\left(\frac{\sqrt{3}}{N}\frac{p_i}{\lambda^{2/N} - 1}\right)^N \left(\frac{r_o}{r}\right)^2 \quad \text{and} \quad \dot{\varepsilon}_z = 0$$

$$= B\left(\frac{2}{N}\right)^N \left(\frac{\sqrt{3}}{2}\right)^{N+1} \left(\frac{p_i}{\lambda^{2/N} - 1}\right)^N \left(\frac{r_o}{r}\right)^2 \quad \text{and} \quad \dot{\varepsilon}_z^c = 0 \quad (11.117)$$

---

* For instance, see Odqvist.[14]

** For $N = \infty$, $\sigma_r = -(p_i/\ln\lambda)\ln(r_o/r)$ and $\sigma_\theta = (p_i/\ln\lambda)(1 - \ln r_o/r)$, which are the same for a fully yielded cylinder of perfect plasticity as given later in Chapter 12.

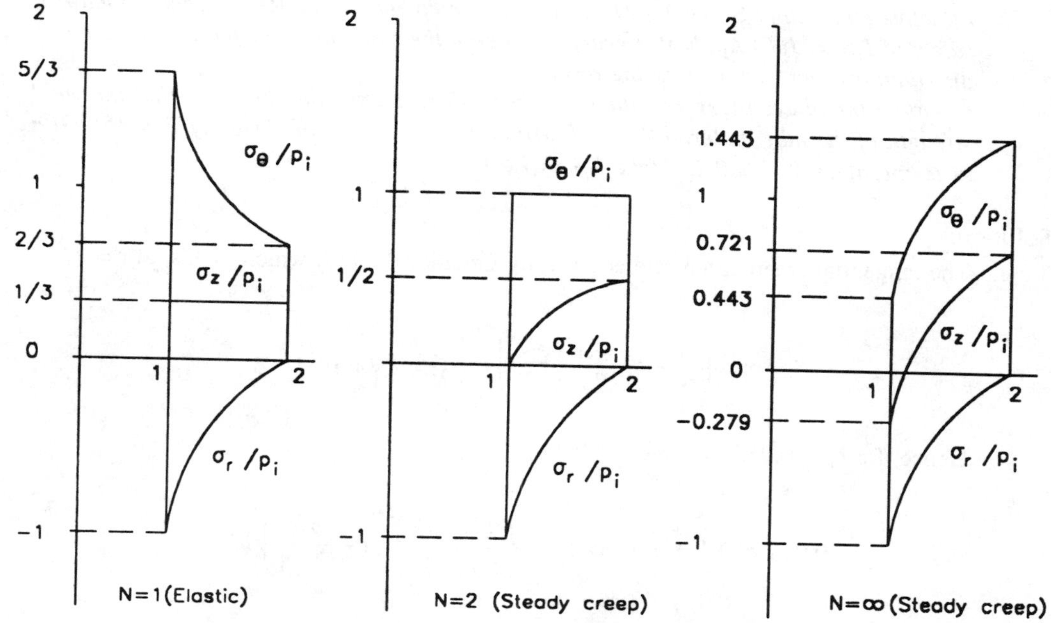

FIGURE 11.19 Stress distributions in a pressurized thick-walled cylinder of $\lambda = 2$: (a) elastic conditions, (b) and (c) steady creep conditions.

In several applications, thick-walled pressure vessels operate under conditions of steady heat conduction for which the radial thermal gradient is given by

$$T = T_i + (T_o - T_i)\ln(r/r_i)/\ln \lambda$$

If the material parameter B in Norton's creep law, Equation 11.89, is replaced by $\beta \exp(-\alpha/T)$ to allow for the material response to varying temperature, the previous solution for the stress remain valid with these new values of

$$B_T = \beta \exp(-\alpha/T_o)\exp\left[-\alpha(T_o - T_i)/T_o^2\right]$$

$$N_T = N\bigg/\left[1 + \frac{\gamma(T_o - T_i)}{2T_o^2 \ln \lambda}\right]$$

replacing B and N, respectively, in Equations 11.116 and 11.117.

Example 11.13:

A thick-walled cylinder of $\lambda = r_o/r_i = 2$ stainless steel (25% Cr, 20% Ni) is subjected to an internal pressure at temperature of 600°C. Determine the service pressure for a life of 10^5 h based on either of the following design criteria:

a. *Permissible tangential creep strain at the bore of 1.5% using a creep equation of state* $\bar{\varepsilon} = B(\bar{\sigma})^N$, *where* $B = 3.232 \times 10^{-16}$ *and* $N = 5$ *at* $T = 600°C$ *(MPa – h units).*

b. *Effective stress limit* S_r = *63.1 MPa inducing creep rupture in* 10^5 *h. (Take a factor of safety of F.S. = 1.4.) Apply this criterion to both the stationary creep and elastic stress distributions and comment on the results.*
c. *Compare the above pressure values with that based on yielding knowing that the yield strength of the tube material* Y = *190 MPa and its modulus of elasticity* E = *155 GPa at temperature* T = *600°C. (Take F.S. = 1.6.)*

Solution:

a. The tangential creep strain rate is given by Equation 11.117, which yields, at $r = r_i$,

$$\left[\dot{\varepsilon}_\theta^c\right]_{r=r_i} = B\left(\frac{2}{N}\right)^N\left(\frac{\sqrt{3}}{2}\right)^{N+1}\left(\frac{p_i}{\lambda^{2/N}-1}\right)^N(\lambda^2)$$

Hence, for $\varepsilon_\theta = 0.015$ for time $t = 10^5$ h,

$$0.015\times10^{-5} = \left(3.232\times10^{-16}\right)\left(\frac{2}{5}\right)^5\left(\frac{\sqrt{3}}{2}\right)^{5+1}\left(\frac{p_i}{2^{2/5}-1}\right)^5(2)^2$$

giving $p_i = 38.93$ MPa.

b. From the stationary creep stress distribution, the effective stress at any radius is given by combining Equations 11.112 and 11.116, as

$$\bar{\sigma} = \frac{\sqrt{3}}{2}(\sigma_\theta - \sigma_r) = \frac{\sqrt{3}}{2}\frac{p_i}{\lambda^{2/N}-1}\left[\frac{2}{N}\left(\frac{r_o}{r}\right)^{2/N}\right]$$

This attains a maximum value at $r = r_i$; hence, the condition that

$$\frac{\sqrt{3}}{2}\left(\frac{p_i}{\lambda^{2/5}-1}\right)\left[\frac{2}{5}\left(2^{2/5}\right)\right] \le \frac{S_r}{\text{F.S.}} \le \frac{63.1}{1.4}$$

gives $p_i = 31.5$ MPa.

Based on the elastic stress distribution, the maximum effective stress occurs at the inner surface. Equating the (S_r/F.S.) value to $\bar{\sigma} = \sqrt{3}\,(\sigma_\theta - \sigma_r)/2$, yields

$$\bar{\sigma} = \frac{\sqrt{3}}{2}\left[\frac{p_i(\lambda^2+1)}{\lambda^2-1} + p_i\right] \le \frac{S_r}{\text{F.S.}} \le \frac{63.1}{1.4}$$

hence, giving $p_i = 19.52$ MPa. This is a smaller value since it does not make use of the beneficial effects of stress redistribution toward stationary creep conditions.

c. At the instant of initial loading $t = 0$, the stress distribution is elastic as given by Equations 6.2. Again, applying the von Mises yield condition at $r = r_i$, where the maximum effective stress occurs, gives

$$\bar{\sigma} = \frac{\sqrt{3}}{2}\left[\sigma_\theta - \sigma_r\right]_{r=r_i} \le \frac{Y}{\text{F.S.}}$$

or

$$\frac{\sqrt{3}}{2}\left[\frac{p_i(\lambda^2+1)}{\lambda^2-1}+p_i\right]\le\frac{190}{1.6}$$

hence, for $\lambda = 2$, $p_i = 51.42$ MPa.

11.4.6 Steady Creep in Rotating Disks

The solution to this problem requires considering the equilibrium differential Equation 1.10c together with the strain compatibility condition. This latter involves the creep rates in terms of the radial displacement rate $\dot{u}$. Assuming that Tresca yield criterion (10.18) and (10.32) applies and using Norton's creep law (11.89), a solution for an annular disk has been obtained* with the boundary conditions $\sigma_r = 0$ at $r = r_i$ and $r = r_o$. The radial and hoop stresses are

$$\sigma_r=\frac{\rho\omega^2}{3}\left[\frac{r_o^3-r_i^3}{r}\left(\frac{r^{(N-1)/N}-r_i^{(N-1)/N}}{r_o^{(N-1)/N}-r_i^{(N-1)/N}}\right)-\frac{r^3-r_i^3}{r}\right]$$

$$\sigma_\theta=\frac{\rho\omega^2}{3}\left[\frac{r_o^3-r_i^3}{r^{1/N}}\frac{(N-1)/N}{r_o^{(N-1)/N}-r_i^{(N-1)/N}}\right] \tag{11.118}$$

The nonuniformity of the stress distribution of σ_θ with r is less for higher values of N. A reduction of about 24% in $(\sigma_\theta)_{N=10}$ compared with $(\sigma_\theta)_{N=3}$ both at $r = r_i$, is found for a disk or $r_o/r_i = 3$. Moreover, the hoop stress distributions for $2 \le N \le 10$ are shown to intersect at a point of almost $r/r_o = 0.64$.

11.4.7 Steady Creep of Circular Shafts under Torsion

Consider a tubular shaft of outer and inner radii r_o and r_i, respectively, and length L, as shown in Figure 11.20a. Under the action of a twisting moment M_t the shaft undergoes an angle of twist per unit length σ and hence a creep shear strain of γ^c at any radius r, expressed as

$$\gamma^c = r\alpha \tag{11.119a}$$

Note that this expression is based on the same assumption of Equation 7.45 of infinitesimal deformations, where plane sections remain plane during twisting.

The rate of shear strain $\dot{\gamma}$ is thus given by

$$\dot{\gamma}^c = r\dot{\alpha} \tag{11.119b}$$

For creep shear deformation, the state of strain and stress are given by ($\dot{\varepsilon}^c$, $-\dot{\varepsilon}^c$, 0) and (τ, $-\tau$, 0), respectively. Hence, the effective stress $\bar{\sigma}$ and the effective strain rate $\dot{\bar{\varepsilon}}$ given by Expressions 10.26 and 11.71, respectively, are

$$\dot{\bar{\varepsilon}} = \dot{\gamma}^c/\sqrt{3} = r\dot{\alpha}/\sqrt{3} \tag{11.120a}$$

* This solution due to Wahl[15] is based on considering $\dot{\varepsilon}_r = 0$, $\dot{\varepsilon}_\theta = \dot{u}/r = -\dot{\varepsilon}_z$; a geometry of deformation different than that of Section 6.2.4. Note that the solution does not reduce to the limiting elastic solution when N = 1.

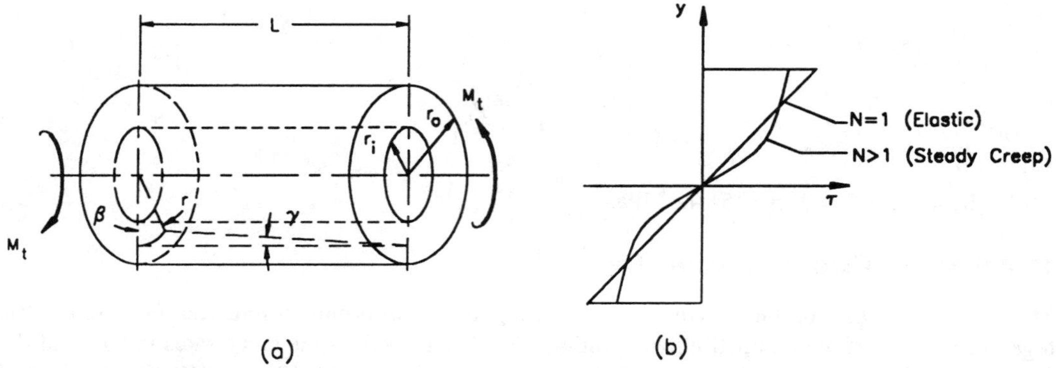

FIGURE 11.20 (a) Torsion of a circular hollow shaft and (b) shear stress distribution in elastic and steady creep conditions.

$$\bar{\sigma} = \sqrt{3}\,\tau \tag{11.120b}$$

These are related under secondary creep conditions by the law (Equation 11.89) as

$$\left(r\dot{\alpha}/\sqrt{3}\right) = B\left(\sqrt{3}\,\tau\right)^N$$

or

$$\tau = \left(\frac{1}{\sqrt{3}}\right)^{(1+1/N)} \left(\frac{r\dot{\alpha}}{B}\right)^{1/N} \tag{11.121}$$

The equilibrium condition is satisfied when

$$M_t = 2\pi \int_{r_i}^{r_o} \tau r^2 dr$$

which gives, after substituting from Equation 11.121 and integrating,

$$M_t = 2\pi \left(\frac{1}{\sqrt{3}}\right)^{(1+1/N)} \left(\frac{N}{1+3N}\right)\left(\frac{\dot{\alpha}}{B}\right)^{1/N} \left[r_o^{(3+1/N)} - r_i^{(3+1/N)}\right] \tag{11.122}$$

Combining Equations 11.121 and 11.122 results in an expression for the shear stress distribution over the shaft cross section as

$$\tau = \frac{1+3N}{2\pi N} \frac{M_t r^{1/N}}{\left[r_o^{(3+1/N)} - r_i^{(3+1/N)}\right]} = \frac{M_t r^{1/N}}{\left(I_o\right)c} \tag{11.123a}$$

where

$$\left(I_o\right)_c = \frac{2\pi N}{1+3N}\left[r_o^{(3+1/N)} - r_i^{(3+1/N)}\right] \tag{11.123b}$$

For $N = 1$, i.e., conditions analogous to elastic behavior, $(I_o)_c$ reduces to $I_o = [\pi(r_o^3 - r_i^3)/2]$, as expected. For a solid shaft $r_i = 0$, the shear stress distribution is

$$\tau = \frac{1+3N}{2\pi N}\frac{M_t r^{1/N}}{r_o^{(3+1/N)}} \tag{11.124}$$

This is shown in Figure 11.20b, where it is seen from the cases corresponding to $N > 1$ and $N = 1$ that τ redistributes during primary creep to be more uniform under steady creep conditions compared with the linear-elastic case ($N = 1$).

Example 11.14:

A solid shaft of 70 mm diameter is to be designed such that the maximum angular twist due to creep between two cross sections separated by length 500 mm is 3°. The shaft operates at 510°C and made of a steel alloy (2.25% Cr–1% Mo) obeying the creep law $(\dot{\bar{\varepsilon}}) = 25 \times 10^{-19} (\bar{\sigma})^{5.31}$ *in MPa-h.*

a. *Determine the maximum twisting moment that can be applied to the shaft without violating the above design criterion over a service life of 30,000 h. Neglect primary creep.*
b. *Compare the results obtained in (a) to the case of a purely elastic response for the same shaft. Take E = 158 GPa and* $\nu = 0.35$.
c. *The stress rupture data for the same material at 510°C is given in the following table. Estimate the margin of safety of the service life using this stress rupture data.*

Stress, MPa	*267*	*200*	*148*	*113.4*	*84*
Rupture time, h	*10*	*10^2*	*10^3*	*10^4*	*10^5*

Solution:

a. The creep shear strain at time $t = 30{,}000$ h is

$$\gamma^c = r_o\alpha = 35\frac{3\times\pi}{180\times 500} = 3.67\times 10^{-3}$$

Hence, the effective creep strain rate (Equation 11.120a) is

$$\dot{\bar{\varepsilon}} = \dot{\gamma}^c/\sqrt{3} = \left(\gamma^c/t\right)/\sqrt{3} = \frac{1}{\sqrt{3}}\frac{3.67\times 10^{-3}}{30{,}000} = 7.06\times 10^{-8}$$

This corresponds — under steady creep condition — to a maximum shear stress τ_c at $r = r_o$:

$$\dot{\bar{\varepsilon}} = B(\bar{\sigma})^N = B\left(\sqrt{3}\,\tau_c\right)^N$$

where $\bar{\sigma}$ is given by Expression 11.120b
Hence,

$$\tau_c = \frac{1}{\sqrt{3}}\left(\dot{\bar{\varepsilon}}/B\right)^{1/N}$$

or

$$\tau_c = \frac{1}{\sqrt{3}}\left(\frac{7.06\times10^{-8}}{25\times10^{-19}}\right)^{1/5.3} = 54.1 \text{ MPa}$$

The twisting moment $(M_t)_c$ corresponding to this shear stress under steady creep conditions is determined from Equation 11.124 as

$$\tau_c = \frac{1+3N}{2\pi N}\frac{(M_t)_c r_o^{1/N}}{r_o^{(3+1/N)}}$$

$$54.1\times10^6 = \frac{1+3\times5.3}{2\pi\times5.3}\frac{M_t\times0.035^{1/5.3}}{0.035^{(3+1/5.3)}}$$

giving $(M_t)_c = 4.57$ kN · m.

b. For the elastic case at a temperature of 510°C, the elastic maximum shear stress τ_e at $r = r_o$ is obtained as

$$\tau_e = G\gamma = \frac{E}{2(1+\nu)}\gamma = \frac{158\times10^3}{2(1+0.35)}\left(3.67\times10^{-3}\right) = 214.76 \text{ MPa}$$

indicating the more uniform stress distribution at conditions of steady creep compared with the initial elastic one. Hence, the corresponding elastic twisting moment $(M_t)_e$ is given by

$$(M_t)_e = \frac{\pi r_o^3}{2}\tau_e = \frac{\pi\times0.035^3}{2}\times214.76\times10^6 = 14.46 \text{ kN}\cdot\text{m}$$

c. Since stress rupture data plots as a straight line on ln σ – ln t_R scales, logarithmic interpolation is used for a stress $\bar{\sigma} = \sqrt{3}\,\tau = \sqrt{3}\times 54.1 = 93.7$ MPa. This results in the rupture time as

$$\ln t_R = \ln 10^4 + \left(\frac{\ln 10^5 - \ln 10^4}{\ln 113.4 - \ln 84}\right)(\ln 113.4 - \ln 93.7)$$

giving $t_R \simeq 43235$ h with a margin of safety for rupture = 43235/30,000 = 1.44.

11.4.8 Creep Buckling of Columns

The phenomenon of buckling of columns occurs instantaneously if the applied compressive stress reaches its critical value within either the elastic or elastoplastic range. However, buckling may develop after some critical time at which the initially applied stress becomes critical.

As a demonstration of the creep buckling phenomenon, a column of length L under a compression load P along its axis z is considered. Only an approximate analytical solution is presented. As derived in Chapter 7, the equilibrium equation for bending moment in terms of the lateral displacement v of the column is given by an expression similar to Equation 7.75a as

$$\frac{dv^2}{dz^2} + \frac{P}{E'I}(v_o + v) = 0 \qquad (11.125)$$

here v_o and v are the initial and creep bending deflection at time $t = 0$ and any time t, respectively. E' is the tangent modulus of the isochronous stress–strain curve at a given temperature, Figure 11.13b. For steady creep, the power law (Equation 11.87) may be applied in an integrated form as

$$\varepsilon = \varepsilon_o + B(\sigma)^N t$$

and, hence,

$$E' = \frac{d\sigma}{d\varepsilon} = \frac{1}{BN(\sigma)^{N-1}t} \qquad (11.126)$$

The solution to Equation 11.125 may be assumed in the form of two Fourier series representing the deflection curve by

$$v_o = \sum_{n=1}^{\infty} C_{on} \sin\frac{n\pi z}{L}$$

$$v_c = \sum_{n=1}^{\infty} C_n \sin\frac{n\pi z}{L} \qquad (11.127)$$

which satisfy the boundary conditions $v = v_o = 0$ at $z = 0$ and $z = L$ for a pinned column.

Substitution of Expressions 11.127 into Equation 11.125 yields

$$\sum_{n=1}^{\infty}\left[-C_n + \frac{PL^2}{n^2\pi^2 E'I}(C_n + C_{on})\right]\sin\frac{n\pi z}{L} = 0 \qquad (11.128)$$

A nontrivial solution is obtained for any value of z by setting the square bracket of Equation 11.128 to zero, thus giving

$$C_n + \frac{C_{on}}{\dfrac{n^2\pi^2 E'I}{L^2 P} - 1}$$

Instability occurs when the denominator goes to zero, i.e.,

$$P_{cr} = \frac{n^2\pi^2 E'I}{L^2}$$

The smallest load value occurs at $n = 1$; hence,

$$P_{cr} = \frac{n^2 E'I}{L^2} \qquad (11.129)$$

For the elastic case $E^t = E$, P_{cr} is thus the Euler critical load previously given by Expression 7.75b. Under creep conditions, E^t is replaced by its value, from Expression 11.126, to give for the critical buckling time t_{cr} as

$$t_{cr} = \frac{\pi^2 I}{AL^2} \frac{1}{BN\sigma^N} \tag{11.130}$$

Example 11.15:

Two pin-ended columns of equal length and circular cross sections of diameters (d) and (2d), respectively, are subjected to compressive loads at an elevated temperature. If the critical time to buckling is the same for both columns, determine the ratio of their buckling loads and compare it with the ratio of Euler buckling loads for the two columns. The creep data for the column material is given by the isochronous curves of Figure 11.21 at 567°C.

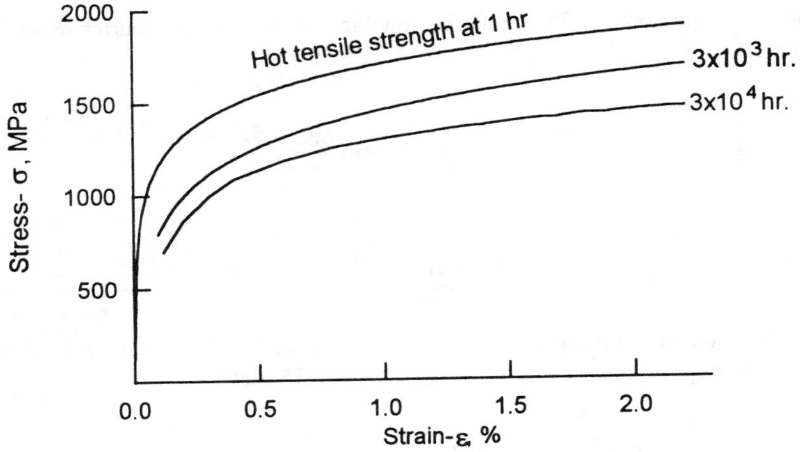

FIGURE 11.21 Example 11.15.

Solution:

The two isochronous creep curves for $t = 3 \times 10^4$ and 3×10^3 h are used to determine the constants in the creep law: $\dot{\varepsilon} = B(\sigma)^N$. Following an inverse procedure, as explained in Section 11.4.1, the following points are read out:

For $\sigma = 142.6$ MPa:	$\varepsilon = 2.3\%$	at	3×10^4 h
	$\varepsilon = 1.02\%$	at	3×10^3 h
For $\sigma = 129.0$ MPa:	$\varepsilon = 1.16\%$	at	3×10^4 h
	$\varepsilon = 0.58\%$	at	3×10^3 h

These correspond to two points $(\sigma, \dot{\varepsilon})$ as $(142.6, 4.07 \times 10^{-7})$ and $(129.0, 2.15 \times 10^{-7})$, respectively. The power N in the creep law is thus

$$N = \ln (4.07/2.15)/\ln (142.6/129) = 6.37$$

$$B = 7.725 \times 10^{-21} \text{ (in MPa-h units)}$$

Since the critical buckling time t_{cr}, as given by Equation 11.130, is the same for both columns; hence,

$$t_{cr} = \frac{\pi^2 I_1}{A_1 L_1^2} \frac{1}{BN\sigma_1^N} = \frac{\pi^2 I_2}{A_2 L_2^2} \frac{1}{BN\sigma_2^N}$$

Knowing that $A_2 = 4A_1$, $I_2 = 16I_1$ and $L_1 = L_2$ gives

$$\left(\frac{\sigma_2}{\sigma 1}\right)^N = \frac{I_2 A_1}{I_1 A_2} \quad \text{or} \quad \frac{P_2}{P_1} = \left(\frac{I_2}{I_1}\right)^{1/N} \left(\frac{A_1}{A_2}\right)^{1-N/N}$$

i.e., $P_2/P_1 = 4.97$ for $N \simeq 6.37$.

The ratio of the elastic Euler buckling loads for the two columns is obtained from Equation 11.129 or by putting $N = 1$ in the above expression, i.e., $P_2/P_1 = I_2/I_1 = 16$. This indicates that the higher stability of a column with larger cross section under elastic conditions is not maintained to the same degree under creep conditions.

11.4.9 The Reference Stress Method

Solutions to stationary creep problems as presented in the preceding sections have been all obtained using the Norton–Baily creep law. This simple power law has been assumed to fit experimental creep data, thus providing an analytical means to analyze stationary creep problems. In reality, materials experimental data under creep conditions do not always conform to this idealized power law. Thus, estimating the material parameters B and N in Equation 11.87 often involves some error because of the relatively high scatter in creep experimental data. Moreover, the nonlinearity of Norton's creep law gives rise to large deviations in calculated creep rates even with small errors in the material parameters, as has been demonstrated in Section 11.4.2 and Example 11.11. To overcome this difficulty, it has been sought to relate the structural creep behavior directly to the relevant uniaxial creep experimental data without too much dependence on the accuracy of fitting a material law. This constitutes the basis of the so-called "reference stress method". The existence of a reference stress in stationary creep problems requires a mathematical proof which can be found in specialized texts.*

The aim of the reference stress method is thus to correlate the creep deformation $\dot{\delta}$ in a structure with an experimentally determined uniaxial creep rate $\dot{\varepsilon}^c_{RF}$ at a selected reference stress σ_{RF}. Hence, it is postulated that

$$\dot{\delta} = \beta \dot{\varepsilon}^c_{RF} \tag{11.131}$$

The factor β is a geometric scaling factor that depends on the structure and boundary conditions but optimistically may be "weakly dependent" on the creep exponent N.

The identification of the reference stress σ_R depends on the simple observation that all stress distributions of stationary creep problems intersect at almost the same point with the initial elastic stress distribution. This observation is valid for beam bending (Figure 11.16), pressurized thick-walled cylinders (Figure 11.19), as well as torsion of circular bars (Figure 11.20). Considering, for instance, the stress distributions in beam bending as shown in Figure 11.22a, the intersection point may be arbitrarly selected by considering the two distributions for $N = 1$ (linear elastic) and $N = \infty$ (perfect plasticity). Hence, the reference stress ($\sigma_{RF} = 4M/bh^2$) occurs at ($y = h/3$), where h is the total depth of the beam. This reference stress is obviously independent of the creep exponent N.

A more-refined method to find a reference stress value may be achieved by assuming that a creep power law holds with an exponent N known within certain bounds, i.e., $N' \leq N \leq N''$. The

* For this proof, see Boyle and Spence,[16] p. 111.

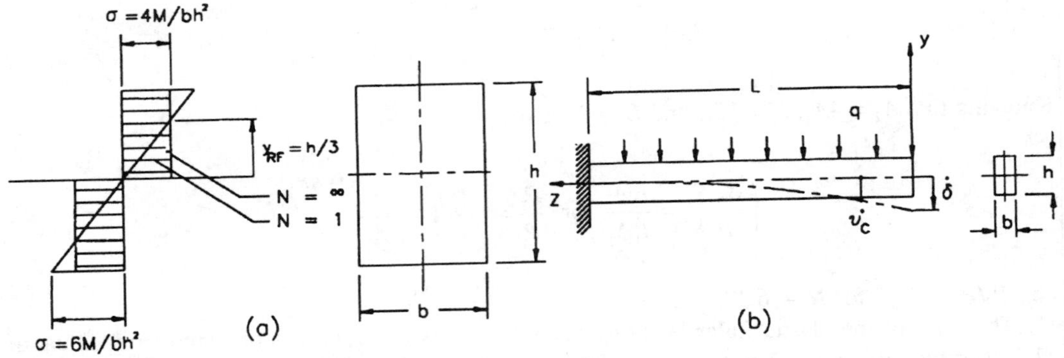

FIGURE 11.22 Development of the reference stress method for a cantilever beam of rectangular cross section subjected to uniformly distributed load.

intersection of the stress distributions corresponding to N' and N'' identifies a more accurate value for σ_{RF}, which is obviously dependent on the bounds selected for N. This dependence, however, is not high and is also often referred to as "weakly dependent" on the creep exponent.

To illustrate this method, consider the case of a cantilever beam of length L loaded by a distributed load of intensity q per unit length as shown in Figure 11.22b. The stress distribution is given by Equation 11.97 as

$$\sigma = \frac{My}{I}\left(\frac{2N+1}{3N}\right)\left(\frac{2y}{h}\right)^{(1/N-1)}$$

For a beam of a rectangular section $I = bh^3/12$, this equation may be rewritten as

$$\left(\frac{\sigma}{M}\frac{bh^2}{4}\right) = \frac{2N+1}{2N}\left(\frac{2y}{h}\right)^{1/N} \tag{11.132}$$

It may be shown numerically that if $(2y/h) = 0.636$, and assuming that the exponent N varies within the range $6 \le N \le 8$, the right-hand side of Equation 11.132 varies between 1.00463 and 1.00406, respectively. Hence, by choosing a mean value of 1.00435, it is sufficiently accurate to identify, for $(2y/h) = 0.636$, the reference stress point as

$$\sigma_{RF} = \frac{1.00435M}{\left(bh^2/4\right)} = 2.00867qL^2/bh^2 \simeq 2\,qL^3/bh^2 \tag{11.133}$$

at

$$y_{RF} = 0.318\text{ h} \quad \text{and} \quad M = M_{\max} = qL^2/2$$

Having determined a reference stress σ_{RF}, hence it is feasible to conduct a uniaxial creep test, thus finding an $\dot{\varepsilon}^c_{RF}$ corresponding to this stress.

To correlate the tip deflection rate $\dot{\delta}$ of the loaded cantilever, recall Equation 11.103b that gives, at $z = 0$,

$$\dot{\delta} = -B\left[\frac{2N+1}{N}\frac{2}{bh^2}\right]^N\left(\frac{2}{h}\right)\left(\frac{q}{2}\right)^N\frac{L^{2N+2}}{(2N+2)} \tag{11.134}$$

Rewriting this expression in the form of Equation 11.131, knowing that $\sigma_{RF} = 2qL^2/bh^2$ results in

$$\dot{\delta} = -\left(\frac{2N+1}{2N}\right)^N \left(\frac{1}{N+1}\right)\frac{L^2}{h}\left(B\sigma_{RF}^N\right) \qquad (11.135a)$$

$$\dot{\delta} = -\beta\dot{\varepsilon}_{RF}^c \qquad (11.135b)$$

where the definition of β is self-evident. For $6 \le N \le 8$, β possesses the values $(0.23093\ L^2/h)$ and $(0.180463\ L^2/h)$, respectively. An average value may be suggested to relate $\dot{\delta}$ to $\dot{\varepsilon}_{RF}^c$ as

$$\dot{\delta} = -0.2057\frac{L^2}{h}\left(\dot{\varepsilon}_{RF}^c\right) \qquad (11.136)$$

Note that in this specific problem, β is not independent or weakly dependent on N and its value is determined as an average in the vicinity of an expected value of N. This is called a local reference stress method as opposed to a global method for any N. This latter method is based on finding a nonzero limiting value of β as $N \to \infty$. Expressions 11.135 indicates clearly that this limit does not exist for this specific cantilever problem.

Now, if a tensile creep test for a material whose $6 \le N \le 8$ is conducted at a stress level corresponding to σ_{RF}, the secondary creep strain rate $\dot{\varepsilon}_{RF}^c$ is determined experimentally. The end deflection rate $\dot{\delta}$ is thus calculated directly from Expression 11.136 without resort to any expressions involving the creep exponent N, thus avoiding the uncertainties in evaluating the creep material parameters. This procedure is demonstrated in Example 11.16.

A further refinement* in identifying the reference stress is based on selecting its value such that to obtain equal tip deflection rates for two arbitrary values of N, say N' and N''. Rewriting Equation 11.135a as

$$\dot{\delta} = \left(\frac{2N+1}{2N}\right)^N \left(\frac{1}{N+1}\right)\frac{L^2}{h}\left(\frac{1}{\alpha}\right)^N B\sigma_{RF}^N \qquad (11.137a)$$

where $\sigma_{RF} = \alpha(2qL^2/bh^2)$, and setting $(\dot{\delta})_{N'} = (\dot{\delta})_{N''}$ gives

$$\alpha = \left[\left(\frac{N'+1}{N''+1}\right)\left(\frac{2N''+1}{2N''}\right)^{N''}\left(\frac{2N'}{2N'+1}\right)^{N'}\right]^{1/(N''-N')} \qquad (11.137b)$$

For practical ranges of stress and temperature, most creep exponents lie in the range $3 \le N \le 9$; hence, this gives for $N' = 3$ and $N'' = 9$, $\alpha = 0.8618$ and $\sigma_R = 1.724(qL^2/bh^3)$ compared with $\sigma_{RF} = 2(qL^2/bh^2)$, as obtained from Equation 11.133.

Throughout the previous analysis, Norton's power law has been applied to obtain an analytical solution, otherwise resort may be made to numerical solutions. The literature on the reference stress method is extensive, and extensions of this method to include combined steady and variable loadings as well as conditions of nonisothermal stress relaxation and creep rupture are found.[16]

Example 11.16:

A wide strip made of superplastic Zn–22%Al sheet of 127 mm thickness and 915 mm width is fixed as a cantilever of length 390.1 mm. It was observed[17] that this strip creeps significantly at

* Other techniques to identify the reference stress are found in the literature. See Krauss,[13] p. 64.

room temperature 28 ± 32°C under its own weight (density of this alloy is 6160 kg/m³). Provided that the uniaxial creep data of the material at room temperature is given by

σ MPa	*8.791*	*9.917*	*11.460*	*12.596*	*14.717*	*21.724*
$\dot{\varepsilon}^c$, h^{-1}	1.844×10^{-5}	5.18×10^{-5}	3.467×10^{-4}	4.023×10^{-4}	1.887×10^{-3}	$0.95\,7 \times 10^{-2}$

Determine the end deflection of the strip:

a. *Using the stationary creep solution based on Norton's creep law as given in Section 11.4.3;*
b. *Using the local reference stress method for 6 ≤ N ≤ 8.*

Solution:

a. To apply the solution of Section 11.4.3, Norton's creep parameters B and N have to be determined by fitting this law to $\sigma-\dot{\varepsilon}^c$ data given above. This yields $B = 8.5959 \times 10^{-12}$ and $N = 6.931$ (in MPa-h units) with correlation coefficient = 0.975.

An expression for the deflection rate of a wide strip fixed as a cantilever may be obtained following the derivation of Section 11.4.3 noting that plane strain condition prevail such that

$$\bar{\sigma} = \sqrt{3}\,\sigma/2 \quad \text{and} \quad \dot{\bar{\varepsilon}} = 2\dot{\varepsilon}^c/\sqrt{3}$$

Hence, it may be shown that the end deflection rate is given by

$$\dot{\delta} = -B\left(\frac{\sqrt{3}}{2}\right)^{N+1}\left(\frac{2N+1}{N}\frac{q}{bh^2}\right)^N\left(\frac{2}{h}\right)\frac{L^{2N+2}}{2N+2}$$

Substituting $q = 6160 \times 9.81 \times 10^{-9} \times 915 \times 1.27 \times 10^{-3} = 0.07022$ N/mm, $b = 915$ mm, $h = 1.27$ mm, and $L = 390.1$ mm yields for the given B and N:

$$\dot{\delta} = -7.459 \text{ mm/h}$$

b. The reference effective stress σ_{RF} is obtained from Equation 11.133, as

$$\bar{\sigma}_{RF} = \frac{\sqrt{3}}{2}\frac{2.00867qL^2}{bh^2} = 12.596 \text{ MPa}$$

If a tensile creep experiment is conducted at this value of stress, a steady creep effective strain rate $\dot{\bar{\varepsilon}}_{RF}$ of about 4.025×10^{-4} h^{-1} is measured as may be read from the given table of $\sigma-\dot{\varepsilon}^c$ for this material.

Therefore, applying Equation 11.136, knowing that $\dot{\varepsilon}^c_{RF} = \sqrt{3}\,\dot{\bar{\varepsilon}}_{RF}/2$, results in the end deflection of the strip as

$$\dot{\delta} = 0.206\frac{L^2}{h}\left(\dot{\varepsilon}^c_{RF}\right)$$

giving $\dot{\delta} = -8.604$ mm/h.

Example 11.17:

A pressurized thick-walled tube of an aspect ratio $\lambda = r_o/r_i = 2$ is made of a material which obeys Norton's creep law $\dot{\bar{\varepsilon}}^c = B(\bar{\sigma})^N$.

a. *Identify a reference stress for this case.*
b. *Derive an expression for the radial expansion rate at the bore using the reference stress method.*
c. *Compare the bore expansion rate according to (b) with the analytical Expression 11.117 for $\lambda = 2$ and N = 5.*

Solution:

a. As pointed out earlier, an approximate but rather conservative method for determining the reference stress is based on the idea that it is independent of the creep exponent. Hence, the reference stress may be identified by the stress solution corresponding to perfect plasticity ($N = \infty$). Accordingly, the effective stress $\bar{\sigma} = \sqrt{3}\,(\sigma_\theta - \sigma_r)/2$, using Equations 11.116 at $N \to \infty$ gives

$$\sigma_{RF} = \frac{\sqrt{3}}{2}\frac{p_i}{\ln\lambda}$$

Another method to define a local reference stress is to assume it as

$$\sigma_{RF} = \alpha\left(\frac{\sqrt{3}}{2}\right)\frac{p_i}{\ln\lambda}$$

The constant α is determined such that equal bore expansion rates are obtained for $3 \le N \le 9$. This yields almost the same value for σ_{RF}, namely,

$$\sigma_{RF} = 1.003\left(\frac{\sqrt{3}}{2}\right)\frac{p_i}{\ln\lambda}$$

b. The bore expansion rate is obtained from Equation 11.117 as

$$(\dot{u})_{r_i} = \frac{\sqrt{3}}{2}\lambda^2 r_i\left(\frac{2\ln\lambda}{N(\lambda^{2/N}-1)}\right)^N B\left(\frac{\sqrt{3}}{2}\frac{p_i}{\ln\lambda}\right)^N$$

In terms of σ_{RF}, this may be written as

$$(\dot{u})_{r_i} = \beta\left(B\sigma_{RF}^N\right) = \beta\dot{\varepsilon}_R^c$$

where the scaling factor is

$$\beta = \frac{\sqrt{3}}{2}\lambda^2 r_i\left(\frac{2\ln\lambda}{N(\lambda^{2/N}-1)}\right)^N$$

The limiting value for β as $N \to \infty$ is found to be

$$(\beta)_{N\to\infty} = \sqrt{3}\, r_i$$

Hence,

$$(\dot{u})_{r_i} = \sqrt{3}\, r_i \left(\dot{\varepsilon}^c_{RF}\right)$$

c. For $\lambda = 2$, $N = 5$, Expression 11.117 and the reference stress method give, respectively,

$$(\dot{u})_{r_i}/r_i = B\left(\frac{2}{N}\right)^N \left(\frac{\sqrt{3}}{2}\right)^{N+1} \lambda^2 r_i \left(\frac{p_i}{\lambda^{2/N}-1}\right)^N$$

$$= 5.189 B p_i^5$$

and

$$(\dot{u})_{r_i}/r_i = \sqrt{3}\, r_i B\left(\frac{\sqrt{3}}{2}\frac{p_i}{\ln\lambda}\right)^N$$

$$= 5.272 B p_i^5$$

11.4.10 Stress Relaxation

Stress relaxation is a creep-related phenomenon observed at elevated temperatures. A body deformed elastically to a fixed initial strain undergoes creep deformations at the expense of a decrease in the initial elastic strain component. This results in stress decrease with time, i.e., stress relaxation. This is observed in several practical applications, e.g., relaxation of tightening loads in bolted flange connections and stress-fitted joints.

A stress relaxation test is conducted by applying an initial fixed strain (say, 0.15%) and recording the relaxed stress (load) with time. A typical record is shown in Figure 11.23 for different initial stresses, the record being a mirror image of creep curves indicating a high initial decrease of stress, which levels off with time.

The analysis of stress relaxation problems assumes that the flow rules of creep apply. To illustrate this, consider a bar initially subjected to a fixed initial elastic strain ε_i associated with an initial stress $\sigma_i = \varepsilon_i E$. As time progresses, creep (plastic) strain ε^c accumulates at the expense of decreasing elastic strain ε^e, such that the following equality is satisfied at all times, $t > 0$,

$$\varepsilon_i = \varepsilon^e + \varepsilon^c \tag{11.138}$$

In terms of time derivatives, this becomes

$$0 = \frac{d\varepsilon^e}{dt} + \frac{d\varepsilon^c}{dt}$$

Substitution from Hooke's law (Equation 3.6a) for ε^e and from the creep law (Equation 11.87) for $\dot{\varepsilon}^c$ yields

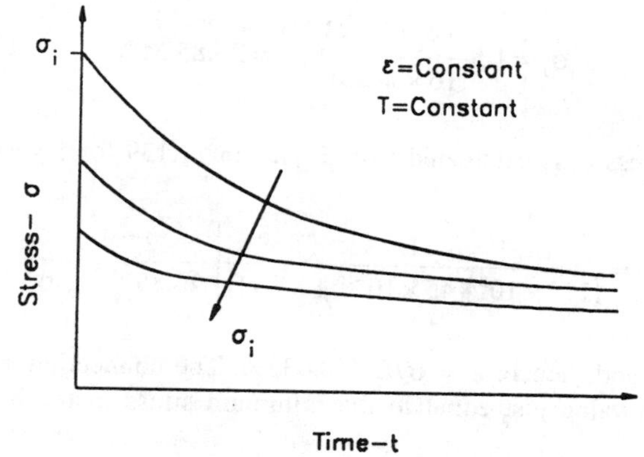

FIGURE 11.23 Typical relaxation behavior at different initial stress levels.

$$\frac{1}{E}\frac{d\sigma}{dt} = -B\sigma^N$$

Rearranging and integrating gives

$$\int dt = -\frac{1}{EB}\int \frac{1}{\sigma^N}\,d\sigma$$

hence,

$$t = -\frac{\sigma^{1-N}}{EB(1-N)} + C$$

From the initial conditions, $\sigma = \sigma_i$ at $t = 0$, the constant C is determined to yield

$$t = -\frac{\sigma^{1-N}}{EB(N-1)}\left(\frac{1}{\sigma_t^{N-1}} - \frac{1}{\sigma_i^{N-1}}\right) \tag{11.139}$$

where σ_t is the relaxed stress after elapsed time t.

In practice, this expression, although it follows the experimental trend, does not fit the rapid initial stress relaxation. This is due to the neglect of primary creep in applying Equation 11.87 into the above analysis.

Example 11.18:

Two flanged cover plates for a pipe connection of a flow area of 0.15 m^2 are held by 16 equally spaced steel bolts of 20 mm diameter. Assuming that the flanges are rigid, what should be the initial tightening stress and strain in the bolts in order that the connection remains tight (with a margin of safety of 1.5) after 1 year? The operating conditions are temperature 450°C and pressure 0.6 MPa. After what time should the bolts be retightened to prevent leakage? Take E = 170 MPa and apply a creep law $\dot{\varepsilon}^c = 45 \times 10^{-16}\,\sigma^{4.1}$ (in MPa-h).

Solution:

The minimum stress in each bolt after 1 year must be greater than the stress resulting from the internal pressure (considering a margin of safety of 1.5); hence,

$$\sigma_t = 1.5\frac{0.6(0.15\times 10^6)}{16\times \pi(20)^2/4} = 26.85 \text{ MPa}$$

The initial tightening stress σ_i is calculated from Equation 11.139 for 1 year, i.e., 8760 h as

$$8760 = \frac{1}{(170\times 10^3)(45\times 10^{-16})(4.1-1)}\left[\frac{1}{26.85^{4.1-1}} - \frac{1}{\sigma_i^{4.1-1}}\right]$$

giving σ_i = 55.66 MPa and, hence, $\varepsilon_i = \sigma_i/E$ = 0.033%. The connection starts to leak when the relaxed stress reaches a value just equal to the minimum stress in the bolt due to the internal pressure; hence,

$$\sigma_t = 26.85/1.5 = 17.9 \text{ MPa}$$

Applying Equation 11.13 again gives for leakage time t_l:

$$t_l = \frac{1}{(170\times 10^3)(45\times 10^{-16})(4.1-1)}\left[\frac{1}{17.9^{4.1-1}} - \frac{1}{55.66^{4.1-1}}\right]$$

$$= 53461.6 \text{ h} = 6.1 \text{ years}$$

11.4.11 Creep under Variable Loading: Time Hardening vs. Strain Hardening

Norton–Baily's creep law as expressed by Equation 11.87 has been used extensively in analyzing creep problems. The simplicity of this law helps in arriving at analytical solutions with acceptable accuracy for creep problems involving steady loadings. For situations when the applied stresses vary with time, either continuously or according to step changes, the use of Norton–Baily's law becomes inaccurate since the phases of primary creep cannot be neglected at every load change. To overcome this, the idea that Equation 11.87 expresses the creep rate as a function of stress σ and current time t, i.e., $\dot{\varepsilon}^c = f(\sigma, t)$ has been replaced by considering $\dot{\varepsilon}^c = f(\sigma, \varepsilon^c)$. The derivation of such functions is as follows:

The primary and secondary creep strains are expressed by Equation 11.86, namely,

$$\varepsilon^c = B\sigma^N t^{N1} \tag{11.140}$$

where $N >> 1$ and $N_1 \leq 1$. The time derivative gives

$$\dot{\varepsilon}^c = BN_1\sigma^N t^{N_1-1} \tag{11.141}$$

This derivation results in the known Time-Hardening rule, where creep strain rate is expressed as a function of the stress σ and time t. Obviously Equation 11.141 reduces to Equation 11.87 by taking $N_1 = 1$, i.e., secondary creep conditions. Note that in differentiating Expression 11.86, the time derivative of stress has been ignored, although the derivation is mainly devoted to variable stress conditions. This shortcoming puts restrictions, at least theoretically, on the application of the law (Equation 11.87) to mainly constant stress or step-changes of stress for long durations.

Another formulation known as "strain hardening" may be derived from Equation 11.141 by eliminating time t, as given by Equation 11.140, namely.

$$t = \left(\frac{\varepsilon^c}{B\sigma^N} \right)^{1/N_1} \tag{11.142}$$

Substitution into Equation 11.141 yields

$$\dot{\varepsilon} = BN_1(\sigma)^N \left(\frac{\varepsilon^c}{B\sigma^N} \right)^{N_1 - 1/N_1} \tag{11.143}$$

or

$$\dot{\varepsilon}^c = B^{1/N_1} N_1 (\sigma)^{N/N_1} \left(\varepsilon^c \right)^{(N_1-1)/N_1}$$

Equation 11.143 expresses the creep rate $\dot{\varepsilon}^c$ as a function of the stress σ and the current creep strain ε^c. Figure 11.24 illustrates schematically the predictions of both time-hardening and strain hardening laws. This figure considers a creep test with a step change in the applied stress from σ_1 at time t_1 to a higher stress σ_2 acting up to time t_2 before another further step change in the stress. Figure 11.24 indicates that, according to the time-hardening rule, the curve ab is a rigid (vertical) translation of the curve a_1b_1. Meanwhile, for strain hardening, the curve ab_2 is a rigid (horizontal) translation of the curve $a_1'b_1'$. For this step change of stress, the time-hardening rule predicts less total creep strain as compared with strain hardening.

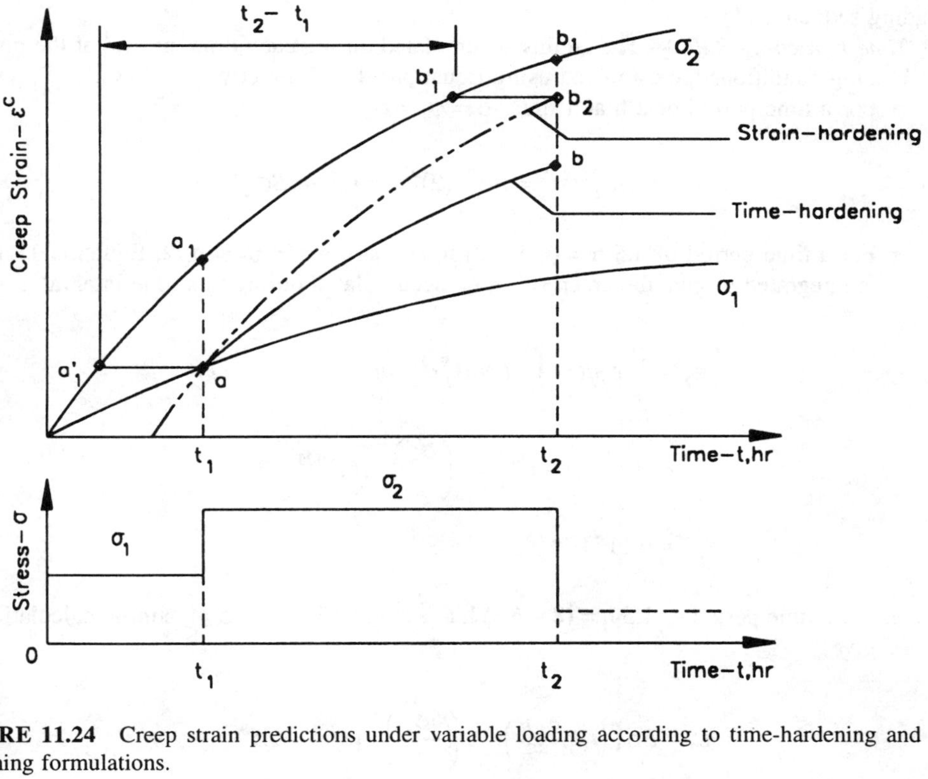

FIGURE 11.24 Creep strain predictions under variable loading according to time-hardening and strain-hardening formulations.

Note that the difference between Equation 11.141 and 11.143 is not phenomenological but procedural as both stem from the same law Equation 11.140. Both laws represent attempts to model mathematically the process of hardening associated with decreasing creep strain rate in the primary creep phase. Obviously, both laws reduce to the same Norton-Baily form for secondary creep when $N_1 = 1$.

Experiments indicate that the strain-hardening formulation is to be favored over time-hardening formulation. However, noting the large scatter in creep data, the use of the simple law of time hardening becomes justifiable in deriving analytical solutions. Evidently, strain hardening offers no difficulty in seeking numerical solutions. Both formulations as given above are applicable only for situations where no stress reversals occur, a situation where modified rules have to be used.* Also both formulations do not account well for the important phenomenon of creep recovery due to unloading or variable cyclic loading.

Example 11.19:

A bar undergoing creep deformation at constant temperature is subjected to step changes in loading. Knowing that the equation of state is given by $\varepsilon^c = B\,\sigma^{2.5}\,t^{0.35}$, *determine the total accumulated creep strain according to the rules of time hardening and strain hardening for the following two loading sequences:*

Sequence (1): stress σ_o *for 2 h, stress* $\sigma_o/2$ *for 4.5 h, and, finally, a stress* $5\sigma_o/4$ *for 1.5 h,*
Sequence (2): Stress σ_o *for 2 h, stress* $5\sigma_o/4$ *for 4.5 h, and finally, a stress* $\sigma_o/2$ *for 1.5 h.*

Note: For simplicity, take $\sigma_o = B = 1$.

Solution:

Loading sequence (1)

a. *Time-hardening Rule* — The strains accumulated during each time interval at the given loading conditions are estimated using Equation 11.140 directly as follows:
   - For a time period of 2 h at a stress $\sigma = \sigma_1 = \sigma_o$,

$$\varepsilon_1^c = B\sigma_1^{2.5}t^{0.35} = B\sigma_1^{2.5}(2)^{0.35} = 1.2746B\sigma_o^{2.5}$$

   - For a time period of 4.5 h = (6.5 – 2) h, at a stress $\sigma = \sigma_2 = \sigma_o/2$, Equation 11.141 is integrated to give the creep strain ε_2^c accumulated during this time interval as

$$\varepsilon_2^c = \int_{t_1}^{t_2} \dot{\varepsilon}_2^c dt = \int_2^{6.5} BN_1\sigma_2^N t^{N_1-1} dt$$

$$= B\sigma_2^N\left(t_2^{N_1} - t_1^{N_1}\right) = B\left(\frac{\sigma_o}{2}\right)^{2.5}\left(6.5^{0.35} - 2^{0.35}\right)$$

$$= 0.1151B\sigma_o^{2.5}$$

   - For a time period of 1.5 h = (8 – 6.5) h at a stress $\sigma = \sigma_3 = 5\sigma_o/4$, similar calculations give

$$\varepsilon_3^c = B\sigma_3^N\left(t_3^{N_1} - t_2^{N_1}\right) = B\left(\frac{5\sigma_o}{4}\right)^{2.5}\left(8^{0.35} - 6.5^{0.35}\right)$$

$$= 0.2535B\sigma_o^{2.5} = 0.2535$$

* See Krause,[13] p. 18.

Hence, the total accumulated strain over the time period (2 + 4.5 + 1.5) = 8 h, for σ_o = B = 1,

$$\left(\varepsilon^c\right)_{\text{total}} = \varepsilon_1^c + \varepsilon_2^c + \varepsilon_3^c = 1.6432$$

b. *Strain-hardening Rule* — The strain accumulated for each loading and time interval is estimated according to the following:

- For a time period of 2 h at a stress $\sigma = \sigma_1 = \sigma_o$, the same strain ε_1^c is the same as found according to the time-hardening rule, i.e.,

$$\varepsilon_1^c = 1.2746 B\sigma_o^{2.5}$$

To determine the accumulated strain $\varepsilon_2{}^c$ at the end of the time interval t_2, Equation 11.143 is integrated as follows:

$$\int \dot{\varepsilon}^c = \int \frac{d\varepsilon^c}{dt} = \int B^{1/N} N_1 \sigma_2{}^{N/N_1} \left(\varepsilon^c\right)^{(N_1-1)/N_1}$$

Separation of variables and integrating between the limits t_1 and t_2 and $\varepsilon_1{}^c$ and $\varepsilon_2{}^c$ for a constant stress σ_2 yields

$$t_2 - t_1 = \int_{\varepsilon_1^c}^{\varepsilon_2^c} \frac{1}{N_1} B^{-1/N_1} \sigma_2^{-N/N_1} \left(\varepsilon^c\right)^{(1-N_1)/N_1} d\varepsilon^c$$

Performing the integration and rearranging to obtain $\varepsilon_2{}^c$ gives

$$\varepsilon_2^c = \left[(t_2 - t_1) B^{1/N_1} \sigma_2^{N/N_1} + \left(\varepsilon_1^c\right)^{1/N_1}\right]^{N_1}$$

- For a time period of 4.5 h = (6.5 – 2) h, t_2 = 6.5 h, t_1 = 2 h, $\sigma_2 = \sigma_o/2$, and ε_1^c = $1.2746B\sigma_o^{2.5}$, the above results in

$$\varepsilon_2^c = \left[(6.5-2) B^{1/N_1} \left(\sigma_o/2\right)^{N/N_1} + \left(1.2746 B\sigma_o^{2.5}\right)^{1/N_1}\right]^{N_1}$$

Recalling that $B = \sigma_o = 1$, $N_1 = 0.35$, and $N_2 = 2.5$ gives $\varepsilon_2{}^c = 1.28166$.

- For a time period of 1.5 h = (8 – 6.5) h, the use of the same expression for ε_3^c and $\sigma_3 = 5\sigma_o/4 = 5/4$ yields

$$\varepsilon_3^c = \left[(8-6.5) B^{1/N_1} (5/4)^{N/N_1} + (1.28166)^{1/N_1}\right]^{N_1}$$

$$\varepsilon_3^c = \left(\varepsilon^c\right)_{\text{total}} = 2.1921$$

which is the total creep strain accumulated at the end of t_3 = 8 h.

This is different from that predicted by the time-hardening rule.

Loading sequence (2)

a. *Time-hardening Rule* — The total strains accumulated during the three time intervals for the given loading conditions are estimated by integrating Equation 11.140 directly as demonstrated in the calculations of the first loading sequence; hence,

$$\left(\varepsilon^c\right)_{\text{total}} = \varepsilon_1^c + \varepsilon_2^c + \varepsilon_3^c$$

$$\left(\varepsilon^c\right)_{\text{total}} = 1.2746B\sigma_o^{2.5} + B\left(5\sigma_o/4\right)^{2.5}\left(6.5^{0.35} - 2^{0.35}\right) + B\left(\sigma_o/2\right)^{2.5}\left(8^{0.35} - 6.5^{0.35}\right)$$

$$\left(\varepsilon^c\right)_{\text{total}} = 2.4372; \text{ for } \sigma_o = B = 1$$

b. *Strain-hardening Rule* — The strain accumulated for each loading and time interval is estimated according to the following:

- For a time period of 2 h at a stress $\sigma = \sigma_o$, the same strain ε_1^c is found according to time-hardening rule, i.e.,

$$\varepsilon_1^c = 1.2746B(\sigma)_o^{2.5}$$

- For a time period of 4.5 h at a stress $\sigma = 5\sigma_o/4$, the same expression used in loading sequence (1) gives

$$\varepsilon_2^c = \left[(6.5-2)B^{1/N_1}\left(5\sigma_o/4\right)^{N/N_1} + \left(1.2746B\sigma_o^{2.5}\right)^{1/N_1}\right]^{N_1}$$

This yields for $\sigma_o = B = 1$, $N = 2.5$, and $N_1 = 0.35$:

$$\varepsilon_2^c = 3.04816$$

- For a time period of 1.5 h = (8 – 6.5) h at a stress $\sigma = \sigma_o/2 = 1/2$, the same procedure is followed to give

$$\varepsilon_2^c = \left[(8-6.5)B^{1/N_1}(1/2)^{N/N_1} + (3.04816)^{1/N_1}\right]^{N_1}$$

$$\varepsilon_3^c = \left(\varepsilon^c\right)_{\text{total}} = 3.04862$$

which is the total creep strain accumulated at the end of $t_3 = 8$ h.

Again, this is different from that predicted by the time-hardening rule. It is observed that for both sequences of loading the time-hardening rule predicts lower strains than these predicted by the strain hardening rule. Note that the difference between the two loading sequences is in the order of applying the last two stresses $\sigma_o/2$ and $5\sigma_o/4$.

11.4.12 Creep Rupture and Damage Concept

In the previous sections, the analysis has been focused on secondary creep deformation to provide the designer with limits against failure due to excessive creep deformation. Now, it is required to consider creep rupture that involves actual breakage of the component. Obviously, this involves the three stages of creep deformation including the more-complicated tertiary creep stage up to rupture point. The analysis in this section is concerned only with the ultimate point of rupture, namely, the analytical determination of creep rupture time for the simple case of uniaxial tensile creep. In this context, two possible creep rupture modes are generally identified; the first is ductile and the second is brittle.

11.4.12.1 Ductile Creep Rupture under Uniaxial Stress

Consider a rod of an initial uniform cross sectional area A_o undergoing creep deformation due to a constant uniaxial tensile load P. To analyze ductile rupture, the rod is assumed to elongate uniformly up to the rupture point at a current value of strain-rate given by

$$\dot{\varepsilon}^c = \frac{1}{L}\frac{dL}{dt} = -\frac{1}{A}\frac{dA}{dt} \tag{11.144}$$

where L and A are the current length and cross sectional area of the bar at any instant. For a process of long duration, primary creep can be ignored since secondary creep constitutes the major part of creep life. Hence, Equation 11.87 may be employed with Equation 11.144 to give, for a current stress value $\sigma = P/A$,

$$-\frac{1}{A}\frac{dA}{dt} = B\left(\frac{P}{A}\right)^N \tag{11.145}$$

Rearranging and integrating give

$$-\frac{1}{N}A^N = BP^N t + C \tag{11.146}$$

The constant C is determined from the initial conditions: at $t = 0$; $A = A_o$ as $C = -A_o^N/L$. Hence, Equation 11.146 results in a relation between current area A and time t as

$$t = \frac{1}{BN}\frac{1-(A/A_o)^N}{\sigma_o^N} \tag{11.147}$$

where $\sigma_o = P/A_o$ is the initial stress. Ductile rupture occurs when $A = 0$ and thus the ductile rupture time t_{RD} (due to Hoff) is given by

$$t_{RD} = \frac{1}{BN}\left(\frac{1}{\sigma_o}\right)^N \tag{11.148}$$

On a plot of log σ_o vs. log t_{RD}, Figure 11.25, Equation 11.145 indicates a descending line representing ductile creep rupture curves qualitatively similar to Figure 11.14. However, in practice, Equation 11.148 does not agree with exprimental creep rupture curves because of neglect of initial elastic and plastic effects, as well as primary creep. The agreement appears to be more adequate for loading causing a long secondary creep stage; hence, Expression 11.148 may be a better fit for the left portion of the creep rupture curves of Figure 11.14. Note that a finite rupture time is predicted for each finite load P and, hence, initial stress σ_o, regardless of whether or not the current area and stress, respectively, go to zero and infinity simultaneously.

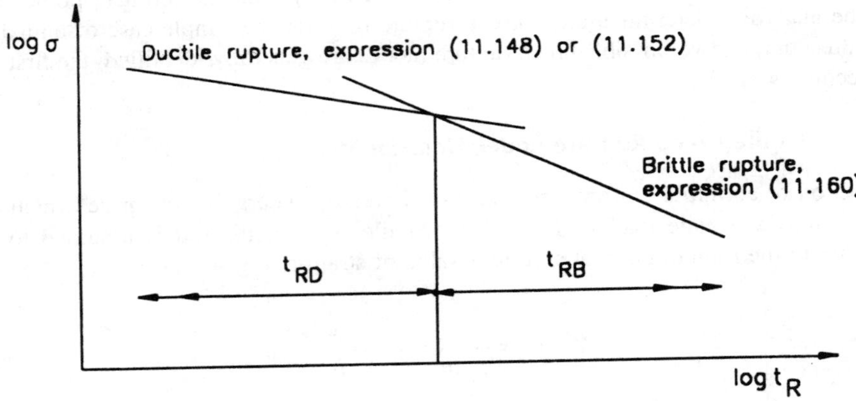

FIGURE 11.25 Prediction of ductile or brittle creep rupture times.

At higher initial stresses, which are associated with the left portion of the creep rupture curve, the initial plastic strain may be of an appreciable value. The analysis can be extended to include the initial plastic strain ε^p characterized by the power law (Equation 10.8b): $\sigma = K(\varepsilon^p)^n$. The total strain-rate at any instant t is thus given by

$$\dot{\varepsilon} = -\frac{1}{A}\frac{dA}{dt} = \dot{\varepsilon}^p + \dot{\varepsilon}^c \tag{11.149}$$

Substitution from Equations 10.8b for $\dot{\varepsilon}^p$ and 11.87 for $\dot{\varepsilon}^c$, respectively, into Equation 11.149 yields

$$-\frac{1}{A}\frac{dA}{dt} = \frac{d}{dt}\left(\frac{\sigma}{K}\right)^{1/n} + B\sigma^N$$

For a current stress $\sigma = P/A$, this becomes

$$-\frac{1}{A}\frac{dA}{dt} = \left[-\frac{1}{n}\left(\frac{P}{K}\right)^{1/n} A^{-1-1/n}\right]\frac{dA}{dt} + \left(\frac{P}{A}\right)^N \tag{11.150}$$

or

$$\left[-\frac{1}{A} + \frac{1}{n}\left(\frac{P}{K}\right)^{1/n} A^{-1-1/n}\right]\frac{dA}{dt} = \left(\frac{P}{A}\right)^N$$

Rearranging and integrating with the initial conditions, at $t = 0$, $A = A_o$ gives

$$A_o^N - A^N + \frac{N}{Nn-1}\left(\frac{P}{K}\right)^{1/n}\left[A^{N-1/n} - A_o^{N-1/n}\right] = NBP^N t \tag{11.151}$$

Ductile rupture time t^*_{RD}, is thus predicted from Expression 11.151 by putting $A = 0$, hence, for $\sigma_o = P/A_o$,

$$t^*_{RD} = \frac{1}{BN}\left(\frac{1}{\sigma_o}\right)^N\left[1 - \frac{N}{Nn-1}\left(\frac{\sigma_o}{K}\right)^{1/n}\right] \tag{11.152}$$

Clearly, the initial plastic strain becomes negligible if σ_o/K approaches zero,a case which reduces to that given by Hoff's rupture time t_{RD} of Expression 11.148. Equation 11.152 predicts infinite rupture time when $N = l/n$ and, hence, ductile rupture would never occur. Since for most materials the strain hardening coefficient n is less than the exponent N in Norton's creep law. In fact, $n < 1$ and $N > 1$ as may be seen from Tables 10.1 and 11.6, respectively.

11.4.12.2 Creep Damage Concept

In reality, creep rupture does not occur when the specimen area diminishes to become a point, i.e., $A = 0$. Generally, at rupture the area remains finite, say, $\alpha_R A_o$, where α_R is a fraction less than unity. This describes brittle fracture where material deterioration or damage in the form of internal microscopic fissuration and cavitation takes place at elevated temperatures. Clearly, for a proper understanding of this phenomenon, its physical description on the microscopic level has to be explored. However, a phenomenological approach describing "damage" may be used to predict creep rupture times by defining a new damage parameter as:[18]

$$\Omega = 1 - A_R/A \quad 0 \le \Omega \le 1 \tag{11.153}$$

Here, A_R (being different from the current area A) is the area after some internal damage has occurred and it represents the actual remaining area that supports the load at a net stress:

$$\sigma_R = \frac{P}{A_R} = \frac{\sigma}{1-\Omega} \tag{11.154}$$

Note that the damage parameter Ω goes from zero initially to unity at rupture, and hence the net stress σ_R is always larger than the current stress σ. The damage rate $\dot{\Omega}$ may be assumed to depend on the net stress σ_R according to the simple power law:

$$\frac{d\Omega}{dt} = \dot{\Omega} = B_1 \sigma_R^{N_2}$$

where B_1 and N_2 are material parameters. Rearranging and integrating between the limits $t = 0$ and $t = t_{RB}$, where t_{RB} stands for creep brittle rupture time, results in

$$\int_0^1 (1-\Omega)^{N_2} d\Omega = B_1 \int_0^{t_{RB}} \sigma^{N_2} dt \tag{11.155}$$

If σ is a known function of time, Expression 11.155 may be used to determine rupture time t_{RB} thus leading to a different form for creep brittle rupture time t_{RB} as

$$B_1 \int_0^{t_{RB}} \sigma^{N_2} dt = \frac{1}{N_2 + 1} \tag{11.156}$$

Introducing t_{RK} as that particular value of t_{RB} for which $\sigma = \sigma_K =$ constant gives

$$B_1 \sigma_K^{N_2} t_{RK} = \frac{1}{N_2 + 1} \tag{11.157}$$

Hence, Expression 11.156 becomes

$$\int_0^{t_{RB}} \left(\frac{\sigma}{\sigma_K} \right)^{N_2} dt = t_{RK} \tag{11.158}$$

Under a complex state of stress, σ in the above criterion is replaced by either σ_{max}, the maximum principal stress, or $\bar{\sigma}$ the effective stress depending on the type of material. This rupture criterion is supported experimentally and used to analyze creep rupture problems such as pressurized cylinders, torsion of tubes and bending of beams.[14] Note that for a constant stress σ_o, Expression (11.157) plots on a log σ–log t_{RB} as a straight line, which slopes down to the right. Again, this has the correct experimental form of a stress rupture curve as shown in Figure 11.25 provided that $N_2 < N$ and $B_1 > B$ are implied as observed experimentally for several materials.

If creep occurs under a stress program, i.e., $\sigma = \sigma(t)$, a linear cumulative creep damage concept* may be applied to give the total rupture time. Combining Equations 11.155 and 11.157 gives

$$\sum_{K=1}^{K} \frac{\Delta t_K}{t_{RK}} = 1 \tag{11.159}$$

where t_{RK} is the rupture time corresponding to a constant stress σ_K and Δt_K is the time spent at this given stress.

Example 11.20:

A rod made of the stainless steel alloy described in Example 11.9 is subjected to a creep testing program composed of the following constant tensile stresses and temperatures:

$$\sigma_1 = 407 \text{ MPa at } T_1 = 450°\text{C for } \Delta t_1 = 25 \text{ h}$$

$$\sigma_2 = 243 \text{ MPa at } T_2 = 505°\text{C for } \Delta t_2 = 150 \text{ h}$$

Determine how many additional hours Δt_3 would elapse before the rod ruptures at a stress $\sigma_3 = 370$ MPa and temperature of $T_3 = 390°$C.

Solution:

Prediction of the additional hours to ruture Δt_3 is found from the law of linear cumulative damage (Equation 11.159):

$$\sum_{K=1}^{3} \frac{\Delta t_K}{t_{RK}} = \frac{\Delta t_1}{t_{R1}} + \frac{\Delta t_2}{t_{R2}} = \frac{\Delta t_3}{t_{R3}} = 1$$

where t_{R1}, t_{R2}, and t_{R3} are the rupture times corresponding to the constant stresses 407, 243, and 370 MPa, respectively. Note that these stresses are applied at constant temperatures 450, 505, and 390°C, respectively.

The creep rupture data of the given stainless steel is fitted in Example 11.9 using the Larson–Miller parameter P_{LM} as

* The similarity of this law with the law of cumulative fatigue damage $\sum_{K=1}^{K} (\Delta N_K/N_{FK})$ is self-evident where the number of cycles N replaces time t.

$$\ln \sigma = 10.729 - 2.9145 \times 10^{-4} P_{LM}$$

where $P_{LM} = T(C + \log_{10} t_R)$; $C = 20$.

Applying these two expressions to the data given above yields the rupture times:

$$t_{R1} = 251.8\ h \text{ at } \sigma_1 = 407 \text{ MPa and } T_1 = 450°C$$

$$t_{R2} = 1233.8\ h \text{ at } \sigma_2 = 243 \text{ MPa and } T_2 = 505°C$$

$$t_{R3} = 8.3403 \times 10^4\ h \text{ at } \sigma_3 = 370 \text{ MPa and } T_3 = 390°C$$

Therefore, direct substitution yields

$$\frac{25}{251.8} + \frac{150}{1233.8} + \frac{\Delta t_3}{83403} = 1$$

or

$$\frac{\Delta t_3}{83,403} = 0.7791$$

hence, $\Delta t_3 = 64,982.6$ h.

Note that subjecting the tensile rod to a loading history of (σ_1, T_1) and (σ_2, T_2) for a total of 175 h reduces the rupture time to about 78% of the total rupture time if it would have been subjected to a single loading (σ_3, T_3).

11.4.12.3 Brittle Creep Rupture under Uniaxial Stress

Equation 11.151 may be used here to establish a brittle rupture time t_{RB} for a rod subjected to a constant tensile stress σ_o. Rupture time is the time elapsed to develop rupture at some cross-sectional area $\alpha_R A_o$ as

$$t_{RB} = \frac{1}{BN}\left(\frac{1}{\sigma_o}\right)^N \left[\left(1-\alpha_R^N\right) - \frac{N}{Nn-1}\left(\frac{\sigma_o}{K}\right)^{1/n}\left(1-\alpha_R^{N-1/n}\right)\right] \tag{11.160}$$

The parameter α_R is not known ahead and resort is made to the damage concept to determine α_R, namely, Expression 11.158:

$$\int_o^{t_{RB}} \left(\frac{\sigma}{\sigma_K}\right)^{N_2} dt = t_{RK}$$

Since, $\sigma = P/A = \sigma_o A_o/A$, then

$$t_{RK} = \left(\frac{\sigma_o}{\sigma_K}\right)^{N_2} A_o^{N_2} \int_{A_o}^{\alpha_R A_o} \frac{dA}{A^{N_2}} \frac{dt}{dA}$$

Substituting for dA/dt from Expression 11.150 and integrating gives

$$t_{RK} = \frac{1}{B}\left(\frac{\sigma_o}{\sigma_K}\right)^{N_2}\left(\frac{1}{\sigma_o}\right)^{N}\left[\frac{1-\alpha_R^{N-N_2}}{N-N_2} - \left(\frac{\sigma_o}{K}\right)^{1/n}\frac{\left(1-\alpha_R^{N-N_2-1/n}\right)}{n\left(N-N_2-1/n\right)}\right] \tag{11.161}$$

In this expression, all terms are known except α_R. Once α_R is determined, the rupture time is obtained from Equation 11.161 for brittle rupture if the condition expressing transition from ductile to brittle rupture is satisfied, $t_{RB} \le t_{RD}^*$. This condition is rewritten using Equations 11.152 and 11.161 as

$$\alpha_R \ge \left(\frac{\sigma_o}{K}\right)\left(\frac{N}{Nn-1}\right)^n \tag{11.162}$$

Example 11.21:

*Investigate the rupture mode whether ductile or brittle for a tensile rod at 600°C made of an alloy with the following properties:**

Plastic behavior: $\sigma = K(\varepsilon)^n$, $K = 833.9$ *MPa, and* $n = 0.222$

Secondary creep: $\dot{\varepsilon}^c = B(\sigma)^N$, $B = 3.232 \times 10^{-16}$, *and* $N = 5$

Damage: $\dot{\Omega} = B_l \sigma_{R^2}^{N}$, $N_2 = 3.5$

Rupture data: $t_{RK} = 10^5$ *h at* $\sigma_K = 87.8$ *MPa*

Consider the two cases when the applied stress is (a) $\sigma_o = 140$ *MPa and (b)* $\sigma_o = 200$ *MPa.*

Solution:

At any applied stress σ_o, brittle rupture occurs at a time t_{RB} as given by Equation 11.160. Substitution into this expression by $B = 3.232 \times 10^{-16}$, $N = 5$, $N_2 = 3.5$, $n = 0.222$, and $K = 833.9$ MPa results in, for any σ_o,

$$t_{RB} = \frac{1}{3.232\times10^{-16}\times5}\left(\frac{1}{\sigma_o}\right)^5\left[\left(1-\alpha_R^5\right) - \frac{5}{5\times0.222-1}\left(\frac{\sigma_o}{833.9}\right)^{1/0.222}\left(1-\alpha_R^{5-1/0.222}\right)\right]$$

$$t_{RB} = 6.1881\times10^{14}\left(\frac{1}{\sigma_o}\right)^5\left[1-\alpha_R^5 - 3.158\times10^{-12}\sigma_0^{4.5}\left(1-\alpha_R^{0.5}\right)\right]$$

In this expression α_R is unknown, and it has to be determined from Expression 11.161 using the specific rupture data $t_{RK} = 10^5$ hr at $\sigma_K = 87.8$ MPa; hence,

$$10^5 = \frac{1}{3.232\times10^{-16}}\left(\frac{\sigma_o}{87.8}\right)^{3.5}\left(\frac{1}{\sigma_o}\right)^5 \left[\frac{\left(1-\alpha_R^{5-3.5}\right)}{5-3.5} - \left(\frac{\sigma_o}{833.9}\right)^{1/0.222}\left(\frac{1}{0.222}\right)\frac{1-\alpha_R^{5-3.5-1/0.222}}{5-3.5-1/0.222}\right] \tag{a}$$

* The given material parameters match closely with those given for T_i, stabilized austenitic steel 25Cr–20Ni. See Table 15.1 in Odqvist.[14]

a. For σ_o = 140 MPa, the above expression may be solved numerically to give

$$\alpha_R = 0.61904$$

For brittle rupture, this value of α_R satisfies the Condition 11.162, namely,

$$\alpha_R \geq \left(\frac{\sigma_o}{K}\right)\left(\frac{N}{Nn-1}\right)^n \geq \left(\frac{\sigma_o}{833.9}\right)\left(\frac{5}{5\times 0.222-1}\right)^{0.222} \geq 2.794\times 10^{-3}\sigma_o$$

or

$$\alpha_R =\geq 0.3912 \text{ for } \sigma_o = 140 \text{ MPa}$$

Substituting with α_R = 0.62 together with σ_o = 140 MPa in the expression for t_{RB} gives

$$t_{RB} = 10{,}424 \text{ h}$$

This rupture time as expected, is less than that for ductile rupture t^*_{RD} = 11,337 h calculated using Equation 11.152.

b. For higher stress σ_o = 200 MPa, α_R is to be determined by substitution into Equation 11.161, or Expression a above to give

$$\alpha_R^{4.5} - 0.134\alpha_R^3 + 0.003646 = 0$$

This equation does not have a real positive root and the brittle rupture mode is not physically anticipated. A ductile rupture mode may occur due to this high stress at time t^*_{RD} given by Expression 11.152 as

$$t^*_{RD} = \frac{1}{3.232\times 10^{-16}\times 5}\left(\frac{1}{200}\right)^5\left[1-\frac{5}{5\times 0.222-1}\left(\frac{200}{833.9}\right)^{1/0.222}\right]$$

$$= 1792.3 \text{ h}$$

PROBLEMS

Unstable Plastic Deformation

Plastic Instability of Uniaxially Loaded Bars

11.1 An aluminum alloy specimen is tested in simple tension. The effective stress–effective strain behavior is given by $\bar{\sigma} = 150(0.12 + \bar{\varepsilon})^{0.25}$ (MPa). Estimate the maximum uniform strain at instability: (a) neglecting elastic effects and (b) considering elastic strains and taking E = 70 GPa.

11.2 In a tensile test on round steel bars the following data are obtained:

Load (kN):	0	103 (max)	81.8 (fracture)
Minimum diameter (mm):	12.8	12.1	8.94

(a) Determine the material constants K and in the law $\bar{\sigma} = K\,\bar{\varepsilon})^n$. (b) Using the Bridgman correction, what are the maximum and average longitudinal stresses in the necked cross section at fracture?

11.3 A tensile test is conducted on a specimen of initial dimensions of 7 mm diameter and 50 mm gauge length. The following load–extension readings were obtained

Load (kN):	13.3	16.0	18.7	20.5(max)	14.7 (fracture)
Extension (mm):	2.1	5.3	10.2	21.8	24

a. Determine the true stress-strain behavior and fit the law $\sigma = K(\varepsilon)^n$ up to the instability point.
b. Apply the Bridgman correction to refit the same law up to fracture using the empirical relations $r_N/R = \sqrt{\varepsilon_N - n}$, knowing that the diameter at the fracture section is 3.74 mm.

11.4 A thin metal strip is subjected to uniaxial tensile loading while maintaining plane strain conditions. Determine the strains and the effective stress at instability. Take a material law of the form $\bar{\sigma} = K(\bar{\varepsilon})^n$.

11.5 Determine the instability strain for a specimen that is subjected to a constant hydrostatic pressure p, then loaded in monotonically increasing uniaxial tension. Take a material law of the form $\bar{\sigma} = K(\bar{\varepsilon})^n$.

11.6 A rolled steel sheet of thickness 1.5 mm ± 0.003 is to be formed by stretching uniaxially. Using the material law $\sigma = 630(\varepsilon)^{0.25}$ (MPa), determine

a. The maximum uniform strains at the onset of diffuse and localized necking while neglecting the initial thickness variation in the sheet.
b. Considering the initial thickness variation, how does this affect the answer obtained in (a)?

Plastic Instability of Biaxially Stretched Sheets

11.7 A circular grid of unit radius is marked on a 70/30 annealed brass sheet of thickness h_o. After biaxial stretching, the circles deform into ellipses described by the relation $3x^2 + 2y^2 = 6$.

a. Determine the principal strains and final thickness at this stage of deformation.
b. If deformation continues proportionally until plastic instability, what are the limiting strains for a material obeying the law $\bar{\sigma} = K(\bar{\varepsilon})^n$?

11.8 A sheet metal is stretched biaxially in the condition of plane stress; $\alpha = \sigma_2/\sigma_1$ and $\sigma_3 = 0$. Determine the limit strains at the onset of plastic instability if the flow curve of this metal is given by $\bar{\sigma} = 883(\bar{\varepsilon})^{0.49}$ MPa.

11.9 A sheet of brass is stretched biaxially in the condition of plane strain. (a) Determine the critical value of the effective strain and stress at which instability occurs. The stress–strain curve of the material is given by $\bar{\sigma} = 730(0.129 + \bar{\varepsilon})^{0.48}$ (MPa). (b) What are the stress and strain components at this instant?

11.10 A thin circular sheet of thickness 1.5 mm is made of an annealed copper obeying the law $\bar{\sigma} = 344\,\bar{\varepsilon})^{0.54}$ (MPa). The sheet is bulged by hydrostatic pressure in a circular die of an aperture of 175 mm diameter. Establish a graph for the bulging pressure p and the bulge height H up to the point of plastic instability.

Plastic Instability of Thin-Walled Containers

11.11 An aluminum sphere 500 mm mean diameter has a wall thickness of 10 mm. If the material behaves according to the law $\bar{\sigma} = 170(\bar{\varepsilon})^{0.25}$ (MPa), calculate the dimensions of the sphere and the internal pressure at the onset of plastic instability.

11.12 It is required to replace a spherical tank for compressed air by a cylindrical one of the same volume and material. Taking a cylinder of length-to-diameter ratio of 4, (a) determine its thickness as a ratio of the spherical tank thickness. (b) Using another material for the cylindrical tank, what would be the required strength for a cylindrical tank having the same thickness as the spherical one? Take the occurrence of plastic instability as a design basis.

11.13 An open-ended, thin-walled cylinder, made from an alloy whose stress–strain curve is given by $\bar{\sigma} = 870\,\bar{\varepsilon})^{0.161}$ (MPa), is subjected to an internal pressure p_i. If the initial tube dimensions are $r_{mo} = 100$ mm and $h_o = 2.5$ mm, (a) determine the dimensions of the cylinder at the instant of plastic instability and the associated maximum pressure. (b) Compare the results with a closed cylinder.

11.14 A pressure vessel is constructed from a thin-walled cylinder and two hemispherical closures, all the same thickness, $h_o/r_{mo} = 0.1$. The vessel is to be tested under internal pressure by a value, the smaller of (a) 0.9 of the pressure required to initiate yielding or (b) 0.6 the pressure to cause plastic instability in any part of the vessel. Determine the test pressure for both conditions. Take $Y = 340$ MPa and $\bar{\sigma} = 780(\bar{\varepsilon})^{0.23}$ (MPa). Neglect discontinuity stresses at the junction of closures.

11.15 A thin-walled pipe with closed ends is subjected to an internal pressure p_i. The pipe is made of stainless steel for which the plastic behavior is represented by $\bar{\sigma} = 1500\,\bar{\varepsilon}\,)^{0.48}$ (MPa).

a. Using the maximum pressure instability criterion, calculate the maximum pressure that the pipe can withstand, knowing that the initial pipe dimensions are 1.2 mm thickness and 130 mm mean radius.
b. Establish a table showing the variation of the pressure with pipe radius for effective strains $\bar{\varepsilon} \geq 0.03$.

11.16 A uniform thin-walled pipe of mean radius r_{mo} just fits over a solid (rigid) rod. Find the tensile load the tube can support before plastic instability occurs. Assume frictionless sliding between the tube and the rod to which no load is applied. Use a law, $\bar{\sigma} = K(\bar{\varepsilon})^n$, for the tube material. Compare the result with that of simple tension. (Apply the results of Section 11.2.5.)

11.17 A thin-walled tube is subjected to an axial load closed internal pressure P and such that the following stress ratio holds throughout the test: $\sigma_z = (1 + \alpha)\sigma_\theta$ where $\alpha = P/(\pi r_m^2 p_i)$. Determine the point of plastic instability for maximum axial load and maximum internal pressure.

11.18 The plastic stress–strain behavior of a steel is given by $\bar{\sigma} = 760\,\bar{\varepsilon}\,)^{0.16}$ MPa. A thin-walled pipe made of this steel with $r_{mo} = 75$ mm and $h_o = 2.5$ mm is subjected to axial tension and internal pressure such that $\sigma_\theta/\sigma_z = 1/2$ where θ and z denote the circumferential and axial directions, respectively. Determine the tube dimensions and the pressure at plastic instability.

11.19 A thin-walled tube of mean radius r_{mo} and thickness h_o subjected to a twisting moment M_t is made of a material obeying the law: $\bar{\sigma} = K(\bar{\varepsilon})^n$. Show that plastic instability cannot occur for $n > 0$.

Strain-Rate-Dependent Behavior — Superplasticity

11.20 In superplastic material testing, the strain-rate-sensitivity exponent m, may be determined by sudden step-changes in strain rate (or simply in the cross-head speed of the testing machine). If the strain rate is suddenly increased by a factor of 5, the flow stress level rises by 80%. What is the value of m? Establish a table indicating how the total elongation of rods made of this superplastic alloy is affected by the presence of an initial cross sectional imperfection f.

11.21 Derive an expression to predict the limit strain in a tensile rod with an initial imperfection f assuming a material behavior given by $\bar{\sigma} = C(\bar{\varepsilon}^n(\dot{\bar{\varepsilon}})^m$. If $n = 0.2$ and $m = 0.1$, what would be the limit strain for rods with $f = 0.98$? Compare the results with that obtained by neglecting either n or m separately.

11.22 In superplastic bulging, the average strain rate has to be controlled within a given range where the material superplastic properties are optimum, i.e., m is maximum. (a) Derive relations between the bulge pressure and time such that the average thickness strain rate $(\dot{\varepsilon}_h)_{av} = \dot{\varepsilon}_o$ as given by Equation 11.81, remains constant throughout the entire bulging process. (b) What would be the bulging pressures and times for H/a = 0.25, 0.5, and 1.

11.23 A circular sheet of thickness-to-radius ratio of 0.025 is to be bulged at constant pressure to form a hemispherical dome. Estimate the required pressure and time for both alloys:

a. Superplastic Zn–22Al: $m = 0.5$, $C = 7.2$ (MPa-min) at $\dot{\bar{\varepsilon}})_{av} = 1.0$ min^{-1} and 270°C.
b. Ti–6Al–4V: $m \simeq 1$, C = 55 (MPa-min) at $(\dot{\bar{\varepsilon}})_{av} = 0.1$ and 980°C.

Adjust the process parameters such that $\dot{\varepsilon})_{av}$ given above occurs at the end of forming.

Creep Deformation

11.24 In short-time creep tensile tests on specimens of 6.5 mm diameter and 75 mm length made of lead–tin alloy, the following data are obtained at room temperature:

- Load = 200 N:

Time, t (min):	40	60	85	123	143	170	200	220	
Elongation (mm):	1.02	1.21	1.35	1.6	1.71	1.86	2.02	2.13	

- Load = 300 N:

Time, t (min):	20	30	62	81	90	120	152	180	200
Elongation (mm):	2.04	2.33	3.1	3.54	3.72	4.39	4.92	5.75	6.19

a. Plot the two creep curves and identify the primary, secondary, and tertiary (if any) creep stages.
b. Assuming that Norton's creep law $\dot{\varepsilon}^c = B(\sigma)^N$ apply for secondary creep, determine the constants B and N.

11.25 The following data were recorded during creep rupture testing for an alloy steel:

Temperature (°C)	Stress (MPa)	Time to Rupture (h)
500	320	4704
600	190	590
700	105	277
800	60	96.3
1000	35	8.1

If a component made from this material is required to last at least 1 year at a stress of 140 MPa, what is its maximum permissible service temperature? Apply the Larson–Miller extrapolation parameter for this alloy.

11.26 Uniaxial creep rupture tests resulted in the following data: 10^3 h at 600°C and 300 MPa and 10^2 h at 500°C and 450 MPa. Using Larson-Miller interpolation parameter, determine:

a. Rupture time at temperature 475°C and stress 365 MPa,
b. Operating temperature for rupture time 20,000 h at 270 MPa.

Creep of Beams

11.27 Two beams made of an aluminum alloy obeying the creep law $\dot{\varepsilon}^c = 11.2 \times 10^{-13}\,\sigma^4$ (MPa-h) at 100°C. The maximum creep deflection at 100°C is to be restricted to 10 mm in 10,000 h. Determine the required dimensions of the square cross sections of each beam knowing that the first beam is simply supported over a span of 800 mm and loaded by 6.5 kN at its midspan. The second beam is supported as a cantilever with an overhang of 750 mm and loaded at its free end.

11.28 A 600-mm-long simply supported beam of 26 mm width and 50 mm height is acted upon at the middle of its span by a concentrated force $P = 5000$ N operating at $T = 500$°C. The beam is made of low-carbon steel having a yield strength of 150 MPa and a modulus of elasticity, $E = 160$ GPa (at $T = 500$°C). The secondary creep equation of state, $\dot{\varepsilon}^c = B(\sigma)$ at $T = 500$°C, has the parameters $B = 1.5 \times 10^{-12}$ (MPa-h) and $N = 3.2$. Neglecting the zone of primary creep, determine the maximum total deflection of the beam after 1 year of loading.

11.29 A 500 mm long cantilever made of lead with rectangular section 50 mm wide × 25 mm deep carries an end load at 300°C. Neglecting elasticity and self-weight, determine the maximum end load such that the end deflection due to secondary creep after 5000 h does not exceed 10 mm. The seconadry creep rate is given by $\dot{\varepsilon} = 3.5 \times 10^{-7}\ \sigma^{1.75}$ (MPa-h). Determine the maximum stress in the beam under stationary conditions.

Creep of Pressure Vessels

11.30 The boiler tubes of the superheater section are 50 mm in diameter by 2 mm wall thickness. They operate at 500°C. It is desired that the radial expansion of the tubes over 10^5/h be less than 2%. Calculate the maximum allowable internal pressure that could be used for tubes of 2.25% Cr, 1% Mo steel with creep behavior N = 5.5 and $B = 1.45 \times 10^{-18}$ (MPa-h).

11.31 For an alloy steel, the following data are obtained from creep tests at 450°C:

Stress s (MPa):	68	105	145	180	210
Steady creep rate $\dot{\varepsilon}^c$ (h^{-1}):	9.83×10^{-8}	2.5×10^{-6}	3.8×10^{-5}	4×10^{-4}	1.34×10^{-3}

a. Represent this data on a log σ – log $\dot{\varepsilon}^c$ plot and determine the constants in the creep law $\dot{\varepsilon}_c = B(\sigma)^N$.
b. Boiler tubes of 60 mm diameter and 1.5 mm thickness made of the same alloy as in (a) operating at 450°C are designed such that the diametral expansion does not exceed 0.5% in 50,000 h. Estimate the maximum allowable internal pressure which could be applied.
c. Compare the pressure determined in (b) to that required for yielding, knowing that the yield strength at 450°C is Y = 350 MPa.
d. What would be the creep life for these tubes if an error of ±3% is made in determining the material parameter N?

11.32 Tensile creep tests at 500°C for a metallic alloy resulted in the following creep data:

σ (MPa):	80	100	120	140	160
$\dot{\varepsilon}^c$ h^{-1} (10^{-9}):	52	156	384	812	1575

Assuming that the Norton's creep $\dot{\varepsilon}^c = B(\sigma)^N$ law applies to this data, (a) determine the constants B and N. (b) A pipline of 40-mm bore and 2-mm wall thickness made of the above alloy is to carry fluid at pressure 0.2 MPa; determine the creep strain after 8 years.

11.33 A pressure vessel is composed of a thin-walled cylinder with hemispherical ends. The vessel operates under secondary creep conditions such that Norton's law, $\dot{\varepsilon}^c = B(\sigma)^{4.5}$, is applicable. Determine the ratio between the thickness of the hemispherical ends to that of the thin-walled cylinder according to either one of the conditions — (a) same hoop stress or (b) same circumferential creep rate — at the joint of the cylinder and its ends.

11.34 A spherical pressure vessel of 700-mm diameter and 8-mm thickness is made of stainless steel of the following properties at the operating temperature 450°C:

- Yield strength = 390 MPa, Young's modulus = 175 GPa, ν = 0.3;
- Strain-hardening exponent 0.12, strength coefficient K = 770 MPa (in the law $\bar{\sigma} = K(\bar{\varepsilon})^n$);
- Creep behavior for the vessel material N = 6.8 and $B = 1.47 \times 10^{-22}$ (in MPa-h) (in the law ($\dot{\bar{\varepsilon}} = B(\bar{\sigma})^N$).

The vessel is designed to operate at service pressure equal to 30% of its burst pressure; determine:

a. The service pressure,
b. The life for a creep diametral deformation of 0.5%.

11.35 A thick-walled cylinder of $r_o/r_i = 3$ stainless steel (25% Cr–20% Ni) is subjected to an internal pressure at a temperature of 600°C. Determine the service pressure for a life of 10^5 h based on either of the following design criteria:

a. Permissible tangential creep strain at the bore of 1.5% using a creep equation of state $\dot{\bar{\varepsilon}} = B(\bar{\sigma})^N$, where $B = 3.232 \times 10^{-16}$ and $N = 5$ at $T = 600°C$ (MPa-h);
b. Effective stress limit $S_r = 63.1$ MPa, inducing creep rupture in 10^5 h (take a factor of safety F.S. = 1.4);
c. Compare the above pressure values with that based on yielding knowing that the yield strength of the tube material $Y = 190$ MPa and its modulus of elasticity $E = 155$ GPa at temperature $T = 600°C$ (take F.S. = 1.6).

11.36 Determine the value of the internal pressure applied to a thick-walled cylinder of $\lambda = r_o/r_i = 2.5$, such that the creep strain at the bore does not exceed 2% under stationary creep conditions at 20,000 h. Also, calculate the stresses at bore and outer surface. Take material creep parameters as $N = 2.75$ and $B = 6.03 \times 10^{-11}$ (MPa-h).

Creep of Shafts under Torsion

11.37 Find the limiting value of σ_θ at $r = r_i$ for a rotating disk using Equation 11.18 for N = 1 and N = ∞. Compare the results for $r_o/r_i = 5$ and $v = 1/2$ with that using Equations 9.19 and 12.109.

11.38 A circular hollow rod of radii $r_o = 90$ mm and $r_i = 60$ mm is subjected to constant twisting moment $M_t = 15$ kN · m. Assuming steady creep conditions, determine the angular displacement after 50,000 h. Take $N = 7$, $B = 1.68 \times 10^{-20}$, and $E = 170$ MPa.

11.39 A solid shaft of 80-mm diameter is to be designed such that the maximum creep angular twist between two cross sections separated by length 600 mm is 2°. The shaft operates at 560°C and is made of a steel alloy (2.25% Cr–1% Mo) obeying the creep law $(\dot{\bar{\varepsilon}}) = 25 \times 10^{-19}\ \bar{\sigma})^{6.37}$ in MPa-h.

a. Determine the maximum twisting moment that can be applied to the shaft without violating the above design criterion over a service life of 50,000 h. Neglect primary creep.
b. Compare the results obtained in (a) to the case of a purely elastic response for the same shaft. Take $E = 152$ GPa and $\nu = 0.35$.
c. The stress rupture data for the same material at 567°C is given in the following table. Estimate the margin of safety of the service life using this stress rupture data.

Stress (MPa):	115	82	60	48	31
Rupture time (h):	10	10^2	10^3	10^4	10^5

11.40 Determine the total angular twist after 20,000, h of a circular hollow shaft of 60 mm and 40 mm outer and inner diameters, respectively. The shaft is 3 m long and carries a steady torque of 500 N · m. The shaft material alloy steel obeys the secondary creep law $\dot{\varepsilon}^c = 30 \times 10^{-10} (\sigma)^{3.2}$ (MPa-h) at 450°C. Take $E = 175$ GPa.

Creep Buckling of Columns

11.41 Two pin-ended columns of equal length and square cross sections of sides h and 2h, respectively, are subjected to compressive loads at an elevated temperature. If the critical time to buckling is the same for both columns, determine the ratio of their buckling loads and compare it with the ratio of Euler buckling loads for the two columns. The creep data for the column material is given by $\dot{\varepsilon}^c = 7.725 \times 10^{-21} (\sigma)^{6.37}$ (MPa-h). What would be the ratio of their buckling times if they were subjected to the same load?

Reference Stress Method

11.42 Identify a reference stress point, based on equal deflection rates within the range of $3 \leq N \leq 6$, for

a. Simply supported beam with a central load P,
b. Cantilever beam with an end load P.

11.43 A bar of a solid circular cross section is subjected to twisting moment M_t. Identify a reference stress by examining the stress distribution for elastic and perfectly plastic conditions. Derive an expression for the reference stress, for $6 \leq N \leq 8$.

Creep Relaxation

11.44 A flanged cover plate for a pipe connection of area 0.2 m^2 is held by 24 equally spaced steel bolts of 18 mm diameter. Assuming that the flanges are rigid, what should be the initial tightening stress and strain in the bolts in order that the connection remains tight (with a margin of safety of 1.5) after 1 year? The operating conditions are temperature 450°C and pressure 0.5 MPa. After what time should the bolts be retightened to prevent leakage? Take $E = 170$ MPa and apply a creep law $\dot{\varepsilon}^c = 45 \times 10^{-16}(\sigma)$ (MPa-h).

11.45 A long pipe of 400-mm diameter and 5-mm thickness is closed by two flanged connections tightened by 20 bolts of 15-mm diameter equally spaced on a pitch circle diameter of 500-mm. The service conditions are such that it operates at an internal pressure of 12 MPa while the temperature is maintained at 200°C. The material properties of all components of the pipe connections are

- Yield strength = 270 MPa, $\nu = 0.3$, $E = 175$ GPa;
- UTS = 450 MPa, $= K(\varepsilon)^{0.3}$;
- $\dot{\bar{\varepsilon}}^c = (50 \times 10^{-16})\ \bar{\sigma}\ )^{3.2}$ (MPa-h).

a. Check the possibility of failure due to bursting.
b. Check the possibility of failure due to leakage. Take an initial bolt tightening strain of 0.1%.

11.46 Derive an expression for uniaxial relaxation of a rod subjected to an initial stress σ_o according to the strain-hardening rule. If the rod relaxes to a stress $\sigma_i/2$, compare the relaxation time to that predicted by the time-hardening rule.

11.47 A spherical pressure vessel (700-mm diameter and 8-mm thickness) is fabricated from two halves which are joined by 36 bolts, each of 24-mm diameter located at a pitch circle diameter of 780 m. the vessel which operates at 450°C, is made of stainless steel of the following properties:

- Yielding strength = 390 MPa, Young's modulus = 175 GPa;
- Strain-hardening exponent $n = 0.12$,
- Strength coefficient $K = 770$ MPa (in the law $\bar{\sigma} = K(\ \bar{\varepsilon}\ )^n$).
- Creep behavior:
  For the vessel material: $N = 7.8$ and $B = 1.47 \times 10^{-22}$
  For the bolt material $N = 6.5$ and $B = 2.34 \times 10^{-22}$
  (in the law $(\ \dot{\bar{\varepsilon}}\ ) = B(\ \bar{\sigma}\ )^N)$ (MPa-h)

The vessel is designed to operate at a service internal pressure equal to 20% of its burst pressure; determine:

a. The service pressure;
b. The life for a creep diametral deformation of 1% (neglect the effect of the flanged connection);
c. The period after which leakage may occur, assuming an initial bolt tightening at 0.2% strain (neglect deformations in the flanges).

Creep Under Variable Loading and Creep Rupture

11.48 A bar made of a material that obeys the creep law $\bar{\varepsilon}^c = B(\sigma)^{3.2}t^{1/3}$. Determine the final strain for both the time-hardening and the strain-hardening rules when

a. A uniaxial stress σ_o acts for 2 h followed by the stress $2\sigma_o$ for another 2 h;
b. A uniaxial stress $2\sigma_o$ acts for 2 h followed by the stress σ_o for another 2 h.

Comment on the results. For simplicity, take $\sigma_o = B = 1$.

11.49 A uniaxial creep specimen is subjected to the following stress history:

- 120 MPa at 580°C for 20 h
- 150 MPa at 650°C for 100 hr
- 60 MPa at 720 C°for 350 hr

Using the data of Example 11.9, estimate how many additional hours at 115 MPa and 520°C the specimen can withstand.

11.50 A bar of an initial cross-sectional area $A_o = 100$ mm^2 is subjected to a tensile load of 15 kN. Estimate the rupture time according to

a. Hoff's ductile creep rupture Equation 11.148 with $N = 3.3$ and $B = 8.08 \times 10^{-13}$;
b. Ductile creep rupture (Equation 11.152) with same N and B, knowing that the bar material obeys the law $\sigma = K(\varepsilon)^n$ within the plastic range, $K = 280$ and $n = 0.18$.

11.51 Investigate the rupture mode whether ductile or brittle, for a tensile rod at 550°C made of an alloy with the following properties:

- Plastic behavior: $\sigma = K(\varepsilon)^n$, $K = 850$ MPa, and $n = 1/3$;
- Secondary creep: $\dot{\varepsilon}^c = B(\sigma)^N$, $B = 2.197 \times 10^{17}$, and $N = 6$;
- Damage: $\dot{\Omega} = B_1(\sigma)_R^{N_1}$, $N_1 = 3.8$,
- Rupture data: $t_{RK} = 10^4$ h at $\sigma_K = 90.8$ MPa

Consider the two cases when the applied stress is (a) $\sigma_o = 150$ MPa and (b) $\sigma_o = 180$ MPa.

REFERENCES

1. Considère, A., *Ann Ponts et Chaussees*, 9(66), 574–775, 1885.
2. Bridgman, P. W., *Rev. Mod. Phys.*, 17, 3, 1945; *Trans. Asm,* 32, 553, 1943.
3. Davidenkov, N. N. and Spiridonova, N. J., *Proc. Am. Soc. Test. Mater.,* 46, 1147, 1946.
4. Alexander, J. M., and Brewer, R. C., *Manufacturing Properties of Materials,* Van Nostrand, London, 1963.
5. Marciniak, Z. and Kuzynski, K., *Int. J. Mech. Sci.*, 9, 609, 1967.
6. Hill, R., A theory of the plastic bulging of a metal diaphragm by lateral pressure, *Philos. Mag.* (Ser.7), 41, 1133, 1950.
7. Chakrabarty, J. and Alexander, J. M., Hydrostatic bulging of circular diaphragms, *J. Strain Anal.*, 5(3), 155–161, 1970.
8. Chakrabarty, J., *Theory of Plasticity,* McGraw-Hill, New York, 1987.
9. Ragab, A. R., Thermoforming of superplastic sheet in shaped dies, *Met. Technol.,* 10, 340, 1983.
10. Cornfield, G. C. and Johnson, R. H., *Int. J. Mech. Sci.,* 12, 479, 1970.
11. Larson, F. R. and Miller, J., A time–temperature relationship for rupture and creep stress, *Trans. ASME,* 174, 765, 1952.
12. Manson, S. S. and Haferd, A. M., A linear time–temperature relation for extrapolation of creep and stress rupture data, NASA TN 2890, NASA, 1953.
13. Krauss, H., *Creep Analysis*, John Wiley & Sons, New York, 1980.
14. Odqvist, F. G., *Mathematical Theory of Creep and Creep Rupture*, Oxford Clarendon Press, London, 1966.
15. Wahl, A. M., Stress distributions in rotating disks subjected to creep at elevated temperatures, *J. Appl. Mech.*, 24(2), 299–305, 1957.

16. Boyle, J. T. and Spence, J., *Stress Analysis for Creep*, Butterworths & Co., London, 1983.
17. Duncan, J. L. and Ragab, A. R, Comparison of bending and tension in creep of sheet metal, *ASTM J. Test. Eval.*, 1(6), 451–456, 1973.
18. Kachanov, L. M., *Izv. Akad. Nauk.*, USSR, 8, 26, 1958.

12 Some Elastic-Plastic Problems

Analytical solutions of elastic–plastic problems have been possible in relatively few cases, in which some simplifying features such as plane strain or axial symmetry prevail. In this chapter analytical solutions to some elastic–plastic problems that have engineering applications are presented. These problems are characterized by direct determination of the elastic–plastic boundary and simple strain–compatibility expressions. The problems presented comprise plane strain and plane stress bending of bars, bending of curved beams, bars subjected to torsion, combined torsion and tension, and buckling of columns. The problem of axisymmetric bending of circular plates, pressurized thick-walled cylinders, and rotating disks of uniform thickness are analyzed. The solutions give the stresses and strains during loading from the onset of yielding to full yielding and plastic collapse and also determine the residual stresses upon unloading. The effect of strain hardening is considered in some applications.

12.1 INTRODUCTION

Analytical solutions of rigid–plastic or elastic–plastic problems, obtained in either an explicit form or through numerical integration, have been possible in relatively few cases in which some simplifying features, such as plane strain or axial symmetry prevail. One of the difficulties involved in the solution is to determine the elastic–plastic boundary along which certain continuity conditions in stresses and displacements have to be satisfied. Another difficulty lies in the application of the elastic–plastic stress–strain relations in which the elastic strain increments are related to the stress increments while the plastic strain increments are related to the total stresses. The integration of these differential stress–strain relations has to follow the history of the deformation from the initiation of yielding. It may be noted that the problems treated in the earlier chapter on plastic deformation, Chapter 10, were characterized by homogeneous states of stress and solutions were obtained basically from equilibrium considerations.

In this chapter analytical solutions to some elastic–plastic problems of practical importance are presented, these comprise plane strain bending of bars, combined bending and tension, bending of curved beams, bars subjected to torsion, combined torsion and tension, buckling of columns, axisymmetric bending of circular plates, thick-walled cylinders, and rotating disks of uniform thickness. These problems are characterized by direct determination of the elastic–plastic boundary and simple strain–compatibility expressions. The solutions give the stresses and strains during loading from initiation of yielding to full yielding. The residual stresses upon unloading are also determined by assuming elastic recovery. These residual stresses may be utilized to improve the load-carrying capacity under subsequent service loads.

The following subscript notation is adopted throughout the analysis: e, elastic state; Y, initial yielding; p, partial yielding; u, full or ultimate yielding; res, residual; T, Tresca yielding criterion; M, von Mises yield criterion, and f, final state. For example M_{YT} is the moment to initiate yielding according to the Tresca criterion, R_p is the radius of curvature at partial yielding, M_u is the moment at which the cross section has fully yielded, and ω_{YT} is the angular velocity at which yielding is initiated according to the Tresca criterion.

12.2 PLANE STRAIN BENDING OF PLATES

For a plate of a small thickness-to-width ratio *h*/*b*, Figure 12.1, the anticlastic curvature in the *y–z* plane is suppressed apart from small regions near the edges. It can be then assumed that the strain along the *z*-direction vanishes. Moreover, the conventional assumption that plane cross sections remain plane during bending results in a linear strain distribution. The stress normal to the plate surface is zero.

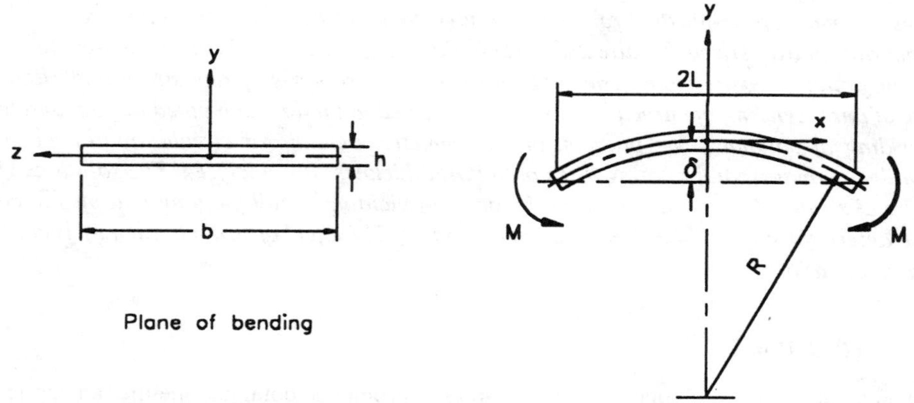

FIGURE 12.1 Plane strain bending of a strip.

According to these assumptions:

$$\varepsilon_z = 0,\ \varepsilon_x = y/R, \text{ and } \sigma_y = 0 \tag{12.1}$$

where *y* is the distance measured from the centroidal (neutral) axis and *R* is the radius of curvature of this axis in the *x–y* plane. Expression 12.1 holds for both elastic and plastic bending provided that the strains are small.

From equilibrium considerations, at any section the axial force *P* and the bending moment *M* are given by

$$P = \int_{-h/2}^{h/2} \sigma_x b dy = 0 \tag{12.2a}$$

and

$$M = \int_{-h/2}^{h/2} \sigma_x b y dy \tag{12.2b}$$

12.2.1 Elastic State

Within the elastic state, Hooke's law, Equations 3.6 applies:

$$\varepsilon_x = \frac{y}{R_e} = \frac{1}{E}\left[\sigma_x - \nu\sigma_z\right]$$

$$\varepsilon_z = 0 = \frac{1}{E}\left[\sigma_z - \nu\sigma_x\right]$$

where $\sigma_y = 0$ and R_e is the radius of curvature within the elastic state.

Simplifying gives

$$\sigma_x = \frac{Ey}{(1-\nu^2)R_e} \tag{12.3a}$$

$$\sigma_z = \frac{\nu Ey}{(1-\nu^2)R_e} \tag{12.3b}$$

Because the centroidal and the neutral axes are coincident, Equation 12.2a is identically satisfied. Equation 12.2b results in an expression for the bending moment and stresses:

$$M = \frac{Ebh^3}{12(1-\nu^2)R_e} = \frac{EI}{(1-\nu^2)R_e} \tag{12.4}$$

$$\sigma_x = \frac{My}{I} \tag{12.5a}$$

$$\sigma_z = \frac{\nu My}{I} \tag{12.5b}$$

where

$$I = \frac{bh^3}{12}$$

Note that for plane strain conditions the term $(1 - \nu^2)$ appears. The stress distributions is shown in Figure 12.2a.

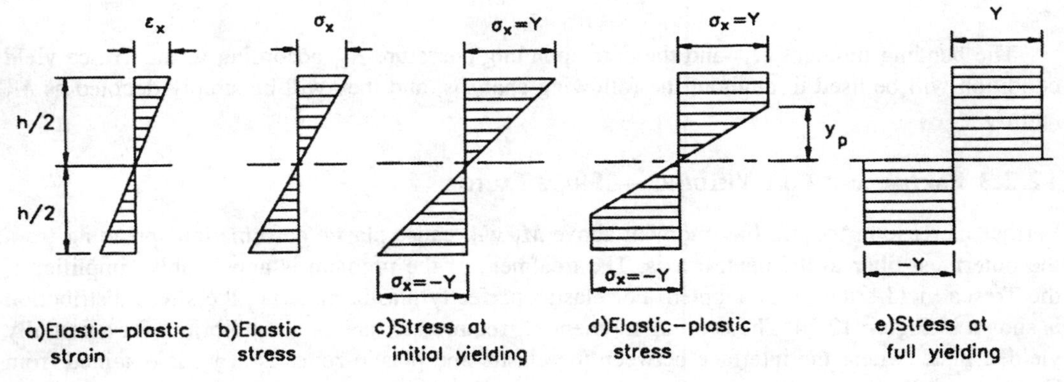

FIGURE 12.2 Distribution of strain and stress through the thickness of the strip of Figure 12.1 for elastic–plastic and fully plastic conditions of an elastic–perfectly plastic material: (a) elastic–plastic strain, (b) elastic stress, (c) stress at initial yielding, (d) elastic–plastic stress, (e) stress at full yielding.

12.2.2 Initial Yielding

As the bending moment is increased, yielding occurs at the outermost fibers of the plate, i.e., at $y = \pm h/2$. The material is assumed to be elastic-perfectly plastic of yield stress Y, the Tresca yield condition, Equation 10.18, applies as

$$\sigma_1 - \sigma_3 = Y$$

where $\sigma_1 = \sigma_x$, $\sigma_2 = \sigma_z$, and $\sigma_3 = \sigma_y = 0$. Hence, $\sigma_x = Y$ at the outermost fibers $y = \pm h/2$, as shown in Figure 12.2c. Substitution into Equation 12.5a and using the definition of $I = bh^3/12$ gives the bending moment required to initiate yielding according to the Tresca condition as

$$M_{YT} = Ybh^2/6 \tag{12.6a}$$

which is independent of Poisson's ratio ν.

If the material obeys the von Mises yield criterion, Equation 10.17, namely,

$$\sigma_x^2 + \sigma_z^2 - \sigma_x \sigma_z = Y^2$$

where $\sigma_z = \nu\sigma_x$; this gives at yielding:

$$\sigma_x = Y/\sqrt{1 - \nu + \nu^2}$$

From Equation 12.5a, the moment required to initiate yielding is thus

$$M_{YM} = Ybh^2 \Big/ \left(6\sqrt{1 - \nu + \nu^2}\right) \tag{12.6b}$$

The moment values given by Equations 12.6a and b differ by 11.8% for $\nu = {}^1/_3$.

The radius of curvature R_{YT} at this instant of yield initiation is obtained from Equation 12.3a at $y = h/2$, and putting $\sigma_x = Y$ according to the Tresca condition gives

$$\frac{1}{R_{YT}} = \frac{2Y\left(1 - \nu^2\right)}{Eh} \tag{12.7}$$

The bending moment M_{YT} and the corresponding curvature R_{YT} according to the Tresca yield condition will be used throughout the following analysis, and they will be simply denoted as M_Y and R_Y.

12.2.3 Partial and Full Yielding — Shape Factor

Further increase in the bending moment above M_Y will cause plastic deformation spreading from the outermost fiber to the neutral axis. The treatment of the problem is appreciably simplified if the Tresca yield criterion is adopted. For elastic–perfectly plastic material, the stress distribution is shown in Figure 12.2d. The bending moment M_p required to maintain equilibrium for a partially yielded plate, where the interface between the elastic and plastic zones is at y_p, is obtained from Equation 12.2b as

$$M_p = 2\int_{y_p}^{h/2} Ybydy + 2\int_0^{y_p} Y\frac{y}{y_p}bydy$$

This gives

$$M_p = Yb\left(3h^2 - 4y_p^2\right)/12 \tag{12.8}$$

where the stresses σ_x and σ_z (Figure 12.2d) are given by

$$\begin{aligned} &\sigma_x = Yy/y_p; \ \sigma_z = \nu\sigma_x \ \text{ elastic } |y| \le y_p \\ &\sigma_x = \pm Y; \ \sigma_z = \sigma_x/2 \ \text{ plastic } |y| \ge y_p \end{aligned} \tag{12.9}$$

where the positive and negative signs applys to the positive and negative values of y respectively. Note that y_p is always positive and that the sign convention of Section 7.1 is to be observed for M and R.

A discontinuity in the calculated stress σ_z along the plastic–elastic boundary exists. This may be avoided by assuming either the elastic value of ν or $\nu = 0.5$ all over. This is justified by the fact that at partial and full yielding the elastic strains are higher than the plastic strains or of the same order of magnitude. This does not affect the relation between the moment and the curvature if the Tresca yield criterion is adopted.

The radius of curvature R_p at this instant of partial yielding depends solely on the remaining elastic core. Hence, it is still given by the compatibility condition (Equation 12.1) as it remains valid for both elastic and plastic small deformations. At the elastic–plastic interface, $y = y_p$; $\sigma_x = \pm Y$ and $\sigma_z = \nu\sigma_x$, then, from Hooke's law,

$$\left(\varepsilon_x\right)_{y=y_p} = \frac{\left(1-\nu^2\right)}{E}Y$$

The radius of curvature is thus

$$\frac{1}{R_p} = \frac{\left(1-\nu^2\right)}{E}\frac{Y}{y_p} \quad \left|y_p\right| \le \frac{h}{2} \tag{12.10}$$

As the bending moment increases, the interface fiber at $y = y_p$ advances toward the neutral axis. Clearly, there always remains a layer of elastic material, no matter what the magnitude of M_p. The thinner this elastic core, the less it will be in error to assume that the material is fully plastic across the whole plate section. However, as this phase is approached the radius of curvature progressively decreases and the initial assumptions of the analysis become more and more in error.

For a perfectly plastic material with a negligible elastic core (i.e., rigid material), the bending moment M_u, which renders the whole plate fully plastic as shown in Figure 12.2e is the collapse moment, obtained by putting $y_p = 0$ in Equation 12.8 giving

$$M_u = Ybh^2/4 \tag{12.11}$$

The ratio M_u/M_{YT} as given by Equations 12.6a and 12.11 is known as the "shape factor" and is equal to 1.5 for the rectangular cross-section. This means that a rectangular beam section is effectively capable of sustaining a bending moment that is 50% greater than that estimated at the onset of yielding.

Table 12.1 lists the shape factor for some cross sections together with the plastic modulus Z_u for each section being defined as $Z_u = M_u/Y$. It is clear that a circular cross section, for instance,

TABLE 12.1
Shape Factors and Plastic Section Modulus for Various Beam Cross Sections

Cross Section	Shape Factor M_u/M_{YT}	Plastic Section Modulus $Z_u = M_u/Y$
Rectangular (b x h)	1.5	$bh^2/4$
Solid circular (r_o)	$16/3\pi$	$4r_o^3/3$
Thin-walled tube (r_m, h)	$4/\pi$	$4r_m^2 h$
Triangular (base b, height h)	$4(2-\sqrt{2})$	$(2-\sqrt{2})bh^2/6$

carries an even greater bending moment than a rectangular section, both loaded at full plasticity of their cross sections.

The curvature of the plate ($1/R$) increases rapidly as full yielding of the cross section is approached at M_u. The moment–curvature relation for elastic–plastic conditions is derived by combining Equations 12.8 and 12.11 for M_p and M_u, respectively, together with Equations 12.7 and 12.10 for R_Y and R_p, respectively. This results in the expression:

$$\frac{M_p}{M_Y} = \frac{1}{2}\left[3 - \left(\frac{R_p}{R_Y}\right)^2\right] \tag{12.12}$$

Expression 12.12 indicates that the radius of curvature R_p is only $^1/_5$ of R_Y when the moment $M_p/M_Y = 1.48$, a value which is very close to 1.5, the limiting collapse moment M_u.

The slope of the deflection curve in elastic–plastic bending of beams is discontinuous at a fully plastic cross section, where the curvature becomes infinitely large. In reality the discontinuouty is the limit of a narrow region through which the slope changes rapidly in a continuous manner.

Example 12.1:

A steel sheet of 1 mm thickness is plastically bent to a radius of 125 mm. The yield stress is 220 MPa.

a. *What fraction of the cross section remains elastic?*
b. *What percent error does neglecting the elastic core cause in the calculation in bending moment?*
c. *What is the radius of curvature under load,* R_p *compared to that at initial yielding,* R_Y*?*

Solution:

a. From Equation 12.10 the elastic–plastic interface y_p is obtained as

$$\frac{1}{R_p} = \frac{(1-\nu^2)}{E}\frac{Y}{y_p} \quad y_p \le \frac{h}{2}$$

$$\frac{1}{125\times10^{-3}} = \frac{(1-0.3^2)}{200\times10^9}\times\frac{220\times10^6}{y_p}$$

hence, $y_p = 0.1251$ mm.

The elastic core fraction is about 25% of the sheet thickness. The bending moment for this partial yielding is given by Equation 12.8 as

$$M_p = Yb\left(3h^2 - 4y_p^2\right)/12$$

$$M_p = 220\times10^6 \times b\left[3\left(1\times10^{-3}\right)^2 - 4\left(0.1251\times10^{-3}\right)^2\right]/12$$

hence, $M_p = 53.852b$ N · m.

b. Neglecting the elastic core, the bending moment is given by assuming full yielding according to Equation 12.11:

$$M_u = Ybh^2/4$$

$$M_u = 220\times10^6 \times b\left(1\times10^{-3}\right)^2/4 = 55b \text{ N}\cdot\text{m}$$

The difference in bending moments is only 2.1% due to the presence of an elastic core.

c. Applying Equations 12.12 together with Equation 12.6a gives

$$M_Y = \frac{Ybh^2}{6} = \frac{220\times10^6\times b\times0.001^2}{6} = 36.667b \text{ N}\cdot\text{m}$$

and

$$\frac{M_p}{M_Y} = \frac{53.852}{36.667} = \frac{1}{2}\left[3 - \left(\frac{R_p}{R_Y}\right)^2\right]$$

giving $R_p/R_Y = 0.251$.

12.2.4 Unloading: Residual Stresses and Springback

In thin sheet forming, bending moment is applied to obtain the required permanent shape through plastic deformation. When the moment is removed after partial or full yielding has occurred, the sheet does not recover completely to its original shape. There will be residual stresses. These are obtained by considering that recovery occurs due to a moment M_p as given by Equation 12.8 applied elastically thus resulting in elastic stresses given by Equations 12.5 as

$$\sigma_x = \frac{M_p y}{I} = \frac{Yb\left(3h^2 - 4y_p^2\right)y}{12I} = Y\left(3h^2 - 4y_p^2\right)\frac{y}{h^3}$$

$$\sigma_z = \nu M_p y/I = \nu Y\left(3h^2 - 4y_p^2\right)\frac{y}{h^3}$$

Subtraction of these stress from the stresses given by Equations 12.9 gives the residual stresses in the strip as

$$\left(\sigma_x\right)_{\text{res}} = \frac{Yy}{y_p} - \frac{Yy}{h^3}\left(3h^2 - 4y_p^2\right) \quad \text{for } |y| \le y_p$$
$$= \pm Y - \frac{Yy}{h^3}\left(3h^2 - 4y_p^2\right) \quad \text{for } |y| \ge y_p \tag{12.13a}$$

$$\left(\sigma_z\right)_{\text{res}} = \nu\left[\frac{Yy}{y_p} - \frac{Yy}{h^3}\left(3h^2 - 4y_p^2\right)\right] \quad \text{for } |y| \le y_p$$
$$= \pm\frac{Y}{2} - \frac{\nu Yy}{h^3}\left(3h^2 - 4y_p^2\right) \quad \text{for } |y| \ge y_p \tag{12.13b}$$

Figure 12.3 shows the residual stress distribution resulting from unloading a partially yielded sheet. The residual stress σ_x attains its greatest magnitude at the outer surface and at the plastic boundary. As the sheet is rendered more plastic, i.e., when $y_p \to 0$ the residual stress at $y = \pm h/2$ approaches the limiting value of $\pm Y/2$ and $\pm Y$ at $y_p = 0$.

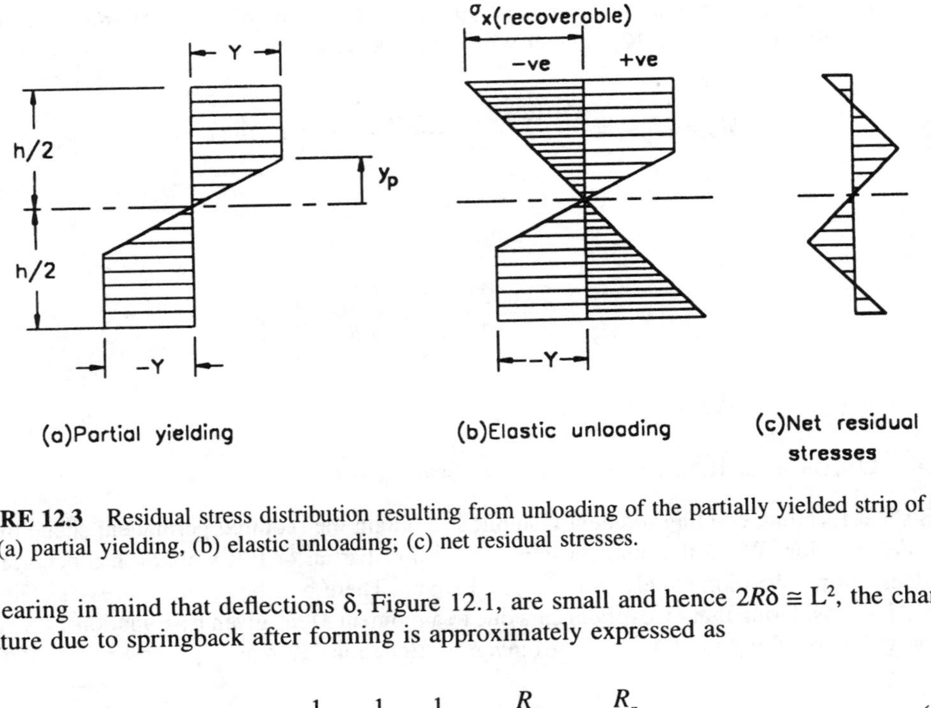

FIGURE 12.3 Residual stress distribution resulting from unloading of the partially yielded strip of Figure 12.1: (a) partial yielding, (b) elastic unloading; (c) net residual stresses.

Bearing in mind that deflections δ, Figure 12.1, are small and hence $2R\delta \cong L^2$, the change in curvature due to springback after forming is approximately expressed as

$$\frac{1}{R_f} = \frac{1}{R_p} - \frac{1}{R_e} \quad \text{or} \quad \frac{R_p}{R_f} = 1 - \frac{R_p}{R_e} \tag{12.14}$$

where R_f, R_p, and R_e are, respectively, the radii of curvature at the final state of springback, plastic bending, and elastic unloading. Eliminating y_p from Equations 12.10 and 12.8 yields

$$M_p = \frac{Yb}{12}\left[3h^2 - \frac{4Y^2}{E^2}\left(1 - \nu^2\right)^2 R_p^2\right] \tag{12.15}$$

The curvature $1/R_e$ upon unloading is also implicitly given by Equation 12.4, where M_p given by Equation 12.8, acts as an elastic moment, hence,

$$M_p = \frac{Ebh^3}{12(1-\nu^2)R_e} \tag{12.16}$$

Equating Expressions 12.15 and 12.16 gives

$$\frac{Ebh^3}{12(1-\nu^2)R_e} = \frac{Yb}{12}\left[3h^2 - \frac{4Y^2}{E^2}(1-\nu^2)^2 R_p^2\right]$$

In view of this expression, Equation 12.14 yields

$$\frac{R_p}{R_f} = 1 - 3\left[\frac{YR_p}{Eh}(1-\nu^2)\right] + 4\left[\frac{YR_p}{Eh}(1-\nu^2)\right]^3 \tag{12.17}$$

When $R_p/R_f = 0$, there is a complete springback, i.e., the bending is wholly elastic. At the other extreme, when $R_p/R_f = 1$, there is no springback at all.

An important feature in forming processes is to determine residual stresses and strains and springback after sheet bending. In such a case, the sheet cross section has fully yielded and the residual stresses are obtained from Equations 12.13a and b by putting $y_p = 0$ for $|y| \geq y_p$

$$\begin{aligned}(\sigma_x)_{\text{res}} &= Y\left(1 - \frac{3y}{h}\right)\\ (\sigma_z)_{\text{res}} &= \frac{1}{2}Y\left(1 - \frac{6\nu y}{h}\right)\end{aligned} \tag{12.18}$$

Springback of a fully plastic bent sheet is determined from Equation 12.14, where R_e is given by Equation 12.4 for $M = M_u$; hence,

$$\frac{R_p}{R_f} = 1 - \frac{3Y(1-\nu^2)}{Eh}R_p \tag{12.19}$$

Residual stresses and springback for sheets of strain-hardening materials may be calculated in the same way, although computations may be tedious. Note that throughout the entire above analysis the Bauschinger effect in reverse yielding has been ignored.

Example 12.2:

Select the tool radius required to produce a final bend radius of 2.5 mm in steel sheets of thickness 1 mm. The yield strength is Y = *300 MPa.*

a. *What would be the tool radius for plates of higher strength with* Y = *500 MPa? Take* E = *200 GPa and* ν = *0.3.*
b. *Calculate the moment* M_p *required to achieve the bending operation.*

Solution:

a. In bending operations, the applied moment produces almost full yielding, and Equation 12.19 may be used where R_p stands for the tool radius as:

$$R_p = R_f\left[1 - \frac{3Y(1-\nu^2)}{Eh}R_p\right]$$

$$R_p = 2.5\times10^{-3}\left[1 - \frac{3\times300\times10^6(1-0.3^2)}{200\times10^9\times0.001}R_p\right]$$

hence,

$$R_p = 2.5\times10^{-3} - 1.0237R_p$$

$$R_p = 0.123 \text{ mm}$$

indicating the importance of considering springback in sheet metal presswork. For steel sheets of $Y = 500$ MPa, the tool radius is $R_p = 0.924$ mm. i.e., a smaller tool radius. Note that all kinds of steel possess almost the same modulus of elasticity and Poisson's ratio irrespective of their strength. The error resulting from assuming full yielding could be checked by applying Equation 12.10 for partial yielding, i.e,

$$\frac{1}{R_p} = \frac{(1-\nu^2)}{E}\frac{Y}{y_p}$$

$$\frac{1}{0.123} = \frac{(1-0.3^2)}{200\times10^9}\times\frac{300\times10^6}{y_p}$$

hence, the elastic core is bounded by $y_p = \pm0.17$ mm, i.e., 0.34% of the sheet thickness.

b. For bending moment calculations, full yielding is assumed and Equation 12.11 is used:

$$M_u = \frac{Ybh^2}{4} = \frac{300\times10^6\times b(1\times10^{-3})^2}{4} = 75b \text{ N}\cdot\text{m}$$

12.3 PLANE STRESS BENDING OF BEAMS

12.3.1 Initial Yielding, Full Yielding, and Springback

A beam of narrow width b compared with its height h under pure bending is analyzed assuming plane stress conditions following the same procedure applied in Section 12.2 for plane strain.

The determination of bending moments required to initiate yielding, at any stage of partial yielding and for full yielding, are based on equilibrium consideration, Expressions 12.2, as derived in Section 12.2.2. Taking into consideration that $\sigma_x = E\varepsilon_x = Ey/R$, while $\sigma_y = \sigma_z = 0$, the following expressions are obtained:

Elastic state: $$M_e = \frac{EI}{R_e} \quad (12.20)$$

Initial yielding: $$M_Y = \frac{Ybh^2}{6} \text{ and } \frac{1}{R_Y} = \frac{2Y}{Eh} \quad (12.21)$$

Partial yielding: $M_p = \frac{Yb}{12}\left[3h^2 - 4y_p^2\right]$ and $\frac{1}{R_p} = \frac{Y}{Ey_p}$ (12.22)

Full yielding: $M_u = \frac{Ybh^2}{4}$ (12.23)

Note that in the above equations E replaces $E/(1 - \nu^2)$ in Expressions 12.4 to 12.11 for plates, respectively. Hence, the determination of springback upon elastic recovery from an elastic–plastic state is directly obtained from Equation 12.17 as

$$\frac{R_p}{R_f} = 1 - 3\frac{YR_p}{Eh} + 4\left(\frac{YR_p}{Eh}\right)^3$$

or

$$\frac{R_p}{R_f} = \left(\frac{YR_p}{Eh} + 1\right)\left(\frac{2YR_p}{Eh} - 1\right)^2 \qquad (12.24)$$

Comparison between Expressions 12.17 and 12.24 indicates that springback is more pronounced for plane stress conditions than for plane strain.

The above analysis has been confined to plates and beams of rectangular cross sections, i.e., sections with two axes of symmetry in which the neutral axis coincides with the centroidal axis. In fact, the analysis may be extended to beams with one axis of symmetry. Consider a beam of a total cross-sectional area A, which is symmetrical about the y-axis, as shown in Figure 12.4. The

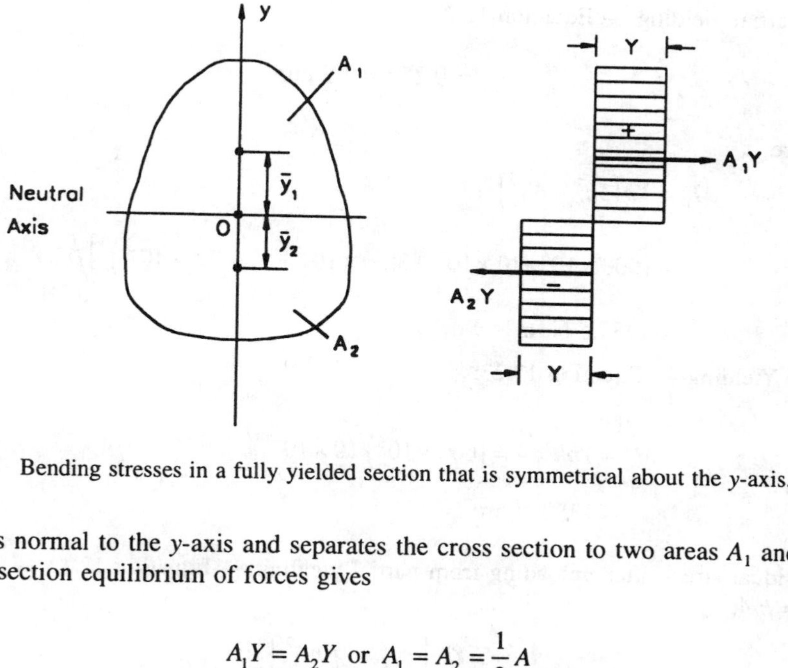

FIGURE 12.4 Bending stresses in a fully yielded section that is symmetrical about the y-axis.

neutral axis is normal to the y-axis and separates the cross section to two areas A_1 and A_2. For a fully yielded section equilibrium of forces gives

$$A_1Y = A_2Y \text{ or } A_1 = A_2 = \frac{1}{2}A$$

Since the neutral axis divides the cross section into two equal areas, the equilibrium of moments is satisfied when the bending moment at full yielding M_u is given by

$$M_u = A_1 Y\bar{y}_1 + A_2 Y\bar{y}_2 \tag{12.25a}$$

where $\bar{y}_1$ and $\bar{y}_2$ are the distances of the centroids of the areas A_1 and A_2 from the neutral axis, respectively.

The plastic section modulus Z_u is thus

$$Z_u = \frac{M_u}{Y} = \frac{1}{2}A(\bar{y}_1 + \bar{y}_2) \tag{12.25b}$$

Note that in this case the neutral axis is not necessarily a centroidal axis.

Example 12.3:

A bar of a rectangular section 10 mm wide and 30 mm high is subjected to a moment M_p. *The bar material is elastic perfectly plastic having* E = *200 GPa and* Y = *600 MPa. Determine:*

a. M_p *at the moments of initial yielding, partial yielding of* $^1/_2$ *the cross section and full yielding.*
b. Residual stress at the outer fiber for the cases of partial and full yielding.

Solution:

a. Initial yielding — Equation 12.21:

$$M_Y = Ybh^2/6 = (600 \times 10^6)\,(10 \times 10^{-3})\,(30 \times 10^{-3})^2/6$$
$$= 900 \text{ N·m}$$

Partial yielding — Equation 12.22:

$$y_p = 0.25h = 7.5 \text{ mm}$$

hence,

$$M_p = Yb\left(3h^2 - 4y_p^2\right)/12$$

$$= \left(600\times10^6\right)\left(10\times10^{-3}\right)\left[3\left(30\times10^{-3}\right)^2 - 4\left(7.5\times10^{-3}\right)^2\right]/12$$

$$= 1237.5 \text{ N}\cdot\text{m}$$

Full Yielding — Equation 12.23:

$$M_u = Ybh^2/4 = \left(600\times10^6\right)\left(10\times10^{-3}\right)\left(30\times10^{-3}\right)^2/4$$

$$= 1350 \text{ N}\cdot\text{m}$$

b. Residual stress after unloading from partial yielding — Equation 12.13 at $y = h/2$ and $y_p = h/4$:

$$\sigma_x = Y - \frac{Yh}{2h^3}\left[3h^2 - 4\left(\frac{h}{4}\right)^2\right] = -3Y/8$$

$$= -225 \text{ MPa (compressive)}$$

Residual stress after full yielding — Equation 12.18 at $y_p = h/2$:

$$\sigma_x = Y\left(1 - \frac{3y}{h}\right) = -\frac{Y}{2} = -300 \text{ MPa (compressive)}$$

12.3.2 Combined Bending and Tension

Consider the case of a long strip subjected to combined bending and tension, Figure 12.5. The

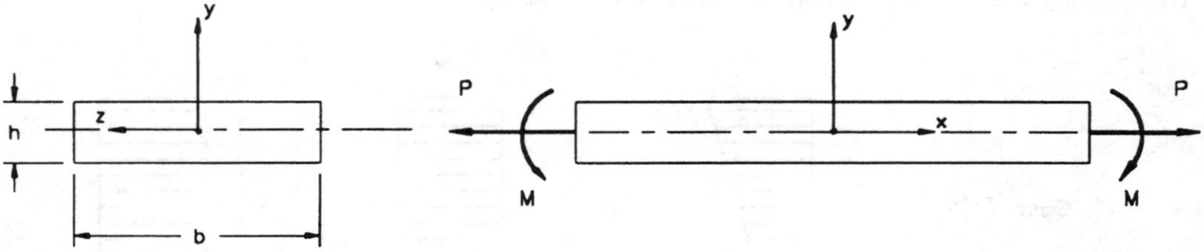

FIGURE 12.5 Strip subjected to combined bending and tension.

material is assumed to be elastic-perfectly plastic during loading and elastic during unloading.

12.3.2.1 Elastic State

Plane cross sections remain plane after deformation. The stress distribution along the plate thickness is obtained directly by superimposing the bending stresses due to M, $\sigma_b = 12My/bh^3$, Figure 12.6a, to the tensile stresses due to P, $\sigma_t = P/bh$, Figure 12.6b, to give the resultant stresses, shown in Figure 12.6c.

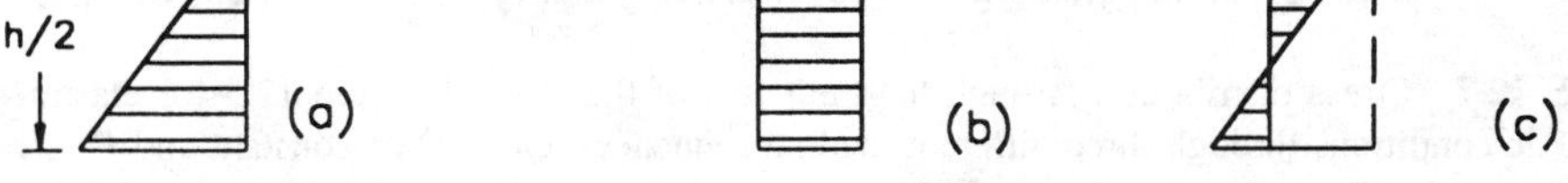

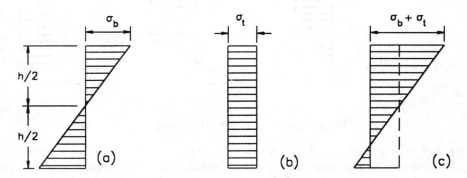

FIGURE 12.6 Elastic stresses in the strip of Figure 12.5: (a) stresses due to bending, (b) stresses due to tensile force, and (c) stress due to combined bending and tension,

12.3.2.2 Elastic–Plastic State

Initial yielding takes place at the top fibers when $\sigma_b + \sigma_t = Y$. This can be achieved either by increasing M or P or by increasing both M and P simultaneously. Further yielding depends upon the loading program. In this respect three cases of loading are considered: (1) M is constant and P is increased, (2) P is constant and M is increased, and (3) both M and P are increased simultaneously. Figure 12.7 shows the stress distribution for each case at yield initiation, partial yielding, and full yielding.

Note that in case 1 full yielding requires that P acts at an offset of $(-M/P)$ away from the centroid. In case 3 the stresses due to increasing M are taken to be higher than the stresses due to increasing P. Figure 12.7 illustrates the dependence of the resulting state of stresses on the history of loading.

12.3.2.3 Unloading and Residual Stresses

Unloading is assumed to be fully elastic. The residual stresses are obtained by superimposing the stresses obtained from the elastic–plastic analysis, Figure 12.7, to the stresses in the elastic state by applying P and M in the opposite direction of loading.

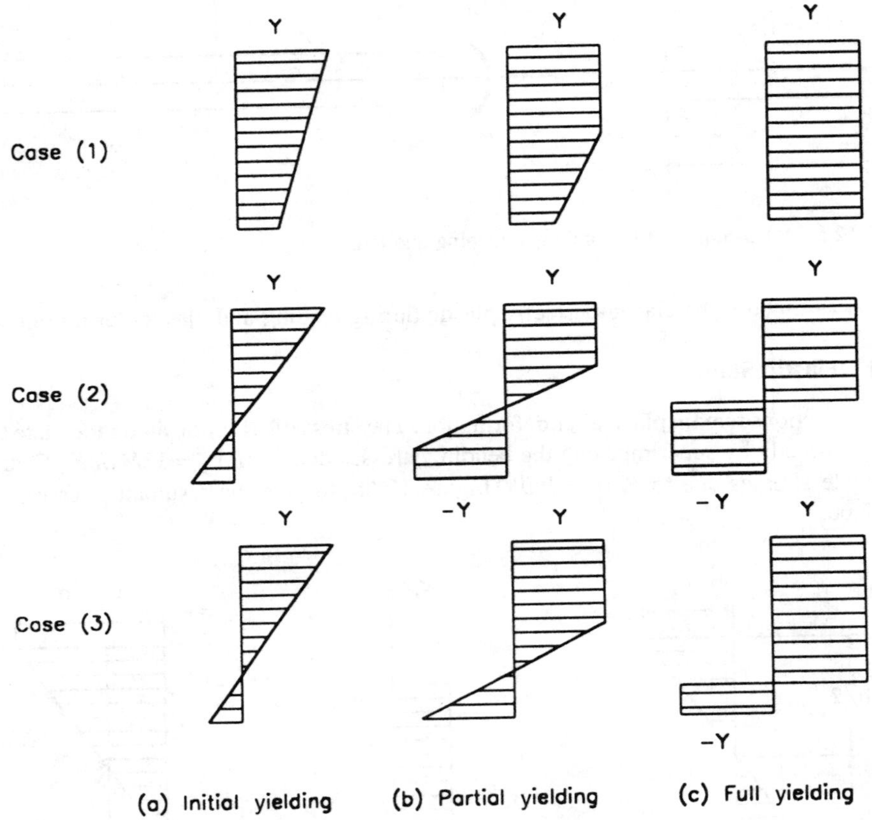

FIGURE 12.7 Stress distribution through the thickness of the strip of Figure 12.5 for elastic–plastic and fully plastic conditions through three different loading sequences, case (1) M constant and P increased, case (2) P constant and M increased, and case (3) M and P increased simultaneously; (a) initial yielding, (b) partial yielding, (c) full yielding.

Example 12.4:

The hanger in Figure 12.8a is made of a steel strip having a cross section of 100 × 24 mm, Y = *240 MPa, to support a load* P *at an offset distance of 50 mm. It is required to determine:*

a. *The load for initial yielding;*
b. *The load for full yielding.*

Plot the stress distributions.

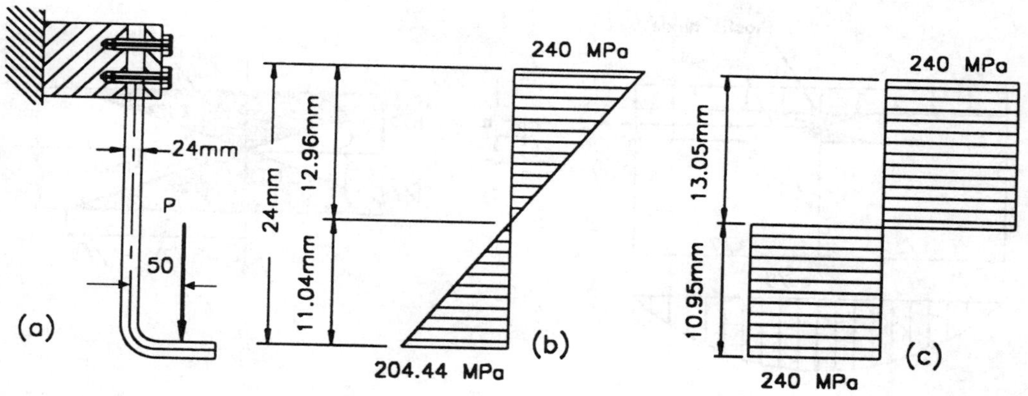

FIGURE 12.8 Example 12.4.

Solution:

This is case 3 of Figure 12.7 since $M = 50P$.

a. For initial yielding:

$$Y = \frac{P}{100 \times 24} + \frac{50P \times 6}{100 \times 24^2} = 0.005625P \text{ MPa}$$

Equating Y to 240 MPa gives P_Y for initial yielding:

$$P_Y = 42.67 \text{ kN}$$

Maximum tensile stress due to M_Y is $(\sigma_b)_Y = 222.24$ MPa, and tensile stress due to P_Y is $(\sigma_t)_Y = 17.76$ MPa. The stress distribution at initial yielding is shown in Figure 12.8b.

b. For full yielding:

$$Y = \frac{P_u}{100 \times 24} + \frac{50P_u \times 4}{100 \times 24^2} = 0.00389P_u \text{ MPa}$$

Taking $Y = 240$ MPa yields:

$$P_u = 61.697 \text{ kN}$$

Maximum tensile bending stress due to M is $\sigma_b = 214.29$ MPa, and tensile stress due to P_u is $\sigma_t = 25.71$ MPa. The stress distribution at full yielding is shown in Figure 12.8c.

12.3.3 Plastic Collapse of Beams — Plastic Hinges

At any section of a beam subjected to pure bending, yielding first occurs at the top and bottom fibers at a bending moment value M_Y. Yielding spreads toward the neutral axis of the beam sections as the bending moment is increased up to its limiting value M_u at which the beam sections are rendered wholly plastic. For a laterally loaded beam, these incidents start to occur at the cross section where the bending moment has its greater value.

At the beam section that has fully yielded, a plastic hinge will form. Under statically determinate conditions, the beam will behave as a mechanism with unlimited deflection, and plastic collapse is hence assumed.

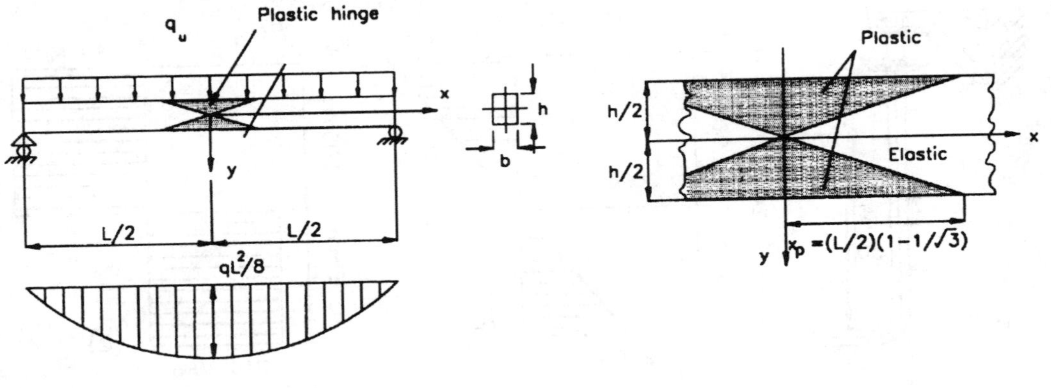

(a) (b)

FIGURE 12.9 Plastic hinge in a simply supported beam subjected to uniformly distributed force: (a) load diagram and (b) plastic hinge detail.

To illustrate the above incidents, consider a simply supported beam of rectangular cross section and subjected to a uniformly distributed load q, as shown in Figure 12.9a. The bending moment attains its maximum $M_{\max}$ at midlength $z = L/2$, while any other section is loaded according to

$$M = M_{\max}\left[1 - 4\left(\frac{x}{L}\right)^2\right] \text{ where } M_{\max} = \frac{qL^2}{8} \tag{12.26}$$

being measured from the beam middle section. The beam cross section at $x = L/2$ will be fully yielded when $M_{\max} = M_u = Ybh^2/4$, as given by Equation 12.23. Obviously, the adjacent sections will be in a state of partial yielding for which the extent is governed by their distances away from the midlength section. For any partially yielded section with an elastic core of depth $2y_p$, the bending moment is given by Equation 12.22 as

$$M_p = \frac{Ybh^2}{4}\left[1 - \frac{4}{3}\left(\frac{y_p}{h}\right)^2\right] = M_u\left[1 - \frac{4}{3}\left(\frac{y_p}{h}\right)^2\right] \tag{12.27}$$

Therefore, limit of the wholly elastic length of the beam and, hence, the extent of plasticity, x_p is determined by equating Expressions 12.26 and 12.27 to give

$$x = \frac{L}{\sqrt{3}}\left(\frac{y_p}{h}\right) \text{ or } \left(\frac{y_p}{h}\right) = \frac{\sqrt{3}}{L}x \tag{12.28a}$$

This indicates that the regions of plasticity near the midsection are triangular shaped ending at x_p $\pm L/2(1-/\sqrt{3})$, as determined from Expression 12.28a by putting $y_p = h/2$. This is shown in Figure 12.9b. The load at this instant of forming a plastic hinge at the midsection is obtained from Equations 12.23 and 12.26 as

$$q_u = 8M_u/L^2 = 8Ybh^2/4L^2 = 2Y(bh^2/L^2) \tag{12.28b}$$

Note that the above analysis neglects the effect of shearing stresses on yielding.

For other conditions of loading,[1] Figure 12.10, a similar analysis may be considered to show that the shape and extent of the plastic zones are

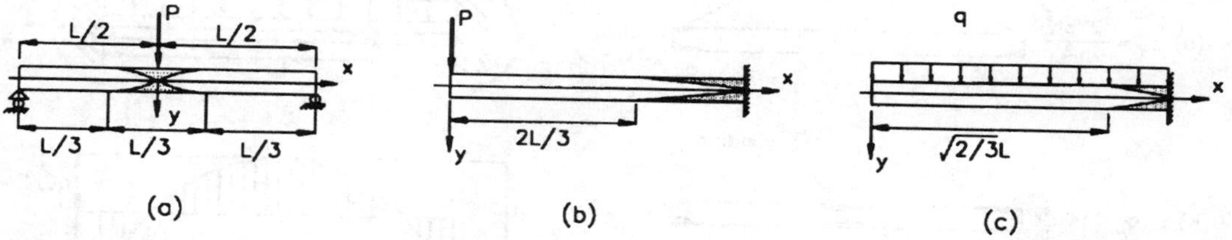

FIGURE 12.10 Plastic hinge for different conditions of loading and supports: (a) simply supported beam subjected to a concentrated force at midspan, (b) cantilever beam subjected to edge force, and (c) cantilever beam subjected to uniformly distributed load.

For a simply supported beam with a central load P:

$$\left(\frac{y_p}{h}\right)^2 = \frac{3}{2L}x; \quad \begin{matrix} \text{elastic for } L/6 \le |x| \le L/2 \\ \text{plastic for } 0 \le |x_p| \le L/6 \end{matrix} \tag{12.29a}$$

For a cantilever beam with an end load P:

$$\left(\frac{y_p}{h}\right)^2 = \frac{3}{4}\left(1-\frac{x}{L}\right); \quad \begin{matrix} \text{elastic for } 0 \le x \le 2L/3 \\ \text{plastic for } 2L/3 \le x_p \le L \end{matrix} \tag{12.29b}$$

For a cantilever beam with uniformly distributed load q:

$$\left(\frac{y_p}{h}\right)^2 = \frac{3}{4}\left[1-\left(\frac{x}{L}\right)^2\right]; \quad \begin{matrix} \text{elastic for } 0 \le x\sqrt{2/3}L \\ \text{plastic for } \sqrt{2/3}L \le x_p \le L \end{matrix} \tag{12.29c}$$

The shapes of the plastic zones are parabolic for the first two cases and elliptical for the third.

The influence of a plastic hinge in a statically indeterminate beam is different from that in a statically determinate one. In the former, a redistribution of moments occurs and unconstrained deformation is postponed until a sufficient number of plastic hinges develop such that the beam behaves as a mechanism. This is illustrated in Example 12.5 for a beam with built-in ends.

Example 12.5:

A beam of rectangular cross section has both ends built-in, as shown in Figure 12.11. If the beam is subjected to a uniformly distributed load q, determine:

a. *The load at which the first plastic hinges are initiated;*
b. *The load at which the beam behaves as a mechanism.*

Solution:

This is a statically indeterminate problem, and the reaction moment at the built-in ends may be obtained by applying the Castigliano method, Expression 3.44, namely,

$$\frac{\partial U}{\partial M_o} = 0 \text{ where } U = \int_0^L \frac{M^2}{2EI}dx$$

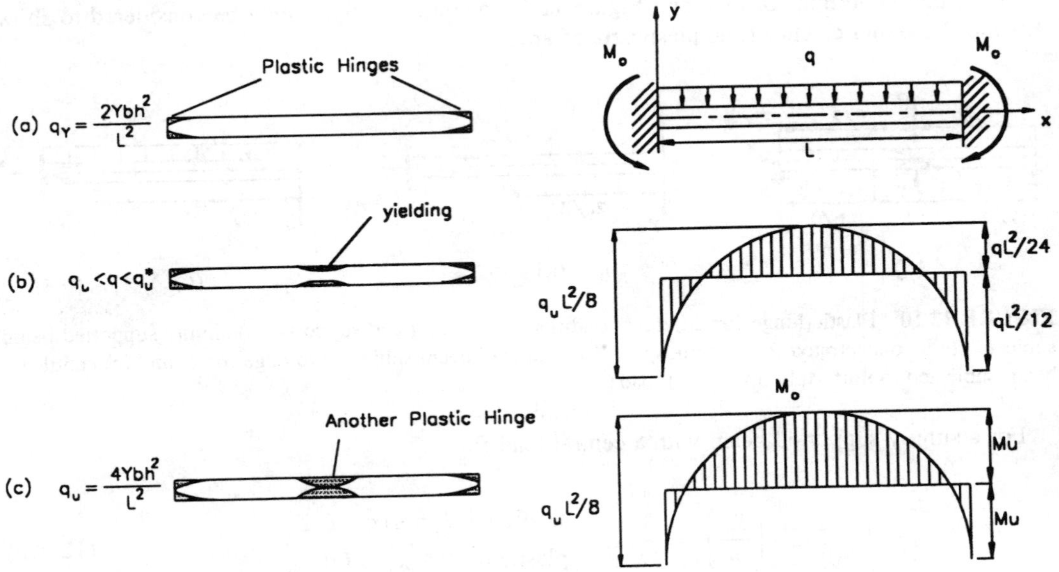

FIGURE 12.11 Example 12.5.

is the bending strain energy. From Figure 12.10 at any x

$$M = M_o - {}^1/_2\, qx^2$$

hence,

$$U = \frac{1}{2EI}\int_0^L \left(\frac{1}{2}qx^2 + M_o - \frac{1}{2}qLx\right)^2 dx$$

Integrating and applying the condition $\partial U/\partial M_o = 0$ results in

$$M_o = qL^2/12$$

At midspan $M = qL^2/24$.

a. The maximum bending moment occurs at the built-in ends and plastic deformation is initiated there when

$$M_o = M_Y \text{ or } qL^2/12 = Ybh^2/6$$

Hence, the load intensity q_Y when the first plastic hinges are initiated at the built-in ends is given by

$$q_Y = 2Y\,(bh^2/L^2)$$

Increasing q beyond q_Y will develop two plastic hinges at the built-in ends, but this will not cause the beam to become a mechanism. A third hinge has to develop for beam collapse.

b. The value of load q_u at which a third plastic hinge is formed is determined by considering the redistribution of moments along the beam length. The equilibrium of a half-span of the beam thus gives the moment at $x = L/2$ as

$$M_u = {}^1/_8\, q_u L^2 - M_u$$

where the value $M_u = Ybh^2/4$ is attained at the midlength and the two built-in ends. This gives

$$q_u = 4Y\,(bh^2/L^2)$$

Clearly, the load causing complete collapse of the beam is twice as much the load that is required to initiate yielding at the built-in ends.

12.3.4 Deflection and Shear Stresses

The determination of elastic–plastic deflections in beams is based on the same assumptions pertaining to small deformation as applied to elastic beams as in Chapter 7, namely,

$$\frac{d^2 v}{dx^2} \cong -\frac{1}{R} \tag{7.10}$$

where v is the deflection, R is the radius of curvature, and x is the coordinate axis along the beam. The basic expressions for moment–curvature are derived from Equations 12.20 through 12.22, such that

$$\frac{R_Y}{R_e} = \frac{M_e}{M_Y} \qquad \text{elastic } |M_e| < M_Y \tag{12.30a}$$

$$\frac{R_Y}{R_p} = \left[3 - 2\left(\frac{M_p}{M_Y}\right)\right]^{-1/2} \qquad \text{elastic} - \text{plastic } |M_p| > M_Y \tag{12.30b}$$

To illustrate the method of derivation, the case of a simply supported beam uniformly loaded as shown in Figure 12.9 is considered.* Due to symmetry, only one-half of the beam is considered by combining Equations 7.10 and 12.30 together with (12.26) to yield:

$$\frac{d^2 v_e}{dx^2} \cong -\frac{(q/q_Y)}{R_Y}\left[1 - \frac{4x^2}{L^2}\right] \qquad x_p < x < L \tag{12.31a}$$

$$\frac{d^2 v_p}{dx^2} \cong -\frac{1}{R_Y}\left[3 - 2\left(\frac{q}{q_Y}\right)\left(1 - \frac{4x^2}{L^2}\right)\right]^{-1/2} \qquad 0 < x < x_p \tag{12.31b}$$

where v_e and v_p denote deflections in the elastic and plastic portions along the beam length, respectively. The above equations are integrated and the following conditions are imposed:

* Several cases of beams loaded in elastic–plastic conditions are found in Chakrabarty.[1]

$$\text{At } x = L/2: \mathrm{v}_e = 0, \ \text{ at } x = 0: d\mathrm{v}_p/dx = 0 \text{ (symmetry)}$$

$$\text{At } x = x_p: \mathrm{v}_e = \mathrm{v}_p \ \text{ and } \ d\mathrm{v}_e/dx = d\mathrm{v}_p/dx \text{ (continuity)}$$

Denoting q/q_Y by α, the deflections for the elastic and plastic portions, respectively, are given by

$$\mathrm{v}_e = \frac{L^2}{4R_e}\left\{\left(1-\frac{2x}{L}\right)\left[\frac{1}{\sqrt{2\alpha}}\sinh^{-1}\sqrt{\frac{2\alpha-2}{3-2\alpha}} - \frac{2\alpha+1}{3}\sqrt{1-\frac{1}{\alpha}}\right]\right.$$
$$\left. + \frac{\alpha}{2}\left[\frac{5}{6} - \frac{4x^2}{L^2}\left(1-\frac{2x^2}{3L^2}\right)\right]\right\} \quad x_p \le x \le L \tag{12.32a}$$

and

$$\mathrm{v}_p = \delta - \frac{L^2}{4R_e\sqrt{2\alpha}}\left\{\frac{2x}{L}\sinh^{-1}\left(\frac{2x}{L}\sqrt{\frac{2\alpha}{3-2\alpha}}\right)\right.$$
$$\left. - \sqrt{\frac{3-2\alpha}{2\alpha}+\frac{4x^2}{L^2}} + \sqrt{\frac{3-2\alpha}{2\alpha}}\right\} \quad 0 \le x \le x_p \tag{12.32b}$$

where the central deflection δ is

$$\delta = \frac{L^2}{4R_e}\left\{\frac{1}{\sqrt{2\alpha}}\sinh^{-1}\sqrt{\frac{2\alpha-2}{3-2\alpha}} + \sqrt{\frac{3-2\alpha}{2\alpha}}\right.$$
$$\left. - \frac{2\alpha+1}{3}\sqrt{1-\frac{1}{\alpha}} + \frac{2\alpha}{3} - \frac{3}{4\alpha}\right\} \tag{12.32c}$$

At initial yielding $\alpha = 1$ and the central deflection is found to be $\delta_y = YL^2/24Eh$ as obtained for a fully elastic beam. As α approaches the collapse value 1.5, the central deflection according to Expression 12.32c tends to infinity as expected, because of the plastic hinge formed at midlength. However, for $\alpha = 1.44$, the ratio δ/δ_y is approximately less than 2, indicating that the deflection remains within an elastic order of magnitude even when loading conditions approach that of full collapse.

To evaluate the deflection v of the beam under elastic–plastic conditions due to transverse loads, it is usually assumed that the shear stress is small in comparison with the normal stress for beams of $L/h > 5$.* For shorter beams the shear stress may affect deflections and collapse load estimations.

The shear stress distribution over any elastic–plastic cross section shows that in the plastic zone, $|y| \le y_p$, the shear stress τ_{xy} vanishes. This follows from the fact that at the cross section top and bottom boundaries $\pm h/2$, $\tau_{xy} = 0$, while $\sigma_x = Y =$ constant due to yielding. Therefore, to satisfy the equilibrium condition (Equation 1.6a), namely,

$$(\partial\sigma_x/dx) + (\partial\tau_{xy}/\partial y) = 0 \text{ at } y = \pm h/2$$

* See Section 5.2.4.

hence, $(\partial \tau_{xy}/\partial y)$ must vanish also. By successive differentiation of the equilibrium and the von Mises yield condition, $\sigma_x^2 + 3\tau_{xy}^2 = Y^2$, it may be shown[2] that the higher order derivatives of the shear stress also vanish, indicating that $\tau_{xy} = 0$ all over the plastic zone of the cross section.

The transverse shearing force within the length $0 \le x \le x_p$ is thus carried entirely by a parabolically distributed shear stress according to the expressions:

$$\tau_{xy} = -\frac{3qx}{4y_p}\left(1 - \frac{y^2}{y_p^2}\right); \quad \text{elastic zone } |y| < y_p$$

$$\tau_{xy} = 0; \quad \text{plastic zone } |y| > y_p$$

Over the beam length, the greatest shear stress occurs at $x = \pm L/2$, but its value for usual L/h ratios is not sufficient to cause yielding at these cross sections.

12.3.5 Effect of Strain Hardening

So far the beam material has been considered to be elastic–perfectly plastic. A more realistic analysis would consider materials with strain-hardening behavior. For simplification, a material with a linear strain-hardening yield stress as given by Expression 10.8h in the plastic range, namely,

$$\sigma = Y_o + E^p \varepsilon^p \quad |y| \ge y_p \tag{10.8h}$$

may be considered to analyze the elastic–plastic problem of pure bending of a beam of a rectangular cross section. Under partial yielding, the stress distribution across the cross section is shown in

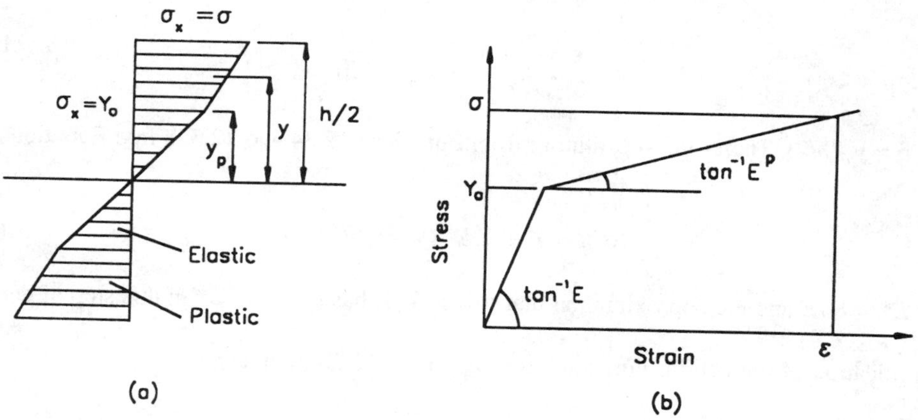

FIGURE 12.12 Effect of strain hardening on the stress distribution through thickness in strip bending: (a) stress distribution and (b) material stress–strain curve.

Figure 12.12a. There exists an elastic core bounded by $\pm y_p$ surrounded by plastic regions $y_p \le |y| \le h/2$. Within the plastic region the stress σ_x varies linearly as a consequence of the linear-hardening behavior of the material, as shown in Figure 12.12b, according to

$$\text{at } y = \pm y_p\text{: } \sigma_x = \pm Y_o \text{ and at } y = \pm h/2\text{: } \sigma_x = \pm\sigma$$

More generally from Figure12.8b, within the elastic region, these expressions are obtained as

$$(\sigma_x)_e = Y_o\left(\frac{y}{y_p}\right) \quad \text{for } |y| \le y_p \tag{12.33a}$$

and, within the plastic region, as

$$(\sigma_x)_p = Y_o + (\sigma - Y_o)(y - y_p)\Big/\left(\frac{h}{2} - y\right) \le y_p \quad \text{for } |y| \ge y_p \tag{12.33b}$$

where $(\sigma_x)_e$ and $(\sigma_x)_p$ denote the stress distribution within the elastic and plastic regions, respectively.

The value of $(\sigma - Y_o)$ is obtained from Expression 10.8i and

$$\sigma - Y_o = \left(\varepsilon - \frac{\sigma}{E}\right)E^p \tag{12.34}$$

where $(\varepsilon - \sigma/E)$ is the plastic strain ε_x^p at $y = \pm h/2$. The strain ε_o at the elastic–plastic boundary $y = y_p$ is given according to Hooke's law and Equation 12.1 as

$$\varepsilon_o = \frac{Y_o}{E} = \frac{y_p}{R_p} \tag{12.35a}$$

Also, because of the linearity of the strain distribution equation (Equation 12.1), the strain ε at the outer fiber is thus

$$\varepsilon = \frac{h/2}{R_p} = \frac{Y_o}{E}\frac{h/2}{y_p} = \frac{Y_o}{E\beta} \tag{12.35b}$$

where $\beta = y_p/(h/2)$. Therefore, substituting from Equations 12.34 and 12.35b into Equation 12.33b gives

$$(\sigma_x)_p = Y_o + Y_o E^* (y - y_p)/(\beta h/2) \tag{12.36}$$

where $E^* = E^p/E$ and the approximation that $\sigma/E \simeq Y_o/E$ has been made* in this step of derivation only.

Application of the equilibrium condition (Equation 12.2b), namely,

$$M_p = \int_{-h/2}^{h/2} \sigma_x b y dy = 2\left[\int_0^{y_p} (\sigma_x)_e b y dy + \int_{y_p}^{h/2} (\sigma_x)_p b y dy\right]$$

Substitution of Equations 12.33a and 12.33b into the above conditions and integrating results in

$$M_p = Y_o b h^2[(3 - \beta^2) + E^*(\beta^2 - 1 + 2/\beta)]/12 \tag{12.37}$$

* This approximation is shown to be equivalent to considering $1 + E^p/E \simeq 1$ for $E^p << E$.

This expression gives the bending moment required to cause partial yielding in a linear-hardening material. Expression 12.37 reduces to Equation 12.22 for an elastic–perfectly plastic material of $E^* = E^p/E = 0$.

Springback occurs upon elastic unloading of the beam from an elastic–plastic state of deformation due to M_p. Recalling that

$$\frac{1}{R_e} = \frac{M_p}{EI} = \frac{12M_p}{Ebh^3} \text{ and } M_Y = \frac{Y_o bh^2}{6}$$

results in

$$\frac{1}{R_e} = \frac{2Y_o}{Eh}\frac{M_p}{M_Y} \tag{12.38}$$

Combining Equations 12.35a and 12.38 with Equation 12.14 gives for $\beta = 2y_p/h$:

$$\begin{aligned} \frac{1}{R_f} &= \frac{1}{R_p} - \frac{1}{R_e} = \frac{2Y_o}{h\beta} - \frac{2Y_o}{Eh}\frac{M_p}{M_Y} \\ \frac{1}{R_f} &= \frac{2Y_o}{Eh}\left[\frac{1}{\beta} - \frac{M_p}{M_Y}\right] \end{aligned} \tag{12.39}$$

As regards to full yielding of the beam cross section, no value can be assigned here for M_u. Due to the assumption of strain hardening, i.e., an ever-increasing yield stress, there will always be an elastic core no matter how high the applied bending moment.

Finally, for a nonlinear strain-hardening material,* the derivations are similar to those given above. For a material that strain hardens according to the law $\sigma = K\varepsilon^n$, where $\varepsilon > (Y/E)$, the moment-curvature relation takes the form:[1]

$$\frac{M_p}{M_Y} = \frac{1}{2+n}\left\{3\left(\frac{R_Y}{R_p}\right)^n - (1-n)\left(\frac{R_p}{R_Y}\right)^2\right\} \tag{12.40}$$

For a nonhardening material, $n = 0$, this reduces to Expression 12.12. Note that the analysis of bending moments and deflections of beams under stationary creep conditions apply directly to beams of materials which strain harden according to the power law $\sigma = K\varepsilon^n$. Hence, expressions of Section 11.4.3 remain valid with $\dot{\varepsilon}$, $\dot{v}$, N, and B replaced by ε, v, $1/n$, and $(1/K^{1/n})$, respectively. For the bending of beams of a material satisfying a nonlinear plastic stress strain relation of the form $\sigma = K\varepsilon^n$, the following equations apply:[4]

$$\frac{M}{I_n} = \frac{K}{R^n} = \frac{\sigma}{y^n}$$

where

* Numerical solutions for beams of materials obeying the Rambnerg–Osgood law (Equation 10.8d) are obtained by Betten.[3]

$$I_n = \frac{bh^{n+2}}{2^{n+1}(n+2)}$$

All reduce to the well-known elastic solution at $n = 1$.

Example 12.6:

A beam of rectangular cross section, b = *25 mm and* h = *40 mm, is loaded gradually so that the plastic zone spreads inward. Determine the values of the bending moments that cause initial yielding, spread of the plastic zone to a depth of 10 mm from the beam top and bottom surfaces, and the springback upon unloading. Consider the following material behaviors:*

a. *Elastic–perfectly plastic;*
b. *Linear strain-hardening with* $\sigma = Y_o + E^p\varepsilon^p$.

Take Y = *250 MPa,* E = *200 GPa, and* E^p = *10 GPa.*

Solution:

a. Elastic–perfectly plastic material: At initial yielding, Equation 12.21 gives

$$M_Y = \frac{Y_o bh^2}{6} = \frac{250\times10^6\times0.025\times0.04^2}{6} = 1666.67\ \text{N}\cdot\text{m}$$

for elastic–plastic conditions at y_p = 10 mm, Equation 12.22 gives

$$M_p = \frac{Y_o b}{12} = \left[13h^2 - 4y_p^2\right] = \frac{250\times10^6\times0.025}{12}\left[3\times0.04^2 - 4\times0.01^2\right]$$

$$= 2291.7\ \text{N}\cdot\text{m}$$

and

$$\frac{1}{R_p} = \frac{Y_o}{Ey_p} = \frac{250\times10^6}{200\times10^9\times0.01} = 0.125\ \text{m}^{-1}$$

Since at recovery

$$\frac{1}{R_e} = \frac{M_p}{EI} = \frac{2291.7}{200\times10^9\left(0.025\times0.04^3/12\right)}$$

giving $1/R_e = 0.0859\ \text{m}^{-1}$.

The final radius of curvature is calculated from Equation 12.14 as

$$\frac{1}{R_f} = \frac{1}{R_p} - \frac{1}{R_e} = 0.125 - 0.0859$$

giving R_f = 25.5 m.

b. Elastic linear strain-hardening material: At initial yielding M_y has the same value: M_Y = 1666.67 N · m. For elastic–plastic conditions at y_p = 10 mm, $\beta = 2y_p/h = 0.5$, and $E^* = E^p/E = 0.05$, Equation 12.37 gives

$$M_p = Y_o bh^2\left[\left(3-\beta^2\right)+E^*\left(\beta^2-1+2/\beta\right)\right]/12$$

$$= \frac{250\times10^6\times0.025\times0.04^2}{12}\left[\left(3-0.5^2\right)+0.05\left(0.5^2-1+2/0.5\right)\right]$$

$$= 2427.1\ \mathrm{N\cdot m}$$

This value is about 6% higher than M_p for perfectly plastic material. The final radius of curvature R_f is obtained from Equation 12.39 as

$$\frac{1}{R_f} = \frac{2Y_o}{Ebh}\left[\frac{1}{\beta}-\frac{M_p}{M_Y}\right] = \frac{2\times250\times10^6}{200\times10^9\times0.04}\left[\frac{1}{0.5}-\frac{2427.1}{2291.7}\right]$$

or

$$R_f = 17\ \mathrm{m}$$

12.4 BIAXIAL BENDING OF FLAT PLATES

Biaxial bending of thin, flat elastic plates has been analyzed in Chapter 8. Provided that the deflections and strains are small, the validity of the assumption of planes normal to the midsurface remaining plane applies in the analysis of elastic–plastic plates.

12.4.1 RECTANGULAR PLATES

The elastic–plastic bending of a thin flat plate, as shown in Figure 12.13a, by bending moments $M_x \neq M_y$ results in the stress distributions σ_x and σ_y of Figure 12.13b.

From Equations 8.2, 8.4, and 8.5, the stresses σ_x and σ_y, respectively, in the elastic state are expressed by

$$\sigma_x = \frac{Ez}{1-\nu^2}\left(\frac{1}{R_{ex}}+\frac{\nu}{R_{ey}}\right)$$

$$\sigma_y = \frac{Ez}{1-\nu^2}\left(\frac{1}{R_{ey}}+\frac{\nu}{R_{ez}}\right) \qquad (12.41a)$$

where R_{ex} and R_{ey} are the elastic radii of curvatures.

If elastic–plastic bending occurs, the elastic core in the material will be different in either direction x or y, i.e., $z_{px} \neq z_{py}$, depending on the values attained by M_x and M_y. For a plate of a material with linear strain-hardening behavior $\sigma = Y_o + E^p\varepsilon^p$, at the elastic–plastic interface, Equations 12.41a gives

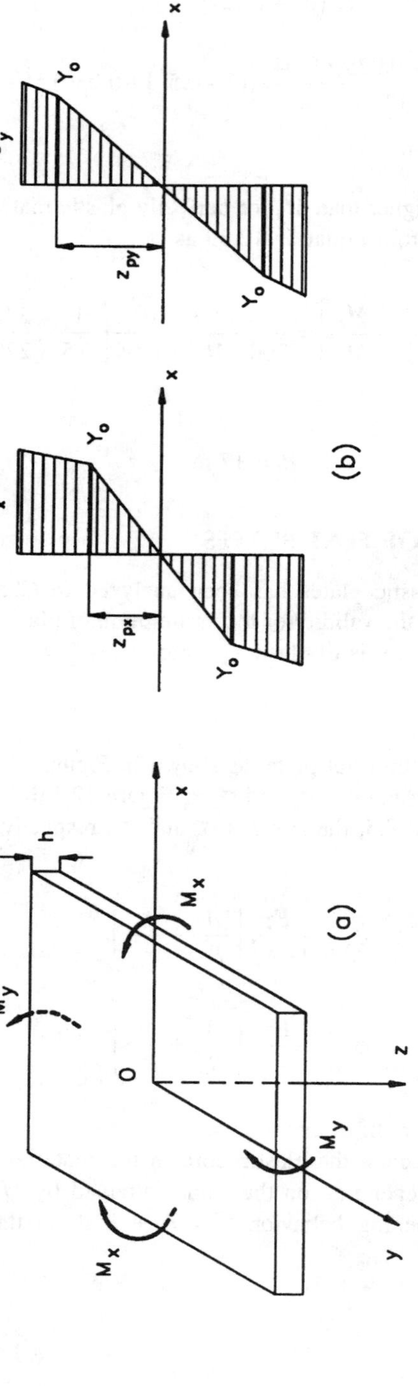

FIGURE 12.13 Elastic–plastic biaxial bending of (a) a rectangular plate and (b) stress distributions.

$$Y_o = \frac{Ez_{px}}{1-\nu^2}\left(\frac{1}{R_{px}} + \frac{\nu}{R_{py}}\right)$$

$$Y_o = \frac{Ez_{py}}{1-\nu^2}\left(\frac{1}{R_{py}} + \frac{\nu}{R_{px}}\right) \tag{12.41b}$$

where R_{px} and R_{py} are the radii of curvature at conditions of yielding in the x- and y-directions, respectively. Equating both Expressions 12.41b and solving for the radii of curvature results in

$$\frac{1}{R_{px}} = \frac{2Y_o}{Eh}\left(\frac{1}{\beta_x} - \frac{\nu}{\beta_y}\right)$$

$$\frac{1}{R_{py}} = \frac{2Y_o}{Eh}\left(\frac{1}{\beta_y} - \frac{\nu}{\beta_x}\right) \tag{12.42}$$

where

$$\beta_x = \frac{z_{px}}{h/2} \text{ and } \beta_y = \frac{z_{py}}{h/2}$$

The bending moment M_{px} and M_{py} per unit length, causing partial yielding in the x- and y-directions, respectively, are obtained from Equation 12.37 where $E^* = E^p/E$, as

$$M_{px} = Y_o h^2\left[\left(3-\beta_x^2\right) + E^*\left(\beta_x^2 - 1 + 2/\beta_x\right)\right]/12$$

$$M_{py} = Y_o h^2\left[\left(3-\beta_y^2\right) + E^*\left(\beta_y^2 - 1 + 2/\beta_y\right)\right]/12 \tag{12.43}$$

The effect of elastic unloading in both directions is determined by combining Equations 8.2 and 8.9 to yield, for the elastic radii of curvature R_{ex} and R_{ey},

$$\frac{1}{R_{ex}} = 12\left(M_{px} - \nu M_{py}\right)/Eh^3$$

$$\frac{1}{R_{ey}} = 12\left(M_{py} - \nu M_{px}\right)/Eh^3 \tag{12.44}$$

The final radii of curvature after recovery are thus determined from Expressions 12.42 and 12.44 according to Equation 12.14.

Example 12.7:

A square, flat plate with clamped edges and thickness h *is subjected to temperatures increase of* T *and decrease of* T *at the top and bottom surface of the plate, respectively. Determine the temperature difference* 2T *sufficient to cause partial yielding to a depth of* h/3.

Solution:

From symmetry of deformation, $R_{px} = R_{py} = R_p$, $z_{px} = z_{py} = z_p$ and $\beta_x = \beta_y = z_p/h/2$. Equation 12.42 thus gives

$$\frac{1}{R_p} = \frac{2Y_o}{Eh}\left(\frac{1}{\beta} - \frac{\nu}{\beta}\right) = \frac{(1-\nu)Y_o}{Ez_p}$$

Since the curvature is due to a temperature difference of $2T$, then

$$\frac{1}{R_p} = \frac{2\alpha T}{h}$$

Therefore,

$$2T = \frac{(1-\nu)Y_o}{\alpha E}\frac{h}{z_p}$$

For $z_p = h/3$:

$$2T = \frac{3(1-\nu)Y_o}{\alpha E}$$

For steel: $Y_o = 200$ MPa, $E = 200$ GPa, $\nu = 0.3$, $\alpha = 12 \times 10^{-6}$/°C, which gives $2T = 175$°C.

The value of bending moment $M_{px} = M_{py} = M_p$ is determined from Equation 12.43 for an elastic–perfectly plastic material, i.e., $E^* = E^p/E = 0$ as

$$M_p = Y_o h^2(3 - \beta^2)/12$$

Upon return to uniform ambient temperature, elastic recovery takes place and Equation 12.44 gives

$$\frac{1}{R_e} = \frac{12(1-\nu)M_p}{Eh^3} = \frac{(1-\nu)Y_o}{Eh}\left(3-\beta^2\right)$$

The final curvature R_f is thus obtained as

$$\frac{1}{R_f} = \frac{1}{R_p} - \frac{1}{R_e} = \frac{3(1-\nu)Y_o}{Eh} - \frac{(1-\nu)Y_o}{Eh}\left(3-\beta^2\right)$$

$$\frac{1}{R_f} = \frac{(1-\nu)Y_o\beta^2}{Eh} = \frac{(1-0.3)\times 200\times 10^6 \times 0.5^2}{200\times 10^9 \times h}$$

or

$$R_f = 5.714 \times 10^3 h \text{ m } (h \text{ in meters})$$

12.4.2 Circular Plates

In Section 8.5, flat circular plates of uniform thickness subjected to axisymmetric bending have been analyzed in the elastic state under different loading and edge conditions. Similar to the case of plate bending, Section 12.2, only the simple case of a circular plate subjected to uniform edge moments, M_o per unit circumference, will be considered here. The plate is simply supported at the outer radius, and, in the case of an annular plate, it will be subjected to edge moments M_o at both the inner and outer radii r_i and r_o, respectively, Figure 12.14. The assumptions of the midsurface remaining unstrained and straight lines normal to the midsurface remaining straight hold in both elastic and plastic bending, provided that the strains are small.

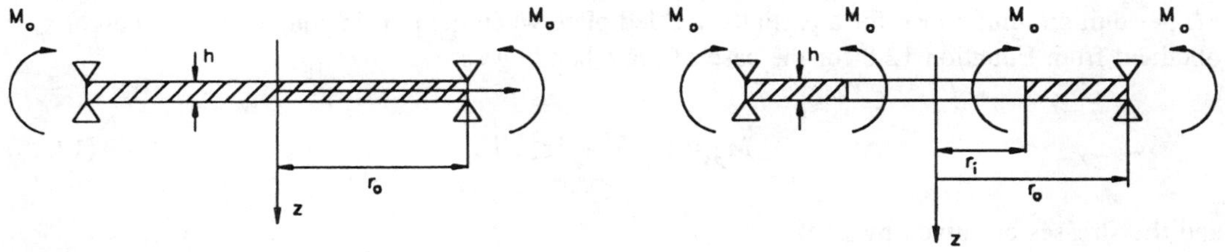

FIGURE 12.14 Axisymmetric elastic–plastic loading of a circular plate subjected to uniform edge bending moment.

The lateral displacement of the midsurface w and the stresses in the elastic state for both plates of Figure 12.14 are given by

$$w = \frac{M_o}{2D(1+\nu)}\left(r_o^2 - r^2\right)$$

$$\sigma_r = \sigma_\theta = \frac{M_o}{h^3/12} z \tag{8.54}$$

and

$$\sigma_z = 0 \tag{8.55}$$

The maximum stress occurs at the plate surface, $z = \pm h/2$, given by

$$\sigma_r = \sigma_\theta = \pm\frac{6M_o}{h^2}$$

The radius of curvature of the deformed plate R is obtained from Equation 8.54 as

$$\frac{1}{R_e} = -\frac{\partial^2 w}{\partial r^2} = \frac{M_o}{D(1+\nu)} = \frac{12(1-\nu)}{Eh^3} M_o$$

The radius of curvature is constant along r, which means that the midsurface deforms into a spherical shape.

As the bending moment is increased yielding occurs at z = $\pm h/2$, when $\sigma_r = \sigma_\theta = Y = 6M_Y/h^2$, which gives

$$M_Y = Y\frac{h^2}{6}$$

and, hence,

$$\frac{1}{R_Y} = Y\frac{h^2}{6D(1+\nu)} = \frac{2Y}{Eh}(1-\nu)$$

Further increase in the bending moment will spread the plastic zone toward the midsurface. The stress distribution for elastic–perfectly plastic material is shown in Figure 12.2b. The moment M_p per unit circumference for a partially yielded plate when the plastic zone reaches a height z_p is obtained from Equation 12.8 for the case of plate bending as

$$M_p = Y\left(3h^2 - 4z_p^2\right)/12 \tag{12.45}$$

and the stresses are given by

$$\begin{aligned} \sigma_r &= \sigma_\theta = Yz/zp \quad \text{for } |z| \le z_p \\ \sigma_r &= \sigma_\theta = Y \qquad \text{for } |z| \le z_p \\ \sigma_z &= 0 \end{aligned} \tag{12.46}$$

The radius of curvature is obtained from Equation 12.1, $\varepsilon_r = z/R$, by substituting $\varepsilon_r = (Y/E)(1 - \nu)$ at $z = z_p$, which yields

$$\frac{1}{R_p} = \frac{Y}{Ez_p}(1-\nu) \tag{12.47}$$

As the bending moment increases, z_p will decrease until the entire plate becomes fully plastic; the bending moment reaches an ultimate value M_u obtained by putting $z_p = 0$ in Equation 12.8 giving

$$M_u = Yh^2/4 \tag{12.48}$$

$$\sigma_r = \sigma_\theta = \pm Y \text{ all over, and } \sigma_z = 0 \tag{12.49}$$

The radius of curvature is indeterminate since for a perfectly plastic material there is no strain hardening; thus as soon as the plate is fully plastic, it will continue to flow ending by collapse.

When the edge moment is removed after partial yielding has occurred, the plate does not recover completely to its original shape. This is due to the occurrence of residual stresses $(\sigma_r)_{res}$ and $(\sigma_\theta)_{res}$, which can be determined by following the same procedure as for the case of plate bending, Equation 12.13a.

$$\begin{aligned} \left(\sigma_r\right)_{res} = \left(\sigma_\theta\right)_{res} &= \frac{Yz}{z_p} - \frac{Yz}{h^3}\left(3h^2 - 4z_p^2\right) \quad \text{for } |z| \le z_p \\ &= \pm Y - \frac{Yz}{h^3}\left(3h^2 - 4z_p^2\right) \quad \text{for } |z| \ge z_p \end{aligned} \tag{12.50}$$

and

$$\frac{1}{R_f} = \frac{(1-\nu)Y}{E}\left[\frac{1}{z_p} - \frac{3h^2 - 4z_p^2}{h^3}\right] \tag{12.51}$$

Example 12.8:
In Example 8.5 a disk of 30 mm diameter and 2 mm thickness is blanked by shearing, Figure 8.20. Material has E = 210 GPa, ν = 0.3, and Y = 230 MPa. It is required to produce a concave blank having radius of curvature of 200 mm. Determine the diametral clearance between punch and die Δ to obtain the required curvature and the maximum values of the residual stresses in the blank. For this calculation a practical value of the ultimate shear strength of 400 MPa is to be used.

Solution:
Substituting in Equation 12.51 the given values of R_f, h, E, Y, and $\nu = 0.3$

$$\frac{1}{200\times10^{-3}} = \frac{230\times10^6\times(1-0.3)}{210\times10^9}\left[\frac{1}{z_p} - \frac{3\times0.002^2 - 4\times z_p^2}{0.002^3}\right]$$

gives $z_p = 0.125$ mm.

Substituting z_p in Equation 12.45 gives

$$M_p = 230\times10^6\,(3\times0.002^2 - 4\times0.000125^2)/12 = 228.8 \text{ N}\cdot\text{mm}$$

In terms of the die clearance Δ and the ultimate shear stress, this bending moment gives

$$M_p = \tau_{\text{ult}} h\frac{\Delta}{2} = 400\times10^6\times0.002\times\frac{\Delta}{2}$$

which yields Δ = 0.572 mm.

The maximum residual stresses occur at $z = z_p$ and are obtained by substituting $z_p = 0.125$ mm in Equation 12.50, which gives max $(\sigma_r)_{\text{res}}$ = max $(\sigma_\theta)_{\text{res}}$ = ±187 MPa at $z = \pm0.125$ mm.

12.5 BENDING OF CIRCULARLY CURVED BEAMS

Bending of a circularly curved beam in the elastic range has been analyzed in Section 6.3.1 using the stress function approach. The stresses are given for a beam of rectangular cross section of unit thickness by

$$\sigma_r = \frac{4M}{Nhr_i^2}\left[\left(\frac{r_o}{r}\right)^2 \ln\left(\frac{r_o}{r_i}\right) + \left(\frac{r_o}{r_i}\right)^2 \ln\left(\frac{r}{r_o}\right) - \ln\left(\frac{r}{r_i}\right)\right]$$

$$\sigma_\theta = \frac{4M}{Nhr_i^2}\left[-\left(\frac{r_o}{r}\right)^2 \ln\left(\frac{r_o}{r_i}\right) + \left(\frac{r_o}{r_i}\right)^2 \ln\left(\frac{r}{r_o}\right) - \ln\left(\frac{r}{r_i}\right) + \left(\frac{r_o}{r_i}\right)^2 - 1\right] \tag{6.43}$$

$$\tau_{r\theta} = 0$$

where

$$N = \left[\left(\frac{r_o}{r_i}\right)^2 - 1\right]^2 - 4\left(\frac{r_o}{r_i}\right)^2\left[\ln\left(\frac{r_o}{r_i}\right)\right]^2$$

Note that the radial stress σ_r is positive throughout the cross section. The hoop stress σ_θ changes from positive to negative when r is increased from r_i to r_o, as shown in Figure 12.15a. The position of the neutral surface corresponding to $\sigma_\theta = 0$, which does not obviously coincide with the centroidal surface, is independent of M as long as the bar is entirely elastic. This neutral surface will change its position as plastic deformation progresses throughout the plate.

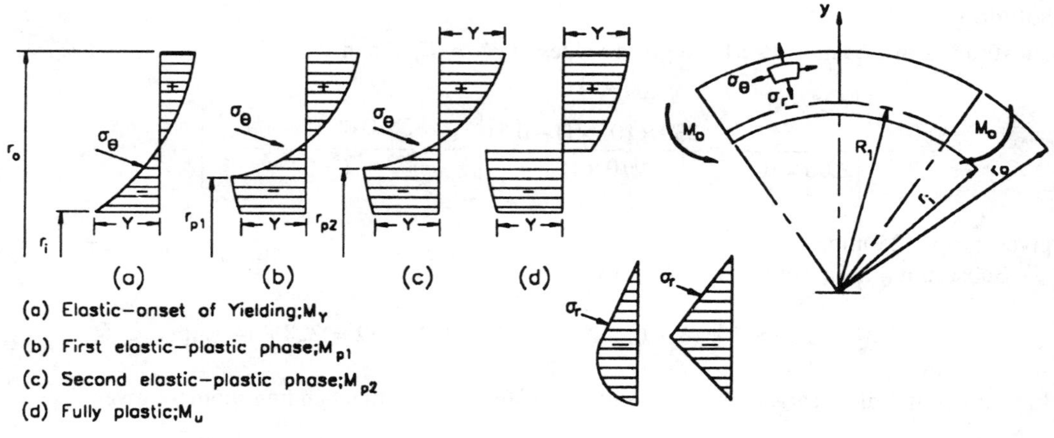

FIGURE 12.15 Stresses in an elastic–plastic circularly curved beam in bending: (a) elastic — onset of yielding, M_Y; (b) first elastic–plastic phase, M_{p1}; (c) second elastic–plastic phase, M_{p2}; (d) fully plastic, M_u.

The stresses $\sigma_\theta > \sigma_z > \sigma_r$ in the plastic region must satisfy the Tresca condition (Equation 10.18); $\sigma_\theta - \sigma_r = Y$.

At the inner concave surface, σ_θ attains its maximum while $\sigma_r = 0$; therefore, at $r = r_i$:

$$\sigma_\theta = -Y \tag{12.52}$$

Combining Equations 6.43 and 12.52 results in an expression for M_Y at which yielding commences as

$$M_Y = \frac{Y}{4}\frac{\left(r_0^2 - r_i^2\right)^2 - 4r_i^2 r_o^2\left(\ln r_o/r_i\right)^2}{\left(r_0^2 - r_i^2\right) + 2r_o^2\left(\ln r_o/r_i\right)} \tag{12.53}$$

As the bending moment is further increased, a plastic zone is formed spreading from $r = r_i$ outward to a radius, say, r_{p1}. The stresses in the elastic region ($r_{p1} < r \leq r_o$) could be determined using the conditions of continuity of σ_r and σ_θ across the elastic–plastic interface radius r_{p1}.

The stresses in the plastic region must satisfy both the equilibrium equation and the yield criterion. Thus, it can be shown that these stresses are given by[5]

$$\sigma_r = -Y \ln(r/r_i)$$
$$\sigma_\theta = -Y[1 + \ln(r/r_i)] \quad r_i < r \le r_{p1} \tag{12.54}$$

By using the expressions for the stresses in the elastic and plastic regions, the applied moment M_{p1} at this phase is:

$$M_{p1} = -\frac{Y}{4}\left[\left(r_{p1}^2 - r_i^2\right) - Lr_{p1}^2\right] \Bigg/ \left[1 + \frac{r_o^2}{r_{p1}^2}\left(\ln\frac{r_o^2}{r_{p1}^2} - 1\right)\right]$$

where

$$L = \left(\frac{r_o^2}{r_{p1}^2} - 1\right)^2 - \frac{r_o^2}{r_{p1}^2}\left[\left(\ln\frac{r_o^2}{r_{p1}^2}\right)\left(\ln\frac{r_o^2}{r_i^2}\right) - \left(\frac{r_o^2}{r_{p1}^2} - 1\right)\left(\ln\frac{r_{pl}^2}{r_i^2}\right)\right] \tag{12.55}$$

When the plastic zone has spread to a sufficient extent, the outer convex boundary becomes stressed to the yield point, i.e., $\sigma_\theta = Y$ at $r = r_o$, as shown in Figure 12.15c. If at this instant the plastic zone has spread to a radius r_{p2}, it can be shown that this occurs when the bending moment reaches its value M_{p2}, as given by

$$M_{p2} = \frac{Y}{4}\left[r_i^2 - r_o^2 + 2r_o^2 \ln\left(\frac{r_i r_o}{r_{p2}}\right)\right] \tag{12.56}$$

At this stage it is said that a second partial yielding has started. The radius r_{p2} is given by

$$\left(r_{p2}/r_o\right)^2 = \left[1 - \ln\left(r_i r_o/r_{p2}^2\right)\right] \Big/ \left[1 + \ln\left(r_o/r_i\right)\right] \tag{12.57}$$

Further increase in M_{p1} causes a progressive spreading of both plastic zones; the elastic zone between them diminishes. Ultimately, the bending moment approaches a value to cause full yielding throughout the plate section, as shown in Figure 12.15d. This moment is given by

$$M_u = \frac{Y}{4}\left(r_o - r_i\right)^2 \tag{12.58}$$

The distribution of stresses within elastic, partial, and full yielding indicates variation in the position of the neutral axis. At full yielding the radius of the neutral axis R_1 and the stresses are given by

$$\begin{aligned}
R_1 &= \sqrt{r_1 r_o} && \\
\sigma_r &= -Y\ \ln(r/r_i) && r_i \le r < R_1 \\
\sigma_\theta &= -Y\ (1 + \ln r/r_i) && \sigma_\theta < \sigma_r \qquad (12.59a)\\
\sigma_r &= Y\ \ln(r_o/r) && R_1 \le r \le r_o \\
\sigma_\theta &= Y(1 - \ln r/r_o) && \sigma_\theta > \sigma_r \qquad (12.59b)
\end{aligned}$$

Note that σ_r is continuous all over, but σ_θ is discontinuous at the neutral axis.

Example 12.9:

For a curved beam of $r_o/r_i = 2$ and unit thickness subjected to pure bending calculate the bending moment for

a. Initial yielding, first partial yielding when $r_{p1}/r_o = 0.6$, and full yielding;
b. Repeat these calculations for a beam of $r_o/r_i = 1.1$ taking the same spread of partial yielding. Compare with the above case and a straight beam of the same depth.

Solution:

a. For $r_o/r_i = 2$

Initial yielding: Equation 12.53

$$M_Y = 0.03224Yr_o^2$$

Partial yielding: Equation 12.55 at $r_{p1}/r_o = 0.6$

$$M_{p1} = 0.05971Yr_o^2$$

Full yielding: Equation 12.58

$$M_u = 0.0625Yr_o^2$$

b. Repeating calculations for $r_o/r_i = 1.1$ and taking the same spread of partial yielding, i.e.,

$$\left(\frac{r_p - r_i}{r_o - r_i}\right)_{r_o/r_i=2} = \left(\frac{r_p - r_i}{r_o - r_i}\right)_{r_o/r_i=1.1}$$

gives $r_p/r_o = 0.9273$. Hence,

$$M_Y = 0.00133Yr_o^2$$

$$M_{p1} = 0.01873Yr_o^2$$

$$M_u = 0.00207Yr_o^2$$

Note that $(r_o - r_i)/r_o = 0.5$ for the first beam, while it is only 0.1 for the second one.

For a straight beam, $h = r_o - r_i = 0.09091r_o$; hence,

$$M_Y = 0.00138Yr_o^2 \text{ Equation 12.21}$$

$$M_u = 0.00207Yr_o^2 \text{ Equation 12.23}$$

Note that for a curved beam with small $(r_o - r_i)/r_o$, computation may be done using the more simpler expressions for bending of a straight beam.

12.6 BUCKLING OF BARS UNDER AXIAL COMPRESSION

In Section 7.7.1 elastic buckling of bars under axial compressive load has been analyzed. For a bar of pinned ends of length L, cross-sectional area A, least second moment of area of the cross section I, and material elastic modulus E, the critical buckling load or Euler load is given by

$$P_{cr} = \frac{\pi^2 EI}{L^2} \tag{7.75}$$

Since the material of the bar is elastic, the compressive stress on the bar cross section P_{cr}/A should be less than the yield strength of the bar material, Y. Substituting this condition in Equation 7.75 gives

$$\frac{\pi^2 EI}{L^2 A} \le Y \text{ or } \lambda = L\Big/\sqrt{\frac{I}{A}} \ge \pi\sqrt{\frac{E}{Y}} \tag{12.60}$$

where λ is the slenderness ratio of the bar.

For an elastic–perfectly plastic material, where Y remains constant so that the bar cannot sustain further increase of axial load, a plastic hinge will develop at the midlength leading to collapse of the bar. This means that the material should be strain hardening to sustain further increase of load. In elastic–plastic analysis of buckling, the material is assumed to be elastic-linear strain hardening, as shown in Figure 12.16a. The stress–strain relation is determined by the elastic modulus E and a plastic modulus E^p. For nonlinear strain hardening E^p is replaced by the tangent modulus corresponding to the slope of the stress–strain curve at the point when the buckling stress is equal to the initial yield stress Y.

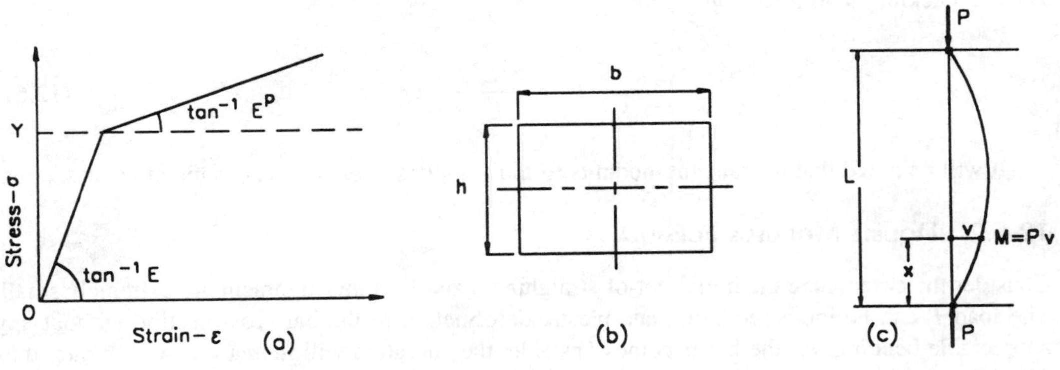

FIGURE 12.16 Elastic–plastic buckling of a bar under axial compression: (a) material stress–strain curve, (b) bar cross section, and (c) bar loading and deformation.

In this analysis, the bar cross section is assumed to be rectangular, Figure12.16b, and the bar deforms as shown in Figure 12.16c. The cross section is thus subjected to combined bending and compression. In this case the stress distribution depends on the history of loading, as has been indicated in Section 12.4. Two cases are considered: (a) increasing both P and M simultaneously, which yields the tangent modulus formula and (b) increasing P first to initiate yielding and then M is applied maintaining P constant, which yields the double-modulus formula.

12.6.1 Tangent Modulus Formula

Consider the case where the initial out-of-straightness and load misalignment are appreciable, Figure12.16c. The application of a load P will induce a bending moment M, which will simultaneously increase with the increase of P. In this respect the summation of the stresses due to P and M will continuously increase. As soon as plastic deformation is initiated, it will increase giving a fully plastic bar section at $P = P_{cr}$. Assuming a linear strain-hardening material, as in Figure 12.16a,

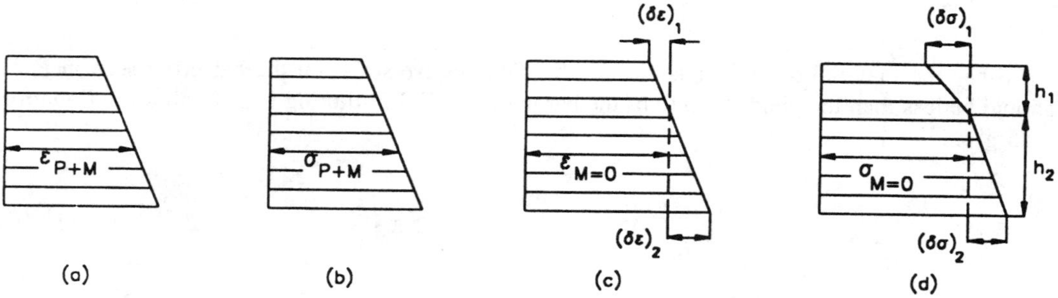

FIGURE 12.17 Strain and stress distributions in elastic–plastic buckling according to initial out-of-straightness: (a), (b) strain and stress for the tangent modulus formula, and (c) and (d) strain and stress distributions for the double-modulus formula.

and a linear strain distribution, Figure 12.17a, the stress distribution will be linear, as shown in Figure 12.17b. Since the Euler formula for elastic analysis, Equation 7.75, is essentially based on a linear strain and stress distributions, it will be possible to express P_{cr} for the case of plastic buckling by the Euler formula if the modulus of elasticity E is replaced by the tangent modulus E^p. The buckling load is thus given by

$$P_{cr} = \frac{\pi^2 E^p I}{L^2} \tag{12.61}$$

It will be noted that the tangent modulus formula applies to bars of any shape of cross section.

12.6.2 Double-Modulus Formula

Consider the case where the initial out-of-straightness and load misalignment are extremely small. The load P can be increased to initiate plastic deformation in the bar cross section without any appreciable bending. As the bar becomes unstable, the curvature will increase at a high rate, thus M increases while P remains virtually constant. As a result of bending, the strain will decrease by $(\delta\varepsilon)_1$ on the convex outer surface and increase by $(\delta\varepsilon)_2$ on the concave inner surface, as shown in Figure 12.17c, corresponding to a decrease and increase in stresses by $(\delta\sigma)_1$ and $(\delta\sigma)_2$, respectively, Figure 12. 17d. The strain recovery along the part of height h_1 is elastic, whereas the strain increase along h_2 continues to be plastic. For an elastic-linear strain hardening material the change in stresses and strains are given by

$$(\delta\sigma)_1 = E(\delta\varepsilon)_1 \text{ and } (\delta\sigma)_2 = E^p(\delta\varepsilon)_2 \tag{12.62a}$$

For a plane cross section remaining plane,

$$(\delta\varepsilon)_1 = \frac{h_1}{R} \text{ and } (\delta\sigma)_2 = \frac{h_2}{R} \tag{12.62b}$$

where R is the radius of curvature of the bent bar. Substituting in Equation 12.62a gives

$$(\delta\sigma)_1 = E\frac{h_1}{R} \text{ and } (\delta\sigma)_2 = E^p\frac{h_2}{R} \tag{12.62c}$$

From the equilibrium of bending stresses,

$$\tfrac{1}{2}(\delta\sigma)_1 h_1 b = \tfrac{1}{2}(\delta\sigma)_2 h_2 b$$

or

$$(\delta\sigma)_1 h_1 = (\delta\sigma)_2 h_2 \tag{12.63a}$$

Substituting in Equation 12.62c yields

$$Eh_1^2 = E^p h_2^2$$

Hence,

$$h_1 = \frac{\sqrt{E^p}}{\sqrt{E}+\sqrt{E^p}}h \text{ and } h_2 = \frac{\sqrt{E}}{\sqrt{E}+\sqrt{E^p}}h \tag{12.63b}$$

The bending moment M is obtained by integration as

$$M = \tfrac{2}{3}h_1[\tfrac{1}{2}(\delta\sigma)_1 h_1 b] + \tfrac{2}{3}h_2[(\tfrac{1}{2}\delta\sigma)_2 h_2 b]$$

Substituting from Equations 12.62c and 12.63b yields

$$M = \frac{bh^3}{12}\frac{4EE^p}{\left(\sqrt{E}+\sqrt{E^p}\right)^2}\frac{1}{R}$$

Putting

$$\frac{bh^3}{12} = I \text{ and } \frac{4EE^p}{\left(\sqrt{E}+\sqrt{E^p}\right)^2} = \bar{E} \tag{12.64}$$

gives

$$M = \frac{\bar{E}I}{R}$$

Again P_{cr} for this case of plastic buckling can be obtained from the Euler formula , Equation 7.81b, if the modulus of elasticity E is replaced by the double-modulus $\overline{E}$, Equations 12.64 to give

$$p_{cr} = \frac{\pi^2 \overline{E} I}{L^2} \tag{12.65}$$

The values for I and $\overline{E}$ given in Equation 12.65 apply to a rectangular cross section. For other cross section shapes I and $\overline{E}$ will have different values. Note that the value of $\overline{E}$ is always greater than E^p since the latter is always smaller than E. This means that P_{cr} obtained from Equation 12.61 is lower than that obtained from Equation 12.65. The tangent modulus formula therefore may be recommended in design, since it is simpler to use for any shape of cross section and yields a more conservative design.

Example 12.10:

Determine the buckling load for a square bar of 50 mm width and 600 mm length made of an aluminum alloy E = *70,* E^p = *42 GPa, and* Y = *80 MPa.*

Solution:

Substituting in Equation 12.60, $A = 50^2 = 2500$ mm², $I = 50^4/12 = 520{,}833$ mm⁴, $E = 70$ GPa, and Y = 80 MPa gives $\lambda = 41.57$ and $\pi\sqrt{E/Y} = 92.93$, so that $\lambda < \pi\sqrt{E/Y}$ and, hence, from Equation 12.60 the deformation is elastic–plastic. Applying the tangent-modulus formula, Equation 12.61, gives

$$P_{cr} = \frac{\pi^2 \times 42 \times 10^3 \times 520{,}833}{600^2} = 599.72 \text{ kN}$$

Applying the double-modulus formula, Equations 12.64 and 12.65,

$$\overline{E} = \frac{4 \times 70 \times 42}{\left(\sqrt{70} + \sqrt{42}\right)^2} = 53.35 \text{ GPa}$$

gives

$$P_{cr} = \frac{\pi^2 \times 53{,}350 \times 520{,}833}{600^2} = 761.8 \text{ kN}$$

Note that if for the same cross section $\lambda = 92.93$, which means a bar length of 1341.3 mm, the critical load is obtained from the Euler formula as 200 kN, which is much less than the actual case when plastic yielding has been initiated.

12.7 BARS SUBJECTED TO TORSION

In Section 7.5, bars subjected to torsion have been analyzed in the elastic state using the Prandtl stress function ϕ from which the shear stress components τ_{zx} and τ_{zy} are determined as

$$\tau_{zx} = \frac{\partial \phi}{\partial y} \quad \text{and} \quad \tau_{zy} = -\frac{\partial \phi}{\partial x}$$

where ϕ satisfies the relations:

$$\nabla^2 \phi = \frac{\partial^2 \phi}{\partial x^2} + \frac{\partial^2 \phi}{\partial y^2} = -2G\alpha \tag{7.54}$$

α is the angle of twist per unit length of bar and

$$\phi = 0 \text{ at the boundary} \tag{7.55}$$

Noting that for a state of free torsion

$$\sigma_x = \sigma_y = \sigma_z = \tau_{zy} = 0$$

The yield condition is expressed as

$$\tau^2 = \tau_{zx}^2 + \tau_{zy}^2 = \left(\frac{\partial \phi}{\partial x}\right)^2 + \left(\frac{\partial \phi}{\partial y}\right)^2 = k^2 \tag{12.66}$$

where k is the yield stress in shear, which is equal to $Y/2$ according to the Tresca criterion (Equation 10.18) or to $Y/\sqrt{3}$, according to the von Mises criterion, Equation 10.17.

A property of Equation 7.54 is that τ is maximum at the boundary, and therefore plastic yielding starts at the boundary at the points satisfying Equations 12.66. These points correspond to the entire boundary in a bar of a circular cross section, minor axes extremities in an elliptical section, the middle points of sides in a square or an equilaterial triangular section. In tubular sections plastic yielding starts at the point of minimum wall thickness. As the applied twisting moment is increased, the plastic zones increase and elastic–plastic boundaries develop. These boundaries are not, in general, easy to determine analytically except in some few cases.* Two cases will be considered here namely, circular sections and thin-walled tubular sections.

12.7.1 Circular Solid and Hollow Sections

For a circular section in torsion the cross section does not warp and remains plane. The shear strain γ at any radius r is expressed by the relation:

$$\gamma = \alpha r \tag{12.67}$$

This relation holds in both the elastic and plastic states. Two cases will be considered: solid section and hollow section. The material is elastic–perfectly plastic, i.e., $k = Y/2$ for Tresca criterion, is constant.

12.7.1.1 Solid Circular Section

For a solid circular section of radius r_o yielding is initiated at $r = r_o$, where $\tau_y = k$ by applying a twisting moment $(M_t)_Y$ given by

$$(M_t)_Y = k\frac{\pi}{2}r_o^3$$

and

$$\alpha_Y = k/(Gr_o) = \gamma_Y/r_o = \tau_Y/Gr_o \tag{12.68a}$$

---

* Extensive discussion of torsion of prismatic bars of arbitrary cross sections is to be found in Chakrabary,[1] Chap. 3. Also, use of membrane and sand-heap analyses to analyze the elastic and plastic torsion problem of an arbitrary cross section is explained in Johnson and Mellor,[4] Chap. 8.

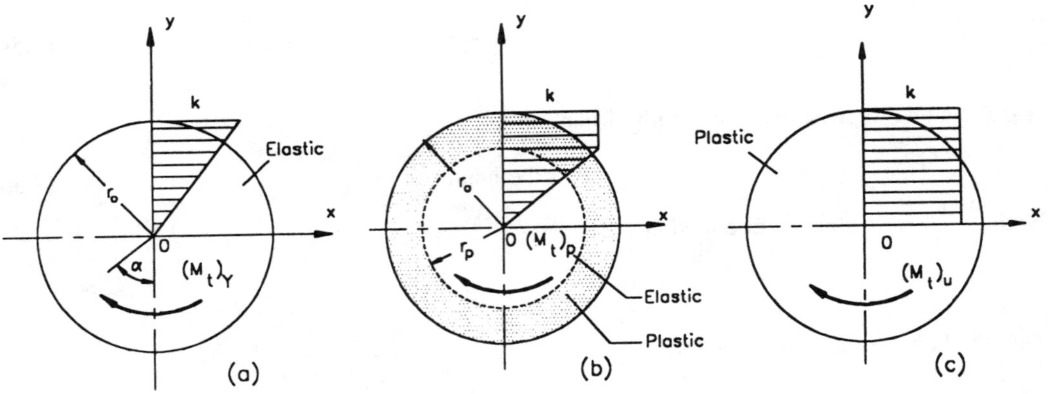

FIGURE 12.18 Shear stress distribution in elastic–perfectly plastic torsion of a solid circular section: (a) elastic state, (b) partial yielding, and (c) full yielding.

The shear stress distribution is shown in Figure 12.18a. Increasing the twisting moment beyond $(M_t)_Y$ will propagate, yielding inwards. Let yielding reach a certain radius r_p. The material included between $r = 0$ and $r = r_p$ will remain elastic while the material included between $r = r_p$ and $r = r_o$ will become plastic. For an elastic–perfectly plastic material, the shear stress distribution for this state of partial yielding is shown in Figure 12.18b. In this case, the twisting moment $(M_t)_p$ is obtained by integration and the angle of twist per unit length α_p is obtained from Expression 12.67 as

$$(M_t)_p = k\frac{\pi}{2}r_p^3 + k\frac{2}{3}\pi r_o^3\left[1-\left(\frac{r_p}{r_o}\right)^3\right]$$

or

$$(M_t)_p = k\frac{2\pi}{3}r_o^3\left[1-\frac{1}{4}\left(\frac{r_p}{r_o}\right)^3\right] \tag{12.68b}$$

and

$$\alpha_p = k/Gr_p$$

Increasing $(M_t)_p$ further will decrease r_p until $r_p = 0$ and the section becomes fully plastic. In this case $(M_t)_u$ and α_u are obtained from Equation 12.68b by substituting $r_p = 0$ giving

$$(M_t)_u = k\frac{2}{3}\pi r_o^3 \tag{12.68c}$$

and $\alpha_u \to \infty$, i.e., full collapse of bar.

12.7.1.2 Hollow Circular Section

For a hollow section of outer radius r_o and inner radius r_i, the corresponding values of $(M_t)_Y$ and α_Y for initial yielding, $(M_t)_p$ and α_p for partial yielding, and $(M_t)_u$ and α_u for full yielding are obtained such as in the case of a solid circular section as

$$(M_t)_Y = k\frac{\pi}{2}r_o^3\left[1-\left(\frac{r_i}{r_o}\right)^4\right] \text{ and } \alpha_Y = K/(Gr_o)$$

(12.69a)

$$(M_t)_p = k\frac{\pi}{2}r_p^3\left[1-\left(\frac{r_i}{r_p}\right)^4\right] + k\frac{2\pi}{3}r_o^3\left[1-\left(\frac{r_p}{r_o}\right)^3\right]$$

$$\alpha_p = k/(Gr_p) \tag{12.69b}$$

The twisting moment for a fully plastic wall is obtained from Equation 12.69b by putting $r_p = r_i$, hence,

$$(M_t)_u = k\frac{2\pi}{3}r_o^3\left[1-\left(\frac{r_i}{r_o}\right)^3\right] \text{ and } \alpha_u = \rightarrow \infty \tag{12.69c}$$

The associated stress distributions are shown in Figure 12.19a, b, and c, respectively.

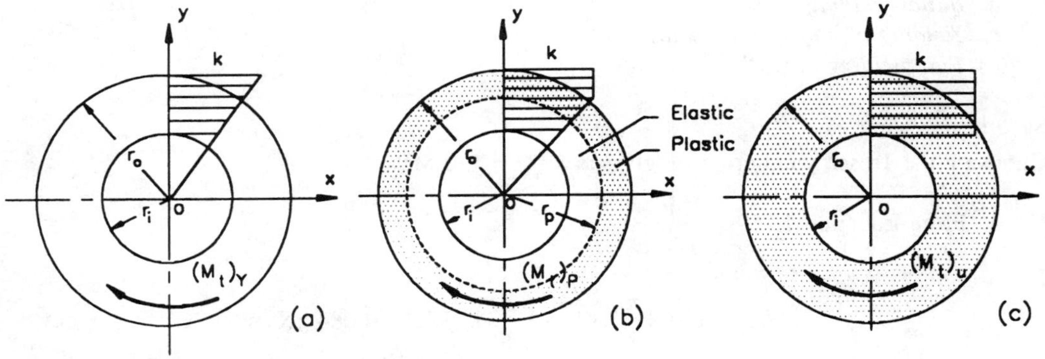

FIGURE 12.19 Shear stress distribution in elastic–perfectly plastic torsion of a hollow circular section: (a) elastic state, (b) partial yielding, and (c) full yielding.

Unloading for solid and hollow sections is fully elastic so that the residual stresses are obtained by superimposing the stresses given in Figures 12.18 and 12.19, respectively, to the stresses in the elastic state resulting from the application of the moment $(M_t)_p$ in the opposite direction.

Consequently, it follows that the residual stress in the previously elastic and plastic zones, respectively, are given by

$$(\tau)_{res} = k\left(\frac{r}{r_p}\right)\left\{1-\frac{4r_p}{3r_o}\left(1-\frac{r_i^4}{r_o^4}\right)\left[1-\frac{1}{4}\left(\frac{r_p}{r_o}\right)^3-\frac{3}{4}\left(\frac{r_o}{r_p}\right)\left(\frac{r_i}{r_o}\right)^4\right]\right\} \quad r_i < r \le r_p \tag{12.70a}$$

$$(\tau)_{res} = k\left\{1-4r\left[1-\frac{1}{4}\left(\frac{r_p}{r_o}\right)^3-\frac{3}{4}\left(\frac{r_o}{r_p}\right)\left(\frac{r_i}{r_o}\right)^4\right]\Big/\left[3r_o\left(1-\frac{r_i^4}{r_o^4}\right)\right]\right. \quad r_p < r \le r_o \tag{12.70b}$$

In the above analysis, the yield shear stress k is taken to be constant. If, however, the material strain-hardens as displayed by the torque-angular displacement curve ($M_t - \alpha$) of a rod tested under torsion, a correction to k has to be made. This correction is suggested[6] to be

$$k = \left(3M_t + \alpha \frac{dM_t}{d\alpha}\right) \Big/ \left(2\pi r_o^3\right) \tag{12.71a}$$

where $dM_t/d\alpha$ is the slope of the M_t–α curve. The accuracy of calculations may be improved specially in the early part of M_t–α curve by considering the equation:

$$k = \left[4M_t + \alpha^2 \frac{d}{d\alpha}\left(\frac{M_t}{\alpha}\right)\right] \Big/ \left(2\pi r_o^3\right) \tag{12.71b}$$

Example 12.11:
A bar of a hollow circular cross section of r_i = *20 mm and* r_o = *30 mm is made of low-carbon steel of* Y = *240 MPa and* G = *80 GPa. Determine the twisting moment and the angle of twist per unit bar length for the following conditions:*

a. *Initial yielding*
b. *Yielding of half wall thickness*
c. *Full yielding of wall thickness*

Solution:
Choosing the Tresca yield criterion gives $k = Y/2 = 120$ MPa

a. From Equation 12.69a,

$$\left(M_t\right)_Y = 120 \times \frac{\pi}{2} 30^3 \left(1 - \left(\frac{20}{30}\right)^4\right) 10^{-6} = 4.084 \text{ kN} \cdot \text{m}$$

$$\alpha_Y = \frac{120}{80 \times 30} \times 10^{-3} = 5 \times 10^{-5} \text{ rad}$$

b. From Equation 12.69b, substituting r_p = 25 mm gives

$$\left(M_t\right)_p = \left[120 \times \frac{\pi}{2} 25^3 \left(1 - \left(\frac{20}{25}\right)^4\right) + 120 \times \frac{2}{3} \pi 30^3 \left(1 - \left(\frac{20}{25}\right)^3\right)\right] 10^{-6}$$

$$= 1.739 + 2.859 = 4.598 \text{ kN} \cdot \text{m}$$

$$\alpha_p = \frac{120}{80 \times 25} \times 10^{-3} = 6 \times 10^{-5} \text{ rad}$$

c. From Equation 12.69c,

$$\left(M_t\right)_u = 120 \times \frac{2\pi}{3} 30^3 \left[1 - \left(\frac{20}{30}\right)^3\right] \times 10^{-6} = 4.775 \text{ kN} \cdot \text{m}$$

$\alpha_u \to \infty$ collapse of bar unless the material develops strain hardening.

12.7.2 Thin-Walled Tubular Sections

In Section 7.5.5 torsion of thin-walled tubular sections has been analyzed based on uniform distribution of shear stresses across the wall thickness h and constant shear flow f along the wall length such that

$$\tau h = f = \text{constant} \tag{7.63}$$

and

$$M_t = 2f A_o \tag{7.64}$$

where A_o is the area enclosed by the midline of the cross section. Equations 7.63 and 7.64 are derived from equilibrium conditions and, hence, are valid for both elastic and plastic states.

Two cases will be considered: uniform wall thickness and nonuniform wall thickness.

12.7.2.1 Uniform Wall Thickness

Since $\tau h = f =$ constant and the thickness is uniform, therefore τ is constant along the perimeter. If M_t is increased to initiate yielding, the entire section will be subjected to the yield shear stress k and become fully plastic. The bar will thus collapse as soon as the twisting moment M_t reaches a value given by

$$M_t = 2khA_o \tag{12.72}$$

unless strain-hardening develops.

12.7.2.2 Nonuniform Wall Thickness

Yielding starts at the location of minimum wall thickness and in this case the shear flow f is equal to kh_{min}. For a constant value of k the shear flow f, which is constant along the wall length, cannot increase beyond this value. This means that the cross section cannot support a further increase in M_t, and hence it will collapse as soon as yielding is initiated unless the material stain-hardens.

12.7.3 Combined Torsion and Tension

12.7.3.1 Solid Circular Section

For a solid circular section of radius r_o, subjected to combined torsion and tension, yielding is initiated at $r = r_o$ according to the von Mises criterion (Equation 10.21a), given by

$$\frac{\sigma^2}{3} + \tau^2 = k^2 \tag{10.21a}$$

Putting at the onset of yielding

$$\sigma = \sigma_Y = \frac{P_Y}{\pi r_o^2} \quad \text{and} \quad \tau = \tau_Y = \frac{2(M_t)_Y}{\pi r_o^2} \tag{12.72}$$

where $0 < \sigma_Y < Y$ and $0 < \tau_Y < k$, and substituting σ_Y and τ_Y from Equations 12.72 into Equation 10.21a gives the relation between P_Y and M_{Y2} to initiate yield as

$$\frac{1}{3}\left(\frac{P_Y}{\pi r_o^3}\right)^2 + \left(\frac{2(M_t)_Y}{\pi r_o^3}\right)^2 = k^2 \tag{12.73}$$

Equation 12.73 gives for $P_Y = 0$ or $M_Y = 0$ the condition of initial yielding for pure torsion or pure tension, respectively.

From Equation 10.21a the shear stress for initial yielding is given by

$$\tau_Y = \sqrt{k^2 - ⅓\sigma_Y^2} \tag{12.74a}$$

By putting $\sqrt{k^2 - ⅓\sigma_Y^2} = k_Y$, then (12.74b)

$$\tau_Y = k_Y,\ (M_t)_Y = k_Y(\pi/2)r_o^3, \text{ and } \alpha_Y = k_Y/(Gr_o). \tag{12.75a}$$

Equations 12.75a are identical to Equation 12.68a if k is replaced by k_Y. Hence, for an elastic–perfectly plastic material, the twisting moment applied to cause partial plastic deformation from r_o to r_p can be obtained from Equation 12.68b as

$$(M_t)_Y = k_p \frac{2}{3}\pi r_o^3\left(1 - \frac{1}{4}\left(\frac{r_p}{r_o}\right)^3\right) \text{ and } \alpha_p = k_p/(Gr_p) \tag{12.75b}$$

where $k_p = \sqrt{k^2 - ⅓\sigma_p^2}$

Here σ_p is the applied axial stress, when partial plastic yielding takes place from $r = r_o$ to $r = r_p$, expressed by the relation $\sigma_p = P_p/\pi r_o^2$, where P_p is the applied axial load.

For a fully plastic cross section, $r_p = 0$, and in this case, similar to Equations 12.68c, $(M_t)_u$ and α_u are given by

$$(M_t)_u = k_u ⅔\pi r_o^3 \text{ and } \alpha_u \to \infty, \text{ i.e., full collapse} \tag{12.75c}$$

where $k_u = \sqrt{k^2 - ⅓\sigma_u^2}$ and $\sigma_u = P_u/(\pi r_o^2)$.

12.7.3.2 Hollow Circular Sections

For a hollow circular section, of outer radius r_o and inner radius r_i subjected to combined torsion and tension, the corresponding values of $(M_t)_Y$, α_Y for initial yielding, $(M_t)_p$, α_p for partial yielding, and $(M_t)_u$, α_u for full yielding are obtained, as in the case of a solid circular section. This requires replacing k in Equations 12.69 for a hollow section in pure torsion by k_Y, k_p, and k_u, respectively, as shown in the next Example 12.12.

Example 12.12:

The hollow circular bar of Example 12.11 is subjected to a combined constant axial load of 100 kN and a twisting moment, M_t. *Determine the twisting moment and angle twist per unit length, taking a yield shear stress* k = *120 MPa, for the following conditions:*

a. *Initial yielding* $(M_t)_Y$, α_Y;
b. *Yielding of half wall thickness* $(M_t)_p$, α_p;
c. *Full yielding* $(M_t)_u$, α_u.

Solution:

From Example 12.11 the results for pure torsion are obtained for k = 120 MPa as

$$(M_t)_Y = 4.084 \text{ N}\cdot\text{m}, \quad \alpha_Y = 0.05 \text{ rad/m}$$

$$(M_t)_p = 4.598 \text{ N}\cdot\text{m}, \quad \alpha_p = 0.06 \text{ rad/m}$$

$$(M_t)_u = 4.775 \text{ N}\cdot\text{m}, \quad \alpha_u \to \infty$$

For the present case of combined torsion and tension

$$\sigma_Y = \sigma_p = \sigma_u = \frac{100 \times 10^3}{\pi\left(30^2 - 20^2\right)} = 63.662 \text{ MPa}$$

Hence, $k_Y = k_p = k_u = \sqrt{120^2 - \tfrac{1}{3}(63.662)^2} = 114.23$ MPa .

The twisting moment and the angles of twist per unit length for pure torsion are reduced by the ratio of 114.23/120, i.e., by 0.952 to give the corresponding values for combined torsion and tension as

$$(M_t)_Y = 3.888 \text{ N}\cdot\text{m}, \quad \alpha_Y = 0.0476 \text{ rad/m}$$

$$(M_t)_p = 4.377 \text{ N}\cdot\text{m}, \quad \alpha_p = 0.0571 \text{ rad/m}$$

$$(M_t)_u = 4.546 \text{ N}\cdot\text{m}, \quad \alpha_u \to \infty$$

12.7.3.3 Thin-Walled Cylinder of Uniform Thickness

1. For an elastic–perfectly plastic thin-walled cylinder of a mean radius r_m and a uniform wall thickness h subjected to combined torsion and tension the three states of initial, partial, and full yielding will occur simultaneously and the twisting moments and angles of twisting per unit length are obtained as

$$\begin{aligned} &(M_t)_Y = (M_t)_p = (M_t)_u = 2k_Y \pi r_m^3 h \\ &\alpha_Y = \alpha_p = \alpha_u \to \infty \end{aligned} \tag{12.76}$$

where

$$k_Y = \sqrt{k^2 - \frac{1}{3}\left(\frac{P}{2\pi r_m h}\right)^2}$$

and P is the axial load at the onset of yielding.

2. For the case of a thin-walled cylinder subjected to combined internal pressure p_i and torsion, yielding takes place according to the von Mises criterion (Equation 10.12b) as

$$\tfrac{1}{2}[(\sigma_r - \sigma_\theta)^2 + (\sigma_\theta - \sigma_z)^2 + (\sigma_z - \sigma_r)^2 + 3\tau^2] = 3k^2$$

Neglecting σ_r with respect to σ_θ and σ_z and putting $\sigma_\theta = 2\sigma_z$ gives

$$\alpha_z^2 + \tau^2 = k^2 \tag{12.77}$$

where

$$\sigma_z = p_i \frac{r_m}{2h}$$

and

$$k_Y = \sqrt{k^2 - \left(p_i \frac{r_m}{2h}\right)^2}$$

The three states of initial, partial, and full yielding will occur simultaneously. Hence,

$$(M_t)_Y = (M_t)_p = (M_t)_u = 2k_Y \pi r_m^2 h$$
$$\alpha_Y \to \infty \tag{12.78}$$

3. For the case of a thin-walled cylinder subjected to combined torsion and bending, yielding takes place according to the von Mises yielding criterion as given by Equation 10.21a. In this case σ_Y at the initiation of yielding is given by

$$\sigma_Y = \frac{(M_b)_Y}{\pi r_m^2 h} \tag{12.79}$$

where $(M_b)_Y$ is the applied bending moment at the onset of yielding. Hence, from Equations 12.72 and 12.74a,

$$\tau_Y = k_Y = \sqrt{k^2 - \tfrac{1}{3}\sigma_Y^2}$$
$$(M_t)_Y = 2k_Y \pi r_m^2 h \text{ and } \alpha_y = k_Y/(Gr_o) \tag{12.80}$$

Material yielding in this case is confined to two points lying at a diameter normal to the moment axis. To spread yielding, the twisting moment M_t should be kept constant at its initial yielding value of $(M_t)_Y$ as given by Equation 12.80. The bending moment can be now increased to spread yielding along the cylinder wall, and, since τ_Y and k are constant, then σ_Y will be also constant. The applied bending moment $(M_b)_p$ to spread yielding to two lengths subtended by two radial angles $2\phi_p$, Figure 12.20, is obtained by integration as

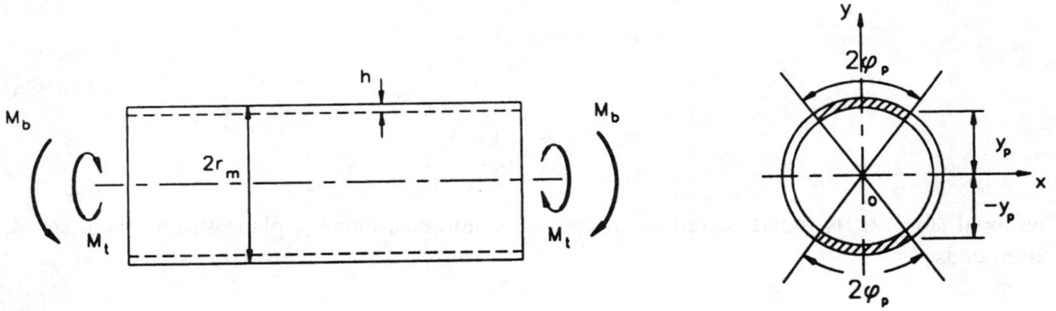

FIGURE 12.20 Elastic and plastic zones in a thin-walled cylinder subjected to combined torsion and bending.

$$\left(M_b\right)_p = 4r_m^2 h\sigma_Y\left[\int_0^{\phi_p}\cos\phi d\phi + \int_{\phi_p}^{\frac{\pi}{2}}\cos^2\phi d\phi\right]$$

$$= 4r_m^2 h\sigma_Y\left[\sin\phi_p + \frac{1}{2}\left(\frac{\pi}{2} - \phi_p\right) - \frac{1}{4}\sin 2\phi_p\right] \tag{12.81a}$$

If $\phi_p = \pi/4$, then half the wall length will be yielding at

$$\left(M_b\right)_p = 3.4r_m^2 h\sigma_Y = 1.082\left(M_b\right)_Y$$

$$\left(M_t\right)_p = \left(M_t\right)_Y \text{ and } \alpha_p = \alpha_Y \tag{12.81b}$$

For full yielding, i.e., $\phi_p = \pi/2$, the applied bending moment $(M_b)_u$ is obtained as

$$\left(M_b\right)_u = 4r_m^2 h\sigma_Y = 1.273\left(M_b\right)_Y$$

$$\left(M_t\right)_u = \left(M_t\right)_Y \text{ and } \alpha_u = \alpha_Y \tag{12.82}$$

12.7.3.4 Remarks

The above analysis of elastic–plastic bars subjected to combined torsion and tension has been limited to circular cross section and elastic–perfectly plastic material obeying the von Mises criterion. The analysis has been quite instructive to show how torsion and tension interact in the initiation and progress of plastic yielding. Treatment of the problem for a strain hardening material is rather involved and would require including incremental strain-stress relations. This problem has received historical interset at the early stages of formulating the mathematical theory of plasticity to establish the basis of an experimental verification of the laws of plasticity,[7,8] as presented in Chapter 10.

12.8 PRESSURIZED THICK-WALLED CYLINDERS

In Section 6.2.1 the stresses in a thick-walled cylinder of outer radius r_o, inner r_i, $\lambda = r_o/r_i$ when subjected to an internal pressure p_i has been obtained in the elastic state as

$$\sigma_r = \frac{p_i}{\lambda^2 - 1}\left(1 - \frac{r_o^2}{r^2}\right)$$

$$\sigma_\theta = \frac{p_i}{\lambda^2 - 1}\left(1 + \frac{r_o^2}{r^2}\right) \tag{12.83a}$$

The axial stress σ_z is found according to the end conditions, namely, plane strain, plane stress, closed ends, as

$$\text{plane stress: } \sigma_z = 0,$$

$$\text{plane stress: } \varepsilon_z = 0, \text{ giving } \sigma_z = \frac{2\nu}{\lambda^2 - 1} p_i \tag{12.83b}$$

$$\text{closed ends: } \sigma_z = \frac{p_i}{\lambda^2 - 1}$$

The values of the above stresses are such that $\sigma_\theta > \sigma_z > \sigma_r$, regardless of end conditions.

These stresses increase as the internal pressure is increased. At a certain value of p_i, the cylinder material will start to yield at the most heavily stressed zone. Upon subsequent increase in p_i, yielding will progress until the cylinder wall becomes fully plastic.

12.8.1 Initial and Partial Yielding

The stress distribution, Figure 6.3, indicates clearly that yielding starts to occur at the inner radius. The internal pressure p_Y at which yielding of the cylinder wall starts is determined by applying an appropriate yield criterion.

1. Tresca yield criterion

Since $\sigma_\theta > \sigma_z > \sigma_r$, yielding according to Equation 10.18, will start at the inner radius if

$$\frac{1}{2}\left[\sigma_\theta - \sigma_r\right]_{r=r_i} = k = \frac{Y}{2}$$

where k and Y are the yield stresses of the material in shear and tension, respectively. Substituting from Equations 12.83a and denoting the pressure at which yielding starts by p_{YT}, gives

$$\frac{1}{2}\left(\frac{2P_{YT}\lambda^2}{\lambda^2 - 1}\right) = \frac{Y}{2}$$

or

$$P_{YT} = \frac{1}{2}Y\left(1 - \frac{1}{\lambda^2}\right) \tag{12.84}$$

Note that Equation 12.84 is valid for all the above-mentioned end conditions.

2. von Mises yield criterion

As explained before, the von Mises yield criterion incorporates the effect of all three components of stresses σ_r, σ_θ, and σ_z. Hence, the internal pressure required to just cause yielding at the inner radius r_i depends on the end conditions, i.e., the value of σ_z as given by Equations 12.83b.

Taking first the plane stress condition $\sigma_z = 0$, yielding will start according to the von Mises criterion, Equation 10.17, if

$$\sigma_r^2 + \sigma_\theta^2 - \sigma_r\sigma_\theta = Y^2$$

Substituting from Equations 12.83a gives the value of the internal pressure at the onset of yielding, denoted as p_{YM}:

$$\text{plane stress}: \; P_{YM} = \frac{Y}{\sqrt{3}}\left(1 - \frac{1}{\lambda^2}\right)\left(\frac{1}{1 + 1/3\lambda^4}\right)^{1/2} \tag{12.85a}$$

The same procedure could be applied for the other two end conditions, namely,

$$\text{plane strain}: \; P_{YM} = \frac{Y}{\sqrt{3}}\left(1 - \frac{1}{\lambda^2}\right)\left(\frac{1}{1 + (1-2\nu)^2/3\lambda^4}\right)^{1/2} \tag{12.85b}$$

$$\text{closed ends}: \; P_{YM} = \frac{Y}{\sqrt{3}}\left(1 - \frac{1}{\lambda^2}\right) \tag{12.85c}$$

The internal pressure at the onset of yielding as predicted by Tresca and von Mises yield criteria are given in Table 12.2 for a cylinder of $\lambda = 2$ and $\nu = 0.3$. From Table 12.2, the maximum difference in the pressure predictions according to the von Mises criterion for the three end conditions is about 14%. A basic difference between the Tresca and von Mises predictions lies in the factor $^1/_2$ and $1/\sqrt{3}$, as seen from Equations 12.84 and 12.85. In view of this and for simplicity, the Tresca yield condition is considered throughout the forthcoming analysis.

TABLE 12.2
Onset Yielding Pressure for a Thick-Walled Cylinder, $\lambda = 2$, $\nu = 0.3$

	End Conditions		
Yield Criterion	**Plane Stress**	**Plane Strain**	**Closed Ends**
Tresca: p_{YT}/Y	0.375	0.375	0.375
von Mises: p_{YM}/Y	0.4375	0.4323	0.433

Once yielding has occurred at $r = r_i$, further increase in the pressure will cause the plastic zone to spread progressively into the wall. The material is assumed to be elastic–perfectly plastic. The radius r_p is taken as the radius of the interface between the plastic and the elastic zones. The radial stress σ_{rp} at this interface, i.e., where yielding is just about to occur, is obtained from Equations 12.83a and 12.84, by putting $p_i = p_{YT}$ and $\lambda = r_p/r_o$, as

$$\sigma_{rp} = -\frac{Y}{2}\left(1 - \frac{r_p^2}{r_o^2}\right) \tag{12.86}$$

Example 12.13:

An open-ended cylinder made of steel of Y = 600 MPa has an inside radius of 40 mm and is subjected to an internal pressure of 140 MPa.

a. *Determine the outside radius of the cylinder at which yielding occurs according to the Tresca criterion.*
b. *Does a cylinder with the same dimensions yield according to the von Mises criterion?*
c. *What would be the outer radius for a cylinder with closed ends under the same conditions?*

Solution:

a. For an open cylinder, a plane stress condition may be assumed. At initial yielding, the Tresca yield criterion, Equation 12.84, gives

$$P_{YT} = \frac{Y}{2}\left(1 - \frac{r_i^2}{r_o^2}\right)$$

$$140 = \frac{600}{2}\left(1 - \frac{40^2}{r_o^2}\right)$$

hence, $r_o = 54.77$ mm.

b. According to the von-Mises yield criterion, Equation 12.85a,

$$P_{YM} = \frac{Y}{\sqrt{3}}\left(1 - \frac{r_i^2}{r_o^2}\right)\left(\frac{1}{1 + r_i^4/3r_o^4}\right)^{1/2}$$

For $r_i = 40$ and $r_o = 54.77$, $p_{YM} = 154.48$ MPa, which is higher than p_{YT}; i.e., yielding will not occur.

c. For a cylinder with closed ends, σ_z is given by

$$\sigma_z = \frac{p_i}{(r_o^2/r_i^2) - 1}$$

Hence, at $r = r_i$, $\sigma_\theta > \sigma_z > \sigma_r$ and the Tresca yield criterion give the same results as for a cylinder with open ends. The von Mises yield criterion applied to a cylinder with closed ends gives, according to Equation 12.84c for $p_{YM} = 1544.48$ MPa, $r_i = 40$ mm, and $Y =$ 600 MPa:

$$P_{YM} = \frac{Y}{\sqrt{3}}\left(1 - \frac{r_i^2}{r_o^2}\right)$$

and $r_o = 53.74$ mm.

Example 12.14:

A long, thick-walled cylinder with diameter ratio $\lambda = 2.5$ is embedded in a rigid wall. Determine the maximum internal pressure at which the cylinder material starts to yield assuming plane strain conditions. Apply both the Tresca and the von Mises yield criteria.

Solution:

This problem has been solved in Example 6.2. The stresses are found to be

$$\sigma_r = -\frac{4p_i}{37}\left[3+\left(\frac{r_o}{r}\right)^2\right]$$

$$\sigma_\theta = -\frac{4}{37}\left[3-\left(\frac{r_o}{r}\right)^2\right]$$

$$\sigma_z = -\frac{8p_i}{37}$$

These stresses are plotted in Figure 12.21.

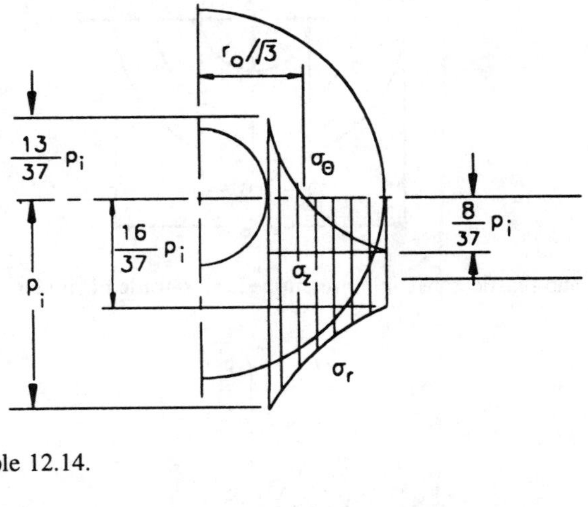

FIGURE 12.21 Example 12.14.

a. According to the Tresca yield criterion, yielding may occur at $r = r_i$; where $\sigma_\theta > \sigma_z > \sigma_r$; hence,

$$\left(\sigma_\theta - \sigma_r\right)_{r=r_i} = \frac{13}{37}p_i - \left(-p_i\right) = Y$$

giving an onset yielding pressure $p_{YT} = 0.74Y$.

b. According to the von Mises criterion, at $r = r_i$,

$$Y = \frac{1}{\sqrt{2}}\left[\left(-p_i - \frac{13}{37}p_i\right)^2 + \left(\frac{13}{37}p_i + \frac{8}{37}p_i\right)^2 + \left(-\frac{8}{37}p_i + p_i\right)^2\right]^{1/2}$$

giving $p_{YM} = 0.851Y$.

The hoop stress σ_θ changes from tensile to compressive at $r = r_o/\sqrt{3}$, at which $\sigma_r = -24/37p$. The yielding condition may be also checked to give

$$Y = \frac{1}{\sqrt{2}}\left[\left(-\frac{24}{37}p_i - 0\right)^2 + \left(0 + \frac{8}{37}p_i\right)^2 + \left(-\frac{8}{37}p_i + \frac{24}{37}p_i\right)^2\right]^{1/2}$$

giving $p_{YM} = 1.283Y$.

Hence, yielding will commence at $r = r_i$ at either $p_{YM} = 0.851Y$ or $p_{YT} = 0.74Y$.

12.8.1.1 Stresses in the Elastic Region $r_p \leq r \leq r_o$

The radial stress given by Equation 12.83 acts as an internal pressure on the elastic annulus of Figure 12.22. The stresses are hence obtained from Equations 12.83 by replacing p_i by $-\sigma_{rp}$ and λ by r_o/r_p; hence,

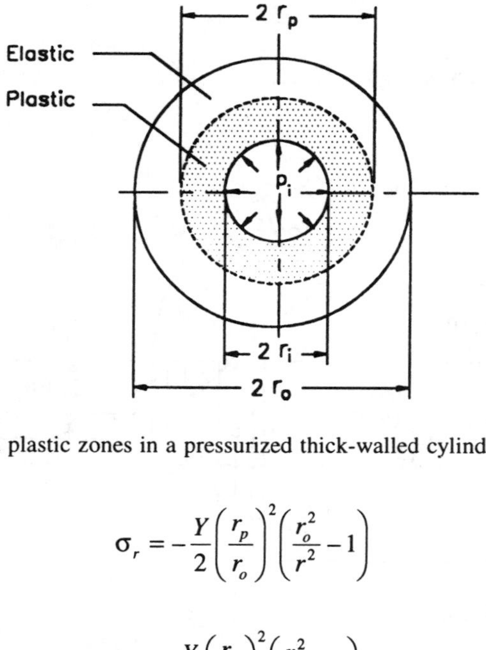

FIGURE 12.22 Elastic and plastic zones in a pressurized thick-walled cylinder.

$$\sigma_r = -\frac{Y}{2}\left(\frac{r_p}{r_o}\right)^2\left(\frac{r_o^2}{r^2} - 1\right) \tag{12.87a}$$

$$\sigma_\theta = \frac{Y}{2}\left(\frac{r_p}{r_o}\right)^2\left(\frac{r_o^2}{r^2} + 1\right) \tag{12.87b}$$

12.8.1.2 Stresses in the Plastic Region $r_i \leq r \leq r_p$

In the plastic region, both the equilibrium equation (Equation 1.10c) and the Tresca yield condition (Equation 10.18) have to be satisfied, i.e.,

$$\frac{d\sigma_r}{dr} + \frac{\sigma_r - \sigma_\theta}{r} = 0 \tag{1.10c}$$

and $\sigma_\theta - \sigma_r = Y$. Substitution gives

$$\frac{d\sigma_r}{dr} = \frac{Y}{r}$$

so that

$$\sigma_r = Y\ln(r) + C_1 \tag{12.88}$$

C_1 is a constant of integration found from the condition that σ_r is continuous at the plastic–elastic interface; that is,

$$r = r_p : \ \sigma_r = -\sigma_{rp} = \frac{Y}{2}\left(1 - \frac{r_p^2}{r_o^2}\right)$$

Application of this condition yields

$$\sigma_r = \frac{Y}{2}\left[\left(\frac{r_p^2}{r_o^2} - 1\right) - \ln\left(\frac{r_p^2}{r^2}\right)\right] \tag{12.89a}$$

and $\sigma_\theta = Y + \sigma_r$, i.e.,

$$\sigma_\theta = \frac{Y}{2}\left[\left(\frac{r_p^2}{r_o^2} + 1\right) - \ln\frac{r_p^2}{r^2}\right] \tag{12.89b}$$

The first of Equations 12.89 gives the radial stress distribution in the plastic region extending from the bore $r = r_i$ to the interface radius $r = r_p$. The value of σ_r at $r = r_i$ represents the internal pressure p_p required to cause partial yielding within the wall up to a radius $r = r_p$. Hence, putting $\sigma_r = -p$ at $r = r_i$ in Equation 12.89a gives

$$p_p = \frac{Y}{2}\left[\ln\frac{r_p^2}{r_i^2} + \left(1 - \frac{r_p^2}{r_o^2}\right)\right] \tag{12.90}$$

which is the relationship between the pressure p_p and the interface radius r_p to which yielding progresses. For any given pressure causing partial yielding, Equation 12.90 is solved to obtain the interface radius r_p. Equation 12.87 and 12.89 are then used to calculate the stresses within the elastic and plastic regions, respectively. A graphical representation of these stresses are shown in Figure 12.23 for a cylinder of $r_o/r_i = \lambda = 2$ at different degrees of partial yielding expressed by r_p/r_o.

Example 12.15:

For a thick-walled cylinder of inner radius r_i *and outer radius* $2r_i$, *assuming the wall material to be elastic–perfectly plastic with yield strength of 250 MPa, calculate:*

a. *The internal pressure at which the elastic–plastic interface radius* $r_p = 1.5r_i$.
b. *The tangential stress distribution due to this internal pressure and plot it.*

Solution:

a. Equation 12.90 gives the internal pressure for $r_p = 1.5\, r_i$ as

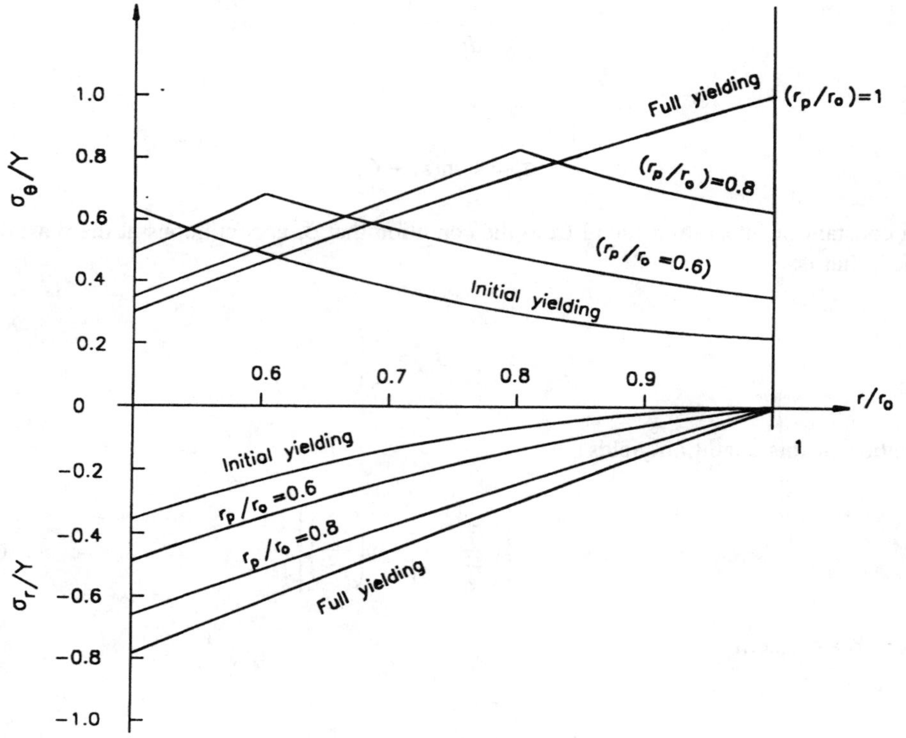

FIGURE 12.23 Distribution of normalized stresses through the wall of a pressurized thick-walled cylinder of $r_o/r_i = 2$ for different values of r_p/r_o.

$$p_p = \frac{Y}{2}\left[\ln\left(\frac{r_p^2}{r_i^2}\right) + \left(1 - \frac{r_p^2}{r_o^2}\right)\right].$$

$$p_p = \frac{250}{2}\left[\ln(1.5)^2 + \left(1 - \frac{1.5}{2}\right)^2\right]$$

$$p_p = 156.1\ \text{MPa}$$

b. The hoop stress in the elastic region $r_p \le r \le r_o$, Equation 12.87:

$$\sigma_\theta = \frac{Y}{2}\left(\frac{r_p}{r_o}\right)^2\left(\frac{r_o^2}{r^2} + 1\right) = \frac{9Y}{32}\left(\frac{r_o^2}{r^2} + 1\right)$$

The hoop stress in the elastic and plastic regions are given by Equations 12.87 and 12.89, respectively, as

$$\sigma_\theta = \frac{Y}{2}\left(\frac{r_p}{r_o}\right)^2\left(\frac{r_o^2}{r^2}+1\right) \qquad r_p \le r \le r_o$$

$$\sigma_\theta = \frac{Y}{2}\left[\left(\frac{r_p^2}{r_o^2}+1\right) - \ln\frac{r_p^2}{r^2}\right] \quad r_i \le r \le r_p$$

are calculated as given in Table 12.3. These stresses are plotted in Figure 12.24.

TABLE 12.3
Tangential Stresses; Example 12.15

Radius, *r*	r_i	r_p	r_o
σ_θ MPa (plastic)	93.9	195.3	—
σ_θ MPa (elastic)	—	195.3	140.6

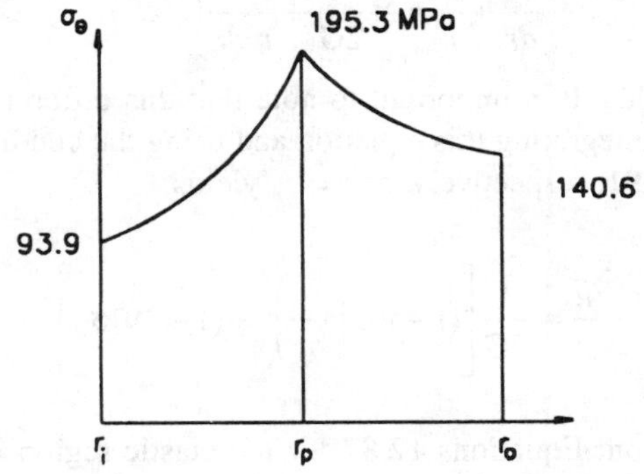

FIGURE 12.24 Example 12.15.

12.8.1.3 Radial Displacement in Partially Yielded Cylinders

To determine the radial displacement in a partially yielded thick-walled cylinder, the end conditions must be specified. For simplicity, plane strain conditions only are considered here.

Within the elastic region, the elastic strains are obtained from Hooke's law (Equation 3.6) using the stresses given by Equations 12.83, knowing that $\sigma_z = \nu(\sigma_r + \sigma_\theta)$ for plane strain conditions. The radial displacement in the elastic region is thus, for $r_p \le r \le r_o$,

$$\frac{u}{r} = \frac{(1+\nu)}{2}\frac{Y}{E}\left(\frac{r_p}{r_o}\right)^2\left[\left(\frac{r_o}{r}\right)^2 + (1-2\nu)\right] \tag{12.91}$$

The total strains are the sum of the elastic and plastic strain components ε^e and ε^p, respectively; hence,

$$\varepsilon_r + \varepsilon_\theta = \left(\varepsilon_r^e + \varepsilon_r^p\right) + \left(\varepsilon_\theta^e + \varepsilon_\theta^p\right)$$

In view of the condition of constancy of volume applied to plastic components of strain, $\varepsilon_r^p = -\varepsilon_\theta^p$ and $\varepsilon_z^p = 0$; hence

$$\varepsilon_r + \varepsilon_\theta = \varepsilon_r^e + \varepsilon_\theta^e$$

The elastic strains are obtained from Hooke's law as above. For cylinders with small and moderate wall ratios, say, $\lambda < 5$, both elastic and plastic strains are small under conditions of partial yielding. Recalling that $\varepsilon_r = du/dr$ and $\varepsilon_\theta = u/r$ gives for the elastic strains:

$$\frac{du}{dr} + \frac{u}{r} = \frac{(1-2\nu)(1+\nu)}{E}\left(\sigma_r + \sigma_\theta\right) \tag{12.92}$$

From the equilibrium equation (Equation 1.10c), namely,

$$\frac{d\sigma_r}{dr} + \frac{\sigma_r - \sigma_\theta}{r} = 0$$

Equation 12.92 reduces to

$$\frac{du}{dr} + \frac{u}{r} = \frac{(1-2\nu)}{2G}\frac{1}{r}\frac{d}{dr}\left(r^2\sigma_r\right)$$

where G is the shear modulus. It is important to note that this differential equation holds in both elastic and plastic regions. Integrating this equation and using the condition that σ_r and u are given by Equations 12.87 and 12.91, respectively, at $r = r_p$ yields

$$\frac{u}{r} = \frac{1}{2G}\left[(1-\nu)Y\left(\frac{r_p}{r}\right)^2 + (1-2\nu)\sigma_r\right] \tag{12.93}$$

Substitution for σ_r either from Equations 12.87 for the elastic region or Equations 12.89 for the plastic region furnishes the radial displacement at any radius.

12.8.2 Full Yielding — Plastic Expansion Process

12.8.2.1 Full Yielding

The pressure p_u required to produce a fully yielded wall is found from Equation 12.88, namely,

$$\sigma_r = Y\ln(r) + C_1$$

Full yielding occurs when

$$\sigma_r = -p_u \quad \text{at } r = r_i$$

$$\sigma_r = 0 \quad \text{at } r = r_o$$

Hence, from Equation 12.88,

$$p_u = Y\ln(r_o/r_i) = Y\ln\lambda \tag{12.94}$$

The same result is obtained from Equation 12.90 by putting $p_p = p_u$ and $r_p = r_o$.

The ratio of the internal pressure to cause full yielding to that which initiates yielding is obtained from Equations 12.84 and 12.94, as

$$\frac{p_u}{p_{YT}} = \frac{2\lambda^2 \ln \lambda}{\lambda^2 - 1} \tag{12.95}$$

This ratio increases as λ becomes larger. For instance, for $\lambda = 2$, $p_u = 1.85p_{YT}$. On the other hand, for a thin-walled tube, say, $\lambda = 1.1$, $p_u = 1.098p_{YT}$, indicating that full yielding occurs shortly after initial yielding. The stresses in the fully yielded cylinder are obtained from Equations 12.89 by putting $r_p = r_o$. These stresses are shown in Figure 12.23 by the curves $r_p/r_o = 1$. Note that under increasing internal pressure, the cylinder expands plastically and the initial inner and outer radii r_i and r_o, respectively, become larger. The current values of these radii must by strictly used in Equations 12.89 and 12.95. However, for elastic–plastic analysis, i.e., contained or restricted plastic flow, the change in dimensions remain infinitesimal and the initial radii could be used without introducing much error.

12.8.2.2 Plastic Expansion Process

When Equation 12.94 is used to calculate the pressure required to expand a thick-walled tube (as a metal-forming process), large strains are encountered. The plastically deformed radii should then be used; hence, Equation 12.94 becomes

$$p_u = Y \ln\left(\frac{r_o'}{r_i'}\right) \tag{12.96}$$

Let r_i' and r_o' be the inner and outer deformed radii at full yielding. Neglecting elastic deformation, the constant volume condition gives

$$r_o^2 - r_i^2 = r_o'^2 - r_i'^2 \tag{12.97}$$

Substitution into Equation 12.96 results in

$$p_u = \frac{Y}{2} \ln\left(1 + \frac{r_o^2 - r_i^2}{r_i'^2}\right) \tag{12.98}$$

Equation 12.98 gives the internal pressure required to maintain the cylinder in a fully plastic state as flow progresses. This pressure is always smaller than that given by Equation 12.94 at the onset of full yielding. This is explained by the fact that the faster expansion of the inner surface of the tube compared with the expansion of the outer surface results in a smaller current r_o'/r_i' ratio. Consequently, a falling pressure–expansion curve is obtained. In reality, this may not be the case since most metallic alloys show strain hardening, thus requiring additional load to continue plastic deformation.[9]

Comparison among the pressures required to cause initial, partial, full yielding, and plastic expansion in a cylinder of $\lambda = 2$ and $\nu = 0.3$ is presented in Table 12.4. The stresses during plastic expansion of a thick-walled cylinder (of inner and outer radii r_i' and r_o', respectively) must satisfy the equilibrium and the yield criterion. Hence, Equations 12.88 apply by putting $r_p = r_o'$ and knowing that

TABLE 12.4
Pressure Required to Deform a Cylinder, $\lambda = 2$ and $\nu = 0.3$

Condition	Pressure/Yield Strength
Initial yielding, Eq. 12.84	$p_{YT}/Y = 0.375$
Partial yielding, Eq. 12.90 ($r_p = 1.5r_i$)	$p_p/Y = 0.624$
Full yielding, Eq. 12.94	$p_u/Y = 0.693$
Plastic expansion, Eq. 12.98 ($r_i' = 1.5r_i$)	$p_u/Y = 0.424$
Plastic expansion, Eq. 12.98 ($r_i' = 2r_i$)	$p_u/Y = 0.280$

$$\text{at } r = r_i': \ \sigma_r = -p_u$$

$$\text{at } r = r_o': \ \sigma_r = 0$$

where p_u is given by Equation 12.98. This gives the stress distribution:

$$\sigma_r = -Y \ln \frac{r_o'}{r'}$$
$$\sigma_\theta = Y\left(1 - \ln \frac{r_o'}{r'}\right) \tag{12.99}$$

where r' is the current radius in the plastically expanded tube, which is simply obtained by applying the constant volume condition as

$$r' = \left[r_i'^2 + \left(r^2 - r_i^2\right)\right]^{1/2} \tag{12.100}$$

Example 12.16:
Three cylinders are subjected to internal pressure. They have wall ratios $\lambda = 2, 3$, and 5, respectively. Compare the radial displacement u at the inner radius for these cylinders at the following conditions:

a. Initial yielding
b. Partial yielding ($r_p = 0.75r_o$)
c. Full yielding

Take $\nu = 0.3$, $G = 78$ GPa, and $Y = 400$ MPa.

Solution:
The radial displacements u for all conditions are given by Equation 12.93 as

$$\frac{u}{r} = (1-\nu)\left(\frac{Y}{2G}\right)\left(\frac{r_p}{r}\right)^2 + (1-2\nu)\left(\frac{\sigma_r}{2G}\right)$$

a. Initial yielding: Equation 12.87 for $r = r_p = r_i$ gives

$$\sigma_r = \frac{Y}{2}\left(\frac{1}{\lambda^2} - 1\right)$$

$$\sigma_r = \frac{400}{2}\left(\frac{1}{\lambda^2} - 1\right)$$

For $Y = 400$ MPa, $\nu = 0.3$, and $G = 78$ GPa, Equation 12.93 gives

$$\frac{u}{r_i} = 1.795 \times 10^{-3} + 5.128 \times 10^{-4}\left(\frac{1}{\lambda^2} - 1\right)$$

b. Partial yielding: Equation 12.89a for $r_p = 0.75 r_o$ gives

$$\sigma_r = \frac{Y}{2}\left[\left(\frac{r_p}{r_o}\right)^2 - 1 - \ln\left(\frac{r_p}{r_i}\right)^2\right] \quad \text{at } r = r_i$$

$$\sigma_r = \frac{400}{2}\left[(0.75)^2 - \ln(0.75\lambda)^2\right] = \frac{400}{2}\left(0.1379 - \ln \lambda^2\right)$$

Hence, Equation 12.93 gives

$$\frac{u}{r_i} = 1.0096 \times 10^{-3}\lambda^2 + 5.128 \times 10^{-4}\left(0.1379 - \ln \lambda^2\right)$$

c. Full yielding: Equation 12.89a for $r_p = r_o$ gives

$$\sigma_r = -Y \ln \frac{r_o}{r_i} \quad \text{at } r = r_i$$

$$\sigma_r = -\frac{400}{2} \ln \lambda$$

Hence, Equation 12.93 gives

$$\frac{u}{r_i} = 1.795 \times 10^{-3}\lambda^2 + 5.128 \times 10^{-4} \ln \lambda$$

The results for different values of λ are calculated from Equation 12.93 and listed as shown in Table 12.5.

Note that the radial displacement at the bore, $r = r_i$, increases appreciably (more than four times for $\lambda = 2$) as the cylinder deforms plastically from the condition of initial yielding to that of full yielding. The magnitude of this displacement is always larger for thicker cylinders, i.e., large radii ratios. However, even at the full yielding condition, the radial displacement at the bore is still negligibly small compared with the cylinder dimensions.

TABLE 12.5
Radial Displacements, Example 12.16

Condition	u/r_i		
	$\lambda = 5$	$\lambda = 3$	$\lambda = 2$
Initial yielding	1.302×10^{-3}	1.340×10^{-3}	1.41×10^{-3}
Partial yielding	23.66×10^{-3}	8.03×10^{-3}	3.40×10^{-3}
Full yielding	44.52×10^{-3}	15.80×10^{-3}	6.82×10^{-3}

Example 12.17:

A steel tube is expanded by applying internal pressure using a tool composed of tapered plug segments. The operation is performed in hot-working condition at a constant yield stress of 45 MPa. If the initial outside and inside diameter are 400 and 200 mm, respectively, and the final inside diameter is 300 mm, assuming plane strain conditions:

a. Calculate the work done in expanding the tube,
b. Show that the radial displacement at the bore is greater than that at the outer surface. How does this affect the pressure required for expansion?

Solution:

a. In thick-walled cylinders, the stresses and strains are not uniformly distributed across the thickness. For an annulus dr, the increment of plastic work done during a bore expansion from r_i' to $(r_i' + dr_i')$, where r_i' is the current expanded bore, is given by

$$dW^p = \sigma_\theta d\varepsilon_\theta + \sigma_r d\varepsilon_r$$

The above expression is valid for both plane stress or plane strain end conditions. Applying the Tresca yield criterion, i.e.,

$$\sigma_\theta - \sigma_r = Y$$

and the constant-volume condition for the plane strain condition,

$$d\varepsilon_\theta = -d\varepsilon_r \quad \text{or} \quad \frac{\delta u}{r} + \frac{d(\delta u)}{dr} = 0$$

where δu is an increment of radial displacement, then

$$dW^p = Y d\varepsilon_\theta = Y\frac{\delta u}{r}$$

Again, from the constant-volume condition,

$$\frac{d(\delta u)}{\delta u} = -\frac{dr}{r}$$

hence, $\delta u = C_1/r$. At $r = r_i'$, $\delta u_i = C_1/r_i'$ and $C_1 = \delta u_i r_i'$; hence,

$$\delta u = \delta u_i \frac{r_i'}{r} = \frac{r_i'}{r} dr_i' \text{ and } dW^p = Y\delta u_i r \frac{1}{r^2}$$

Integrating the plastic work increment over the whole tube cross section for a unit length gives

$$W^p = \int dW^p = \int_{r_i}^{r_o} Y\delta u_i r_i' \frac{1}{r^2} 2\pi r dr$$

or

$$W^p = 2\pi Y r_i' \delta u_i \ln \frac{r_o}{r_i'}$$

The total plastic work W_T^p is done over expanding the bore from 0.1 to 0.15 m; thus,

$$W_T^p = \int_{r_i'=0.1}^{r_i'=0.15} W^p dr_i' = 2\pi Y \int_{0.1}^{0.15} r_i \ln\left(\frac{r_o}{r_i'}\right) dr_i'$$

where $dr_i' = \delta u_i$.

Recalling that at any instant $r_o^2 - r_i^2 = 0.2^2 - 0.1^2 = C_2^2$, then

$$W_T^p = 2\pi Y \int_{0.1}^{0.15} r_i' \ln\left(1 + \frac{C_2^2}{r_i'^2}\right) dr_i'$$

Integrating by parts gives per unit length of the cylinder for $Y = 45$ MPa:

$$W_T^p = 1.889 \times 10^6 \text{ N} \cdot \text{m/m}$$

Note that this work is equivalent to

$$W_T^p = \int_{r_i'=0.1}^{r_i'=0.15} p_u dV$$

where $dV = 2\pi r_i' dr_i'$ per unit length of the cylinder. The pressure p_u is given by Equation 12.98.

b. Applying Equation 12.100 at $r = r_o$ gives

$$r_o' = \left[r_i'^2 + \left(r_o^2 - r_i^2\right)\right]^{1/2} = \left[150^2 + \left(200^2 - 100^2\right)\right]^{1/2}$$

$$r_o' = 229.13 \text{ mm}$$

Hence, $u|_{r=ro} = 29.13$ mm, which is less than $u|_{r=ri} = 50$ mm. Consequently, the wall ratio initially equal to 2 decreases to (229.13/150) = 1.526 at the end of the expansion process. The expansion pressure at this instant is obtained from Equation 12.96 as 0.4226*Y*, i.e., less than 0.6932Y, which was attained at the beginning of the process.

12.8.3 Residual Stresses — The Autofrettage Process

For an elastically deforming thick-walled cylinder, i.e., loaded such that $p_i < p_{YT}$, the removal of the internal pressure brings back the cylinder to its initial dimensions with no stresses left in the cylinder walls. This is not the case when unloading occurs in a partially yielded cylinder under a pressure $p_i < p_p$. Upon removal of the applied pressure, only the elastic strains will disappear while plastic strains remain. As a result, residual stresses are left in the cylinder walls. These stresses are obtained by considering the unloading process to occur elastically. Thus, Equations 12.83 are employed to find the stresses σ_r and σ_θ due to a pressure p_p as determined from Equation 12.90:

$$\sigma_r = \frac{Y}{2(\lambda^2-1)}\left[\ln\frac{r_p^2}{r_i^2}+\left(1-\frac{r_p^2}{r_o^2}\right)\right]\left(1-\frac{r_o^2}{r^2}\right)$$

$$\sigma_\theta = \frac{Y}{2(\lambda^2-1)}\left[\ln\frac{r_p^2}{r_i^2}+\left(1-\frac{r_p^2}{r_o^2}\right)\right]\left(1+\frac{r_o^2}{r^2}\right) \qquad (12.101)$$

The residual stresses are thus obtained by subtracting the above stresses from those existing before unloading, as given by Equations 12.87 and 12.89 for the elastic and plastic regions, respectively. These are given by

In the plastic region $r_i \le r \le r_p$

$$\sigma_r = -\frac{Y}{2}\left[\frac{p_p}{p_{YT}}\left(1-\frac{r_i^2}{r^2}\right)-\ln\left(\frac{r^2}{r_i^2}\right)\right]$$

$$\sigma_\theta = -\frac{Y}{2}\left[\frac{p_p}{p_{YT}}\left(1+\frac{r_i^2}{r^2}\right)-\left[2+\ln\left(\frac{r^2}{r_i^2}\right)\right]\right] \qquad (12.102)$$

In the elastic region $r_p \le r \le r_o$

$$\sigma_r = -\frac{Y}{2}\left(\frac{r_p^2}{r_i^2}-\frac{p_p}{p_{YT}}\right)\left(\frac{r_i^2}{r^2}-\frac{r_i^2}{r_o^2}\right)$$

$$\sigma_\theta = -\frac{Y}{2}\left(\frac{r_p^2}{r_i^2}-\frac{p_p}{p_{YT}}\right)\left(\frac{r_i^2}{r^2}+\frac{r_i^2}{r_o^2}\right) \qquad (12.103)$$

where p_{YT} and p_p are given by Equations 12.84 and 12.90, respectively.

To demonstrate this calculation procedure, consider a cylinder of $\lambda = 2$, which has partially yielded to a radius $r_p/r_o = 0.6$. Stresses in the elastic–plastic regions as well as residual stresses at the radii $r = r_i$, $r = r_p$, and $r = r_o$ are given in Table 12.5 and these residual stresses are shown in Figure 12.25a and b for the data of Table 12.6. It is seen that σ_r is everywhere compressive and σ_θ is compressive in the inner part and tensile in the outer region of the tube.

Table 12.6 indicates that the residual stress σ_r is very small and may be neglected. The hoop stress σ_θ at the inner radius is negative, i.e., compressive. This has a beneficial effect if the cylinder is reloaded again. Upon reloading by internal pressure inducing only elastic deformations, the stresses given by Equations 6.2 act together with the residual stresses. Notably, the tensile stress σ_θ at the inner radius is reduced by the compressive residual value, as shown in Figure 12.25c. It is thus clear that the resulting residual stresses provide favorable conditions when the cylinder is

TABLE 12.6
Residual Stresses in a Pressurized Cylinder, $\lambda = 2$ and $r_p/r_o = 0.6$

Condition	Equations Employed	Stress Ratio	Radius r_i	r_p	r_o	Curves in Figure 12.25
Initial partial	12.87	σ_r/Y	–0.5	–0.32	0	1
yielding	12.89	σ_θ/Y	–0.5	0.68	0.36	2
Elastic	12.101	σ_r/Y	–0.5	–0.3	0	3
unloading		σ_θ/Y	0.84	0.63	0.33	4
Residual stresses	12.102	σ_r/Y	0	–0.02	0	5
	12.103	σ_θ/Y	–0.34	0.05	0.03	6

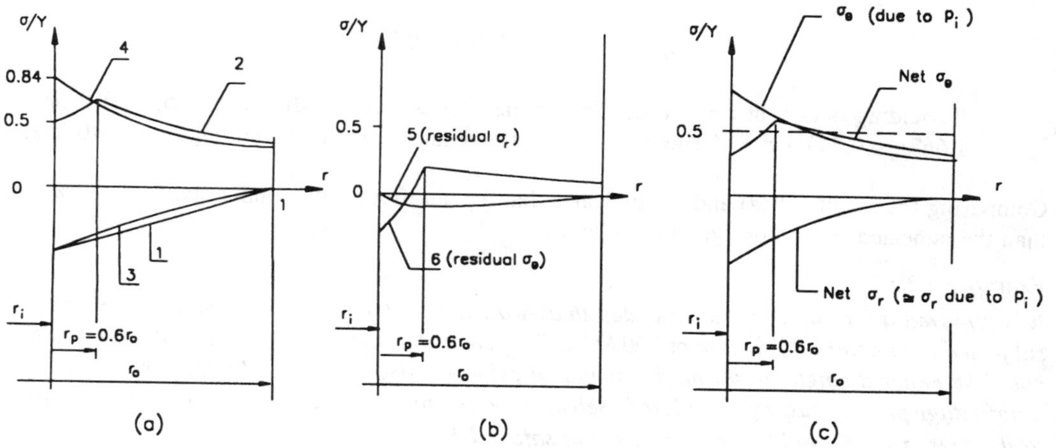

FIGURE 12.25 Distribution of normalized stresses through the wall of a thick-walled cylinder subjected to autofrettage according to Table 12.4: (a) initial partial yielding and elastic unloading, (b) residual stresses, and (c) reloading stresses.

loaded again. Such a process based on inducing residual stresses by plastic deformation is called *autofrettage.*

The autofrettage process is often used in strengthening high-pressure cylinders, such as gun barrels and extrusion containers. A non-hardening material cannot be strengthened theoretically by more than a factor of 2. However, one should not consider the values of the residual stresses as totally reliable. In particular, all metals behave inelastically, and the Bauschinger effect is observed when the cylinder is unloaded. This results in a decrease in the beneficial effects of the residual stresses calculated on the assumption of elastic unloading.

Example 12.18:

For a cylinder of $\lambda = 2$, determine the pressure required to initiate yielding if

a. *The cylinder is to be pressurized after being fully annealed (no residual stresses);*
b. *The cylinder is prestressed by an autofrettage process inducing plastic region of* r_p/r_o = *0.6 (use the results of Table 12.6).*

Solution:

a. Annealed cylinder (without residual stresses): Using Equation 12.84,

$$p_{yT}/Y = 0.375$$

b. Autofrettaged cylinder (with residual stresses): Using the results of Table 12.6 gives for residual stresses at $r = r_i$: $\sigma_r = 0$ and $\sigma_\theta = -0.34Y$. Upon elastic reloading by an internal pressure p_i, the stresses at $r = r_i$ are given by Equation 12.83 as

$$\sigma_r = -p_i$$

$$\sigma_\theta = 1.667 p_i$$

These stresses are added to the residual stresses, giving

$$\sigma_r = -p_i,\ \sigma_\theta = 1.667p_i - 0.34Y$$

If yielding occurs at $r = r_i$ according to the Tresca yield condition $\sigma_\theta - \sigma_r = Y$; hence, $1.667p_i - 0.34Y + p_i = Y$. The internal pressure required to initiate yielding is $p_i/Y = 0.502$.

Comparing the results of (a) and (b) indicates that the autofrettaged cylinder stands 33.87% more than the annealed one before yielding initiates.

Example 12.19:

It is required to design an open-ended thick-walled cylinder of $r_i = 100$ *mm and* $r_o = 200$ *mm, subjected to an internal pressure of 200 MPa. The selected design which should not fail by yielding could be either a single or shrink-fit compound cylinder. Also a single cylinder prestressed by an autofrettage process may by considered. Select a steel of minimum strength requirement according to the Tresca yield condition and a factor of safety, F.S., of 1.75.*

For the compound tube take $r_c = 141.4$ *mm and* $\Delta = 0.1414$ *mm. The autofrettage process is conducted such that the elastic–plastic interface radius* $r_p = 0.6r_o$.

Tabulate your calculations for σ_r, σ_θ *at the inner, common, elastic–plastic interface, and outer radii, respectively.*

Solution:

a. Single cylinder: Equation 12.83 gives the stresses due to internal pressure (see Table 12.7A. Failure occurs when $\sigma_\theta - \sigma_r$ attains a maximum value equal to Y/F.S. This may occur at $r = 100$ mm:

TABLE 12.7A
Stresses in a Single Cylinder, Example 12.19

Stress (MPa)	$r = 100$ mm	$r = 120$ mm	$r = 141.4$ mm	$r = 200$ mm
σ_r	−200.0	−118.5	−66.7	0.0
σ_θ	333.3	251.9	200.0	133.3
$\sigma_\theta - \sigma_r$	533.3	370.4	266.7	133.3

$$\left(\sigma_\theta - \sigma_r\right)_{max} = 333.3 - (-200) = \frac{Y}{1.75}$$

Thus a high strength steel of yield stress 933.3 MPa is required.

b. Compound cylinder (shrink-fit): The stresses due to shrinkage are calculated from Equations 6.16 and 6.17 as shown in Table 12.7b.* The net stresses due to both shrinkage

TABLE 12.7B
Stresses in a Compound Cylinder, Example 12.19

Shrinkage Stresses (MPa)	Inner Cylinder		Outer Cylinder	
	$r = 100$ mm	$r = 14.1$ mm	$r = 141.4$ mm	$r = 200$ mm
σ_r	0	–33.3	–33.3	0.0
σ_θ	–133.3	–100.0	100.0	66.7

and applied internal pressure are calculated by adding the stresses calculated for a single cylinder in (a) to the above stresses, thus resulting in the stresses shown in Table 12.7c.

TABLE 12.7C
Net Stresses in a Compound Cylinder, Example 12.19

Net Stresses (Shrinkage + Pressure) MPa	Inner Tube		Outer Tube	
	$r = 100$ mm	$r = 141.1$ mm	$r = 141.4$ mm	$r = 200$ mm
σ_r	–200	–100	–100	0.0
σ_θ	200	100	300	200
$\sigma_r - \sigma_\theta$	400	200	400	200

Failure may occur at the inner surface of the inside cylinder and the outside cylinder. To prevent this,

$$(\sigma_\theta - \sigma_r)_{max} = 200 - (-200) = \frac{Y}{1.75}$$

Thus a steel of yield strength 700 MPa is required.

c. Autofrettaged cylinder: The residual stresses due to plastic deformation up to $r_p/r_o = 0.6$, prior to loading are obtained from Table 12.6, as shown in Table 12.7d. The net stresses

TABLE 12.7D
Stresses in Autofrettaged Cylinder, Example 12.19

Prestresses (Autofrettage) (MPa)	$r = 100$ mm	$r = r_p = 120$ mm	$r = 200$ mm
σ_r	0	$-0.022Y$	0
σ_θ	$-0.339Y$	$0.047Y$	$0.025Y$

* These results were previously determined in Example 6.7.

in the autofrettaged cylinder are calculated by adding the stresses calculated for a single cylinder as in (a) to the above stresses, thus giving the stresses in Table 12.7e. Failure

TABLE 12.7E
Net Stresses in Autofrettaged Cylinder, Example 12.19

Net Stresses (Autofrettage + Pressure) MPa	r = 100 mm	r = 120 mm	r = 200 mm
σ_r	–200	–118.5 – 0.022Y	0.0
σ_θ	333.3 – 0.339Y	251.9 + 0.047Y	133.3 + 0.025Y
$\sigma_\theta - \sigma_r$	533.3 – 0.339Y	370.4 + 0.069Y	133.3 + 0.025Y

is prevented if $\sigma_\theta - \sigma_r \leq$ F.S. Hence,

at r = 100 mm: 533.3 – 0.339Y = Y/1.75, giving Y = 585.8 MPa, or
at r = 120 mm: 370.4 + 0.069Y = Y/1.75, giving Y = 737.2 MPa.

Thus, the autofrettaged cylinder must have a minimum strength of 738.7 MPa.

Comparing the strength of steels required for the three cases compounded by shrink-fit requires a steel of a lower yield strength than that for autofrettage. However, for long cylinders, such as gun barrels, shrink-fit is not a technologically feasible process, and autofrettage is used in production by drawing a lubricated, oversized tapered plug through the bore of the cylinder to be autofrettaged.

12.8.4 Effect of Strain Hardening and Temperature Gradient

12.8.4.1 Strain Hardening

The analysis of elastic–plastic deformation of pressurized thick-walled cylinders made of strain hardening material does not generally end up with a closed-form solution. For a linear strain hardening material obeying the law (Equation 10.8i), namely, $\bar{\sigma} = Y_o + E^p\,\bar{\varepsilon}^p$, a solution may be obtained as follows. According to the Tresca yield criterion for plane strain conditions (Equation 10.32),

$$\sigma_\theta - \sigma_r = Y = (\bar{\sigma})_T \quad \text{for} \quad \sigma_\theta > \sigma_z > \sigma_r \tag{10.32}$$

and

$$\left(\varepsilon^{-p}\right)_T = \varepsilon_\theta^p \ \text{ for } \ \varepsilon_\theta > \left(\varepsilon_r = -\varepsilon_\theta/2\right) > \left(\varepsilon_z^p = 0\right)$$

where Y is the current yield stress.

Substitution of $(\bar{\sigma})_T$ and $\left(\bar{\varepsilon}^p\right)_T$ into Equation 10.8i results in

$$\bar{\sigma} = \sigma_\theta - \sigma_r = Y_o + E^p\varepsilon_\theta^p \tag{12.104}$$

The total strain component $\varepsilon_\theta = u/r$ is given by Equation 12.93, which holds throughout the whole cylinder. Subtracting the elastic strain component expressed by Equation 12.91 yields the net plastic strain component:

$$\varepsilon_\theta^p = \left(1 - \nu^2\right)\left[\frac{Y_o r_p^2}{E r^2} - \frac{\bar{\sigma}}{E}\right] \quad r_i \le r \le r_p$$

Combining Equations 12.104 and 10.32 to eliminate ε_θ^p gives

$$\sigma_\theta - \sigma_r = Y_o\left[1 + \left(1 - \nu^2\right)\frac{E^p}{E}\left(\frac{r_p}{r}\right)^2\right] \Big/ \left[1 + \left(1 - \nu^2\right)\frac{E^p}{E}\right] \tag{12.105}$$

Inserting in the equilibrium equation (Equation 1.10c) and integrating results in, for $r_i \le r \le r_p$;

$$\sigma_r = -\frac{Y_o}{2}\left[\left(1 - \frac{r_p^2}{r_o^2}\right) + \ln\frac{r_p^2}{r^2} + \left(1 - \nu^2\right)\frac{E^p}{E}\left(\frac{r_p^2}{r^2} - \frac{r_p^2}{r_o^2}\right)\right] \Big/ \left[1 + \left(1 - \nu^2\right)\frac{E^p}{E}\right] \tag{12.106a}$$

$$\sigma_\theta = \frac{Y_o}{2}\left[\left(1 + \frac{r_p^2}{r_o^2}\right) - \ln\frac{r_p^2}{r^2} + \left(1 - \nu^2\right)\frac{E^p}{E}\left(\frac{r_p^2}{r^2} + \frac{r_p^2}{r_o^2}\right)\right] \Big/ \left[1 + \left(1 - \nu^2\right)\frac{E^p}{E}\right] \tag{12.106b}$$

The internal pressure required to cause yielding up to any radius r_p is thus obtained as $p_p = -\sigma_r$ at $r = r_i$. Expressions 12.106 for stresses reduce to Equation 12.89 for elastic–perfectly plastic material by setting $E^p/E = 0$.

The effect of strain hardening is to increase the magnitude of the stresses in the plastic region. The internal pressure also steadily increases without attaining a maximum as the elastic–plastic interface progresses. For a cylinder of $\lambda = 2$ and $E^p/E = 0.1$, the pressure p_p causing plasticity up to $r_p/r_i = 1.5$ is found to be $0.6425Y_o$ using Equations 12.106a. This compares to $0.6242Y_o$ as found in Example 12.15, a value which is about 2.8% less.

For a fully yielded, thick-walled cylinder made of a rigid–perfectly plastic material obeying the law, $\bar{\sigma} = K(\bar{\varepsilon})^n$, the solution given in Section 11.4.5 of stationary creep is applicable when $\dot{\varepsilon}^c$, N, and B are replaced by ε^p, $1/n$, and $(1/K)^{1/n}$, respectively.*

12.8.4.2 Radial Temperature Gradient

In Section 6.2.2 the elastic solution to the problem of a thick-walled cylinder subjected to a steady-state temperature distribution has been given. Combined overloading due to an internal pressure and a temperature gradient may initiate yielding or even elastic–plastic deformation in the cylinder wall. This problem has been solved[11] and the internal pressure necessary to render the tube plastic within a radius r_p is given by

$$p_i = \frac{1}{2}\left[Y - \frac{\beta E}{\ln\left(r_o/r_i\right)}\right]\left[1 - \left(\frac{r_p}{r_o}\right)^2 + \ln\left(\frac{r_p}{r_i}\right)^2\right] + \beta E \tag{12.107}$$

where

* This solution is, in fact, the same as may be obtained by solving the problem using the total deformation theory, which has been referred to in Section 10.8.5. For this, see Skrzypek and Hetnarski,[10] Chap. 7.

$$\beta = \frac{\alpha(T_i - T_o)}{2(1-\nu)}$$

During elastic–plastic deformation, the pressure is augmented if $\beta > 0$ and reduced if $\beta < 0$. If $\beta = Y \ln \lambda/E$, the pressure attains a value $Y \ln \lambda$, a value which is equal to the collapse pressure, Equation 12.94, without temperature gradient. For instance, for a steel cylinder with $\lambda = 2$, $Y =$ 400 MPa, $E = 200$ GPa, $\nu = 0.3$, and $\alpha = 12 \times 10^{-6}$, the temperature difference of 161.7°C combined with a pressure of 277 MPa renders the cylinder fully plastic. Also, the pressure reaches its maximum value of $Y \ln \lambda$ when the cylinder becomes fully plastic, i.e., at $r_p = r_o$.

12.9 ANNULAR ROTATING DISKS OF UNIFORM THICKNESS

The stresses in thin annular disks of uniform thickness undergoing elastic deformation are given by the equation:

$$\sigma_r = \frac{3+\nu}{8}\rho\omega^2\left(r_o^2 + r_i^2 - \frac{r_o^2 r_i^2}{r^2} - r^2\right)$$

$$\sigma_\theta = \frac{3+\nu}{8}\rho\omega^2\left(r_o^2 + r_i^2 + \frac{r_o^2 r_i^2}{r^2} - \frac{1+3\nu}{3+\nu}r^2\right) \tag{6.18}$$

$$\sigma_z = 0$$

As the speed of rotation is increased, yielding will initiate at the inner radius $r = r_i$. Upon further increase in rotational speed, an inner plastic annulus will form surrounded by the rest of the disk material which remains elastic. This elastic region will diminish as the plasticity progresses because of further increase in the rotational speed and will end by full yielding of the disk material.

12.9.1 Initial Yielding

In this section it is required to determine the angular speed at which yielding starts. From the elastic stress distribution given in Chapter 6, both σ_r and σ_θ are positive, while $\sigma_z = 0$, for plane stress conditions. Thus, the stresses are such that $\sigma_\theta > \sigma_r > \sigma_z$.

12.9.1.1 Tresca Yield Criterion

Application of the Tresca yield condition gives

$$\sigma_\theta - \sigma_z = Y \text{ or } \sigma_\theta = Y$$

The maximum value of σ_θ occurs at $r = r_i$ as given by Equation 6.19b, and yielding starts when

$$(\sigma_\theta)_{r=r_i} = \frac{3+\nu}{4}\rho\omega^2\left(r_o^2 + \frac{1-\nu}{3+\nu}r_i^2\right) = Y$$

i.e., yield will start at ω_{YT}:

$$\omega_{YT}^2 = \frac{4Y/\rho r_o^2}{(2+\nu)+(1-\nu)(r_i^2/r_o^2)} \tag{12.108}$$

12.9.1.2 Von Mises Yield Criterion

According to the von Mises yield criterion, the three stress components σ_θ, σ_r, and σ_z must be considered. By referring to Figure 6.26b, there may be two possible locations where yielding may occur, namely, at $r = r_i$ and at $r = \sqrt{r_i r_o}$. However, yielding for plane stress conditions takes place when the maximum value of ($\sigma_\theta^2 + \sigma_r^2 - \sigma_r\sigma_\theta$) attains Y. It can be easily shown that this occurs first at $r = r_i$ where $\sigma_r = 0$. Thus, yielding will initiate always at the inner radius of an annular disk and both the Tresca and von Mises yield criteria give at this instant the same rotational speed according to Equation 12.108. For simplicity, solutions applying the Tresca condition will be adopted hereafter.

The value of ω_{YT} at $r = r_i$ as calculated from Equation 12.108 for $r_o = 5r_i$ and $\nu = 0.3$ shows that

$$\omega_{YT}^2 = 1.202Y/\rho r_o^2$$

12.9.2 Partial and Full Yielding

For an angular speed $\omega_p > \omega_{YT}$, the plastic region will spread out from the inner radius. The disk will then consist of an inner plastic region extending to a radius r_p surrounded by an outer elastic region of $r_p \le r \le r_o$.

12.9.2.1 Stresses in the Plastic Region $r_i \le r \le r_p$

Within the plastic region, the stresses are required to satisfy both the equilibrium equation and the yield criterion. The equation of equilibrium (Equation 1.11b) is given by

$$\frac{d\sigma_r}{dr} + \frac{\sigma_r - \sigma_\theta}{r} = -\rho\omega_p^2 r$$

Substituting $\sigma_\theta = Y$ according to the Tresca yield criterion gives

$$\frac{d}{dr}(r\sigma_r) = Y - \rho\omega_p^2 r^2$$

Integrating results in

$$\sigma_r = Y - \frac{\rho\omega_p^2 r^2}{3} + \frac{D}{r}$$

where D is a constant of integration obtained from the condition that $\sigma_r = 0$ at the inner radius $r = r_i$. Thus, the stresses in the plastic region, $r_i \le r \le r_p$, are given by

$$\sigma_r = Y\left(1 - \frac{r_i}{r}\right) - \frac{1}{3}\rho\omega_p^2 r^2\left(1 - \frac{r_i^3}{r^3}\right)$$

and

$$\sigma_\theta = Y \tag{12.109}$$

At the interface radius the radial stress σ_r is given by

$$\sigma_{r_p} = Y\left(1 - \frac{r_i}{r_p}\right) - \frac{1}{3}\rho\omega_p^2 r_p^2\left(1 - \frac{r_i^3}{r_p^3}\right) \tag{12.110}$$

This stress must be continuous across the elastic–plastic interface boundary and, hence, is used as a boundary condition to determine the stresses in the elastic region.

12.9.2.2 Stresses in the Elastic Region $r_p \le r \le r_o$

These stresses are given by the same expressions originally derived in Chapter 6, using the stress function approach, namely,

$$\sigma_r = \frac{A}{r^2} + 2C - \frac{3+\nu}{8}\rho\omega_p^2 r^2$$

$$\sigma_\theta = -\frac{A}{r^2} + 2C - \frac{1+3\nu}{8}\rho\omega_p^2 r^2$$

The constants A and C are determined from the two conditions:

At $r = r_p$: $\sigma_r = \sigma_{rp}$,
At $r = b$: $\sigma_r = 0$.

Applying these conditions gives, within the elastic region, $r_p \le r < r_o$:

$$\sigma_r = \left[\frac{r_i r_p}{2r^2}\left(Y - \frac{1}{3}\rho\omega_p^2 r_i^2\right) + \frac{1+3\nu}{24}\rho\omega_p^2\frac{r_p^4}{r_o^2}\right]\left[1 - \frac{r_o^2}{r^2}\right] + \left[\frac{1+3\nu}{8}\rho\omega_p^2 r_o^2\left(1 - \frac{r^2}{r_o^2}\right)\right]$$

$$\sigma_\theta = -\left[\frac{r_i r_p}{2r_o^2}\left(Y - \frac{1}{3}\rho\omega_p^2 r_i^2\right) + \frac{1+3\nu}{24}\rho\omega_p^2\frac{r_p^4}{r_o^2}\right]\left[1 + \frac{r_o^2}{r^2}\right] + \left[\frac{1+3\nu}{8}\rho\omega_p^2 r_o^2\left(\frac{3+\nu}{1+3\nu} - \frac{r^2}{r_o^2}\right)\right] \tag{12.111}$$

For a given radius r_p of the elastic–plastic interface, ω_p can be determined from the second of Equations 12.111 by putting $\sigma_\theta = Y$ and noting that σ_r is continuous across $r = r_p$:

$$\omega_p^2 = \frac{3Y}{\rho r_o^2}\left[2 - \frac{r_i}{r_p}\left(1 + \frac{r_p^2}{r_o^2}\right)\right] \Big/ \left[2 + \frac{1+3\nu}{4}\left(1 - \frac{r_p^2}{r_o^2}\right)^2 - \frac{r_i^3}{r_p r_o^2}\left(1 + \frac{r_p^2}{r_o^2}\right)\right] \tag{12.112}$$

As the rotational speed increases, the disk is rendered more and more plastic and ultimately full yielding occurs at $\omega = \omega_u$. This is obtained by putting $r_p = r_o$ in Equation 12.112 to give

$$\omega_u^2 = \frac{3Y/\rho r_o^2}{1 + (r_i/r_o) + (r_i^2/r_o^2)} \tag{12.113}$$

which is independent of ν.

As an example, consider a disk of $r_o = 5r_i$ and $\nu = 0.3$. From Equation 12.113 the angular velocity $\omega_u = 2.4193Y/\pi\ \omega_u^2$, which shows that ω_u is about 2.013 times ω_{YT}. It may be noted that for a ring, $r_i \cong r_o$, Equations 12.108 and 12.113 give $\omega_{YT}^2 = \omega_u^2 \cong Y/pr_o^2$.

Example 12.20:

A flat turbine disk of uniform thickness has 1000-mm outside diameter and a 400-mm inside diameter. The tensile rim loading due to blades and shrouding is $\sigma_o = 6$ MPa. Assuming that the inside periphery is unloaded, calculate the speeds at which

a. Yield initiates at the inside radius,
b. The disk becomes fully plastic.

Assume Y = *300 MPa, $\nu = 0.3$, and a specific weight of 78330 N/m³.*

Solution:

Due to the applied tensile load at the outer rim, the stresses in the disk are the sum of stresses due to rotation, as given by Equations 6.18 and the rim tensile loading according to Equation 6.3, where $p_o = 6$ MPa and $\lambda = 2.5$. Hence,

$$\sigma_r = \frac{6\times10^6}{2.5^2-1}\left(2.5^2 - \frac{0.5^2}{r^2}\right) + \frac{3.3}{8}\frac{78330}{9.81}\left(0.5^2 + 0.2^2 - \frac{0.5\times0.2}{r^2} - r^2\right)\omega^2$$

$$\sigma_\theta = \frac{6\times10^6}{2.5^2-1}\left(2.5^2 + \frac{0.5^2}{r^2}\right) + \frac{3.3}{8}\frac{78330}{9.81}\left(0.5^2 + 0.2^2 + \frac{0.5\times0.2}{r^2} - \frac{1.9}{3.3}r^2\right)\omega^2$$

The stress distributions due to rotation and all-around tensile loading are shown in Figure 12.26a and b, respectively.

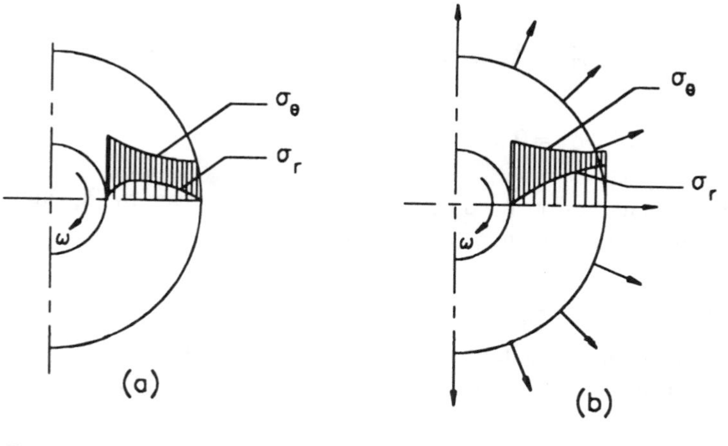

FIGURE 12.26 Example 12.20.

a. Initial yielding — According to the Tresca criterion, yielding will commence at the inner surface. By noting that $\sigma_\theta > \sigma_r > (\sigma_z = 0)$, then, at $r = r_i$, $\sigma_\theta = Y = 300$ MPa, which gives $\omega = 177$ rad/s or $N = 1690$ rpm.
b. Full yielding — This occurs when the Tresca criterion and the equilibrium equation are satisfied throughout the entire disk, i.e.,

$$\frac{d\sigma_r}{dr}+\frac{\sigma_r-\sigma_\theta}{r}=-\rho\omega_u^2 r$$

and $\sigma_\theta = Y$, where $\sigma_\theta > \sigma_r > (\sigma_z = 0)$; hence,

$$\frac{d}{dr}(r\,\sigma_r)=Y-\rho\omega_u^2 r^2 \quad \text{or} \quad \sigma_r = Y-\frac{1}{3}\rho\omega_u^2 r^2+\frac{C_1}{r}$$

The constant C_1 and the full yielding angular velocity ω_u are determined by applying the boundary conditions:

At $r = r_i$: $\sigma_r = 0$, and
At $r = r_o$: $\sigma_r = \sigma_o = 6$ MPa.

Hence, $\omega_u^2 = 3\dfrac{Y(r_o - r_i) - \sigma_o r_o}{\rho(r_o^3 - r_i^3)}$

For $\sigma_o = 6$ MPa, $r_i = 0.2$ m, $r_o = 0.5$ m, and $Y = 300$ MPa, $\omega_u = 528.6$ rad/s or $N =$ 5047.4 rpm, almost three times the rotational speed at initial yielding.

12.9.3 Residual Stresses at Stoppage

At stoppage, the disk is partially released of stresses as a result of elastic recovery. These released stresses are determined from Equation 6.18 for elastic deformation by substituting ω_p from Equation 12.112 to give

$$\sigma_r = \frac{3+\nu}{8}\rho\omega_p^2\left(r_o^2 + r_i^2 - \frac{r_o^2 r_i^2}{2} - r^2\right)$$

$$\sigma_\theta = \frac{3+\nu}{8}\rho\omega_p^2\left(r_o^2 + r_i^2 + \frac{r_o^2 r_i^2}{2} - \frac{1+3\nu}{3+\nu}r^2\right) \tag{12.114}$$

$$\sigma_z = 0$$

The residual stresses are thus obtained by subtracting stresses given by Equation 12.114 from the corresponding stresses of Equations 12.109 and 12.111.

12.9.4 Shrink-Fitted Disks

In many high-speed applications, a rotating disk is shrunk-mounted on the shaft. In this case, the disk is subjected to an internal pressure p_c at the inner radius given by Equation 6.14 as

$$p_c = E\frac{\Delta}{r_i}\frac{\lambda^2 - 1}{2\lambda^2} \tag{6.14}$$

where $\lambda = r_o/r_i$ and Δ is the current radial interference given by the expression

$$\Delta = \Delta_o - u$$

where Δ_o is the initial interference and u is the radial displacement at $r = r_i$ due to rotation, obtained from $(\sigma_r, \sigma_\theta)$ of Equation 6.18 by substituting into Hooke's law as

$$u = \frac{r_i}{E}\left(\sigma_\theta - \nu\sigma_r\right)$$

Yielding initiates at the inner radius of the disk when σ_θ at $r = r_i$ is equal to Y. In this case, the stress σ_θ is the summation of σ_θ due to shrinkage, Equation 6.11, and σ_θ due to rotation, Equation 6.18, as given by

$$\sigma_\theta = Y = p_c\frac{\lambda^2+1}{\lambda^2-1} + \frac{3+\nu}{4}\rho\left(\omega_{YT}^2\right)\left(r_o^2 + \frac{1-\nu}{3+\nu}r_i^2\right) \tag{12.115}$$

from which $(\omega_{YT})_s$, the angular velocity at the onset of yielding for the shrunk-mounted disk is obtained as

$$\left(\omega_{YT}^2\right)_s - \frac{4}{\rho r_o^2}\left[Y = p_c\frac{\lambda^2+1}{\lambda^2-1}\right]\Big/\left[(3+\nu)+(1-\nu)\frac{1}{\lambda^2}\right] \tag{12.116}$$

Inspection of Equations 12.116 and 12.108 shows that $(\omega_{YT}^2)_s < \omega_{YT}^2$ as long as u is less than Δ_o. By increasing the angular velocity, u will increase until reaching a value equal to Δ_o. At this instant, the interference between the disk and shaft is lost and the disk gets loose. It may be assumed that during this period in which p_c decreases there is no spread of plastic deformation such that $r_p = r_i$ and hence $(\omega_{YT}^2)_s = \omega_{YT}^2$, as given by Equation 12.108.

Example 12.21:

A thin annular disk of $\lambda = r_o/r_i = 5$ is shrunk on a shaft made of the same material. The assembly is rotating at an angular velocity ω. If the radial interference is $\Delta_o = 0.25$ mm and $r_o = 500$ mm, find the maximum angular speed such that the disk neither gets loose nor starts to yield.

Take $Y = 400$ MPa, $E = 200$ GPa, $\nu = 0.3$, and $\rho = 8000$ kg/m³.

Solution:

The angular velocity at which yielding starts is obtained from Equation 12.116 in which p_c is determined from the current radial interference $\Delta = \Delta_o - u$, i.e,

$$u = \frac{r_i}{E}\left(\sigma_\theta - \nu\sigma_r\right)$$

Substituting from Equation 6.18 for $r = r_i$ and $\sigma_r = 0$ yields

$$u = \frac{r_i}{E}\left(\frac{3+\nu}{4}\right)\rho\omega^2\left(r_o^2 + \frac{1-\nu}{3+\nu}r_i^2\right)$$

$$= 0.0208\frac{\rho\omega^2}{E} = 8.32\times10^{-7}\omega^2 \text{ mm}$$

From Equation 6.14, namely,

$$\frac{\Delta}{r_i} = \frac{2p_c}{E}\frac{\lambda^2}{\lambda^2-1}$$

hence,

$$p_c = \frac{E}{2r_i}\left(0.25 - 8.32\times10^{-7}\omega^2\right)\frac{24}{25}$$

$$= \left(240 - 7.987\times10^{-4}\omega^2\right)\ \text{MPa}$$

Substituting p_c in Equation 12.116 yields

$$\omega^2 = \left[\frac{4}{8000\times0.5^2}\right]\left[400\times10^6 - \left(240 - 7.987\times10^{-4}\omega^2\right)\times10^6\times\frac{26}{24}\right]$$

$$/\left[3 + 0.3 + (1-0.3)/25\right]$$

which gives the angular velocity at which yielding starts — $(\omega_{YT})_S = 418.7$ rad/s.

The angular velocity at which the disk gets loose is obtained by setting either $p_c = 0$ or $u = \Delta_o$; both conditions give the same results as $\omega = 548.2$ rad/s.

The disk will yield at lower angular velocity, almost 76% of that required to loosen the assembly. Hence, the disk neither gets loose nor starts to yield if $\omega < 418.7$ rad/s.

12.10 SOLID ROTATING DISKS OF UNIFORM THICKNESS

The stresses in a solid rotating disk undergoing elastic deformations are given by Equations 6.23 as

$$\sigma_r = \frac{3+\nu}{8}\rho\omega^2\left(r_o^2 - r^2\right)$$

$$\sigma_\theta = \frac{3+\nu}{8}\rho\omega^2\left(r_o^2 - \frac{1+3\nu}{3+\nu}r^2\right) \tag{6.23}$$

Both σ_r and σ_θ attain their maximum values at the disk center $r = 0$, as

$$\left(\sigma_r\right)_{\max} = \left(\sigma_\theta\right)_{\max} = \frac{3+\nu}{8}\rho\omega^2 r_o^2$$

12.10.1 Initial, Partial and Full Yielding

12.10.1.1. Initial Yielding

As the rotational speed is increased, yielding will commence at the center.

Tresca yield criterion

Since $\sigma_\theta > \sigma_r > (\sigma_z = 0)$, the Tresca condition is satisfied as

$$\sigma_\theta = Y$$

At the center:

$$\left(\sigma_\theta\right)_{\max} = \frac{3+\nu}{8}\rho\omega^2 r_o^2 = Y$$

which gives the angular velocity at the onset of yielding as

$$\omega_{YT}^2 = \frac{8Y/\rho r_o^2}{3+\nu} \tag{12.117}$$

Note that for an annular disk of a radius r_i approaching zero, Equation 12.108 gives a value for $\omega_{YT}^2/\rho r_o^2$ only one half that obtained for a solid disk by Equation 12.117.

von Mises yield criterion

At the disk center, $(\sigma_r)_{max} = (\sigma_\theta)_{max}$. This substituted into the von Mises yield criterion for plane stress, Equation 10.17, gives

$$(\sigma_\theta)_{max} = Y$$

Thus, yielding initiates at the same angular velocity as obtained by the Tresca yield criterion.

12.10.1.2 Partial Yielding

Here, a procedure similar to that of an annular disk may be followed. The conditions of $\sigma_r = 0$ at $r = r_o$ and continuity of σ_r across the elastic–elastic interface boundary r_p must be satisfied.

Stresses in the elastic region $r_p \le r \le r_o$

$$\sigma_r = \frac{\rho\omega_p^2 r_o^2}{8}\left[(3+\nu) - \frac{1}{3}(1+3\nu)\frac{r_p^4}{r^2 r_o^2}\right]\left(1-\frac{r^2}{r_o^2}\right)$$

$$\sigma_\theta = \frac{\rho\omega_p^2 r_o^2}{8}\left[(3+\nu) + (1+3\nu)\left[\frac{r_p^4}{3r_o^4}\left(\frac{r_o^2}{r^2}+1\right) - \frac{r^2}{r_o^2}\right]\right] \tag{12.118}$$

Stresses in the plastic region $0 \le r \le r_p$

$$\sigma_r = Y - \frac{1}{3}\rho\omega_p^2 r^2$$

$$\sigma_\theta = Y \tag{12.119}$$

The angular velocity ω_p corresponding to any elastic–plastic interface radius r_p is obtained from Equations 12.118 with $\sigma_\theta = Y$, noting that σ_r is continuous across $r = r_p$; hence,

$$\omega_p^2 = \frac{3Y}{\rho r_o^2}\Bigg/\left[1+\frac{1+3\nu}{8}\left(1-\frac{r_p^2}{r_o^2}\right)^2\right] \tag{12.120}$$

12.10.1.3 Full Yielding

This occurs when the speed attains a value ω_u obtained by putting $r_p = r_o$ in Equation 12.120 as

$$\omega_u^2 = \frac{3Y}{\rho r_o^2} \tag{12.121}$$

In both annular disks and solid disks the displacements in the elastic regions are obtained by direct substitution of the stresses into Hooke's law. The displacements within the plastic region are more elaborate to determine.[1]

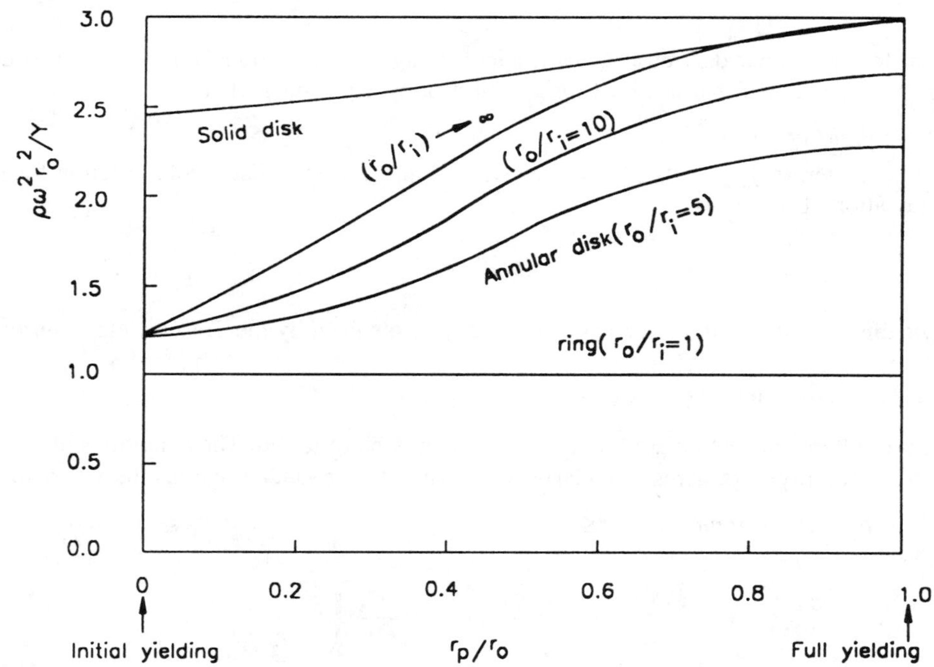

FIGURE 12.27 Variation of the angular velocity in relation to the elastic–plastic interface radius for elastic–plastic rotating disks of different geometry.

Figure 12.27 shows the variation of the angular velocity ω_p required to spread yielding with an elastic–plastic interface radius r_p for annular disks ($r_o/r_i = 5$ and 10), a ring ($r_o/r_i \cong 1$), and a solid disk, all of $\nu = 0.3$. As seen from Figure 12.27, the presence of an extremely small hole at the center of a disk will result in early yielding at about $^1/_2$ the speed which a solid disk can sustain before yield initiates at the center.

Example 12.22:

An annular disk of $r_o = 4r_i$ *and a solid disk of the same* r_o *are made of the same material,* $\nu = 0.25$. *Calculate for both disks the angular velocities required for*

a. *Initial yielding,*
b. *Partial yielding* $r_p = 0.3r_o$, *and*
c. *Full yielding.*

Tabulate the results and comment.

Solution:

a. Annular disk:
   - Initial yielding — Equation 12.108:

$$\omega_{YT}^2 = 1.213Y/\rho r_o^2$$

TABLE 12.8
Angular Speed, Example 12.22

Condition	Initial Yielding $\omega_{YT}^2/(Y/\rho r_o^2)$	Partial Yielding $\omega_p^2/(Y/\rho r_o^2)$ $(r_p = 0.3r_o)$	Full Yielding $\omega_u^2(Y/\rho r_o^2)$
Annular disk	1.213	1.421	2.286
Solid disk	2.46	2.54	3.0

- Partial yielding — Equation 12.112:

$$\omega_p^2 = 1.421Y/\rho r_o^2$$

- Full yielding — Equation 12.113:

$$\omega_u^2 = 2.286Y/\rho r_o^2$$

b. Solid disk:
- Initial yielding — Equation 12.117:

$$\omega_{YT}^2 = 2.462Y/\rho r_o^2$$

- Partial yielding — Equation 12.120:

$$\omega_p^2 = 2.540Y/\rho r_o^2$$

- Full yielding — Equation 12.121:

$$\omega_u^2 = 3Y/\rho r_o^2$$

The above results are tabulated in Table 12.8.

For the annular disk, the ratio $\omega_u/\omega_{YT} = 1.885$, while it is only 1.219 for the solid disk. Once initial yielding occurs in a solid disk, a relatively slight increase in the rotational speed renders the disk fully plastic. However, a solid disk with the same outer radius as an annular one can rotate at a speed 2.03 times higher than the latter without yielding at any point. The superiority of a solid disk compared with an annular one has been indicated before in Chapter 6, Section 6.3.2.

12.10.2 Residual Stresses at Stoppage

When the disk is stopped, the recoverable elastic deformations are those due to the stresses obtained from Equations 6.23 by substituting the value of ω_p for partial yielding obtained from Equation 12.120. The residual stresses are obtained by following the same procedure adopted in the case of an annular rotating disk, Section 12.9.3.

12.11 SHAKEDOWN LIMIT: APPLICATION TO PRESSURIZED CYLINDERS

Throughout this chapter, the analysis has demonstrated that a state of residual stresses exists in structures or machine components which have been unloaded after exceeding the elastic limit. The existence of residual stresses has been shown to be favorable as in the autofrettage processes of pressurized thick-walled cylinders in which the cylinder is strengthed by an initial overstrain in the plastic range.

Limitations, however, have to be set on the structure, such that reloading must keep the total stresses, the residual added to the current service stresses, well within the elastic range. To meet this objective, two basic theorems are involved, namely, limit analysis and shakedown. Limit analysis is concerned with the determination of ultimate loads at which the structure undergoes unrestricted plastic flow, i.e., behaves like a mechanism. These limit loads or, say, collapse loads, have been determined in this chapter for several components, e.g., bent bars and plates, twisted bars, pressurized tubes, and rotating disks.

The shakedown theorem* is concerned with structures having a residual stress distribution resulting from unloading from the plastic state. The theorem defines the extreme values of loads that lead nowhere to total stresses beyond the yield limit when both the residual and current stresses are superimposed. This elastic response of the structure — upon reloading — is viewed as if it has "settled-down" or "shaked-down" to safe working limits against fatigue failure.

The determination of shakedown limits is best illustrated by considering the case of a pressurized thick-walled cylinder. The residual stresses left over within the walls of the cylinder after unloading from the condition of partial yielding are given by Equations 12.102 and 12.103.

Upon unloading, there is a possibility that the resulting residual stresses may exceed those required for initiating a secondary or reversed yielding at the bore of the unloaded cylinder. Upon unloading the residual stresses for a plane case are $\sigma_r = \sigma_z = 0 > \sigma_\theta$ at the inner radius of the cylinder. Hence, reversed yielding, neglecting the Bauschinger effect, may occur when $\sigma_\theta = -Y$ at $r = r_i$. Substituting into the second of Equations 12.102 gives for $r = r_i$:

$$-Y = -\frac{Y}{2}\left[\frac{p_p}{p_{YT}}\left(1+\frac{r_i^2}{r_i^2}\right)-\left(2+\ln\frac{r_i^2}{r_i^2}\right)\right]$$

hence giving

$$p_p/p_{YT} \leq 2 \qquad (12.122)$$

Thus, if yielding is to be prevented upon unloading, the pressure causing partial yielding p_p must not exceed twice the pressure required to initiate yielding for this cylinder. Note that such a limit is derived disregarding any Bauschinger effect due to repeated loading and unloading.

Now the limit for p_p is p_u, i.e., the pressure required for full yielding. Hence, from Equation 12.94 for p_u and Equation 12.84 for p_{YT}, Condition 12.122 gives

$$Y\ln\frac{r_o}{r_i} \leq Y\left(1-\frac{r_i^2}{r_o^2}\right)$$

which sets a lower limit on $\lambda = (r_o/r_i)$ as 2.22.

For cylinders of $\lambda > 2.22$, where $2p_{YT} < p_i < p_u$, reverse yielding occurs upon complete unloading. However, for cylinders of any λ, subsequent reloading by $p_i < 2p_{YT}$ results only in elastic response.

* For a proof for the shakedown theorem, see Chakabarty,[1] p. 99.

Therefore, the shakedown condition for a thick-walled cylinder, which is repeatedly loaded, is that the internal pressure must be less than p_u when $\lambda \le 2.22$ and less than $2p_{YT}$ when $\lambda \ge 2.2$.

Figure 12.28 represents the shakedown limit pressures for thick-walled cylinders of various radii ratios $1.5 \le \lambda \le 4$. For cylinders of radii ratios $\lambda \ge 2.22$, reloading with a pressure less than $2p_{YT} = Y(1 - 1/\lambda^2)$ results in an elastic stress distribution. Internal pressures larger than $2p_{YT}$ and less than $p_u = Y \ln \lambda$ cause an elastoplastic state of deformation. For cylinders of $\lambda < 2.22$, the shakedown limit is the full yielding pressure p_u, i.e., both the shakedown limit and collapse pressure coincide. Note that the above limits have been determined for a perfectly plastic material, i.e., one without strain hardening, hence offering conservative shakedown limits. On the other hand, neglect of the Bauschingher effect may aggravate the situation when stress reversals due to repeated loading and unloading lower the yield strength of the material in compression compared with its value in tension.

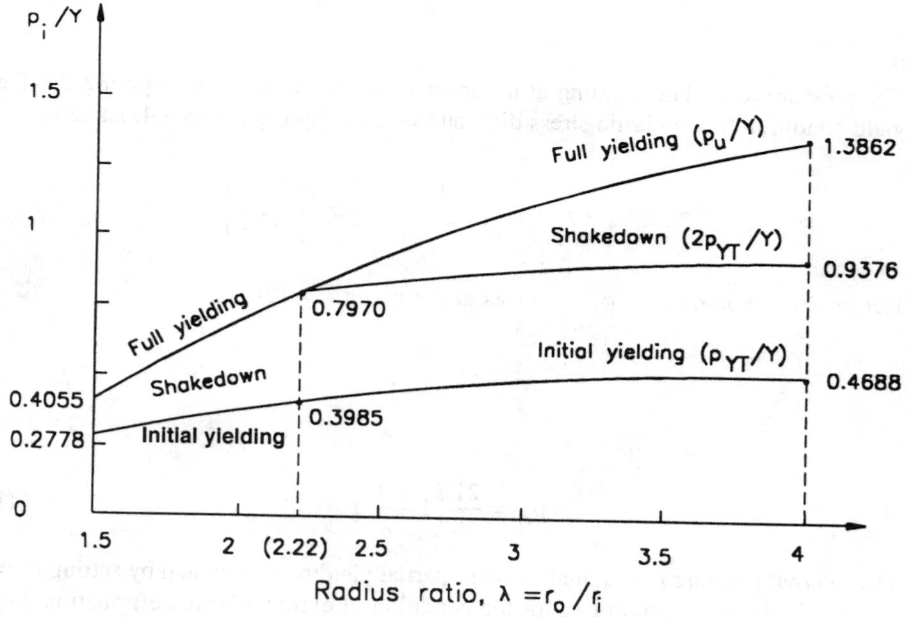

FIGURE 12.28 Shakedown limit pressures for thick-walled cylinders of different outer-to-inner radii ratios.

Loading and unloading beyond the shakedown limit and below the collapse limit may incur alternating plastic strains and eventually failure by low-cycle fatigue (10 to 10^4 cycles). Another possibility of failure may be due to "ratcheting" resulting from the occurrence of plastic strain increments during each loading cycle, eventually leading to unacceptable changes (growth or shrinkage) in the dimensions of the body. Both types of failure prevail in structures where cyclic thermal loads are superimposed on steady mechanical loads.

Example 12.23:

For a pressurized spherical vessel of outer and inner radii r_o *and* r_i*, respectively, determine:*

a. *The internal pressure* p_{YT} *required to initiate yielding,*
b. *The internal pressure* p_p *required to cause partial yielding,*
c. *The shakedown limit.*

Apply the Tresca yield criterion and make use of the following:

- *The full yielding pressure:*

$$p_u = 2Y\, ln(r_o/r_i) \tag{12.123}$$

- *The stress distribution within the plastic region of a partially yielded sphere,* $r_i < r < r_p$:

$$\begin{aligned} \sigma_r &= -\frac{2Y}{3}\left[1-\left(\frac{r_p}{r_o}\right)^3 + ln\left(\frac{r_p}{r}\right)^3\right] \\ \sigma_\theta = \sigma_\phi &= \frac{2Y}{3}\left[\frac{1}{2}+\left(\frac{r_p}{r_o}\right)^3 - ln\left(\frac{r_p}{r}\right)^3\right] \end{aligned} \tag{12.124}$$

Solution:

a. The pressure to initiate yielding at the inner radius is obtained by applying the Tresca yield condition to the elastic stress distribution given by Equations 4.4, namely,

$$\sigma_\theta = \sigma_\phi = p_i\frac{r_o^3/2r^3+1}{r_o^3/r_i^3-1} \quad \sigma_r = -p_i\frac{r_o^3/r^3-1}{r_o^3/r_i^3-1}$$

Hence, the condition $\sigma_\theta - \sigma_r = Y$ gives at $r = r_i$ and $\lambda = r_o/r_i$:

$$Y = p_i\frac{\lambda^3}{\lambda^3-1}$$

or

$$p_{YT} = \frac{2Y}{3}\left(1-\frac{1}{\lambda^3}\right) \tag{12.125}$$

b. The internal pressure p_p required to cause partial yielding is obtained by setting $\sigma_r = -p_p$ at $r = r_i$ in the expression of σ_r for the condition of elastic–plastic deformation, i.e.,

$$p_p = \frac{2Y}{3}\left[1-\left(\frac{r_p}{r_o}\right)^3 + \ln\left(\frac{r_p}{r_i}\right)^3\right] \tag{12.126}$$

when r_p is set equal to r_o, i.e., full yielding of the wall, this gives $p_u = 2Y\ \ln(r_o/r_i)$ as given before.

c. The shakedown limit is derived from the expression for the residual stress of σ_θ. This expression is found by subtracting an elastic stress σ_θ due to p_p from the plastic stress distribution of σ_θ within the plastic region; $r_i \le r \le r_p$; hence,

$$(\sigma_\theta)_{res} = \frac{2Y}{3}\left[\frac{1}{2}+\left(\frac{r_p}{r_o}\right)^3 - \ln\left(\frac{r_p}{r}\right)\right] - \frac{p_p}{\lambda^3-1}\left[\frac{1}{2}\left(\frac{r_o}{r}\right)^3+1\right]$$

In terms of the ratio of p_p/p_{YT}, this reduces to

$$(\sigma_\theta) = Y - \frac{2Y}{3}\left[\frac{p_p}{p_{YT}}\left(1 + \frac{r_i^3}{2r^3}\right) - \ln\left(\frac{r}{r_i}\right)^3\right] \tag{12.127}$$

For secondary yielding to occur, this residual stress which attains its maximum at $r = r_i$ has to satisfy the Tresca yield condition with $\sigma_z = \sigma_r = 0$, i.e.,

$$\sigma_\theta = -Y$$

$$-Y = Y - \frac{2Y}{3}\left[\frac{p_p}{p_{YT}}\left(1 + \frac{r_i^3}{2r_i^3}\right) - \ln\left(\frac{r_i}{r_i}\right)^3\right]$$

or $p_p/p_{YT} \le 2$. Now the limiting value for p_p is p_u: hence, the above condition gives

$$p_u \le 2p_{YT}$$

or, according to Equations 12.123 and 12.125,

$$2Y \ln \lambda = \frac{4Y}{3}\left(1 - \frac{1}{\lambda^3}\right)$$

This sets a limit of $\lambda \ge 1.7$, which means that reverse yielding would occur when p_p is greater than $2p_{YT}$ for a sphere with a wall ratio $\lambda \ge 1.7$. Otherwise, the sphere wall material behaves elastically if reloaded by a pressure $p_p < 2p_{YT}$. For spheres with $\lambda <$ 1.7, reverse yielding occurs only when p_p is greater than p_u. To summarize, the shakedown limits are

$$\text{For } \lambda \ge 1.7: \quad p_p \le 2\text{pYT}, \quad \text{and}$$

$$\text{For } \lambda < 1.7: \quad p_p \le p_u \tag{12.128}$$

PROBLEMS

Elastic–plastic Flat Plates and Beams

12.1 A plate of rectangular cross section is subjected to a bending moment M, which is greater than M_Y but less than M_u, so that the outer fibers are yielded to a depth $(h/2 - y_p)$, where h is the height of the cross section. If the material of the beam is elastic–perfectly plastic, show that

$$y_p = 3\left[\left(\frac{h}{2}\right)^2 - \frac{M}{bY}\right]^{1/2}$$

where b is the width of the section and Y is the yield stress of the material.

12.2 It is necessary to bend an aluminum alloy sheet (1 mm thick) to a final radius of curvature of 75 mm. The stress–strain curve may be approximated as elastic perfectly plastic, $Y = 150$ MPa, $\nu = 0.25$. Accounting for springback, what radius of curvature must the tool have if loading is pure bending? What fraction of the cross section remains elastic?

12.3 A plate of rectangular cross section 500 mm wide and 5 mm thick, is subjected to bending produced by equal and opposite moments of magnitude $0.9M_u$ along the plate width. Calculate and sketch the bending stress distribution across the plate.

12.4 A bar having a rectangular section 20 × 60 mm; the material is elastic–perfectly plastic having $E = 200$ GPa, $Y = 400$ MPa and $\nu = 0.3$. Determine the following:

a. M_p and R_p when $^1/_3$ of beam height, 60 mm has yielded. Determine the residual stresses and radius of curvature when the load is removed.
b. Determine the fiber stress and radius of curvature if the bar is loaded again, by a moment = $^1/_3 M_p$. Show that the stresses in the bar have been improved.

12.5 A helical spring is produced by cold winding of a wire around a cylindrical mandrel. The wire has a square cross section 3 × 3 mm and the coil mean diameter = 30 mm. The wire material is assumed to be elastic–perfectly plastic having $Y = 600$ MPa, $E = 200$ GPa. Considering elastic recovery, determine the diameter of the mandrel that gives the required coil diameter.

12.6 A leaf spring is given a circular shape by cold bending of a straight strip. The strip has a cross section $b \times h$ and length L. Material is elastic–perfectly plastic. If the required final camber = c, determine the displacement under load at the middle of the leaf to obtain this camber.

12.7 Determine the "shape factor" for a beam of circular tubular section subjected to pure bending. Take mean radius r_{mo} and thickness $h = 0.5r_{mo}$.

12.8 Determine the shape factor for a beam of a triangular cross section of base b and height h.

12.9 A steel bar 50-mm-square section is subjected to a compression load of 250 kN acting in an eccentric way parallel to the bar axis and in the principal plane of bending. Determine the eccentricity that will render the whole section to be become plastic if $Y = 270$ MPa.

12.10 A strip 0.6 mm thick, 18 mm wide, and 250 mm long first takes a permanent set when bent to a radius of curvature of 250 mm. Estimate its yield strength and the applied bending moment if $E = 210$ GPa and $\nu = 0.33$.

12.11 A coiler is being designed for a cold-rolling line of a steel mill. The coil diameter should allow plastic bending. The sheet is 2 m wide and 0.8 mm thick and has a yield strength of 300 MPa. What is the maximum and minimum diameters of the coil assuming initial yielding for the former and $y_p = h/4$ for the latter? Determine also the maximum moment for bending the sheet taking $E = 210$ GPA and $\nu = 0.3$.

12.12 A uniformly distributed load q is acting laterally on a beam of rectangular cross section b × h. If the load is gradually increased to $0.96q_u$, the collapse load, and then gradually removed, determine the stresses at the outer fibers of the beam and the deflection (a) under load and (b) after the load is removed.

12.13 Determine the collapse concentrated load acting in the midspan and length of the plastic zone of a simply supported beam.

Elastic–Plastic Rectangular and Circular Plates

12.14 A square plate of 1 m side and 40 mm thickness is subjected to pure bending $M_x = 2M_y$.

a. Determine the values of these moments at the onset of yielding.
b. If these moments are increased to 1.25 of their values, what is the depth of the remaining elastic boundary?
c. Determine the final radii of curvature upon complete unloading. Take $Y = 300$ MPa, $\nu = 0.3$, and $E = 200$ GPa and neglect strain hardening.

12.15 A solid, circular, flat plate of diameter 200 and thickness 5 mm is subjected to an edge moment M per unit circumference. Determine the value of M for (a) initial yielding, (b) for 50% of the thickness yielding, and (c) for full thickness yielding. Determine the residual stresses at the outer fibers and radius of curvature when the moment is removed. Take $E = 78$, and $Y = 110$ MPa and $\nu = 0.25$.

12.16 A spring washer shaped as a dished disk, 25ϕ and 50ϕ inner and outer diameter, respectively, is made of 1-mm spring steel strip by punching. The dish total height is 1.5 mm. Determine the necessary diametral clearance between punch and die to produce the dished disk. Take $E = 210$ GPa, $Y = 480$ MPa, $\nu = 0.3$ and ultimate shear strength = 540 MPa. Note that punching the inner hole takes place after blanking the disk solid.

12.17 It is proposed to produce dishes by applying uniform edge bending moment to a freely supported circular flat disk. Determine the moment to reach a maximum depth of the dish as 250 mm for a dish of 2.4-m diameter and 3-mm thickness made of aluminum, $Y = 90$ MPa, $E = 72$ GPa and $\nu = 0.33$. Note that the depth obtained by this process is small and other methods such as bulging or spinning have to be used for the production of deeper dishes.

Curved Beams

12.18 The hanger in Figure 12.8 is subjected to an edge bending moment M instead of the load P. If the radius of the bend is 50 mm, determine the value of M at which yielding is initiated considering the bend to be a circularly curved beam. Compare the results with those of Example 12.4.

12.19 A square solid bar of a side of 25-mm is shaped into a right angle bend having a mean radius of 75 mm. The bar material is steel having $E = 210$ GPa, $Y = 240$ MPa, and $\nu = 0.3$. It is required to decrease the bar curvature so that the mean radius of curvature is increased to 80 mm by applying a bending moment M at each edge. Determine the value of M by considering the bar to be (a) a straight beam and (b) a curved beam.

Buckling

12.20 Determine the buckling load for a bar, 800 mm long, of a rectangular cross section 40 × 60 made of an aluminum alloy, $E = 72$ GPa, $Y = 100$ MPa and $E^p = 42$ GPa. Apply (a) tangent-modulus formula, and (b) double modulus formula.

12.21 A steel pipe, 220-mm outside diameter, 5-mm thickness, $E = 210$ GPa, $E^p = 10$ GPa and $Y = 280$ MPa, is used as a column of length 2 m. Find the maximum allowable compressive load that can be carried by the column for a factor of safety = 2.5.

Torsion

12.22 A hollow shaft of circular cross section with outer and inner diameters 75 and 50-mm, respectively, is subjected to a twisting moment M_t. (a) Determine the twisting moment and the angle of twist at which yielding is initiated. (b) If the twisting moment is increased to spread the plastic zone to a diameter of 60-mm, determine the residual stresses and the residual angle of twist. The material is steel having $E = 210$ GPa, $\nu = 0.3$ and $Y = 320$ MPa.

12.23 A solid circular bar of 40-mm diameter and 200-mm length is used as a torsion bar to absorb explosion energy through elastic-plastic deformation. Determine the maximum dissipated strain energy for $r_p/r_i = 0.1$ if the bar material has a yield stress in shear = 120 MPa and G = 82 GPa.

12.24 For the hollow circular shaft of Problem 12.22, an axial force is applied instead of increasing the twisting moment initiating yielding. Determine the axial force required to (a) spread yielding to a diameter of 60 mm and (b) produce full yielding.

12.25 A thin-walled cylinder of mean radius r_m and wall thickness h is subjected to a twisting moment M_t and a bending moment M_b. If the material is elastic–perfectly plastic of yield shear stress k, (a) derive an expression for M_b to initiate yielding. (b) If M_b is kept constant, determine the ratio of M_b to M_t to make the cylinder wall fully plastic.

Elastic–Plastic Pressurized Cylinders

12.26 A thick-walled cylinder has an inside diameter of 180 mm and an outside diameter of 420 mm. It is made of steel having a yield stress of $Y = 460$ MPa. Determine the fully plastic internal pressure for cylinder according to the Tresca criterion.

12.27 Determine the working pressure p_i for the thick-walled cylinder in Problem 12.26 if

a. It is designed with a factor of safety of F.S. = 3.0 based on the fully plastic pressure.
b. What is the factor of safety based on the maximum elastic pressure p_Y?

12.28 Two cylinders are shrunk-fit together to form a compound, open cylinder. Both cylinders are made of a brittle material with a linear stress–strain curve up to the ultimate strength $\sigma_u = 480$ MPa. The inside cylinder has inner and outer radii of 50 and 75, respectively. The outside cylinder has inner and outer radii of 75 mm and 150 mm, respectively. Determine the shrinkage pressure p_c and the maximum internal pressure p_i that results in initiation of fracture simultaneously at the inner radii of both cylinders. Use the Tresca yield criterion.

12.29 A steel ring is expanded by applying internal pressure using a tool composed of a tapered, segmented plug. The operation is a hot-working one at a constant yield stress of 45 MPa. If the initial outside and inside diameters of the ring are 200 and 500 mm, respectively, and the final inside diameter is 300 mm, calculate the work done in the operation assuming plane strain conditions.

12.30 A thick-walled circular cylinder has an internal radius r_i and an external radius $2r_i$. The cylinder is subjected to an internal pressure sufficient to cause yielding throughout half the wall thickness. Determine the residual stress component at the inside and outside surfaces of the cylinder after the release of the pressure. Assume plane stress conditions. The material is (a) elastic–perfectly plastic, (b) elastic–linear hardening with $E^p/E = 0.1$, $E = 200$ GPa, and $\nu = 0.3$.

12.31 For a thick-walled cylinder subjected to internal pressure, calculate the thickness of the yielded portion of the cylinder for an internal pressure of 93 MPa. The bore of the cylinder is 12 mm, and the outside diameter is 20 mm. What is the maximum internal pressure that may be applied to produce full yielding of the wall? Solve the problem for

a. An elastic–perfectly plastic material $Y = 308$ MPa;
b. Elastic-linear strain-hardening with $E^p/E = 0.15$, $E = 200$ GPa, and $\nu = 0.3$.

12.32 A thick-walled cylinder, 50-mm inside diameter and 150-mm outside diameter, is subjected to an internal pressure such that the cylinder yields at a diameter of 60 mm. Sketch the stress distribution in the cylinder wall for this condition. If the internal pressure is released, calculate the permanent change in the inside diameter. Assume elastic–perfectly plastic material with a yield stress of 200 MPa and take $E = 200$ GPa and $\nu = 0.3$.

12.33 A thick-walled cylinder $\lambda = 2$, material elastic–perfectly plastic is designed according to the maximum shear theory using a factor of safety F.S. = 1.5 based on yielding. The working internal pressure is $0.3Y$.

a. Show that the cylinder cannot support this pressure.
b. If the cylinder has been treated by an autofrettage process to produce favorable residual stresses, in this process $r_p/r_o = 0.6$, determine the distribution of the residual radial and tangential stresses in the cylinder wall. Show that the cylinder after this treatment can support the working pressure safely.
c. Repeat for $r_p/r_o = 0.9$ and comment on the results obtained.

12.34 The steel container for cold extrusion is a prestressed thick-walled cylinder having an inner diameter of 100 mm and an outer diameter 200 mm. The maximum internal pressure is 100 MPa. The container is designed according to the maximum shear theory with yield shear stress $k = 125$ MPa. It is required to make calculations for the following two possible designs:

a. The container is built up of two cylinders shrunk together. Determine the fitted intermediate diameter and the amount of interference for optimum strength conditions.
b. The container is made of one autofrettaged cylinder. Determine the required autofrettage pressure to obtain the maximum allowable stress when the service pressure is applied.

12.35 A steel shaft 50 mm in diameter is shrunk-fit in a hub of 75-mm diameter of the same steel of $Y = 300$ MPa. Determine the diametral interference at which yielding starts assuming Y remains constant.

12.36 Two cylinders are shrunk-fit together to form a composite, open cylinder. Both cylinders are made of steel having a yield stress of $Y = 700$ MPa. The inner cylinder has inside and outside diameters of 100 and 150 mm, respectively. The outer cylinder has inside and outside diameters of 150 and 300 mm, respectively.

a. Determine the shrinkage pressure p_c and the maximum internal pressure p_i that can be applied to the cylinder if it has been designed with a factor of safety of F.S. = 1.85 based on simultaneous initiation of yielding at the inner radii of the inner and outer cylinders, respectively. Use the Tresca yield criterion.
b. Determine the outer diameter of the inner cylinder required for this design. For the steel, $E = 200$ GPa and $\nu = 0.29$.

Elastic–Plastic Rotating Disks

12.37 Plot the stress ratios σ_θ/Y and σ_r/Y against r/r_o for values of $r_p/r_o = 0.5, 0.75, 1.0$ for a rotating disk of uniform thickness:

a. Annular disk with a central hole $r_o/r_i = 2$;
b. Solid disk.

Take $\nu = 0.3$.

12.38 A rotating solid disk of uniform thickness, 1000 mm in diameter, is made of steel having $Y = 300$ MPa, $\rho = 7800$ kg/m^3, $\nu = 0.28$. Determine the value of ω at which

a. Yield starts;
b. Yield extends to middle radius;
c. Disk is fully plastic.

Compare with a disk made of an aluminum alloy having $Y = 160$ MPa, $\rho = 2700$ kg/cm^3, $\nu = 0.32$.

12.39 A flat turbine disk of uniform thickness has 900-mm outside diameter and 200-mm inside diameter; the tensile radial rim loading due to blades and shrouding is 7 MPa at 2800 rpm. Calculate the speeds at which (a) yield first occurs at the inside radius and (b) the disk just becomes fully plastic. Take $Y = 450$ MPa, $P = 7800$ kg/m^3, and $\nu = 0.3$.

Shakedown Limit

12.40 A thick-walled cylinder is made of a material with a yield strength of 600 MPa. The outer and inner diameters of the cylinder are 300 and 100, respectively. Determine the required pressure for (a) initial yielding, (b) shakedown, and (c) full yielding. For the same shakedown pressure, what would be the radius ratio of r_o/r_i for a thick-walled sphere?

12.41 An annular rotating disk is rendered fully plastic by increasing its speed of rotation. At stoppage, show that the distribution of residual hoop stress upon elastic unloading is given by

$$\sigma_\theta = Y\left\{\frac{r_i}{r_o} - \frac{1+3\nu}{8}\left(1+\frac{r_i^2}{r_o^2}-\frac{3r^2}{r_o^2}\right) - \frac{3}{8}(3+\nu)\frac{r_i^2}{r^2}\right\} \Big/ \left(1+\frac{r_i}{r_o}+\frac{r_i^2}{r_o^2}\right)$$

Show that secondary yielding occurs upon unloading if the radius ratio r_o/r_i is greater than 3.414 for $\nu = 1/3$.

REFERENCES

1. Chakrabarty, J., *Theory of Plasticity*, McGraw-Hill, New York, 1987, Chap. 3.
2. Venkatraman, B. and Patel, S. A., *Structural Mechanics with Introductions to Elasticity and Plasticity*, McGraw-Hill, New York, 1970, Chap. 21.
3. Betten, J., *Eng. Arch.*, 44, 199, 1975.
4. Johnson, W. and Mellor, P. B., *Engineering Plasticity*, Van Nostrand, London, 1973, Chap. 8.
5. Shaffer, B. W. and House, R.N., *J. Appl. Mech.*, 24, 305, 1955.
6. Nadai, A., *Theory of Flow and Fracture of Solids*, McGraw-Hill, New York, 1950.
7. Morrison, J. L. and Shepherd, W. M., *Proc. Inst. Mech. Eng.*, 163, 1, 1950.
8. Hill, R. and Siebel, M. P. L., *Philos. Mag.*, 42, 722, 1951.
9. Ford, H. and Alexander, J. M., *Advanced Mechanics of Materials*, Longman, London, 1963.
10. Skrzypek, J. J. and Hetnarski, R. B., *Plasticity and Creep*, CRC Press, Boca Raton, FL, 1993, Chap. 7.
11. Bland, D. R., *J. Mech. Phys. Solids*, 4, 209, 1956.

Index

A

Airy stress function, 172, 174
Angle of rotation, 51, 360
Angle of twist, 360, 371
Anisotropic materials
  elasticity of, 135–140
  plastic deformation
    effective strain increment, 691–692
    effective stress, 691–692
    flow rule, 688
    normal anisotropy, 689–690
    planar isotropy, nonquadratic criterion, 697–700
    planar isotropy, rotational symmetry, 692–697
    strain ratio, 689
    through-thickness yield strength, 688, 689–690
    yield criterion, 687–688
Annular circular plates
  bending of flat
    simply supported, edge moments, 475–476
    simply supported, shearing force at inner edge, 477–478
    superposition for load conditions, 478–479
  disks, rotating
    initial yielding, 886–887
    partial and full yielding, 887–890
    residual stresses, 890
    shrink-fitted, 890–892
    uniform thickness, 262–264
Autofrettage process, 881–884
Axial loads, *see* Bars; Columns
Axially nonsymmetric problems, polar coordinates
  beams, concentrated load, 301–303
  curved beam, bending
    end moment and normal force, 282
    end shearing force, 274–280
    inclined end force, 282–283
    pure bending, 280–281
  curved beams, thermal stresses, 283–286
  disk, solid circular, diametral loads, 299–301
  straight boundaries
    concentrated line load, 295–297
    uniformly distributed line load, 297–299
  wedges subjected to loads
    bending moment at vertex, 292–295
    force acting along axis, 286–289
    force inclined to axis, 290–292
    force perpendicular to axis, 289–290
Axial symmetry
  state of strain, 72
  state of stress, 9
Axisymmetric problems, polar coordinates
  cylinder, steady-state radial thermal gradient
    open or closed end, 254–255
    plane strain, 250–253
    plane stress, 253–254
  cylinder, thick-walled, uniform pressure
    external only, 247–248
    internal and/or external, 243–246
    internal only, 246, 248–249
  cylinder, with shrink fits, 255–262
  disk, rotating solid, uniform strength, 266–269
  disk, rotating, uniform thickness
    annular, 262–264
    solid, 264–266
    thermal gradients, 270–273
  drums and rotors, rotating, 269–270

B

Bars, *see also* Rods
  axial loads, stress function solution, 233–234
  bending moment stresses, 320
  buckling, axial compression, elastic-plastic
    discussion, 853–854
    double-modulus formula, 854–856
    tangent modulus formula, 854, 856
  definition, 319
  free torsion, elastic
    comparison of types, 367–368
    discussion, 359–360
    end constraint effects, 374–376
    internal stiffening webs, 372–373
    Saint-Venant's, 360–363
    solid circular section, 363–364, 374–376
    solid elliptical section, 365–366
    solid rectangular section, 366–367
    state of stress, 9
    stress tensor, 7, 9, 72–73
    thin-walled closed sections, 370–372
    thin-walled open sections, 368–369
    twisting moment and warping, 359–363
  limit tensile strain, 740–741, 764–765
  necking, 734–738, 742–746
  open vs. closed sections, 372
  steady creep, circular shafts, 785–788
  torsion, of elastic-plastic
    combined with tension, 861–865
    example problems, 860–861
    hollow circular section, 858–860
    shear stress components, 856–857
    solid circular section, 857–858
    thin-walled tubes, 861
  twisting moment
    discussion, 359–360
    end constraint effects, 374–376

internal stiffening webs, 372–373
Saint-Venant's free torsion, 360–363
solid circular section, 363–364
solid elliptical section, 365–366
solid rectangular section, 366–367
thin-walled closed sections, 370–372
thin-walled open sections, 368–369
Bauschinger effect
in autofrettage process, 881
behavior of plastically deformed solids, 643
effective plastic strain and, 670
ignored in reverse yielding, 827
isotropic-hardening rule and, 679
Beam-columns, maximum deflection, 410–412
Beams, *see also* Rods
bending of, *see* Beams, bending of
buckling of, lateral, 412–415
concentrated load, polar coordinates, 301–303
creep deformation under bending, 773–776
definition, 319
deflection of curved, 387–391
elastic curve example, 328–330
on elastic foundations
discussion, 415–416
infinitely long, 416–422
semi-infinite, 422–425
short, 425
elementary theory of, 216–217
lateral loads in middle, 232–233
stress function
cantilever beam, end load, 220–229
simply supported, lateral loads, 232–233
simply supported, sinusoidal load, 230–232
uniformly distributed load, 213–220
thermal stresses in curved, 283–286
Beams, bending of
combined with tension
elastic-plastic state, 831–832
elastic state, 831
unloading and residual stresses, 832–833
creep deformation, 773–776
curved
determination of neutral axis, 336
elastic-plastic, 849–853
end moment and normal force, 282
end shearing force, 274–280
errors from approximating neutral axis, 336–337
inclined end force, 282–283
lateral force, plane of axis, 337
location of neutral axis, 334–336
maximum stresses, 337
pure bending, 280–281
solution comparisons, 339
strain energy, 338–339
tangential stresses, 330–334
moment in, 320
plane stress, elastic-plastic
combined bending and tension, 831–833
curved, 849–853
deflection and shear stresses, 837–839
plastic hinges, 833–837
springback, 829–830
strain-hardening effects, 839–843
yielding, 828–830
Beltrami-Michell stress-compatibility equations, 156
Bending moment
bar stresses, 320
beam, curved, 278
in deflection of curved beams, 390
elastic curve in rods, 326–300
membrane stresses in pressure vessels, 527–531
non-linear strain-hardening material, 841
plastic deformation, 822–823
plastic hinge formation, 834
and plastic zone, 850–851
plates
circular, 847–848
rectangular, 843–845
during recovery, 825–827
shape factor and, 823–824
at wedge vertex, 292–295
and yield initiation, 822, 828–830
Bending of plates, *see* Plates, bending of flat
Biaxial stretching, plastic instability
bulging, circular sheet, 749–754
flat sheet, 746–749
Biharmonic equation, 173, 195, 205–206
Body forces, definition of, 1
Bore expansion rate, 795
Boundary-value problems, *see* Elastic boundary-value problems
Bridgman's formula, 743, 744, 745–746
Brittle rupture, 805–808
Brittle vs. ductile materials, 561
Buckling
bars, axial compression, elastic-plastic
discussion, 853–854
double-modulus formula, 854–856
tangent modulus formula, 854, 856
columns
clamped end/free end, 400
clamped end/pinned end, 400
and creep, 788–791
ends clamped, 401
energy method, minimum potential, 403–410
equilibrium approach, 398–403
example problems, 402–403
pinned ends, 398–400
similarity to plate buckling, 540–542
columns vs. plates, 540–542
critical pressure, cylindrical vessels, 550
plate, uniformly compressed
all-around clamped, 542–543
circular, axisymmetric, 544–545
simply supported, 543–544
strain energy approach, 540–542
rods, *see* Rods, buckling of
shells
critical pressure, 550
out-of-roundness effects, 550
thin-walled cylinders, external uniform pressure, 546–550

Bulging
biaxial stretching, circular sheet, 749–754
with planar isotropy, 698
for strain-rate dependent material, 765–767
Bulk modulus of elasticity, 108

C

Cantilever beam, end load example
displacements, 222–225
example problems, 225–229
stresses, 220–221
Cartesian coordinate elastic plane problems
algebraic polynomials used to solve
discussion, 205–208
end load, cantilever beam, 220–229
hydrostatic pressure, retaining wall, 209–213
uniformly distributed load, simply supported, 213–220
example problems, 235–237
plane strain, Hooke's law, 167
sign convention, 8
static equilibrium, 10–14
strain compatibility equations, 62–64
strain-displacement relations, 54–59
stress function
derivation, 182–183
example problems, 176–182
glossary of, 195–197
plane strain, 171–174
plane stress, 174
thermoelastic plane strain, 174–175
thermoelastic plane stress, 175–176
trigonometric series used to solve
axial loads, bar, 233–234
discussion, 229–230
equal lateral loads in middle, 232–233
lateral loads, simply supported beam, 232–233
sinusoidal loads, simply supported beam, 230–232
Castigliano's theorems
applied to displacement in rods, 376–377
stress-strain relations, 132–134
Cauchy's stress formulas, 23
CCF (configuration correction factor), 574–575
Centrifugal body forces, axisymmetric plane problems, 189–191
Circular flat plates
bending of annular
simply supported, edge moments, 475–476
simply supported, shearing force, inner edge, 477–478
superposition for load conditions, 478–479
bending of solid
clamped, center force, 472–474
clamped, uniform pressure, 469–472
discussion, 466–468
simply supported, center force, 474–475
simply supported, uniform pressure, 468–469
buckling of, 544–545
Circular holes, stress concentration in
biaxial loading, 308
example problems, 308–311
formula derivation, 303–308
local yielding situations, 303
COD (Crack opening displacement)
application to structural design, 630–632
discussion, 627–629
experimental determination of, 629–630
Coefficient of linear thermal expansion, 114
Columns
buckling of
clamped end/free end, 400
clamped end/pinned end, 400
ends clamped, 401
energy method, minimum potential, 403–410
equilibrium approach, 398–403
example problems, 402–403
pinned ends, 398–400
similarity to plate buckling, 540–542
creep buckling and, 788–791
definition of, 319
Compatibility, strain
Beltrami-Michell equations, 156
conditions
Cartesian coordinates, 62–64
cylindrical polar coordinates, 64–65
spherical polar coordinates, 65–67
equations
plane strain, 167
plane stress, 169
polar coordinates, plane strain, 184–185
fields, 62
Complementary energy, 119, 124
Conditions of integrability, *see* Strain, compatibility equations
Conditions of strain continuity, *see* Strain, compatibility equations
Configuration correction factor (CCF), 574–575
Constitutive law, 101
Contact problems, straight boundary
force acting along axis, 296–297
force acting normal, 295–296
force inclined to axis, 297
Cracking, *see* Fracture mechanics
Crack opening displacement (COD)
application to structural design, 630–632
discussion, 627–629
experimental determination of, 629–630
Crack propagation, *see also* Fracture mechanics
dependence on cyclic loading, 597–599
example problems, 601–603
mixed-mode loading, 589–590
nonpropagating, region (i), 599
and plane strain, 564
steady, region (ii), 599–601
stress intensity factor, 577–579, 599
unstable growth rate, region (iii), 601
Crack tip opening displacement (CTOD), 627–629, 630–632
Creep deformation
bending of beams, 773–776
buckling of columns, 788–791
effective strain rate for, 781
elastic analogue, 782

equations of state
creep parameters for alloys, 772
multiaxial behavior, 773
uniaxial behavior, 771–773
extrapolation parameter, 770
isochronous stress-strain curves, 768
reference stress method, 791–796
rupture
brittle, uniaxial stress, 807–808
damage concept, 805–807
ductile, uniaxial stress, 803–805
example problems, 808–809
testing and data, 769
time, 803–808
stages of creep, 767–768
stationary, in thick-walled cylinders, 780–784
steady
beams under bending, 773–776
circular shafts under torsion, 785–788
rotating disks, 785
thin-walled cylinders, 777–780
strain vs. time curves, 768
stress relaxation, 796–798
temperature effects, 796–798
time hardening vs. strain hardening, 798–802
variable loading, 798–802
Critical energy release rate, 564, 570
CTOD (Crack tip opening displacement), 627–629, 630–632
Curved beams
bending of
determination of neutral axis, 336
elastic-plastic, 849–853
end moment and normal force, 282
end shearing force, 274–280
errors from approximating neutral axis, 336–337
inclined end force, 282–283
lateral force, plane of axis, 337
location of neutral axis, 334–336
maximum stresses, 337
pure bending, 280–281
solution comparisons, 339
strain energy, 338–339
tangential stresses, 330–334
deflection of, 387–391
bending moments comparison, 390
displacement solutions comparison, 389
force comparison, 391
thermal stresses, 283–286
Curvilinear coordinates, 18–21
Cyclic loading and crack growth, *see* Crack propagation; Safe-life prediction
Cylinders
critical buckling pressure, 550
elastic plane problems, polar coordinates
open or closed end, 254–255
with shrink fits, 255–262
steady-state radial thermal gradient, 250–254
uniform pressure, 243–249
thick-walled, *see* Thick-walled cylinders, pressurized
thin-walled, *see* Thin-walled cylinders, pressurized
Cylindrical polar coordinates, 14–17
strain compatibility equations, 64–65
strain-displacement relations, 59–61

D

Damage tolerance requirement, 604
Deflection
of circular plates
due to shear, 482–483
large, 483–486
of curved beams
bending moments comparison, 390
discussion, 387–391
displacement solutions comparison, 389
force comparison, 391
elastic-plastic, in beams, 837–839
Deformable bodies, definition of, 1
Deformations
from crack modes, 570–571, 589–590
creep, *see* Creep deformation
extension ratio, 84
natural (logarithmic) strain, 84–85
patterns, in plates, 442–444
plastic, *see* Plastic deformation
strain-displacement relations for large, 83
strain rate, 87
stress-strain relation for large, 146–150
warping of plane sections, 225
De Laval disk, 266–269
Deviatoric strain, 77
Deviatoric stress, 39–41
Dilatation and distortion, 76–78, 112–114
Direct integration method, 448–449
Direct strain, *see* Normal strain
Disks
annular rotating, elastic-plastic
initial yielding, 886–887
partial and full yielding, 887–890
residual stresses, 890
shrink-fitted, 890–892
annular rotating, uniform thickness, 262–264
solid rotating, elastic-plastic
diametral loads, 299–301
initial yielding, 892–893
partial and full yielding, 893–895
residual stresses, 895
solid rotating, uniform strength, 266–269
solid rotating, uniform thickness, 264–266
steady creep in rotating, 785
stress determination
radial thermal gradients, 270–273
uniform strength, solid, 266–269
uniform thickness, annular, 262–264
uniform thickness, solid, 264–266
stress intensity/CCF factors for cracks, 582–583
thermal stresses, 270–273, 479–480
Displacement relations
axisymmetric shells, 493–494
Cartesian coordinates, 54–59
infinitely long beams

concentrated force, 416–418
concentrated moment, 419–420
uniform load, 420–422
large deformations, 83
plane strain, 166, 183
plane stress, 168
polar coordinates
cylindrical, 59–61
plane strain, 183
spherical, 61–62
in rods
Castigliano's theorem, 376–377
deflection of curved, 387–391
Mohr's unit load method, 384–387
spring applications, 391–397
strain energy and, 377
Distortion energy, 656–657
Double-modulus formula for bars, 854–856
Drums and rotors
plane stress determination, 269–270
polar coordinate problems, 269–280
stress intensity/CCF factors for cracks, 582–583
Ductile failure, *see* Plastic deformation
Ductile rupture, *see* Creep deformation, rupture
Ductile vs. brittle materials, 561

E

Edge moment, 475–476
Effective strain
plastic deformation, 691–692
plastic work, 669–674, 679
rate, creep deformation, 781
Effective strain rate, creep deformation, 781
Effective stress
plastic deformation, 691–692
plastic work, 669–674
Elastic analogue and creep analysis, 782
Elastic boundary-value problems
boundary conditions, 156–159
discussion, 155–156
planes, elastic
discussion, 164–165
strain formulation for, 165–167
stress deduction from strain equations, 169–170
stress formulation for, 167–169
stress function, *see* Stress function
Saint-Venant's principle, 159–160
stress solution for pressurized sphere, 161–164
uniqueness needed for solution, 160
Elastic constants
composite, 146–147
generalized Hooke's law, 135–140
relations between, 108–110
Elastic curve due to bending moments, 326–330
Elastic energy theorems
Castigliano's theorems, 132–134, 376–377
complementary energy, 119, 124
stationary potential energy, principle of, 131–132
virtual work, principle of, 129–131
work, principle of, 128–129
Elastic equivalence, 159
Elastic foundations, beams on
discussion, 415–416
infinitely long
concentrated force, 416–418
concentrated moment, 419–420
uniform load, 420–422
semi-infinite, 422–425
short, 425
Elasticity, 101–102
anisotropic, 135–140
elastic limit, 102
modulus of, 108
Elastic limit, 102
Elastic plane problems
algebraic polynomials, Cartesian coordinates
discussion, 205–208
end load, cantilever beam, 220–229
hydrostatic pressure, retaining wall, 209–213
uniformly distributed load, simply supported, 213–220
axially nonsymmetric, polar coordinates
beams, concentrated load, 301–303
curved beams, bending, 274–283
curved beams, thermal stresses, 283–286
disk, solid circular, 299–301
straight boundaries, 295–299
wedges subjected to loads, 286–295
axisymmetric, polar coordinates
cylinders, with shrink fits, 255–262
cylinder, thermal gradient, 250–255
disks, 262–269
drums and rotors, 269–270
thick-walled cylinders, 243–249
trigonometric series, Cartesian coordinates
axial loads, bar, 233–234
discussion, 229–230
equal lateral loads in middle, 232–233
lateral loads, simply supported beam, 232–233
sinusoidal loads, simply supported beam, 230–232
Elastic planes, Cartesian coordinates
boundary-value problem
discussion, 164–165
strain formulation for, 165–167
stress deduction from strain equations, 169–170
stress formulation for, 167–169
stress function, *see* Stress function
discussion, 164–165
problem solving methods, *see* Elastic plane problems
strain formulation, 165–167
stress concentration, small hole
biaxial loading, 308
example problems, 308–311
formula derivation, 303–308
local yielding situations, 303
stress deduction from strain equations, 110–112, 169–170
stress function
derivation, 170–171, 182–183
example problems, 172–176
plane strain, 171–174
plane stress, 174

thermoelastic plane strain, 174–175
thermoelastic plane stress, 175–182
Elastic planes, polar coordinates
beams, curved
bending of, 274–283
thermal stresses in, 283–286
cylinders
compounding by shrink fits, 255–262
pressurized thick-walled, 243–249
thermal stresses, 250–255
disks, rotating
annular, uniform thickness, 262–264
solid, diametral loads, 299–301
solid, uniform strength, 266–269
solid, uniform thickness, 264–266
drums, rotating, 269–270
problem solving methods, *see* Elastic plane problems
rotors, 269–270
thermal stresses
beams, 283–286
cylinders, 250–255
wedges under loads
bending moment at vertex, 292–295
force acting along axis, 286–289
force inclined to axis, 290–292
force perpendicular to axis, 289–290
Elastic-plastic fracture mechanics, nonlinear
crack opening displacement, 627–629
CTOD and structural design, 630–632
experimental determination of COD, 629–630
experimental determination of *J*, 619–622
fracture estimation using *J*, 623–627
J-integral, 616–619
Elastic-plastic problems
bending of beams, curved, 849–853
bending of beams, plane stress
combined bending and tension, 831–833
deflection and shear stresses, 837–839
plastic hinges, 833–837
springback, 829–830
strain-hardening effects, 839–843
yielding, 828–830
bending of plates
biaxial, 843–849
plane strain, 820–828
buckling of bars, axial compression, 853–856
rotating disk
annular, 886–892
solid circular, 892–896
thick-walled pressurized cylinders, 865–886, 896–899
torsion in bars, 856–865
Elastic shakedown, 641, 896–899
Elastic strain energy, *see* Strain energy
Elastic stress-strain relations, *see* Stress-strain relations
Electric resistance strain gauge, 80
Elementary beam theory
longitudinal normal stress, 216–217
wedge bending stress, 290
Elliptical holes, stress concentration in
example problems, 568–570
polar coordinates, 568
stress distribution, 565–568
Energy approach, *see also* Strain energy
bending of plates
direct integration solution, 448–449
strain energy expression, 452
buckling of columns, 403–410
buckling of plates, 540–542
complementary energy, 119, 124
displacements in rods
Castigliano's theorems, 376–377
deflection of curved beams, 387–391
Mohr's unit load method, 384–387
spring applications, 391–397
minimum potential energy, 131–132
Engineering strain, 53, 84
EPRI scheme, 624, 626
Equilibrium approach
buckling of columns, 398–403
plane strain
Cartesian coordinates, 166
polar coordinates, 183
plane stress, 168
plastic zone equations, 590–591
plates
bending of, 448–449, 452, 466–468
stresses in, 445–448
shear stresses in rods, 345–347
shells, axisymmetric, 499–507
static, 2–3
Cartesian coordinates, 10–14
curvilinear coordinates, 18–21
cylindrical polar coordinates, 14–17
moment, 12
spherical polar coordinates, 17–18
stress differential, 12
tetrahedron element, 21–24
Euler load, 399, 401, 406
Euler method, 82, 83
Extension ratio, 84

F

Factor of safety, *see* Safe-life prediction
Fatigue life, *see* Safe-life prediction
Fiber-reinforced composites, 140–145
Flat compression spring, 394
Flat torsion spring, 395–396
Flow curve determination, 674–678
Flow rule, *see* Levy-Mises flow rule
Forces, internal, 3–6
Forman equations, 600–601
Fourier's series, 231–232
Fracture mechanics
during bending moment, 850–851
brittle vs. ductile, 561
crack growth
crack propagation, *see* Crack propagation
safe-life prediction, *see* Safe-life prediction
elastic stress field at crack tip, 566–570
Griffith energy criterion

critical energy release rate, 564
microcrack effects, 562
plastic work added, 565
strain energy per unit thickness, 563
mixed-mode loading, crack deformation, 589–590
nonlinear elastic-plastic
crack opening displacement, 627–629
CTOD and structural design, 630–632
experimental determination of COD, 629–630
experimental determination of *J*, 619–622
fracture estimation using *J*, 623–627
J-integral, 616–619
plastic zone
equilibrium equations, 590–591
example problems, 594–595
plastic radii, 593
shape of, 591–592
specimen thickness, 592
safety factor, *see* Safe-life prediction
stress concentration, elliptical holes, 565–566
example problems, 568–570
polar coordinates, 568
stress distribution, 565–568
stress corrosion cracking, 613–616
stress intensity factor
circular rods and tubes, 579–581
configuration correction factor, 574
crack propagation, 599
cracks from circular holes, 577–579
disks and drums, solid rotating, 582–583
example problems, 571–574, 583–587
fracture criterion equation, 574
fracture properties, 572
interim, in toughness testing, 596
modes of crack deformation, 570–571
plane strain fracture toughness, 571
plates under bending, 577–579
plates under tensile loading, 575–578
relation to critical energy release rate, 570
thick-walled cylinders, pressurized, 581–582
superposition for load conditions, 587–589
toughness testing
interim stress intensity factor, 596
load-displacement records, 596
methodology, 595–596
Free torsion
state of stress, 9
stress tensor, 7, 9, 72–73
twisting moment and warping, 359–363

G

Gauge rosettes, strain, 80–81
Glossary, stress functions
Cartesian coordinates, 195–197
polar coordinates, 197–200
Gravity membrane stresses
conical container, top edge free, 500–501
example problems, 501–504
hemispherical tank, top edge free, 499–500
spherical container, skirt support, 504–507
Green strains, 83, 86
Green yield function, porous solids, 716, 717
Griffith energy criterion
critical energy release rate, 564
microcrack effects, 562
plastic work added, 565
strain energy per unit thickness, 563
Gun barrels
autofrettage process, 881
shrink fit compounding, 255
Gurson yield function, porous solids, 716–718

H

Helical compression spring, 391–393
Hill's yield criterion, 693, 694, 697
Hoff's rupture time, 803, 805
Holes, stress concentration in
circular
biaxial loading, 308
example problems, 308–311
formula derivation, 303–308
local yielding situations, 303
elliptical
example problems, 568–570
polar coordinates, 568
stress distribution, 565–568
Homogeneity, 102–103
Hooke's law
generalized
anisotropic elasticity, 135–140
fiber-reinforced composites, 140–145
inverse form, 110–112, 158
plane strain
Cartesian coordinates, 167
polar coordinates, 183–184
plane stress, 168
sample problems, 105–108
shear loading, 103
strain energy for a solid
complementary energy and, 124
energy components, 24
example problems, 125–128
total density derivation, 122–123
stress in terms of strain, 110–112
stress state, uniaxial, 647
superposition, principle of, 105, 123
tension, uniaxial, 102–103
triaxial loading, 104–105
Hoop strain, 67
Hoop stress, 303, 535
Hydrostatic stress
definition of, 39–41
plane strain, 76, 209–213
volumetric strain and, 108

I

Inelastic behavior, *see* Plastic deformation
Infinitely long beams, displacements of
concentrated force, 416–418

concentrated moment, 419–420
uniform load, 420–422
Internal stiffening webs, 372–373
Isochronous stress-strain curves
column creep buckling application, 789
strain vs. time, 768
Isotropic hardening
vs. kinematic hardening, 700–703
yield surface representation, 678–679
Isotropic solid and strain energy, 116–122
Isotropy, definition of, 103

J

J-integral
discussion, 616–619
experimental determination of, 619–622
fracture estimation using, 623–627

K

Kinematic hardening rule
discussion, 700
flow rule, 709–715
uniaxial behavior, cyclic loading, 701–709
vs. isotropic hardening, 700–703
yield function, 709–715

L

Lagrangian method, 82, 83
Lame's constants, 111, 136, 137
Larson-Miller parameter, 770
LEFM (Linear Elastic Fracture Mechanics), *see* Fracture mechanics
Levy-Mises flow rule, 667–668
effective stress/strain, plastic work, 669–670
kinematic hardening and, 709–715
planar isotropy and, 697
plastic deformation, 665, 688, 710–711
and porous solids, 719–720
for strain-rate sensitivity, 761–763
Limit plastic strains, 748
Linear Elastic Fracture Mechanics (LEFM), *see* Fracture mechanics
Linear strain, 49–50, 69
Linear thermal expansion, coefficient of, 114
Load cycles to failure, *see* Safe-life prediction
Ludwik's law, 647, 649

M

Maximum shear strain, 74–76
Maximum shear stress, 34–37
Maximum stress criterion, *see* Tresca yield criterion
Mean strain, 77
Mean stress, 39–41
Measurement of strain, 80–81
Membrane stresses
gravity
conical container, top edge free, 500–501
example problems, 501–504
hemispherical tank, top edge free, 499–500
spherical container, skirt support, 504–507
pressure vessels
curved end, 527–529
discussion, 523–524
example problems, 531–534
flat end, 524–527
hemispherical end, 529–534
maximum, table of, 531
in shells
in circular cylindrical, 491
in conical, 491–492
in cylindrical storage tanks, 534–537
gravity loading, 499–507
in pressure vessels, 523–534
in pressurized containers, 490–493
radial displacement in, 493–499
thermal gradient effects, 537–540
toroidal, 492
from uniform internal pressure, 486–490
Metal powder compacts, plastic deformation, 722–723
Minimum potential energy
columns, buckling of, 403–410
plates, bending of, 448–449, 452
principle of, 131–132
Modulus
bulk, 108
of elasticity, 108, 147
of rigidity, 103, 140
shear, 103, 105, 108
tangent, formula for bars, 854, 856
Young's, 103, 140
Mohr's circle
sign convention, 79
of strain, 78–79
of stress, 27–28
Mohr's unit load method
displacements in rods, 384–387
spring applications, 391–393
Moment, bending
bar stresses, 320
beam, curved, 278, 282
in deflection of curved beams, 390
elastic curve in rods, 326–333
membrane stresses in pressure vessels, 527–531
nonlinear strain-hardening material, 841
plastic deformation, 822–823
plastic hinge formation, 834
and plastic zone, 850–851
plates
circular, 847–848
rectangular, 843–845
during recovery, 825–827
shape factor and, 823–824
at wedge vertex, 292–295
and yield initiation, 822, 828–830
Moment, edge, 475–476
Moment equilibrium, 12
Moment, twisting, *see* Twisting moment

N

Natural (logarithmic) strain, 84–85, 644, 645
Navier equations, 156, 170
Necking
local
biaxially stretched sheet, 748
wide strip, 738–739
neck-free elongations, 763
stresses, tensile bar, 734–738
round bar, 742–745
wide strip, 745–746
Neutral axis
bending stresses, 323
determination of, 336
errors from approximating, 336–337
location of, 334–336
Nonlinear elastic-plastic fracture mechanics
crack opening displacement, 627–629
CTOD and structural design, 630–632
experimental determination of COD, 629–630
experimental determination of *J*, 619–622
fracture estimation using *J*, 623–627
J-integral, 616–619
Nonsymmetric axial loads, *see* Axially nonsymmetric problems, polar coordinates
Normal anisotropy, plastic deformation, 689–690
Normal strain, 49–50, 69
Normal stresses, definition of, 4
Norton-Baily creep law, 771, 772, 781, 783

O

Oblique bending, 324
Octahedral shear strains, 75
Octahedral shear stress, 38–39, 656
Out-of-roundness effects, 550

P

Paris law, 600, 604, 609, 611
Pipes, *see* Rods; Shells
Planar isotropy, plastic deformation
nonquadratic criterion, 697–700
rotational symmetry, 692–697
Planes, elastic, *see* Elastic plane problems; Elastic planes, Cartesian coordinates; Elastic planes, polar coordinates
Plane strain, 9, 49, 72
bending of plates
discussion, 820
elastic state, 820–821
initial yielding, 822
partial and full yielding, 822–825
residual stresses, 825–828
springback, 826–828
and crack propagation, 564
fracture toughness, 571–572
Hooke's law
Cartesian coordinates, 167
polar coordinates, 183–184
hydrostatic pressure example, 209–213
plastic zone size, 591
and stress function, 174
thermal gradient, cylinder, 250–253
thermoelastic, 174–175
Plane stress, 8, 72; *see also* Stress function
bending of beams
combined bending and tension, 831–833
deflection and shear stresses, 837–839
plastic hinges, 833–837
springback, 829–830
strain-hardening effects, 839–843
yielding, 828–830
and crack propagation, 564
at a cut, 5–6
Hooke's law, 168
maximum shear, 35
plastic zone size, 591
polar coordinates equations, 185–186
thermal gradient, cylinder, 253–254
thermoelastic, 175–176
transformation law, 27–28
uniformly distributed load example, 213–220
Plastic deformation
anisotropic materials
effective strain increment, 691–692
effective stress, 691–692
flow rule, 688
normal anisotropy, 689–690
planar isotropy, 692–697
planar isotropy, nonquadratic criterion, 697–700
strain ratio, 689
through-thickness yield strength, 688, 689–690
yield criterion, 687–688
basic assumptions, 642–643
Bauschinger effect, 643, 670
creep, *see* Creep deformation
discussion, 641–642
flow curve determination, 674–678
flow rule, 665, 688
isotropic hardening, 678–679
kinematic hardening rule
discussion, 700
flow rule, 709–715
triaxial behavior, 709–715
uniaxial behavior, cyclic loading, 701–709
vs. isotropic hardening, 700–703
modes of, *see* Creep deformation; Plastic instability; Superplasticity
necking, *see* Necking
normality, principle of, 667–668
path independence, 679–683
plastic potential, principle of, 668
plastic work
effective strain increment, 669–674
effective stress, 669–674
porous solids
discussion, 715–716
flow rule, 719–720
metal powder compacts, 722–723

void growth, 720–722
yield functions, 716–719
strain definition for large, 644–646
strain hardening, simple tension, 646–647
stress-strain curves
example problems, 650–652
idealized, 652–653
power laws, 647–648
Ramberg-Osgood law, 649
rise in flow stress, 650
strain vs. time, 768
stress-strain relations
basic assumptions, 642–643
elastic-plastic, 683–686
flow rule, 664–667
uniqueness, 679–683
yield criteria, anisotropic materials, 687–688
and elastic distortion energy, 656–657
example problems, 657–658
experimental verification of, 662–664
state of deformation, 653–654
Tresca yield criterion, 658–659
von Mises vs. Tresca, 659–662, 663
von Mises yield criterion, 654–657
yield function, 654
Plastic expansion process, thick-walled cylinders, 875–876
Plastic hinges, 833–837
Plastic instability, *see also* Plastic deformation
biaxial stretching
bulging, circular sheet, 749–754
flat sheet, 746–749
discussion, 733–734
limit tensile strain, imperfect bar, 740–741
necking
local, biaxially stretched sheets, 748
local, wide strip, 738–739
neck-free elongations, 763
stresses, round bar, 742–745
stresses, tensile bar, 734–738
stresses, wide strip, 745–746
thin-walled pressurized containers
cylinder, closed end, 756–758
example problems, 758–760
spherical shell, 754–756
Plastic strain, effective, 669–674, 679
Plastic work
effective strain increment, 669–674, 679
effective stress, 669–674
Griffith energy criterion, 565
Plastic zone
during bending moment, 850–851
equilibrium equations, 590–591
example problems, 594–595
plastic radii, 593
shape of, 591–592
specimen thickness, 592
Plates
bending of, *see* Plates, bending of flat
buckling of
circular, axisymmetric, 544–545
rectangular, uniformly compressed, 540–544
deflection of circular
due to shear, 482–483
large, 483–486
deformation pattern, 442–444
direct integration method, 448–449
equilibrium equations, 445–448
fracture stress from cracks, 571
state of stress, 441–442
stress intensity/CCF factors
under bending, 579
cracks from circular holes, 577–579
tensile loading, 575–578
stress resultants, 444–445
thermal stresses in circular
axisymmetric radial gradient, 480–482
disk with clamped edges, gradient across, 480
disk with free edges, gradient across, 479
Plates, bending of flat
annular circular
simply supported, edge moments, 475–476
simply supported, shearing force at inner edge, 477–478
superposition for load conditions, 478–479
biaxial, elastic-plastic
circular, 847–849
rectangular, 843–846
energy approach
direct integration method, 448–449
strain energy expression, 452
plane strain, elastic-plastic
discussion, 820
elastic state, 820–821
initial yielding, 822
partial and full yielding, 822–825
residual stresses, 825–828
springback, 826–828
radius of curvature, 820, 843–845, 847
rectangular
pure bending, 448–449
strip method, approximate, 465–466
uniformly loaded, all edges clamped, 458–464
uniformly loaded, all edges supported, 452–458
solid circular
clamped, center force, 472–474
clamped, uniform pressure, 469–472
discussion, 466–468
simply supported, center force, 474–475
simply supported, uniform pressure, 468–469
stress intensity factor, 579
thermal gradient
clamped edges, 450
free edges, 449–450
simply supported edges, 451–452
thermal radial gradient effects, 449–452
Poisson's ratio, 103, 108, 140, 147
Polar coordinate problems
axially nonsymmetric, elastic plane
beams, concentrated load, 301–303
curved beams, bending, 274–280
curved beams, thermal stresses, 283–286
disk, solid circular, 299–301

straight boundaries, 295–299
wedges subjected to loads, 286–289
axisymmetric, elastic plane
cylinders, with shrink fits, 255–262
disks, 262–273
thick-walled cylinders, 246–255
circular plates, 466–468, 544–545
glossary of equations, 197–200
plane strain, 183–184
stresses at a crack tip, 568
stress function derivation
axisymmetric plane, 187–195
plane strain, 183–185
plane stress, 185–186
Porous solids, plastic deformation
discussion, 715–716
flow rule, 719–720
metal powder compacts, 722–723
void growth, 720–722
yield functions, 716–719
Power laws, plastic deformation, 647–648
Prandtl-Reuss relations, 683
Prandtl stress function, 363
Pressure vessels, *see* Pressurized containers
Pressurized containers, *see also* Thick-walled cylinders, pressurized; Thin-walled cylinders, pressurized
bending and membrane stresses
curved end, 527–529
discussion, 523–524
example problems, 531–534
flat end, 524–527
hemispherical end, 529–534
maximum, table of, 531
in shells, 491–492
bending of shells
discussion, 507–513
long pipe, rigid flange at end, 513–517
long pipe, uniform circumferential load, 521–523
long pipe, uniform radial compression, 518–521
pressure vessel, end closures, 523–534
short pipe, rigid flanges both ends, 517–518
critical buckling pressure, 550
cylindrical storage tanks, 534–537
shakedown limit, 896–899
sphere, boundary-value in terms of stresses, 161–164
Prestressed compound cylinder, 255
Principal strains, 74–76
Principal stresses, 29–34
Principle of minimum potential energy, 131, 403–410
Principle of normality, 667–668
Principle of plastic potential, 668
Principle of superposition
for annular circular plate loading conditions, 478–479
applied to Hooke's law, 105, 123
fracture mechanics combined loading, 587–589
principle of, 105
Proportionality constant, 103
Proportional limit, 102
Pure shear, 38–39

R

Radial stress, 332
Radial thermal gradient, *see* Thermal gradient
Radius of curvature
circular plate in bending, 847
equilibrium equations, curvilinear coordinates, 18–20
and initial yielding, 822
for partial yielding, 823
plates, bending of, 820, 843–845
and springback, 826, 829
Rainflow counting, 609
Ramberg-Osgood law, 649
Rayleigh-Ritz method, 131, 403–410, 448–449
Reference stress method, 791–796
Regions of crack propagation
example problems, 601–603
nonpropagating, (i), 599
steady, (ii), 599–601
unstable growth rate, (iii), 601
Residual stresses
autofrettage process, 880–884
disks, elastic-plastic
annular rotating, 890
solid rotating, 895
plane strain bending of plates, 825–828
unloading and, 832–833
Rigid bodies, definition of, 1
Rigidity, modulus of, 103, 140
Rods, *see also* Bars; Beams; Columns
beams on elastic foundations
discussion, 415–416
infinitely long, 416–422
semi-infinite, 422–425
short beams, 425
bending of
curved beams, 330–341
elastic curve, 326–330
stresses, 322–326
thermoelastic stresses, straight bars, 341–345
unsymmetrical, 324
buckling of, *see* Rods, buckling of
definition and types, 319
displacements in, energy approach
Castigliano's theorem, 376–377
deflection of curved beams, 387–391
Mohr's unit load method, 384–387
spring applications, 391–397
shear stresses in
circular solid section, 349–352
equilibrium equations derivation, 345–347
rectangular solid section, 347–349
thin-walled closed sections, 357–359
thin-walled open sections, 352–357
springs, *see* Springs
stress intensity/CCF factors for circular cracks, 579–581
stress resultants
discussion, 319–320
sign convention, 321–322
torsion of bars
angle of twist, 360, 371

comparison of types, 367–368
discussion, 359–360
end constraint effects, 374–376
example problems, 860–861
free torsion, 360–363
hollow circular section, 858–860
internal stiffening webs, 372–373
open vs. closed, 372
shear stress components, 856–857
solid circular section, 363–364, 374–376, 857–858
solid elliptical section, 365–366
solid rectangular section, 366–367
steady creep in shafts, 785–788
and tension, combined, 861–865
thin-walled closed sections, 370–372
thin-walled open sections, 368–369, 375–376
thin-walled tubes, 861
Rods, buckling of
beam-columns, 410–412
columns
clamped end/free end, 400
clamped end/pinned end, 400
creep, 788–791
discussion, 398
ends clamped, 401
energy method, minimum potential, 403–410
example problems, 402–403
pinned ends, 398–400
lateral, of beams, 412–415
Rosettes, strain gauge, 80–81
Rotating disks, *see* Disks
Rotational symmetry, *see* Planar isotropy, plastic deformation
Rotation, angle of, 51
Rotors and drums
plane stress determination, 269–270
polar coordinate problems, 269–280
stress intensity/CCF factors for cracks, 582–583
Rubber, stress-strain relations, 146–150
Rupture, creep
brittle, uniaxial stress, 807–808
damage concept, 805–807
ductile, uniaxial stress, 803–805
example problems, 808–809
testing and data, 769
time
brittle, 805–808
ductile, 803–805

S

Safe-life prediction
crack closure, 612
example problems, 604–608, 612–613
growth of short cracks, 611–612
margin of safety, 609
mixed-mode crack growth, 611
S-N curve correlation, 611
stable crack growth, 603–604
stages of fatigue life, 603
variable amplitude loading, 609–611
Saint-Venant's principle
applied to end loads, 221
boundary condition modifications, 159–160
free torsion, 360–363
hole stresses, 303
in simple supported beam, 216
warping function, 361, 366
SCC (Stress corrosion cracking), 613–616
Semi-infinite beams, displacement in, 422–425
Semi-inverse method, 160
Shakedown limit, 641, 896–899
Shape factor, bending moment and, 823–824
Shear center, thin-walled open sections, 355–357
Shear flow lines, 346
Shearing force, bar stresses, 320
Shear modulus, 103, 105, 108
Shear strain, 50–51
finite, 83
maximum, 74–76
octahedral, 75
principal, 74–76
Shear stress, 4
cantilever beam, end load example
displacements, 222–225
stresses, 220–221
and deflection, 482–483, 837–839
loading, Hooke's law, 103
maximum, 34–37
octahedral, 38–39, 656
in rods
circular solid section, 349–352
equilibrium equations derivation, 345–347
rectangular solid section, 347–349
thin-walled closed sections, 357–359
thin-walled open sections, 352–355
trajectories, 346
twisting moment in bars
discussion, 359–360
end constraint effects, 374–376
internal stiffening webs, 372–373
Saint-Venant's free torsion, 360–363
solid circular section, 363–364
solid elliptical section, 365–366
solid rectangular section, 366–367
thin-walled closed sections, 370–372
thin-walled open sections, 368–369
Shells
bending of thin-walled pressurized cylinders
discussion, 507–513
long pipe, rigid flange at end, 513–517
long pipe, uniform circumferential load, 521–523
long pipe, uniform radial compression, 518–521
pressure vessel, end closures, 523–534
short pipe, rigid flanges both ends, 517–518
buckling of
critical pressure, 550
out-of-roundness effects, 550
thin-walled cylinders, external uniform pressure, 546–550
membrane stresses
in circular cylindrical, 491

in conical, 491–492
in cylindrical storage tanks, 534–537
gravity loading, 499–507
in pressure vessels, 523–534
in pressurized containers, 490–493
radial displacement in, 493–499
thermal gradient effects, 537–540
in toroidal, 492
from uniform internal pressure, 486–490
state of stress, 441–442
Shrink fits for cylinders
disks, 890–892
elastic plane stress problems, 255–262
gun barrels, 255
Sign convention
Cartesian coordinates, 8
Mohr's circle, 79
for strain, 51
stress resultants, 321–322
Sinusoidal loads, stress function example, 230–232
Skew bending, 324
Skirt supports, 504–507
S-N curves, safe-life prediction correlation, 611
Spherical polar coordinates, 17–18
strain compatibility equations, 65–67
strain-displacement relations, 61–62
Spiral helical compression spring, 393–394
Springback
plane strain bending of plates, 826–828
plane stress bending of beams, 829–830
radius of curvature and, 826, 829
Springs
displacements in rods, energy approach
example problems, 396–397
flat compression, 394
flat torsion, 395–396
helical compression, 391–393
spiral helical compression, 393–394
stiffness
flat compression, 394
flat torsion, 395–396
helical compression, 392–393
spiral helical compression, 393–394
Statically indeterminate conditions, *see* Compatibility, strain
Static equilibrium, 2–3
Cartesian coordinates, 10–14
curvilinear coordinates, 18–21
cylindrical polar coordinates, 14–17
moment, 12
spherical polar coordinates, 17–18
stress differential, 12
Stationary creep, 780–784
Stationary potential energy, principle of, 131–132
Steady creep
beams under bending, 773–776
circular shafts under torsion, 785–788
rotating disks, 785
thin-walled cylinders, 777–780
Stiffening webs, internal, 372–373
Straight boundaries, polar coordinates
concentrated line load
force acting along boundary, 296–297
force acting inclined to boundary, 297
force acting normal, 295–296
uniformly distributed line load, 297–299
Strain
compatibility conditions
Cartesian coordinates, 62–64
cylindrical polar coordinates, 64–65
spherical polar coordinates, 65–67
compatibility equations
plane strain, 167
plane stress, 169
polar coordinates, plane strain, 184–185
compatibility fields, 62
definition, 2
infinitesimal, 2
large, 644–646
deviatoric matrices, 77
dilatation and deviations, 76–78
displacement relations
Cartesian coordinates, 54–59
cylindrical polar coordinates, 59–61
for large deformations, 83
spherical polar coordinates, 61–62
effective
plastic deformation, 691–692
plastic work, 669–674, 679
rate, creep, 781
elastic plane formulation, 165–167
energy, *see* Strain energy
finite, 82–87
gauge rosettes, 80–81
hardening, *see* Strain-hardening
hoop, 67
under hydrostatic pressure, 76
mean, 77
measurement of, 80–81
Mohr's circle of, 78–79
natural (logarithmic), 84–85, 644, 645
normal, 49–50, 69
plane, *see* Plane strain
principal
maximum, 74–76
octahedral, 75
rate during deformation, 87–90
ratio, plastic deformation, 689
shear, 50–51
finite, 83
maximum, 74–76
octahedral, 75
principal, 74–76
state of
axial symmetry, 72
example problems, 73–74
free torsion, 72–73
plane, 9, 72
stress determined from, 110–112, 169–170
tensor of the second rank, 24, 67–71

transformation law, 69–70, 74
volumetric, 52–54
Strain-displacement relations
axisymmetric shells, 493–494
plane strain, 166
plane stress, 168
polar coordinates, plane strain, 183
Strain energy
alternative forms, 123
and angle of twist, 371
bending of rectangular plates
pure, 448–449
strip method, approximate, 465–466
uniformly loaded, all edges clamped, 458–464
uniformly loaded, all edges supported, 452–458
buckling, column, critical load, 403–410
buckling, plates
circular, polar coordinates, 544–545
clamped rectangular, 542–543
discussion, 540–542
simply supported, 543–544
in curved beams, 338–339
density, 118
and displacement, 377
Griffith energy criterion, 563
Hooke's law for a solid
complementary energy and, 124
energy components, 24
example problems, 125–128
total density derivation, 122–123
per unit thickness, in fracture, 563
stress-strain relations
complementary energy, 119–120
density, 118
due to shear stresses only, 117–118
equations for, 116–117
example problems, 120–122
uniaxial stress state, 647
Strain-hardening
bending moment in nonlinear materials, 841
isotropic hardening rule, 679
plane stress bending of beams, 839–843
plastic deformation, 646–647
pressurized thick-walled cylinder, 884–885
vs. time hardening, 798–802
Strain-Hardening rule, 798–802
Strain-rate-dependent material, *see* Superplasticity
Strain ratio, plastic deformation, 689
Stress
bending
in pressure vessels, 523–524
in rods, 322–326, 337
boundary-value problems in terms of, 161–164
Cartesian coordinates, 6–8
concentration
circular holes, 303–311
elliptical holes, 565–570
corrosion cracking, 613–616
definition of, 2
deviatoric, 39–41
differential equations for, 12
effective
plastic deformation, 691–692
plastic work, 669–674
elastic limit, 102
equilibrium equations for, 10–14, 445–448
field, at crack tip, 566–570
function, for elastic plane problems, *see* Stress function
hoop, 303, 535
hydrostatic
definition of, 39–41
plane strain, 76, 209–213
volumetric strain and, 108
intensity factor
circular rods and tubes, 579–581
configuration correction factor, 574
crack propagation, 599
cracks from circular holes, 577–579
disks and drums, solid rotating, 582–583
example problems, 571–574, 583–587
fracture criterion equation, 574
fracture properties, 572
interim, in toughness testing, 596
modes of crack deformation, 570–571
plane strain fracture toughness, 571
plates under bending, 577–579
plates under tensile loading, 575–578
relation to critical energy release rate, 570
thick-walled cylinders, pressurized, 581–582
maximum shear, 34–37
mean, 39–41
membrane, in shells
in circular cylindrical, 491
in conical, 491–492
in cylindrical storage tanks, 534–537
gravity loading, 499–507
in pressurized containers, 490–493
radial displacement in, 493–499
thermal gradient effects, 537–540
toroidal, 492
from uniform internal pressure, 486–490
in metal plasticity, 642–643
Mohr's circle of, 27–28, 78–79
octahedral shear, 38–39, 656
plane, *see* Plane stress; Stress function
principal, 29–34
proportional limit, 102
radial, 332
residual
autofrettage process, 880–884
disks, elastic-plastic, 890, 895
plane strain bending of plates, 825–828
unloading and, 832–833
shear, *see* Shear stress
state at a point
axial symmetry, 9
curvilinear coordinates, 18–21
cylindrical polar coordinates, 14–17
determining for a cut plane, 5–6
free torsion, 9
internal forces, 3
plane, 8

spherical polar coordinates, 17–18
stress tensor, 7, 9
stress vector, 4
static equilibrium equations
Cartesian coordinates, 10–14
curvilinear coordinates, 18–21
cylindrical polar coordinates, 14–17
moment, 12
spherical polar coordinates, 17–18
stress differential, 12
tangential, 330–334
tensor, 7, 9, 72–73
in terms of strain, 110–112, 169–170
thermal
curved beams, 283–286
cylinders, 250–255
plates, circular, 479–482
transformation law, 21–28
vector, 4–5
Stress function
algebraic polynomials used to solve
discussion, 205–208
end load, cantilever beam, 220–229
hydrostatic pressure, retaining wall, 209–213
uniformly distributed load, simply supported, 213–220
axially nonsymmetric, polar coordinates
beams, concentrated load, 301–303
curved beams, bending, 274–283
curved beams, thermal stresses, 283–286
disk, solid circular, 299–301
straight boundaries, 295–299
wedges subjected to loads, 286–295
axisymmetric, polar coordinates
cylinders, with shrink fits, 255–262
cylinder, thermal gradient, 250–255
disks, 262–269
drums and rotors, 269–270
thick-walled cylinders, 243–249
derivation, 170–171, 182–183
example problems, 176–182
glossary of equations
Cartesian coordinates, 195–197
polar coordinates, 197–200
planes, axisymmetric
with centrifugal body forces, 189–191
with radial temperature gradient, 191–195
without body forces, 187–188
plane strain, 171–174, 183–185
plane stress, 174, 185–186
thermoelastic formulation
plane strain, 174–175
plane stress, 175–176
trigonometric series used to solve
axial loads, bar, 233–234
discussion, 229–230
equal lateral loads in middle, 232–233
lateral loads, simply supported beam, 232–233
sinusoidal loads, simply supported beam, 230–232
Stress matrix, *see* Stress, tensor
Stress relaxation, creep deformation, 796–798
Stress resultant vector, *see* Stress, vector
Stress-strain relations
curves, plastic deformation
example problems, 650–652
idealized, 652–653
power laws, 647–648
Ramberg-Osgood law, 649
rise in flow stress, 650
strain vs. time, 768
deformation, large elastic, 146–150
dilatation and distortion, 112–114
elastic constants
anisotropic materials, 135–140
composite, 146
relations between, 108–110
elastic energy theorems
Castigliano's, 132–134, 376–377
complementary energy, 119–120
least work, principle of, 134
stationary potential energy, principle of, 131–132
virtual work, principle of, 129–131
work, principle of, 128–129
elasticity, 101–102
homogeneity, 102–103
Hooke's law
example problems, 105–108
generalized, 135–145
inverse form, 110–112, 158
plane strain, 167, 183–184
plane stress, 168
shear loading, 103
strain energy for a solid, 122–128
superposition, principle of, 105, 123
triaxial loading, 104–105
uniaxial stress state, 647
uniaxial tension, 102–103
isotropy, 103
plastic deformation
basic assumptions, 642–643
elastic-plastic, 683–686
flow rule, 664–667
reloading, 647
strain energy
complementary energy, 119–120
density, 118
due to shear stresses only, 117–118
equations for, 116–117
example problems, 120–122
thermoelastic, 114–116, 537–540
Stress tensor, 7, 9, 72–73
Stress vector, 4–5
Strut, definition of, 319
Superplasticity, *see also* Plastic deformation
forming time, bulged circular sheet, 765–767
limit tensile strains, 764–765
neck-free elongations, 763
strain-rate dependent behavior, 760–763
Superposition
for annular circular plate loading conditions, 478–479
applied to Hooke's law, 105, 123
fracture mechanics combined loading, 587–589
principle of, 105

Surface free energy, 564–565
Surface tractions, definition of, 1

T

Tangential stresses, 330–334
Tangent modulus formula for bars, 854, 856
Temperature effects, *see* Thermal gradient
Tension and bending, combined in beams, 831–833
Tension and torsion, combined in bars
  hollow circular section, 862–863
  solid circular section, 861–862
  thin-walled cylinder, uniform thickness, 863–865
Tensor of the second rank, 24, 67–71
Tetrahedron element, equilibrium equation for, 21–24
Thermal expansion, coefficient of linear, 114
Thermal gradient
  axial, in cylinder walls, 537–540
  bending of bars, 341–345
  and creep analysis, 796–798
  cylinders, polar coordinates
    open or closed end, 254–255
    plane strain, 250–253
    plane stress, 253–254
  effect on plates in bending
    clamped edges, 450
    free edges, 449–450
    simply supported edges, 451–452
  membrane stresses in shells, 537–540
  pressurized thick-walled cylinder, 250–255, 885–886
  stresses in circular plates
    axisymmetric radial gradient, 480–482
    disk, rotating, 270–273
    disk with clamped edges, gradient across, 480
    disk with free edges, gradient across, 479
  stress relaxation, 796–798
  stress-strain problems, 191–195
Thermal stresses
  curved beams, 283–286
  cylinders, 250–255
  plates, circular
    axisymmetric radial gradient, 480–482
    disk with clamped edges, gradient across, 480
    disk with free edges, gradient across, 479
Thermoelastic effects
  plane strain, 174–175
  plane stress, 175–182
  stresses in bending of bars, 341–345
  stress-strain relations, 114–116
Thick-walled cylinders, pressurized
  discussion, 865–866
  elastic problem solution, sphere, 161–164
  example problems, 876–879
  residual stresses, 880–884
  shakedown limit, 896–899
  stationary creep, 780–784
  strain-hardening effects, 884–885
  stress intensity/CCF factors, 581–582
  thermal gradient, 250–255, 885–886
  uniform pressure
    external only, 247–248
    internal and/or external, 243–246
    internal only, 246, 248–249
  uniform pressure, polar coordinates
    external only, 247–248
    internal and/or external, 243–246
    internal only, 246, 248–249
  yielding
    elastic region stresses, 870
    example problems, 868–870, 871–873
    full, 874–875
    plastic expansion process, 875–876
    plastic region stresses, 870–871
    radial displacement, 873–874
    Tresca yield criterion, 866, 867
    von Mises yield criterion, 867
Thin-walled closed sections in torsion, 370–372
Thin-walled cylinders, pressurized
  bending of
    long pipe, rigid flange at end, 513–517
    long pipe, uniform circumferential load, 521–523
    long pipe, uniform radial compression, 518–521
    pressure vessel, end closures, 523–534
    short pressurized pipe, rigid flanges both ends, 517–518
  buckling of shells, 546–560
  plastic instability
    closed end, 756–758
    example problems, 758–760
    spherical shell, 754–756
  steady creep in, 777–780
  stress-strain relations, 509
  torsion and tension combined, 863–865
Thin-walled open sections
  equations, 352–355
  shear center, 355–357
  torsion, 368–369
Tie, definition of, 319
Time-Hardening rule, 798–802
Time rate of strain, 88
Torsion of bars and rods
  angle of twist, 360, 371
  comparison of types, 367–368
  discussion, 359–360
  end constraint effects
    discussion, 373–374
    solid sections, 374–375
    thin-walled sections, 375–376
  example problems, 860–861
  free torsion
    state of stress, 9
    stress tensor, 7, 9, 72–73
    twisting moment and warping, 359–363
  hollow circular section, 858–860
  internal stiffening webs, 372–373
  open vs. closed, 372
  Prandtl stress function, 363
  shear stress components, 856–857
  solid circular section, 363–364, 374–375, 857–858
  solid elliptical section, 365–366
  solid rectangular section, 366–367
  steady creep in circular shafts, 785–788

tension and, combined
hollow circular section, 862–863
solid circular section, 861–862
thin-walled cylinder, uniform thickness, 863–865
thin-walled closed sections, 370–372
thin-walled open sections, 368–369, 375–376
thin-walled tubes, 861
Torus, 492
Toughness testing, fracture
interim stress intensity factor, 596
load-displacement records, 596
methodology, 595–596
Transformation law
plane stress, 27–28
shear strain, 69–70, 74
stress, 21–27
Tresca yield criterion
disks
annular rotating, 886
solid rotating, 892–893
plastic deformation, 658–659
thickness strain at instability, 752–753
thick-walled pressurized cylinders, 866, 867
vs. von Mises yield criterion, 659–663
Twist, angle of, 360, 371
Twisting moment, 190
bar stresses, 320
discussion, 359–360
end constraint effects, 374–376
internal stiffening webs, 372–373
Saint-Venant's free torsion, 360–363
solid circular section, 363–364
solid elliptical section, 365–366
solid rectangular section, 366–367
thin-walled closed sections, 370–372
thin-walled open sections, 368–369

U

Ultimate tensile strength (UTS), 735
Unloading, 825–828
Unstable plastic deformation, *see* Plastic instability
Unsymmetrical bending, 324
UTS (Ultimate tensile strength), 735

V

Variable amplitude loading, 609–611
Velocity relations for strain, 87–88
Virtual work, principle of, 129–131
Void growth, porous solids, 720–722
Void volume fraction, 716
Volumetric strain, 52–54
von Mises yield criterion
annular rotating disks, 887
plastic deformation, 654–657
and porous solids, 716
relation to Hill's yield criterion, 697
solid rotating disk, 893
thick-walled pressurized cylinders, 867
vs. Tresca criterion, 659–662, 663

W

Warping, *see also* Torsion of bars and rods
deformation of plane sections, 225
Saint-Venant's function, 361, 366
twisting moment and, 359–363
Webs, internal stiffening, 372–373
Wedges subjected to loads
bending moment at vertex, 292–295
bending stresses, 290
force acting along axis, 286–289
force inclined to axis, 290–292
force perpendicular to axis, 289–290
Williams stress function, 568
Work
during deformation, 116
principle of, 128–129
principle of least, 134
stationary potential energy, principle of, 31–132
virtual, principle of, 129–131

Y

Yield criteria, plastic deformation
anisotropic materials, 687–688
and elastic distortion energy, 656–657
example problems, 657–658
Hill's yield criterion, 693, 694, 697
kinematic hardening rule, 709–715
porous solids, 716–719
state of deformation, 653–654
Tresca yield criterion, 658–659
von Mises vs. Tresca, 659–662, 663
von Mises yield criterion, 654–657
yield function, 654
Yielding
initial, elastic-plastic
disk, annular rotating, 886–887
disk, solid rotating, 892–893
plane strain bending of plates, 822
radius of curvature and, 822
initiation and bending moment, 822, 828–830
kinematic hardening function, 709–715
partial and full, elastic-plastic
disk, annular rotating, 887–890
disk, solid rotating, 893–895
plane strain bending of plates, 822–825
radius of curvature and, 823
plane stress bending of beams, 828–830
stress concentration in small holes, 303
Yield strength, anisotropic materials, 688, 689–690
Yield surface, isotropic hardening, 678–679
Young's modulus, 103, 140